Microwave Circuit Design

Microwave Circuit Design

A Practical Approach Using ADS

Kyung-Whan Yeom

Pearson

New York • Boston • Indianapolis • San Francisco
Toronto • Montreal • London • Munich • Paris • Madrid
Capetown • Sydney • Tokyo • Singapore • Mexico City

For information about buying this title in bulk quantities, or for special sales opportunities (which may include electronic versions; custom cover designs; and content particular to your business, training goals, marketing focus, or branding interests), please contact our corporate sales department at corpsales@pearsoned.com or (800) 382-3419.

For government sales inquiries, please contact governmentsales@pearsoned.com.

For questions about sales outside the United States, please contact intlcs@pearson.com.

Visit us on the Web: informit.com

Library of Congress Cataloging-in-Publication Data

Yeom, Kyung-Whan.

Microwave circuit design : a practical approach using ADS / Kyung-Whan Yeom.

pages cm

Includes index.

ISBN 978-0-13-408678-1 (hardcover : alk. paper)—ISBN 0-13-408678-3

1. Microwave circuits—Computer-aided design. 2. Advanced design system. I. Title.

TK7876.Y46 2015

621.381'32—dc23

2015004174

ISBN-13: 978-0-13-408678-1
ISBN-10: 0-13-408678-3

ScoutAutomatedPrintCode

To my wife Cho Soon-Duk (조순덕)
and son Yeom Jae-Hyung (염재형)

Contents

Preface

This book primarily examines **active microwave circuit design**, an important part of microwave engineering. This subject has worldwide appeal given the incredible growth in mobile and satellite communications. In the past, the use of microwaves was limited to radars and weapon systems, and to remote sensing and relay systems. However, due to the rapid expansion of mobile and satellite communication systems in recent years, systems that use radio waves or microwaves can be found in almost every sphere of our lives. Thus, it is clear there is an increased need for educational materials about active microwave circuit designs. This text is intended as a guide for graduate students who have majored in electronic engineering and its related fields. It should also be useful to engineers and professionals working in these fields who want to update their knowledge through independent study.

In writing this guide, I make the assumption that readers have majored in electronics or related fields as undergraduate students. In particular, readers are assumed to have the **prerequisite knowledge of circuit theory, electronic circuits, and electromagnetics,** which are usually covered in mandatory courses at the undergraduate level.

Numerous books have been published on the subject of active microwave circuit design. However, many of these works do not present the hands-on approach required in modern curricula, making it difficult for readers who only have the basic prerequisite knowledge mentioned above, to understand and follow such texts. For these readers, practical design skills may be hard to acquire by simply reading a text that presents only theory based primarily on mathematical explanations. On the other hand, most people working in this field have become familiar with the prevalence of **design software** employed in active microwave circuit designs, such as the **Advanced Design System (ADS)** from Agilent Technologies and **Advancing the Wireless Revolution (AWR)** from AWR Corporation. The design environment for active microwave circuits has changed drastically with the continuous expansion of microwave applications into our daily lives. Recently, a variety of software design tools applicable to circuit design, system design, and electromagnetic analysis of passive structures has emerged. This has significantly reduced the need for analytical methods and specific design-oriented, in-house programs for the design of circuits and systems. With these advances, the rapid exchange of results between designers has facilitated independent study and experimentation with basic concepts using software tools and practical designs. Clearly, innovations in the field underscore the necessity for advanced education in active microwave circuit design and improvements to relevant software tools. The practical design skills for active microwave circuit designers can be effectively improved through hands-on practice with design software. More than ever, the importance of ongoing education to an engineer in this field cannot be overemphasized.

Given this perspective, it is my view that an education incorporating these features has become imperative. With more than 17 years of experience educating graduate students, I have written this guide to address the critical importance of this subject. With this book, readers will acquire **the practical skills required for active microwave circuit design** using the design software. The popular Advanced Design System (ADS) from Agilent Technologies is the design tool used in the book as it has the longest proven track record compared to other design software. However, since most features of ADS are also available in other, similar design software, I believe that selecting ADS as the design tool will not present any critical limitations to readers.

This book is primarily composed of two parts: **basic concepts for active microwave circuit designs**, and **practical design examples** such as low-noise amplifiers (LNA), power amplifiers (PA), microwave oscillators, phase-locked loops (PLL), and mixers. The designs of LNAs, PAs, oscillators, and mixers are essential in building various communication systems, radars, and other microwave transmitting and receiving systems. Additional components such as phase shifters, variable attenuators, and switches, although important, appear only in limited applications and are not used as frequently when compared to the previously mentioned set of components.

The **basic concepts** are **concisely** and **clearly explained based on their physical characteristics**. These concepts, essential in an introduction to an active microwave circuit design course, include passive devices, transmission-line theory, high-frequency measurement, and an introduction to active devices. For these basic concepts, this book focuses more on physical concepts and on understanding the meaning of calculated results rather than on exhaustive mathematical calculations. This is achieved by presenting critical concepts as clearly and succinctly as possible. In addition, complex calculations are avoided whenever possible and Agilent's ADS is employed to replace them. **The software is used to analyze or verify the basic concepts**, enabling readers to achieve a deeper and more thorough understanding of them. Pertinent, real-world examples facilitate comprehension and independent study.

For the design of LNAs, PAs, oscillators, and mixers, readers are provided with **practical design examples** using ADS that they can subsequently use to design similar active microwave circuits. I am confident this book will provide readers with the practical skills necessary for active microwave circuit design. Finally, although the book is designed for graduate students, it can also be very helpful as source material for **independent study or as a reference book** for professionals.

The text is composed of materials that provide a two-semester course curriculum. Depending on the students, this can be reduced to a one-semester course when the foundation topics in the first part of the book are skipped or covered only briefly. For the design of LNAs, PAs, oscillators, and mixers, a project-style lecture may be useful. (After a brief explanation of the basic design components, students establish a lecture style and present their design.) A solution manual is available for instructors at Pearson's Instructor's Resource Center (IRC). I welcome and appreciate any corrections or suggestions for improvement to this content.

Register your book at informit.com/title/9780134086781 to access this book's ADS examples and problems.

Kyung-Whan Yeom
April 2015

Acknowledgments

I am very grateful to my lab students for their efforts editing many of the figures and the contents of this textbook. Also, I would like to thank the Prentice Hall staff, and particularly Bernard Goodwin, for their support. I would especially like to express my deep appreciation to the reviewers for evaluating this text and offering valuable suggestions. Finally, I would like to thank my wife, Soon-Duk Cho, and son, Jae-Hyung Yeom, for their tolerance and support, which allowed me to write this book.

Reviewers
Professor Changzhi Li (Texas Tech University); Full text
Professor Jin Wang (University of South Florida); Full text
Dong-Wook Kim (Chungnam National University, Korea); Chapter 9, 10
Professor Matthew Radmanesh (California State University, Northridge); Chapter 9
Anonymous reviewer; Chapter 9

Translation
Abdul-Rahman Ahmed (Chungnam National University, Korea)

LNA Design Example
Hyun-Seok Oh (Agency for Defense Department, Korea) and Dong-Hyun Lee

Oscillator Design Example
Hae-Chang Jeong and Beom-Ik Son (Chungnam National University, Korea)

Mixer Simulation
Seong-Sik Yang (Samsung Thales, Korea)

Preparation of Manuscript and Figures
Hyun-Mi Kim(KOMSCO, Korea)
Man-Hee Lee (LIG Nex1, Korea)
Hyun-Seok Oh (Agency for Defense Department, Korea)
Yun-Seong Heo (Hanhwa, Korea)
Hae-Chang Jeong (Chungnam National University, Korea)
Beom-Ik Son (Chungnam National University, Korea)
Seok-Jeong Lee (Chungnam National University, Korea)

About the Author

Kyung-Whan Yeom was born in Seoul, Korea, in 1957. He received a B.S. degree in electronics from Seoul National University in 1980 and M.S. and Ph.D. degrees in electrical engineering from the Korea Advanced Institute of Science and Technology (KAIST), Daejeon, Korea, in 1982 and 1988, respectively.

From 1985 to 1991, he worked at LG Precision as a principal engineer. He worked on the MIC team as a team leader and was later involved in the military electronics division for EW Equipment. When he was at LG Precision, he received a technical achievement award for the ABEK program from Teledyne Microelectronics.

From 1991 to 1995, he worked at LTI on power amplifier modules for analog cellular phones. He joined the Chungnam National University as assistant professor in 1995 and is currently a professor in the Department of Radio Science and Engineering, Chungnam National University, Daejeon, Korea. His research interests are in the design of hybrid and monolithic microwave circuits and microwave systems.

Professor Yeom has been a member of the Korean Institute of Electromagnetic Engineering and Science (KIEES) and the Institute of Electrical and Electronics Engineers (IEEE) since 1995. He was the editor-in-chief of KIEES from 2004 to 2006. He received the IR-52 Jang Youg-Sil Prize from the Ministry of Science and Technology (MOST) of Korea for his work on cell phone power amplifiers in 1994. He received an academic award from KIEES for the design and fabrication of a novel 60 GHz GaAs pHEMT resistive double balanced star MMIC mixer in 2004. He also received the best paper award from the Korean Federation of Science and Technology Societies (KOFST) for his work, "A Novel Design Method of Direct Coupled Bandpass Filter Based on EM Simulation of Individual Resonator."

CHAPTER

1

Microwave Integrated Circuits

CHAPTER OUTLINE

1.1 CLASSIFICATION OF MICROWAVE INTEGRATED CIRCUITS

An active microwave circuit can be defined as a circuit in which active and passive microwave devices such as resistors, capacitors, and inductors are interconnected by transmission lines. At low frequencies, the transmission lines are a simple connection; however, at microwave frequencies they are no longer just simple connections and their operation becomes a complicated distributed circuit element. As a result, a microwave integrated circuit's classification is based on the fabrication method of the transmission lines used for interconnection.

There are various types of transmission lines in microwave integrated circuits; some common examples are waveguides, coaxial, and microstrip lines. Figure 1.1 shows the transmission lines used in microwave circuits. Although there are special cases of microwave integrated circuits that are composed of coaxial lines and waveguides, in most cases the microwave integrated circuits are formed using planar transmission lines. Therefore, the content of this book is restricted to microwave integrated circuits formed using planar transmission lines, examples of which are microstrip, slot line, and co-planar waveguide (CPW), as shown in Figure 1.2. These planar transmission lines are frequently used in the large-scale production of microwave circuits and generally form the basic transmission lines for microwave circuits.

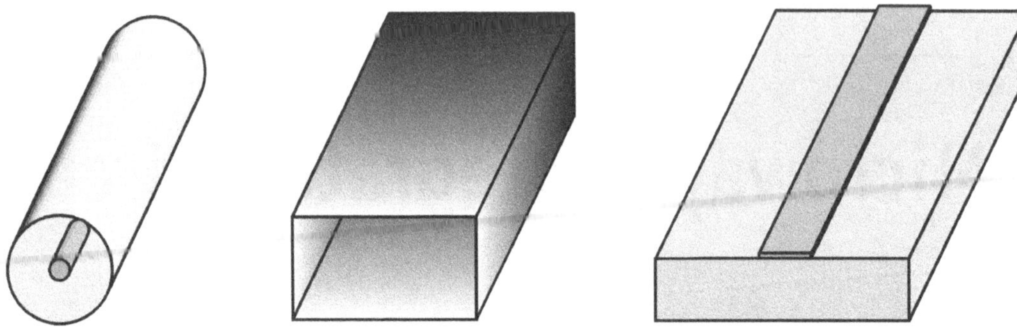

(a) (b) (c)

Figure 1.1 Some common transmission lines used in microwave circuits: (a) coaxial line, (b) rectangular waveguide, and (c) microstrip line

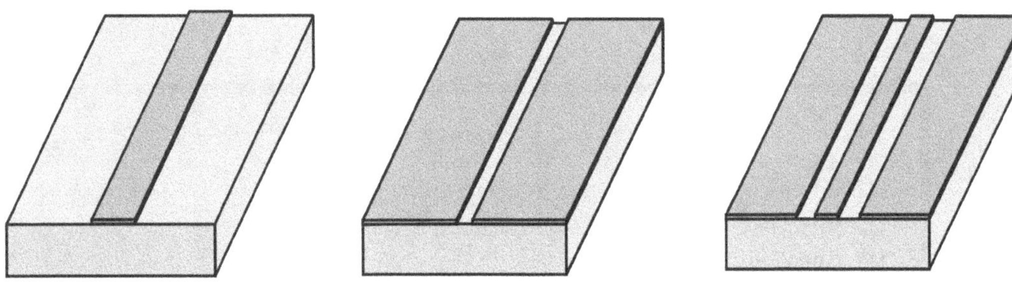

(a) (b) (c)

Figure 1.2 Some common planar transmission lines used in microwave circuits: (a) microstrip, (b) slot line, and (c) CPW (co-planar waveguide). They are explained in Chapter 3.

The implementation of planar transmission lines on substrates can be classified into two basic groups: *monolithic* and *hybrid integrated circuits*. In monolithic integration, the active and passive devices as well as the planar transmission lines are grown *in situ* on one planar substrate that is usually made from a semiconductor material called a *wafer*.

Figure 1.3 shows an example of monolithic integration. Figure 1.3(a) is a photograph of the top side of a wafer and Figure 1.3(b) shows a single monolithic microwave integrated

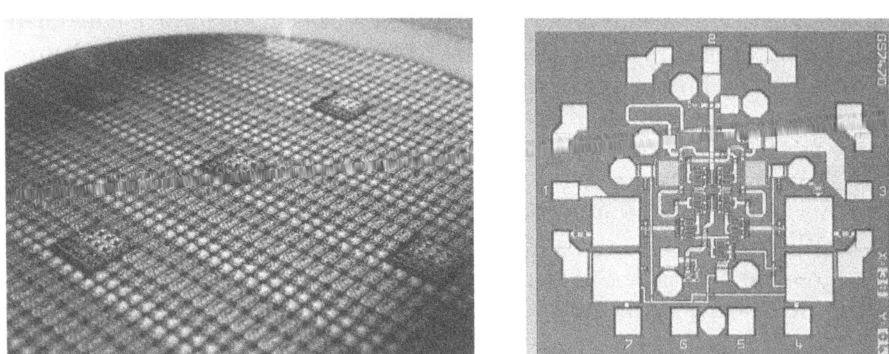

(a) (b)

Figure 1.3 Monolithic integration: (a) a wafer and (b) a monolithic microwave integrated circuit on the wafer (28 GHz GaAs pHEMT Gilbert cell up-converting mixer; refer to Chapter 12).

circuit; the identical circuits are repeatedly produced on the wafer in Figure 1.3(a). The monolithic microwave integrated circuit in Figure 1.3(b) is found to contain active and passive devices, and planar transmission lines. The monolithic integration provides a compact-sized circuit and eliminates a significant amount of assembly when building a component or a system. Especially because size is of critical importance in most recent RF systems, monolithic integration is frequently employed to provide a compact component. An advantage of monolithic integration is that it is well suited for large-scale production, which results in lower costs. A disadvantage is that monolithic integration takes a long time to develop and fabricate, and small-scale production results in highly prohibitive costs.

Hybrid integration is a fabrication method in which the transmission lines are implemented by conductor patterns on a selected substrate with either *printing* or *etching*, and active and passive devices are assembled on the patterned substrate by either soldering or wire bonding. When implementing transmission lines by conductor patterns on a substrate, careful consideration must be given to the substrate material and the conductor material for the transmission lines because these materials can have significant effects on the characteristics of transmission lines. Hybrid integration is thus classified into three types based on the method by which the lines are formed on the substrate: a *printed circuit board* (PCB), a *thick-film* substrate, and a *thin-film* substrate.

Figure 1.4 shows an example of how connection lines are formed on a PCB substrate. Both sides of the dielectric material are attached with copper cladding that is then etched to obtain the desired conductor patterns. For PCB substrate materials, *epoxy fiberglass* (FR4), *teflon*, and *duroid* are widely used. FR4 substrate (a kind of epoxy fiberglass) can be used from lower frequencies to approximately 4 GHz, while teflon or duroid can be used up to the millimeter wave frequencies, depending on their formation. Generally, all these materials lend themselves to soldering while wire bonding for an integrated circuit assembly is typically not widely used. Furthermore, compared with other methods that will be explained later, a PCB can result in lower costs; its fabrication is easy and requires less time to produce. In addition, production on a small scale is possible without the use of expensive assembly machines; it is easy to fix and could also be used in large-scale production, and is thus widely used.

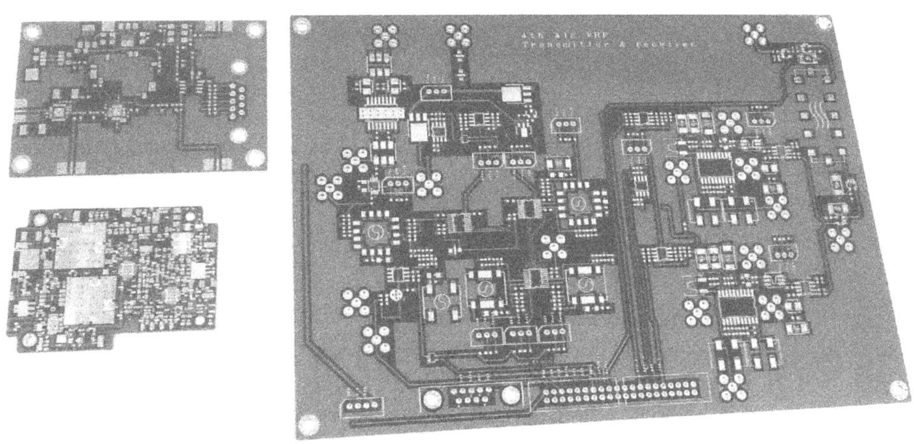

Figure 1.4 A photograph of epoxy fiberglass PCBs. The PCBs on the left are for the X-band and 2 GHz frequency synthesizers using the phase locked loop. The PCB on the right is for the VHF automatic identification system, which has a similar block diagram shown in Figure 1.7. The power amplifier is implemented in a separate block.

Thick-film substrates are produced by screen-printing techniques in which conductor patterns are formed by pushing conductive paste on a ceramic substrate through a patterned screen and then firing printed conductor patterns. The substrate is called thick film because the patterns formed by such techniques are generally much thicker than those formed using thin-film techniques. As a benefit of using screen-printing techniques, multiple screen printings are possible. Dielectric or resistor patterns can also be formed by similar screen-printing techniques using dielectric or resistor pastes. Using an appropriate order of multiple screen printings, it is also possible to form capacitors and resistors on the ceramic substrate. Since the ceramic substrate is more tolerant of heat, it is easy to assemble active devices in the form of chips. On the other hand, considering the lines and patterns formed by this process, the pattern accuracy of thick film is somewhat inferior compared to that of thin film. The costs and development time, on a case-by-case basis, are somewhere between those of the PCB and thin-film processes. Recently, however, the integration based on thick-film technology has become rare because its cost and pattern accuracy are between the PCB and thin-film technology, while thick film is widely used to build multifunction components. A typical example is the package based on LTCC (low-temperature co-fired ceramics) technology.[1] Multilayer ceramics and structuring are possible in LTCC technologies. Figure 1.5 shows a photograph of thick-film patterned substrates fabricated using the thick-film process.

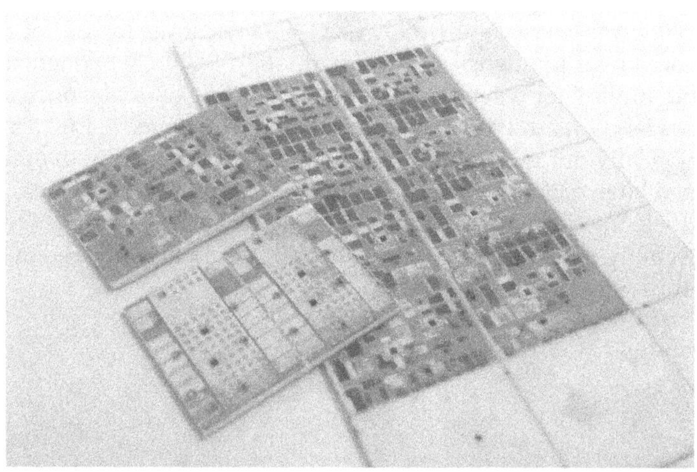

Figure 1.5 A photograph of substrates fabricated by the thick-film process. Identical circuits can be arrayed for efficient production. This circuit is for the mobile communication VCO presented in Chapter 10.

The thin-film technique is very widely used in the fabrication of microwave circuits for military and microwave communication systems. In the case of the thin-film process, a similar ceramic substrate material used in thick film is employed, but compared to the thick-film substrate, a fine surface-finish substrate is used. The most widely used substrate is 99% alumina (Al_2O_3). Other substrates such as fused silica, quartz, and so on are possible for conductor-pattern generation based on thin-film technologies. The pattern

1. Refer to Barry LTCC, LTCC Surface Mount T/R Module, at www.barryind.com.

formation on the substrate is created with a photolithographic process that can produce fine tracks of conductor patterns similar to those in a semiconductor process. Since the thin-film substrate is also alumina as in the case of a thick-film substrate, the assembly of semiconductor chips using wire bonding is possible. Thin film compared with PCB and thick film is more expensive, and due to the requirement of fine tracks, a mask fabrication is necessary and the process generally takes longer. Passive components such as resistors and air-bridge capacitors can be implemented using this process. In addition, integrated circuits produced by the thin-film process require special wire bonders and microwelding equipment for assembly. Compared to the monolithic integration process, the thin-film process tends to be cheaper in terms of cost, but compared to MMIC, the assembled circuit using the thin-film patterned substrate is difficult to characterize precisely because of unknown or poorly described parasitic circuit elements associated with the assembly methods such as wire bonding and die attach. Before the emergence of MMICs (monolithic microwave integrated circuits), thin-film technology was the conventional method for building microwave-integrated circuits (MICs). Figure 1.6 is a photograph of thin-film circuits fabricated with the thin-film technique.

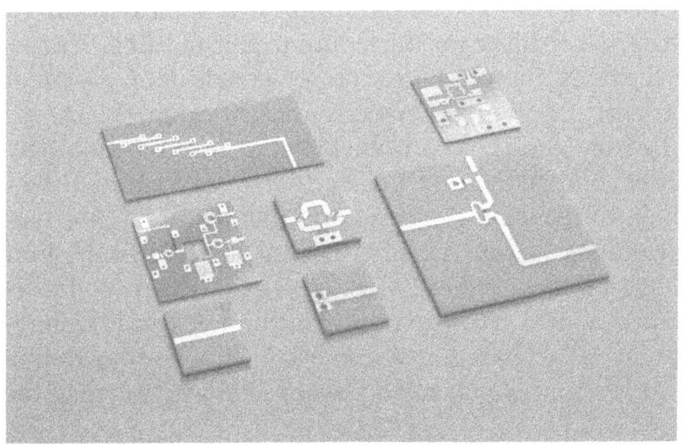

Figure 1.6 A photograph of substrates produced by the thin-film process. From top left to bottom right, they are filter, phase shifter, power amplifier (presented in Chapter 9), path-switching circuit by assembly, power divider, and 50 Ω lines.

The choice of integration method depends on the application and situation, taking into account several factors mentioned previously, such as the operating frequency of the integrated circuit, the types of semiconductor components (chip or packaged), the forms of the passive components, large-scale fabrication costs, and method of assembly. These factors should all be considered when selecting the optimum method of integration. For a description of microwave-patterned substrate fabrication, assembly with wire bonding and soldering, and packaging, see reference 1 at the end of this chapter. The book provides general information about microwave-circuit fabrications. Table 1.1 provides a comparison of the hybrid integrations described previously.

Table 1.1 Comparison of hybrid integration

Technology	Cost	Fabrication Time	Pattern Accuracy	Assembly
PCB	Low	Short	Low	Soldering
Thick film	Middle	Middle	Low	Soldering and wire bonding
Thin film	High	Long	Fine	Soldering and wire bonding

Now we will consider the application of the planar transmission lines such as microstrip, slot, and CPW to the monolithic and hybrid integration technologies. Microstrip lines are the most widely used transmission lines for both monolithic and hybrid integration technologies. In microstrip lines, the top conductor pattern is usually connected to the ground by a through hole or a via hole. Thus, the back-side process for the through-hole or via-hole fabrication is essential to building a circuit based on microstrip lines. This back-side process is inconvenient especially in the monolithic integration. In hybrid integration, the holes can be fabricated through simple mechanical drilling for a PCB case and through laser or ultrasonic drilling for thick- and thin-film cases. Then, plating the fabricated holes completes the fabrication of a through or via hole. However, to fabricate via holes in monolithic integration, a wafer that typically has a normal thickness of about 600 μm should be polished down to about 100 μm thickness. Current technology does not support via-hole fabrication beyond 100 μm. In Figure 1.2, we can see that the CPW and slot lines do not need the back-side metallic ground and they eliminate the need for any additional back-side metallization process. The CPW is very helpful in monolithic integration and is widely used to build MMICs without vias. However, the discontinuities of CPWs are not well understood compared to those of microstrip lines and the integration based on a CPW is not as popular as that based on a microstrip. The various discontinuities of microstrip and slot lines, CPWs, and planar transmission lines are covered in reference 2 at the end of this chapter.

1.2 MICROWAVE CIRCUITS IN A COMMUNICATION SYSTEM

Microwave integrated circuit classification has been discussed previously. The microwave integrated circuit was classified according to the method of implementing the planar transmission lines for the purpose of connecting active and passive devices. The functions of microwave integrated circuits vary greatly and we will now consider several important microwave integrated circuits, the designs of which will be discussed in later chapters. Some examples of these circuits are low noise amplifiers (LNA), power amplifiers (PA), oscillators, mixers, directional couplers, switches, attenuators, and filters, among a host of other microwave-integrated circuits. Among these, directional couplers, switches, attenuators, filters, and so on, are basically passive microwave circuits, although they are very widely used. Thus, they are not covered in this book because they are considered to be outside its scope. In addition, although components such as switches, variable attenuators, phase shifters, and other control circuits are important and are composed of semiconductor devices, they are generally not regarded as the basic building blocks of a wireless communication system. Therefore, this book will only cover low-noise amplifiers, power amplifiers, oscillators, and mixers, which are the most widely used circuits in the construction of wireless communication systems. The basic design theory of these circuits as well as the devices related to them will be explained in this book.

As an example of a wireless communication system, Figure 1.7 shows a block diagram of an analog cellular phone handset (Rx frequency is 869–894 MHz and Tx frequency is 824–849 MHz).[2] A general transceiver used for the transmission and reception of analog signals (usually voice) has a similar block diagram that is shown in Figure 1.7. A weak RF signal with a typical power level of about –100 dBm (0.1 nW) received from an antenna first goes through a filter called a diplexer and the signal is received only in the receiver frequency band. The filtered signal is too weak for direct demodulation or signal processing, and a low-noise amplifier (LNA) with a gain of 20–30 dB is required to amplify the received signal. Too much gain may cause distortion and an LNA with a gain of 20–30 dB is usually employed. Chapter 8 provides a detailed explanation of the design of an LNA.

Next, because the received signal frequency is so high, the first mixer shown in Figure 1.7 translates the carrier frequency to a lower frequency band called first IF (intermediate frequency). A double-conversion superheterodyne receiver is more widely used than a single-conversion super heterodyne receiver in a communication system. The filter in front of the first mixer again suppresses both the image frequency signal and other signals at the outside of the receiving frequency band. Since multiple users in service are using the same frequency band, multiples of other user signals generally coexist with the signal in the first

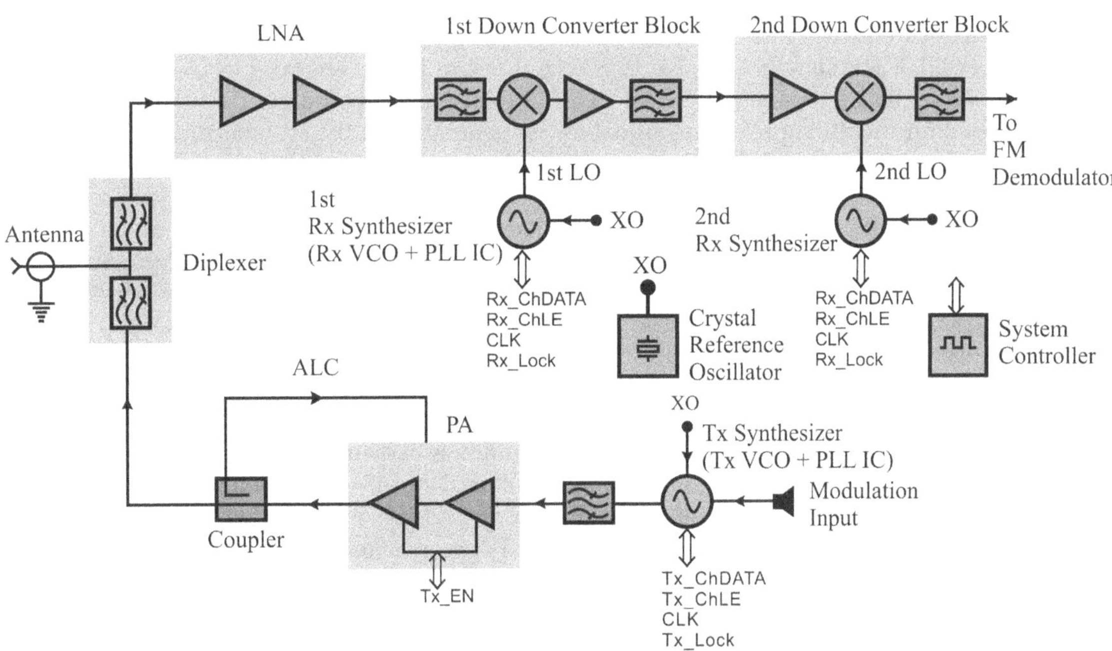

Figure 1.7 A block diagram of an analog mobile phone handset (AMPS standard). Tx_EN stands for Tx enable and ALC stands for automatic level control. Tx_ and Rx_data are required to set the programmable frequency dividers in Tx and Rx synthesizers. LE stands for Load Enable. When LE is high, the digital channel data are loaded to the corresponding programmable frequency divider in PLL IC. Synthesizers are explained in Chapter 11. Lock signal indicates that the synthesizer using PLL is in a locked state.

2. Refer to AMPS (Advanced Mobile Phone System) standard.

IF. Intermodulations among the multiple signals are one of the crucial issues in mixer design. Chapter 12 describes the typical topologies of various mixers for suppressing such spurious signals. In order to filter out possible spurious signals that appear at the first mixer output, the signal is passed through a narrow bandpass filter that has a bandwidth of about the signal bandwidth. The first IF filter removes many unwanted spurious signals although it may not be completely sufficient. The first IF output is converted again through the second mixing. Now the center frequency of the second IF is low enough, the highly selective filter is available, and the spurious signals can be sufficiently suppressed through the second IF filter. In addition, the signal frequency is low enough and can be demodulated for the recovery of the original signal. The demodulator is an FM demodulator and is almost the same as the FM demodulator that is commercially popular.

Note that the mixer requires the input signal from a *local oscillator* (LO) for the translation of the signal frequency to the IF. The two LO signals are supplied from the two Rx-synthesizers and each Rx-synthesizer consists of a voltage-controlled oscillator (VCO) and a commercial PLL (phase-locked loop) IC (integrated circuit). Since the frequency of most VCOs is not stable enough to be used in such communication systems, the frequency of a VCO must be stabilized using a stable crystal oscillator (XO in Figure 1.7) with a typical temperature stability of 2 ppm (parts per million) and a phase-locked loop (PLL). Furthermore, the LO frequency should be moved up and down according to the base station commands. Such frequency synthesis and stabilization can be achieved by a *phase-locked loop* (PLL). To build a frequency synthesizer using PLL, the VCO frequency as well as the crystal oscillator frequency must be divided by appropriate programmable frequency dividers in the PLL IC. The signals CLK, Rx_ChDATA, Rx_ChLE, and Rx_Lock, shown in Figure 1.7, are the digital signals between the PLL IC and the system controller. The clock signal CLK is used for the timing reference signal that is generated by the system controller using the crystal oscillator. Rx_ChDATA sent from the controller represents the digital data to set the programmable frequency dividers. The signal Rx_ChLE selects the corresponding programmable divider for Rx_ChDATA to be loaded among several frequency dividers in the PLL IC. When phase lock is achieved, the PLL IC sends the signal Rx_Lock to the system controller to inform the phase lock completion. The two Rx synthesizers are necessary for the double-conversion superheterodyne receiver. The commercial PLL IC generally includes the necessary components to achieve the phase lock for two VCOs in a single PLL IC. Thus, the LO signal for the second conversion is similarly synthesized using a single PLL IC. The design of the Tx and Rx VCOs in Figure 1.7 as well as the other microwave VCOs are described in Chapter 10, while the PLL's operation is explained in Chapter 11.

In the transmission operation, the modulation input signal (usually voice) goes to the modulation input of a Tx synthesizer. The Tx synthesizer is similarly composed of a VCO and a PLL IC. Through the PLL IC, the desired carrier center frequency is similarly synthesized as in the Rx synthesizer. The digital signals CLK, Tx_ChDATA, Tx_ChLE, and Tx_Lock are similarly interpreted as in the Rx synthesizer. The modulation signal has a generally higher frequency than the PLL loop bandwidth and thus can modulate a VCO without the effects of a PLL. Therefore, the *frequency-modulated* (FM) signal appears at the Tx synthesizer output with the synthesized carrier frequency. The modulated signal then passes through the bandpass filter that removes unnecessary or spurious signals. The average output power level of the modulated signal is generally low; thus, in order to obtain the desired RF power output level, the signal must be amplified by a power amplifier (PA) whose typical maximum output power level is about 1W. The function ALC (Automatic Level Control) is generally built in to control the transmitting power level. When a user is close to the base station, the transmitting power level is set to low;

otherwise, it is set to high for a better quality of communication. The PA output signal is then passed through a diplexer without affecting the receiver and radiated via the antenna. A power amplifier is important in this type of communication system because it consumes most of the DC power supplied from a battery. Furthermore, because a power amplifier operates in large-signal conditions, significant distortion arises. In Chapter 9, we will discuss the design and linearity evaluation of a power amplifier.

Given the preceding discussion, the key circuits in building a communication system are a low-noise amplifier, a power amplifier, oscillators, and mixers. With that in mind, this book will discuss in detail the design and evaluation method of these circuits.

1.3 SUMMARY

- Microwave integrated circuits can be classified according to the fabrication method of the patterned substrate and in terms of monolithic and hybrid integration. Hybrid integration can be further classified into integrations based on PCB, thick film, and thin film. In the selection of integration, one type cannot be said to be superior to the other; the choice is made depending on the application and given situation, and by taking into consideration several factors such as cost, time, pattern accuracy, and assembly.
- Among active microwave circuits, the most commonly used building blocks for wireless communication systems or other systems, such as repeaters, transponders, and radars, are amplifiers, oscillators, and mixers.

REFERENCES

1. T. S. Lavergetta, *Microwave Materials and Fabrication Techniques,* Dedham, MA: Artech House, Inc., 1984.
2. K. C. Gupta, *Microstrip Lines and Slot Lines*, 2nd ed. Dedham, MA: Artech House, Inc., 1996.

PROBLEMS

1.1 A waveguide generally has lower line loss than a microstrip. An SIW (substrate integrated waveguide) can be considered as the planar version of a waveguide. How is an SIW configured using a substrate?

1.2 Find the TR (transmission and receiving) module example built using a LTCC on the Web site www. barryind.com.

1.3 How is the ALC in Figure 1.7 constructed?

1.4 Refer to the FM demodulator IC SA605, which is used to demodulate an FM signal. Explain how the FM signal is demodulated using its block diagram.

1.5 Refer to the Web site of vendors of PLL IC such as Analog Devices Inc. or other companies. Explain the synthesizer data bus shown in Figure 1.7.

1.6 How can the PLL be modulated? Explain how to set the PLL loop bandwidth by taking the bandwidth of a bandlimited modulation signal into consideration.

CHAPTER

2

Passive Devices

CHAPTER OUTLINE

2.1 IMPEDANCES

Passive devices include resistors, capacitors, and inductors. When these devices are used in a circuit, their most fundamental property is their impedance. The impedance Z of a resistor, capacitor, and inductor are expressed, respectively, below in Equations (2.1)–(2.3).

$$Z_R = R \tag{2.1}$$

$$Z_C = \frac{1}{j\omega C} \tag{2.2}$$

$$Z_L = j\omega L \tag{2.3}$$

Figure 2.1 shows the log-log scale plot of $|Z|$ in the equations above. In the case of a resistor, it exhibits constant impedance independent of frequency, while the impedance of a capacitor linearly decreases, and that of an inductor increases with frequency, as shown in Figure 2.1. There are various ways of fabricating resistors, capacitors, and inductors, which produce similar impedance characteristics for some limited ranges of frequency. In this chapter, we will examine how the passive devices are classified based on their fabrication methods, and we will look at the impedance characteristic dependence on fabrication

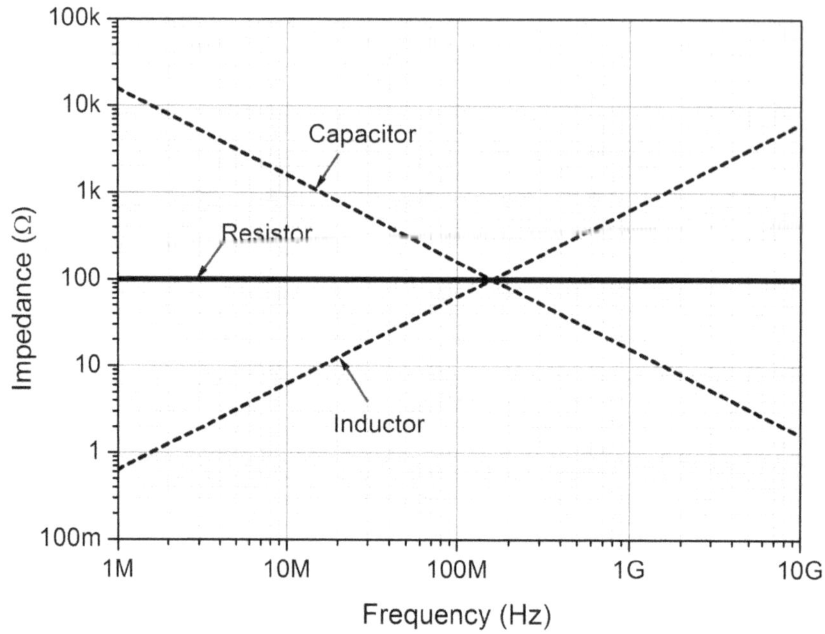

Figure 2.1 Impedance plots of a 100 Ω resistor, a 10 pF capacitor, and a 100 nH inductor

methods. Commercially available resistors, capacitors, and inductors generally exhibit the impedance characteristics shown in Figure 2.1 at low frequencies; however, they generally do not have the simple and ideal characteristics described above as the frequency increases. The impedance characteristic typically becomes more complicated at higher frequencies and it is necessary to examine the characteristics before circuit design. In practice, complicated equivalent circuits are needed for their representations as the frequency increases, which adds to the difficulty and complexity of circuit design at high frequencies.

In such circumstances, the equivalent circuit that properly reflects the impedance characteristics in the frequency range of operation should be used in circuit design. The critical factors will be the *understanding of the typical equivalent circuit of such commercially available resistors, capacitors and inductors*, and *the technique to determine the equivalent circuit values with given data*. Typical forms of data are available as a datasheet and library in design software. When data are not available, measurements or EM simulations can be carried out to determine them. Therefore, in this chapter, detailed data analysis techniques and appropriate measurement techniques in the absence of data will be discussed.

2.2 CLASSIFICATION

Largely based on their fabrication method, passive devices are classified into *lead-type* and *chip-type* components, and *pattern-type* passive components. Lead-type and chip-type components are widely available commercially. Photographs of them are shown in Figure 2.2 and Figure 2.3, respectively. Pattern-type passive components are formed by patterns on substrate.

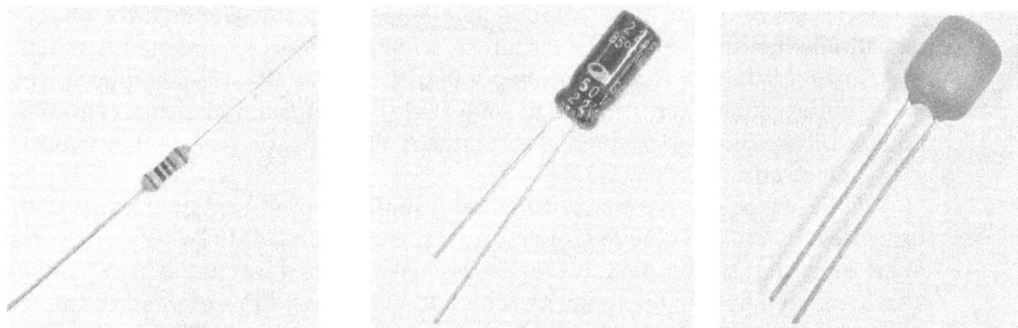

(a) (b) (c)

Figure 2.2 Lead-type components: (a) resistor, (b) capacitor, and (c) inductor

(a) (b) (c)

Figure 2.3 Chip-type components: (a) resistor, (b) capacitor, and (c) inductor

The assembly method of lead-type and chip-type components differs. While lead-type components are basically assembled using an *insertion technique*, chip-type components are assembled using a *surface mounting technique*. In assembling a lead-type component, the lead terminals are first bent and inserted into through holes formed in the printed circuit board. The unnecessary lead terminals are then cut and the printed circuit board with lead-type components is dipped into melted solder, and finally the lead terminals are soldered to the round conductor patterns formed around the through holes on the bottom side of the PCB. In the case of chip-type components, *solder creams* are first printed to *land patterns* formed on a printed circuit board or substrate, and chip-type components are mounted on the land patterns manually or automatically. Finally, passing through a *reflow machine* with an appropriate temperature profile, the soldering of chip-type components to the land patterns is completed. In the case of lead-type components, it should be noted that the lead terminals are supposed to be used only for connections. However, they also give rise to parasitic inductance at high frequencies, and consequently the impedance of the remaining lead terminal adds to the impedance of the passive device. In the past, these lead-type components were mostly used as passive devices before chip-type components became commercially popular. But due to the fluctuations arising from the length of the lead terminals when assembled, lead-type components are seldom used at high frequency although they are still used for low-frequency applications. In addition, lead-type components are generally bigger than chip-type components.

Compared to lead-type components, chip-type components have relatively fewer parasitic elements caused by the terminals, which becomes advantageous at high frequencies. However, the key reason for the popularity of these devices is miniaturization rather than their advantage for use at high frequencies. In addition, the surface-mounting technique, which provides an advantage in large-scale fabrication, is another reason why they are widely used.

Pattern-type passive components can usually be produced on the substrate by using the process of a monolithic microwave integrated circuit (MMIC) or thin film in situations where a circuit designer directly creates passive devices through the use of patterns. However, there are limits to the range of values attainable with this method compared to chip-type or lead-type components, which is a disadvantage.

In the case of a resistor, as shown in Figure 2.4(c), a thin film of a resistive material (usually NiCr and TaN) is used as a resistor pattern on which the conductor pattern (for the purpose of connection) is formed by the appropriate technique. The resistor in Figure 2.4(c) has a uniform thickness and the resistance is thus proportional to the length (L) and inversely proportional to the width (W). The proportionality constant is defined as the *sheet resistivity* R_S, and the resistance R is given by Equation (2.4).

$$R = R_S \frac{L}{W} \tag{2.4}$$

Note that R_S has the dimension of Ω/square.

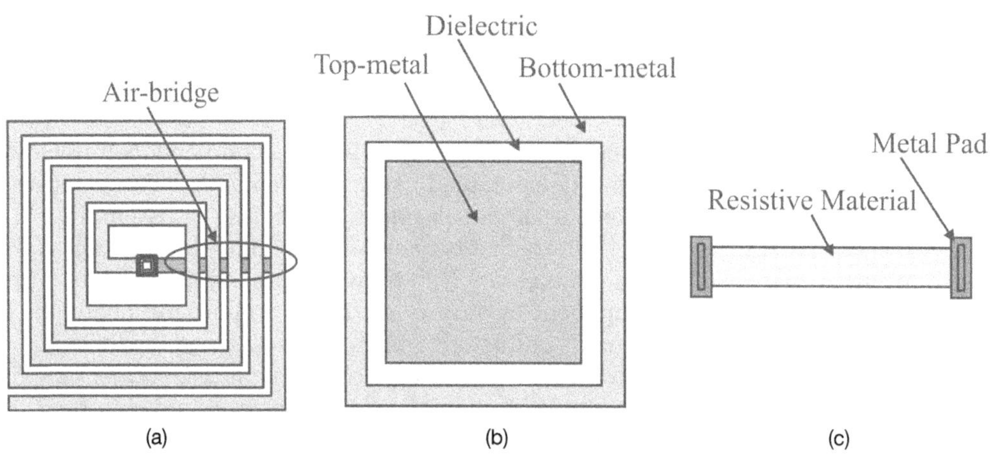

Figure 2.4 Passive components produced by pattern formation: (a) inductor, (b) capacitor, and (c) resistor

In the case of the capacitor in Figure 2.4(b), a uniform-thickness dielectric sheet is formed on the bottom conductor pattern, a top conductor pattern is again formed on the dielectric, and a metal-insulator-metal (MIM) capacitor is formed. The dielectric material's thickness, generally determined by the process, is constant and once the *sheet capacitance* C_S has been determined, the capacitance is then directly proportional to the surface area (A) and is expressed in Equation (2.5).

$$C = C_S A \tag{2.5}$$

Of course, as the frequency increases, the parasitic elements inherent in the MIM capacitor will appear, and will need appropriate modeling in the frequency range of operation.

In the case of an inductor, it is usually implemented as a spiral type, as shown in Figure 2.4(a). Its design is not as simple as that of a capacitor or resistor. Foundry-service companies usually provide measured results or data on several configurations of inductors. In the absence of this information, the values of inductors are determined through electro-magnetic (EM) simulation. In addition, although the assessment of these pattern-type components is important, when viewed from a design point of view, the construction of their equivalent circuits is the same as that of chip-type or lead-type components, and so its specific discussion will be omitted here.

Example 2.1

(1) Given that the sheet resistance R_S is 50 Ω/square, calculate the resistance of the component shown in Figure 2E.1 below.
(2) The permittivity of an MIM capacitor is 7.2 and its thickness is 0.4 μm. Determine the sheet capacitance in pF/mm^2 and then calculate the capacitance of a 50 μm^2 capacitor.

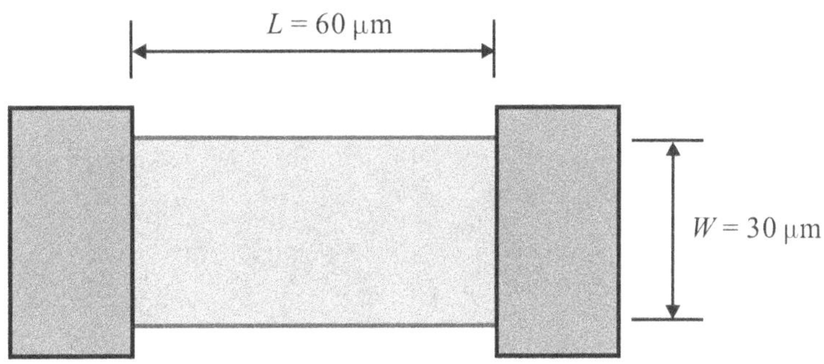

Figure 2E.1 Resistor pattern

Solution

(1) In the case of the resistor, since the value of the sheet resistance R_S is 50 Ω/square, then

$$R = R_S \frac{L}{W} = 50 \times \frac{60}{30} = 100\ \Omega$$

(2) In the case of the capacitor, the sheet capacitance per square mm is

$$C_S = \varepsilon_r \varepsilon_0 \frac{A}{t} = 8.854\ \text{pF/m} \times 7.2 \times \frac{1\ \text{mm}^2}{0.4\ \mu\text{m}}$$
$$= 8.854\ \text{pF/m} \times 7.2 \times \frac{1\ \text{m}}{0.4} = 159\ \text{pF}$$

Thus, $C_S = 159$ pF/mm^2 and for a capacitor having an area of 50 μm × 50 μm, the capacitance is

$$C = C_S \times 0.05^2 = 159 \times 0.05^2\ \text{pF} = 0.398\ \text{pF}$$

2.3 EQUIVALENT CIRCUITS

As mentioned earlier, pattern-type passive components require the thin-film or MMIC process. When pattern-type components are not used, the most likely choice for passive components in high-frequency applications is chip-type components because lead-type components are not appropriate for higher-frequency applications. In this section, the methods for manufacturing and evaluating chip-type components will be explained. In addition, the method for extracting the equivalent circuit from given data will be described.

2.3.1 Chip-Type Capacitors

A chip-type capacitor is usually constructed using a multilayered structure, as shown in Figure 2.5. Each terminal consists of a number of parallel conducting plates (in Figure 2.5, they are labeled as internal electrodes), and the sum of the capacitance formed between the parallel conducting plates appears at the terminals of the chip-type capacitor. The dielectric fills the space between the conducting plates and its dielectric constant further increases the capacitance between the terminals.

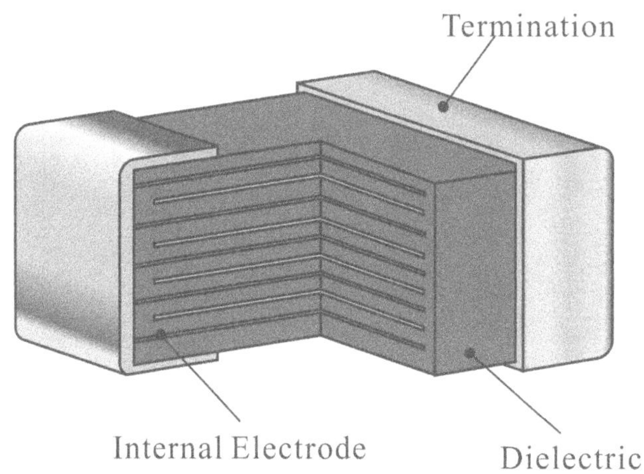

Figure 2.5 Structure of a chip capacitor. The capacitance appears at the two terminals and is the sum of the capacitance appearing between two nearby conducting plates. Using high-permittivity ceramic material, the capacitance can be increased.

The classification of chip-type capacitors is based on the geometrical parameters shown in Figure 2.6: length L between the terminals and terminal width W. Based on a standard unit of mm, a capacitor having a length of 1.0 mm and a width of 0.5 mm is called type 1005; following a similar definition, type 1608 is a capacitor that has a length of 1.6 mm and a width of 0.8 mm. This classification is not limited to capacitors; resistors and inductors are classified in the same way. Thus, a 1608 resistor represents a chip resistor with a length of 1.6 mm and a width of 0.8 mm.

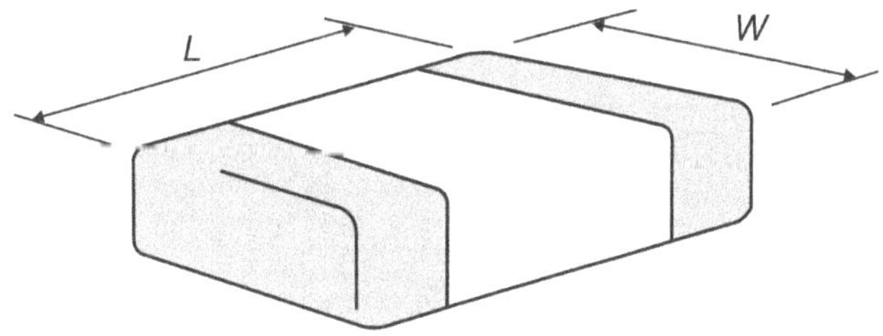

Figure 2.6 Dimensions of a chip-type component (*L*: length, *W*: width)

The impedance of such a capacitor depends on the frequency and the general equivalent circuit as shown in Figure 2.7. In that equivalent circuit, *C* represents the capacitance of a chip capacitor, *L* is the parasitic inductance that appears due to the multilayer structure, and *R* represents the loss in the dielectric material used in the capacitor. Therefore, from a structural point of view, as the size of the capacitor becomes smaller, inductance *L* generally becomes smaller. It should also be noted that the inductance generally tends to be constant regardless of the value of the capacitance when the size of the chip-type capacitor is equal.

C: capacitance of the capacitor
L: parasitic inductance (0.7 nH–1 nH)
R: represents the loss of the conductor plate

Figure 2.7 The equivalent circuit of a chip-type capacitor. *C* is the value of the capacitor and *L* occurs due to conductor patterns used to form the capacitor. *R* is due to the conductor and dielectric losses.

Figure 2.8 shows the impedances of chip-type capacitors with respect to frequency. Looking at the 100 pF curve in the figure, a frequency of about 100 MHz, the impedance decreases linearly with frequency. This is obvious because at low frequency the chip capacitor behaves as an ideal capacitor. Since it shows an impedance of about 30 Ω at 50 MHz, it can be seen to have a capacitance of

$$C = \frac{1}{2\pi f X_C} \cong \frac{1}{2\pi \times 50 \times 10^6 \times 30} = \frac{1000}{2\pi \times 1.5}\ \text{pF} \cong 106\ \text{pF}$$

The error in the calculated capacitance value is due to the approximate reading of the impedance value from the graph. From Figure 2.8, as the frequency increases, the impedance, after reaching a minimum, rises again. This can be understood from the equivalent circuit of the chip-type capacitor in Figure 2.7. While the impedance of capacitor *C* in Figure 2.7 is decreasing, that of inductor *L* is increasing with frequency. Thus, as the frequency increases, inductor *L* in Figure 2.7 becomes dominant, which explains why the impedance of the chip-type capacitors shown in Figure 2.8 increases with frequency.

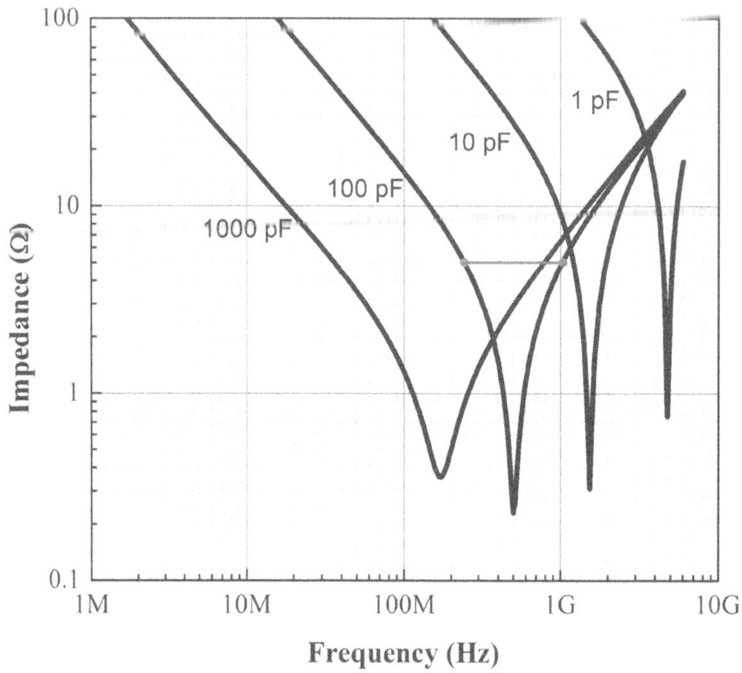

Figure 2.8 Impedance characteristics of GRM36 series chip-type capacitors (refer to the datasheet[1])

From the minimum point of the 100 pF curve in Figure 2.8, series resistance R in the equivalent circuit of Figure 2.7 is found to be approximately 0.2 Ω. Furthermore, at a frequency of 1 GHz, assuming that the impedance is mainly determined by an inductor, the approximate value of the inductor can be obtained from the curve. Since its impedance is approximately 5 Ω at 1 GHz, the approximate inductance value is found to be

$$L = \frac{X_L}{2\pi f} \cong \frac{5}{2\pi \times 1 \times 10^9} = \frac{5}{2\pi}\ \text{nH} \cong 0.8\ \text{nH}$$

More accurate values of the equivalent circuit can be determined by curve fitting, which is widely used to fit such data.

Thus, in the case of the 100 pF capacitor, when used at a frequency of over 1 GHz, it acts as an inductor rather than a capacitor. In order to use the 100 pF capacitor as a DC block, it must have, at most, an impedance below 5 Ω (which is estimated as 1/10 of a standard impedance value of 50 Ω). From the graph in Figure 2.8, it is possible to use the 100 pF capacitor as a DC block or a bypass capacitor in the frequency range of 300 MHz to 900 MHz. Its use as a DC block at a higher frequency poses a difficulty due to the effect of the parasitic inductor.

1. Murat Manufacturing Co. Ltd., Chip Monolithic Ceramic Capacitor, 1999.

2.3.2 Chip-Type Inductors

The manufacturing methods of chip-type inductors are somewhat more diverse when compared to those of chip-type capacitors. The structures of chip inductors vary, depending not only on their manufacturing methods but also on their values, DC current limits, and frequency coverage. Thus, the structures are not shown here and readers should refer to the datasheet. Basically, an inductor is formed by winding *enamel-coated copper wire* on either a ferrite core or dielectric materials. The enamel-wire winding is sometimes replaced by thick-film printing technologies. After the inductor is formed by winding, appropriate solder terminals for connections are made. More winding is possible with a wire of thinner diameter. The inductance can be significantly increased by increasing the number of windings. However, it should be noted that this will lead to increasing series resistance and decreasing current capacity. In addition, this will also lead to a proportional increase of parasitic capacitance. Therefore, the inductor with a larger inductance value usually cannot be used at high frequencies. Consequently, when inductors are employed in designing a circuit, the datasheet should be carefully consulted in order to determine the frequency range of operation. Since the inductor generally shows a parallel resonance due to parasitic capacitance, the inductor can be used as a resonator. However, the Q of the parallel resonance is generally low. Thus, when applied as a resonator to an oscillator, for example, a close examination of the datasheet for the Q value is required.

Figure 2.9 shows a typical equivalent circuit of a chip-type inductor where L represents the inductance arising from the winding, R represents the winding resistance, and C represents the sum of parasitic capacitances appearing between the windings.

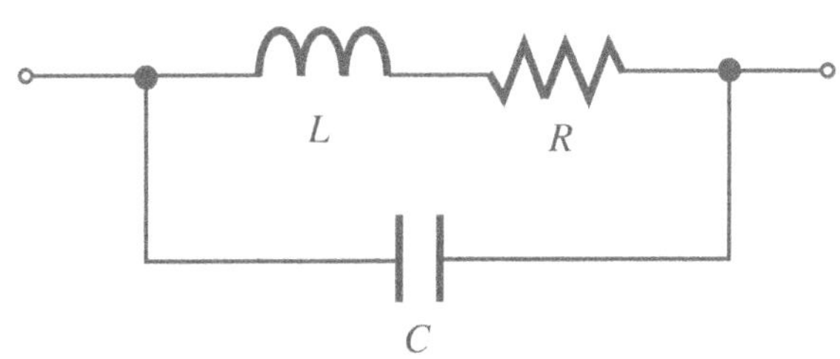

Figure 2.9 Equivalent circuit of an inductor. L is the value of inductor and C represents stray capacitance between the winding. R represents the winding-coil loss.

In Figure 2.9, at an extremely low frequency the chip-type inductor acts like a resistor R. As the frequency increases, the impedance of L in the figure becomes dominant and it acts as an inductor. As the frequency increases further, the impedance of C in the figure becomes smaller and then the inductor behaves as a capacitor. Thus, the approximate frequency range for the chip-type inductor in Figure 2.9 to be used as an inductor is shown in Equation (2.6).

$$\frac{R}{L} < \omega < \frac{1}{\sqrt{LC}} \tag{2.6}$$

Figure 2.10 shows the impedances of chip-type inductors with respect to frequency. In this figure, examining the 100 μH curve, it is found that the impedance increases linearly with frequencies up to 10 MHz. Thus, below 10 MHz, the chip-type inductor behaves as an

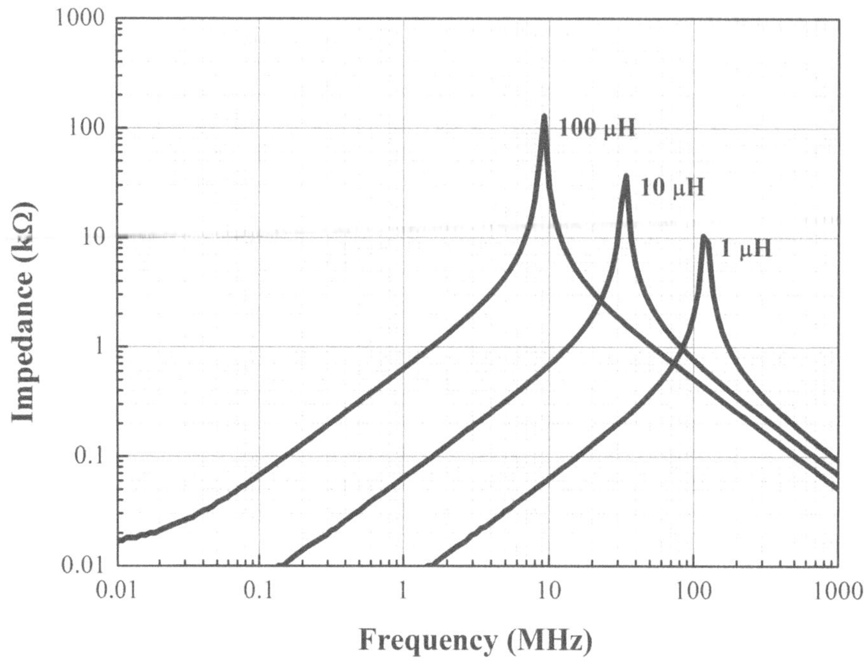

Figure 2.10 Impedance characteristics of IMC1210 series chip-type inductors (refer to the datasheet[2])

inductor. Since the impedance value at 1 MHz appears to be approximately 0.6 kΩ, the inductance value is computed to be

$$L = \frac{X_L}{2\pi f} \cong \frac{0.6 \times 10^3}{2\pi \times 1 \times 10^6} = \frac{0.6}{2\pi} \text{ mH} = 95.4 \text{ μH}$$

Due to the approximate reading of the impedance value from the graph, an error arises in the calculated inductance value. As the frequency increases, the impedance begins to fall after reaching a maximum point, which is due to the influence of the parasitic capacitor in the equivalent circuit in Figure 2.9. Thus, as the frequency becomes much higher, the inductor acts as a capacitor.

Assuming that at a frequency of 100 MHz the impedance is mainly due to a capacitor, then because its impedance is approximately 0.5 kΩ at this frequency, the value of the capacitor is computed to be

$$C = \frac{1}{2\pi f X_C} \cong \frac{1}{2\pi \times 0.5 \times 10^3 \times 100 \times 10^6} = \frac{10}{2\pi \times 0.5} \text{ pF} \cong 3.2 \text{ pF}$$

From Figure 2.10, the resistance at the maximum point (or parallel resonance) is 100 kΩ. The impedance value at this frequency becomes, approximately,

$$\frac{(\omega_o L)^2}{R} = 100 \text{ K}$$

2. Vishay Intertechnology, Inc., http://www.vishay.com/docs/34043/imc1210.pdf, 2012.

and the value of the resistor in Figure 2.10 is

$$R = \frac{(\omega_o L)^2}{100\ \text{k}} = \frac{\left(2\pi \times 10 \times 10^6 \times 100 \times 10^{-6}\right)^2}{100 \times 10^3} = 4\pi^2 \times 10 \cong 390\ \Omega$$

2.3.3 Chip-Type Resistors

The structure of a chip-type resistor is shown in Figure 2.11(a), where the resistor is manufactured by printing a resistive material (RuO_2) on a ceramic substrate. In addition, the terminals for connection are formed using a similar thick-film printed conductor pattern as shown in the figure. The thick-film printed terminals are plated to make soldering possible. Furthermore, in order to prevent oxidation or damage to the resistive material, a glassy coating is added to the resistive material in a post-processing application.

A similar classification to that of capacitors (1005, 1608, and 2010) is also used in the classification of chip resistors. As the size of chip-type resistors gets smaller, similar to a capacitor the impedance characteristics generally become more ideal, that is, they can be applied to higher frequencies. Another factor to note is power consumption. The power consumption that a chip-type resistor can withstand is usually listed in the manufacturer's datasheet. The power consumption capability becomes generally smaller as the size of the resistor shrinks.

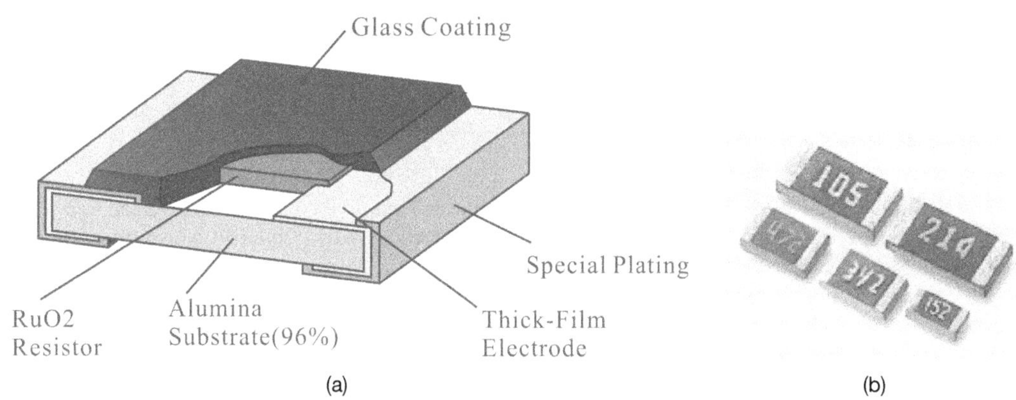

Figure 2.11 Chip-type resistor; (a) structure and (b) photo. A chip-type resistor is formed by printing the resistive material using thick-film technology. A glass coating is applied for protection. The resistor's value is sometimes marked on the glass coating. Two digits represent the value of the resistor and the remaining digits represent the exponent. Thus, the three digits 342 stand for 3.4 kΩ.

Depending on the manufacturer, the value of the resistor is sometimes marked on its surface, as shown in Figure 2.11(b), which makes the identification of that value much easier. In Figure 2.11(b), following a general notation, the first two digits represent the value of the resistance and the remaining digits represent the exponent. Thus, a 3.4 kΩ resistor is denoted as

$$342 = 34 \times 10^2 = 3.4 \times 10^3 = 3.4\ \text{k}$$

The impedance characteristic with respect to frequency and the equivalent circuit of a chip-type resistor are not generally known. The method explained in the following section can be used for measuring the unknown impedance characteristic of a chip-type resistor. Using the measured impedance characteristic, the equivalent circuit of chip-type resistor may be found to be similar to chip-type capacitors and inductors.

2.4 IMPEDANCE MEASUREMENTS

Chip-type passive components can all be considered as one-port components. Thus, using an *impedance analyzer* or a *network analyzer*, their impedance characteristics can be measured. Here, the method of measuring the impedance characteristics using the widely available network analyzer is presented. It should be noted that this method may be accompanied by significant errors when used with high frequencies.

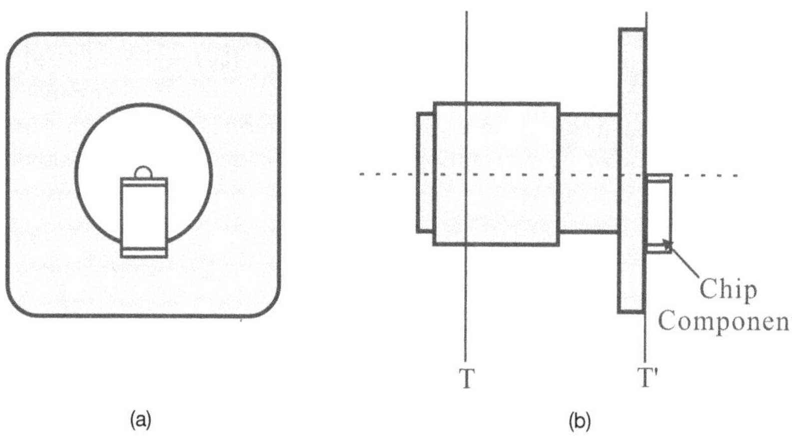

Figure 2.12 Assembly for the measurement of a passive component; (a) back and (b) side views. Plane T is the conventional reference plane that appears after the completion of the network-analyzer calibration. The reference plane T can be moved to a new reference plane T′ using electrical delay or port extension utilities.

First, in order to perform measurements using the network analyzer, a coaxial *small miniature assembly* (SMA) connector must be prepared and its soldering tab (if present) should be removed, as shown in Figure 2.12. Next, the network analyzer should be calibrated with the one-port calibration procedure that is installed in the network analyzer. The calibrated reference plane appears at plane T in Figure 2.12. The calibrated reference plane T means that the measured impedance using the network analyzer includes all the effects of components connected after plane T. Here, for the time being, the calibration can be understood as the elimination of the non-ideal characteristics of the network analyzer and the movement of the measurement plane to plane T. In practical measurement, cables and adapters can be included for connecting the network analyzer to the assembly shown in Figure 2.12. Also, the network analyzer itself has the non-ideal characteristics. Without the calibration procedure, both the characteristics of cables and adapters and the non-ideal characteristics of the network analyzer are included in the measured impedance. Both of these effects can be removed through the calibration. After the calibration, the impedance can be correctly measured without the effects of cables and adapters and the non-ideal characteristics of the network analyzer. After completing the one-port calibration, the calibrated reference plane appears normally at plane T of the connector.

The measured results at the calibrated plane T obviously include the length effect of the SMA connector, but the correct impedance of the passive component is defined at plane T′. Therefore, the length effect of the SMA connector should be removed from the measured impedance, which can usually be done by using the *electrical delay* or *port extension* function installed in the network analyzer. Since the measured impedance at plane T′ in Figure 2.12 without the chip component should be ∞, it should be corrected to be open through the adjustment of the electrical delay. Now, the calibrated reference plane is moved to T′. The measured impedance based on this method may include some errors because such an open is not an ideal open.

Next, the passive *device under test* (DUT) is connected by soldering the chip component to a connector, as shown in Figure 2.12(b). The measured impedance at plane T′ corresponds to the impedance value of the chip component. Figure 2.13 shows an example of the measured impedance following the method presented. The DUT is a zero-biased varactor diode. The series resistance of the varactor diode is found to be approximately 0.5 Ω. At low frequency, the measured reactance is negative and this implies that the varactor diode acts as a capacitor. However, as frequency further increases, the diode behaves as an inductor. Thus, the varactor diode can be modeled by the series resonant equivalent circuit in Figure 2.7. The values of R, L, and C can be determined through optimization.

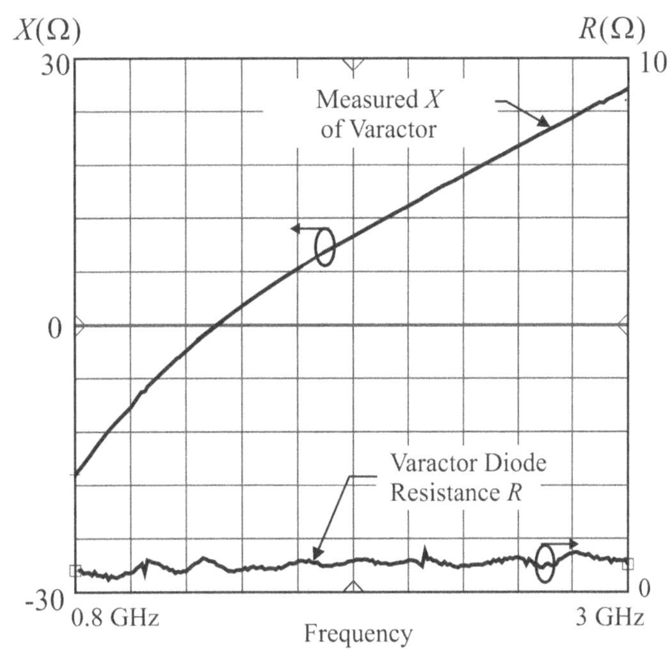

Figure 2.13 Measured impedance for a chip varactor diode using the method presented. The varactor diode can be modeled by a *RLC* series resonant circuit with a resonance frequency of about 1.4 GHz.

The limitations of the method presented will primarily be in the connector. Typical SMA connectors can be used up to 18 GHz. Thus, the measurement of impedance characteristics is possible only below 18 GHz using an SMA connector. However, even the use of other high-frequency coaxial connectors (2.9 mm or 2.4 mm coaxial connector, to be described later) instead of the SMA does not improve the accuracy of the measurements at a higher frequency. The reason for this is because an open circuit created by the open coaxial connector is not truly open. In practice, the open end of the connector shows the fringing capacitance and the radiation resistance. In addition, no matter how these factors are corrected, when a device is attached, the field shape of the coaxial connector is distorted, which is the reason for the difficulty in obtaining accurate impedance characteristics for a device at a higher frequency.

Example 2.2

Open the Murata capacitor library in ADS. After computing the 10 pF impedance by simulation, obtain its equivalent circuit.

Solution

Figure 2E.2 shows the simulation schematic for computing the impedance of a 10 pF capacitor. In the schematic of Figure 2E.2, **Vout** becomes the impedance since the AC current source is set to 1 A. Plotting the real and imaginary parts of **Vout** separately, the graphs in Figure 2E.3 are obtained. From Figure 2E.3(a), we can obtain $R = 0.17\ \Omega$.

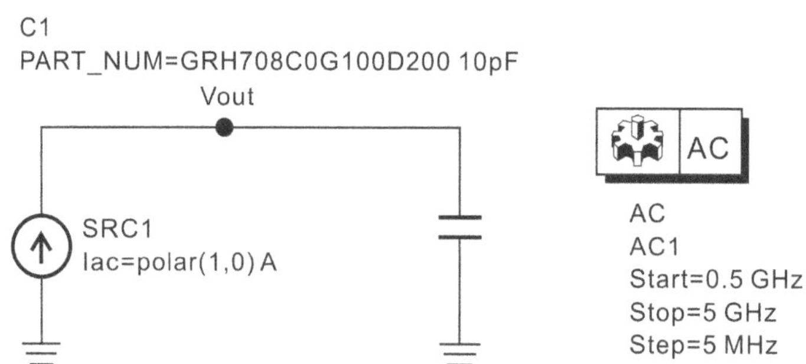

Figure 2E.2 Simulation schematic for the Murata 10pF capacitor impedance. Since the AC current source is set to 1 A, the voltage **Vout** becomes the impedance of the Murata 10 pF capacitor.

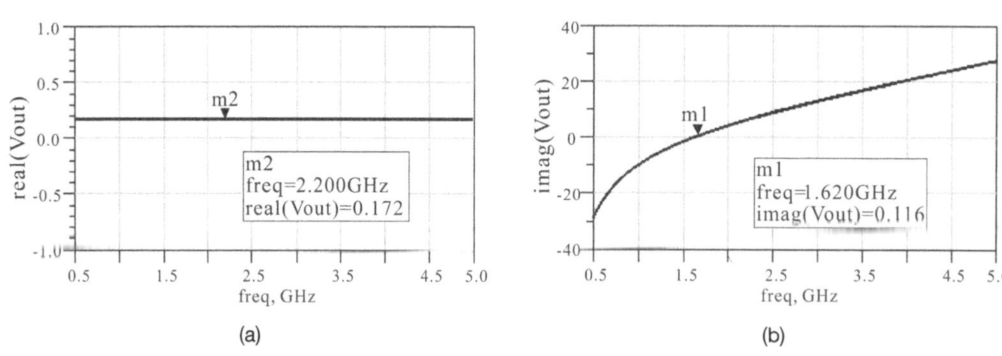

Figure 2E.3 (a) Real and (b) imaginary parts of the impedance. The real part is constant for frequency change and the imaginary part shows a series resonant circuit with a resonance frequency of about 1.62 GHz. Thus, the Murata 10 pF capacitor can be modeled as a series *RLC* circuit.

From Figure 2E.3(b), the imaginary part of the impedance shows a series resonance. The slope near the resonant frequency becomes

$$\left.\frac{\partial X}{\partial f}\right|_{f_o} = 2\pi\left.\frac{\partial X}{\partial \omega}\right|_{f_o} = 2\pi\frac{\partial}{\partial \omega}\left.\left(\omega L - \frac{1}{\omega C}\right)\right|_{f_o} = 4\pi L$$

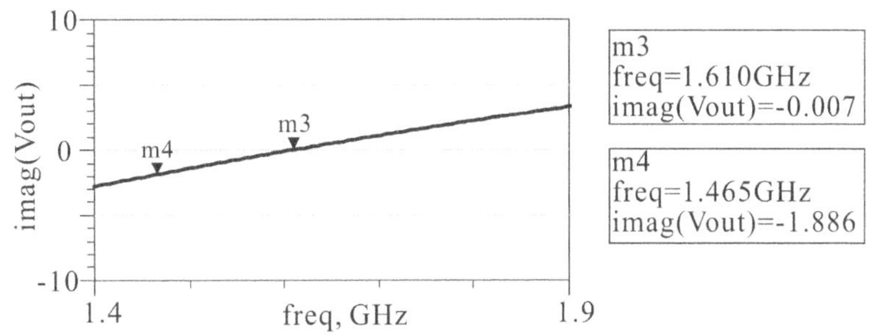

Figure 2E.4 The imaginary part of the impedance in Figure 2E.3(b) is redrawn around the resonant frequency.

This means the inductance of the series resonant circuit can be obtained using the slope of the reactance at the resonance frequency. A plot of the imaginary part near the resonant frequency is presented in Figure 2E.4. To compute the slope, markers are inserted in the plot. The equations to calculate the slope at the resonance frequency using the marker values are shown in Measurement Expression 2E.1, where **m**3 and **m**4 represent the y values (vertical) of the markers. In addition, **indep**(**m**3) and **indep**(**m**4) are the x (horizontal) values of the markers. Thus, the first equation in Measurement Expression 2E.1 becomes the slope divided by 4π, which corresponds to the inductance calculated from $L=\partial X/\partial f/(4\pi)$. Using the determined value of L and the resonant frequency, the capacitance can be calculated from the relation $C = 1/(\omega_o)^2 L$.

Eqn L=(m3-m4)/(indep(m3)-indep(m4))/(4*pi)

Eqn C=1/((2*pi*indep(m3))**2*L)

Eqn X=2*pi*freq*L-1/(2*pi*freq*C)

Measurement Expression 2E.1 Equations for the calculation of the equivalent circuit values in the display window

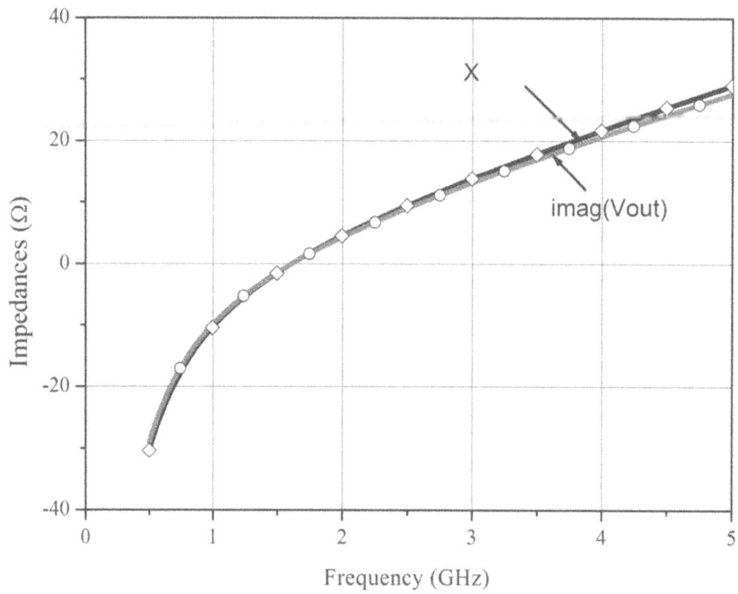

Figure 2E.5 Comparison of the two impedances obtained from calculation and from simulation. The imag(**Vout**) is the imaginary part of the impedance of the Murata 10 pF capacitor shown in Figure 2E.3(b) and **X** represents the computed impedance of the series *LC* circuit for frequency. The values of *LC* are computed using equations in Measurement Expression 2E.1.

The computed values are $R = 0.17\ \Omega$, $L = 0.977$ nH, and $C = 10$ pF. Thus, the equivalent circuit of the 10 pF Murata chip capacitor is obtained. Finally, the last equation represents the reactance **X** calculated from the obtained values of *L* and *C*. The reactance **X** gives the verification of the computed *L* and *C* values. If the computed *L* and *C* values are close enough to fit the imaginary part, **X** will show good agreement. The comparison is shown in Figure 2E.5, where a very good agreement can be seen. Therefore, it can be found that the equivalent circuit produces the reactance close to the imaginary part obtained from the simulation.

2.5 SUMMARY

- Passive devices are largely classified based on their fabrication method and are classified into lead-type and chip type components, and pattern-type passive components. Chip-type components are widely used in hybrid microwave integrated circuits.
- The extraction of the equivalent circuit of a passive component from the data can be achieved using the impedance plot. The commercial passive component shows a combination of the impedance plot of the ideal resistor, inductor, and capacitor.
- The impedance of a passive component can be measured using an impedance analyzer or network analyzer. In this section, we showed how to obtain the equivalent circuit from the measured data.

REFERENCES

1. L. Young, ed., *Advances in Microwaves*, vol. 7. New York and London: Academic Press, 1971.
2. GEC-Marconi, *GaAs IC Foundry Design Manual,* Chelsmford, UK: GEC-Marconi Company, October 1997.
3. I. D. Robertson, ed., *MMIC Design,* London: The Institution of Electrical Engineers, 1995.

PROBLEMS

2.1 Company A fabricates a thin-film resistor whose sheet resistivity is 50 Ω/square; in the case of a company B, an identical thin-film resistor has a sheet resistivity of 100 Ω/square. What is the difference between the processes of the two companies? Further, given that the material's volume resistivity is ρ and its thickness is t, find its sheet resistivity.

2.2 Given that the sheet resistivity is 50 Ω/square and we want to design a 100-Ω resistor, if the current flowing in this resistor is 2 mA, find its minimum width. Its rated current per width is 0.5 mA/μm.

2.3 The dielectric material of an MIM capacitor is silicon nitride and its dielectric constant is 7.2. If the area is 50 μm^2, find the thickness of a 0.53 pF capacitor. In addition, find the capacitance per unit area in (F/μm^2).

2.4 In the equivalent circuit of an inductor such as that in Figure 2.9, show that the approximate magnitude of the impedance at resonance is

$$\left|Z_{max}\right| = \frac{\left(\omega_o L\right)^2}{R}$$

$$\omega_o = 1/\sqrt{LC}$$

2.5 In this chapter, we have covered the method for extracting the equivalent circuit of a capacitor from $|Z(\omega)|$ characteristics obtained from an impedance analyzer such as that shown in Figure 2.8. The impedance of the equivalent circuit of a capacitor is

$$Z(\omega) = R + jX = R + j\left(\omega L - \frac{1}{\omega C}\right)$$

Using this equation, the real and imaginary parts for a frequency can be obtained by graphical representation. With that in mind, show that the approximate impedance near the resonant frequency ω_o is

$$Z(\omega) \cong R + j2L\left(\omega - \omega_o\right)$$

2.6 Similar to problem 2.5 above, for a parallel resonant circuit, show that the admittance near the resonant frequency $Y(\omega)$ is given by

$$Y(\omega) \cong G + j2C(\omega - \omega_o)$$

2.7 **(ADS Problem)** In Example 2.2, the equivalent circuit of the 10 pF capacitor from Murata is obtained using the slope of the reactance versus the frequency and resonance frequency such that

$$\left.\frac{\partial X}{\partial f}\right|_{f_o} = 4\pi L \text{ and } \omega_o = 1/\sqrt{LC}$$

Alternatively, the equivalent circuit can be obtained using the impedance plot in the log-log scale. Compute the equivalent circuit values using ADS.

CHAPTER

3

Transmission Lines

CHAPTER OUTLINE

3.1 INTRODUCTION

Figure 3.1 shows an example of a high-frequency circuit fabricated on a 100-μm-thick GaAs substrate. In the circuit (a three-stage 28 GHz GaAs pHEMT LNA), the voltages and currents in the line shown in the figure generally are functions of position at a high frequency. The pair of voltages (V_1 and V_2) and currents (I_1 and I_2) are simply considered to be equal at a low frequency, as the line acts as a simple connection. However, the pair of voltages and currents at these two points becomes unequal as the frequency increases. The shape of the lines in the circuit appears too complex. As a first approximation to treat and understand the lines' behavior, we will therefore consider an isolated straight line (or transmission line) filled with homogenous media along the direction of propagation. The principles and applications of uniform transmission lines will be presented in this chapter.

As explained in Chapter 2, the characteristics of passive devices are represented by their impedance. As in the case of a resistor, for example, it can be characterized by the parameter of resistance. Similarly, in the case of transmission lines, there are two parameters required for their characterization. We first discuss these two parameters. Then, we will describe a qualitative investigation of the transmission lines most widely used at high

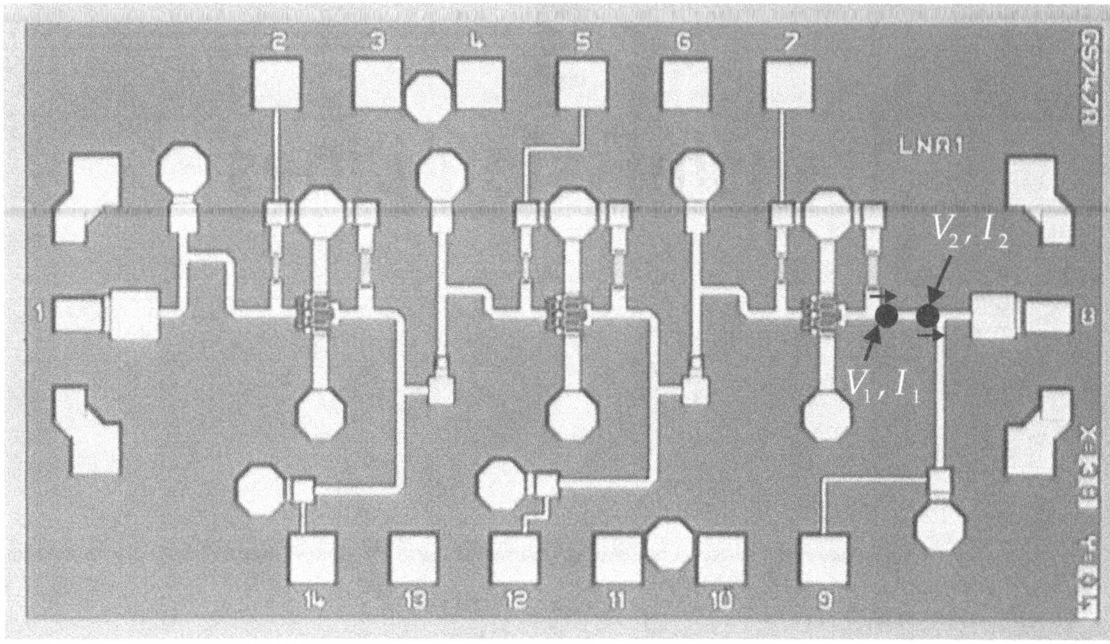

Figure 3.1 An example of a high-frequency circuit. The circuit is a three-stage 28 GHz GaAs pHEMT LNA. The V_1 and I_1 and V_2 and I_2 at different positions are not equal, although there is no discontinuity between them.

frequencies: coaxial and microstrip. In addition, when a transmission line is terminated by a passive device, incident and reflected signals appear in the transmission line. We will therefore learn about the defining parameters in such a situation, namely reflection coefficient and return loss.

The applications of transmission lines to circuits are various, and examples of typical applications will also be explored in this chapter. These examples include short-length transmission lines, integer multiples of quarter-wavelength transmission lines, and two-port circuit applications of transmission lines. Furthermore, in practice, when one end of a transmission line is open, or a number of transmission lines are connected at a point, many undesirable discontinuities arise and we will therefore consider their effects and how to treat them.

3.2 PARAMETERS

3.2.1 Phase Velocity

Figure 3.2(a) shows a uniform transmission line aligned along the positive z-direction. Since the current and voltage are functions of position z and time t, they are expressed as $i(z,t)$, and $v(z,t)$, respectively. The equivalent circuit of an infinitesimal section of the transmission line of length Δz from position z is shown in Figure 3.2(b). The magnetic field arising due to the flow of current in the conductors is represented as an inductance, while the electric field arising between the lines is represented as a capacitance. Since the transmission line is uniform in the propagation direction, the transmission line of length Δz can be

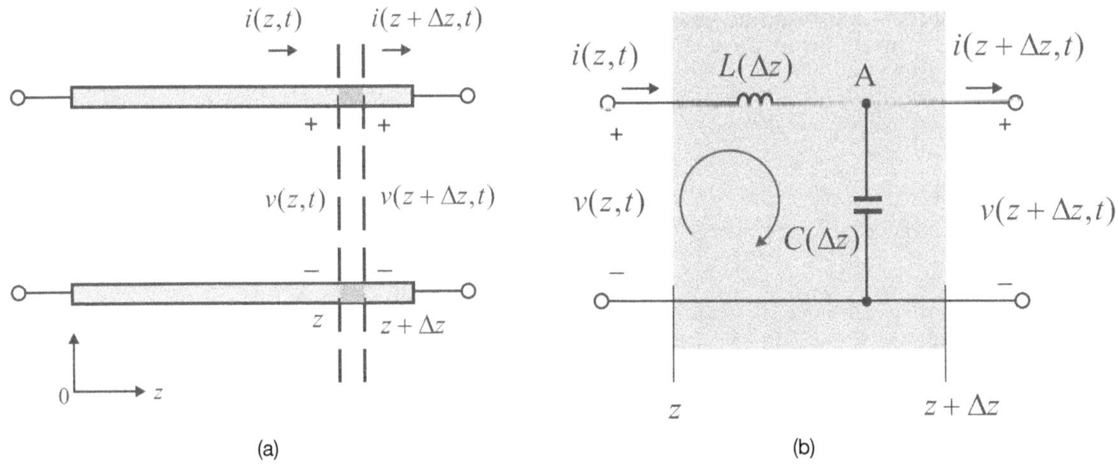

Figure 3.2 (a) Transmission line and (b) its lumped equivalent circuit for the infinitesimal length Δz. Various forms of lumped equivalent circuits for the infinitesimal length Δz are possible, but they yield the same transmission-line equations.

represented by a lumped equivalent circuit using the following unit-length inductance and capacitance respectively:

L : Inductance per unit length [H/m]

C : Capacitance per unit length [F/m]

Then, $L\Delta z$ and $C\Delta z$ become the corresponding inductance and capacitance for the transmission line of infinitesimal length Δz. Generally, the conductors have losses as well as the dielectric material that fills the conductors. These losses are also represented using the resistance R and conductance G per unit length. The resistance R represents the conductor loss per unit length, while conductance G represents the dielectric loss per unit length. Similarly, the resistance and conductance for a line of length Δz becomes $R\Delta z$ and $G\Delta z$.

Here, we assume the transmission line has no loss ($R = G = 0$). However, the results including loss can be similarly obtained. Applying KVL to the circuit in Figure 3.2(b) in the direction indicated by the arrow, we obtain Equation (3.1).

$$v(z,t) = L\Delta z\frac{\partial i(z,t)}{\partial t} + v(z + \Delta z,t) \tag{3.1}$$

Thus,

$$\Delta v = v(z + \Delta z,t) - v(z,t) = -L\Delta z\frac{\partial i(z,t)}{\partial t}$$

Dividing both sides of this equation by Δz, we have

$$\frac{\partial v(z,t)}{\partial z} = -L\frac{\partial i(z,t)}{\partial t} \tag{3.2}$$

Equation (3.2) implies that *voltage v(z,t) decreases due to the voltage drop by the inductor* as the voltage propagates in the z-direction. Similarly, the result of applying KCL at node A in Figure 3.2(b) is

$$\Delta i = -C(\Delta z)\frac{\partial v(z,t)}{\partial t}$$

Similarly, by dividing both sides of this equation by Δz, we obtain

$$\frac{\partial i(z,t)}{\partial z} = -C\frac{\partial v(z,t)}{\partial t} \tag{3.3}$$

Again, Equation (3.3) implies that the *current i(z,t) decreases due to the leakage current through the capacitor* as the current propagates in the z-direction. Equations (3.2) and (3.3) are coupled partial differential equations for the voltage and current. Differentiating Equation (3.2) with respect to z and using Equation (3.3), the result is

$$\frac{\partial^2 v(z,t)}{\partial z^2} = LC\frac{\partial^2 v(z,t)}{\partial t^2} = \frac{1}{v_p^2}\frac{\partial^2 v(z,t)}{\partial t^2} \tag{3.4}$$

This is a one-dimensional *wave equation* having a phase velocity of v_p and the solution $v(z,t)$ of Equation (3.4) is a voltage wave propagating in the +z and −z directions. Similarly, for the current we obtain Equation (3.5).

$$\frac{\partial^2 i(z,t)}{\partial z^2} = LC\frac{\partial^2 i(z,t)}{\partial t^2} = \frac{1}{v_p^2}\frac{\partial^2 i(z,t)}{\partial t^2} \tag{3.5}$$

In conclusion, both the voltage and current have a phase velocity of v_p and the equations represent a wave propagating in the +z and −z directions.

$$\text{Phase velocity, } v_p = \frac{1}{\sqrt{LC}} \tag{3.6}$$

The general solution of Equation (3.4) is

$$v(z,t) = C_1 f\left(t - \frac{z}{v_p}\right) + C_2 g\left(t + \frac{z}{v_p}\right) \tag{3.7}$$

where C_1 and C_2 are constants and $f(t)$ and $g(t)$ represent arbitrary waveforms. These $f(t)$ and $g(t)$ waveforms will be determined by the boundary conditions at the point of $z = 0$, where the voltage source is connected. The left part of Equation (3.7) represents a wave traveling in the +z direction and the right part represents a wave traveling in the −z direction.

Let us first consider only the wave traveling in the + z direction. The voltage at $z = 0$ is

$$v(0,t) = C_1 f(t)$$

while the wave at an arbitrary position z becomes

$$v(z,t) = C_1 f\left(t - \frac{z}{v_p}\right)$$

Let $C_1 = 1$. Setting $C_1 = 1$ does not lose the generality. Here, $f(t)$ can be considered to be an external applied voltage to the transmission line. Two duration-limited waveforms, $v(0,t)$ and $v(z,t)$, are shown in Figure 3.3 with $C_1 = 1$. Since the value of $f(t - t_1)$ at $t = t_1$ is equal to the value of $f(t)$ at $t = 0$, $f(t)$ at $t = 0$ appears at a delayed time $t = t_1$ in $f(t - t_1)$. If similar logic is applied to a time other than $t = 0$, then $f(t - t_1)$ represents a function that $f(t)$ is shifted by t_1 in the forward direction along the time axis. Thus, the change at $z = 0$ with time is transferred to a point z with a delay of t_1. As another interpretation of this delay t_1, when the time-varying waveform at the point $z = 0$ is transmitted, the waveform arrives at a distance of z with a time elapse of $t_1 = z/v_p$ because the phase velocity is v_p. Thus, the voltage at a distance z appears with a time delay of $t_1 = z/v_p$. Note that the voltage propagates without distortion and damping in the lossless transmission line.

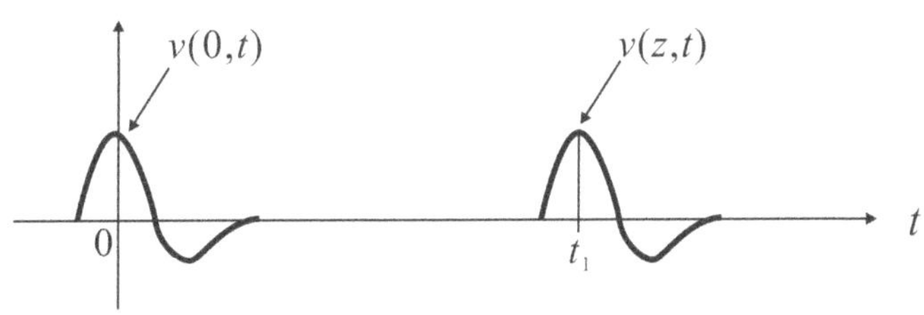

Figure 3.3 Solution of the wave equation. The observer at $z = 0$ observes the externally applied voltage waveform $v(0,t)$ at $z = 0$, while the observer located at z observes $v(z,t)$. The waveform $v(z,t)$ is delayed in time by $t_1 = z/v_p$.

Example 3.1

The voltages $\delta(t)$, $\cos(\omega t)$, and $e^{j\omega t}$ are applied to a transmission line at $z = 0$, respectively, that extends from position $z = 0$ to infinity. Determine the waveform at a distance z.

Solutions

(1) When $v(0,t) = \delta(t)$, at a distance z, the voltage is

$$v(z,t) = \delta\left(t - \frac{z}{v_p}\right)$$

As shown in Figure 3E.1, this represents $v(0,t)$ shifted by $t_1 = z/v_p$ when observed from the time axis.

(2) When $v(0,t) = \cos(\omega t)$, the waveform at a distance z is

$$v(z,t) = \cos\left\{\omega\left(t - \frac{z}{v_p}\right)\right\} = \cos\left(\omega t - \omega\frac{z}{v_p}\right) = \cos(\omega t - \beta z)$$

Comparing the voltage $\cos(\omega t)$ at a distance $z = 0$, $\cos(\omega t - \beta z)$ has a phase delay of βz. Note that the phase delay per unit length is β, and β is called the *propagation constant*.

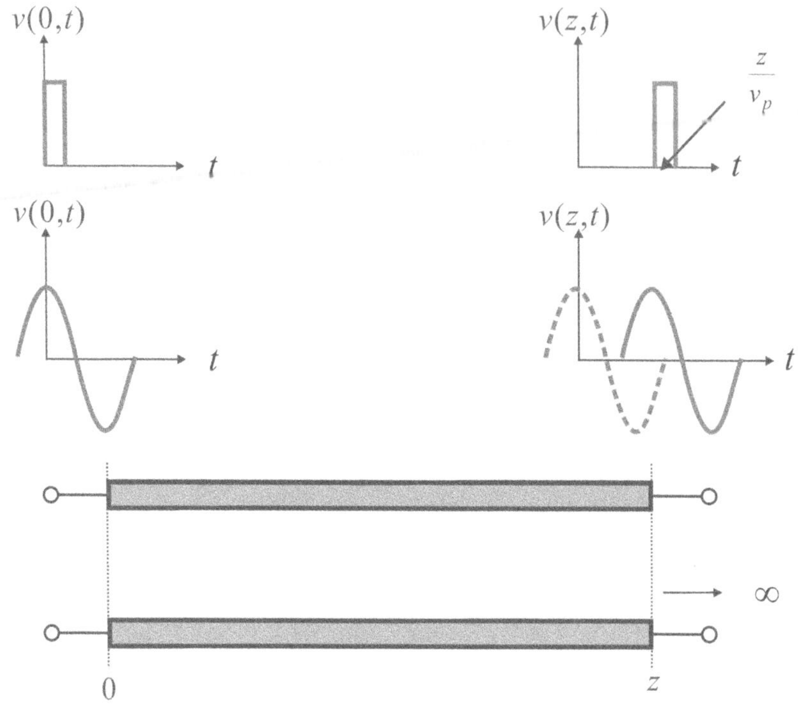

Figure 3E.1 Waveform propagation in a transmission line. The pulse is time delayed, while the phase delay is observed in the case of a sinusoidal voltage input.

(3) When $v(0,t) = e^{j\omega t}$, the waveform at the distance z is

$$v(z,t) = e^{j(\omega t - \beta z)} = e^{j\omega t} e^{-j\beta z}$$

The complex waveform $e^{j\omega t}$ in Example 3.1 is often used to determine sinusoidal response. When the wave $v(z,t)$ is expressed using a *phasor* $V(z)$, it becomes

$$V(z) = e^{-j\beta z}$$

because the rest of the complex waveform without the $e^{j\omega t}$ is defined as the phasor. $V(z)$ represents the phase delay at a distance z. When the time-domain waveform is required from the phasor, the waveform can be found by multiplying the phasor by $e^{j\omega t}$ and taking the real part of the result as

$$v(z,t) = \mathrm{Re}\left(V(z)e^{j\omega t}\right) = \cos(\omega t - \beta z)$$

3.2.2 Wavelength

As we have seen previously, the time delay in a transmission line depends on both phase velocity and the length of the transmission line. The time delay t_d through the transmission line is expressed as

$$t_d = \frac{l}{v_p} \tag{3.8}$$

Note that given two physically different transmission lines of different lengths, they may give the same time delay but have different physical parameters. Thus, in practice, when indicating the length of a transmission line, it can be specified by the time delay t_d given in Equation (3.8). Alternatively, when both the length and phase velocity are simultaneously specified, the effect of the length of the transmission line (in terms of time delay) can be determined.

As another expression for the length effect of a transmission line, wavelength is used. When a sinusoidal source is applied at the input (for example, a voltage of $\cos(\omega t)$ applied at the point $z = 0$), the spatial waveform at a fixed time t_1 is shown in Equation (3.9).

$$v(z,t) = \cos(\omega t_1 - \beta z) = \cos(\beta z + \phi) \tag{3.9}$$

Its graph is shown in Figure 3.4.

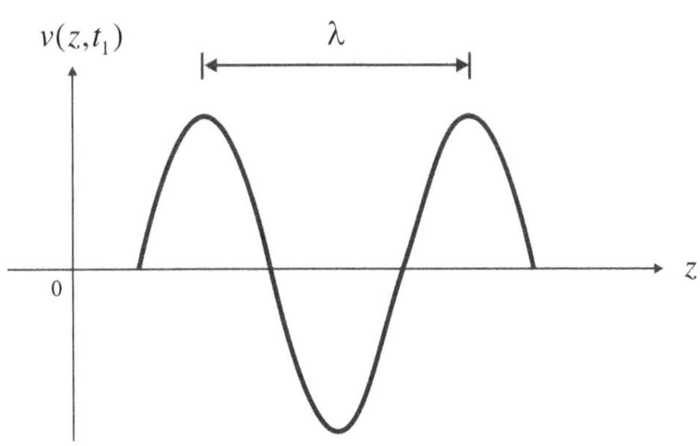

Figure 3.4 Definition of wavelength

A spatial sinusoidal waveform appears along the transmission line. The period of the sine wave occurring in space is called wavelength λ, and is defined as the distance between two successive *troughs* or the distance between two successive *crests* of the waveform. This is expressed mathematically in Equation (3.10)

$$\lambda = \frac{2\pi}{\beta} = \frac{2\pi v_p}{\omega} \tag{3.10}$$

and the relationship $v_p = f\lambda$ results. In addition, the wavelength can be seen to vary with frequency. Therefore, another way to express the length of the transmission line is to represent it by the frequency and corresponding number of wavelengths.

Another way to express the length of the transmission line at a given frequency is *electrical length θ*. The electrical length is expressed as the phase delay of a transmission line at a given frequency. Thus,

$$\theta = \beta l = \frac{\omega l}{v_p} \tag{3.11}$$

Usually expressed in degrees, one wavelength corresponds to 360° from Equation (3.11), half a wavelength is 180°, and a quarter wavelength becomes 90°. In summary, to represent the length of a transmission line, three conventions are used that are expressed in terms of the following parameters: (length, phase velocity), (wavelength, frequency), and (electrical length, frequency). Figure 3.5 shows the methods for expressing the length of the transmission line.

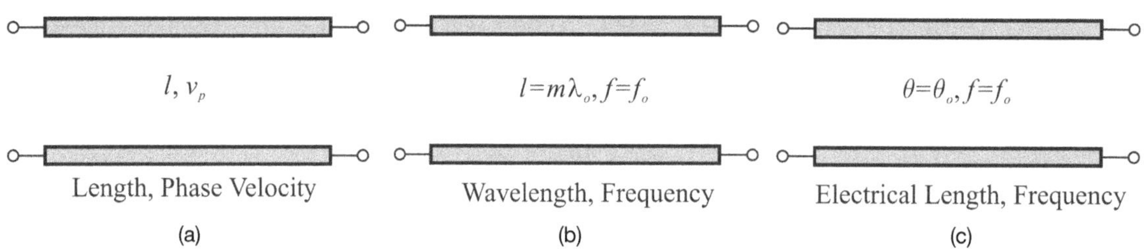

Figure 3.5 Methods for expressing the length of a transmission line. In (a), the time delay of the transmission line is specified with length and phase velocity. In (b), the length is specified based on the wavelength and the frequency. In (c), the length is specified using the phase delay, which is equal to the electrical length of the transmission line at a given frequency.

Example 3.2

For a particular transmission line, the phase velocity is equal to the speed of light. What is the wavelength at a frequency of 1 GHz? In that transmission line, if the time delay due to the length of the transmission line is 1 nsec, determine the corresponding length of the transmission line using the wavelength at a frequency of 1 GHz.

Solutions

From Equation (3.10),

$$\lambda_0 = \frac{c}{f} = \frac{3\times 10^8}{1\times 10^9}\,[m] = 0.3 \text{ m} = 30 \text{ cm}$$

and

$$t_d = \frac{l}{v_p} = \frac{l}{c} \Rightarrow l = ct_d = 3\times 10^8 \cdot 1\times 10^{-9} = 0.3 \text{ m}$$

In addition, because $\lambda_o = 0.3$ m at a frequency of 1 GHz, it can be seen from the result above that

$$l = \lambda_o$$

Example 3.3

Figure 3E.2 shows an ideal transmission line in the ADS schematic window. The **E** in the figure represents the electrical length explained earlier.

(1) If the wavelength at 1 GHz is denoted as λ_o, what is the length of the transmission line?
(2) What is the electrical length **E** at a frequency of 3 GHz?

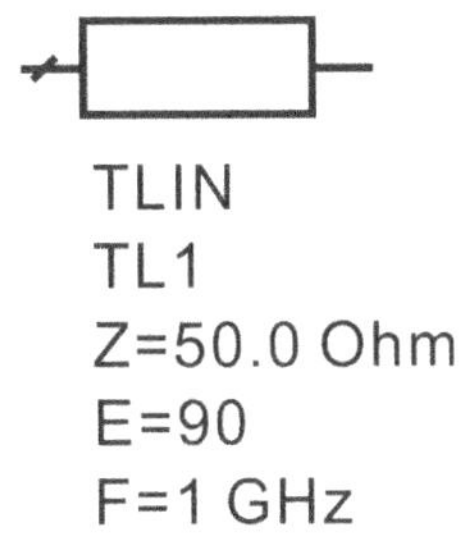

Figure 3E.2 An ideal transmission line component in ADS

Solutions

(1) Since

$$\theta = \beta l = \frac{2\pi l}{\lambda_o} = \frac{\pi}{2}$$

then the length is found to be $l = 0.25\ \lambda_o$.

(2) The electrical length is

$$\theta = \beta l = \frac{\omega l}{v_p}$$

Now, keeping the phase velocity and length fixed, the electrical length is proportional to the frequency. Thus, at 3 GHz, **E** = 270°.

3.2.3 Characteristic Impedance

As discussed earlier, both the current and voltage waves travel with the same phase velocity and there are three different ways to specify the time delay that is due to the length of the transmission line. It should be noted that the traveling current and voltage waveforms are not independent. They have the same waveforms and can be related by a constant of proportionality. The constant is defined as the *characteristic impedance* of the transmission line or simply *impedance* for short. In order to examine this relationship, we assume that there exist only the current and voltage waves traveling in the positive z direction, and there are no waves in the opposite direction. Then, the waveforms can be written as

$$v(z,t) = v^+\left(t - \frac{z}{v_p}\right) \tag{3.12a}$$

$$i(z,t) = i^+\left(t - \frac{z}{v_p}\right) \tag{3.12b}$$

Now, substituting Equations (3.12a) and (3.12b) into Equation (3.2),

$$\frac{\partial v(z,t)}{\partial z} = -\frac{1}{v_p}\frac{dv^+}{du} = -L\frac{di^+}{du}$$

where $u = t - z/v_p$. Thus, we obtain

$$\frac{d}{du}\left(v^+(u) - v_p L i^+(u)\right) = 0 \tag{3.13}$$

For Equation (3.13) to be 0, then the variables in the parentheses should have no dependence on u. Again, if the equation has to be 0, both $v^+(u)$ and $i^+(u)$ must have the same waveform, which means v_pL is then a constant of proportionality. The same result will be obtained using Equation (3.3). Therefore, the current and voltage waves traveling on the transmission line have the same waveforms and are related to each other by the constant of proportionality that has the same dimension as resistance and is called the *characteristic impedance* of the transmission line, as shown in Equation (3.14).

$$Z_o = v_p L = \frac{1}{v_p C} = \sqrt{\frac{L}{C}} \tag{3.14}$$

The same result can be obtained by considering the wave traveling in the $-z$ direction. However, a negative constant of proportionality appears between the voltage and the current waves. The voltage has a plus sign regardless of the direction of propagation, which is measured with respect to a ground line. But, in the case of the current, it becomes negative when the direction is changed. As a result, the voltages and current waves in a transmission line are respectively expressed as Equations (3.15) and (3.16).

$$v(z,t) = v^+\left(t - \frac{z}{v_p}\right) + v^-\left(t + \frac{z}{v_p}\right) \tag{3.15}$$

$$i(z,t) = \frac{1}{Z_o}\left(v^+\left(t - \frac{z}{v_p}\right) - v^-\left(t + \frac{z}{v_p}\right)\right) \tag{3.16}$$

To summarize the previous results, a transmission line is characterized by two parameters: the characteristic impedance defines the relationship between the current and voltage waves in a transmission line, and the time delay is specified by the parameters of length and phase velocity. These transmission line characterizations are shown in Figure 3.6.

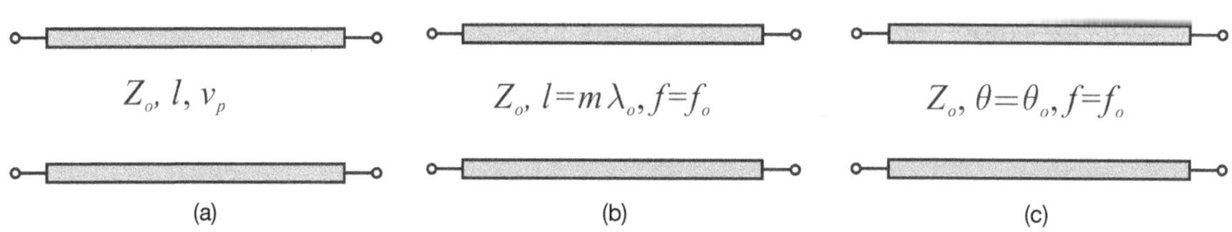

Figure 3.6 Ways of specifying the transmission-line parameters by (a) Z_o, l, v_p; (b) Z_o and the multiple of wavelength; and (c) Z_o and the electrical length. The specifications of (b) and (c) require frequency.

3.2.4 Measurements

The characteristic impedance and phase velocity of a lossless transmission line, expressed in terms of capacitance per unit length and inductance per unit length, are given by Equations (3.17) and (3.18).

$$Z_o = \sqrt{\frac{L}{C}} \tag{3.17}$$

$$v_p = \frac{1}{\sqrt{LC}} \tag{3.18}$$

Thus, once the capacitance per unit length and the inductance per unit length have been determined or obtained by measurement, the transmission line parameters, such as characteristic impedance and phase velocity, can be obtained. Structurally, there are various types of transmission lines that include microstrip, strip line, coaxial, CPW, and waveguide. The parameters Z_o and v_p of these transmission lines are often required in circuit design. In order to obtain the unit length capacitance C, one end of the short-length transmission line ($l<<\lambda$) is opened, as shown in Figure 3.7. When the impedance of an open-end transmission line is measured at the other end of the line, the current through the inductor is close to 0, and the voltage drop due to the inductor can be ignored. The impedance Z_{OPEN} will be due to the sum of the distributed capacitance per unit length, as shown in Equation (3.19).

$$Z_{OPEN} = \frac{1}{j\omega l \cdot C} \tag{3.19}$$

The transmission line is then shorted and the impedance Z_{SHORT} is measured. Considering that the length of the transmission line is short, the voltage across the capacitance of the line is small and the leakage current due to the capacitor is close to 0. The impedance then becomes

$$Z_{SHORT} = j\omega l \cdot L \tag{3.20}$$

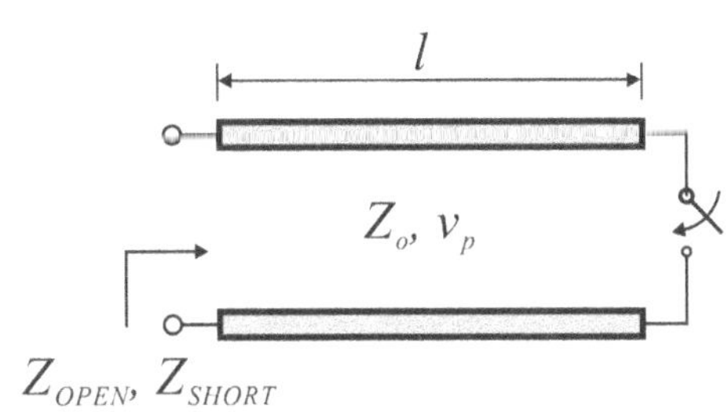

Figure 3.7 Measurement of the parameters of a transmission line. Using the impedances Z_{OPEN} and Z_{SHORT}, the transmission-line parameters, such as the characteristic impedance and electrical length, can be computed.

Combining Equations (3.19) and (3.20), the Equations (3.21) and (3.22) are obtained.

$$Z_o = \sqrt{\frac{L}{C}} = \sqrt{Z_{OPEN} \cdot Z_{SHORT}} \tag{3.21}$$

$$v_p = \frac{1}{\sqrt{LC}} = \frac{1}{\omega l}\sqrt{\left|\frac{Z_{OPEN}}{Z_{SHORT}}\right|} \tag{3.22}$$

The measurement is assumed to be conducted at a low frequency, but even at a higher frequency Equation (3.21) can be used to obtain the characteristic impedance. However, Equation (3.22) is similar in principle for a higher frequency, but the result becomes a little different. The exact equation can be obtained by using the results explained in section 3.4.2. Equation (3.22) is applicable if the frequency is low enough. Here, a low-enough frequency implies

$$f << \frac{v_p}{l} \tag{3.23}$$

Hence, Equations (3.21) and (3.22) are applicable up to the frequency usually equal to 1/10 of the right-hand side of Equation (3.23).

3.3 COAXIAL AND MICROSTRIP LINES

Structurally, there are various types of transmission lines, among which special considerations are given to coaxial and microstrip lines. The coaxial line can be manufactured to be a nearly ideal transmission line with broadband characteristics. Due to these advantages, the coaxial line has been adopted in the past for the high-frequency measurement of devices and it is still widely used. On the other hand, the microstrip line has a planar structure that facilitates the construction of smaller microwave-integrated circuits. Here, we present a brief explanation of the two lines—coaxial and microstrip.

3.3.1 Coaxial Line

Figure 3.8 shows the structure of a coaxial line and its electric and magnetic field lines in cross-section. The characteristic impedance Z_o and phase velocity v_p of the coaxial line with inner and outer radii of a and b become Equations (3.24) and (3.25).

$$Z_o = \frac{60}{\sqrt{\varepsilon_r}} \ln\frac{b}{a} \tag{3.24}$$

$$v_p = \frac{c}{\sqrt{\varepsilon_r}} \tag{3.25}$$

Thus, in constructing a coaxial line, the desired characteristics of impedance can be obtained by appropriately adjusting the inner radius a and the outer radius b.

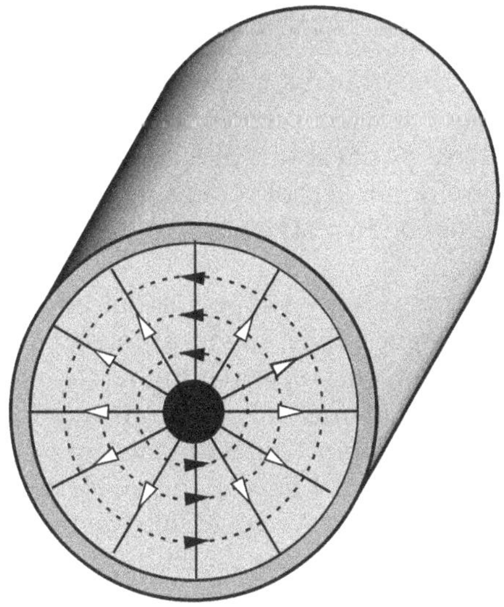

Figure 3.8 Structure of a coaxial line and the field patterns of the cross-section (solid lines represent the electric field, and dotted lines represent the magnetic field).

The real coaxial line is constructed in the form of a cable and a connector. As shown in Figure 3.9, the cable has the form of the coaxial line shown in Figure 3.8 and the connector appears at the end of the cable for convenient connection to other devices.

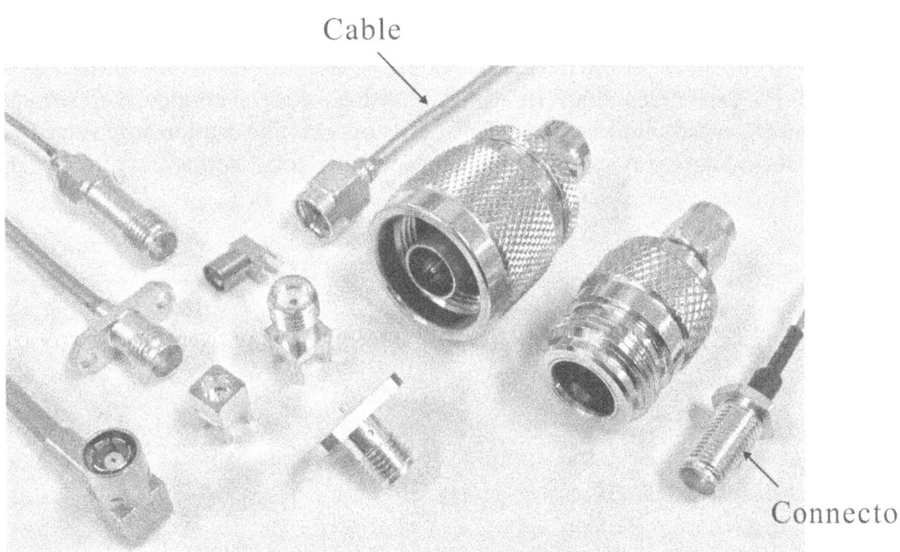

Figure 3.9 Photograph of various coaxial lines and connectors

Depending on the outer conductor material of the cable, coaxial cables are classified as flexible, semirigid, and handy-reformable. Flexible cable refers to a cable that bends easily, and the outer conductor usually consists of braids for flexibility. Semirigid cables are jacketed with copper to provide a certain degree of flexibility, but the semirigid cable is not easily deformed. This type of cable is often used for connecting microwave components in equipment. Handy-reformable cable is also metal jacketed, but the outer conductor is bellow jacketed for easier bending. The extent of reformability is between that of flexible and semirigid cables.

These coaxial lines usually support the electromagnetic wave characterized by the transverse electromagnetic (TEM) mode. However, higher modes can propagate within the coaxial line as the frequency becomes higher. To prevent higher modes, the coaxial line must have a smaller outer diameter. For a 7 mm outer-diameter air-coaxial line, TEM mode can be propagated alone without higher modes up to 18 GHz; for 3.5 mm up to 26.5 GHz; for 2.9 mm up to 40 GHz; for 2.4 mm up to 50 GHz; and for 1 mm up to 110 GHz. Thus, as the outer diameter decreases, the coaxial line can be used in higher frequencies, but there is a disadvantage that is due to the difficulty in manufacturing coaxial lines with such small diameters. Similar to cables, the connectors that facilitate connection at the end of the cables are also classified according to the size of their outer diameter. Precision air-line connectors used for measurement are available for 7 mm and 3.5 mm diameters (for use up to 18 GHz and 26.5 GHz, respectively); 2.9 mm (called K-connector, for use up to 40 GHz); 2.4 mm (for use up to 50 GHz); and 1 mm (for use up to 110 GHz). In addition, at microwave frequencies, one of the most widely used connectors is the small miniature assembly (SMA) connector. The inside dielectric material is usually Teflon and it can be mated with a 3.5 mm coaxial connector. Also, the K or 2.9 mm connector can be mated with a 3.5 mm coaxial connector.

In general, different connector types are incompatible. In order to solve this problem of incompatibility, *coaxial adaptors* (as shown in Figure 3.10) are used to connect two incompatible connectors. Their frequency range of operation is thus limited by the connector used at the lower frequency. In addition to these coaxial connectors, there are other kinds of connectors classified as BNC, TNC, N-type, etc. The reader may refer to the numerous materials available on the subject for these connectors' frequency range of operation.

Figure 3.10 Photograph of various adaptors[1]

1. Agilent, Agilent 83059 Precision 3.5mm Coaxial Adapters, 5952-2836E, available at http://www.agilent.com/ 2006.

Example 3.4

A coaxial cable has an inner-radius a inch and an outer-radius b inch, as shown in Figure 3E.3. By using ADS, determine the characteristic impedance of the coaxial cable in terms of the ratio a/b.

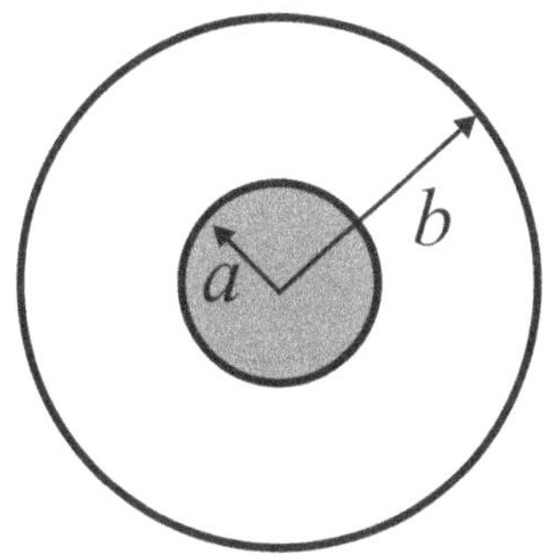

Figure 3E.3 Cross-section of a coaxial cable

Solutions

Figure 3E.4 is a simulation schematic for the determination of the characteristic impedance using ADS. The radius of b is fixed at 0.116 inch and that of a is varied to change the value of the characteristic impedance.

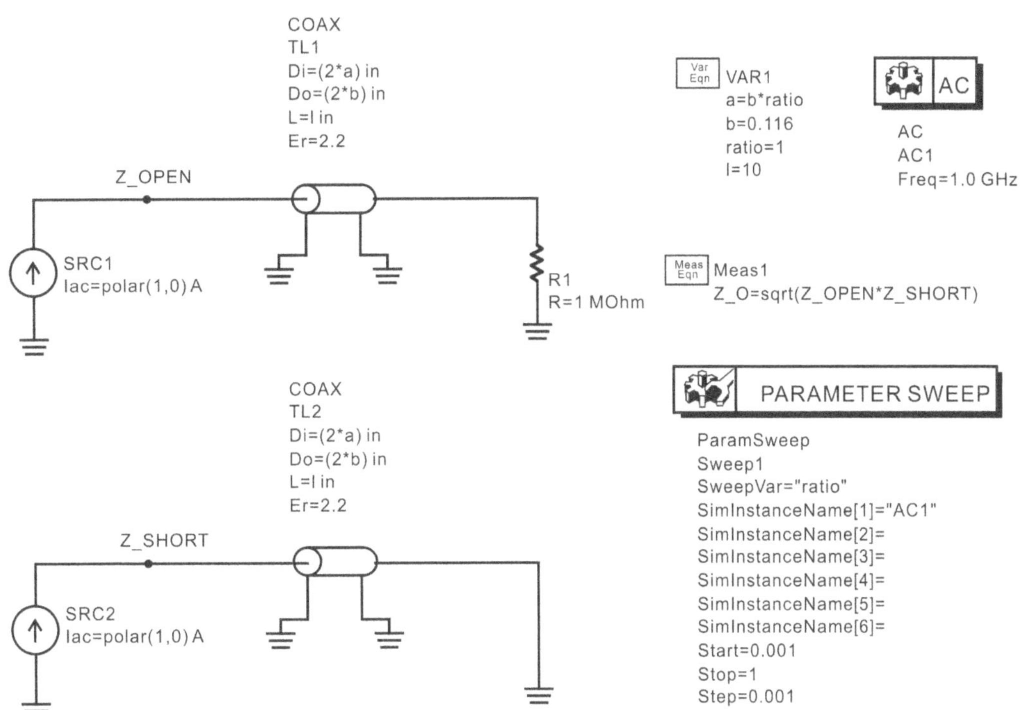

Figure 3E.4 Circuit setup for calculating characteristic impedance. The length and permittivity of the coaxial line are fixed at 1 inch and 2.2 inches, respectively. The outer radius is fixed at 0.116 inch, while the inner radius a is varied by equation **a** = **b*ratio**. To avoid numerical discontinuity, the open end is implemented using a resistor of 1 MΩ. The voltages **Z_OPEN** and **Z_SHORT** correspond to the impedances because the current source has 1 A. The characteristic impedance given by Equation (3.21) is implemented in **Meas**1 in Figure 3E.4. Thus, the output **Z_O** appears at the output dataset.

The radius a is varied to make the ratio a/b vary between 0.001–1. Now, since the computed **Z_OPEN** and **Z_SHORT** in Figure 3E.4 correspond to the Z_{OPEN} and Z_{SHORT} of Equation (3.21), the graph of the characteristic impedance with respect to a/b can be obtained. This can be done by entering Equation (3.21) in the ADS display window. Alternatively, by inserting the **MeasEqn** utility shown in Figure 3E.5, the characteristic impedance can be directly obtained.

Theoretically, the characteristic impedance of a coaxial cable can be obtained from the equation given by

$$Z_o = 60\sqrt{\frac{\mu_r}{\varepsilon_r}}\ln\left(\frac{b}{a}\right)$$

Thus, for the purpose of the comparison, the equation for the theoretical characteristic impedance is entered in the display window as shown in Measurement Expression 3E.1.

```
Eqn z=60*sqrt(1/2.2)ln(1/ratio)
```

Measurement Expression 3E.1 Theoretical impedance of a coaxial cable

Figure 3E.5 shows the plot of these two characteristic impedances, which are found to show good agreement.

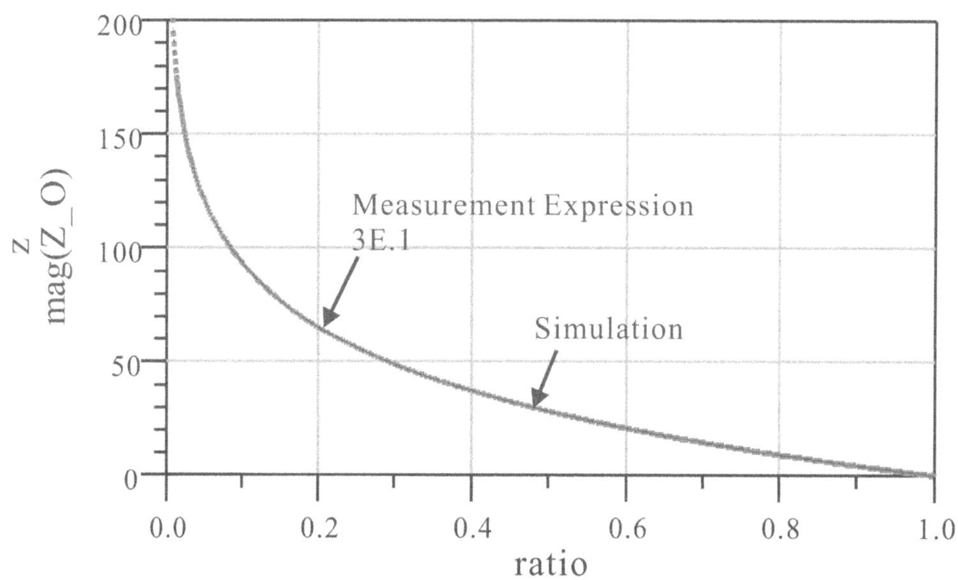

Figure 3E.5 Theoretical and simulated characteristic impedances of a coaxial cable with respect to b/a. The impedance Z is the impedance computed using Measurement Expression 3E.1, while the impedance **Z_O** is computed using the short-open method.

3.3.2 Microstrip Line

Planar transmission lines can be microstrip lines, strip lines, and CPWs. Among the planar transmission lines mentioned, the most widely used type for microwave-integrated circuit designs is the microstrip line. Figure 3.11 shows the cross-sectional geometry and electromagnetic field patterns of a microstrip line.

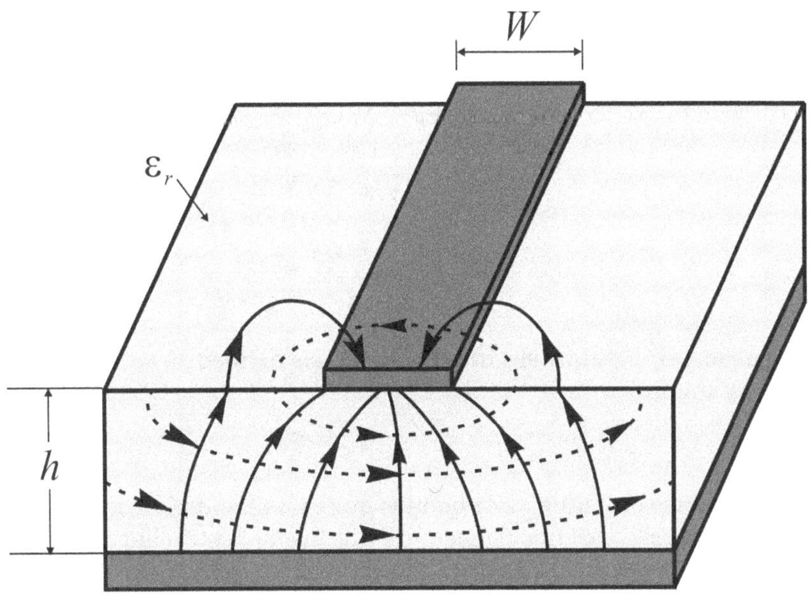

Figure 3.11 Cross-section and field pattern of a microstrip line (solid lines represent the electric field and dotted lines represent the magnetic field; fundamental propagation mode is quasi-TEM).

The fundamental propagation mode in a microstrip line, unlike a coaxial line, is not a TEM mode, but because its propagation mode is close to a TEM mode, it is called a quasi-TEM mode. The characteristic impedance and phase velocity of a microstrip line depend on its filling dielectric and cross-sectional geometry. Here, we shall qualitatively look at how the characteristic impedance and phase velocity of a microstrip line depend on the dielectric constant and the cross-sectional geometry. The exact values for a given dielectric constant and cross-sectional geometry can be calculated using software like ADS. The phase velocity and the characteristic impedance can be similarly represented using the capacitance per unit length and the inductance per unit length, as explained in section 3.2.1. Thus,

$$v_p = \frac{1}{\sqrt{LC}}$$

$$Z_o = \frac{1}{v_p C} = \sqrt{\frac{L}{C}}$$

Rather than Z_o and v_p in the equations above, the more physically meaningful Z_o and v_p can be obtained by comparing the microstrip line with an air-filled microstrip line. When the dielectric material is replaced by air, the microstrip's line structure is homogeneous and the capacitance per unit length C^a can be obtained analytically. In the case of the air-filled microstrip line, the phase velocity is

$$c = \frac{1}{\sqrt{LC^a}} \tag{3.26}$$

Here, c is the speed of light. Note that the inductance per unit length L obtained from Equation (3.26) is the same as that of the dielectric-filled microstrip in Figure 3.11. Using C^a, Z_o, and v_p can be expressed as shown in Equations (3.27) and (3.28).

$$v_p = \frac{1}{\sqrt{LC}} = c\sqrt{\frac{C^a}{C}} = \frac{c}{\sqrt{\varepsilon_{eff}}} \tag{3.27}$$

$$Z_o = \frac{1}{v_p C} = \frac{1}{c\sqrt{CC^a}} \tag{3.28}$$

In Equation (3.27), the dielectric constant ε_{eff} is defined as the ratio of C to C^a and is given by Equation (3.29).

$$C = \varepsilon_{eff} C^a \tag{3.29}$$

The dielectric constant ε_{eff} can be interpreted as the dielectric constant of an equivalent dielectric material that homogeneously fills the microstrip line in Figure 3.11. Note that the geometry of the equivalent microstrip is the same as the microstrip in Figure 3.11 and the only difference is in dielectric filling. The microstrip line homogeneously filled with an equivalent dielectric material can give the same phase velocity as the inhomogenously filled microstrip in Figure 3.11. The equivalent dielectric constant for the homogeneously filled microstrip is called an *effective permittivity* or an *effective dielectric constant*.

The effective permittivity is a function of a dielectric constant ε_r and the geometrical parameters such as strip width W and dielectric thickness h shown in Figure 3.11. The field pattern will be evenly distributed between the air and the dielectric material when the width W is extremely narrow; when the width is very large, the electromagnetic field becomes mainly concentrated in the dielectric material, so the effective permittivity is given by Equation (3.30).

$$\varepsilon_{eff} = \begin{cases} \dfrac{\varepsilon_r + 1}{2} & w \to 0 \\ \varepsilon_r & w \to \infty \end{cases} \tag{3.30}$$

For ε_{eff}, the phase velocity is

$$v_p = \frac{c}{\sqrt{\varepsilon_{eff}}}$$

Thus, the phase velocity increases as the width of the microstrip line becomes narrower, and as the width widens, the phase velocity decreases. For Z_o, as the width narrows, the sectional capacitance becomes smaller and consequently Z_o becomes higher. In contrast, as the width widens, the capacitance becomes larger, which leads to lower Z_o.

Next, we will look at the capacitance per unit length of the microstrip line that is equal to the cross-sectional capacitance. It can be found that the capacitance consists of two components: the capacitance formed beneath the width of the transmission line, called *plate capacitance*, C_p, and the capacitance formed at the edge of the transmission line called *fringing capacitance*, C_f. Thus, the capacitance of a microstrip line can be expressed using Equation (3.31)

$$C = C_p + 2C_f \tag{3.31}$$

and the plate capacitance can be expressed using Equation (3.32).

$$C_p = \varepsilon_r \varepsilon_o \frac{W}{h} \tag{3.32}$$

Here, ε_o represents the dielectric constant of free space.

Example 3.5

A microstrip is implemented on a 25-mil-thick alumina substrate with a dielectric constant of 9.8. Find the effective dielectric constant ε_{eff} and the characteristic impedance Z_o of the microstrip line using the open- and short-circuit method previously discussed.

Solutions

Figure 3E.6 shows the microstrip line's structure with a width w, dielectric constant ε_r, and thickness h.

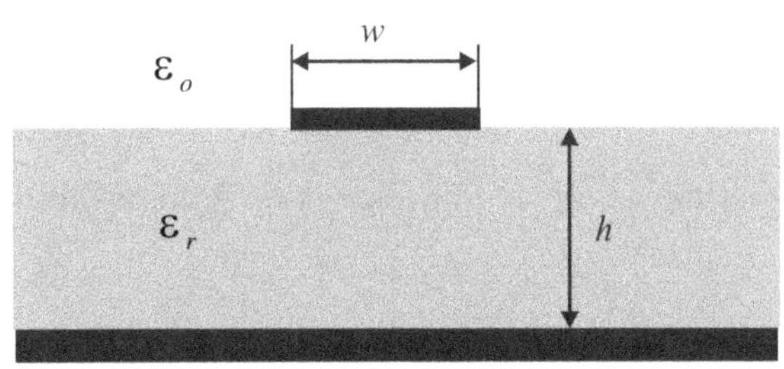

Figure 3E.6 Sectional view of a microstrip

Figure 3E.7 is an ADS schematic setup for determining the characteristic impedance and the effective permittivity of two microstrip lines. **H** and **er** in **MSUB** are set to 25 mil and 9.8, which correspond to h and ε_r, respectively. The variable **w** in **MLIN**, which corresponds to w, is set to vary. Here, w is varied from 0.001 to 100 mil. The impedances of the short-circuited **TL1** and open-circuited **TL2** are used to compute Z_o of the microstrip line from the equation $Z_o = (Z_{OPEN}\, Z_{SHORT})^{1/2}$. Here, a 1-MΩ resistor is used for achieving an open condition. The effective permittivity is computed by comparing the impedances of **TL2** with $\varepsilon_r = 9.8$ and of **TL3** with $\varepsilon_r = 1$ (air) as

$$\varepsilon_{eff} = \frac{C}{C_a} = \frac{\text{Z\_OPEN\_air}}{\text{Z\_OPEN}}$$

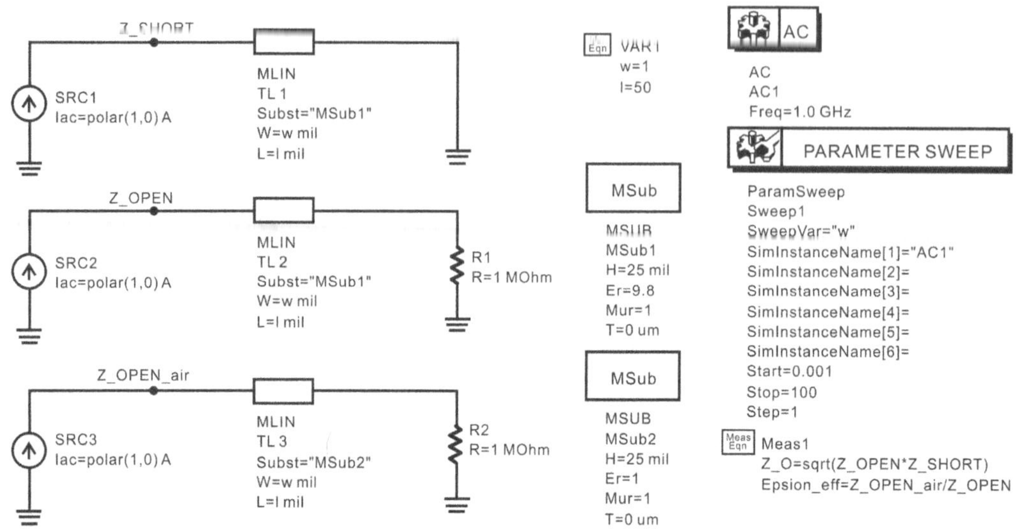

Figure 3E.7 Circuit setup for ADS simulation. The lengths of all microstrip lines are fixed at 50 mil and the widths are varied by the parameter sweep controller from 0.001 to 100 mil. The two microstrip lines in the figure share the common substrate **Msub**1, whereas the bottom microstrip line has the substrate **Msub2**; the only difference between them is in the permittivity set to 1.Thus, the characteristic impedance of the microstrip can be computed by **Z_SHORT** and **Z_OPEN**, while the effective permittivity can be computed using **Z_OPEN** and **Z_OPEN_air**. The equations necessary for the characteristic impedance and effective permittivity are in **Meas**1 in the figure.

Figure 3E.8 shows the values of Z_o with respect to w. As the width increases, Z_o decreases as expected.

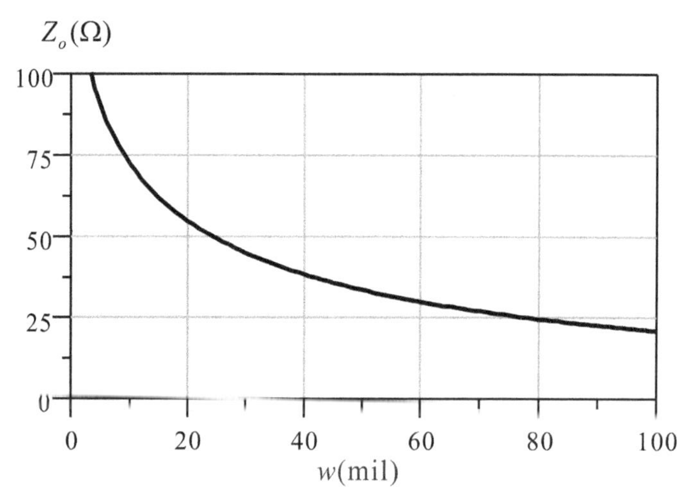

Figure 3E.8 Characteristic impedance of a microstrip line with respect to width (h = 25mil, ε_r = 9.8)

The effective permittivity ε_{eff} can be approximately expressed as

$$\varepsilon_{eff} = \frac{\varepsilon_r + 1}{2} + \frac{\varepsilon_r - 1}{2} \frac{1}{\sqrt{1 + 12h/w}}$$

In order to compare the computed ε_{eff} from the ADS with the ε_{eff} in the equation above, the following equations are entered in the display window:

Eqn er=9.8

Eqn h=25

Eqn e_eff=(er+1)/2 + (er-1)/2/sqrt(1+12*h/w)

Measurement Expression 3E.2 Calculation of the effective permittivity of a microstrip line

Now, the accuracy in terms of the agreement can be found by comparing the computed ε_{eff} from ADS with that given by the approximate equation. Figure 3E.9 displays the comparison of the two results. The computed ε_{eff} from ADS shows good agreement with the approximate equation of the effective permittivity.

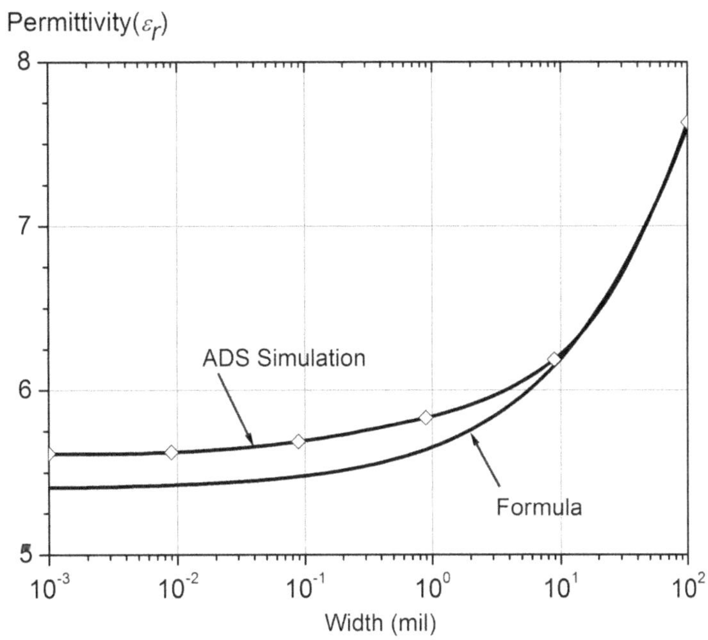

Figure 3E.9 Simulated effective permittivity of a microstrip line with respect to width. The effective permittivity given by the empirical formula in Measurement Expression 3E.2 is also plotted for comparison.

3.4 SINUSOIDAL RESPONSES

3.4.1 Phasor Analysis

The response to a sinusoidal source $v_s(t) = V_m\cos(\omega t + \phi)$ frequently appears in electrical-circuit analysis. Instead of computing the response directly to the sinusoidal source, a complex waveform source of $Ve^{j\omega t}$ is preferred due to its mathematical convenience. Here, the complex number $V = V_m e^{j\phi}$, which represents the phase and amplitude of the sinusoidal source, is called a *phasor*. Conversely, for a given phasor, the corresponding time-domain waveform can be obtained by multiplying the phasor by $e^{j\omega t}$ and taking the real part of the result. That is,

$$v_s(t) = \text{Re}\left(Ve^{j\omega t}\right) = \text{Re}\left(V_m e^{j\omega t + j\phi}\right) = V_m \cos(\omega t + \phi)$$

Example 3.6

(1) What is the phasor that corresponds to the sinusoidal waveform $v(t) = 2\cos(\omega t + 30°)$?

(2) A phasor for the free-space Green's function is given by $e^{-jkr}/4\pi r$. What is the corresponding waveform in the time domain?

Solutions

(1) The phasor is $V = 2\angle 30°$ and the corresponding time domain wave is

$$v_s(t) = \text{Re}\left(Ve^{j\omega t}\right) = \text{Re}\left(2e^{j30°}e^{j\omega t}\right) = 2\cos(\omega t + 30°)$$

(2) $v(t) = \text{Re}\left(\dfrac{e^{-jkr}}{4\pi r}e^{j\omega t}\right) = \dfrac{1}{4\pi r}\cos(\omega t - kr)$

From the solutions to the wave equations given by Equations (3.15) and (3.16), the voltage and current at an arbitrary position z on a transmission line for a sinusoidal input can be expressed as the sum of two traveling waves given by

$$v(z,t) = V_m^+ \cos(\omega t - \beta z + \phi^+) + V_m^- \cos(\omega t + \beta z + \phi^-)$$

$$i(z,t) = \frac{1}{Z_o}\left\{V_m^+ \cos(\omega t - \beta z + \phi^+) - V_m^- \cos(\omega t + \beta z + \phi^-)\right\}$$

The corresponding complex waveform can be written as Equations (3.33a) and (3.33b).

$$\begin{aligned} v_c(z,t) &= V_m^+ e^{j(\omega t - \beta z + \phi^+)} + V_m^- e^{j(\omega t + \beta z + \phi^-)} \\ &= e^{j\omega t}\left(V^+ e^{-j\beta z} + V^- e^{j\beta z}\right) \end{aligned} \tag{3.33a}$$

$$\begin{aligned} i_c(z,t) &= \frac{1}{Z_o}\left(V_m^+ e^{j(\omega t - \beta z + \phi^+)} - V_m^- e^{j(\omega t + \beta z + \phi^-)}\right) \\ &= \frac{1}{Z_o}e^{j\omega t}\left(V^+ e^{-j\beta z} - V^- e^{+j\beta z}\right) \end{aligned} \tag{3.33b}$$

Here,

$$V^+ = V_m^+ e^{j\phi^+} \tag{3.34a}$$

$$V^- = V_m^- e^{j\phi^-} \tag{3.34b}$$

Therefore, the sinusoidal voltage waveforms above, propagating in the +*z* and −*z* directions, can be expressed using complex phasors. The voltage phasor $V^+ e^{-j\beta z}$ represents a sinusoidal voltage waveform traveling in the z–direction, while the voltage phasor $V^- e^{j\beta z}$ represents a sinusoidal voltage waveform traveling in the −z direction. At $z = 0$, the voltage and current in Equations (3.33a) and (3.33b) become Equations (3.35a) and (3.35b).

$$v_c(0,t) = e^{j\omega t}\left(V^+ + V^-\right) \tag{3.35a}$$

$$i_c(0,t) = \frac{1}{Z_o} e^{j\omega t}\left(V^+ - V^-\right) \tag{3.35b}$$

This means that voltage phasor V^+ in Equation (3.34a) is the phasor for the sinusoidal voltage waveform at $z = 0$, which travels in the *z*–direction. On the other hand, V^- in Equation (3.34b) is the phasor at $z = 0$ for the traveling waveform in the −z direction. Thus, the voltage phasor at z in Equations (3.34a) and (3.34b) can be physically interpreted using the two traveling waves. The phasor at $z = 0$ given by V^+ travels in the +*z* direction to position *z* and the voltage phasor at *z* has a phase delay of $e^{-j\beta z}$ compared with that at $z = 0$. As a result, the voltage phasor at *z* is given by $V^+ e^{-j\beta z}$. In contrast, in the case of the voltage wave traveling in the −*z* direction, the voltage phasor at *z* is given by $V^- e^{j\beta z}$. As the voltage wave is transmitted at position z and arrives at $z = 0$, the phase at $z = 0$ is delayed by an amount of $e^{-j\beta z}$. Thus, it results in the phasor $V^- e^{j\beta z} e^{-j\beta z} = V^-$ at $z = 0$. This is shown in Figure 3.12. Therefore, the phase of the voltage wave traveling in the right direction is delayed while the phase of the voltage wave at position *z* shows phase advance for the voltage traveling in the left direction. Furthermore, in the case of the current wave, it is worth noting that the current wave traveling in the +z direction is denoted as + while that traveling in the −*z* direction is denoted as −, but the phase relation is the same as that of the voltage wave.

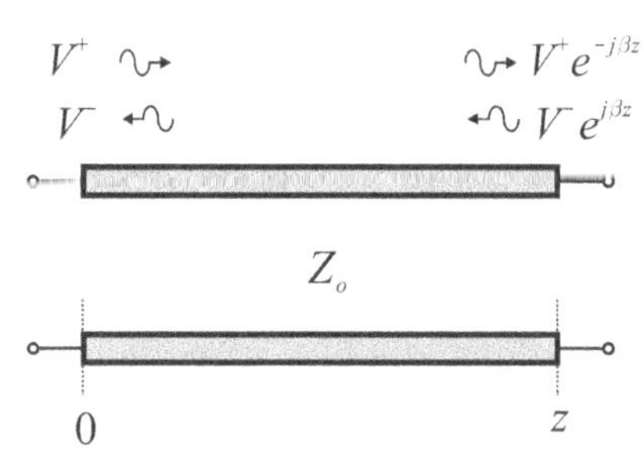

Figure 3.12 Phasor expressions of voltage waves in a transmission line. Note that the phase delay of $e^{-j\beta z}$ occurs along the propagation direction irrespective of the wave-propagation direction.

It is worth noting that *both waves have phase delay in their direction of propagation*. The interpretation of the phasor above is frequently used in the analysis of reflections in transmission lines as well as impedance analysis based on reflection. Therefore, a thorough understanding of the phasor is necessary.

3.4.2 Reflection and Return Loss

When a transmission line is terminated by a load, the incident wave is reflected by the load. In contrast to the case of an infinite-length transmission line, current and voltage waves traveling in both directions appear in the transmission line terminated by the load. The wave traveling toward the load is called the *incident wave*, while the wave traveling in the opposite direction to the incident wave is called the *reflected wave*. In this section, we will learn about reflection and the related return loss on a transmission line.

Figure 3.13 shows a transmission line whose length is z, terminated by a load Z_L. With the coordinate shown in Figure 3.13, the voltage incident at $z = 0$ toward the load is denoted as phasor $V+$ and the reflected wave from the load is denoted as phasor V^-. With these phasor notations, the incident and reflected voltages at z can be expressed as shown in Figure 3.13.

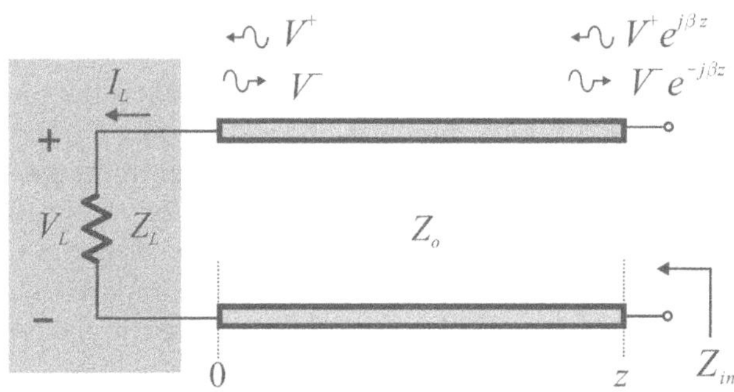

Figure 3.13 A transmission line terminated by a load. V^+ and V^- are defined as the incident and reflected voltage waves at $z = 0$. The incident and reflected voltages at z are accordingly represented as shown in the figure.

The reflection coefficient Γ_L is defined in Equation (3.36).

$$\Gamma_L = \frac{V^-}{V^+} \tag{3.36}$$

Thus, it can be seen that the reflection coefficient is a complex number. The voltage drop V_L by the load is given by $V_L = Z_L I_L$. At the same time, V_L is the voltage of the transmission line at 0. Thus, V_L becomes

$$V_L = V^+ + V^- \tag{3.37}$$

and similarly I_L becomes

$$I_L = I^+ - I^- \tag{3.38}$$

Substituting Equations (3.37) and (3.38) into $V_L = Z_L I_L$, Equation (3.39) is

$$Z_L I_L = Z_L\left(I^+ - I^-\right) = Z_L\left(\frac{V^+}{Z_o} - \frac{V^-}{Z_o}\right) = V_L = V^+ + V^- \tag{3.39}$$

Solving for V^-/V^+, the reflection coefficient then becomes Equation (3.40).

$$\Gamma_L = \frac{V^-}{V^+} = \frac{Z_L - Z_o}{Z_L + Z_o} \tag{3.40}$$

From the reflection coefficient given by Equation (3.40), it can be found that the magnitude of Γ_L is less than 1, for example, $|\Gamma_L| \le 1$ for a passive load Z_L that has the positive real part. In addition, the reflection coefficient can be defined at an arbitrary position on the transmission line. The reflection coefficient at z, shown in Equation (3.41), becomes

$$\Gamma(z) = \frac{V^-(z)}{V^+(z)} = \frac{V^- e^{-j\beta z}}{V^+ e^{+j\beta z}} = \Gamma_L e^{-2j\beta z} \tag{3.41}$$

It can be seen in Equation (3.42) that the reflection coefficient has the same magnitude as that at the load plane. That is,

$$|\Gamma(z)| = |\Gamma_L| \tag{3.42}$$

but the phase is delayed by an amount of $2\beta z$. Since the magnitude of the reflection coefficient is independent of position, it can be determined by means of *scalar* measurement without resorting to *vector* measurement. The scalar measurement is measuring only the magnitude of a complex quantity, while the vector measurement is the measurement of both the magnitude and phase of the complex quantity. When the reflection coefficient is expressed in decibels, it is called the *return loss* and is defined as shown in Equation (3.43).

$$\text{RL(Return Loss)} = 20\log|\Gamma_L| \tag{3.43}$$

From Equation (3.40), when the load impedance is equal to Z_o,

$$Z_L = Z_o, \quad \Gamma_L = 0$$

and there is no reflection; on the other hand, when the load is short-circuited,

$$Z_L = 0, \quad \Gamma_L = -1$$

which leads to the total reflection. Finally, when the load is open-circuited, it can be seen that

$$Z_L = \infty, \quad \Gamma_L = 1$$

By using the reflection coefficient obtained at position z, the input impedance Z_{in} looking into the load from the transmission line can be determined. That is, since the voltage and current at position z are as shown in Equations (3.44) and (3.55),

$$V(z) = V^+(z) + V^-(z) = V^+ e^{j\beta z} + V^- e^{-j\beta z} \tag{3.44}$$

$$I(z) = I^+(z) + I^-(z) = I^+ e^{j\beta z} - I^- e^{j\beta z} \tag{3.45}$$

then input impedance is computed as

$$Z_{in} = \frac{V(z)}{I(z)} = \frac{V^+(z)+V^-(z)}{\frac{V^+(z)}{Z_o} - \frac{V^-(z)}{Z_o}} = Z_o\frac{1+\Gamma(z)}{1-\Gamma(z)}$$

$$= Z_o\frac{Z_L + jZ_o\tan(\beta z)}{Z_o + jZ_L\tan(\beta z)} \tag{3.46}$$

From Equation (3.46), when the load is short-circuited ($Z_L = 0$), the input impedance

$$Z_{in} = jZ_o\tan(\beta z) \tag{3.47}$$

is obtained in Equation (3.47). When the load is open-circuited ($Z_L = \infty$), the input impedance is shown in Equation (3.48).

$$Z_{in} = \frac{Z_o}{j\tan(\beta z)} \tag{3.48}$$

Also, when the load impedance is equal to the characteristic impedance, that is, $Z_L = Z_o$, Z_{in} is independent of the position, and $Z_{in} = Z_o$. Furthermore, when the transmission line is a quarter wavelength long, $\tan(\beta z)$ approaches infinity and the input impedance is given by Equation (3.49).

$$Z_{in} = \frac{Z_o^2}{Z_L} \tag{3.49}$$

This implies the inversion of the load impedance. Thus, a quarter-wavelength transmission line is often called an *impedance inverter*. Finally, when the transmission line is a half wavelength long, $\tan(\beta z) = 0$ and the input impedance is found to be equal to the load impedance, as shown in Equation (3.50).

$$Z_{in} = Z_L \tag{3.50}$$

3.4.3 Voltage Standing Wave Ratio (VSWR)

The previously described reflection coefficient can be measured using a slotted line. The slotted line is a coaxial air line that has a slot on the outer conductor along the propagation direction. When a probe is inserted in the slot, the DC voltage proportional to the internal electric field of the transmission line at the point where the probe is inserted appears at the probe output. Thus, the voltage of the transmission line at the arbitrary position can be measured. Moving the probe along the propagation direction, it is possible to observe the change of the voltage with respect to position. The voltage along the propagation direction shows a standing-wave pattern due to reflection, which will be explained later in this section. Using the standing-wave pattern, the reflection coefficient of the unknown load can be determined and it is also possible to determine its impedance. The term *voltage standing wave ratio* (VSWR) originates from the slotted-line method. Recently, as network analyzers have become popular, the slotted-line method is not used as often as in the past. However,

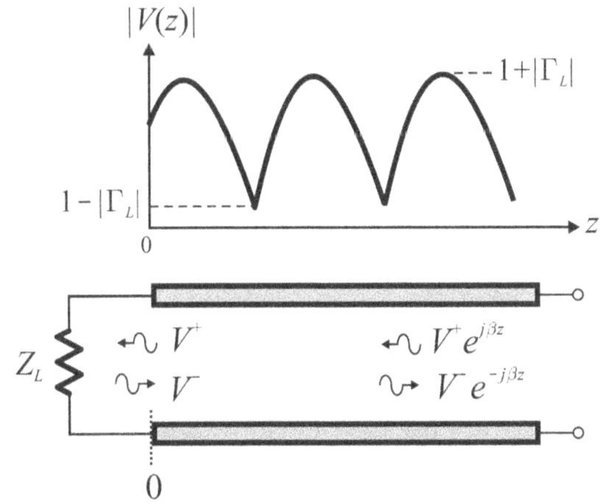

Figure 3.14 Standing-wave pattern. When reflection occurs, the voltage amplitude varies along z.

because it has been used extensively to determine the reflection coefficient for an unknown load impedance, the term VSWR is still widely used for specifying the reflection phenomenon together with the return loss or reflection coefficient.

When the load Z_L is connected to the end of the transmission line, as shown in Figure 3.14, the voltage at the location z is expressed as shown in Equation (3.51)

$$\begin{aligned} V(z) &= V^+(z) + V^-(z) \\ &= V^+e^{j\beta z} + V^-e^{-j\beta z} \\ &= V^+e^{j\beta z}\left\{1 + \Gamma_L e^{-2j\beta z}\right\} \end{aligned} \tag{3.51}$$

and the waveform in the time domain becomes

$$\begin{aligned} v(z,t) &= \mathrm{Re}\left(V(z)e^{+j\omega t}\right) \\ &= \left|V(z)\right|\cos\left(\omega t + \angle V(z)\right) \end{aligned} \tag{3.52}$$

In order to examine the meaning of Equation (3.52), we substitute $\Gamma_L = -1$ into Equation (3.51) and rewrite it to obtain Equation (3.53).

$$V(z) = 2jV^+\sin(\beta z) \tag{3.53}$$

Thus, the time-domain waveform of the transmission line voltage is

$$\begin{aligned} v(z,t) &= \mathrm{Re}\left(V(z)e^{+j\omega t}\right) \\ &= 2\left|V^+\right|\sin(\beta z)\cos\left(\omega t + 90^\circ\right) \end{aligned} \tag{3.54}$$

The time-domain waveform in Equation (3.54) is a simple harmonic oscillation with time, and its amplitude is given by $2|V^+|\sin(\beta z)$. The waveform does not propagate but oscillates up and down with time. Also, the peak and null positions remain fixed and do not move. Thus, the waveform is called a *standing wave*. In contrast, for a traveling wave given by $\cos(\omega t - \beta z)$ in Example 3.1, the peak value moves in time with phase velocity.

Moving the probe along the propagating axis, the value measured by the probe is proportional to $|V(z)|$, which is depicted in Figure 3.14. If Γ_L is denoted in Equation (3.55)

$$\Gamma_L = |\Gamma_L| e^{j\theta} = \rho e^{j\theta} \tag{3.55}$$

then

$$|V(z)| = |V^+| |1 + \rho e^{j(\theta - 2\beta z)}| \tag{3.56}$$

Here, the $1 + \rho e^{j(\theta - 2\beta z)}$ represents a circle with the center of (1, 0) and a radius ρ, as shown in Figure 3.15, as z varies. Therefore, when the probe is moved along the distance z of the transmission line, the minimum value of the amplitude is $1 - \rho$ and the maximum value is $1 + \rho$, as shown in Figure 3.15.

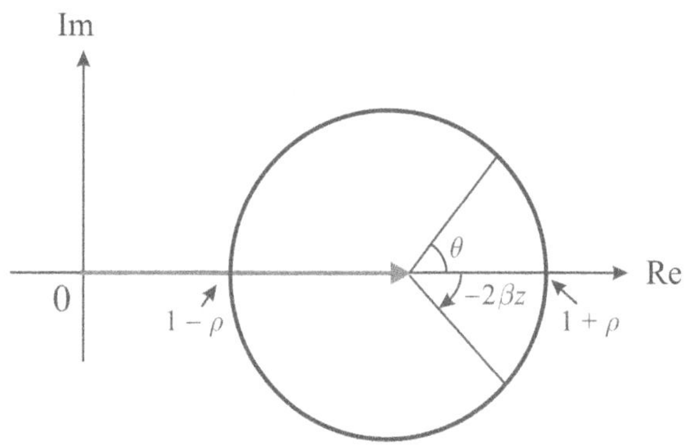

Figure 3.15 Trajectory of $1 + \rho e^{j(\theta - 2\beta z)}$ in the complex plane. The locus is the circle: the center is at (1, 0) and the radius is ρ.

The ratio of this maximum value to the minimum value is referred to as the voltage standing wave ratio (VSWR) and is defined as shown in Equation (3.57).

$$\text{VSWR} = \frac{1 + \rho}{1 - \rho} \tag{3.57}$$

As $0 \le \rho \le 1$ in this equation, the minimum value of the VSWR is 1 and, in most cases, it is greater than 1. When the VSWR = 1, it indicates that $\rho = 0$. This means the load impedance is equal to the characteristic impedance $Z_L = Z_o$ from the equation of the reflection coefficient. From VSWR measurements, the magnitude of the reflection coefficient ρ can be obtained. However, even though not discussed here, ρ and the phase of the reflection coefficient θ can be determined using the first minimum position of $|V(z)|$ from the load.

Example 3.7

Determine the magnitude of the reflection coefficient and the return loss for a load with VSWR = 2.

Solutions

$$\text{VSWR} = \frac{1+\rho}{1-\rho} = 2$$
$$\rho = |\Gamma| = \frac{\text{VSWR} - 1}{\text{VSWR} + 1} = \frac{1}{3}$$

The return loss is

$$\text{RL} = 20\log(|\Gamma|) = 20\log\frac{1}{3} = -9.5 \text{ dB}$$

3.4.4 Smith Chart and Polar Chart

The reflection coefficient defined previously is a complex number whose magnitude lies between 0 and 1. That is, $0 \le |\Gamma_L| \le 1$ and it can be plotted on a complex plane. Since the reflection coefficient can be expressed as shown in Equation (3.58),

$$\Gamma_L = |\Gamma_L| e^{j\theta} \tag{3.58}$$

it can be plotted on a polar chart in terms of its magnitude and angle, as shown in Figure 3.16(a). However, when the reflection coefficient is plotted on the polar chart, the corresponding load impedance cannot be found from the chart. In this case, the load impedance has to be determined again from the reflection coefficient using Equation (3.59).

$$Z_L = Z_o \frac{1+\Gamma_L}{1-\Gamma_L} \tag{3.59}$$

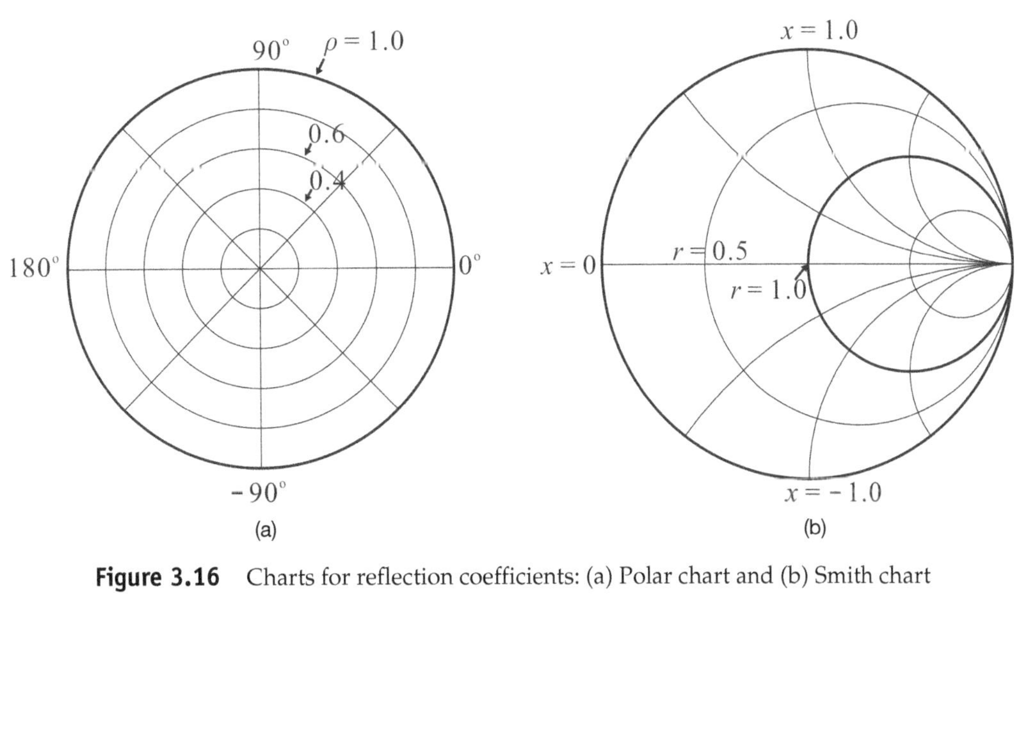

Figure 3.16 Charts for reflection coefficients: (a) Polar chart and (b) Smith chart

This cumbersome calculation can be avoided when the scale of Z_L is plotted. The reflection coefficient is expressed in terms of this normalized load impedance, $z = Z_L/Z_o = r + jx$, as shown in Equation (3.60).

$$\Gamma_L = \frac{z-1}{z+1} \tag{3.60}$$

The trajectories of Γ_L can be plotted on the same complex plane for the change of x with r kept constant, and also for the change of r with x kept constant. The resulting plot is called a Smith chart. Therefore, when Γ_L is plotted on the complex plane, the normalized load impedance by Z_o can be directly read from the chart without requiring any calculation.

Figure 3.16(b) shows the commonly used impedance based on a Smith chart. A Smith chart with both admittance and impedance scales represented by different colors is also available. Due to the frequent use of a ruler and compass in working with a Smith chart, a radial scale is usually provided at the bottom of the chart, as shown Figure 3.17. The scales for the reflection coefficient, VSWR, and return loss are shown at the bottom of the chart.

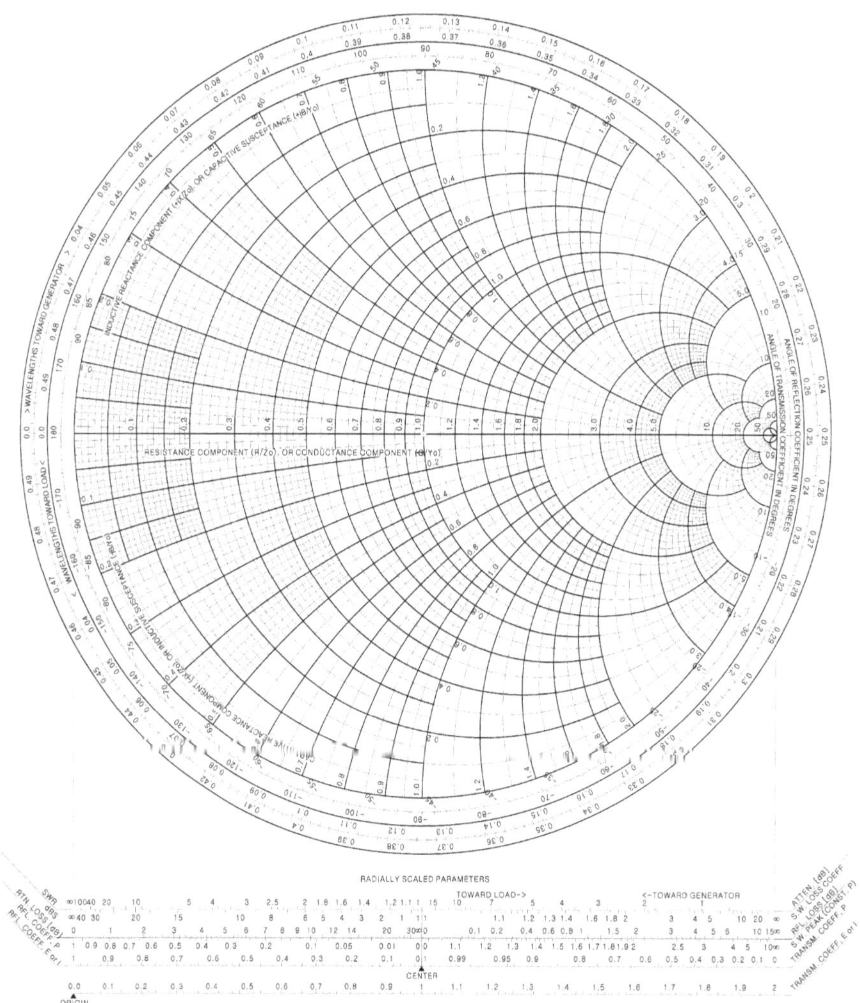

Figure 3.17 A Smith chart. The scale for the reflection coefficient, VSWR, and return loss are at the bottom of the chart. The scale can be used to set a compass.

Using those radial scales, VSWR, the magnitude of reflection coefficient, and return loss can easily be read with a compass, which does not require any calculation. In addition, the angles are provided on the circumference of the Smith chart, as well as the circumference scale corresponding to wavelength.

Example 3.8

For a load reflection coefficient $\Gamma_L = 0.6\angle{-30°}$ with a reference impedance of $Z_o = 50\ \Omega$, determine the load impedance using a Smith chart.

Solutions

Find the point corresponding to the angle −30° on the circumference scale of the Smith chart and draw a straight line with a ruler from the center of the chart to this point as shown in Figure 3E.10. Measure a radius of 0.6 using the radial scale at the bottom of the chart with a compass. With the one end of the compass fixed at the center of the chart, draw the $|\Gamma_L| = 0.6$ circle. The intersection point of the line and the $|\Gamma_L| = 0.6$ circle will appear at A. The normalized impedance at this point is read as $z = 2 - j1.8$. Then, de-normalizing z by the 50-Ω reference impedance results in $Z_L = 100 - j90$.

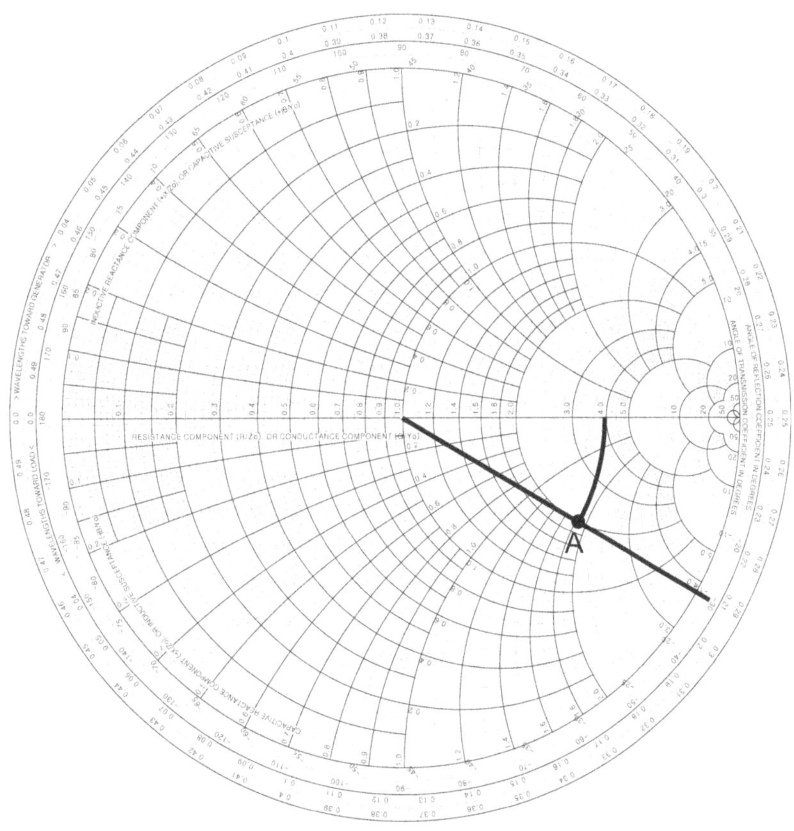

Figure 3E.10 Impedance calculation using a Smith chart. The compass is set using the radial scale of 0.6 shown in Figure 3.17 below the Smith chart. Locate one leg of the compass at the origin of the Smith chart and draw an $r = 0.6$ circle. Draw the straight line corresponding to −30° using the circumferential angle scale along the Smith-chart circle. The intersection point of the $r = 0.6$ circle and the straight line A is shown above. Reading the value of the normalized impedance is about $z = 2 - j1.8$.

3.5 APPLICATIONS

The application of transmission lines to a circuit is a vast subject and so, in this book, we will limit the discussion to a few of the many practical applications. The applications for these lines can be broadly categorized into that of a short-length transmission line and that of a quarter- and a half-wavelength-long transmission line. Another important application is the use of a quarter-wavelength transmission line as an impedance inverter. In a situation where the transmission line in a circuit requires a very large area at the frequency of operation, this is often implemented as lumped elements to save space. In this case, the lumped-equivalent two-port circuit of the transmission line is used. This will be discussed briefly in this section.

3.5.1 Short-Length Transmission Line

The inductors or capacitors described in Chapter 2 can be implemented in a limited-frequency range using transmission lines. A transmission line having electrical length θ is shown in Equation (3.61).

$$\theta = \beta l = \frac{\omega l}{v_p} \tag{3.61}$$

The impedance of a short-circuited transmission line is $jZ_o \tan(\theta)$. As the length θ is small, it gives a positive reactance that is proportional to ω, so it can be used as an inductor. The equivalent inductance can then be calculated using Equation (3.62).

$$j\omega L = jZ_o \tan(\theta) \tag{3.62}$$

When Z_o is large and θ is small, the shorted transmission line functions more like an inductor.

In contrast, when the end of a short-length transmission line is open, it gives a negative reactance, so it can be considered as a capacitor and the equivalent capacitance can be determined from Equation (3.63).

$$\frac{1}{j\omega C} = \frac{Z_o}{j\tan(\theta)} \tag{3.63}$$

In this case, when Z_o and θ become smaller, the open-end short-length transmission line functions more like a capacitor.

3.5.2 Resonant Transmission Line

A resonator acts as the main component in filter design. Figure 3.18 shows both series and parallel resonant LC circuits. At the resonant frequency,

$$\omega_o = \frac{1}{\sqrt{LC}}$$

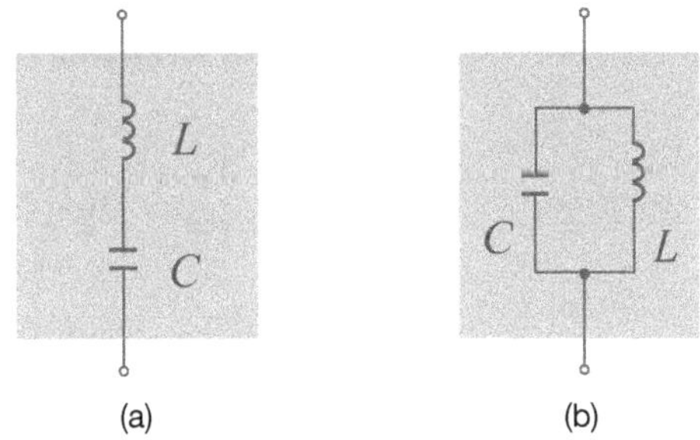

Figure 3.18 Resonant circuits: (a) *LC* series resonant circuit and (b) *LC* parallel resonant circuit

the series and parallel resonant *LC* circuits become short- and open-circuited, respectively. That is, at ω_o, the impedance in the case of the series resonant circuit becomes 0, which means a short circuit, while for the parallel resonant circuit, the admittance becomes 0 and the impedance approaches infinity, which means an open circuit.

These series and parallel resonant circuits can be constructed using transmission lines with one end that is either shorted or opened. In this section, we will examine the series or parallel resonances of the transmission line with one end shorted or opened. When the transmission line with one end shorted or open functions as both series and parallel resonant circuits, we can also determine the values of *LC* for both those circuits. The series resonance or parallel resonance can be found using the voltage and current standing-wave patterns. Through the standing-wave patterns, we can easily determine whether the transmission line functions as series *LC* or parallel *LC* resonant circuits.

The voltage at *z* of the one-end shorted transmission line, as shown in Figure 3.19, is

$$V(z) = V^+(z) + V^-(z) = V^+e^{+j\beta z} + V^-e^{-j\beta z}$$

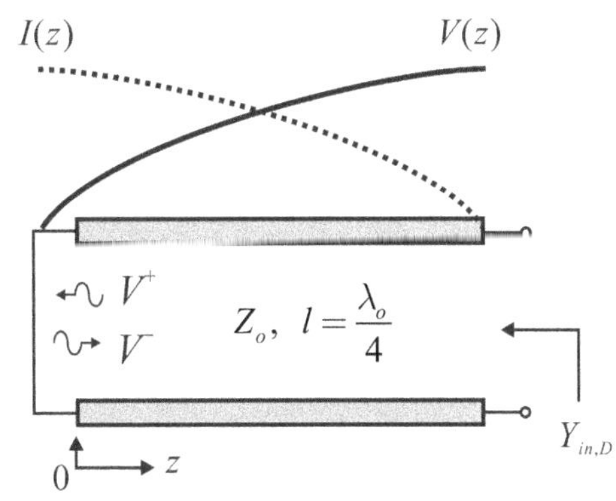

Figure 3.19 Standing waveforms in a quarter-wavelength shorted transmission line. *V*(*z*) and *I*(*z*) represent the voltage and current standing waveforms.

Since

$$\Gamma_L = \frac{V^-}{V^+} = -1$$

$V(z)$ results in

$$V(z) = 2jV^+ \sin(\beta z) \tag{3.64}$$

Note that the voltage is 0 at $z = 0$, which is shorted. Since the line is a quarter wavelength long, $\beta z = \pi/2$, we know that the maximum voltage occurs at the input. Similarly, for the current,

$$\begin{aligned} I(z) &= I^+(z) - I^-(z) = \frac{1}{Z_o}\left(V^+e^{+j\beta z} - V^-e^{-j\beta z}\right) \\ &= \frac{2V^+}{Z_o}\cos(\beta z) \end{aligned} \tag{3.65}$$

This means the maximum current flows at the plane of $z = 0$, but the current at the input of the transmission line can be found to be 0. Intuitively, because the transmission line is short-circuited at $z = 0$, maximum current flows while the voltage, on the other hand, is 0. Also, because the input of the transmission line corresponds to the quarter cycle of the sinusoidal waveform, it can easily be determined that the current will be 0 and the voltage peak will be at the input plane. Therefore, the impedance at the input terminal of a shorted quarter-wavelength transmission line is ∞ and so its behavior is similar to the parallel resonant circuit. In contrast, when one end of the line is open-circuited, the voltage is maximum and the current is 0 at $z = 0$. When the standing wave reaches the position of a quarter cycle, the opposite occurs. The voltage is now 0 and the current is maximum at the position of a quarter cycle. This means the impedance becomes 0. The open-end quarter-wavelength transmission line can thus be found to behave similarly to a series-resonant circuit.

We will now try to compute the equivalent LC values of the short-circuited transmission line. Dividing Equation (3.65) by Equation (3.64), the input admittance of the circuit in Figure 3.19 is shown in Equation (3.66).

$$Y_{in,D} = -jY_o \cot \beta l = -jY_o \cot\left(\frac{\pi}{2}\frac{\omega_o + \Delta\omega}{\omega_o}\right) = -jY_o \cot\left(\frac{\pi}{2} + \frac{\pi}{2}\frac{\Delta\omega}{\omega_o}\right) \tag{3.66}$$

Here, $\Delta\omega$ represents the deviation from the resonant frequency ω_o. Around the resonant frequency ω_o, $Y_{in,D}$ can be approximately expressed as shown in Equation (3.67)

$$Y_{in,D} = -jY_o \cot\left(\frac{\pi}{2} + \frac{\pi}{2}\frac{\Delta\omega}{\omega_o}\right) \cong jY_o \frac{\pi}{2}\frac{\Delta\omega}{\omega_o} \tag{3.67}$$

which can be seen to increase linearly with frequency. On the other hand, the input admittance of the parallel resonant LC circuit, shown in Figure 3.20, is given by Equation (3.68).

$$Y_{in,L} = j\left(\omega C - \frac{1}{\omega L}\right) \tag{3.68}$$

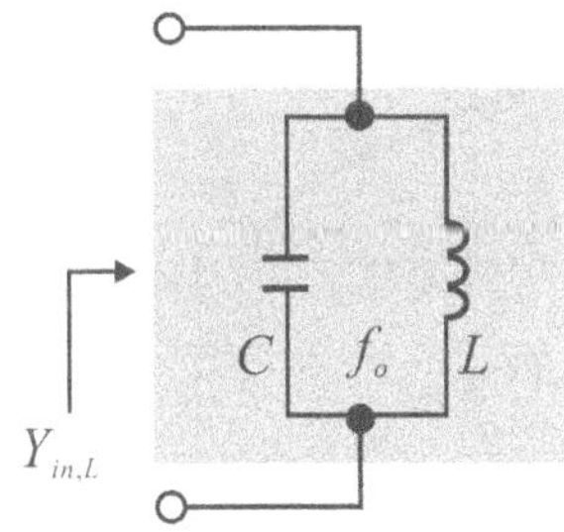

Figure 3.20 Lumped parallel resonant circuit

Around $\omega_o = 1/(LC)^{1/2}$, $Y_{in,L}$ can be approximated as

$$\begin{aligned} Y_{in}^{L} &= j\left(\omega C - \frac{1}{\omega L}\right) = j\left\{(\omega_0 + \Delta\omega)C - \frac{1}{(\omega_0 + \Delta\omega)L}\right\} \\ &\cong j\left(\Delta\omega \cdot C + \frac{\Delta\omega}{\omega_o^2 L}\right) = j2C\Delta\omega \end{aligned} \tag{3.69}$$

Comparing Equation (3.69) with Equation (3.67), we obtain Equation (3.70).

$$C = \frac{\pi Y_o}{4\omega_o} = \frac{1}{8 f_o Z_o} \tag{3.70}$$

Now, the value for the inductance L, which is related to the resonant frequency, is determined as shown in Equation (3.71).

$$L = \frac{1}{\omega_o^2 C} \tag{3.71}$$

The equivalent L and C values for these short-circuited quarter-wavelength and half-wavelength transmission lines can be similarly determined and each is shown in Figure 3.21. Note that the half-wavelength transmission line shows a series resonance and the value of L is first determined similarly to the previous procedure. Also, because the length is twice that of the quarter-wavelength transmission line, the factor 4 instead of 8 appears. In Figure 3.22, the summary for open-circuited quarter- and half-wavelength transmission lines is shown. In the open-circuited case, series resonance appears at the quarter wavelength and parallel resonance appears at the half wavelength.

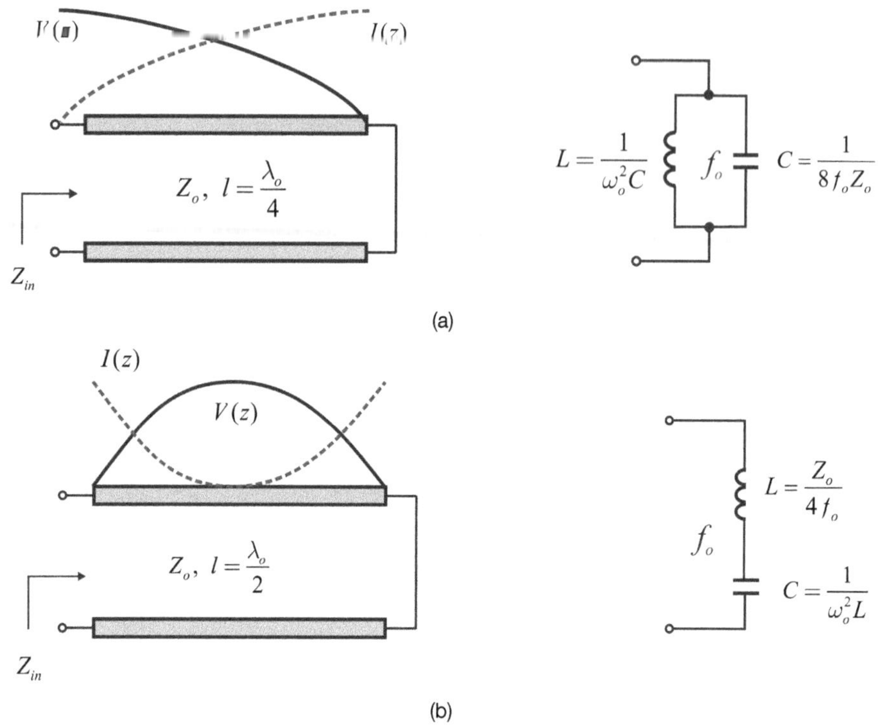

Figure 3.21 Equivalent circuits for (a) shorted quarter-wavelength and (b) half-wavelength transmission lines

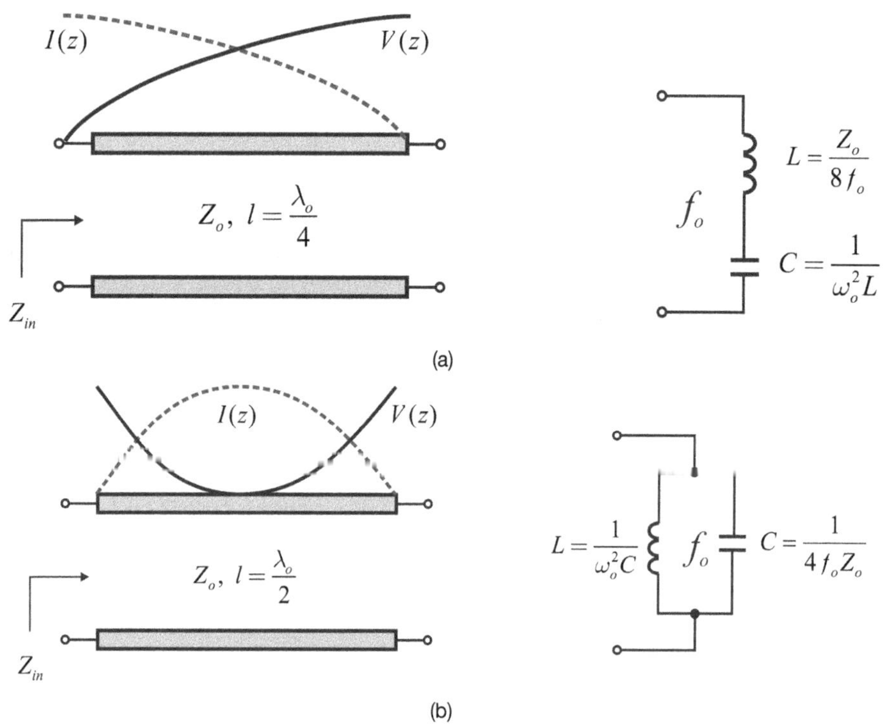

Figure 3.22 Equivalent circuit for (a) open-end quarter-wavelength and (b) half-wavelength transmission lines

Example 3.9

For the shorted quarter-wavelength transmission line with $Z_o = 50\ \Omega$ shown in Figure 3.19, compute the lumped LC-equivalent circuit values at a frequency of 1 GHz.

Solutions

From Figure 3.21(a), the value of C

$$C = \frac{1}{8 f_o Z_o} = \frac{1}{8 \times 1 \times 10^9 \times 50} = 2.5\ \text{pF}$$

is obtained, and the corresponding L value is

$$L = \frac{1}{\omega_o^2 C} = \frac{1}{\left(2\pi \times 10^9\right)^2 \times 2.5 \times 10^{-12}} = 10.13\ \text{nH}$$

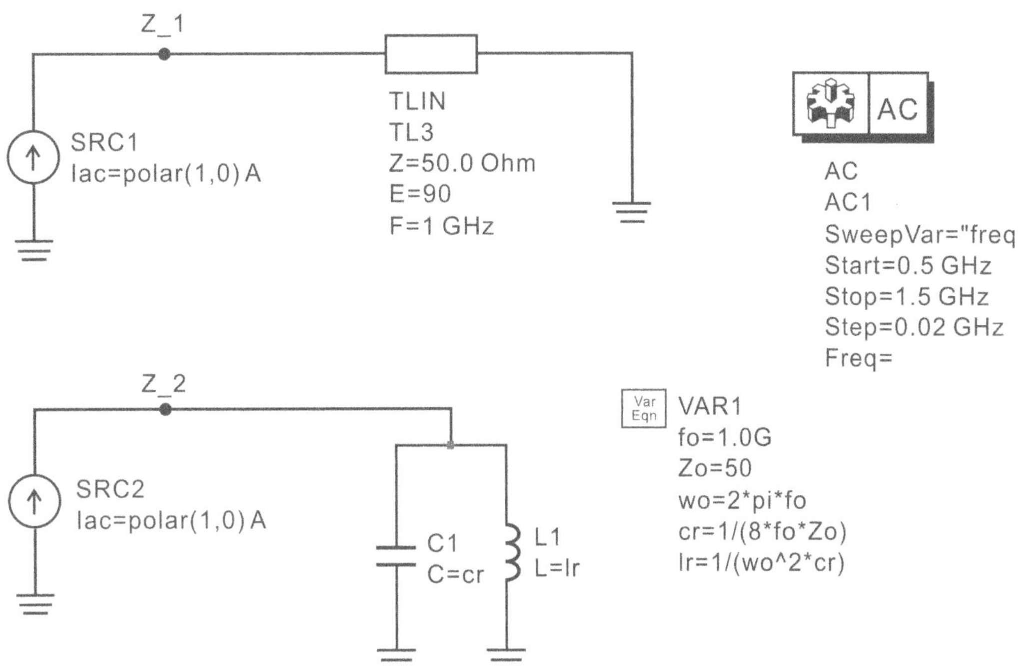

Figure 3E.11 Impedance computation for a parallel resonant circuit and the shorted quarter-wavelength transmission line. The values of LC for the parallel resonant circuit are computed in **VAR**1 at a frequency of 1 GHz. Since unit current sources are used, the voltages **Z**_1 and **Z**_2 become the impedances.

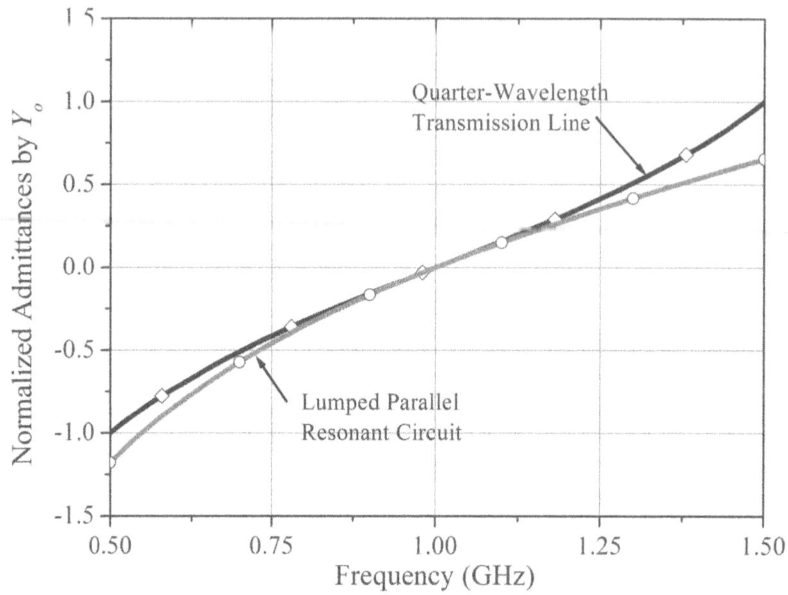

Figure 3E.12 Comparison of the admittances of a quarter-wavelength transmission line and a lumped parallel resonant circuit. The admittances of the quarter-wavelength transmission line and the parallel resonant circuit are computed by imag(1/**Z**_1) and imag(1/**Z**_2), respectively.

In order to examine the frequency range at which the lumped *LC*-equivalent circuit can approximate the transmission line, we set up the circuit shown in Figure 3E.11 in ADS and compare the impedances.

Figure 3E.12 also shows the simulated admittances. The admittance of the lumped *LC* parallel resonant circuit, 1/**Z**_2, is compared to that of the shorted, quarter-wavelength transmission-line resonator 1/**Z**_1. The two admittances show relatively close agreement near the resonant frequency, but differences can be found as the frequency moves away from the resonant frequency.

Example 3.10

Figure 3E.13 shows an RF circuit connected to a quarter-wavelength transmission line, the end of which is shorted with a bypass capacitor. Explain why the transmission line does not affect the RF circuit at the center frequency f_o.

Solutions

Since the bypass capacitor *C* acts as a short, the transmission line seen from the RF circuit acts as a shorted quarter-wavelength transmission line. As a result, a parallel resonant circuit appears at the input of the RF circuit, which becomes an open circuit at the resonant frequency. Thus, the transmission line circuit has no effect on the RF circuit and acts as an RF choke (RFC). Notably, DC voltage can be supplied to the RF circuit by applying the DC voltage to the top of the bypass capacitor. The transmission line circuit can operate as an RFC for a broader bandwidth as the characteristic impedance of the transmission line becomes higher.

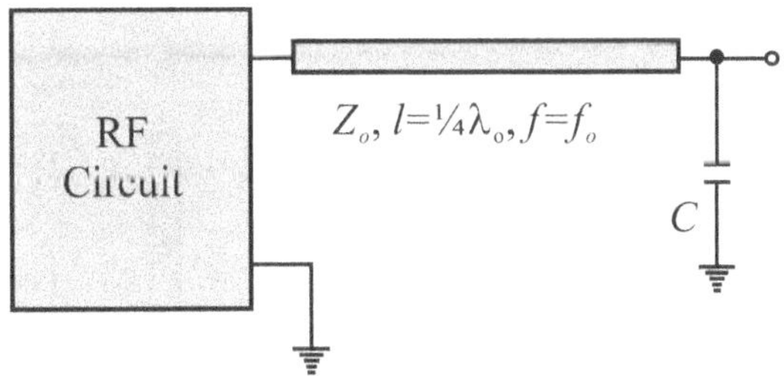

Figure 3E.13 RF choke circuit. The capacitor C at the end of the quarter-wavelength transmission line is sufficiently large and operates as a bypass capacitor. The impedance seen in the capacitor-loaded quarter-wavelength transmission line is a parallel resonant circuit and it approaches infinity at the resonant frequency. Thus, the capacitor-loaded quarter-wavelength transmission line does not affect the RF circuit at the resonance frequency. It is called an RF choke (RFC).

Example 3.11

Figure 3E.14 is a photograph of an oscillator that employs a dielectric resonator oscillator.

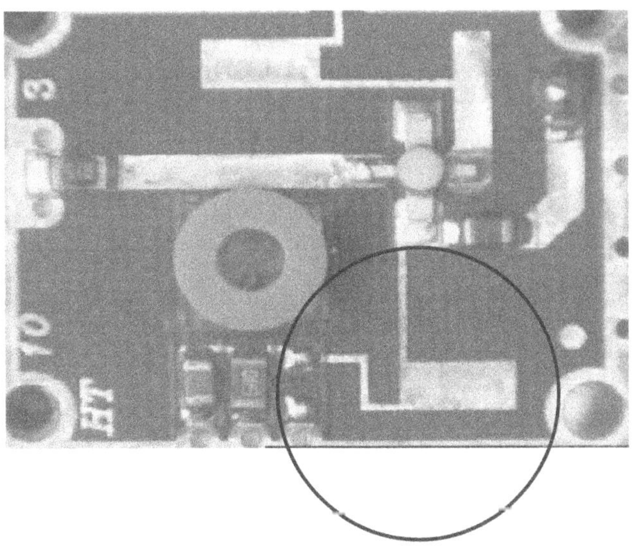

Figure 3E.14 An RF choke used in a dielectric resonator oscillator. The circled area is the RF choke that is used to bias the transistor.

In the oscillator circuit, an RFC circuit similar to the RFC circuit explained in the previous example is also used. Explain the function of the area circled in Figure 3E.14.

Solutions

Figure 3E.15 shows an equivalent circuit for the dielectric-resonator oscillator in Figure 3E.14. Transmission lines A, B, and C represent the equivalent circuit of the circled area in Figure 3E14. The characteristic impedance of narrow microstrip lines is denoted as Z_H, while that of the wide microstrip is denoted as Z_L. Here, the impedance of transmission line C seen from the connection point behaves as

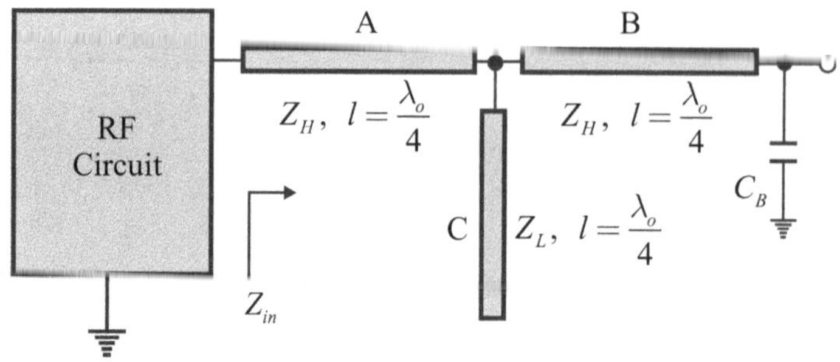

Figure 3E.15 Equivalent circuit of the circled area shown in Figure 3E.14

a series resonant circuit, as shown in Figure 3.22(a). Consequently, the point where transmission lines A, B, and C are connected is actually shorted to the ground. Since the connection point is grounded, transmission line A presents an impedance $Z_{in} = \infty$ when viewed from the RF circuit, and the circled circuit acts as an open circuit to the RF circuit at the resonant frequency.

Note that the one-end-open, quarter-wavelength transmission line such as transmission line C in the figure presents a short-circuit impedance when seen from the other end of the line. In contrast, the one-end-shorted, quarter-wavelength transmission line provides an open-circuit impedance when seen from the other end of the line. Through similar reasoning, transmission line B provides an open-circuit impedance at the connection point because its end is shorted to the ground by the bypass capacitor C_B. However, its effect does not appear because the connection point is shorted to the ground. The role of transmission line B is to widen the bandwidth of RFC. At a lower frequency, the transmission lines A and B function as inductors, and transmission line C functions as a capacitor, and together they function as a low-pass filter. The circuit in Figure 3E.15 behaves as a broader-band RFC than the RFC circuit in Example 3.10.

Example 3.12

Figure 3E.16 is an oscillator employing a quarter-wavelength coaxial-line resonator. Explain the principle of the oscillator's operation.

Solutions

The active device in Figure 3E.16 sees the impedance of a shorted quarter-wavelength coaxial line that appears as a parallel resonant circuit. Therefore, the equivalent circuit is configured as shown in Figure 3E.17.

Since the active device is biased to have negative resistance, the total resistance of the resulting parallel resonant circuit is negative. Consequently, an exponentially increasing voltage appears across the active device, which at equilibrium becomes a sine wave with constant amplitude. Accordingly, the oscillation frequency becomes approximately the same as the resonant frequency of the quarter-wavelength coaxial line. Generally, the oscillation frequency differs slightly from the quarter-wavelength resonant frequency due to the parasitic elements arising from the active device. The screw in Figure 3E.16 is inserted to obtain the desired oscillation frequency by means of its adjustment.

The loop probe is inserted for sensing magnetic fields. The reason for the placement of the loop probe is that the current is at its maximum at the shorted end. As the current is proportional to the magnetic field, the magnetic field is strongest near the shorted end. Therefore, in order to couple the oscillation power from the point where the magnetic field is strongest, the loop probe is inserted as illustrated. If a dipole probe was inserted instead of the loop probe, the electric field near the shorted

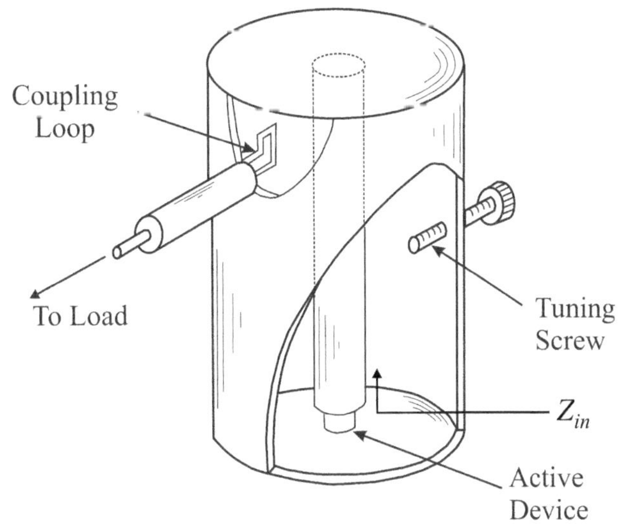

Figure 3E.16 An oscillator employing a coaxial line. The active device is a one-port device with negative resistance. The coupling loop is used to couple the power inside the oscillator. Since the magnetic field is strong at the shorted end of the coaxial line, the loop probe is inserted. The tuning screw is used to tune the oscillation frequency.

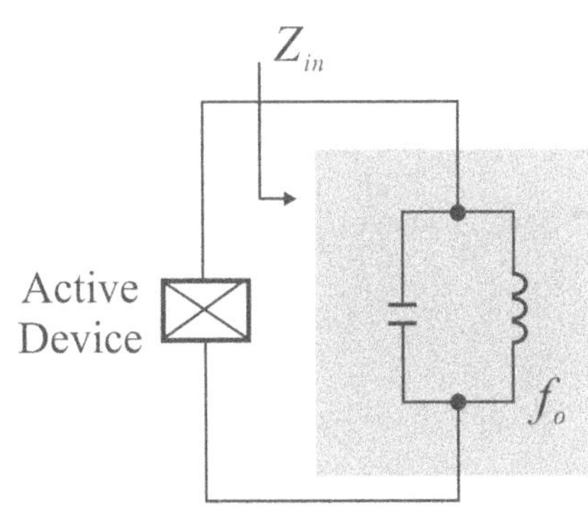

Figure 3E.17 Equivalent circuit of the oscillator in Figure 3E.16. The shorted quarter-wavelength coaxial line in Figure 3E.16 can be modeled as a parallel resonant circuit with a resonance frequency of f_o.

end would be almost nonexistent and there would be almost no coupled output power. Taking the resonance property of the shorted quarter-wavelength coaxial line into consideration, the dipole probe should be inserted near the active device.

3.5.3 Two-Port Circuit Application

We have previously seen the one-port circuit application of a short-length transmission line to a capacitor or an inductor. We have also seen how quarter- or half-wavelength transmission lines can be used to replace lumped *LC*-resonant circuits. In addition to these applications, transmission lines are frequently used as two-port circuit components. The application of a quarter-wavelength transmission line as an impedance inverter is a typical example. In this section, we will learn the widely used two-port circuit representation of a transmission line.

3.5.3.1 Impedance Inverters

When a load Z_L is connected to one end of a quarter-wavelength transmission line with the characteristic impedance Z_o, the input impedance Z_{in} is shown in Equation (3.72)

$$Z_{in} = Z_o \frac{Z_L + Z_o \tan(\beta z)}{Z_o + Z_L \tan(\beta z)} = \frac{Z_o^2}{Z_L} \tag{3.72}$$

which behaves as an impedance inverter. This is a typical application in power dividers, the structure of which is shown in Figure 3.23. The purpose of the power divider is to divide an input power equally into two output ports without reflection. When the two output ports are directly connected to the input port without transmission lines, the input impedance Z_{in} becomes $Z_o/2$ and there will be reflection at the input due to mismatch. Thus, impedance matching is required. To match the input, the impedance Z_1 shown in Figure 3.23 must be $2Z_o$ so that power can be delivered to the two output ports without reflection. When a quarter-wavelength transmission line of Z_x is inserted for the purpose of impedance matching, the impedance Z_1 seen from the input power into the transmission line then becomes

$$Z_1 = 2Z_o = \frac{Z_x^2}{Z_o}$$

The impedance of the transmission line Z_x is then determined as shown in Equation (3.73)

$$Z_x = \sqrt{2}Z_o \tag{3.73}$$

and the input power will be equally divided into the two output ports without reflection.

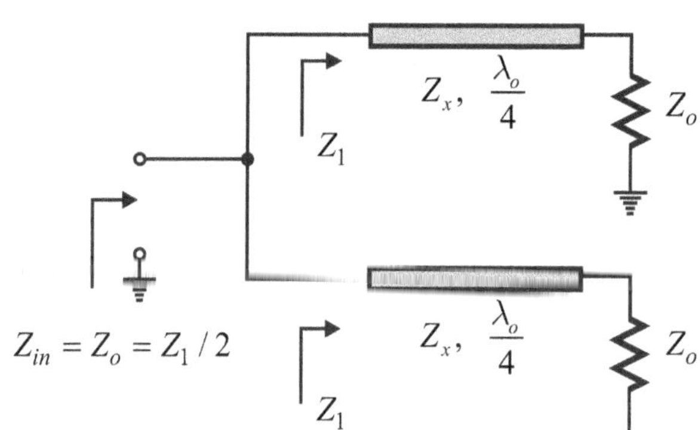

Figure 3.23 Schematic of a power divider. The quarter-wavelength transmission line matches the 50 Ω to 100 Ω. As a result, Z_{in} becomes 50 Ω.

Example 3.13

To satisfy the condition $Z_1 = 2Z_o$, it can be achieved with a resistor instead of the transmission line, as shown in Figure 3E.18(a). When $Z_o = 50\ \Omega$, Z_1 will be $100\ \Omega$. Then, the power from the input port can be equally divided and delivered to the two output ports. This type of divider is shown in Figure 3E.18(b) and it is called a *resistive power divider.* Determine the difference in the power delivered to the two ports when the resistive power divider is used instead of the power divider shown in Figure 3.23.

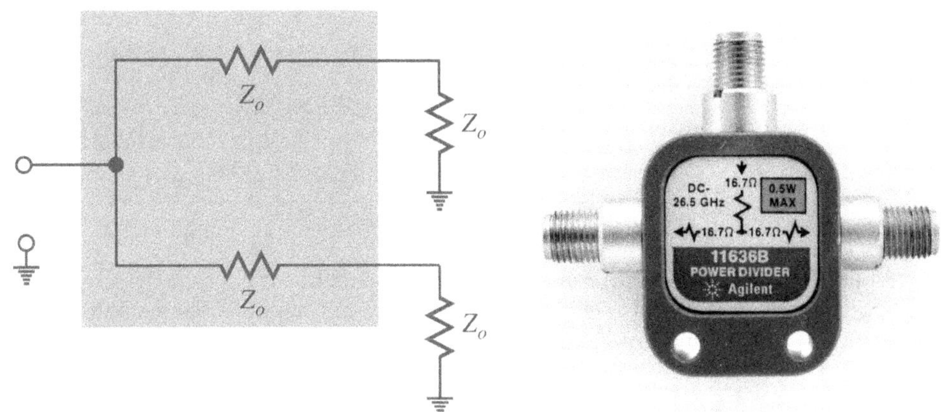

(a) (b)

Figure 3E.18 (a) A resistive power divider and (b) a photo of a resistive power divider.[2] Note that the power delivered to the load becomes the half of the power divider in Figure 3.23. However, the resistive power divider provides a broadband operation.

Solutions

In the power divider shown in Figure 3.23, all the input power is delivered to the load with none consumed, whereas in the case of the resistive power divider, as power is delivered only to Z_0 out of the total resistance $2Z_o$, the delivered power is half when compared with the divider in Figure 3.23. Thus, the output power is reduced by 6 dB when compared with the input power. The resistive power divider is widely used in instruments due to their broadband operation despite the loss of the delivered power.

3.5.3.2 Two-Port Lumped-Element Equivalent Circuit

A transmission line is a passive two-port circuit and it can be represented by a two-port lumped-element equivalent circuit within a limited frequency range. The two-port circuit representation of a transmission line allows the lumped-component implementation of circuits such as delay line, branch-line coupler, rat-race ring, or power dividers, which are easily implemented using transmission lines. In this section, we will learn how to construct such transmission-line two-port equivalent circuits as well as how they are used.

2. Agilent Technologies, Differences in Application Between Power Dividers and Power Splitters, Application Note 5989-6699EN, 2007.

From circuit theory, a general, passive two-port circuit can be represented by T- or pi-shaped equivalent circuits, as shown in Figure 3.24. In the case of symmetry, the circuit in Figure 3.24 satisfies (a) $Z_a = Z_c$ and (b) $Y_1 = Y_3$. Since the transmission line of electrical length θ, shown in Figure 3.25(a), is a symmetrical, passive two-port circuit, it can be represented by the symmetrical cases of T- and π-equivalent circuits shown in Figure 3.24(a) and Figure 3.24(b). The values of the symmetrical T- and π-equivalent circuits can be obtained by making their two-port parameters equal. A bisection method can be used to determine the values of the symmetrical T- and π-equivalent circuits.

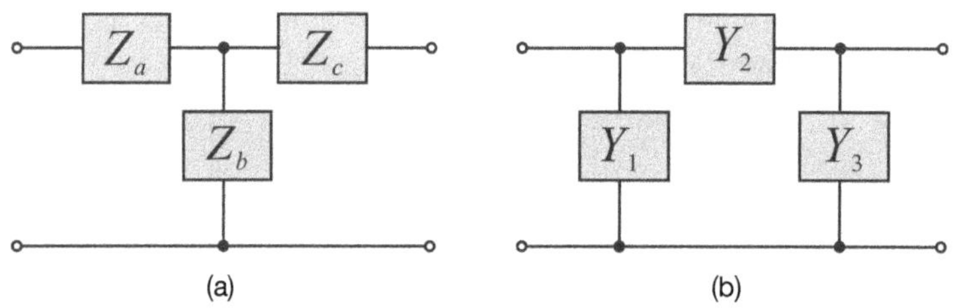

Figure 3.24 (a) T- and (b) π-shaped equivalent circuits for a general passive two-port network. The derivation of an equivalent circuit is explained in Chapter 5. For a symmetrical two-port network $Z_a = Z_c$ in (a) and $Y_1 = Y_3$ in (b).

When the T-shaped and π-shaped equivalent circuits in Figures 3.25(b) and (c) are opened along the line of symmetry, the open-circuit impedances of T- and π-shaped circuits should match the impedance of the bisected transmission line, and this can be written as Equation (3.74).

$$Z_a + 2Z_b = \frac{1}{Y_1} = \frac{Z_o}{j\tan\left(\dfrac{\theta}{2}\right)} \tag{3.74}$$

Also, when the T- and π-shaped circuits are shorted along the line of symmetry, the short-circuit impedances of the T- and π-shaped circuits should match that of the bisected transmission line, and this results in

$$Z_a = \frac{1}{Y_1 + 2Y_2} = jZ_o\tan\left(\frac{\theta}{2}\right) \tag{3.75}$$

Therefore, Z_a in Equation (3.75) for the T-shaped circuit can be determined, and when it is substituted in Equation (3.74), it will yield the value of Z_b. Thus, the values for the T-shaped equivalent circuit are shown in Equations (3.76a) and (3.76b).

$$Z_a = Z_c = jZ_o\tan\left(\frac{\theta}{2}\right) \tag{3.76a}$$

$$Z_b = \frac{1}{2}\left[-jZ_o\tan\left(\frac{\theta}{2}\right) + \frac{Z_o}{j\tan\left(\dfrac{\theta}{2}\right)}\right] = -jZ_o\csc(\theta) \tag{3.76b}$$

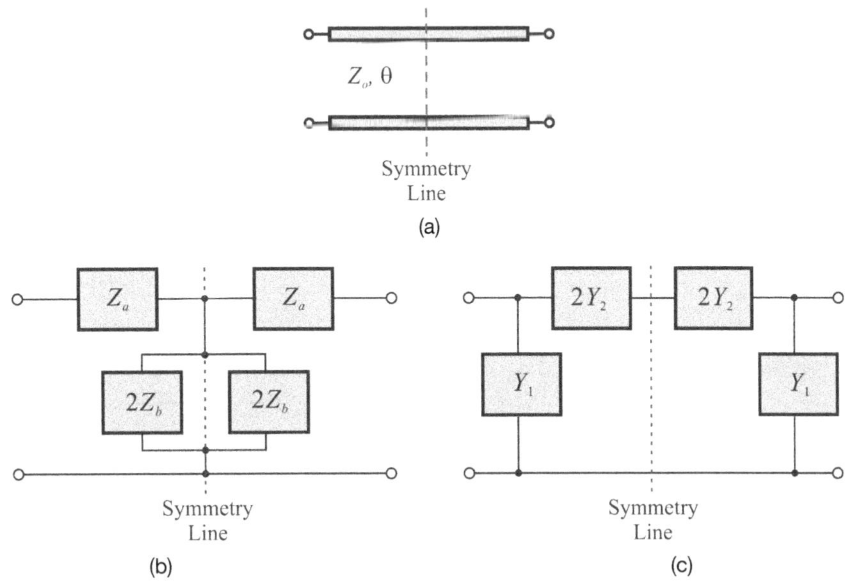

Figure 3.25 Lumped-element equivalent circuit of (a) a θ-long transmission line; (b) a T-shaped equivalent circuit; and (c) a π-shaped equivalent circuit. The values of T- and π-shaped equivalent circuits can be derived by bisection along the symmetry line. $Z_a = jZ_o\tan(\theta/2)$ and $Z_b = -jZ_o\csc(\theta)$ in (a) while $Y_1 = jY_o\tan(\theta/2)$ and $Y_2 = -jY_o\csc(\theta)$ in (b).

Furthermore, Y_1 in Equation (3.74) for the pi-shaped circuit can be determined, which when substituted in Equation (3.75), will give Y_2. The circuit values for the π-shaped equivalent circuit are shown in Equations (3.77a) and (3.77b).

$$Y_1 = Y_3 = jY_o \tan\left(\frac{\theta}{2}\right) \tag{3.77a}$$

$$Y_2 = -jY_o \csc(\theta) \tag{3.77b}$$

For a quarter-wavelength line, $\tan(\theta/2) = \csc(\theta) = 1$ and the equivalent circuit shown in Figure 3.26 can be obtained. Note that all the reactance of inductors and capacitors is equal to the characteristic impedance of the transmission line Z_o.

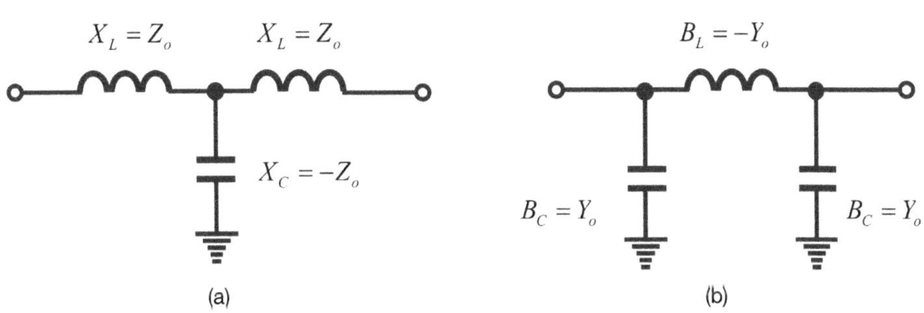

Figure 3.26 (a) T- and (b) π-equivalent circuits for a quarter-wavelength transmission line. Note that the values of the three elements have the same magnitude and are determined by Z_o or Y_o.

Example 3.14

By using the lumped-element equivalent circuit in Figure 3.26(b), design the power divider in Figure 3.23 and verify its operation using ADS simulation.

Solutions

The π-equivalent circuit of a quarter-wavelength transmission line in Figure 3.26(b) is used to replace the two quarter-wavelength transmission lines in the power divider and the power divider can then be configured as shown in Figure 3E.19.

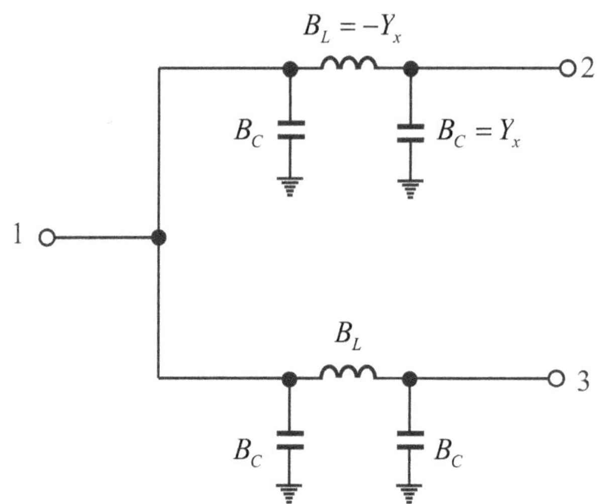

Figure 3E.19 A power divider employing lumped components. The two quarter-wavelength transmission lines in Figure 3.23 are replaced by π-shaped equivalent circuits.

The circuit shown in Figure 3E.20 is set up in ADS to verify its operation. In this circuit, both ports 2 and 3 are terminated by 50-Ω resistors and the voltages appearing at the ports are represented by **vout1** and **vout2**. The input is represented by a Norton equivalent circuit with a current source of 1 A and a 50-Ω resistor in parallel. Both the values of L and C are assigned to have an impedance of $Z_x = (2)^{½} \cdot 50$ using **VAR** (variables and equations components). The computed voltages are shown in Table 3E.1.

Table 3E.1 Calculated voltages (note that since the impedance seen from the current source is 25 Ω, vin1=25 V)

freq	vout1	vout2	vin1
1.0 GHz	17.678/−90°	17.678/−90°	25/0°

First, since the impedance when looking into the power-divider circuit is 50 Ω, the input resistance seen from the current source **SRC2** will be 25 Ω because the input resistance is a parallel combination of **R1** and the power-divider input resistance. A current of 1 A flowing into the parallel

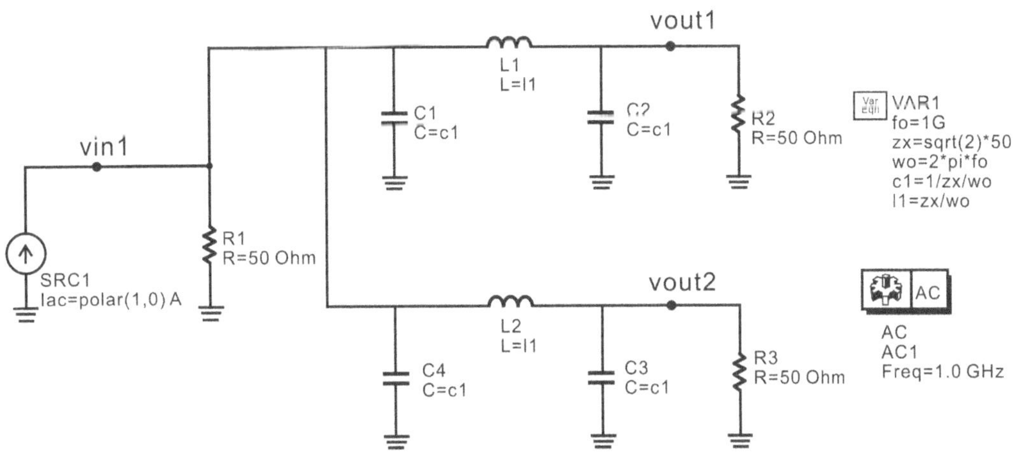

Figure 3E.20 Simulated circuit for a power divider using lumped components at a frequency of 1 GHz. First, **zx** in **VAR1** is defined to match 50 Ω to 100 Ω. Then, the values of an inductor and capacitor that give the impedance **zx** at a frequency of 1 GHz are computed. The source is represented by the Norton equivalent circuit with the impedance 50 Ω.

combined input resistance of 25 Ω will develop a voltage 25 V for **vin**1. Second, in order to know whether the available power has been delivered into the two output ports without reflection, the following equations are entered in the display window.

Eqn pa=1/8*50
Eqn p1=1/2*mag(vout1)**2/50
Eqn r=p1/pa

Measurement Expression 3E.3 Calculation for available and delivered power in the display window

Here, power **pa** means the maximum available power from the Norton source and **p**1 means the power delivered to the 50-Ω load. Therefore, comparing the two, the ratio **r** should be half if the input power is equally divided and delivered to the load without reflection. Now we can find whether the available power is delivered without reflection through **r**. Table 3E.2 shows the result of the **r** value displayed using the listing box. The result will be that the available power is delivered without reflection.

Table 3E.2 Calculation results using Measurement Expression 3E.3 (the variable *r* is the ratio of available power to delivered power to the load)

freq	r
1.0 GHz	0.5

3.6 DISCONTINUITIES

In the practical implementation of a transmission-line circuit, multiple transmission lines cannot be connected to a single point and often require an area, which is called *discontinuity*. These discontinuities store a part of the energy around them and the stored energy cannot propagate through the transmission lines. Consequently, the discontinuities can be modeled as energy-storing inductors or capacitors. In this section, typical examples of such discontinuities will be discussed and their effects will be explained qualitatively. Approximate circuit models of the discontinuities are already built in as components in ADS, which makes it possible to simulate their effects without specific knowledge of them. Thus, a qualitative understanding may be sufficient to help understand the circuit-simulation results. In addition, we will also learn how to evaluate such discontinuities by example.

3.6.1 Open-End Microstrip

Figure 3.27(a) shows a top view of an open-end microstrip line, while Figure 3.27(b) shows its equivalent circuit. Additional fringing fields arise at the end of the open-end microstrip. The effect of the open-end capacitance can be represented by an additional capacitance C_{oc}.

The capacitance C_{oc} is usually called a microstrip *open-end capacitance*, the approximate effect of which looks like an extension of the microstrip line's length. The length extension from the reference plane defined at T–T′ in Figure 3.27 is given by

$$\Delta l = \frac{C_{oc}}{C} \tag{3.78}$$

The C in Equation (3.78) represents the capacitance per unit length of the microstrip line and C_{oc} represents the microstrip's open-end capacitance.

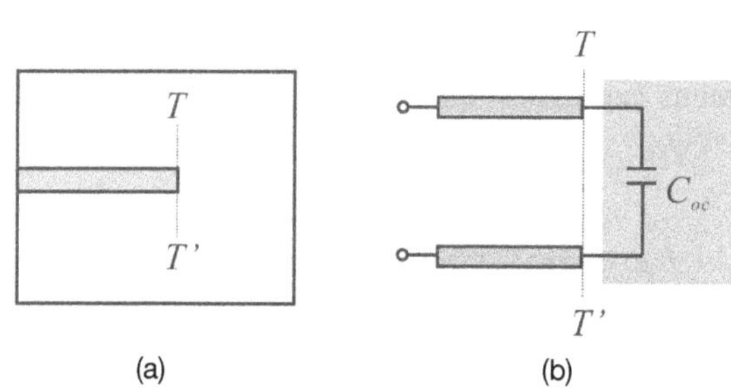

Figure 3.27 Open-end microstrip: (a) top view and (b) its equivalent circuit. The capacitor C_{oc} reflects the extra fringing capacitance due to the open end.

Example 3.15

For microstrip components in ADS, there are two open stubs; one component, **MLEF**, takes an open-end capacitance into account whereas the other component, **MLOC**, does not. For a 10-mil-thick alumina substrate, compare the reflection coefficients of these two open stubs at a frequency of 10 GHz. In addition, calculate the length extension due to the open-end capacitance.

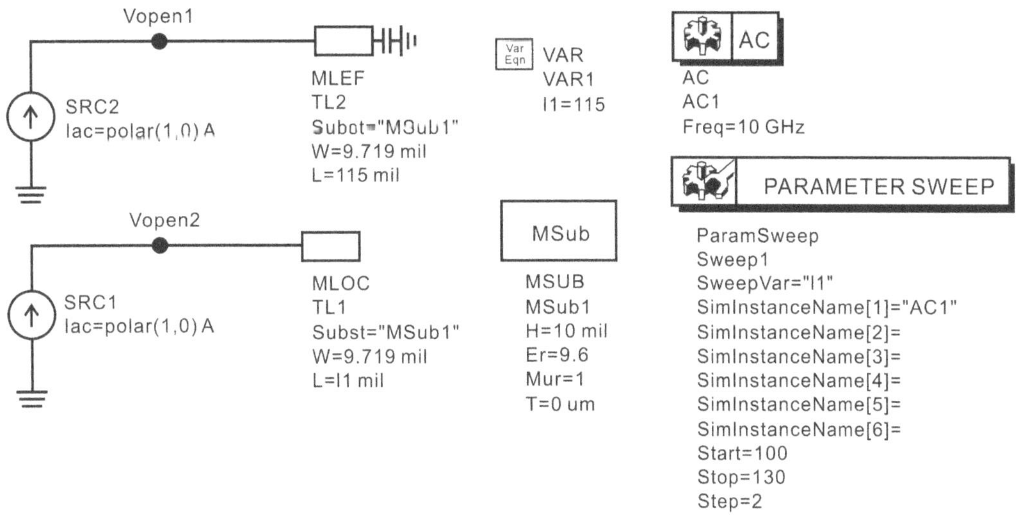

Figure 3E.21 Evaluation of a microstrip open-end effect by simulation. The component **MLEF** is the microstrip circuit element taking the open-end effect into account, whereas **MLOC** is just the microstrip line whose end is open. **Vopen**1 and **Vopen**2 become the impedances because the AC current sources are both set to 1 A. To find the length extension of **MLEF**, the length l_1 of **MLOC** is varied.

Solutions

Set up the circuit as shown in Figure 3E.21. The **MLEF** is a microstrip open stub with the open-end effect and **MLOC** is a microstrip open stub without the open-end effect. The width and length for a 50-Ω quarter-wavelength line are calculated using **LINECALC**. The values of width and length are 9.719 mil and 115 mil, respectively. In order to find the length extension of **MLEF** as a consequence of the open-end effect, the length l1 of the **MLOC** is set as a variable.

In Figure 3E.21, **Vopen**1 and **Vopen**2 directly become the impedances. In order to convert the impedances to reflection coefficients, the following equations in Measurement Expression 3E.4 are entered in the display window:

Eqn s1=(Vopen1-50)/ (Vopen1+50)

Eqn s2=(Vopen2-50)/ (Vopen2+50)

Measurement Expression 3E.4 Calculation of reflection coefficients

Next, the result is displayed in Figure 3E.22. Since the length of **MLOC** is set to vary, the reflection coefficient of **MLOC** will increase in a clockwise direction. Comparing the reflection coefficient of the two lines, when the length of the **MLOC** increases by approximately 3 mil from 115 mil, **MLOC** with the increased length gives the same reflection coefficient as the **MLEF**. Therefore, through this example it can be seen that the open-end capacitance corresponds to the length extension of approximately 3 mil at 10 GHz.

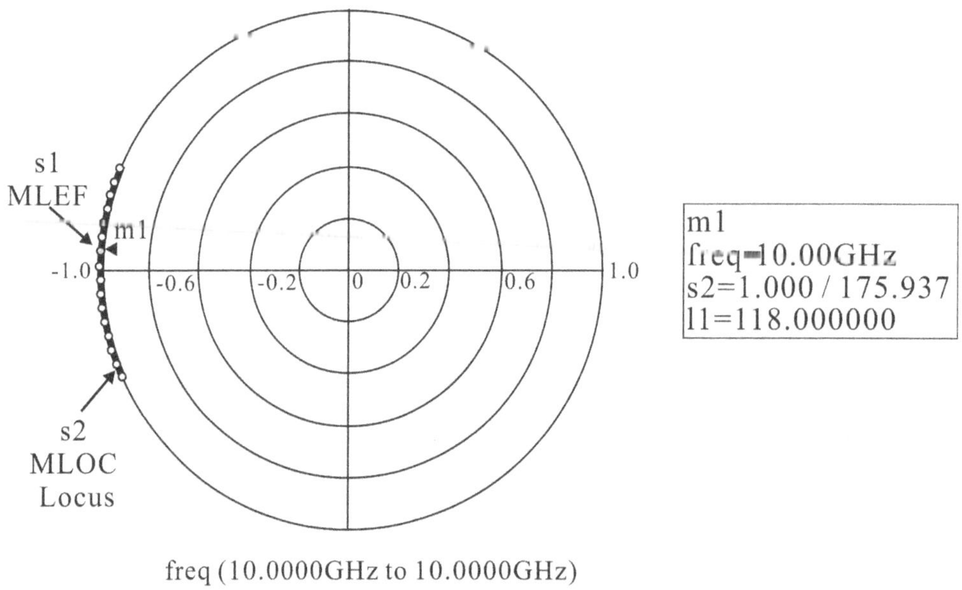

Figure 3E.22 Simulated reflection coefficients of the two open-end microstrip lines in Figure 3E.21. The reflection coefficient **s1** is for **MLEF** and **s2** is for **MLOC**. The length l_1 of **MLOC** should be 118 mil to give the same reflection coefficient as that of **MLEF**, which is approximately 3-mil long compared with the length of **MLEF**.

3.6.2 Step and Corner Discontinuities

Figure 3.28 shows step and corner discontinuities. The equivalent circuit of the step and corner discontinuities is shown in Figure 3.28(c). Due to the current discontinuities at the step and corner, there will be energy stored around these discontinuities, which can be represented as an inductor in the equivalent circuit. Two inductors appear in both connecting lines to form a step and corner discontinuity. These inductors are usually small and their effects can be approximately represented by a small extension of the length of both microstrip lines. Defining L_{o1} and L_{o2} as the inductances per unit length of the lines on the left- and right-hand sides, and L_1 and L_2 as the inductances arising from the discontinuity, length extension l_1 and l_2, similar to the open-end case, are approximately given by Equation (3.79).

$$l_1 = \frac{L_1}{L_{o1}}, \qquad l_2 = \frac{L_2}{L_{o2}} \tag{3.79}$$

In Figure 3.28(c), the change in length of the lines from reference planes T and T' is denoted as l_1 and l_2. In the figure, a capacitor C_B is shown that represents an additional fringing capacitance or a capacitance caused by the area of the discontinuity.

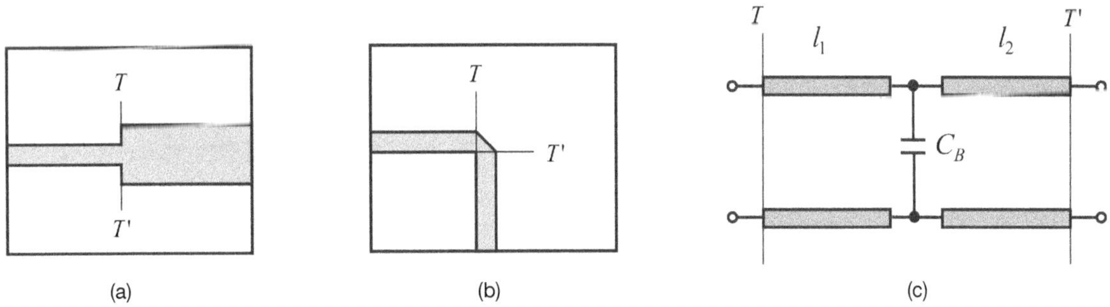

Figure 3.28 Microstrip (a) step, (b) corner, and (c) their equivalent circuit. In the case of the step discontinuity, the discontinuity has no length. In the case of the corner discontinuity, the discontinuity region is defined by the triangle with reference planes T and T'.

3.6.3 T-Junction and Cross Junction

As shown in Figures 3.29 and 3.30, when three or four microstrip lines are connected, the connection requires a common area that is approximated to a point in circuit analysis. The effect of such discontinuities is generally not small. However, similar to step and corner discontinuities, they are modeled as a length extension of the lines and a discontinuity capacitor. It is noteworthy that the equivalent circuit appears to be connected in shunt because the electric field at the junction area is common to all the connection lines. Approximating the common junction area as a point, the voltage at the connection point can be approximately equal and the shunt connection is physically plausible. With planar transmission lines, in the case of a slot line or CPW, care should be taken in building up the equivalent circuit because the electric field at T and the cross junction are not common to all the connection lines. Refer to reference 3 at the end of this chapter for detailed information.

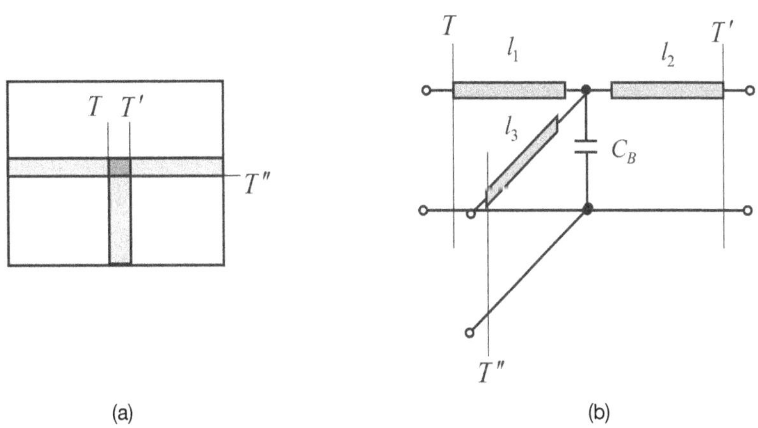

Figure 3.29 (a) A T-junction and (b) its equivalent circuit. The T-junction equivalent circuit in Figure 3.29(b) describes the discontinuity of the rectangle defined by reference planes T, T', and T''. The capacitor C_B is physically interpreted as the capacitance due to the area of the discontinuity rectangle and the lengths l_1, l_2, and l_3 of the transmission lines are due to the length that reaches the center point of the discontinuity rectangle.

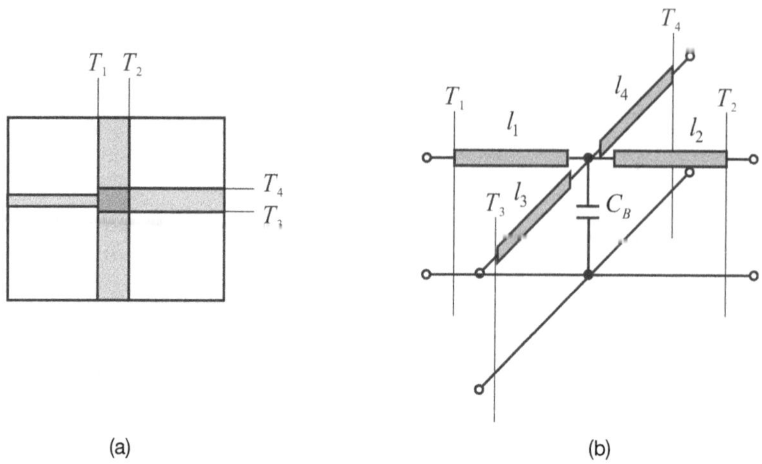

Figure 3.30 (a) Cross-junction and (b) its equivalent circuit. The cross-junction equivalent circuit describes the discontinuity rectangle defined by reference planes T_1, T_2, T_3, and T_4. The physical interpretation of the equivalent circuit is similar to T-junction discontinuity.

3.7 SUMMARY

- The voltage and current of a transmission line are waves propagating along the propagation axis z with a phase velocity v_p and their waveforms have the same shape. The ratio of the voltage to the current is defined as the characteristic impedance Z_o.
- The time delay appears in the transmission line due to v_p and thus the transmission line of a finite length is specified in various ways such as (v_p, length), (fraction of wavelength, frequency), and (electrical length, frequency).
- The transmission-line parameters Z_o and v_p can be measured using open- and short-circuit impedances.
- Coaxial and microstrip lines are the most widely used transmission lines. As the outer diameter of a coaxial line becomes smaller, the line can be used as a single-mode transmission line up to higher frequencies.
- When a mismatched load is connected to the end of a transmission line, a reflection occurs. As a result, a standing wave is formed. The reflection can be described by the reflection coefficient, VSWR, and the return loss. VSWR is the index representing the standing waveform.
- Series or parallel resonances occur for a transmission line with a length of $n\lambda/4$. The application of the transmission-line resonator and the extraction of the values of LC at resonance are explained.
- A transmission line can be represented by T-shaped, π-shaped equivalent circuits, or a lumped two-port network because the transmission line is a passive device. Also, the applications for the two-port representation are shown.
- Physical implementation of a circuit using a transmission line accompanies discontinuities. The method for characterizing the discontinuities using the approximate discontinuity models built in to ADS is explained.

REFERENCES

1. R. G. Brown, R. A. Sharpe, W. L. Hughes, and R. E. Post, *Lines, Waves, and Antennas*, 2nd ed. New York: John Wiley & Sons, Inc., 1973.
2. J. C. Freeman, *Fundamentals of Microwave Transmission Lines*, New York: John Wiley & Sons, Inc., 1996.
3. K. C. Gupta, R. Garg, I. Bahl, and P. Bhartia, *Microstrip Lines and Slot Lines*, 2nd ed. Dedham, MA: Artech House, Inc., 1995.

PROBLEMS

3.1 In Example 3.1, a traveling sine wave propagating along z direction is given by $\cos(\omega t - \beta z)$. For a constant C, if $\omega t - \beta z = C$ is chosen, what is the meaning? Also, when this constant is taken as the peak of the sine wave, prove that its velocity is v_p.

3.2 In the special case of the transmission line when θ is small, find the values of L and C using the results in Figure 3.25. In addition, show that the equivalent circuits in Figure 3P.1 yield the same wave equation as $\theta \to 0$:

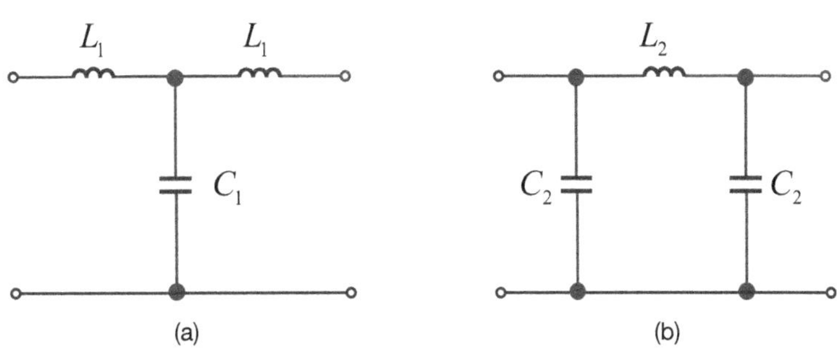

Figure 3P.1 Equivalent circuit of a short-length transmission line: (a) T- and (b) π-shaped

3.3 Considering the two-port circuit in Figure 3P.2 as a transmission line, determine its equivalent characteristic impedance using the open- and short-circuit methods presented in section 3.2.4.

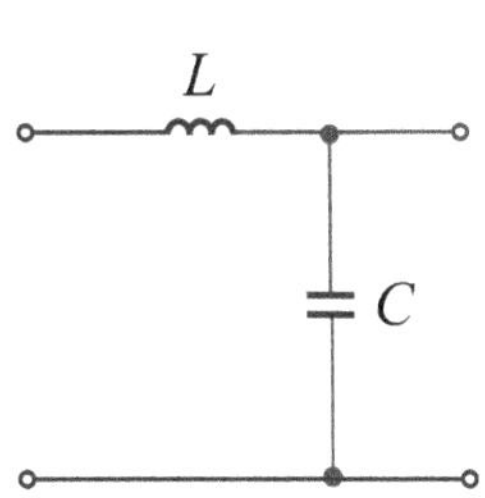

Figure 3P.2 *LC* circuit

3.4 (1) If the electrical length at a frequency of 1 GHz is 90°, what is the electrical length at a frequency of 3 GHz?
(2) Also, if the phase velocity is equal to the speed of light, what is its length?
(3) Finally, what is its propagation constant?

3.5 In the definition of the reflection coefficient shown below, when the real part of Z_L is positive, prove that $|\Gamma_L| \leq 1$.

$$\Gamma_L = \frac{V^-}{V^+} = \frac{Z_L - Z_o}{Z_L + Z_o}$$

3.6 Figure 3P.3 shows a slot-line measurement. Voltage standing-wave minima location is shown by the numbers for two measurements. When the load is replaced by a short, the first minimum occurs at 90 cm. The distance $d_{\min}$ from the first minimum to the short is 15 cm (= 90 – 75 cm in Figure 3P.3) at a frequency of 375 MHz. Find the load impedance.

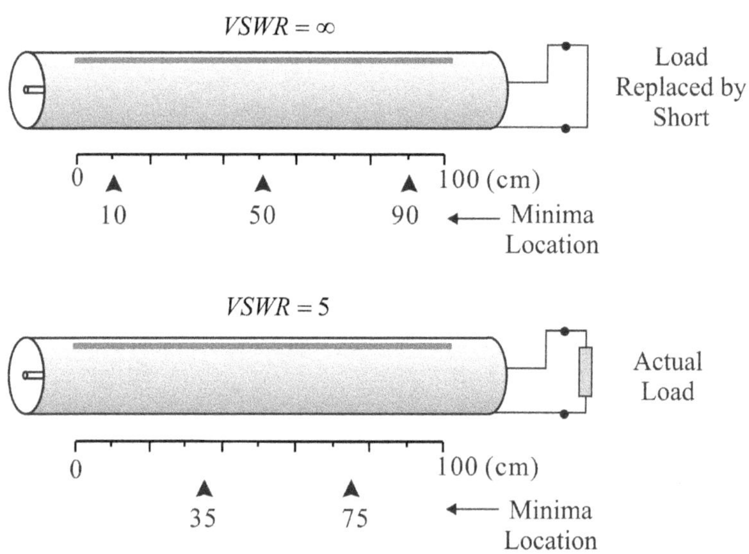

Figure 3P.3 Slot-line measurements

3.7 In Figure 3P.4, V^+ and V^- represent incident and reflected voltage phasors at load plane. Write the voltage V_1 using V^+ and V^-. When the VSWR at the input is 2, find the magnitude of Γ_{in} and the corresponding return loss. In the condition above, setting $Z_o = 50\ \Omega$, find the value of resistor Z_L.

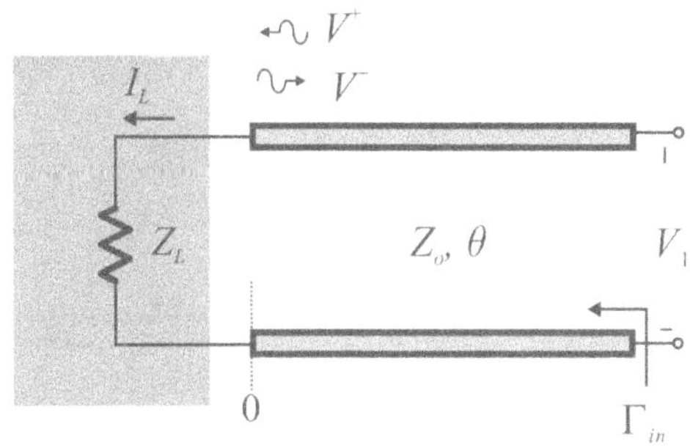

Figure 3P.4 Transmission line terminated by load

3.8 Using the scale at the bottom of the Smith chart in Figure 3.17 and a compass, find the VSWR and the return loss corresponding to $\Gamma = 0.6$.

3.9 For a transmission line with Z_o and θ, derive the *ABCD* parameters defined as

$$\begin{pmatrix} V_1 \\ I_1 \end{pmatrix} = \begin{pmatrix} A & B \\ C & D \end{pmatrix} \begin{pmatrix} V_2 \\ I_2 \end{pmatrix}$$

3.10 Figure 3P.5 shows an equivalent circuit of a distribution amplifier similar to that in Figure 3.1. Find V_1 and V_2.

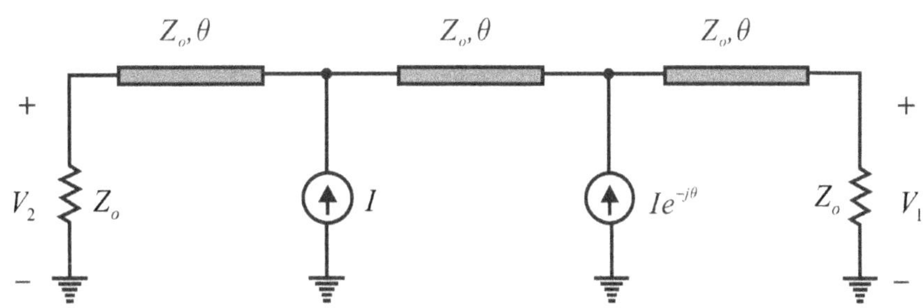

Figure 3P.5 Equivalent circuit of the output of a distribution amplifier

3.11 If a 50-Ω transmission line has a quarter-wavelength length at 1 GHz and its end is short-circuited, can this be considered a parallel resonant circuit near the 1 GHz frequency? What, then, is the capacitance of the parallel resonant circuit?

3.12 In Figure 3P.6, find the characteristic impedance Z_x for a maximum power transfer to load $2Z_o$. For this, find the time-domain voltage waveform $v_1(t)$. Also find the time-domain voltage waveform $v_2(t)$.

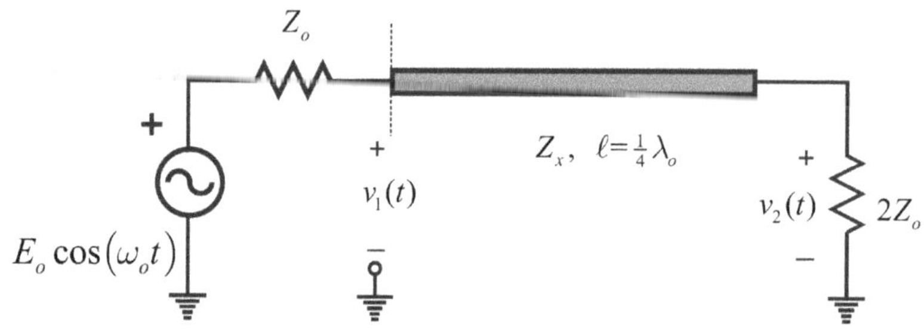

Figure 3P.6 Problem 3.12 circuit

3.13 **(ADS Problem)** The characteristic impedance of a transmission line can be found by applying a pulse signal to the line. Figure 3P.7 shows a simulation setup for computing the characteristic impedance of the transmission line. By computing the current and voltage pulses, and the ratio of the pulse, show that the ratio is the characteristic impedance given by 50 Ω.

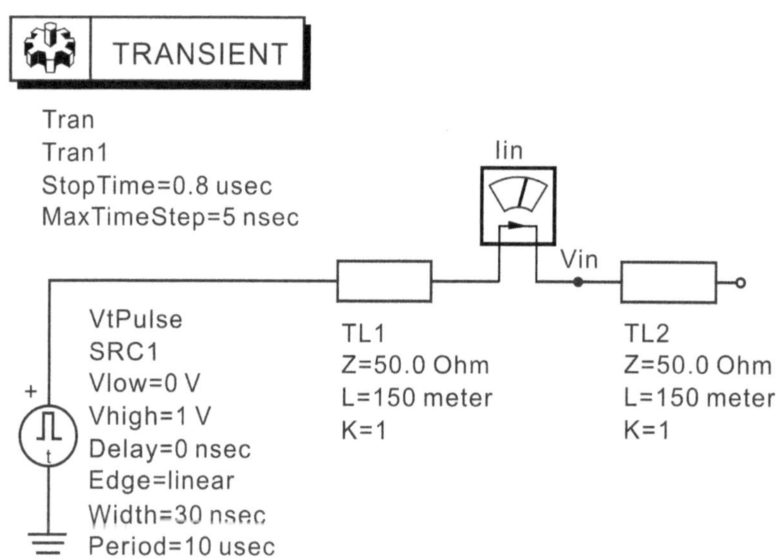

Figure 3P.7 Simulation setup to compute the characteristic impedance using a pulse

3.14 **(ADS Problem)** The standing wave can be simulated by sweeping the length of the transmission line as shown in Figure 3P.8. Plot the mag(**v**1) vs. l1 results in the standing-wave pattern. Also, plot the standing waveform with time as a parameter. It will be apparent that the minimum point does not move despite the time change, which is why it is called a "standing wave."

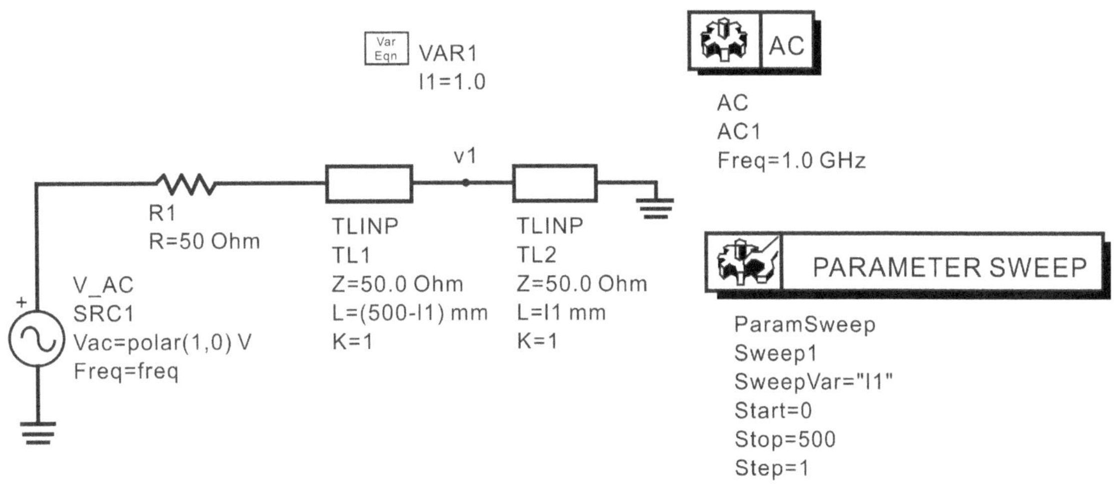

Figure 3P.8 Standing waveform simulation

3.15 **(ADS Problem)** Using ADS, extract the equivalent circuit of corner discontinuity **MCORN**, which has a width of a 50-Ω-line microstrip. Note that it can be represented by a T-equivalent circuit because the corner is a passive circuit. Here, the substrate is a 10-mil-thick alumina with $\varepsilon_r = 9.6$.

CHAPTER

4

S-Parameters and Noise Parameters

CHAPTER OUTLINE

4.1 S-PARAMETERS

A two-port network is a circuit that has two accessible ports. A port is a pair of terminals that satisfies a *port condition*. The port condition requires that the current flowing into one port's terminal should be equal to that flowing out of the other port's terminal. Among multiport networks, a two-port network may be the basic building block for composite functional circuits. Reference directions for the port variables, such as port currents and voltages, will usually be defined as shown in Figure 4.1(a). Generally, this kind of two-port network is represented by two-port parameters that use the relationship between the port voltages and currents. Figure 4.1 shows Z- and Y-parameters defined in terms of port currents and voltages. Of the port variables, when the port currents are chosen as independent variables or sources, it results in the definition of Z-parameters as shown in Equation (4.1).

$$\begin{pmatrix} V_1 \\ V_2 \end{pmatrix} = \begin{pmatrix} z_{11} & z_{12} \\ z_{21} & z_{22} \end{pmatrix} \begin{pmatrix} I_1 \\ I_2 \end{pmatrix} \tag{4.1}$$

Alternatively, when the port voltages are chosen as independent variables or sources, it results in the definition of Y-parameters as shown in Equation (4.2).

$$\begin{pmatrix} I_1 \\ I_2 \end{pmatrix} = \begin{pmatrix} y_{11} & y_{12} \\ y_{21} & y_{22} \end{pmatrix} \begin{pmatrix} V_1 \\ V_2 \end{pmatrix} \tag{4.2}$$

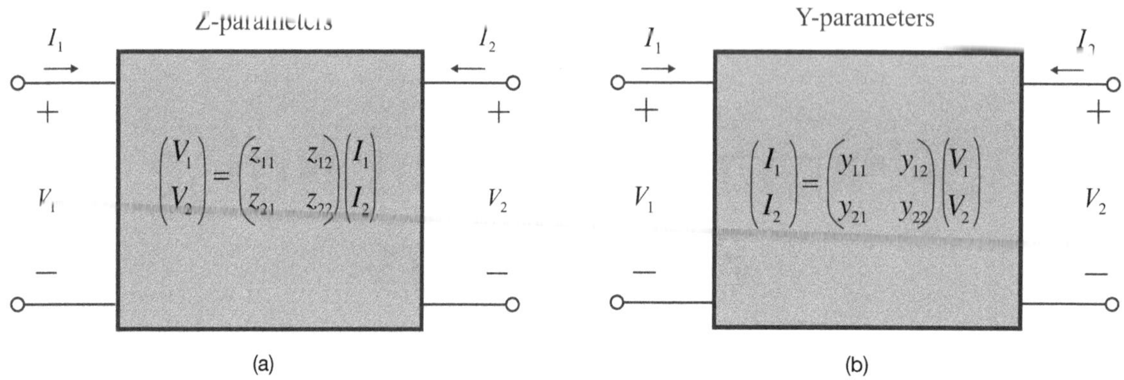

Figure 4.1 Two-port parameters: (a) Z-parameters and (b) Y-parameters

However, the measurement of the Z- and Y-parameters requires open- and short-circuiting of ports, and the measurement of the port voltage or current for the open or shorted port, respectively. Creating an open or short-circuit at the port planes or *reference planes* of the device under test (DUT) requires certain modifications of the DUT, which can be problematic. Furthermore, it is usually difficult to measure port voltages and currents at the port planes. In addition, such open- or short-circuited ports can sometimes cause an oscillation in the case of a sensitive active device, which could cause measurement problems.

On the other hand, in contrast to the Z- and Y-parameters that are directly defined by the port voltages and currents, S-parameters are defined by the ratio between the incident and reflected voltages. This definition does not require any significant modification of the DUT. Measurement can easily be conducted by employing a *directional coupler*. Thus, S-parameters are obviously more convenient when measuring and are thus widely used. In addition, they can also be converted to other parameters. The S-parameters are commonly measured using the network analyzer, a photograph of which is shown in Figure 4.2. In that figure, the network analyzer is composed of a dedicated source, an S-parameter test set, and the main frame. Here, an S-parameter test set is configured with a directional coupler and other microwave components for measuring the incident and reflected voltages that are used to obtain S-parameters. The main frame provides a user interface for controlling the test set and source. In addition, the main frame displays the measured S-parameter results through the test set.

In this chapter, we will discuss S-parameters defined by incident and reflected voltages and we will introduce the concept of *normalization*. We will also consider the concept of a conventional S-parameter and its conversion to other parameters, as well as the problem of shifting reference planes.

4.1.1 Voltage S-Parameter Definition

Figure 4.3 shows the measurement of S-parameters for a two-port network. Two sinusoidal voltage sources, E_{o1} and E_{o2}, with internal source resistances Z_{o1} and Z_{o2}, are applied to ports 1 and 2, respectively. The internal source resistance is called the *reference impedance*. For the purpose of defining the S-parameters, transmission lines are inserted at the input and output of the two-port network, as shown in Figure 4.3. They are assumed to have

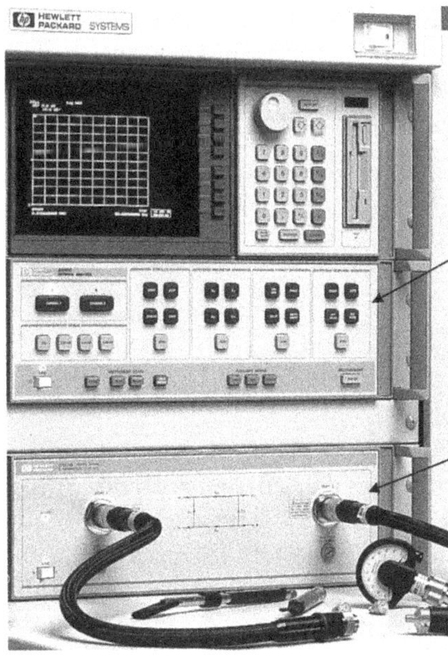

Figure 4.2 8510 Network Analyzer.[1] The S-parameters are measured using the S-parameter test set and the measured data are delivered to the main frame. The main frame provides the user interface and displays the measured data.

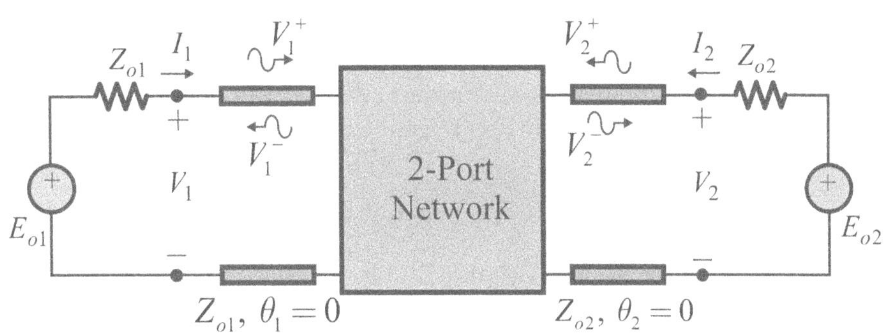

Figure 4.3 A two-port network. V_1, V_2, and I_1, I_2 are port voltages and currents; Z_{o1} and Z_{o2} are reference impedances; V_1^+, V_2^+, and V_1^-, V_2^- are incident and reflected voltages.

0 length, and their characteristic impedances are equal to the input and output source impedances Z_{o1} and Z_{o2}, respectively. These transmission lines are inserted to distinguish between the incident and the reflected voltages at the input and output ports.

Note that these characteristic impedances are, for the time being, assumed to be real and to be equal to the source internal resistances of port 1 and port 2, respectively.

1. Agilent Technologies, HP 85106D Millimeter-Wave Network Analyzer System, 1998.

Denoting each of the incident and reflected voltages of ports 1 and 2 as V_1^+, V_1^-, V_2^+, and V_2^-, respectively, and port voltage and current at ports 1 and 2 as V_1, V_2, I_1, and I_2, respectively, we can express this as shown in Equations (4.3)–(4.6).

$$V_1 = V_1^+ + V_1^- \tag{4.3}$$

$$I_1 = I_1^+ - I_1^- = \frac{1}{Z_{o1}}\left(V_1^+ - V_1^-\right) \tag{4.4}$$

$$V_2 = V_2^+ + V_2^- \tag{4.5}$$

$$I_2 = I_2^+ - I_2^- = \frac{1}{Z_{o2}}\left(V_2^+ - V_2^-\right) \tag{4.6}$$

We can see that all the port voltages and currents can be expressed in terms of the incident and reflected voltages V_1^+, V_1^-, and V_2^+, V_2^-. Therefore, instead of representing a two-port network by Z- and Y-parameters defined by the port voltages and currents, the two-port network can be represented by two-port parameters defined by the incident and reflected voltages. If the incident voltages V_1^+, $V_2^+$are selected as the independent variables, and the reflected voltages V_1^-, V_2^- are selected as the dependent variables, the newly defined two-port parameters are shown in Equation (4.7).

$$\begin{pmatrix} V_1^- \\ V_2^- \end{pmatrix} = \begin{pmatrix} S_{11}^V & S_{12}^V \\ S_{21}^V & S_{22}^V \end{pmatrix} \begin{pmatrix} V_1^+ \\ V_2^+ \end{pmatrix} \tag{4.7}$$

Such a voltage-defined S^V_{ij} is called an S-parameter. Also, from Equations (4.3) and (4.5), by applying KVL at the input and output ports, we can obtain

$$E_{o1} - Z_{o1}I_1 = V_1 = V_1^+ + V_1^- \tag{4.8}$$

$$E_{o2} - Z_{o2}I_2 = V_2 = V_2^+ + V_2^- \tag{4.9}$$

If we substitute Equations (4.4) and (4.6) into Equations (4.8) and (4.9), respectively, we obtain

$$E_{o1} - Z_{o1} \cdot \frac{1}{Z_{o1}}\left(V_1^+ - V_1^-\right) = V_1^+ + V_1^- \tag{4.10}$$

$$E_{o2} - Z_{o2} \cdot \frac{1}{Z_{o2}}\left(V_2^+ - V_2^-\right) = V_2^+ + V_2^- \tag{4.11}$$

From Equations (4.10) and (4.11), we can see that the incident waves V_1^+ and V_2^+ are given by, respectively,

$$V_1^+ = \frac{E_{o1}}{2}, \quad V_2^+ = \frac{E_{o2}}{2} \tag{4.12}$$

This means the incident voltage to the two-port network is independent of the load-matching conditions and is determined only by the presence of the source. In other words, there is no incident voltage at the input port if $E_{o1} = 0$; that is, the source is turned off.

When the source is turned on, that is, $E_{o1} \neq 0$, there is incident voltage at the input port given by Equation (4.12). In addition, the incident voltages can be measured by directly connecting the loads with the same value as the reference impedances to the sources. Therefore, we can see that the incident voltage depends only on the presence of the source and its value, regardless of the port's condition.

The simultaneous application of sources E_{o1} and E_{o2} can also be decomposed into two separate cases using the superposition principle. One case would be that E_{o1} alone is active and the other case would be that E_{o2} alone is active. When $E_{o2} = 0$, then from Equation (4.12), $V_2^+ = 0$ and the reflected voltage from port 2, V_2^- is shown in Equation (4.13).

$$V_2^-\Big|_{E_{o2}=0} = V_2 - V_2^+ = V_2 - 0 = V_2 \tag{4.13}$$

Also, the reflected voltage from port 1, V_1^- becomes Equation (4.14).

$$V_1^-\Big|_{E_{o2}=0} = V_1 - V_1^+ = V_1 - \frac{E_{o1}}{2} \tag{4.14}$$

Similarly, by making $E_{o1} = 0$ and determining the reflected voltage for this case, we get Equations (4.15) and (4.16).

$$V_2^-\Big|_{E_{o1}=0} = V_2 - \frac{E_{o2}}{2} \tag{4.15}$$

$$V_1^-\Big|_{E_{o1}=0} = V_1 - V_1^+ = V_1 - 0 = V_1 \tag{4.16}$$

With these voltages determined, the voltage-defined S-parameters can be rewritten as Equations (4.17a)–(4.17d).

$$S_{11}^V = \frac{V_1^-}{V_1^+}\Bigg|_{V_2^+=0} = \frac{V_1 - \frac{E_{o1}}{2}}{\frac{E_{o1}}{2}} \tag{4.17a}$$

$$S_{21}^V = \frac{V_2^-}{V_1^+}\Bigg|_{V_2^+=0} = \frac{V_2}{\frac{E_{o1}}{2}} \tag{4.17b}$$

$$S_{22}^V = \frac{V_2^-}{V_2^+}\Bigg|_{V_1^+=0} = \frac{V_2 - \frac{E_{o2}}{2}}{\frac{E_{o2}}{2}} \tag{4.17c}$$

$$S_{12}^V = \frac{V_1^-}{V_2^+}\Bigg|_{V_1^+=0} = \frac{V_1}{\frac{E_{o2}}{2}} \tag{4.17d}$$

$S^V{}_{11}$ and $S^V{}_{22}$ represent the reflection coefficients at ports 1 and 2, respectively. To confirm this fact, suppose that the input voltage at port 1 is E_{o1} and at port 2 it is terminated by the load impedance Z_{o2}. Defining the input impedance looking from port 1 as Z_{in}, the voltage across port 1 becomes

$$V_1 = \frac{Z_{in}}{Z_{o1} + Z_{in}} E_{o1} \tag{4.18}$$

Substituting Equation (4.18) into Equation (4.17a) and then rewriting, we get Equation (4.19).

$$S_{11}^V = \frac{\frac{Z_{in}}{Z_{o1} + Z_{in}} E_{o1} - \frac{E_{o1}}{2}}{\frac{E_{o1}}{2}} = \frac{Z_{in} - Z_{o1}}{Z_{in} + Z_{o1}} \tag{4.19}$$

Therefore, $S^V{}_{11}$ is the reflection coefficient looking into port 1 when port 2 is terminated by Z_{o2}. Similarly, $S^V{}_{22}$ is the reflection coefficient looking into port 2 when port 1 is terminated by Z_{o1}.

Example 4.1

Determine the voltage-defined S-parameter of the two-port networks when the reference impedances of ports 1 and 2 are Z_o.

Solution

(1) Series Impedance
From Figure 4E.1, when port 2 is terminated by Z_o, the impedance looking into port 1 is $Z_s + Z_o$. The reflection coefficient looking into port 1 is

$$S_{11} = \frac{Z_s + Z_o - Z_o}{Z_s + Z_o + Z_o} = \frac{Z_s}{2Z_o + Z_s}$$

Also, since the two-port network is symmetrical,

$$S_{11} = S_{22}$$

In addition, the voltage across port 2's termination Z_o is

$$V_2 = \frac{Z_o E_{o1}}{Z_o + Z_o + Z_s}$$

Therefore,

$$S_{21} = \frac{\frac{Z_o}{2Z_o + Z_s} E_o}{\frac{E_o}{2}} = \frac{2Z_o}{2Z_o + Z_s} = S_{12}$$

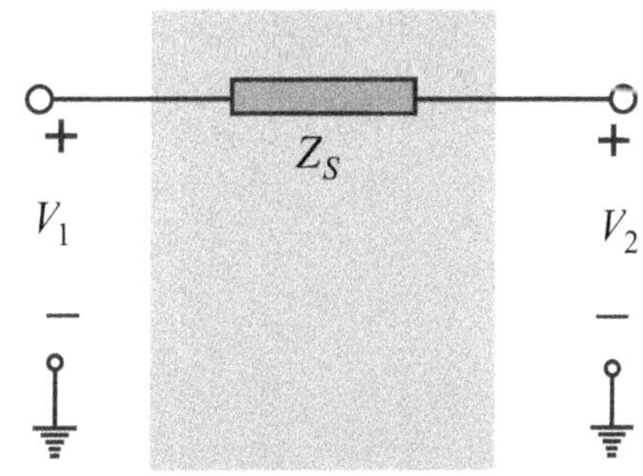

Figure 4E.1 A two-port circuit (series)

(2) Parallel Impedance
Similarly, the reflection coefficient for Figure 4E.2 is

$$S_{11} = \frac{Z_p \parallel Z_o - Z_o}{Z_o + Z_o \parallel Z_p} = S_{22}$$

Also, from the voltage across port 2,

$$S_{21} = \frac{\dfrac{Z_p \parallel Z_o}{Z_o + Z_o \parallel Z_p} E_o}{\dfrac{E_o}{2}} = \frac{2(Z_p \parallel Z_o)}{Z_o + Z_o \parallel Z_p} = S_{12}$$

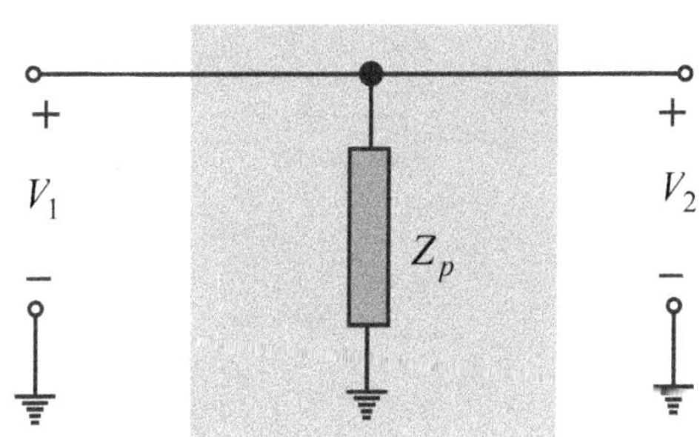

Figure 4E.2 A two-port circuit (shunt)

(3) Transmission Line (Figure 4E.3)
The reflection coefficient when port 2 is terminated by Z_o is

$$S_{11} = \frac{Z_o - Z_o}{Z_o + Z_o} = 0 = S_{22}$$

Also, as the voltage at port 1 is

$$V_1 = \frac{E_o}{2}$$

The voltage V_1 is delayed by the electrical length θ and appears at port 2. Thus,

$$S_{21} = \frac{\frac{E_o}{2} e^{-j\theta}}{\frac{E_o}{2}} = e^{-j\theta} = S_{12}$$

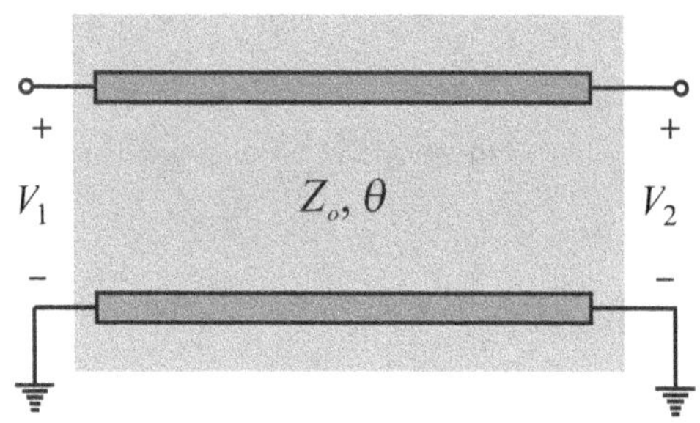

Figure 4E.3 A two-port circuit (transmission line)

Example 4.2

Verify the S-parameter calculation of the transmission line in Example 4.1(3) using AC simulation.

Solution

Set up the schematic shown in Figure 4E.4 below:

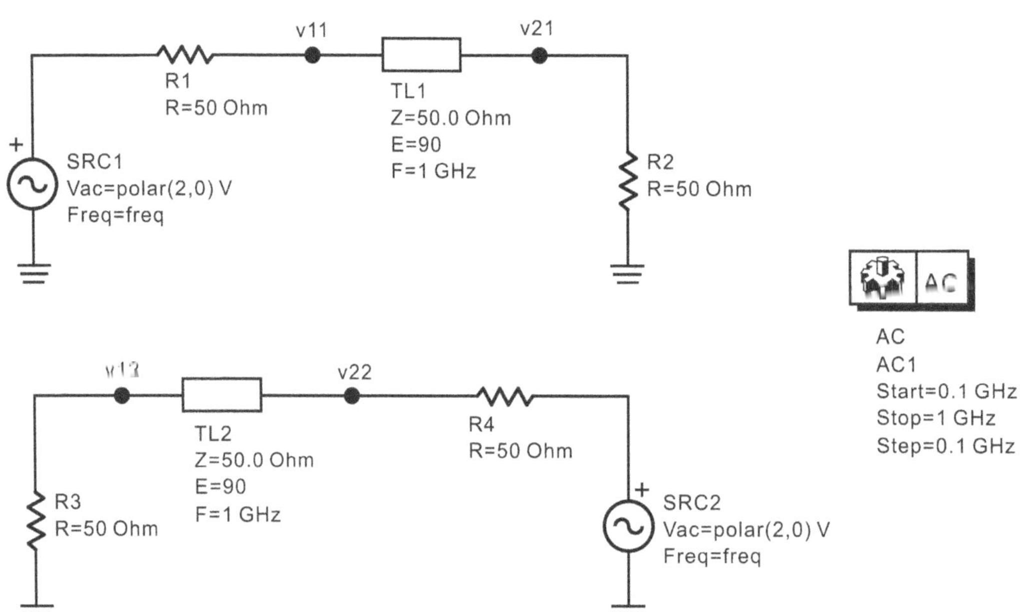

Figure 4E.4 Circuit for the verification of transmission-line S-parameters. The values of the two AC voltage sources are set to 2, S_{11} = **v11** – 1, S_{21} = **v21**, S_{12} = **v12**, and S_{22} = **v22** – 1.

From Equation (4.17), the S-parameters become S_{11} = **v**11 – 1, S_{21} = **v**21, S_{12} = **v**12, and S_{22} = **v**22 – 1. These equations are then entered in the display window as shown in Measurement Expression 4E.1:

Eqn n11=v11–1 Eqn n12=v12
Eqn n21=v21 Eqn n22=v22–1

Measurement Expression 4E.1 Equations written in the display window

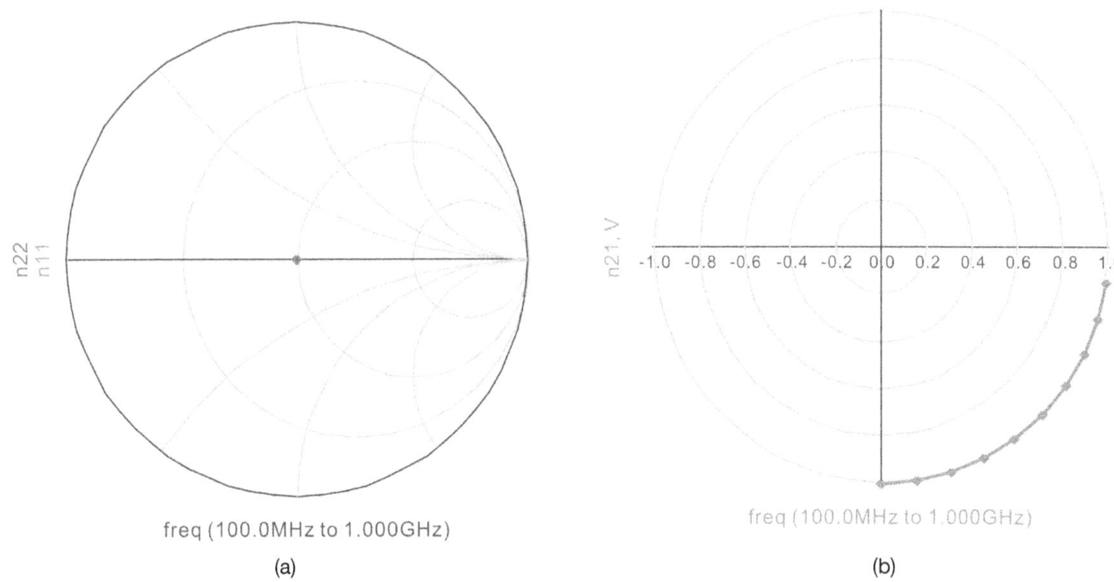

Figure 4E.5 Smith- and polar-chart plots for (a) **n**11, **n**22, and (b) **n**21

The computed **n**11 and **n**22 are plotted on a Smith chart and **n**21 is plotted on a polar chart. The traces appear as shown in Figure 4E.5. Thus, we can see that, $S_{11} = S_{22} = 0$, and $S_{12} = S_{21}$ rotates the unit circle clockwise with respect to the frequency as explained in Example 4.1(3).

4.1.2 Definitions and Properties of S-Parameters

As we saw in the previous section, the voltage S-parameters have been defined in terms of incident and reflected voltages, but we can also define current-based S-parameters in terms of the incident and reflected currents. To avoid a variety of definitions that depend on the choice of voltages or currents, S-parameters are defined in terms of *standardized* or *normalized* incident and reflected voltages that, by squaring directly, give the power as follows:

$$\begin{pmatrix} a_1 \\ a_2 \end{pmatrix} = \begin{pmatrix} \dfrac{V_1^+}{\sqrt{2Z_{o1}}} \\ \dfrac{V_2^+}{\sqrt{2Z_{o2}}} \end{pmatrix}, \quad \begin{pmatrix} b_1 \\ b_2 \end{pmatrix} = \begin{pmatrix} \dfrac{V_1^-}{\sqrt{2Z_{o1}}} \\ \dfrac{V_2^-}{\sqrt{2Z_{o2}}} \end{pmatrix} \tag{4.20}$$

In Equation (4.20), the division by $(2)^{½}$ converts the peak value to the *root mean square* (RMS) value. In a case where the phasors V_i^+, V_i^- are already defined as RMS values, $(2)^{½}$ is not required. Now, for the normalized incident and reflected voltages a, b, respectively, the normalized S-parameters or simply *S-parameters* are then defined as shown in Equation (4.21).

$$\begin{pmatrix} b_1 \\ b_2 \end{pmatrix} = \begin{pmatrix} S_{11} & S_{12} \\ S_{21} & S_{22} \end{pmatrix} \begin{pmatrix} a_1 \\ a_2 \end{pmatrix} \tag{4.21}$$

The relationship between the S-parameters and the previously derived voltage S-parameters becomes Equations (4.22a)–(4.22d).

$$S_{11} = \left.\frac{b_1}{a_1}\right|_{a_2=0} = \left.\frac{V_1^-}{V_1^+}\right|_{V_2^+=0} = S_{11}^V \tag{4.22a}$$

$$S_{21} = \left.\frac{b_2}{a_1}\right|_{a_2=0} = \left.\sqrt{\frac{Z_{o1}}{Z_{o2}}}\frac{V_2^-}{V_1^+}\right|_{V_2^+=0} = \sqrt{\frac{Z_{o1}}{Z_{o2}}}S_{21}^V \tag{4.22b}$$

$$S_{22} = \left.\frac{b_2}{a_2}\right|_{a_1=0} = \left.\frac{V_2^-}{V_2^+}\right|_{V_1^+=0} = S_{22}^V \tag{4.22c}$$

$$S_{12} = \left.\frac{b_1}{a_2}\right|_{a_1=0} = \left.\sqrt{\frac{Z_{o2}}{Z_{o1}}}\frac{V_1^-}{V_2^+}\right|_{V_1^+=0} = \sqrt{\frac{Z_{o2}}{Z_{o1}}}S_{12}^V \tag{4.22d}$$

Therefore, when the chosen input and output reference impedances are equal, the voltage-defined S-parameters are equal to the S-parameters. Here, the S-parameters' definition was based on the real-valued reference impedances. However, the S-parameter can also be defined for the complex-valued reference impedances. The details can be found in reference 2 at the end of this chapter.

To see the meaning of the standardized incident and reflected voltages a, b defined above, first consider $|a_1|^2$

$$|a_1|^2 = \left|\frac{V_1^+}{\sqrt{2Z_{o1}}}\right|^2 = \frac{|E_{o1}|^2}{8Z_{o1}} \tag{4.23}$$

Equation (4.23) represents the power delivered to the load when the load resistance is matched to the reference impedance Z_{o1}. The power given by Equation (4.23) is the maximum power that can be derived from the source, which will be explained later in Chapter 6, and is called the *maximum available power* or *shortly available power*. Thus, $|a_1|^2$ is the maximum available power that can be derived from the source connected to port 1, and $|a_2|^2$ similarly represents the maximum available power from the source connected to port 2.

Also, $|b_2|^2$ for $E_{o2} = 0$ is

$$|b_2|^2 = \frac{|V_2|^2}{2Z_{o2}} \tag{4.24}$$

The power given by Equation (4.24) represents the *delivered* power to the load of Z_{o2} connected to port 2 from the source at port 1. In addition, the power delivered to the input port, or port 1 is shown in Equation (4.25)

$$\frac{1}{2}\text{Re}\left(V_1^* I_1\right) = \frac{1}{2}\text{Re}\left(\sqrt{2Z_{o1}}\left(a_1 + b_1\right)^* \sqrt{2}\frac{\left(a_1 - b_1\right)}{\sqrt{Z_{o1}}}\right) = \left|a_1\right|^2 - \left|b_1\right|^2 \tag{4.25}$$

where $|a_1|^2$ is the incident power and $|b_1|^2$ is the reflected power. This means the available power is incident at the input port and a part of the available power that corresponds to $|b_1|^2$ is reflected back to the source. Thus, the net delivered power to the input is equal to the incident power minus the reflected power.

4.1.3 Ports and S-Parameter Simulation

As we saw in Example 4.2, S-parameters can be obtained through AC simulation. However, some problems arise when trying to obtain S-parameters through the AC simulation. First, two identical circuits should be drawn to obtain the S-parameters. This is significant when the port number increases to n. A number of n identical circuits should be drawn to obtain n-port S-parameters. Another issue is that voltage labels are necessary to distinguish a two-port network from its sources. Ports in ADS eliminate the problems, although in principle the computation may be the same as an AC simulation. These ports primarily provide a clear distinction between a two-port network and its sources. An example of the circuit used to determine the S-parameters for a two-port network is shown in Figure 4.4.

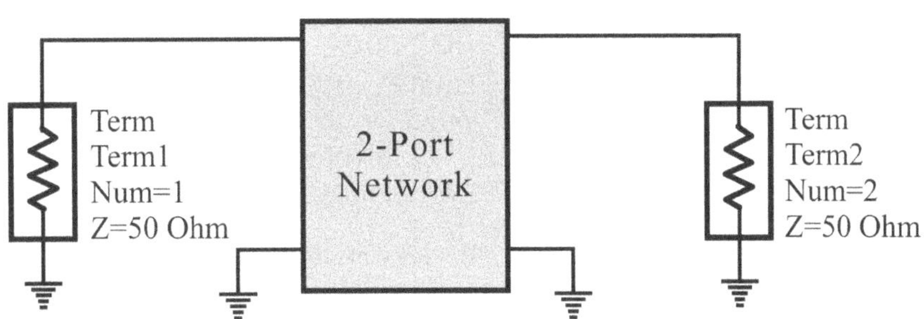

Figure 4.4 A two-port network connected to the ports

To understand the ports in the circuit shown in Figure 4.4, when port 1 is active, it becomes the series connection of source and internal resistance, as shown in Figure 4.5(a); otherwise, the deactivated ports act only as simple load resistors, as shown in Figure 4.5(b). In this case, the voltage appearing across all the ports can be computed, and the S-parameters for the excitation at port 1 can also be computed, as explained in Example 4.2. After calculation, the computed results are saved. By repeating this n times for all the ports by activating a port in the order of the port's number, the voltages appearing at all the ports can be computed and saved. With this method, n-port S-parameters can be obtained.

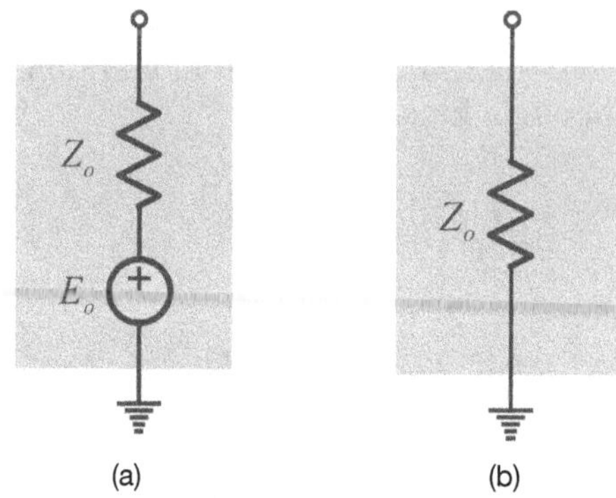

Figure 4.5 Equivalent circuit for (a) activated and (b) deactivated ports

The necessary parameters to specify for the port are the port number and the reference impedance. The port number corresponds to **Num** in Figure 4.4. The reference impedance is specified by entering the number at parameter **Z**, which can be a complex number.

In addition, ports can be used in nonlinear circuit analysis. Figure 4.6 shows a large signal port. The port number and reference impedance are the same as in the case of linear ports. However, the large-signal port requires the power level of the signal, which is specified in terms of power. In Figure 4.6, the power level is specified by **P** and it represents *the delivered power to the load when the load impedance is conjugate matched to the source impedance.*

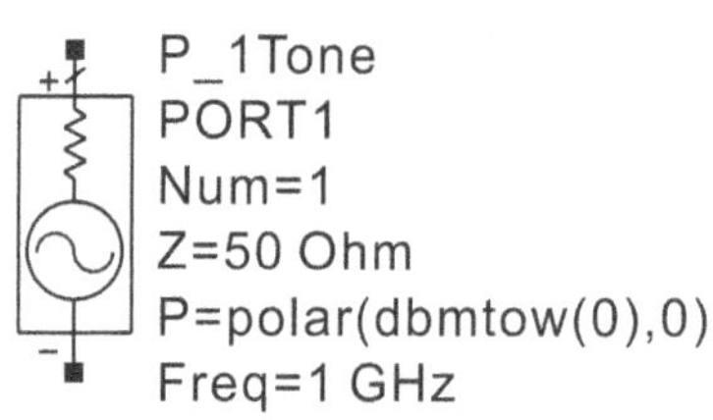

Figure 4.6 Large-signal port in ADS. **P** represents the power. When the port is conjugate matched to the load, the power **P** is delivered to the load. The other parameters can be similarly interpreted as is done in the S-parameter port.

Example 4.3

Use S-parameter simulation in ADS to determine the S-parameters of the transmission line in Example 4.2.

Solution

Set up the circuit as shown in Figure 4E.6 and perform the S-parameter simulation.

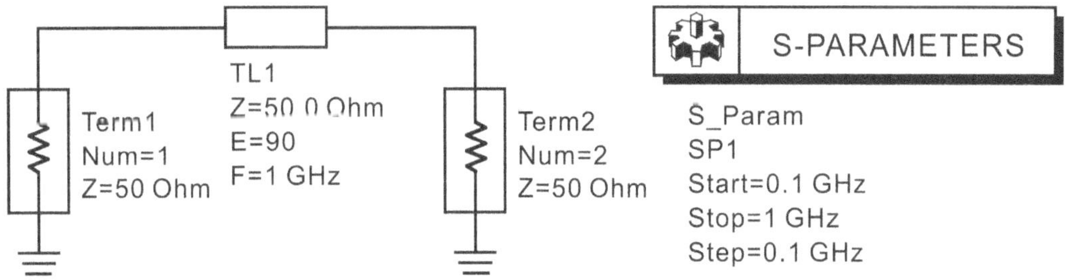

Figure 4E.6 An S-parameter simulation circuit. The S-parameter ports and S-parameter simulation controller specify the S-parameter simulation.

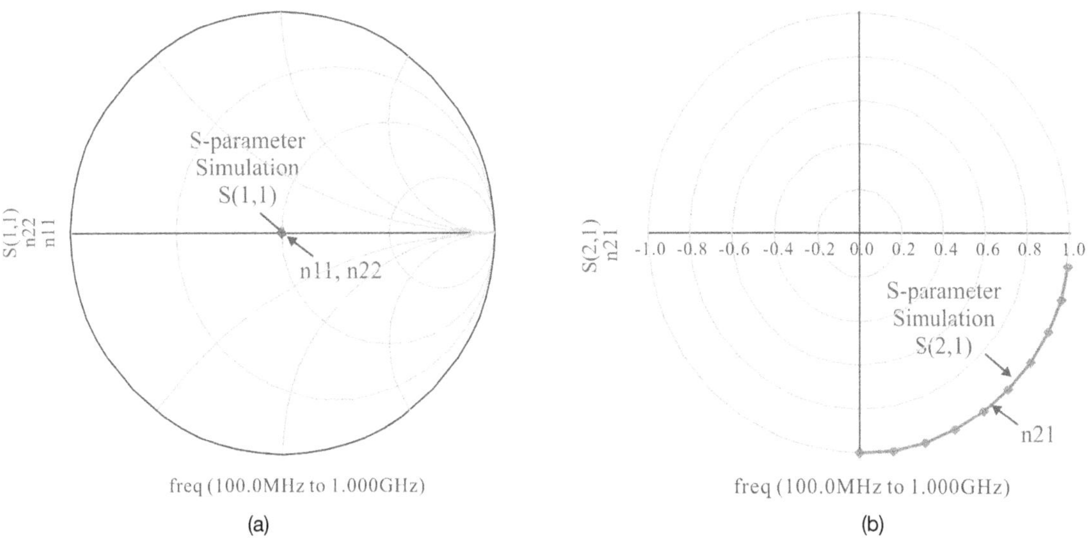

Figure 4E.7 Comparison of S-parameters obtained with the result of Example 4.2(a) S_{11} and (b) S_{21}

The S-parameter simulation results are then obtained. For the purpose of comparison with the earlier results, both results are simultaneously plotted, as shown in Figure 4E.7. From that figure, we can see that the S-parameter simulation gives the same S-parameters as those in Example 4.2. It should be noted that the voltage S-parameters in Example 4.2 are equal to the S-parameters. Here, **n**11, **n**22, and **n**21 are the results from Example 4.2.

4.1.4 S-Parameter Conversion

S-parameters defined in terms of incident and reflected voltages are convenient for measurement purposes, but understanding a circuit by means of S-parameters is usually difficult. On the other hand, Y- and Z-parameters defined in terms of port voltages and currents can present measurement difficulties but, because they evolve directly from circuit theory, they offer an easy way to understand an unknown circuit. In this section, we will see how to convert S-parameters to other parameters defined by port voltages and currents. We will also describe how this conversion is handled in ADS. In addition, we will study what happens to the S-parameters in a situation where the transmission-line lengths are not 0. In the previous definition of S-parameters, the transmission-line lengths inserted in the input and output of a two-port network were intentionally set to 0. We will also find out how the effects of the non-zero-length transmission lines can be removed.

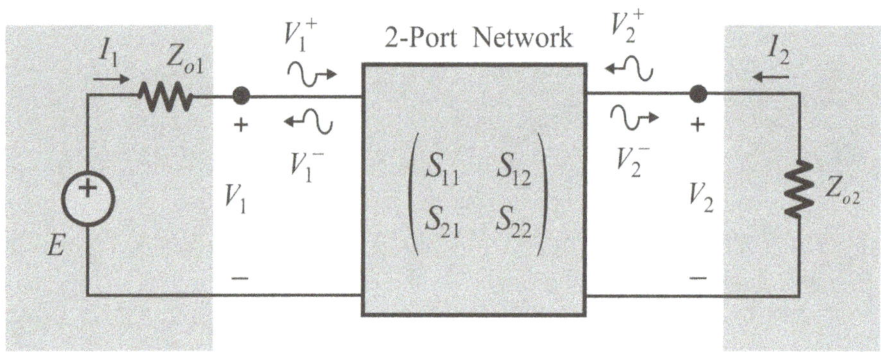

Figure 4.7 Port voltages and currents, as well as incident and reflected voltages. The incident and reflected voltages can be expressed using port voltages and currents.

Figure 4.7 shows a two-port network. In order to find the conversion relation between S-parameters and other parameters, the relationship between the incident and reflected voltages and the port voltages and currents is the first to be found. The relationship is shown in Equations (4.26a)–(4.26d).

$$V_1 = V_1^+ + V_1^- \tag{4.26a}$$

$$I_1 = I_1^+ - I_1^- = \frac{1}{Z_{o1}}\left(V_1^+ - V_1^-\right) \tag{4.26b}$$

$$V_2 = V_2^+ + V_2^- \tag{4.26c}$$

$$I_2 = I_2^+ - I_2^- = \frac{1}{Z_{o2}}\left(V_2^+ - V_2^-\right) \tag{4.26d}$$

Using the equations above, the normalized incident and reflected voltages in terms of the port voltages and currents can be expressed as follows:

$$\begin{pmatrix} a_1 \\ a_2 \end{pmatrix} = \begin{pmatrix} \dfrac{V_1^+}{\sqrt{2Z_{o1}}} \\ \dfrac{V_2^+}{\sqrt{2Z_{o2}}} \end{pmatrix} = \begin{pmatrix} \dfrac{V_1 + Z_{o1}I_1}{2\sqrt{2Z_{o1}}} \\ \dfrac{V_2 + Z_{o2}I_2}{2\sqrt{2Z_{o2}}} \end{pmatrix} \tag{4.27}$$

$$\begin{pmatrix} b_1 \\ b_2 \end{pmatrix} = \begin{pmatrix} \dfrac{V_1^-}{\sqrt{2Z_{o1}}} \\ \dfrac{V_2^-}{\sqrt{2Z_{o2}}} \end{pmatrix} = \begin{pmatrix} \dfrac{V_1 - Z_{o1}I_1}{2\sqrt{2Z_{o1}}} \\ \dfrac{V_2 - Z_{o2}I_2}{2\sqrt{2Z_{o2}}} \end{pmatrix} \tag{4.28}$$

Now, when the reference impedance $Z_{o1} = Z_{o2} = Z_o$, Equations (4.27) and (4.28) can then be expressed as

$$\begin{pmatrix} a_1 \\ a_2 \end{pmatrix} = \frac{1}{\sqrt{2Z_o}}\begin{pmatrix} V_1^+ \\ V_2^+ \end{pmatrix} = \frac{1}{2\sqrt{2Z_o}}\begin{pmatrix} V_1 \\ V_2 \end{pmatrix} + \frac{1}{2}\sqrt{\frac{Z_o}{2}}\begin{pmatrix} I_1 \\ I_2 \end{pmatrix} \tag{4.29a}$$

$$\begin{pmatrix} b_1 \\ b_2 \end{pmatrix} = \frac{1}{\sqrt{2Z_o}}\begin{pmatrix} V_1^- \\ V_2^- \end{pmatrix} = \frac{1}{2\sqrt{2Z_o}}\begin{pmatrix} V_1 \\ V_2 \end{pmatrix} - \frac{1}{2}\sqrt{\frac{Z_o}{2}}\begin{pmatrix} I_1 \\ I_2 \end{pmatrix} \tag{4.29b}$$

Expressing Equation (4.29) as a vector and substituting the Z-parameter relationship $\mathbf{V} = \mathbf{ZI}$ into the equation above, we obtain

$$\mathbf{a} = \frac{1}{2\sqrt{2Z_o}}\mathbf{V} + \frac{1}{2}\sqrt{\frac{Z_o}{2}}\mathbf{I} = \frac{1}{2\sqrt{2Z_o}}\left(\mathbf{Z} + Z_o\mathbf{U}\right)\mathbf{I} \tag{4.30a}$$

$$\mathbf{b} = \frac{1}{2\sqrt{2Z_o}}\mathbf{V} - \frac{1}{2}\sqrt{\frac{Z_o}{2}}\mathbf{I} = \frac{1}{2\sqrt{2Z_o}}\left(\mathbf{Z} - Z_o\mathbf{U}\right)\mathbf{I} \tag{4.30b}$$

Using Equation (4.30), we can then obtain the following relationship between Z-parameters and S-parameters shown in Equation (4.31).

$$\mathbf{S} = \left(\mathbf{Z} - Z_o\mathbf{U}\right)\left(\mathbf{Z} + Z_o\mathbf{U}\right)^{-1} \tag{4.31}$$

In addition, substituting the Y-parameter relationship $\mathbf{I} = \mathbf{YV}$, the relationship between Y-parameters and S-parameters is obtained as Equation (4.32).

$$\mathbf{S} = \left(Y_o\mathbf{U} - \mathbf{Y}\right)\left(Y_o\mathbf{U} + \mathbf{Y}\right)^{-1} \tag{4.32}$$

Here, $Y_o = 1/Z_o$. Conversely, from Equation (4.31), representing $\mathbf{Z}$ in terms of $\mathbf{S}$, we obtain Equation (4.33).

$$\mathbf{Z} = Z_o\left(\mathbf{U} + \mathbf{S}\right)\left(\mathbf{U} - \mathbf{S}\right)^{-1} \tag{4.33}$$

Similarly, from Equation (4.32), expressing $\mathbf{Y}$ in terms of $\mathbf{S}$, we obtain Equation (4.34).

$$\mathbf{Y} = Y_o\left(\mathbf{U} - \mathbf{S}\right)\left(\mathbf{U} + \mathbf{S}\right)^{-1} \tag{4.34}$$

The results of Equations (4.33) and (4.34) are shown in Table 4.1. Also, S-parameters can be similarly expressed using Y- and Z-parameters, and those results are shown in Table 4.2. It is also possible to convert S-parameters to other parameters such as ABCD and H-parameters. Using Equation (4.30), which shows the relationship between the port currents and voltages and the normalized incident and reflected voltages, all the conversions are possible.

Table 4.1 S-parameters converted to Z-parameters and Y-parameters

Z-parameters	Y-parameters
$z_{11} = Z_o \dfrac{(1+S_{11})(1-S_{22})+S_{21}S_{12}}{(1-S_{11})(1-S_{22})-S_{21}S_{12}}$	$y_{11} = Y_o \dfrac{(1-S_{11})(1+S_{22})+S_{21}S_{12}}{(1+S_{11})(1+S_{22})-S_{21}S_{12}}$
$z_{12} = Z_o \dfrac{2S_{12}}{(1-S_{11})(1-S_{22})-S_{21}S_{12}}$	$y_{12} = Y_o \dfrac{-2S_{12}}{(1+S_{11})(1+S_{22})-S_{21}S_{12}}$
$z_{21} = Z_o \dfrac{2S_{21}}{(1-S_{11})(1-S_{22})-S_{21}S_{12}}$	$y_{21} = Y_o \dfrac{-2S_{21}}{(1+S_{11})(1+S_{22})-S_{21}S_{12}}$
$z_{22} = Z_o \dfrac{(1-S_{11})(1+S_{22})+S_{21}S_{12}}{(1-S_{11})(1-S_{22})-S_{21}S_{12}}$	$y_{22} = Y_o \dfrac{(1+S_{11})(1-S_{22})+S_{21}S_{12}}{(1+S_{11})(1+S_{22})-S_{21}S_{12}}$

Table 4.2 S-parameters converted from Z-parameters and Y-parameters

Z-parameters	Y-parameters
$S_{11} = \dfrac{(z_{11}-Z_o)(z_{22}+Z_o)-z_{21}z_{12}}{z_{11}z_{22}-z_{12}z_{21}}$	$S_{11} = \dfrac{(Y_o-y_{11})(Y_o+y_{22})+y_{21}y_{12}}{y_{11}y_{22}-y_{12}y_{21}}$
$S_{12} = \dfrac{2Z_o z_{12}}{z_{11}z_{22}-z_{12}z_{21}}$	$S_{12} = -\dfrac{2Y_o y_{12}}{y_{11}y_{22}-y_{12}y_{21}}$
$S_{21} = \dfrac{2Z_o z_{21}}{z_{11}z_{22}-z_{12}z_{21}}$	$S_{21} = -\dfrac{2Y_o y_{21}}{y_{11}y_{22}-y_{12}y_{21}}$
$S_{22} = \dfrac{(z_{11}+Z_o)(z_{22}-Z_o)-z_{21}z_{12}}{z_{11}z_{22}-z_{12}z_{21}}$	$S_{22} = \dfrac{(Y_o+y_{11})(Y_o-y_{22})+y_{21}y_{12}}{y_{11}y_{22}-y_{12}y_{21}}$

Example 4.4

Using the results in Table 4.1, determine the input impedance Z_{in} of the two-port network in Figure 4E.8 with S-parameters for the load impedance Z_L, (1) $Z_L = \infty$, (2) $Z_L = 0$, and (3) $Z_L = Z_o$.

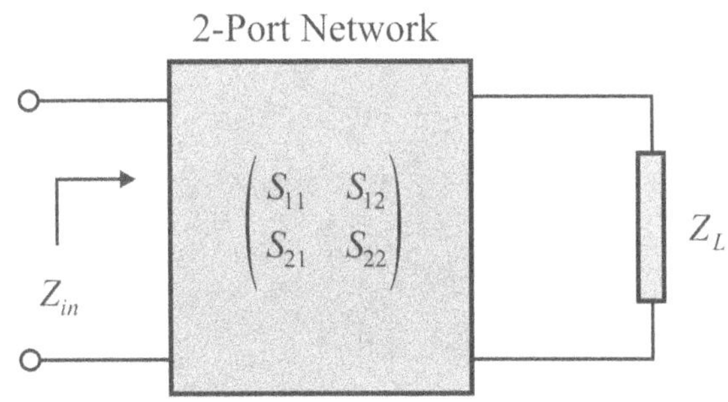

Figure 4E.8 A two-port network terminated by a load

Solution

When $Z_L = \infty$, from the definition of the Z-parameters, the input impedance is $Z_{in} = z_{11}$. Thus, expressing z_{11} in terms of the S-parameters in Table 4.1,

$$Z_{in} = Z_o \frac{(1+S_{11})(1-S_{22}) + S_{12}S_{21}}{(1-S_{11})(1-S_{22}) - S_{12}S_{21}}$$

When $Z_L = 0$, $Z_{in} = 1/y_{11}$ and expressing this in terms of the S-parameters of Table 4.1,

$$Z_{in} = \frac{1}{y_{11}} = Z_o \frac{(1+S_{11})(1+S_{22}) - S_{12}S_{21}}{(1-S_{11})(1+S_{22}) + S_{12}S_{21}}$$

When $Z_L = Z_o$, the reflection coefficient at the input is S_{11} and thus

$$Z_{in} = Z_o \frac{1+S_{11}}{1-S_{11}}$$

4.1.5 Shift of Reference Planes

Figure 4.8 shows a two-port network that includes non-zero transmission lines at the input and output ports. We will now examine the S-parameters for this situation. In practice, when conducting S-parameter measurement, finite-length transmission lines are unavoidably inserted in the setup. These transmission lines can be considered as the two cables inserted between the S-parameter test set in Figure 4.2 and a DUT. The lines provide an easy connection to the DUT. Because we want to obtain only the S-parameters of the DUT, we need to remove the effects of the transmission lines inserted during measurement. This process is called *calibration*. In practice, calibration is a complex procedure that involves more than simply removing the effects of transmission lines from the S-parameters. However, the removal of the transmission-line effects from S-parameters, which will be described later in this chapter, is very close to the practical network analyzer calibration.

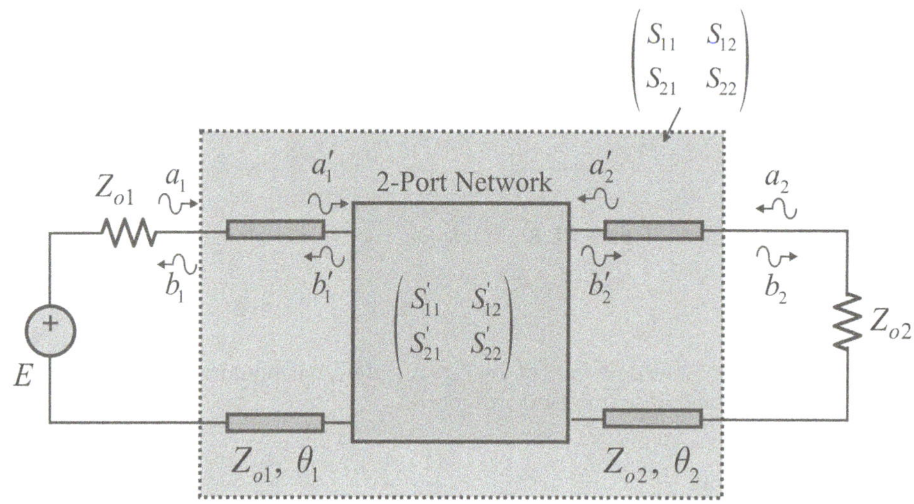

Figure 4.8 S-parameter measurements with non-zero transmission lines included

Let the electrical length of each of the transmission lines inserted in the input and output ports of the DUT be θ_1 and θ_2, respectively. In addition, let the reference impedance for S-parameters be equal to the characteristic impedances of the transmission lines. In this case, there is absolutely no reflection between the source and transmission lines. Also, let the normalized incident and reflected voltages at the ports of the DUT be a_1', b_1' and a_2', b_2', respectively. Finally, let the normalized incident and reflected voltages at the ports of the two-port network defined by the dotted-line rectangle in the figure be a_1, b_1 and a_2, b_2, respectively. Then, the following relationship in Equation (4.35) holds:

$$\begin{pmatrix} a_1' \\ a_2' \end{pmatrix} = \begin{pmatrix} a_1 e^{-j\theta_1} \\ a_2 e^{-j\theta_2} \end{pmatrix} \tag{4.35a}$$

$$\begin{pmatrix} b_1' \\ b_2' \end{pmatrix} = \begin{pmatrix} b_1 e^{j\theta_1} \\ b_2 e^{j\theta_2} \end{pmatrix} \tag{4.35b}$$

The S-parameters of the DUT are shown in Equation (4.36).

$$\begin{pmatrix} b_1' \\ b_2' \end{pmatrix} = \begin{pmatrix} S_{11}' & S_{12}' \\ S_{21}' & S_{22}' \end{pmatrix} \begin{pmatrix} a_1' \\ a_2' \end{pmatrix} \tag{4.36}$$

Now, if we substitute Equation (4.35) into Equation (4.36), we obtain Equation (4.37).

$$\begin{pmatrix} b_1' \\ b_2' \end{pmatrix} = \begin{pmatrix} b_1 e^{j\theta_1} \\ b_2 e^{j\theta_2} \end{pmatrix} = \begin{pmatrix} S_{11}' & S_{12}' \\ S_{21}' & S_{22}' \end{pmatrix} \begin{pmatrix} a_1 e^{-j\theta_1} \\ a_2 e^{-j\theta_2} \end{pmatrix} \tag{4.37}$$

Rewriting Equation (4.37), the S-parameters of the DUT that includes transmission lines result in Equation (4.38).

$$\begin{pmatrix} b_1 \\ b_2 \end{pmatrix} = \begin{pmatrix} S_{11}' e^{-2j\theta_1} & S_{12}' e^{-j(\theta_1+\theta_2)} \\ S_{21}' e^{-j(\theta_1+\theta_2)} & S_{22}' e^{-2j\theta_2} \end{pmatrix} \begin{pmatrix} a_1 \\ a_2 \end{pmatrix} \tag{4.38}$$

Therefore,

$$\begin{pmatrix} S_{11} & S_{12} \\ S_{21} & S_{22} \end{pmatrix} = \begin{pmatrix} S_{11}' e^{-2j\theta_1} & S_{12}' e^{-j(\theta_1+\theta_2)} \\ S_{21}' e^{-j(\theta_1+\theta_2)} & S_{22}' e^{-2j\theta_2} \end{pmatrix} \tag{4.39}$$

We can provide a physical interpretation for the result of S_{11} from Equation (4.39). The incident voltage a_1 traverses the input transmission line of length θ_1, reflects from the DUT, and returns through the input transmission line. Consequently, S_{11} has a phase delay equal to $e^{-2j\theta_1}$ with respect to S'_{11} due to the phase delay from the round trip, and S_{11} can be expressed as $S'_{11}e^{-2j\theta_1}$. Similarly, S_{21} also experiences a phase delay equal to traversing both the input and output transmission lines, and consequently S_{21} lags $S'_{21}$by a phase delay equal to $e^{-j(\theta_1+\theta_2)}$. Therefore, if we know the electrical lengths of the input and output transmission lines in advance, we can determine the S-parameters of the DUT from the measured S-parameters of the DUT that includes the transmission lines. In addition, the S-parameters of the DUT that includes the transmission lines have the same magnitudes and $|S_{ij}| = |S'_{ij}|$. This means the magnitudes of the S-parameters of the DUT that includes the transmission lines do not change; only the phase changes. Therefore, when the only concern is to measure the magnitudes, the property of $|S_{ij}| = |S'_{ij}|$ is useful. For example, in measuring the gain $|S_{21}|^2$, the transmission lines will not have any effect on the measured results if they have no losses.

In practice, to remove the effect of the inserted transmission lines, we need to know the electrical lengths and impedances of those lines. These can be determined from the *open-* and *short-circuit method* discussed in Chapter 3. However, the inserted transmission lines are not ideal transmission lines as described above. For example, they do not have the same impedances as the reference impedances (source mismatch), they also have some losses that cannot be neglected (loss from source to port 1), and so on. Thus, removing the effects of the transmission lines may not be as simple as described, but it can still be done through a similar method called *calibration*. Typical calibration methods include OSL (open-short-load) as well as the calibration method suitable for microstrip lines called TRL (thru-reflect-line). Please refer to references such as 3 at the end of this chapter for detailed descriptions of these calibration methods.

Example 4.5

The circuit in Figure 4E.9 is a series resonant circuit in which the input and output transmission lines are inserted both at the input and the output. Given that the input and output reference impedances for the S-parameters of the circuit in Figure 4E.9 are equal to the characteristic impedances of the input and output transmission lines, compute the S-parameters and draw $|S_{11}|$. Using the measured $|S_{11}|$, the circuit-element values of L and C can be determined. This method is frequently used to determine the circuit-element values. Explain how to extract the circuit-element values from the measured $|S_{11}|$.

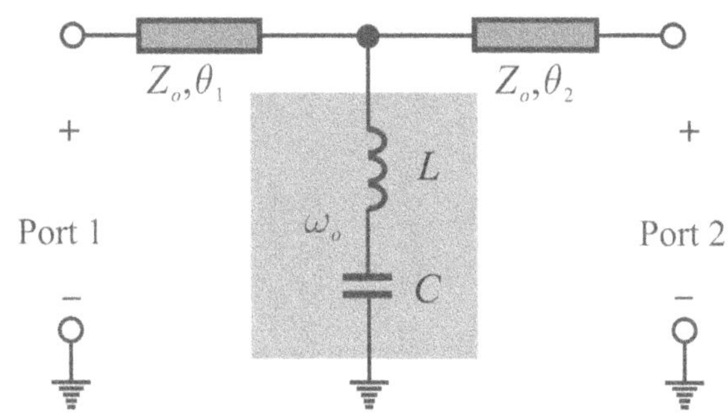

Figure 4E.9 Series resonant circuit with two input and output transmission lines

Solution

We can find the S-parameters of the series resonant circuit in the shaded area without the transmission lines. Then, the desired S-parameters are found by shifting the input and output reference planes by θ_1 and θ_2, respectively. The S-parameters of the circuit without the transmission lines, from Example 4.1(2), are

$$S_{11} = S_{22} = \frac{Z_p \parallel Z_o - Z_o}{Z_o + Z_o \parallel Z_p}$$

$$S_{21} = S_{12} = \frac{2(Z_p \parallel Z_o)}{Z_o + Z_o \parallel Z_p}$$

$$Z_p = j\left(\omega L - \frac{1}{\omega C}\right)$$

Z_p can be rewritten as follows:

$$Z_p = j\left(\omega L - \frac{1}{\omega C}\right) = \frac{jQZ_o}{2}\left(\frac{\omega}{\omega_o} - \frac{\omega_o}{\omega}\right) = jQZ_o\delta$$

$$\delta = \frac{1}{2}\left(\frac{\omega}{\omega_o} - \frac{\omega_o}{\omega}\right) \cong \frac{(\omega - \omega_o)}{\omega_o}$$

$$Q = \frac{2\omega_o L}{Z_o}$$

Substituting S_{11} and S_{21}, and then rearranging,

$$S_{11} = S_{22} = \frac{-1}{1 + 2jQ\delta}$$

$$S_{21} = S_{12} = \frac{2jQ\delta}{1 + 2jQ\delta}$$

Next, applying the shift of the reference planes, the S-parameters become

$$S_{11} = \frac{-1}{1 + 2jQ\delta} e^{-2j\theta_1}, \; S_{22} = \frac{-1}{1 + 2jQ\delta} e^{-2j\theta_2}$$

$$S_{21} = S_{12} = \frac{2jQ\delta}{1 + 2jQ\delta} e^{-j(\theta_1 + \theta_2)}$$

Both S_{11} and S_{22} give the same results when their magnitudes are plotted. The transmission lines yield only simple phase delays from the shift of the reference planes. The $|S_{11}|$ is sketched in Figure 4E.10. When S_{11} and S_{22} are measured, the Q can be determined from the 3-dB bandwidth, as shown in Figure 4E.10.

Since the 3-dB bandwidth here becomes $2Q \cdot \delta = \pm 1$, we can see that Q is

$$Q = \frac{f_o}{f_2 - f_1} = \frac{f_o}{BW}$$

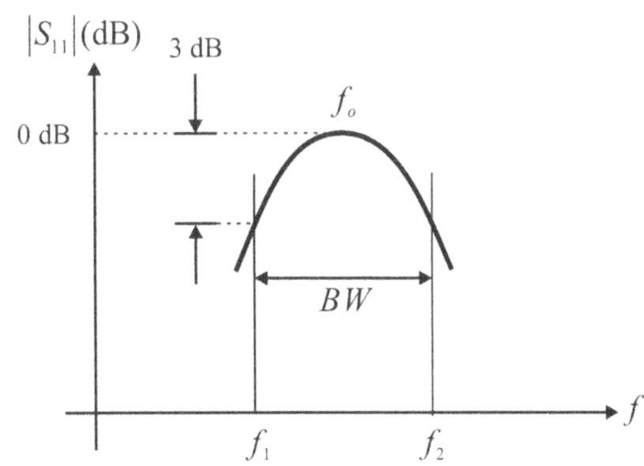

Figure 4E.10 Measurement of Q. BW is the 3-dB bandwidth and Q can be computed by f_o/BW.

Now, using the measured Q and the resonant frequency, the circuit's L and C values can be determined from the following equations:

$$L = \frac{QZ_o}{2\omega_o}, \quad C = \frac{1}{\omega_o^2 L}$$

Therefore, the equivalent circuit values can be obtained using only the scalar measurement of $|S_{11}|$ without resorting to a vector measurement.

4.1.6 Insertion Loss and Return Loss

In this section, we will discuss the insertion loss and return loss of the two-port network. Previously, we found that the maximum available power can be delivered to the matched load. When a two-port network, represented by S-parameters, is inserted between a matched source and a load, as shown in Figure 4.9, the power delivered to the load P_L is less than the maximum available power P_A. The ratio of P_L and P_A is called the *insertion gain* and the inverse of the ratio is called the *insertion loss*. The insertion gain is defined in Equation (4.40).

$$IG = \frac{P_L}{P_A} \tag{4.40}$$

Suppose that the S-parameters of a two-port network are measured based on the reference impedances Z_o and the source and load impedances are also Z_o. Then, the insertion gain becomes Equation (4.41).

$$IG = \frac{P_L}{P_A} = \frac{\dfrac{|V_2|^2}{2Z_o}}{|a_1|^2} = \frac{|b_2|^2}{|a_1|^2} = |S_{21}|^2 \tag{4.41}$$

Therefore, $|S_{21}|^2$ is equal to the insertion gain.

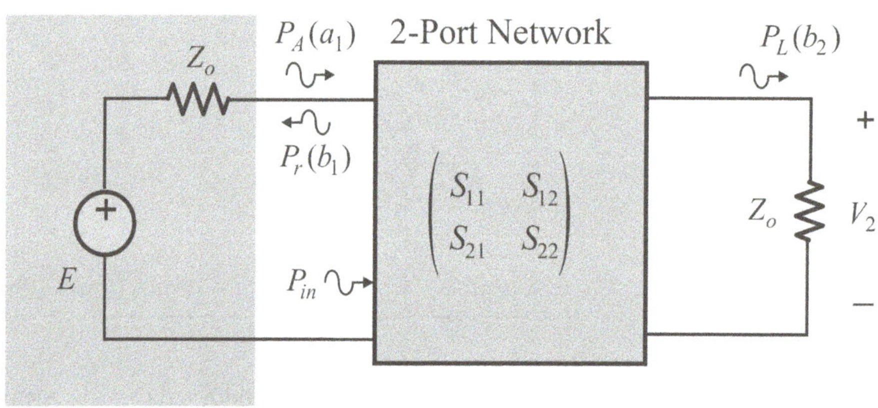

Figure 4.9 Insertion and return losses

On the other hand, for maximum power delivery to the load, the reflected voltage should be 0; that is, $b_1 = 0$, as shown in Figure 4.9. However, when the two-port network is inserted, the maximum available power from the source is generally not delivered to the input port, and a part of the maximum available power is reflected due to mismatch. Thus, the reflected power P_r divided by the maximum available power P_A is defined as the *return gain*, the inverse of which is called the *return loss*. Thus, the return gain is defined in Equation (4.42).

$$RG = \frac{1}{RL} = \frac{P_r}{P_A} \tag{4.42}$$

In addition, the return gain using S-parameters is shown in Equation (4.43).

$$RG = \frac{P_r}{P_A} = \frac{|b_1|^2}{|a_1|^2} = |S_{11}|^2 \quad (4.43)$$

Note that S_{11} represents the reflection coefficient that appears in Equation (4.19). The reflection causes a standing wave and VSWR can be used as the metric for the reflection phenomenon in place of S_{11} or the return loss. As defined in the transmission line, VSWR can be defined as

$$VSWR = \frac{1 + |S_{11}|}{1 - |S_{11}|}$$

In summary, the insertion of a two-port network causes the input reflection and $|S_{11}|^2$ represents the return loss. Similarly, the insertion of the two-port network prevents the maximum power delivery and $|S_{21}|^2$ represents the insertion loss. If a lossless, passive, two-port network is inserted, there is no loss in the two-port network. A portion of the supplied power is reflected and the remainder is delivered to the load from the supplied power. As a result, the sum of the two powers should be equal to the available power due to power conservation. Thus, in Equation (4.44), we can see that the following relationship is satisfied:

$$|S_{11}|^2 + |S_{21}|^2 = 1 \quad (4.44)$$

Example 4.6

When a lossless, passive, two-port network is terminated by Z_o, as shown in Figure 4E.11, VSWR = 2 is measured at the input side at frequency f_o.

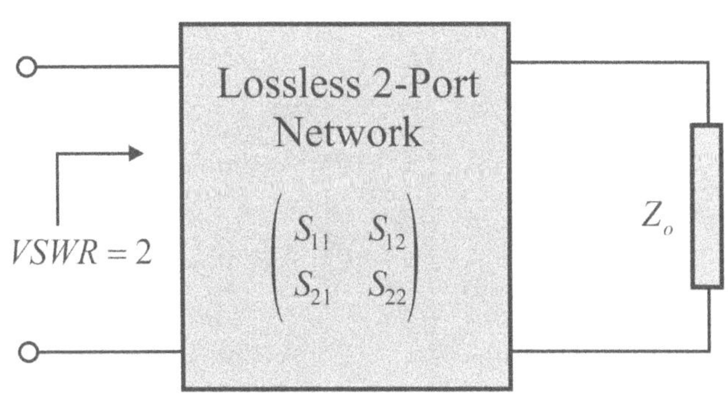

Figure 4E.11 A lossless two-port network terminated by a load

What is the VSWR at the output side when the input is terminated by Z_0? Also, what is the insertion loss? The S-parameters of the two-port network are measured using the reference impedance Z_0.

Solution

From the relationship between the VSWR and the reflection coefficient,

$$|S_{11}| = \frac{VSWR - 1}{VSWR + 1} = \frac{1}{3}$$

Since it is a lossless network, Equation (4.44) should be satisfied. Therefore, the insertion gain is

$$|S_{21}|^2 = 1 - |S_{11}|^2 = \frac{8}{9}$$

This is expressed in decibels as 0.5 dB. In addition, when the source is placed at the output, the following relationship is satisfied at the output:

$$|S_{22}|^2 + |S_{12}|^2 = |S_{11}|^2 + |S_{21}|^2 = 1$$

Because the two-port network is passive, reciprocity holds, $S_{12} = S_{21}$, and we obtain

$$|S_{22}| = |S_{11}| = \frac{1}{3}$$

Thus, the output VSWR is 2, which is the same as the input VSWR.

4.1.7 Input Reflection Coefficient

Figure 4.10 shows a two-port network terminated by a general load. The load impedance is different from the two-port, S-parameter reference impedance. In this case, the input reflection coefficient is different from S_{11} due to the presence of a_2 as a consequence of the load mismatch at the output. The input reflection coefficient depends on the load impedance, as shown in Example 4.4. In this section, as a generalization, we will present the input reflection coefficient when a two-port network is terminated by an arbitrary load Γ_L.

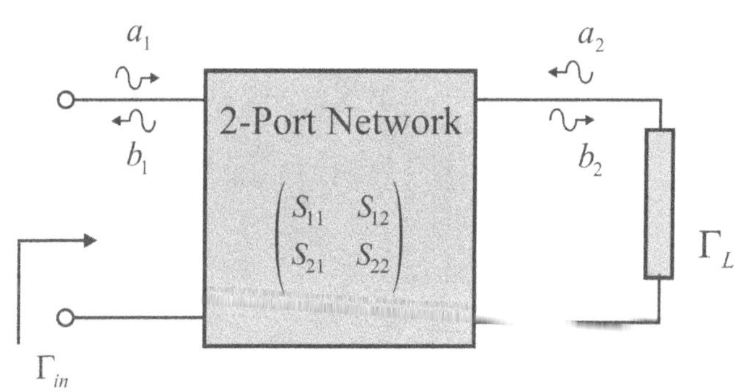

Figure 4.10 Input reflection coefficient for a mismatched load at port 2

From the load condition, we obtain $a_2 = \Gamma_L b_2$. From the relationship at the output given by S-parameters, a_2 becomes

$$b_2 = \frac{1}{\Gamma_L} a_2 = S_{21} a_1 + S_{22} a_2 \rightarrow a_2 = \frac{\Gamma_L S_{21}}{1 - S_{22}\Gamma_L} a_1 \tag{4.45}$$

Substituting Equation (4.45) into the input S-parameter equation, the reflection coefficient looking into the input, Γ_{in} is obtained as

$$\Gamma_{in} = \frac{b_1}{a_1} = \frac{S_{11} a_1 + S_{12} a_2}{a_1} = S_{11} + \frac{\Gamma_L S_{12} S_{21}}{1 - S_{22}\Gamma_L} \tag{4.46}$$

When $\Gamma_L = 0$ (the case of a matched load), we see that Γ_{in} in Equation (4.46) becomes equal to S_{11}. To interpret Equation (4.46) in another way, we can write that equation into an infinite series as

$$\Gamma_{in} = S_{11} + \Gamma_L S_{12} S_{21} \left(1 + \left(S_{22}\Gamma_L\right) + \left(S_{22}\Gamma_L\right)^2 + \ldots\right) \tag{4.47}$$

The interpretation of Equation (4.47) is shown in Figure 4.11, where S_{11} represents the direct reflection at the input. The term $\Gamma_L S_{12} S_{21}$ is the result of the multiplication of a forward transmission to the load S_{21}, followed by a reflection from load Γ_L, and a back-transmission to the input S_{12}. The terms in () represent the multiple reflections between port 2 and the load Γ_L, and they leak out to the input through multiple reflections that occur as a consequence of the mismatch at the output.

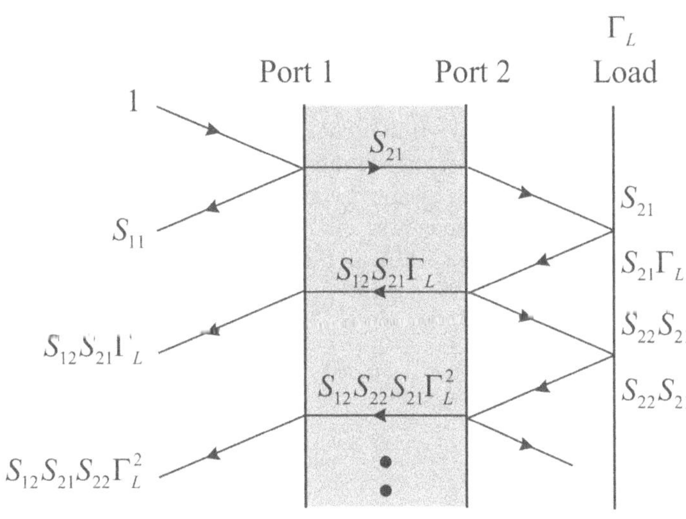

Figure 4.11 Analysis of the input reflection coefficient due to multiple reflections

Similarly, when a mismatched source impedance is connected to the input, the reflection coefficient viewed from the output is given by Equation (4.48).

$$\Gamma_{out} = \frac{b_2}{a_2} = S_{22} + \frac{\Gamma_s S_{12} S_{21}}{1 - S_{11}\Gamma_s} \tag{4.48}$$

4.2 NOISE PARAMETERS

4.2.1 Expression of Internal Noise

A two-port network generally has internal noise sources. As a result, both the externally applied signal sources and the internal noise sources appear together at the input and output of a two-port network. Due to the internal noise sources, the externally applied signal sometimes gets buried under the noise power level if it is weak, which makes it difficult to detect the externally applied signal. For the expression of the internal noises, we must understand the physical characteristics of those noise sources and model the two-port network that includes the internal noise sources. The modeling method of the internal noise sources is to represent the internal noise sources by two equivalent external sources applied to the two-port network. In this section, we first explain how to represent the internal noise sources as external equivalent sources. To do this, the sources need not be noise sources because the noise sources are also a kind of voltage or current source. We will consider a general two-port network that has internal sources.

Figure 4.12(a) shows a two-port network with internal sources. Suppose that the external voltage sources V_1 and V_2 are applied to the input and output of the two-port network with internal sources. Let the currents flowing in the input and output ports be I_1 and I_2. Then, the currents I_1 and I_2 are the *superposition* of the currents due to both the internal sources and the external voltage sources V_1 and V_2. Thus,

$$\begin{pmatrix} I_1 \\ I_2 \end{pmatrix} = \begin{pmatrix} y_{11} & y_{12} \\ y_{21} & y_{22} \end{pmatrix} \begin{pmatrix} V_1 \\ V_2 \end{pmatrix} + \begin{pmatrix} i_{n1} \\ i_{n2} \end{pmatrix} \tag{4.49}$$

The first term on the right-hand side of Equation (4.49) represents the current due to the external voltage sources V_1 and V_2, while the second term, i_{n1}, i_{n2}, represents the currents due to the internal sources. Although written in lowercase letters, it must be noted that the second term represents phasors.

Figure 4.12(b) shows the equivalent circuit of Figure 4.12(a). Using Equation (4.49), we can determine the currents due to the internal sources alone by removing the external voltage sources. This can be done by short-circuiting the ports and then measuring the currents flowing into those ports. If the short-circuited currents are placed outside of the two-port network, as shown in Figure 4.12(b), all the internal sources in the two-port network should be removed, which is also shown in Figure 4.12(b), because the short-circuited currents

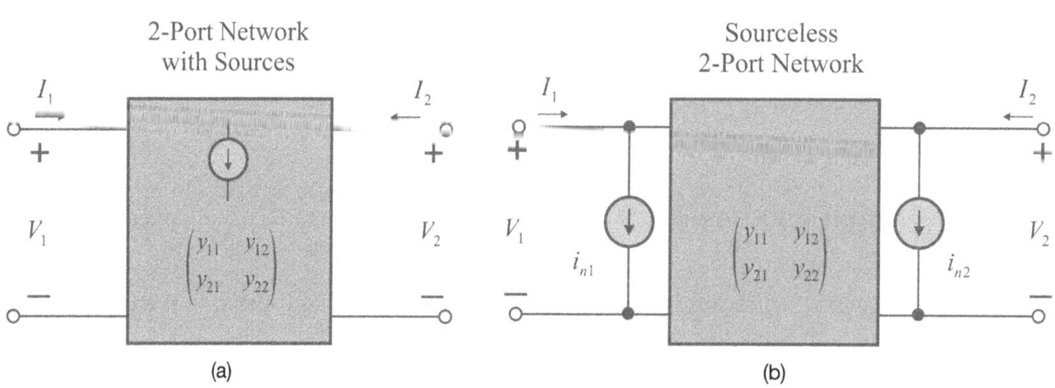

Figure 4.12 Treatment of internal noise sources of a two-port network: (a) two-port network having internal sources and (b) internal sources extracted to the outside of the network

reflect the contributions of all the internal sources. As a result, all the internal sources can be represented by the external current sources outside the two-port network. Thus, the circuit in Figure 4.12(b) can be considered as a general two-port network with the internal sources. In this way, although the two networks in Figures 4.12(a) and 4.12(b) are different, they provide the same voltages and currents to the external circuit connected to the two-port network. In addition, it is worth noting that an internal source can contribute to both i_{n1} and i_{n2}. In other words, some portions of i_{n1} and i_{n2} come from the same internal source.

Similarly, if the externally applied sources are current sources I_1 and I_2, the Z-matrix representation can be used instead of the Y-matrix representation. Then, the internal sources can be represented by the two equivalent external voltage sources v_{n1} and v_{n2} connected in series to the ports. Using the same method, there are various ways to represent the internal sources by the external equivalent sources placed outside of the two-port network, all of which depend on the expressions of the two-port network. Of these expressions, the ABCD-matrix representation (depicted in Figure 4.13, with internal sources extracted to the input of the network) is useful in noise calculation.

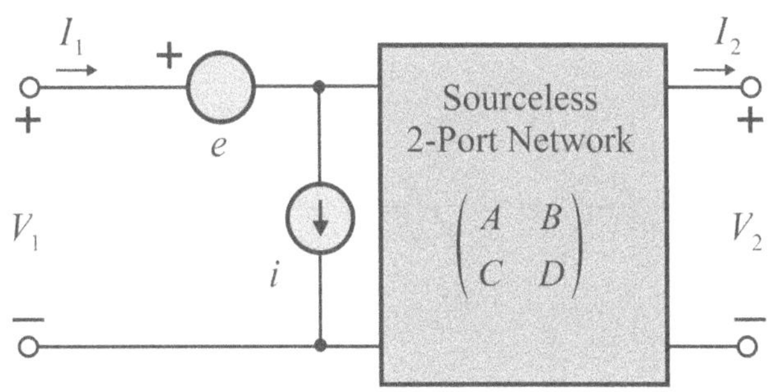

Figure 4.13 Internal source extraction using ABCD parameters

The relationship between the extracted external sources can be obtained similarly to the two-port parameter conversion. Basically, the extracted external sources should have the same effect on the rest of the connected circuit irrespective of their representations. Given this fact, the relationship between the externally extracted sources given by Y-parameters and those given by the ABCD parameters can be obtained. Thus, when the input and output are shorted, both representations should yield i_{n1} and i_{n2} at ports 1 and 2, respectively, and they are expressed as Equations (4.50) and (4.51).

$$i_{n1} = i - y_{11}e \tag{4.50}$$

$$i_{n2} = -y_{21}e \tag{4.51}$$

As shown previously, the relationship between the other external equivalent sources that is expressed using different two-port parameters can be similarly derived.

4.2.2 Representation of Noise Signals

4.2.2.1 Basic Concept

Now let us consider a situation in which the extracted external voltage and current sources are noise sources. When the voltage of a noise source is measured in the time domain, a random waveform such as that shown in Figure 4.14(a) is observed. When this voltage is remeasured, a waveform different from the previous waveform will be displayed. A noise waveform obtained from one measurement trial is called a *sample function* and a collection or set of all the sample functions is called an *ensemble*.

Let the value of a sample function for a noise source at time t_1 be $v(t_1)$, as shown in Figure 4.14(a). The repeated measurement of $v(t_1)$ at t_1 for other sample functions will give a different value. Repeating such a process will lead to a probability distribution described as a *probability density function* (PDF), $f(v)$. Function $f(v)$ can be interpreted as the probability that the voltage $v(t_1)$ is observed. An example of such a PDF is shown in Figure 4.14(b). Now, at the measurement time t_1, the mean voltage and mean power can be expressed in Equations (4.52) and (4.53) using a probability density function as follows:

$$E\left(v\left(t_1\right)\right) = \int_{-\infty}^{\infty} v\left(t_1\right) f\left(v\right) dv \tag{4.52}$$

$$E\left(v^2\left(t_1\right)\right) = \int_{-\infty}^{\infty} v^2\left(t_1\right) f\left(v\right) dv \tag{4.53}$$

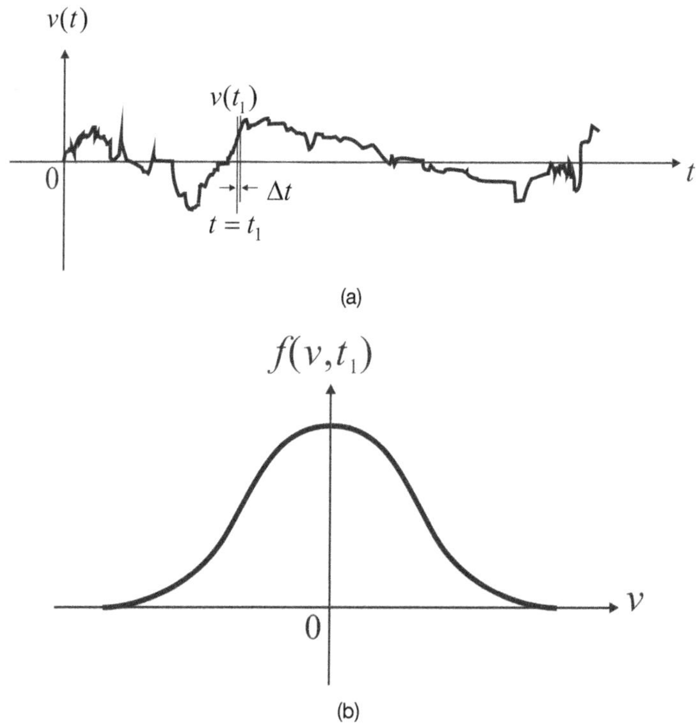

Figure 4.14 (a) Noise signal with respect to time and (b) probability distribution function at t_1

Note that the mean voltage and mean power are measured in the open-circuit condition. The mean power given by Equation (4.43) does not represent either the delivered power to the load or the available power. Since the squared voltage is proportional to power, it is called the mean power.

It should be noted that the mean voltage and mean power are based on the statistical average of sample function values at t_1, called the *ensemble average*. In general, noise sources have a PDF of the Gaussian distribution function with a 0 mean and a standard deviation v_s, and can be described as shown in Equation (4.54).

$$f(v) = \frac{e^{\frac{-v^2}{2v_s^2}}}{v_s\sqrt{2\pi}} \tag{4.54}$$

In most cases, of the noises that we treat, the PDF will be the same even if a measurement time is changed from t_1 to t_2. In addition, noise sources, regardless of measurement time, have the same mean and standard deviation. If the PDF is measured at two different times, t_1 and t_2 are the same, and the correlation between the two measured voltages depends only on $t_2 - t_1$, these noises are called a *stationary process*. If we consider the noise sources we usually encounter, we find that they are mostly stationary processes.

Figure 4.15 shows various kinds of noise waveforms. Figure 4.15(a) is a stationary process noise whose mean voltage and mean power at any time do not vary over time, whereas Figure 4.15(b) represents a noise waveform where the mean voltage varies with respect to time. Figure 4.15(c) shows a noise signal whose mean voltage does not vary with time but the mean power does vary over time. Therefore, (b) and (c) are nonstationary processes because their mean voltage and mean power vary over time.

Among the stationary processes, the concept of an ergodic process is useful when measuring the mean voltage and mean power of a noise signal. To explain the ergodic process, suppose that a certain value of the voltage v is observed at t_2. If the occurrence of v at t_2 is not influenced by the voltage occurring at t_1, the probability of observing a voltage of v at an arbitrary time t will be equal to the probability of observing the voltage v at t_1. In this case, the PDF for repeated trials over time will be the same as the PDF from the ensemble and the following relation in Equation (4.55) holds:

$$f(v)\Delta v = \lim_{T\to\infty}\frac{1}{T}\sum(\Delta t)_v \tag{4.55}$$

Here, T is the total measurement time interval and $(\Delta t)_v$ is the time interval in which voltage value v is observed. This kind of process is called an *ergodic process*.

In the case of an ergodic process, the time averages for a sample function can replace the statistical averages. The statistical mean voltage and mean power given in Equations (4.52) and (4.53) can therefore be obtained by computing their time averages as shown in Equations (4.56) and (4.57).

$$E\{v(t_1)\} = \int_{-\infty}^{\infty} v(t_1)f(v)dv = \lim_{T\to\infty}\frac{1}{T}\int_{-T/2}^{T/2} v(t)dt = \langle v(t)\rangle \tag{4.56}$$

$$E\{v^2(t_1)\} = \int_{-\infty}^{\infty} v^2(t_1)f(v)dv = \lim_{T\to\infty}\frac{1}{T}\int_{-T/2}^{T/2} v^2(t)dt = \langle v^2(t)\rangle \tag{4.57}$$

Here, $<\bullet>$ represents the time average. Generally, the ergodic process becomes a stationary process. However, in many stationary processes, within a very short time the occurrence of the voltage value at t_1 influences the occurrence of the voltage value at t_2 and most of the

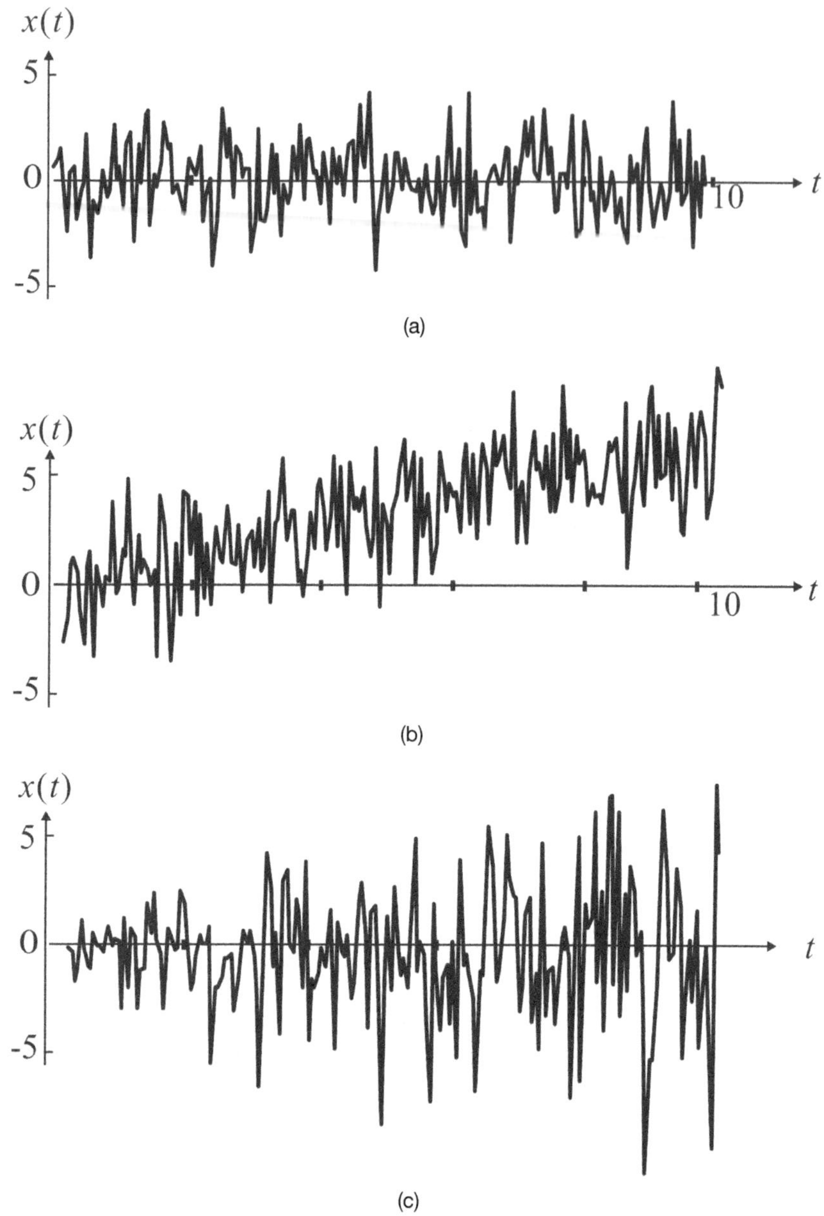

Figure 4.15 Noise signal with time: (a) stationary process and nonstationary process, (b) mean voltage changes, and (c) mean power changes with time

noise signals are not ergodic processes. In other words, there is some correlation between them. However, if the voltages $v(t_1)$ and $v(t_1 + \tau)$ occur with a sufficiently large time difference $\tau = t_2 - t_1$, there is no correlation between them. Thus, the two voltages widely separated in time can be seen to be independent, and this frequently happens in measurement. Consider two sets of sample functions: One is a noise sample function that is measured over a long period T. The other sample function is measured over the same period of T but the period's starting time is sufficiently separated from the previous measurement period. Similarly, multiple sets of the waveforms over the period T can be built. These sets for the period T can be thought of as independent or as an ergodic process because they have no correlation. Consequently, these sets are equivalent to the collection of sample functions obtained from independent trials and these sets can be used in place of an ensemble. As a result, the statistical averages necessary for mean voltage and mean power can be obtained using these sets. The statistical average is actually computed using one sample function. This is the case for most noise signals and it is used to measure the ensemble averages, such as mean voltage and mean power.

4.2.2.2 Spectrum Analysis of Noise Signals

In circuit analysis, spectrum analysis is preferred to time-domain analysis. When the time-domain noise waveform is truncated between $[-T/2, T/2]$, as shown in Figure 4.16(a), the truncated waveform can be expressed as shown in Equation (4.58).

$$e_T(t) = \begin{cases} e(t) & -\frac{T}{2} < t < \frac{T}{2} \\ 0 & |t| > \frac{T}{2} \end{cases} \tag{4.58}$$

The spectrum for $e_T(t)$ can be obtained using either of two approaches. One is to take the Fourier transform of $e_T(t)$ and T is made to approach infinity; that is, $T \to \infty$. The other is from the Fourier series of $e_T(t)$, in which the periodic waveform is defined using $e_T(t)$ as shown in Equations (4.59a) and (4.59b).

$$e_T(t) = e(t) \text{ for } -\frac{T}{2} < t < \frac{T}{2} \tag{4.59a}$$

$$e_T(t + nT) = e_T(t) \tag{4.59b}$$

This can be expanded in the Fourier series and the spectrum of $e_T(t)$ can be obtained by making T approach infinity.

Mathematically, the Fourier transform is considered to be a rigorous approach; however, the second approach is much closer to actual measurement. Thus, the Fourier-series method has been adopted by the IRE committee.[2] Now, the Fourier series of $e_T(t)$ can be given by Equation (4.60)

$$e_T(t) = \sum_{m=-\infty}^{\infty} E_T(\omega) e^{j\omega t} \tag{4.60}$$

where Equations (4.61a)–(4.61c) are

$$\omega = 2\pi \cdot m\Delta f \tag{4.61a}$$

$$\Delta f = \frac{1}{T} \tag{4.61b}$$

$$E_T(\omega) = \frac{1}{T} \int_{-\frac{T}{2}}^{\frac{T}{2}} e_T(t) e^{-j\omega t} dt \tag{4.61c}$$

2. IRE Subcommittee on Noise, "IRE Standards on Methods of Measuring Noise in Linear Two Ports, 1959," *Proceedings of the IRE* 48, no. 1 (January 1960): 60–68.

Figure 4.16 shows the $E_T(\omega)$ of the Fourier series.

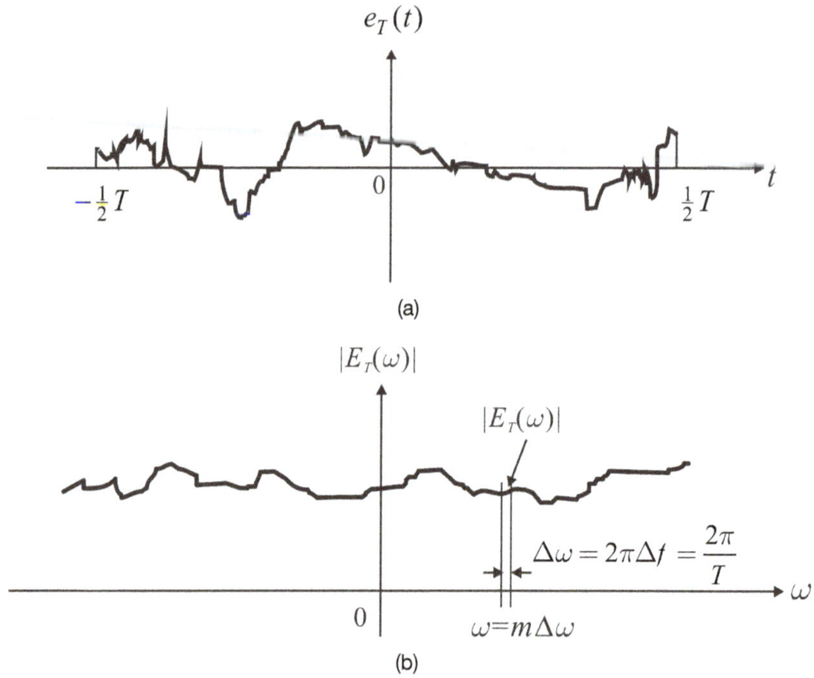

Figure 4.16 (a) Truncated noise signal and (b) its Fourier series

Thus, $2E_T(\omega)$ can be interpreted as the phasor of frequency $m\Delta f$. Since the period of $e_T(t)$ is T, note that the frequency resolution is given by Δf in Equation (4.61b). Thus, the *resolution frequency* Δf is determined by the measurement time T. To make Δf small, a long measurement time is necessary.

Since the noise has been expanded in the Fourier series, as given by Equation (4.60), note that $E_T(\omega)$ becomes a random variable. A sample waveform of $e_T(t)$ for a period T will determine a sample $E_T(\omega)$. Another $e_T(t)$ measured for another period T will yield a different value of $E_T(\omega)$. Thus, every measurement for T, $E_T(\omega)$ obtained from the Fourier series of $e_T(t)$ yields different values. Therefore, $E_T(\omega)$ is the *random phasor* of a noise at a frequency $\omega = 2\pi(m\Delta f)$ and its mean voltage and mean power should be determined by the statistical average of the Fourier series of $e_T(t)$. In addition, as explained previously, every sampled $e_T(t)$ after a sufficiently separated time has elapsed has no correlation. All the waveforms of $e_T(t)$ measured over period T can be seen to be an ergodic process and can be thought of as the sets of the sample functions of $e_T(t)$ from independent trials. Therefore, *$E_T(\omega)$ can be determined by the time average of the Fourier series of $e_T(t)$* for each widely separated time.

In practice, the measurement hardware based on the Fourier series is difficult to implement. A practical way to measure $E_T(\omega)$ is to pass a noise signal through the bandpass filter of bandwidth Δf centered at $\omega = 2\pi f$, as shown in Figure 4.17. Note that the filter output differs from the previously defined $E_T(\omega)$ by a factor of 2 because the Fourier series in Equation (4.60) is two-sided, while the spectrum measurement in Figure 4.17 is one-sided. Also, the filter output is continuous and it is denoted by $E'(\omega)$. The $E'_T(\omega)$ is the filter output $E'(\omega)$ sampled on the time interval T. The value of $E'_T(\omega)$ can be determined from the amplitude and phase of the sampled filter output. The time averages of $E'_T(\omega)$ can be obtained from the successive samples defined over T, which are sufficiently separated and they correspond to the statistical average values of $E'_T(\omega)$. The dependence of $E'_T(\omega)$ on ω can be found by sweeping the center frequency of the filter. However, the phase of $E'_T(\omega)$ is often difficult to measure because the filter output is usually connected to a power detector that measures an amplitude of $E'_T(\omega)$. Generally, the mean value of $|E'_T(\omega)|$ can be measured. Now, because the spectrum is defined as the power per unit bandwidth, the one-sided power spectral density $W_e(\omega)$ can be expressed as Equation (4.62).

$$W_e(\omega) = \frac{\left\langle \left|E'(\omega)\right|^2 \right\rangle}{\Delta f} = \frac{\left\langle \left|E'_T(\omega)\right|^2 \right\rangle}{\Delta f} \text{ [W/Hz]} \tag{4.62}$$

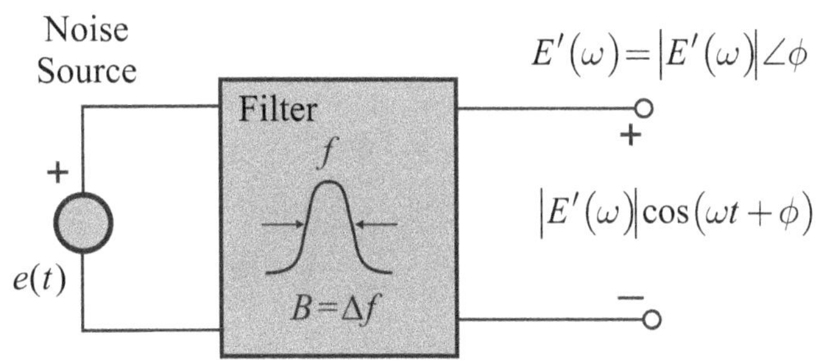

Figure 4.17 Noise spectrum measurement

In conclusion, the noise signal can be spectrally decomposed and we can analyze a circuit with noise sources using noise phasors in a frequency domain. The spectral average noise power can be measured using a bandpass filter and its power spectral density is given by Equation (4.62). This measurement is illustrated in Figure 4.17.

4.2.2.3 Thermal Noise

Thermal noise arises as a result of the random motion of electrons in a resistive device. An example of a thermal noise waveform would be the waveform shown in Figure 4.15(a). Thermal noise was first discovered by Johnson from experiments he conducted in 1926 and was later explained theoretically by Nyquist in 1928.

Thermal noise is a stationary process and is known to have a zero mean voltage and Gauss distribution. In addition, for a time difference τ not equal to 0, the voltages $v(t)$ and $v(t+\tau)$ of the thermal noise are completely uncorrelated and they can be categorized into an ergordic process. In addition, its mean power spectral density is known to be constant regardless of frequency. The mean power of an open-circuited resistor R is expressed in Equation (4.63)

$$\left\langle v^2 \right\rangle = 4kT_oRB \tag{4.63}$$

where B is the bandwidth of the instrument measuring the thermal noise voltage and T_o represents the temperature of the resistor in Kelvin. As B approaches ∞, the measured thermal noise power also approaches ∞.

Now consider the mean of the thermal noise power. After the thermal noise passes through the bandpass filter of bandwidth Δf, the mean power of the filtered output is equal to

$$\frac{\left\langle \left| E'(\omega) \right|^2 \right\rangle}{\Delta f} = 4kT_o R \tag{4.64}$$

Note that $E'(\omega)$ is the rms value and only positive frequency power is accounted for in Equation (4.64). Conversely, using the mean spectral noise power, the *equivalent thermal noise resistance* R_n can be defined in Equation (4.65).

$$R_n = \frac{\left\langle \left| E'(\omega) \right|^2 \right\rangle}{4kT_o \Delta f} \tag{4.65}$$

Now, taking into account the negative frequency spectral density, the mean spectral density can be expressed as shown in Equation (4.66).

$$W_e(f) = \frac{\left\langle \left| E'(\omega) \right|^2 \right\rangle}{2\Delta f} = 2kT_o R \quad [\text{W/Hz}] \tag{4.66}$$

Since the thermal noise has a constant power density regardless of frequency, it is called *white noise*. In addition, the mean available power within a bandwidth B for a matched load is expressed in Equation (4.67).

$$P_n = \frac{4kT_o RB}{4R} = kT_o B \; [\text{W}] \tag{4.67}$$

Example 4.7

Figure 4E.12 shows a circuit in which two noisy resistors are connected in parallel. The voltage sources v_1 and v_2 represent the noise voltages appearing at resistors R_1 and R_2, respectively. As a result, the two resistors in Figure 4E.12 are noiseless resistors. The equipment of the bandwidth B is connected to the output to measure the mean noise power. What is the mean power of v_o within a bandwidth B?

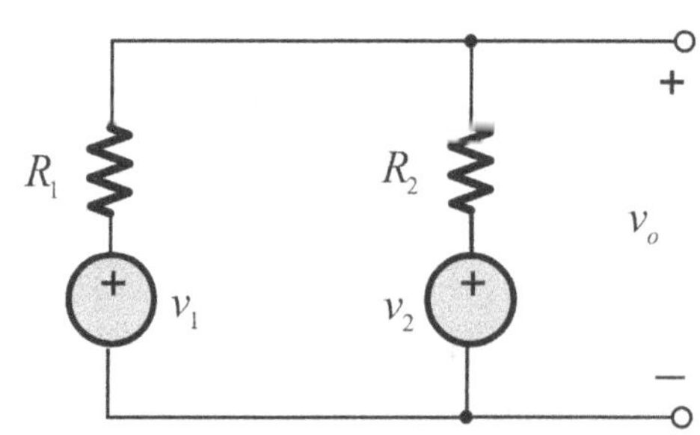

Figure 4E.12 Parallel resistance circuit

Solution

Since there are two sources, we can calculate the output voltage using superposition. The output voltage due to each source is

$$v_{o1} = \frac{R_2}{R_1 + R_2} v_1, \quad v_{o2} = \frac{R_1}{R_1 + R_2} v_2$$

The output voltage then becomes $v_o = v_{o1} + v_{o2}$. Since the two sources are independent,

$$E(v_o^2) = E(v_{o1}^2) + E(v_{o2}^2) = \frac{R_2^2}{(R_1 + R_2)^2} E(v_1^2) + \frac{R_1^2}{(R_1 + R_2)^2} E(v_2^2)$$

Substituting $E(v_1^2) = 4kT_oR_1B$ and $E(v_2^2) = 4kT_oR_2B$,

$$E(v_o^2) = 4kT_oB\frac{R_1R_2}{(R_1 + R_2)}$$

is obtained. Note that the result can be obtained by inspection. Since the resistance looking into the two resistors is $R_1||R_2$, we see that the power above becomes the mean thermal noise power of $R_1||R_2$.

Example 4.8

An attenuator is a device that reduces the input power and delivers it to the load. The attenuator also provides the impedance $Z_{in} = 50\ \Omega$ when the attenuator's input or output is terminated by a resistor $R_o = 50\ \Omega$, as shown in Figure 4E.13. In that figure, the noise voltage of R_o is extracted as noise voltage source v_1. What is the maximum available noise power at the output?

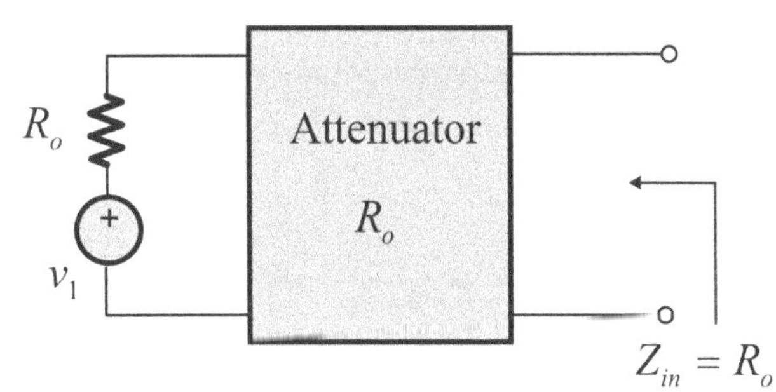

Figure 4E.13 Attenuator connected to termination R_o

Solution

Since the resistance looking into the terminals is still 50 ohm, the available power is achieved when a 50-Ω load is connected. From Example 4.7, it can be found that the attenuator generates the same thermal noise as a 50-Ω resistor. Thus, the available power within a bandwidth B becomes

$$P_o = kT_oB$$

This is equal to the available noise power from a 50-Ω resistor. As a result, the available noise power kT_oB applied to the attenuator from the 50-Ω termination resistor is reduced, but because the attenuator itself is made from resistors, the internal noise contributes to the available output power. The internal noise power is exactly equal to the amount of the input noise power decreased by the attenuator, and the resulting available power is the same as the input available noise power.

4.2.3 Noise Figure

Noise factor is a figure of merit that represents the noise performance of a two-port network. As previously mentioned, a two-port network usually has internal noises. When a load is connected to the two-port network, both the external source and the internal noise sources contribute to the power delivered to the load. A definition for a noise factor would be *the ratio of the actual delivered power to a load to the delivered power to the load from an external source alone.* As an external source, *a thermal noise source of room temperature,* $T_0 = 290$ °K is used. Note that if a high-temperature external noise source is applied, the ratio becomes close to 1. Since the temperature of the two-port network is fixed, the internal noise sources provide only a small amount of the actual delivered power to the load compared with the high-temperature external noise source, and the ratio is close to 1. Thus, a definition of the noise factor without specifying the external noise source temperature depends on temperature. As a result, the noise factor is defined in Equation (4.68).

$$\text{Noise factor} = \frac{\text{Actual noise power output to load}}{\text{Power output due to noise source of } T_o = 290°K} \tag{4.68}$$

The defined noise factor obviously depends on frequency because the internal noise sources generally deliver a frequency-dependent noise power to the load. To measure the frequency response of the noise factor, a narrowband filter that can sweep the center frequency is inserted, as shown in Figure 4.18. Thus, the frequency response of the noise factor can be obtained. It is worth noting that $kT_0\Delta f$ in Figure 4.18 represents the available power and does not represent the magnitude of the noise's current source. Such an expression is often used because it enhances intuitive understanding.

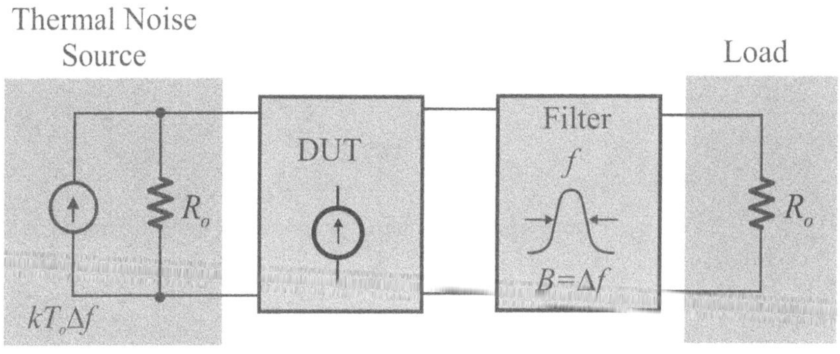

Figure 4.18 Noise factor measurement

The noise figure is the noise factor expressed in decibels and is thus defined as shown in Equation (4.69).

$$F = 10\log(\text{noise factor}) \tag{4.69}$$

Based on the definition in Equation (4.68), alternative expressions that are useful in the computation of a noise factor can be derived. Because Equation (4.68) is the ratio of powers delivered to the same load, another expression can be derived by representing the circuit with a Norton equivalent circuit. When the circuit seen from the load is transformed into the Norton equivalent circuit, the Norton impedance is not affected by the internal noise sources or by the external source. The Norton impedance represents only the function of the source impedance. The change due to the internal sources appears only in the Norton current source. *The external and internal noise sources contribute to the value of the Norton current source but have no effects on Norton impedance.* Thus, the noise figure is independent of the load impedance; it depends only on the impedance looking into the output of a two-port network, which in turn is a function of the external source impedance alone. Since it is independent of the load impedance, we pick a load that is matched to the equivalent Norton impedance. Then, the noise factor can be expressed as Equation (4.70).

$$\text{Noise factor} = \frac{\text{Available noise power from the output of device}}{\text{Available output power due to noise source of } T_o} \tag{4.70}$$

Let available power gain G be defined as shown in Equation (4.71).

$$G = \frac{\text{Available power from the output}}{\text{Available power from the input}} \tag{4.71}$$

Since the available power that can be obtained from the thermal noise is $kT_o\Delta f$, the denominator of Equation (4.70) becomes $(kT_o\Delta f)G$. Therefore, the noise factor is also expressed as

$$F = \frac{P_{no}}{kT_o\Delta fG} = \frac{\frac{P_{no}}{G}}{kT_o\Delta f} \tag{4.72}$$

Here, P_{no} is the available power at the output. The $kT_o\Delta f$ in Equation (4.72) is the available thermal noise power at the input. Since P_{no}/G is the equivalent available noise power at the input, the noise factor can also be expressed as Equation (4.73).

$$\text{Noise factor} = \frac{\text{Equivalent available noise power at the input}}{\text{Availabe power from noise source of } T_o} \tag{4.73}$$

Therefore, the equivalent noise power available at the input becomes $FkT_o\Delta f$. In addition, since the available power from the thermal noise source is $kT_o\Delta f$, the available input noise power due to the DUT can be expressed as $(F - 1)kT_o\Delta f$ because the two sources are independent or uncorrelated. This is illustrated in Figure 4.19, where the two current sources are shown. One is the current source from the thermal noise source and the other comes from the internal noises of the two-port network. This type of representation is useful for the noise-figure computation of cascaded two-port networks.

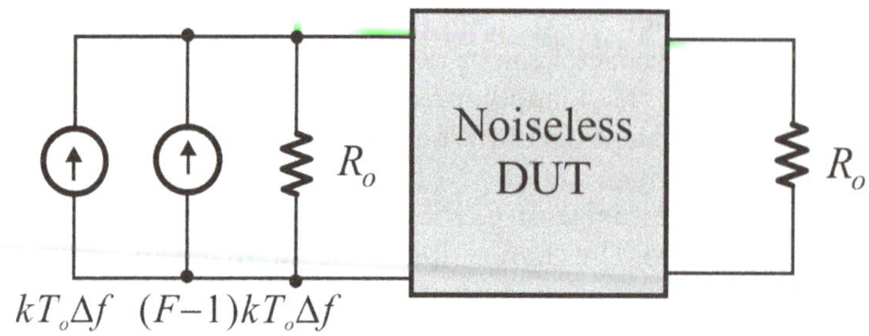

Figure 4.19 Representation of the DUT noise using the input equivalent noise

Now we will consider what happens when two signals, signal and noise, are applied to the input of the two-port network. The two signals at the input are shown in Figure 4.20. The available power of the input signal is S_i and the available noise spectrum density is N_i ($= kT_o$). Similarly, the available signal power at the output is S_o and the output noise spectrum density is N_o. From the definition of available gain in Equation (4.71), $G = S_o/S_i$. Then, the noise factor in Equation (4.72) can be seen to be the ratio of the signal to the noise ratio of the output to that of the input as in Equation (4.74).

$$F = \frac{P_{no}}{kT_o \Delta f G} = \frac{P_{no} S_i}{kT_o \Delta f S_o} = \frac{\dfrac{N_o}{S_o}}{\dfrac{N_i}{S_i}} \tag{4.74}$$

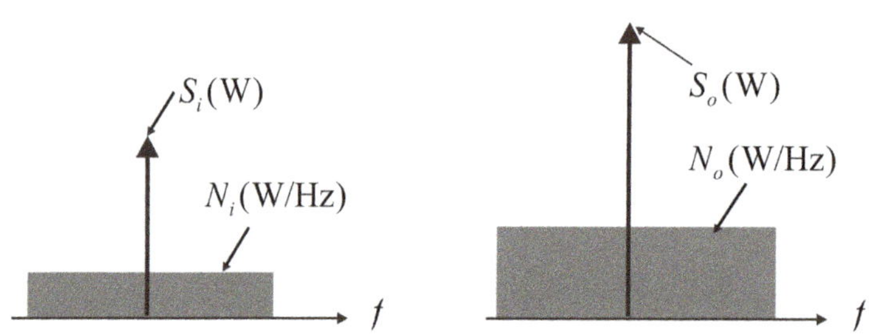

Figure 4.20 The input and output spectrums of a two-port network

4.2.4 Expression of Noise Parameters

The noise figure represents the noise contribution of a two port network to a load compared with an external noise source at T_o. However, the noise figure changes when the source impedance changes because the Norton equivalent circuit to the load depends on the source impedance. We need the noise parameters in order to be able to find the dependence of the noise figure on the source impedance. To achieve this, the internal noise sources of the two-port network are extracted to the outside of that network, as shown in Figure 4.21(a), a procedure that was previously explained in Figure 4.13. From Figure 4.21(a), two noise sources, e and i, appear together with the external noise source i_S. The two noise sources and the external noise source can be combined into a single Norton current source, as shown in Figure 4.21(b). Using the resulting circuit in Figure 4.21(b), the noise power at the output can be calculated. It is worth noting that the directions of the voltage and current in Figure 4.13 are changed. However, the result is same.

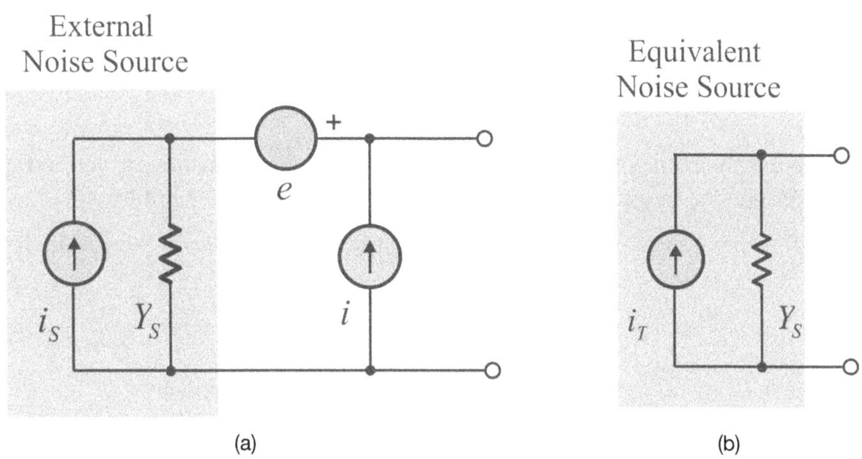

Figure 4.21 (a) Equivalent circuits at the input including the noise sources of the two-port network and (b) its Norton equivalent circuit

The noise sources e and i are random noise phasors and these sources are not completely independent. Some internal noise sources in the two-port network commonly appear at both e and i, while others appear only at either e or i. Therefore, these two noise sources have some correlations. In order to express the two noise sources, we require four parameters in total. Two parameters are required for the mean powers of e and i and two are required for the correlation between e and i.

When the three noise sources of the input shown in Figure 4.21 are transformed into the Norton equivalent circuit in Figure 4.21(b), the Norton admittance is equal to the external source admittance given by $Y_s = G_s + jB_s$. The Norton current source i_T becomes Equation (4.75).

$$i_T = i_s + i + Y_s e \tag{4.75}$$

Suppose that i_u denotes a part of the noise current i, which has no correlation at all with noise voltage e. Then, the remaining part of the current i becomes $i - i_u$. Thus, the current $i - i_u$ is fully correlated to e as a result. The Norton current source can be rewritten as Equation (4.76).

$$i_T = i_s + \left(i - i_u\right) + i_u + Y_s e \tag{4.76}$$

Let the correlation admittance between $i - i_u$ and e be $Y_\gamma = G_\gamma + jB_\gamma$; then the equation above becomes Equation (4.77).

$$i_T = i_s + i_u + \left(Y_s + Y_\gamma\right)e \tag{4.77}$$

Since the three noise sources are independent, their mean power becomes equal to Equation (4.78).

$$E(i_T^2) = E(i_s^2) + E(i_u^2) + |Y_s + Y_\gamma|^2 E(e^2) \tag{4.78}$$

The power of the thermal noise source in the bandwidth Δf is expressed in Equation (4.79).

$$E(i_s^2) = 4kT_oG_s\Delta f \tag{4.79}$$

If we use the concept of the equivalent thermal noise resistance given in Equation (4.67) for the remaining two noise sources, they can be expressed as Equations (4.80) and (4.81).

$$E(i_u^2) = 4kT_oG_u\Delta f \tag{4.80}$$

$$E(e^2) = 4kT_oR_n\Delta f \tag{4.81}$$

From the definition of the noise factor, F can be expressed as

$$F = \frac{E(i_T^2)}{E(i_s^2)} = 1 + \frac{G_u}{G_s} + \frac{R_n}{G_s}|Y_s + Y_\gamma|^2 \tag{4.82}$$

The noise factor in Equation (4.82) depends on the real part of the source admittance G_s. As G_s approaches zero, the noise factor becomes larger and approaches infinity. Also, as G_s approaches infinity, the noise factor approaches infinity, too. The noise factor shows a minimum for appropriate Y_s. Thus, the admittance that gives this minimum value of F is defined as $Y_{opt} = G_{opt} + jB_{opt}$. Equation (4.82) can be rewritten using the definition of minimum noise factor, F_{min}. It can then be expressed as shown in Equations (4.83a)–(4.83d).

$$F = F_{min} + \frac{R_n}{G_s}\left\{(G_s - G_{opt})^2 + (B_s - B_{opt})^2\right\} \tag{4.83a}$$

$$G_{opt} = \left[\frac{G_u + R_nG_\gamma^2}{R_n}\right]^{1/2} \tag{4.83b}$$

$$B_{opt} = B_\gamma \tag{4.83c}$$

$$F_{min} = 1 + 2R_n(G_\gamma + G_{opt}) \tag{4.83d}$$

Therefore, the two-port noise parameters can be defined in terms of F_{min}, R_n, and Y_{opt} instead of their mean powers, and the correlation of the two sources e and i, which becomes another definition for the two-port noise parameters.

In addition, by using the reflection coefficient, the equation above can be rewritten as Equations (4.84a)–(4.84c).

$$F = F_{min} + \frac{4r_n|\Gamma_s - \Gamma_{opt}|^2}{\left(1-|\Gamma_s|^2\right)|1+\Gamma_{opt}|^2} \tag{4.84a}$$

$$\Gamma_s = \frac{Y_s + Y_o}{Y_s - Y_o} \tag{4.84b}$$

$$\Gamma_{opt} = \frac{Y_{opt} + Y_o}{Y_{opt} - Y_o} \tag{4.84c}$$

Here, r_n is R_n/Z_o. Generally, because F_{min}, r_n, and Γ_{opt} defined in the equations above can be directly measured with a noise figure meter, they are the most widely used as two-port noise parameters.

Example 4.9

Using ADS, determine the noise parameters F_{min}, R_n, and Γ_{opt} for the circuit shown in Figure 4E.14, for frequencies of 1 GHz to 10 GHz.

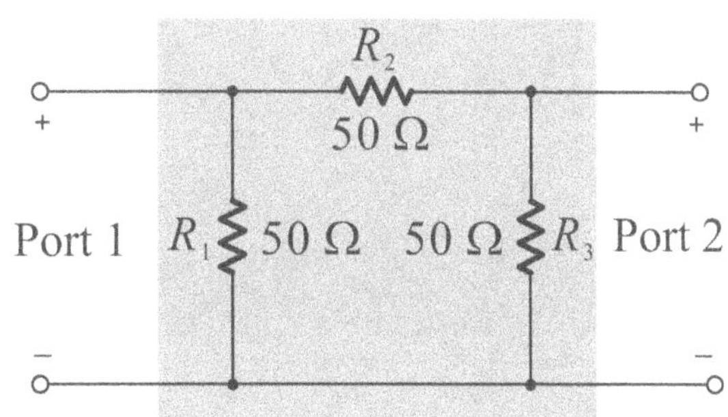

Figure 4E.14 A two-port circuit with noisy resistors

Solution

Figure 4E.15 shows the simulation circuit set up to calculate the noise parameters. As mentioned earlier, the noise factor is defined at the temperature of 290 °K. However, the default temperature of ADS is set to 25 °C. Thus, the noise factor computed by ADS is slightly different from the exact noise factor. Therefore, the **option controller** shown in Figure 4E.15 is inserted to set the temperature at 16.85 °C. ADS does not have a separate noise simulation controller and the noise simulation is carried out by S-parameter simulation. However, the **S-parameter simulation controller** should be set for noise simulation to get the noise simulation results from S-parameter simulation. Figure 4E.16 shows the pop-up window of the **S-parameter simulation controller**. The box next to **Calculate noise** should be checked for noise simulation. In the port fields, the input is assigned to 1 and the output is assigned to 2. This assignment of the ports reflects the port numbering in the schematic shown in Figure 4E.15. Thus, the port has **Num**=1 is the input port and **Num**=2 as the output port. After the simulation, the noise figure and noise parameters such as $F_{\min}$, R_n, and Γ_{opt} are stored in a dataset as **nf**, **NFmin**, **Rn**, and **Sopt**, respectively.

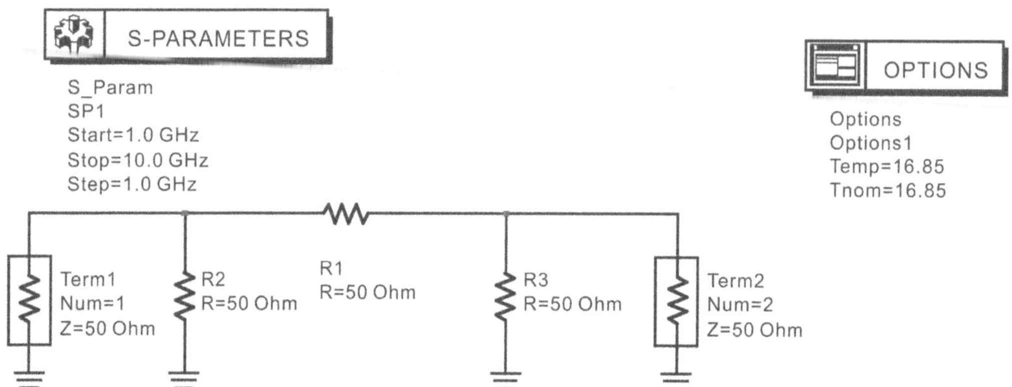

Figure 4E.15 Simulation schematic. The temperature in the ADS simulation is set to 25 °C by default. However, the noise figure is defined at 290 °K. Thus, the option controller is included to set the simulation temperature at 290 °K, which is 16.85 °C.

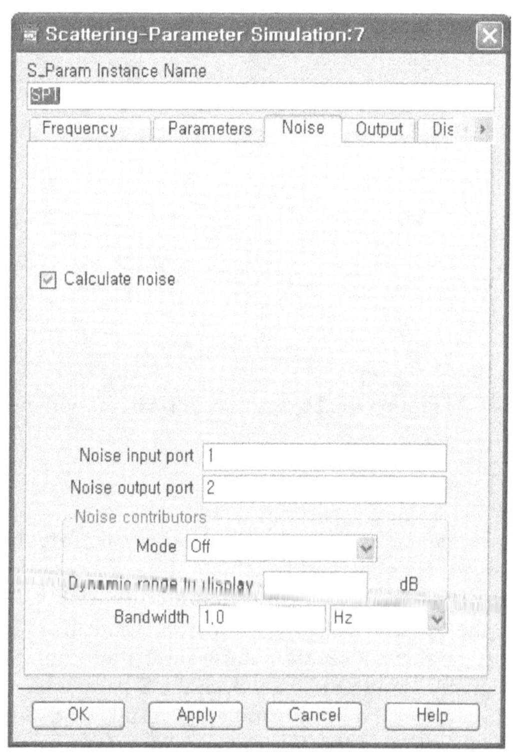

Figure 4E.16 Pop-up window of S-parameter simulation controller. When the Calculate noise check box is not checked, noise simulation is not performed. Noise input and output ports are the port numbers for noise sources and are usually set to be equal to the S-parameter port numbers. Using **Mode**, we can find the contributions of various noise sources in a two-port network from the dataset after the simulation.

The simulation results are shown in Figure 4E.17 and Figure 4E.18. In Figure 4E.17, the noise figure appears as **nf**(2). The **nf**(2) represents the noise figure at the output port (i.e., port 2) when a noise source is connected to the input (i.e., port 1). On the other hand, **nf**(1) represents the noise figure at the input port (port 1) when a noise source is connected to the output, but **nf**(1) is not frequently used. It can be seen in Figure 4E.17 that the noise figure is 11.76 dB and $F_{\min}$ is 11.439 dB. Figure 4E.18 shows the reflection coefficient that gives the minimum noise, that is, Γ_{opt} (**Sopt**). It can be seen that **Sopt** is independent of frequency and has a constant value of 28.86 Ω.

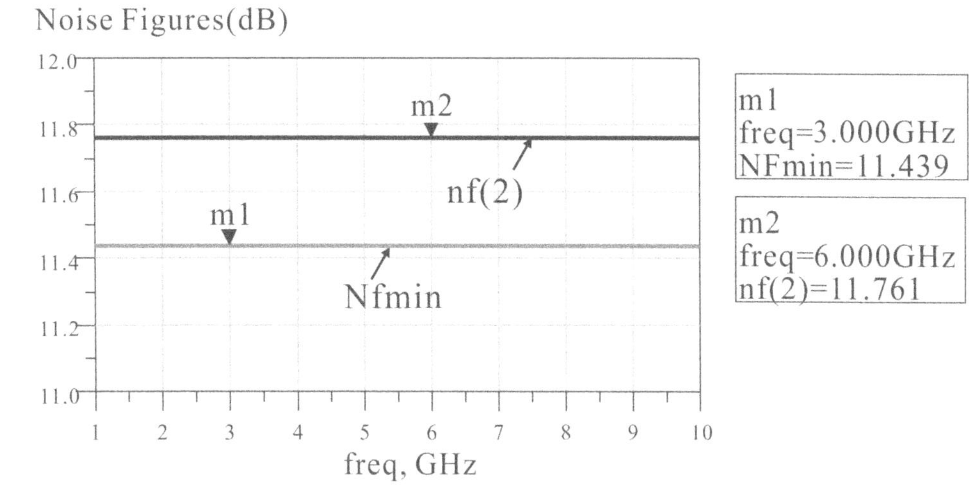

Figure 4E.17 Simulation results of noise figure and $NF_{\min}$ given as decibel values

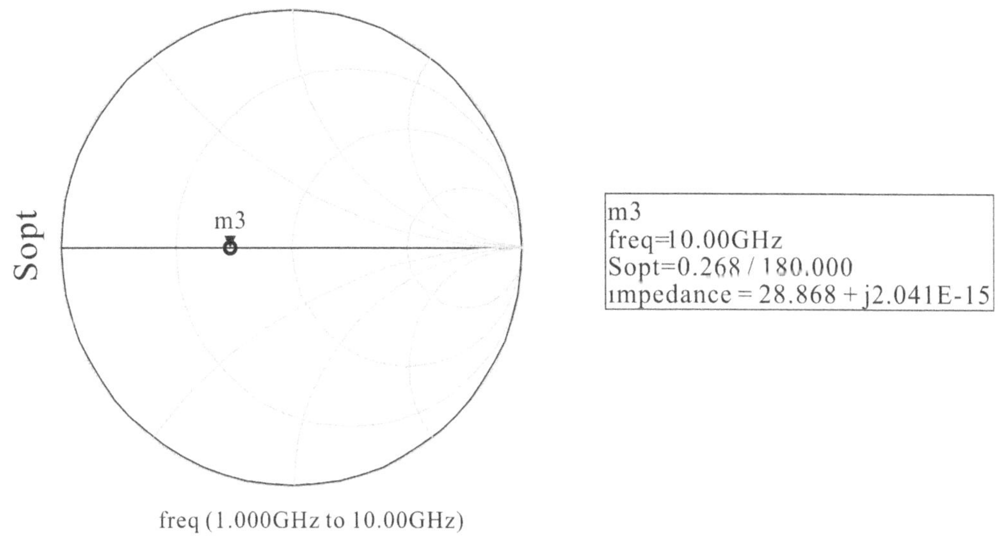

Figure 4E.18 Plot of S_{opt}

Example 4.10

The noise simulation results **icor**(1,1), **icor**(1,2), and **icor**(2,2) after the S-parameter noise simulation represent the noise correlations $E(|i_{n1}|^2)$, $E(|i_{n2}|^2)$, and $E(i_{n1}^{*}i_{n2})$ in Figure 4.12, which are in an ADS dataset appearing as a result of the simulation performed in Example 4.9. Determine them by calculation and also verify them using ADS.

Solution

The Y-parameters of Figure 4E.14 are

$$y_{11} = 2Y_o = (50 \parallel 50)^{-1} = y_{22}$$

$$y_{12} = -Y_o = -(50)^{-1} = y_{21}$$

The i_{n1} and i_{n2} correspond to the short-circuit noise currents that appear at ports 1 and 2 and are due to the three noisy resistors. The short-circuit noise current at port 1 is due to resistors R_1 and R_2, and that at port 2 is due to R_2 and R_3. The noise current in parallel with a 50-Ω resistor per 1 Hz bandwidth is

$$E(|i_n|^2) = 4kTY_o$$

Thus, $E(|i_{n1}|^2)$ and $E(|i_{n2}|^2)$ that correspond to **icor**(1,1) and **icor**(2,2) in the ADS dataset are computed to be

$$E\left(\left|i_{n1}\right|^2\right) = 2E\left(\left|i_n\right|^2\right) = 8kTY_o = \textbf{icor}(1,1) = \textbf{icor}(2,2)$$

The common noise current due to R_2 appears at ports 1 and 2. Since this current appears in the opposite direction, we obtain

$$E(i_{n1}^{*}i_{n2}) = E(i_{n1}i_{n2}) = -E\left(i_n^2\right) = -4kTY_o = \textbf{icor}(1,2) = \textbf{icor}(2,1)$$

The following equations are entered in the display window to compute $E(|i_{n1}|^2)$, $E(|i_{n2}|^2)$, and $E(i_{n1}^{*}i_{n2})$.

```
Eqn T=290
Eqn i1=4*boltzmann*T/50
Eqn i_port1_sq=2*i1
Eqn icor12=i1
```

Measurement Expression 4E.2 Equations

The **boltzmann** in Measurement Expression 4E.2 is a built-in constant that represents the Boltzmann constant, while **i1** represents $E(|i_n|^2)$ of a 50-Ω resistor. Also, **i_port1_sq** and **icor12** represent the computed values for **icor**(1,1) and **icor**(1,2), respectively. Figure 4E.19 shows the comparison and we can see that the computed and simulated results show an exact match.

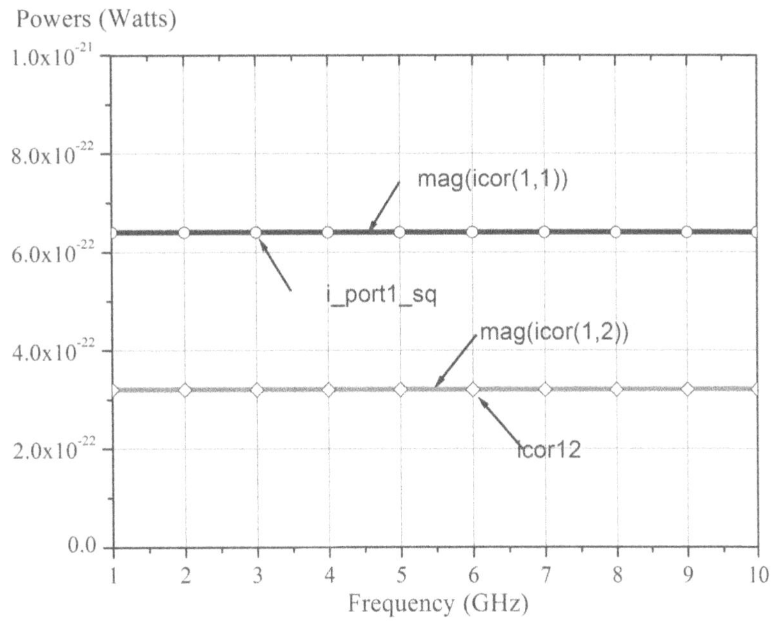

Figure 4E.19 Comparison of the calculation results. The simulated outputs **icor**(i,j) after the noise simulation represent the short-circuit noise currents appearing at ports 1 and 2 that are verified by comparing them with the directly computed values.

Example 4.11

Calculate the noise parameters of the circuit in Figure 4E.14 and compare the results with the previously simulated ADS results in Example 4.9.

Solution

The short-circuit noise currents at ports 1 and 2 can be represented by e and i as

$$i_{n1} = i + y_{11}e$$

$$i_{n2} = y_{12}e$$

Using the equations above, e and i are

$$e = \frac{i_{n2}}{y_{12}}$$

$$i = i_{n1} - \frac{y_{11}}{y_{12}}i_{n2}$$

Their ensemble averages are

$$E(e^2) = \frac{1}{y_{12}^2} E(|i_{n2}|^2) = \frac{1}{y_{12}^2} \cdot 8kTY_o = 8kTZ_o$$

$$E(|i|^2) = E(|i_{n1}|^2) + \left(\frac{y_{11}}{y_{12}}\right)^2 E(|i_{n2}|^2) - 2\frac{y_{11}}{y_{12}} E(i_{n1} i_{n2})$$

$$= 8kTY_o + 4(8kTY_o) - 4(4kTY_o) = 24kTY_o$$

In addition, the correlation between i and e is

$$E(e^* \cdot i) = E(e \cdot i) = \frac{1}{y_{12}} E\left\{ i_{n2}^* \left(i_{n1} - \frac{y_{11}}{y_{12}} i_{n2} \right) \right\} = \frac{1}{y_{12}} E(i_{n1} i_{n2}) - \frac{y_{11}}{y_{12}^2} E(|i_{n2}|^2)$$

Keeping the correlation of i and e in mind, the results from Example 4.10 can be rewritten as

$$E(e \cdot i) = 4kT - \frac{y_{11}}{y_{12}^2} 8kTY_o = -12kT$$

When a thermal noise source is applied to the input, the total noise current of the input becomes

$$i_T = i_s + i - G_s e$$

Therefore, the average of the total current noise at the input is

$$E(|i_T|^2) = E(|i_s|^2) + E\left(|i - G_s e|^2\right) = E(|i_s|^2) + E(|i|^2) + G_s^2 E(|e|^2) - 2G_s E(e \cdot i)$$

$$= 4kTG_s + 24kTY_o + \frac{G_s^2}{Y_o}(8kT) + 2G_s(12kT)$$

Hence, the noise factor is

$$F = \frac{E(i_T^2)}{E(i_s^2)} = 7 + 6\frac{Y_o}{G_s} + \frac{2G_s}{Y_o}$$

Since $G_s = Y_o$, the noise figure is

$$\mathbf{nf}(2) = F = 7 + 6 + 2 = 15 = 11.76 \text{ dB}$$

The minimum value of F occurs when

$$6\frac{Y_o}{G_s} = \frac{2G_s}{Y_o} \rightarrow G_s = \sqrt{3}Y_o$$

Therefore,

$$Z_{opt} = \sqrt{\frac{1}{3}} Z_o = 28.6\ \Omega$$

The minimum noise factor is then

$$F_{\min} = 7 + 2\sqrt{12} = 13.92 = 11.43 \text{ dB}$$

By comparing these results, we can see that the computed $F_{\min}$ and Γ_{opt} agree exactly with the previously simulated results in ADS. Alternatively, by using the definition of the noise factor in Equation (4.68), we can directly calculate the noise factor, but that is not presented here.

4.2.5 Frii's Formula

Figure 4.22 shows an n-stage cascaded amplifier. The available power gain and noise figure of each stage is denoted as G_{Ai} and F_i, respectively. Then, the equivalent available input noise power at each stage will be $(F_i - 1)kT_o\Delta f$, as shown in Figure 4.22. It is also worth noting that each amplifier is assumed to be matched.

Thus, the output noise power N_T is

$$N_T = F_1 kT_o \Delta f G_1 G_2 \cdots G_n + (F_2 - 1)kT_o \Delta f G_2 \cdots G_n + \cdots + (F_n - 1)kT_o \Delta f G_n \tag{4.85}$$

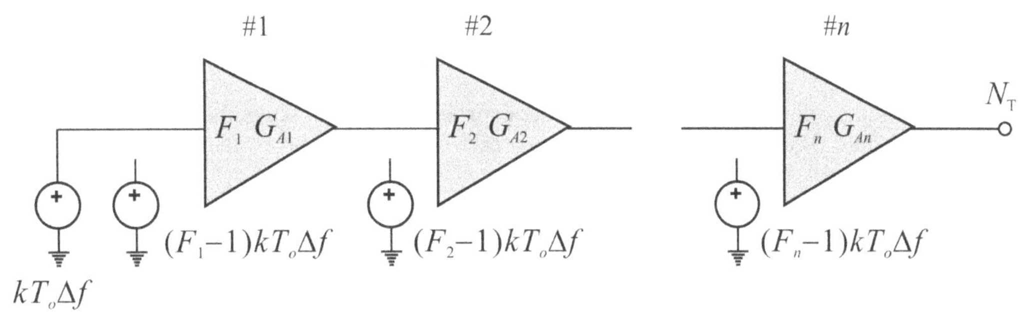

Figure 4.22 An n-stage cascaded amplifier

Suppose that the overall noise factor of the cascaded amplifier is F_{tot}. Then, the equivalent available input noise power of the cascaded amplifier is $F_{tot}kT_o\Delta f$. The equivalent input noise power $F_{tot}kT_o\Delta f$ will be amplified by the cascaded amplifier and delivered to the load. Thus, N_T can be expressed as

$$N_T = F_{tot} kT_o \Delta f G_1 G_2 \cdots G_n \tag{4.86}$$

Comparing Equation (4.85) with Equation (4.86), the overall noise factor F_{tot} can be expressed as Equation (4.87).

$$F_{tot} = F_1 + \frac{F_2 - 1}{G_1} + \frac{F_3 - 1}{G_1 G_2} \cdots \tag{4.87}$$

Also, we can see that the overall power gain G_t is

$$G_t = G_1 G_2 \cdots G_n \tag{4.88}$$

Equation (4.87) is known as Frii's formula, which can be used to obtain the overall noise figure of cascaded stages when the noise factor and available power gain of each stage are known. Also, from Frii's formula, when the gain G_1 of the first stage is large enough, it can be found that *the overall noise figure is mainly determined by the noise figure of the first-stage amplifier*.

Example 4.12

In the block diagram of the cascaded three-stage amplifier shown in Figure 4E.20, calculate the overall noise figure and gain.

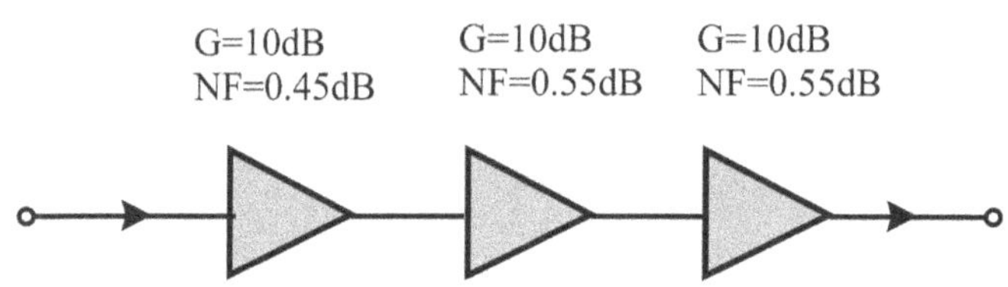

Figure 4E.20 Block diagram of a cascaded three-stage amplifier

Solution

Since Equations (4.87) and (4.88) are not expressed in decibel scale, the decibel values in the second and third columns of Table 4E.1 are first converted into linear-scale numbers in columns 4 and 5, respectively. The values are calculated using

$$G = 10^{\frac{G[dB]}{10}}, F = 10^{\frac{F[dB]}{10}}$$

Table 4E.1 Calculated noise figures

G and F are the converted gain and noise figures from their decibel values. NFt (dB) is the noise figure computed at the input of the corresponding stage.

	G(dB)	F(dB)	G	F	F_t	NF_t (dB)
Amp1	10	0.45	10	1.11	1.12	0.51
Amp2	10	0.55	10	1.14	1.15	0.60
Amp3	10	0.55	10	1.14	1.14	0.55

The conversion from decibel to linear scale must be carried out before applying Frii's formula. Rather than the direct application of Equation (4.87), the overall noise figure can be computed by repeating the computation of the two-stage noise figure. Denoting the second-stage noise factor as F_2 and the gain and noise factor of the first stage as G_1 and F_1, respectively, the overall noise factor F_t of the two stages becomes

$$F_t = F_1 + \frac{F_2 - 1}{G_1}$$

Applying the formula above iteratively and moving from the last to the first stage of the cascaded stages, the noise factor can be calculated. The noise figure of the third-stage amplifier alone becomes its own noise figure, as shown in the fourth row and **Ft** column. Then, the noise figure at the second-stage amplifier input can be calculated using the formula above, and the result is shown in the third row. The total noise figure seen from the first-stage amplifier can be calculated similarly. The values in the **Ft** column are then converted to decibels using

$$NF_t[dB] = 10\log F_t$$

The results in Table 4E.1 are calculated using Microsoft's Excel spreadsheet program. It can be seen that the overall noise figure is 0.51 dB. Considering the noise figure for the first stage is 0.45 dB, the overall noise figure appears to be slightly increased. However, when the gain of the first stage is high enough, the overall noise figure approaches the first stage's noise figure.

Example 4.13

For the block diagram of the cascaded receiver shown in Figure 4E.21, calculate the overall noise figure and gain.

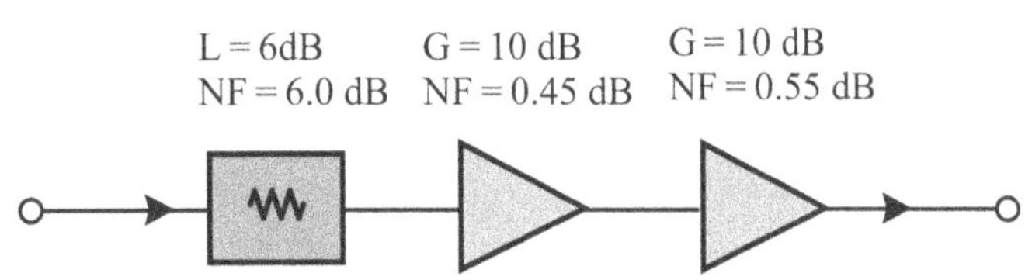

Figure 4E.21 Block diagram of a cascaded receiver

Solution

The overall noise figure and gain can be solved in the same way as in Example 4.12 and their values are shown in Table 4E.2. It is worth noting that if an attenuator with an attenuation of L appears first and its output is connected to the next stage of a noise factor of F_2, then the overall noise factor becomes approximately

$$F_t = \frac{1}{L} + L(F_2 - 1) \simeq LF_2$$

Thus, the noise figure becomes

$$F_t[dB] = L[dB] + F_2[dB]$$

Table 4E.2 Result of calculated noise figures

The designation is the same as in Table 4E.1.

·	G(dB)	F(dB)	G	F	F_t	NF_t (dB)
Att1	–6	6	0.25	3.98	4.57	6.60
Amp2	10	0.55	10	1.14	1.15	0.60
Amp3	10	0.55	10	1.14	1.14	0.55

That is, the overall noise figure is obtained by adding the attenuator losses to the next-stage noise figure in decibels. Looking at Table 4E.2, the results can be seen to reflect this fact. Therefore, when constructing a low-noise amplifier, the loss in the input matching network is directly associated with the noise figure. Thus, the reduction of the input matching network loss is crucial in the low-noise amplifier design.

4.2.6 Measurement of Noise Figure and Noise Parameters

4.2.6.1 Noise Figure Measurement

The noise figure of a two-port network can be measured using the setup shown Figure 4.18. Generally, the noise figure depends on frequency and the measurement of a *spot noise figure* for a given frequency requires the measurement of the spectral noise power. The spectral noise power can be measured using a tunable bandpass filter that moves its center frequency within a specified narrow bandwidth. The equipment that can measure the spectral noise power is a *noise figure meter* shown in Figure 4.23. An early-stage noise figure meter can measure the noise figure up to a frequency of 1.5 GHz due to the difficulty of the hardware's implementation.

Recently, new equipment that seems to perform better than the former noise figure meter has been developed. However, the explanation we will present is based on that former noise figure meter. Although the meter can directly measure noise power only up to a frequency 1.5 GHz, extending it to higher frequencies is possible through a frequency down-conversion. The required equipment for the frequency down-conversion includes a noise figure test set and an RF signal source. The noise figure test set can be viewed as a mixer. The RF signal source plays the role of a local signal source. This combination of equipment is known as a *noise figure measurement system* and is shown in Figure 4.23.

We now turn our attention to a noise source, an example of which is shown in Figure 4.24. When a DC voltage of 0 V is applied to the 28 V pulse input shown in Figure 4.24, the noise source behaves like a thermal noise source at room temperature, and the noise source is said to be in the *cold state*. When a DC voltage of 28 V is applied, a larger thermal noise than room temperature appears at the coaxial connector output, and the noise source is said to be in the *hot state*. The figure of merit for such noise sources is *Excessive Noise Ratio*. Excessive Noise Ratio (ENR) is defined as the ratio of the excessive noise power in the hot state to the thermal noise in a cold state. Defining the thermal noise at room temperature or a cold state as $kT_o\Delta f$, then the excessive noise in a hot state can be expressed as $k(T_1 - T_o)\,\Delta f$, where T_1 is the equivalent noise temperature in the hot state. Thus, ENR can be expressed as Equation (4.89).

$$ENR = \frac{T_1 - T_o}{T_o} \tag{4.89}$$

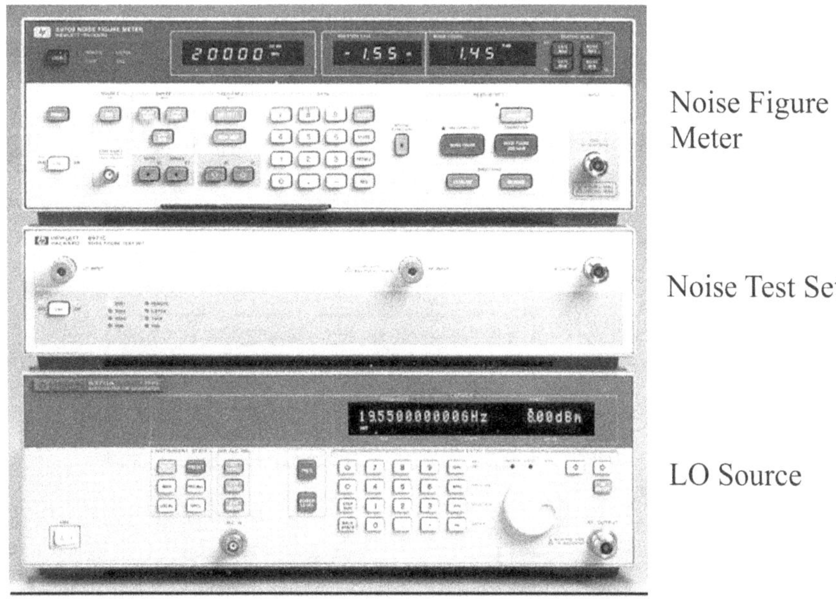

Figure 4.23 Noise figure measurement system.[3] The noise figure meter measures the noise figure of a DUT, but its frequency range is limited to 1.5 GHz. The noise test set is the down-converter for the noise figure meter. In the down conversion, an LO source is necessary. The signal generator at the bottom acts as the LO source for the down-conversion.

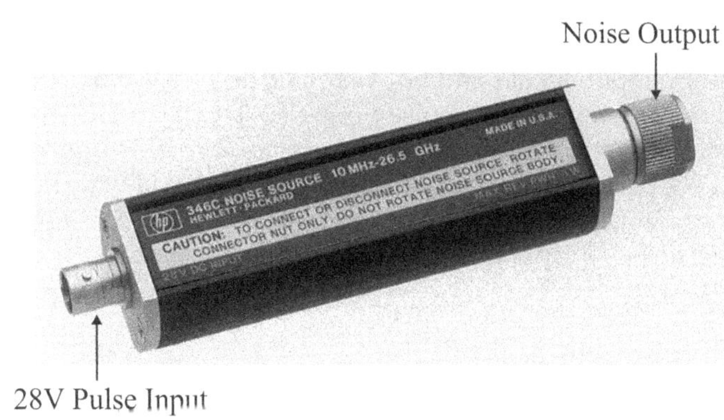

Figure 4.24 Noise signal source.[4] At the BNC connector labeled 28 V pulse input, the pulse signal from the noise figure meter is applied. The pulse signal periodically changes from 0 to 28 V. At the connector labeled noise output, broadband noise signals appear. The noise power at 0 V is close to the thermal noise power level, while at 28 V, an increased noise power from the thermal noise power appears at the output.

3. Agilent technologies, HP8970 Series, Noise Figure Measurement Products, 5091-6049E.pdf, May, 1999.

4. Agilent technologies, 346A/B/C Noise sources 10 MHz to 26.5 GHz, 5953-6452.pdf, 2000.

Figure 4.25 shows the setup for the measurement of the noise figure of a DUT. In Figure 4.25, the noise figure meter drives the noise source using the periodical pulse voltage that changes from 0 to 28 V. To measure the noise figure of the DUT, the cold noise source is first applied to the DUT, and the resulting output noise power of the DUT is measured. This corresponds to the noise output power of the DUT for the thermal noise input power at room temperature. Then, the hot noise is applied to the DUT and the output noise power of the DUT is measured. From these two measurements, the noise figure of the DUT can be determined.

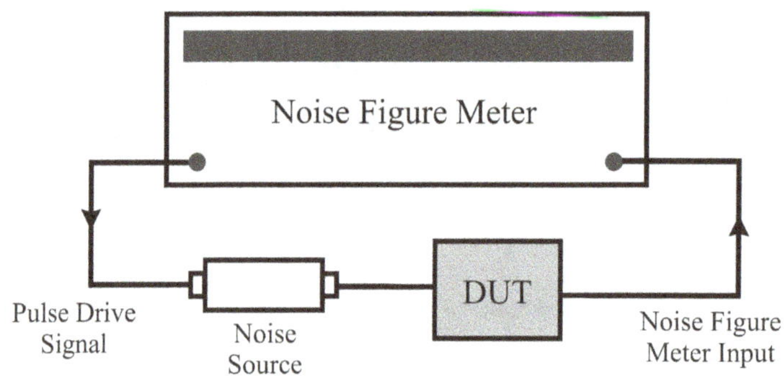

Figure 4.25 Noise figure measurement setup using noise figure meter and noise source

Figure 4.26 shows the concept of the noise figure measurement shown in Figure 4.25. In Figure 4.26, the noise powers delivered to the load in the noise figure meter are denoted by N_o and N_1. Noise powers N_o and N_1 represent the DUT output noise powers for the thermal noise input at cold and hot states, respectively. Now, denoting the noise figure of the DUT as F, the equivalent noise power $(F-1)kT_o\Delta f$ of DUT can be placed at the input. Consequently, the two-port DUT becomes a noise-free two-port network because all internal noises are placed at the input. The hot-state and cold-state noise inputs are represented by switches. Thus, in the cold state, thermal noise of $kT_o\Delta f$ is applied to the DUT, whereas the hot noise $kT_1\Delta f$ is applied to the DUT in the hot state.

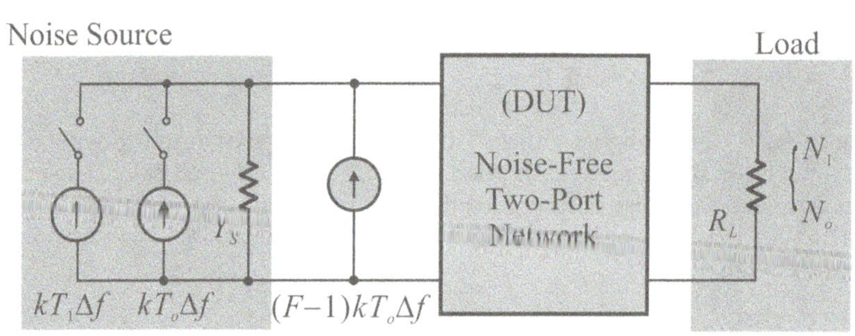

Figure 4.26 Representation of the noise figure measurement in Figure 4.25

When the input is thermal noise at room temperature, the output becomes $N_o = FkT_oG\Delta f$. Here, G represents the DUT power gain. When the input is in the hot state, because there is no change in the noise temperature of the DUT, N_1 becomes Equation (4.90).

$$N_1 = kT_1\Delta fG + (F - 1)kT_o\Delta fG \tag{4.90}$$

If we take the ratio of the two powers, which is called the *Y-factor*, we obtain

$$Y = \frac{N_1}{N_o} = \frac{kT_1\Delta fG + (F - 1)kT_o\Delta fG}{FkT_o\Delta fG} \tag{4.91}$$

Using Equation (4.91), the noise factor can be determined as

$$F = \frac{T_1 - T_o}{T_o}(Y - 1)^{-1} \tag{4.92}$$

Equation (4.92) is rewritten using the definition of ENR and the noise figure is thus expressed in Equation (4.93).

$$F(dB) = 10\log(ENR) - 10\log(Y - 1) \tag{4.93}$$

From Equation (4.92), the noise figure can be determined using ENR and the Y-factor. Thus, changing the center frequency of the bandpass filter with fixed bandwidth Δf in the noise figure meter, the frequency-dependent noise figure can be measured. The bandwidth Δf can also be controlled by manipulating the front panel keys of the noise figure meter.

4.2.6.2 Noise Parameter Measurement

The setup to measure the noise parameters using the noise figure meter is shown in Figure 4.27. With this setup, the noise parameters can be indirectly measured. In the setup shown in Figure 4.27, the source impedance seen from the DUT can be adjusted through the tuner.

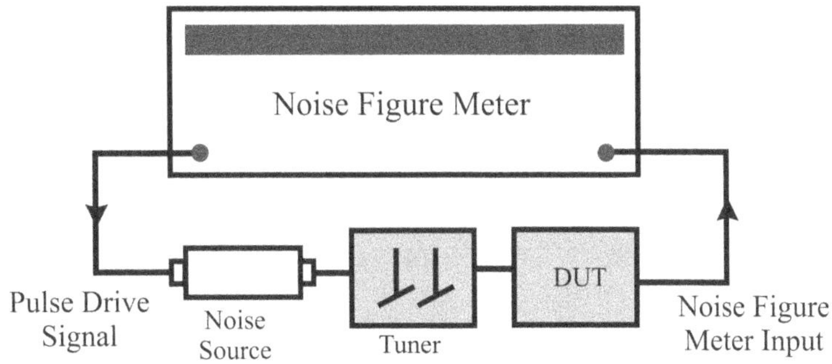

Figure 4.27 Noise parameter measurement setup

Here, the tuner is the commercially available mechanical impedance tuner. Thus, by adjusting the tuner, the point that shows the minimum noise figure can be found. The minimum noise figure is F_{min}. The impedance Γ_{opt} that yields the minimum noise figure can now be obtained by measuring the tuner impedance using a network analyzer. To measure Γ_{opt}, the measurement can be made by replacing the noise source with a 50-Ω load. For the remaining parameter r_n, the tuner is removed and the noise figure of the DUT is measured again. This is the noise figure when the input and output impedances are 50 Ω. Denoting the measured noise figure as F_{50} and then substituting $\Gamma_s = 0$ into Equation (4.84a), the normalized noise resistance r_n can be calculated as shown in Equation (4.94).

$$r_n = F_{50} - F_{min} \frac{\left|1 + \Gamma_{opt}\right|^2}{4\left|\Gamma_{opt}\right|^2} \tag{4.94}$$

Thus, the noise parameters can be measured using a noise figure meter and tuner.

4.3 FILE FORMATS

Generally, the suppliers of high-frequency devices provide the measured S-parameters. If the S-parameters are absent, they can be obtained by direct measurement. These S-parameters are necessary for designing circuits or performing circuit simulations. Conversely, the performance of a designed circuit is also evaluated in terms of the S-parameters. The S-parameters are written as a normal ASCII text file and the data is delivered to a **data device** for simulation in ADS. This text file exists in various formats and will be described in this section. The file formats for recording two-port parameters can be generally classified into four types: a **Touchstone** file, a **CITI** file, an **IC-CAP** file, and an **MDIF** file. Among these, the most commonly used formats are **Touchstone** files and **CITI** files, which we will describe in this section.

A **Touchstone** file is an ASCII text file and a **Touchstone** format is also commonly called an **SnP** format. The letter S in **SnP** stands for S-parameters, while n stands for the number of ports. The two-port parameters such as G-, H-, Y-, and Z-parameters instead of the S-parameters can also be written in the **Touchstone** file format. This is summarized in Table 4.3. In addition to the S-parameters, two-port noise parameters may also be included in the **Touchstone** file format.

Table 4.3 Parameter abbreviations in the Touchstone format

S	=	Scattering parameters
Y	=	Admittance parameters
Z	=	Impedance parameters
H	=	Hybrid-h parameters
G	=	Hybrid-g parameters

Touchstone file formats are composed of an **Option line**, a **Data line**, and **Comment lines**. The data structure of the Touchstone file format is shown in Table 4.4. Here, # represents a delimiter character of the **Option line** that specifies the data format of the following data lines. The <frequency unit> in Table 4.4 represents the unit of frequency and the <parameter> represents the type of two-port parameters. The <format> defines how the two-port parameter is written since it is a complex number. **DB** indicates that the complex number is written in dB and angle format, while **MA** means that the complex number is written in a magnitude and angle format. Finally, the <Rn> represents the reference impedance. Examples of the **Option line** are shown in the Table 4.5.

Table 4.4 Touchstone file structure

Option line ▶ # <frequency unit><parameter><format><Rn>
<data line>
.
.
<data line>

Table 4.5 Example of the Option line

# GHZ S RI R 100	Frequency in GHz, S-parameters in real-imaginary format, normalized by 100 ohm
# HZ Z MA R 1	Frequency in Hz, Z-parameters in magnitude-angle format, normalized by 1 ohm
# MHZ S DB R 50	Frequency in MHz, S-parameters in dB-angle format, normalized by 50 ohm

The example of the **S2P** file is shown in Table 4.6. All the data are arranged with frequency as a parameter and in order from the left as S11 S21 S12 S22. Note that the format follows the definition of the **Option line**.

Table 4.6 S-parameters example

```
# GHZ S MA R 50
0.100000   0.218 171.4   4.487 171.3   0.136 -6.8    0.117 172.9 ! freq S11, S21, S12, S22
0.144750   0.221 167.6   4.472 167.8   0.133 -10.7   0.120 165.5
0.189500   0.215 163.2   4.452 163.7   0.132 -14.9   0.125 163.0
2.00000   8.359 0.3819 179.99  0.75 ! NFmin[dB] S-opt Rn/Zo
```

CITI stands for Common Instrumentation Transfer and Interchange and it is a widely used file format in measurement instruments. A **CITI** file is composed of **Header** and **Data**. **Header** consists of **Keyword** and **Setup** information, while **Data** mostly consists of one or several data arrays. Table 4.7 provides an example of a typical **CITI** file. **Header** indicates the information about the data listed on the following header as well as information about the setup of the measuring equipment. **CITIFILE A.01.00** is the version number and **NAME** is the name of the data, which is specified as **Memory** in this example. **VAR** specifies the independent variable. In this example, the name of the variable is **FREQ**; it has magnitude (**MAG**) and the number of variables is 3. The values of the variables are listed in **DATA** and are found to be 2, 3, and 4 GHz. Next, the **DATA** line in **Header** specifies the data. The name of **DATA** is **S** (this means S-parameters) and is presented in the form of **RI** (the complex number representation of the real and imaginary parts). **VAR_LIST_BEGIN** indicates the start of the previously defined **VAR** data and the **VAR_LIST_END** indicates the end of the **VAR** data. The **DATA** is listed starting with the **BEGIN** keyword and ends with the keyword **END**.

Table 4.7 CITI file example

Header	CITIFILE A.01.00 NAME MEMORY VAR FREQ MAG 3 DATA S RI
DATA	VAR_LIST_BEGIN 2000000000 3000000000 4000000000 VAR_LIST_END BEGIN –35.4545E-2, –1.38601E-3 0.23491E-2, –3.9883E-3 2.00382E-3, –1.40022E-3 END

This is an example of one-port S-parameters, which are listed as RI-format complex numbers. In the case of two-port S-parameters, **Header** should include four data lines such as **DATA S**[1,1] **RI**, **DATA S**[2,1] **RI**, **DATA S**[1,2] **RI**, **DATA S**[2,2] **RI**, and the corresponding **BEGIN** and **END** keywords in the data field will also appear four times

4.4 SUMMARY

- The S-parameters are the two-port network parameters that are the most conveniently measured and can be converted to other two-port network parameters.
- The independent variable for the voltage S-parameters are incident voltages, and the reflected voltages are the dependent variables determined by the S-parameters and the incident voltages.

- The a- and b- waves are defined using the incident voltages and reflected voltages, and their squares correspond to powers, especially the a-waves that can be physically interpreted as the available power from the sources. The conventional S-parameters are defined using the a- and b-waves.
- The S-parameters can be computed using S-parameter simulation in ADS. Terms are used to distinguish ports from the two-port network in ADS.
- When transmission lines are connected to a cascaded two-port network, the effects of the transmission lines appear as the phase shift in the S-parameters proportional to the electrical lengths of the transmission lines. The effects can be removed through calibration.
- The return loss and insertion loss of a two-port network can be related to the two-port S-parameters.
- Internal noise sources inside a two-port network can be equally placed outside of the input and output of a two-port network, which yields the same response to the external networks.
- Noise spectrums can be measured using multiple time samples, which are obtained from a sample function and defined on a sufficiently long time interval.
- Noise figure is defined as

$$\text{Noise factor} = \frac{\text{Actual noise power output to load}}{\text{Power output due to noise source of } T_o = 290°K}$$

  and it is the function of the source impedance.
- The dependence of the noise figure can be described using the noise parameters $NF_{\min}$, S_{opt}, and R_n.
- Noise simulation can be performed using an S-parameter simulation.
- Touchstone and CITI file formats that describe the S-parameters and noise parameters as text files are introduced.

REFERENCES

1. D. M. Pozar, *Microwave Engineering,* 2nd ed. New York: John Wiley & Sons, Inc., 1998.
2. H. Fukui, ed., *Low-Noise Microwave Transistor & Amplifiers*, New York: IEEE Press, 1981.
3. Agilent Technologies, *Agilent Network Analyzer Basics*, 5965-7917E, pdf, 2004.
4. R. E. Ziemer and W. H. Tranter, *Principles of Communications*, New York: John Wiley & Sons, Inc., 2002.
5. H. W. Ott, *Noise Reduction Techniques in Electric Systems*, New York: John Wiley & Sons, Inc., 1976.
6. H. Hildebrand and P. H. Russer, "An Efficient Method for Computer-Aided Noise Analysis of Linear Amplifier Networks," *IEEE Transactions on Circuits and Systems* 23, no. 4 (April 1976): 235–238.

PROBLEMS

4.1 For a circuit in Figure 4P.1, calculate the two-port S-parameters normalized by the reference impedance Z_o.

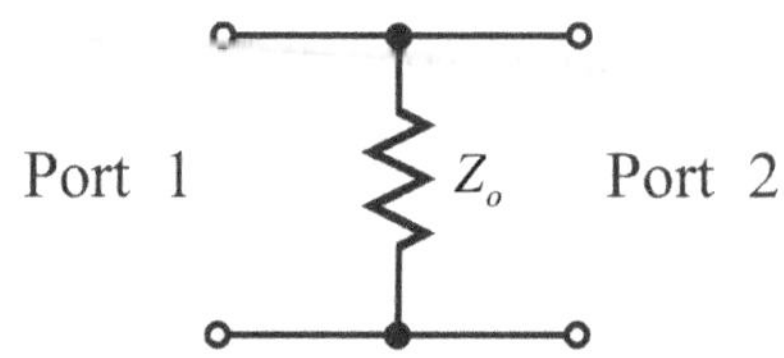

Figure 4P.1 A two-port circuit

Using the results of the previous calculation, now calculate the two-port S-parameters for the circuit in which two transmission lines of electrical lengths θ_1 and θ_2 are added to the circuit in Figure 4P.1 as shown in Figure 4P.2.

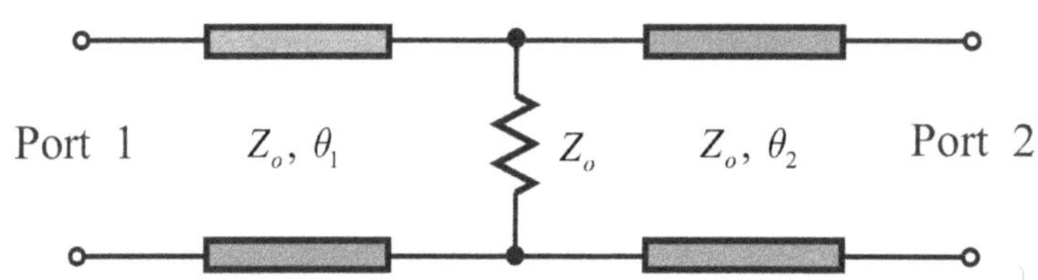

Figure 4P.2 A two-port circuit with transmission lines

4.2 Calculate the S-parameters of the parallel and serial resonance circuits in Figure 4P.3.

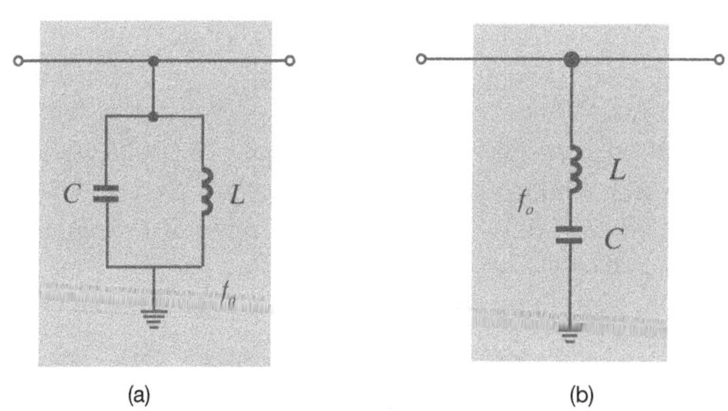

Figure 4P.3 (a) A parallel resonator and (b) a serial resonator

4.3 Explain how to obtain the circuit values of the parallel resonant or series resonant circuits in Figure 4P.3 using $|S_{21}|$ and $|S_{11}|$ (refer to Example 4.5).

4.4 Find the maximum value of $|S_{21}|$ for the circuit shown in Figure 4P.4 and sketch the frequency response ($\omega_o = 2\pi f_o = 2\pi \times 10$ GHz).

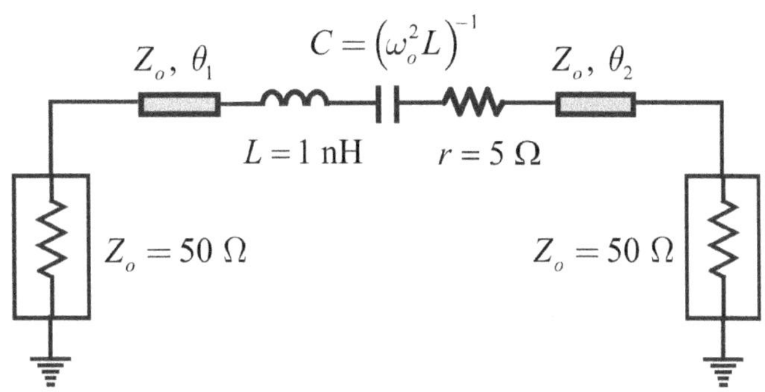

Figure 4P.4 A series resonant circuit

4.5 Find voltage V_{out} for the circuit in Figure 4P.5.

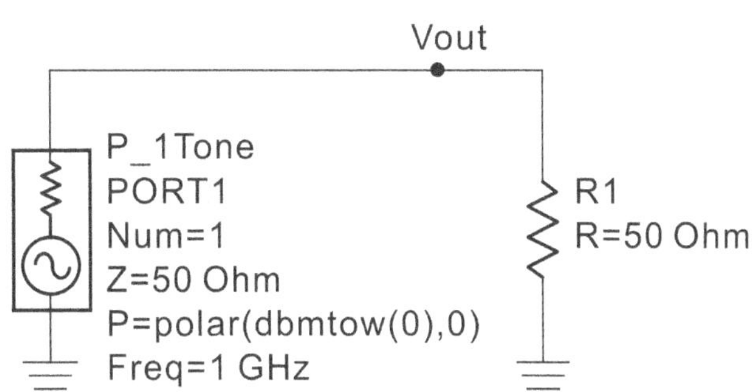

Figure 4P.5 Circuit for problem 4.5

4.6 Given that the insertion loss in a passive lossless two-port network is 3 dB, calculate the return loss.

4.7 Calculate the S-parameters of the attenuator circuit shown in Figure 4P.6. Determine whether $|S_{21}|^2 + |S_{11}|^2 = 1$ is satisfied.

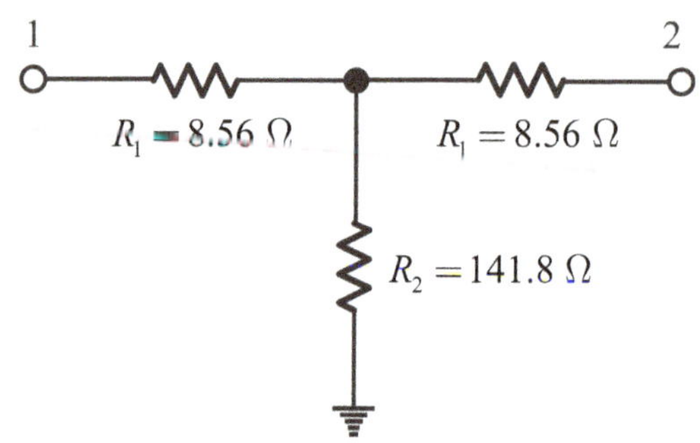

Figure 4P.6 T-type 3 dB attenuator

4.8 Derive the results in Equation (4.83).

4.9 Derive the results in Equation (4.84).

4.10 For the circuit in Figure 4P.7, find $F_{\min}$ and Γ_{opt}. Use the definition in Equation (4.68) after transforming the circuit into a Norton equivalent circuit at port 2.

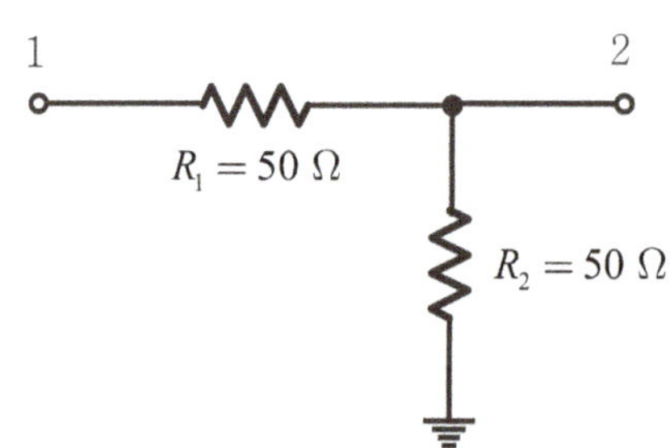

Figure 4P.7 A two-port resistor circuit

4.11 The noise temperature T_e is defined in terms of the noise factor as $T_e = (F - 1)T_o$. Using this relationship, with the noise temperature of each stage known, derive the expression below by expressing the noise temperature in terms of Frii's formula.

$$T_e = T_{e1} + \frac{T_{e2}}{G_1} + \frac{T_{e3}}{G_1 G_2} + \cdots$$

4.12 The attenuation with an attenuator is L (L is not in decibel scale); given that its temperature is T_o, determine the noise factor of the attenuator.

4.13 When the attenuator in problem 4.12 is connected next to a noise source having ENR_1, what is the resulting ENR_2? Assume the cold temperature of the noise source is T_o.

4.14 **(ADS problem)** Assume that **icor**(1,1) and **icor**(2,2), appearing after an S-parameter noise analysis, represent the short-circuit noise current squares at ports 1 and 2, respectively. The short-circuit noise currents can be obtained through AC simulation in ADS. Figure 4P.8 shows AC noise simulation to obtain the short-circuit noise currents. Since the noise currents cannot be obtained directly, the short-circuit noise currents are computed using the node voltages across small-valued resistors. Compute the short-circuit noise currents through the AC noise simulation and show that the results are in good agreement with the **icor**(1,1) and **icor**(2,2) obtained through the S-parameter noise simulation.

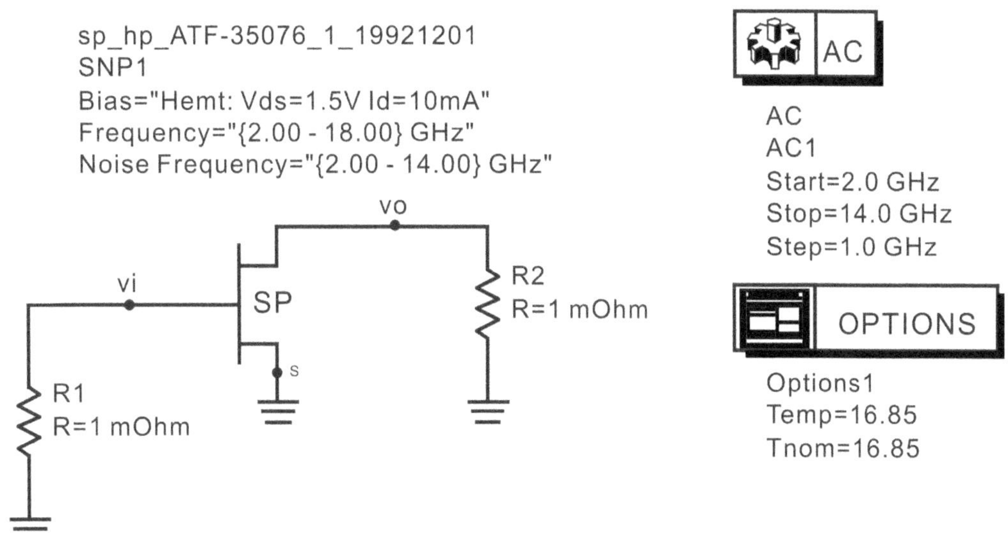

Figure 4P.8 AC noise simulation to compute the short-circuit noise currents

CHAPTER

5

Introduction to Microwave Active Devices

CHAPTER OUTLINE

5.1 INTRODUCTION

Before 1970, available microwave semiconductor devices were largely diodes and Si BJTs (bipolar junction transistors). Gunn, IMPATT, varactor, PIN, and Schottky diodes were frequently used in microwave applications. A reverse-biased *pn* junction diode has a depletion capacitance that changes when the reverse-biased DC voltage changes. A varactor diode is a variable capacitor that utilizes this depletion capacitance property. The varactor diode is often used to adjust the oscillation frequency and it can also amplify a weak signal. A parametric amplifier employs the varactor diode and its operation resembles that of a mixer. In the past, the parametric amplifier played an important role as a low-noise amplifier because there were no suitable active devices for amplifier applications, especially for low-noise amplifiers. Another diode using the *pn* junction is the PIN diode. By creating an intrinsic-region (I-region) in the *pn* junction, the PIN diode can be formed, and the resistance of this I-region depends on the DC voltage. Based on this property, electronic switches can be implemented in the microwave region. In addition, by combining the PIN diodes with transmission lines of the appropriate length, they can also be utilized as digital phase shifters. A PIN diode can also function as an analog-type variable attenuator.

A Schottky diode has, for a long time, been used in detectors and mixers due to its rectifying property. The Schottky diode uses majority carrier diffusion while the *pn* diode uses minority carrier diffusion. Consequently, the Schottky diode has no diffusion capacitance that is associated with minority carriers. Thus, the Schottky diode provides a number of benefits when used in mixers at high frequencies.

Until the 1970s, Gunn diodes and IMPATT diodes were primarily used as active components in oscillators and amplifiers. In those days, it was difficult to utilize transistors at frequencies higher than 4 GHz. The DC characteristics of these diodes showed a negative resistance when they were biased appropriately. By using negative resistance, it was much easier to design oscillators and they could also be used as amplifiers. Since the reflection coefficient of devices with a negative resistance is greater than 1, the Gunn and IMPATT diodes could be configured as reflection amplifiers in combination with a circulator. However, one problem with these diodes was that they were not very efficient. As a result, heat dissipation always had to be considered. Therefore, they were used in the construction of circuits with waveguides that easily adapted to thermal design. These heat problems become important limiting factors in circuit integration. Another disadvantage was that these diodes could not be integrated with other devices in a single process, making it intrinsically difficult to build up complex-functioning integrated circuits that need other devices such as Schottky or varactor diodes used for mixer or frequency control.

Based on their operation, transistors can be generally categorized into two types: the bipolar junction transistor (BJT) and the field effect transistor (FET). BJTs use two carriers, holes and electrons, and they use a diffusion mechanism in current flow. BJTs control current flow by raising or lowering the barrier height formed at junctions. The number of diffusing carriers depends on the barrier height, which is altered by the voltage across the base-emitter junction. In contrast, FETs use one majority carrier, an electron, and they use a drift mechanism in current flow. An FET forms a channel through which electrons can flow. The number of electrons flowing through the channel can be controlled by narrowing or widening the channel's thickness, which is achieved by controlling the gate voltage. FETs are also classified according to their channel formation. There are two kinds of channel formation: *enhancement type* and *depletion type*. In the enhancement type, the channel is not formed under the gate when the applied gate voltage is below a threshold voltage. The channel under the gate is formed by applying the adequate gate voltage. The gate voltage collects the carriers with opposite polarity and the accumulated carriers under the gate form the channel. In contrast, in the depletion type, the channel with carriers is formed in advance and the gate voltage is used to deplete the channel by repelling the carriers in the channel with the same polarity. Thus, the gate voltage is used to control the channel conductivity in both types of FETs.

The silicon process was the only available process technology for the fabrication of transistors until the 1970s. Microwave Si BJTs, which improved low-frequency BJTs in many ways, were exclusively used up to the range of microwave frequencies. However, these Si BJTs were not suitable for applications at frequencies higher than 4 GHz. In the early 1970s, the GaAs (gallium arsenide) MESFET (MEtal Semiconductor FET) or simply GaAs FET was developed as a result of the advance of the GaAs compound semiconductor process technology. The electron mobility in the GaAs is six times faster than in the Si. With this electron mobility advantage, GaAs FET has far better performance than the Si transistor. When GaAs FETs are fabricated using similar technology to the Si process, their performance is also as much as six times better. With the emergence of GaAs FETs, Gunn diodes, which had been popular until that time, became obsolete for amplifiers and oscillators in microwave frequencies up to the Ku band, and their application is now limited to the millimeter-wave frequencies. However, a further improvement in the characteristics of GaAs FETs led to the emergence of the HEMT (high-electron-mobility transistor) and the pHEMT (pseudomorphic HEMT). It is now possible to construct an integrated circuit up to a frequency of 200 GHz with these HEMTs. Thus, the future use of Gunn or IMPATT diodes in the millimeter wave remains unclear.

Three-terminal devices such as BJTs and GaAs MESFETs have several advantages when compared to diodes. They provide a flexibility in circuit design that can be applied to various components such as amplifiers, oscillators, and so on. In terms of efficiency, they are superior to the Gunn or other diodes. They do not require heat dissipation structures if they are not operated at significantly high power. They can also be used as switches by controlling the gate or base voltage. In addition, the characteristics of the varactor diode are inherently included in these devices, and they can be utilized in planar circuits as well, making them superior to other active components in terms of their advantages for integration. Every year, the high-frequency characteristics and performance of BJTs and GaAs MESFETs have been steadily improved by many researchers and even to date, their performance has been improved every year by applying new materials or new operating principles.

In this chapter, we will briefly investigate the operation and characteristics of these transistors used in microwave applications. Among diodes, we will also look at the varactor diode in oscillator design. The varactor diode is discussed further in Chapter 10. The operating principles and performance of Schottky diodes will also be discussed in the mixer design found in Chapter 12.

5.2 FIELD EFFECT TRANSISTOR (FET)

As explained in this chapter's introduction, there are two major types of three-terminal semiconductor devices: FETs and BJTs. The MOSFET (metal-oxide semiconductor FET), based on the silicon process, is the most widely used three-terminal device for implementing low-frequency integrated circuits, including digital circuits. However, during the 1970s, the performance of the MOSFETs fabricated with the Si process were not adequate for microwave applications, and their high-frequency performance was inferior to that of BJTs fabricated with the Si process. The GaAs MESFET (metal semiconductor FET) fabricated using the GaAs compound semiconductor technology was the first type of FET device applied to microwave circuits. Since the electron mobility in the GaAs is about six times faster than the Si, it was possible to fabricate an FET-type device using concurrent semiconductor process technologies.

The performance of GaAs MESFETs has been further improved with recent advances in process technologies that have made the fabrication of complicated epi-layer structures possible. As a result, GaAs pHEMTs can now be used at frequencies reaching several hundred GHz. In addition, the Si complementary MOS (CMOS) fabrication technology has demonstrated remarkable advances in various mobile-communication services. Recently, many CMOS circuits operating at millimeter wave frequencies have been reported. However, the basic operation of CMOS circuits is the same as it was in the past, a subject that is extensively covered in many textbooks related to electronic circuits. Thus, in this section, we will consider the operation and equivalent circuits of FETs based on GaAs MESFETs among the many other types of FETs.

5.2.1 GaAs MESFET

A top planar view of a GaAs FET with a gate length of 0.3 μm and a gate width of 250 μm is shown in Figure 5.1. The cross-section at A-A′ in that figure is also shown in Figure 5.2. In the GaAs FET shown in Figure 5.2, an electron-rich *n*-type epitaxial layer (epi-layer) is grown on the semi-insulating GaAs substrate. On the epitaxial layer, two ohmic contacts are formed for the drain and source terminals, while a Schottky contact is formed for the gate terminal. The Schottky diode formed between the gate and the epitaxial layer is reverse-biased, applying a negative gate voltage, and no gate current flows. However, a depletion region with no carriers occurs in the epitaxial layer due to the reverse-biasing gate voltage. The further increase of the negative gate voltage widens the depletion region and makes the channel narrow. This causes the drain-source current to decrease. Therefore, the drain-source current can be controlled using the gate voltage.

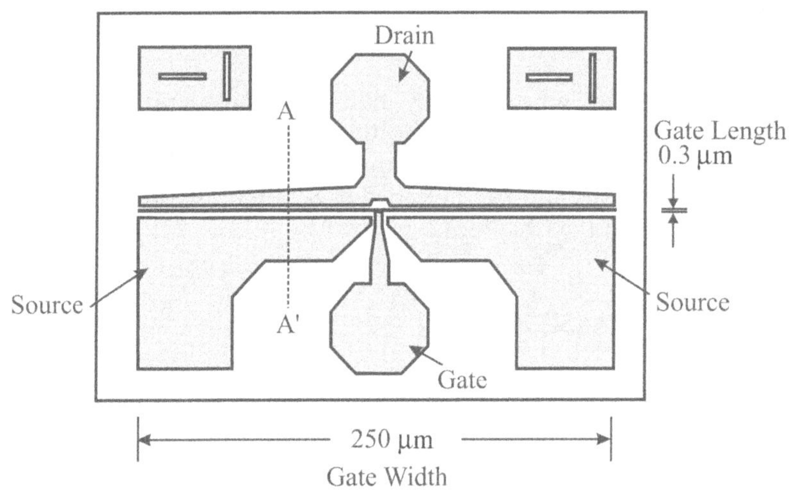

Figure 5.1 A top planar view of a GaAs FET

The equivalent circuit of a GaAs FET is shown in Figure 5.2. Resistors R_s and R_d represent the ohmic resistances that occur from the source and drain ohmic contacts. Resistor R_g represents the gate metallization resistance deposited to form the Schottky diode. The dependence of the drain current on the gate voltage is represented by the transconductance g_m, and that of the drain current on v_{DS} is represented by the resistance R_{ds}. In contrast with the R_g, R_s, and R_d shown in the figure, g_m and R_{ds} are nonlinear components and they represent the FET's DC characteristics. Resistor R_i is called channel resistance and represents the resistance between the depletion region and source terminals. The depletion region that occurs in the gate region is not directly connected to the source terminal but is connected to the source through a channel region. R_i represents the resistance of such a channel resistance to the source terminal.

Capacitors C_{gs} and C_{gd} represent the capacitances caused by the depletion region between the gate and the source, and between the gate and the drain terminals. The DC-bias dependence of these capacitors has characteristics similar to those of a depletion capacitance. On the other hand, C_{ds} represents the capacitance occurring between the terminals of the source drain. Capacitor C_{ds} between the drain and source terminals results

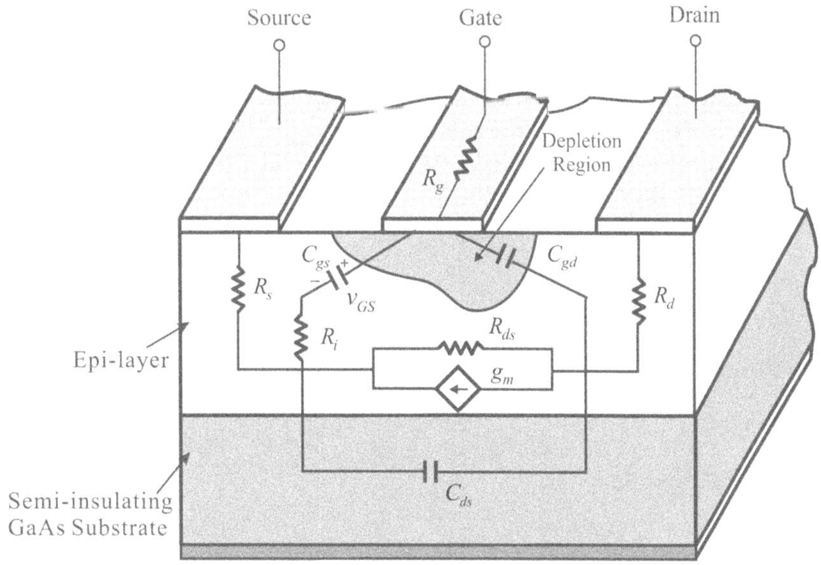

Figure 5.2 Cross-section of a GaAs FET and its equivalent circuit. C_{gs} and C_{gd} represent the depletion capacitances of the Schottky diode. The transconductance g_m represents the current controlled by the voltage across C_{gs}. R_d and R_s come from the ohmic contact resistances. R_g comes from the gate metallization resistance. R_{ds} and C_{ds} represent the resistance and capacitance between the drain and source terminals.

from the capacitances through the channel and air regions. Capacitor C_{ds} is also considered to be nonlinear and its characteristics are different from those of the depletion capacitance. The equivalent circuit in Figure 5.2 is redrawn in Figure 5.3. The transconductance is expressed as $y_m = g_m e^{-j\omega\tau}$ because the drain current flows with a time delay of τ compared to the gate-source control voltage.

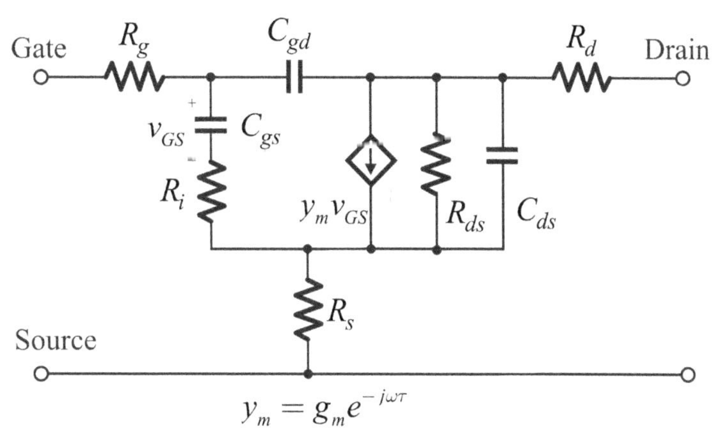

Figure 5.3 Small-signal GaAs MESFET equivalent circuit redrawn from Figure 5.2. Here, τ represents delay because the drain current is not instantly controlled by the gate-source voltage.

5.2.2 Large-Signal Equivalent Circuit

Figure 5.4 shows a large-signal equivalent circuit for a GaAs FET. Since the physical origin of R_g, R_s, and R_d is itself linear, these circuit elements can be treated as linear components. The physical operations of the gate-source and the gate-drain junctions are originated from Schottky diodes, and i_{GD} and i_{GS} can be described by the Schottky diode I–V characteristics shown in Equations (5.1) and (5.2).

$$i_{GS} = I_{s1}\left(e^{\frac{v_{GS}}{\eta V_T}} - 1\right) \tag{5.1}$$

$$i_{GD} = I_{s2}\left(e^{\frac{v_{GD}}{\eta V_T}} - 1\right) \tag{5.2}$$

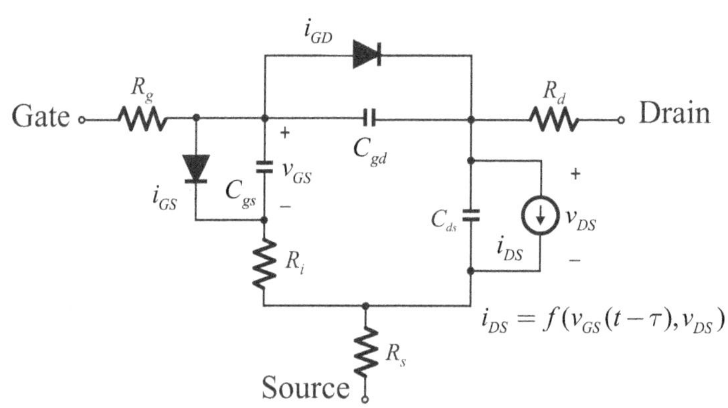

Figure 5.4 Large-signal equivalent circuit of a GaAs MESFET: i_{GD} and i_{GS} represent the Schottky diode formed between the gate drain and the gate source, respectively. C_{gs} and C_{gd} are depletion capacitances of the Schottky diode. The current source i_{DS} represents the current source, depending on the gate-source and drain-source voltages. Other circuit elements are assumed to be linear elements.

By measuring the forward DC current characteristics of the Schottky diode, the parameters such as ideality factor η and saturation current I_s can be determined. Also, since C_{gs} and C_{gd} originate from the depletion capacitance, they can be determined by measuring the C–V characteristics given by Equations (5.3) and (5.4).

$$C_{gs} = \frac{C_{gs0}}{\left(1 - \frac{v_{GS}}{\phi_{GS}}\right)^{m_S}} \tag{5.3}$$

$$C_{gd} = \frac{C_{gd0}}{\left(1 - \frac{v_{GD}}{\phi_{GD}}\right)^{m_D}} \tag{5.4}$$

In contrast, the nonlinear characteristics of C_{ds} and R_i are not so significant, and the small-signal values of C_{ds} and R_i can be used in the large-signal model.

However, the I_{DS}–V_{DS} characteristics of the GaAs FET are significantly different from those of the low-frequency FET devices. Therefore, a new mathematical model is required to represent its I_{DS}–V_{DS} characteristics, which are shown in Figure 5.5. There can be various mathematical models representing the I_{DS}–V_{DS} characteristics, but of these, there are three that are well known.

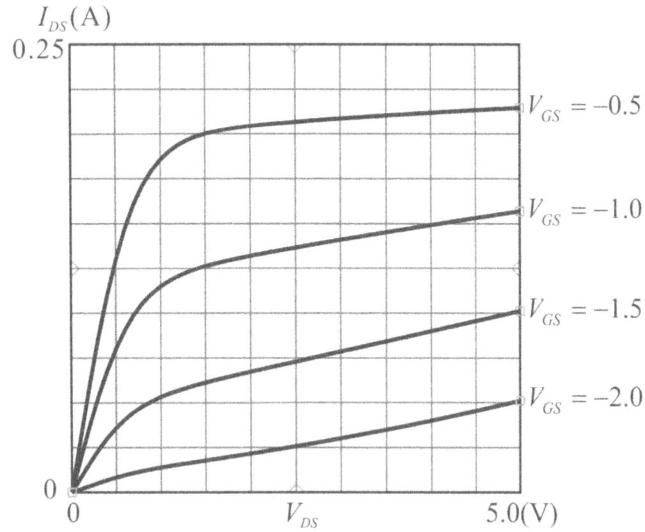

Figure 5.5 An example of I_{DS}–V_{DS} characteristics

The Curtice model describes the characteristics shown in Figure 5.5 using the equation $(1 + \lambda V_{DS})\tanh(\alpha V_{DS})$. The function $\tanh(x)$ can be approximated by x for small x, and approaches 1 as $x \to \pm\infty$. Thus, for a small V_{DS}, $(1 + \lambda V_{DS})\tanh(\alpha V_{DS})$ behaves as a straight line for V_{DS} because $(1 + \lambda V_{DS})\tanh(\alpha V_{DS}) \cong \alpha V_{DS}$, which steeply increases for V_{DS}. In contrast, for a sufficiently large V_{DS}, the equation can be approximated as $(1 + \lambda V_{DS})$, which increases slightly for V_{DS}. Thus, the I_{DS}–V_{DS} curves can be qualitatively represented by $(1 + \lambda V_{DS})\tanh(\alpha V_{DS})$. Now, we can determine the given equation through curve fitting of the measured I_{DS}–V_{DS} characteristics with constants α and λ.

However, since the dependence of the drain current on V_{GS} is close to the parabola shown in Figure 5.6, the I_{DS}–V_{GS} characteristics can be modeled using a quadratic equation. In reality, the use of the quadratic equation does not closely approximate the measured results. The cubic equation given in Equation (5.5) may be a better way to describe the I_{DS}–V_{GS} characteristics in Figure 5.6 because this equation has three degrees of freedom. In addition, the I_{DS}–V_{GS} characteristics depend to some degree on V_{DS}. In order to describe this, V_1 is used instead of V_{GS}, which depends linearly on V_{DS} as in Equation (5.5b). Thus, the I_{DS}–V_{DS} characteristics shown in Figure 5.5 are expressed mathematically in Equations (5.5a) and (5.5b).

$$I_{DS} = \left(A_0 + A_1V_1 + A_2V_1^2 + A_3V_1^3\right)\left(1 + \lambda V_{DS}\right)\tanh\left(\alpha V_{DS}\right) \tag{5.5a}$$

$$V_1 = V_{GS}\left[1 + \beta\left(V_{DS} - V_{to}\right)\right] \tag{5.5b}$$

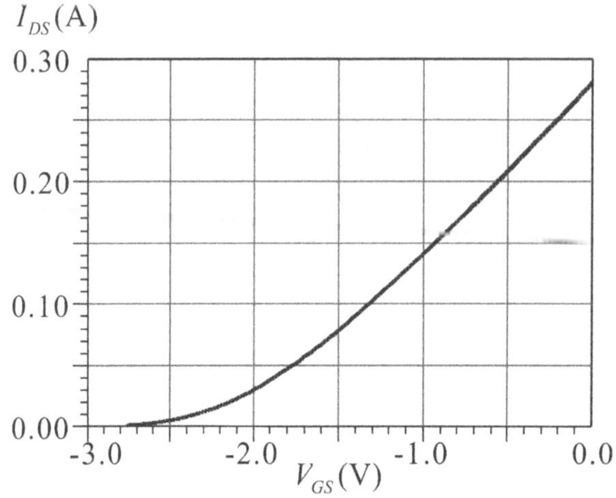

Figure 5.6 An example of I_{DS}–V_{GS} characteristics

For the Materka model, the relationship between the drain current and V_{GS} is described in a well-known parabolic relationship, and the slope for V_{DS} is expressed by modifying tanhx in the I_{DS}–V_{DS} characteristics as shown in Equations (5.6a) and (5.6b).

$$I_{DS} = I_{dss}\left(1 - \frac{V_{GS}}{V_p}\right)^2 \tanh\left(\frac{\alpha V_{DS}}{V_{GS} - V_p}\right) \tag{5.6a}$$

$$V_p = V_{po} + \gamma V_{DS} \tag{5.6b}$$

Note that the pinch-off voltage V_p also depends linearly on V_{DS}.

The Raytheon model is a mathematical model developed by the Raytheon Corporation and is expressed in Equation (5.7).

$$I_{DS} = \frac{\beta\left(V_{GS} - V_{to}\right)^2}{1 + \Theta\left(V_{GS} - V_{to}\right)}\left(1 + \lambda V_{DS}\right)\left[1 - \left(1 - \alpha\frac{V_{DS}}{3}\right)^3\right] \tag{5.7}$$

In addition, there are various other mathematical models that are difficult to distinguish in terms of their pros and cons, but they all describe the DC drain-current dependences on V_{GS} and V_{DS}, and all of them can be used in performing large-signal simulations. In addition, it is worth noting that all the equations from (5.5) to (5.7) represent a mathematical relationship for DC characteristics. Since the RF drain current i_{DS} depends on RF gate voltage v_{GS} with a time delay of τ, the large-signal microwave operations of i_{DS} can be modeled by replacing V_{GS} by $v_{GS}(t-\tau)$ in the equations from (5.5) to (5.7) using Equation (5.8):

$$i_{DS} = f\left(v_{GS}(t - \tau), v_{DS}\right) \tag{5.8}$$

This equation can be used as a mathematical model for microwave operation.

5.2.3 Simplified Small-Signal Equivalent Circuit and S-Parameters

The equivalent circuit shown in Figure 5.3 clearly explains the device's physical origin, but it is rather complicated. Resistors R_g, R_s, and R_d are the resistances that occur due to the contacts, and the values of the elements are generally small. Also, these resistors are not regarded as intrinsic parts of FETs, and so they are called extrinsic elements. Thus, their values are usually approximated to zero, and the approximate, simplified, small-signal equivalent circuit that results is shown in Figure 5.7. This simplified equivalent circuit is used to explain qualitatively the measured S-parameters of the GaAs FET.

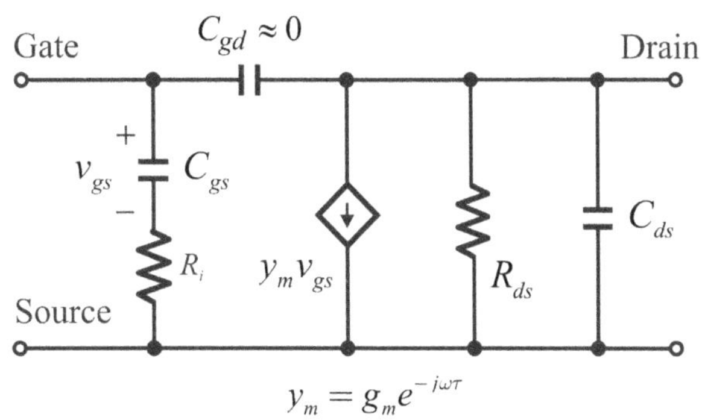

Figure 5.7 A simplified equivalent circuit of an FET. The extrinsic circuit elements R_g, R_d, and R_s shown in Figure 5.3's equivalent circuit are omitted here.

For increased simplification, further approximation is possible because the depletion capacitance C_{gd} is generally small compared to C_{gs} and sometimes it is also assumed to be 0. Thus, the FET's input and output are isolated. Such an approximation is called a *unilateral approximation*.

Figure 5.8 shows the measured S-parameters of a typical chip-state GaAs FET. After the GaAs FET chip was assembled by wire bonding on the carrier, as shown in Figures 5.9(b) and 5.9(c), the carrier assembly was mounted on a jig, as shown in Figure 5.9(a). In Figure 5.9(b), the reference planes are shown. Thus, the measured S-parameters usually include the inductance of bonding wires.

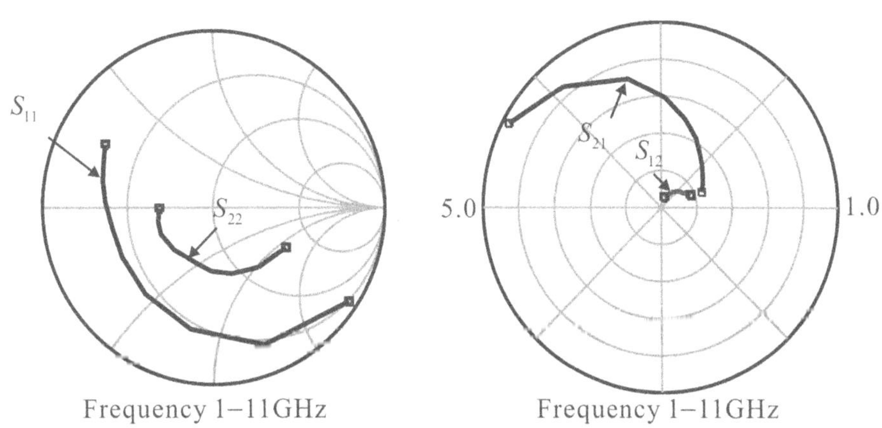

Figure 5.8 Typical measured GaAs FET S-parameters

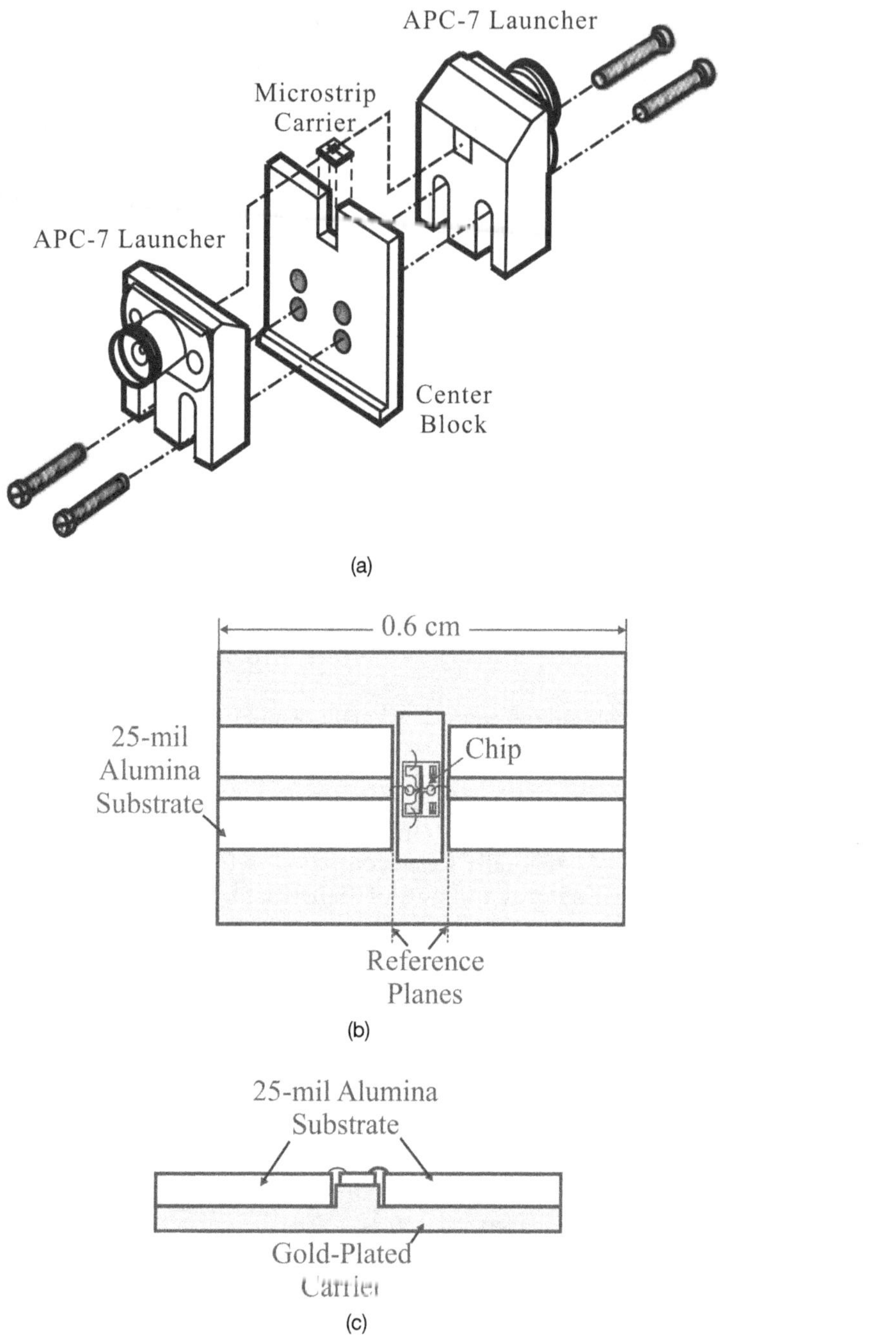

Figure 5.9 Illustration of the assembly for S-parameter measurement of a GaAs FET: (a) assembly of the test jig, (b) top view of the microstrip carrier, and (c) cross-sectional view of the microstrip carrier[1]

1. Agilent Technologies, *High-Frequency Transistor Primer Part IV, GaAs FET Characteristics,* 1999. Available at http://paginas.fe.up.pt/~hmiranda/etele/trans_primer4.pdf.

In Figure 5.8, S_{11} and S_{22} are related to input and output impedances. Thus, they are usually plotted on the Smith chart, as shown on the left side of Figure 5.8. In contrast, S_{12} and S_{21} are the transfer functions and so they are plotted on the polar chart on the right side of Figure 5.8 to show their magnitude and phase. On the polar chart, the radius is set to 5.0 for S_{21} and the radial division scale is found to be 1.0. However, as S_{12} is small, the radius is set to 1.0 and the radial division scale becomes 0.2. The frequency responses of the S-parameters can be explained using the simplified equivalent circuit shown in Figure 5.7.

From Figure 5.9, the measured S_{11} includes the bonding-wire inductance. The input impedance of the FET is then found to be a series *R-L-C* circuit approximating $C_{gd} = 0$ in the simplified equivalent circuit. For a frequency change, the reactance of *L* and *C* changes but R_i is constant. Thus, the locus of S_{11} lies in a constant resistance circle for a frequency change. At low frequency, the circuit behaves like a series *R-C* circuit. However, as the frequency increases, the inductance of the bonding wire becomes dominant, and so the circuit behaves as a series *R-L* circuit. From this, the resistance at the series resonance corresponds to the channel resistance R_i in the simplified equivalent circuit, but if the ignored contact resistance were to be considered, it would approximate the value of $R_i + R_g + R_s$. In addition, applying the method explained in Example 2.2 of Chapter 2, the bonding-wire inductance and capacitance can be obtained by calculating the *L* and *C* values at resonance. The resulting capacitance and inductance correspond approximately to C_{gs} and the bonding-wire inductance.

In the case of S_{22}, the drain-source impedance can be approximated by a circuit of R_{ds} and C_{ds} connected in parallel. Therefore, as the frequency approaches 0, the drain-source impedance approaches R_{ds}. As the frequency increases, S_{22} moves along a constant conductance circle and the trajectory appears in the capacitive region due to capacitor C_{ds}. As the frequency further increases, S_{22} is observed to move away from the constant conductance circle and to approximate a constant resistance circle similar to S_{11} due to the effects of bonding-wire inductor and C_{gd}.

In the case of S_{21}, at an extremely low frequency, S_{21} can be computed as

$$S_{21} = -2g_m Z_o = 2g_m Z_o \angle 180^\circ \tag{5.9}$$

Using Equation (5.9), the approximate value of g_m can be found from the low-frequency measurement data. As the frequency increases, the voltage across C_{gs} decreases. As a result, $|S_{21}|$ is reduced because it corresponds to the voltage across the termination Z_o. In addition, the influence of C_{ds} further reduces $|S_{21}|$. The phase of S_{21} increases in a clockwise direction due to these capacitors.

In the case of S_{12}, when the frequency is extremely low, we can see that S_{12} approximates

$$S_{12} = 2j\omega C_{gd} Z_o = 2\omega C_{gd} Z_o \angle 90^\circ \tag{5.10}$$

From Equation (5.10), we see that $|S_{12}|$ increases as the frequency increases. Similar to the phase of S_{21}, the phase of S_{12} also increases in a clockwise direction as the frequency increases.

5.2.4 Package

When a chip-type active device is used in a circuit, the circuit can be constructed without further performance degradation of the active device. However, active devices can easily be damaged by external environmental factors. In particular, the assembly of chips on a PCB is bothersome and inconvenient because components are mainly attached by soldering on the PCB. Thus, for convenience of handling, chips are sometimes used in packaged form, albeit with some degree of performance degradation. An assembly with a packaged chip can be soldered and does not require techniques such as wire bonding and the attachment of dies. In addition, packaged chips offer the advantage of good protection against external environmental factors.

Figure 5.10 shows a GaAs FET chip assembled using a commercial ceramic package. Figure 5.10(a) shows the top view with the lid removed and (b) shows the back view. The source is commonly used as the ground terminal. To minimize the inductance arising from the assembly of the source terminal, the terminal is frequently wire bonded in two places, as shown in Figure 5.10(a). The inductance resulting from the assembly of the source terminal, however small, causes feedback from output to input and creates the strong possibility of oscillation or device instability. Thus, to minimize the inductance, the package terminals where the source is connected are usually made wider when compared to the other terminals. The chip is mounted directly on the lead terminal, and the two source pads in the chip are wire bonded twice in each of two places on the lead terminal. The heat is thus dissipated through these lead terminals but it is worth noting that this heat dissipation may not be sufficient in certain cases.

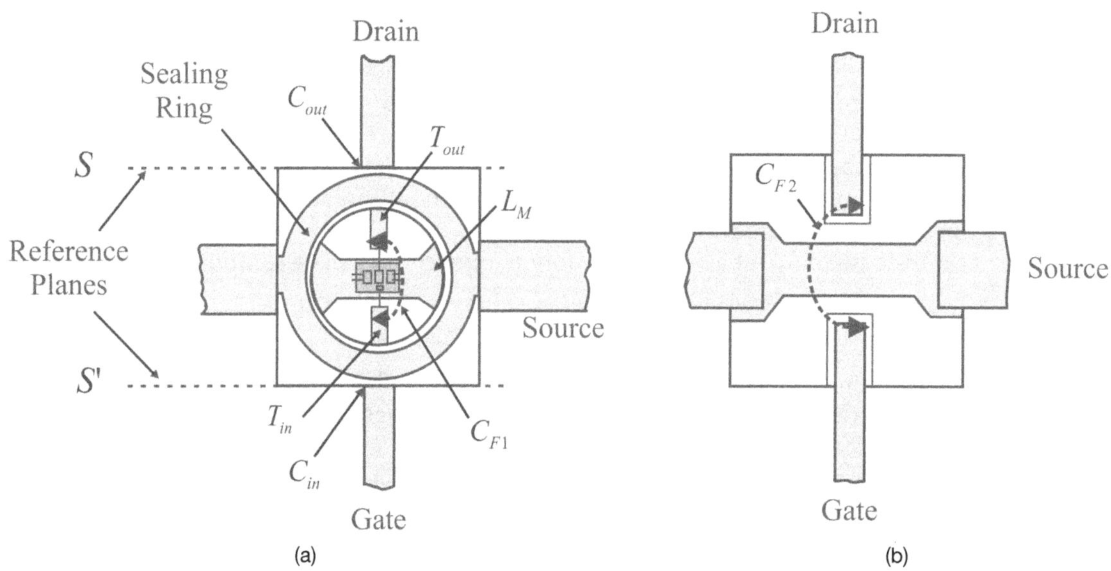

Figure 5.10 GaAs FET package assembly: (a) top view and (b) bottom view. S-parameters are usually defined at the reference planes S and S′.

The planes S and S' shown in Figure 5.10(a) become the reference planes of the S-parameter measurement and the measured S-parameters are usually provided with these reference planes. Thus, many parasitic elements are added to the chip's equivalent circuit in the packaged device. This is shown in Figure 5.11. The bonding-wire inductances L_g, L_s, and L_d occur as a consequence of wire bonding. In addition, since lines with a finite length are inserted into the package, they thus appear as transmission lines or inductance. Transmission lines T_{in}, T_{out}, and inductor L_M in Figure 5.11 represent these transmission lines

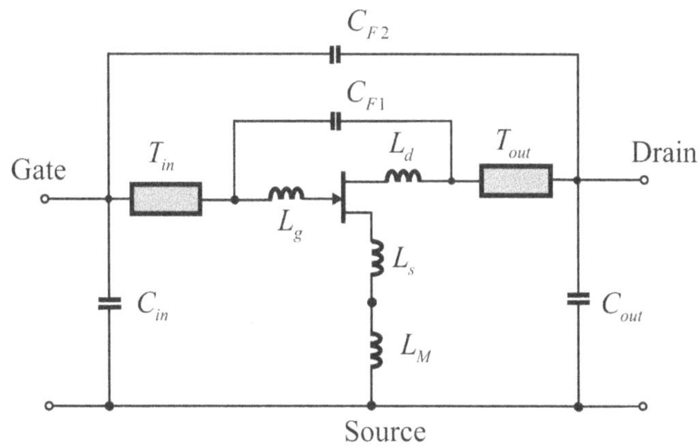

Figure 5.11 Equivalent circuit of the packaged device shown in Figure 5.10

as inductance. Various parasitic capacitances also occur. Capacitances C_{in} and C_{out} occur due to the discontinuity of the transmission lines. The capacitances C_{F1} and C_{F2} represent feedback capacitances occurring as a result of the coupling between the lines in the package.

These packaged-circuit parasitic elements prevent the accurate determination of the values of the active device's equivalent circuit. Thus, it is common to obtain first the value of the chip's equivalent circuit and then the value of the package's equivalent. These are combined to yield the overall value of the equivalent circuit.

5.2.5 GaAs pHEMT

The electron mobility in the channel of an FET is closely related to its high-frequency performance. As previously explained, the GaAs FET has better high-frequency performance than the Si because the electron mobility in the GaAs is about six times faster than the Si. However, this electron mobility is usually degraded by the doped impurity atoms that are doped to generate electrons. After the generation of electrons, the impurity atoms take on + polarity while the electrons take on – polarity. The moving electrons are thus attracted and scattered by the fixed impurity atoms with + polarity. As a result, the electron mobility is significantly degraded by the *impurity scattering*. This electron mobility degradation can be avoided by employing a complex epitaxial layer structure instead of a single epitaxial layer. Figure 5.12 shows the complex epitaxial layer structure that improves electron mobility.

In Figure 5.12, the electrons are generated in the *n*-type AlGaAs layer, where impurity atoms are richly doped. The heterojunction is formed between the undoped AlGaAs and the undoped GaAs. As a result, an *electron well* is formed as a result of the heterojunction. The generated electrons in the *n*-type AlGaAs layer are then easily trapped and gathered in the electron well that is formed beneath the undoped GaAs layer. Since the electron well is very thin, the trapped electrons in the well can be considered to be a sheet of electron gas that is called 2DEG (2-dimensional electron gas). In addition, note that the undoped GaAs layer is almost intrinsic because there are no impurity atoms. Thus, the electrons can move according to the applied electric field without the influence of the impurity atoms. This results in an electron velocity that is much faster than in the GaAs FET. Therefore, the high-frequency gain performance can be improved when compared with a conventional GaAs FET because the electrons move much faster in the undoped channel. For this reason, the device is called an HEMT (high-electron-mobility transistor).

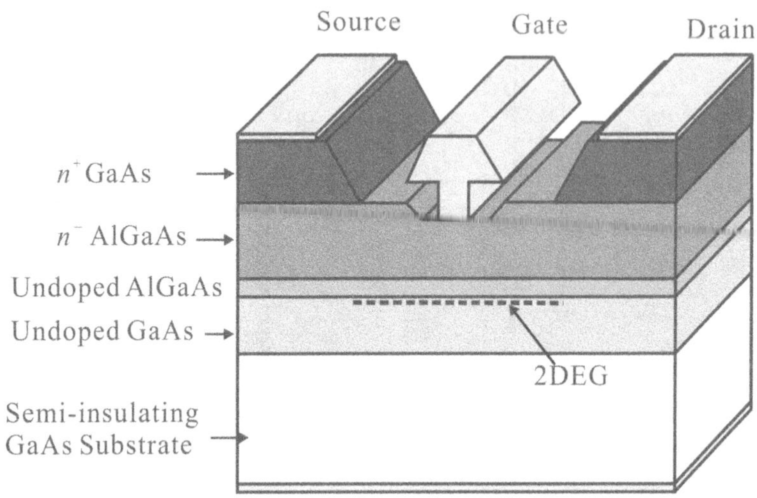

Figure 5.12 Cross-sectional structure of a GaAs HEMT

In the fabrication of a GaAs pHEMT, the lattice constants of the AlGaAs and GaAs differ significantly, making it difficult to grow a stable AlGaAs layer on the GaAs. The problem is solved by using the recently developed *pseudomorphic* technology. Inserting an extremely thin, undoped InGaAs layer between the undoped GaAs and the undoped AlGaAs layers, the stable AlGaAs layer can be grown, which makes it possible to manufacture these high-performance devices. Since the pseudomorphic technology is used to fabricate these devices, they are often called pHEMTs.

5.3 BIPOLAR JUNCTION TRANSISTOR (BJT)

5.3.1 Operation of an Si BJT

The structure of a BJT is shown in Figure 5.13, in which (a) shows the top view and (b) shows the cross-sectional view through line S–S′. As seen in the figure, first an *n*-type epitaxial layer is grown on the *n*+ substrate. Then, on the *n*-type epitaxial layer, the base region is formed by locally doping a *p*-type material. The emitter area is then formed by locally doping an n^+-type material on the base region. Using this procedure, an *npn* BJT is formed.

The principles of operation for an *npn* transistor are usually explained using the dotted-line area in Figure 5.13(b), which is also shown in Figure 5.14. Two *pn* junctions, the base-emitter (BE) junction and the collector-base (CB) junction, appear in the transistor. The two *pn* junction diodes are connected back-to-back. The BJT can operate in various modes such as active, cutoff, inverse, and saturation. In the active mode, the base-emitter (BE) junction is forward-biased while the collector-base (CB) junction is reverse-biased. As a result, the barrier height of the BE junction is lowered while the barrier height of the CB junction is raised. Due to the lowered BE-junction barrier height, the majority carriers, which are the electrons in the emitter region, are diffused into the base region while the holes in the base region are diffused into the emitter region. However, because the CB junction is reverse-biased, the diffusion between the collector and base do not occur due to the increased barrier height. Normally, the emitter region is more heavily doped than the base region. Thus, more electrons will be diffused from the emitter to the base than will holes from the base be diffused to the emitter. Consequently, the current contribution from the diffusion of holes can be ignored.

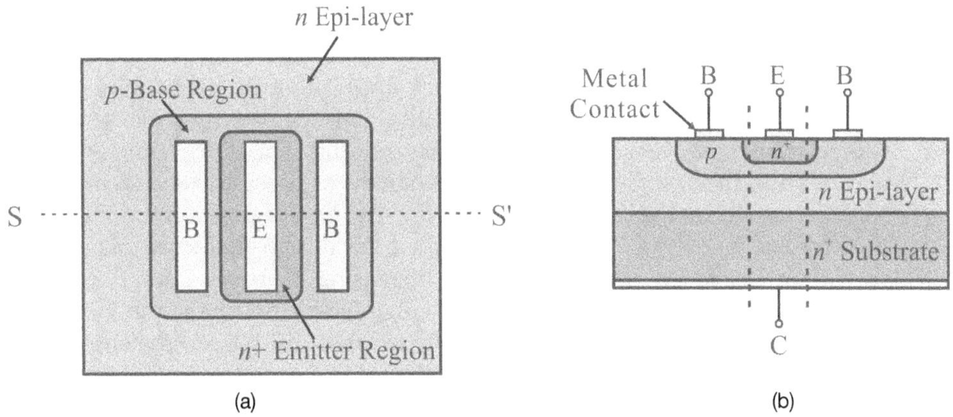

Figure 5.13 Structure of a BJT: (a) top view and (b) cross-section through line S–S′

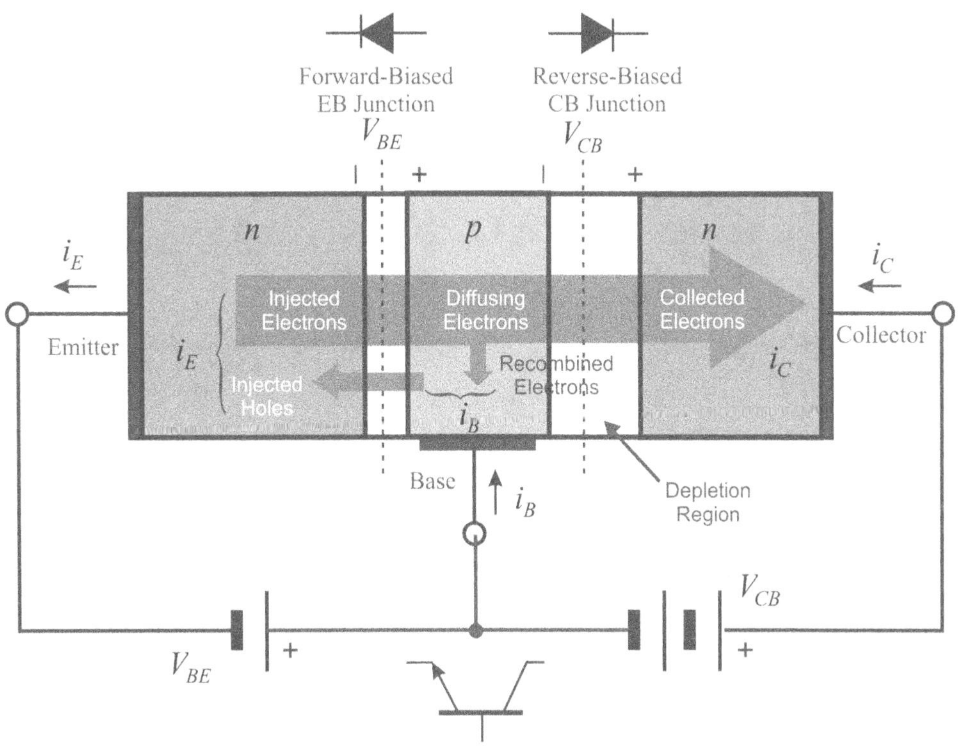

Figure 5.14 Current flow of an *npn* transistor biased to operate in the active mode

A small number of the diffused electrons from the emitter recombine and thereby disappear in the base region, while most of the electrons are collected in the collector region. Thus, the collector current i_C is almost equal to the emitter current i_E. Note that the emitter current i_E depends on the BE junction barrier height, which in turn is controlled by a small voltage applied to the BE junction, V_{BE}. Therefore, the large emitter current flow of the BJT can be controlled by a small-voltage V_{BE}, which thus acts as an amplifier. Therefore, the BE-junction voltage plays a similar role to the gate voltage in an FET.

However, the structure of the BJT shown in Figure 5.13(b) needs some improvements before it can be used in a high-frequency application. Basically, high-frequency performance is closely related to the base width. First, the electrons injected from the emitter should transit the base region to reach the collector. The transit time is related to the base width. Thus, the base width should be as narrow as possible. Second, a *base-spreading resistance* occurs due to the distance between the true base region and the actual base terminals, as shown in Figure 5.13(b). As a result, the gain at high frequencies is reduced. The base-spreading resistance can be reduced by increasing the base-region doping. However, the significantly increased doping of the base region increases the number of holes and, consequently, the number of holes that are diffused from the base to the emitter increases. As a result, the base current increases, which is desirable. As a way of reducing the base-spreading resistance, the base and emitter are implemented using the interdigital structure shown in Figure 5.15. When the finger width and spacing of the interdigital structure are narrowed, the base-spreading resistance can be significantly reduced due to the short length between the true base region and the base terminal.

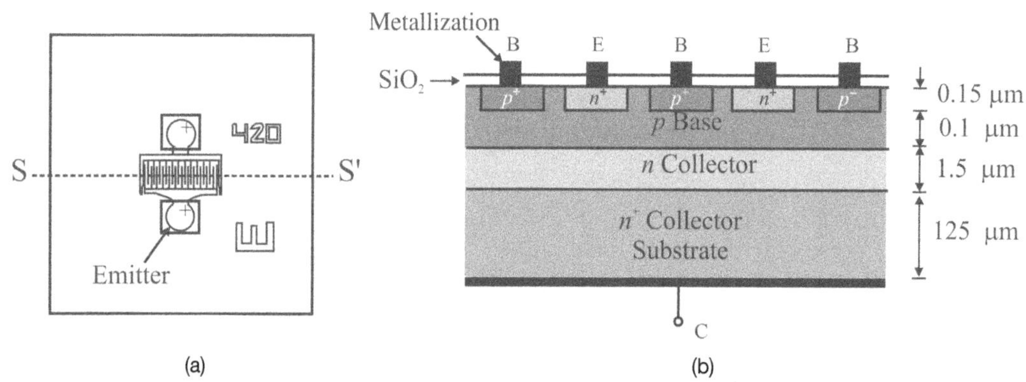

Figure 5.15 A high-frequency BJT's (a) top and (b) cross-sectional views. The base-spreading resistance that appears between the base terminal and the true base region is reduced by using the interdigital configuration for the base and emitter. Also, the cut-off frequency f_T can be made higher in order to have a shallow base region using the process control.

In addition, because the electrons should pass through the n-type epitaxial layer in order to reach the collector terminal, the epitaxial layer is made sufficiently thin in order for the electrons to arrive at the collector terminal at a much faster rate. This BJT structure is shown in Figure 5.15 and the cross-sectional structure along S–S′ is shown in Figure 5.15(b). In that figure, the base thickness typically has the order of 0.1 μm, and the n-type collector thickness typically has the order of 1.5 μm, which is extremely thin. The base and emitter spacing is found to be about 1 μm in order to reduce the base-spreading resistance.

5.3.2 Large-Signal Model of a BJT

A large-signal model of a BJT is shown in Figure 5.16. From Figure 5.14, because the BE and BC junctions can be represented by diodes, they are represented as diodes I_{BR} and I_{BF} in Figure 5.16. The subscript B represents the base, while the subscripts F and R stand for forward and reverse. In addition, the space-charge-region diode is known to occur in the pn junction when the junction currents are at a low level. Thus, these space-charge-region diodes appear in the BE and BC junctions, and they are connected in parallel. In Figure 5.16, they are represented by diodes I_{LE} and I_{LC}. It is worth noting that the collector current is not generated due to the space-charge-region diode currents, and the currents that result from these diodes can be treated as leakage currents, which is why the subscript L is added.

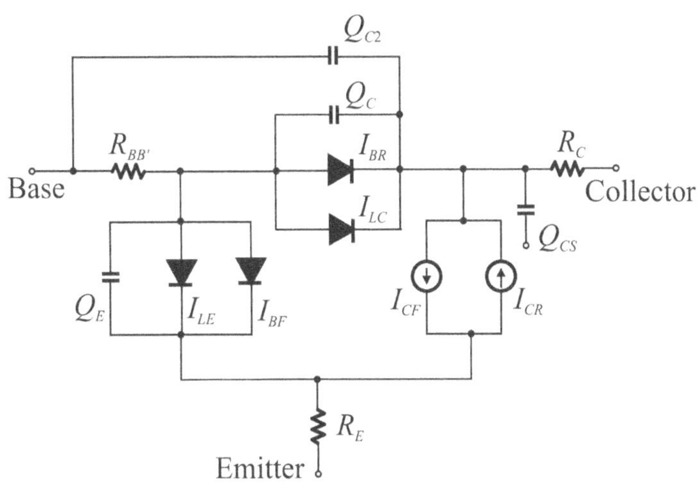

Figure 5.16 Large-signal model of a BJT. The charges Q_E and Q_C include the diffusion and depletion capacitances. I_{LE} and I_{LC} represent space-charge-region diodes and do not contribute to the collector current I_C. I_{BF} and I_{CF} are the base and collector currents in the active mode, while I_{BR} and I_{CR} are the base and collector currents in the inverse mode. R_E and R_C are contact resistances, while $R_{BB'}$ is the base-spreading resistance. The charge models Q_{C2} and Q_{CS} are device dependent.

Therefore, the base current can be expressed in terms of the collector saturation current I_s as follows:

$$I_B = \frac{I_S}{\beta_F}\left(e^{\frac{qV_{BE}}{n_F kT}} - 1\right) + \frac{I_S}{\beta_R}\left(e^{\frac{qV_{BC}}{n_R kT}} - 1\right) + C_2 I_S\left(e^{\frac{qV_{BE}}{n_{EL} kT}} - 1\right) + C_4 I_S\left(e^{\frac{qV_{BC}}{n_{CL} kT}} - 1\right) \quad (5.11)$$

The first two terms of Equation (5.11) are related to the collector current flowing when each of the BE and BC junctions are forward-biased; each of them is divided by its respective current gain, β_F and β_R. The last two terms represent the current of the space-charge-region diode, which is independent of the collector current.

It can be seen that the current due to the BE and BC junction diodes constitutes the total collector current $I_{CT} = I_{CF} - I_{CR}$. The current is caused by two modes: the first occurs when the BE junction is forward-biased and the BC junction is reverse-biased (active mode); the second occurs when the BC junction is forward-biased and the BE junction is reverse-biased (inverse mode). In this case, the current is in the reverse direction. Thus, the collector current I_{CT} can be expressed as shown in Equation (5.12).

$$I_{CT} = \frac{I_S}{q_b}\left(e^{\frac{qV_{BE}}{n_F kT}} - 1\right) - \frac{I_S}{q_b}\left(e^{\frac{qV_{BC}}{n_R kT}} - 1\right) \tag{5.12}$$

The q_b in the expression is a factor that represents the Early effect and the Kirk effect that appear in a large collector current and they are expressed as follows:

$$q_b = \frac{1}{1 - \frac{V_{BC}}{V_A} - \frac{V_{BE}}{V_B}}\left(1 + \sqrt{1 + 4q_2}\right) \tag{5.13}$$

$$q_2 = \frac{I_S}{I_{KF}}\left(e^{\frac{qV_{BE}}{kT}} - 1\right) + \frac{I_S}{I_{KR}}\left(e^{\frac{qV_{BC}}{kT}} - 1\right) \tag{5.14}$$

V_A and V_B in Equation (5.13) represent the forward and reverse Early voltages, respectively, while I_{KF} and I_{KR} in Equation (5.14) represent the forward and reverse knee currents of the collector current. Equations (5.11) through (5.14) represent the DC characteristics of the BJT.

Figure 5.17 illustrates how these parameters are determined. Plotting I_B and I_C with respect to V_{BE} enables us to extract the BE junction-related parameters given by Equations (5.11) through (5.14). This extraction is usually called a *Gummel plot*. From Figure 5.17,

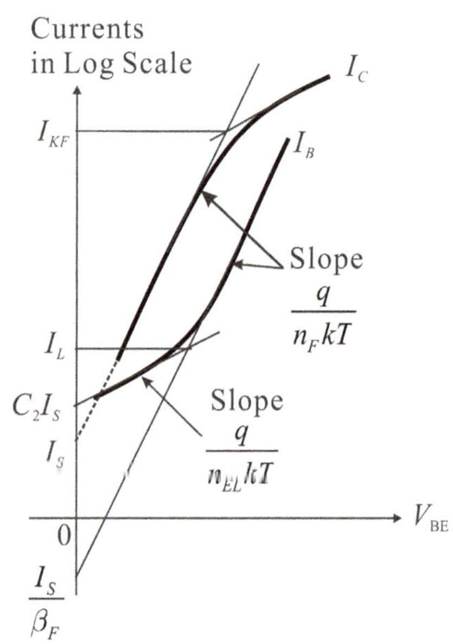

Figure 5.17 Gummel plot for V_{BE}. The space-charge-region diode current I_{LE} is dominant in I_B for a small V_{BE}, while I_{BF} is dominant in I_B for a large V_{BE}. Using the slopes and intercepts, the parameters of the two diodes can be determined. The collector current I_C saturates as I_C increases due to the Kirk effect. Similarly, the parameters of the collector current can be determined using its slopes and intercepts. Repeating for V_{BC}, the DC parameters for I_C and I_B can be completely determined.

the knee current I_{KF} of the collector current can be determined from the turning point and from the slopes where the current begins to show. For the base current, on the other hand, the space-charge-region diode and the BE junction related parameters can be determined from the turning point and the slopes in a small base current. Similarly, by plotting the Gummel plot for V_{BC} the BC junction-related parameters can be extracted. The parameters are grouped and shown in Table 5.1.

Next, we consider the resistances caused by contacts. These are R_E, R_B, and R_C, but of these, R_B is not a simple contact resistance. It varies according to the current and so two

Table 5.1 Gummel Poon model parameters

Parameter	Meaning	Parameter	Meaning
IS	Transport saturation current	NR	Reverse ideality factor
NF	Forward ideality factor	IKR	Reverse knee current
IKF	Forward knee current	NC	BC ideality factor
NE	BE ideality factor	ISC	BC saturation current
ISE	Base emitter saturation current	BR	Reverse beta
BF	Forward beta	VAR	Reverse early voltage
VAF	Forward early voltage		
RE	Emitter resistance		
RC	Collector resistance		
RB	Zero bias resistance		
RBM	Minimum base resistance		
IRB	Current where base resistance is halfway between RB and RBM		
CJE	BE zero bias capacitance	CJC	BC zero bias capacitance
VJE	BE built-in potential	VJC	BC built-in potential
MJE	BE grading coefficient	MJC	BC grading coefficient
FC	Models transition from junction to diffusion capacitance		
XCJC	Models distributed nature of base		
TF	Forward transit time	TR	Reverse transit time
ITF	Models TF dependence in ic		
VTF	Models TF dependence on Vbc		
PTF	Excess phase of TF		
XTF	Coefficient of TF bias dependence		

additional parameters (RBM, IRB) are required to describe it. These parameters are summarized in Table 5.1 and are classified as groups.

In addition, depletion and diffusion capacitances appear at the BE and BC junctions. The depletion capacitance is given in Equation (5.15) as

$$C_{JE} = \frac{C_{je}(0)}{\left(1 - \frac{V_{BE}}{\phi_{BE}}\right)^{m_E}} \tag{5.15}$$

The depletion capacitance has the parameters $C_{je}(0)$, m_E, and ϕ_{BE}. They refer to the capacitance at 0 V, the grading coefficient, and the built-in potential, respectively. The parameters for the BE junction's depletion capacitance can be determined experimentally from the C–V measurement for V_{BE}. Through curve fitting of the measured results in Equation (5.15), the parameters can be determined. Similarly, the same parameter group can be defined for the BC-junction's depletion capacitance and those parameters can be experimentally extracted using the C–V measurement for V_{BC}. The parameters for the BE- and BC-depletion capacitances are also grouped and shown in Table 5.1. However, because these expressions for the BE- and BC-junction depletion capacitances have singularity at $V_{BE} = \phi_{BE}$ and $V_{BC} = \phi_{BC}$, their applications are usually limited to $V_{BE} \le FC \cdot \phi_{BE}$ and $V_{BC} \le FC \cdot \phi_{BC}$. For V_{BE} or V_{BC} values greater than these, a straight line that is given by the tangent at $FC \cdot \phi_{BE}$ is used instead of Equation (5.15). Thus, the value FC in Table 5.1 represents the range of V_{BE} and V_{BC}.

The diffusion capacitance is usually characterized by the transit time, which appears in the active and inverse mode operations. The transit time in the active and inverse mode operations is represented by TF and TR in Table 5.1, respectively. In addition, since the transit time is not constant but depends on several parameters, the parameters that represent this dependence (ITF, VTF, PTF, and XTF) form a group. The capacitors Q_E and Q_C in Figure 5.16 represent the capacitances that are the result of the previously explained depletion and diffusion capacitances. The capacitors Q_{C2} and Q_{CS} depend on the fabrication method, and the reader may refer to reference 7 at the end of this chapter for details. Also, some additional circuit elements can be added to the BJT equivalent circuit described in this book according to the fabrication process. The reader is encouraged to consult other relevant references for details.

5.3.3 Simplified Equivalent Circuit and S-Parameters

Figure 5.18 shows a BJT small-signal equivalent circuit derived from the large-signal model in Figure 5.16. Since the BC junction is generally reverse-biased in active mode, the BC junction can be represented by the parallel RC circuit. The value of g_μ comes from a reverse-biased diode and its value is usually quite small. The capacitor C_μ represents a depletion capacitance. Similarly, the forward-biased BE junction can also be represented by the parallel RC circuit. The resistor g_π represents the small-signal conductance of the forward-biased diode. Capacitor C_π represents the combined capacitances from the diffusion and depletion capacitances. Generally, the diffusion capacitance is larger than the depletion capacitance and the value of C_π is primarily determined by the diffusion capacitance. In a low-frequency application, the diffusion capacitance C_π is typically small compared to g_π and so is ignored. However, in a high-frequency application, C_π becomes dominant and the effects of both elements appear. Especially in microwave applications, the occurrence of g_π can be ignored when compared to C_π. Transconductance g_m represents

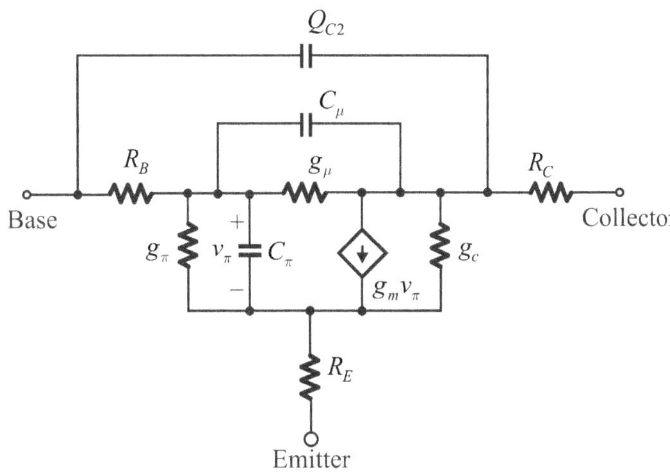

Figure 5.18 A small-signal equivalent circuit. The circuit elements g_π and C_π represent the small-signal equivalent circuit for the BE-junction diode while g_μ and C_μ represent those of the BC-junction diode. R_E and R_C are contact resistances, and R_B is the base-spreading resistance. Q_{C2} is the capacitance, which is device dependent.

the collector current controlled by the BE-junction voltage, while g_c represents the resistance from the Early effect. Resistors R_E, R_B, and R_C represent the contact resistances.

The circuit shown in Figure 5.18 is complex and can be difficult to understand. Therefore, the simplified equivalent circuit shown in Figure 5.19 is often used to explain the qualities of the measured S-parameters. The resistor g_π in the simplified equivalent circuit is ignored in a high-frequency application. Since the value of g_μ is also small, it is also ignored and even C_μ is sometimes ignored. The values of R_E and R_C are typically small, and because they are the resistances caused by contacts, they are also ignored. The equivalent circuit thus obtained is shown in Figure 5.19.

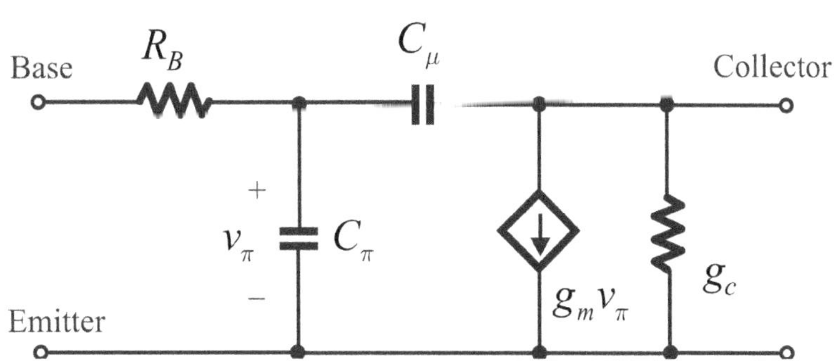

Figure 5.19 Simplified small-signal equivalent circuit

In Figure 5.20, similar to the FET, S_{11} and S_{22} are plotted on the Smith chart since they are related to impedance. In contrast, S_{12} and S_{21} are the transfer functions and so they are plotted on the polar chart to show their magnitude and phase. The radius of S_{21} is set to 10.0, and the radial scale corresponds to a division of 2.0 per grid. However, as S_{12} is small, the radial scale is selected to be 0.2 per grid.

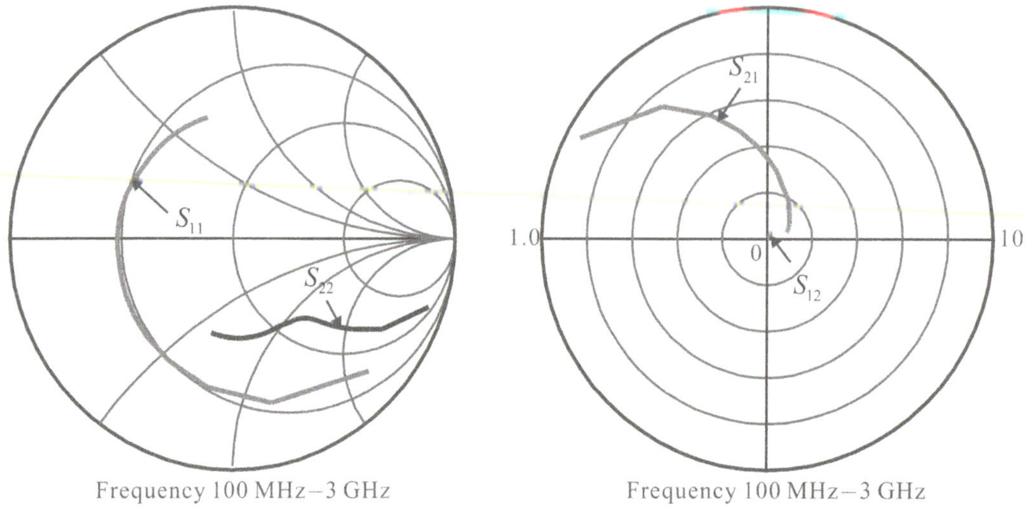

Figure 5.20 Typical measured S-parameters of a BJT

Here, as in a GaAs FET, the input impedance includes the bonding wire in the measurement. Thus, the input approximates a series *RLC* circuit from the simplified equivalent circuit shown in Figure 5.19. As a result, the locus of the S-parameters follows a constant resistance circle, as shown in Figure 5.20. At low frequency, the locus lies in the capacitive region of the Smith chart. However, as the frequency increases, the inductance of the bonding wire becomes dominant and so the locus lies in the inductive region. The resistance value at resonance approximately corresponds to the base resistance R_B in the simplified equivalent circuit. In addition, by using the method described in Chapter 2 to compute the L and C values at resonance, both the inductance and the capacitance of the bonding wire can be obtained. The capacitance obtained can be interpreted as C_π.

In the case of S_{22}, because the collector resistance g_c is small, the effect of g_c is seldom observed in S_{22}. Rather than g_c, the output circuit can be approximated as the series connection of C_μ and $r_\pi \parallel (Z_o + R_B)$ for an extremely low frequency. In this case, because r_π is generally large, the output circuit appears to approximate the series connection of C_μ and $(Z_o + R_B)$. Therefore, the locus moves, following the constant resistance circle. As the frequency becomes higher, the approximate output circuit appears to be g_c in parallel with a parasitic capacitor C_c. Thus, with increasing frequency, the locus moves along the constant conductance circle. Due to the effect of C_c, the trajectory appears in the capacitive region. The resulting locus is the combined locus that moves along the constant resistance circle at low frequencies, and moves along the constant conductance circle at high frequencies. Even though it is not shown here, with a further increase in frequency, S_{22} follows a trajectory similar to that of S_{11} due to the effects of bonding-wire inductors and C_μ.

From the simplified equivalent circuit of Figure 5.19, S_{21} in Equation (5.16) is

$$S_{21} = -2g_m Z_o = 2g_m Z_o \angle 180^\circ \tag{5.16}$$

Therefore, by using the low-frequency measurement data, the approximate value of g_m can be determined. Since the voltage across C_π decreases as the frequency increases, its magnitude decreases. It can also be seen that the phase increases in a clockwise direction. Similar to the FET explanation, we can also see in Equation (5.17) that for the low-frequency limit S_{12} becomes

$$S_{12} = 2j\omega C_\mu Z_o = 2\omega C_\mu Z_o \angle 90° \tag{5.17}$$

The phase of S_{12} can also be seen to increase in a clockwise direction as the frequency increases. From the equation above, the magnitude also increases as the frequency increases.

5.3.4 Package

Figure 5.21 shows a typical example of a low-power BJT package.

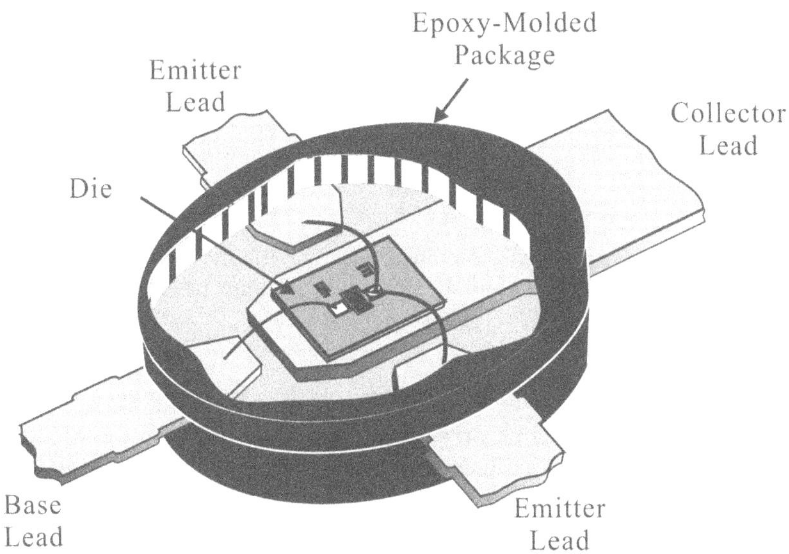

Figure 5.21 Example of a low-power BJT package. On the lead frame, the BJT chip is attached and wire-bonded. Then, the assembled BJT is molded using an epoxy material and any unnecessary lead frames are cut.

In Figure 5.21, the BJT is first attached to the lead frame. It is worth noting that the bottom of the chip generally becomes the collector, as shown in Figure 5.13(b). Then, the base and emitter terminals are wire-bonded to the corresponding lead-frame terminals. Since the parasitic elements appearing at the emitter terminal from the assembly provide feedback from output to input, they have a significant effect on device performance. In order to minimize these parasitic emitter inductances due to the assembly, it is common to assign two terminals to the emitter.

After the assembly, the wire-bonded chip is molded using an epoxy material. Then, the lead terminals are appropriately cut from the lead frame to be used as a packaged device. This packaging for a BJT will result in performance degradation at high frequencies as in the case of an FET.

5.3.5 GaAs/AlGaAs HBT

Figure 5.22 shows the cross-sectional view of a GaAs heterojunction bipolar transistor (HBT). The advantages of a compound semiconductor GaAs over an Si and the characteristics of a heterojunction are used in the GaAs HBT to improve the performance of the Si BJT. However, note that the base and emitter have the same interdigital structure that is used in the Si BJT.

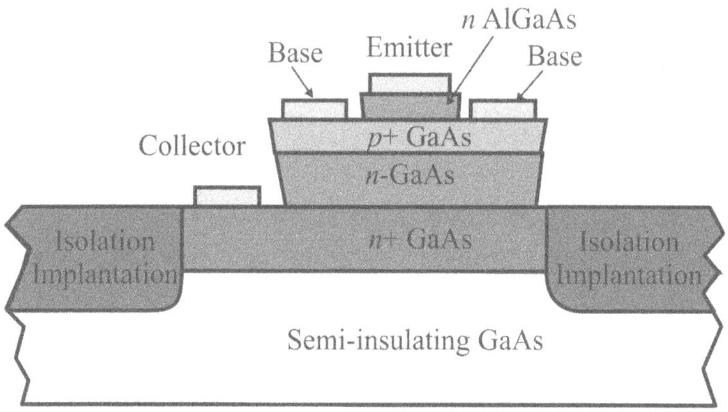

Figure 5.22 Cross-sectional structure of a GaAs HBT. The base-emitter junction is composed of the heterojunction of the AlGaAs and the GaAs, which suppresses the diffusion of the base-region holes into the emitter.

The BE junction of the GaAs HBT is formed by using the *n*-type AlGaAs emitter and the *p*-type GaAs base. Thus, an energy trap occurs between the AlGaAs and GaAs heterojunction. The energy trap is useful for suppressing the diffusing holes from the base toward the emitter that appears for a forward-biased BE junction. The diffusing holes are easily trapped and therefore the holes' diffusion is significantly suppressed. Due to the energy trap, the doping of the base region can be increased. As a result, the base resistance at high frequencies that limits device performance can be decreased. Therefore, the unit-gain frequency given in Equation (5.18) is increased.

$$f_{max} = \sqrt{\frac{f_T}{8\pi R_B C_\mu}} \tag{5.18}$$

Here, f_T is the cut-off frequency at which the short-circuit current gain becomes unity. The frequency f_{max} represents the maximum oscillation frequency at which the maximum power gain becomes unity.

As previously explained, the injected electrons from the emitter to the base arrive at the collector by diffusion. Thus, the base transit time τ_b is determined by the diffusion constant. The diffusion constant of electrons in the GaAs, D_n is four times larger when compared to that in the Si, and a smaller base transit time results. The base transit time τ_b is directly related to the cutoff frequency f_T. The smaller τ_b results in a further increase in f_{max} given by Equation (5.18).

The electrons arrive at the collector then move from the collector region by a drift mechanism and finally arrive at the collector terminal. Suppose that the devices fabricated using the Si and the GaAs have the same collector thickness. The drift velocity v_d of electrons in the GaAs is approximately six times faster than in the Si. As a result, the faster v_d will reduce the collector transit time, τ_d. The reduced τ_d also leads to a rise in f_T. Thus, due to such improvements, the GaAs HBT can be used up to 10 GHz. With more advanced processes, the GaAs HBT can be used up to the millimeter wave frequency.

5.4 DC BIAS CIRCUITS

For the use of a BJT or an FET in amplifier or oscillator circuits, it is necessary to bias it with appropriate DC sources, such as DC voltage or current sources. The DC bias circuit can determine the operating point of the BJT or FET. However, the DC bias circuit should not affect the RF signal flow. Consequently, decoupling the DC bias circuit from an RF circuit is necessary. In this section, we will discuss the DC bias circuit of microwave transistors as well as the decoupling method.

5.4.1 BJT DC Bias Circuits

5.4.1.1 DC Bias Circuit

Figure 5.23 shows two typical DC bias circuits for a BJT. In the case of the DC bias circuits in Figure 5.23(a), the approximate base current is $I_B = 0$, and supply voltage V_{CC} divided by the resistors R_1 and R_2 appears at the base. As shown in Equation (5.19), the base voltage V_B is

$$V_B = \frac{R_2}{R_1 + R_2} V_{CC} \tag{5.19}$$

Then, Equation (5.20) finds the emitter current I_E to be

$$I_E = \frac{V_B - V_{BE}}{R_E} \tag{5.20}$$

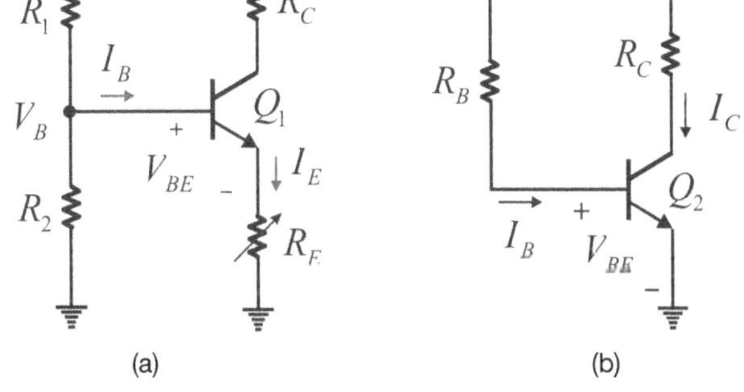

Figure 5.23 DC bias circuits; (a) circuit using emitter resistor, (b) circuit with emitter resistor removed. The circuit in (a) provides a stable I_C, while the circuit in (b) provides a stable ground.

Thus, the desired value of the emitter current I_E is obtained by varying the resistor R_E. In addition, since the RF output is usually taken from the collector, the voltage V_{CE} will be limited by the resistor R_C. In order to overcome the RF signal-swing limitation by the resistor R_C, an RF choke (RFC) is often employed instead of resistor R_C.

In the case of Figure 5.23(b), the base current I_B is obtained from Equation (5.21) as

$$I_B = \frac{V_{CC} - V_{BE}}{R_B} \tag{5.21}$$

Thus, the collector current I_C is subsequently given by Equation (5.22) as

$$I_C = \beta I_B \tag{5.22}$$

The typically large variations in the value of β among units of the same device type appear. This will result in the large variations of the collector current I_C determined by Equation (5.22) and I_C may generally deviate from the design value. Consequently, the value of the resistor R_B should be adjusted to obtain the desired value of the collector current I_C. Compared with the DC bias circuit shown in Figure 5.23(b), the DC bias circuit that uses the emitter resistor shown in Figure 5.23(a) can yield a designed collector current insensitive to the variations in the values of β. Therefore, the DC bias circuit in Figure 5.23(b) is not generally used at low frequencies, while the DC bias circuit that uses the emitter resistor shown in Figure 5.23(a) is preferred for low-frequency applications.

However, the emitter terminal is usually grounded in an RF application, and a stable ground is needed for an RF application. When the bypass capacitor is added in parallel to the emitter resistor R_E in Figure 5.23(a) for the AC ground, many problems arise due to the unstable AC ground. Practically, this type of configuration is accompanied by unpredictable, small, parasitic elements between the emitter terminal and the ground. Even assuming an ideal bypass capacitor, in order to connect the capacitor in parallel with the resistor R_E, inevitably some land patterns and connection lines are required, which make it very difficult to correctly predict the impedance attached to the emitter terminal. In a worst-case scenario, the emitter's parasitic impedance may cause oscillations or it may even significantly change the S-parameters of the BJT, which often leads to failure in obtaining the desired gain. Thus, although the circuit shown in Figure 5.23(a) supplies a stable DC collector current, this type of DC bias circuit should be avoided in an RF amplifier circuit, especially when using packaged BJTs. On the other hand, despite the flaws of the bias circuit shown in Figure 5.23(b), such as the device-to-device DC collector current change and the necessity for the adjustment of R_B, the DC bias circuit is preferred in RF amplifier applications because it provides a stable RF ground.

5.4.1.2 RF Decoupling

Figure 5.24 shows the method for decoupling a designed DC bias circuit from an RF circuit. The two capacitors, C_B, block the DC current flowing out to the RF input and output, and are called DC block capacitors. These capacitors are usually set sufficiently large so as to behave as shorts at the operating frequency. However, the DC block capacitors are not pure capacitors. As described in Chapter 2, parasitic elements, such as a series inductor, are inherent in the DC block capacitors. As a result, a large-valued chip capacitor may not function as a DC block at higher frequencies and can cause significant insertion loss due to the parasitic series inductor. In general, up to a frequency of a few hundred MHz, the chip capacitor operates well as a DC block because the value of the series inductor is small.

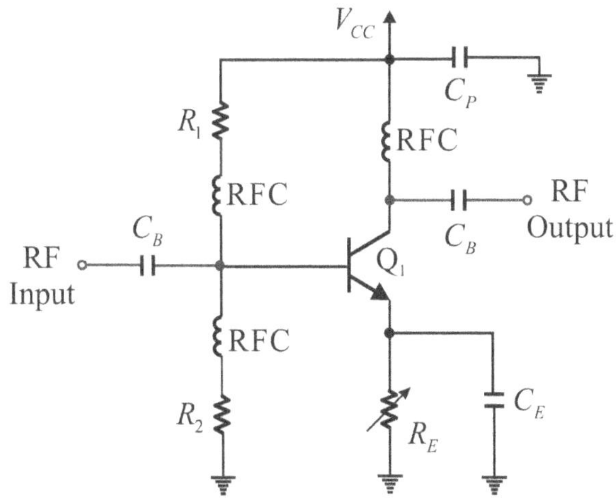

Figure 5.24 An example of an RF decoupled circuit. The RFCs connected to R_1 and R_2 can be removed when the values of R_1 and R_2 are sufficiently large. For the selection of the DC block capacitors C_B, refer to Chapter 2. Bypass capacitor C_P should not only provide the circuit ground but also block the incoming DC supply noise.

However, beyond this frequency, the capacitor does not work adequately as a DC block due to the influence of the series inductor. The chip capacitors should be replaced by other capacitors to behave as a true DC block. In the case of thin-film implementation, MIM capacitors can be also used for a better DC block. The MIM capacitors can usually be connected with wire bonding. Since the bonding wires also produce inductances, the length and number of the bonding wires should be set to produce only minimal parasitic inductances. In PCB assembly, the DC block is often constructed with coupled transmission lines. However, the coupled transmission lines can usually be used as a narrowband DC block when a broadband DC block is not required. Using coupled-transmission-line DC block will avoid the effects of the imprecise parasitic inductances.

The next problem is the selection of the bypass capacitor C_P for isolating the RF circuit from the external bias circuitry. The bypass capacitor C_P, similar to the DC block capacitor, should be sufficiently large enough to behave as a short at the operating frequency. It is critical that the C_P should be placed so as not to affect the RF circuit. Careful consideration should be given in advance as to whether the flow of the RF signal could be affected. Obviously, when C_P is sufficiently large, the lower-frequency AC noise from the DC supply that may flow into the internal RF circuitry can be effectively blocked. However, a bypass capacitor chosen to block the low-frequency AC noise from the DC supply does not act as a short at the RF as discussed in section 2.3.1 of Chapter 2. Thus, to make a short circuit at the RF, two parallel capacitors or, occasionally, *multiple capacitors in parallel* are used. In this case, a small-value capacitor operating as a short at the RF and a large-valued capacitor to block the AC noise from the DC supply are selected.

In addition, the capacitor C_E is inserted to provide the RF ground. Since this capacitor bypasses the emitter resistor R_E, it is called a bypass capacitor. However, the bypass capacitor causes significant problems at high frequencies. A capacitor that can be a short at the RF must be chosen for C_E. However, in reality it is difficult to choose such capacitor. Although such a bypass capacitor can be chosen, extra connecting lines and land patterns are necessary to connect the chosen bypass capacitor in parallel to the emitter resistor R_E.

As a result, the effects of the connecting lines and land patterns appear at RF and these effects should be also minimized. These effects are more pronounced as the operation frequency becomes higher and so the ground by the bypass capacitor as shown in Figure 5.24 is usually not recommended, except for oscillators and for amplifiers operating at frequencies below a few hundred MHz.

The RFC shown in Figure 5.24 acts as an open circuit at the operating frequency and opens the collector terminal. In the case of RFCs connected to the base, they open the base terminal from the bias resistors R_1 and R_2. In general, however, the RFC bandwidth is narrow and other resonance phenomena can appear. Note that *RFCs are not necessary where a resistor can sufficiently ensure an open circuit for an RF*. Thus, when resistors R_1 and R_2 are sufficiently large compared with the surrounding impedances, it is usually a good idea to use resistors alone without the addition of RFCs due to the broadband characteristic of the resistors. In contrast, the replacement of the collector RFC by a resistor is usually not recommended due to the DC collector current. Thus, it is necessary to use the RFC in spite of resonance and its narrowband property. In this case, the RFC is selected to resonate at the operating frequency since it opens at the resonance frequency.

5.4.1.3 Active DC Bias Circuit

Figure 5.25 shows an active DC bias circuit. Transistor Q_1 is a DC-biasing low-frequency operating *pnp* transistor and transistor Q_2 represents the RF transistor. From the circuit shown in Figure 5.25, the supply voltage V_{CC} divided by the resistors R_2 and R_3 appears at the base of Q_1. Thus, the base voltage V_B of Q_1, as shown in Equation (5.23), becomes

$$V_B = \frac{R_3}{R_2 + R_3} V_{CC} \tag{5.23}$$

Therefore, using Equation (5.24), a current I_1

$$I_1 = \frac{V_B - V_{BE}}{R_1} \tag{5.24}$$

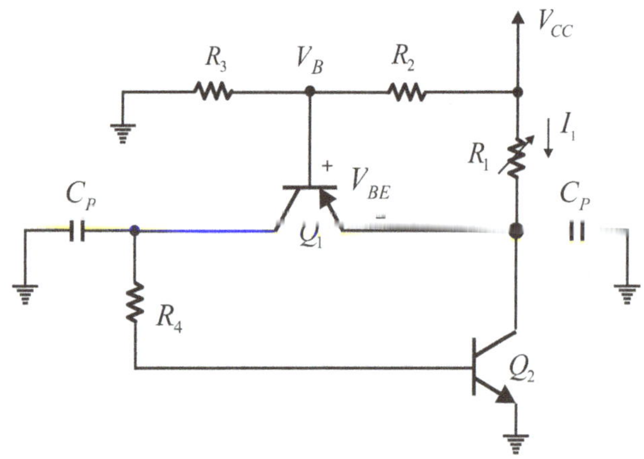

Figure 5.25 An active DC bias circuit. Q_2 is an RF transistor while Q_1 is a low-frequency transistor for biasing Q_2.

will flow through resistor R_1. Current I_1 is the sum of the collector current of Q_2 and the emitter current of Q_1. The emitter current of Q_1 is equal to the base current of Q_2, which is negligible compared to the collector current of Q_2. Thus, I_1 is approximately equal to the collector current of Q_2. Therefore, we can see that the current flowing in the transistor Q_2 is determined by the resistor R_1 as in Equation (5.24), regardless of the β of the transistor. Note that the emitter terminal of transistor Q_2 is also directly grounded and the circuit retains the two advantages of the DC bias circuits shown in Figure 5.23. The disadvantage is that the supply voltage V_{CC} must be greater than the collector voltage of transistor Q_2. This becomes a serious problem when the collector current is large because the power loss due to the collector resistor R_1 is high. Bypass capacitors C_P are inserted to isolate the DC bias circuit from the RF circuit. The criteria for the selection of such capacitors are the same as for the selection of the bypass capacitor C_P previously described.

5.4.2 FET DC Bias Circuit Design

The unique feature of the DC bias circuit for a depletion-mode FET is that a negative voltage is necessary for the gate bias. Figure 5.26 shows two types of DC bias circuits for the depletion-mode FET.

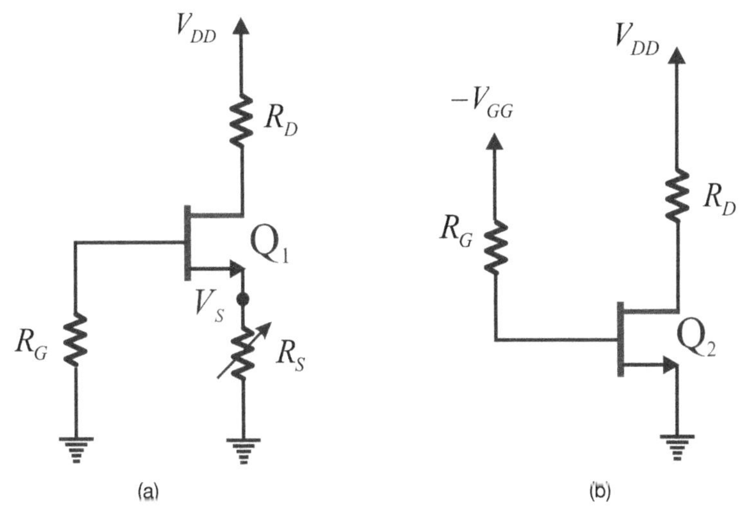

Figure 5.26 FET DC bias circuit: (a) a self-bias circuit and (b) a DC bias circuit using two DC sources. The circuit in (a) gives a stable I_D while that in (b) provides the stable ground.

In Figure 5.26(a), the source resistor R_S is inserted, so that when the drain current flows, a voltage drop across resistor R_S appears, which is the voltage at the source V_S. Since the DC current does not flow through the gate, the gate DC voltage is 0. Therefore, the voltage V_{GS} between the gate and source is $-V_S$. This V_{GS} determines the drain current, which can be adjusted by changing the value of the resistor R_S. If a large value is selected for R_S, a large negative V_{GS} appears and the drain current becomes small. On the other hand, when R_S is small, a large drain current flows. As previously explained, since there is usually no DC current flowing in the gate, gate resistor R_G has no effect. The reason for inserting the gate resistor in the circuit is to protect the device. In abnormal operation, a positive voltage may appear across the gate and a large gate current flows, which can damage the FET. To prevent this, a resistor is often inserted in the gate for protection. The circuit in Figure 5.26(a) is called a self-bias circuit.

In Figure 5.26(b), two DC voltage sources are used. A negative DC voltage is applied to the gate while a positive voltage is applied to the drain. Thus, the drain current is adjusted by voltage $-V_{GG}$. The reason for inserting resistor R_G in the gate is the same as for that shown in Figure 5.26(a). The resistor inserted in the drain sets the drain-source DC voltage. Resistor R_D may not be needed in RF circuit applications. In addition, the pros and cons of two DC bias circuits are the same as those for the BJT DC bias circuits. Figure 5.26(a) shows a stable DC drain current, while Figure 5.26(b) shows a stable RF ground.

Figure 5.27 illustrates an active DC bias circuit of an FET. The transistor Q_1 is a low-frequency DC-biasing *pnp* transistor, while Q_2 represents an RF FET. The resistors R_2 and R_3, as in the case of the BJT active DC bias circuit, divide the supply voltage. The difference from the active DC bias circuit shown in Figure 5.25 is that both negative and positive voltages are used in the FET active DC bias circuit. The base voltage determined by the voltage division then determines the current through the resistor R_1. The current through R_1 is equal to the sum of the currents flowing through the transistors Q_1 and Q_2. Because most of the emitter current of Q_1 appears at the collector, the negative voltage that results from the voltage division appears at the gate of the FET, and this determines the DC current flowing in the drain of the FET. This method for isolating the DC bias circuit is similar to that of the BJT. Also, the resistor R_6 inserted in the gate is intended to protect the FET from damage when, in an abnormal operation, a large current flows as a result of possible positive voltage appearing at the gate.

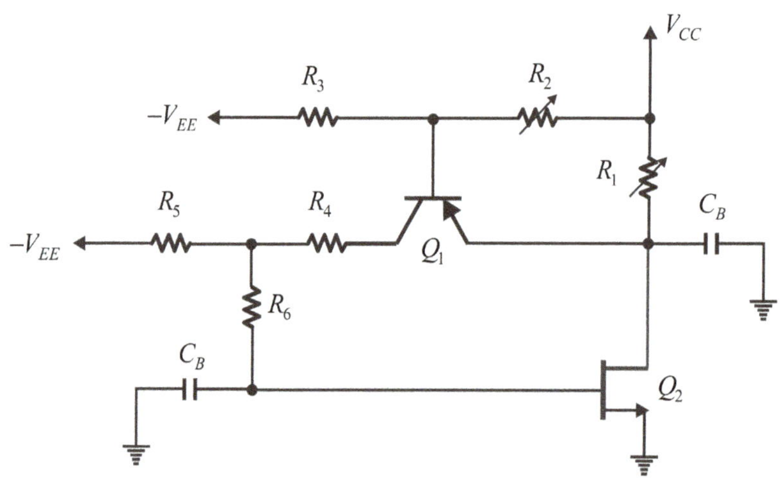

Figure 5.27 An active DC bias circuit. Q_1 is a low-frequency *pnp* transistor for biasing Q_2.

5.4.3 S-Parameter Simulation

Figure 5.28 shows an example of an S-parameter simulation for an FET.

A real measurement using test equipment is similar to the same S-parameter simulation shown in Figure 5.28. The DC feed in Figure 5.28 is a component representing an RFC that becomes a short for the DC and an open for the RF. The DC block represents a component that is open at the DC and short for all the RF frequencies. The gate and drain voltages are set to $V_{GS} = -1$ V and $V_{DS} = 3$ V. Note that the bypass capacitors for DC voltage supplies are not necessary because the DC voltage source provides a complete short at AC.

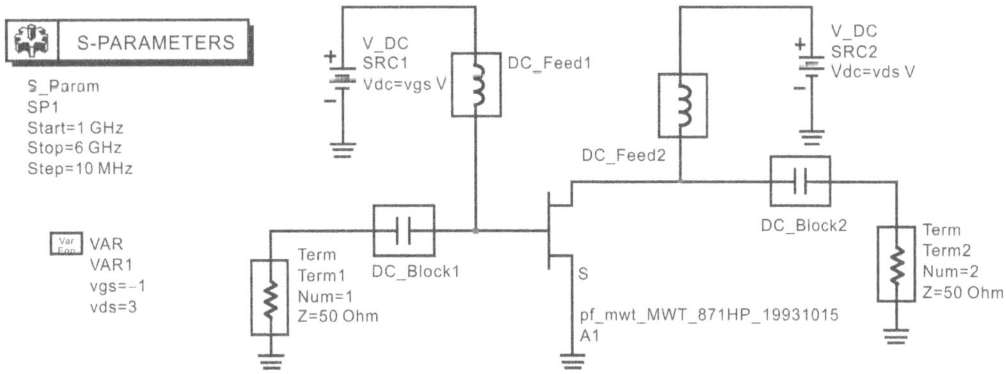

Figure 5.28 An S-parameter simulation of an FET. The component **DC_Block** is an ideal DC block. The **DC_Block** is short at AC, and open at DC. In contrast, the component **DC_Feed** is an ideal RFC. In the simulation, bypass capacitors for DC voltage sources are not necessary because the DC voltage sources are ideally short at AC.

Figure 5.29 shows the S-parameter simulation results. When the S-parameter simulation is performed with this setup, DC analysis will automatically be performed in advance without the need to specify it separately. At the established DC operating point, the FET will be converted automatically into the corresponding small-signal equivalent circuit, and the S-parameters of the FET will be computed. The computed S-parameters will show the previously explained frequency dependencies.

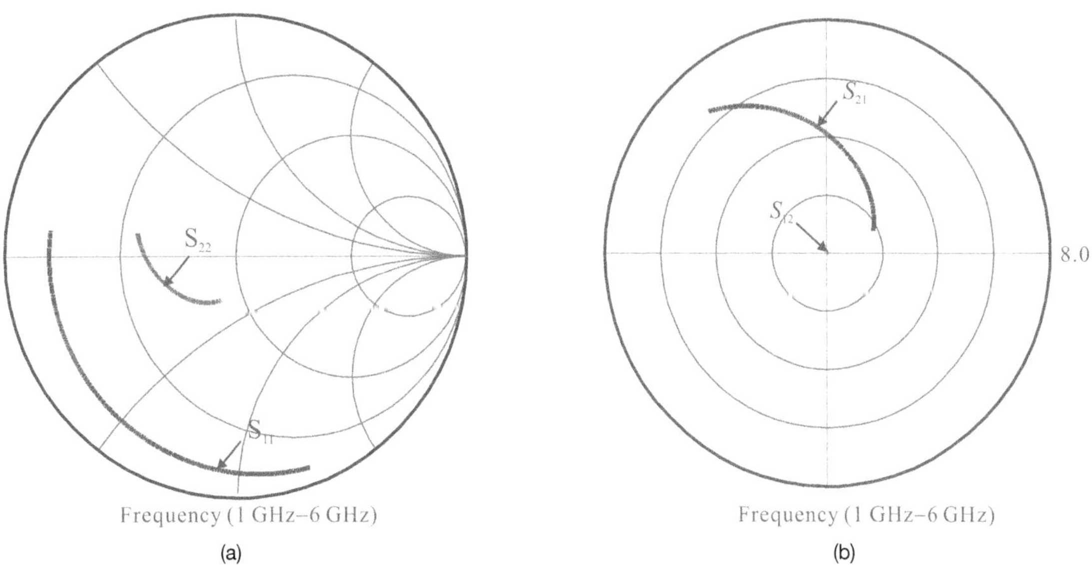

Figure 5.29 S-parameter simulation results: (a) S_{11} and S_{22}, (b) S_{21} and S_{12}

Figure 5.30 shows a BJT S-parameter simulation setup. It is similar to an FET S-parameter simulation. The only difference is that a DC current source is used for DC biasing the base. Since the operating point of the BJT is usually determined by the

DC collector current, the corresponding base current is supplied by the current source. In addition, because the DC current source is open at the AC, an RFC will not be needed. The calculated S-parameters of this set up are shown in Figure 5.31.

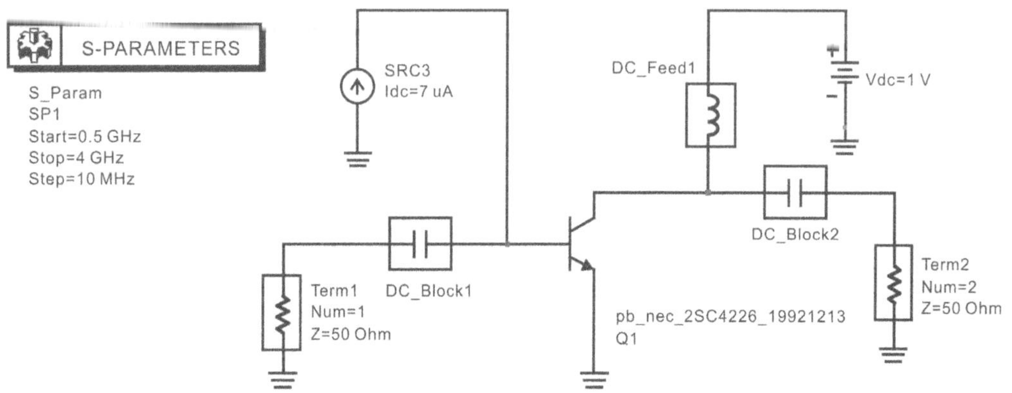

Figure 5.30 A BJT S-parameter simulation setup. Since the BJT collector current is controlled by the base current, the current source is used to bias the BJT. Note that the DC current source acts as open at the AC and the RFC in series is not necessary. The explanations for other elements are similar to those for an FET bias circuit.

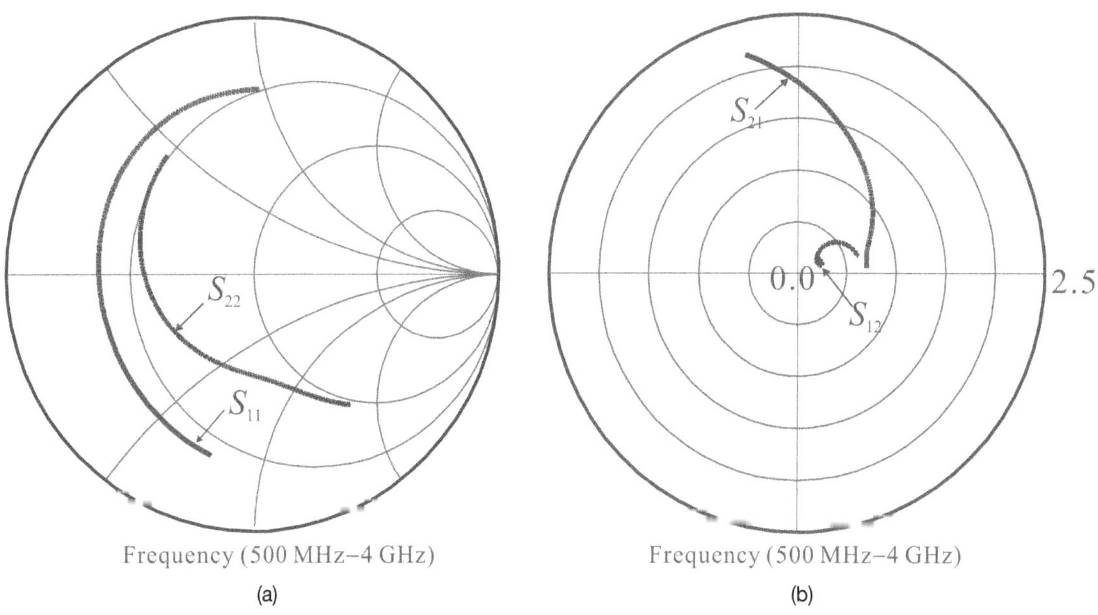

Figure 5.31 S-parameter simulation results: (a) S_{11} and S_{22}, (b) S_{21} and S_{12}

5.5 EXTRACTION OF EQUIVALENT CIRCUITS

Physically modeled equivalent circuits rather than measured S-parameters for passive or active microwave devices sometimes provide designers with more insight for understanding those devices' operations. Many microwave devices can be physically modeled as either T-type or π-type equivalent circuits, and their equivalent circuit element values often need to be extracted from the measured S-parameters for further analysis and design. For devices that can be modeled by T- and π-type equivalent circuits, their equivalent circuit element values can easily be extracted by converting their S-parameters to Z- and Y-parameters. Since the Z- and Y-parameters themselves can be naturally represented by T- or π-type equivalent circuits respectively, the T- and π-type equivalent circuit element values can be easily obtained from the converted Z- or Y-parameters.

Figure 5.32 shows a T-type equivalent circuit. We will show that the derivation of the T-type equivalent circuit from the Z-parameters is converted from the measured S-parameters. From the definition of the Z-parameters, the port voltages can be expressed as

$$V_1 = z_{11}I_1 + z_{12}I_2 \tag{5.25}$$

$$V_2 = z_{21}I_1 + z_{22}I_2 \tag{5.26}$$

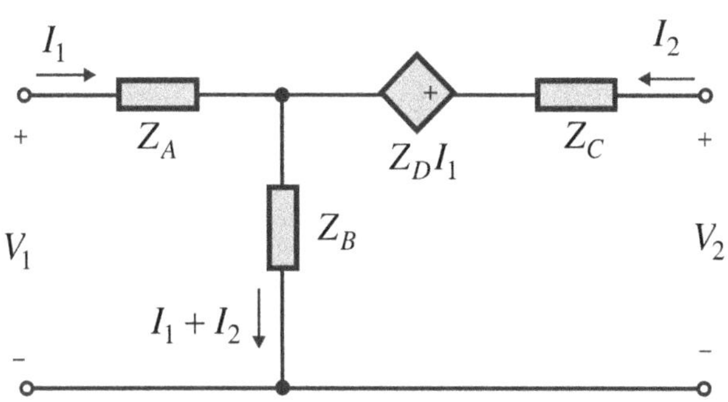

Figure 5.32 T-type equivalent circuit

Since the voltage V_1 at port 1 in Figure 5.32 depends on the currents I_1 and $I_1 + I_2$, by rewriting Equation (5.25), we obtain Equation (5.27):

$$V_1 = (z_{11} - z_{12})I_1 + z_{12}(I_1 + I_2) \tag{5.27}$$

Thus, comparing Equation (5.27) with V_1 of Figure 5.32, Z_A and Z_B are expressed in Equations (5.28) and (5.29).

$$Z_A = z_{11} - z_{12} \tag{5.28}$$

$$Z_B = z_{12} \tag{5.29}$$

Also, by arranging voltage V_2 at port 2 as determined by I_2 and $I_1 + I_2$, and arranging the remaining terms as determined by the current I_1, we obtain

$$V_2 = z_{21}I_1 + z_{22}I_2 = z_{12}(I_1 + I_2) + (z_{22} - z_{12})I_2 + (z_{21} - z_{12})I_1$$

Comparing the equation above with V_2 derived from the circuit in Figure 5.32 yields Equations (5.30) and (5.31).

$$Z_C = z_{22} - z_{12} \tag{5.30}$$

$$Z_D = z_{21} - z_{12} \tag{5.31}$$

Using Equations (5.28) through (5.31), the circuit in Figure 5.32 can be represented using Z-parameters as shown in Figure 5.33.

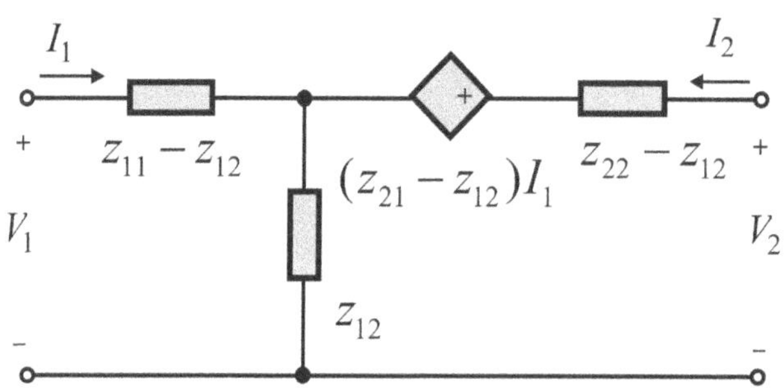

Figure 5.33 A Z-parameter equivalent circuit

Similarly, Y-parameters can be represented by a π-type equivalent circuit. Figure 5.34 shows a π-type equivalent circuit.

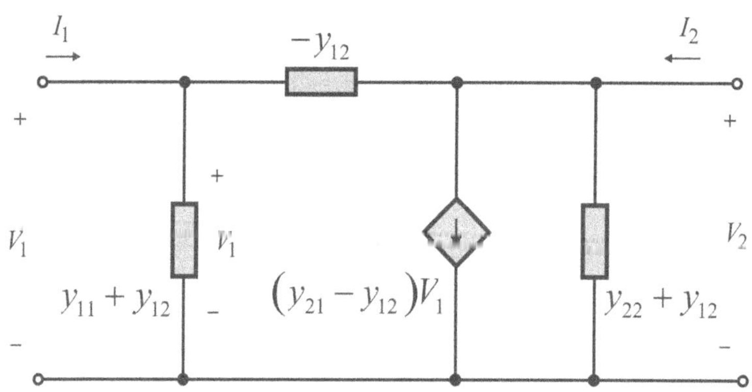

Figure 5.34 A π-type equivalent circuit

Note that when two port voltages are present simultaneously, the current flowing from port 1 to port 2 depends on the voltage difference $V_1 - V_2$. Thus, current I_1 at port 1 can be expressed as

$$I_1 = y_{11}V_1 + y_{12}V_2 = -y_{12}(V_1 - V_2) + (y_{11} + y_{12})V_1$$

Similarly, current I_2 at port 2 can be expressed as

$$I_2 = y_{21}V_1 + y_{22}V_2 = -y_{12}(V_2 - V_1) + (y_{22} + y_{12})V_2 + (y_{21} - y_{12})V_1$$

Using the two equations above, we can obtain the π-type equivalent circuit shown in Figure 5.34.

The equivalent circuits in Figure 5.33 and Figure 5.34 are general examples and can be further simplified for passive devices. The equivalent circuit for a passive device is shown in Figures 5.35(a) and (b), respectively. By the reciprocity $z_{12} = z_{21}$ and $y_{12} = y_{21}$ in the Z- and Y- parameters, the controlled sources disappear in the passive device's equivalent circuits.

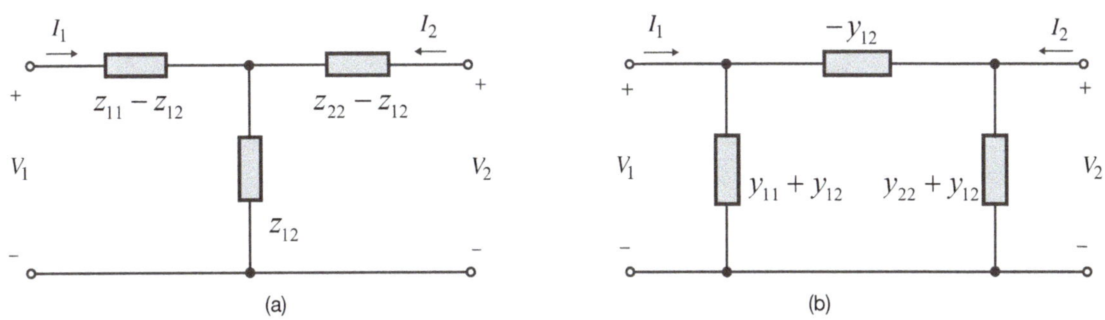

Figure 5.35 The equivalent circuit for a passive device (a) in Z-parameters and (b) in Y-parameters

The equivalent circuits in Figures 5.35(a) and (b) are useful for analyzing unknown inductors or capacitors. Practically speaking, $(z_{11} - z_{12})$, z_{12}, and $(z_{22} - z_{12})$ in the T-type equivalent circuit, and $(y_{11} + y_{12})$, y_{12}, and $(y_{22} + y_{12})$ in the π-type equivalent circuit may not be represented by a single element such as an inductor or a capacitor, and they may show a frequency dependence. In this case, the frequency dependence of each element can be further decomposed and represented by the complex circuit that consists of frequency-independent elements, as explained in Chapter 2. In addition, using a similar technique, a multiport passive device can be represented in a similar way. The reader may wish to consult reference 8 at the end of this chapter for details.

Example 5.1

For a 10-mil-thick alumina substrate with a permittivity of 9.6, a microstrip ring inductor that has 2 turns, width and spacing of 10 mil, and an inner radius of 50 mil can be represented by the equivalent circuit in Figure 5E.1 for frequencies up to 2 GHz. Calculate the values of the equivalent circuit. Here, L_s represents the inductance that results from the microstrip ring-type inductor, and C_{p1} and C_{p2} represent the parasitic capacitances.

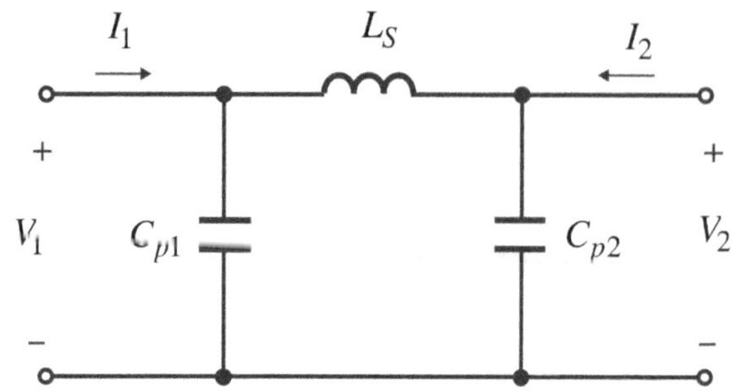

Figure 5E.1 The equivalent circuit for a microstrip ring-type inductor

Solution

Figure 5E.2 shows the S-parameter simulation setup for the microstrip ring-type inductor.

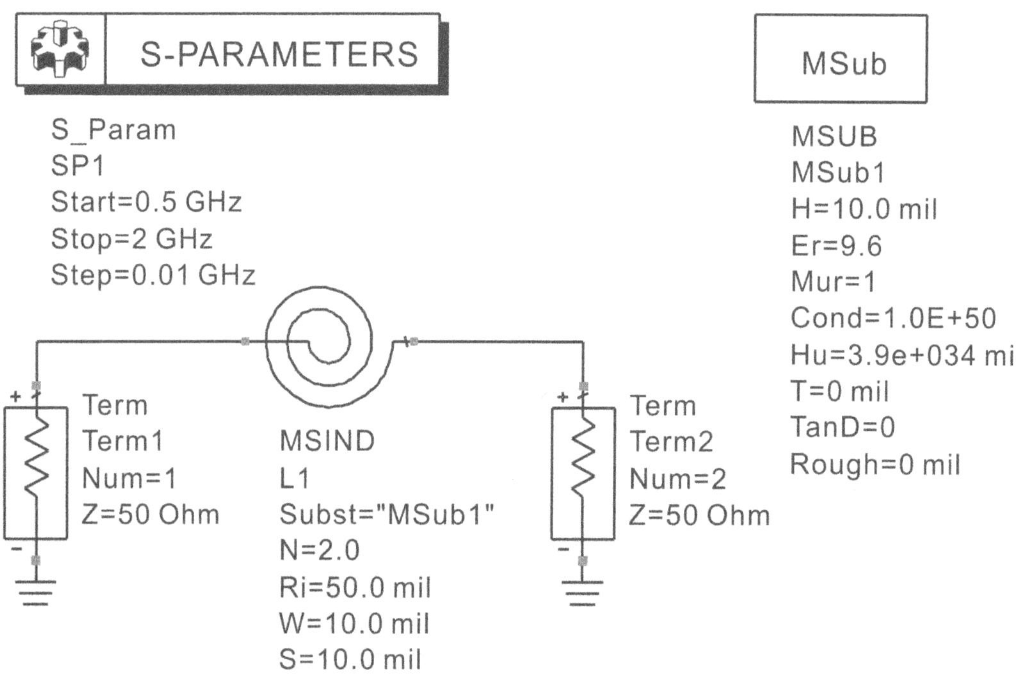

Figure 5E.2 S-parameter simulation of a microstrip ring-type inductor. The S-parameters as well as the Y- or Z-parameters can be obtained simultaneously after the S-parameter simulation. The parameters **N**, **Ri**, **W**, and **S** represent the number of turns, the inner radius, and the width and spacing of the spiral inductor **MSIND**.

Before calculating the S-parameters, open the S-parameter simulation controller in ADS, and check the Y-parameter calculation in the **Parameters** tab. This setting provides the S-parameters as well as the Y-parameters to be stored in the dataset after the simulation. The following equations shown in Measurement Expression 5E.1 are then entered in the display window. The values of **Ls**, **Cp**1, and **Cp**2 are plotted with respect to frequency and are shown in Figure 5E.3.

Eqn w=2*pi*freq

Eqn Cp1=(imag(Y(1,1)+Y(1,2)))/w

Eqn Cp2=(imag(Y(2,2)+Y(1,2)))/w

Eqn Ls=-1/imag(-Y(1,2))/w

Measurement Expression 5E.1 Equation for calculating the value of the equivalent circuit values in Figure 5E.1

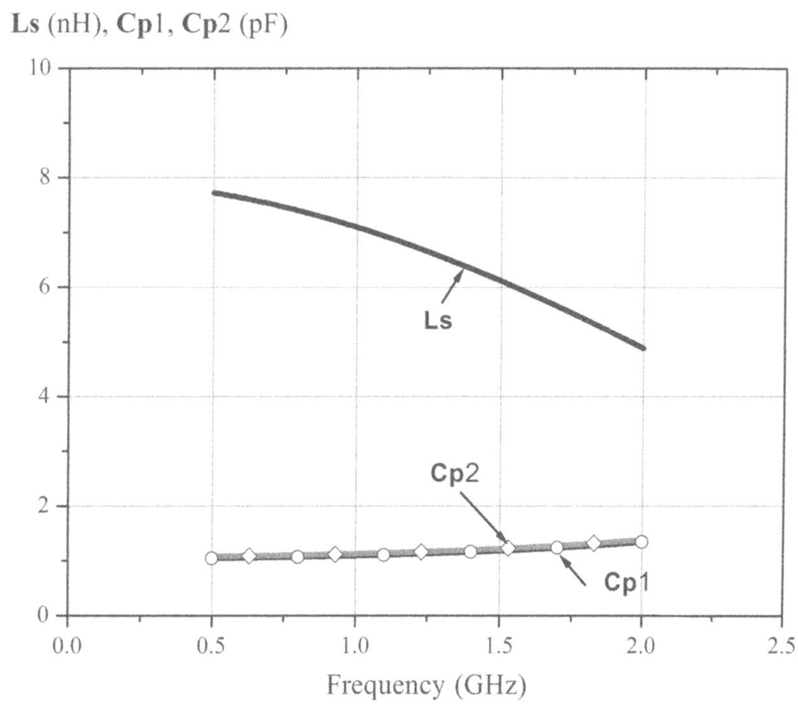

Figure 5E.3 Capacitances **Cp1** and **Cp2**, and the series inductance of a microstrip ring-type inductor. In most cases, the number of significant digits is two, so the scale of the y-axis is set to vary from 0–10.

Cp1 in Measurement Expression 5E.1 represents the shunt capacitor at port1, while **Cp2** represents the shunt capacitor at port 2. The inductor **Ls** represents the series inductor that connects port 1 and port 2. Note that **Cp1** and **Cp2** are almost frequency independent. However, Ls shows some frequency dependence. Thus, Ls cannot be represented by a single inductor and should be represented by a more complex circuit.

We now present the method for obtaining the values of the simplified small-signal equivalent circuit shown again in Figure 5.36(a) from the measured S-parameters.[2] The π-type equivalent circuit in Figure 5.36(b) is similar to the simplified equivalent circuit of

2. K. W. Yeom, T. S. Ha, and J. W. Ra, "Frequency Dependence of GaAs FET Equivalent Circuit Elements Extracted from the Measured Two-Port S-parameters," *Proceedings of the IEEE*, vol. 76, no. 7, pp. 843–845, July 1988.

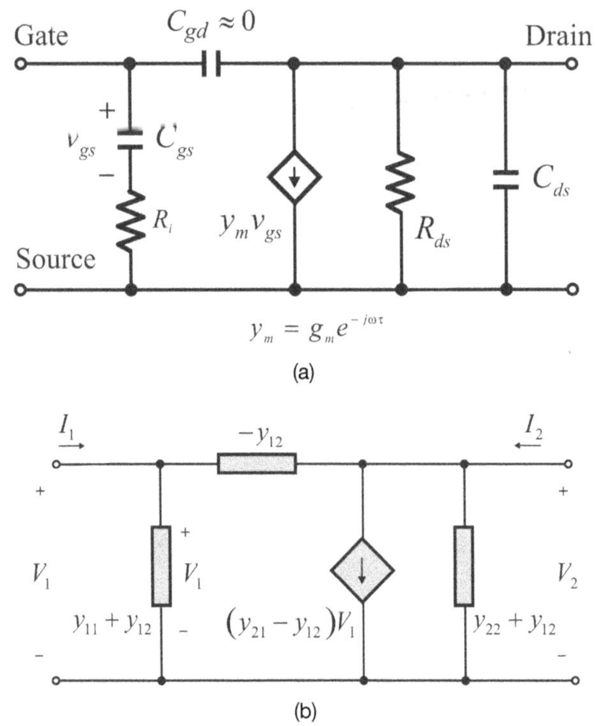

Figure 5.36 (a) A GaAs FET simplified equivalent circuit and (b) a Y-parameter equivalent circuit. The two representations look similar and the only difference is in the control voltage: in (a), the control voltage appears across C_{gs}, while in (b) V_1 is the control voltage.

an FET. Therefore, by directly matching the simplified FET equivalent circuit to the π-type equivalent circuit, the values of the simplified equivalent circuit can be directly obtained from the measured S-parameters. First, comparing the output impedances of the two circuits in Figures 5.36(a) and (b), the admittance of $R_{ds} \| C_{ds}$ should be equal to $y_{22} + y_{12}$. From this comparison, we obtain Equations (5.32) and (5.33).

$$R_{ds}^{-1} = \mathrm{Re}(y_{22} + y_{12}) \tag{5.32}$$

$$C_{ds} = \frac{1}{\omega}\mathrm{Im}(y_{22} + y_{12}) \tag{5.33}$$

Similarly, comparing y_{12} and C_{gd}, we obtain Equation (5.34).

$$C_{gd} = \frac{1}{\omega}\mathrm{Im}(-y_{12}) \tag{5.34}$$

Comparing the input impedances of the two circuits, we obtain Equation (5.35).

$$\frac{1}{y_{11} + y_{12}} = R_i + \frac{1}{j\omega C_{gs}} \tag{5.35}$$

Rewriting Equation (5.35), R_i and C_{gs} can be determined using Equations (5.36) and (5.37).

$$R_i = \text{Re}\left(\frac{1}{y_{11} + y_{12}}\right) \tag{5.36}$$

$$\frac{1}{\omega C_{gs}} = -\text{Im}\left(\frac{1}{y_{11} + y_{12}}\right) \tag{5.37}$$

However, the control voltages are different in the dependent sources shown in Figures 5.36(a) and 5.36(b). In the case of Figure 5.36(a), voltage V_{gs} across C_{gs} is the control voltage, while in the case of Figure 5.36(b), the control voltage is V_1. The relationship between V_1 and V_{gs} is

$$\frac{V_{gs}}{V_1} = \frac{1}{1 + j\omega C_{gs} R_i} = \frac{\text{Im}\left(\frac{1}{y_{11} + y_{12}}\right)}{\frac{1}{y_{11} + y_{12}}}$$

Also,

$$(y_{21} - y_{12})V_1 = y_m V_{gs}$$

Using these two equations, g_m can be obtained with Equation (5.38).

$$g_m = |y_{21} - y_{12}| \cdot \left|\frac{V_1}{V_{gs}}\right| = \frac{|y_{21} - y_{12}|}{|y_{11} + y_{12}|}\left\{\text{Im}\left(\frac{1}{y_{11} + y_{12}}\right)\right\}^{-1} \tag{5.38}$$

And τ is determined with Equation (5.39).

$$\tau = -\frac{1}{\omega}\angle\left(j\frac{y_{21} - y_{12}}{y_{11} + y_{12}}\right) \tag{5.39}$$

Using Equations (5.32) through (5.39), we find that the values of the simplified equivalent circuit of FET can be directly determined from the measured S-parameters.

Example 5.2

This example shows how to obtain the simplified FET equivalent circuit. Open the model of the chip pHEMT FHX15 in the ADS library and compute the S-parameters at $V_{DS} = 2$ V and $V_{GS} = -0.2$ V. Using the S-parameters, compute the simplified FET equivalent circuit element values.

Solution

Figure 5E.4 shows a DC simulation setup. The I_D–V_{DS} characteristics are plotted in Figure 5E.5. From the I_D–V_{DS} characteristics, the drain current can be found to be $I_{DS} = 18$ mA at $V_{DS} = 2$ V and $V_{GS} = -0.2$ V. Setting V_{GS} and V_{DS} for $I_{DS} = 18$ mA, the S-parameter simulation is performed as shown in Figure 5E.6. To obtain the simplified FET equivalent circuit element values using the obtained S-parameters, the following equations shown in Measurement Expression 5E.2 are entered in the display window.

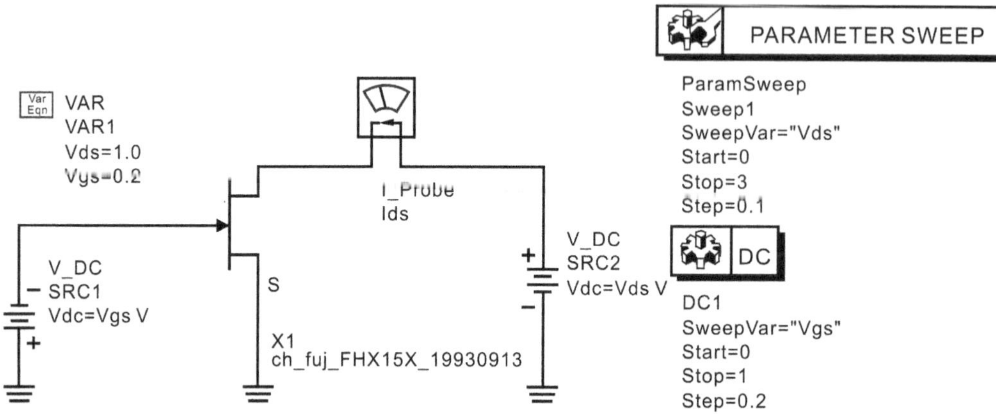

Figure 5E.4 DC simulation circuit to determine the drain current I_{DS}. The current probe **Ids** measures the simulated drain current. The **ParamSweep** sweeps **Vds** from 0 to 3 V and the DC simulation component sweeps **Vgs** from 0 to –1 V.

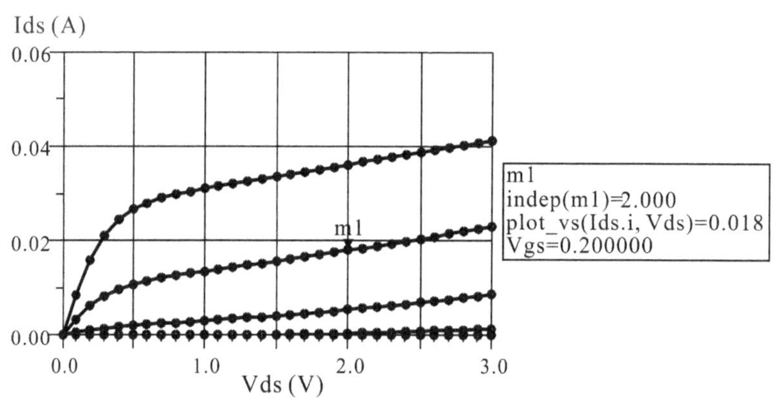

Figure 5E.5 DC characteristics of pHEMT FHX15. I_{DS} = 18mA at **Vgs** = -0.2 V, **Vds** = 2 V

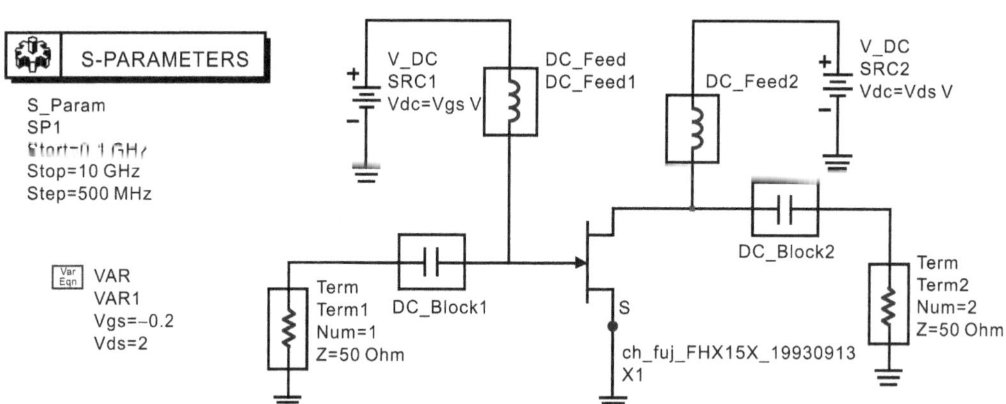

Figure 5E.6 S-parameter simulation setup. To yield I_{DS} = 18 mA, **Vgs** = –0.2 V, **Vds** = 2 V are applied. Note that the Y-parameters are computed simultaneously using the S-parameter simulation.

```
Eqn w=2*pi*freq
Eqn Cgd=imag(-Y(1,2))/w
Eqn Rds=1/real(Y(2,2)+Y(1,2))
Eqn Cgs=-1/(w*(imag(1/Y(1,1)+Y(1,2))))
Eqn Cds=1/w*imag(Y(2,2)+Y(1,2))
Eqn Ri=real(1/Y(1,1)+Y(1,2)))
Eqn gm=-mag((Y(2,1)-Y(1,2))/(Y(1,1)+Y(1,2)))/imag(1/(Y(1,1)+Y(1,2)))
Eqn tau=(-1/w)*phase(j*(Y(2,1)-Y(1,2))/(Y(1,1)+Y(1,2)))
```

Measurement Expression 5E.2 Equations for obtaining the values of the simplified equivalent circuit

Figures 5E.7 through 5E.9 show plots for the equivalent circuit element values with respect to frequency. From these figures, we find that $g_m = 77$ mA/V, $\tau = 0.26$ ns, $R_{ds} = 158\ \Omega$, $C_{gs} = 0.25$ pF, $C_{gd} = 31$ fF, and $C_{ds} = 0.1$ pF.

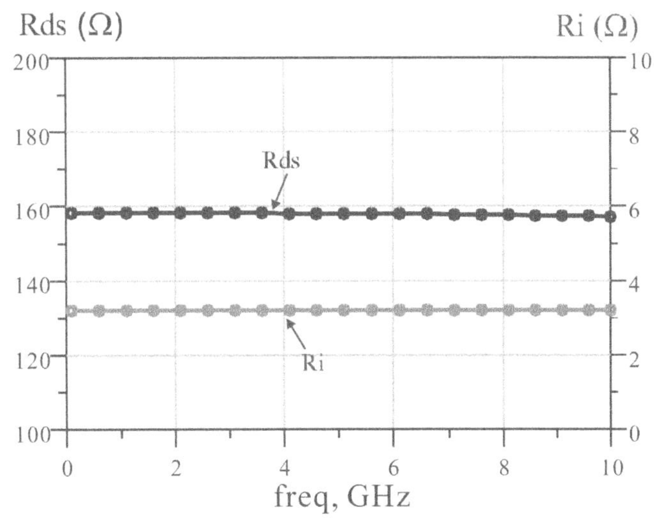

Figure 5E.7 Computed **Rds** and **Ri**

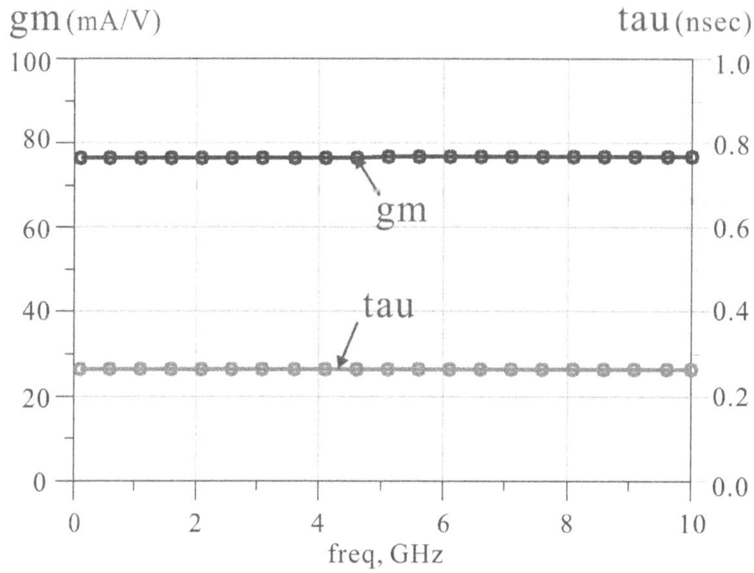

Figure 5E.8 Computed **gm** and **tau**

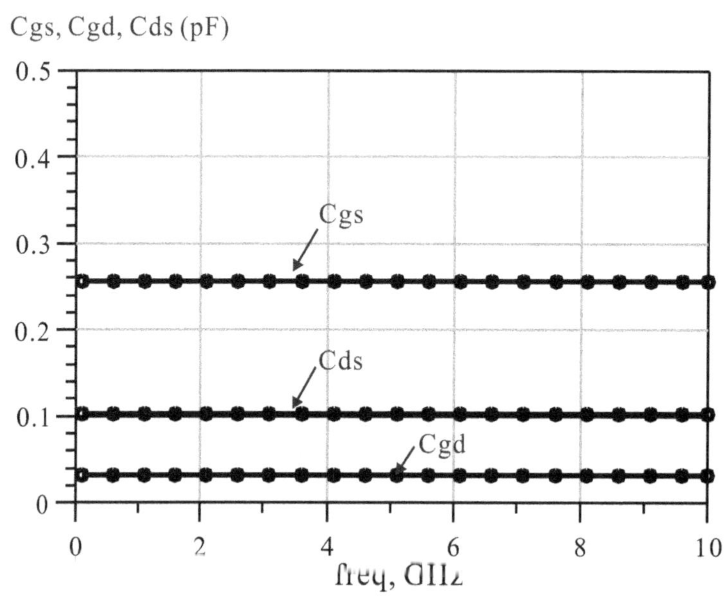

Figure 5E.9 Computed **Cgs**, **Cgd**, and **Cds**

From Example 5.2, we can determine the simplified FET equivalent circuit element values. Figure 5.37 again shows the equivalent circuit of a chip FET. The equivalent circuit is known to predict quite closely the measured S-parameters of a chip FET.

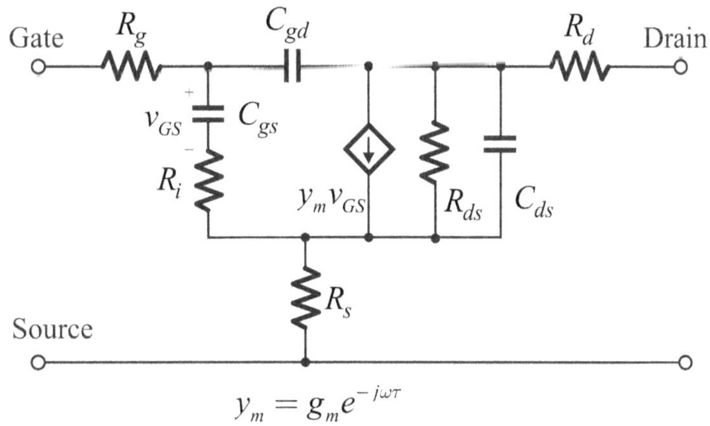

Figure 5.37 Small-signal equivalent circuit of the chip FET

The difference between the chip FET and the simplified equivalent circuits is in R_g, R_s, and R_d, which are called extrinsic elements. A method for determining the chip FET equivalent circuit element values, which is similar to the extraction of the simplified equivalent circuit, does not exist at present. The values of R_g, R_s, and R_d were previously determined by optimization, which yields the best fit to the measured S-parameters. However, because resistors R_g, R_s, and R_d in the active state of a FET make only a small contribution to the S-parameters, there is significant uncertainty about their values when determined by this method, and obtaining their exact values is difficult.[3] If we assume that R_g, R_s, and R_d do not change even when the DC operating point changes, their values may be determined more accurately using the measured S-parameters at other DC operating points where their effects are dominant. Thus, such operating points should be found first, after which their values can be extracted more accurately.

The operating points of $V_{GS} = 0$ V and $V_{DS} = 0$ V are thought to be suitable for the extraction of R_g, R_s, and R_d. At this bias point, the FET completely becomes a passive device, and the equivalent circuit appears, as shown in Figure 5.38. The zero-bias depletion capacitance C_b appears, and will appear on both the source and drain sides. Because there is no change in the values of R_g, R_s, and R_d, their values are the same as in active mode. The measurement at $V_{GS} = 0$ V and $V_{DS} = 0$ V is often referred to as a *cold-FET measurement*.[4] Even in the cold FET condition, because the values of R_g, R_s, and R_d are typically small, and the impedance of the depletion capacitance C_b is usually very large, it is difficult to determine their values accurately when measurement errors occur.

3. R. L. Vaitkus, "Uncertainty in the Values of GaAs MESFET Equivalent Circuit Elements Extracted from the Measured Two-Port Scattering Parameters," presented at the *1983 IEEE Cornell Conference on High Speed Semiconductor Devices and Circuits*, Cornell University, 1983.

4. G. Dambrine, A. Cappy, F. Helidore, and E. Playez, "A New Method for Determining the FET Small-Signal Equivalent Circuit," *IEEE Transactions on Microwave Theory and Techniques*, vol. 36, no. 7, pp. 1151–1159, July 1988.

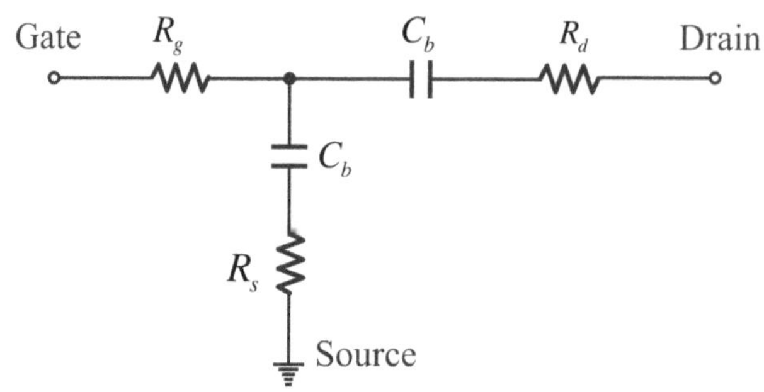

Figure 5.38 FET small-signal equivalent circuit for $V_{GS} = 0$ V and $V_{DS} = 0$ V. Since $V_{DS} = 0$ V, the same depletion capacitance appears between the gate drain and the gate source.

To overcome this problem, a small positive voltage is applied to the gate terminal, which makes the gate drain and gate source behave like Schottky diodes. The Schottky diodes then begin to conduct for a small positive voltage and so the contribution of C_b disappears. Instead, a small-signal resistance of a Schottky diode appears. In addition, the channel resistance is formed between the drain and source terminals. Since the effect of the channel resistance appears in a distributed form, it must be considered as a distributed circuit. However, the circuit's distributed effect is not pronounced and so it can be regarded as a lumped element. With the channel resistor considered, the Z-parameters are expressed in Equations (5.40)–(5.42).

$$z_{11} = R_s + R_g + \frac{R_c}{3} + \frac{nkT}{qI_g} \tag{5.40}$$

$$z_{12} = R_s + \frac{R_c}{2} \tag{5.41}$$

$$z_{22} = R_s + R_d + R_c \tag{5.42}$$

Resistor R_c represents the channel resistance, and nkT/qI_g of z_{11} represents the small-signal Schottky diode resistance that is dependent on the DC gate current I_g. The different contributions of R_c to z_{11}, z_{12}, and z_{22} are because the effect of the distributed channel resistance yields different contributions to z_{11}, z_{12}, and z_{22}. From Equations (5.40) through (5.42), we find that only z_{11} depends on the small-signal Schottky diode resistance. Thus, the term that excludes the small-signal Schottky diode resistance can be found by plotting Re(z_{11}) versus Schottky diode current I_g. Figure 5.39 shows a plot of Re(z_{11}) versus I_g. Using this plot, the term that excludes the small-signal Schottky diode resistance in Equation (5.40) can be found from the intercept value obtained by extrapolating the straight line.

However, using Equations (5.40) through (5.42), the values of R_g, R_s, and R_d cannot be determined because there are four unknowns but only three available equations. Therefore, a separate independent measurement is necessary. There are two methods: The first is to use the Fukui method, with which the value of $R_s + R_d$ can be determined.[5] The second is to directly measure the DC gate resistance in a device. Then, the values of R_g, R_s, and R_d can be determined.

5. H. Fukui, "Determination of the Basic Device Parameters of a GaAs MESFET," *Bell System Technical Journal*, vol. 58, no. 3, pp. 771–795, March 1979.

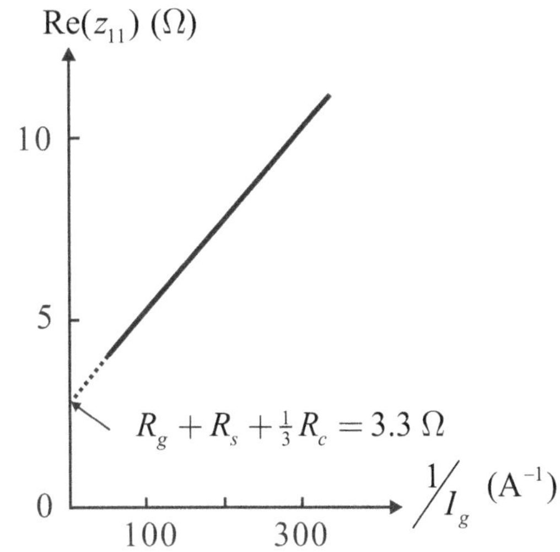

Figure 5.39 Plot for obtaining $R_g + R_s + 1/3{\cdot}R_c$ (drawn after the reference).[6] The z_{11} represents the Z-parameters of the forward-biased cold FET.

Using the obtained values of R_g, R_s, and R_d, the remaining values of the chip FET equivalent circuit in Figure 5.37 can be determined from the measured S-parameters in active mode. Figure 5.40 shows the method for removing the contribution of resistors R_g, R_s, and R_d from the measured S-parameters in the active mode. Then, the remaining circuit is a simplified FET equivalent circuit (or intrinsic equivalent circuit of an FET). By converting the resulting S-parameters into Y-parameters, the values of the simplified equivalent circuit can be determined, as shown in Example 5.2.

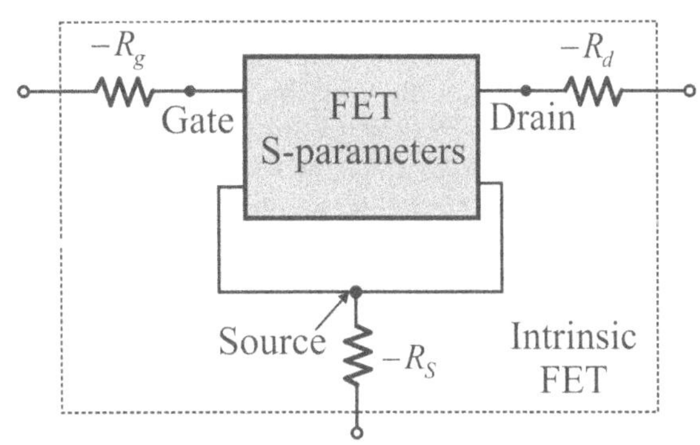

Figure 5.40 The method for removing the effects of resistors R_g, R_s, and R_d in the measured S-parameters. The added three resistors with – values remove the effects of the resistors R_g, R_s, and R_d on the S-parameters and the S-parameters of the intrinsic FET can then be obtained.

6. G. Dambrine, A. Cappy, F. Helidore, and E. Playez, "A New Method for Determining the FET Small-Signal Equivalent Circuit," *IEEE Transactions on Microwave Theory and Techniques*, vol. 36, no. 7, pp. 1151–1159, July 1988.

Example 5.3

In Example 5.2, the values of R_g, R_s, and R_d were approximated to 0 in order to extract the simplified FET equivalent circuit of the pHEMT FHX15 at $V_{DS} = 2$ V and $V_{GS} = -0.2$ V. When $R_g = 1.4$, $R_s = 1.5$, and $R_d = 1.5\ \Omega$, determine the values of the simplified FET equivalent circuit element values with the effects of resistors R_g, R_s, and R_d removed.

Solution

Figure 5E.10 shows a circuit setup to remove the effects of resistors R_g, R_s, and R_d. In Figure 5E.10, the S-parameter data component **SNP**1 represents the S-parameters used for computing the simplified equivalent circuit element values in Example 5.2. Similar to Example 5.2, the check box of the calculation of Y-parameters in the S-parameter simulation controller **SP**1 is checked.

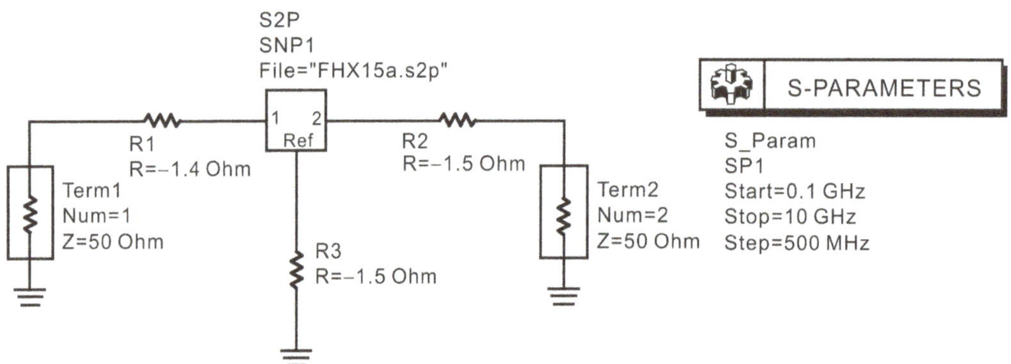

Figure 5E.10 Circuit for removing the effects of the resistances R_g, R_s, and R_d

The formulas in Measurement Expression 5E.2 are entered again in the display window and then the changes in the values of the simplified FET equivalent circuit element values can be obtained. The computed values of the simplified equivalent circuit are $g_{mo} = 88$ mA/V, $\tau_o = 0.25$ ns, $R_{dso} = 137\ \Omega$, $C_{gso} = 0.28$ pF, $C_{gdo} = 29$ fF, and $C_{dso} = 0.12$ pF. The general trend is that the resistance values decrease while the capacitor values increase. The frequency dependences of the computed results are not appreciable, and the values simply move up and down. The change in the transconductance is especially noteworthy as it is directly related to the gain. It can be seen that the change $\Delta g_m = g_{mo} - g_m = 11$ mA/V. The comparison is shown in Figure 5E.11.

In Figure 5E.11, g_{mo} represents the compensated result for R_g, R_s, and R_d, while g_m represents the value in the previous example, which does not take into consideration the effects of resistors R_g, R_s, and R_d. Here, it is generally known that R_s is the main reason for the reduction in g_{mo} and g_{mo} is reduced by R_s as in

$$g_m = \frac{g_{mo}}{1 + g_{mo}R_s} = \frac{88}{1 + 0.088 \times 1.5} = 77\ \text{(mA/V)}$$

We can see that Δg_m is the same as was calculated in Example 5.2. Therefore, in order to prevent the reduction of g_{mo} by R_s in a large transconductance FET, special attention must be taken to make R_s smaller.

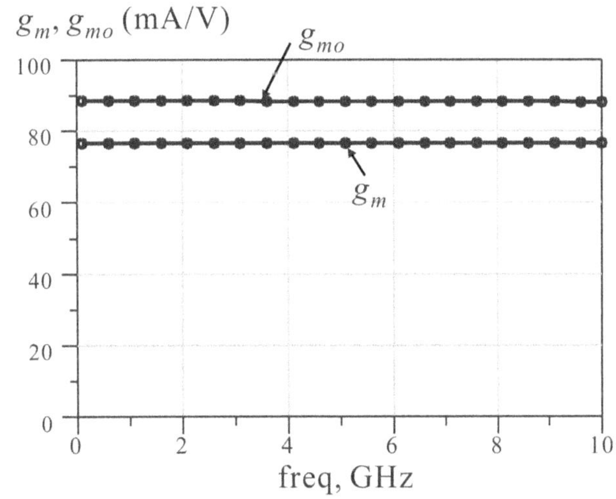

Figure 5E.11 Changes in g_m due to the effects of R_g, R_s, and R_d

5.6 SUMMARY

- Three-terminal devices, FETs and BJTs, have advantages over various one-port devices in the implementation of complex-functioning integrated circuits.
- The principles of operation for small- and large-signal models, and for the typical S-parameters of GaAs MESFET are introduced.
- A GaAs pHEMT has multiple epi-layers and its electron mobility in a channel is improved by removing the impurity scattering that occurs in a GaAs MESFET.
- The base width of a high-frequency Si BJT is minimized to the limit that the process allows and the interdigital structure is employed to implement the base and the emitter in order to reduce the base-spreading resistance.
- The operations of small- and large-signal models, as well as the typical S-parameters of BJT, are explained.
- In a GaAs HBT, the base-emitter junction is implemented using a heterostructure. As a result, the current gain β is not degraded for high base-region doping, which makes it possible to implement small base-spreading resistance. The diffusion in p-type GaAs is faster than that in an Si, which results in the improvement of f_T. Also, the collector drift is reduced because the electron mobility is faster than that in an Si. Combining these facts, the high-frequency performances of the GaAs HBT are superior to those of the Si BJT.
- Self-bias and two-supply DC bias circuits are introduced. The two-supply DC bias circuit has an advantage over the self-bias circuit in providing a stable ground.
- The extraction of a GaAs FET equivalent circuit from S-parameters is explained.

REFERENCES

1. G. D Vendelin, A. M. Pavio, and U. L. Rohode, *Microwave Circuit Design Using Linear and Nonlinear Techniques*, Hoboken, NJ: John Wiley & Sons, Inc., 1990.
2. G. Gonzalez, *Microwave Transistor Amplifiers: Analysis and Design*, 2nd ed. Englewood Cliffs, NJ: Prentice Hall, 1997.
3. R. Soares, J. Graffeuil, and J. Obregon, *Applications of GaAs MESFETs*, Dedham, MA: Artech House, Inc., 1983.
4. J. M. Golio, *Microwave MESFETs and HEMTs*, Dedham, MA: Artech House, Inc., 1991.
5. J. Mun, ed., *GaAs Integrated Circuits*, Oxford: BSP Professional Books, 1988.
6. H. Fukui, ed., *Low-Noise Microwave Transistors and Amplifiers*, New York: IEEE Press, 1981.
7. G. Massobrio and P. Antognetti, *Semiconductor Device Modeling with SPICE*, 2nd ed. New York: McGraw-Hill, Inc., 1993.
8. N. Marcuvitz, *Waveguide Handbook*, Vol. 2 of the MIT Radiation Laboratory Series. New York: Dover Publications, Inc., 1965.

PROBLEMS

5.1 Figure 5P.1 shows a simplified equivalent circuit of a low-frequency GaAs MESFET.

(1) Calculate the S-parameters setting the reference impedance to Z_o.

(2) Setting $Z_o = 50\ \Omega$, compute g_m when $S_{21} = 5\angle 180°$.

(3) When resistor R_f is connected in parallel, as shown in Figure 5P.2, find the value of R_f that makes $S_{11} = 0$.

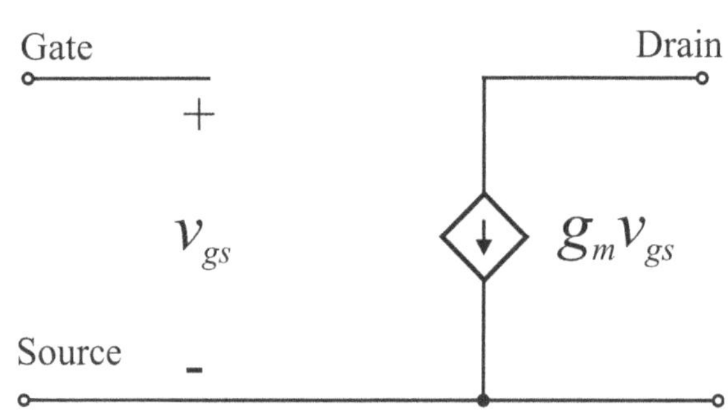

Figure 5P.1 A simplified FET equivalent circuit

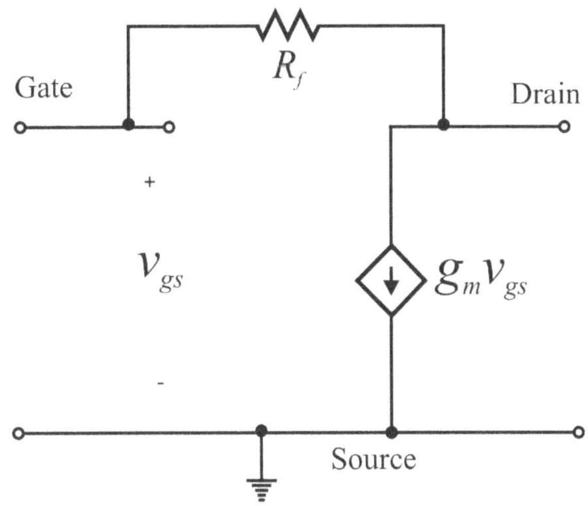

Figure 5P.2 A simplified FET equivalent circuit with a parallel feedback resistor

5.2 Figure 5P.3 shows a simplified equivalent circuit of an FET with a source resistance R_s at low frequency. Due to R_s, the equivalent transconductance g_{me} is lowered from g_m, as explained in Example 5.3. Defining the equivalent transconductance g_{me} as

$$g_{me} = \left.\frac{i_d}{v_{gs}}\right|_{\text{Drain Shorted}}$$

show that

$$g_{me} = \frac{g_m}{1 + g_m R_S}$$

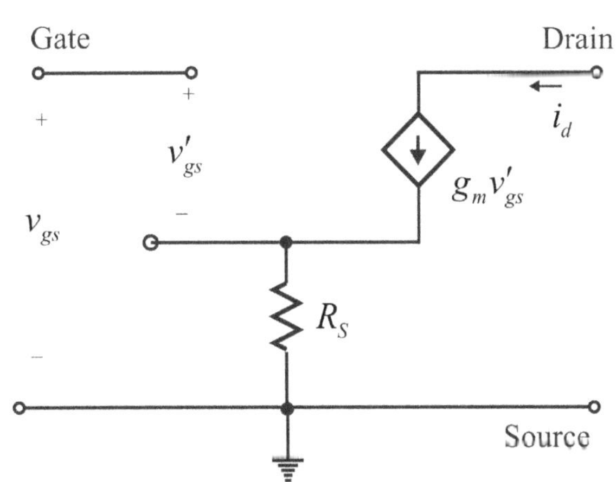

Figure 5P.3 An FET equivalent with a source resistor

5.3 The small-signal equivalent circuit of a GaAs MESFET, including wire bonding, can be approximately represented by Figure 5P.4; given the following S_{11}, determine the values of R_i and L_g. ($C_{gs} = 0.5$ pF).

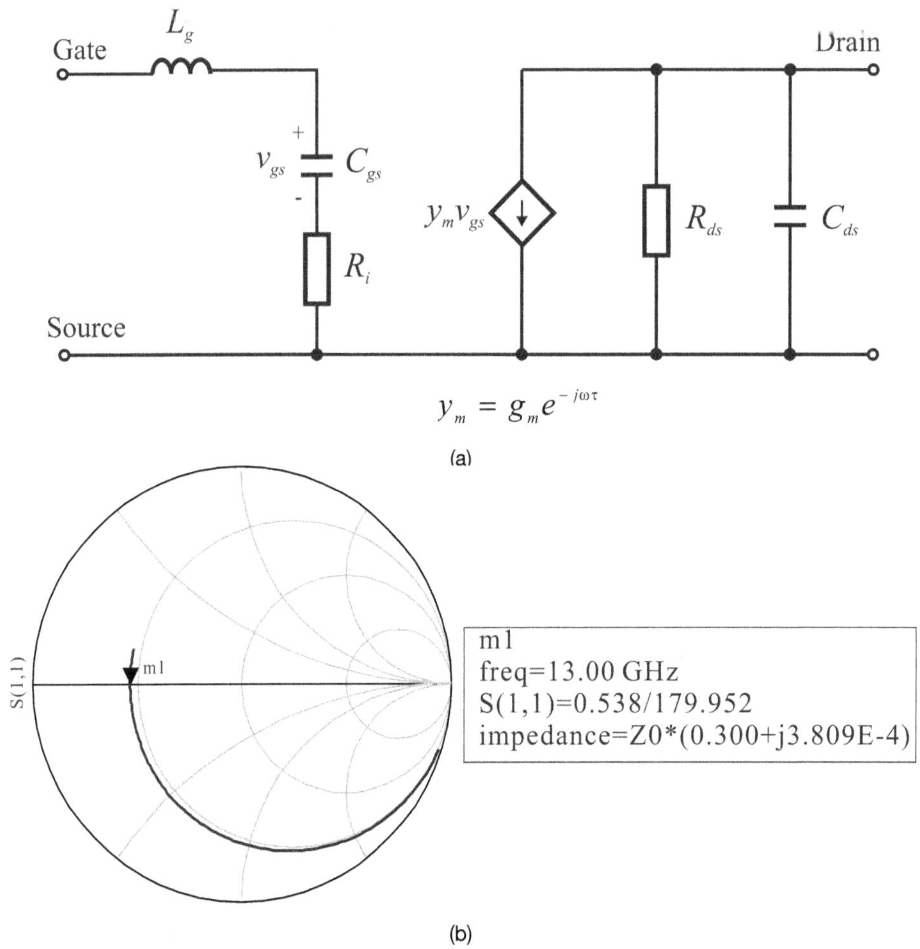

Figure 5P.4 (a) A GaAs FET equivalent circuit and (b) S_{11}

5.4 In the following active DC bias circuit (Figure 5P.5), determine the value of resistor R_1 for a 20 mA current to flow into the collector of transistor Q_2.

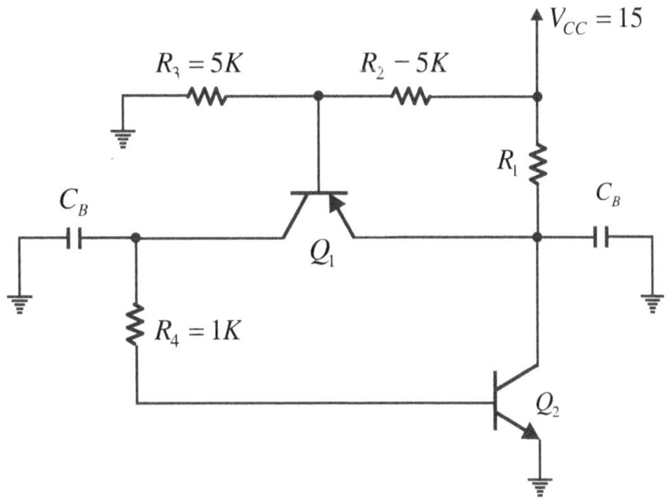

Figure 5P.5 Active DC bias circuit

5.5 (ADS problem) Using the 2SC4226 in the ADS library, set up the circuit shown in Figure 5P.6 below.

(1) Determine its I_C–V_{CE} characteristics as shown in Figure 5P.7. ($I_B = 20\ \mu\text{A} - 160\ \mu\text{A}$, with a step of 20 μA.)

(2) Compute the S-parameters at $V_{CE} = 5$ V and $I_C = 5$ mA for a frequency range of 1–4 GHz with a step of 0.05 GHz.

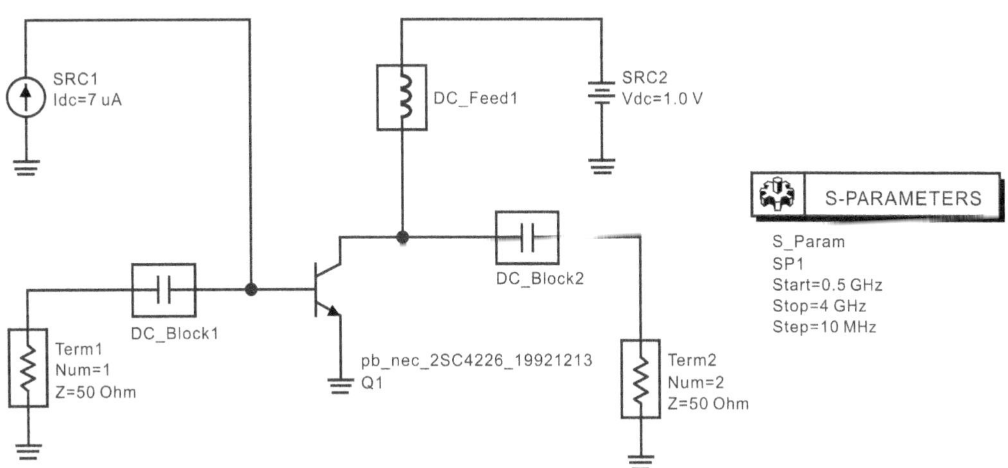

Figure 5P.6 Circuit of problem 5.5

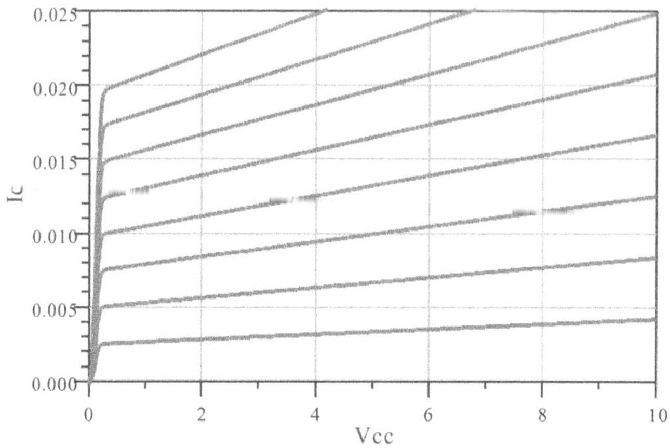

Figure 5P.7 DC characteristics of 2SC4226

CHAPTER

6

Impedance Matching

CHAPTER OUTLINE

6.1 INTRODUCTION

When there is a mismatch between a source and a load, the available power from the source is not delivered to the load and this results in a loss of power. The mismatch can be resolved by using a matching circuit. The function of the matching circuit is to transform the load impedance into an impedance that matches the source impedance. As noted above, a matching circuit is usually required for maximum power transfer, but it is also sometimes used for other purposes. For example, in low-noise amplifier design, the source impedance is transformed to an impedance that gives a minimum noise figure. In oscillator design, the load impedance is transformed into a particular impedance that may be suitable for purposes other than maximum power transfer. Thus, in a broad sense, the matching circuit can be defined as a circuit that transforms a given impedance into a desired impedance.

As a first condition for a matching circuit, an impedance match is necessary and, in addition to the impedance matching, the matching of field shapes should also be considered. For example, a successful matching between a coaxial line and a widely used two wire line for TV reception cannot be achieved solely by a simple impedance matching. A field-shape matching of the two lines is also required because the two field shapes of the lines differ significantly and a simple impedance matching will not yield a successful match. Thus, if the shapes of the electromagnetic fields are different, the field shapes should be matched. To accomplish this, an understanding of circuit theory and electromagnetic field theory is necessary. However, these complex theories are beyond the scope of this discussion, which only covers transmission-line circuits, so we will limit our focus to impedance matching.

There are various types of matching circuits and in this chapter we will first describe the method for constructing a matching circuit using lumped elements. The design concepts of this type of circuit can also be applied to a matching circuit that uses transmission lines. In addition, both single- and double-stub tuners as well as a quarter-wavelength, transmission-line-impedance matching circuit will be discussed in this chapter.

6.2 MAXIMUM POWER TRANSFER THEOREM

Figure 6.1 shows a circuit in which a source with an impedance of $Z_S = R_S + jX_S$ is connected to a load with an impedance of $Z_L = R_L + jX_L$. Suppose that the source impedance is fixed and the load impedance can vary for maximum power transfer to the load.

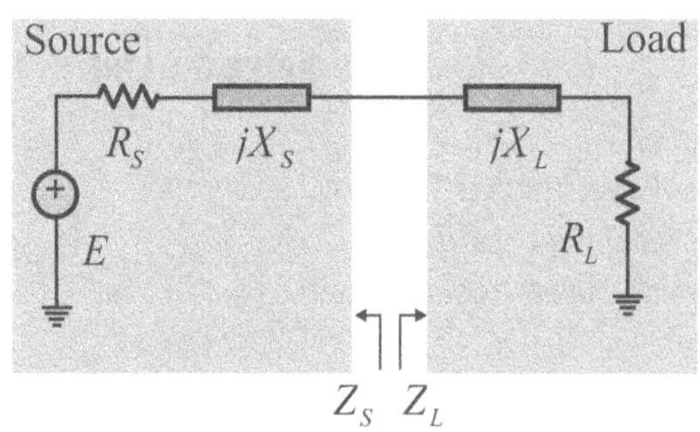

Figure 6.1 Source connected to a load circuit. The source impedance $Z_S = R_S + jX_S$ is fixed while the load impedance $Z_L = R_L + jX_L$ can be varied for the maximum power delivery.

The power delivered to the load is shown in Equation (6.1) as

$$P_L = \frac{1}{2}\mathrm{Re}\left(V_L^* I_L\right) = \frac{1}{2}R_L\left|I_L\right|^2 \tag{6.1}$$

The current I_L is represented by Equation (6.2).

$$I_L = \frac{E}{\left(R_L + R_S\right) + j\left(X_S + X_L\right)} \tag{6.2}$$

Substituting I_L into Equation (6.1), we obtain Equation (6.3).

$$P_L = \frac{1}{2}\frac{|E|^2 R_L}{\left(R_S + R_L\right)^2 + \left(X_S + X_L\right)^2} \tag{6.3}$$

Therefore, for maximum P_L, $X_S = -X_L$ should be satisfied first. The resulting P_L is then

$$P_L = \frac{1}{2}\frac{|E|^2 R_L}{(R_S + R_L)^2} \tag{6.4}$$

P_L in Equation (6.4) is plotted in Figure 6.2 and the maximum power transfer occurs at $R_S = R_L$. The maximum power is then given by Equation (6.5) as

$$P_{L,\max} = \frac{|E|^2}{8R_S} \tag{6.5}$$

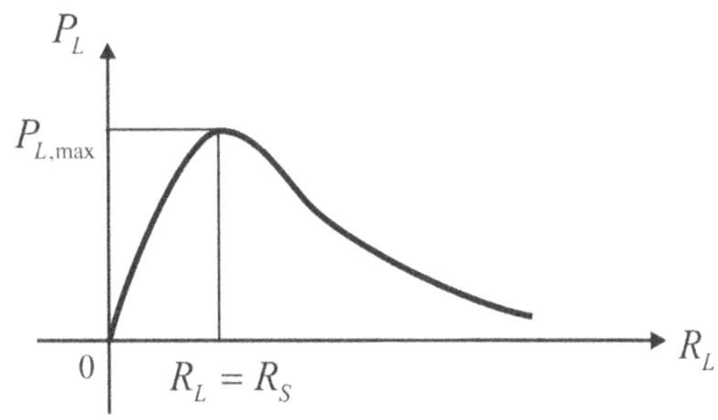

Figure 6.2 Delivered power P_L with respect to R_L

This is referred to as *maximum available power*, or simply *available power*. Note that the maximum available power is delivered to the load when the source and load impedance have the relationship shown in Equation (6.6).

$$Z_S^* = Z_L \tag{6.6}$$

That is, maximum power is delivered when the source impedance is equal to the conjugate of the load impedance. This is called *conjugate matching*.

Usually, the load is not the variable impedance shown in Figure 6.1, but it has a fixed impedance Z_L as shown in Figure 6.3. From Figure 6.3, the conjugate matching to the source impedance is achieved through a lossless two-port matching circuit composed of, for example, transmission lines or lossless components such as inductors and capacitors. Suppose that the Z-parameters of such a lossless two-port matching network are known. For maximum power delivery from the source to the input of the two-port matching circuit, the following condition given by Equation (6.7) should be satisfied:

$$Z_S = Z_{in}^* = \left(z_{11} - \frac{z_{12}^2}{z_{22} + Z_L}\right)^* \tag{6.7}$$

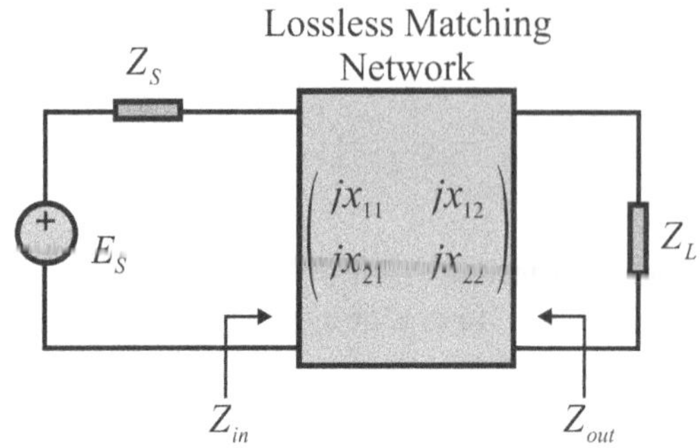

Figure 6.3 Source and load impedances Z_s and Z_L are conjugate matched through a lossless matching circuit. In the case where $Z_{in} = Z_s^*$, then $Z_{out} = Z_L^*$.

Since the matching circuit is passive, by reciprocity, $z_{12} = z_{21}$. Also, since there are no losses, z_{ij} is purely imaginary. Therefore, Equation (6.7) can be expressed as

$$Z_S^* = jx_{11} + \frac{x_{12}^2}{jx_{22} + Z_L} \tag{6.8}$$

From Equation (6.8), Z_L can be obtained as shown in Equations (6.9) and (6.10).

$$Z_L = -jx_{22} + \frac{x_{12}^2}{-jx_{11} + Z_S^*} \tag{6.9}$$

$$Z_L^* = jx_{22} - \frac{x_{12}^2}{jx_{11} + Z_S} \tag{6.10}$$

The right side of Equation (6.10) is the output impedance seen from the load. Thus, when the conjugate match is achieved at the input side through the lossless matching circuit in Equation (6.7), the output side is also naturally conjugate matched. Therefore, when the arbitrary load impedance is conjugate matched to the source impedance through a lossless matching circuit, the impedance looking into the source from the load will automatically be the complex conjugate of the load impedance.

In other words, if the input side is matched through a lossless matching circuit, as shown in Figure 6.3, then $Z_S = Z_{in}^*$; under this condition, the output side will also be automatically matched by $Z_L = Z_{out}^*$. Rewriting the relation in terms of the reflection coefficients, Equation (6.11) gives us

$$\Gamma_{in} = \Gamma_S^* \rightarrow \Gamma_L = \Gamma_{out}^* \tag{6.11}$$

6.3 DISCRETE MATCHING CIRCUITS

6.3.1 Series-to-Parallel Conversion

The matching circuit design uses the basic concept of parallel-to-series conversion, and its inverse, series-to-parallel conversion. When a parallel reactance jX_P is added to a given resistor R_P, as shown in Figure 6.4, this can be converted to the series circuit shown on the right-hand side of Figure 6.4. In this figure, the real part R_S is less than R_P due to the added jX_P. Therefore, when it is necessary to lower a large resistance R_P to a certain desired value, that resistance value can be achieved by adding jX_P in parallel, which will also produce an additional series reactance jX_S. Note that the sign of jX_S is equal to that of jX_P; that is, when a capacitor jX_P is added, jX_S has the sign of the capacitor, while an added parallel inductor will result in a series inductance.

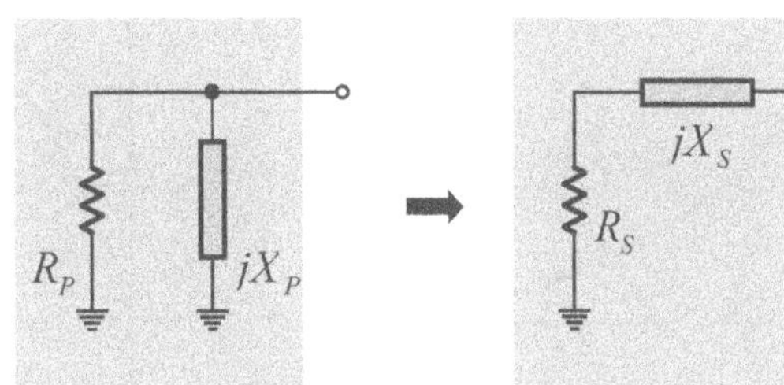

Figure 6.4 Parallel-to-series conversion. The impedance connected in parallel can be converted to the impedance connected in series.

To verify the properties mentioned above, the impedance of a parallel connection is computed and the impedance is separated into its real and imaginary parts. As a result, we obtain

$$\frac{jX_P R_P}{R_P + jX_P} = \frac{jX_P R_P (R_P - jX_P)}{R_P^2 + X_P^2} = \frac{X_P^2 R_P}{R_P^2 + X_P^2} + \frac{jX_P R_P^2}{R_P^2 + X_P^2}$$

Therefore, using Equations (6.12) and (6.13), R_S and X_S are

$$R_S = \frac{X_P^2 R_P}{R_P^2 + X_P^2} = \frac{R_P}{1 + b_p^2} \tag{6.12}$$

$$X_S = \frac{X_P R_P^2}{R_P^2 + X_P^2} = \frac{b_p R_P}{1 + b_p^2} \tag{6.13}$$

where

$$b_p = \frac{R_P}{X_P} = \frac{B_P}{G_P} \tag{6.14}$$

represents the normalized susceptance shown in Equation (6.14). In addition, the Q of the two circuits should be equal. Thus, Equation (6.15) results in

$$\frac{X_S}{R_S} = \frac{R_P}{X_P} \tag{6.15}$$

Figures 6.5 (a) and (b) show the plots of R_S and X_S. Since the variable in Equations (6.12) and (6.13) is b_p, the plots are drawn with respect to b_p. From Figure 6.5, as expected, R_S decreases monotonically as b_p increases, while X_S does not show monotonic increase or decrease. Note that the sign of b_p is equal to X_S, which means that X_P and X_S have the same signs.

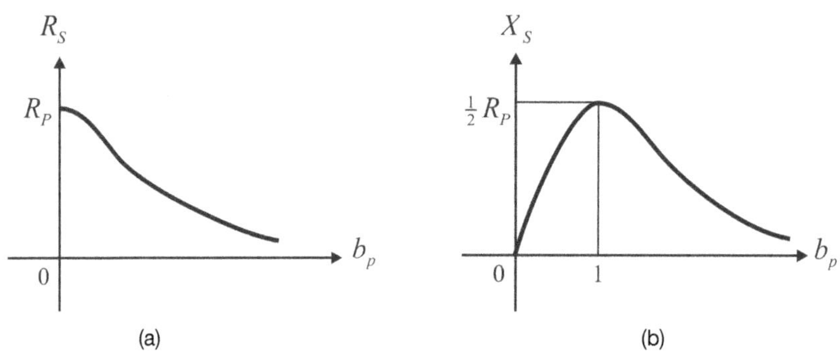

Figure 6.5 Variation of (a) R_S and (b) X_S with respect to b_p

Conversely, a series-to-parallel conversion is used to increase the value of a given resistor. When jX_S is added to a given resistor R_S in series, as shown on the left-hand side of Figure 6.6, this can be transformed to the parallel circuit shown on the right-hand side of Figure 6.6. The real part R_P becomes larger than R_S due to the added jX_S. Therefore, to transform a small resistance to a larger desired value, a series jX_S is added, as shown on the left of the figure. As a result of the transform, an additional parallel reactance jX_P is produced. The sign of X_P is the same as the added X_S.

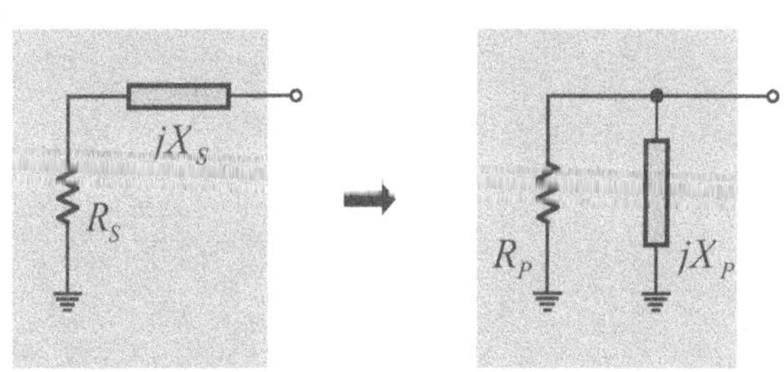

Figure 6.6 Series-to-parallel conversion. Similar to Figure 6.4, the impedance connected in series can be converted into the impedance connected in parallel.

The values of R_P are shown in Equation (6.16).

$$\frac{1}{R_P} = \text{Re}\left(\frac{1}{R_S + jX_S}\right) = \frac{R_S}{R_S^2 + X_S^2} \tag{6.16}$$

Then, R_P is given in Equation (6.17) as

$$R_P = \frac{R_S^2 + X_S^2}{R_S} = R_S + \frac{X_S^2}{R_S} \tag{6.17}$$

Also, since the Q of the two circuits is equal, X_P in Equation (6.18) is

$$X_P = R_S \frac{R_P}{X_S} = \frac{R_S\left(R_S + \frac{X_S^2}{R_S}\right)}{X_S} \tag{6.18}$$

Figure 6.7 shows the plots of Equations (6.17) and (6.18) with respect to X_s.

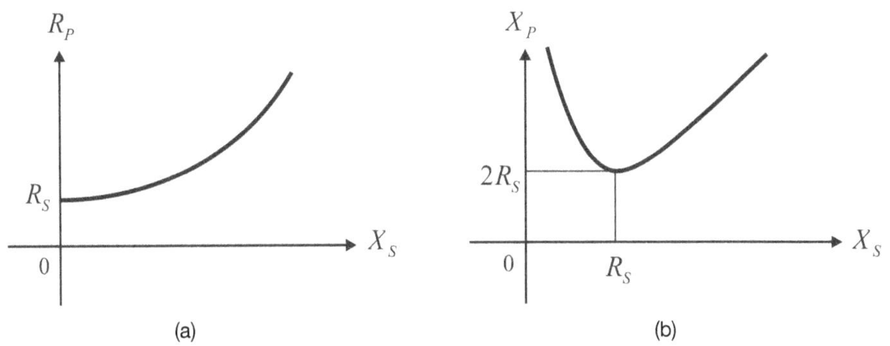

Figure 6.7 (a) R_P and (b) X_P with respect to X_s

In summary, a parallel reactance is added to a resistor to decrease the value of a given resistance while a series reactance is added to increase the value of a given resistance. By inserting the appropriate reactance, the desired value of resistance can be achieved.

6.3.2 L-Type Matching Circuit

Figure 6.8 shows an L-type matching circuit. In that figure, R_L is assumed to be larger than R_S. The inserted L-type matching circuit in the shaded area will transform the resistance R_L into R_S. The parallel reactance X_2 is added to reduce R_L to the value of the source resistor R_S. Therefore, the added parallel jX_2 acts as a narrowband impedance transformer. For an appropriately determined X_2, the series reactance with the same sign as X_2 is produced as a result of the series-to-parallel conversion. The series reactance produced as a result of that the conversion can be tuned out by inserting X_1, which has the opposite sign from that

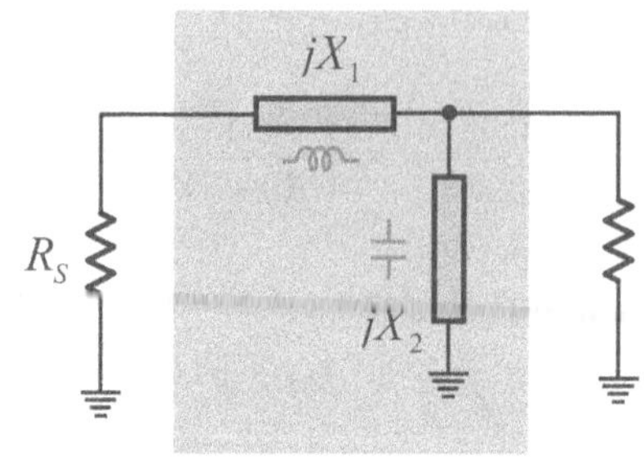

Figure 6.8 L-type matching circuit. X_2 acts as a narrow-band impedance transformer and X_1 tunes out the reactance produced by the parallel-to-series conversion of R_L and jX_2.

series reactance. If X_2 is implemented with a capacitor, an inductor should be used as X_1 and, conversely, if X_2 is implemented with an inductor, a capacitor should be used as X_1. Thus, the impedance matching is achieved for the desired frequency. Note that if a capacitor is used as X_2, the matching circuit behaves as a low-pass filter, while if an inductor is chosen, the matching circuit acts as a high-pass filter.

For matching, from Equation (6.12), the value of X_2 should satisfy Equation (6.19).

$$R_S = \frac{X_2^2 R_L}{R_L^2 + X_2^2} \tag{6.19}$$

When the ratio n is defined as shown in Equation (6.20),

$$n = \frac{R_L}{R_S} \tag{6.20}$$

Equation (6.19) can then be rewritten as Equation (6.21).

$$(n-1)X_2^2 = n^2 R_S^2 \tag{6.21}$$

Suppose that X_2 is implemented using a capacitor. Since X_2 is a negative value, Equation (6.22) is obtained.

$$X_2 = -\frac{nR_S}{\sqrt{n-1}} \tag{6.22}$$

Also, since jX_1 should eliminate the reactance produced after the parallel-to-series conversion, X_1 should be an inductor. Using Equation (6.13), X_1 then can be rewritten as Equation (6.23).

$$X_1 = \frac{X_2 R_L^2}{R_L^2 + X_2^2} = R_S\sqrt{n-1} \tag{6.23}$$

In addition, Q at a given frequency is the ratio of the real and imaginary parts, and Q becomes

$$Q = \frac{X_1}{2R_S} = \frac{1}{2}\sqrt{n-1} = \frac{1}{2}\sqrt{\frac{R_L}{R_S} - 1}$$

The Q is related to the bandwidth: the higher the Q, the narrower the bandwidth becomes. From the Q of the L-type matching circuit above, it can be found that *the bandwidth becomes narrower as the ratio of the source and load resistances becomes larger.*

Example 6.1

For the circuit in Figure 6E.1, design an L-type matching circuit whose source and load resistances are 50 Ω and 100 Ω, respectively.

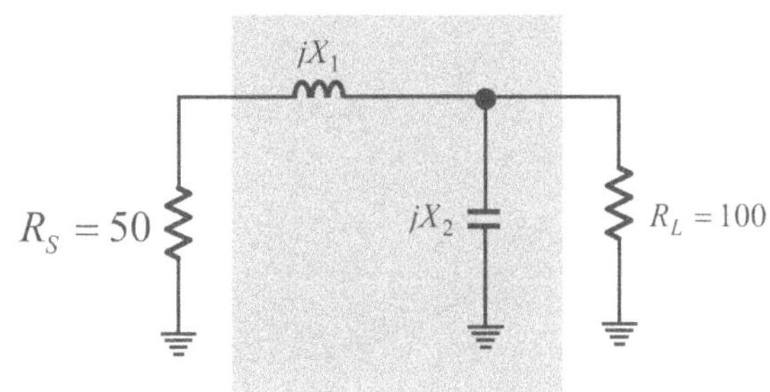

Figure 6E.1 Example of an L-type matching circuit

Solution

From the circuit,

$$n = \frac{R_L}{R_S} = 2$$

Therefore, if X_2 is implemented with a capacitor, then

$$X_2 = -\frac{nR_S}{\sqrt{n-1}} = -2 \times 50 = -100\ \Omega$$

Then, X_1 is

$$X_1 = R_S\sqrt{n-1} = 50\ \Omega$$

The L-type matching circuit shown in Figure 6E.1 can thus be constructed using the values X_1 and X_2.

The solution for Example 6.1 can also be obtained using a Smith chart, which provides a clearer understanding of the operation of each element. Figure 6.9 shows how the impedance matching is achieved through the L-type matching circuit. The value of load resistor R_L appears at point A and the value of source resistor R_S appears at point B. In order to reduce R_L, a parallel reactance jX_2 is added. Since the resulting conductance is constant as $1/R_L$, irrespective of the added jX_2, the impedance of $R_L \| jX_2$ moves along the constant-conductance circle of $1/R_L$, as shown in Figure 6.9. The locus of $R_L \| jX_2$ cuts several constant-resistance circles as B_2 ($jB_2 = 1/jX_2$) increases.

Now the value of X_2 should be determined. Let the real part of impedance $R_L \| jX_2$ be R. Note that the series jX_1 in the L-type matching circuit does not change the value of R.

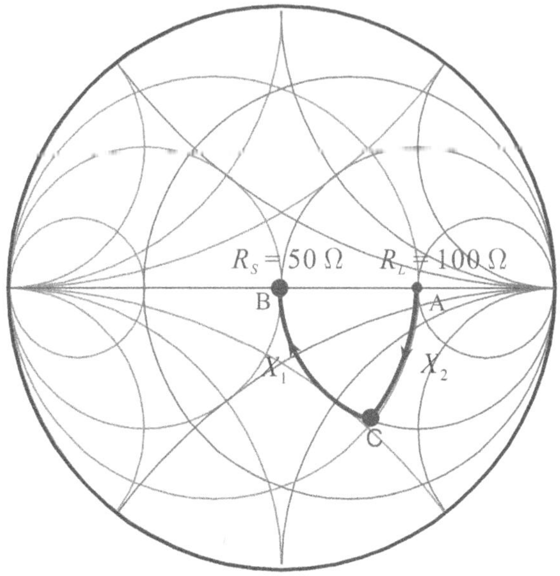

Figure 6.9 L-type matching circuit on a Smith chart. X_2 moves the impedance at point A to point C along the constant-conductance circle. X_1 moves the impedance at point C to point B along the constant-resistance circle.

Thus, as jX_1 changes, the locus moves along the constant-resistance circle given by R. From this reasoning, jX_2 $(=1/jB_2)$ should be chosen for point A to move to point C, which has the real-part impedance of 50 Ω. Reading the corresponding susceptance values of point C from the Smith chart, jB_2 can be obtained. Then, the value of X_2 is $-1/B_2$. For the determined X_2, the added jX_1 in series moves the impedance at point C to the origin. Reading the corresponding reactance of point C, the value of X_1 can be found. Thus, the L-type matching circuit values can be obtained using the Smith chart.

Example 6.2

Using the principle of operation of the L-type matching circuit in the Smith chart shown in Figure 6.9, design the L-type matching circuit of Example 6.1 at a frequency of 1 GHz using ADS.

Solution

The L-type matching circuit is set up in ADS, as shown in Figure 6E.2. To obtain the value of c1, the series inductor is deactivated and shorted. The S-parameter simulation is then performed, sweeping the value of **c**1. The resulting S_{11} is plotted on the Smith chart as shown in Figure 6E.3. The capacitor value that crosses the 50-Ω resistance circle can be found by the moving marker **m**1. The value of **c**1 is found to be about 1.6 pF, as shown in Figure 6E.3.

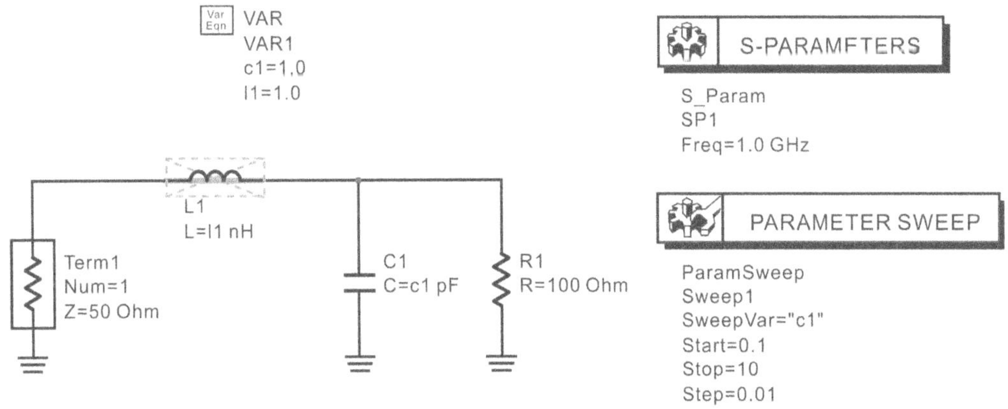

Figure 6E.2 L-section matching circuit design using a Smith chart. The series **L1** is deactivated and shorted. Varying the value of **c1**, the impedance seen from port 1 will move along the constant-conductance circle.

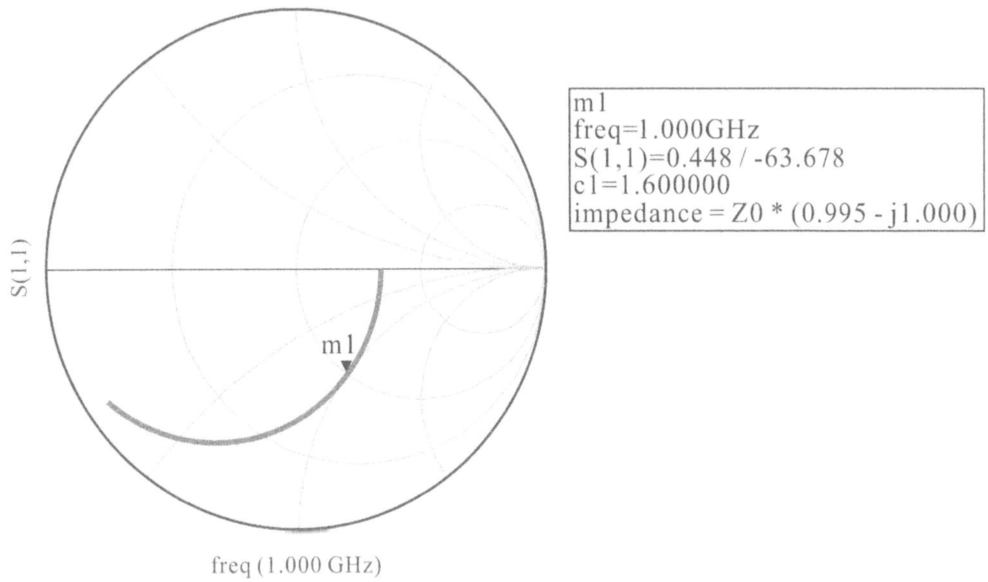

Figure 6E.3 The locus with respect to the capacitance **c1** change. The impedance moves along the constant-conductance circle. The value of **c1** can be determined by reading the point crossing the 50-Ω resistance circle.

The value of this capacitor is fixed and the inductor in Figure 6E.2 is activated. The inductor value l1 is swept from 0.1 nH to 10 nH with a step of 0.05 nH. S_{11} is plotted after simulation and the result is shown in Figure 6E.4. From the plot, we can find l1 = 7.95 nH.

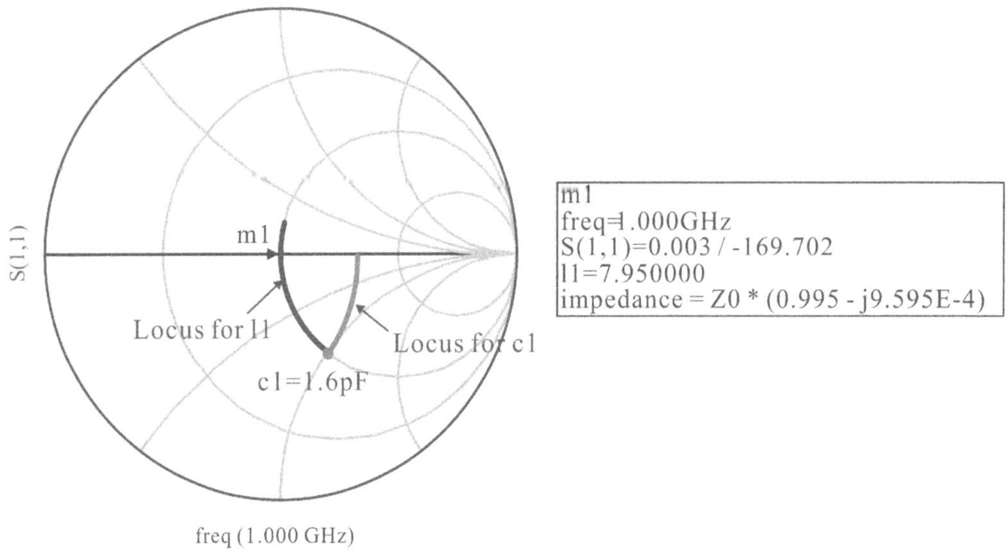

Figure 6E.4 Locus with the inductance as a parameter. For the fixed value of c1 = 1.6 pF, the impedance moves along the 50-Ω resistance circle as the inductance changes. The value of the inductor is found by reading the value at the origin of the Smith chart.

6.3.3 A π-Type Matching Circuit

Figure 6.10 shows a π-type matching circuit. By appropriately dividing one series element into two elements, as shown in Figure 6.10, the π-type matching circuit can be viewed as two L-type matching circuits that face each other. Denoting the impedances looking into the source and load from the reference plane as R_π, the value of R_π turns out to be smaller than both the source and load resistances R_S and R_L. In Figure 6.10, the load impedance R_L is lowered by added reactance jX_2, and this impedance is matched to R_π using the series reactance jX_1. Thus, the value of R_π is less than that of R_L. Using the same reasoning, the value of R_π is smaller than R_S.

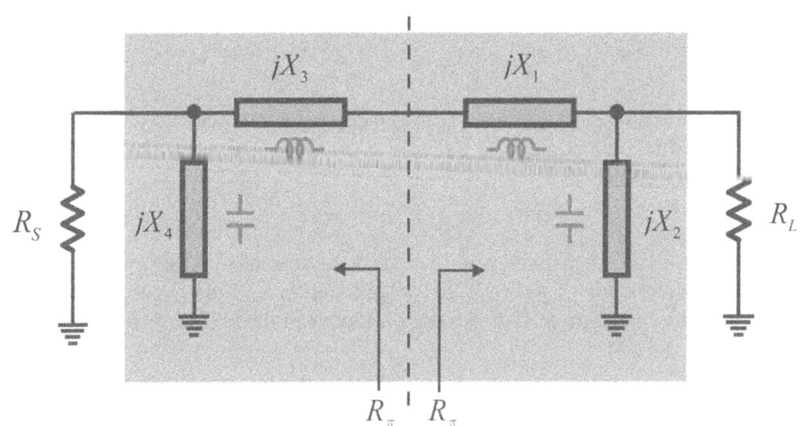

Figure 6.10 A π-type matching circuit, which is composed of two mirroring L-type matching circuits. When the two L-type matching circuits that are to match the chosen R_π are combined, a π-type matching circuit can be designed.

Thus, the value of R_π smaller than the source and load resistances should be selected first to design a π-type matching circuit. Then, the source and load impedances can be matched to the selected value of R_π using two L-type matching circuits. The π-type matching circuit will be complete by combining the resulting series elements of the two L-type circuits.

Choose a value of R_π that satisfies $R_\pi < R_S$, and R_L. The values of jX_4 and jX_2, which act as impedance transformers, can be determined from Equation (6.22). The impedance ratios $n_1$and n_2 are defined as shown in Equation (6.24).

$$n_1 = \frac{R_S}{R_\pi}, \qquad n_2 = \frac{R_L}{R_\pi} \tag{6.24}$$

If jX_4 and jX_2 are implemented with a capacitor, the values for X_4 and X_2 can be obtained as shown in Equations (6.25a) and (6.25b).

$$X_4 = -\frac{n_1 R_\pi}{\sqrt{n_1 - 1}} \tag{6.25a}$$

$$X_2 = -\frac{n_2 R_\pi}{\sqrt{n_2 - 1}} \tag{6.25b}$$

For the determined X_4 and X_2, the values of the series elements X_3 and X_1 can be obtained using Equations (6.26a) and (6.26b).

$$X_3 = \omega L_3 = R_\pi \sqrt{n_1 - 1} \tag{6.26a}$$

$$X_1 = \omega L_1 = R_\pi \sqrt{n_2 - 1} \tag{6.26b}$$

The Q of the π-type matching circuit is given by

$$Q \cong \frac{X_1 + X_3}{2R_\pi} = \frac{1}{2}\left(\sqrt{\frac{R_S}{R_\pi} - 1} + \sqrt{\frac{R_L}{R_\pi} - 1}\right) \tag{6.27}$$

The Q given by Equation (6.27) is typically larger than that of the L-type matching circuit and, as a result, the matching bandwidth is reduced.

Figure 6.11 shows how the π-type matching circuit works in a Smith chart. The load $R_L = 100\ \Omega$ and the source resistor $R_S = 50\ \Omega$ are selected and appear at points A and B, respectively. The value of R_π is set to 25 Ω. The added jB_2 (=$1/jX_2$) moves the load R_L (point A) along the constant-conductance circle, as shown in Figure 6.11. Here, the jX_2 is selected to be a capacitor. The value of jB_2 should be selected such that the load resistance R_L moves up to point C, which yields the series resistance value of R_π. After fixing jB_2, adding jX_1 in series makes the locus move along the constant R_π resistance circle. Note that jX_1 is an inductor. Thus, point C moves along the direction of increasing reactance shown in Figure 6.11. Therefore, reading the added reactance from the Smith chart, which gives point D, will give the precise value of X_1.

Adding jX_3 in series moves the resulting impedance at point D further along the constant R_π resistance circle in the direction of increasing reactance. Considering that $R_S \| jX_4$ lies in the constant-conductance circle $1/R_S$, the added jX_3 should move the resistor R_π

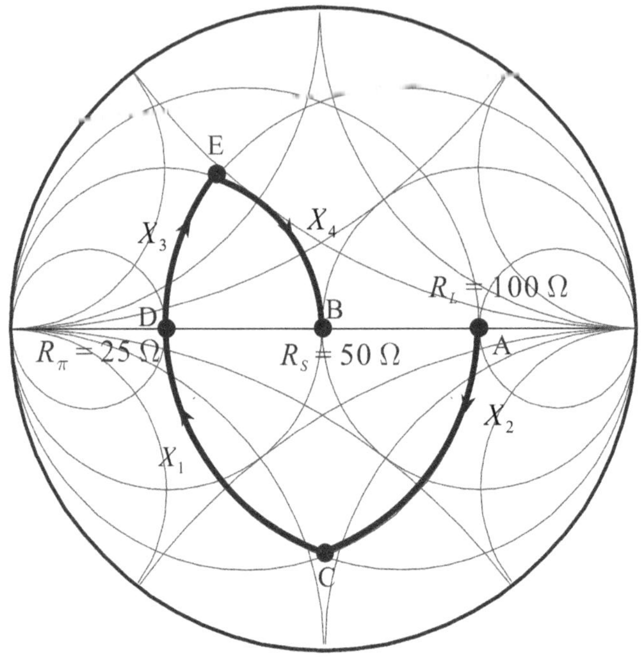

Figure 6.11 A π-type matching circuit viewed in a Smith chart. X_2 in Figure 6.10 moves R_L to point C; then the series X_1 moves point C to point D (R_π). The series X_3 moves point D to point E and the shunt X_4 finally moves point E to point B and matching is achieved.

(point D) up to point E. Then, jX_4 added in parallel can move point E up to the desired point B. This way, the matching from points A to B can be achieved.

6.3.4 T-Type Matching Circuit

Figure 6.12 shows a T-type matching circuit. As in the π-type matching circuit, by appropriately dividing one shunt element into two shunt elements (jX_2 and jX_4), as shown in the figure, the circuit can then be viewed as two L-type sections facing each other,

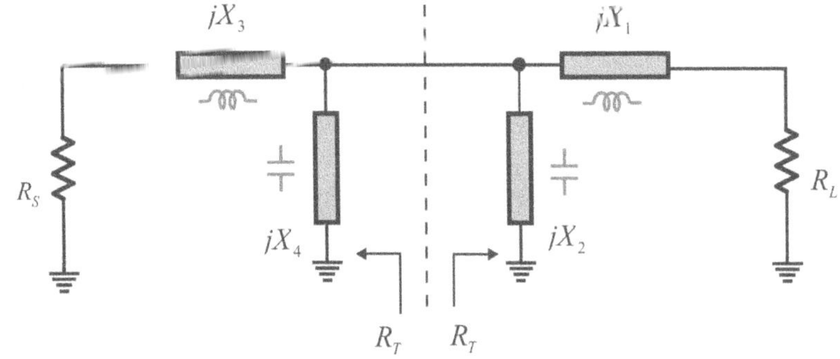

Figure 6.12 A T-type matching circuit that is composed of two mirroring L-type matching circuits; however, the location of shunt elements differs from that of a π-type matching circuit. As a result, R_T should be chosen to satisfy $R_T > R_L$ and R_S.

The impedances looking into the source and load from the reference plane are denoted as R_T. Using similar reasoning, the value of R_T is thus greater than both the source and load resistances, as explained in the π-type matching circuit. For a design with a T-type matching circuit, a value of R_T greater than the source and load resistances should be selected first, and the source and load can be matched using two L-type matching circuits. The design of the T-type matching circuit will be completed by combining the two parallel elements X_4 and X_2 that constitute the two L-type matching circuits.

Let $R_T > R_S$, R_L be selected. The impedance ratios n_1 and n_2 are defined in Equation (6.28) as

$$n_1 = \frac{R_T}{R_S}, \quad n_2 = \frac{R_T}{R_L} \tag{6.28}$$

Suppose that jX_4 and jX_2, which act as impedance transformers in the T–type matching circuit, are implemented with capacitors. Then, the following values can be obtained with Equations (6.29a) and (6.29b) for X_4 and X_2, respectively, as

$$X_4 = -\frac{n_1 R_S}{\sqrt{n_1 - 1}} \tag{6.29a}$$

$$X_2 = -\frac{n_2 R_L}{\sqrt{n_2 - 1}} \tag{6.29b}$$

As in the π-type matching circuit, X_3 and X_1 can be similarly determined with Equations (6.30a) and (6.30b) as

$$X_3 = \omega L_3 = R_S\sqrt{n_1 - 1} \tag{6.30a}$$

$$X_1 = \omega L_1 = R_L\sqrt{n_2 - 1} \tag{6.30b}$$

The T-type matching circuit can be implemented using the values given by Equations (6.29a) and (6.29b) and (6.30a) and (6.30b). The Q of the T-type section, similar to the p-type section, can be determined as shown in Equation (6.31).

$$Q \cong \frac{B}{G} = \frac{\left|\frac{1}{X_2} + \frac{1}{X_4}\right|}{\frac{2}{R_T}} = \frac{1}{2}\left(\sqrt{\frac{R_T}{R_S} - 1} + \sqrt{\frac{R_T}{R_L} - 1}\right) \tag{6.31}$$

Note that Q in Equation (6.31) is also larger than the Q of the L-type matching circuit and the matching bandwidth is also narrower than that of the L-type section. The larger R_T is set, the narrower the matching bandwidth becomes. The π- or T-type matching circuits are often used where a narrow matching bandwidth is required, such as a tuner.

The behavior of each element of the T-type matching circuit can be found using the Smith chart shown in Figure 6.13. The load $R_L = 50\ \Omega$ and the source $R_S = 25\ \Omega$ are selected and appear at points A and E, respectively. The value of R_T in Figure 6.13 is set to 100 Ω.

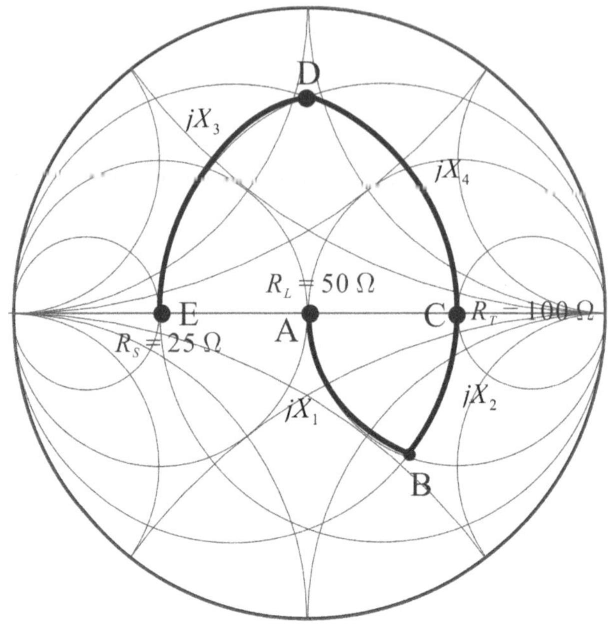

Figure 6.13 T-type matching circuit viewed in a Smith chart. The behaviors of the series and shunt elements are similar to those in the π-type matching circuit.

The added jX_3 in series moves the source impedance (point E) along the constant-resistance circle to point D. Here, jX_3 is selected as an inductor. Since an inductor is added, it moves along the direction of increasing reactance, as shown in the figure.

Point D yields the conductance value of R_T. Then, the added jB_4 (=$1/jX_4$) in parallel moves point D to point C in the direction of increasing susceptance because B_4 is a capacitor. Reading the value of the added susceptance will give the precise value of B_4.

Adding jX_2 in parallel moves point C further along the constant R_T conductance circle in the direction of increasing susceptance. Note that the impedance $R_L + jX_1$ has the constant resistance R_L irrespective of X_1. Thus, the value of X_2 should be increased to give the resistance value of R_L that corresponds to point B. Finally, adding X_1 in series moves the impedance in the direction of increasing reactance, which moves point B up to point A along the constant-resistance circle. As a result, the matching from points E to A can be determined.

6.3.5 Double L-Type Matching Circuit

A double L-type matching circuit is shown in Figure 6.14. This circuit is formed by connecting two cascading L-type matching circuits. The impedance R_D is defined as the impedance looking into the load side of Figure 6.14. Thus, R_D is smaller than the load resistance R_L and bigger than the source resistance R_S. Therefore, to determine the double-L matching circuit, the value of R_D should be selected such that $R_L > R_D > R_S$ and the double L-type section is completed by designing the two L-type matching circuits separately using the value of R_D.

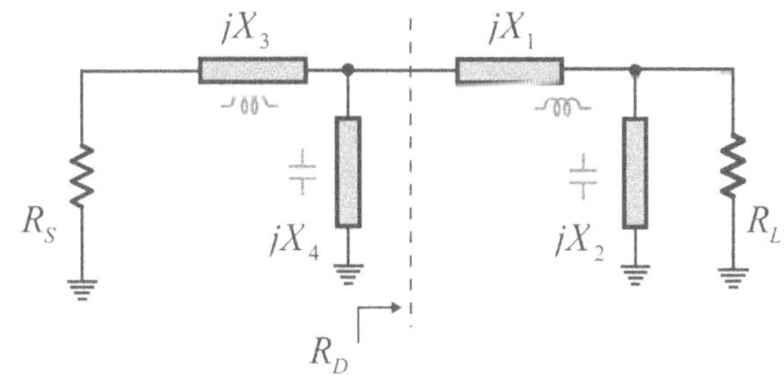

Figure 6.14 Double L-type matching circuit that is composed of two cascading L-type matching circuits. The value R_D should be chosen to satisfy $R_L > R_D > R_S$.

The exact Q of the double L-type matching circuit is complex. However, by multiplying the Q of the two L-type matching circuits, the approximate value of Q can be obtained as

$$
\begin{aligned}
Q &\cong \frac{X_1}{R_S}\frac{X_3}{R_D} = \sqrt{\frac{R_L}{R_D} - 1}\sqrt{\frac{R_D}{R_S} - 1} \\
&= \sqrt{\frac{R_L}{R_S} + 1 - \left(\frac{R_L}{R_D} + \frac{R_D}{R_S}\right)} \le \sqrt{\frac{R_L}{R_S} - 1}
\end{aligned}
\tag{6.32}
$$

We see from Equation (6.32) that the Q of the double L-type matching circuit is lower than that of the L-type matching circuit. Thus, the bandwidth of the double L-type is found to be wider than that of the L-type. Usually, the objective of broadband matching can be achieved by cascading the connection of the L-type matching circuits. Figure 6.15 shows a Smith chart of the double L-type matching circuit. The load R_L is set to 100 Ω (point A) and the source R_S is set to 25 Ω (point E). The value of R_D for the double L-type matching circuit is set to 50 ohm.

6.3.6 Matching Circuit Design for a General Source Impedance

All the source and load impedances described so far had only the real part impedances. However, we will now consider the general case of the source or load of complex impedances.

In the case of a complex-valued source or load, we first consider the real-valued source and load that include only the real part of the complex impedances. Then, a matching circuit, such as an L-type or double L-type matching circuit, can be designed for the real-valued source and load using the previously described method. As shown in Figure 6.16, the design of the matching circuit includes the imaginary parts of the source or load. The desired matching circuit for the complex-valued source or load can be obtained after its imaginary part contributions are removed from the designed matching circuit. This design method is quite general and applicable for matching the load and source with complex impedances.

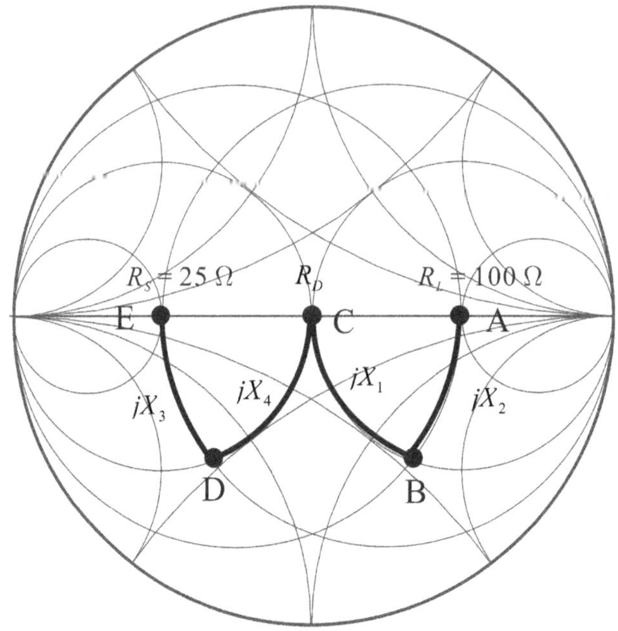

Figure 6.15 Double L-type matching circuit viewed in a Smith chart. The behaviors of X_1 to X_4 are similar to those of two separate L-type matching circuits. X_1 and X_2 of the first L-type matching circuit match R_L to R_D while X_3 and X_4 match R_D to R_S.

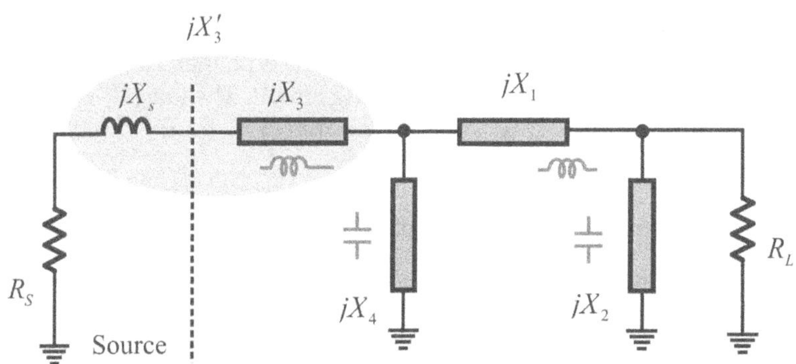

Figure 6.16 Matching circuit configuration for a complex impedance source

Example 6.3

Design an L-type matching circuit for a circuit having the source and load impedances given by 50 + j25 Ω and 100 Ω, respectively.

Solution

The imaginary part of the source impedance is set to 0 for the time being. Let X_2 and X_1 of the L-type matching circuit be implemented with a capacitor and inductor, respectively. The following values were obtained from Example 6.1:

$$X_2 = -100\ \Omega,\ X_1 = 50\ \Omega$$

Since the imaginary part of the source impedance is 25 Ω, the L-type matching circuit for the given source and load impedances should be:

$$X_2 = -100\ \Omega,\ X_1 = 25\ \Omega$$

The designed L-type matching circuit is shown in the shaded rectangle of Figure 6E.5.

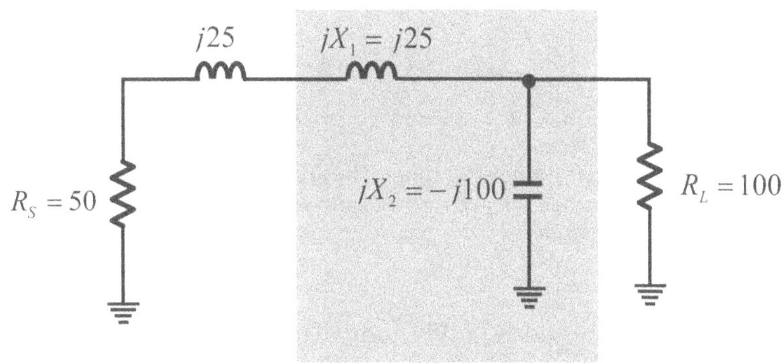

Figure 6E.5 An example of an L-type matching circuit that matches the impedance $R_L = 100\ \Omega$ to $Z_S = 50 + j25\ \Omega$.

The method presented can be applicable to the design of a broadband matching circuit. The reactive elements of the source and load are removed, and a filter that can match the real valued source and load is designed first, after which the reactance of the load and source is then removed from the matching circuit. It is worth noting that the configuration of the matching circuit must be selected in such a way as to have the same sign as that of the reactance of the load and source. The reactance of the designed matching circuit without the consideration of the source and load reactive elements must be sufficiently large to remain positive even after removing the reactance contributions from the source and load. Because of this, the bandwidth of the matching circuit for such a source and load with reactive elements is limited to some extent.

6.4 TRANSMISSION-LINE MATCHING CIRCUITS

6.4.1 Single-Stub Tuner

Figure 6.17 shows a single-stub tuner. The short- (or in some cases, open-) circuited transmission line connected in shunt is called a *stub*. It is called single-stub tuner because it only has one stub. Arbitrary characteristic impedance for the transmission lines is possible for a single-stub tuner but for convenience, all the characteristic impedances are made to be the same as $Z_0 = 50\ \Omega$. The parameters that are allowed to change are thus the length of the stub and the length of the transmission line. We will explain the matching method in this situation.

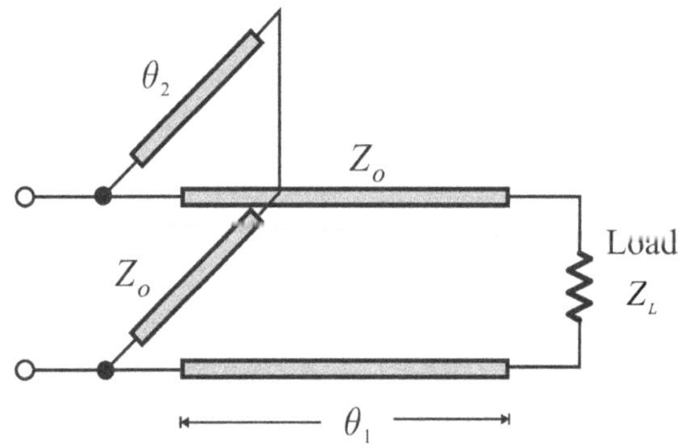

Figure 6.17 Circuit of single-stub tuner. The characteristic impedances Z_o of the two transmission lines are fixed while the electrical lengths θ_1 and θ_2 are varied to match arbitrary chosen Z_L to Z_o.

First, an arbitrary load (Z_L) is normalized by the characteristic impedance Z_o. Using the value of the reflection coefficient Γ_L that corresponds to the normalized load impedance, the load can be represented on a Smith chart as shown in Figure 6.18. Then, when the length of the transmission line is adjusted (θ_1 in Figure 6.17), the reflection coefficient looking from just before the stub will be $\Gamma(\theta_1) = \Gamma_L e^{-2j\theta_1}$, which is a circle rotating clockwise with the origin as the center. When θ_1 is varied, the locus will pass through two points of the constant-conductance circle $G = 1$ (unit conductance circle). Taking into consideration that the stub has a negative susceptance (as an inductor) for a length shorter than a quarter wavelength, the point A in Figure 6.18 is chosen. Let the susceptance of point A be B_d, the length of the stub is then determined as expressed in Equation (6.33).

$$B_d = \frac{Y_o}{\tan(\theta_2)} \tag{6.33}$$

Given that the stub length is allowed up to a half wavelength, both points A and B in Figure 6.18 are possible matching points. Thus, arbitrary load impedance can be matched to 50 Ω using the single-stub tuner as shown in Figure 6.18. The previous L-type matching circuit resembles the single-stub tuner. When the impedance of the transmission line connected to the load is high, it approximates the behavior of the inductor in the L-type matching circuit. Also, if the open stub is used instead of the shorted stub, the stub acts as a shunt capacitor. Since the single-stub tuner can match arbitrary load impedance, it is frequently used in matching. For example, by moving the chip capacitor along the 50-Ω microstrip, matching to the unknown load impedance can be achieved.

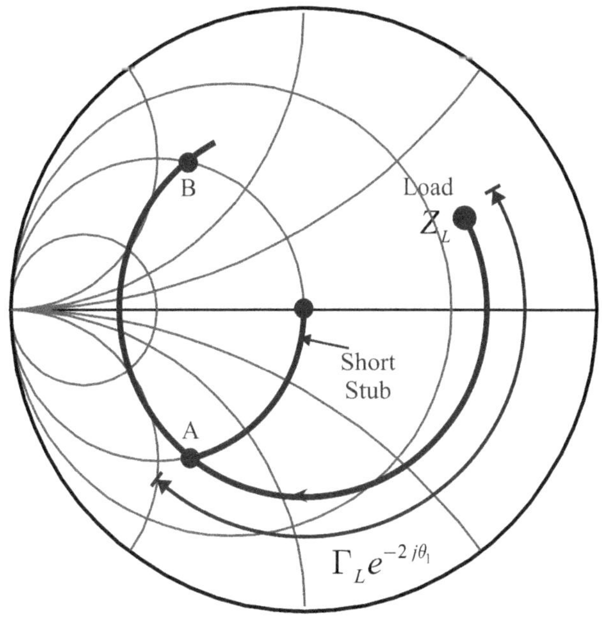

Figure 6.18 Single-stub tuner operation. Varying the electrical length θ_1 moves the load impedance Z_L along the circle centered at the origin with a radius of $|\Gamma_L|$. The circle will intersect 50-Ω conductance circle at two points, A and B. It is possible to match the two points to 50 Ω. For the chosen point A, the stub moves it to the origin and matching is achieved.

Example 6.4

Design the single-stub matching circuit using ADS, which matches a load impedance $Z = 100 + j100\ \Omega$ to 50 Ω at a frequency of 1 GHz. Use the single-stub tuner operation in the Smith chart explained in Figure 6.18.

Solution

The simulation circuit is set up as shown in Figure 6E.6.

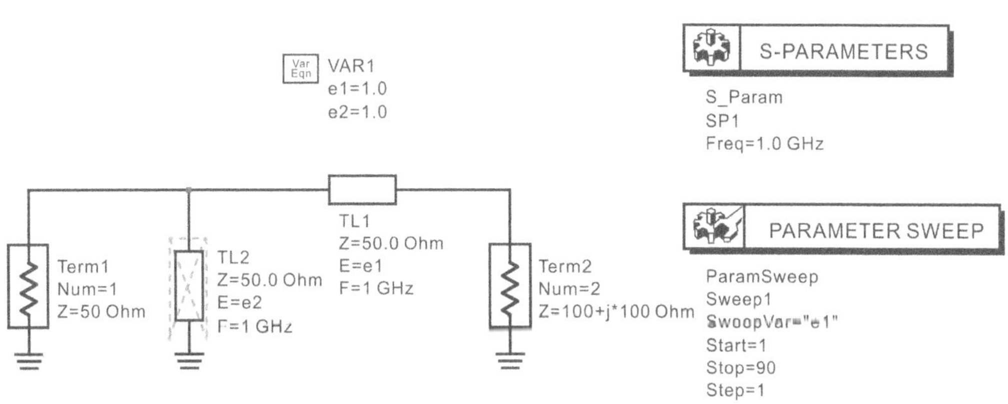

Figure 6E.6 Setup for a single-stub matching circuit simulation. First, the stub is deactivated and then the electrical length **e**1 that crosses the 50-Ω conductance circle is found by sweeping the value of **e**1.

The transmission-line stub is deactivated and the simulation is performed by varying the electrical length **e**1 of the transmission line **TL**1. The simulated S_{11} is shown in Figure 6E.7 where the S_{11} is plotted on the admittance-based Smith chart. The admittance-based Smith chart can be set using the *Plot Option* of the Smith chart in ADS. We can see that the result **e**1 = 79° intersects with the 50-Ω constant-conductance circle, and the electric length of the transmission line **TL**1 is determined as **e**1 = 79°.

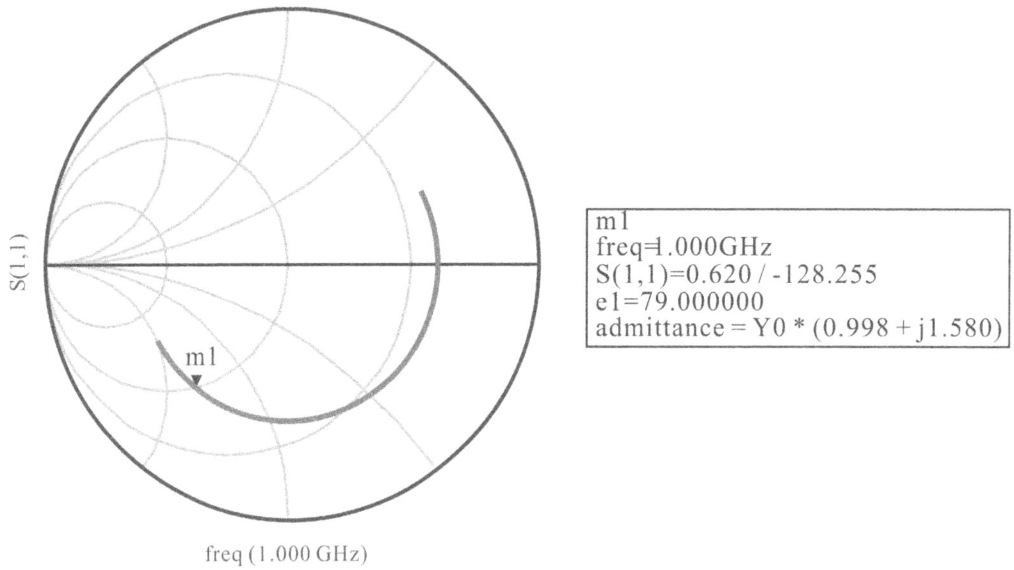

Figure 6E.7 Simulation result as electrical length **e**1 is varied. The value of **m**1 gives the impedance at the 50 Ω conductance circle.

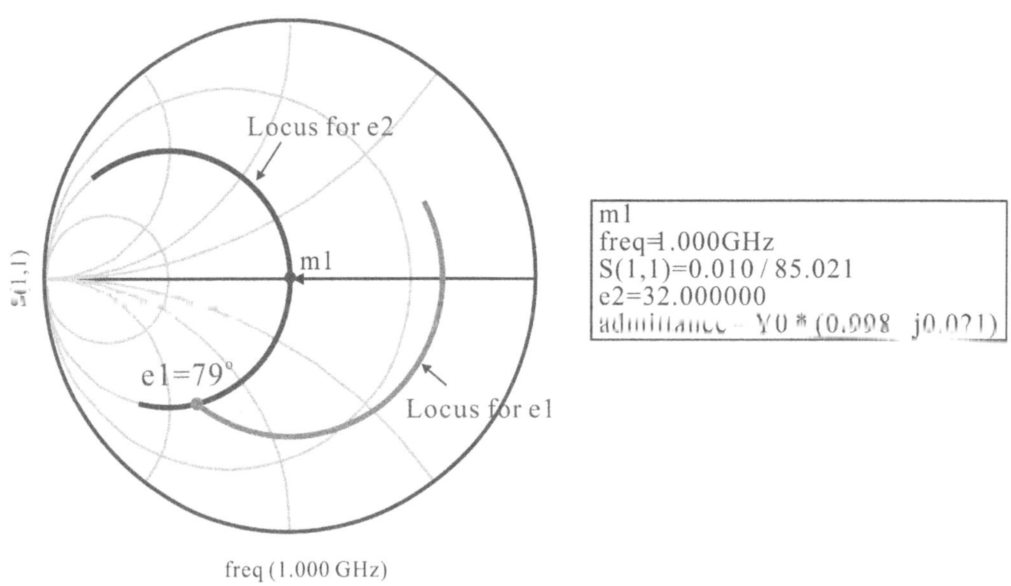

Figure 6E.8 Simulation result as electrical length **e2** is varied. Fixing **e**1 = 79°, the electrical length **e**2 of the stub is swept. The value of **e**2 can be found by reading the value of **e**2. The locus for **e**1 is the previously simulated trace.

Next, fixing **e1** to 79°, the transmission-line stub **TL2** is activated. The electrical length **e2** of **TL2** is varied from 10° to 135°. The simulated S_{11} is shown in Figure 6E.8 together with that of the previous simulation. We can observe in the figure that the load is matched to the 50-Ω source at e2 = 32°. Therefore, matching to the 50-Ω source is achieved when **e1** = 79° and **e2** = 32°.

Note that all the load impedances in the Smith chart can be matched to 50 Ω using the single-stub tuner matching circuit. However, in some applications the electrical length θ_1 of the transmission line cannot be allowed to vary, which can be a problems because the single-stub tuner matching circuit is hard to implement using coaxial lines. The double- or triple-stub tuners that employ a fixed-length transmission line can be used in such cases. The *double-stub tuner matching circuit* uses a single transmission line and two stubs.[1] The transmission-line length used for the double-stub tuner is set normally to a one-eighth wavelength at the matching frequency. This structure can be easily implemented using coaxial lines and is commercially available as a component. The operation of the double-stub tuner matching circuit is considered in problem 6.7 at the end of this chapter. The disadvantage of the double-stub tuner matching circuit is that it does not match all the impedances in the Smith chart, which can be overcome using a *triple-stub tuner* that can provide matching for all the impedances in the Smith chart; refer to reference 5 at the end of this chapter for more information.

6.4.2 Impedance Inverter

In the single-stub tuner, the load is matched to the source using transmission lines with fixed characteristic impedance Z_o. However, matching is possible using a quarter-wavelength transmission line, as explained in Chapter 3. Figure 6.19 shows a quarter-wavelength transmission line connected to a load. Matching can be achieved by adjusting only the characteristic impedance.

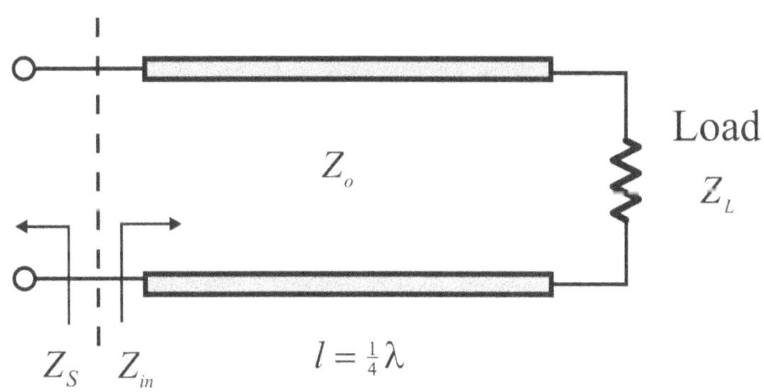

Figure 6.19 Matching circuits using a quarter-wavelength transmission line

1. Maury microwave stub tuners, www.maurymw.com.

Equation (6.34) describes the input impedance looking into the transmission line Z_{in}.

$$Z_{in} = \frac{Z_o^2}{Z_L} \tag{6.34}$$

Therefore, for a source with impedance Z_S, the load and source can be matched when the characteristic impedance Z_o of a quarter-wavelength transmission line is determined as

$$Z_o = \sqrt{Z_S Z_L}$$

In the case of a complex load impedance, a shunt stub is added and the imaginary part of Z_L can be tuned out. Then, the problem is the same as the real load problem previously explained. A single stage of an impedance inverter may not yield a satisfactory VSWR in the matching bandwidth. For a broader bandwidth, the number of stages is often increased and a rather complex filter design method is often employed. Consult reference 6 at the end of this chapter for more information.

6.5 SUMMARY

- Maximum power delivery occurs when the source and load impedances are conjugate matched and the available power from the source is delivered to the load.
- When the source and load impedances are conjugate matched using a lossless two-port network, a conjugate match at the source plane results in an automatic conjugate match at the load plane.
- Discrete matching circuits such as L-, π-, T-, and double L-type matching circuits are explained. The L-type matching circuit is the basic matching circuit from which the other matching circuits can be derived.
- The bandwidth of the π- and T-type matching circuits is narrower than that of the L-type matching circuit, while the bandwidth of the double L-type matching circuit is broader than the L-type matching circuit. The π- and T-type matching circuits are generally used for tuners.
- There are single-stub, double-stub, and triple-stub tuner matching circuits in transmission-line matching circuits. Matching circuits in single-stub and triple-stub tuners can match all the impedances in the Smith chart.

REFERENCES

1. R. G. Brown, R. A. Sharpe, W. L. Hughes, and R. E. Post, *Lines, Waves, and Antennas*, 2nd ed. Hoboken, NJ: John Wiley & Sons, Inc., 1973.
2. K. K. Clarke and D. T. Hess, *Communication Circuits: Analysis and Design*, Reading, MA: Addison-Wesley Publishing Company, 1971.
3. R. Soares, J. Graffeuil, and J. Obregon, *Applications of GaAs MESFETs*, Dedham, MA: Artech House, Inc., 1983.

4. T. T. Ha, *Solid-State Microwave Amplifier Design*, Hoboken, NJ: John Wiley & Sons, Inc., 1981.
5. R. E. Collin, *Foundations for Microwave Engineering*, New York: McGraw Hill, 1966.
6. G. Matthaei, L. Youg, and E. M. T. Jones, *Microwave Filters, Impedance-Matching Networks, and Structures*, Dedham, MA: Artech House, Inc., 1964.

PROBLEMS

6.1 The values of capacitors and inductors in matching circuits such as L-, π-, T-, and double L-type sections can be determined using a Smith chart. For 50- and 100-Ω source and load impedances, design the following matching circuits:
(1) For an L-type, determine the shunt susceptance and series reactance using only a Smith chart.
(2) For a π-type, using $R_\pi = 25\ \Omega$, determine the element values in susceptance or reactance using only a Smith chart.
(3) Design double L-type matching circuit with $R_D = (50 \times 100)^{1/2}\ \Omega$.

6.2 Prove that the Q of the π-type matching circuit in problem 6.1.2 is larger than that of the L-type matching circuit. This can be shown by proving that the value given by Equation (6.27) is always greater than the value given by the following equation:

$$Q = \frac{1}{2}\sqrt{\frac{R_L}{R_S} - 1}$$

6.3 In the circuit shown in Figure 6P.1, the source and load impedances are 9 and 1 kΩ. Determine the values of the matching circuit, which matches at $\omega_o = 10^8$ [rad/sec]. The value of C_2 is set at 100 pF. Determine the other element values.

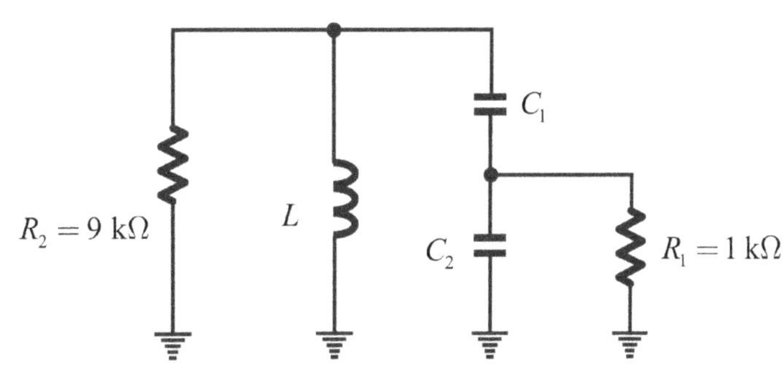

Figure 6P.1 High Q matching circuit

6.4 Prove that the Q of the double L-type matching circuit given by Equation (6.32) is less than the Q of the L-type matching circuit.

6.5 **(ADS Problem)** Design a double L-shape section matching circuit to match 100 Ω to 50 Ω for $R_D = 60\ \Omega$, 70.7 Ω, and 80 Ω. In the design, what value of R_D yields the widest matching bandwidth? Prove your answer using Equation (6.32).

6.6 Design an L-type section matching circuit for the source and load impedances shown in Figure 6P.2.

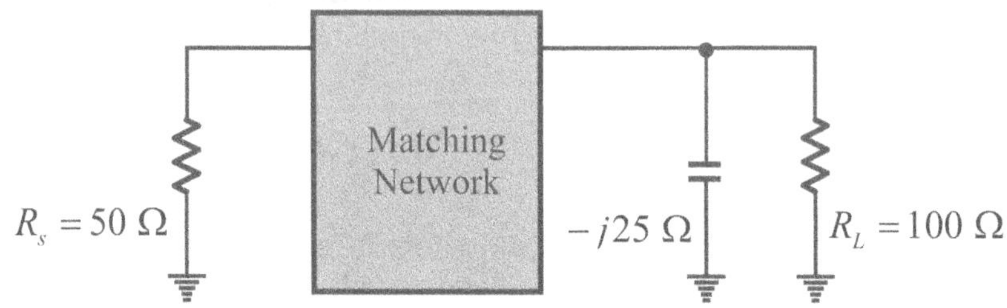

Figure 6P.2 Problem 6.6 matching circuit

6.7 (ADS Problem) Using an open-circuited single-stub tuner, design a matching circuit to match a 100 Ω to 50 Ω. Assume the impedance of all the transmission lines to be 50 Ω and the electrical length of open stub is shorter than a one-quarter wavelength.

6.8 (ADS Problem) Using a short-circuited double-stub tuner shown in Figure 6P.3, determine **theta1** and **theta2** in the matching circuit to match a 100 Ω to 50 Ω at a frequency of 1GHz. The impedance of all the transmission lines is 50 Ω.

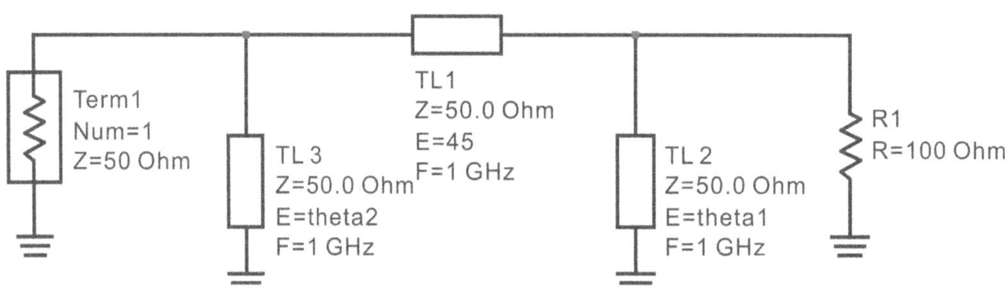

Figure 6P.3 Double-stub tuner matching circuit

CHAPTER

7

Simulation and Layout

CHAPTER OUTLINE

7.1 SIMULATION IN ADS

Circuit simulation in ADS is basically performed using a window framework such as the *Schematic/Layout window* and the *Display window* shown in Figure 7.1.[1] Picking up a component from the Component palette, placing the component, and wiring that component for a circuit can be carried out in the Schematic/Layout window. After wiring, an appropriate simulation-control component, such as DC, AC, transient simulation, etc., which defines the circuit-simulation method, is placed in the same window. Then, the simulation for a wired circuit can be performed and the simulation results are stored in a *dataset*. The results in the dataset can be opened and further processed in the display window, which provides various plots and listings for viewing and examining the results. Thus, the Schematic/Layout and Display windows provide basic tools for circuit simulation in ADS. The flow of the simulation is shown in Figure 7.1.

The work that a user performs in the Schematic/Layout window is first saved to a specific file with the pre-assigned file extension **.dsn* (design file) and the file is stored in a pre-assigned directory named *networks*. Next, the simulation results are also saved to a file with the pre-assigned file extension **.ds* (dataset) and the file is stored in the pre-assigned directory named *data*. The simulation results can be shown in the display window in

1. Agilent Technologies, Advanced Design System 2009U1, 2009.

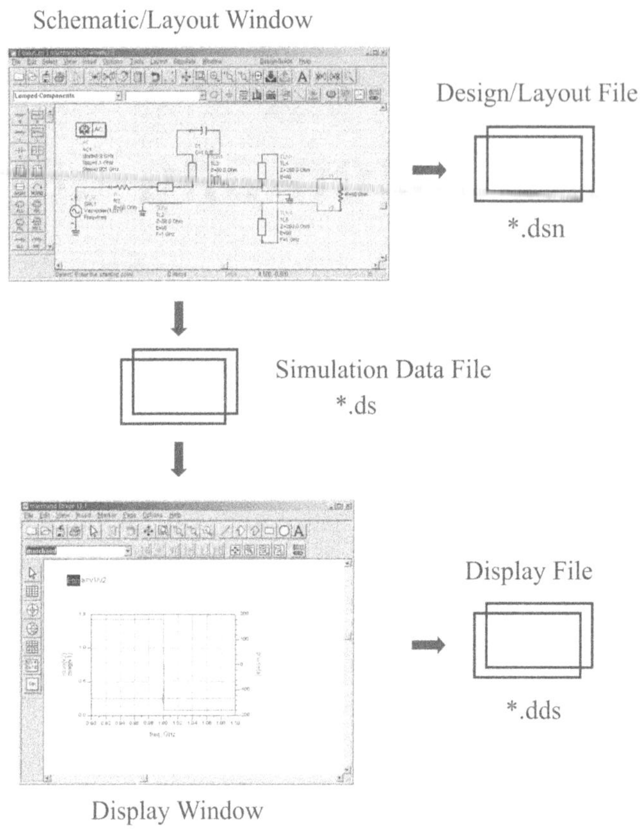

Figure 7.1 Simulation flow in ADS. Simulation is performed in the Schematic/Layout window and the work is stored as a file having extension *.dsn in the network directory. The simulated data is stored in a file with the extension *.ds in the data directory. The simulated data file can be viewed and further processed in the display window and the work is stored with the file extension *.dds in the project directory.

various ways to accommodate the user's objectives, as shown in Figure 7.1. Finally, the work in the display window is also stored in a file with the pre-assigned file extension **.dds* (display file) and the file is stored in the working project directory. The default directory is a user-named *project* directory. Note that the results thus stored can be opened, viewed, and edited in the specific windows such as Schematic/Layout and display.

Therefore, it is critical to know how to perform tasks in the Schematic/Layout window to be able to simulate a circuit. It is also necessary to know how to perform tasks in the display window in order to view stored dataset files generated from simulations.

A given task may be composed of various circuit-simulations. For example, if a circuit is composed of several blocks, the blocks are simulated separately as a first step. Based on those separate simulations, the entire circuit is again simulated as one large circuit. In performing each block simulation, the resulting files (mentioned above as *.dsn, *.ds, and *.dds, etc.) are stored in a pre-assigned directory such as network, data, and a default project directory. The combined directories form a folder or directory named a *project*. The name of the project directory is specified in ADS by the user, and *_prj directory extension is created in the default working directory. Most users perform multiple tasks and, as a result, create multiple projects. A *file manager* that organizes and edits the multiple projects is the *Main window*, which is shown in Figure 7.2. In Figure 7.2, many project directories with the extension _prj are in the directory named MW_Ckts. In the Main window, many file-management functions can be performed such as creating and deleting a project, as well as adding or removing other projects from a project directory.

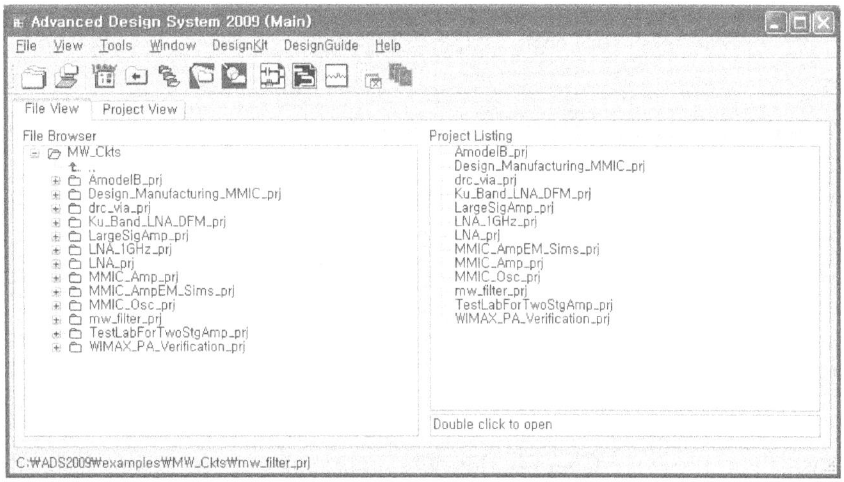

Figure 7.2 Main window and projects. The Main window shows current projects with the file extension *_prj. The Main window functions as a file manager.

When the project named mw_filter_prj is opened, the directories in the project are shown in Figure 7.3. There are data and network directories among others. As mentioned earlier, the design file (*.dsn), dataset (*.ds), and display data file (*.dds) are created as a result of a circuit simulation. The files are stored automatically in the corresponding pre-assigned directories in a named current working project directory; (1) design files are stored under networks, (2) the simulation result files are stored under data, and (3) the plotted results in the display window are saved directly under the named project directory.

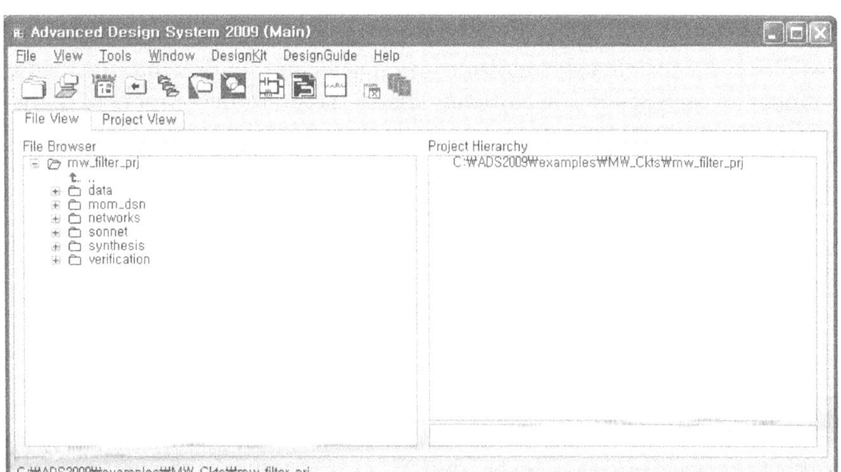

Figure 7.3 Directories autogenerated in the Project directory. The simulated data are stored in the data directory, the work for the simulation circuits is stored in the network directory, and the work done in the Display window is saved in the *_prj directory.

The autogenerated files can also be viewed in the Window environment without the aid of ADS. In that environment, even specially created ADS files are displayed. The files can be copied, deleted, and moved in the Window environment but these types of actions by a user could change or remove files protected for ADS operation. Thus, file management tasks such as copy, move, delete, etc., should be done using the available ADS file manager so as not to destroy an entire project.

7.2 CIRCUIT SIMULATIONS

7.2.1 Classification of Circuit Simulations

The circuit simulation of a wired circuit should be specified to be simulated in ADS. The circuit simulations are classified according to the sources used in the circuit, such as voltage and current. Depending on the current and voltage sources used, different circuit-simulation methods and procedures are required to compute the voltage and current responses in the circuit.

Circuit simulations in ADS are categorized into *DC, AC, S-parameter, large-signal S-parameter, transient, harmonic balance*, and *envelope simulations.* S-parameter simulation is basically a kind of AC simulation while large-signal S-parameter simulation, in a broad sense, belongs to the harmonic balance category. Envelope simulation is also generally considered to be a modification of harmonic balance simulation. To summarize, the basic simulations thus include *DC, AC, transient,* and *harmonic balance simulations.* In the sections that follow, we will briefly discuss the principles of circuit analysis for the four kinds of simulation.

7.2.2 DC Simulation

In DC simulation, DC sources are applied to a circuit and the responses are computed for those sources. All outputs will be DC voltages or DC currents. In ADS, DC simulation is performed by the *DC simulation control component*, shown in Figure 7.4, which can be used for DC simulation of a linear circuit containing resistors, capacitors, inductors, and other linear passive devices, as well as for a nonlinear circuit containing nonlinear devices such as diodes and transistors without any distinction. The DC simulation control component is usually set to the default condition and it is recommended that the reader refer to the ADS manual for instructions on adjusting the DC simulation component to the desired accuracy in nonlinear circuit analysis.

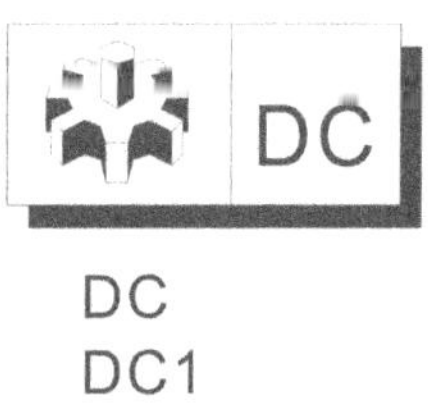

Figure 7.4 DC simulation control component. **DC** is the name of the component and **DC**1 is the attributed name. Thus, DC1 can be edited by the user while DC cannot.

We will first discuss the principles of linear DC simulation with a simple example, after which we will discuss a nonlinear DC simulation method. Figure 7.5 shows an example circuit to illustrate DC simulation. As shown in the figure, ADS automatically generates node numbers for the wired circuit and builds a node equation. From circuit theory, it is well known that the node equation for a given circuit can be automatically built using the assigned node numbers. The admittance of a one-port device, such as a resistor, capacitor, or inductor, appears at the matrix element with its node numbers in the node matrix **A**. For example, if the one-port device is connected to the i and j nodes, then the admittance of the one-port device appears both at a_{ii} and a_{jj}. In addition, the negative value of the admittance appears at a_{ij} and a_{ji}. In the case of the transconductance g_m, suppose that (i,j) are the node numbers where the control voltage appears and (k,l) are the node numbers where the controlled current appears. Then, the transconductance g_m appears at the elements with (i,j) columns and (k,l) rows of the node matrix with the appropriate signs. Repeating for each element, the node matrix can be automatically built. Thus, the node equation for the circuit shown in Figure 7.5 is

$$\begin{pmatrix} \frac{1}{R_s}+\frac{1}{r_\pi} & -\frac{1}{r_\pi} & 0 \\ -\frac{1}{r_\pi}-g_m & \frac{1}{R_e}+\frac{1}{r_\pi}+\frac{1}{r_o}+g_m & \frac{1}{r_o} \\ g_m & -\frac{1}{r_o}-g_m & \frac{1}{R_L}+\frac{1}{r_o} \end{pmatrix} \begin{pmatrix} V_1 \\ V_2 \\ V_3 \end{pmatrix} = \begin{pmatrix} I_s \\ 0 \\ 0 \end{pmatrix} \tag{7.1}$$

Equation (7.1) can be considered in terms of a linear matrix equation $\mathbf{Ax} = \mathbf{b}$. For a given source vector **b**, the unknown output voltage vector **x** can be determined. Many numerical methods are available for solving the linear matrix equation and providing the desired node voltages.

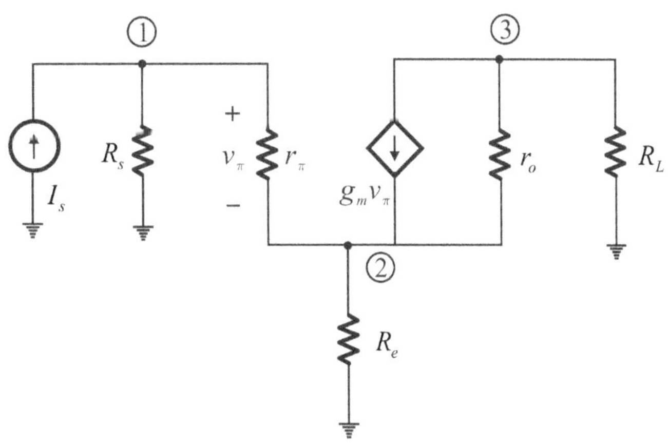

Figure 7.5 Circuit example for linear DC analysis

Nonlinear DC circuits, which include transistors or diodes, can be analyzed by the repetition of the DC analysis of linear circuits. Here, nonlinear DC circuit analysis will be illustrated using a simple circuit example rather than the rigorous analysis method. To illustrate the principle of nonlinear DC analysis, a simple diode circuit and its Norton equivalent circuit are shown in Figure 7.6.

The nonlinear diode is assumed to have the relationship $i = f(v)$. Expanding this relationship in a Taylor series around a given initial voltage v_n gives

$$\begin{aligned} i &= f(v_n) + f'(v_n)(v - v_n) \\ &= f(v_n) - v_n f'(v_n) + f'(v_n)v \end{aligned} \tag{7.2}$$

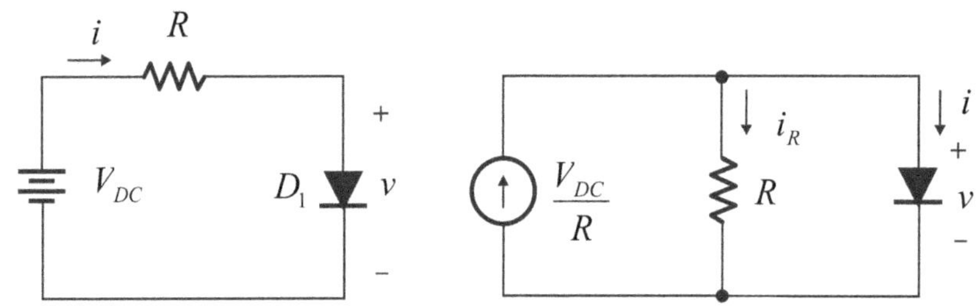

Figure 7.6 A simple diode circuit example and its Norton equivalent circuit

Since the value of v_n is given, the first two terms in Equation (7.2) are constant and represent the DC current. The last term can be considered as a resistor having conductance $f'(v_n)$. Therefore, near the given initial voltage v_n, the nonlinear device can be considered as a parallel connection with the DC current source and resistor, as shown in Figure 7.7.

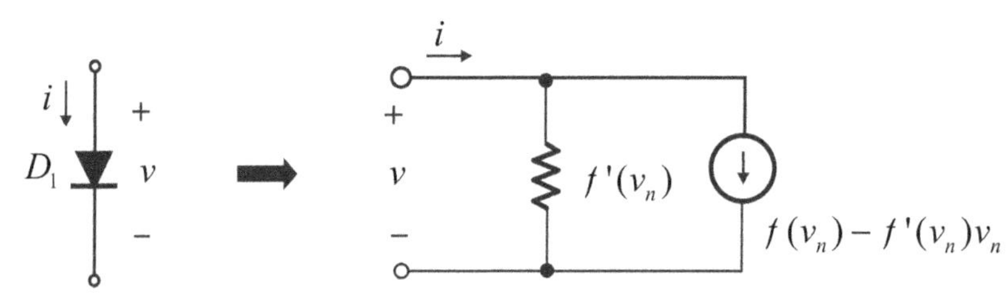

Figure 7.7 Equivalent circuit of a nonlinear device near the voltage v_n. The nonlinear device can be approximated as a resistor and current source in parallel near the voltage v_n.

Therefore, for a given nonlinear circuit, after replacing the nonlinear devices with the linear equivalent circuits composed of a resistor and current source, as shown in Figure 7.7, the nonlinear circuit then becomes a linear circuit, and the method of linear circuit analysis explained previously can be applied to obtain the new voltage of the circuit, v_{n+1}. When the resulting voltage v_{n+1} is close to v_n, the initial guess voltage v_n becomes the solution of the circuit in Figure 7.6. If the difference is large, then because the initial selection is erroneous, the initial value is updated by the newly obtained voltage v_{n+1} and the new voltage can be computed again. Thus, through this iteration, the voltage of the nonlinear circuit shown in Figure 7.6 can be determined within an allowable error. This is an example for a simple diode circuit. However, any nonlinear device can be generally approximated by a linear resistor and a DC current source near an initial voltage. By replacing the nonlinear device in the circuit with the linear resistor and current source, an exact solution can be reached through iteration. This process is shown in Figure 7.8 as a flow chart.

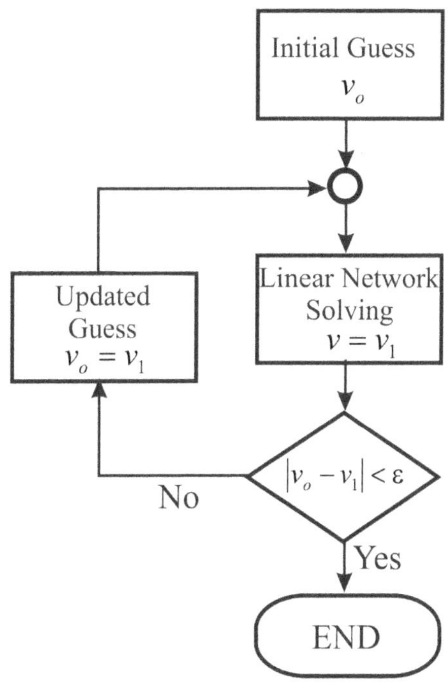

Figure 7.8 Flow chart of DC analysis. The voltages in the circuit, including a nonlinear resistive device, can be solved by iterative circuit analysis.

Example 7.1

The I–V characteristic of the diode shown in Figure 7.6 is given by

$$i = I_s(e^{\frac{v}{V_T}} - 1)$$

When $I_s = 10^{-12}$ A, $V_{DC} = 5$ V, $V_T = 25$ mV, and $R = 1$ kΩ for the circuit shown in Figure 7.6, obtain the voltage across the diode by iteration.

Solution

The circuit in Figure 7.6 is replaced by the approximate linear circuit near the voltage v_n, which is redrawn in Figure 7E.1.

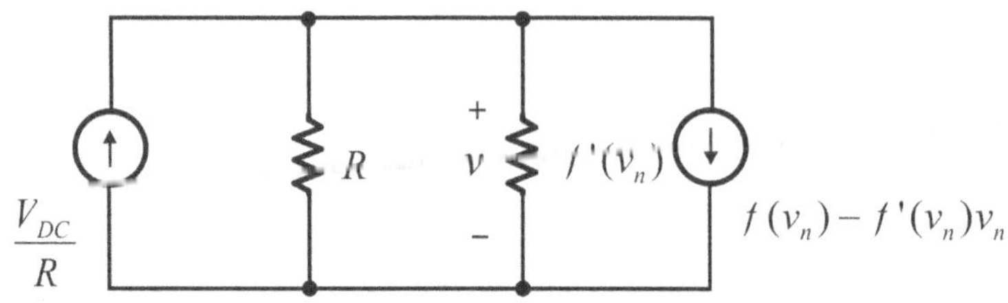

Figure 7E.1 The circuit is redrawn by replacing the diode with the linear equivalent circuit near voltage v_n.

Computing $v = v_{n+1}$ for a given v_n, the relationship shown below is obtained.

$$v = v_{n+1} = \frac{\frac{V_{DC}}{R} - \left[f(v_n) - f'(v_n)v_n\right]}{f'(v_n) + \frac{1}{R}}$$

The nominator represents the total current flowing into $R\,||\,f'(v_n)$ and the denominator represents the conductance of $R\,||\,f'(v_n)$. This equation can be used for iteration. Setting the initial voltage v_n to 0.6 V and iterating, the result shown in Table 7E.1 is obtained. The results in the table are computed using Microsoft Excel. From those results, the voltage across the diode is approximately found to be v = 0.5554 V.

Table 7E.1 Iteration results for the circuit in Figure 7E.1

v_n	$f'(v_n)$	$f(v_n) - f'(v_n)v_n$	v_{n+1}	$\varepsilon = \lvert v_{n+1} - v_n \rvert$
0.6000	1.0596E+00	−6.0925E-01	0.579172305	0.020827695
0.5792	4.6059E-01	−2.5525E-01	0.56380388	0.015368425
0.5638	2.4908E-01	−1.3420E-01	0.55664305	0.007160829
0.5566	1.8704E-01	−9.9440E-02	**0.55540554**	0.00123751

7.2.3 Transient Simulation

Transient simulation is the circuit analysis for a time-varying voltage source $v_s(t)$ or a current source $i_s(t)$ applied to a circuit. Transient simulation can be performed using the *transient simulation component* in ADS that is shown in Figure 7.9. This component can be used to analyze both linear and nonlinear circuits without any distinction as in the DC simulation component. To set the analysis, it is necessary to specify the start and stop times, and the time step. In the transient simulation component shown in Figure 7.9, the **StartTime**, unless otherwise specified, is 0 sec by default. The **StopTime** in this component is set to 100.0 nsec while the **MaxTimeStep** is set to 1.0 nsec.

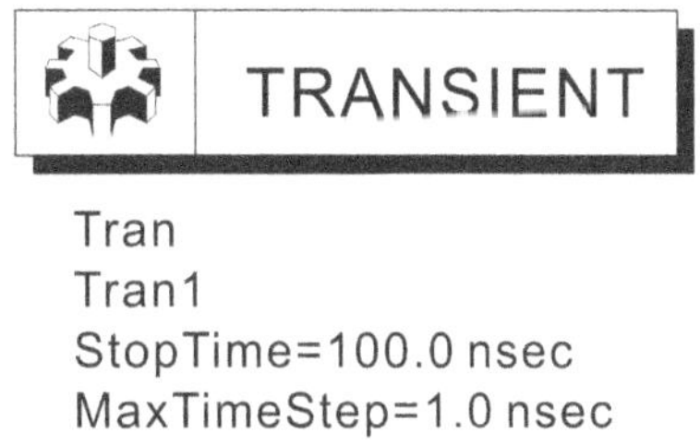

Figure 7.9 Transient simulation component; by default, start time is set to 0.

Figure 7.10 shows a simple *RC* circuit as an illustration. The circuit is used to illustrate how to calculate a time-varying voltage response using the previously explained linear DC circuit analysis. The current–voltage relationship for the capacitor in Figure 7.10 is given by Equation (7.3).

$$i = C\frac{dv}{dt} \tag{7.3}$$

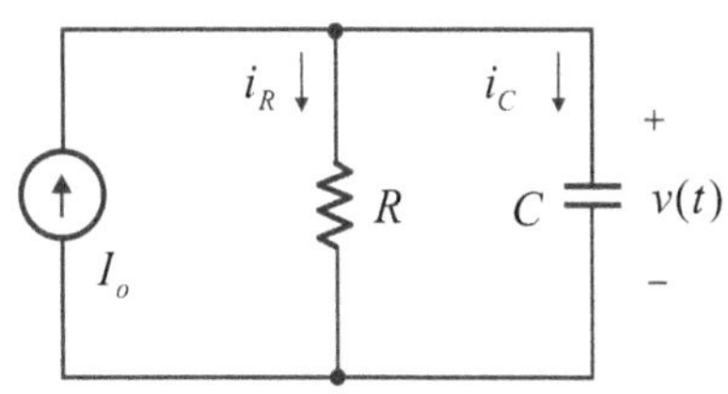

Figure 7.10 RC circuit example

As shown in Figure 7.11, suppose that voltage $v(T_n)$ at time T_n is known. The voltage at a neighboring time $T_{n+1} = T_n + T$ is unknown and denoted as $v(T_{n+1})$. Here, T is a fixed time step, and is assumed to be small enough. Near T_n, the voltage and current of the capacitor in Equation (7.3) can be written as

$$i = i_{n+1} = C\frac{dv}{dt} \cong C\frac{v - v_n}{T} = C\frac{v}{T} - C\frac{v_n}{T} \tag{7.4}$$

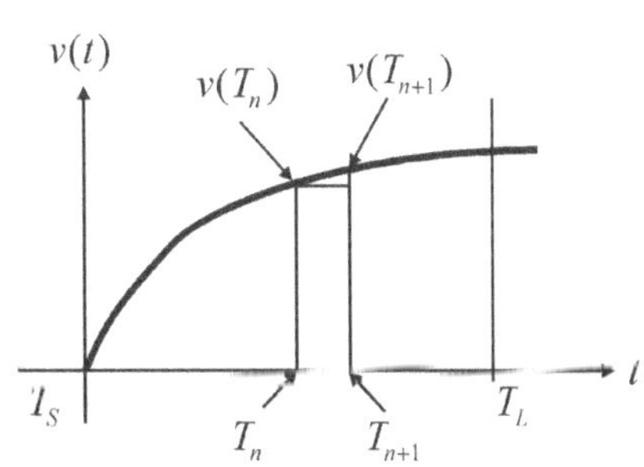

Figure 7.11 The voltage across a capacitor with respect to time

The first term on the right-hand side of Equation (7.4) can be considered as a resistor that is proportional to the applied voltage. The last term, including v_n, is a constant that can be considered as a current source. Therefore, the capacitor can be approximated as a parallel connection of a resistor and current source in a small time step T in the neighborhood of T_n. Figure 7.12 shows the equivalent circuit of the capacitor near $t = T_n$. Replacing the capacitor in Figure 7.10 with the equivalent circuit in Figure 7.12, voltage $v(T_{n+1})$ can be found through linear DC circuit analysis. Repeating the calculation to compute the voltage at the next time step with the determined voltage at the present time, the overall voltage waveform with respect to time can be determined. Even though this example uses a capacitor, inductors and other energy-storing linear elements like transmission lines usually can be similarly represented by a resistor and current source as shown in Figure 7.12.

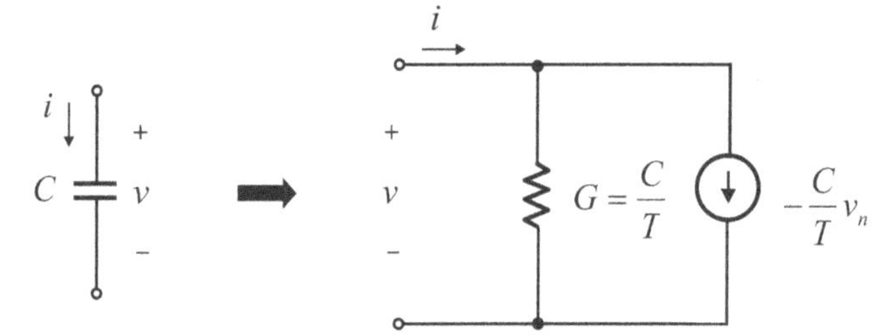

Figure 7.12 The equivalent circuit of a capacitor in a small time step near v_n. Here, v_n is the known value and the voltage v and current i are the unknown voltage and current at the next time step.

Example 7.2

For the parallel RC circuit shown in Figure 7.10, determine the voltage across the capacitor for a time step of $T = 1$ msec using the iteration method explained previously. The values in the RC circuit in Figure 7.10 are $I_o = 1$ A, $R = 1\ \Omega$, $C = 1$ F, and $v_c(0) = 0$ V. Compare the computed results with the theoretical value.

Solution

The theoretically computed voltage across the capacitor is

$$v(t) = I_o R\left(1 - e^{\frac{-t}{RC}}\right)$$

In order to perform the calculation iteratively, the capacitor is replaced by the equivalent circuit of Figure 7.12 and is redrawn, as shown in Figure 7E.2.

Solving the circuit of Figure 7E.2,

$$v = v_{n+1} = \frac{I_o + \frac{C}{T}v_n}{\frac{1}{R} + \frac{C}{T}} = R\frac{I_oT + Cv_n}{T + RC}$$

The calculated voltage $v(t)$ with respect to time, shown in Table 7E.2, is also computed using Excel. The error between computed and theoretical values is also shown in the last column of the table. Some errors, although small, are found to occur but, as the time step T is reduced, the errors are reduced and the voltage computed by the iteration becomes closer to that computed by the theoretical equation.

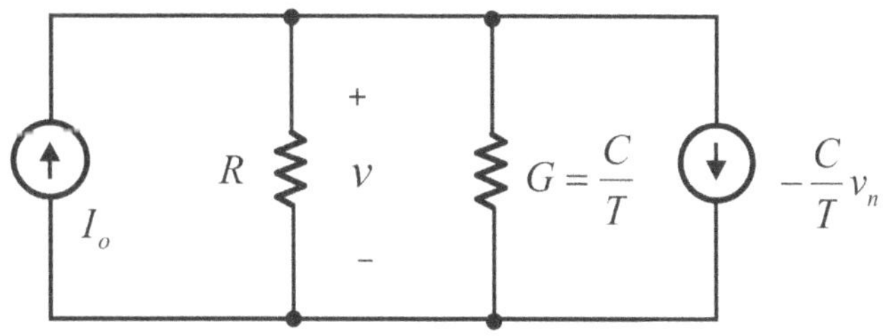

Figure 7E.2 The circuit in which the capacitor is replaced by the small time-step equivalent circuit

Table 7E.2 The voltage across the capacitor computed by an iterative method

T[msec]	v_n	v_{n+1}	$v(t)$	$\varepsilon=\lvert v_n - v(t)\rvert$
0	0.000 E+00	9.091E-02	0.000E+00	0.000E+00
1	9.091E-02	1.736E-01	9.561E-02	4.253E-03
2	1.736E-01	2.487E-01	1.813E-01	7.3716E-03
3	2.487E-01	3.170E-02	2.592E-01	1.050E-02
4	3.170E-01	3.791E-01	3.297E-01	1.269E-02
5	3.791E-01	4.355E-01	3.935E-01	1.439E-02
6	4.355E-01	4.868E-01	4.512E-01	1.566E-02
7	4.868E-01	5.335E-01	5.034E-01	1.657E-02
8	5.335E-01	5.759E-01	5.507E-01	1.718E-02
9	5.759E-01	6.145E-01	5.934E-01	1.753E-02
10	6.145E-01	6.495E-01	6.321E-01	1.766E-02

The transient circuit analysis flow chart of the previous example is shown in Figure 7.13. When nonlinear devices are included in a circuit, the solution for the next neighboring time, T_{n+1}, is not obtained by direct calculation as in Example 7.2. The computation of the voltage at the next time requires an iteration similar to that noted in Example 7.1.

7.2.4 AC Simulation

AC simulation is a circuit analysis for sinusoidal sources. As explained in the discussion of transient simulation, a transient waveform generally appears because the sinusoidal source is a time-varying source. However, after the transient waveform disappears, all the voltages and currents in the circuit become the sine waves having the same frequency as the source. This is called *AC steady state*. Thus, all the voltages and currents in the circuit differ only in amplitude and phase. For the purpose of analyzing the circuit in AC steady state, the sine wave source is usually expressed using a complex number with magnitude and phase, as shown on the right-hand side of Figure 7.14. The complex number representing a sinusoidal waveform is called a *phasor*. The magnitude and phase of the phasor can

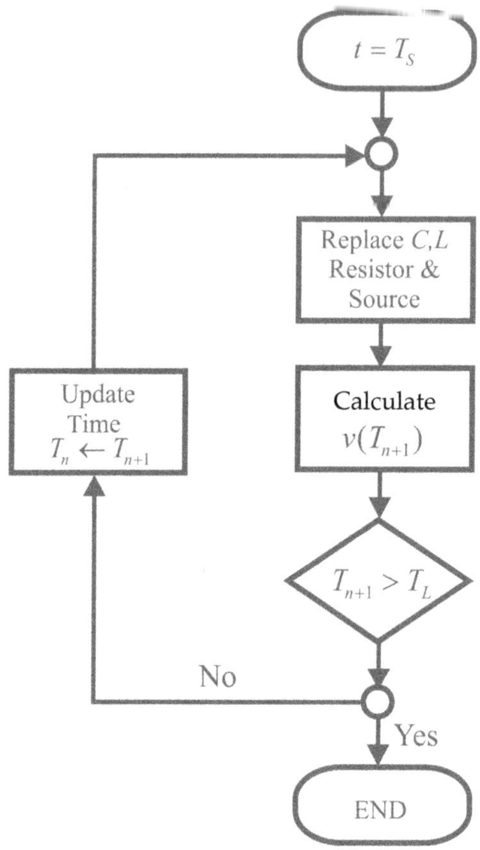

Figure 7.13 Transient simulation flow chart

be interpreted as the amplitudes and phase of a sinusoidal waveform. As a result, in AC steady state, all the currents and voltages can be represented by phasors. In addition, since all the currents and voltages are represented by phasors, the current and voltage relationship of all elements can be considered as a kind of resistor with the resistance represented by a complex number. Consequently, the circuit can be composed of complex-valued DC sources and resistors and, as a result, AC analysis can be thought of as a linear DC circuit analysis in the complex number domain. After the AC analysis of the circuit, the voltage and current phasors are obtained rather than direct sinusoidal waveforms.

The AC analysis is possible only for linear circuits. In situations with nonlinear elements, harmonic voltages and currents are generated in response to sine wave sources. Thus, a nonlinear circuit cannot be analyzed using the AC analysis previously explained. However, when AC analysis is employed for a nonlinear circuit in ADS, the software performs a *small-signal AC analysis*. In that case, ADS automatically performs a nonlinear DC analysis even when there is no DC simulation component in the schematic. Then, ADS internally constructs a small-signal AC equivalent circuit. Nonlinear devices such as transistors are replaced by their *small-signal equivalent circuits* determined from nonlinear DC analysis. AC analysis is then carried out for the resulting small-signal linear circuit. Thus, AC simulation for a nonlinear circuit in ADS is basically a small-signal AC analysis.

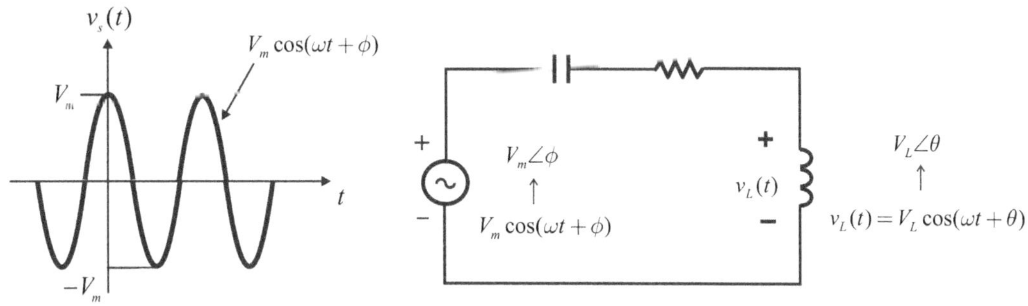

Figure 7.14 The concept of AC analysis. The sinusoidal time-domain voltage is represented by a phasor. The magnitude of the phasor represents the amplitude of the sinusoidal time-domain voltage, while its phase represents the phase of the sinusoidal time-domain voltage.

Figure 7.15 shows an *AC simulation component* in ADS for performing AC analysis. Basically, the frequency response can be obtained in AC circuit analysis as a default. The variable **freq** is defined as a *global variable*. This means that the variable name **freq** cannot be used to represent a variable for any other purpose. The variable **Start** in Figure 7.15 indicates the start frequency for AC analysis, while the variable **Stop** indicates the end frequency, and the variable **Step** indicates the frequency step for AC analysis.

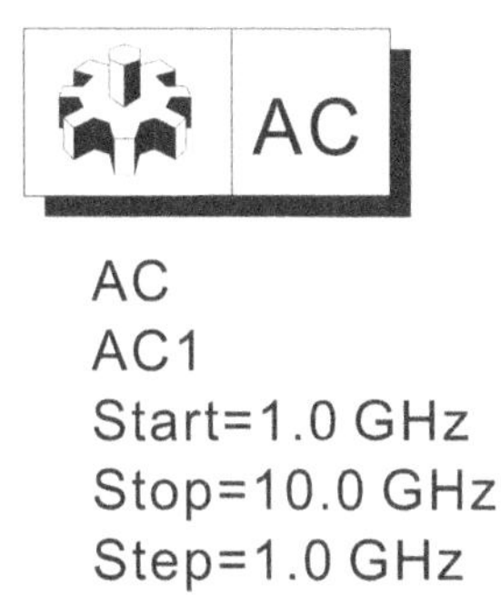

Figure 7.15 AC simulation component

7.2.5 Harmonic Balance Simulation

Harmonic balance simulation is a true nonlinear AC simulation. When a sinusoidal source is applied to the nonlinear device in a circuit, the harmonics of the sinusoidal source will occur. Thus, in AC steady state, the harmonic frequency nf_o will be generated in the circuit for the sinusoidal source of frequency f_o. As a result, not only the fundamental frequency f_o but also the harmonic frequency nf_o appear at each node voltage.

In this case, since an orthogonal relationship exists for each harmonic, a separate circuit for each harmonic frequency can be built and analyzed. However, this is not so simple because the current-voltage relationship, as $i = f(v)$, for a nonlinear device is generally defined in a time domain. In building a separate circuit for each harmonic frequency, the relationship between the k-th harmonic voltage V_k and the k-th harmonic current I_k is required. Thus, the harmonic relationship must be obtained using the time domain relationship of $i = f(v)$.

Assuming the k-th harmonic voltage is V_k, the time-domain waveform $v(t)$ can be expressed as the Fourier series given by

$$v(t) = \sum_{k=-N}^{N} V_k e^{jk\omega_o t} \tag{7.5}$$

Note that since $v(t)$ is real, $V_k = V_{-k}^*$. It is worth noting that V_k in Equation (7.5) is not a phasor. The phasor voltage V_k^p can be defined using V_k as expressed in Equation (7.6).

$$V_k^p = \begin{cases} V_k & k = 0 \\ 2V_k & k \neq 0 \end{cases} \tag{7.6}$$

Using Equation (7.5), the k-th harmonic current I_k, shown in Equation (7.7), is

$$I_k = \frac{1}{T}\int_0^T f(v(t)) e^{-jk\omega_o t} dt = F_k(V_o, V_1, \ldots, V_N) \tag{7.7}$$

This represents the I_k–V_k relationship of a nonlinear device. If $f(v(t))$ is expanded in a Taylor series near $v(t) = v_o(t)$, it can be written as Equation (7.8),

$$f(v(t)) = f(v_o(t)) + f'(v_o(t))\delta v(t) = f(v_o(t)) + g(t)\delta v(t) \tag{7.8}$$

where $\delta v(t) = v(t) - v_o(t)$. The δ is added to represent the small signal. When the initial voltage $v_o(t)$ is close to the solution, $\delta v(t)$ can be treated as a small signal. In addition, note that $g(t)$ can also be expressed in a Fourier series because it is also periodic with the same period T. The Fourier series of $g(t)$ is defined by Equations (7.9a) and (7.9b) as

$$g(t) = f'(v_o(t)) = \sum_{k=-N}^{N} G_k e^{jk\omega_o t} \tag{7.9a}$$

$$G_k = \frac{1}{T}\int_0^T g(t) e^{-jk\omega_o t} dt \tag{7.9b}$$

Note that $g(t)$ is a real waveform and G_k has the following relationship:

$$G_{-k} = G_k^*$$

Substituting Equation (7.8) into Equation (7.7), Equation (7.10) shows that

$$\begin{aligned} I_k &= \frac{1}{T}\int_0^T f(v(t)) e^{-jk\omega_o t} dt \\ &= \frac{1}{T}\int_0^T f(v_o(t)) e^{-jk\omega_o t} dt + \frac{1}{T}\int_0^T g(t)\delta v(t) e^{-jk\omega_o t} dt \end{aligned} \tag{7.10}$$

The first term is the k-th phasor current that is determined by the initial voltage $v_0(t)$, while the second term represents the k-th harmonic current obtained by applying the small-signal voltage $\delta v(t)$ to the linear time-varying conductance $g(t)$.

Also expressing $\delta v(t)$ in a Fourier series,

$$\delta v(t) = \sum_{m=-N}^{N} \delta V_m e^{jm\omega_o t} \tag{7.11}$$

from the real condition of $\delta v(t)$, $\delta V_m = (\delta V_{-m})^*$. Using Equations (7.11) and (7.9), the k-th harmonic of the second term in Equation (7.10) is expressed in Equation (7.12).

$$k\big(g(t)\delta v(t)\big) = \sum_{m=-N}^{N} G_{k-m}\delta V_m \tag{7.12}$$

Here, $k(\bullet)$ represents the function that takes the k-th harmonic from the Fourier series. Using Equations (7.7) and (7.12), Equation (7.10) can be rewritten as

$$\begin{aligned} I_k &= F_k\left(V_0^{old}, V_1^{old}, \dots V_N^{old}\right) + \sum_{m=-N}^{N} G_{k-m}\delta V_m \\ &= F_k\left(V_0^{old}, V_1^{old}, \dots V_N^{old}\right) + \sum_{m=-N}^{N} G_{k-m}\left(V_m - V_m^{old}\right) \\ &= F_k\left(V_0^{old}, V_1^{old}, \dots V_N^{old}\right) - \sum_{m=-N}^{N} G_{k-m}V_m^{old} + \sum_{m=-N}^{N} G_{k-m}V_m \end{aligned} \tag{7.13}$$

Here, V_m^{old} and V_m are the Fourier coefficients of $v_o(t)$ and $v(t)$, respectively. The first and second terms in Equation (7.13) represent the current source that depends on the harmonics of the initially selected $v_o(t)$. The right-hand term in Equation (7.13) shows the linear relationship between I_k and $(V_0, V_1, \cdot\cdot, V_N)$. Among them, the relationship of I_k–V_k can be seen as a resistor and it can be expressed as $I_k = G_0V_k$. The dependence of I_k on the remaining harmonic voltages V_k can be interpreted as the dependent current sources. Thus, separately built harmonic circuits for each harmonic frequency are coupled to each other by these dependent sources. This is how these circuits differ from the conventional linear harmonic circuits built up for each harmonic frequency. Using the relationship near the initial guess voltage $v_o(t)$ in Equation (7.13), the nonlinear device can be approximated by a resistor, a current source, and dependent current sources coupled by other harmonic voltages. This is shown in Figure 7.16.

The circuit in Figure 7.16 can be used to build a separate harmonic circuit that includes nonlinear devices for each harmonic frequency. The resulting circuit then becomes many separate harmonic circuits but they are coupled to each other by the dependent sources. Note that these coupled harmonic circuits are linear and the harmonic voltage V_m can be obtained from the solutions for those coupled circuits. When the computed voltages do not converge, the initial values are updated with the newly computed voltages and reiterated until the desired convergence is achieved.

The harmonic balance simulation example is performed by an *HB simulation component* in ADS, as shown in Figure 7.17, where **Freq** indicates the fundamental frequency while **Order** indicates the number of harmonics.

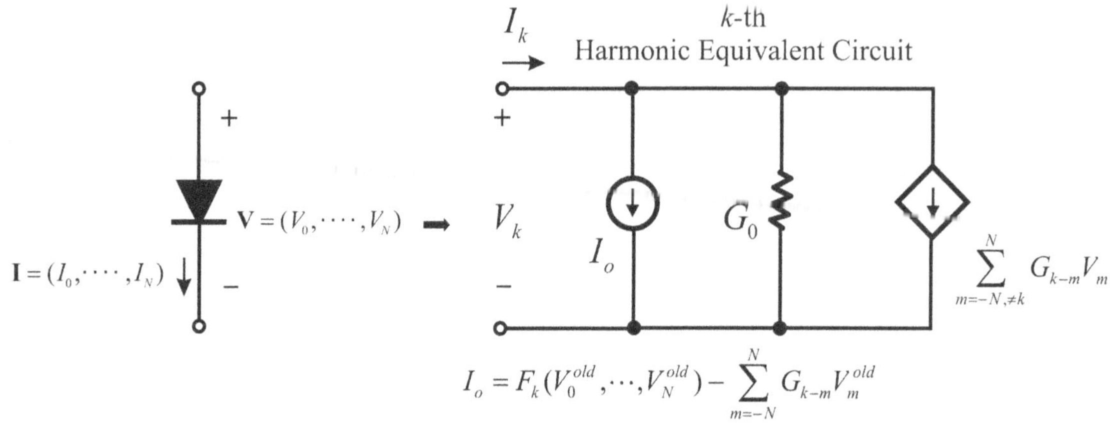

Figure 7.16 The equivalent circuit of the nonlinear device for the k-th phasor near $v_o(t)$. Due to the dependent current sources, the separately built-up k-th harmonic circuit is coupled to other harmonic circuits.

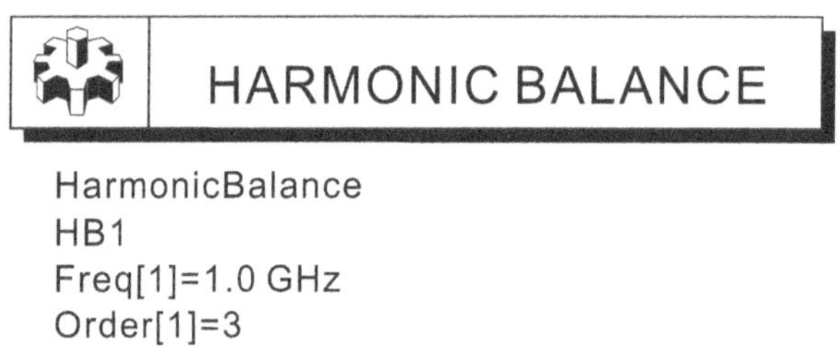

Figure 7.17 HB simulation component

Example 7.3

Figure 7E.3 is a half-wave rectifier circuit. The values of the elements are $R = 1\ \text{k}\Omega$, $I_1 = 5$ mA, and $\omega/(2\pi) = f = 1$ kHz, and the diode has this relationship:

$$i = I_s\left(e^{\frac{v}{V_T}} - 1\right) \text{ and } I_s = 10^{-14} \text{ A}, V_T = 25 \text{ mV}$$

Set the number of harmonics to 3. Calculate the voltage across the diode $v(t)$ using the method of harmonic balance.

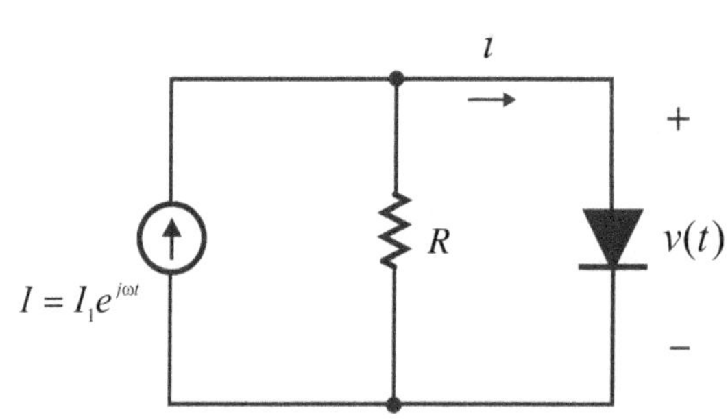

Figure 7E.3 Half-wave rectifier circuit

Solution

In the circuit shown in Figure 7E.3, the diode turns on and off according to the applied sinusoidal signal. When the diode is conducting, a voltage of 0.7 V appears across the diode, while a voltage of about 5 V ($=I_1R$) peak sine wave appears when the diode is off. The result analyzed by ADS is shown in Figure 7E.4. A time-domain voltage waveform is shown in Figure 7E.4(a), while Figure 7E.4(b) shows the current waveform. The computation is rather complex. This was written in MathCad, which is discussed in Appendix C. The $v(t)$ calculated using ADS is represented by the harmonic phasors (V_0, $2V_1$, $2V_2$, $2V_3$). The values of the harmonic phasors are −1.288, 2.883, −1.008, and 0.1. The calculation results using MathCad that follow the given algorithm are found to yield the exact harmonic phasor obtained by ADS and shown in Appendix C.

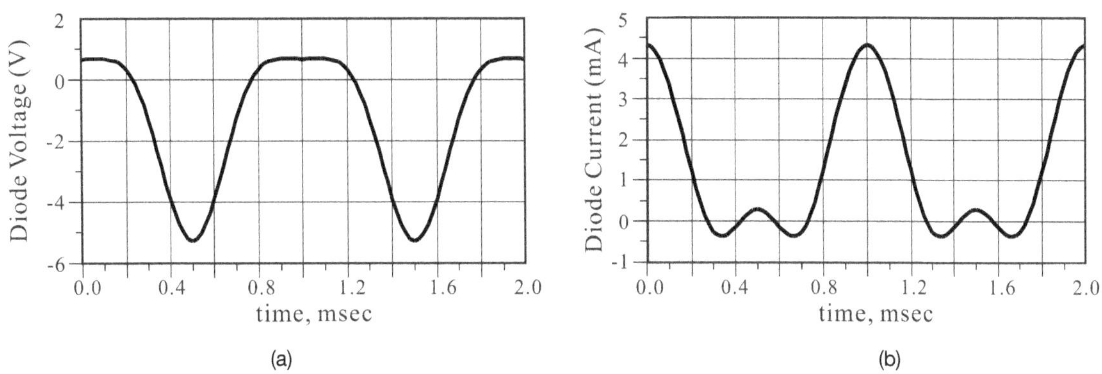

Figure 7E.4 (a) Time-domain voltage and (b) current waveforms across the diode

7.2.6 Multi-Tone Harmonic Balance

When a sinusoidal source with two different frequencies, f_1 and f_2, is applied to a circuit that includes nonlinear devices, those devices will give rise not only to the harmonic frequencies mf_1 and nf_2, but also to the intermodulation frequency components $mf_1 \pm nf_2$. Thus, when two different sinusoidal sources are applied, the voltage of the nonlinear device $i = f(v)$ can be expressed using a Fourier series as

$$v(t) = \sum_{m=-N}^{N} \sum_{n=-M}^{M} V_{m,n} e^{jm\omega_1 t + jn\omega_2 t} \tag{7.14}$$

However, in most cases, $v(t)$ given by Equation (7.14) may not be periodic. The simplest idea is to reduce the quasi-periodic regime to a strictly periodic regime by taking the greatest common divider of the intermodulating tones as the fundamental frequency of operation, and then to use the fast Fourier transform. Although this approach has the obvious advantage of immediacy, the main limitation arises from the fact that low values of the fundamental may lead to huge storage and CPU time requirements that can easily exceed the available resources of even large supercomputers. This results in considerable restrictions on the combinations of input frequencies that are practically usable, which makes this choice unsuitable for a general-purpose simulator. One way of significantly reducing such redundancy is to resort to a multidimensional Fourier transformation, as shown in Equations (7.15a) and (7.15b).[2]

$$v(t_1,t_2) = \sum_{m=-N}^{N} \sum_{n=-M}^{M} V_{m,n} e^{jm\omega_1 t_1 + jn\omega_2 t_2} \tag{7.15a}$$

$$V_{m,n} = \frac{1}{T_1}\frac{1}{T_2}\int_0^{T_1}\int_0^{T_2} v(t_1,t_2) e^{-j(m\omega_1 t_1 + n\omega_2 t_2)} dt_1 dt_2 \tag{7.15b}$$

Time-domain sampling thus requires a multidimensional grid of sampling points, and conversion to the frequency domain can be performed by multiple Fourier transformations. When voltages and currents are transformed into the multidimensional frequency domain, the coupled frequency-domain circuits for each intermodulation frequency are produced. Using the solved $V_{m,n}$, the $v(t)$ can be obtained from Equation (7.14). The nice feature of this approach is that it can be implemented in a conventional harmonic balance program without any major change to the original program structure, and thus requires little programming effort.

The harmonic balance simulation component for a two-tone or more is the same as the harmonic balance simulation component shown in Figure 7.17. In the two-tone case, two frequencies appear as the variables, **Freq**[1] and **Freq**[2], and their number of harmonics appears as **Order**[1] and **Order**[2] in the harmonic balance simulation component. The number of the intermodulation frequencies is specified using **MaxOrder** in the simulation component. Thus, the number of frequencies depends on the variables **Freq**[1], **Freq**[2], and **MaxOrder**. Figure 7.18 shows an example. From that figure, the number of harmonics for f_1 is $N = 5$ and the number for f_2 is $M = 4$. **MaxOrder**, which specifies the number of intermodulation frequencies, is set to 3. This means the intermodulation frequencies in the rectangle are $|m \pm n| = 3$, as shown in Figure 7.18. Thus, the harmonic and intermodulation frequencies marked as • in the figure are simulated in the harmonic balance simulation.

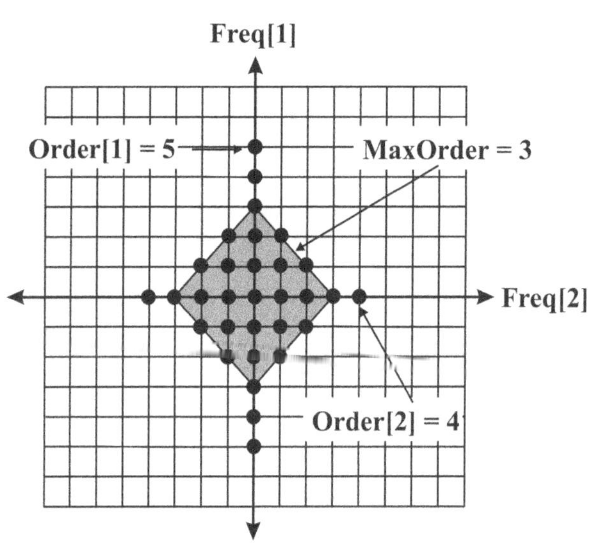

Figure 7.18 Example specifying the number of harmonics

2. V. Rizzoli, C. Cecchetti, A. Lipparini, and F. Mastri, "General-Purpose Harmonic Balance Analysis of Nonlinear Microwave Circuits under Multitone Excitation," *IEEE Transactions on Microwave Theory and Techniques*, vol. 36, no. 12, pp. 1650–1660, December 1988.

Example 7.4

List the harmonic frequencies and their number when the order of frequency f_1 is 5, that of f_2 is 4, and the **MaxOrder** is 2.

Solution

It can be seen from Figure 7.18 that there are 13 harmonics and they are listed as follows:

$$0, f_1, 2f_1, 3f_1, 4f_1, 5f_1;\ \text{6 harmonics}$$
$$0, f_2, 2f_2, 3f_2, 4f_2;\ \text{5 harmonics}$$
$$f_1 \pm f_2; 2\ \text{(intermodulated frequency when the MaxOrder is 2)}$$

Next, to determine the voltage phasor that corresponds to a specific frequency from among the mixed frequency components, the phasor voltage that corresponds to the specific frequency must be specified. However, ADS sets the single index in an ascending order from DC to higher frequencies. For Example 7.4, $f_1 = 1.5$ GHz and $f_2 = 1.15$ GHz, and **MaxOrder** is set to 3. In this case, all the frequencies are listed as well as their corresponding frequency index. The frequency index is shown in the third column in Table 7.1. From that table, the frequency index for $f_1 = 1.5$ GHz is found to be 4, while that of $f_2 = 1.15$ GHz is 3.

Table 7.1 Example of mixed frequency index

Frequency	Intermodulation [*m*,*n*]	Frequency Index
0	[0, 0]	freq[0]
350 MHz	[1, -1]	freq[1]
800 MHz	[1, -2]	freq[2]
1.15 GHz	[0, 1]	freq[3]
1.5 GHz	[1, 0]	freq[4]
1.85 GHz	[2, -1]	freq[5]
2.30 GHz	[0, 2]	freq[6]
2.65 GHz	[1, 1]	freq[7]
3.00 GHz	[2, 0]	freq[8]
3.45 GHz	[0, 3]	freq[9]
3.8 GHz	[1, 2]	freq[10]
4.15 GHz	[2, 1]	freq[11]
4.5 GHz	[3, 0]	freq[11]
4.6 GHz	[0, 4]	freq[12]
6.0 GHz	[4, 0]	freq[13]
7.5 GHz	[5, 0]	freq[14]

However, rather than the single index specification, it may be more convenient to specify the frequency using the [*m*, *n*] index. Here, *m* corresponds to the index of **Freq**[1] and *n* corresponds to the index of **Freq**[2]. The frequency specification that uses two independent frequency indices, **Freq**[1] and **Freq**[2], is possible using the built-in **mix(•)** function in ADS. For example, when the node voltage **vOut** is calculated as a result of harmonic balance simulation, the following **mix(•)** function is used to specify the voltage of the frequency component $2f_1 - f_2$:

$$y = \mathbf{mix}\left(\mathbf{vOut}, \{2, -1\}\right)$$

7.2.7 Optimization

Optimization is a numerical search technique that can be used to determine the element values in a circuit that satisfy given goals, which makes a design possible without the need for complex calculations. To perform the optimization, (1) the optimization *algorithm* must be selected, (2) *goals* for a circuit should then be selected, and (3) circuit simulation methods must finally be selected. The design goal is thus defined based on circuit simulation results. Many design goals can appear in optimization. In this case, the *optimization controller* in ADS evaluates those goals after circuit simulation and updates the circuit element values to satisfy the goals using selected optimization algorithms such as gradient, random, and so on. The procedure is repeated until all the design goals are satisfied or no significant changes are found in any of the goals.

The optimization technique can be used in various situations such as a matching circuit design or the circuit modeling mentioned in Chapter 5. Hence, from the designer's perspective, a circuit can be designed using optimization without the need for complex design formulas or calculations. It is worth noting that for a complex circuit, it may not be possible to obtain the desired values through optimization if the chosen initial guess values are not close to the desired values. Therefore, the user should select initial values as close as possible to the desired values for a successful optimization.

Figure 7.19 shows an optimization example for a matching circuit design. In the example, a matching circuit is configured as an L-type matching circuit to match a 100-Ω source to 50-Ω load. The goals are defined on S-parameter simulation results and the circuit element values are optimized to deliver maximum power to the load. Note that because the port impedances are 50 Ω and 100 Ω respectively, the S-parameters are calculated based on 50 Ω and 100 Ω as the reference impedances.

The maximum power transfer conditions can be stated as $|S_{21}| = 1$. The insertion gain $|S_{21}|^2$ is defined as the ratio of the delivered power to port 2 to the maximum available power from port 1. Since the matching network is lossless, the maximum available power from port 1 can be delivered to the load without loss. Therefore, under the maximum power transfer condition, $|S_{21}|^2 = 1$. Thus, the goal must be set such that $|S_{21}| = 1$ to achieve matching for a given frequency. Furthermore, since the matching network is lossless, $|S_{11}| = 0$. Consequently, the goal can be set to achieve the condition $|S_{11}| = 0$, or to achieve both conditions concurrently. All the conditions are the same. However, in the optimization process, because the convergence may be slow, depending on the goal, it is comparatively useful to set concurrent conditions. Figure 7.20 shows the optimized results; as expected, $|S_{21}| = 1$ and $|S_{11}| = |S_{22}| = 0$. When S_{11} and S_{22} are plotted on the Smith chart, the impedance looking into the input or output can be determined. However, it should be noted that the input is normalized by 50 Ω while the output is normalized by 100 Ω. Therefore, since $|S_{11}| = |S_{22}| = 0$, it represents an input impedance of 50 Ω and an output impedance of 100 Ω. In addition, the element values that yield the S-parameters in Figure 7.20 are shown in Figure 7.19. The values of **ls** and **cm** are found to be 7.96 nH and 1.59 pF, respectively.

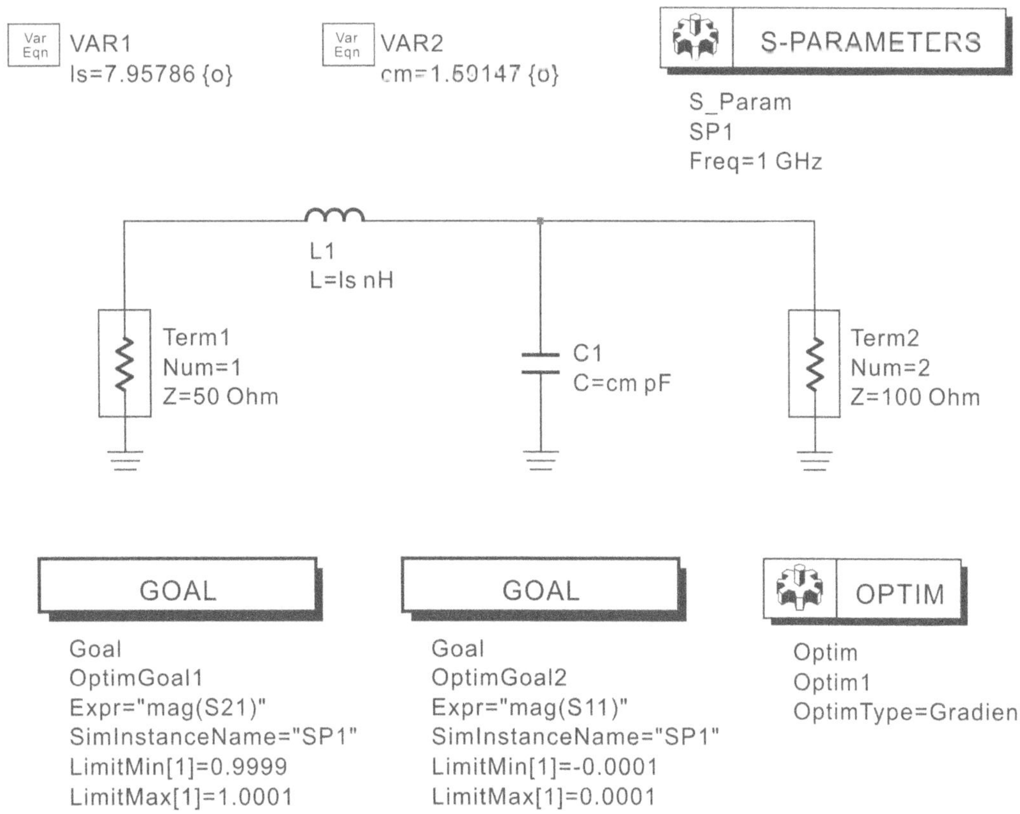

Figure 7.19 Matching circuit design using optimization. The optimization controller **Optim** first calculates the cost function given by the goals, which compute the errors from the desired values using the simulated results. It then determines the new circuit element values based on the given optimization algorithm specified by the user. This is iterated until the goals are satisfied or until the given number of iterations is reached.

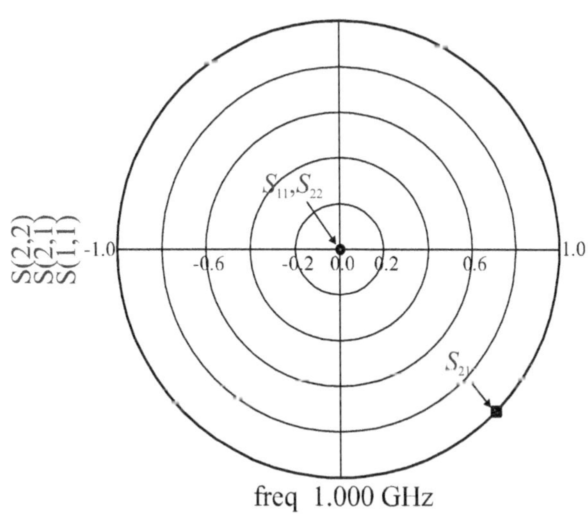

Figure 7.20 Optimized S-parameters. For the given source and load impedances, the maximum power delivery is achieved when $|S_{21}| = 1$, which means $|S_{11}| = 0$ at the same time.

7.3 LAYOUT

A *layout* is necessary to fabricate a designed circuit on a PCB or semiconductor wafer. The methods for creating the layout can be classified into two types: *auto-layout* and *manual layout*. In this section, we will present the basic concept of a *manual layout* such as *layer*, *grid*, and *component* (also called *block* and *instance* in other software). To illustrate the manual layout concept, we will use an example layout of a simple circuit.

In the example, the fabrication of a two-sided PCB is shown in Figure 7.21. The two-sided PCB is usually fabricated with a substrate copper plated on both sides. First, holes are drilled and then they are copper plated to connect the top and bottom copper plates. Next, the conductor patterns are etched on the top and bottom copper plates. After etching, both sides of the patterned PCB are selectively coated with an SR (solder resist) material for soldering convenience during assembly; when components are soldered to the land patterns, the SR prevents the formation of unwanted connections between the solders. Finally, the components are placed on the land patterns that do not have the SR coating and the desired circuit is formed by soldering.

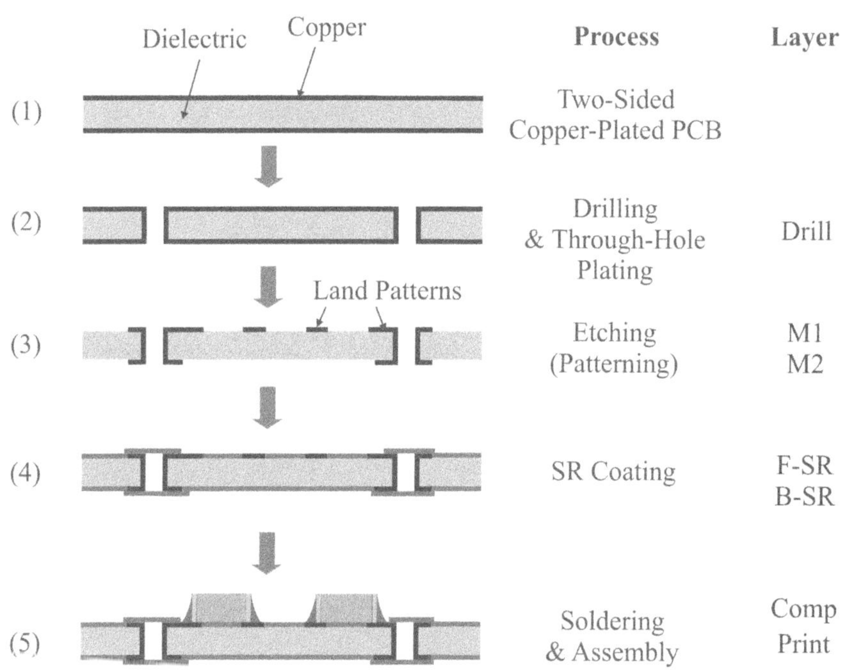

Figure 7.21 Two-sided PCB fabrication process

Each process shown in Figure 7.21 requires a mask or film: for drilling, film **Drill**; for top and bottom conductor pattern etching, films **M1** and **M2**; and for SR coating, films **F-SR** and **B-SR**. Thus, the film or mask required by each process should be fabricated prior to the PCB fabrication. The films are fabricated using a *Gerber format* file converted from the layout. The designer must draw a layout for the film required for each process. The layout for the film or mask corresponding to each process becomes a *layer*. In other words, a layer can be considered as a film drawing for one process among the many processes involved in the PCB fabrication. Therefore, the layout for a PCB consists of the many layers necessary for PCB fabrication.

In Figure 7.21, for example, in order to perform a drilling operation, a drawing that specifies the position to be drilled as well as the diameter of the drill must be created first. This drawing represents the **Drill** layer. Then, drawings for the conductor patterns on the top and bottom sides of the substrate are required in order to form the desired wiring; these drawings represent the layers **M1** and **M2**. Next, the top and bottom SR processes must also be drawn for the **F-SR** and **B-SR** layers, respectively. Then, the sizes and shapes of the components to be mounted on top of the PCB must also be drawn by the designer on the **Comp** layer (component layer). In addition, on the **Print** layer, the shapes to be printed with white ink are drawn for classifying the PCB or to assist with understanding its fabrication. Therefore, each layer can be considered as a decomposed plot for each process of a PCB fabrication.

The next design concept includes the use of an instance, or component, layout. To explain the concept of component, an example layout is shown in Figure 7.22 where components with the identical size and shape repeatedly appear at different locations. Drawing these components one by one is a cumbersome task. In addition, the component layout can be used in designing other circuits besides the example layout. Thus, each component that repeatedly appears is drawn separately and each drawing for the component is prepared in advance as a template. As a result, whenever the component is needed, it can be accessed and used for other layouts.

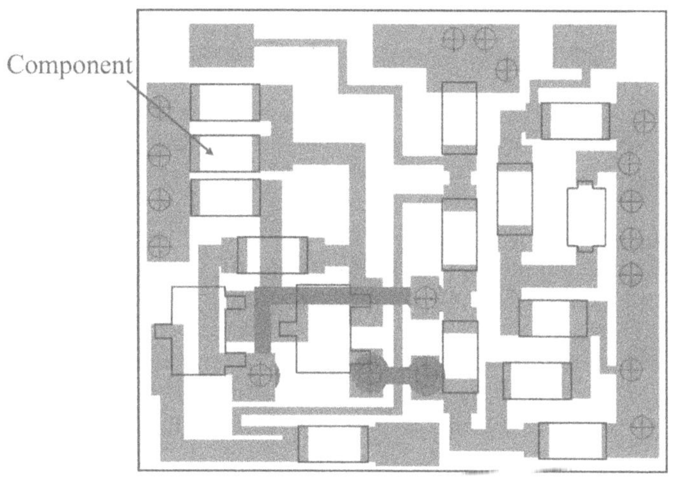

Figure 7.22 PCB layout example. Many identical elements appear in the layout. For efficiency, these identical components are usually prepared as a component layout.

The concept of a grid can easily be understood with a graph paper. The thin, solid lines are marked on the graph paper every 1 mm and the thick, solid lines every 1 cm. These grids enable a user to read the length between two points easily and facilitate the drawings for shapes such as rectangles, polygons, and circles that have mm-unit-based dimensions. These marked solid lines on the graph paper become the *minor* and *major* grids in layout. In ADS, the *layout unit* is first defined. It specifies a basic unit length, such as 1 mm, 1 mil, 1 inch, and so on. Based on the layout unit, the user defines the *minor* and *major* grids. In addition, as it is often difficult to place and click the mouse precisely on a grid, by entering a specific radius of snap, the software recognizes the approximate mouse-click position as the center when the mouse is clicked within the radius of snap. This is called *snap*. Using snap, the user can easily draw a shape that has vertices precisely located on the grid. The concepts of layer, component, and grid will be illustrated using an example.

7.3.1 Layout Example

Figure 7.23 shows the circuit of a common emitter amplifier that we will use to illustrate how to create a layout.

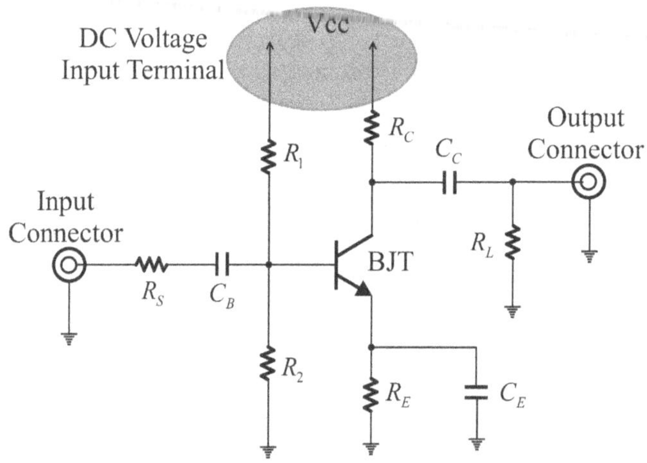

Figure 7.23 Common emitter amplifier circuit for an example layout

As shown in the figure, one BJT as well as 9-chip components such as resistors and capacitors are used in the circuit. Note that the 1608-size chip resistors and capacitors have the same size and so it is possible to use a single component for both. Thus, a layout for the 1608 chip components and a layout for the BJT are prepared for the circuit layout. In addition, the layout of the via hole is also needed as a component and it can be used to provide the ground for the 1608-chip components. Other components such as a DC voltage terminal and two AC input and output connectors are necessary. The AC input connector is for applying the AC signal and the AC output connector is for measuring the output signal. Note that since the AC input and output connectors are the same components, it is also possible to use a single component for both. Therefore, a total of five kinds of components should be prepared.

7.3.2 Layer Preparation for Layout

First, create a new project in ADS and name it **Manual_Layout**_prj. The **Manual_Layout_**prj window is shown in Figure 7.24. Then, click the *New Layout Window* icon and the layout window shown in Figure 7.25 will appear.

On the menu bar, select *File* > *Save Design* and save the layout as **CE_Amp_Layout**. As discussed previously, a layer definition is required for films and this can be defined using the *Layer Editor*. Select *Options* > *Layers* on the menu bar. The *Layer Editor* is displayed, as shown in Figure 7.26, which also shows a modified layer definition from the layer definition autogenerated by the Layer Editor.

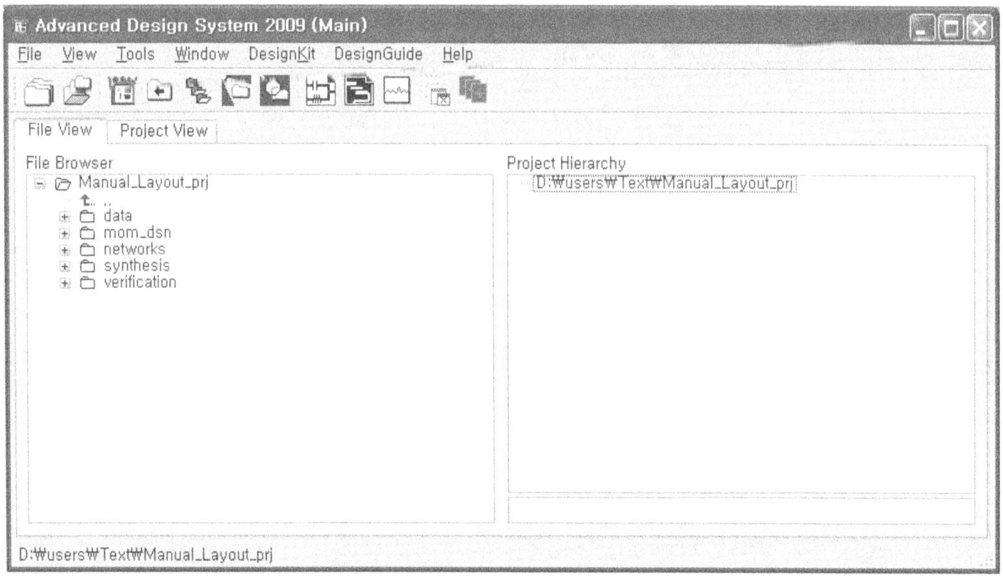

Figure 7.24 Project created for layout

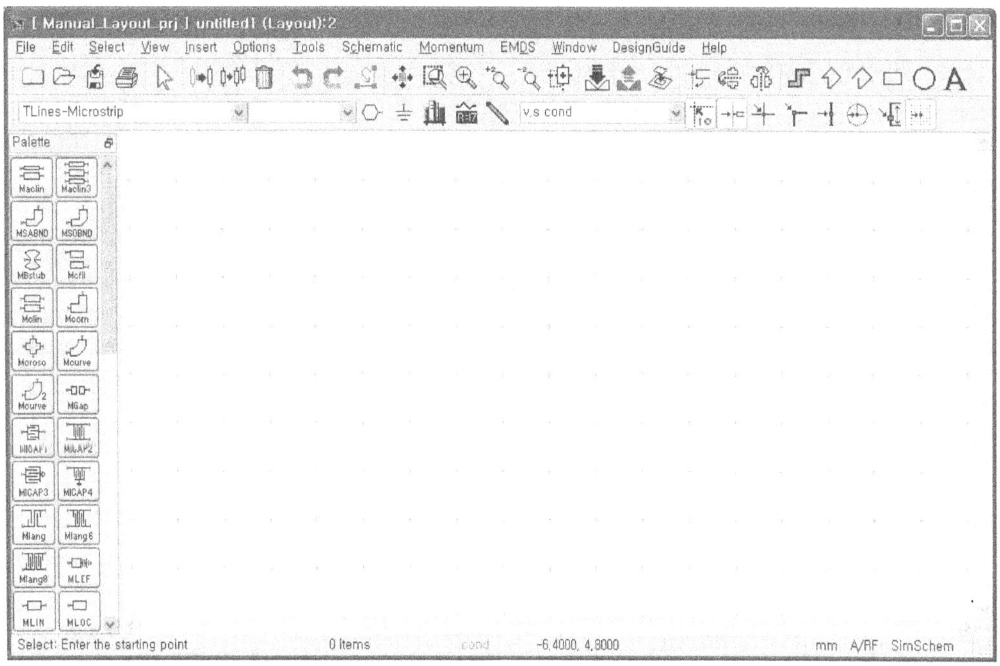

Figure 7.25 Layout window

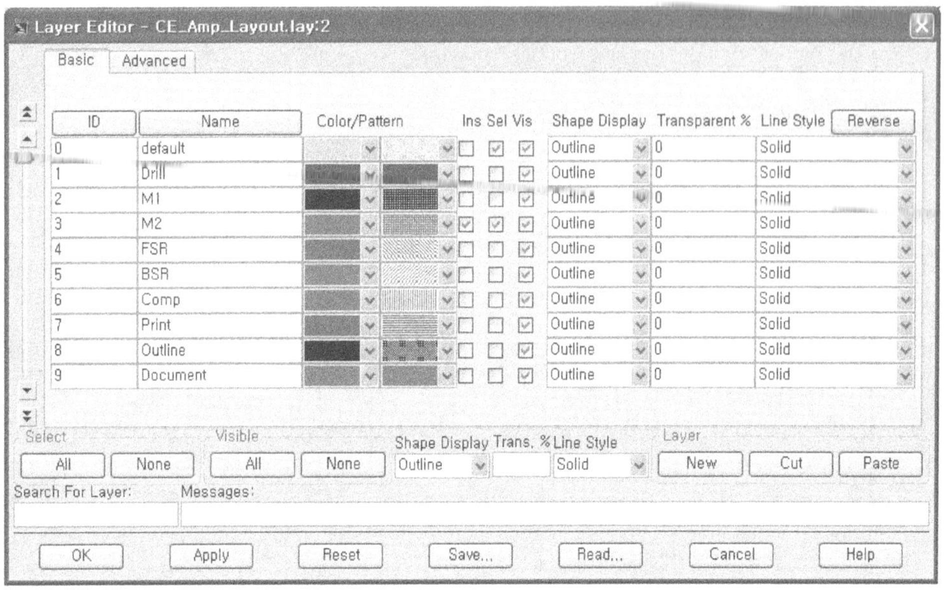

Figure 7.26 **Layer Editor** setup. Layers from ID 1 to 5 are for the two-sided PCB process shown in Figure 7.21. **Print** (ID 7) is the layer where the patterns for assembly convenience are drawn. **Comp** (ID 6) is the layer for component drawings, while **Outline** (ID 8) is the layer where the size of the PCB circuit is drawn. **Document** (ID 9) is the layer where the text labels for the components are drawn.

The Layer Editor's respective functions are as follows:

- *Name*: defines the name of each layer.
- *Color/Pattern*: represents the colors and fill patterns of each layer.
- *Ins, Sel, Vis*: represent the state of the layer. When *Vis* (Visible) is first selected, the drawings on this layer become visible, while anything on the unselected layer is not visible. Only the layers in *Sel* (Select) can be selected or edited. One layer selected in *Ins* (Insert) is available for inserting and drawing, and the drawing that has been worked on appears in this layer.
- *Shape Display*: shows the shape of a specified layer. *Outline* shows the shape by outline that is not filled, while *Filled* shows a filled shape.
- *Line Style*: specifies the line shape of the outline.
- *New, Cut, Paste*: *New* is used to create a new layer. Selecting a layer's *ID* or the *Name* of an unused layer and clicking *Cut* removes that layer. The previously cut layer can be returned to the desired location using *Paste.*
- *Save..., Read...*: used to save a newly created layer definition or recall a previously used layer definition, respectively.

A layer is defined as shown in Figure 7.26. Since a two-sided substrate is used in the figure, the seven basic layers described in Figure 7.21 are required. In addition to the seven layers, the layer **Outline** is added for drawing the size of the circuit, which also becomes the substrate size. The layer **Document** is added for the document of the circuit layout, which helps to clarify the roles of each component in the layout. Generally, the layer **M2** is used as the ground for the **M1** layer that uses a plated through hole and has no patterns for component mounting. Thus, the **M2** layer is used for the circuit ground and most of **M2** layer is considered to be filled with a conductor for the circuit ground. However some component terminals, such as the DC voltage terminal and connectors, should not be grounded and they should be soldered to the patterns on the **M2** layer. Thus, the drawings on the **M2** layer are interpreted as *slots* in the area where the conductor on the **M2** layer is removed. Similarly, the drawings on the layers **FSR** and **BSR** are also considered as slots as on the **M2** layer. All other layers automatically generated are removed using *Cut* except the designated layers and **default** layer with **ID** 0. The resulting layers are defined as shown in Figure 7.26. When the layer definitions are completed, click *Save...* and save the layout as **CE_Amp_Layout**. The saved file has the extension **.lay*.

7.3.3 Layout Units and Grid Set

On the Layout Window menu bar, click *Options* > *Preference* and the *Preferences for Layout* window shown in Figure 7.27 appears. Click the *Layout Units* tab and select *mm units*.

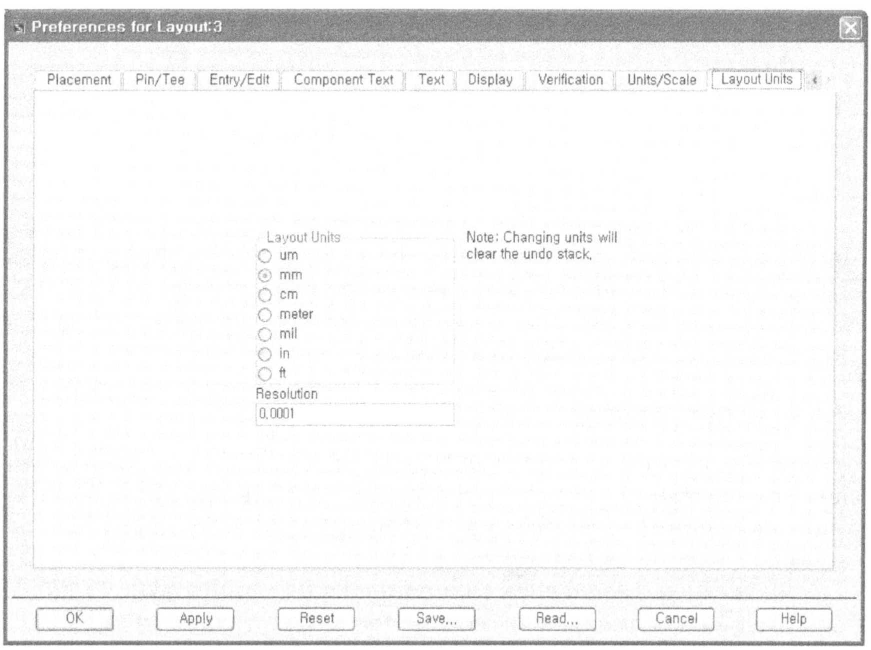

Figure 7.27 **Layout Units** settings tab in the **Preferences for Layout** window

After clicking *Apply*, select the *Grid/Snap* tab, which is shown in Figure 7.28. First, select the check boxes for the *Minor Grid* and *Major Grid* options in the *Display* area on the left-hand side of the window and then select the *Dots* option in the *Type* area. Next, set the *Spacing* as shown in Figure 7.28. *Snap Grid Distance* is the spacing for snap. Minor grid spacing is set by the *Snap Grid Per Minor Display Grid* values. Major grid spacing is set by *Minor Grid Per Major Display Grid* values. For the grid thus set, the spacing between two adjacent minor grids is 0.02 mm and the spacing between two adjacent major grids is 0.1 mm.

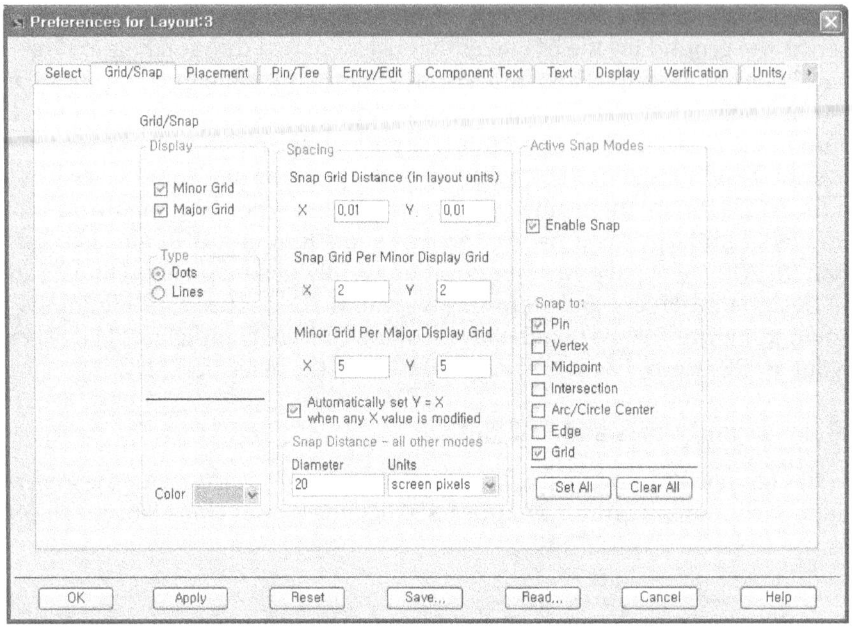

Figure 7.28 **Grid/Snap** settings in the Preferences for Layout tab. As a result of the settings shown, the minor grid has 0.02 mm distance and the major grid has 0.1 mm distance. Pointing with the mouse below 0.01 mm is not allowed because **Snap Grid** is set to 0.01 mm.

To use the snap mode, the *Enable Snap* option is checked in the *Active Snap Modes* area on the right-hand side of the window in Figure 7.28. With this setting, an object's shape will be placed only on *Snap Grid*. Thus, a polygon's vertices or a circle's center or radius can only be placed using *Snap Grid* when inserting, moving, and resizing objects. After all the settings have been made, click *Save.*

7.3.4 Outline Setting

In order to insert a shape on the **Outline** layer, specify the layer to be drawn as the **Outline** layer by selecting the *Ins* check box for the **Outline** layer in the Layer Editor window shown in Figure 7.26 or select **v,s Outline** from the drop-down menu on the right-hand side of the window in Figure 7.25.

In addition, click the icon on the toolbar in the layout window shown in Figure 7.25 to set the origin at the center of the screen. If the origin does not appear, click the icon until it does.

Basic shapes can be drawn by selecting these toolbar icons: . Their functions are as follows:

- can be used to draw traces (lines with width), and simultaneously set line widths and bend shapes. In the layout window, it can connect devices like wires.
- is used when drawing a polygon.
- is used to draw a polyline.
- is used to draw a rectangle.
- is used to draw a circle.
- s used to insert text.

Select *Options* > *Preference* on the menu bar in Figure 7.25 and then select the *Entry/Edit* tab, as shown in Figure 7.29. Then, check the box for *Show Coordinate Entry Dialog for* ... shown in Figure 7.29.

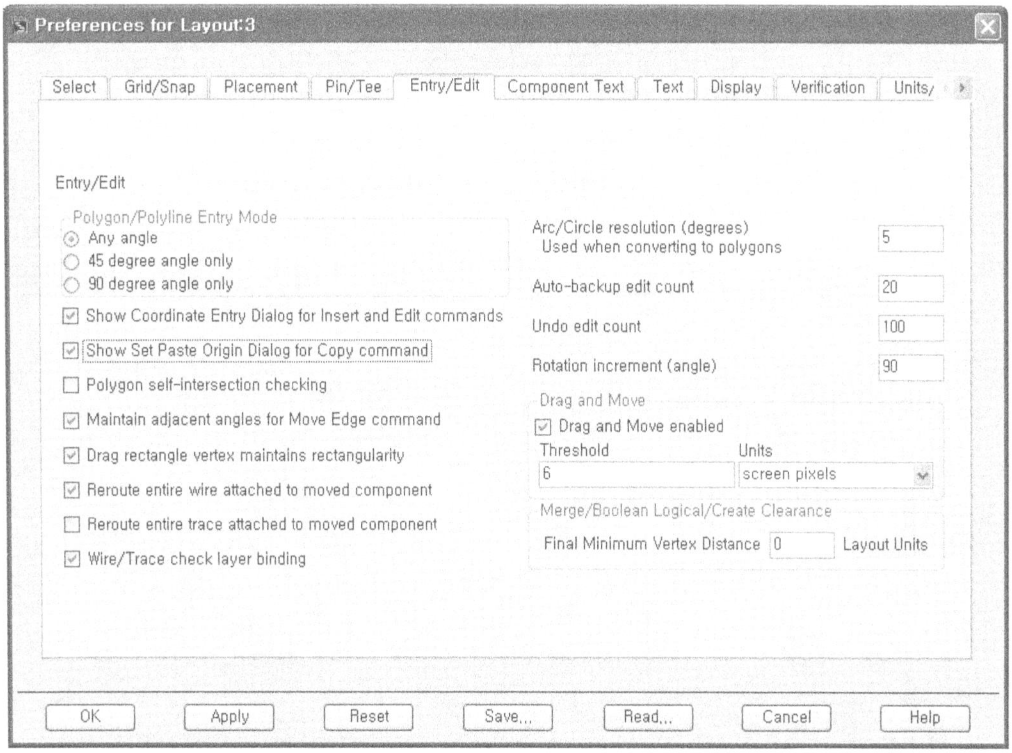

Figure 7.29 **Entry/Edit** tab of the Preference for Layout window. By selecting **Show Coordinate Entry Dialog for Insert and Edit commands**, the coordinate entry by number is enabled. Sometimes the number entry is convenient.

In order to specify the size of the circuit, click the *rectangle icon* from among the previously described toolbar icons. The *Coordinate Entry* window shown in Figure 7.30 appears as a result of the *Show Coordinate Entry Dialog for...* option that was checked previously. The coordinate values of the starting points X and Y are entered as **0** and **0** and, after clicking Apply again, enter **60** for X and **40** for Y and then click *OK*.

At the end of this process, a rectangle 60 mm in horizontal length and 40 mm in vertical length is created, as shown in Figure 7.31. Note that the bottom-left corner of the rectangle is at the origin (0, 0). If you choose to draw without selecting the *Show Coordinate Entry Dialog*... option, then after clicking , it is possible to move the mouse to the desired position by left clicking and then dragging.

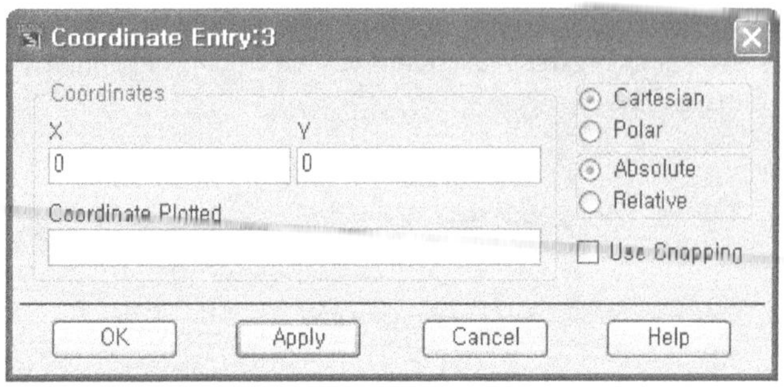

Figure 7.30 **Coordinate Entry** window

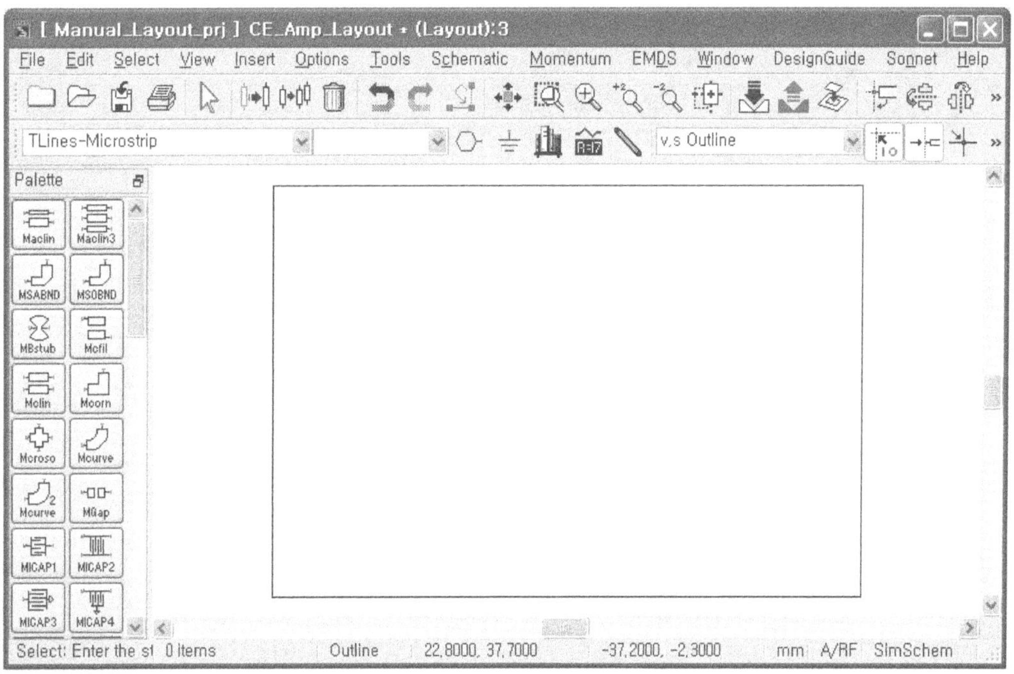

Figure 7.31 The completed outline

7.3.5 Component Layout

We will first discuss how to draw the component layout of the 1608 chip.

1. Click *File* > *New Design* in the window shown in Figure 7.31 and a *New Design* pop-up window appears. Enter the name **Chip1608** and click *OK*. A new layout window named **Chip1608** opens. Specify the inserting layer as **Comp** and draw a rectangle 1.6 × 0.8 mm by inputting from (0, 0) to (0.8, 1.6) in the *Coordinate Entry* pop-up window. This represents the size of the 1608 chips.
2. Specify the inserting layer as **M1**, click the rectangle icon, and draw a rectangle at the *Coordinate Entry* from (0, –0.4) to (0.8, 0.3) and from (0, 1.3) to (0.8, 2). These rectangles will be used for land patterns.

3. Specify the inserting layer as **FSR** in the same way and draw a rectangle with the same size and at the same position as **M1**. **FSR** means that the SR material will be coated outside the rectangle drawn on the FSR layer. Figure 7.32 shows such a drawing and only one color can be seen because of the overlap of **M1** and **FSR**. Click and click Save... to close the window.

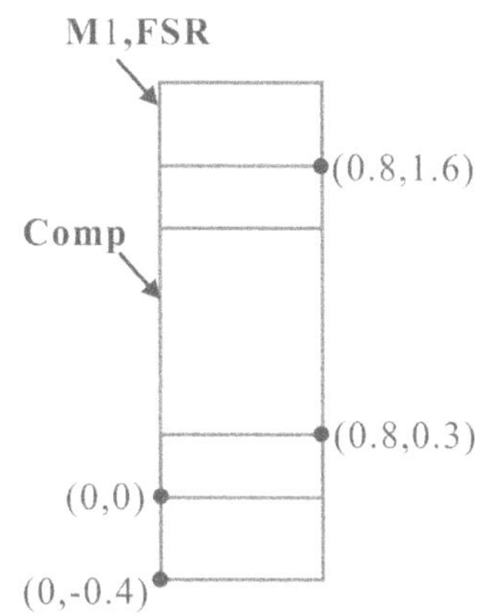

Figure 7.32 1608 chip component layout

4. Open a new layout window and save it as **BCX**19 to create the BJT component layout, as shown in Figure 7.33. Use the coordinates shown in that figure and draw the shapes on the **M**1, **FSR**, and **Comp** layers in the same way as was done in Steps 2 and 3. Select the **Print** layer as the inserting layer and draw on the **Print** layer by clicking the icon. The shape drawn on this layer will be marked on the PCB board in white to indicate the position of the BJT component. Select the **Document** layer as the inserting layer and write the name of the **BCX**19 chip as well as the name of each terminal, as shown in Figure 7.33. Save the file after completing these tasks and close the window

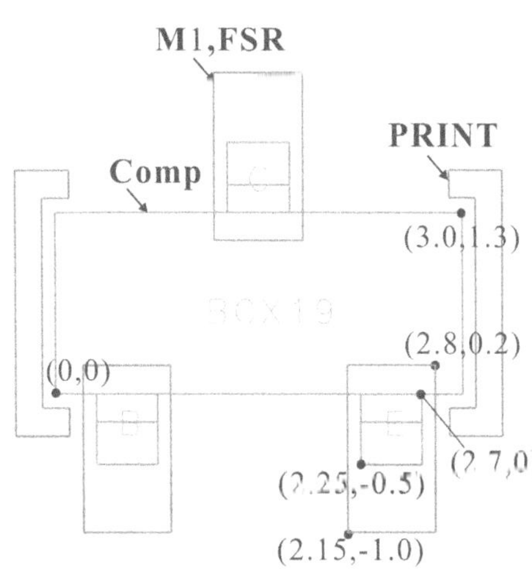

Figure 7.33 BJT (BCX19) chip component layout

5. Open a new window and save it as **GRD**. For the purpose of connecting the device terminal to the ground, the configuration shown in Figure 7.34 is required. Drawing a circle on the **Drill** layer is required to indicate the size and position of the hole to be drilled. A rectangular pattern also must be drawn on the **M1** layer to connect the top conductor patterns. It should be noted that no shape needs to be drawn on the **M2** layer because that layer is filled with the conductor.

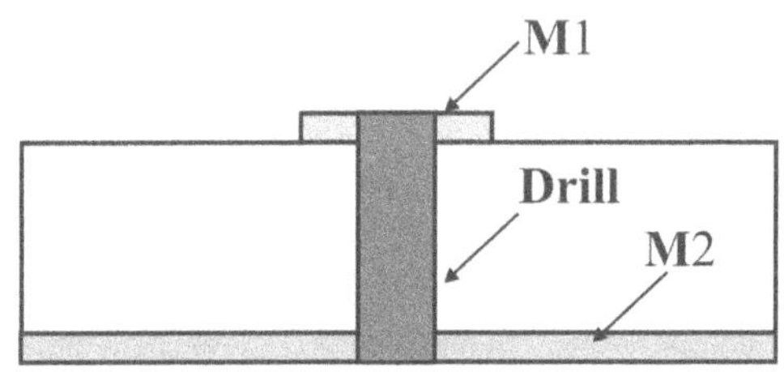

Figure 7.34 Sectional view of through hole to the ground

6. Specify the insertion layer as the **M1** layer and click the rectangle icon. Input (–0.4, –0.4) in the bottom-left corner in the *Coordinate Entry* box and, after clicking Apply, input (0.4, 0.4) in the top-right corner of the rectangle and click OK. This draws a 0.8-mm square rectangle centered at the origin. Select the insertion layer as the **Drill** layer and, in the same way, create a **0.25**-mm radius circle centered at the origin.
7. Then, click the polyline icon. Enter (0, –0.25) for the starting point of the polyline in the *Coordinate Entry* box and then click Apply; enter (0, 0.25) again and click OK; the line then follows the mouse, which you can move to the end of the solid line. By clicking the left mouse button at the end of the line, a solid line of length 0.5 mm is created and, by creating a similar line from (–0.25, 0) to (0.25, 0) on the **Drill** layer, the cross is created as shown in Figure 7.35. These lines will assist in determining the center of the circle.

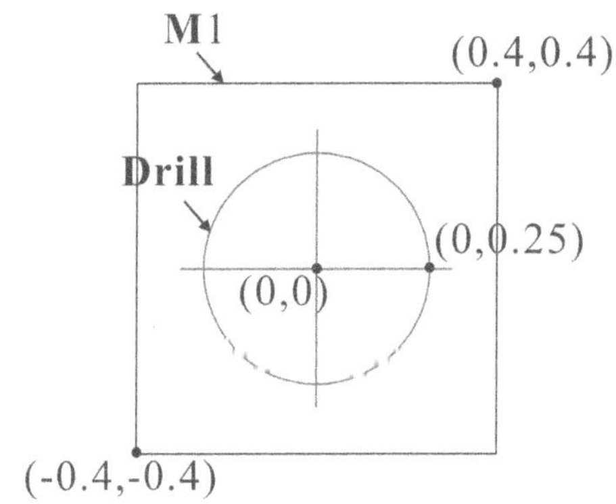

Figure 7.35 Ground component layout

8. When all the processes are finished, the **GRD** layout shown in Figure 7.35 is created. Save and close the window.

9. Open a new window and save it as **Bias_pgp**. Specify the insertion layer as the **Comp** layer and draw a rectangle from (0, 0) to (5.2, 10.2); also draw a line from (5.2, 8.77) to (3.76, 10.2) as shown in Figure 7.36. On the **Drill** layer, draw a circle with a radius of **0.5** mm centered at (3.25, 2.55). Also draw a 0.9-mm radius circle centered at (3.25, 2.55) as **M1**, **FSR**, and **BSR** layers, respectively. Writing **GND** in the **Document** layer will complete the ground terminal of the DC voltage supply terminal. Note that nothing must be drawn in the **M2** layer because this hole should be connected to the ground of the **M2** layer.

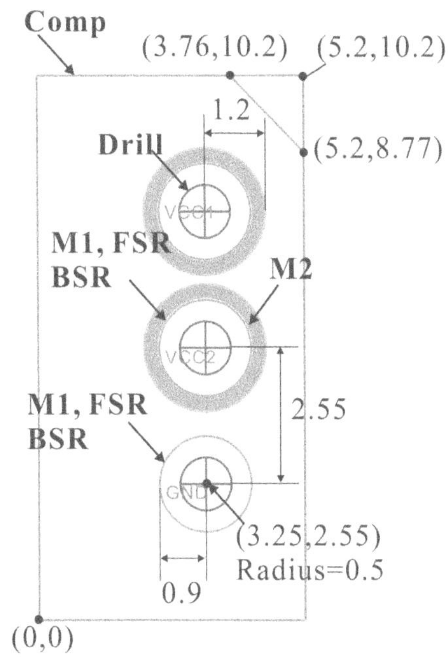

Figure 7.36 DC voltage supply terminal layout. Note that the ring patterns in the **M2** layer represent slot patterns and the terminals for **VCC1** and **VCC2** are not grounded.

10. Click *Edit* > *Advanced Copy/Paste* > *Copy Relative* in the menu bar to display the *Copy Relative* pop-up window shown in Figure 7.37. Then, enter (0, 2.55) and click Apply to form a circle with the same size and description as that centered at (3.25, 5.1). Draw a 0.9-mm radius circle centered at (3.25, 5.1) for **M2**.

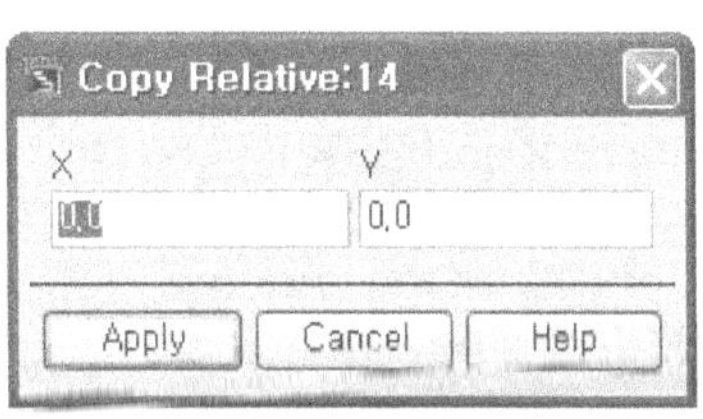

Figure 7.37 **Copy Relative** pop-up window

To prevent the terminal from being connected to the ground, a ring-shaped pattern is necessary. This pattern is interpreted as a slot on the **M2** layer. To draw the ring-shaped pattern, first draw a concentric circle with a radius of 1.2 mm on the **default** layer. The ring-shaped pattern can be generated by subtracting the circle drawn on the **M2** layer from the **default** layer. This can be performed by selecting the *Edit/Boolean Logical* command on the menu bar. The pop-up window appears; enter the settings shown in Figure 7.38.

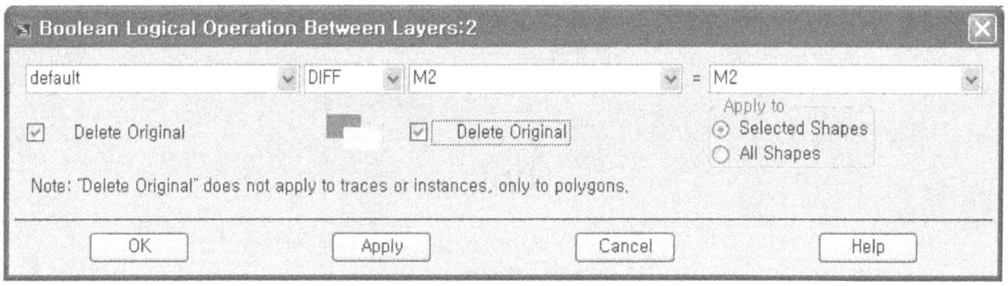

Figure 7.38 **Boolean Logical Operation Between Layers** window

Then, the ring pattern appears in the **M2** layer. This prevents the + source terminal from being shorted to the ground's bottom conductor during soldering. Write **VCC1** in the **Document** layer. Then, copy the drawings of the **VCC1** terminals to a relative position (0, 2.55) and modify **VCC1** as a **VCC2** terminal. Finally, save and close the window.

11. Open a new window and save it as **Connector**. Draw a circle with a radius of 0.8 mm and a cross centered at (–2.55, –2.55) on the **Drill** layer; then, draw circles with the same center of a 1.5-mm radius on the **M1**, **FSR**, and **BSR** layers.
12. Using *Copy Relative*, copy the drawings in step 11 centered at (2.55, –2.55), (2.55, 2.55), (–2.55, 2.55) and (0, 0), in that order. Draw circles with a 1.5-mm radius and centered at (0, 0) on the **M2** layer. The circles centered at (0, 0) are for the signal while the remaining circles are for the ground. In order that the circle centered at (0, 0) is not shorted by the bottom conductor, draw a circle with a radius of 1.9 mm and a center (0, 0) on the **default** layer and create the ring-shaped pattern on the **M2** layer using the Boolean logical operation shown in Figure 7.38.
13. Finally, draw a square from (3.15, 3.15) to (–3.15, –3.15) on the **Comp** layer, and this will complete the AC input and output connecter layouts shown in Figure 7.39. Save and then close the window.
14. Open a new window and save it as **PCB_Stand**. Draw a circle with a 1.75-mm radius centered at (0, 0) on the **Drill** layer; also draw a circle with a 3-mm radius and the same center on the **FSR** and **BSR** layers. Draw a rectangle from (5.08, 5.08) to (–5.08, –5.08) on the **M1** layer. This will create the **PCB Stand** component layout shown in Figure 7.40. Save and close the window.

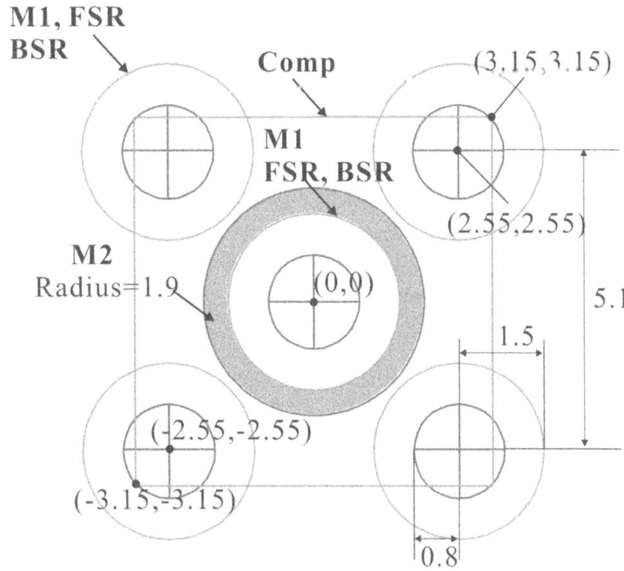

Figure 7.39 AC input and output connector layout

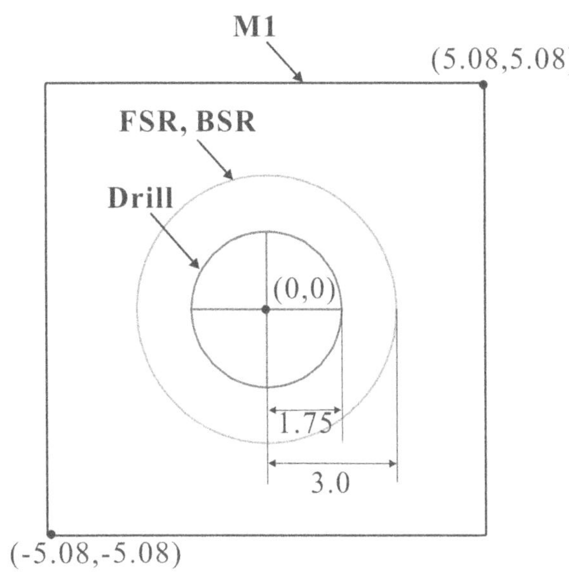

Figure 7.40 PCB Stand component layout

7.3.6 Layout Using Components

1. Open the **CE_Amp_Layout**, where the outline of the circuit is drawn and click the icon on the toolbar to display the *Component Library* pop-up window. Click *Projects* on the left-hand side of the *Component Library* window to display, on the right-hand side of the window, all the *Components* created so far, as shown in Figure 7.41.
2. Insert the **PCB_Stand** by clicking the **PCB_Stand** in the *Component Library* window, as shown in Figure 7.41. Return to the current **CE_Amp_Layout** window and click the **PCB_Stand** component to place it in the layout window. Now, place four **PCB_Stand** components in the layout window. Since the **PCB_Stand** is intended as a support for the PCB board, the outline of the square is positioned as shown in Figure 7.42.

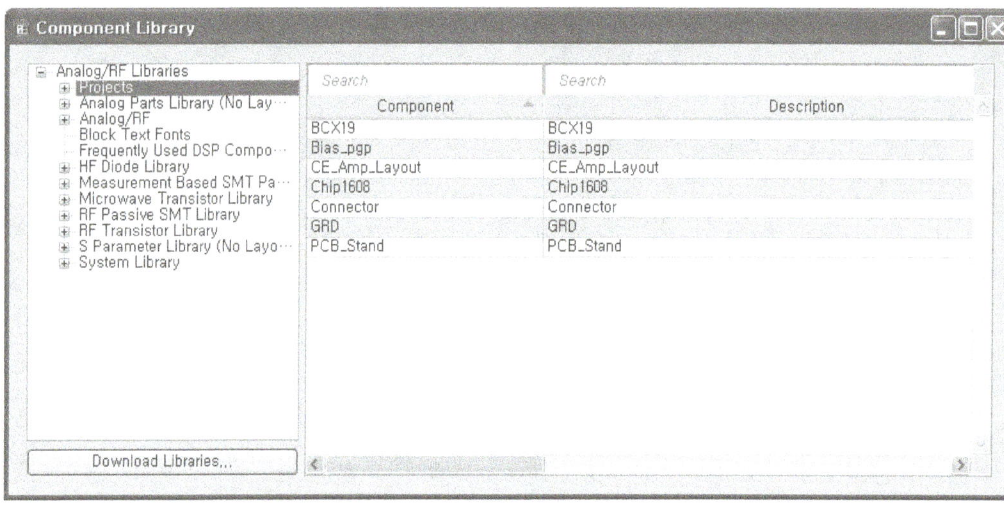

Figure 7.41 **Component Library** pop-up window

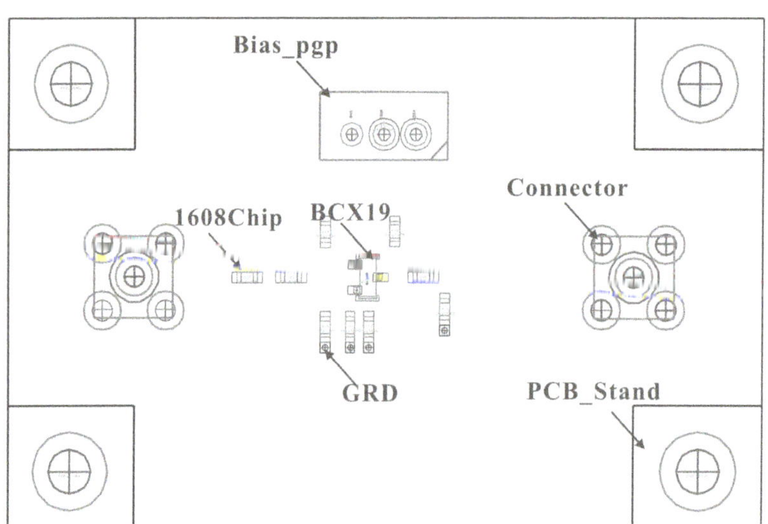

Figure 7.42 Components placement using the Component Library

3. All the other components are also placed in the outline of the **CE_Amp_Layout** window shown in Figure 7.42. As mentioned previously, by creating each component separately, they will not need to be redrawn each time and can be used repeatedly.
4. If you need to edit the component being used, click the component and then click the icon on the toolbar to move into the component window; then, after editing, move back to the previous task window by clicking Save... and . Note that although a single component has been edited, the other components can also be automatically edited at the same time.
5. To connect each device in the circuit configuration, the **M1** layer must be set to *Ins*. From the *Entry/Edit* tab of the *Preference for Layout* window, remove the check box selection for the *Show Coordinate Entry Dialog*. Click the icon to display the *Path* pop-up window shown in Figure 7.43. Set the *Width* to 0.5 and select the *Square* corner type.

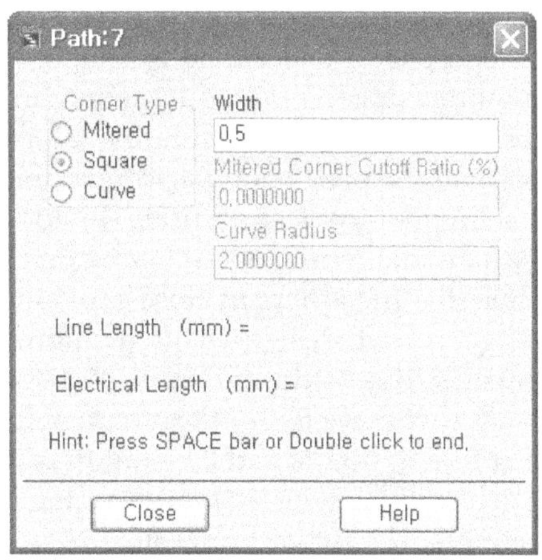

Figure 7.43 **Path** settings window

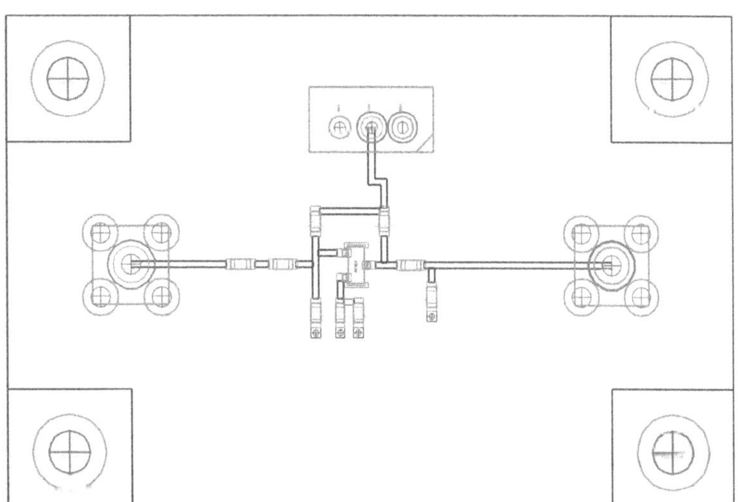

Figure 7.44 The completed **CE_Amp_Layout**

6. Move the pointer to one component's shape on the M1 layer and click the mouse. The trace line begins. Move the pointer to the other component's shape on the M1 layer to connect them and, by double clicking, a straight connecting line is drawn between the two components. Routing a line with many segments also can be done using a similar method. The line begins by clicking. Then, move the pointer to the line segment's endpoint and click to complete the segment. By repeating this procedure, a line with many segments can be drawn and double clicking at the end point of the line will complete the line's routing. Connecting all the components with routing lines, complete the layout as shown in Figure 7.44.

7.4 MOMENTUM

A circuit designed with simulation is usually implemented on a planar substrate by using microstrip lines. However, in a practical layout, as discussed previously in Chapter 3, discontinuities frequently occur when a circuit is designed using microstrip lines. Some of the discontinuities are modeled as circuit components in ADS and their effects can be predicted to some extent through circuit simulation. However, these circuit models are basically approximations and the prediction performed with a circuit simulation that includes the discontinuities is, in practice, less accurate. In addition, a situation sometimes occurs for economic reasons in which only a small area on the board is reserved for the circuit area. A microstrip line that meanders in close proximity to other lines is inevitable in the layout. While creating a layout, designers often ignore most of the effects that are not precisely modeled. However, due to the existence of these effects, the circuit's behavior may be different from what was expected in the circuit simulation. To overcome this problem, designers often insert tuning points in the layout to adjust its geometry; this can help them achieve the desired goal of the layout, but these tuning points are undesirable and, depending on the circumstances, it may be impossible to insert them, especially in the case of MMIC. The trial-and-error method of inserting tuning points can be avoided if the circuit behavior can be accurately predicted before fabrication.

To calculate passive structures, methods based on electromagnetic theory have been developed in the past and are still being studied but the method of calculation varies considerably depending on the structures. A general numerical method for solving general structures remains an extremely difficult problem even to date. However, in our examples, most of the microstrip line circuits are implemented on planar structures and, in most cases, the substrate structure is vertically uniform. We will present an efficient calculation method for these cases that uses the integral equation of Green's function (impulse response that satisfies the boundary conditions in space).

The structure analysis that uses Maxwell equations for electromagnetics is known as EM simulation (electromagnetic simulation). The EM-simulation tool for a planar structure is built into ADS and is called Momentum. In this section, the principles of Momentum will be briefly explained, after which an example of an S-parameters calculation method for a passive structure will be discussed.

7.4.1 Theory

A simple structure that illustrates the Momentum calculation principle is shown in Figure 7.45, where a conductor pattern S of an arbitrary geometry is placed on the dielectric substrate. All the surfaces that surround the structure are considered to be the perfect conductors for simplicity. Note that the cross-sectional structure is uniform everywhere when viewed vertically. The problem is to find the two-port S-parameters of the structure shown in Figure 7.45. The various formulations for the structure shown in the figure have already been presented in the literature on electromagnetics and we will explain the computation method based on that research.[3,4]

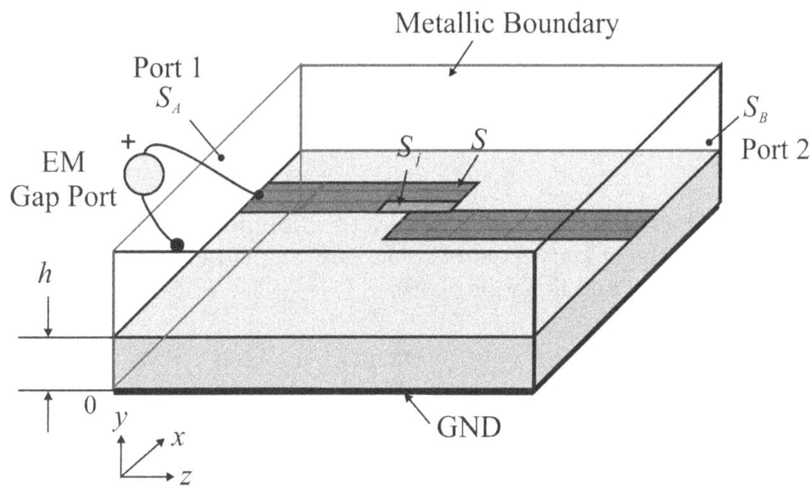

Figure 7.45 An example of a two-layer planar structure. The top conductor pattern is divided into small subsections called a mesh. In the mesh, the field is assumed to be uniform.

Denoting the unknown current-density vector **J** flowing on the conductor surface S, the tangential electric field **E** on the conductor surface is given in Equation (7.16) as

$$\mathbf{E}_t(\mathbf{r}) = \iint_S \mathbf{G}(\mathbf{r} \mid \mathbf{r}')\mathbf{J}(\mathbf{r}')dS' \tag{7.16}$$

Here, **r** represents the position vector at the point of observation, while **r′** represents the position vector of the source. The surface S represents the conductor pattern on which the unknown current flows. The current density **J** is a vector with two spatial components, x and z. The electric field **E** can have three spatial components, x, y, and z. The y component of **E** can also be expressed with a similar equation; however, the y component is not required in order to solve the structure in Figure 7.45. Thus, the function $\mathbf{G}(\mathbf{r} \mid \mathbf{r}')$ is a dyadic Green's function expressed in Equation (7.17).

$$\mathbf{G}(\mathbf{r} \mid \mathbf{r}') = \begin{pmatrix} G_{xx} & G_{xz} \\ G_{zx} & G_{zz} \end{pmatrix} \tag{7.17}$$

---

3. J.C. Rautio and R.F. Harrington, "An Electromagnetic Time-Harmonic Analysis of Shielded Microstrip Circuits," *IEEE Transactions on Microwave Theory and Techniques*, vol. 35, no. 8, pp. 726–730, August 1987.

4. J. Mosig, "Arbitrarily Shaped Microstrip Structures and Their Analysis with a Mixed Potential Integral Equation," *IEEE Transactions on Microwave Theory and Techniques*, vol. 36, no. 2, pp. 314–323, February 1988.

Equation (7.16) represents the electric field **E** for the unknown current **J**. Due to the uniform vertical structure, the Green's function can be obtained as various forms that use electromagnetic theory; for more information, refer to the reference in footnote 4. Thus, the electric field for all points in space can be calculated if the current density is known.

In order to solve the integral Equation (7.16), the conductor shape S above is divided into small subsections called a *mesh* or a *cell*, as shown in Figure 7.45. The size of the mesh is assumed small enough for the fields to be considered uniform. Thus, the fields can be expanded using a *basis function*. Denoting the basis function as $\mathbf{B}_i(\mathbf{r})$ defined on the mesh-i in S_i, then $\mathbf{J}(x, h, z)$ in $(x, h, z) \in S$ can be written as shown in Equation (7.18).

$$\mathbf{J}(\mathbf{r}) = \sum_{i=1}^{N} I_i \mathbf{B}_i(\mathbf{r}) \tag{7.18}$$

The basis function $\mathbf{B}_i(\mathbf{r})$ is a vector basis function defined on $(x, h, z) \in S_i$ of mesh i, and I_i then becomes the unknown quantity. Substituting Equation (7.18) into Equation (7.16) and integrating, we obtain Equation (7.19).

$$\mathbf{E}_t(\mathbf{r}) = \sum_{i=1}^{N} I_i \iint_{S_i} \mathbf{G}(\mathbf{r} \mid \mathbf{r}')\mathbf{B}_i(\mathbf{r}')dS' \tag{7.19}$$

Here, $\mathrm{E}_t(\mathbf{r})$ is the tangential electric field generated by the mesh S_i. Similarly, the tangential electric field $\mathbf{E}_t(\mathbf{r})$ can be expressed using the same basis function $\mathbf{B}_i(\mathbf{r})$. The tangential electric field appearing at mesh i can be expressed as shown in Equation (7.20).

$$\mathbf{E}_t(\mathbf{r}) = \sum_{i=1}^{N} V_i \mathbf{B}_i(\mathbf{r}) \tag{7.20}$$

Substituting Equation (7.20) into Equation (7.19) and multiplying $\mathbf{B}_i(\mathbf{r})$ on both sides of Equation (7.19) results in

$$V_i = \sum_{j=1}^{N} Z_{ij} I_j \tag{7.21}$$

In Equation (7.21), Z_{ij} is defined by Equation (7.22).

$$Z_{ij} = \iint_{S_i} \mathbf{B}_i(\mathbf{r}) \bullet \left(\iint_{S_j} \mathbf{G}(\mathbf{r} \mid \mathbf{r}')\mathbf{B}_j(\mathbf{r}')dS' \right) dS \tag{7.22}$$

Now, Equation (7.19) becomes a matrix equation as shown in Equation (7.21) that can be numerically solved.

To obtain the two-port S-parameters, the exciting fields at the input and output ports are necessary. However, the exciting fields are seldom known directly. Imagine an infinitesimal gap between the port surface S_A and the conductor pattern. This gap obviously does not significantly distort the computed $\mathbf{G}(\mathbf{r} \mid \mathbf{r}')$. Now, assume that a unit gap voltage source is connected between the edge of the conductor pattern S and port surface S_A. The other port surface S_B still remains short. In Equation (7.21), due to the unit gap voltage, the value of V_i in the port cell can be set to 1, while all other V_i can be set to 0. All the voltages in other cells are 0 because they are a tangential electric field on the perfect conductor. I_j in

Equation (7.22) can then be solved using a conventional numerical algorithm. Thus, the unknown **J** can be found. The port currents at ports 1 and 2 then become the summation of the current densities of the port cells. This defines the two-port Y-parameters for the structure shown in Figure 7.45. The Y-parameters y_{11} and y_{21} can be determined as a result. By interchanging the roles of the ports, the remaining Y-parameters can be found. Thus, the complete S-parameters for the structure shown in Figure 7.45 can be determined through Y- to S-conversion. Such an excitation obviously generates port discontinuity, which can be calibrated and the S-parameters can be obtained without port discontinuity. See reference 2 at the end of this chapter.

Sometimes the S-parameters inside a conductor pattern are required rather than the box shown in Figure 7.45. For example, to minimize the effects of discontinuities on ports, the discontinuities are set to occur sufficiently far away from the ports. As a result, higher-order modes will not be present at the ports. In this case, the S-parameters defined on the inside of the geometry become the designers' real interest. In this case, the electric fields at the ports S_A and S_B can be considered as those that appear in the geometries where the port surface is infinitely extended. Therefore, the electric field shape can be numerically determined as a two-dimensional problem for the ports, and the transmission-line parameters, such as characteristic impedance and propagation constant, can be obtained from the two-dimensional solutions. As a result, the S-parameters inside the geometry can be obtained by the shift of the reference plane using the computed transmission-line parameters. The transmission-line parameters for the ports are generally included with the S-parameter solutions for the structure shown in Figure 7.45.

7.4.2 Settings and EM Simulation

7.4.2.1 Substrate and Layout Layers

As explained earlier, because Momentum provides electromagnetic solutions for a planar structure (a vertically uniform dielectric structure), the substrate structure (the vertical structure) composed of multiple substrate layers must first be specified. Also, users should map multiple metallization layers on the substrate layers. For the specified substrate and metallization structures, Momentum calculates the Green's function and stores it for calculation efficiency.

Figure 7.46 shows a window for determining the substrate structure. On the menu bar of the ADS Layout window, select *Momentum > Substrate > Create/Modify* to display the window shown in Figure 7.46. The *Substrate Layers* tab in the figure defines the structure of the substrate. The dielectric structure in this example consists of *FreeSpace*, *Alumina*, and *GND*. When multilayers are required, a new substrate layer can be added by clicking *Add*; layers can also be deleted by clicking *Cut*. The boundary conditions for *FreeSpace* are the same as the properties in free space. *FreeSpace* can be set to a shielded structure, as shown in Figure 7.45, where a conductor plate is placed at the ceiling. By clicking the layers in the dielectric structure, its parameters can be modified. For example, click the dielectric layer *Alumina* and activate the window that defines the attributes of that dielectric layer. The *Dielectric Thickness* and *Substrate Layer Name* can be specified. The specifications for each layer are completed by entering the *Thickness*, *Dielectric Permittivity*, and *Dielectric Loss Tangent* information for each dielectric. Its permittivity *Er* and permeability *Mu* can be defined in a variety of formats. The permittivity and permeability are defined in terms of the general *Re*, *Loss Tangent* form and it is also possible to define them in other ways.

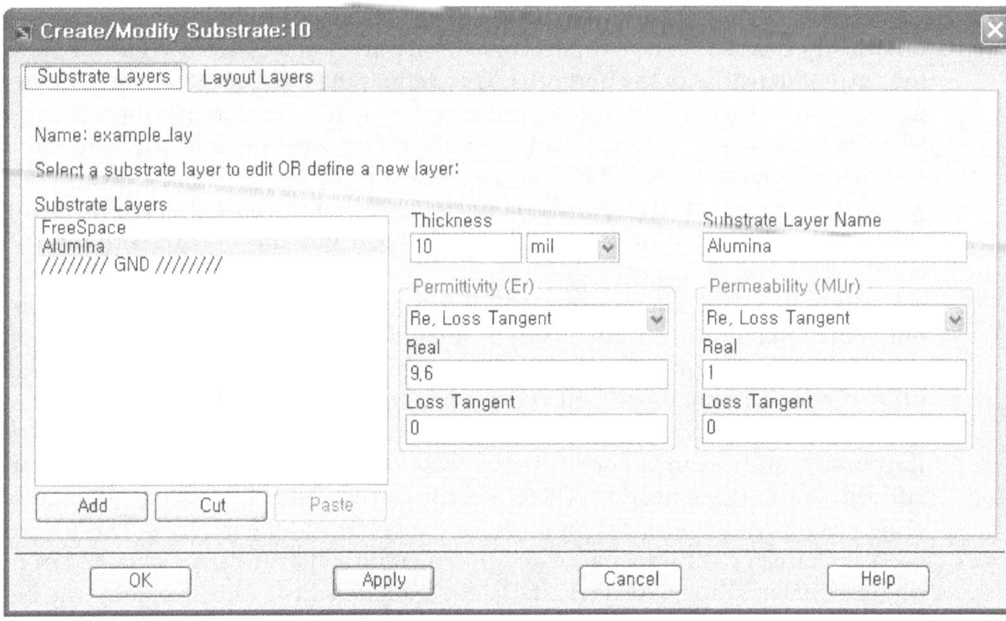

Figure 7.46 **Substrate Layer** tab window

Next, the location of the metallization must be specified for the calculation of the Green's function, which is done by opening the *Layout Layers* tab and specifying the metallization on which the substrate layer is located. The window for the *Layout Layers* tab is shown in Figure 7.47. This is marked "........." in Figure 7.47. The shapes in the specified layer can be interpreted as *Strip*, *Slot*, and *Via*. *Strip* means that the drawing in the layout layer is interpreted as a conductor pattern on the mapped substrate layer. If *Slot* is selected, the drawing in the layout layer is interpreted as a slot on the mapped substrate layer; that is, the negative image of the drawing is interpreted as a conductor on the mapped substrate layer. *Via* means the drawing in the layout layer is interpreted as a via through the mapped substrate layer. Also, *Unmap* is used to cancel the layer mapping. In Figure 7.47, the layer named **cond** is specified on top of the **Alumina** substrate layer, and the drawing on the **cond** layer is interpreted as conductor patterns on the **Alumina** substrate. The conductor's conductivity can be specified by adjusting the Layout Layer conductivity parameters. Finite conductivity as well as a perfect conductor can be specified. In addition, the resistive material with sheet resistance can also be used. Finally, *Overlap Precedence* is used to give priority to a layer when there are more than two layers mapped on the same substrate.

This process is absolutely necessary if you do not want to use the substrate provided in ADS, but intend to enter the substrate information directly. Once specified as described above, the Green's function required to perform the simulation can be computed. To do this, click *Momentum* > *Substrate* > *Precompute Substrate Function*. This computes the Green's function for the desired frequency range and, because the Green's function will be used repeatedly, it does not matter even if the calculation is done for a wider frequency range.

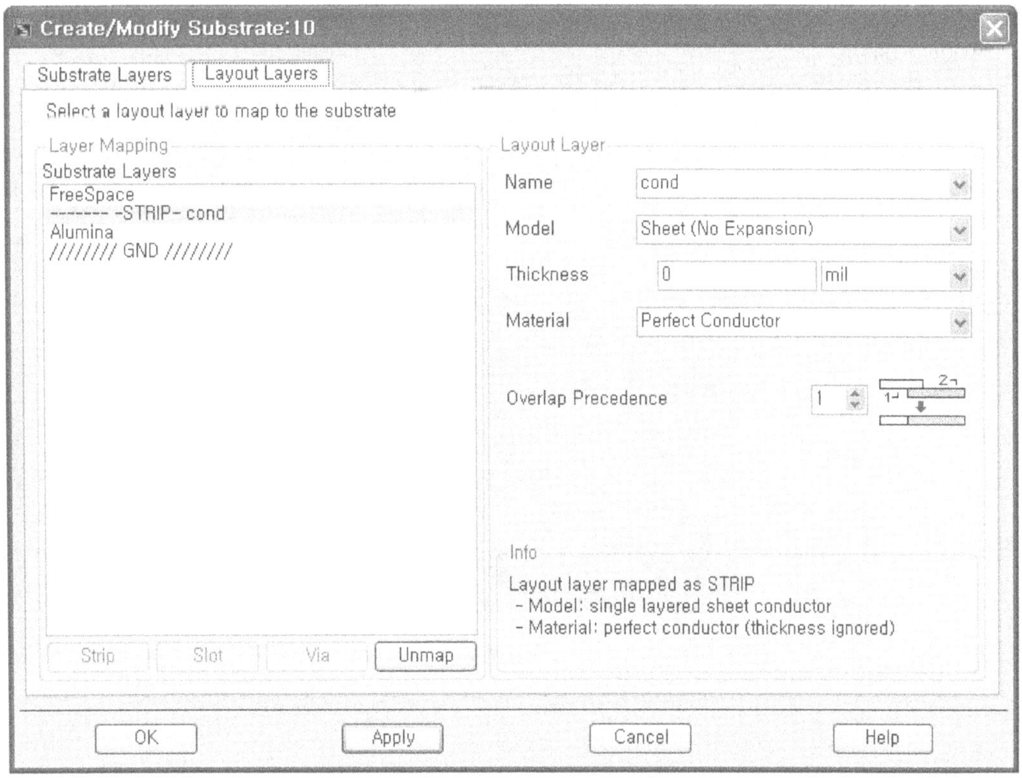

Figure 7.47 **Layout Layers** tab window

7.4.2.2 Ports and Reference Planes

Figure 7.48 is a conductor shape drawn on the layer named **cond**. A microstrip line having a 25-mil width and 100-mil length is inserted on the substrate shown in Figure 7.46 and 7.47. The microstrip line is drawn first as a circuit in the Schematic window; the drawing is generated using the *Layout > Generate/Update Layout* commands on the menu bar of the Schematic window.

The input and output ports must be set up in order to perform Momentum simulations. To this end, the layer where the ports are connected must be specified as the inserting layer. After selecting the **cond** layer as the inserting layer, the ports can be attached by clicking the icon. Then, the drawn shape appears, as shown in Figure 7.49.

To set the parameters of the port, select *Momentum > Ports > Editor* on the menu bar. Click the port to display the Port Properties Editor window shown in Figure 7.50. Port modes include *Single*, *Internal*, *Differential*, *Coplanar*, *Common*, and *Ground Reference*. A *Single Mode* port is assigned by default, which can be widely used for a general microstrip-structure simulation. The port impedance can be set in the *Impedance* field and it can be set to a complex value using the *Real* and *Imaginary* fields. The port impedances become the reference impedances for the S-parameters. *Reference Offset* moves the reference line of the computed S parameters. It is used to remove the effect of the length of the microstrip lines. When the *Port Properties Editor* settings are completed, new perpendicular lines are created that indicate the position of the reference plane.

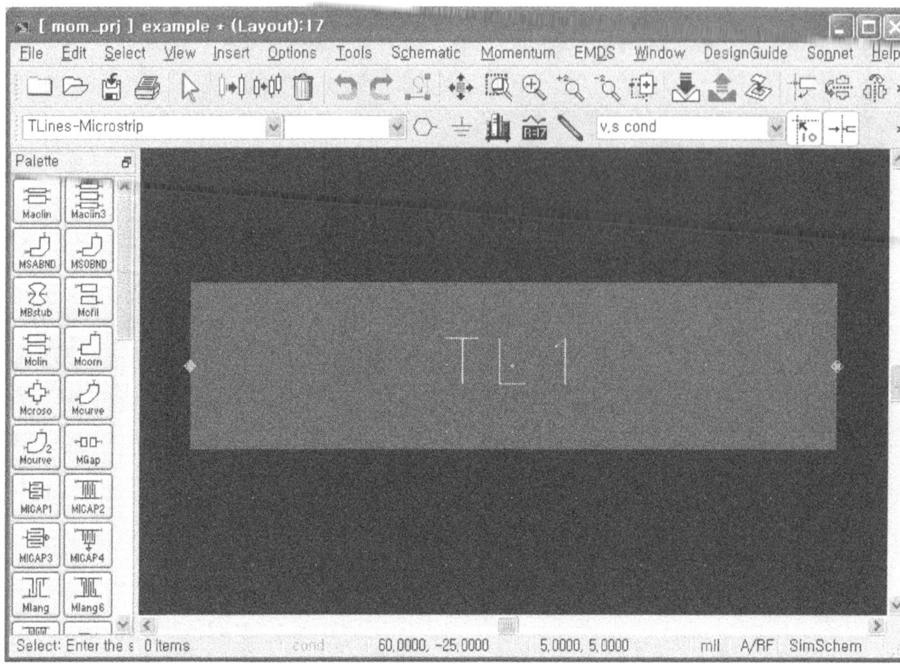

Figure 7.48 25-mil-wide, 100-mil-long microstrip line

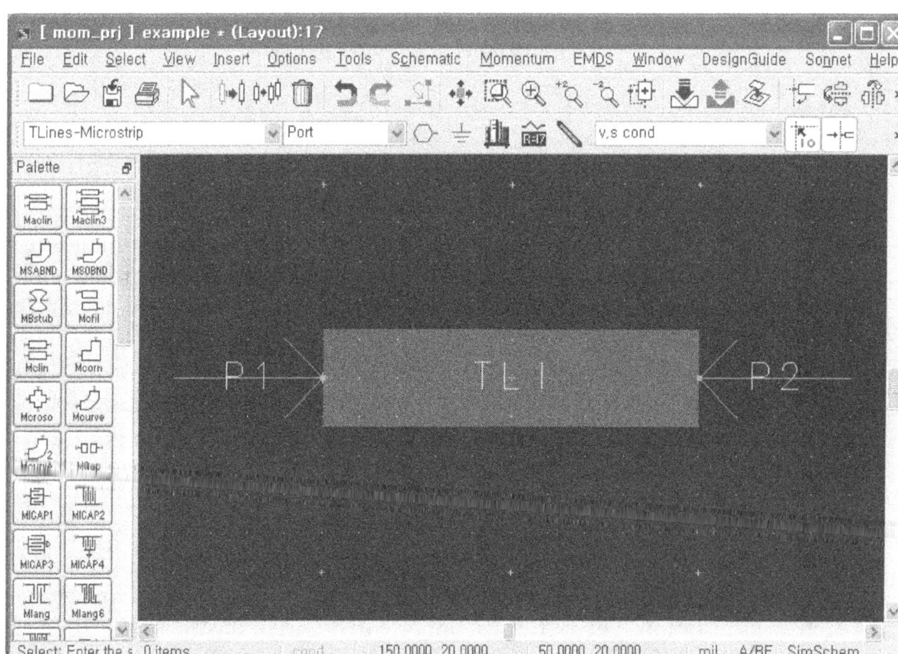

Figure 7.49 The drawing after the ports are inserted

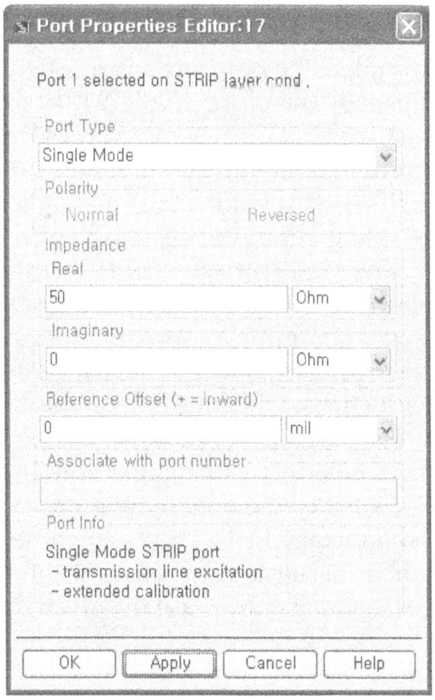

Figure 7.50 Port Properties Editor window

Figure 7.51 shows the window after completing the port setup. The reference lines can be moved by entering the *Reference Offset* parameter in the *Port Properties Editor* window and can also be moved using the mouse by selecting the reference line and dragging it.

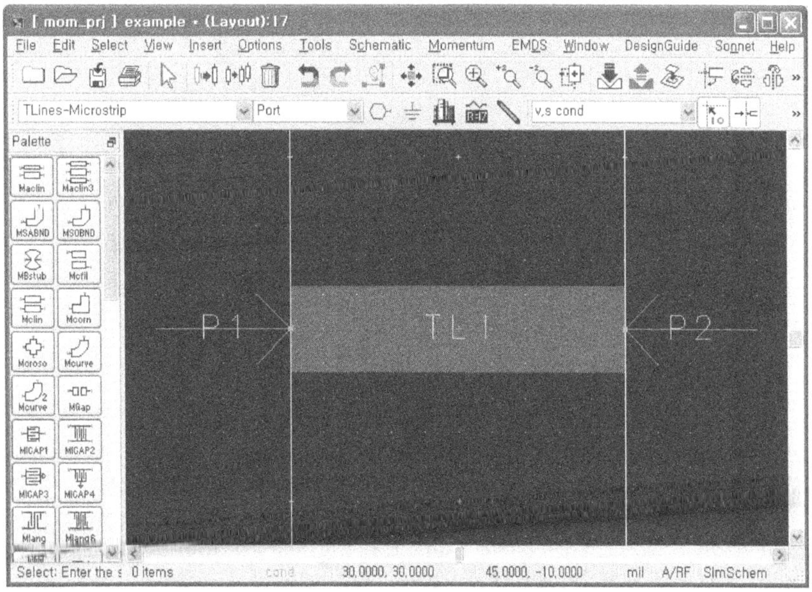

Figure 7.51 The window after completing the Port Properties Editor settings

7.4.2.3 Meshes

When the work to complete the port setup is finished, the electromagnetic problem setup of a structure is completed. Therefore, after creating a mesh in Momentum, the previously described matrix Equation (7.21) is generated and solved. Even when a user does not specify the mesh, Momentum automatically generates the mesh after which the calculation is performed; however, the user can adjust the generation parameters of the mesh.

The *Mesh Frequency* is set in the *Global* tab of the Mesh Setup Controls window, as shown in Figure 7.52. The *Mesh Frequency* field determines the reference frequency for mesh generation and is normally set to the highest frequency of simulation. The *Mesh Density* field determines the cells per unit wavelength. In the case of a circuit that is 3-wavelengths long, setting the *Mesh Density* to 20 will create a total of 60 cells. A dense mesh is required in the case of complex structures, while the default mesh value is adequate for simple circuits such as microstrips. The curve portions of the circuit are divided into triangular meshes. The *Arc Resolution* field determines the angle of triangular meshes and can be set up to a maximum of 45°. The smaller the angle, the more exact the analysis, but smaller angles should be avoided because they make processing time too long. The *Transmission Line Mesh* field defines the number of mesh per unit width of a microstrip line, and is disabled by default. Usually, the remaining entries use the given default values.

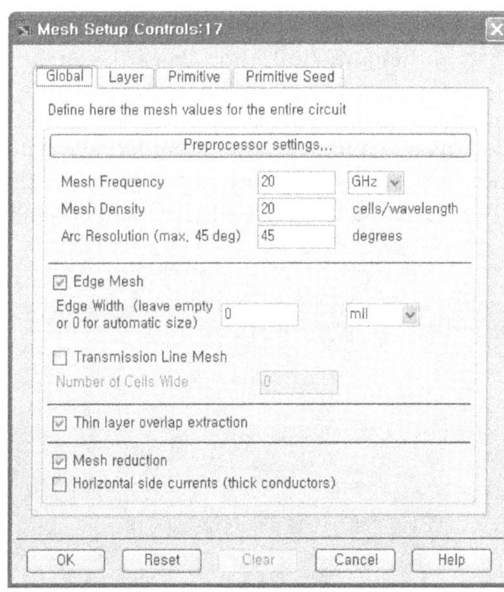

Figure 7.52 Mesh Setup Controls window

The conductor surface specified for Momentum calculations is divided into triangular or rectangular meshes. To set the mesh parameters, click *Momentum* > *Mesh* > *Setup* and the Mesh Setup Controls window shown in Figure 7.52 appears.

After the mesh setting is finished, the meshes can be shown in advance by clicking *Momentum* > *Mesh* > *Preview*; then, the highest frequency of the simulation frequency range, (20 GHz as shown in Figure 7.52) is entered. Click OK and the layout window will display the mesh, as shown in Figure 7.53.

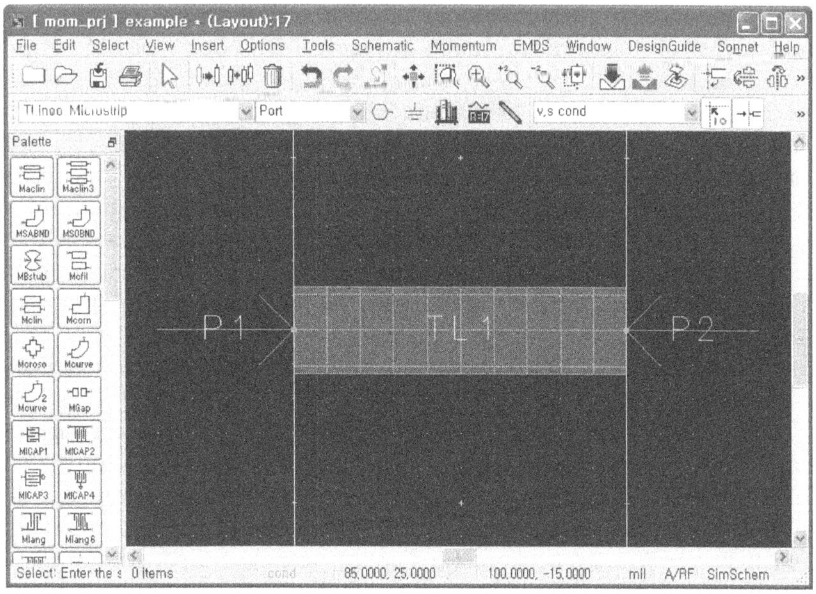

Figure 7.53 Microstrip lines are divided by mesh

7.4.2.4 Simulation and Data

Now the settings for Momentum simulation are completed. For the simulation, when you select *Momentum > Simulation > S-parameters*, the window shown in Figure 7.54 appears.

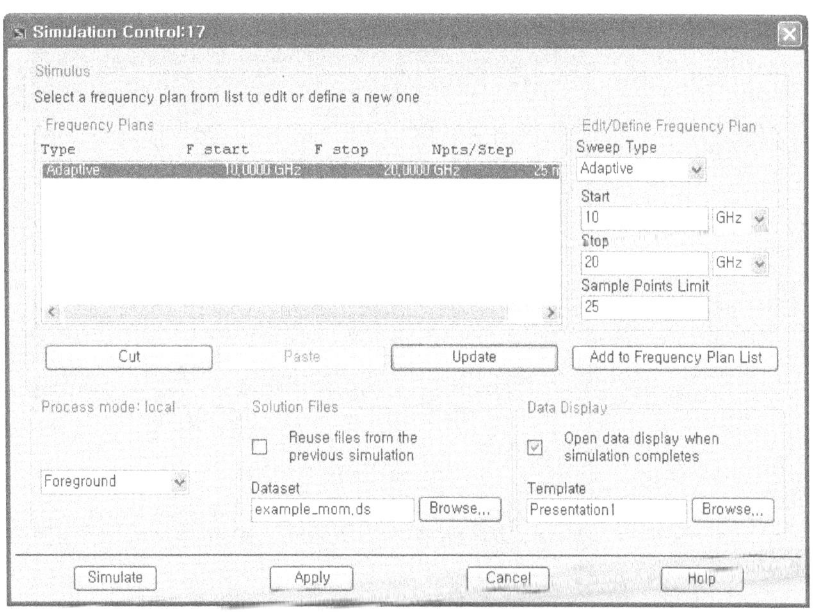

Figure 7.54 Simulation Control window

Simulation frequencies can be selected using the *Edit/Define Frequency Plan* in the top-right corner of the window shown in Figure 7.54. Four types of frequency selections are provided: *Adaptive, Logarithmic, Linear* and *Single Point*. Adaptive simulation frequencies are computed by adaptive algorithm. Since the S-parameters of the given frequency range can be obtained by interpolating a small number of S-parameter samples, the algorithm determines the simulation frequencies so as to eliminate the uncertainty of the interpolated S-parameters. The Adaptive method is used because Momentum simulation requires a long computation time. The *Sample Points Limit* field specifies the maximum number of samples. The Logarithmic selection sets the simulation frequencies to increase logarithmically. The simulation frequencies are specified by start and stop frequencies, and the numbers per decade. The Linear selection sets the simulation frequencies linearly to increase from the start to the stop frequencies. The Single Point selection is the simulation for one frequency.

The calculation results are stored as a dataset in the directory *data*. A user can identify the dataset by entering its name in the *Dataset* field; the default name is **layout name_mom**. There is a check box for whether or not previously simulated data will be reused; this, however, is usually not selected by default. In Figure 7.54, since *Open data display when simulation completes* is checked, the display window is opened automatically after simulation. The simulation frequency range is set to 10–20 GHz and the *Sweep Type* in Figure 7.54 is set to *Adaptive* and *Sample Points Limit* is set to 25.

When the simulation is finished, the display window will automatically open due to the selection of that option in Figure 7.54. In order to view the results, click the icon *Rectangular plot* in the display window. The Plot Traces & Attributes window in Figure 7.55 opens and it will contain the computed S-parameters shown in Figure 7.55. The **S** here represents the S-parameters normalized by the port impedances set by the Port Properties Editor during Momentum set up. The port impedances were set to 50 Ω. **S**_50 represents

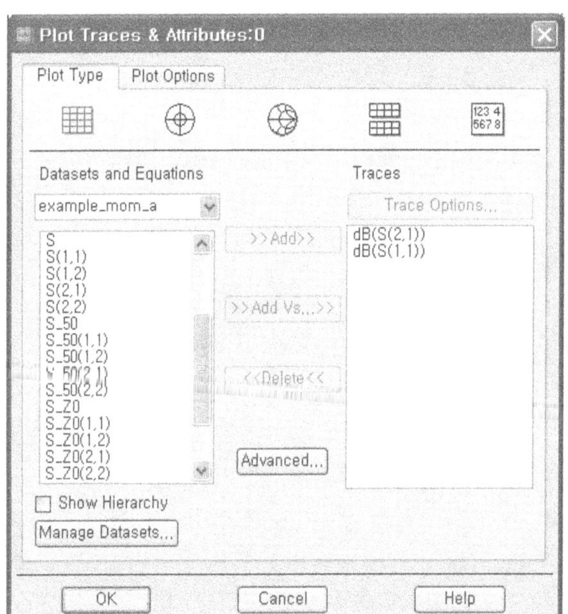

Figure 7.55 Plot Traces & Attributes window

the S-parameters normalized by 50 Ω. Therefore, **S_50** and **S** are the same S-parameters in this case. However, if the port impedances set by the Port Properties Editor are different, the two are different S-parameters. Similarly, **S_Z0** represents the S-parameters normalized by impedance **Z0**. Momentum first solves the port impedances and propagation constants based on the cross-sectional shape of the ports, as explained previously. The results are stored in **Z0** and **Gamma**, where **Z0** represents the characteristic impedance and **Gamma** represents the propagation constant of the ports. Thus, **S_Z0** represents the S-parameters normalized by impedance **Z0** that is determined by Momentum.

Figures 7.56 and 7.57 show **S** and **S_Z0**. The previously defined 100-mil-long by 25-mil-wide microstrip line has a characteristic impedance of approximately 29.3 Ω and an

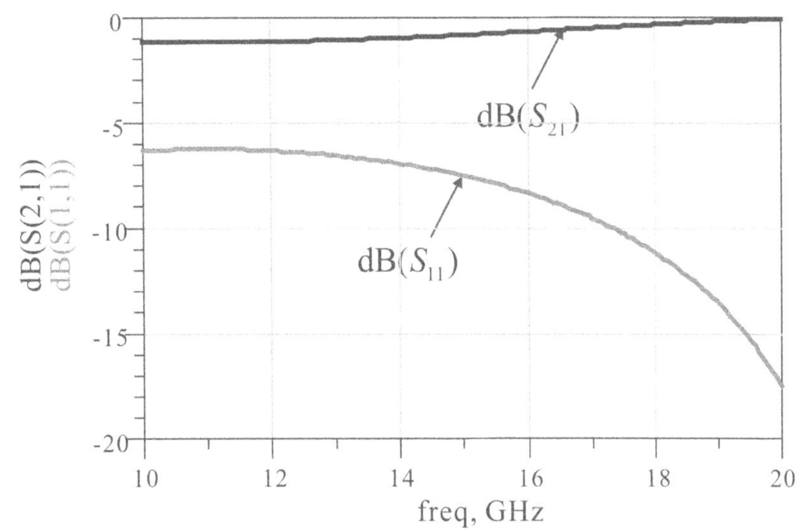

Figure 7.56 Momentum simulation results. The S-parameters represented by **S** in the dataset are the S-parameters with the reference impedances specified by the Port Properties Editor.

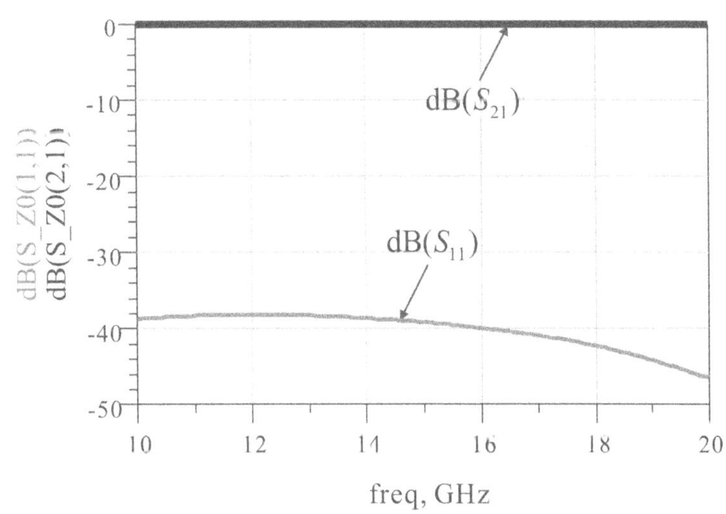

Figure 7.57 S_Z0 results. The **S_Z0** represents the S-parameters with the reference impedances Z0, which represents the port impedances computed in the Momentum simulation.

electrical length of 82° at 10 GHz when computed using **LineCalc**. The Momentum calculation results of **Z0** show a 29.7-Ω characteristic impedance that is close to that of the **LineCalc** computation. The impedance **Z**0 is used as a reference impedance for the normalized S-parameters **S_Z**0. Since the analyzed structure is a simple microstrip line, **S_Z**0 will show the small insertion loss with respect to frequency. This is shown in Figure 7.57, while the S-parameters shown in Figure 7.56 exhibit a significant insertion loss and return loss due to mismatch.

However, in Figure 7.57 the return loss still shows an approximate value of −40 dB, which should be infinity. The exact reason for this is unknown, but may be due to dispersion in the microstrip line and cross-coupling of the port.

7.5 SUMMARY

- The simulation in ADS is performed in the ADS Schematic/Layout window and the work in this window is saved in the network directory of the working project. The simulated data is saved in the data directory of the working project. The data can be viewed in the display window and the display window data is saved in the working project directory.
- Circuit simulations can be generally classified into DC, transient, AC, and harmonic balance. The S-parameter simulation is a kind of AC simulation.
- Layout is classified into two types, manual layout and auto layout. The layers, grid, and components are the basic design concepts in drawing a layout.
- Momentum is the tool for EM solving of a planar structure. The substrate structure, layer metallization, and assignment of ports are necessary for EM simulation of a planar structure.
- After the Momentum simulation, the S-parameters **S**, **S**_50, **S_Z**0 appear in the dataset and their physical meanings are explained. Also, port data **Z**0 and **Gamma** appear in the dataset and they represent the port impedances and propagation constants, respectively.

REFERENCES

1. K. W. Yeom, *Electronic Circuits and Microwave Circuit Design Lab*, Daejeon, South Korea: Chungnam National University Press, 2004.
2. J. C. Rautio, "EM-Component-Based-Design of Planar Circuits," *IEEE Microwave Magazine*, vol. 8, no. 4, pp. 79–90, August 2007.
3. R. Gilmore and L. Besser, *Practical RF Circuit Design for Modern Wireless Systems*, vol. 2. Boston: Artech House, Inc., 2003.

PROBLEMS

7.1 In transient simulation, a capacitor can be represented by a resistor and a current source in a sufficiently short time, as shown in Figure 7P.1:

$$i = C\frac{dv}{dt} \cong C\frac{v - v_n}{T} = C\frac{v}{T} - C\frac{v_n}{T}$$

(1) Using this expression, the time response of the circuit below can be expressed as follows:

$$v_{n+1} = \frac{I_o + \frac{C}{T}v_n}{\frac{1}{R} + \frac{C}{T}} = R\frac{I_oT + Cv_n}{T + RC}$$

(2) In addition, using the expressions above for $I_o = 1$ A, $R = 1\ \Omega$, $C = 1$ F, and $v_C(0) = 0$ V, determine for a parallel circuit the voltage across the capacitor from 0 to $t = 10$ msec.

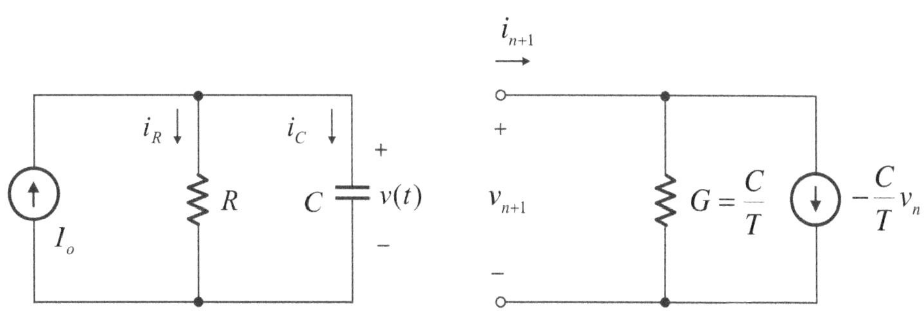

Figure 7P.1 Circuit of problem 7.1

7.2 (ADS problem) Determine the lumped-element matching circuit that matches a $10 - j3\ \Omega$ to $50\ \Omega$ at a frequency of 1 GHz using optimization.

7.3 Explain the *Gerber*, *DXF*, and *GDS II* formats.

7.4 (ADS problem)

(1) A four-layer PCB board is shown in Figure 7P.2. Construct a layer structure for PCB fabrication. A through hole for grounding the patterns on layers 1 and 3 is also required. Assuming a blind hole is not supported in the process of the PCB fabrication, draw a ground-hole component. In addition, construct a component to interconnect the patterns on layers 1 and 3.

(2) Create a PCB layout of the following circuit in Figure 7P.3: MAG83563 GaAs RFIC was used in this circuit, and the size and configuration are in the datasheet. Other passive components are all 1608-size chip components. First, look at the actual size and then draw the MAG83563 component (SOT-363 package) using the dimensions in the datasheet. Use the components created in this lab for the other components.

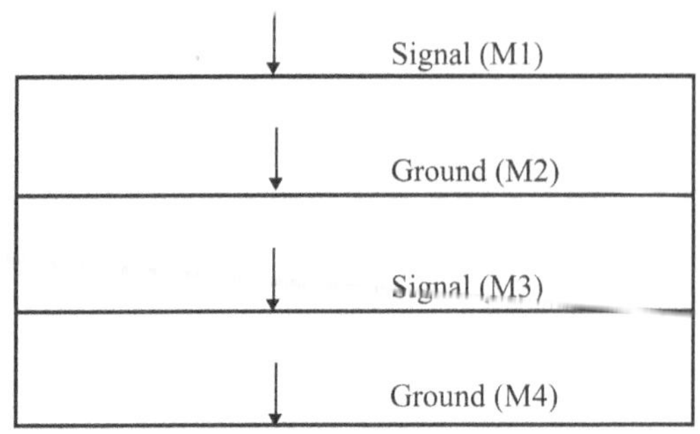

Figure 7P.2 PCB board cross-section of problem 7.4

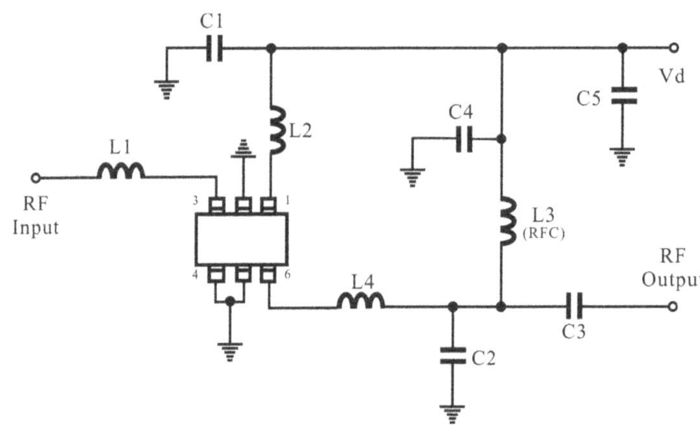

Figure 7P.3 Schematic of problem 7.5

7.5 A momentum simulation is performed for a 30-Ω microstrip line with an electrical length of 90° at 1 GHz. The port 1 and 2 impedances are set to 50 Ω and 30 Ω, respectively, using the Port Properties Editor, and Momentum simulated. For the simulated results in the dataset, answer the approximate values of **S**_Z0(2,1) and **S**(2,1).

CHAPTER

8

Low-Noise Amplifiers

CHAPTER OUTLINE

8.1 INTRODUCTION

A low-noise amplifier is generally placed at the front end of a receiver and plays the role of amplifying received signals that are weak. As shown in Figure 8.1, the input signal of the low-noise amplifier consists of both signal P_s and noise N_s that are amplified together. The impedance Z_o at the source and load has a typical value of 50 Ω, which is determined by a coaxial connector. In addition, the internal noise sources of the low-noise amplifier also contribute to output noise power N_o. As a result, the signal-to-noise ratio at the output is worse than that at the input. Thus, the low-noise amplifier should be designed so that its internal noise sources minimally contribute to the output noise power, and it should amplify the received signal to the extent that the next-stage signal processing components can process the signal.

The amplifier shown in Figure 8.1 can be generally represented as shown in Figure 8.2. Thus, the design of a typical low-noise amplifier must address the problem of determining the impedance looking into the source and load from the active device. In this chapter, we will first explain the gain that is an important measure of a low-noise amplifier as a function of the source and load reflection coefficients, Γ_S and Γ_L, shown in Figure 8.2,

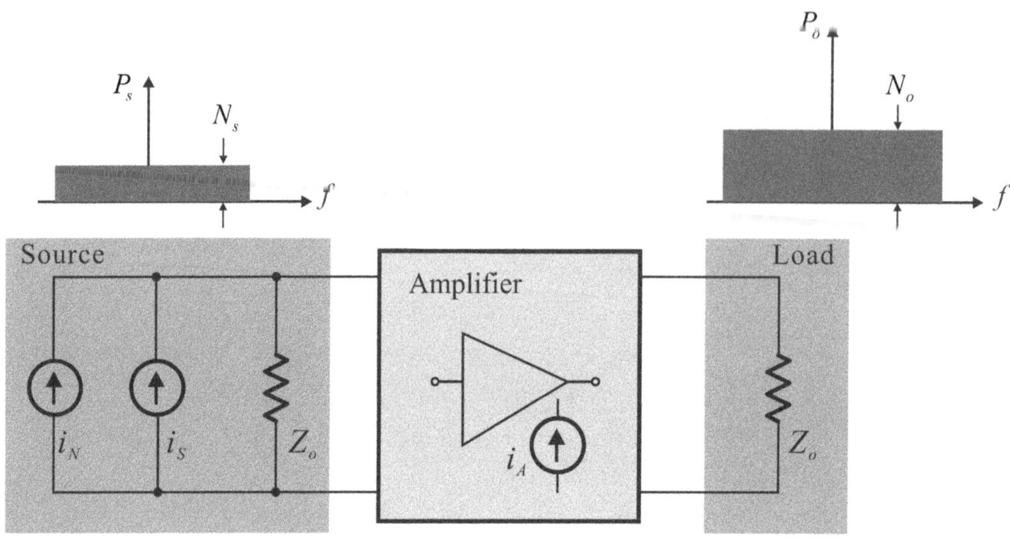

Figure 8.1 Low-noise amplifier design concept. The signal and noise powers P_s and N_s are represented by the current sources i_S and i_N, and P_o and N_o are the delivered signal and noise powers to the load. The current source i_A in the amplifier represents the amplifier's internal noise source. The impedance Z_o at the source and load side has a typical value of 50 Ω.

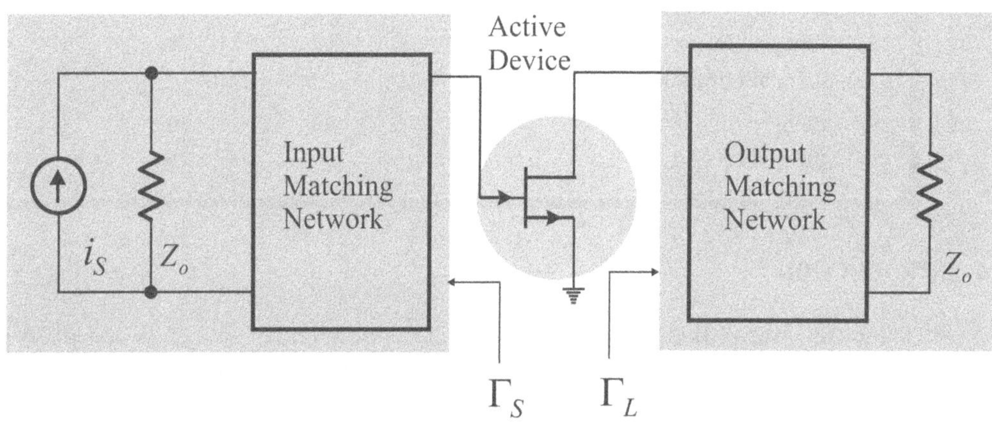

Figure 8.2 The definition of the source and load reflection coefficients Γ_S and Γ_L. Usually Γ_S and Γ_L are defined using the same reference impedance as those used for the S-parameters of the active device.

and we will also look at the Γ_S and Γ_L that give maximum gain from a design perspective. However, in designing a low-noise amplifier, not only the gain but also the noise figure must be taken into consideration. We will discuss how to select Γ_S and Γ_L by considering the noise figure and gain.

In addition, because the active device has some inherent degree of feedback, it shows a negative resistance for specific Γ_S and Γ_L, and because of this, the amplifier can turn into an oscillator. Therefore, the method for examining whether or not the active device yields a negative resistance at a given frequency will be presented, and then the method of stabilizing the active device in that situation will be explored.

8.2 GAINS

8.2.1 Definition of Input and Output Reflection Coefficients

The source and load impedances are generally fixed as $R_S = R_L = Z_o$. However, the optimum gain and noise figure of an active device are rarely obtained for the fixed source and load impedances $R_S = R_L = Z_o$. The gain and noise figure of the active device depend significantly on the load and source impedances. Thus, the fixed-value source and load impedances should be converted into the appropriate impedances using the matching circuits. Due to the inclusion of the matching circuits, the active device generally sees the converted source and load impedances that are different from Z_o. This is shown in Figure 8.3. Here, Γ_S represents the reflection coefficient looking into the input matching circuit from the active device, while the reflection coefficient looking into the output matching circuit from the active device is denoted as Γ_L. Note that the reference impedance is required in order to measure Γ_S and Γ_L, and it is assumed to be the same as the reference impedance used to measure the S-parameters of the active device.

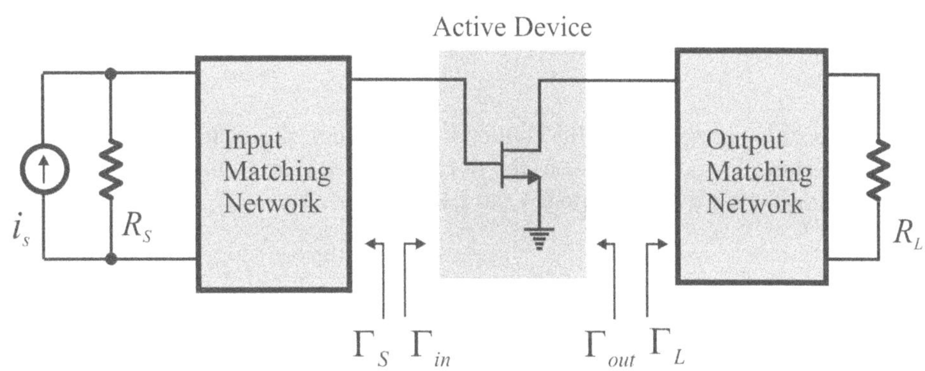

Figure 8.3 Definition of reflection coefficients Γ_S, Γ_L, Γ_{in}, and Γ_{out}

The reflection coefficient seen from the active device input when a load Γ_L is connected to the active device output is denoted as Γ_{in}; similarly, the reflection coefficient seen from the active device output when a source Γ_S is connected to the active device input is denoted as Γ_{out}. Therefore, using the S-parameters of the active device, the input and output reflection coefficients are expressed in Equations (8.1) and (8.2).

$$\Gamma_{in} = S_{11} + \frac{S_{12}S_{21}\Gamma_L}{1 - S_{22}\Gamma_L} \tag{8.1}$$

$$\Gamma_{out} = S_{22} + \frac{S_{12}S_{21}\Gamma_S}{1 - S_{11}\Gamma_S} \tag{8.2}$$

8.2.2 Thevenin Equivalent Circuit

Figure 8.4 shows a circuit for measuring the reflection coefficient of a Thevenin equivalent circuit, which is often used to represent a one-port network containing internal sources. In Figure 8.4, the Thevenin equivalent circuit is shown in the shaded area and has an open-circuit voltage E_T and impedance Z_T.

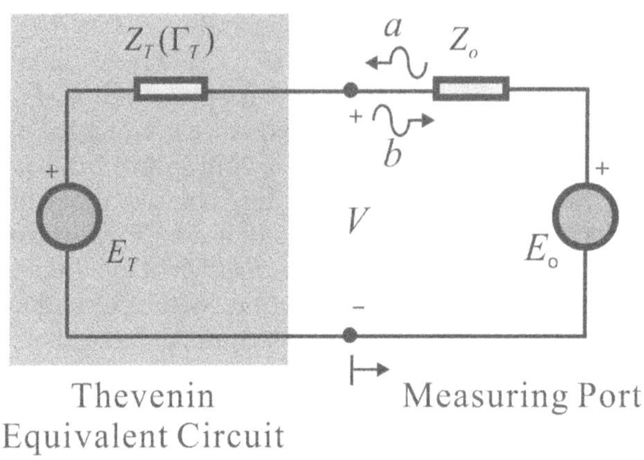

Figure 8.4 Wave representation of Thevenin equivalent circuit that has an open-circuit voltage of E_T and impedance Z_T. The measurement port has a voltage E_o and Z_o.

The port is connected to the Thevenin equivalent circuit to measure its reflection coefficient. Here, E_o and Z_o represent the port voltage and impedance, respectively. Using superposition, voltage V in Figure 8.4 can be determined as

$$V = E_o \frac{Z_T}{Z_o + Z_T} + E_T \frac{Z_o}{Z_o + Z_T} = V^+ + V^- \tag{8.3}$$

Voltage V is the sum of the incident and reflected voltages as shown in Equation (8.3). The incident voltage is given by

$$V^+ = \frac{E_o}{2} \tag{8.4}$$

Therefore, substituting Equation (8.4) into Equation (8.3), reflected voltage V^- from Equation (8.3) is given by

$$V^- = V^+ \frac{Z_T - Z_o}{Z_T + Z_o} + E_T \frac{Z_o}{Z_o + Z_T} \tag{8.5}$$

Expressing reflected voltage V^- in Equation (8.5) in terms of a normalized incident and reflected voltages a and b, we obtain Equations (8.6) and (8.7).

$$b = \frac{V^-}{\sqrt{2Z_o}} = \Gamma_T a + b_T \tag{8.6}$$

$$b_T = \frac{E_T}{\sqrt{2}} \frac{\sqrt{Z_o}}{Z_o + Z_T} \tag{8.7}$$

Reflected voltage b in Equation (8.6) consists of two terms. The first term is the reflection coefficient looking into the Thevenin circuit when $E_T = 0$. The second term is the reflection coefficient that depends on the voltage source E_T that can be interpreted as the voltage appearing across the port impedance Z_o when $E_o = 0$. In circuit theory, E_T is determined by measuring the open-circuit voltage of a one-port network. Similarly, impedance Z_T is obtained by measuring the impedance of the one-port network with all the internal sources in that network turned off. The reflection coefficient given by Equation (8.6) corresponds to this circuit theory. The reflection coefficient Γ_T is obtained by measuring the reflection coefficient with source E_T turned off. Similarly, the reflection coefficient due to source E_T is found by measuring the voltage delivered to the port impedance Z_o instead of measuring the open-circuit voltage. The sum of the two reflections constitutes the total reflection in Equation (8.6). The difference is in measuring the internal source contribution by using a termination Z_o instead of an open circuit.

Example 8.1

Prove that, $|b_T|^2$, obtained from Equation (8.7), is the power delivered to the port resistance Z_o when $E_o = 0$.

Solution

$$|b_T|^2 = \frac{1}{2} Z_o \frac{|E_T|^2}{|Z_o + Z_T|^2}$$

Because

$$P_L = \frac{1}{2} Z_o |I_L|^2 = \frac{1}{2} Z_o \left| \frac{E_T}{Z_o + Z_T} \right|^2 = |b_T|^2$$

it can be seen that $P_L = |b_T|^2$.

Now, the amplifier in Figure 8.3 can be represented by the simple schematic that uses the Thevenin equivalent circuit shown in Figure 8.5. The input can be represented by the Thevenin equivalent circuit, while the output can be represented by a simple load.

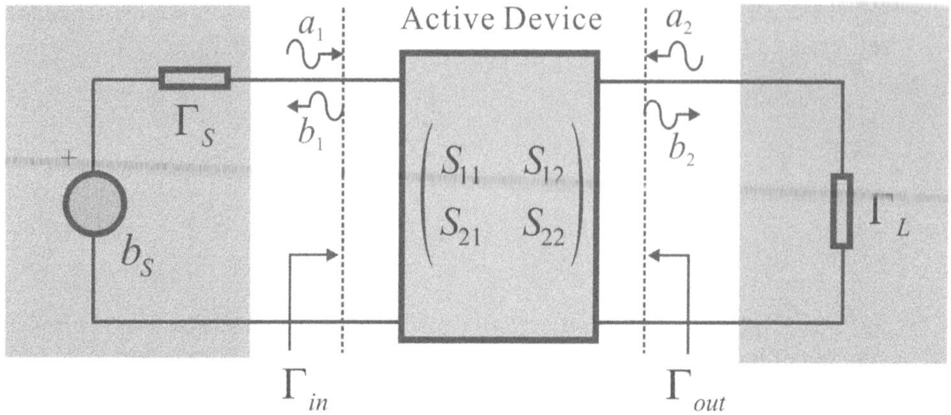

Figure 8.5 The simple schematic representation of the amplifier at the device planes

The active device is represented by the S-parameters, and the incident and reflected voltages (a_1, a_2) and (b_1, b_2) are defined using the reference impedances that measured the S-parameters of the active device.

Next, we examine the power delivered to the load when a source represented by a Thevenin equivalent circuit is connected to an arbitrary load Γ_L, as shown in Figure 8.6. Note that both the reflection coefficients of the load and source represented by the Thevenin equivalent circuit are measured based on the same reference impedance Z_o.

From Equation (8.6), the reflected voltage b_g toward the load in Figure 8.6 is expressed in Equation (8.8)

$$b_g = \Gamma_S a_g + b_S \tag{8.8}$$

while the reflected voltage from the load is shown in Equation (8.9).

$$b_L = \Gamma_L a_L \tag{8.9}$$

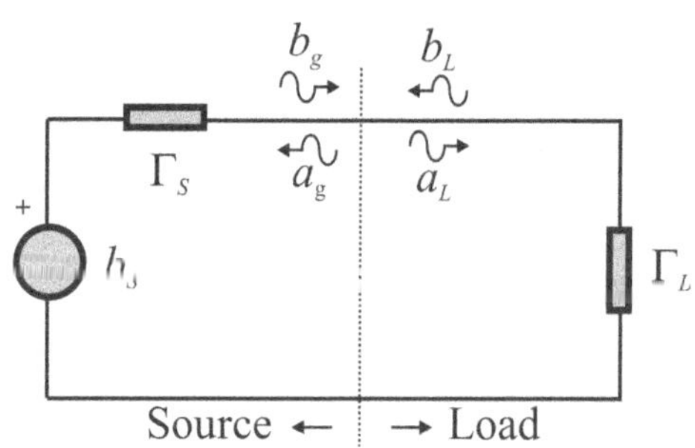

Figure 8.6 The circuit where the source is connected to the load. The source and load reflections are measured using the same reference impedances.

Since $b_g = a_L$, and $a_g = b_L$, the incident voltage toward the load a_L can be obtained from Equations (8.8) and (8.9), as shown in Equation (8.10).

$$a_L = \frac{b_S}{1 - \Gamma_S \Gamma_L} \tag{8.10}$$

Thus, the power delivered to the load is

$$P_L = |a_L|^2 - |b_L|^2 = |a_L|^2 \left(1 - |\Gamma_L|^2\right) \tag{8.11}$$

Substituting Equation (8.10) into Equation (8.11), P_L can be expressed with Equation (8.12).

$$P_L = \frac{|b_S|^2 \left(1 - |\Gamma_L|^2\right)}{|1 - \Gamma_S \Gamma_L|^2} \tag{8.12}$$

In addition, the available power from the source is delivered to the load when the source and the load impedances are conjugate matched, that is, when $\Gamma_S = (\Gamma_L)^*$. Substituting $\Gamma_S = (\Gamma_L)^*$ into Equation (8.12),

$$P_A = \frac{|b_S|^2}{1 - |\Gamma_S|^2} \tag{8.13}$$

To verify that the result given by Equation (8.13) is the available power, Equation (8.7) is substituted into Equation (8.13), and the manipulation of P_A results in

$$P_A = \frac{|b_S|^2}{1 - |\Gamma_S|^2} = \frac{\left(\frac{E_S}{\sqrt{2}} \frac{\sqrt{Z_o}}{Z_o + Z_S}\right)^2}{1 - \left(\frac{Z_S - Z_o}{Z_S + Z_o}\right)^2} = \frac{E_S^2}{8Z_S} \tag{8.14}$$

We can see that P_A in Equation (8.14) is equal to the available power that can be obtained from the source having internal resistance Z_S. This is expressed in Equation (6.5) of Chapter 6. In addition, using Equations (8.12) and (8.13), the ratio of power delivered to the load to available power can be computed as

$$G = \frac{P_L}{P_A} = \frac{\left(1 - |\Gamma_S|^2\right)\left(1 - |\Gamma_L|^2\right)}{|1 - \Gamma_S \Gamma_L|^2} \tag{8.15}$$

8.2.3 Power Gains

Gain is one of the key performance indicators for amplifiers and is defined in various ways depending on measurement and design purposes. Among these gains is the *transducer power gain* that is frequently used in measurements and is defined in Equation (8.16).

$$G_T = \frac{\text{Delivered power to load}}{\text{Available power from source}} \tag{8.16}$$

In measuring the transducer gain, the available power from a source with a 50-Ω internal resistance is measured by connecting a power meter (with a 50-Ω internal resistance) to the source. Then, inserting the amplifier between the meter and the source, and measuring the delivered power to the power meter, the ratio of power delivered to the load to available power can be obtained and the ratio can be interpreted as the transducer gain given by Equation (8.16).

On the other hand, from a design perspective, when the output impedance of the amplifier is conjugate matched by varying the load impedance, the available power at the output can be obtained. The ratio of the available output power to the available power from the source is called *available power gain* and is defined in Equation (8.17).

$$G_A = \frac{\text{Available power from output}}{\text{Available power from source}} \tag{8.17}$$

Using the simplified schematic shown in Figure 8.7, the available power gain G_A is found to be equal to the ratio of P_L/P_A. Since the input matching network is assumed to be lossless, the available power from the active device input is equal to that from the source. In addition, the tuning of the load impedance for the maximum power delivery is equivalent to the tuning of Γ_L. Thus, the available power gain is equal to P_L/P_A.

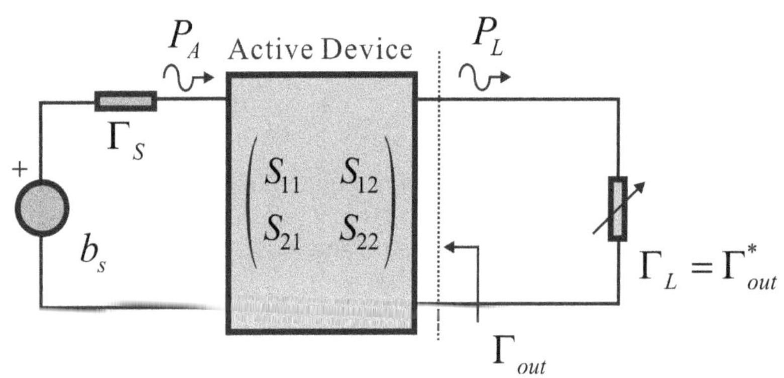

Figure 8.7 The concept of available power gain. The load Γ_L is tuned for the maximum power delivery and the available power gain is then the ratio of P_L/P_A at the maximum power delivery.

The available power gain is the maximum gain derived from the output for a given source impedance, which is a function of the source impedance only. Therefore, the available power gain represents the degradation of the gain due to the fixed source impedance.

Based on the symmetrical concept for a fixed load, when the source impedance is adjusted to deliver maximum power to the load as shown in Figure 8.8, then the ratio of the input power to the delivered load power is called *power gain*. Using the simplified schematic, power gain can be defined as

$$G_p = \frac{\text{Delivered power to load}}{\text{Input power at device input}} = \frac{P_L}{P_{in}}$$

The power gain, as an indicator of the degradation of the gain due to a chosen fixed load, is commonly used in design. The previous three power gains can be expressed in terms of the two-port S-parameters and the source and load reflection coefficients or impedances at the device plane. This will be discussed further in the next section.

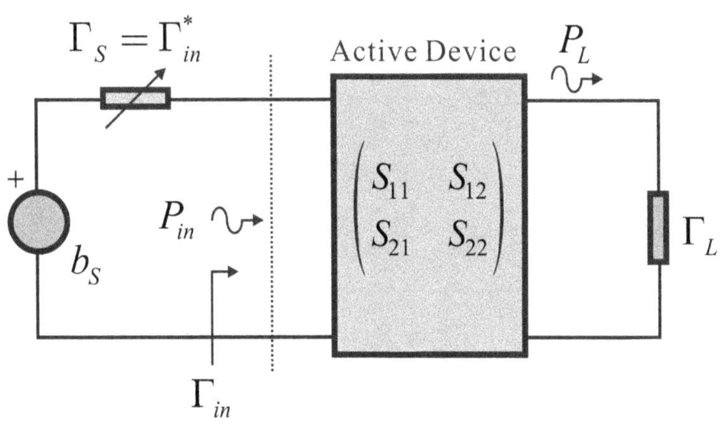

Figure 8.8 The concept of power gain. The source is tuned for the maximum P_L. The power gain is the ratio of P_L/P_{in} at the maximum P_L.

8.2.3.1 Transducer Power Gain

Figure 8.9 shows again a simplified schematic representation of the amplifier.

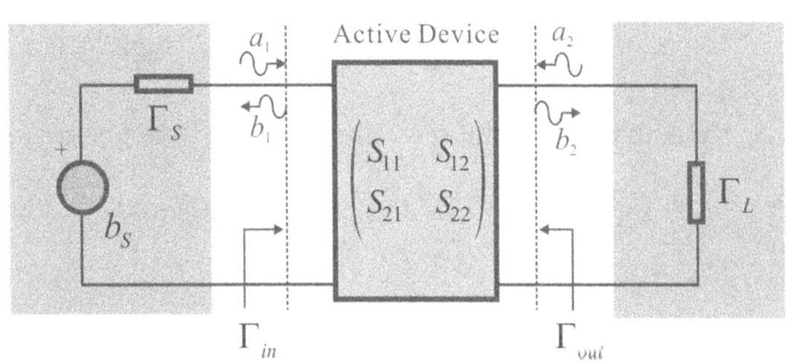

Figure 8.9 Simplified schematic representation of the amplifier

In Figure 8.9, the one-port circuit seen from the load Γ_L can be represented by the Thevenin equivalent circuit that includes the active device and the input circuit, and the previously developed power relationship in Equations (8.13) and (8.15) can be applied. At the device output plane, the Thevenin reflection coefficient Γ_T is shown in Equation (8.18).

$$\Gamma_T = \Gamma_{out} \tag{8.18}$$

The Thevenin reflected voltage b_T is the reflected voltage that appears across the reference impedance Z_o when the load Γ_L is replaced by Z_o. From Figure 8.10, b_T is given by Equation (8.19).

$$b_T = b_2\big|_{a_2=0} = S_{21}a_1 \tag{8.19}$$

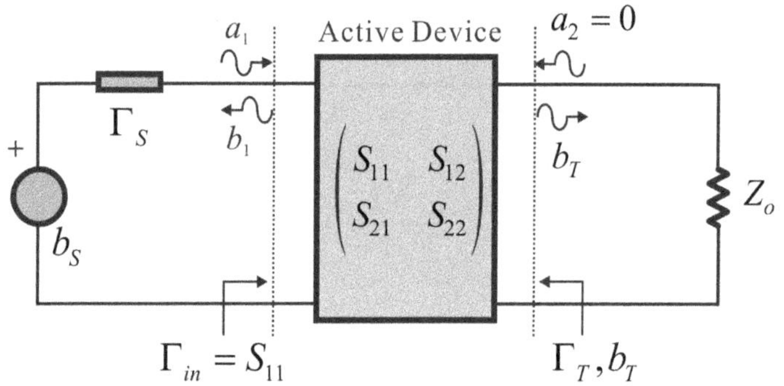

Figure 8.10 The circuit for the extraction of the Thevenin equivalent at the load plane

Note that $\Gamma_{in} = S_{11}$ because $a_2 = 0$. Applying Equation (8.10), a_1 is given by

$$a_1 = \frac{b_S}{1 - S_{11}\Gamma_S} \tag{8.20}$$

Thus, from Equation (8.20), the Thevenin reflected voltage b_T is shown in Equation (8.21).

$$b_T = S_{21}\frac{b_S}{1 - S_{11}\Gamma_S} \tag{8.21}$$

Therefore, the equivalent circuit seen from the load Γ_L can be represented as shown in Figure 8.6.

Applying Equation (8.12), the power delivered to the load Γ_L is given by Equation (8.22).

$$P_L = \frac{|b_T|^2\left(1 - |\Gamma_L|^2\right)}{|1 - \Gamma_{out}\Gamma_L|^2} = |b_S|^2 \frac{|S_{21}|^2}{|1 - S_{11}\Gamma_S|^2} \frac{\left(1 - |\Gamma_L|^2\right)}{|1 - \Gamma_{out}\Gamma_L|^2} \tag{8.22}$$

Since the available power from the source is the same as in Equation (8.13), the transducer power gain G_T is expressed as

$$G_T = \frac{P_L}{P_A} = \frac{1-|\Gamma_S|^2}{|1-S_{11}\Gamma_S|^2}|S_{21}|^2\frac{1-|\Gamma_L|^2}{|1-\Gamma_{out}\Gamma_L|^2} \tag{8.23}$$

The transducer gain in Equation (8.23) is derived at the device output plane, but it can also be derived at the device input plane. The derived transducer gain is expressed as

$$G_T = \frac{P_L}{P_A} = \frac{1-|\Gamma_S|^2}{|1-\Gamma_{in}\Gamma_S|^2}|S_{21}|^2\frac{1-|\Gamma_L|^2}{|1-S_{22}\Gamma_L|^2} \tag{8.24}$$

Although G_T in Equation (8.24) is different from that in Equation (8.23), G_T in Equation (8.24) is the same as that in Equation (8.23).

8.2.3.2 Available Power Gain

Available power gain is defined as the ratio of available power at the device input to available power at the output. Figure 8.11 again shows how to derive the available power gain, which is accomplished by using the previously derived transducer power gain.

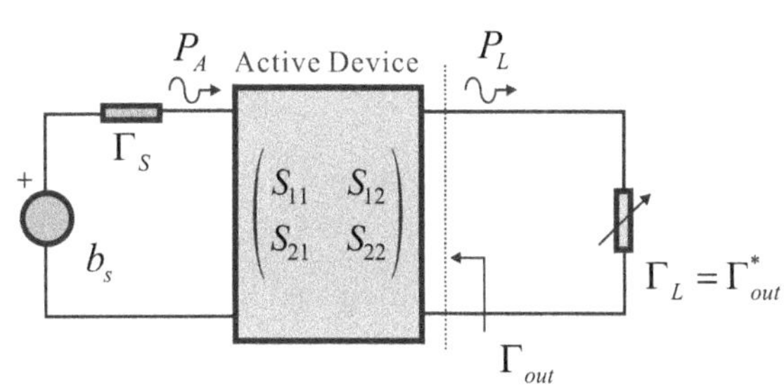

Figure 8.11 Circuit for calculating available power gain

Note that the transducer power gain is defined in terms of the available power at the input and the delivered power to the load. The available power at the output is the power when the load impedance is conjugate matched to Γ_{out}, as shown in Figure 8.11. This condition is $\Gamma_L = (\Gamma_{out})^*$. Then, the available power gain can be obtained by substituting $\Gamma_L = (\Gamma_{out})^*$ into the transducer gain in Equation (8.23). Therefore, the available power gain is given by Equation (8.25).

$$G_T = \frac{P_L}{P_A} = \left.\frac{\left(1-|\Gamma_S|^2\right)|S_{21}|^2\left(1-|\Gamma_L|^2\right)}{|1-\Gamma_{out}\Gamma_L|^2|1-S_{11}\Gamma_S|^2}\right|_{\Gamma_L=\Gamma^*_{out}} = \frac{\left(1-|\Gamma_S|^2\right)|S_{21}|^2}{|1-S_{11}\Gamma_S|^2\left(1-|\Gamma_{out}|^2\right)} \tag{8.25}$$

The available power gain can also be interpreted as the maximum gain for a fixed Γ_S.

8.2.3.3 Power Gain

The power P_{in} delivered to the input of the device shown in Figure 8.12 is given by Equation (8.26).

$$P_{in} = |a_1|^2 - |b_1|^2 = |a_1|^2\left(1 - |\Gamma_{in}|^2\right) \tag{8.26}$$

Using Equation (8.10), P_{in} results in Equation (8.27).

$$P_{in} = \frac{|b_S|^2\left(1 - |\Gamma_{in}|^2\right)}{|1 - \Gamma_{in}\Gamma_S|^2} \tag{8.27}$$

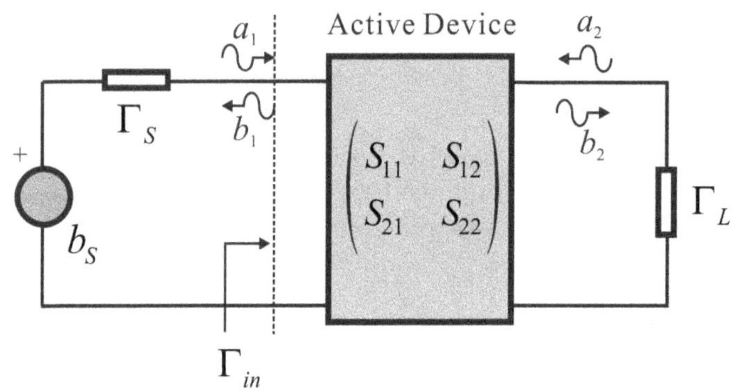

Figure 8.12 Circuit for the calculation of power gain

In addition, a_1 and a_2 in Figure 8.12 are given by

$$a_1 = \frac{b_S}{1 - \Gamma_{in}\Gamma_S} \tag{8.28}$$

$$a_2 = \Gamma_L b_2 \tag{8.29}$$

Substituting a_1 and a_2 of Equations (8.28) and (8.29) into b_2 gives Equation (8.30).

$$b_2 = S_{21}a_1 + S_{22}a_2 = S_{21}\frac{b_S}{1 - \Gamma_{in}\Gamma_S} + S_{22}\Gamma_L b_2 \tag{8.30}$$

Then, b_2 can be computed as Equation (8.31).

$$b_2 = \frac{S_{21}b_S}{(1 - \Gamma_{in}\Gamma_S)(1 - S_{22}\Gamma_L)} \tag{8.31}$$

Thus, the power delivered to the load is given by Equation (8.32).

$$P_L = |b_2|^2\left(1-|\Gamma_L|^2\right) = |b_S|^2 \frac{|S_{21}|^2}{|1-\Gamma_{in}\Gamma_S|^2} \frac{\left(1-|\Gamma_L|^2\right)}{|1-S_{22}\Gamma_L|^2} \tag{8.32}$$

Therefore, based on the definition of power gain, the ratio of the input power P_{in} to the power delivered to the load P_L is given by Equation (8.33).

$$G_p = \frac{P_L}{P_{in}} = \frac{|S_{21}|^2\left(1-|\Gamma_L|^2\right)}{\left(1-|\Gamma_{in}|^2\right)|1-S_{22}\Gamma_L|^2} \tag{8.33}$$

The power gain is the ratio of input power to the power delivered to the load. However, the power gain can be interpreted in another way. The maximum power is delivered to the device's input when the source impedance is conjugate matched to the device's input impedance. The ratio of the input power at the conjugate matching condition to the power delivered to the load leads to the same result given by Equation (8.33). This is found in problem 8.4 at the end of this chapter. Thus, another way of interpreting the power gain is that it represents the maximum gain of the active device for a specific fixed load Γ_L.

8.2.3.4 Unilateral Gain

In general, $S_{12} \neq 0$ and the active device has some feedback. However, S_{12} can generally be made 0 by adding the appropriate feedback circuit. In this case, there will be no feedback from output to input. This condition is then said to be *unilateral*. The resulting gain is called Mason's gain, U, which is expressed in Equations (8.34a) and (8.34b).

$$U = \frac{\left|\frac{S_{21}}{S_{12}} - 1\right|^2}{2k\left|\frac{S_{21}}{S_{12}}\right| - 2\,\mathrm{Re}\left(\frac{S_{21}}{S_{12}}\right)} \tag{8.34a}$$

$$k = \frac{1-|S_{11}|^2-|S_{22}|^2+|D|^2}{2|S_{12}S_{21}|} \tag{8.34b}$$

Since Mason's gain is the gain measured after the active device has been completely stabilized by removing feedback, it can be considered as the true gain of the active device. Therefore, it is used as the criterion for determining whether a measured device at an arbitrary frequency is active or passive.

Since S_{12}, which represents the feedback from output to input, is generally less than 1, a unilateral approximate expression for the transducer gain is often used in amplifier design and active device assessment. To obtain the unilateral transducer power gain,

$S_{12} = 0$ is substituted into Equation (8.25). In this case, since $\Gamma_{in} = S_{11}$ and $\Gamma_{out} = S_{22}$, the transducer power gain is expressed in Equation (8.35).

$$G_{TU} = G_T\Big|_{S_{12}=0} = \frac{\left(1-\left|\Gamma_S\right|^2\right)\left|S_{21}\right|^2\left(1-\left|\Gamma_L\right|^2\right)}{\left|1-S_{11}\Gamma_S\right|^2\left|1-S_{22}\Gamma_L\right|^2} \tag{8.35}$$

When the input and output reflection coefficients in Figure 8.13 are each conjugate matched ($\Gamma_S = (\Gamma_{in})^*$, $\Gamma_L = (\Gamma_{out})^*$), maximum gain is then obtained and the maximum unilateral gain $G_{TU,\max}$ is shown in Equation (8.36).

$$G_{TU,\max} = \frac{\left|S_{12}\right|^2}{\left(1-\left|S_{11}\right|^2\right)\left(1-\left|S_{22}\right|^2\right)} \tag{8.36}$$

The calculation of the unilateral gain in Equation (8.35) is simple compared to the computations for other gains and it was commonly used in the past to evaluate the maximum gain of active devices when computers were not readily available.

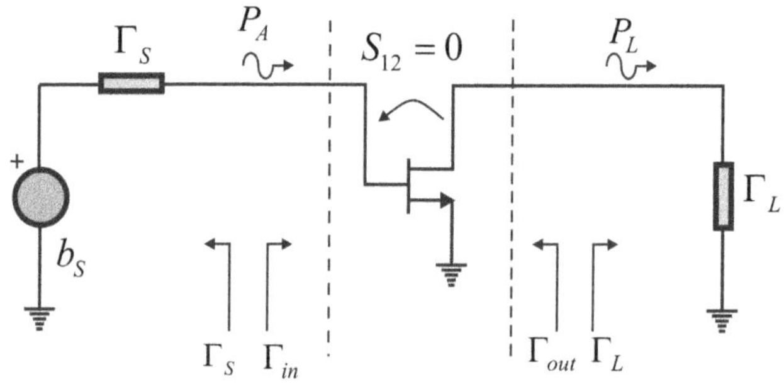

Figure 8.13 Unilateral gain

8.3 STABILITY AND CONJUGATE MATCHING

Active devices generally have feedback from output to input, however small. Due to this feedback, some load impedance values yield a negative resistance at the input. Similarly, a negative resistance can appear at the output for some values of the source impedance. When the negative resistance at the input or output appears at certain out-of-band frequencies, a designed amplifier may turn into an oscillator. In the case that the negative resistance appears at in-band frequencies, it limits the matching of an amplifier and must be eliminated through a stabilization method. For a stabilized device, the maximum gain occurs at the source and load impedances of the conjugate matching. In this section, we will discuss how to find the stable regions of the source and load impedances, and the mathematical derivation of the conjugate matching source and load impedances.

8.3.1 Load and Source Stability Regions

The input or output impedances of an active device may have a negative resistance for a particular load or source impedance, so the region of the load and source impedances that induce a negative resistance must first be found. When a negative resistance is induced at the input or output of an active device, the magnitude of the input or output reflection coefficients are mathematically greater than 1. That is, $|\Gamma_{in}| > 1$ and $|\Gamma_{out}| > 1$. Therefore, the problem of finding the stability regions requires determining two conditions:

1. A region of Γ_L that gives $|\Gamma_{in}| \le 1$; and
2. A region of Γ_S that gives $|\Gamma_{out}| \le 1$.

Condition 1 limits the range of load selection and is called the *load stability region*, and condition 2 limits the range of source selection and is thus called a *source stability region*. After determining the locus of the boundary ($|\Gamma_{in}| = 1$ and $|\Gamma_{out}| = 1$), the locus divides a region of Γ_L or Γ_S into two regions. Once a stable region is found between those two regions, the remaining region becomes automatically unstable, which gives rise to a negative resistance. From Figure 8.14, the input reflection coefficient is expressed by Equation (8.37)

$$\Gamma_{in} = S_{11} + \frac{S_{12}S_{21}\Gamma_L}{1 - S_{22}\Gamma_L} = \frac{S_{11} - D\Gamma_L}{1 - S_{22}\Gamma_L} \tag{8.37}$$

where $D = S_{11}S_{22} - S_{21}S_{12}$.

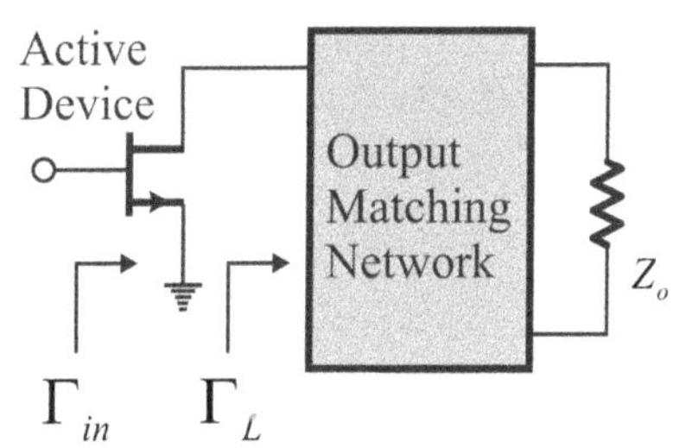

Figure 8.14 Load stability or the locus of Γ_L that gives $|\Gamma_{in}| = 1$

The reflection coefficient Γ_{in} for the impedance of a passive device satisfies $|\Gamma_{in}| \le 1$. However, for the impedance that has a negative resistance, $|\Gamma_{in}| > 1$. From Equation (8.37), Γ_L that satisfies $|\Gamma_{in}| = 1$ is

$$|\Gamma_{in}| = \left|\frac{S_{11} - D\Gamma_L}{1 - S_{22}\Gamma_L}\right| = 1 \tag{8.38}$$

Multiplying both sides of Equation (8.38) by $|S_{22}/D|$ and rewriting, the following expression is obtained:

$$\left|\frac{\Gamma_L - \dfrac{S_{11}}{D}}{\Gamma_L - \dfrac{1}{S_{22}}}\right| = \left|\frac{S_{22}}{D}\right| \tag{8.39}$$

Equation (8.39) can also be represented as Equation (8.40).

$$\left|\frac{\Gamma_L - A}{\Gamma_L - B}\right| = m \tag{8.40}$$

Constants A and B represent two points on the Γ_L plane and are expressed as follows:

$$A = \frac{S_{11}}{D} = \frac{S_{11}D^*}{|D|^2}$$

$$B = \frac{1}{S_{22}} = \frac{S_{22}^*}{|S_{22}|^2}$$

From Equation (8.40), m represents the ratio of the distances from a point Γ_L to two points, A and B. The locus satisfying Equation (8.40) is well-known and is represented by the circle shown in Figure 8.15. The internal (I) and external (E) division points with the division ratio m:1 can be obtained from points A and B. The center of the circle becomes their midpoint between I and E. Half of the distance between points I and E becomes the radius of the circle.

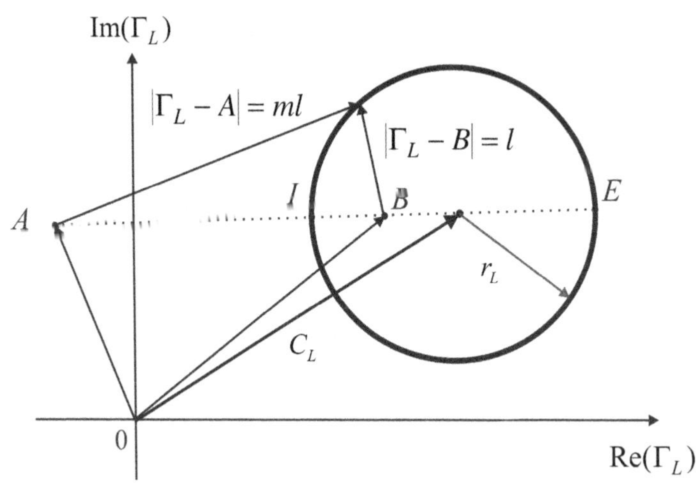

Figure 8.15 Locus of $|\Gamma_{in}| = 1$ in the Γ_L plane

Points I and E are determined as

$$I = \frac{mB + A}{m + 1} \qquad E = \frac{mB - A}{m - 1}$$

The center of the circle C_L and the radius r_L are then determined from points I and E as

$$C_L = \frac{I + E}{2} = \frac{m^2 B - A}{m^2 - 1} \tag{8.41a}$$

$$r_L = \left|\frac{E - I}{2}\right| = \left|\frac{m(B - A)}{m^2 - 1}\right| \tag{8.41b}$$

Substituting A, B, and m into Equations (8.41a) and (8.41b), the center C_L and the radius r_L in the Γ_L plane are given by

$$C_L = \frac{S_{22}^* - D^* S_{11}}{|S_{22}|^2 - |D|^2} \tag{8.42a}$$

$$r_L = \frac{|S_{12} S_{21}|}{\left||S_{22}|^2 - |D|^2\right|} \tag{8.42b}$$

Load stability circles drawn using Equations (8.42a) and (8.42b) are shown in Figure 8.16. In order to determine the region of stability between the two regions divided by the load stability circle, one test point is necessary. The load of $\Gamma_L = 0$ provides a good test point. Setting $\Gamma_L = 0$ results in the input reflection coefficient $\Gamma_{in} = S_{11}$. If $|S_{11}| < 1$, it implies that $|\Gamma_{in}| < 1$ for $\Gamma_L = 0$. Since $|\Gamma_{in}| < 1$ for $\Gamma_L = 0$, the stable region is located inside the unit Smith-chart circle that includes the origin, as shown in Figure 8.16(a). Otherwise, the region that includes the origin is unstable.

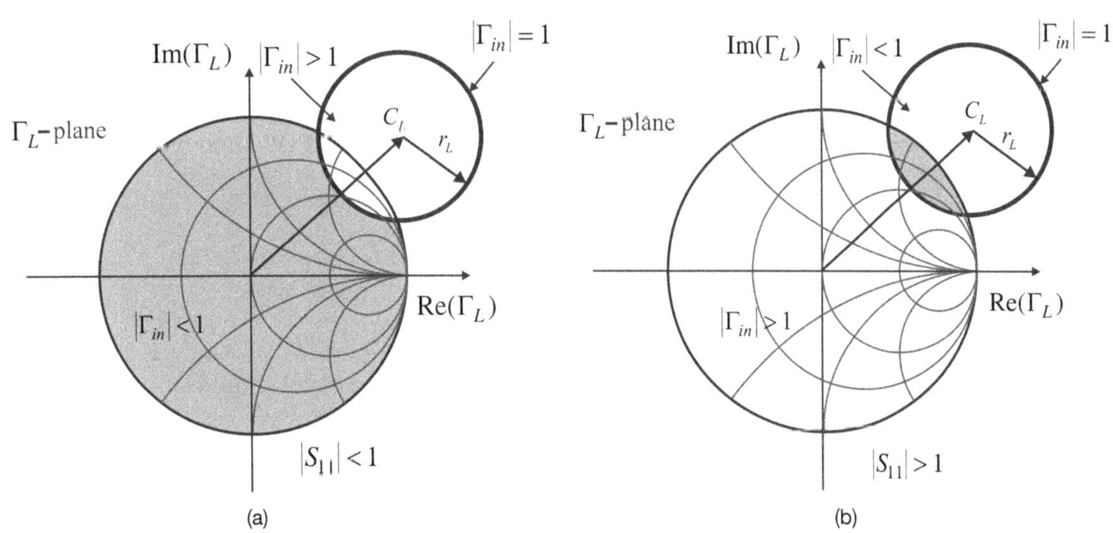

Figure 8.16 Stable region in the Γ_L plane: (a) when $|S_{11}| < 1$ and (b) when $|S_{11}| > 1$

Similarly, the region of Γ_S giving $|\Gamma_{out}| < 1$ can be determined. The output reflection coefficient Γ_{out} in Figure 8.17 is expressed as

$$\Gamma_{out} = S_{22} + \frac{S_{12}S_{21}}{1 - S_{11}\Gamma_S} = \frac{S_{22} - D\Gamma_S}{1 - S_{11}\Gamma_S} \tag{8.43}$$

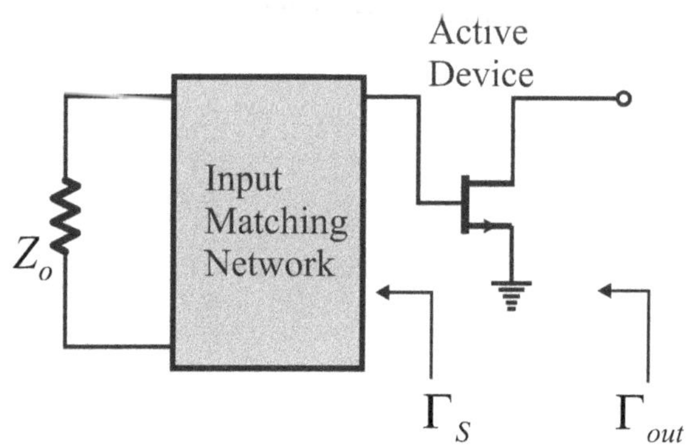

Figure 8.17 Source stability where the locus of Γ_S gives $|\Gamma_{out}| = 1$

If Γ_{out} is set to $|\Gamma_{out}| = 1$, the source stability circles can be determined. Equation (8.43) can be obtained when the subscripts 1 and 2 in Equation (8.37) are interchanged. Therefore, by interchanging those subscripts in Equation (8.42), the center C_S and radius r_S of the source stability circle giving $|\Gamma_{out}| = 1$ in the Γ_S–plane are given by Equations (8.44a) and (8.44b).

$$C_S = \frac{S_{11}^* - D^* S_{22}}{|S_{11}|^2 - |D|^2} \tag{8.44a}$$

$$r_S = \frac{|S_{12}S_{21}|}{\left||S_{11}|^2 - |D|^2\right|} \tag{8.44b}$$

The stability circles drawn with the center C_S and radius r_S are shown in Figure 8.18 and, similar to our reasoning in the case of the load stability circle, when $|S_{22}| < 1$, the region inside the unit Smith chart that includes the origin is the stable region shown in Figure 8.18(a). Otherwise, the region that includes the origin is the unstable region. In Figures 8.18(a) and (b), the resulting stable regions inside the unit Smith chart are represented by the shaded area.

8.3.2 Stability Factor

From Figures 8.16(a) and 8.18(a), in order for the entire region $|\Gamma_S| \leq 1$and $|\Gamma_L| \leq 1$ on the Smith chart to be the region of stability, Equations (8.45) and (8.46) should be satisfied from Figure 8.19.

$$\left||C_L| - r_L\right| > 1 \quad \text{for } |S_{11}| < 1 \tag{8.45}$$

$$\left||C_S| - r_S\right| > 1 \quad \text{for } |S_{22}| < 1 \tag{8.46}$$

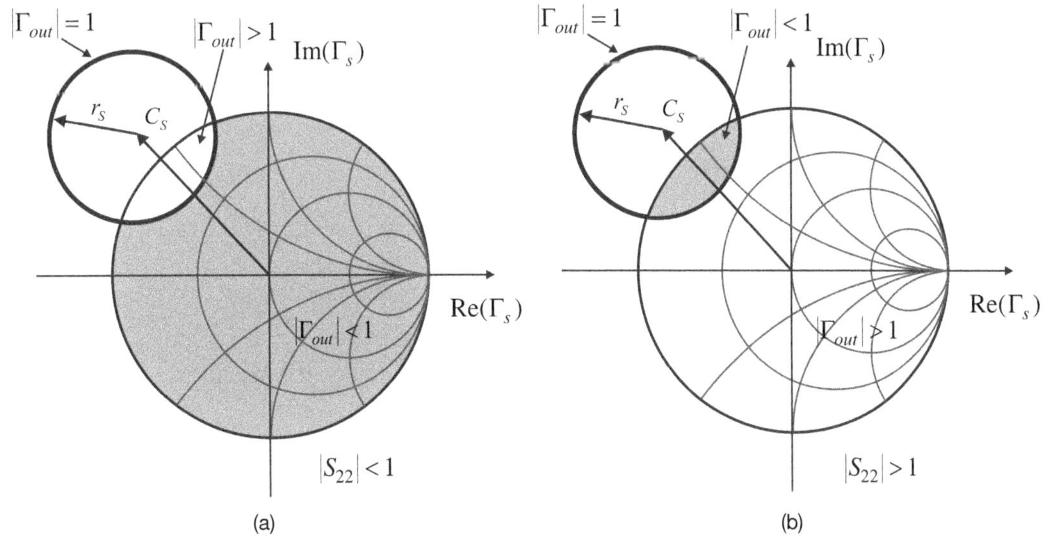

Figure 8.18 Stable region in the Γ_S plane: (a) when $|S_{22}| < 1$ and (b) when $|S_{22}| > 1$

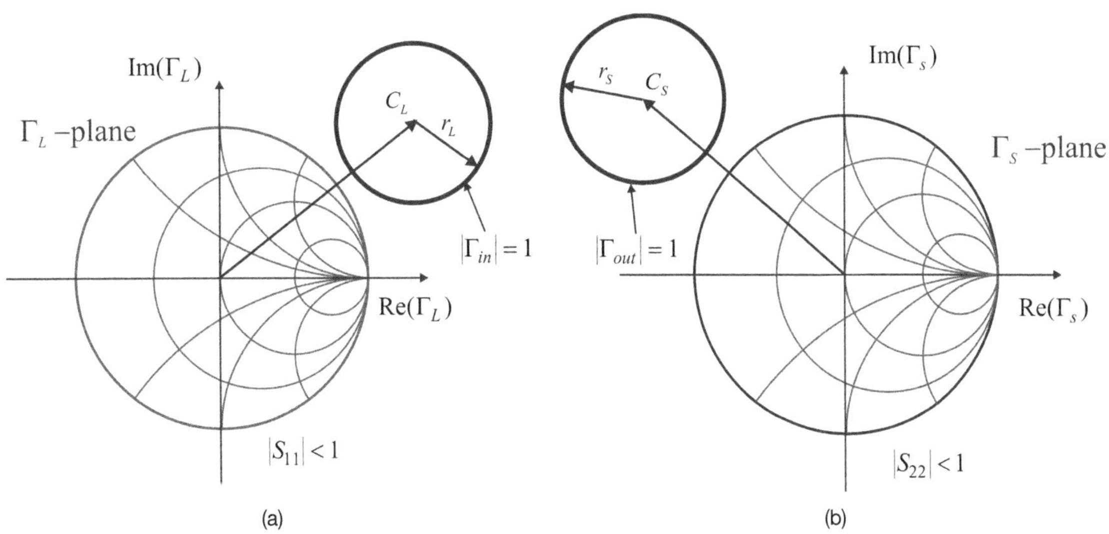

Figure 8.19 Unconditional stability: (a) Γ_L plane, (b) Γ_S plane

Substituting the center and radius of the load stability circle obtained from Equation (8.42) into Equation (8.45), we obtain

$$\left|\frac{|S_{22} - DS_{11}^*| - |S_{12}S_{21}|}{|S_{22}|^2 - |D|^2}\right| > 1 \tag{8.47}$$

Equation (8.47) can be rewritten as Equation (8.48).

$$\left|\left|S_{22} - DS_{11}^*\right| - \left|S_{12}S_{21}\right|\right|^2 > \left|\left|S_{22}\right|^2 - \left|D\right|^2\right|^2 \tag{8.48}$$

Expanding Equation (8.48), we obtain Equation (8.49).

$$2\left|S_{12}S_{21}\right|\left|S_{22} - DS_{11}^*\right| < \left|S_{22} - DS_{11}^*\right|^2 + \left|S_{12}S_{21}\right|^2 - \left|\left|S_{22}\right|^2 - \left|D\right|^2\right| \tag{8.49}$$

Note that the following identity holds:

$$\left|S_{22} - DS_{11}^*\right|^2 = \left|S_{12}S_{21}\right|^2 + \left(1 - \left|S_{11}\right|^2\right)\left(\left|S_{22}\right|^2 - \left|D\right|^2\right) \tag{8.50}$$

Squaring each term of Equation (8.49) and substituting Equation (8.50) into Equation (8.49), and then rewriting, Equation (8.51) is obtained.

$$\left(\left|S_{22}\right|^2 - \left|D\right|^2\right)^2\left[\left\{\left(1 - \left|S_{11}\right|^2\right) - \left(\left|S_{22}\right|^2 - \left|D\right|^2\right)\right\}^2 - 4\left|S_{12}S_{21}\right|^2\right] > 0 \tag{8.51}$$

Thus, the necessary condition for Equation (8.51) is

$$1 - \left|S_{11}\right|^2 - \left|S_{22}\right|^2 + \left|D\right|^2 > 2\left|S_{12}S_{21}\right| \tag{8.52}$$

Dividing both sides of Equation (8.52) by $2\,|\,S_{12}S_{21}\,|$ yields the stability factor k, which is used to examine the unconditional stability of the two-port network, as shown in Equation (8.53).

$$k = \frac{1 - \left|S_{11}\right|^2 - \left|S_{22}\right|^2 + \left|D\right|^2}{2\left|S_{12}S_{21}\right|} > 1 \tag{8.53}$$

The other required condition can be obtained from Figures 8.16(b) and 8.18(b). The radius is large enough to include the Smith chart. The condition is $r_L - |\,C_L\,| > 1$. Substituting the center and radius results from Equation (8.42) into this condition gives Equation (8.54).

$$\left|\frac{S_{22} - DS_{11}^*}{\left|S_{22}\right|^2 - \left|D\right|^2}\right| < \frac{\left|S_{12}S_{21}\right|}{\left|S_{22}\right|^2 - \left|D\right|^2} - 1 \tag{8.54}$$

This requires that the right-hand side of the equation be positive, which is expressed in Equation (8.55).

$$\frac{\left|S_{12}S_{21}\right|}{\left|S_{22}\right|^2 - \left|D\right|^2} > 1 \tag{8.55}$$

From the stability factor,

$$2k = \frac{1-|S_{11}|^2}{|S_{12}S_{21}|} + \frac{|D|^2-|S_{22}|^2}{|S_{12}S_{21}|} > 2 \tag{8.56}$$

is obtained. However, since the second term of Equation (8.56) is less than 1, from Equation (8.55), the first term must be greater than 1. Therefore,

$$1-|S_{11}|^2 > |S_{12}S_{21}| \tag{8.57}$$

Equation (8.57) is an additional condition that must be satisfied. In addition, from the source stability circles, the following condition is obtained with Equation (8.58):

$$1-|S_{22}|^2 > |S_{12}S_{21}| \tag{8.58}$$

In conclusion, the necessary and sufficient conditions for the unconditional stability of a two-port circuit are expressed in Equations (8.59a)–(8.59c).

$$k > 1 \tag{8.59a}$$

$$1-|S_{11}|^2 > |S_{12}S_{21}| \tag{8.59b}$$

$$1-|S_{22}|^2 > |S_{12}S_{21}|. \tag{8.59c}$$

These are the three conditions that must be simultaneously satisfied.

Example 8.2

Using the 3 V, 10 mA S-parameters of the FHX35LG/LP in ADS, plot the stability factor k and the following terms for the frequency range of 1–18 GHz:

$$\frac{1-|S_{11}|^2}{|S_{12}S_{21}|}, \quad \frac{1-|S_{22}|^2}{|S_{12}S_{21}|}$$

Draw the source and load stability circles at 1 GHz.

Solution

Set up the circuit as shown in Figure 8E.1 using the two-port S-parameters of the ADS FHX35LG/LP. Perform the simulation and enter the equations in the display window to plot the stability factor and the two terms as shown in Measurement Expression 8E.1. The plot is shown in Figure 8E.2.

In Figure 8E.2, **S**1 and **L**1 are observed to be greater than **k**. Since the two terms are mostly greater than k, the stability factor k is critical to determining the stability of the circuit. The relationship that **S**1 and **L**1> k is generally true. In addition, as can be seen from the figure, above the 12-GHz frequency, $k > 1$ and the device is found to be stable while it is unstable at frequencies lower than 12 GHz.

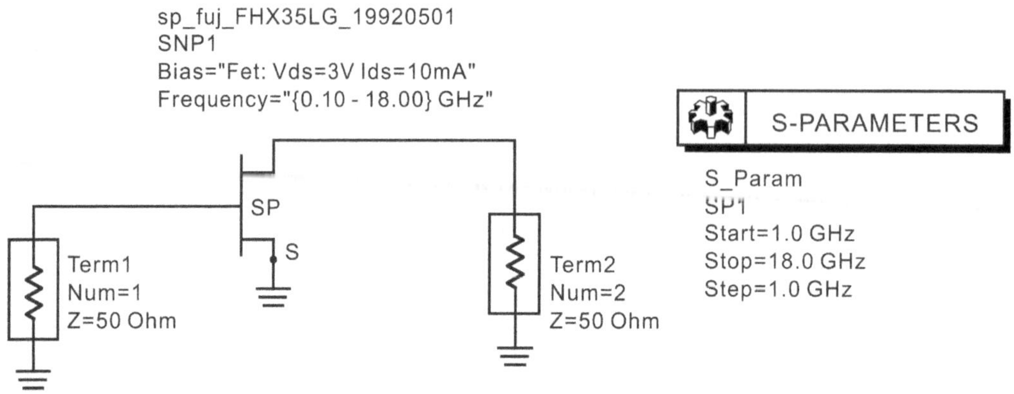

Figure 8E.1 Circuit for determining the stability of FHX35LG/LP. SNP1 is the data component and the DC bias circuit is not necessary.

Eqn k=stab_fact(S)

Eqn S1=(1-(mag(S11))**2)/(mag(S12*S21))

Eqn L1=(1-(mag(S22))**2)/(mag(S12*S21))

Measurement Expression 8E.1 Equations for stability in the display window

Since $k < 1$ at a frequency of 1 GHz, the device is found to be unstable at 1 GHz. In order to find the stable region, the equations in Measurement Expression 8E.2 are also entered in the display window to plot the load and source stability circles at 1 GHz.

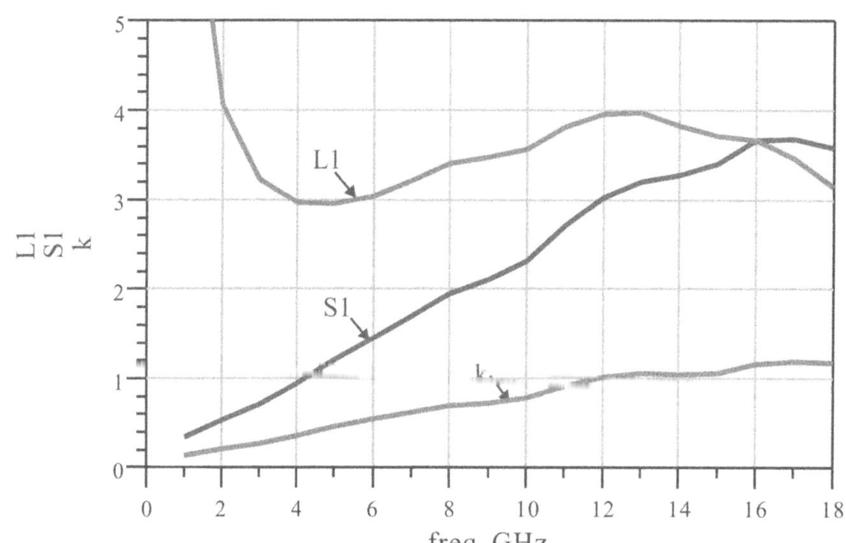

Figure 8E.2 Simulation results. The stability factor k is the smallest of the stability factors **L1**, **S1**, and k. Thus, the stability can be determined using k alone.

Eqn Source_Stability_Circle=s_stab_circle(S[0], 51)

Eqn Load_Stability_Circle=l_stab_circle(S[0],51)

Measurement Expression 8E.2 Equations for drawing the stability circles

The function **s_stab_circle**(**S**[0], 51) is the function for drawing the source stability circle given by Equation (8.44). The S-parameter index is set to [0] for the selection of S-parameters at a frequency of 1 GHz. The number 51 at the end represents the number of points used in plotting the circle. Similarly, **l_stab_circle**(**S**[0],51) is a function for plotting the load stability circle. The circles plotted on the Smith chart using these functions are shown in Figure 8E.3, where it can be seen that since $|S_{11}| < 1$ and $|S_{22}| < 1$, the region including the origin of the Smith chart is the region of stability.

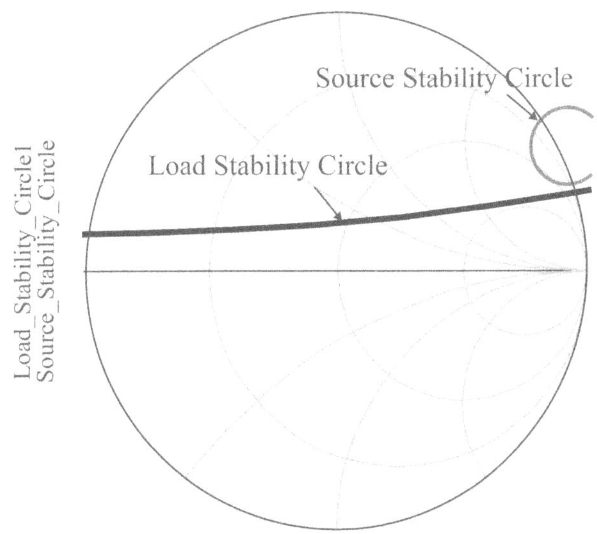

Figure 8E.3 Source and load stability circles

8.3.3 Conjugate Matching

Figure 8.20 shows the amplifier whose input and output are conjugate matched. The objective of conjugate matching is to obtain maximum gain from the active device. When the input is conjugate matched, maximum available power is transferred from the source to the input of the active device. Simultaneously conjugate matching the output ensures that the maximum power that is amplified by the active device is delivered to the load. Therefore, maximum gain is obtained when the input and output are conjugate matched as shown in Equations (8.60a) and (8.60b).

$$\Gamma_{in} = S_{11} + \frac{S_{12}S_{21}\Gamma_L}{1 - S_{22}\Gamma_L} = \Gamma_S^* \tag{8.60a}$$

$$\Gamma_{out} = S_{22} + \frac{S_{12}S_{21}\Gamma_S}{1 - S_{11}\Gamma_S} = \Gamma_L^* \tag{8.60b}$$

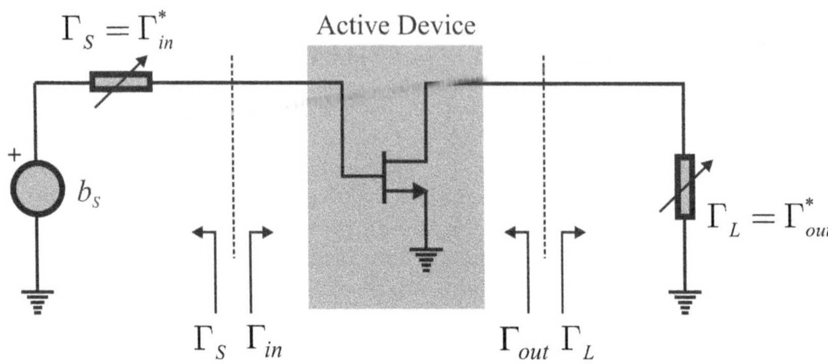

Figure 8.20 Simultaneous conjugate matching conditions. When $\Gamma_S = (\Gamma_{in})^*$ and $\Gamma_L = (\Gamma_{out})^*$, the maximum gain occurs.

Equations (8.61a) and (8.61b) are obtained by rewriting Equations (8.60a) and (8.60b).

$$\left(1 - \Gamma_L S_{22}\right)\left(S_{11} - \Gamma_S^*\right) + \Gamma_L S_{12} S_{21} = 0 \tag{8.61a}$$

$$\left(1 - \Gamma_S S_{11}\right)\left(S_{22} - \Gamma_L^*\right) + \Gamma_S S_{12} S_{21} = 0 \tag{8.61b}$$

To solve the simultaneous equations, Γ_L is expressed in terms of Γ_S using Equation (8.60a) and vice versa, which yields

$$\Gamma_L = \frac{\Gamma_S^* - S_{11}}{\Gamma_S^* S_{22} - D} \tag{8.62a}$$

$$\Gamma_S = \frac{\Gamma_L^* - S_{22}}{\Gamma_L^* S_{11} - D} \tag{8.62b}$$

Substituting Equation (8.62a) into Equation (8.61a) and manipulating the results in the quadratic equation of Γ_S as shown in Equation (8.63),

$$\Gamma_S^2 - \Gamma_S \frac{B_1}{C_1} + \frac{C_1^*}{C_1} = 0 \tag{8.63}$$

where

$$C_1 = S_{11} - D S_{22}^* \tag{8.64}$$

$$B_1 = 1 - \left|S_{22}\right|^2 + \left|S_{11}\right|^2 - \left|D\right|^2 \tag{8.65}$$

The solution to this quadratic equation is obtained in Equation (8.66).

$$\Gamma_{SM} = \frac{B_1}{2C_1} \pm \frac{1}{2}\sqrt{\left(\frac{B_1}{C_1}\right)^2 - 4\frac{C_1^*}{C_1}} = \frac{C_1^*}{|C_1|}\left[\frac{B_1}{2|C_1|} \pm \sqrt{\frac{B_1^2}{|2C_1|^2} - 1}\right] \tag{8.66}$$

Here, $B_1/(2|C_1|)$ is a real number and can be expressed as shown in Equation (8.67).

$$\frac{B_1}{2|C_1|} = \frac{1 + |S_{11}|^2 - |S_{22}|^2 - |D|^2}{2|S_{11} - S_{22}^* D|} > 1 \tag{8.67}$$

The term $B_1/(2|C_1|)$ is greater than 1 when $k > 1$. To prove this, the denominator is expressed as shown in Equation (8.68).

$$|S_{11} - S_{22}^* D|^2 = |S_{12}S_{21}|^2 + \left(1 - |S_{22}|^2\right)\left(|S_{11}|^2 - |D|^2\right) \tag{8.68}$$

Squaring the right-hand side of Equation (8.67) and using Equation (8.68) gives Equation (8.69).

$$\left(1 + |S_{11}|^2 - |S_{22}|^2 - |D|^2\right)^2 > 4|S_{12}S_{21}|^2 + 4\left(1 - |S_{22}|^2\right)\left(|S_{11}|^2 - |D|^2\right) \tag{8.69}$$

This can be simplified into Equation (8.70).

$$\left[\left(1 - |S_{22}|^2\right) + \left(|S_{11}|^2 - |D|^2\right)\right]^2 - 4\left(1 - |S_{22}|^2\right)\left(|S_{11}|^2 - |D|^2\right) > 4|S_{12}S_{21}|^2 \tag{8.70}$$

Making use of the identity $(a + b)^2 - 4ab = (a - b)^2$ and rewriting Equation (8.70) yields Equation (8.71).

$$\left(1 - |S_{11}|^2 - |S_{22}|^2 + |D|^2\right)^2 > 4|S_{12}S_{21}|^2 \tag{8.71}$$

Thus, $B_1/(2|C_1|)$ is greater than 1 when $k > 1$.

This implies that when the transistor is stable, that is, when $k > 1$, the term inside the square root is positive. In the selection of the sign of the square root in Equation (8.66), the sign must be selected such that $|\Gamma_{SM}| \le 1$ because Γ_{SM} cannot be implemented using a passive circuit for $|\Gamma_{SM}| \ge 1$. Since the magnitude of $(C_1)^*/C_1$ outside the bracket [•] in Equation (8.66) is 1, the terms in the bracket [•] must therefore be less than 1. The terms in the bracket [•] can be considered in the form $x - (x^2 - 1)^{½}$ and $x + (x^2 - 1)^{½}$. The term $x + (x^2 - 1)^{½}$ becomes greater than 1 for $x \ge 1$. Therefore, the realizable solution of Γ_{SM} with a passive circuit is expressed in Equation (8.72).

$$\Gamma_{SM} = \frac{C_1^*}{|C_1|}\left[\frac{B_1}{2|C_1|} \quad \sqrt{\frac{B_1^2}{|2C_1|^2} - 1}\right] \tag{8.72}$$

Similarly, Γ_{LM} can be obtained as shown in Equation (8.73)

$$\Gamma_{LM} = \frac{C_2^*}{|C_2|}\left[\frac{B_2}{2|C_2|} - \sqrt{\frac{B_2^2}{|2C_2|^2} - 1}\right] \tag{8.73}$$

where

$$C_2 = S_{22} - DS_{11}^*$$

$$B_2 = 1 - |S_{11}|^2 + |S_{22}|^2 - |D|^2$$

When $k < 1$, $|\Gamma_{SM}| = 1$ and $|\Gamma_{LM}| = 1$. The source and load impedances have purely imaginary values, and real power is not delivered. That is, when $k < 1$, conjugate matching with Equations (8.72) and (8.73) is not possible.

The maximum gain is obtained by substituting Γ_{SM} and Γ_{LM}, given in Equations (8.72) and (8.73), into the transducer power gain, but the calculation is a fairly complicated process and reference 1 at the end of this chapter can be consulted for details. The gain thus obtained is known as the maximum available gain (MAG), G_{max} and is expressed in Equation (8.74).

$$MAG = G_{max} = \left|\frac{S_{21}}{S_{12}}\right|\left(k - \sqrt{k^2 - 1}\right) \tag{8.74}$$

In the case of instability, $k < 1$, meaningful gain can be derived up to the boundary of the stability condition. Thus, substituting $k = 1$ into Equation (8.74), the maximum meaningful gain is derived while maintaining stability. This is also known as the maximum stable gain (MSG) and is expressed as

$$MSG = \left|\frac{S_{21}}{S_{12}}\right|$$

Example 8.3

Using the previous S-parameters in Example 8.2, plot the MSG, the MAG, and $|S_{21}|^2$ for FHX35LG/LP in the 1 GHz–20 GHz band. Also, find Γ_{SM} and Γ_{LM} at 12 GHz.

Solution

Enter the following equation in the display window to obtain the MSG and the MAG.

```
Eqn MaxG=max_gain(S)
```

Measurement Expression 8E.3 Equation for MSG and MAG

The function **max_gain(S)** returns MSG for $k < 1$ and returns the MAG for $k > 1$. Using the values of k that depend on frequency, it is possible to know whether **max_gain(•)** represents the MAG or the MSG. Thus, by simultaneously plotting k values together, the **max_gain(•)** values can be distinguished. The plot is shown in Figure 8E.4. As can be seen from the plot, above 12 GHz, the **max_gain** plot represents the MAG, while below 12 GHz, it represents the MSG. $|S_{21}|^2$ is also plotted for frequency. By comparing the MAG with $|S_{21}|^2$, the gain improvement by simultaneous conjugate matching can be found.

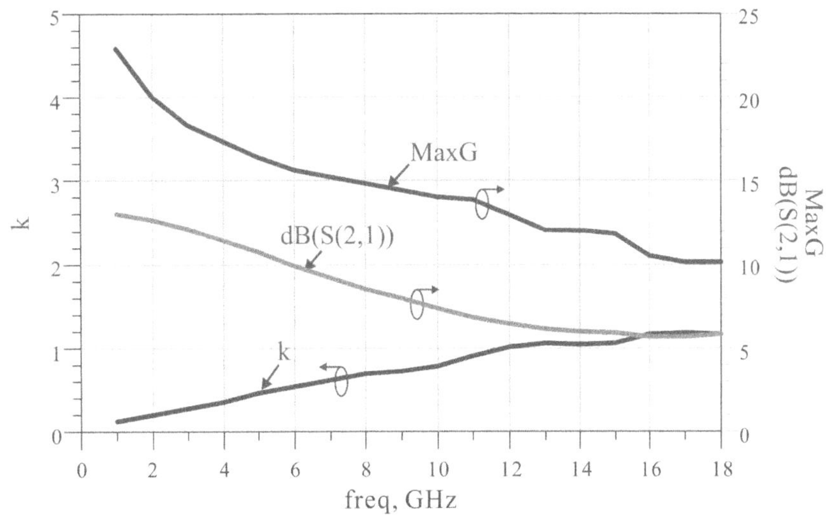

Figure 8E.4 Graph of **Max_gain**, stability factor, and $|S_{21}|^2$ with respect to frequency

Now we can find Γ_{SM} and Γ_{LM} at a frequency of 12 GHz where the selected device is stable. The ADS functions **sm_gamma1(S)** and **sm_gamma2(S)** are used to find Γ_{SM} and Γ_{LM}. The function **sm_gamma1(S)** gives Γ_{SM} in Equation (8.72), while the function **sm_gamma2(S)** gives Γ_{LM} in Equation (8.73). In addition, these two functions give 0 when $k < 1$, and they give the correct values only when $k > 1$. The function **find_index(•)** is added to find the corresponding index number to the frequency 12 GHz.

```
Eqn m=find_index(freq, 12G)
Eqn g1=sm_gamma1(S[m])
Eqn g2=sm_gamma2(S[m])
```

Measurement Expression 8E.4 Equations for obtaining the conjugate matching points

Figure 8E.5 shows **g1** and **g2** in Measurement Expression 8E.4. The conjugate matching reflection coefficients Γ_{SM} and Γ_{LM} at 12 GHz are read as

$$\Gamma_{SM} = 0.944\angle -168.5^\circ$$

$$\Gamma_{LM} = 0.919\angle 179.3^\circ$$

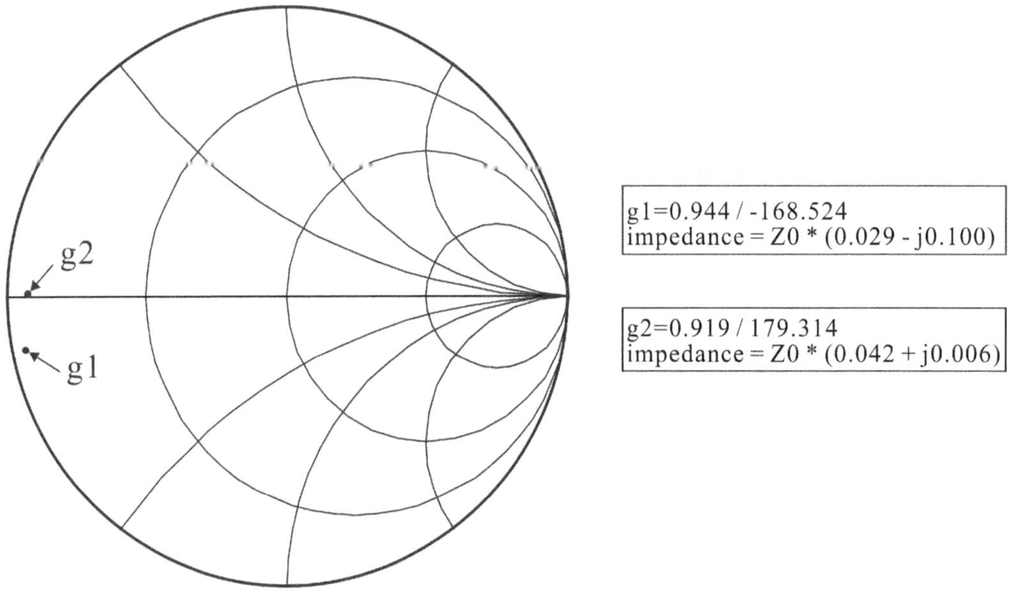

Figure 8E.5 Conjugate matching reflection coefficients at 12 GHz

8.4 GAIN AND NOISE CIRCLES

The previously explained Γ_{SM} and Γ_{LM} for the conjugate matching of source and load reflection will yield a maximum gain. However, from a design perspective, other aspects of performance must also be considered, and the source and load reflection coefficients are often set to values other than Γ_{SM} and Γ_{LM}, which can result in decreased gain. We will consider two cases; one when the source reflection coefficient cannot be selected as Γ_{SM} and the other when the load reflection coefficient cannot be selected as Γ_{LM}. In these two cases, the remaining reflection coefficient is usually set to meet the conjugate matching condition. We often need to estimate the subsequent decrease in gain for the two cases. In this section, we will examine constant gain circles as well as constant noise-figure circles.

8.4.1 Gain Circles

The previously determined power gain in Equation (8.33) can be expressed as Equation (8.34)

$$G_p = \frac{\left(1-|\Gamma_L|^2\right)|S_{21}|^2}{1-|S_{11}|^2+|\Gamma_L|^2\left(|S_{22}|^2-|D|^2\right)-2\,\mathrm{Re}\left(C_2\Gamma_L\right)} \tag{8.75}$$

where C_2 is defined in Equation (8.76).

$$C_2 = S_{22} - DS_{11}^* \tag{8.76}$$

In addition, for ease of mathematical computation, a normalized gain parameter g_p is defined as in Equation (8.77).

$$g_p = \frac{G_P}{|S_{21}|^2} \tag{8.77}$$

The locus of Γ_L giving the same power gain implies that g_p is a constant. For a constant g_p, Equation (8.75) can be rewritten as Equation (8.78).

$$g_p = \frac{\left(1-|\Gamma_L|^2\right)}{1-|S_{11}|^2+|\Gamma_L|^2\left(|S_{22}|^2-|D|^2\right)-2\,\mathrm{Re}\left(C_2\Gamma_L\right)} \tag{8.78}$$

With the following definitions expressed in Equations (8.79) and (8.80),

$$D_2 = |S_{22}|^2 - |D|^2 \tag{8.79}$$

$$X = \frac{1+g_pD_2}{g_p} \tag{8.80}$$

Equation (8.78) can be rewritten as

$$X^2|\Gamma_L|^2 - 2X\,\mathrm{Re}\left(C_2\Gamma_L\right)+|C_2|^2 = X^2 - XB_2 + |C_2|^2 \tag{8.81}$$

Then, rewriting Equation (8.81), we obtain

$$\left|X\Gamma_L - C_2^*\right|^2 = X^2 - XB_2 + |C_2|^2 \tag{8.82}$$

Using Equation (8.82), we can then obtain Equation (8.83) for a circle.

$$\left|\Gamma_L - \frac{C_2^*}{X}\right|^2 = \frac{X^2 - XB_2 + |C_2|^2}{X^2} \tag{8.83}$$

Therefore, the center C_p and radius r_p of the circle giving the same power gain are expressed in Equations (8.84) and (8.85).

$$C_p = \frac{C_2^*}{X} = \frac{g_p\left(S_{22}^* - D^*S_{11}\right)}{1+g_p\left(|S_{22}|^2-|D|^2\right)} \tag{8.84}$$

$$r_p = \frac{\sqrt{1-2k|S_{12}S_{21}|g_p + |S_{12}S_{21}|^2 g_p^2}}{\left|1+g_p\left(|S_{22}|^2-|D|^2\right)\right|} \tag{8.85}$$

The circle thus obtained is called a *power gain circle.*

In addition, when Γ_L is changed to Γ_S and the subscripts 1 and 2 are interchanged in Equation (8.75), it becomes the equation for the available power gain expressed in Equations (8.86) and (8.87).

$$G_A = \frac{\left(1 - |\Gamma_S|^2\right)|S_{21}|^2}{1 - |S_{22}|^2 + |\Gamma_S|^2\left(|S_{11}|^2 - |D|^2\right) - 2\,\mathrm{Re}\left(C_1\Gamma_S\right)} \tag{8.86}$$

$$C_1 = S_{11} - DS_{22}^* \tag{8.87}$$

Similarly, to obtain the locus of Γ_S with the same available power gain, the following normalized gain is defined in Equation (8.88).

$$g_a = \frac{G_A}{|S_{21}|^2} \tag{8.88}$$

Interchanging the subscripts 1 and 2 in Equations (8.84) and (8.85) will yield the radius r_a and center C_a for the available gain circle expressed in Equation (8.89).

$$C_a = \frac{g_a\left(S_{11}^* - D^*S_{22}\right)}{1 + g_a\left(|S_{11}|^2 - |D|^2\right)} \tag{8.89}$$

$$r_a = \frac{\sqrt{1 - 2k|S_{12}S_{21}|g_a + |S_{12}S_{21}|^2 g_a^2}}{\left|1 + g_a\left(|S_{11}|^2 - |D|^2\right)\right|} \tag{8.90}$$

The circles thus obtained are called *available gain circles* and they have the same available power gain.

8.4.2 Noise Circles

The noise figure obtained in Chapter 4 in terms of the noise parameters depends only on the source reflection coefficient Γ_S that is expressed in Equation (8.91).

$$F = F_{min} + 4r_n\frac{|\Gamma_S - \Gamma_{opt}|^2}{|1 + \Gamma_{opt}|^2\left(1 - |\Gamma_S|^2\right)} \tag{8.91}$$

To obtain the locus of Γ_S having the same noise figure, a new noise figure is defined for computational simplicity, as shown in Equation (8.92).

$$N_i = \frac{F_i - F_{min}}{4r_n}|1 + \Gamma_{opt}|^2 \tag{8.92}$$

The noise-figure equation can then be written as Equation (8.93).

$$N_i = \frac{\left|\Gamma_S - \Gamma_{opt}\right|}{1 - \left|\Gamma_S\right|^2} \tag{8.93}$$

Expanding that equation, it can be written as Equation (8.94).

$$\left|\Gamma_S - \Gamma_{opt}\right|^2 = \left|\Gamma_S\right|^2 + \left|\Gamma_{opt}\right|^2 - 2\,\mathrm{Re}\left(\Gamma_S \Gamma_{opt}^*\right) = N_i - N_i\left|\Gamma_S\right|^2 \tag{8.94}$$

Equation (8.94) can then be rewritten as

$$\left|\Gamma_S\right|^2 (1 + N_i) + \left|\Gamma_{opt}\right|^2 - 2\,\mathrm{Re}\left(\Gamma_S \Gamma_{opt}^*\right) = N_i \tag{8.95}$$

Multiplying both sides of Equation (8.95) by $1 + N_i$ results in Equation (8.96).

$$\left|\Gamma_S\right|^2 (1 + N_i)^2 + \left|\Gamma_{opt}\right|^2 - 2(1 + N_i)\mathrm{Re}\left(\Gamma_S \Gamma_{opt}^*\right) = N_i^2 + N_i\left(1 - \left|\Gamma_{opt}\right|^2\right) \tag{8.96}$$

Thus, the locus of Γ_S having the same N_i is given in Equation (8.97).

$$\left|\Gamma_S - \frac{\Gamma_{opt}}{1 + N_i}\right|^2 = \frac{N_i^2 + N_i\left(1 - \left|\Gamma_{opt}\right|^2\right)}{(1 + N_i)^2} \tag{8.97}$$

Therefore, the radius r_F and center C_F of the *noise circle* are expressed in Equations (8.98a) and (8.98b).

$$C_F = \frac{\Gamma_{opt}}{1 + N_i} \tag{8.98a}$$

$$r_F = \frac{\sqrt{N_i^2 + N_i\left(1 - \left|\Gamma_{opt}\right|^2\right)}}{1 + N_i} \tag{8.98b}$$

Example 8.4

Using the FHX35LG/LP, draw the noise circles and the available power gain circles at 12 GHz.

Solution

To calculate the noise parameters, open the S-parameter controller, check *Calculate noise*, and then specify the noise input and output ports as shown in Figure 8E.6. Simulate and then enter the following equations in Measurement Expression 8E.5 in the display window to draw the circles.

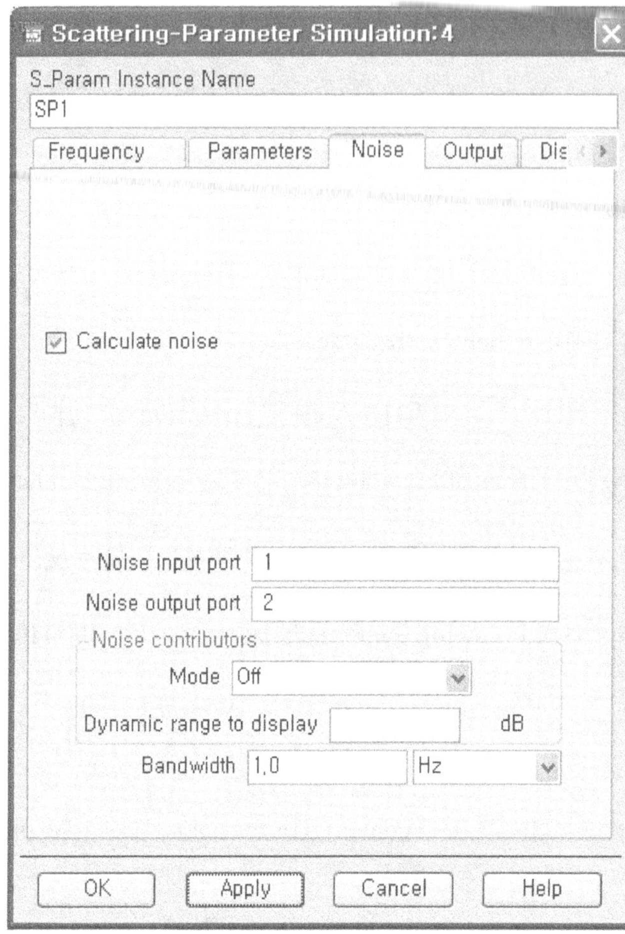

Figure 8E.6 Settings for the S-parameter controller to calculate noise parameters

Eqn ga1=ga_circle(S[m])

Eqn ns1=ns_circle({0,1,2,3}+NFmin[m], NFmin[m], Sopt[m], Rn[m]/50, 51)

Measurement Expression 8E.5 Equations for drawing the locus of the available power gain circles and the noise circles

The function **ns_circle(•)** is used for drawing the noise circles given by Equation (8.98), while the function **ga_circle(•)** is used for drawing the available power gain circles given by Equations (8.89) and (8.90). The function **ga_circle(•)** draws the available power gain circles lower than the MAG by {1,2,3} dB as the default. The available gain circles plotted by the function **ga_circle(•)** are shown on the Smith chart in Figure 8E.7. The maximum point of the available power gain circle in the figure is Γ_{SM}.

In the variables of the function **ns_circle**(), {0,1,2,3}+**NFmin[m]** represents a noise figure lower than the minimum noise figure by {1,2,3} dB. As previously mentioned, **m** represents the frequency index. To plot the noise circles, the corresponding noise parameters **NFmin[m]**, **Sopt[m]** and **Rn[m]**/50 are required. It must be noted that the function **ns_circle(•)** uses r_n, the normalized noise resistance by 50 Ω. The number 51 represents the number of points used for drawing a circle.

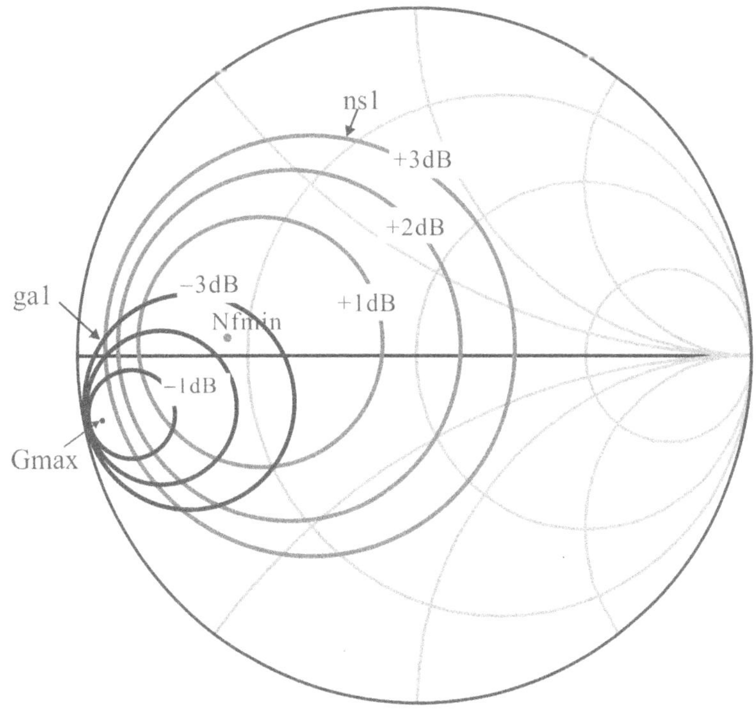

Figure 8E.7 Noise and available power gain circles of FHX35LG/LP at 12 GHz

8.5 SUMMARY OF GAINS AND CIRCLES

8.5.1 Summary of Gains

Transducer power gain G_T is defined as the ratio of the power delivered to the load to the available power from the source, and corresponds to the gain of an amplifier by the usual measurement. In the case of a 50-Ω source and load, $\Gamma_S = \Gamma_L = 0$, the transducer power gain becomes $|S_{21}|^2$. Thus, $|S_{21}|^2$ is the transducer power gain for a 50-Ω source and load.

First, the available power gain is the gain for a given source reflection coefficient with a conjugate matched load. The available power gain represents the gain degradation due to the specific source reflection coefficient and it is used to evaluate the gain decrease by the source mismatch. Similarly, the power gain is the gain for the selected load with the conjugate matched source, and it represents the gain degradation for selected load reflection coefficient.

Next, the maximum gain appears when the simultaneous conjugate matching at both the input and output of an active device is achieved. The maximum available power gain, MAG, can be obtained if the active device is stable. When it is unstable, the maximum stable gain, MSG, can be obtained at the stability boundary $k = 1$. The unilateral approximation is applied to estimate the gain. The maximum unilateral gain is an approximate maximum power gain. This clearly shows the extent of the improvement in the transducer power gain $|S_{21}|^2$ by conjugate matching. The passivity or activity of an active device can also be found using the Mason's gain, which determines whether an active device can amplify or not. The Mason's gain is the gain that is obtained after the removal of feedback from the output to the input of an active device. The previously described gains are summarized in Table 8.1.

Table 8.1 Gain formulas given in terms of S-parameters

Gain	Formula
Transducer power gain in 50-Ω source and load	$G_T = \|S_{21}\|^2$
Transducer power gain	$G_T = \dfrac{(1-\|\Gamma_L\|^2)\|S_{21}\|^2(1-\|\Gamma_S\|^2)}{\|(1-S_{22}\Gamma_L)(1-S_{11}\Gamma_S)-S_{12}S_{21}\Gamma_L\Gamma_S\|^2}$
Power gain	$G_P = \dfrac{(1-\|\Gamma_L\|^2)\|S_{21}\|^2}{\|1-S_{22}\Gamma_L\|^2(1-\|\Gamma_{in}\|^2)}$
Available power gain	$G_A = \dfrac{(1-\|\Gamma_S\|^2)\|S_{21}\|^2}{\|1-S_{11}\Gamma_S\|^2(1-\|\Gamma_{out}\|^2)}$
Mason's unilateral gain	$U = \dfrac{\|S_{21}/S_{12}-1\|^2}{2k\|S_{21}/S_{12}\|-2\mathrm{Re}(S_{21}/S_{12})}$
Maximum unilateral power gain	$G_{TU,\max} = \dfrac{\|S_{12}\|^2}{(1-\|S_{11}\|^2)(1-\|S_{22}\|^2)}$
Maximum stable gain	$MSG = \left\|\dfrac{S_{21}}{S_{12}}\right\|$
Maximum available gain	$MAG = G_{\max} = \left\|\dfrac{S_{21}}{S_{12}}\right\|(k-\sqrt{k^2-1})$

$$\text{Stability factor, } k = \frac{1-|S_{11}|^2-|S_{22}|^2+|D|^2}{2|S_{12}S_{21}|}$$

8.5.2 Summary of Circles

The previously derived center and radius for drawing circles on the Smith chart are summarized in Table 8.2. The available power gain circle is the locus of the source reflection coefficient Γ_S yielding the same available power gain. Note that the load is assumed to be conjugate matched to the output reflection coefficient Γ_{out} given by Γ_S. Conversely, the power gain circle is the locus of the load reflection coefficient Γ_L yielding the same power gain. In plotting the circles, it is assumed that the source reflection coefficient Γ_S is conjugate matched to the input reflection coefficient Γ_{in} given by Γ_L. The power gain circle shows a gain degradation for a selected load reflection coefficient. Together with the available power gain circles, the power gain circles are commonly used to evaluate gain

Table 8.2 Formulas of the various circles given in S-parameters

Circles	Center	Radius
Available gain	$C_a = \dfrac{g_a(S_{11}^* - D^* S_{22})}{1 + g_a(\lvert S_{11}\rvert^2 - \lvert D\rvert^2)}$	$r_a = \dfrac{\sqrt{1 - 2k\lvert S_{12}S_{21}\rvert g_a + \lvert S_{12}S_{21}\rvert^2 g_a^2}}{\lvert 1 + g_a(\lvert S_{11}\rvert^2 - \lvert D\rvert^2)\rvert}$
Power gain	$C_p = \dfrac{g_p(S_{22}^* - D^* S_{11})}{1 + g_p(\lvert S_{22}\rvert^2 - \lvert D\rvert^2)}$	$r_p = \dfrac{\sqrt{1 - 2k\lvert S_{12}S_{21}\rvert g_p + \lvert S_{12}S_{21}\rvert^2 g_p^2}}{\lvert 1 + g_p(\lvert S_{22}\rvert^2 - \lvert D\rvert^2)\rvert}$
Noise mismatch	$C_F = \dfrac{\Gamma_{opt}}{1 + N_i}$	$r_F = \dfrac{\sqrt{N_i^2 + N_i(1 - \lvert\Gamma_{opt}\rvert^2)}}{(1 + N_i)}$
Source stability	$C_S = \dfrac{S_{11}^* - D^* S_{22}}{\lvert S_{11}\rvert^2 - \lvert D\rvert^2}$	$r_S = \dfrac{\lvert S_{12}S_{21}\rvert}{\lvert\,\lvert S_{11}\rvert^2 - \lvert D\rvert^2\rvert}$
Load stability	$C_L = \dfrac{S_{22}^* - D^* S_{11}}{\lvert S_{22}\rvert^2 - \lvert D\rvert^2}$	$r_L = \dfrac{\lvert S_{12}S_{21}\rvert}{\lvert\,\lvert S_{22}\rvert^2 - \lvert D\rvert^2\rvert}$

$$\text{Normalized power gain, } g_p = \frac{G_P}{|S_{21}|^2}$$

$$\text{Normalized available power gain, } g_a = \frac{G_A}{|S_{21}|^2}$$

$$\text{Normalized noise figure, } N_i = \frac{F_i - F_{min}}{4r_n}|1 + \Gamma_{opt}|^2$$

degradation from the MAG for the selected load or source reflection coefficients in the course of design. Noise circles are used to evaluate the degradation of the noise figure when the source reflection coefficient is not selected for the minimum noise figure, and they are commonly used together with the available gain circles mentioned previously in low-noise amplifier design.

The stability circles help us to determine whether negative resistance appears at the input or output for the selected load or source reflection coefficients, and to know whether or not there is the possibility of oscillation. Accordingly, we are able to identify the region of stability where the load or source reflection coefficients do not induce negative resistance at the input or output. Source stability circles determine whether or not the selected source reflection coefficient is in the region of stability, while the load stability circles determine whether or not the selected load reflection coefficient is in the region of stability.

8.6 DESIGN EXAMPLE

A low-noise amplifier is an amplifier that has a low noise figure and it is usually located at the front end of a receiver. According to the Frii's formula, the overall noise figure not only depends on the noise figure of the front-end amplifier, but is also related to the noise figure of subsequent stages. Thus, the overall noise figure depends on both the noise figure and gain of the low-noise amplifier. Because of these facts, the design of the low-noise amplifier can be viewed as a trade-off between the noise figure and the gain. The design involves a process of 1) the appropriate active device selection, 2) the selection of optimal source and load impedances, 3) a design of input and output matching circuits, 4) the addition of DC bias circuits to the matching circuits, and 5) the determination of stability. The selected active device is sometimes unstable at the design frequency. In this case, stabilization is achieved through feedback or by adding small losses to the selected active device. Reference 1 at the end of this chapter provides details of the stabilization method. In addition, considerable attention to the last step of the stability check is recommended since a designed low-noise amplifier often oscillates at frequencies other than the design frequency. This section explains low-noise amplifier design using ADS and is based on the theories previously developed in this book.

8.6.1 Design Goal

This section presents the design goal of a single-stage low-noise amplifier. As an example, the design specifications for the low-noise amplifier operating in the X-band (8–12 GHz frequency band) are selected, as shown in Table 8.3. The appropriate active device that can satisfy the design specifications must then be selected. It is important that an active device has a minimum noise figure, F_{min}, lower than the design goal at the design frequency. Also, the selected active device should have a maximum gain, MAG, greater than the desired gain. An appropriate active device satisfying the design goal may not be available. In that case, the gain can be increased to meet the design goal by cascading a number of stages. Therefore, the key parameter is the noise figure. The active device ATF-36077 is selected to meet the required gain and noise figure and the S-parameters of this device are provided in ADS.

Table 8.3 Low-noise amplifier design specifications

No.	Spec Item	Value
1	Frequency range	9.7–10.3 GHz
2	DC supply voltage	< 3.3 V
3	DC current consumption	< 30 mA
4	Gain	> 10 dB
5	Noise figure	< 1.5 dB

8.6.2 Active Device Model

The active device models can be categorized as a large-signal model and a measured S-parameter data model at a given DC bias, and they can be used in the low-noise amplifier design. The large-signal model of a chip-type active device generally shows a good agreement; however, the S-parameters generated using the large-signal model of a packaged active device at a given DC bias do not generally show a good agreement with the measured S-parameters due to the inaccuracy of the large-signal model. As explained in Chapter 5, the large-signal model of a packaged active device generally does not show sufficient accuracy due to the complexity of the equivalent circuit. Thus, the design of a low-noise amplifier using a packaged device is generally carried out using the measured S-parameter data model. The data models are sometimes already available in the ADS library; otherwise, a user needs to construct the specific data model using given measured data.

Figure 8.21 shows the S-parameter data model of the ATF-36077 provided in ADS. The model has S-parameter values for 0.5 GHz to 18 GHz at $V_{ds} = 1.5$ V and $I_d = 20$ mA. No separate DC bias circuit is required to operate the data model and it can be simulated just for the RF signal. In the case of the design that uses a large-signal model, the S-parameters at a given DC bias should be computed by first using the large-signal model, and a similar design can then be carried out using the computed S-parameters following the design shown below.

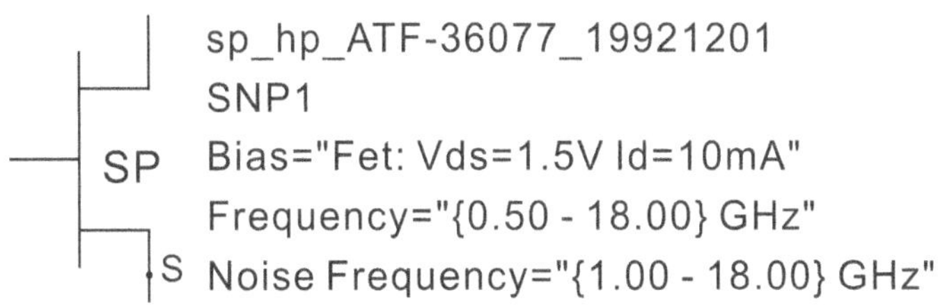

Figure 8.21 Small-signal data model of the ATF-36077

8.6.3 Device Performance

Figure 8.22 shows an S-parameter simulation set up for determining the stability and maximum gain (MSG or MAG) as well as the minimum noise figure with respect to frequency. In the previous Examples, the determination of the MAG and stability was done in the display window using the built-in measurement expressions. It is also possible to do this in the schematic window, as shown in Figure 8.22. After the simulation, the MAG and stability factor are directly available as variables **MaxGain**1 and **StabFact**1 in the dataset.

The simulated stability factor, MAG, and the minimum noise figure are shown in Figure 8.23. From Figure 8.23, k is greater than 1 above 16 GHz but k is less than 1 in the design frequency band. The **MaxGain**1 corresponding to MAG or MSG shows a gain above 15 dB at the design frequency of 10 GHz, as shown in Figure 8.23. Note that **MaxGain**1 in the design frequency band becomes the MSG because k is less than 1. The minimum noise figure, **NFmin** is constant as 0.3 dB below 6 GHz. This is not a simulation error. The device data itself is 0.3 dB. In the author's opinion, because a noise figure below 0.3 dB could not be accurately measured, the data are entered as constant 0.3 dB. However, the minimum noise figure is found to be about 0.5 dB at the design frequency of 10 GHz. Thus, the low-noise amplifier that meets the design specifications can be designed using the selected ATF-36077.

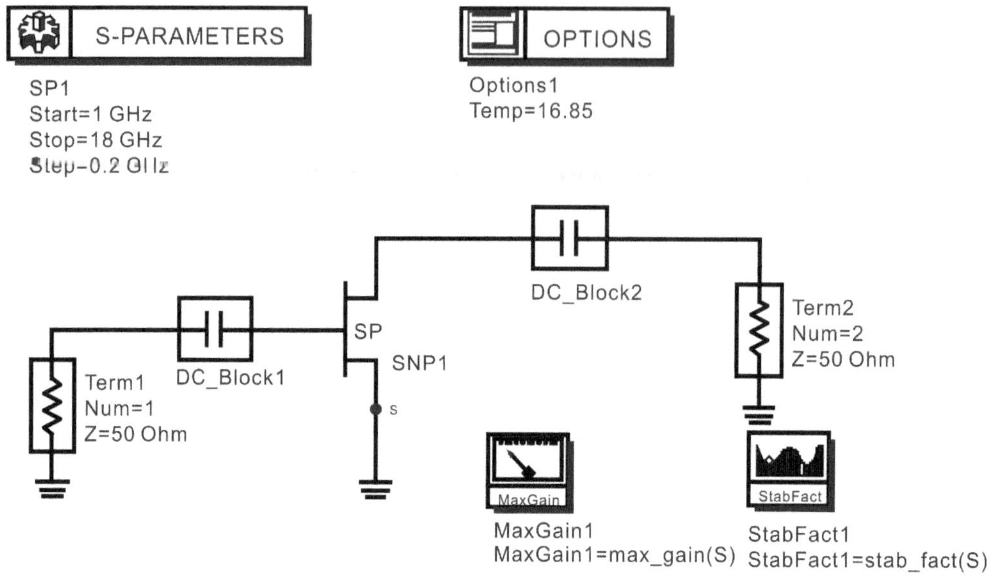

Figure 8.22 Determination of the gain and stability of the ATF-36077. The components **MaxGain** and **StabFact** directly yield the MAG and the stability factor in the dataset after the simulation.

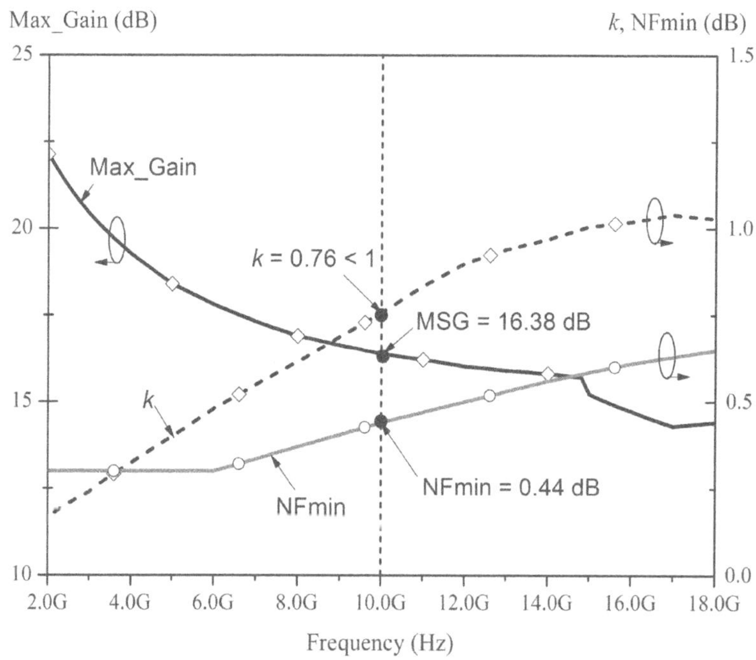

Figure 8.23 Stability factor k, maximum gain, and NF_{min} with respect to frequency. At 10 GHz, $k = 0.76 < 1$ and is unstable. The MSG and NF_{min} at 10 GHz are 16.38 dB and 0.44 dB, respectively.

Since the selected device is unstable at the design frequency band, stabilization is necessary to make k greater than 1 at 10 GHz. One way of improving the stability is shown in Figure 8.24. A series feedback inductor **Le** is connected to the source terminal. To some extent, the inductor added to the source terminal generally increases the stability factor. Figure 8.24 is a circuit setup in which the inductance is swept from 0 to 0.3 nH at a fixed frequency of 10 GHz to evaluate the change in k.

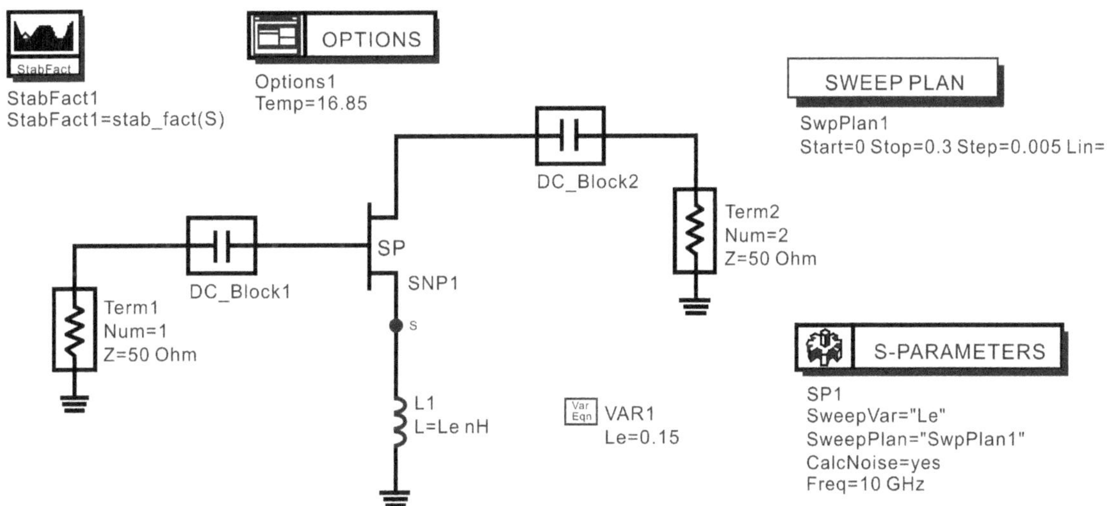

Figure 8.24 Schematic for evaluating the variation of k with the series feedback inductor as a parameter

Figures 8.25(a) and (b) show the simulated stability factor. Figure 8.25(a) shows the stability factor change for **Le**. From Figure 8.25(a), k has a maximum value of 1.031 at 10 GHz when **Le** is approximately 0.05 nH, and the stability is thus achieved. Therefore, **Le** is set to 0.05 nH and k is plotted with respect to frequency as shown in Figure 8.25(b), which shows k greater than 1 at the design frequency. The improvement in the k value is achieved by the addition of **Le**. However, the bandwidth is narrow and the k value is too close to 1. A slight error in the implementation of **Le** may cause the active device to be unstable even at 10 GHz and further stabilization would be necessary. There are several other ways to improve the stability of the circuit further. The addition of resistors is one of the methods commonly used (see reference 1 at the end of this chapter). The method of improving k by the addition of a shunt resistor to the drain terminal is used in this book. The reason for the connection to the drain terminal is to minimize the impact of the amplifier on the noise figure.

Figure 8.26 shows a shunt stabilizing circuit added to the drain terminal. An **R1** = 35 Ω chip resistor (for improving the value of k) is connected in series to a 100 nF DC block capacitor. The microstrip lines represent connecting lines and land patterns for mounting the chip resistor and DC block capacitor. The shunt stabilizing circuit in Figure 8.26 is configured as a subcircuit that is named **R_stab**. Thus, the amplifier circuit can be represented by a simple schematic, which avoids the necessity of a complex schematic during the design.

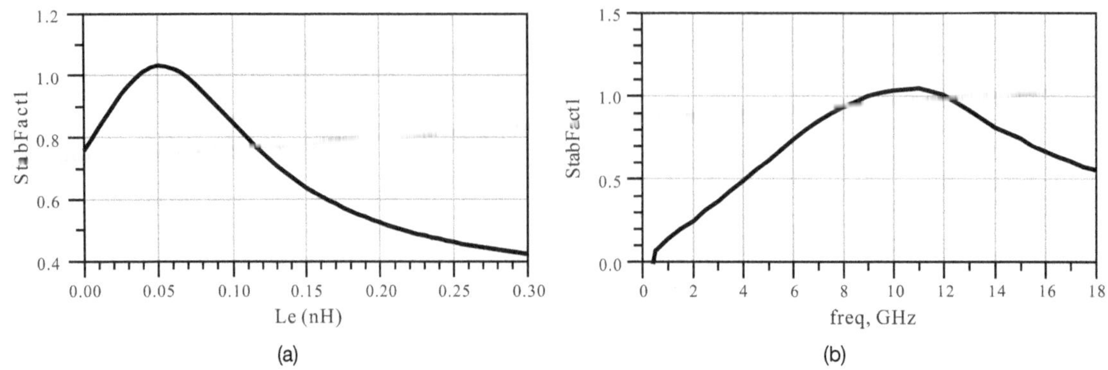

Figure 8.25 (a) Stability factor with respect to the value of **Le** at 10 GHz and (b) the frequency response of k for **Le** = 0.05 nH

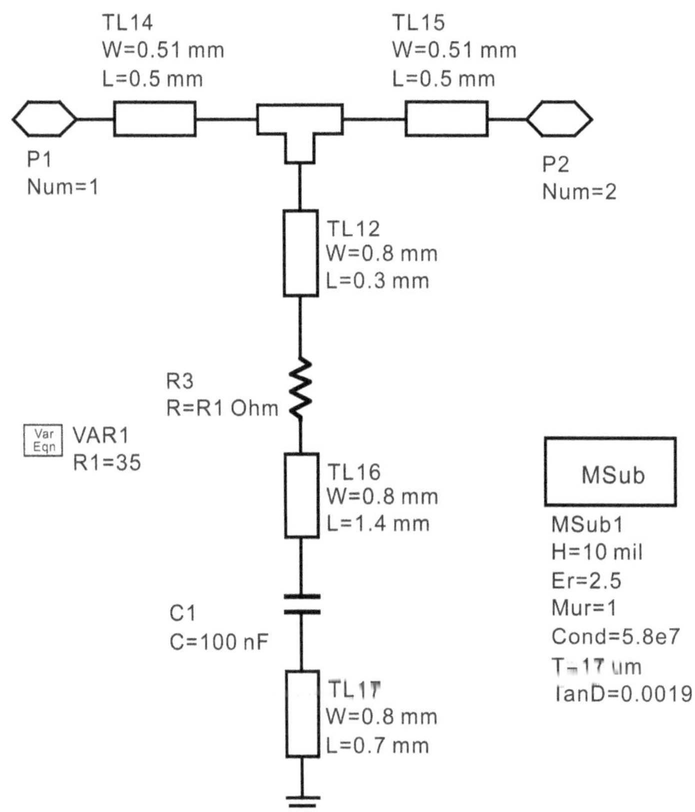

Figure 8.26 Stabilization circuit added for increasing k; subcircuit (**R_stab**)

The **R_stab** subcircuit is added to the stability evaluation circuit in Figure 8.24 and a parameter-swept S-parameter simulation for the resistor value **R**1 in Figure 8.26 is performed to determine the change in k. The value of **R**1 is swept from 0 to 50 Ω. Figure 8.27(a) shows the k for **R**1. The value of resistor **R**1 is selected as 36 Ω. For the value of 36 Ω,

Figure 8.27(b) shows the plot of the stability factor for a frequency range of 1–18 GHz. The stability factor k is found to be greater than 1 for wider frequency range. However, the gain slightly decreases due to the resistor added for stabilization, and the effect is slightly compensated by modifying **Le** = 0.04 nH.

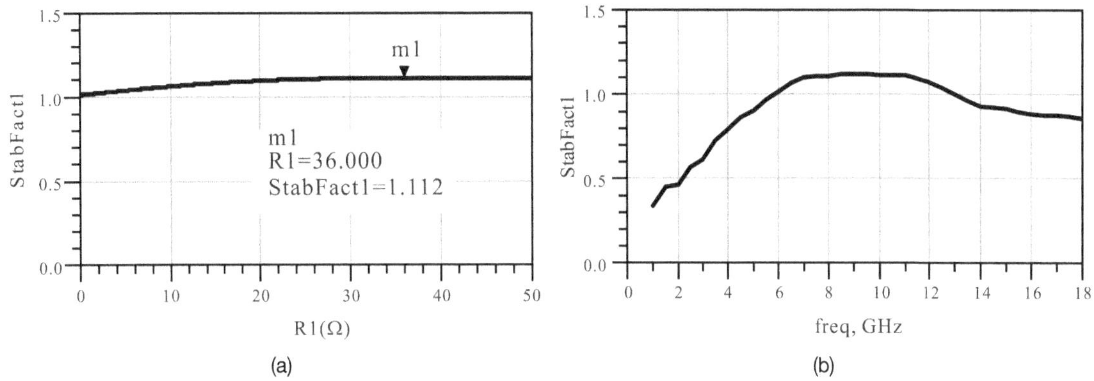

Figure 8.27 The change of k due to the added **R_stab** subcircuit: (a) the stability factor with respect to resistance **R1** at 10 GHz; (b) frequency response of k when the resistance **R1** is 36 Ω

8.6.4 Selection of Source and Load Impedances

In this section, we will examine the selection of the source and load impedances to satisfy the design goals. Figure 8.28 is a circuit setup in which the feedback inductor **Le** and the **R_stab** subcircuit are added to the active device. To plot the available power gain and the

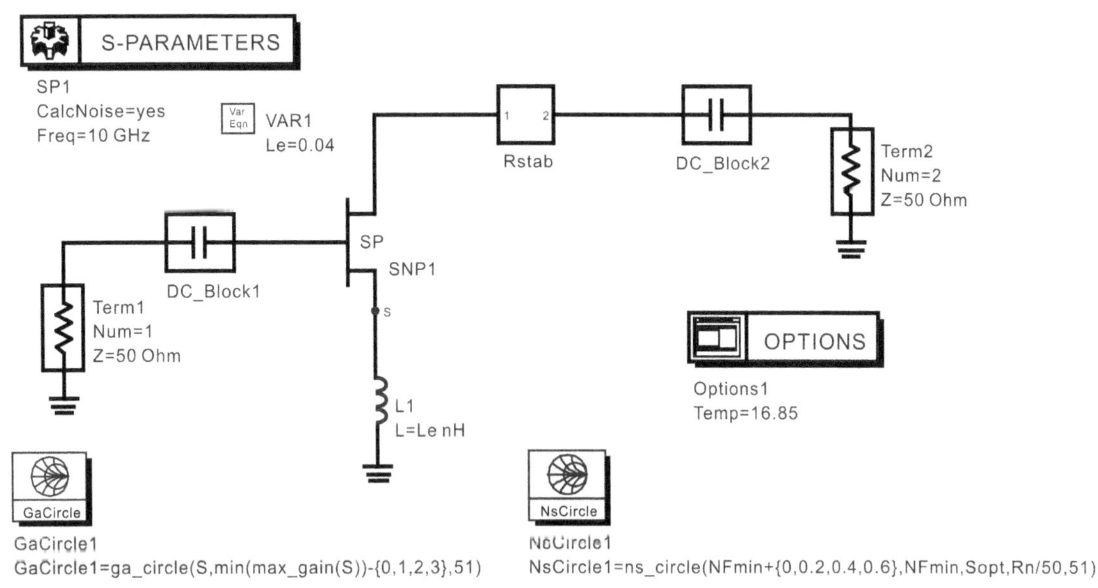

Figure 8.28 Circuit setup for measuring available power gain and noise circles. The available gain circles and noise circles can be directly obtained through the circuit simulation.

noise circles at 10 GHz, we use the Measurement Expressions in the schematic window shown in Figure 8.28. As previously explained, this can also be done in the display window. In Figure 8.28, three available gain circles whose gain reduces by a 1-dB step from the MAG and three noise circles whose noise figure increases by a 0.2-dB step from the minimum noise figure **NFmin** are specified. After the simulation, the available gain and noise circles are stored as **GaCircle1** and **NsCircle1** in the dataset, respectively. The available gain and noise circles are shown in Figure 8.29.

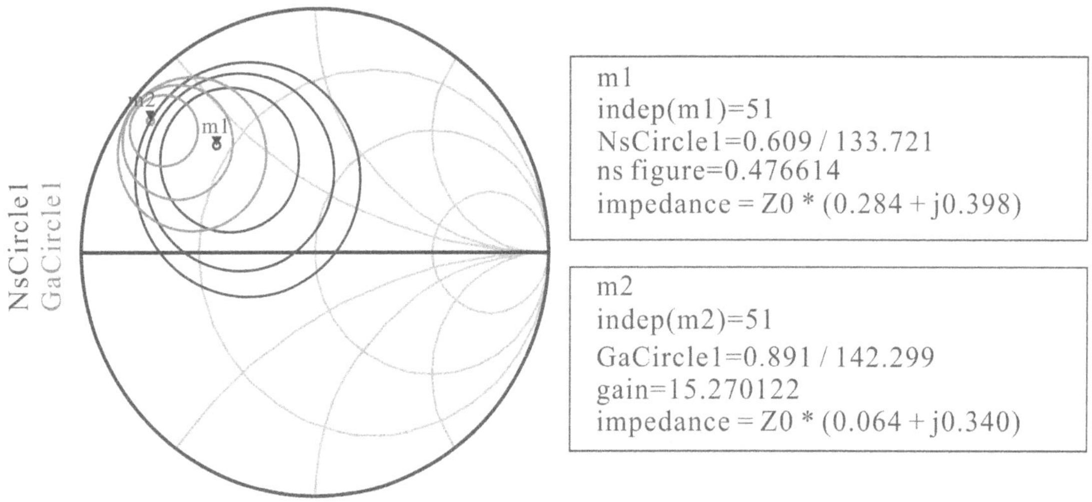

Figure 8.29 Available power gain and noise circles

Marker **m1** reads the minimum noise figure of 0.47dB, while marker **m2** reads the maximum available gain of 15.27 dB. The gain and noise figure performances depend on the choice of the source reflection coefficient, which must be properly selected in accordance with the design goal. Considering the given specifications of a 10-dB gain and a 1.5-dB noise figure, the source reflection coefficient Γ_S is selected as marker **m1** that has a minimum noise figure of 0.47 dB. The selected source reflection coefficient in Figure 8.29 will give a noise figure of 0.47 dB and a gain above 13 dB.

The source reflection coefficient thus selected is Γ_S = **m1** = 0.609∠133.7°. The corresponding conjugate matching load reflection coefficient is given by

$$\Gamma_L = \left(S_{22} + \frac{S_{12}S_{21}\Gamma_S}{1 - S_{11}\Gamma_S} \right)^* = 0.548\angle -159.05^\circ$$

The determined Γ_S and Γ_L are converted to the impedances, and they are given by

$$Z_{source} = \left(Z_{in}\right)^* = 50 * \left(0.285 + j0.398\right)$$

$$Z_{load} = \left(Z_{out}\right)^* = 50 * \left(0.301 - j0.169\right)$$

To confirm that the selected impedances give the estimated gain and noise figure at the design frequency, the circuit in Figure 8.30 is set up. Since the source and load impedances have the selected impedances, the gain and noise figure should show a gain above 13 dB and a noise figure of 0.47 dB.

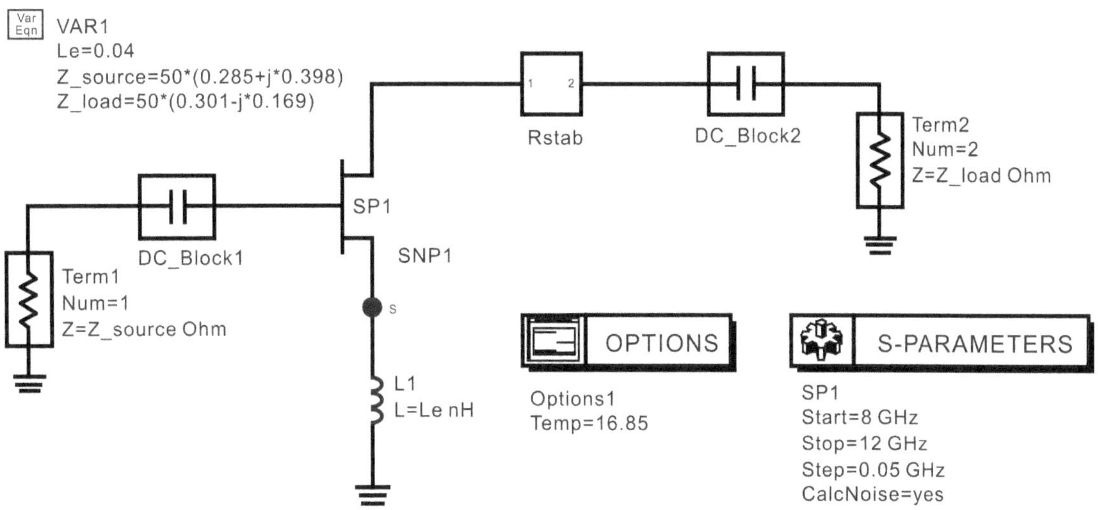

Figure 8.30 Schematic for confirming the selected source and load impedances. The computed impedances are **Z_source** and **Z_load** in the variable block named **VAR**1.

The simulated gain and noise figure are shown in Figure 8.31. As expected, the gain is found to be more than 13 dB and the noise figure is about 0.47 dB at 10 GHz. Thus, the selected impedances yield the gain and noise figure that satisfy the design specifications in the design frequency band. Therefore, the selected source and load impedances are found to be appropriate for the desired low-noise amplifier.

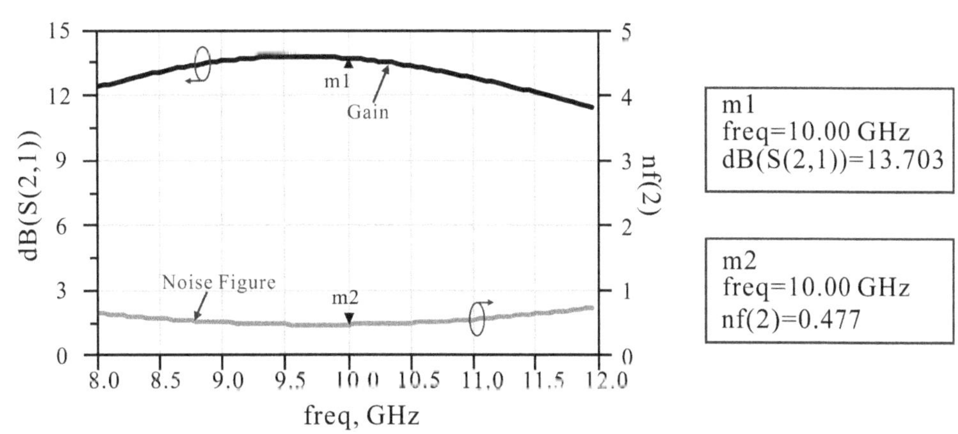

Figure 8.31 Gain and noise figure results for the selected source and load impedances. As expected from Figure 8.29, the gain is above 13 dB and the noise figure is 0.47 dB.

8.6.5 Matching Circuit Design

We found that the desired gain and noise figure can be obtained when the previously selected source and load impedances are connected to the active device. However, actual source and load impedances are 50 Ω, and the matching circuits are necessary to convert the 50 Ω source and load impedances to the previously selected impedances. The lumped-element matching circuits can be designed using the theories previously explained in Chapter 6. However, the optimization is available in ADS and the lumped-element matching circuits can be conveniently designed by the optimization technique. In this section, the lumped-element matching circuits will be primarily designed using the optimization technique. However, for the matching circuits that employ distributed elements such as transmission lines, it is more realistic to implement them on a PCB-like substrate at a frequency of 10 GHz. When microstrip lines are used, impedance matching may be limited by discontinuities, which make it difficult to determine whether or not the matching actually works. Therefore, in this section we will first present the lumped-element matching circuit designs; then, the distributed matching circuits will be designed by replacing the lumped elements with microstrip lines.

8.6.5.1 Lumped-Element Matching Circuits

Figure 8.32 shows a lumped-element input matching circuit. The circuit is set up to determine the values of the lumped-element input matching circuit at 10 GHz using optimization. Port 1 represents the source with a 50-Ω impedance while port 2 represents the input impedance of the active device. The active device's input impedance is the complex conjugate of the previously selected source impedance. When the maximum power transfer condition is achieved, the impedance seen from port 2 must be the complex conjugate of the port 2 impedance, which is equal to the selected source impedance. Therefore, the matching circuit that gives the desired source-impedance value can be determined.

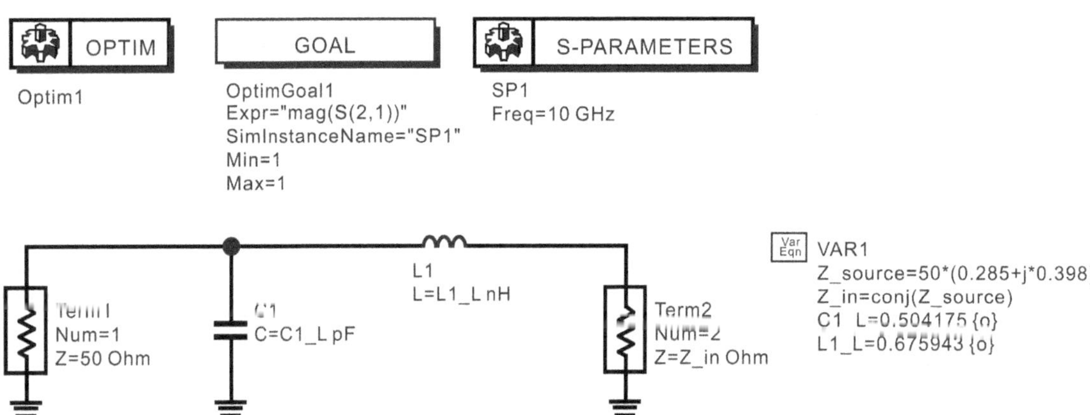

Figure 8.32 Lumped-element input matching circuit setup. Since the impedance seen from port 2 should be the selected **Z_source**, the impedance of **Term2** is set to the conjugate of the **Z-source**.

In Figure 8.32, the values of capacitor **C1_L** and inductor **L1_L** are optimized for maximum power transfer. The goal is set using $|S_{21}|$ and the maximum power transfer is implemented by setting $|S_{21}| = 1$. Note that $|S_{21}| = 1$ is a sufficient condition for conjugate matching. Since the matching circuit consists of lossless elements, the S-parameters satisfy the relation $|S_{11}|^2 + |S_{21}|^2 = 1$. Thus, $S_{11} = 0$ when $|S_{21}|^2 = 1$. In addition, as the lumped-element matching circuit is a passive circuit, $S_{21} = S_{12}$, by invoking the lossless condition it can also be seen that $S_{22} = 0$. After the optimization, the values of the lumped-element matching circuit in Figure 8.32 are updated by the optimized values. The value of **C1_L** is 0.504 pF and that of **L1_L** is 0.675 nH. Figure 8.33 shows the S-parameters of the lumped-element matching circuit with updated values. It can be seen that $|S_{21}|$ is close to 1 while $|S_{22}|$ is close to 0, which indicates successful impedance matching.

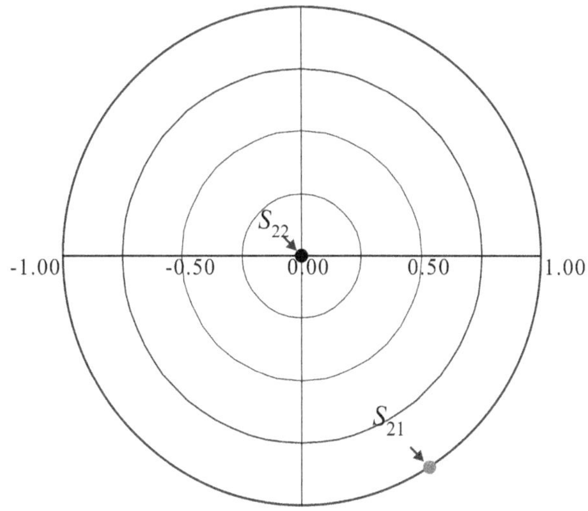

Figure 8.33 Simulation results of the input matching circuit at a frequency of 10 GHz. S_{11} and S_{22} are close to 0, while S_{21} is close to 1. This indicates that the matching is successful.

Similarly, the lumped-element output matching circuit is set up as shown in Figure 8.34. After optimization, the values of the lumped-element output matching circuit are determined as **C1_L** = 0.478 pF and **L1_L** = 0.231 nH. The simulated S-parameters for the determined values are shown in Figure 8.35, where $|S_{21}| = 1$, while $|S_{11}| = |S_{22}| = 0$ and the successful impedance matching is achieved.

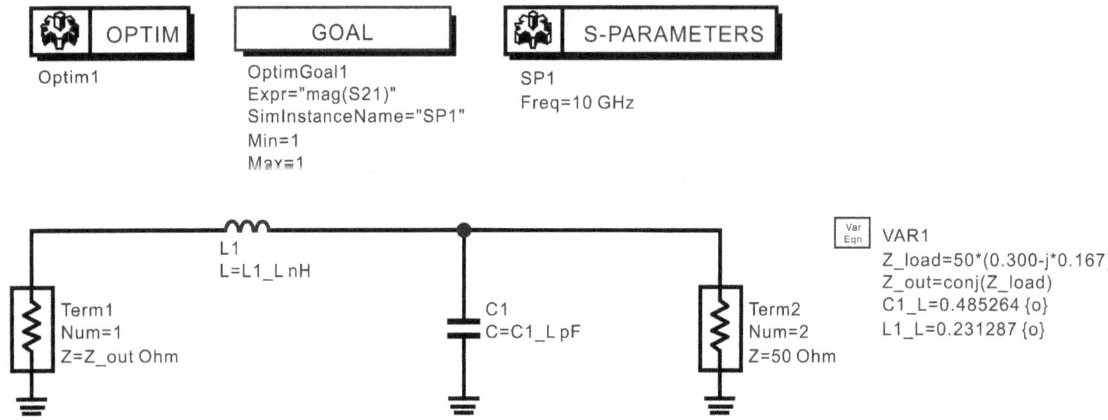

Figure 8.34 Schematic for the lumped–element output matching circuit simulation. Similar to the lumped-element input matching circuit, the impedance of **Term**1 is set to the conjugate of **Z_load**.

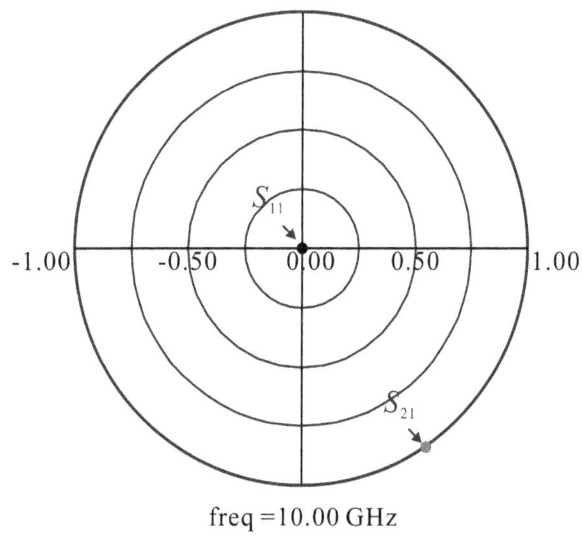

Figure 8.35 Optimization results of the output matching circuits at a frequency of 10 GHz

The lumped-element input and output matching circuits are configured as subcircuits. Using these subcircuits, the low-noise amplifier circuit is configured as shown in Figure 8.36 to confirm the designed matching circuits.

Figure 8.37 is the simulated gain and noise figure. The gain from Figure 8.37 is found to be more than 13 dB, while the noise figure is below 1.5 dB in the desired frequency band. Due to the frequency-dependent matching circuits, the difference arises as the frequency moves away from the design frequency of 10 GHz.

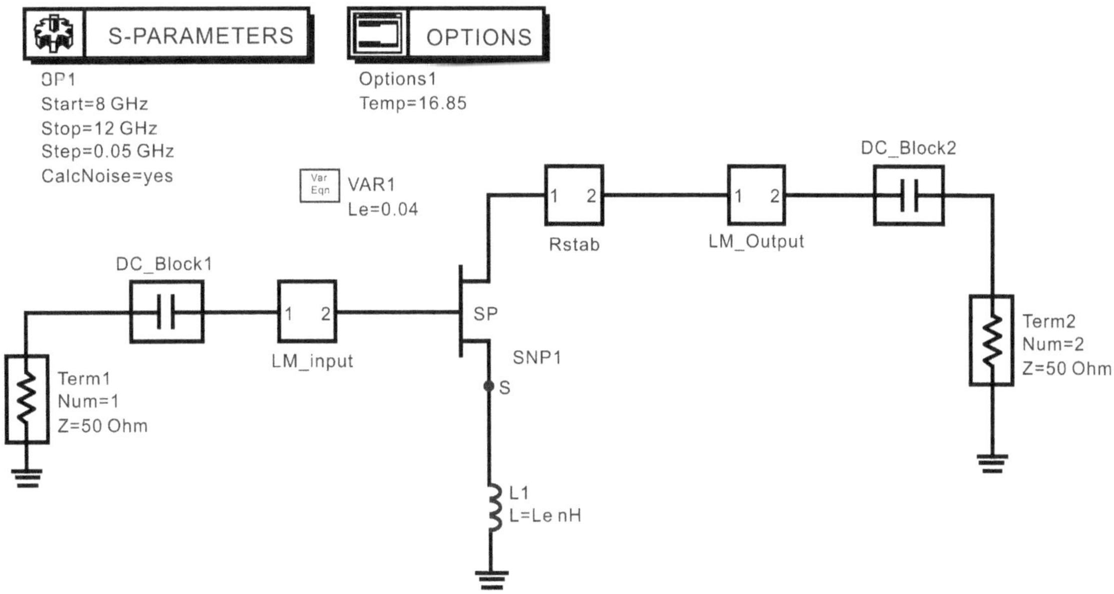

Figure 8.36 The low-noise amplifier circuit composed of the designed lumped-element input and output matching circuits

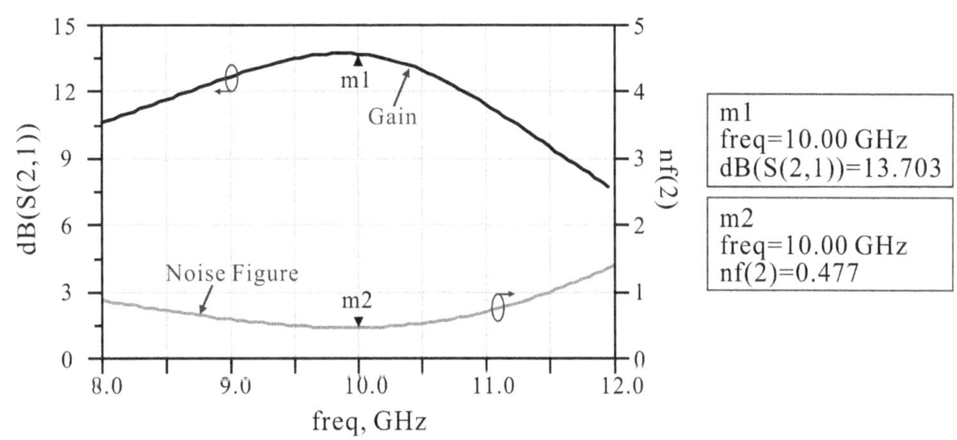

Figure 8.37 Simulation results of the low-noise amplifier composed of lumped-element input and output matching circuits. The gain and the noise figure have almost same values as those in the impedance confirmation.

8.6.5.2 Transmission-Line Matching Circuits

The lumped-element matching circuit discussed in the previous section is intuitive and provides numerous options for design. However, excepting an MMIC, it is generally difficult to implement on a PCB. On the other hand, a matching circuit that uses transmission lines is easy to implement on a PCB-like substrate at high frequency. We will first present the approximate transformation of the previously designed lumped-element matching circuits into the transmission-line matching circuits, and then we will examine a matching circuit design that uses transmission lines.

The approximate transformation can be done by replacing the lumped elements with transmission lines. Figure 8.38 shows the transmission-line circuit that can replace the designed lumped-element matching circuit. The transmission line with characteristic impedance Z_1 and length θ_1 can act as a series inductor, while the two parallel-connected open transmission-line stubs having characteristic impedance Z_2 and length θ_2 can act as the shunt capacitor. The values of Z_1, θ_1, Z_2, and θ_2 can be approximately determined using Equations (8.99a) and (8.99b).

$$X_1 = Z_1 \tan(\theta_1) \approx \omega L \tag{8.99a}$$

$$X_2 = \frac{Z_2}{2\tan(\theta_2)} \approx \frac{1}{\omega C} \tag{8.99b}$$

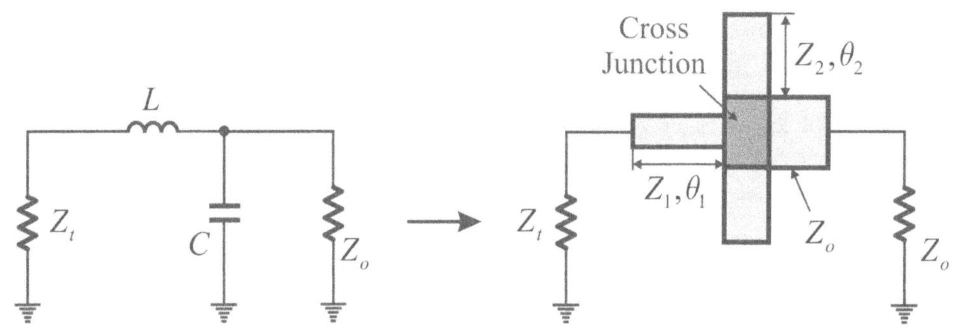

Figure 8.38 Conversion of a lumped-element matching circuit to a transmission-line matching circuit

It is worth noting that because the variables in Equation (8.99a) are Z_1 and θ_1, there are two degrees of freedom in implementing the value of an inductor. The higher Z_1 is, the closer it can be implemented as a series inductor. However, considering that the packaged active device is mounted and soldered on top of the inductor transmission line, there should be no problem soldering the active device. Thus, the inductor transmission line has a width wider than that of the active device's terminal. In addition, the two degrees of freedom are also true for Equation (8.99b), and the lower Z_2 is, the closer to a capacitor it will be. However, because increasing the transmission line's width shortens its length, the implemented transmission-line circuit is close to two cascaded transmission lines with step-discontinuity rather than a cross-junction connection. To avoid this, the lengths of the transmission lines connected at the cross-junction are made greater than their widths.

In addition, when an external circuit is connected directly to the matching circuit, the field of the cross-junction shown in Figure 8.38 may be affected, especially when a coaxial connector is directly connected to the matching circuit. To prevent this, the transmission line with the same value of characteristic impedance Z_o is inserted. The output transmission line has no function when viewed from the load side. In addition, the length must be sufficiently long to adequately attenuate the higher-order modes arising in the cross-junction.

Figure 8.39 shows a schematic to determine through optimization the input matching circuit composed of microstrip lines. As previously explained, the microstrip line provides two degrees of freedom, such as the width and length. Of these two, the width is fixed as previously described. The matching circuit is configured by simply changing the length. The width of the transmission line for the shunt capacitor is set slightly wider than the width of a 50-Ω line as **W_C** = 1.0 mm, while the width of the transmission line for the series inductor is set slightly smaller than the width of a 50-Ω line as **W_L** = 0.5 mm. With the widths thus fixed, the lengths of the two transmission lines, **l_L** and **l_C** are adjusted in a way similar to that of the lumped-element matching circuit until maximum power is delivered. The optimized dimensions are also shown in Figure 8.39.

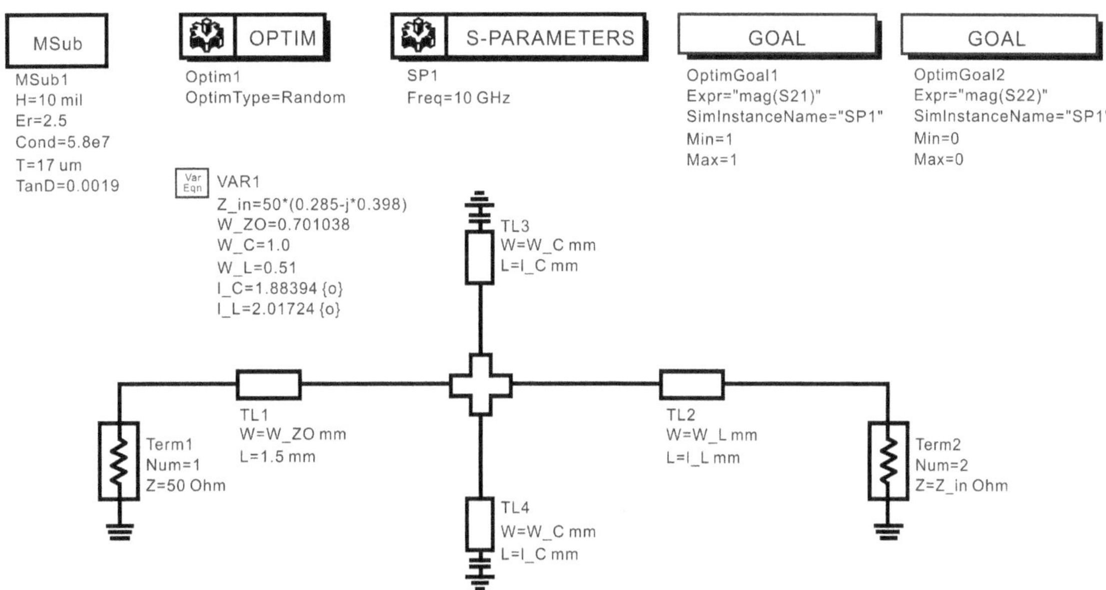

Figure 8.39 Input matching circuit employing microstrip lines. The variable **W_ZO** represents the 50-Ω width of the microstrip line. **W_C** and **W_L** represent the widths of the open-stub and series matching line, respectively. The widths of microstrip lines are fixed and only the lengths are optimized for conjugate matching.

The optimized S-parameters are shown in Figure 8.40, where $|S_{21}| = 1$ while $|S_{22}| = 0$, and the successful impedance matching is achieved. Thus, the microstrip-line matching circuit with the optimized dimensions can be used in place of the lumped-element input matching circuit. Similarly, the microstrip-line output matching circuit is set up as shown in Figure 8.41. The length of the microstrip lines **l_L** and **l_C** are optimized for the maximum power transfer. The impedance of port 1 is the complex conjugate of the load impedance, while the impedance of port 2 is 50 Ω.

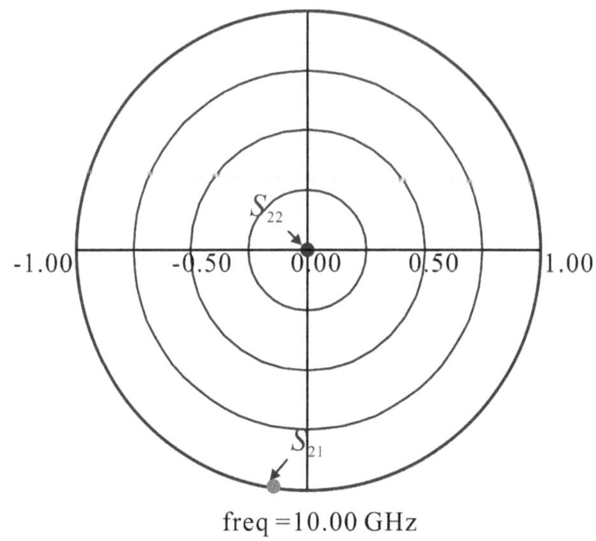

Figure 8.40 Optimization results of microstrip-line input matching circuit at a frequency of 10 GHz

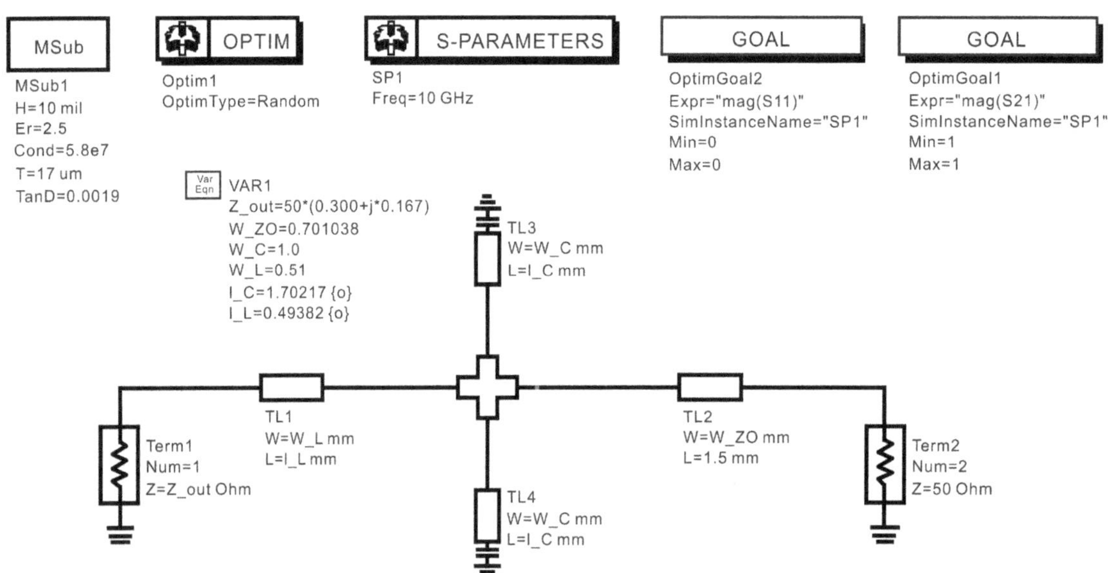

Figure 8.41 Output matching circuit employing microstrip lines. The widths are similarly defined and only the lengths are optimized for conjugate matching.

The S-parameters for the optimized dimensions are shown in Figure 8.42. From the results, $|S_{21}|$ is approximately 1 while $|S_{11}|$ and $|S_{22}|$ are approximately 0, which indicates successful matching. Thus, the microstrip-line matching circuit with the optimized dimensions can be used in place of the lumped-element output matching circuit.

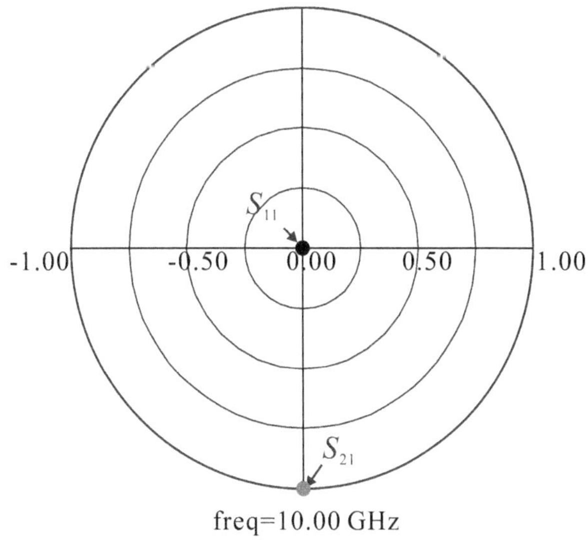

Figure 8.42 Optimization results of microstrip-line output matching circuit at a frequency of 10 GHz

Figure 8.43 shows the simulation results for the low-noise amplifier with the microstrip-line input and output matching circuits. It can be seen that the gain of 13.6 dB and the noise figure of 0.54 dB have been achieved. The values of the gain and the noise figure are close to those that result from using the lumped-element matching circuits. However, the gain is found to be reduced by about 0.1 dB and the noise figure increases by 0.07 dB. This is caused by the losses of the input and output microstrip-line matching circuits.

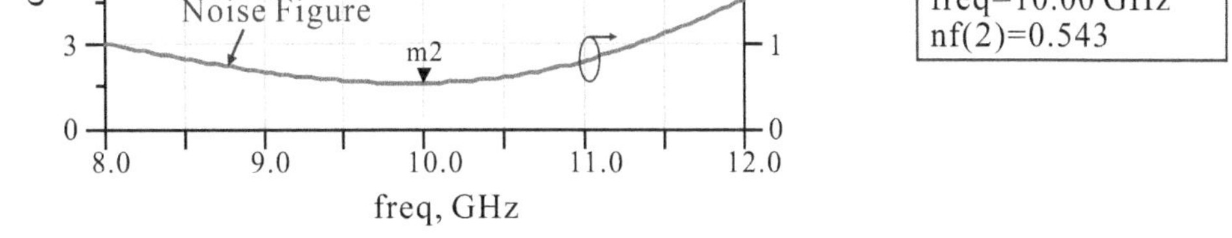

Figure 8.43 Simulation results of the low-noise amplifier composed of microstrip-line matching circuits. The gain is reduced by about 0.1 dB and the noise figure increases by 0.07 dB compared with the gain and the noise figure of the amplifier using the lumped-element matching circuits.

8.6.6 DC Supply Circuit

For the designed low-noise amplifier, an appropriate DC voltage must be applied to the active device through DC supply circuits. These supply circuits must have as minimal an effect on the designed RF matching circuits as possible. Generally, the DC supply circuits will have effects on the matching circuits, and those circuits will have to be redesigned to take into consideration the effects of the DC supply circuits. In this section, we will present a matching circuit design that includes DC supply circuits.

A DC block and an RFC are usually required for DC biasing. Taking into consideration the parasitic inductance discussed in Chapter 2, it is difficult to use chip capacitors as a DC block in the current design's frequency band. Therefore, the coupled lines shown in Figure 8.44 are usually used as a DC block. However, the spacing between the coupled lines generally tends to be too narrow to be fabricated. In addition, the bandwidth of the coupled-line DC block can also be too narrow. As a result, the values of the previously determined matching circuits can sometimes change significantly due to the coupled-line DC block. Although the redesign of the matching circuits due to the inclusion of the coupled-line DC block is similar to the previous design of the matching circuit, the discussion of the coupled-line DC block will not be covered here.

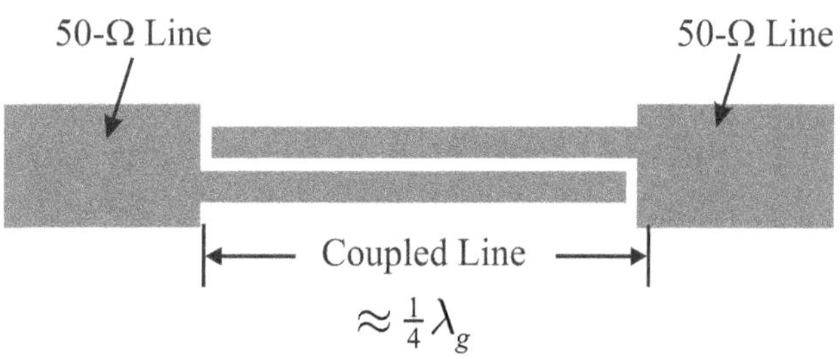

Figure 8.44 DC block using coupled transmission lines

Recently, a high-frequency chip DC block capacitor has been developed and widely commercialized. The advantage of this capacitor is that it generally does not cause any significant changes to the designed matching circuit compared with the coupled-line DC block shown in Figure 8.44. In this section, we will illustrate a process to redesign the matching circuits, including the chip DC block capacitor. ATC's 1-pF chip DC block capacitor is used as a DC block capacitor.[1] The S-parameters of the chip DC block capacitor are provided by the manufacturer and are shown in Figure 8.45, where the 1-pF chip DC block capacitor operates as a DC block from 5 GHz to 26.5 GHz. Using the S-parameters of the chip DC block capacitor that are provided, the chip DC block capacitor can be configured using the S2P data item of ADS. Refer to the datasheet for more details on its frequency characteristics.

Figure 8.46 shows the redesigned input matching circuit. The chip DC block capacitor is configured as a data item, and the RFC is composed of a quarter-wavelength microstrip line (see Chapter 3) and a radial stub. The dimensions of the radial stub

1. American Technical Ceramics Corporation, ATC 500S Series BMC Broadband Microwave Millimeter-Wave 0603 NPO SMT Capacitors, available at www.atceramics.com.

MRSTUB are determined through a separate simulation. The parameters **Wi** and **Angle** are fixed, and the length is determined to yield its input impedance 0. The DC gate voltage is applied to the DC bias point shown in Figure 8.46. The radial open stub and DC block capacitor have effects on the designed input matching circuit but they are not significant. Thus, it is necessary to carry out a new optimization. The goals are similar to the previous optimization. The microstrip line lengths **l_C** and **l_L** are selected as the optimizing parameters of the input matching circuit, as in the case of the previous optimization process. The other variables such as the line width and length of the RF choke, except length

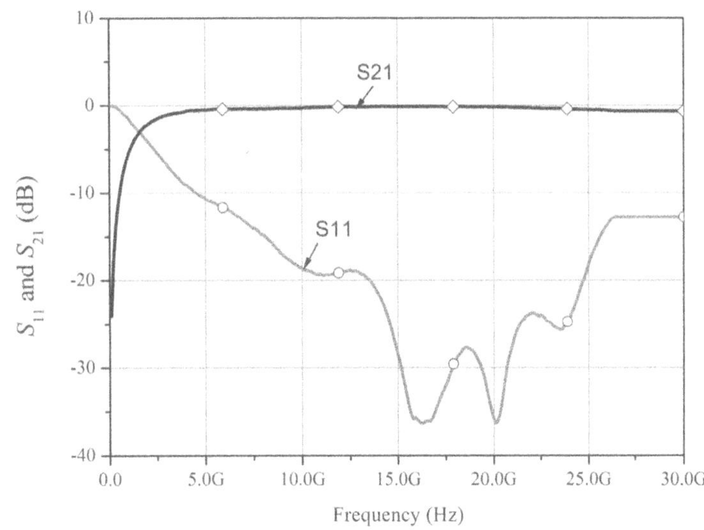

Figure 8.45 The frequency characteristics of a 1-pF DC block provided by the manufacturer

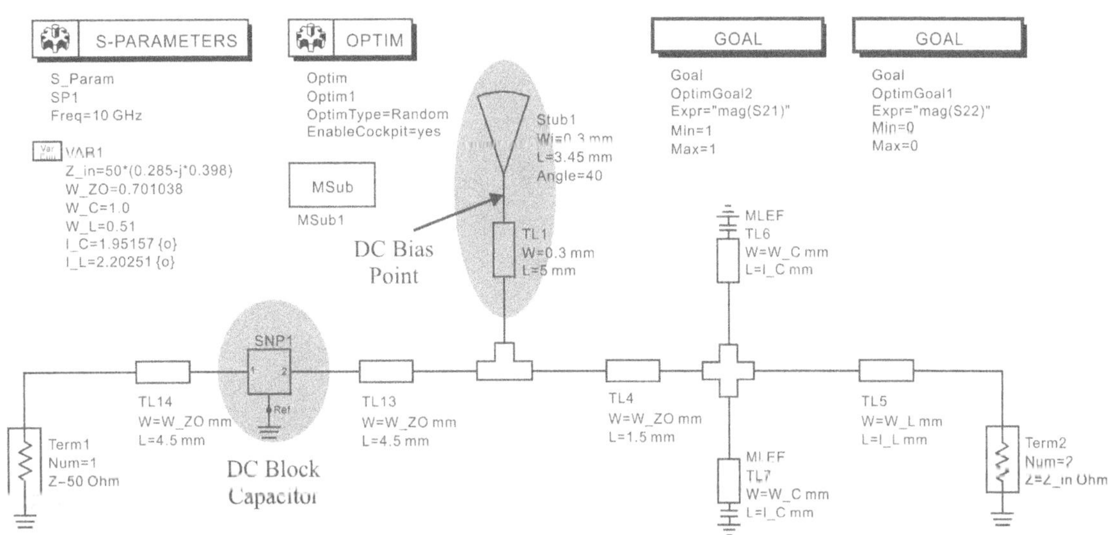

Figure 8.46 Input matching circuit that includes the DC supply circuit. The S-parameter data component is employed for the DC block capacitor. The parameters of the radial stub, **Stub**1, and choke line, **TL**1, are separately simulated and fixed. The DC bias point represents the point at which the DC voltage is supplied.

l_C and l_L, are set to be constant and the optimization is carried out. The optimized S-parameters are shown in Figure 8.47. As in the previous optimization, S_{11} and S_{22} are both 0, while $|S_{21}|$ is close to 1, indicating successful matching. Note that S_{21} is slightly less than 1 due to the loss of the microstrip lines and the DC block capacitor.

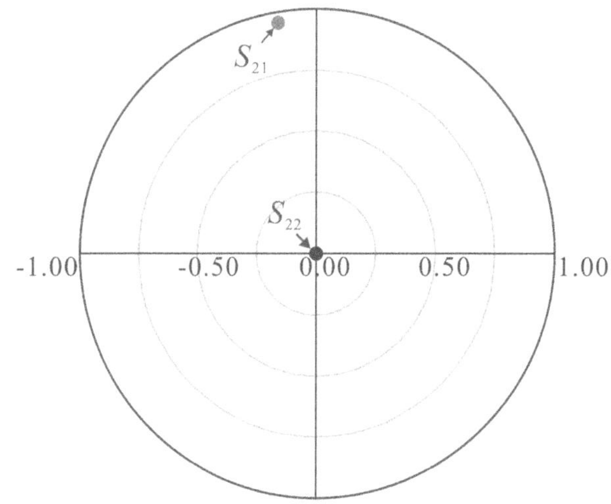

Figure 8.47 Optimization results for the input matching circuit that includes the DC supply circuit. S_{11} is not shown but it is as close to 0 as S_{22}. Note that S_{21} has some loss although S_{11} and S_{22} are matched. This is due to the loss of the microstrip lines and the DC block capacitor.

Similarly, the output matching circuits are set up as shown in Figure 8.48. The DC block is configured using the data item and the same RFC as in the input matching circuit. The DC drain voltage is supplied at the DC bias point shown in Figure 8.48. The simulation results are shown in Figure 8.49. Similar to the input matching circuit, S_{11} and S_{22} are both 0, while $|S_{21}|$ is 1, indicating successful matching.

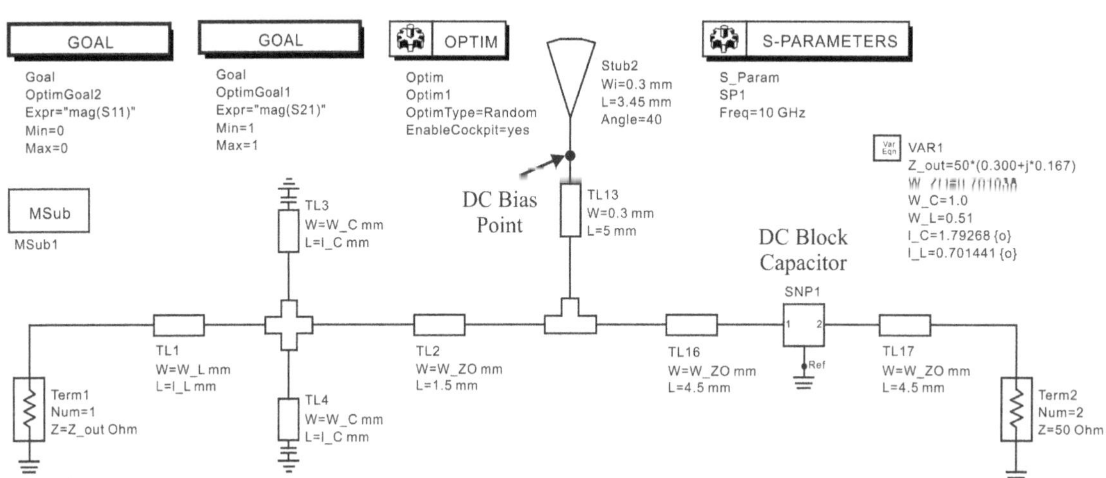

Figure 8.48 Output matching circuit that includes the DC supply circuit. The circuit is similar to the input microstrip matching circuit.

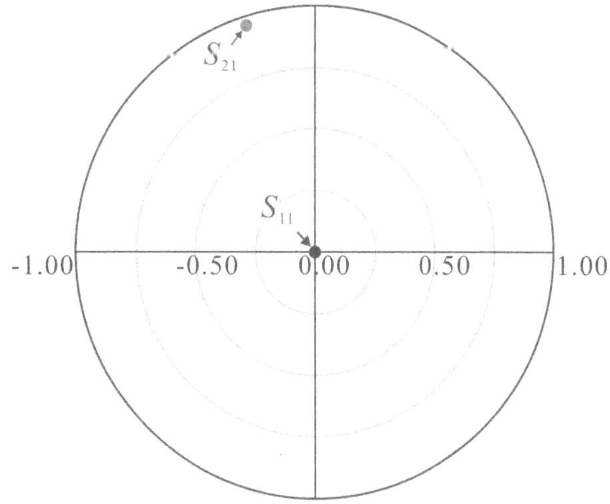

Figure 8.49 Optimization results for the microstrip output matching circuit that includes the DC supply circuit

Figure 8.50 shows the gain and noise figure of the low-noise amplifier with the determined input and output matching circuits that include the DC supply circuits. The gain at 10 GHz is about 13.7 dB, which is about 0.1 dB less than that of the previous low-noise amplifier. The noise figure increases by approximately 0.3 dB from the previous value of about 0.48 dB. This is believed to be due to the loss of the DC block capacitor.

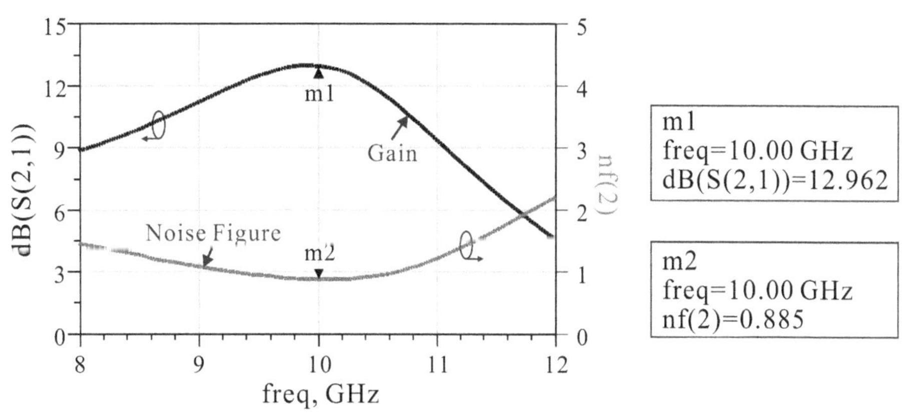

Figure 8.50 Simulation results for the low-noise amplifier that includes the DC supply circuit

8.6.7 Stability

The designed low-noise amplifier is obviously stable at the design frequency but the stability at low frequencies is not certain and should be checked. To determine the stability, the source and load stability circles of the active device should first be plotted on the Smith chart with frequency as a parameter. Then, the loci of Γ_S and Γ_L of the designed matching circuits with frequency as a parameter are plotted together with the corresponding stability circles. The stability can then be determined as to whether or not Γ_S and Γ_L for each frequency lie in the stable region. Although the process is not difficult, it is very cumbersome.

Alternatively, the stability at low frequency can be checked using the S-parameters of the low-noise amplifier rather than those of the active device. The frequency-dependent stability circles of the low-noise amplifier can be similarly plotted on the Smith chart. In this case, the source and load impedances Γ_S and Γ_L become the port impedances that result in $\Gamma_S = \Gamma_L = 0$. Therefore, the stability at low frequency can be checked by determining whether or not the origin of the Smith chart is in the stable region at each frequency. In the latter method, since the source and load stability locus is fixed as $\Gamma_S = \Gamma_L = 0$, the origin of the Smith chart that repeatedly determines whether or not the origin is in the stable region is visually much easier compared to the former method that checks the stability at the active device plane.

Sometimes, the unstable frequency can be missed because the source and load stability circles are drawn at many discrete frequency samples. The stability of the designed low-noise amplifier means that Γ_{in} and Γ_{out} for the selected Γ_S and Γ_L do not show negative resistances. Therefore, $|\Gamma_{in}| < 1$ and $|\Gamma_{out}| < 1$ must be satisfied for the selected Γ_S and Γ_L. In the latter method, $\Gamma_{in} = S_{11}$ and $\Gamma_{out} = S_{22}$ and so $|S_{11}| < 1$ and $|S_{22}| < 1$ must also be satisfied. Therefore, the determination of low-frequency stability based on the S-parameters of the low-noise amplifier is dependent on:

1. Checking the frequency range in which $|S_{11}| > 1$ and $|S_{22}| > 1$.
2. Plotting the stability circles and checking whether their origins are stable or unstable for the frequency range in which $|S_{11}|>1$ and $|S_{22}|>1$.

This procedure significantly reduces the work required to check the stability of the designed low-noise amplifier. When the designed matching circuits for the low-noise amplifier are unstable at some low frequencies, the matching circuits should be readjusted until they move out of the unstable region. The circuit shown in Figure 8.51 is set up to determine stability at frequencies other than the design frequency. Note that the matching circuits include the DC bias supply circuits shown in Figures 8.46 and 8.48.

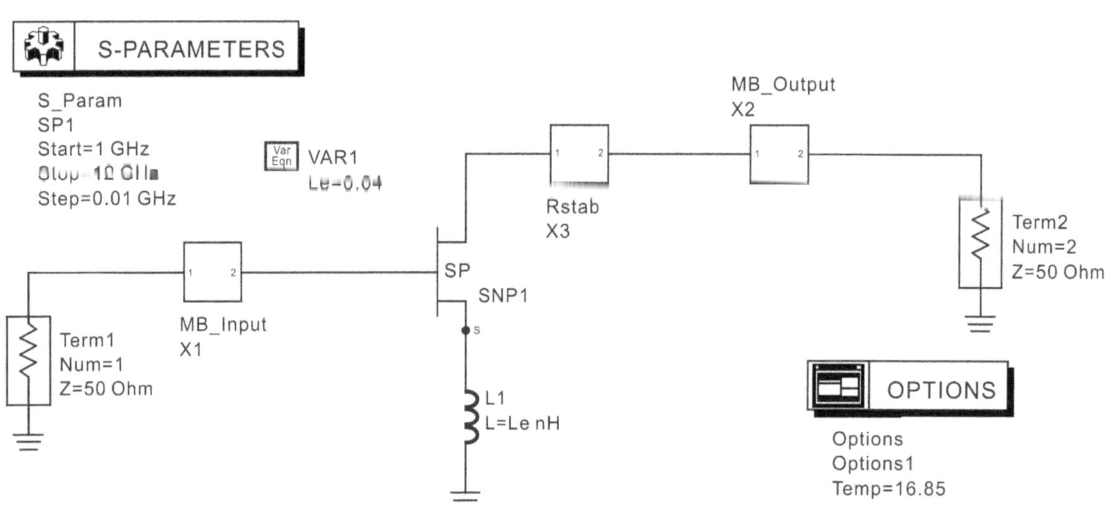

Figure 8.51 Circuit setup for determining low-frequency stability

After simulating the S-parameters, $|S_{11}|$ and $|S_{22}|$ are plotted. Fortunately, $|S_{22}| < 1$ for all frequencies; however, at around 1 GHz, $|S_{11}| > 1$, as shown in Figure 8.52. Note that $|S_{11}|$ can be greater than 1 because the load Γ_L lies in the unstable region. Thus, the load stability circle at 1 GHz is also drawn with S_{11}. The stable region can be determined by using the ADS function **l_stab_region(S)**, which identifies whether or not the inside of the stability circle is stable. The load's stable regions at 1 GHz are inside the stability circle. This means that the load at the origin of the Smith chart is unstable at 1 GHz. As a result, $|S_{11}|$ is greater than 1 at a frequency of 1 GHz. Therefore, the designed amplifier is unstable at 1 GHz.

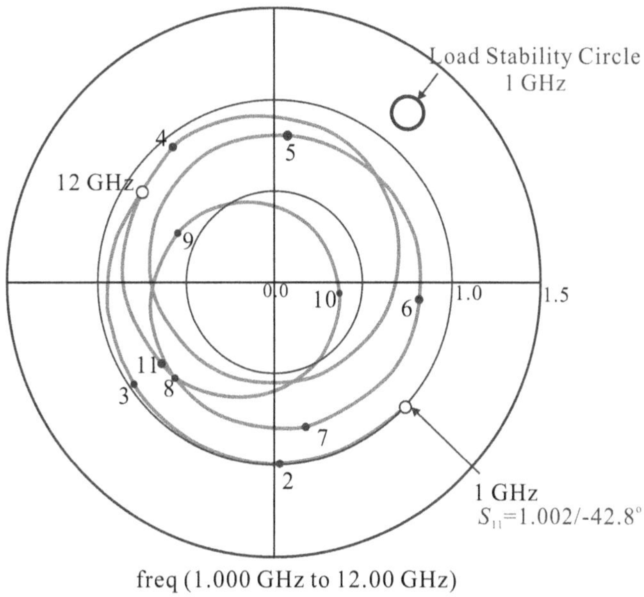

Figure 8.52 Frequency dependence of S_{11} and load stability circle at 1 GHz. The load is placed at the origin, irrespective of frequency changes. The stable region at 1 GHz is inside the stability circle. As a result, the load is in the unstable region at 1 GHz. The load is stable at other frequencies and the stability circles are not drawn. Also, because $|S_{22}| < 1$ and the source placed at the origin is stable for all frequencies, they are not drawn.

The reason for the appearance of $|S_{11}| > 1$ at low frequencies could be attributed to the DC block capacitors inserted in the matching circuits, which open at low frequencies. Therefore, to solve the stability problem, the impedance of the input or output matching circuit seen from the active device is set close to 50 Ω at low frequencies. This looks somewhat like a trial-and-error approach but it is the commonly used method. Usually, the S-parameters of most active devices are measured using 50-Ω reference impedances and the S_{11} or S_{22} of active devices is usually less than 1 for most cases; consequently, the devices are stable in these situations. Therefore, for any frequency at which the active device is unstable, setting the impedance seen from the active device into the source or load close to 50 Ω at that frequency will achieve stability. In this sense, the impedance seen from the active device is set close to 50 Ω at low frequencies. Most stabilizing circuits can essentially be regarded as the application of this concept. It is worth noting that the insertion of the low-frequency stabilizing resistors can affect the DC bias voltage or the flow of current, and this must be prevented when inserting the resistors.

In the case of the instability in the designed low-noise amplifier, the modification of the input matching network may be efficient for solving the stability problem. The change in Γ_S at 1 GHz will cause changes in the load stability circle's center and radius at this frequency. Thus, the load at the origin of the Smith chart will be stable. To change the load stability circle's center and radius, a 50-Ω resistor is inserted in the input matching circuit, as shown in Figure 8.53. The capacitor **C1** acts as a bypass capacitor. For the modified input matching circuit, the S-parameters of the amplifier are simulated again with the output matching circuit remaining unchanged. The stability circles S_{11}, and S_{22} are shown in Figure 8.54.

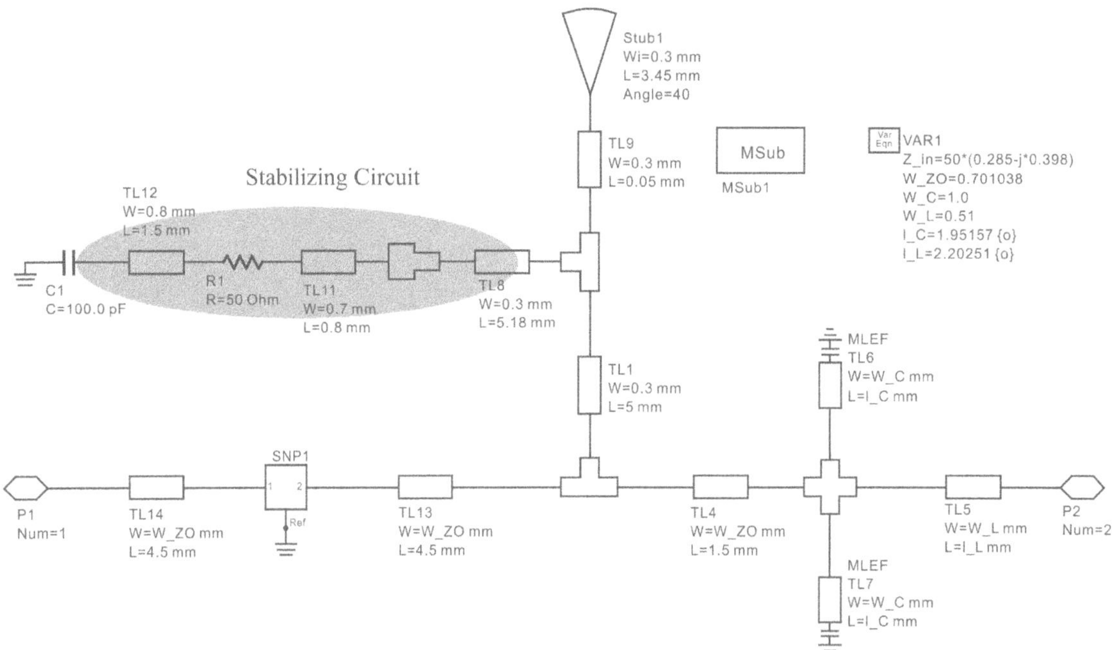

Figure 8.53 Modified input matching circuit for solving a low-frequency source stability problem

As can be seen in that figure, both S_{11} and S_{22} lie in the unit circle. In conclusion, the low-frequency instability is solved by inserting a low-frequency stabilizing resistor in the DC supply circuit; this must be done carefully so as not to affect the matching circuit at the design frequency. In this way, the instability at low frequencies can be largely eliminated.

8.6.8 Fabrication and Measurements

A layout for the previously designed low-noise amplifier is required in order to fabricate that amplifier. A direct layout for the low-noise amplifier circuit is possible using the *auto-layout utility* in ADS. Note that the layout of the matching circuit generally includes numerous discontinuities that are approximately modeled in the circuit simulation. Consequently, the measured results of the designed low-noise amplifier may differ from the circuit simulation results. A final validation of the amplifier's layout through EM simulation is necessary for the first pass of the design. In this section, we will begin by describing the implementation of the source inductor at the design frequency. Then, we will discuss the tuning of the matching circuit layouts in Momentum simulation.

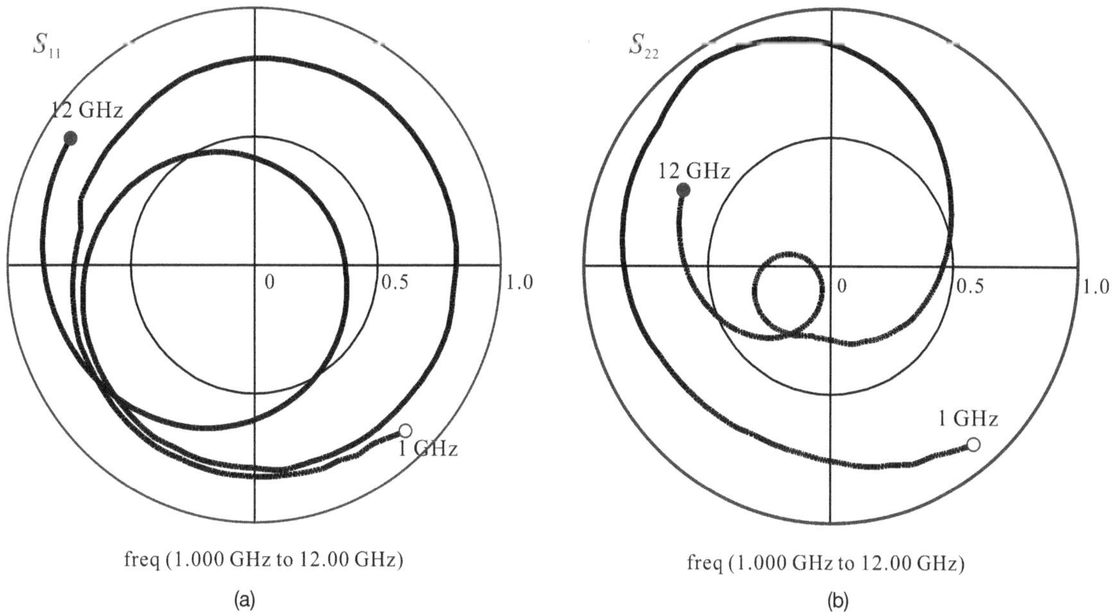

Figure 8.54 S_{11} and S_{22} after stabilization. Both S_{11} and S_{22} are inside the unit circle and the stabilization is successful.

The substrate used for the implementation of the microstrip matching circuits has an inherent dielectric loss as well as a conductor loss due to the finite conductivity of the conductor material. Table 8.4 shows the loss-related parameters of the selected substrate, which is Taconic's TLX-9;[2] the reader may refer to the corresponding datasheet. Note that the substrate losses are not considered in the previously designed low-noise amplifier. The substrate losses can be included in the ADS circuit simulation. It is also possible to include the substrate losses in the settings of the **MSUB** in the circuit simulation.

Table 8.4 PCB substrate parameters

No.	Item	MSUB_menu	Value
1	Conductor thickness	T	17 mm
2	Conductivity	Cond	5.8×10^7 S/m (copper)
3	Loss tangent	TanD	0.0019

Careful consideration of the substrate loss is particularly important because the loss degrades the noise figure, especially the substrate loss at the input matching circuit because it degrades the noise figure of the designed low-noise amplifier. Similar optimization techniques can be applied to further refine the values of the matching circuits and take into consideration the substrate loss.

2. Taconic Advanced Dielectric Division, TLX, available at www.taconic-add.com/en--index.php.

8.6.8.1 Source Inductor

So far, the circuit simulations have been carried out by treating the source inductor as an ideal inductor. However, it is necessary to replace the source inductor by more easily implemented circuit elements such as vias. Since the source inductor is an important element that determines the overall stability of the low-noise amplifier, a careful implementation is required. Figure 8.55 shows the via dimension that is compliant with the manufacturer's design guidelines. The via dimensions may change depending on those guidelines.

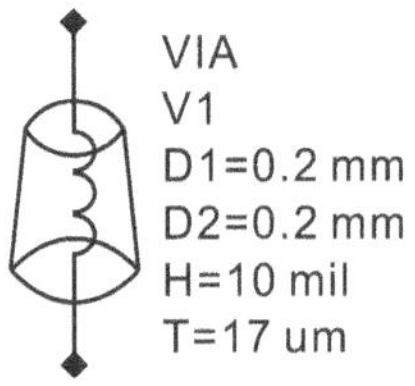

Figure 8.55 Via used in the design

Figure 8.56 shows the distributed inductor circuit for the source inductor. This inductor is implemented using a microstrip line and four vias. The ATF-36077 transistor has two source terminals. Thus, the value of the distributed inductor circuit in Figure 8.56 must have twice the inductance value of the source inductor, so its value is 0.08 nH.

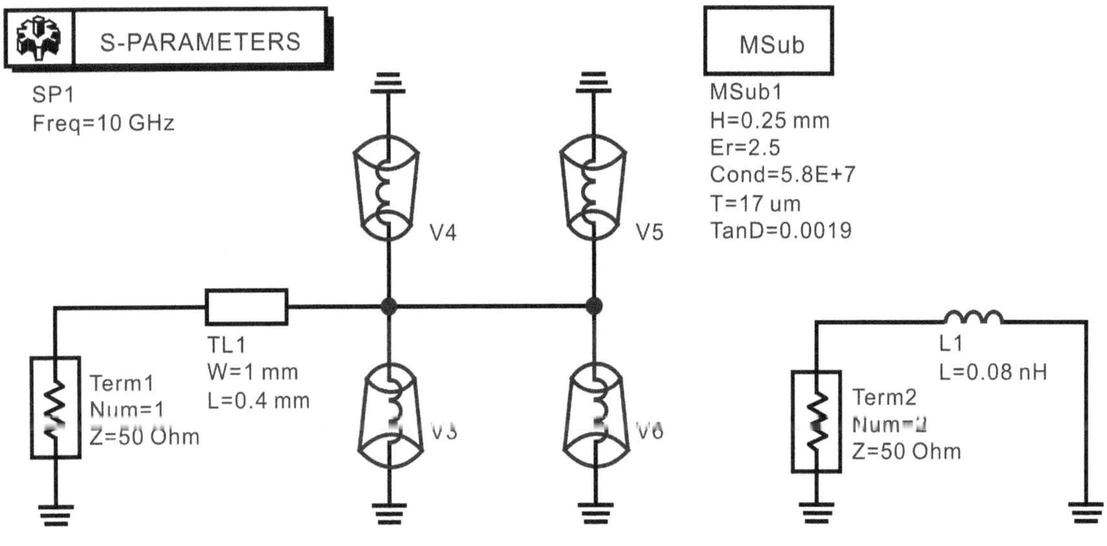

Figure 8.56 Distributed inductor circuit implemented with a microstrip line and four vias. The right-side circuit is the reference inductor circuit.

In Figure 8.56, the reference inductor circuit is shown on the right side and it has an inductance value of 0.08 nH. The reference inductor circuit provides an impedance reference of 0.08 nH. The inductance of a single via is about 0.08 nH. To implement the source inductor, the number of vias is 4, and the inductance of 0.08 nH can be obtained adjusting the microstrip-line length. Sweeping the length of the microstrip line **TL1**, the appropriate microstrip-line length can be found that makes the impedance of the distributed inductor circuit equal to that of the reference inductor circuit. The value of the microstrip-line length thus obtained through the circuit simulation is shown in Figure 8.56. Then, the layout for the distributed inductor circuit in that figure is generated using the autolayout utility of ADS and modified as shown in Figure 8.57. The layer settings for Momentum simulation are configured by importing the **MSUB** information in Figure 8.56.

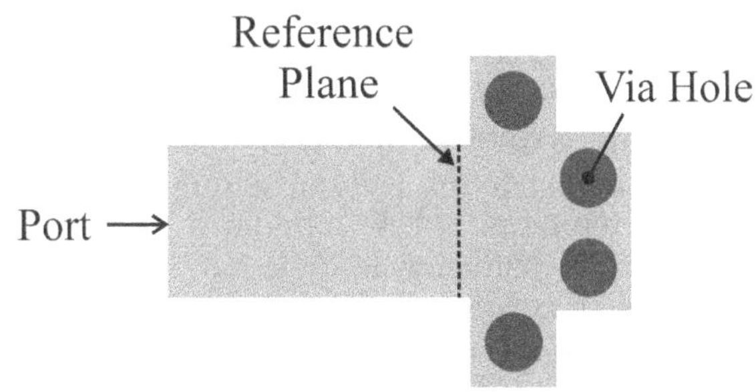

Figure 8.57 Layout implementation of the distributed inductor circuit using a microstrip line and four vias

A slight difference between the Momentum- and circuit-simulated impedances is observed at 10 GHz. In Momentum simulation, by reducing the microstrip-line length that is determined in the circuit simulation, the inductor value of 0.08 nH can be achieved. Figure 8.58 shows the comparison of Momentum- and circuit-simulated impedances with the reference impedance, which is the impedance computed from the reference inductor circuit. It can be seen that the three impedances show good agreement.

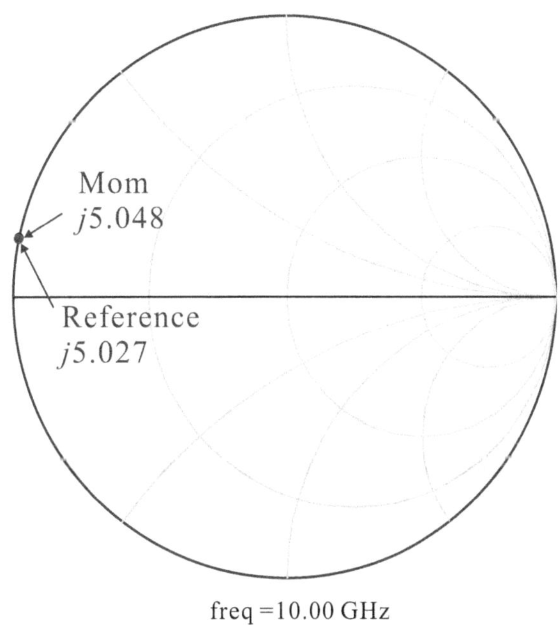

Figure 8.58 Comparison of the impedances obtained from circuit and Momentum simulations

8.6.8.2 Matching Circuits

The matching circuit configurations are the same as the previously designed matching circuit configurations. As a result, the lengths of the matching circuits are slightly modified, but these matching circuits are still not accurate even though the substrate parameters are included in the circuit simulation. Since the approximate models for discontinuities such as T, cross junctions, and so on are used in the circuit simulation, the accurate matching circuits cannot be obtained using circuit simulation alone; thus, they should be verified using Momentum simulation.

The Momentum simulation of the whole matching circuit is still difficult to perform due to its limitations. In particular, Momentum cannot handle the DC block that has an unknown three-dimensional structure. Thus, the DC block and the load-side 50-Ω microstrip line are removed and the layout is prepared for the Momentum simulation. The input matching circuit layout for Momentum simulation is shown in Figure 8.59. Figure 8.60 compares the Momentum-simulated impedances to the circuit-simulated impedances. Here, the circuit-simulated impedance is computed using the circuit that corresponds to the input matching circuit layout in Figure 8.59. Figure 8.60(a) shows the comparison of the two source impedances seen from the active device. Similarly, the output matching circuit layout is prepared and Momentum simulated. Figure 8.60(b) shows the comparison of the load impedances seen from the active device.

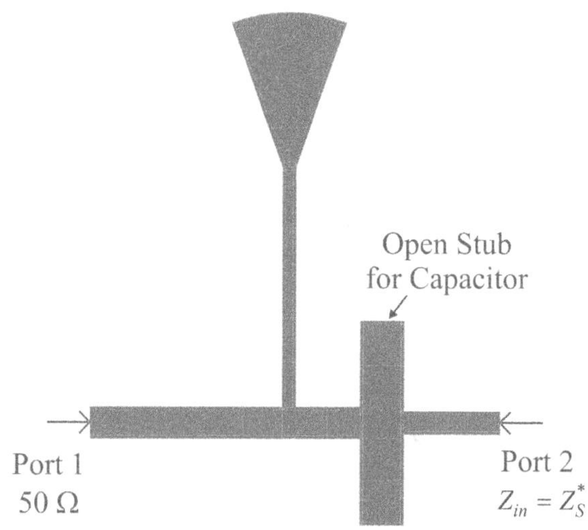

Figure 8.59 Layout configuration of the input matching circuit for Momentum simulation

Due to the approximate models for discontinuities in the circuit simulation, the Momentum-simulated impedances are slightly different from the circuit-simulated impedances. Figure 8.61 shows the comparison of the two impedances for the input and output matching circuits after the length adjustments. They are found to show good agreements. The Momentum-simulated impedance is tuned by adjusting the lengths of both the inductor line and the open stub in the matching circuit. The length reduction by about 0.15 mm of the two open stubs in parallel and the reduction by about 0.12 mm for the input matching circuit make the circuit-simulated and Momentum-simulated input impedances coincide, as shown Figure 8.61(a). Similarly, the length reduction by about 0.07 mm of the two open stubs and the reduction by about 0.08 mm of the inductor line for the output matching circuit make the circuit-simulated and Momentum-simulated output impedances coincide, as shown Figure 8.61(b).

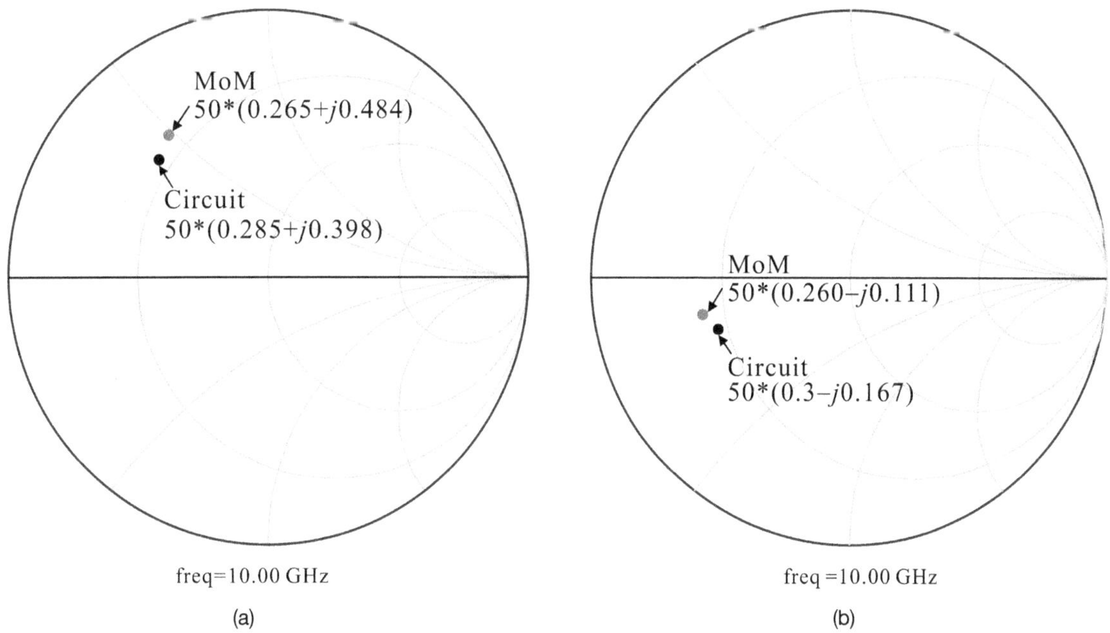

Figure 8.60 Circuit simulation results compared with Momentum simulation results for (a) input and (b) output matching circuit at frequency 10 GHz. They show some differences.

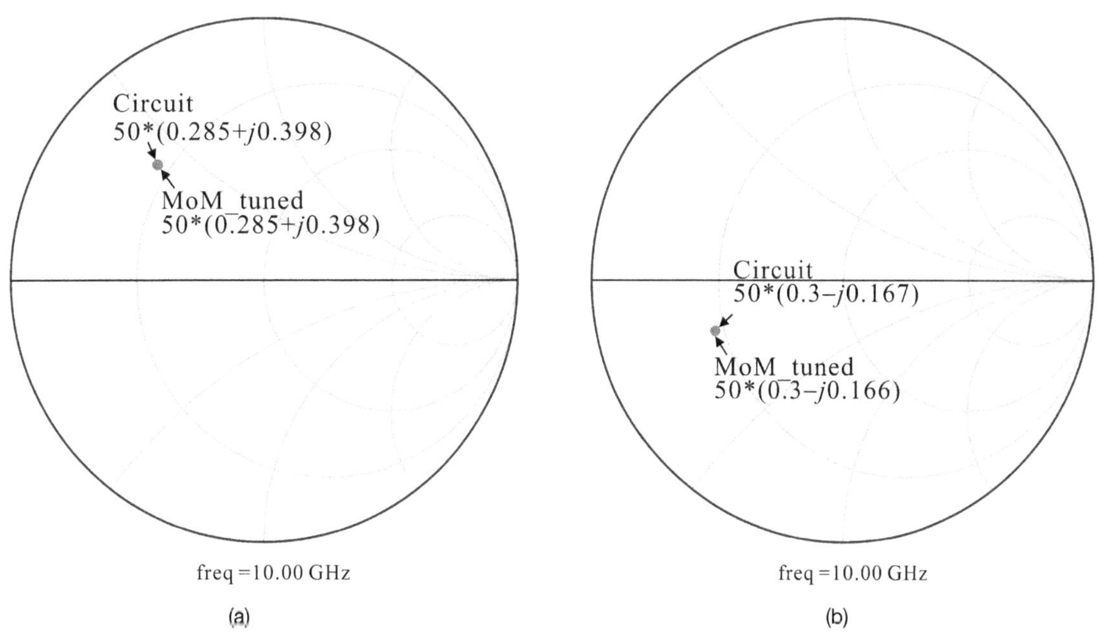

Figure 8.61 Comparison of results for (a) input and (b) output matching circuit after length adjustments are made at 10 GHz. For matching to circuit-simulated impedances, the lengths are adjusted in the Momentum simulation. In the case of the input matching circuit, two open-stub lengths and an inductor-line length are reduced by 0.15 mm and 0.12 mm, respectively. In the case of the output matching circuit, two open-stub lengths and an inductor-line length are reduced by 0.07 mm and 0.08 mm, respectively.

It is necessary to verify whether or not the Momentum-simulated impedances provide adequate performances for the low-noise amplifier. The matching circuits as well as the source inductor obtained from Momentum simulation are represented by the S-parameter data shown in Figure 8.62. Note that the DC block is the same as the circuit simulation and it is also included as a data item. Figure 8.62 shows the circuit setup to confirm the Momentum-simulated results.

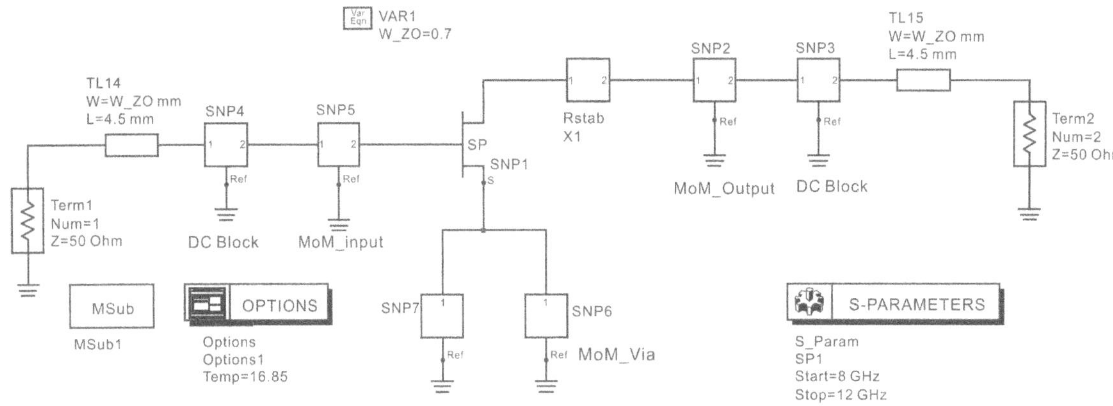

Figure 8.62 Circuit for a low-noise amplifier confirmation using data components. Momentum-simulated results are represented by S-parameter data items.

Figure 8.63 shows the simulated gain and noise figure. The gain is about 13 dB and the noise figure is about 0.9 dB. Comparing the gain and noise figure with those of the circuit-simulation results, it can be seen that the gain has decreased while the noise figure has increased; in particular, the noise figure has increased by 0.2 dB. This is thought to be due to the substrate loss included in the Momentum simulation. However, the gain is found to be more than 12 dB and the noise figure is found to be less than 1.5 dB, which satisfies the design specifications.

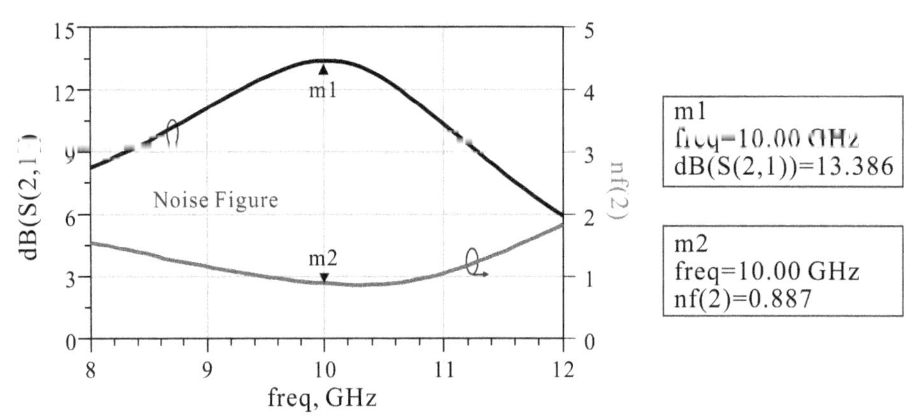

Figure 8.63 Simulated gain and noise figure using Momentum-simulated data components

8.6.8.3 Fabrication and Measurement

Figure 8.64 is a photograph of the actual low-noise amplifier fabricated with the settings and simulations described above. The substrate used in the fabrication is Taconic's 10-mil-thick TLX-9 substrate, which has a dielectric constant of 2.5. ATC's 1 pF broadband 0603 type 500S was used for the DC block capacitors. The gate bias was applied to the connection point of a 100-pF capacitor and a 51-Ω chip resistor. A 100-pF chip capacitor was also attached in shunt for the drain bias. Next, to prevent the AC noise of the DC power supply, a 3.3 μF tantal capacitor was added to the gate and drain bias.

Figure 8.64 The fabricated 10 GHz low-noise amplifier

The gate bias voltage was adjusted from the pinch-off voltage until a specified drain current of 10 mA flowed. A gate voltage of $V_{GS} = -0.33$ V provided the drain current close to 10 mA. The measured drain current was then 10.7 mA. After DC bias adjustment, the noise figure and gain were measured using a noise figure analyzer (NFA). In measuring the noise figure, an attenuator was inserted at the input of the low-noise amplifier. The internal function of the NFA was able to calibrate the inserted attenuator. This was done to reduce the mismatch between the low-noise amplifier and the noise source. After the calibration, the noise figure and gain of the low-noise amplifier was measured.

Figure 8.65 shows the measured and simulated noise figure and gain for the fabricated low-noise amplifier. The measured gain peak shifts downward about 300 MHz and shows the narrower bandwidth compared to the simulated gain. The measured noise figure does not show frequency shift but the noise figure is generally higher than the simulated noise figure. The measured noise figure at 10 GHz is about 1.1 dB, while the simulated noise figure is about 0.89 dB. The noise figure increases by about 0.2 dB. It can be seen that the measured results satisfy the design specifications. However, the measured noise figure and gain show a difference from the simulated noise figure and gain.

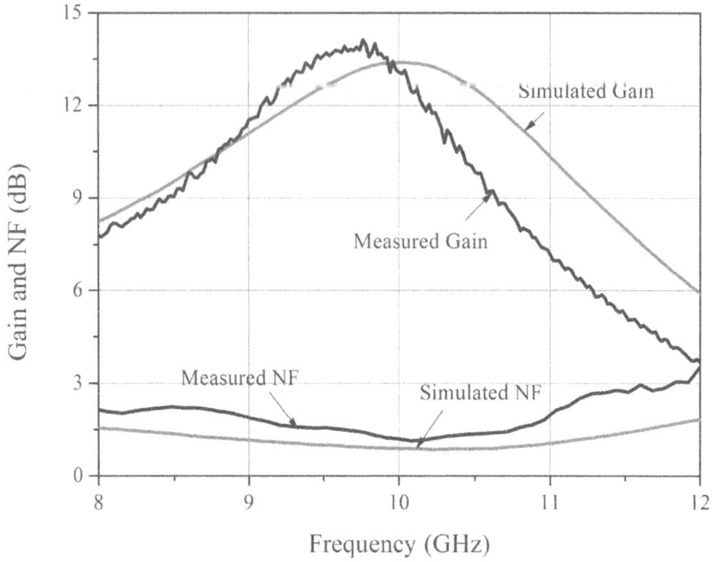

Figure 8.65 Measured gain and noise figure of 10 GHz LNA. The simulated and measured noise figures at 10 GHz are 0.89 dB and 1.1 dB, respectively.

8.6.8.4 Discussion

The reasons for the difference between the measured and simulated gain and noise figure can be a result of the following factors:

1. Substrate parameter variations such as permittivity and thickness.
2. Pattern tolerance in the PCB fabrication.
3. Inaccuracy in the simulation; the S-parameters and noise parameters of the active device may be different from the ADS data.

Factors 1 and 2 cannot explain the shift of the measured gain peak. Since the permittivity change is 2.45–2.55 in the datasheet, the shift of the gain peak is estimated at about

$$\Delta f = f_o \times \frac{\Delta \varepsilon_r}{2\varepsilon_r} = 10G \times \frac{0.05}{5} = 100 \text{ MHz}$$

Since Δf is about 300 MHz, Δf cannot be explained using the permittivity change. Also, Δf cannot be explained by the thickness change. Similarly, the pattern tolerance is determined but it cannot explain Δf.

To determine the difference between the measured and simulated gain and noise figure, the S-parameters of the input and output matching circuits are measured. The S-parameters of both the active device with the series feedback inductor and the DC block capacitor are also measured. The measured S-parameters of the input and output matching circuits are close to the simulated S-parameters; however, the S-parameters of the active device show a significant difference compared with the generated S-parameters using the ADS data model. Thus, it can be concluded that the difference in the S-parameters

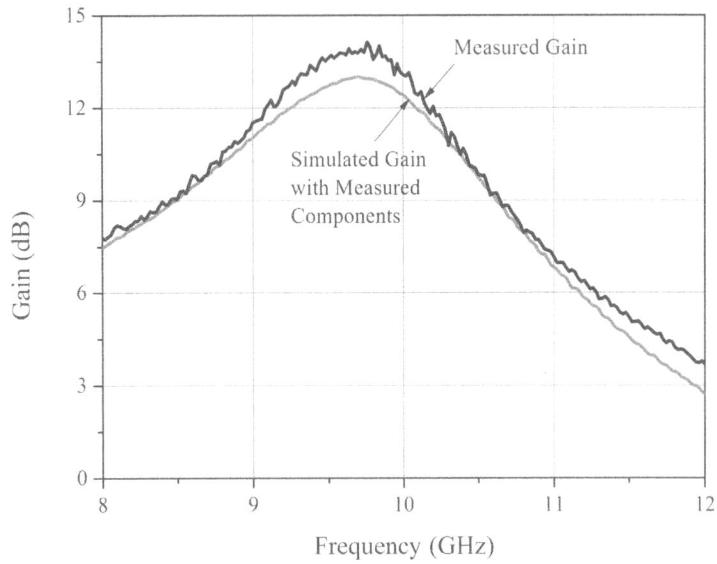

Figure 8.66 Comparison of the measured gain with that computed using the measured parameters. Instead of the ADS data model, the measured S-parameters of the active device are used to compute the gain.

of the active device is the key reason for the difference between the measured and simulated gain and noise figure. This may be caused by differences in the S-parameter measurement environment. In order to verify that the difference of the gains in Figure 8.65 is caused by the difference of the active device's S-parameters, the gain is computed using the measured S-parameters of the active device and then compared with the measured gain. Figure 8.66 shows the comparison of the computed gain with the measured gain. Although there is still a slight difference, they show good agreement. In conclusion, before creating the design, it is better to measure the S-parameters and noise parameter in advance; the use of measured parameters in the design is also recommended.

8.7 SUMMARY

- The gain of an amplifier is a function of Γ_S, Γ_L, and the S-parameters of active devices. Three expressions for gains are derived. The transducer power gain is the ratio of the delivered power to the load to the available power from the source, which represents the conventionally measured gain. The available power gain and power gain are derived for the purposes of the design. The available power gain represents the gain degradation by source mismatch, while the power gain represents the gain degradation by load mismatch.
- A stability check is required to prevent the designed amplifier from becoming an oscillator. A source stability circle is the locus of Γ_S, which yields $|\Gamma_{out}| = 1$. The load stability is the locus of Γ_L, which yields $|\Gamma_{in}| = 1$. Using the stability circles, the stable regions of Γ_S and Γ_L can be found.

- Stability is determined by using the stability factor k. In the case of unconditional stability, simultaneous conjugate matching is possible.
- Available gain and noise circles provide a convenient way to find the gain and noise figure degradations. Using these circles, we can determine the optimum source and load impedances that satisfy the design objective.
- A low-noise amplifier that satisfies the design objective can be fabricated by following the procedures described below:
  1. Select an active device.
  2. Perform a stability check at the design frequency. If unstable, stabilize the active device using stabilization circuits.
  3. Draw the available gain circles and noise circles.
     Determine the optimum source-load impedances.
  4. Design input and output matching circuits that give the determined impedances.
  5. Add DC bias circuits to the designed matching circuits. Modify the designed matching circuits to compensate for the effects of the DC bias circuit.
  6. Perform a stability check at low frequencies. If unstable, modify the matching circuits to make them stable.
  7. Perform a layout check with Momentum.
  8. Complete fabrication and measurement.

REFERENCES

1. G. Gonzalez, *Microwave Transistor Amplifiers*, 2nd ed. Englewood Cliffs, NJ: Prentice Hall, 1997.
2. D. M. Pozar, *Microwave Engineering*, 2nd ed. New York: John Wiley & Sons, Inc., 1998.
3. G. D. Vendelin, A. M. Pavio, and U. L. Rohde, *Microwave Circuit Design Using Linear and Nonlinear Techniques*, New York: John Wiley & Sons, Inc., 1990.

PROBLEMS

8.1 Referring to Figure 8P.1, determine the Thevenin equivalent circuit based on the reflection coefficient Γ_T and b_T of the circuit for a 50-Ω measurement reference impedance.

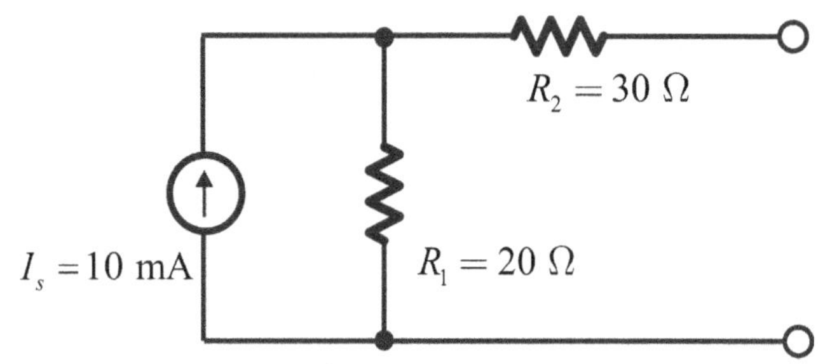

Figure 8P.1 Circuit with source and resistors

8.2 Prove the transducer power gain in Equation (8.24) using the input reflection coefficient of the active device.

8.3 Available power gain can be directly derived using the Thevenin equivalent circuit at the device output plane rather than using the conjugate matching condition described in Section 8.1.3.2. Derive the available power gain using the Thevenin equivalent circuit at the device output plane shown in Figure 8P.2.

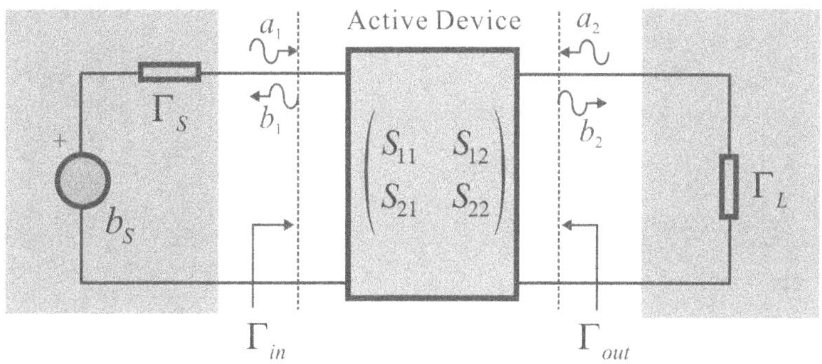

Figure 8P.2 Low-noise amplifier block diagram

8.4 The power gain is the fraction of the input power delivered to the load. Using this definition, the power gain is derived in the text.

However, power gain can also be thought of as the ratio of the power delivered to the load to the input power when the input is matched for maximum power transfer. Referring to Figure 8P.3, prove the power gain using this definition.

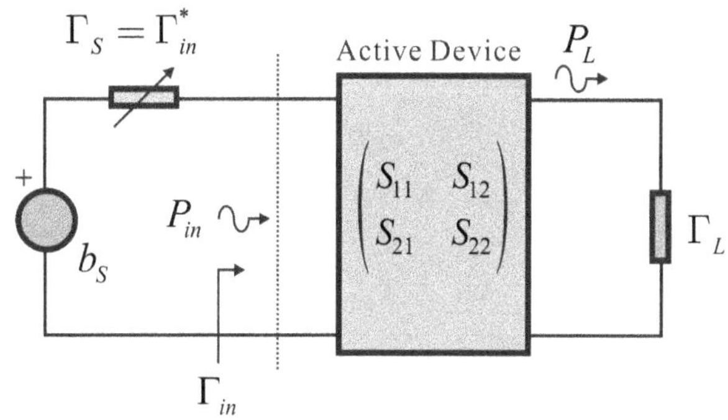

Figure 8P.3 Another interpretation of power gain

8.5 Solve the ten problems below by using the S-parameters of ATF-35076 in the S-parameter library in ADS shown in Figure 8P.4.

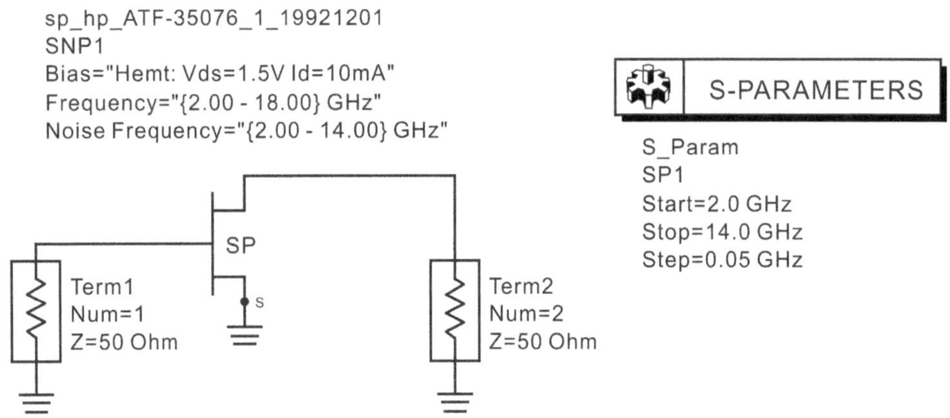

Figure 8P.4 Circuit for Problem 8.5

(1) Plot the stability factor and maximum gain for the frequency range of 2–14 GHz. Find the maximum gain G_{max} at 8 GHz and identify this gain as either MSG or MAG.

(2) Find the minimum noise figure F_{min} (dB) at 8 GHz. Also, determine if this device is adequate for amplifier design; does it satisfy the noise figure < 1 dB and the gain > 10 dB at 8 GHz?

(3) Figure 8P.5 is a schematic for the stabilization of a transistor using series inductive feedback. Find the value of inductor L_s that yields a maximum stability factor.

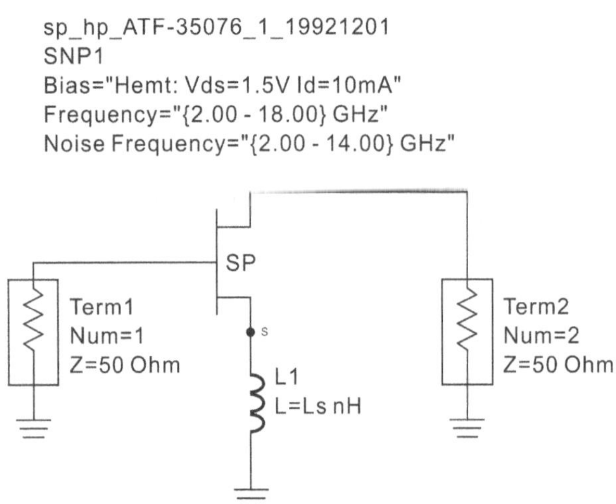

Figure 8P.5 Stabilization using series feedback inductor

(4) For the selected L_s in Figure 8P.5, find the maximum gain and minimum noise figure at 8 GHz.

(5) Draw the available gain circles degraded by {1, 2, 3} dB from the maximum gain; then draw the noise-figure circles degraded by {0.1, 0.2, 0.3} dB from minimum noise figure.

(6) When source reflection coefficient Γ_S is selected to yield the minimum noise-figure point, find the load reflection coefficient Γ_L that yields the maximum gain for the selected Γ_S. Also, find the corresponding impedances Z_S and Z_L in real and imaginary formats and find the expected gain G_{ex} and noise figure F_{ex} for the selected impedances.

(7) Plot the frequency response for the frequency range 2–14 GHz with a 0.05-GHz step. Verify the expected gain and noise figure that appear at 8 GHz.

(8) Design the low-pass input and output matching networks with lumped-element inductors and capacitors. Find the values of input and output inductors and capacitors C_i, L_i, C_o, and L_o.

(9) Using the matching circuits that have the values of inductors and capacitors determined in the preceding step, check the desired gain and noise figure that appear with the frequency response plots for the frequency range of 2–14 GHz with a 0.05-GHz step.

(10) Using **DCFEED** and **DC_Block** in ADS, add the bias circuit to the previous amplifier circuit. Find the frequency where the amplifier is unstable using the ADS function for the frequency range 2–14 GHz with a 1-GHz step.

CHAPTER

9

Power Amplifiers

CHAPTER OUTLINE

9.1 INTRODUCTION

A power amplifier is generally located just before the antenna, and determines the transmitting power level of a communication system. The design of a power amplifier is significantly different from that of the low-noise amplifier presented in Chapter 8. To demonstrate the difference, suppose that two amplifiers are designed with the same active device; one is based on the small-signal design presented in Chapter 8 and has the maximum small-signal gain; the other produces the maximum output power based on the large-signal design that will be presented in this chapter. Figure 9.1 shows a plot of the delivered power P_L of the two amplifiers versus the input power P_{in}.

As shown in Figure 9.1, at a low input power P_{in}, the output power P_L increases in proportion to the input power. However, at a high input power, the output power shows a marginal increase. A further increase in the input power drives the output power of the amplifier into saturation. A point with a 1-dB deviation from the linear output power is referred to as the 1-dB compression point of an amplifier and the output power level at saturation is called the saturated power of an amplifier. As discussed in Chapter 8, an amplifier with maximum small-signal gain can be designed by selecting appropriate

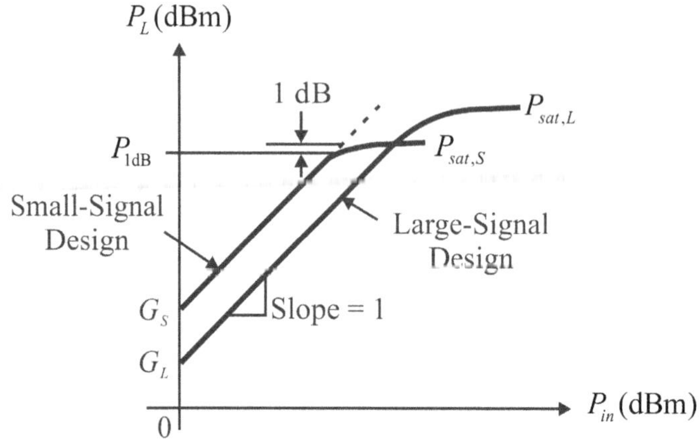

Figure 9.1 Comparison of $P_L - P_{in}$ characteristics of low-noise and power amplifiers. The small-signal design has a higher gain G_S than the gain G_L of the large-signal design; however, the small-signal design saturates early at a lower input power than does the large-signal design.

source and load impedances computed from the measured small-signal S-parameters at a given DC bias. The designed maximum small-signal gain amplifier generally gives a high gain but also has a lower 1-dB compression point and saturated power compared to an amplifier designed to give maximum output power as shown in Figure 9.1. This is not a special case but a general fact.

To explain this qualitatively, a simplified large-signal FET equivalent circuit is shown in Figure 9.2. Here, the current source I_{DS}, which depends on the gate-source and drain-source voltages, has the characteristics shown in Figure 9.3. In addition, excluding the current source I_{DS}, all the other elements in the equivalent circuit of Figure 9.2 are assumed to be linear elements since they typically have low nonlinearity.

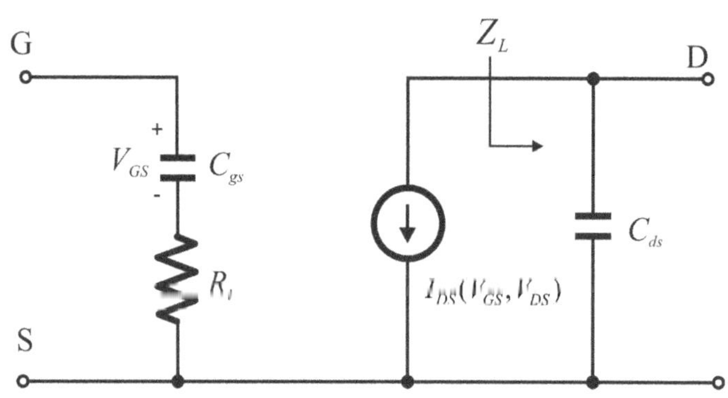

Figure 9.2 Simplified large-signal FET equivalent circuit. Here, the current source I_{DS} represents the DC I_{DS}–V_{DS} characteristics of an FET.

The maximum small-signal gain at the operating point Q can be achieved when the load is conjugate matched to the output impedance. Thus, the impedance Z_L of Figure 9.2 must satisfy $Z_L = r_{DS} = (\partial I_{DS}/\partial V_{DS})^{-1}$. This causes the load line to have a slope of $(r_{DS})^{-1}$, as

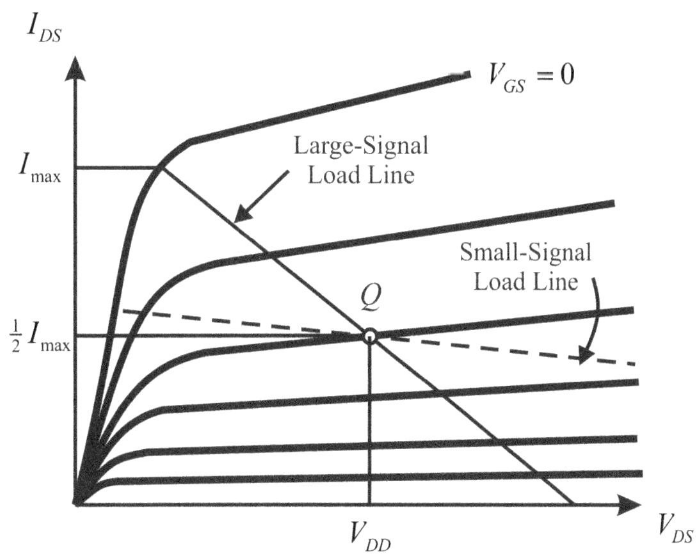

Figure 9.3 Optimum load lines for small- and large-signal power levels

shown in Figure 9.3. However, the load line that gives maximum output power is obviously the load line represented by the solid line in Figure 9.3. Since the gain is approximately proportional to $g_m R_L$, the gain of the amplifier based on the small-signal design is clearly higher than that based on the large-signal design. Yet the amplifier based on the small-signal design has a low saturated power and it generally reaches saturation early as shown in Figure 9.1. The small-signal design clearly does not exploit the maximum output power capability that the active device can supply.

Figure 9.4 shows a power amplifier configuration. The power amplifier can be represented by a block diagram that is similar to a low-noise amplifier. Therefore, the source and load reflection coefficients Γ_S and $\Gamma_{L,opt}$ must be determined by experimental or theoretical means in order to design the power amplifier. When Γ_S and $\Gamma_{L,opt}$ have

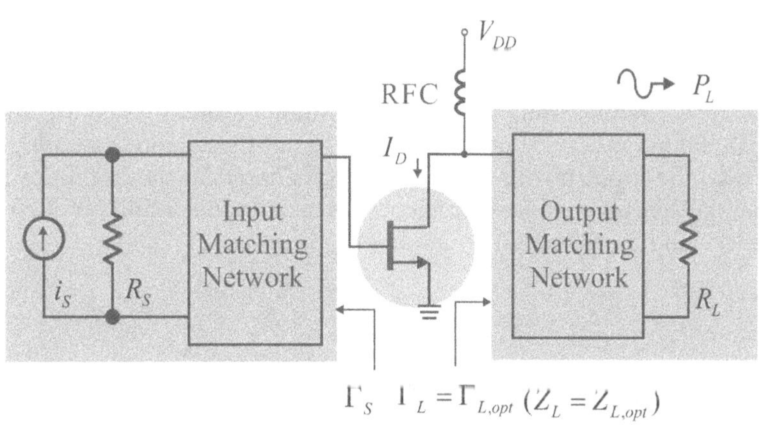

Figure 9.4 Power amplifier block diagram and its design concept. $\Gamma_{L,opt}$ is chosen for optimum output power and then Γ_S is determined for the chosen $\Gamma_{L,opt}$ for conjugate matching at the input.

been determined, the matching circuit can be constructed similar to that of the low-noise amplifier design. The slight difference between the power amplifier and the low-noise amplifier designs is that the input matching circuit losses must be minimized in the case of the low-noise amplifier design, whereas in the power amplifier design, it is the output matching circuit losses that must be minimized. The input matching circuit losses in the low-noise amplifier are directly associated with noise figure degradation. By denoting the loss at the input matching circuit as L(dB) and the noise figure of the active device as F(dB), then the noise figure of the low-noise amplifier as explained in Chapter 4 is approximately given by $L + F$(dB). Thus, for a minimum noise figure, the input matching circuit must be constructed so as to have minimum losses. On the other hand, because the loss in the output matching circuit is directly related to the loss in the output power, the loss in a given output matching circuit must be minimized in the case of the power amplifier.

The impedance $Z_{L,opt}$ in Figure 9.4 that gives maximum output power can easily be inferred from the DC characteristics in Figure 9.3 at low frequency; however, at high frequency, due to the complexity of the equivalent circuit of the active device and the parasitic elements of its package, it is not easy to derive $Z_{L,opt}$ analytically. In a case where the large-signal equivalent circuit is known, the optimum impedance $Z_{L,opt}$ can be obtained at the reference plane of the current source I_{DS} in Figure 9.2. Using $Z_{L,opt}$, the optimum impedance at the output terminals of the package can also be obtained by taking the external parasitic elements into consideration. However, unlike at low frequency, the reference plane of the current source I_{DS} in the large-signal equivalent circuit is generally not obvious at high frequency. Thus, $Z_{L,opt}$ is usually determined experimentally through a load-pull measurement or with simulation; that is, load-pull simulation using software when the large-signal model is available. We will cover the determination of $Z_{L,opt}$ using load-pull measurement and simulation in this chapter.

The next important issue in the implementation of a power amplifier is efficiency, which can be expressed in terms of RF output power for a given supplied DC power, and two definitions are widely used. The first is the *collector* or *drain efficiency*. Denoting the DC power consumed in the output terminal as P_{DC} and the RF output delivered to the load as P_L, the drain efficiency η_D is defined as shown in Equation (9.1).

$$\eta_D = \frac{P_L}{P_{DC}} = \frac{P_L}{V_{DD} I_D} \tag{9.1}$$

Here, V_{DD} represents the DC supply voltage and I_D is the DC supply current when the delivered power to the load at RF is P_L. To some extent, the drain efficiency may be a good measure for representing the efficiency of an amplifier when the amplifier gain is sufficiently high; however, the gain is usually not high enough at a high frequency. Consequently, the input power P_{in} must be considered in the definition of the efficiency. In general, PAE (power-added efficiency) is used in the definition of efficiency at a high frequency and it is defined in Equation (9.2).

$$PAE = \frac{P_L - P_{in}}{P_{DC}} = \frac{P_L - P_{in}}{V_{DD} I_D} \tag{9.2}$$

This definition of PAE can be interpreted as the ratio of the output power added by the active device to the DC power supplied to the active device.

Example 9.1

Given that $V_{DD} = 5.8$ V, $I_D = 400$ mA, $P_{in} = 23$ dBm, and $P_{out} = 33$ dBm, determine the drain efficiency and the PAE.

Solution

Since 33 dBm corresponds to 2 W, the drain efficiency is

$$\eta_D = \frac{2}{5.8 \times 0.4} \times 100 = 86.2\%$$

An input power of 23 dBm corresponds to 0.2 W; therefore the PAE is

$$PAE = \frac{2 - 0.2}{5.8 \times 0.4} \times 100 = 77.5\%$$

The DC biased amplifier shown in Figure 9.3 is usually classified as a class-A amplifier; the problem with this type of amplifier is that its maximum efficiency does not exceed 50%; that is, for the maximum output as shown in Figure 9.3, the approximate drain efficiency can be seen to be

$$\eta_D = \frac{P_L}{V_{DD}I_D} \cong \frac{\frac{1}{2} \times \frac{1}{2} I_{\max} V_{DD}}{\frac{1}{2} I_{\max} V_{DD}} = 0.5$$

In order to understand the efficiency problem in a class-A amplifier, consider an example of this amplifier with a 20 W RF output power. Even if its efficiency is given as the maximum value of 50%, out of the total 40 W power supplied, 20 W is consumed while 20 W is delivered to the load. This will be more problematic in the absence of any RF input power. The total supplied DC power of 40 W to the active device would be consumed by that device in the absence of the RF input power, and all the supplied DC power would be converted into heat. Therefore, power consumption of the class-A amplifier is most problematic in the case where there is no RF input power. This problem can be solved by changing the active device's operation and there are various operation techniques for improving efficiency, which we will discuss in the next sections. However, designing the amplifier to improve its efficiency generally gives rise to the problem of distortion. Therefore, for a communication system that employs both amplitude- and phase-modulated signals, it is necessary to reduce the distortion to improve the amplifier's linearity. The methods for improving linearity will also be explained in this chapter.

9.2 ACTIVE DEVICES FOR POWER AMPLIFIERS

Microwave active devices have already been discussed in Chapter 5. Of these, pHEMTs and HBTs were noted as having excellent high-frequency characteristics. To employ these devices in power amplifiers, their structures must be expanded to handle a large output power. They can be used as power devices by expanding the gate width in the case of a pHEMT or the emitter area in the case of an HBT. This will increase their output power capability, but their breakdown voltage is basically low, which limits their application as high-power devices. In addition to these devices, there are other types of high-power active devices that are widely used today. Among these, GaN (gallium nitride) HEMTs and LDMOSFETs (laterally diffused MOSFETs), simply referred to as LDMOS, have attracted a lot of attention as active devices for power amplifiers, and they will also be introduced in this chapter.

9.2.1 GaN HEMT

The important parameters of semiconductor properties for a high-power active device are electron mobility, energy band-gap (band-gap), and thermal conductivity. Among these parameters, *electron mobility* can be used as a criterion to estimate the high-frequency applicability of an active device fabricated using a given semiconductor process. In other words, for two active devices fabricated using the same processing method but on different semiconductors, the active device using the semiconductor with a higher electron mobility generally yields a higher gain and it can be used in applications up to higher frequencies in contrast to a semiconductor with a lower electron mobility. *Thermal conductivity* is an important parameter of semiconductor properties for power amplifier application. The higher the thermal conductivity, the greater the advantage is in heat dissipation. A device fabricated with a higher thermal conductivity will have a higher power consumption capability. The *energy band-gap*, which is simply referred to as the band-gap parameter, is closely related to the breakdown voltage of an active device. The larger the band-gap is, the higher the breakdown voltage will be. Thus, the higher band-gap semiconductor is advantageous in the fabrication of power devices.

The energy band-gap is defined as the difference in energy between the conduction band and the valence band. Thus, a higher-energy band-gap means the valence-band electrons can seldom move into the conduction band because those electrons require energy sufficiently higher than the energy band-gap to move into the conduction band. On the other hand, when the energy band-gap is low, the valence-band electrons can easily move into the conduction band as they can readily attain the required energy. The energy band-gap is thus an important measure for estimating the breakdown voltage of an active device for a given semiconductor material. The breakdown voltage refers to drain or collector voltages where the drain or collector current increases very rapidly. Therefore, the useful range of an active device's drain voltage is usually limited by the breakdown voltage.

The phenomenon of breakdown can be explained using electron collisions. The electrons accelerated by an applied drain voltage collide with atoms at the drain, thereby generating electrons. When a drain voltage higher than the breakdown voltage is applied, the generated electrons thus attain sufficient energy to move into the conduction band and contribute to the drain current. As a result, the drain current increases rapidly due to the contribution of electrons generated by the collision. However, when the energy band-gap is high, the electrons generated by the collision with the accelerated electrons cannot attain sufficient energy to move into the conduction band. Thus, a high-energy band-gap semiconductor is typically associated with a high breakdown voltage. Due to this fact, when an active device is fabricated with a high-energy band-gap semiconductor, the active device shows a high breakdown voltage; a high DC bias voltage can then be applied and a large output power will be obtained. Therefore, a high-energy band-gap semiconductor is suitable for use as a high-power active device.

Thermal conductivity represents a material's ability to conduct heat. A higher thermal conductivity means the heat generated in an active device due to power consumption is easily dissipated. As we have seen previously, no matter how well a power amplifier is designed, a portion of the supplied power is consumed as heat. The heat generated increases the temperature of the active device, which can damage the device if the heat exceeds the absolute maximum temperature for the device. A well-designed heat-dissipation structure such as a heat sink may help an active device endure a certain temperature increase, but the amount of heat dissipated will be fundamentally limited by the thermal conductivity of a given semiconductor. Therefore, thermal conductivity is an important parameter for the semiconductor materials used in power amplifiers.

The physical properties of various semiconductors are shown in Table 9.1. In that table, SiC and GaN have an excellent thermal conductivity compared with Si. Because the thermal conductivity of Si is comparable to that of a metal, SiC and GaN semiconductors are known to have a fairly good thermal conductivity. In contrast, GaAs has a lower thermal conductivity than Si by a factor of about one third. Thus, in terms of thermal conductivity, Si, GaN, and SiC semiconductors are suitable for the fabrication of active devices for a power amplifier. Next, in terms of an energy band-gap, SiC and GaN have a higher energy band-gap than Si by a factor of about 3. As a result, the fabricated active devices that use SiC and GaN show a breakdown voltage 3 times higher than that of Si. As a direct measure of breakdown voltage, the breakdown voltage in Table 9.1 is directly related to the breakdown voltage that is approximately 6 times larger in SiC and GaN compared to Si. Thus, a high DC supply voltage is possible with SiC and GaN, which leads to an increase in output power capability.

Table 9.1 Comparison of the properties of various semiconductors

Property	Si	GaAs	SiC	GaN
Energy band-gap [eV]	1.11	1.43	3.2	3.4
Breakdown voltage [V/cm]	6.0×10^5	6.5×10^5	3.5×10^6	3.5×10^6
Saturation velocity [cm/sec]	1.0×10^7	2.0×10^7	2.0×10^7	2.5×10^7
Electron mobility [$cm^2/V{\cdot}s$]	1350	6000	800	1600*
Thermal conductivity [W/cm°K]	1.5	0.46	3.5	1.7
Heterojunction structure	SiGe/Si	AlGaAs/GaAs InGaP/GaAs AlGaAs/InGaAs	None	AlGaN/GaN InGaN/GaN

*The electron mobility of a GaN is the electron mobility of an AlGaN/GaN at the heterojunction.

Therefore, it can be seen that SiC and GaN are suitable semiconductors for a power amplifier's active devices, considering their thermal conductivity and energy band-gap. However, the electron mobility of SiC is lower compared to that of Si, and thus SiC active devices are inferior to Si active devices in high-frequency applications. Fortunately, the electron mobility in GaN depends on the condition of the heterojunction formation and this mobility can be made to exceed that in Si by using the heterojunction. An active device suitable for a high-frequency power amplifier can therefore be fabricated using a GaN process. Despite the GaN advantages, the late emergence of GaN devices is due to the lack of a suitable substrate for GaN growth. Recently, the technology for epitaxial-layer growth of GaN on SiC has seen significant advances, and the study of GaN-active devices continues to attract great interest.

Figure 9.5 shows the structure of a GaN HEMT formed on SiC. The reason for selecting an HEMT structure is primarily related to the electron mobility of GaN. As this mobility is not much superior to that of Si, its maximum exploitation is required. By employing an HEMT structure, maximum electron mobility can be obtained because that mobility in the channel can be achieved in the absence of impurities, as discussed in Chapter 5. The frequency characteristics of GaN active devices can therefore be improved although the electron mobility of GaN is not greatly superior to that of Si.

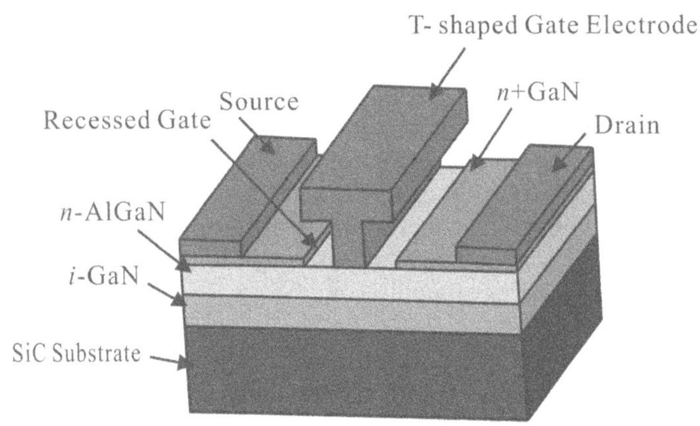

Figure 9.5 Structure of a GaN HEMT. To reduce the gate metallization resistance and gate length, a T-shaped gate metallization is used. An AlGaN/GaN heterojunction is formed for the maximum exploitation of the electron mobility of GaN.

In Figure 9.5, an intrinsic GaN epitaxial layer is grown on SiC, on top of which an n–type AlGaN layer is formed for the heterojunction. The electron well created by the heterojunction is formed in the epitaxial layer of the intrinsic GaN, and the doped electrons in the AlGaN are collected in the electron well and thus move through a channel free from impurity atoms. Therefore, great improvement in the frequency characteristics is possible. The T-shaped gate terminal on the n–AlGaN layer in Figure 9.5 minimizes the resistance of the gate terminal and decreases the gate length, and thus improves the frequency characteristics. Note that the drain and source terminals are formed on the n+ GaN layer for ohmic contact.

Figure 9.6(a) is a photograph of a GaN HEMT. The breakdown drain voltage V_{DS} is reported to be 142 V. This is an advantage that results from the physical properties of a GaN semiconductor. Figure 9.6(b) shows the measured output characteristics of the GaN HEMT at a frequency of 9.3 GHz. The device shows a gain of about 14 dB, and a saturated power of about 35.7 dBm. The PAE is found to be about 40% at 35.7 dBm. The output power per unit gate width is estimated to be 3.7 W/mm.

The frequency characteristics of a 0.25-μm GaN HEMT from Cree, Inc., are shown in Figure 9.7. In that figure, the frequency f_{max} is approximately 50 GHz where the maximum power gain U in Chapter 8 is 1. The frequency f_T is near 20 GHz where the short-circuit current gain h_{21} is 1. The f_{max} and f_T are comparable to the frequency characteristics of a GaAs MESFET. Considering the performances given in Figures 9.6 and 9.7, which were obtained from the initial stages of the GaN process, significant improvements are expected in the future.

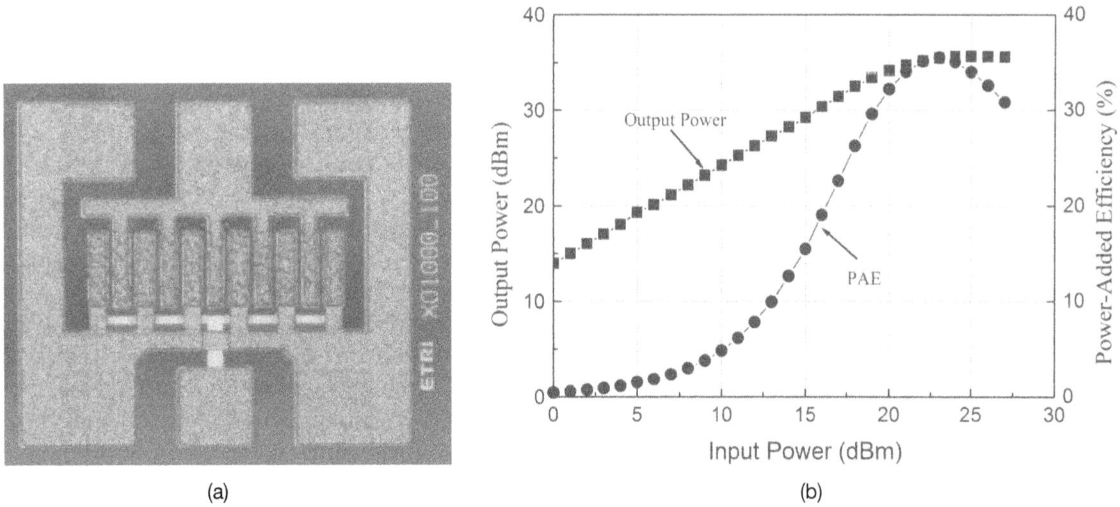

Figure 9.6 (a) Photograph of a GaN HEMT XO1000_100. The gate length and width are 0.25 μm and 1000 μm, respectively. The back side is polished to a thickness of 100 μm. (b) Output power characteristics are at 9.3 GHz. Source: (private communication): ETRI IT Convergence and Component Technology Research Section, 218 Gajeongro Yuseonggu, Daejeon, 305-700, Korea, www.etri.re.kr/kor/main/main.etri.

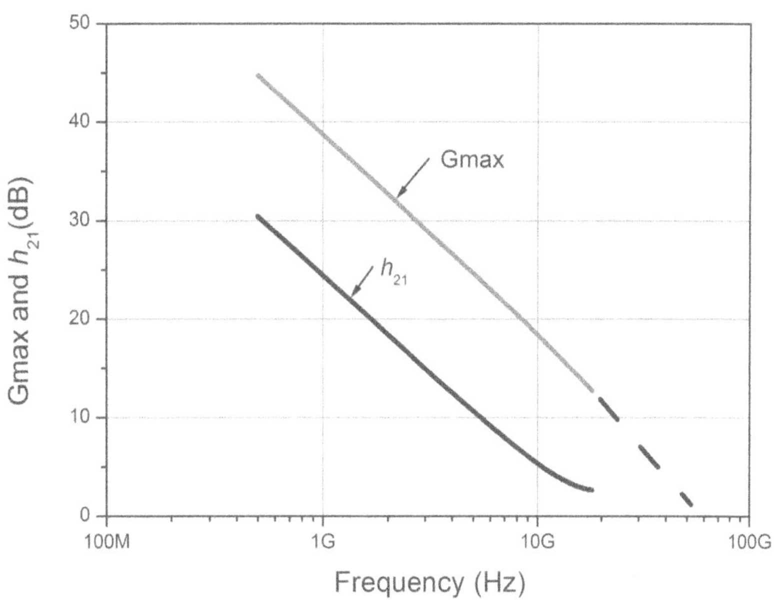

Figure 9.7 GaN HEMT frequency characteristics (refer to the datasheet).[1]

1. Cree, Inc., datasheet, CGHV1J006D (the S-parameters are at V_{DS} = 40 V, I_{DS} = 60 mA); available at www.cree.com/RF/Products/General-Purpose-Broadband-40-V/Discrete-Bare-Die/CGHV1J006.

9.2.2 LDMOSFET

An LDMOSFET (laterally diffused MOSFET) is an active power device fabricated using Si process technology and it has a structure designed to improve the breakdown voltage of an MOSFET. As explained earlier, the channel electrons attain sufficient energy at a high drain voltage V_{DS} and they collide with the atoms in the drain region with high impurity doping for ohmic contact. Numerous electrons are generated due to the collision and these electrons gain the energy sufficient to move into the conduction band taking the relatively small Si band-gap energy into consideration. Thus, there is a sudden increase in the drain current and, as a result, the lower breakdown voltage appears. This lower breakdown voltage can be improved by adjusting a doping profile in the drain region.

Figure 9.8 shows a cross-section of the LDMOSFET. In that figure, the drain region is laterally divided into a lightly doped region (NHV) and highly doped region (n + Drain) for the ohmic contact. In addition, the drain terminal is located at a considerable distance from the channel, which causes the accelerated electrons in the channel to primarily collide in the drain region with a low impurity concentration; as a consequence, the electrons lose energy when they reach the n+ drain region. Thus, this leads to a higher breakdown voltage compared with the case in which the electrons collide directly in the highly doped drain region.

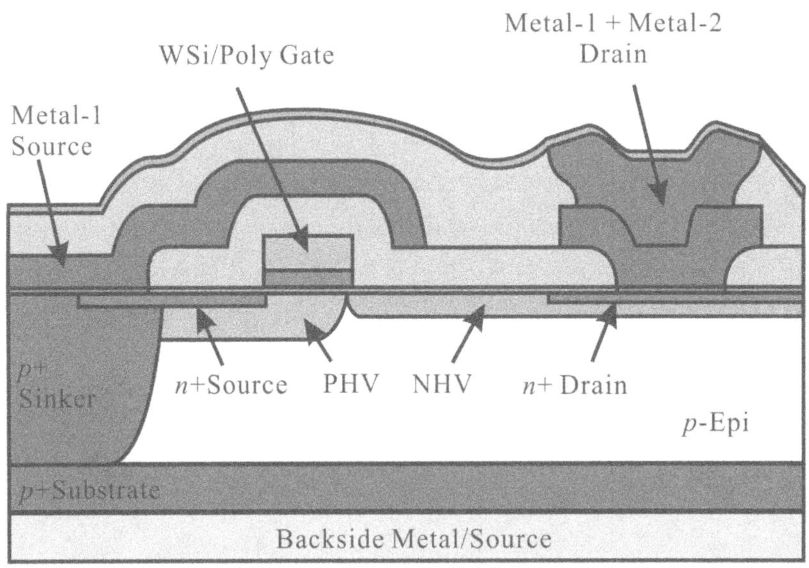

Figure 9.8 Cross-section of an LDMOSFET.[2] PHV is a high-voltage p-region; NHV is a high-voltage n-region with a low n type doping concentration (drawn per the literature identified in the footnote).

The next thing to note is that the source terminal is attached to the bottom of the device, which means that wire bonding is not required to connect the source terminal to the ground. A simple die attachment on the metal carrier is sufficient for the connection of the source terminal to the ground. Thus, the inductance that would otherwise arise from

2. Motorola, Semiconductor Technologies for RF Power F701, 2002.9.24.

the wire bonding during packaging assembly is eliminated, which is an advantage during assembly. Figure 9.9 shows the electric field simulation around the drain terminal that has a strong electric field across the NHV region, while the $n+$ drain region can be seen to have a weak electric field. As a result, the generation of electrons by collision occurs primarily in the lightly doped NHV region, where the number of those electrons is relatively smaller. This leads to a higher breakdown voltage than in the case of direct collision.

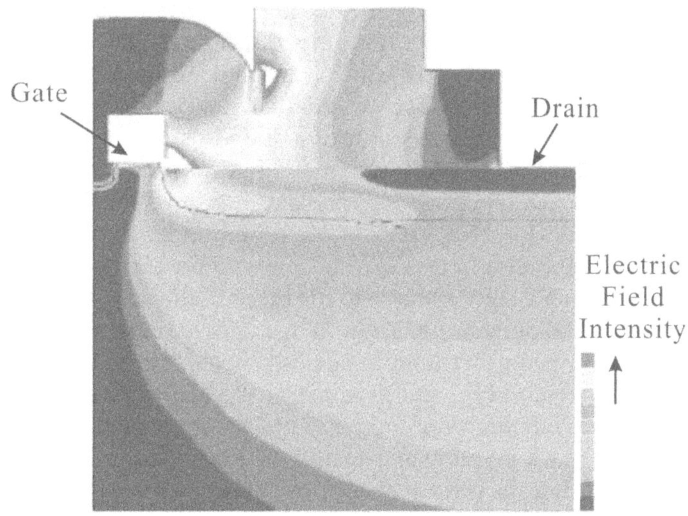

Figure 9.9 Electric-field simulation at the drain terminal of the LDMOSFET shown in Figure 9.8 (drawn per the literature.[3]) Note that the simulated electric-field intensity at the drain terminal becomes weak.

The manufacturing technology for LDMOSFETs is well-established. An LDMOSFET device that can supply up to several hundred W of RF power has been reported in the literature. However, the physical properties of Si limit its operation below the 4-GHz frequency region as the gain becomes low.

9.3 OPTIMUM LOAD IMPEDANCES

In designing the power amplifier, the first problem is determining optimum load impedance. The reason for naming the optimum impedance, rather than the maximum output power impedance, is that the design goal can be set for maximum efficiency or other parameters, depending on the design conditions, and the optimum load impedance does not necessarily give the maximum output power.

In designing a low-noise amplifier, the optimum source and load impedances can be calculated directly from the measured small-signal S-parameters at a given DC bias. However, because the power amplifier basically operates in the large-signal region, the optimum load impedance cannot be determined using the small-signal S-parameters.

3. Ibid.

Therefore, the optimum load impedance is determined either from direct experimental measurement or computed from the large-signal model prior to creating the power amplifier design. In this section, we will describe the method of obtaining the optimum load impedance by the experimental method as well as by a computational method that uses software.

9.3.1 Experimental Load-Pull Method

Load-pull measurement refers to measuring an active device's efficiency or output power by varying the load impedance. An impedance tuner is generally used to vary the load impedance. As mentioned earlier, the optimum load impedance of a power amplifier cannot be obtained from the measured small-signal parameters. Thus, the load impedance connected to the active device's output is tuned using an impedance tuner, and the optimum load impedance is obtained. This is referred to as a load-pull measurement. Figure 9.10 shows two impedance tuners that are inserted before and after the active device, and the source and load impedances are tuned under a given input power to obtain the optimum output power. In this method, the input tuner is first adjusted to deliver maximum power from the signal source to the input of the active device, after which the output tuner is adjusted to deliver maximum output power to the load. Once the adjustments are completed, the impedance tuners are disassembled and the desired load and source reflection coefficients Γ_S and $\Gamma_{L,opt}$ are obtained by measuring the impedances of the impedance tuners. The spectrum analyzer and power meter connected to the output impedance tuner in Figure 9.10 make it possible to measure the exact value of the output power while simultaneously observing the output spectrum in the spectrum analyzer. The power meter can accurately measure the output power but it cannot detect the presence of spurious signals, which necessitates the use of the auxiliary spectrum analyzer. Nonharmonic spurious signals often occur at the output of the power amplifier but the power meter cannot distinguish whether or not the nonharmonic spurious signals occur, which is why the spectrum analyzer is required. Thus, the spectrum analyzer is included for observing the spurious signals.

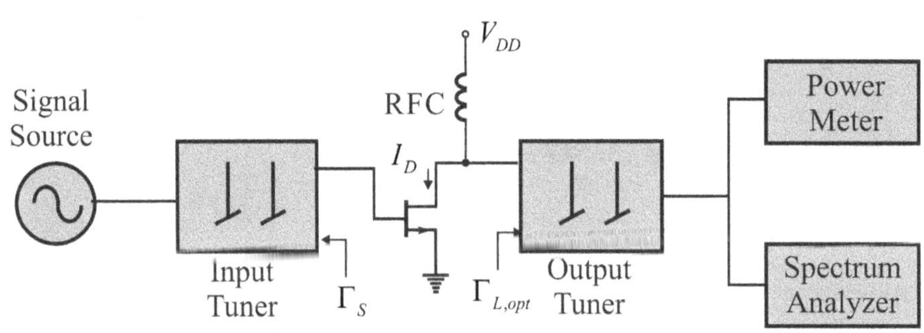

Figure 9.10 Load-pull setup. The output power is accurately measured by the power meter and the spectrum analyzer checks for the occurrence of spurious signals in the output signal.

In the case of low harmonics, the optimum load impedances can easily be obtained from the previous load-pull measurement. However, a problem does occur when the harmonic impedances must be considered. In general, the maximum output power is determined from the load impedance of the fundamental frequency, but the efficiency varies according to the harmonic load impedances. Recently, an impedance tuner that can independently tune the harmonic load impedances in order to resolve the harmonic tuning problem has been reported in the literature. A photograph of a harmonic impedance tuner is shown in Figure 9.11. Most recent impedance tuners can also be controlled by a PC, and their impedances can be read without disassembling them from the load-pull setup for the impedance measurement. The PC-controlled impedance tuners also provide utilities that draw a contour plot of the impedance, yielding the same output power or efficiency from measured data. Measuring the optimum source and load impedances using the load-pull method assists in the convenient and reliable design of power amplifiers.

Figure 9.11 Photograph of an impedance tuner with three carriages. The three harmonic impedances can be separately controlled using this impedance tuner. Source: Focus Microwaves Inc., Three Carriage Three Harmonic Tuner, January 2012.

There is another method for designing and fabricating power amplifiers experimentally without relying on load-pull measurements. An example of this method is shown in Figure 9.12. Using this method, a power amplifier's design allows for adjustable input and output matching circuits that use the optimum impedances in the datasheet, and the power amplifier is designed by adjusting the input and output matching circuits with a given RF input power and DC bias. Once the required specifications such as output power, efficiency, harmonic characteristics, and intermodulation are met, then the adjusted power amplifier can be used as a power amplifier if there are no other problems in size and mass production. However, the size of the power amplifier may be too large for some applications and further miniaturization may be required for those applications. Unlike the power amplifier shown in Figure 9.12, a power amplifier without tuning points may also be necessary for mass production. In those cases, the power amplifier shown in Figure 9.12 is disassembled and the source and load impedances are measured with a network analyzer as in a load-pull measurement. A miniaturized power amplifier can then be redesigned using the measured impedances.

In Figure 9.12, the chip capacitor in the input matching circuit is for input tuning and its position and value can be moved by soldering. Thus, the input matching circuit can be

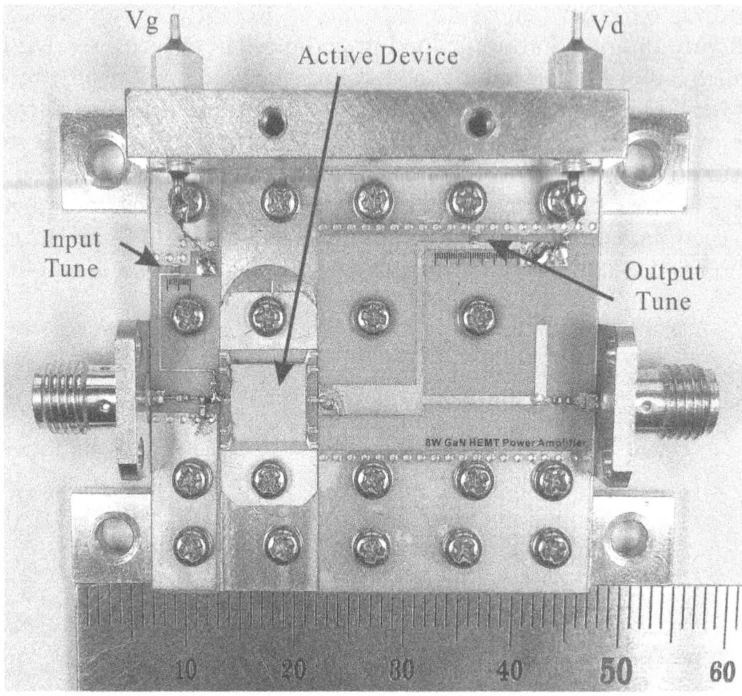

Figure 9.12 Experimental design of an adjustable power amplifier. The location and value of the chip component can be varied for the input and output matchings.

adjusted by trial and error until the optimum input impedance is obtained. The output matching circuit can be similarly adjusted. Thus, the output matching can be achieved through these adjustments.

The advantage of the adjustable circuit impedances is that the power amplifier can be designed without the cumbersome task of modeling the active devices, so the characteristics of the fabricated power amplifier can be determined experimentally, which otherwise would be difficult to predict accurately with the large-signal equivalent circuit. The disadvantage is that because the adjustable range of the input and output matching circuit impedances is limited compared to that in the prescribed load-pull measurement, the input and output matching circuits must be designed to provide the desired input and output matching impedances. Otherwise, the designated goal cannot be achieved. In addition, when changes occur in the DC operating point or input power for a new power amplifier, the obtained input and output matching impedances are not generally optimal for those changed conditions. Therefore, to obtain the optimum input and output impedances when those conditions change, the adjustable power amplifier in Figure 9.12 should be retested using the method discussed above that includes adjusting component values and positions. In the case of mass production, tolerance analysis may also be necessary; however, this method can present difficulties in terms of predicting the effects of parameter changes in an active device or in power amplifier's matching circuits.

9.3.2 Load-Pull Simulation

An alternative method for obtaining the optimum source and load impedances is the load-pull simulation based on the large-signal model of an active device.

When the large-signal model of an active device is available, this method involves setting up a load-pull simulation circuit in ADS, as shown in Figure 9.13, and calculating the desired optimum source and output load impedances; see reference 1 at the end of this chapter. This approach is frequently used in the design of an MMIC (monolithic microwave integrated circuit) power amplifier.

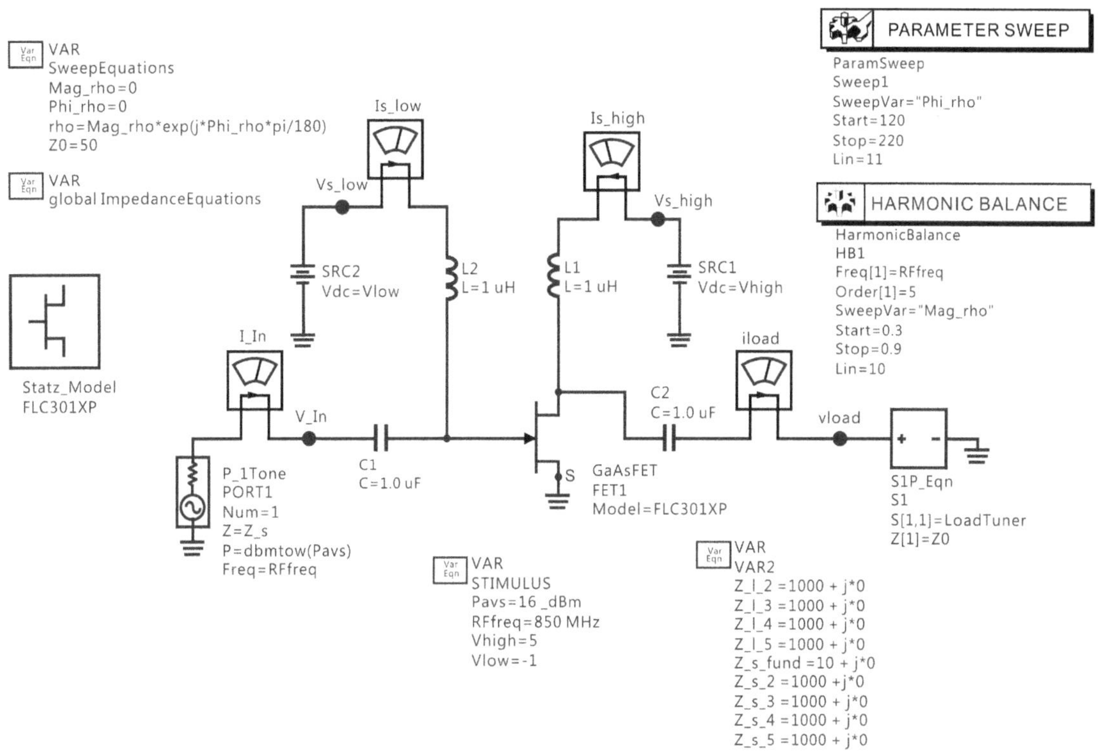

Figure 9.13 Load-pull simulation in ADS. The load impedance is adjusted using the **S1P Equation** component. The value **LoadTuner** of the **S1P Equation** component is determined using the variable block named **globalImpedanceEquation**, which is hidden. The **LoadTuner** value is two-dimensionally swept by the **Parmeter Sweep** and **Harmonic Balance** simulation components.

However, in the case of a packaged active device, the equivalent circuit should take into account the parasitic elements that occur in packaging and may cause the equivalent circuit to become extremely complex and may degrade its accuracy. The advantage that the load-pull simulation method has over the experimental load-pull method is that the optimum source and load impedances can be obtained simply through simulation that reflects the changes that may occur in the power amplifier's circuit conditions, such as frequency and the DC bias operating point. In contrast, with the experimental load-pull method, each time the power amplifier's condition changes—for example, if the frequency is changed—a new measurement must be taken to obtain the optimum source and load impedances. The disadvantage is that because it is not easy to obtain the accurate large-signal model, the large-signal model itself includes a certain degree of inaccuracy, which makes the accuracy of the results problematic.

9.3.2.1 Load Impedance

In Figure 9.13, the input frequency **RFfreq** is 850 MHz and input power **Pavs** is 10 dBm. Inductors **L1** and **L2** function as an RFC because their values are set to 1 μH. The values have been set sufficiently large for the chosen frequency. The DC-block capacitors **C1** and **C2** have values of 1 μF, which is also sufficiently large. Port 1 is the input port that has an impedance of **Z_s** and a power level of **Pavs**. The function **dbmtow(·)** converts the power level expressed in dBm to W. An impedance tuner implemented as a one-port circuit is connected to the output of the active device. In Figure 9.13, **Z_l_2**, **Z_l_3**, **Z_l_4**, and **Z_l_5** of **VAR2** are the load impedances at the harmonic frequencies of $2f_o$, $3f_o$, $4f_o$, and $5f_o$, respectively. By altering the harmonic load impedances, the efficiency or output power of the power amplifier can be tuned. Similarly, **Z_s_2**, **Z_s_3**, **Z_s_4**, and **Z_s_5** are the source impedances at the harmonic frequencies of $2f_o$, $3f_o$, $4f_o$, and $5f_o$, respectively. The harmonic source impedances are also adjustable. The value of the source impedance at the fundamental frequency **Z_s_fund** is set to 10 Ω. The frequency-dependent source and load reflection coefficients are defined using the variable block **globalImpedanceEquations** shown in Measurement Expression 9.1, which is not shown in the schematic window because of its complexity.

```
VAR Eqn VAR
global Impedance Equations
;Tuner reflection coefficient
LoadTuner=LoadArray[iload]
LoadArray=list(0,rho,fg(Z_l_2),fg(Z_l_3),fg(Z_l_4),fg(Z_l_5))
iload=int(min(abs(freq)/RFfreq+1.5,length(LoadArray)))
fg(x)=(x-Z0)/(x+Z0)
;Source impedances
Z_s=SrcArray[isrc]
SrcArray=list(Z0,Z_s_fund, Z_s_2, Z_s_3,Z_s_4, Z_s_5)
isrc=min(iload,length(SrcArray))
```

Measurement Expression 9.1 Source and load impedance setup

Measurement Expression 9.1 is the opened view of the **globalImpedanceEquations**. The value of **LoadTuner** here represents the resulting load impedance. The function **fg**(·) is defined to convert the given harmonic load impedances into the corresponding reflection coefficients. The row vector, **LoadArray**, is formed using the converted reflection coefficients at each harmonic frequency. The function **list**(·) is used to form a vector. Here, the first value of the **LoadArray** is the reflection coefficient at DC and the second value is the reflection coefficient at the fundamental frequency. All the values are set this way up to the fifth harmonic. It should be noted that the load impedance is open at DC because the reflection coefficient at DC is set to 0. The reflection coefficient of the fundamental frequency in **LoadArray** is set to **rho**, which is defined by the variable block named **SweepEquations** in Figure 9.13.

Figure 9.14 shows the desired frequency-dependent load reflection coefficient. Thus, the load reflection coefficient is $\Gamma_L = \rho e^{j\phi}$ for frequency $0.5f_o < f < 1.5f_o$. The ranges of other harmonic frequencies are defined similarly to the fundamental frequency. The variable **LoadTuner** in Measurement Expression 9.1 represents the frequency-dependent load reflection coefficient shown in Figure 9.14 and it is synthesized using the row vector **LoadArray.**

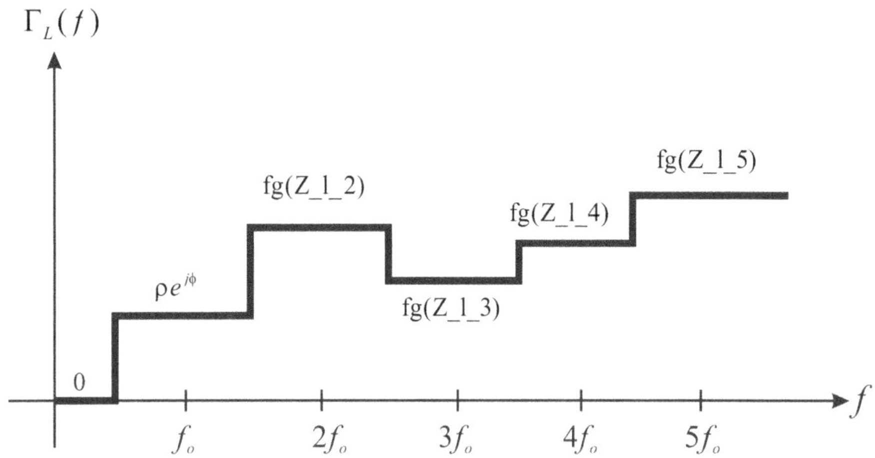

Figure 9.14 Frequency-dependent load reflection coefficient

In Measurement Expression 9.1, the frequency-dependent load reflection coefficient **LoadTuner** is implemented using the index of the row vector **LoadArray**. The variable **iload** represents the index. Define k as expressed in Equation (9.3).

$$k = abs(\text{freq})/\text{RFfreq} + 1.5 \tag{9.3}$$

When $0.5 \times$ **RFfreq < freq <** $1.5 \times$ **RFfreq**, the range of k is $2 < k < 3$ from Equation (9.3). Also, it can be found that **length(LoadArray)** = 6. A smaller value between the two values for **length(LoadArray)** and k is selected using the **min**(·) function. For $2 < k < 3$, the smaller value becomes k. Thus, taking the integer part of k and using the **int**(·) function gives a value of 2. Therefore, the value of **iload** becomes 2. The value of **LoadTuner** = **LoadArray[iload]** is the value corresponding to the index 2, which is $\Gamma_L = \rho e^{j\phi}$. When $4.5 \times$ **RFfreq < freq**, it becomes $k > 6$ from Equation (9.3). The smaller value thus becomes **length(LoadArray)** = 6. The corresponding reflection coefficient to index 6 is found to be **fg(Z_l_5)**. With similar reasoning, it can be found that the frequency-dependent reflection coefficient **LoadTuner** shown in Figure 9.14 is synthesized.

It can be seen that the source reflection coefficient is determined in a similar way in Measurement Expression 9.1. The variable **Z_s** represents the frequency-dependent source reflection coefficient. The variable **SrcArray** is the row vector of the source reflection coefficients at each harmonic. The index that determines the value of the source reflection coefficient is **isrc**. It should be noted that the value of **SrcArray** at DC is set to the reference impedance **Z**0.

9.3.2.2 Sweep

In Figure 9.13, the sweep of the load reflection coefficient is performed using harmonic balance simulation and parameter sweep. The variable **rho** that represents the load reflection coefficient at the fundamental frequency consists of **Mag_rho** and **Phi_rho**. The magnitude **Mag_rho** varies from 0.1 to 0.9, while the phase **Phi_rho** varies from 120° to 220°. This variation in the reflection coefficient is shown in Figure 9.15(a). However, the density of samples decreases as the radius approaches 1, thereby reducing the accuracy in the output power and efficiency contour plots. Instead, even though Figure 9.15(b) is somewhat complex in setting the sweep parameters, a region of the circle in the Smith chart can be uniformly sampled, which is advantageous when drawing more accurate contour plots for output power and efficiency. The simulation schematic for the samples shown in Figure 9.15(b) is shown in Figure 9.16.

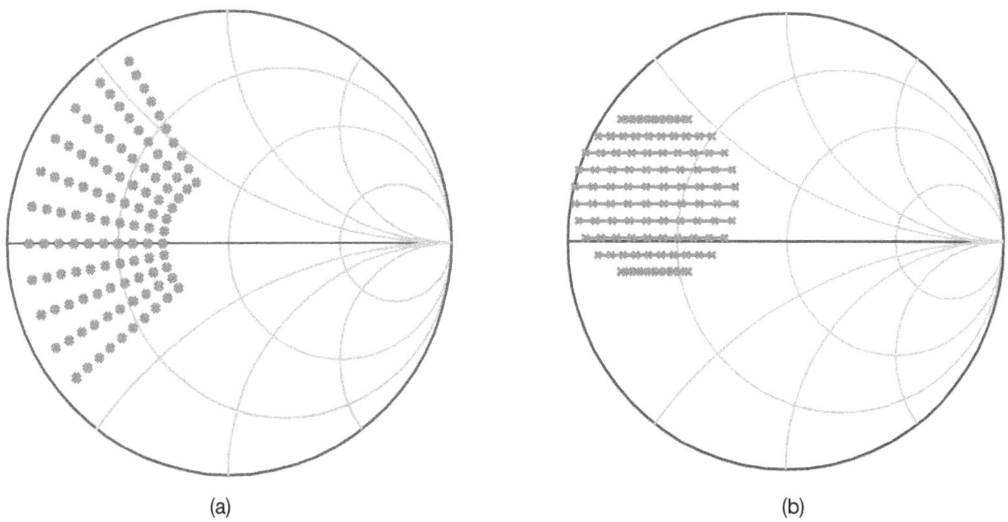

Figure 9.15 Reflection coefficient sweeping methods: (a) samples obtained using the magnitude and angle method (polar sweep) and (b) uniform sampling of a region of the circle (uniform sample sweep)

The simulation for the samples in Figure 9.15(b) can be carried out by sweeping the value of Γ_x for a given Γ_y when the load reflection coefficient is defined as $\Gamma_L = \Gamma_x + j\Gamma_y$. In Figure 9.16, the sweep variables **real_indexs**11 and **imag_indexs**11 are defined to represent Γ_x and Γ_y. In the simulation schematic of Figure 9.16, harmonic balance simulation sweeps the value of Γ_x for a given Γ_y in the parameter sweep. The user must specify a number of samples and a region of the circle to be sampled. This is done by entering the number of samples and the center and radius of the circle. The variable block named **SweepEquations** in Figure 9.16 includes the necessary variables. The variables **s11_center**,

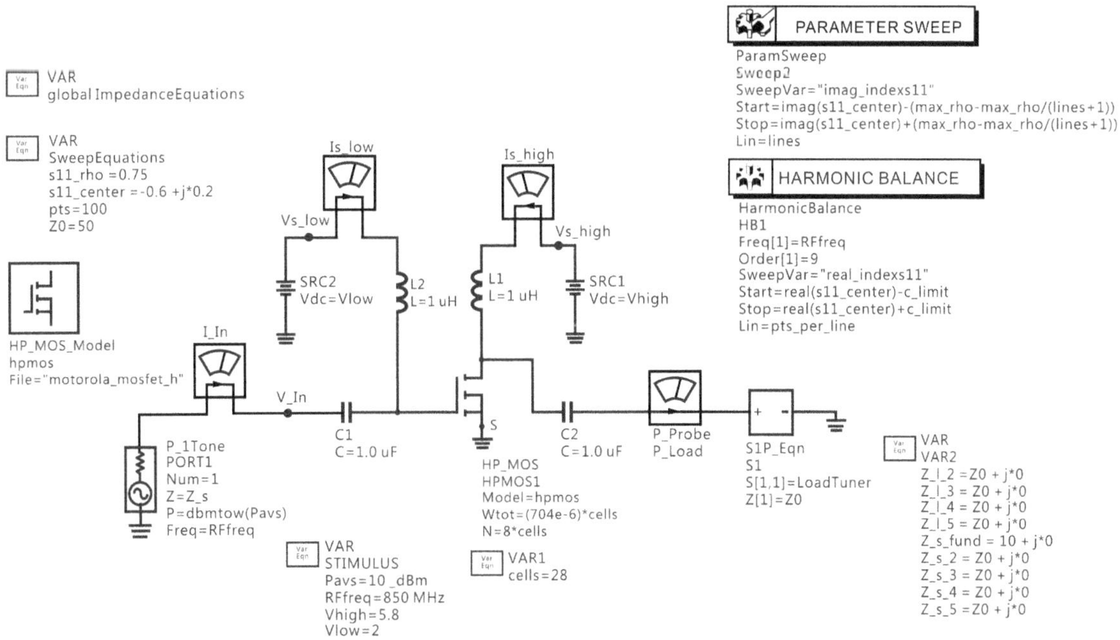

Figure 9.16 Load-pull simulation for the newly specified sample method shown in Figure 9.15(b).

s11_rho, and **pts** represent the center, the radius, and the number of samples. Using the user-specified variables, the number of lines in the direction of the y-axis, **lines**, and the number of samples per line, **pts_per_line**, are calculated and set in the variable **SweepEquations** VAR, which is hidden. Measurement Expression 9.2 is an opened view of **SweepEquations**.

```
real_indexs11=0
imag_indexs11=0
Z0=50
index_s11=real_indexs11+j*imag_indexs11
s11_rho=0.75
s11_center=-0.6+j*0.2
max_rho=min(1.0-mag(s11_center),mag(s11_rho))
pts=100
lines=max(int(sqrt(pts)),1)
pts_per_line=int(pts/lines)
argument=max_rho^2-(imag(s11_center)-imag_indexs11)^2
c_limit=sqrt(if(argument<0) then 0 else argument endif)
```

Measurement Expression 9.2 Opened view of the variable block named **SweepEquations**

The first two lines in the expression are for declaring the sweep variables **real_indexs**11 and **imag_indexs**11. Note that although they are set to 0, their values are newly defined through the parameter sweep. Variable **index_s**11 in the next line is the definition of Γ_L at the fundamental frequency using the variables **real_indexs**11 and **imag_indexs**11. The value of **index_s**11 is used as Γ_L instead of **rho** in the **globalImpedanceEquations** shown in Measurement Expression 9.1. Variables **s**11**_rho** and **s**11**_center** were described earlier and **max_rho** sets the new radius using the user-specified radius. This is necessary when the center and radius settings entered by the user are out of the unit Smith chart. Thus, the newly defined **max_rho** replaces the user-specified **s**11**_rho**.

Next, the number of lines in the direction of the y-axis, **lines**, is set using the total number of sample points, **pts**. Since the total number of sample points is proportional to the area, the variable **lines** is determined from **sqrt(pts)**. The number of points per line, **pts_per_line**, then becomes the total number sample points divided by the number of lines, **lines**. Since **lines** and **pts_per_line** must be integers, the function **int**(·) converts them all into integers. Due to such definitions, the sample points at both ends of the sample lines will be tightly spaced, while they will be widely sampled around the center line.

After the settings are made, in the case of the x-axis, the sweep range from the left to the right should be specified. However, the sweep range varies according to the y-axis values. That is, given y, from the equation of a circle with radius r, the possible values of x in the circle are represented by $-(r^2 - y^2)^{½} \leq x \leq (r^2 - y^2)^{½}$. The value $(r^2 - y^2)^{½}$ is represented by **argument**. To prevent the argument from being 0, the variable **c_limit** is newly defined. Therefore, the variable **real_indexs**11 corresponding to the x-axis reflection coefficient is swept from **real(s**11**_center)–c_limit** to **real(s**11**_center)+c_limit**, as shown in Figure 9.16. The y-axis reflection coefficient could have been swept in the range **imag(s**11**_center)-max_rho<y<imag(s**11**_center)+max_rho** for the number of lines. However, when **lines** = 2, y sweeps the two points at the end of the radius, an undesirable situation. Thus, at the end of the radius, the range is set to sweep above and below the line by a sweep range of **max_rho/(lines** + 1**)**, as shown in Figure 9.17.

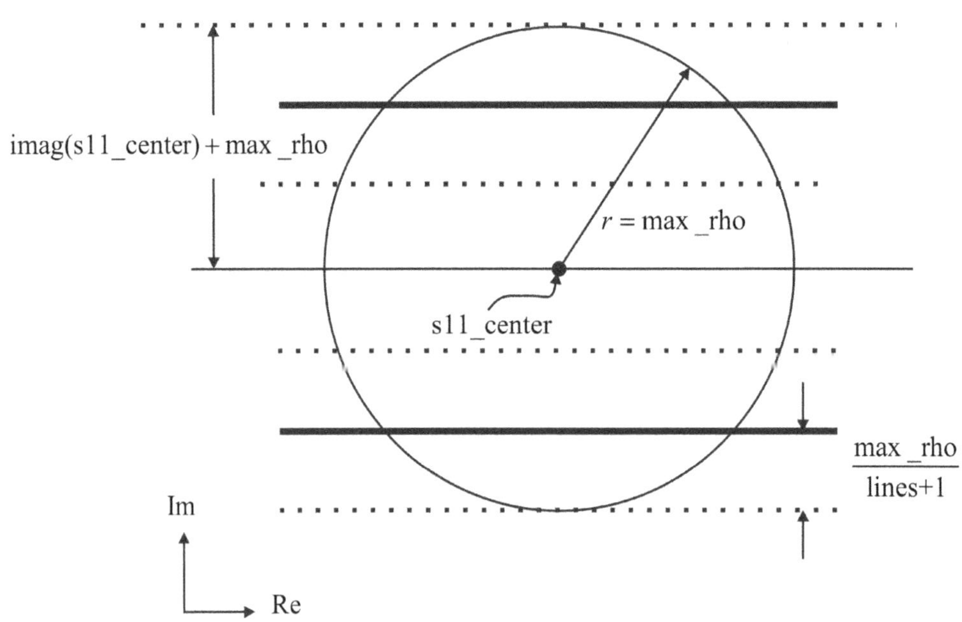

Figure 9.17 Sweep range setting of **imag_indexs**11 (**lines** = 2 case)

9.3.2.3 Display

The simulated results for the previously defined settings are the voltages and currents for the load reflection coefficient change. In order to calculate the output power and efficiency, the equations shown in Measurement Expression 9.3 are entered in the display window to calculate the DC power consumption. The **index**[0] in this expression represents the DC component and the **exist**(·) function determines whether or not the variable calculated from the simulation exists or not. The function **exist**(·) gives a value of 0 when the value specified by the expression does not exist. Therefore, during simulation, when the DC source and DC current probe are not defined by the names shown in Measurement Expression 9.3, the calculation results in the wrong values. The value 1e-20 is added to the DC power consumption **Pdc**. When the power consumption is 0, the division by zero occurs in the efficiency calculation. Then, 1e-20 is added in order to prevent division by zero without affecting the calculation results.

```
Eqn Vs_l=exist("real(Vs_low[0])")
Eqn Vs_h=exist("real(Vs_high[0])")
Eqn Is_l=exist("real(Is_low.i[0])")
Eqn Is_h=exist("real(Is_high.i[0])")
Eqn Pdc=Is_h*Vs_h+Is_l*Vs_l+1e-20
```

Measurement Expression 9.3 Equations in the display window for calculating the DC power consumption

```
Eqn Pdel_Watts=real(0.5*vload[1]*conj(Iload.i[1]))
Eqn Pavs_Watts=10**((Pavs[0,0]-30)/10)
Eqn PAE=100*(Pdel_Watts-Pavs_Watts)/Pdc
Eqn Pdel_dbm=10*log10(Pdel_Watts)+30
```

Measurement Expression 9.4 Equations for calculating the power delivered to the load

Measurement Expression 9.4 shows the equations for calculating the power delivered to the load and the efficiency of using DC power consumption obtained from the load-pull simulation. The first line of this expression calculates the delivered power at the fundamental frequency. However, when a power probe **P_Load** is used, as shown in Figure 9.16, the first line changes as shown in Measurement Expression 9.5.

```
Eqn Pdel_Watts=P_Load.p[1]
```

Measurement Expression 9.5 Equation for the delivered power using the power probe

The second line in Measurement Expression 9.4 calculates the input power using a variable stored in the dataset in the simulation shown in Figure 9.16. The variable **Pavs** becomes a two-dimensional variable because the parameter-swept simulation was carried out for two independent variables. Thus, the constant value from the two-dimensional variable **Pavs** is obtained by specifying **Pavs** as **Pavs**[0,0]. In addition, because **Pavs** is specified in dBm, the unit is changed to W by the equation in the second line. The third line calculates the PAE and the last line calculates the delivered output power in dBm. Once **PAE** and **Pdel_dbm** are calculated, they can be used to plot the contours.

```
Eqn PAE_step=2
Eqn NumPAE_lines=5
Eqn PAEmax=max(max(PAE))
Eqn PAE_contours=contour(PAE,PAEmax-0.1-[0::(NumPAE_lines-1)]*PAE_step)
```

Measurement Expression 9.6 Equations for drawing the contour plot

It is common practice to plot the contours from the maximum point at equally spaced step changes in a descending order. To do this, we first need to obtain the maximum value. The method for doing this is identical to that for finding the maximum of the delivered power, so only the plot of **PAE** contours will be discussed. Since the **max**(·) function used to obtain the maximum value gives the maximum for a single sweep variable, the maximum can be obtained by using the equation **max(max(PAE))** shown in Measurement Expression 9.6. The **NumPAE_lines** in this expression represents the number of contour plots to be drawn. The interval is set by **PAE_step**. Since **PAE** in Measurement Expression 9.4 is calculated as a percentage, this corresponds to a step of 2%. The function **contour**(·) in Measurement Expression 9.6 is a function that gives the coordinates (x, y) as the output. For the coordinates (x, y), x is the independent variable of the function **contour**(·) and y is the value of the function **contour**(·). In addition, x is the primary sweep variable, while y is a secondary sweep variable. In the case of the example in Figure 9.16, **real_indexs11** is the independent variable of the **PAE_contours** defined in Measurement Expression 9.6. In the case of the load-pull simulation for the polar sweep in Figure 9.13, **Mag_rho** is the independent variable of **PAE_contours**. The y coordinate value is stored in **PAE_contours**. Thus, to explain Measurement Expression 9.6, for example, the (x, y) coordinates can be represented as (**indep(PAE_contours)**, **PAE_contours**) and the corresponding value of the contour level is stored as a separate independent variable. The obtained **PAE_contours** should be converted to the complex numbers in the Smith chart using the equations shown in Measurement Expression 9.7. The first two equations in that expression are used to convert the uniformly sampled swept **PAE_contours** into the complex numbers in the Smith chart, while the last two equations are used to convert the polar swept **PAE_contour** into the complex numbers in the Smith chart.

```
Eqn PAE_contours=contour(PAE, PAEmax-0.1-[0::(NumPAE_lines-1)]*PAE_step)
Eqn PAE_contours_p=[indep(PAE_contours)+j*PAE_contours]
Eqn PAE_contours=contour(PAE, PAEmax-0.1-[0::(NumPAE_lines-1)]*PAE_step)
Eqn PAE_conts_forSmithCh=indep(PAE_contours)*exp(j*PAE_contours*pi/180)
```

Measurement Expression 9.7 Conversion of the contour plot values into complex numbers for plotting on the Smith chart

---

Example 9.2

Open the **HB1Tone_LoadPull.dsn** of **examples/RF_board/NADC_PA_prj** in ADS and modify the original simulation schematic as shown in Figure 9E.1.

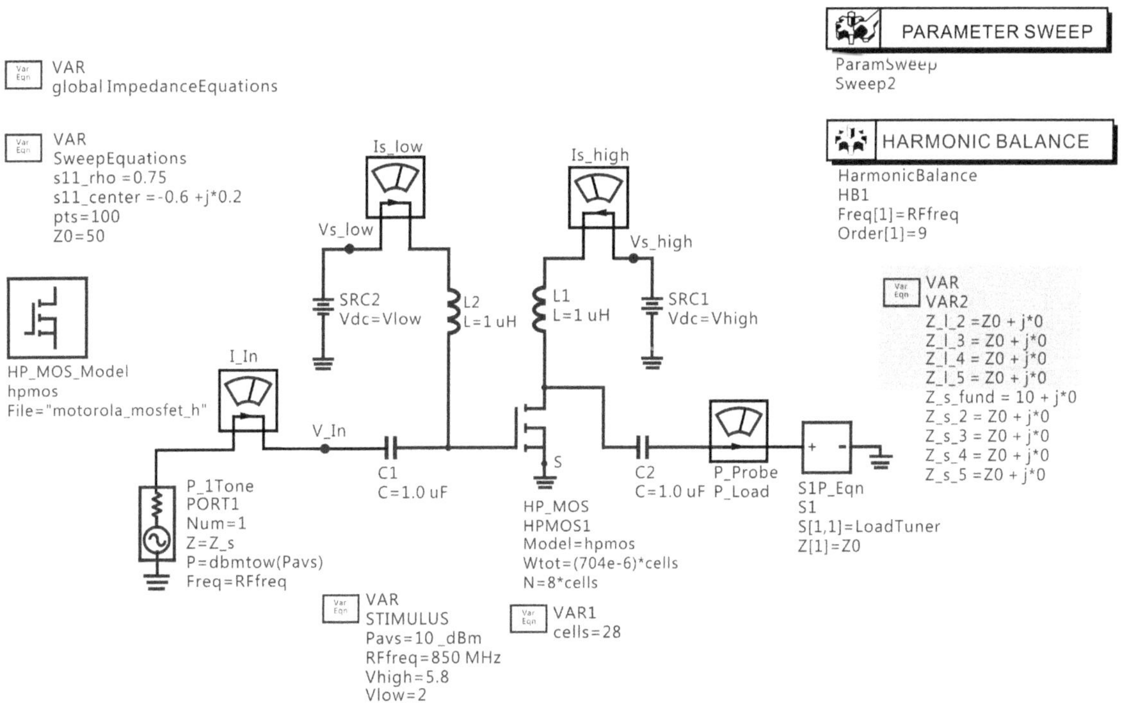

Figure 9E.1 **HB1Tone_LoadPull.dsn** in **examples/RF_board/ NADC_PA_prj**. The load impedance is swept by a uniform sample sweep. The values in the shaded area are changed from the values in Figure 9.16 for comparison.

Perform the simulation and plot the PAE and delivered power contours. Then, set the even harmonic impedance of the **LoadTuner** to 0 Ω and the odd harmonic impedance of the **LoadTuner** to 1 kΩ, and investigate the variation in the contour plot. The values of the **LoadTuner** can be specified using the numbers in the shaded area of Figure 9E.1.

Solution

A schematic that is different from that shown in Figure 9E.1 is displayed when the file mentioned in the specified directory is opened. After deleting the explanations in the original schematic, the schematic in Figure 9E.1 can be obtained. It can be seen that all the load impedances in Figure 9E.1 at all harmonics, except at the fundamental frequency, are set to 50 Ω. In addition, the sweep of the fundamental impedance is performed using the method shown in Figure 9.15(b). Also, the power probe **P_Probe** was used to calculate the power delivered to the load. Therefore, the power for the fundamental frequency becomes **P_Load.p**[1].

Figure 9E.2 is a comparison of the results obtained from the simulation with all harmonic load impedances set to 50 Ω (broadband load) with the simulation that has the even harmonic load impedances set to 0 and the odd harmonic impedances set to 1 kΩ (harmonic load). The comparison of contour plots for the delivered power is shown in Figure 9E.2(a) and the PAE contour plot comparison is shown in Figure 9E.2(b). It can be seen in Figure 9E.2(a) that the variations of the maximum point of the delivered power and the contours do not significantly depend on the harmonic impedance variation. On the other hand, unlike the contour plot of the delivered power, the PAE shows, to some degree, a change in response to changes in harmonic impedances. This will be explained in a later section of this chapter on class-F amplifiers. In addition, the results of the

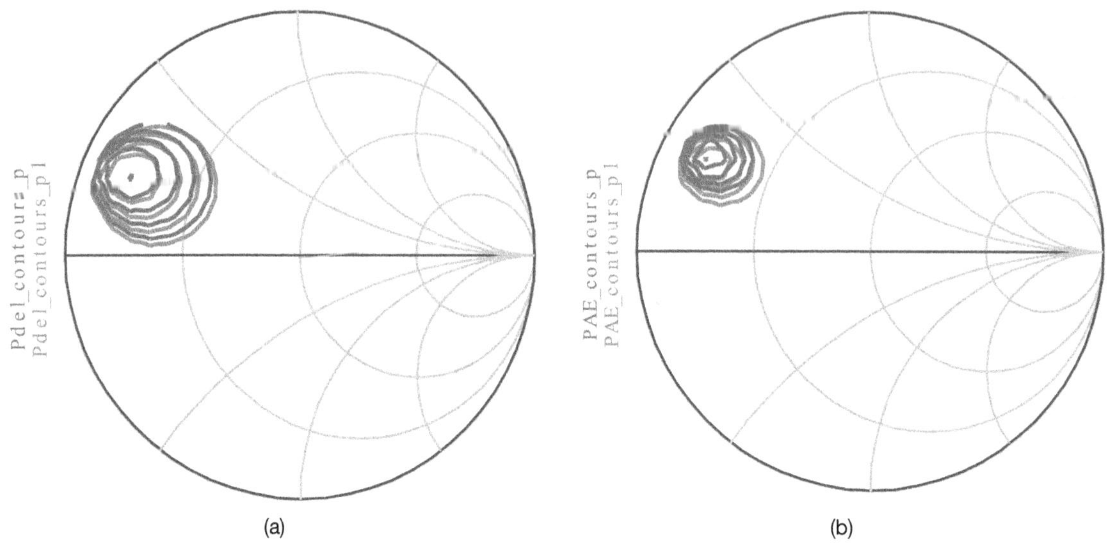

Figure 9E.2 Comparison of (a) contour plots of the output powers and (b) PAEs with broadband load and harmonic load. **Pdel_contours_p** and **Pdel_contours_p1** are the contour plots for delivered power for broadband and harmonic loads, respectively. PAE contours are similarly drawn.

Table 9E.1 Output and efficiency variations due to harmonic impedance variation

	Maximum Power Condition		Maximum Efficiency Condition	
Load	**Power(dBm)**	**Impedance(ohm)**	**Efficiency (%)**	**Impedance(ohm)**
Broadband load	25.67	$6.409 + j10.180$	42.32	$5.743 + j12.756$
Harmonic load	25.52	$6.415 + j10.159$	39.12	$5.762 + j12.667$

comparison are summarized as shown in Table 9E.1. The maximum powers in Table 9E.1 do not change significantly with respect to the change in the harmonic impedances but the maximum PAEs show significant changes. However, it can be seen that the fundamental impedance that gives maximum output power and maximum PAE does not change significantly.

9.4 CLASSIFICATION

Power amplifiers are classified according to the DC bias point of the active device and the characteristics of the load network. This classification has a long history and is based on understanding the operation of the active device's output behavior. Figure 9.18 shows the simplified I_{DS}–V_{GS} and I_{DS}–V_{DS} characteristics of an FET in order to classify the power amplifiers based on the DC bias point.

In the figure, when the amplifier is biased at $V_{GS} = V_{GS,A}$, $V_{DS} = V_{DD}$ (operating point Q_A), considerable drain current $I_{q,A}$ (quiescent drain current) flows in the amplifier even in the absence of RF input. A power amplifier that operates at this DC bias point is called a class-A power amplifier. When a sinusoidal input is applied to the class-A power

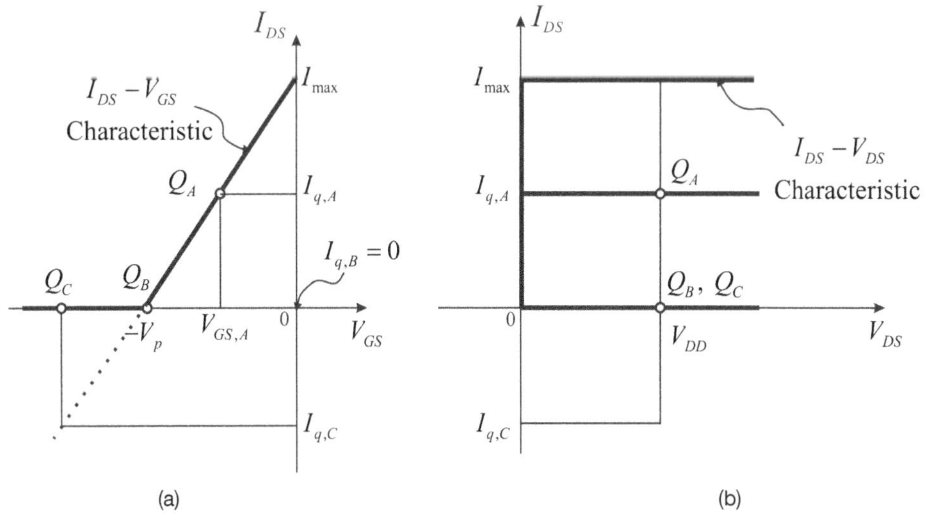

Figure 9.18 Simplified (a) I_{DS}–V_{GS} and (b) I_{DS}–V_{DS} and characteristics of an FET. Points Q_A, Q_B, and Q_C represent class-A, -B, and -C operating points, respectively.

amplifier, the output drain current $i_{DS}(t)$ conducts for a whole cycle of the sinusoidal input, as shown in Figure 9.19. Therefore, the input signal is faithfully amplified without distortion and shows a satisfactory linearity. However, as explained in section 9.1, the class-A power amplifier will have a maximum efficiency of 50%. In addition, the drain current flows even in the absence of RF input, which results in DC power consumption. Note that in the absence of RF input, the active device's DC power consumption is maximum, which requires a large heat sink for sufficient heat dissipation.

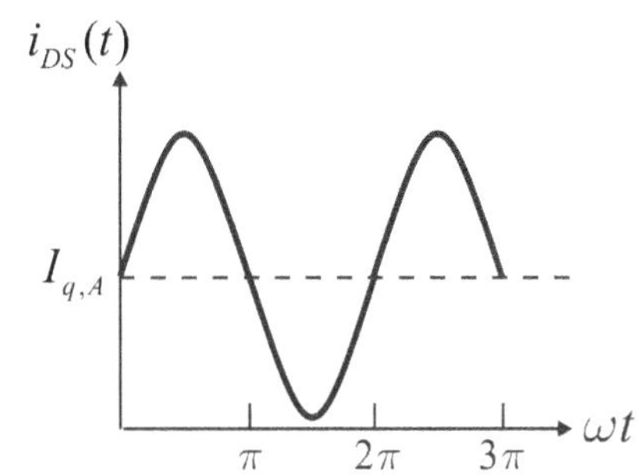

Figure 9.19 Drain-current waveform in a class-A power amplifier

The power consumption in the absence of RF input can be improved by moving the active device's DC bias point. Suppose that the FET is DC biased at $V_{DS} = V_{DD}$ and V_{GS} is set at the FET's pinch-off voltage $V_{GS} = -V_p$, as shown in Figure 9.18. The DC bias point is called a class-B operating point and the power amplifier operating at this point is called a class-B power amplifier. In the class-B case, the drain current is $I_{q,B} = 0$ and will not flow in the absence of RF input. Thus, unlike the class-A amplifier, there is no DC power consumption in the absence of RF input. As another form of bias, V_{GS} can be set smaller than the pinch-off voltage V_p of the FET. This DC bias point is called a class-C operating point and the power amplifier operating at the class-C DC bias point is called a class-C power amplifier. It should be noted that the drain current of the class-C power amplifier is also zero in the absence of RF input. On the other hand, the operating point is often mathematically interpreted as the DC bias for a negative drain current $I_{q,C}$ in the absence of RF input, as shown in Figure 9.18. However, the real DC drain current is zero. The $I_{q,C}$ is the projected value from the I_{DS}–V_{GS} characteristics curve. The drain current $I_{q,C}$ at the class-C DC bias point indicates the required amplitude for the sinusoidal RF input applied to the gate to reach the pinch-off voltage. As such, the class-B and class-C power amplifiers have no DC power consumption in the absence of RF input, which is an advantage.

For a sinusoidal RF input applied to the class-B and C power amplifiers, the drain current $i_{DS}(t)$ flows for a conduction time interval determined according to their DC bias points, while it is zero for the remaining time interval, as shown in Figure 9.20. The time interval in which the drain current flows is called the *conduction angle*; the conduction angle of class-B is 180° while that of class-C is less than 180°. In addition, for an improvement in linearity and an increase in gain, a small DC current is made to flow in the absence of RF input, in which case the conduction angle becomes higher than 180°, an operation called class-AB.

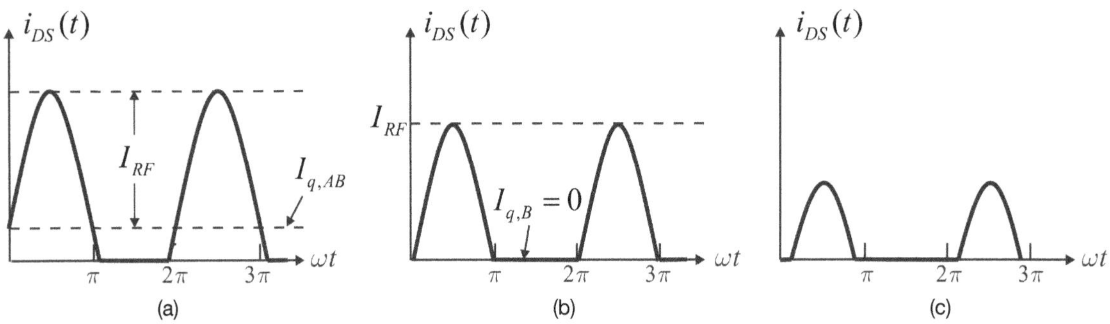

Figure 9.20 Drain-current waveforms according to an operating point; (a) class-AB, (b) class-B, and (c) class-C

As shown in Figure 9.20, the drain current waveform shows the shape of a sinusoidal tip in the conduction angle; otherwise, the RF input waveform is cut off. Distortion can naturally be expected when the RF input signal is amplified through the drain current with the sinusoidal-tip shape. Suppose that a class-B or class-C power amplifier is employed in a communication system. The amplitude and phase of the waveform in Figure 9.20 vary according to the modulated input signal. Therefore, when the RF signal is amplified by the class-B or class-C power amplifier, as shown in Figure 9.20, a loss of information may result. However, if expanded in a Fourier series, the waveform of Figure 9.20 will result in harmonics of nf_o. The harmonic signals can be removed using a filter after the amplification with the sinusoidal-tip-shaped waveforms. Both the amplitude and phase of the filtered waveform at the fundamental frequency are obviously proportional to the modulated RF input because the amplitude and phase of the sinusoidal-tip-shaped waveform in Figure 9.20 is proportional to the input signal. Thus, when the I_{DS}–V_{GS} characteristic is linear, as shown in Figure 9.18, a relatively faithful amplified signal can be obtained. However, the transfer characteristic, such as I_{DS}–V_{GS}, has some nonlinearity and results in distortion. Notably, the distortion becomes more significant as the operation point moves into a deeper class-C.

Previous explanations assume that the output of the active device is considered to be a transconductance-type current source. The output resistance of the current source is ignored. In a case where the output resistance cannot be ignored, it can be included in the load network. Thus, the necessary assumption for a transconductance-type current source is that the amplitude of the input is not set sufficiently large to drive the active device into the saturation region.

In the case of the class-D and class-E power amplifiers that are introduced in this section, the active device does not operate on the basis of the transconductance-type current source. The active device in class-D and class-E operations is assumed to be a switch. That is, when the input signal is sufficiently large, the active device no longer acts as a transconductance-type current source but instead functions as a switch. The modeling of an active device as a switch becomes more appropriate when the active device is driven by a square wave. Using the square wave can significantly increase efficiency, an operation that will be explained in this section. The classification of a class-F amplifier is somewhat ambiguous. Most of these amplifiers are classified according to the load network's characteristics.

9.4.1 Class-B and Class-C Power Amplifiers

Figure 9.21 shows class-B and class-C power amplifier configurations. In that figure, the transistor can operate in class-AB, class-B, or class-C, depending on the DC bias voltage V_{GS}. Here, $v_G(t)$ represents the RF input signal and the drain of the FET is biased at a DC voltage V_{DD}. Capacitor C_B at the output is a DC block capacitor. Inductor L forms a parallel resonant circuit with the capacitor C at the fundamental frequency f_o.

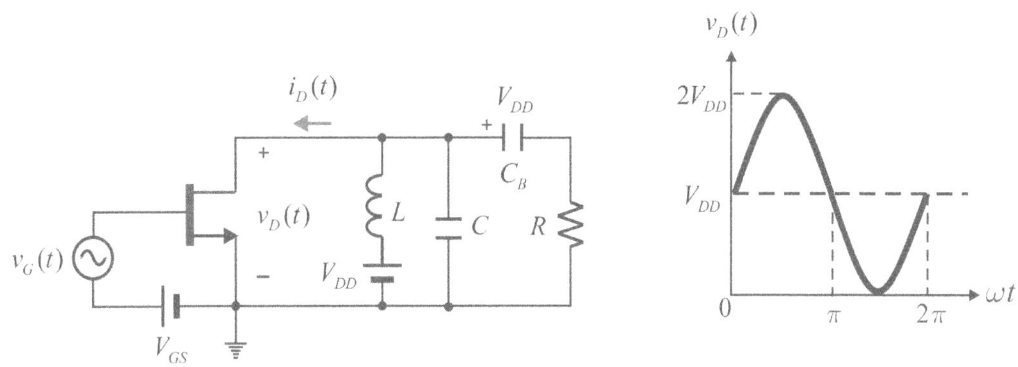

Figure 9.21 Class-B and class-C power amplifier configuration and drain-voltage waveform. The gate bias voltage V_{GS} sets the operating point of the class-B or -C accordingly. Capacitor C_B is a DC block capacitor and the parallel resonant circuit at the fundamental frequency is formed by RLC. Due to the parallel resonance, the sinusoidal waveform appears at the drain and load.

The $Q = \omega_0 CR$ of the parallel resonant circuit is assumed to be sufficiently large. Thus, the impedance of the parallel resonant circuit is 0 at all harmonics except at the fundamental frequency. Consequently, the voltage due to the harmonic currents is 0 and does not appear across the load resistor R. Only the sinusoidal voltage due to the fundamental current appears across R. The voltage across R is sinusoidal and the drain voltage $v_D(t)$ is thus

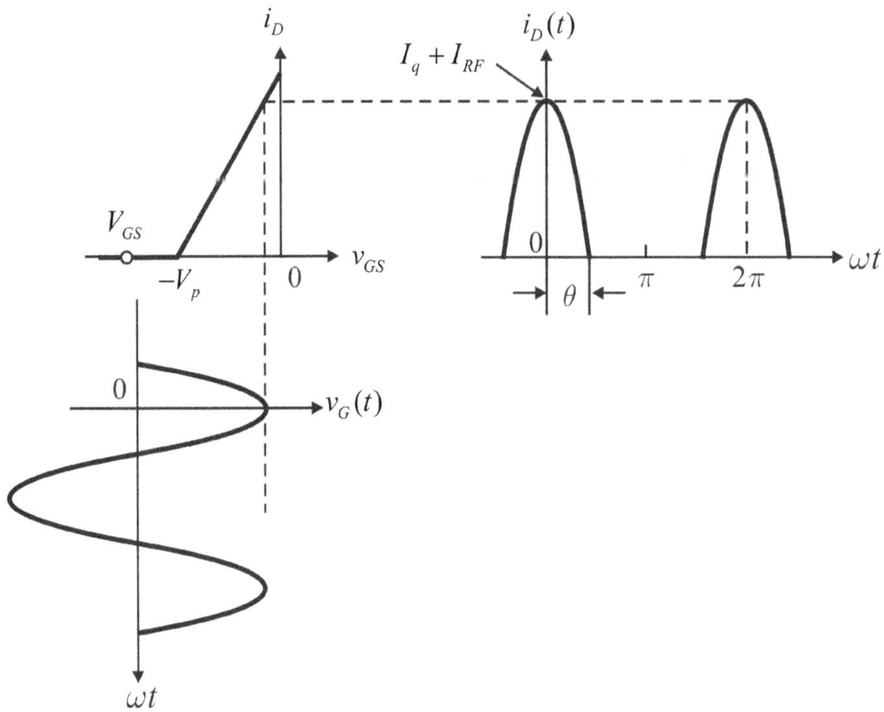

Figure 9.22 Drain-current waveform that depends on the DC bias voltage V_{GS}. According to V_{GS}, the drain current may have a class-B or -C sinusoidal-tip-shaped waveform.

a sinusoidal voltage waveform raised by supply voltage V_{DD} due to the DC voltage across the DC block capacitor, as shown in Figure 9.21. Since the supply voltage V_{DD} is fixed, by selecting an appropriate load resistor R, the maximum peak voltage can reach V_{DD}. It must be noted that a negative voltage does not appear at the drain. When an RF input $V_G\cos(\omega t)$ is applied, as shown in Figure 9.22, the output waveform of the drain current can be written as

$$i_D(t) = \begin{cases} I_q + I_{RF}\cos\omega t & -\theta < \omega t < \theta \\ 0 & \theta < \omega t < 2\pi - \theta \end{cases} \tag{9.4}$$

The I_q in Equation (9.4) becomes the DC quiescent current that flows in the absence of RF input. That is, $I_q = 0$ when the amplifier operates in class B and in class C when $I_q < 0$. In addition, I_{RF} is the amplitude of the fundamental drain current. Using Equation (9.4), the conduction angle θ can be determined as expressed in Equation (9.5).

$$\begin{gathered} i_D = 0 = I_q + I_{RF}\cos\omega t \\ \theta = \pm\cos^{-1}\frac{I_q}{I_{RF}} \end{gathered} \tag{9.5}$$

Now, using θ, Equation (9.4) can be rewritten and expressed as Equation (9.6).

$$i_D(t) = I_{RF}(\cos\omega t - \cos\theta) \tag{9.6}$$

Thus, the maximum current is expressed in Equation (9.7).

$$I_{\max} = I_{RF}(1 - \cos\theta) \tag{9.7}$$

Expanding $i_D(t)$ in a Fourier series,

$$i_D(t) = I_0 + I_1\cos\omega t + I_2\cos 2\omega t + \cdots$$

where the DC current component can be obtained as shown in Equation (9.8)

$$I_0 = \frac{1}{2\pi}\int_{-\theta}^{\theta} I_{RF}(\cos x - \cos\theta)dx = \frac{I_{RF}}{\pi}(\sin\theta - \theta\cos\theta) = I_{RF}\gamma_0 \tag{9.8}$$

and the fundamental wave component I_1 and harmonic components I_n can be expressed as Equations (9.9a) and (9.9b),

$$I_1 = \frac{1}{\pi}\int_{-\theta}^{\theta} I_{RF}(\cos x - \cos\theta)\cos x dx = \frac{I_{RF}}{2\pi}(2\theta - \sin 2\theta) = I_{RF}\gamma_1 \tag{9.9a}$$

$$I_n = \frac{1}{\pi}\int_{-\theta}^{\theta} I_{RF}(\cos x - \cos\theta)\cos nx dx = \frac{I_{RF}}{\pi}\left\{\frac{\sin(n-1)\theta}{n(n-1)} - \frac{\sin(n+1)\theta}{n(n+1)}\right\} = I_{RF}\gamma_n \tag{9.9b}$$

where γ_0 and γ_1 are functions of the conduction angle. Thus, denoting the voltage at fundamental frequency as V_1, the drain efficiency can be computed as shown in Equation (9.10).

$$\eta = \frac{P_1}{P_{DC}} = \frac{V_1 I_1}{2V_{DD}I_0} = \frac{1}{2}\xi\frac{\gamma_1}{\gamma_0} \tag{9.10}$$

Here, ξ is defined as

$$\xi = \frac{V_1}{V_{DD}}$$

and, since $V_1 < V_{DD}$, generally $\xi \leq 1$. Therefore, by substituting $\xi = 1$ and $\theta = \frac{1}{2}\pi$ into Equation (9.10), the maximum efficiency of a class-B amplifier, $\eta_{B,\max}$ is expressed in Equation (9.11).

$$\eta_{B,\max} = \frac{\pi}{4} = 78.5\% \tag{9.11}$$

From I_{max} in Equation (9.7), the value of the load giving maximum output power is given by Equation (9.12).

$$R_{opt} = \frac{V_{DD}}{I_{RF}\gamma_1} = \frac{V_{DD}(1-\cos\theta)}{I_{max}\gamma_1} \tag{9.12}$$

Then, the maximum output power is expressed in Equation (9.13).

$$P_{max} = \frac{V_{DD}^2}{2R_{opt}} \tag{9.13}$$

Example 9.3

Calculate the optimum impedance R_{opt} of a 20-W class-B power amplifier for $V_{DD} = 20$ V.

Solution

From Equation (9.13),

$$R_{opt} = \frac{V_{DD}^2}{2P_{max}} = \frac{20^2}{2\times 20} = 10\ \Omega$$

The meaning of the optimum load in Equation (9.12) can be understood by using a load line at the device output plane, which represents the trajectory of the voltage $v_D(t)$ and current $i_D(t)$. Using $v_D(t)$ and $i_D(t)$ in Figure 9.22, the load line can be drawn as shown in Figure 9.23. In that figure, the optimum load value that gives maximum output power

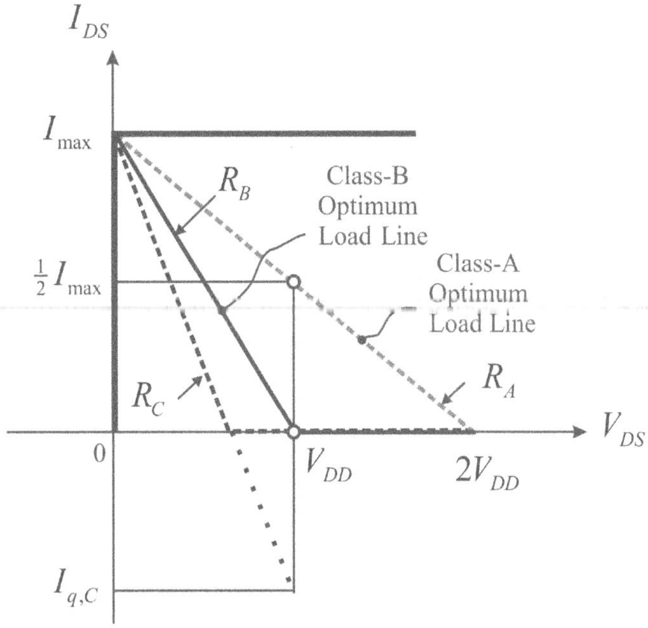

Figure 9.23 Load lines of power amplifiers operating with class-A, -B, and -C. The optimum resistance is increasing as the class moves from C to A.

in a class-B amplifier in Equation (9.12) can be easily inferred from Figure 9.23. In addition, the load resistance of a class-A amplifier that gives maximum output power is double that of the class-B amplifier. Also, from the load line of the class-C amplifier, as the conduction angle decreases, the optimum load resistance that gives maximum output power can also be seen to decrease, as shown in Figure 9.23.

Example 9.4

Set up a 1-A half-wave current source in ADS and configure the output circuit of a class-B amplifier. Set the load value from Equation (9.12) and confirm the voltage waveform $v_D(t)$ and drain efficiency. Also, plot the load line.

Solution

A half-wave current source can be configured in ADS using the n–tone current source, as shown in Figure 9E.3. The Fourier series of the half-wave current source,

$$i(t) = \begin{cases} \cos\omega t & -\dfrac{\pi}{2} < \omega t < \dfrac{\pi}{2} \\ 0 & \dfrac{\pi}{2} < \omega t < \dfrac{3\pi}{2} \end{cases}$$

can be expressed as

$$i(t) = \sum_{n=-\infty}^{\infty} I_n e^{jn\omega t}$$

$$I_0 = \frac{1}{\pi}$$

$$I_1 = \frac{1}{4}$$

$$I_n = \frac{1}{\pi(1-n^2)}, \quad n = 3,5,\cdots$$

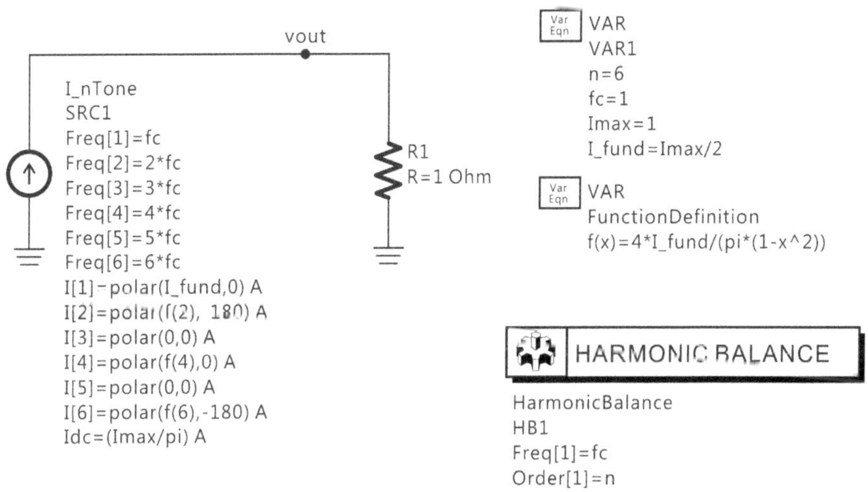

Figure 9E.3 Half-wave current source configuration. The half-wave current source is implemented using a Fourier series.

It should be noted that the Fourier series uses the two-sided spectra while the phasor uses only the positive frequency, that is, the one-sided spectrum. As a result, when expressed as a phasor, and for the harmonics $n \neq 0$, the doubling of the value of I_n is required. In addition, since harmonics are, by default, represented by phasors in ADS, the harmonic phasor is represented by doubling the coefficient of the Fourier series given in the equations above. Thus, the half-wave current source can be configured as shown in Figure 9E.3. Note that I_1 (**I_fund**) in Figure 9E.3 represents the phasor value and is set to $I_{max}/2$. The values of the other harmonics are accordingly set using I_1. The output voltage **vout** across 1 Ω is used to check whether the expanded harmonic current of the Fourier series correctly produces the half-wave waveform in the time domain.

After the verification of the half-wave waveform, the output circuit of the class-B power amplifier can be configured as shown in Figure 9E.4. The current source of Figure 9E.3 is employed in the circuit of Figure 9E.4. The value of **Imax** in Figure 9E.4 is set to 1 A and the maximum value of the half-wave current source is thus 1 A. Using the maximum current value, the value of the optimum load **Ropt** can be computed to be **2Vcc** from Equation (9.12). Also, in order to filter out harmonics, the parallel resonant circuit must have a high Q. The Q value of the parallel resonant circuit is set to $Q = 10$. Using **Q** and **Ropt**, the values of the inductor and the capacitor that constitute the parallel resonant circuit can be calculated.

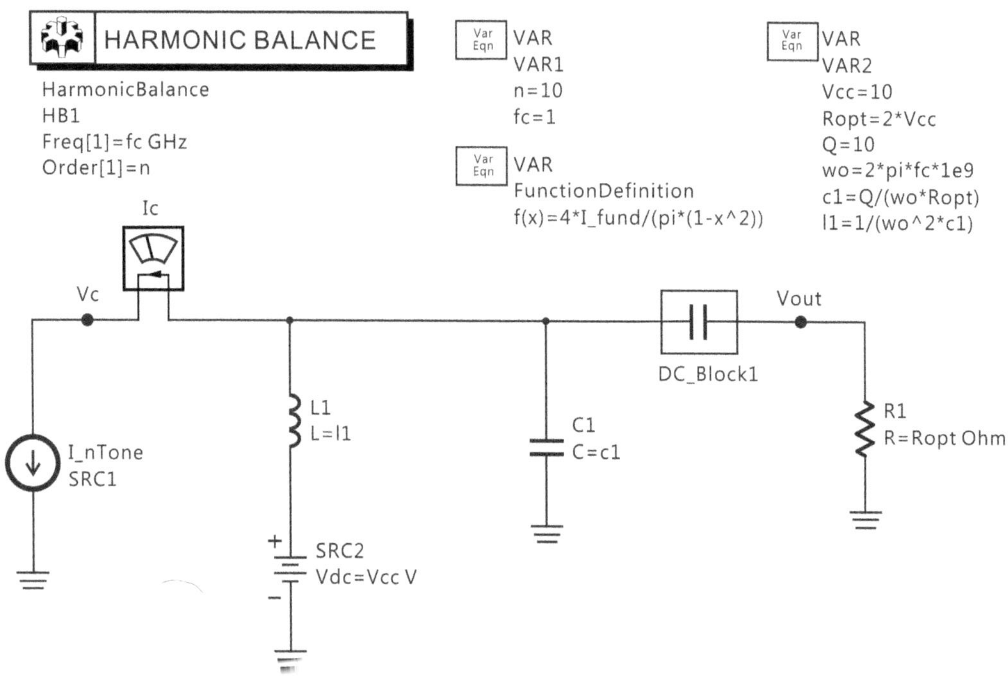

Figure 9E.4 Output circuit of a class-B power amplifier. The current source is a half-wave and **L1** and **C1** are resonant at the fundamental frequency. The Q value of the parallel resonant circuit is set to 10.

With the current and voltage waveforms shown in Figure 9E.5, plotting the current versus the voltage gives the load line, which is shown in Figure 9E.6. Since the voltage is not an exact sinusoidal waveform and has a slight distortion, the load line in Figure 9E.6 is not an exact straight line and shows a slight disparity.

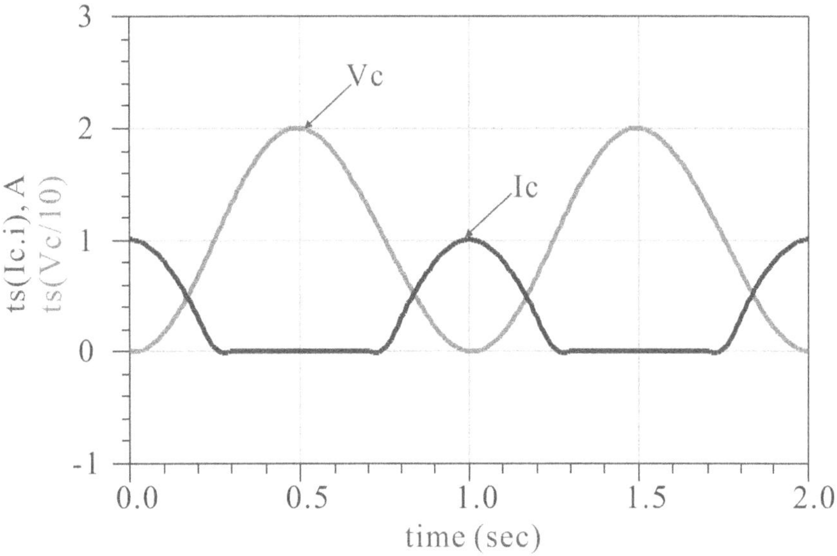

Figure 9E.5 Voltage $V_c(t)$ and current $I_c(t)$ waveforms. $V_c(t)$ is almost sinusoidal while $I_c(t)$ is a half-wave current. The function **ts**(·) generates the time-domain waveform using the set of harmonic voltages or currents.

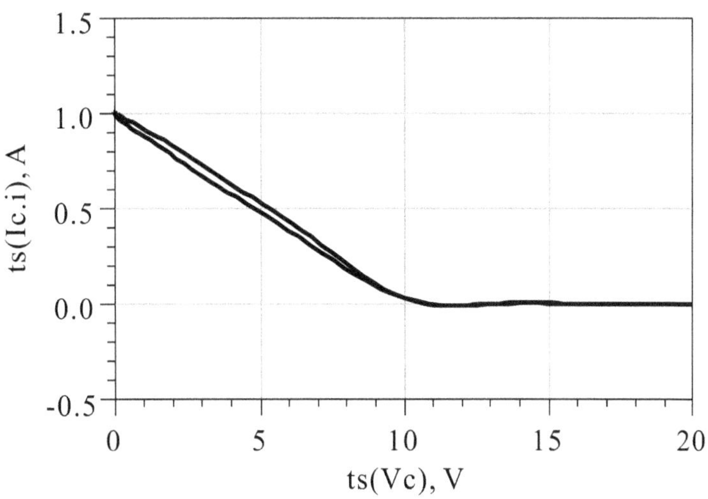

Figure 9E.6 Load line. Note that the load line is the locus of voltage $V_c(t)$ with respect to current $I_c(t)$.

In order to calculate the efficiency, the equations in Measurement Expression 9E.1 are entered in the display window.

```
Eqn Vcc=10
Eqn Ropt=2*Vcc
Eqn Idc=real(Ic.i[0])
Eqn eff=(mag(Vout[1]))**2/(2*Ropt)/(Vcc*Idc)*100
```

Measurement Expression 9E.1 Equations entered in the display window to calculate the efficiency

The calculated efficiency is 78.55%, which is close to the maximum efficiency of $\pi/4$ of a class-B power amplifier.

From Equation (9.10), the maximum efficiency is obtained when $\xi = 1$. The maximum efficiency due to variation in the conduction angle can be expressed as Equation (9.14).

$$\eta = \frac{1}{2}\frac{\gamma_1}{\gamma_0} \tag{9.14}$$

In addition, normalizing the maximum output power P, which varies with respect to the conduction angle, by the maximum power of the class-A power amplifier P_A, gives Equation (9.15).

$$\frac{P}{P_A} = \frac{2\gamma_1}{1 - \cos\theta} = \frac{2\theta - \sin 2\theta}{\pi\left(1 - \cos\theta\right)} \tag{9.15}$$

This is shown in Figure 9.24. As the conduction angle in that figure gets closer to 0, the efficiency gets closer to 100%. Also, as the conduction angle decreases, the output power decreases compared with that of the class-A power amplifier. Therefore, even though the efficiency is good, the maximum output power of the active device cannot be fully utilized. In addition, because the gain approaches 0, a significant increase in the input power is also required. Therefore, power amplifiers below class-B are not practically applicable even though they have increased efficiency.

In the previous explanation using Figure 9.22, a class-B power amplifier is described as using a half-wave current source. It is noteworthy that a sinusoidal voltage appears across the load even though the output current of the active device shows a distorted half-wave current waveform. In addition, the phase and amplitude of the sinusoidal voltage across the load are proportional to those of the sinusoidal input. This indicates that the input signal is linearly amplified. In general, however, a class-B amplifier shows the P_L–P_{in} characteristic shown in Figure 9.25. The P_L–P_{in} characteristic has an S-shape for input power that causes distortion to appear in the output. This is primarily because the I_{DS}–V_{GS} characteristic is not linear. As a result, the P_L–P_{in} characteristic has a distorted S-shape. This S-shaped P_L–P_{in} characteristic can, to some extent, be corrected to a straight line by operating the amplifier in class-AB. An accurate measure of the distortion can be achieved by measuring the distortion for a modulated input signal, which will be described later in this chapter.

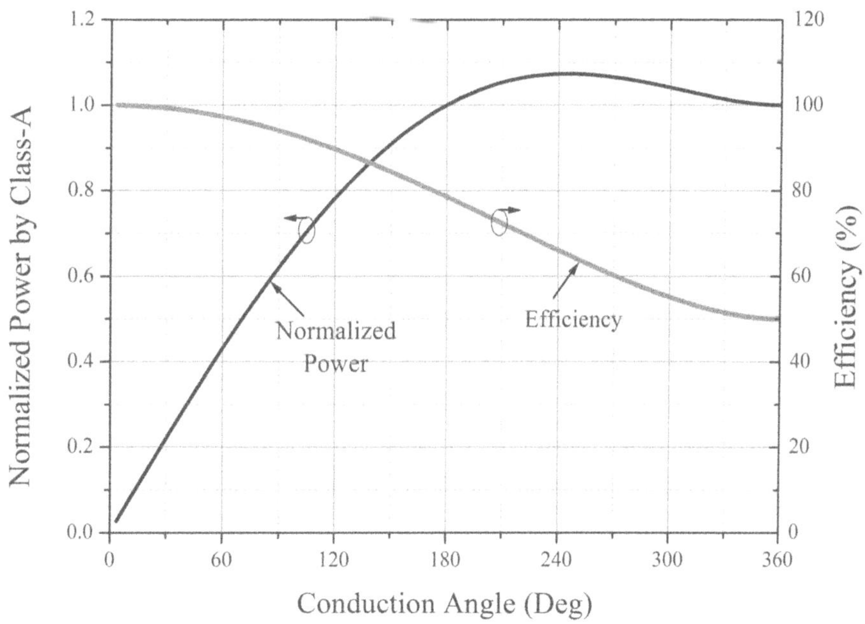

Figure 9.24 The output power and efficiency variations vs. conduction angle. As the conduction angle decreases, the efficiency increases but the output power decreases.

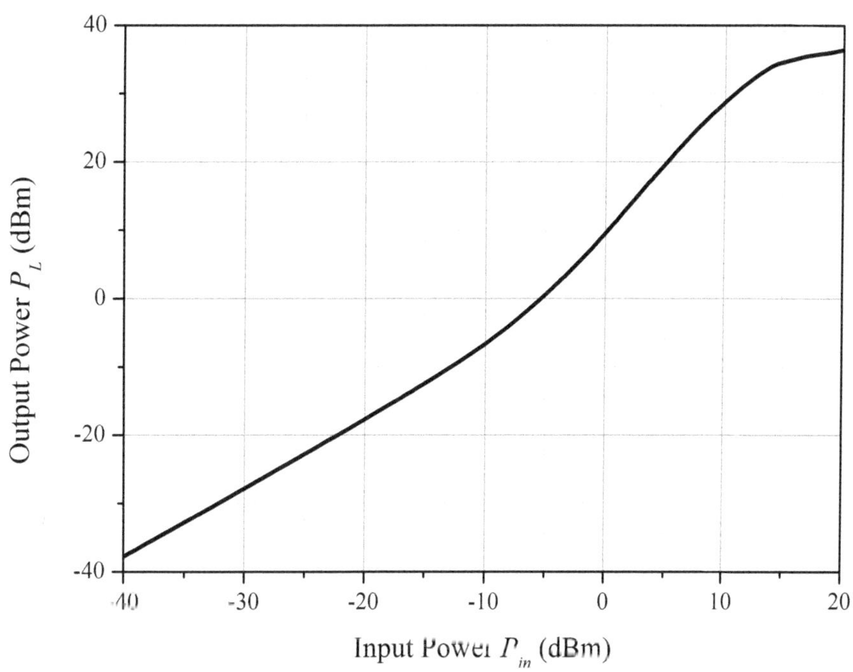

Figure 9.25 Typical output-power characteristic of a class-B amplifier for input power. Due to the nonlinearity of the I_D–V_{GS} characteristic, the S-shape curve appears.

9.4.2 Class-D Power Amplifiers

A class-A amplifier's efficiency can be significantly improved using a class-B operation. The collector or drain efficiency of the class-B operation is, however, still under 100%. A further improvement in efficiency that is closer to 100% can be achieved by operating the active device as a switch. In the class-B power amplifier, the active device operates as a transconductance-type current source that has a sinusoidal-tip-shaped waveform. When the sinusoidal input power is further increased, the active device's output can be modeled as a switch that turns on and off according to the input signal. As the input increases, the output of the active device is shorted in the positive half cycle, while that output is opened in the negative half cycle. Note that because the active device's output operates as a switch, there is no DC power consumption at that output. However, delivering a large input power to the active device can lead to real problems such as damaging the active device. Therefore, applying a square-wave input instead of a larger sinusoidal input makes the active device operate more like a switch. This can be implemented by inserting a drive circuit to convert the sine wave to a square wave. The active device then operates similarly to a switch according to the input signal. This operation is called class-D. Through the class-D operation, the drain or collector efficiency can be 100%. Table 9.2 shows the difference between the two operations when the active device is operated as a switch and as a transconductance-type current source.

Table 9.2 Power amplifier design factors when an active device is operated as a current source or switch

Characteristic	Current-Source Operation	Switch Operation
Output resistance	Always high impedance	"ON" low, "OFF" high
Active device operating in saturation	Not permitted	When "ON"
Voltage across active device while conducting current	Drain or collector voltage is higher than the specified minimum voltage	As low as can be obtained
What determines output voltage while conducting current	Depends on load circuit	0 because switch is closed
What determines the current of active device	Input voltage	Load circuit
Load-circuit design factors	Output-current waveform according to input voltage	Load-circuit response according to short- or open-circuit states of active device

An example of the class-D power amplifier circuit is shown in Figure 9.26, where inductor L operates as an RFC and allows only DC current. The DC current value is assumed to be I_o. Capacitor C_∞ acts as a DC block and allows only DC voltage, which is intuitively determined to be V_{DD}. Thus, when the active device is driven into the off state by the input signal, the voltage across the drain v_D is given by

$$v_D = V_{DD} + I_L R_L$$

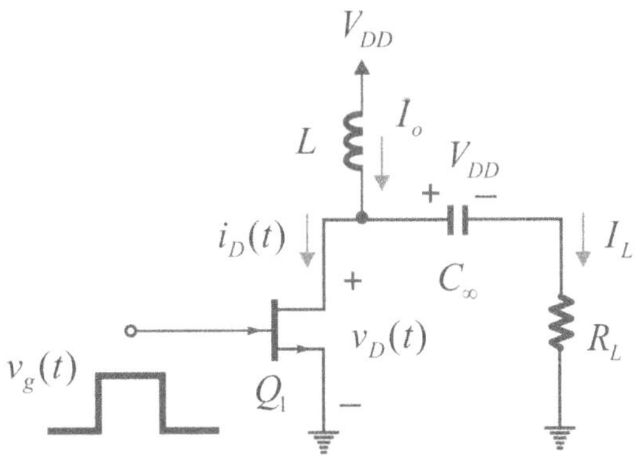

Figure 9.26 Class-D power amplifier circuit. Here, L and C_∞ are an RFC and a DC block. The application of the square wave makes the transistor operate as a switch.

In addition, when the active device is on, v_D is 0 and

$$v_D = 0 = V_{DD} + I_L R_L$$

Thus, the drain voltage alternates from 0 to $V_{DD} + I_L R_L$ according to the active device's state, which changes from on to off. In contrast, when the active device is on, I_L the current flowing through load R_L is obtained as

$$I_L = -\frac{V_{DD}}{R_L}$$

and the drain current becomes

$$i_D = I_o - I_L$$

Note that when the device is off, the drain current is 0, and when the device is on, the drain current has a value of $I_o - I_L$. Since the DC component must be I_o, $I_o - I_L = 2I_o$, then

$$I_o = -I_L = \frac{V_{DD}}{R_L}$$

The resulting voltage and current waveforms are shown in Figure 9.27.

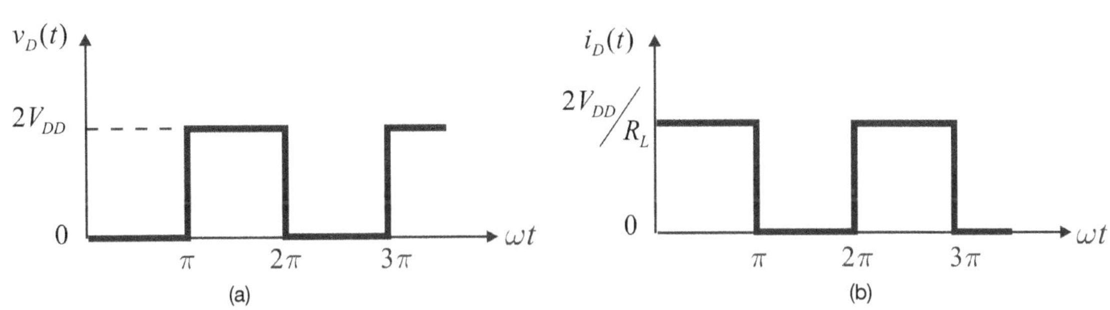

Figure 9.27 The output waveform of a class-D amplifier; (a) drain voltage and (b) drain current

Example 9.5

Using the switch provided in ADS, verify the class-D operation of the circuit in Figure 9.26.

Solution

A class-D power amplifier using the switch is as shown in Figure 9E.7. The drive signal is a square wave that varies between 1 and –1 with respect to time as the **Vdc** = 0 of the square-wave source. The load value is set to **Ropt** and the resistance of the switch for –1 V is set to 1 MΩ and for 1 V to **Ropt*R_load_factor**. Note that **R_load_factor** is set to 0.01 Ω, which is sufficiently small compared with **Ropt**.

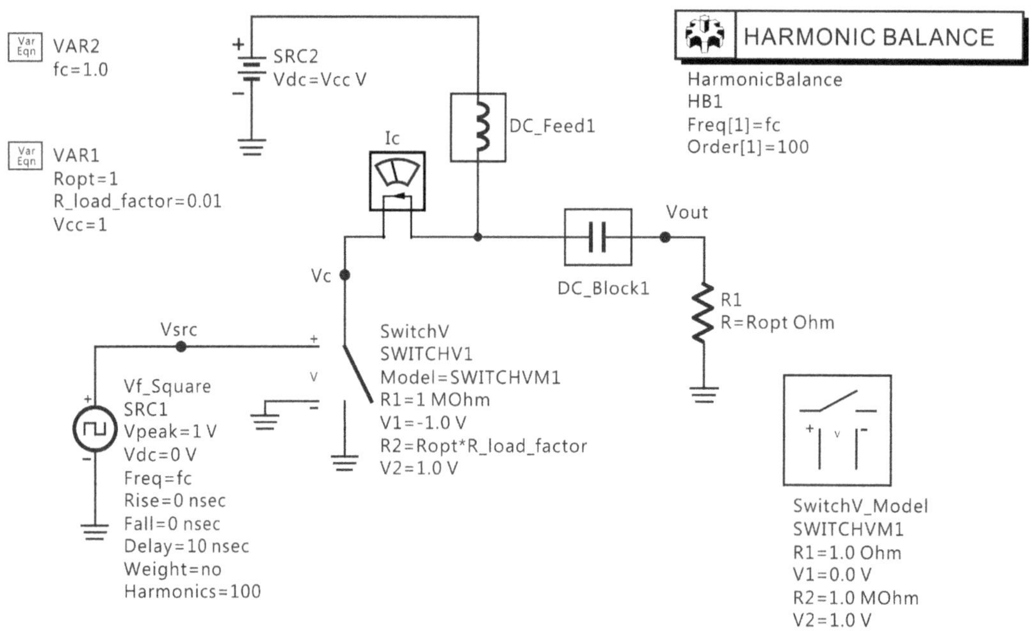

Figure 9E.7 A class-D power amplifier simulation. The switch is modeled by **SwitchV**, which has a resistance 1 MΩ at –1 V **Vsrc** input and has an on-resistance of **Ropt*R_load_factor** at 1 V **Vsrc** input.

The simulated waveforms of this setup are shown in Figure 9E.8. In that figure, when the voltage is maximum, the current is 0 and vice versa. The power consumption of the DC voltage source **SRC2** can be seen to be 1 W. Since the output power is 1 W, the efficiency can be seen to be 100%.

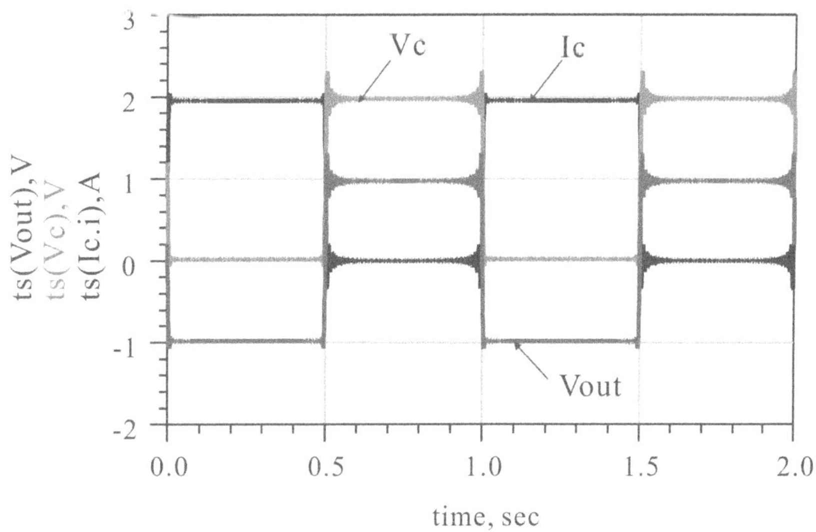

Figure 9E.8 Simulation results; (a) drain current and voltage, and (b) output voltage. Note that the average DC value of the load voltage **Vout** is 0. At the edge of the waveforms, ringing occurs due to Gibb's phenomenon.

With the class-D operation, the active device's output switches on/off according to the phase of the input signal. However, even though the phase information of the input signal is preserved at the amplifier output, the amplifier shows some losses to the amplitude information. The technique for solving this problem is that the amplitude and phase of the input voltage are separated using signal processing, and a square wave that varies according to the phase of the input signal is used as the driving input for the amplifier. The amplitude information can be restored by varying the supply voltage V_{DD} in Figure 9.26 according to the amplitude of the input signal. Since the amplitude of the output signal of the class-D amplifier is related only to the supply voltage, the amplitude and phase information of the input signal can be restored at the amplifier output without any loss.

Even after solving the problem of the loss of the amplitude information, a significant amount of harmonics appears at the output, which presents another problem. A separate filter is required to remove the harmonics. There are several efficient ways of dealing with the harmonic problem of the class-D power amplifier. The voltage and current switching technique shown in Figure 9.28 is one of the ways to deal with the harmonic problem; see reference 3 at the end of this chapter for more information.

In general, the efficiency of a practical class-D power amplifier seldom reaches 100%. One reason for the degraded efficiency is that the output voltage and current of the active device have finite rising and falling times, and a non-zero voltage appears when the device is conducting. Similarly, a non-zero current appears when the device is off. In addition, the charge on the drain capacitor, which is charged to maximum during the device's off state, flows instantly to the drain when the device enters the on state. This sometimes causes permanent damage to the device and loss in the efficiency.

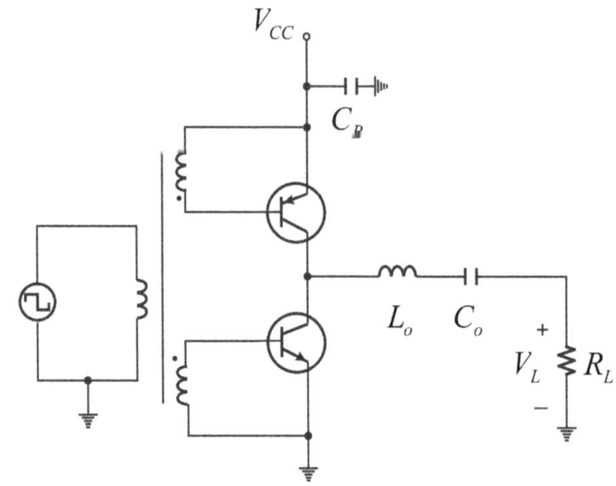

Figure 9.28 Voltage switching in a class-D power amplifier with a harmonics-eliminating filter. For high-state input, the *npn* transistor turns on and the current flows from the load R_L to the *npn* transistor. For low-state input, the *pnp* transistor turns on the current flows from the *pnp* transistor to the load. The series resonant circuit eliminates the harmonics of the current and the sinusoidal voltage appears at the load.

9.4.3 Class-E Power Amplifiers

The active device for a class-E amplifier, just as in a class-D amplifier, is operated as a switch. A patent for a class-E amplifier was applied for at the time when the class-E operation principle was still under development and was not yet a well-defined concept. The correct operation principle for the class-E amplifier was first analyzed by N. Sokal and A. Sokal in 1975, when they also presented its design approach.[4] In addition, as there was a clear difference between the operation of the class-D and class-E amplifiers, the Sokals defined that operation as a class-E operation. The class-E power amplifier circuit developed by the Sokals is shown in Figure 9.29. However, there are various class-E amplifiers in addition to the class-E amplifier shown in that figure.

The configuration of the class-E input circuit shown in Figure 9.29 is the same as that for a class-D amplifier and the active device operates as a switch, which periodically turns on and off. Because of this, a transient analysis that depends on the on/off states is necessary to explain the class-E amplifier. Collector capacitance may exist in the collector terminal. In this case, the collector capacitance can be included in the capacitor C. DC current is supplied to the collector through the RFC, which is denoted as I_o. The series resonator L_o–C_o eliminates harmonics and is set to series resonate at the fundamental frequency. The Q of the series resonant branch is assumed to be sufficiently high, and harmonic currents other than those at the fundamental frequency cannot flow. Inductor L is added in series to the resonator; when the inductor is included, it makes the series resonant circuit resonate at a frequency lower than the fundamental (usually the resonator is said to be *detuned*). The class-E power amplifier generally has an optimized efficiency for the detuned series resonator, and it is analyzed by placing a separate L in series with the series resonator

4. N. O. Sokal and A. D. Sokal, "Class E-A New Class of High-Efficiency Tuned Single-Ended Switching Power Amplifiers," *IEEE Journal of Solid-State Circuits* 10, no. 3 (June 1975): 168–176.

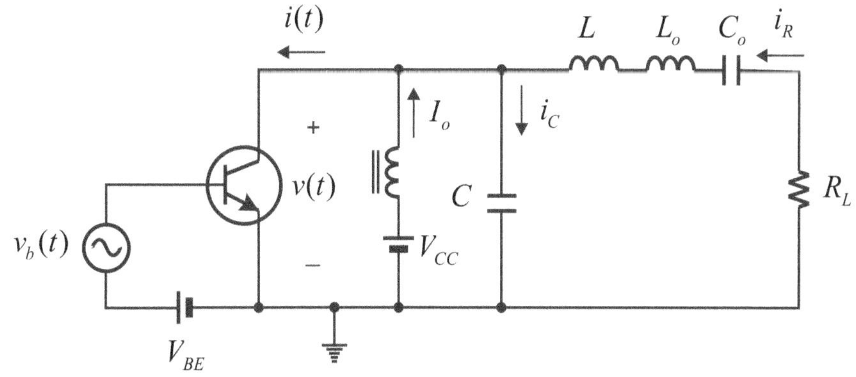

Figure 9.29 Class-E power amplifier configuration. The input $v_b(t)$ is a square wave and causes the transistor to operate as a switch. L_o and C_o are resonant at the fundamental frequency. Note that the load circuit is not resonant at the fundamental frequency.

resonating at the fundamental frequency. It must be noted that the contribution of the inductor L appears only at the fundamental frequency and has no effects on the harmonics.

The output circuit of Figure 9.29 can be analyzed by considering the active device's output as a switch. When the switch is opened, because a current $I_o + i_R$ flows through capacitor C, the capacitor voltage rises and thus the collector voltage increases. On the other hand, when the switch is closed, the accumulated charge in the capacitor instantaneously flows through the switch in the form of an impulse current. In this case, even if there is a small voltage on the switch, power loss in the switch increases. This condition is not desirable for optimized efficiency. If the switch is open for the interval of $0.5T \leq t \leq T$, the following must be satisfied at $t = T$ for optimum efficiency, as expressed in Equation (9.16):

$$v(T) = v_C(T) = 0 \tag{9.16}$$

In addition to the impulse current due to the capacitor, a current $I_o + i_R$ flows into the collector when the switch is closed. Although this is not an impulse-type current as in the case of the capacitor, it can cause a current with a step discontinuity. As this current flows instantaneously into the collector, it also leads to a loss of efficiency. In order to make this loss 0, Equation (9.17) must be satisfied.

$$\left.\frac{dv_C(t)}{dt}\right|_{t=T} = \left.\frac{dv(t)}{dt}\right|_{t=T} = 0 \tag{9.17}$$

The conditions of this equation mean that the capacitor current $i_C = I_o + i_R$ should be zero just before the switch is turned on. Thus, at the instant when the switch is closed, the current does not flow and the discontinuity of the current does not appear. These two conditions become the optimum conditions of the class E power amplifier.

As a first step, the values of L and C that satisfy the optimum conditions for a given load R and supply voltage V_{CC} should be determined through analysis. Then, the power delivered to the load P_L and the DC current I_o can be computed. Other parameters that must be determined include the maximum value of $v(t)$, V_{max}, and the maximum value of $i(t)$, I_{max}. Using the determined V_{max} and I_{max}, the active device appropriate for this class-E power amplifier can be selected.

Equivalent circuits based on the switch states are shown in Figure 9.30. In Figure 9.30(a), because the series resonator is assumed to have a high Q, the current through load R can be represented by $i_R(t) = I_R \sin(\omega t + \phi)$. Therefore $i(t)$ can be expressed as

$$i(t) = I_o + i_R = I_o + I_R \sin(\omega t + \phi) \tag{9.18}$$

Since $i(0) = 0$, Equation (9.18) can be rewritten as

$$i(t) = I_R\left(\sin(\omega t + \phi) - \sin\phi\right) \tag{9.19a}$$

$$I_o = -I_R \sin\phi \tag{9.19b}$$

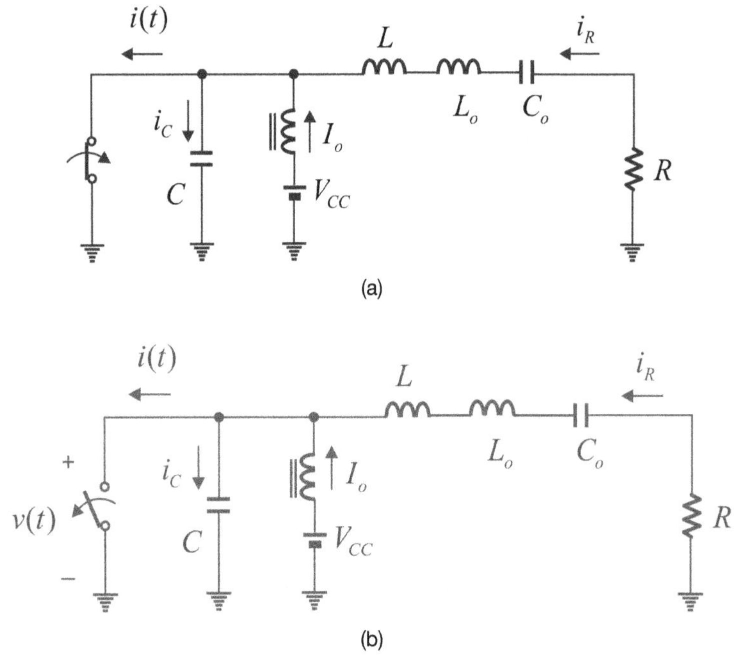

Figure 9.30 Equivalent circuit based on switch states: (a) switch is closed and (b) switch is open

The current given by Equation (9.19) for the closed switch flows into the capacitor when the switch is open. Therefore, the voltage across the capacitor is obtained as shown in Equation (9.20).

$$\begin{aligned} v_C(t) = v(t) &= \frac{1}{\omega C}\int_{\pi}^{\omega t} i(\omega t)\, d\omega t \\ &= -\frac{I_R}{\omega C}\left\{\cos(\omega t + \phi) + \cos\phi + (\omega t - \pi)\sin\phi\right\} \end{aligned} \tag{9.20}$$

Applying $v_C(T) = 0$,

$$\phi = -\tan^{-1}\left(\frac{2}{\pi}\right) = -32.482° \tag{9.21}$$

Using tan(ϕ) given in Equation (9.21), sinϕ and cosϕ can be obtained as expressed in Equation (9.22).

$$\sin\phi = \frac{-2}{\sqrt{\pi^2 + 4}}, \qquad \cos\phi = \frac{\pi}{\sqrt{\pi^2 + 4}} \tag{9.22}$$

Then, substituting sinϕ and cosϕ into Equation (9.20) and rewriting, we obtain Equation (9.23).

$$v(t) = \frac{I_o}{\omega C}\left\{\omega t - \frac{3}{2}\pi - \frac{\pi}{2}\cos\omega t - \sin\omega t\right\} \tag{9.23}$$

Since the average voltage of $v(t)$ should be V_{CC}, the following relationship is obtained with Equation (9.24):

$$V_{CC} = \frac{1}{2\pi}\int_0^{2\pi} v(\omega t)d\omega t = \frac{I_o}{\pi\omega C} \tag{9.24}$$

Therefore, normalizing $v(t)$ by V_{CC}, we get Equation (9.25).

$$v_n(t) = \frac{v(t)}{V_{CC}} = \pi\left\{\omega t - \frac{3}{2}\pi - \frac{\pi}{2}\cos\omega t - \sin\omega t\right\} \tag{9.25}$$

In addition, normalizing $i(t)$ by I_o and rewriting Equation (9.18), we obtain Equation (9.26).

$$i_n(t) = \frac{i(t)}{I_o} = \frac{\pi}{2}\sin\omega t - \sin\omega t + 1 \tag{9.26}$$

The normalized waveforms of $v_n(t)$ and $i_n(t)$ are shown in Figure 9.31.

By differentiating Equation (9.25), the maximum voltage is obtained and expressed in Equation (9.27).

$$V_{max} = -2\pi\phi V_{CC} = 3.562V_{CC} \tag{9.27}$$

Also, by using the relationships shown in Equations (9.19b) and (9.22) where the efficiency is 100%, the power delivered to the load and DC current I_o is obtained as expressed in Equations (9.28) and (9.29).

$$P_L = V_{CC}I_o = \frac{1}{2}I_R^2 R \tag{9.28}$$

$$I_o = \frac{V_{CC}}{R}\frac{8}{\pi^2 + 4} = 0.577\frac{V_{CC}}{R} \tag{9.29}$$

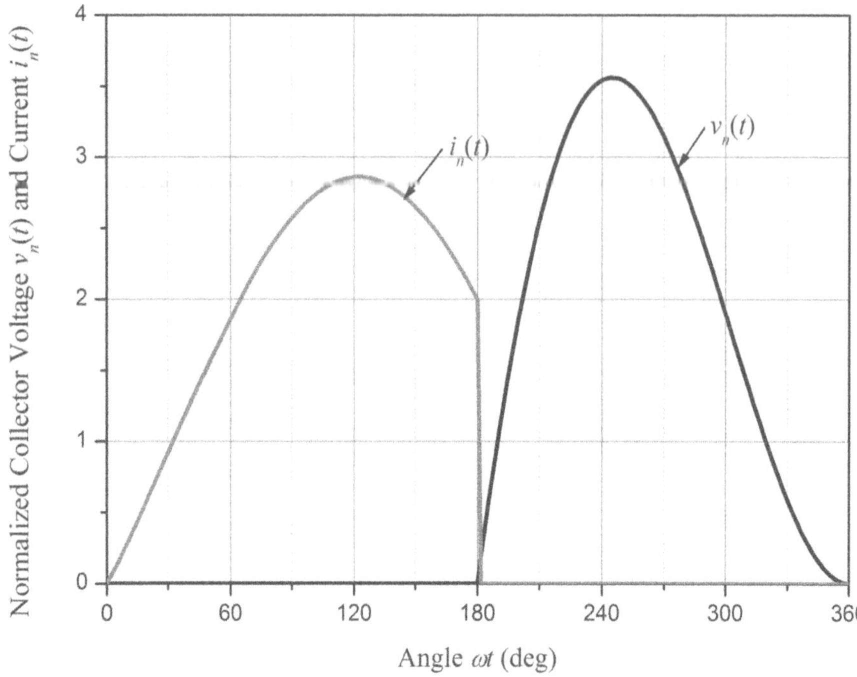

Figure 9.31 Normalized drain current and voltage waveforms. The drain voltage is continuous while the drain current shows an abrupt transition. When $i_n \neq 0$, $v_n = 0$, and when $v_n \neq 0$, $i_n = 0$. Thus, the transistor does not dissipate power and the resulting efficiency is 100%.

The maximum current $I_{\max}$ is obtained by differentiating Equation (9.26) and rewriting as Equation (9.30).

$$I_{\max} = \left(\frac{\sqrt{\pi^2 + 4}}{2} + 1 \right) I_o = 2.8621 I_o \tag{9.30}$$

Also, applying Equations (9.19b) and (9.22) to Equation (9.28), we obtain Equation (9.31).

$$R = \frac{8}{\pi^2 + 4} \frac{V_{CC}^2}{P_L} = 0.5768 \frac{V_{CC}^2}{P_L} \tag{9.31}$$

However, the values of L and C are still not determined. From the voltage in Equation (9.25), the fundamental component of the voltage is the sum of the voltage drops across the resistor R and inductor L, V_R, and V_L. Therefore V_R and V_L can be determined as

$$V_R = -\frac{1}{\pi} \int_0^{2\pi} v(\omega t) \sin(\omega t + \phi) \, d\omega t$$

$$V_L = -\frac{1}{\pi} \int_0^{2\pi} v(\omega t) \cos(\omega t + \phi) \, d\omega t$$

The ratio of these results is expressed in Equation (9.32).

$$\frac{\omega L}{R} = \frac{V_L}{V_R} = 1.1525 \tag{9.32}$$

In addition, by using Equations (9.24) and (9.29) we obtain Equation (9.33).

$$\omega CR = 0.1836 \tag{9.33}$$

A class-E amplifier can be designed using the equations obtained above. Table 9.3 summarizes the formulas required for the design.

Table 9.3 Design formulas for class-E power amplifier

Item	Equation
Active device's maximum voltage	$V_{max} = 3.562V_{CC}$
Active device's maximum current	$I_{max} = 1.651\frac{V_{CC}}{R}$
Optimum load value	$R = 0.5768\frac{V_{CC}^2}{P_L}$
DC current	$I_o = 0.577\frac{V_{CC}}{R}$
Inductor	$\frac{\omega L}{R} = 1.1525$
Capacitor	$\omega CR = 0.1836$

Example 9.6

Verify the waveform of the class-E amplifier shown in Figure 9.31 by using a switch that varies according to the drive signal, and determine the load line and efficiency.

Solution

The configuration of the equivalent circuit diagram of the class-E power amplifier output is shown in Figure 9E.9. In the figure, **R_load, L_main** and **C_main** are set using the equations in Table 9.3, and the Q of the series resonant circuit is set to 20.

Simulation of the circuit in Figure 9E.9 gives the collector voltage and current shown in Figure 9E.10. It can be seen that the waveforms are the same as those calculated theoretically. Note that the voltage appearing at the load in Figure 9E.10 is close to a sinusoidal voltage, although the collector voltage and current are significantly distorted.

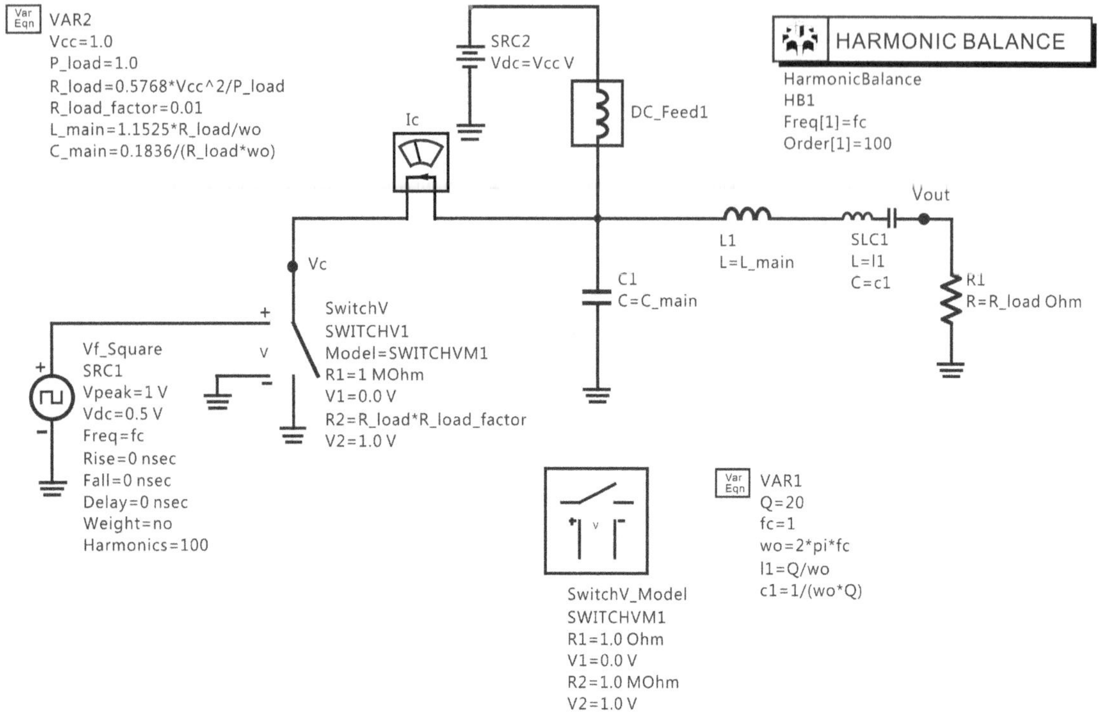

Figure 9E.9 Class-E amplifier simulation schematic. The component **SLC1** is a series resonant circuit at the fundamental frequency. The value of **L_main** and **C_main** are determined using the equations in Table 9.3. The component **SwitchV** acts similarly to its function in the class-D power amplifier.

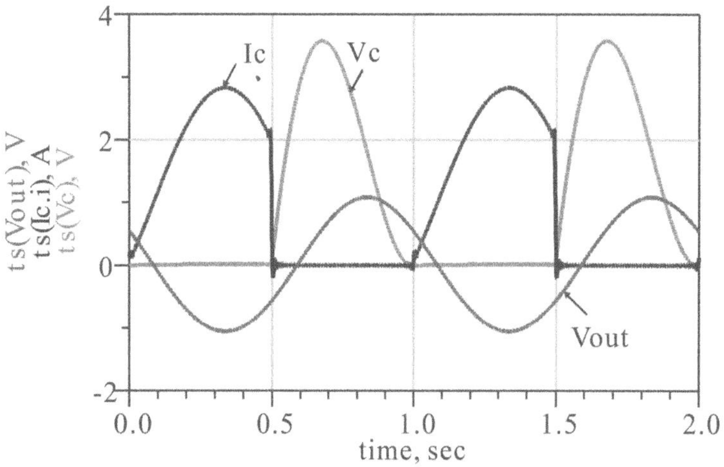

Figure 9E.10 Simulated collector voltage $V_c(t)$, current $I_c(t)$, and the load voltage $V_{out}(t)$. Note that due to the series resonant circuit, an almost sinusoidal voltage appears at the load.

Next, the load line is shown in Figure 9E.11. The load line is considered to be the trajectory between time domain current *i*(*t*) and voltage *v*(*t*). From Figure 9E.11, the collector voltage appears when the collector current is 0 and vice versa. Thus, the active device operates as a switch and the efficiency can be seen to be 100%.

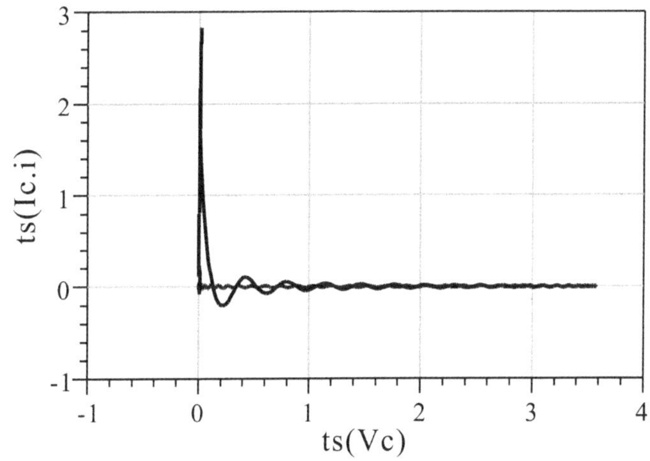

Figure 9E.11 Load line of a class-E load circuit. The load line is the locus of $I_C(t)$ versus $V_c(t)$.

The first characteristic feature of the load circuit of the class-E amplifier is that the load circuit does not resonate at the fundamental frequency as previously explained. However, due to the series resonator in the load circuit, only sinusoidal output appears at the load. The second feature is that the current flows only when the voltage is zero, while the voltage appears only when the current is zero due to the switch operation of the active device. An amplifier with these features is a class-E amplifier and therefore the class-E amplifier is very diverse. The load circuit of the class-E amplifier explained earlier is simple and easy to implement. However, for other types of class-E load circuits that differ from that in the example, the collector voltage and current waveforms should be computed for a new optimum load circuit. The computation of the optimum load circuit becomes more complex when components such as transmission lines are used. In this case, *the class-E load circuit can be approximately designed by setting the impedance of the load network at each harmonic frequency to have the same impedance as that of the class-E amplifier* in Figure 9.29. For more details, refer to reference 3 at the end of this chapter.

Another feature of the class-E amplifier is that the ratio between the average and the peak values of the collector voltage or current waveform, as shown in the amplifier example, is greater than 2. Therefore, the active device used for a class-E amplifier should have a very high breakdown voltage and a maximum current. That is the disadvantage of the class E amplifier: the active device's maximum output capacity is not efficiently exploited.

In addition, in the calculations for the class-E amplifier, the active device was assumed to be a switch. Thus, in order to operate the active device as a switch for a class-E operation, either a driving circuit must be inserted or sufficiently high power must be applied to the input. When the input is not sufficiently high, the active device begins to operate as a half-wave current source similar to a class-B operation. Since the class-E operation is achieved when the device operates as a switch, the class-E operation is not optimum for the transconductance-type current source operation in class-B. Thus, the amplifier designed for a class-E operation does not show an optimum class-B operation. When the input drive is not appropriate, the class-E amplifier will show degradation in terms of efficiency compared to the class-B operation previously described.

9.4.4 Class-F Power Amplifiers

A class-F power amplifier is categorized by its load circuit characteristics. The load circuit is composed of multiple parallel resonators that resonate at odd harmonics in series with a class-B load circuit. The multiple parallel resonators in the output load circuit improve the efficiency of the power amplifier. Class-F power amplifiers have a longer history than do class-E amplifiers. When the parallel resonator that resonates at the third harmonic frequency is inserted in series with the load circuit of a class-B power amplifier, as shown in Figure 9.32, the collector voltage waveform can include the third harmonic frequency components. The inclusion of the third harmonic frequency makes the collector voltage waveform increasingly flat around the maximum and minimum values. A flatter waveform leads to increased efficiency.

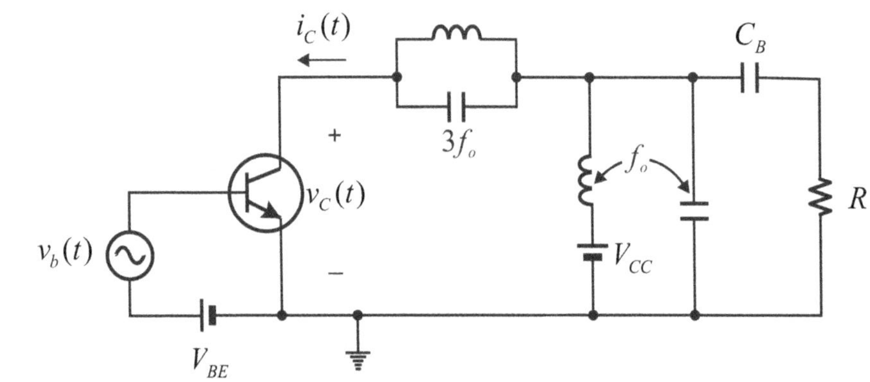

Figure 9.32 Class-B power amplifier with the third-order harmonic peaking circuit

Inserting an infinite number of odd harmonic parallel resonators in series, the collector voltage waveform becomes closer to a square wave and the efficiency becomes 100%. Instead of these multiple-harmonic resonators, Tyler presented a class-F power amplifier that uses a quarter-wavelength transmission line at the fundamental frequency, as shown in Figure 9.33. All of the amplifier's harmonics in that figure can be ideally controlled and the class-F operation is achieved. The input drive circuit in Figure 9.33 acts as it does in a class-B power amplifier. The capacitor C_B in the load side is a DC block

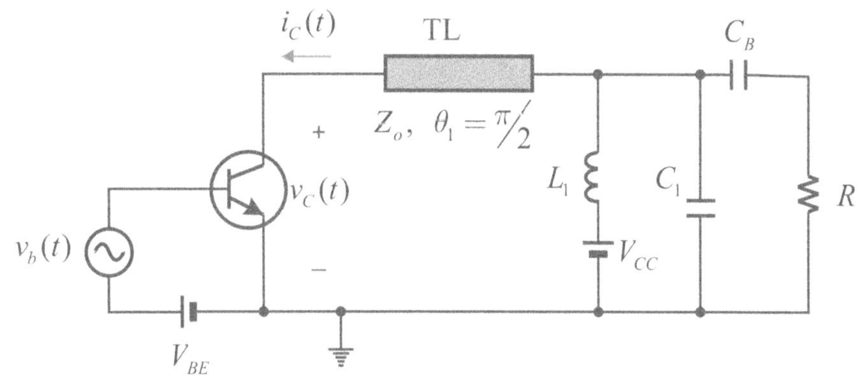

Figure 9.33 Class-F power amplifier implemented using a one-quarter-wavelength transmission line TL. L_1 and C_1 are parallel resonant at the fundamental frequency. C_B is a DC block capacitor.

capacitor. Inductor L_1 and capacitor C_1 form a parallel resonant circuit that resonates at the fundamental frequency. Thus, assuming a high Q, only the fundamental voltage appears across the load. The transmission line is one-quarter-wavelength long and the impedance seen from the collector toward the load at the fundamental frequency is $(Z_o)^2/R$. At even-order harmonics, since the transmission line becomes an integer multiple of a half wavelength, the impedance seen from the collector toward the load appears similar to that of a short circuit, while at odd-order harmonics, it appears similar to an open circuit due to the characteristic of the one-quarter-wavelength transmission line. Therefore, the one-quarter-wavelength transmission line provides a short-circuit impedance at the collector for even-order harmonics, and it provides an open-circuit impedance at odd-order harmonics. Thus, when the circuit in Figure 9.33 is compared to that shown in Figure 9.32, it acts as the infinite number of parallel resonators inserted in series to resonate at odd-order harmonics.

Since the impedance is an open circuit at odd harmonics, voltages of odd harmonic components can appear at the collector; however, because the impedance is a short circuit for even harmonics, voltages of even-order harmonic components cannot appear at the collector. In contrast, the opposite is the case for the current. Thus, setting $\theta = \omega t$ and expressing the collector voltage and current waveforms in a Fourier series can be represented by Equations (9.34a) and (9.34b).

$$v_C(\theta) = V_{CC} + V_1 \sin\theta + V_3 \sin 3\theta + \cdots \tag{9.34a}$$

$$i_C(\theta) = I_{DC} - I_1 \sin\theta - I_2 \cos 2\theta - I_4 \cos 4\theta - \cdots \tag{9.34b}$$

Figure 9.34 shows the expected collector voltage and current waveforms because, as mentioned earlier, the collector voltage waveform of a class-B power amplifier around the minimum and maximum is expected to be flattened due to the parallel resonators.

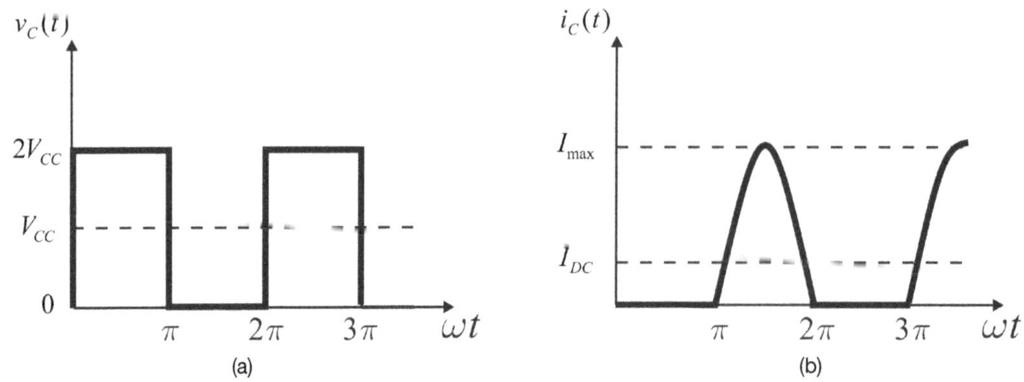

Figure 9.34 Output waveforms of a class-F power amplifier; (a) collector voltage and (b) collector current

---

Example 9.7

Model the active device as a switch and obtain the collector voltage, the current waveforms, and the load voltage waveform of the class-F amplifier shown in Figure 9.33. Also, show the trajectory of the load line.

Solution

When the active device is modeled as a switch, the class-F amplifier of Figure 9.33 can be represented as shown in Figure 9E.12.

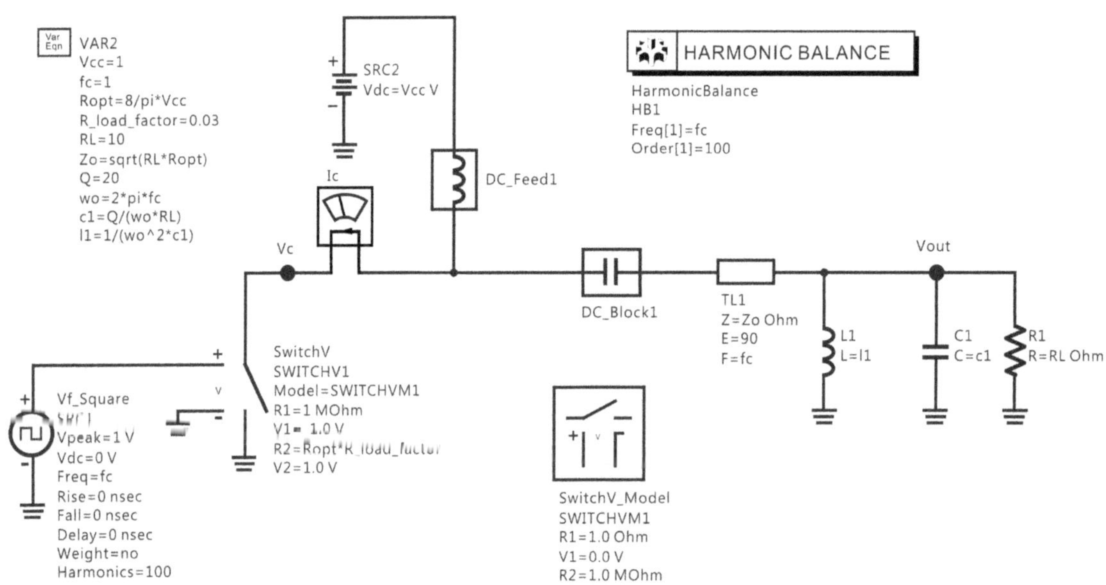

Figure 9E.12 A class-F power amplifier simulation schematic. **L1** and **C1** are parallel resonant at the fundamental frequency. The optimum load resistance **Ropt** is set to $8V_{CC}/\pi$. The transmission line has a quarter wavelength and its **Zo** is set to satisfy the relationship **Ropt** = **Zo**^{2}/**RL**.

Figure 9E.13 shows the collector voltage and current waveforms, and the load voltage waveform obtained from the simulation. The collector voltage waveform in that figure can be found to approximate a square wave, while the collector current has the shape of a half wave. Some distortion in the collector voltage waveform appears as a result of the approximate switch implementation; this is due to the on-resistance of the switch. See reference 3 at the end of this chapter for more information. However, the load voltage shows an almost sinusoidal waveform due to the high Q parallel resonator connected to the load, which is set to 20. Similar to a class-E amplifier, when the voltage is not 0, the current is 0; when the current is not 0, the voltage is approximately 0 and the efficiency is close to 100%. If the on-resistance of the switch is 0, the efficiency ideally becomes 100%.

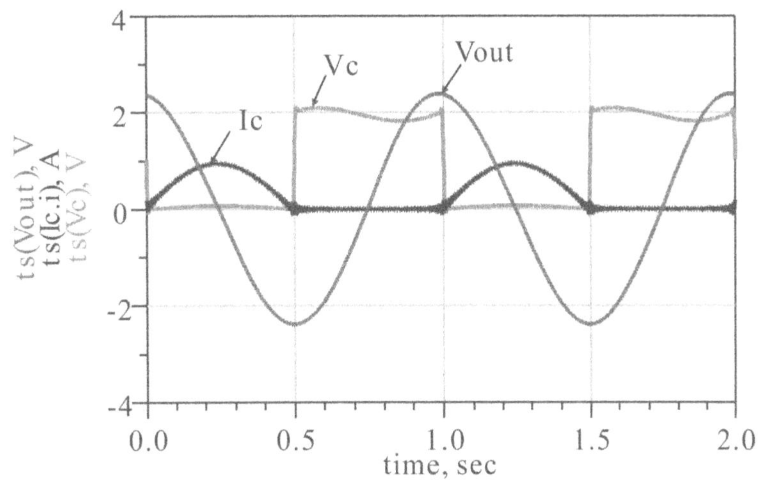

Figure 9E.13 $V_c(t)$, $I_c(t)$, and $V_{out}(t)$ obtained from the simulation

The trajectory of the load line is shown in Figure 9E.14. The load line is similar to that of class-E amplifier. However, due to the on-resistance, a slight slope occurs when the switch is on. The load line also shows distortion to some degree in the switch's transition from off to on.

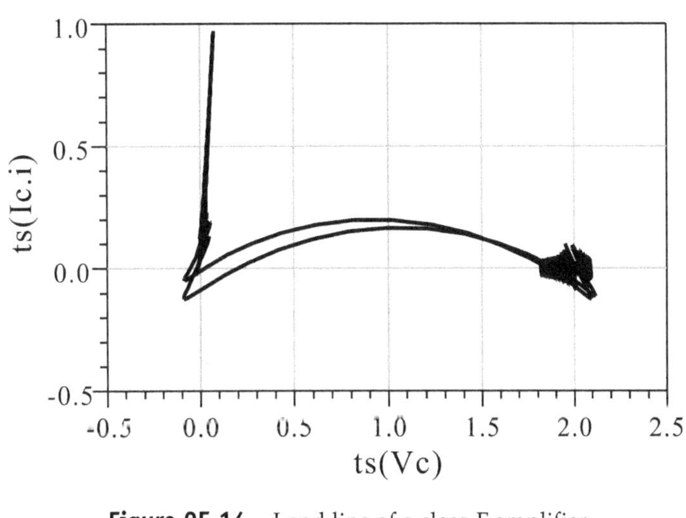

Figure 9E.14 Load line of a class-F amplifier

In order to design a class-F power amplifier, it is necessary to take into account the impedance of the load circuit at each harmonic. This can be determined by dividing the harmonic voltage by the corresponding harmonic current shown in Equation (9.34). Then, the load circuit of the ideal class-F power amplifier should be designed to provide the computed load impedances. The load impedance must be designed to have the optimum resistance R_{opt} at the fundamental frequency. For harmonics, the impedance value must be 0 at even harmonic frequencies and ∞ at odd harmonic frequencies.

An ideal class-F load circuit can be constructed using a transmission line and a parallel resonator that resonates at the fundamental frequency shown in Figure 9.33. However, because the impedance of the transmission line used to implement the class-F load circuit is generally too low, problems can arise when implementing an ideal class-F load circuit. Thus, instead of the load circuit in Figure 9.33, the class-F load circuit is generally made to satisfy the class-F load impedance condition for a limited number of harmonics, as it is generally difficult to satisfy the class-F load conditions for all harmonics. In the case of the harmonic-limited class-F load circuit, the collector voltage and current waveforms given by Equation (9.34) are reduced to those with limited harmonics. Raab has summarized the limited-harmonics waveforms and their efficiencies.[5,6] This is not a simple truncation of the Fourier series in Equation (9.34) that retains only a limited number of harmonics. The simple truncation of the Fourier series results in the well-known Gibbs phenomenon and the resulting waveform with the Gibbs phenomenon is not acceptable for the output waveform of a power amplifier. Thus, the waveform with a limited number of harmonics will differ from that obtained by truncating higher-order harmonics using Equation (9.34). Raab proposed two waveforms: maximally flat waveforms and maximum efficiency waveforms. In this section, we will review the maximum efficiency waveforms. To obtain these waveforms, the drain voltage and current are assumed not to take on negative values. To demonstrate this process, a voltage waveform example up to the third-order harmonic can be written as shown in Equation (9.35).

$$v(\theta) = V_{CC} + \sin\theta + b\sin 3\theta \tag{9.35}$$

Here, the amplitude of the fundamental component has been normalized to 1. The minimum and maximum values can be obtained by taking the derivative of the equation above.

$$\frac{dv(\theta)}{d\theta} = \cos\theta + 3b\cos 3\theta = \cos\theta\left(1 + 12b\cos^2\theta - 9b\right) \tag{9.36}$$

Solving Equation (9.36), we obtain Equation (9.37).

$$\cos^2\theta_m = \frac{3}{4} - \frac{1}{12b} \tag{9.37}$$

5. F. H. Raab, "Maximum Efficiency and Output of Class-F Power Amplifiers," *IEEE Transactions on Microwave Theory and Techniques* 49, no. 6 (June 2001): 1162–1166.

6. F. H. Raab, "Class-F Power Amplifiers with Maximally Flat Waveforms," *IEEE Transactions on Microwave Theory and Techniques* 45, no. 11 (November 1997): 2007–2012.

Substituting θ_m into Equation (9.35), the value of $v(\theta)$ at the minimum is obtained. The value should be $v(\theta) \geq 0$. Thus $v(\theta_m) = 0$ for the maximum swing, and it results in

$$V_{CC} = -\sin\theta_m - b\sin 3\theta_m \tag{9.38}$$

V_{CC} in Equation (9.38) becomes a function of b. When the value of b is adjusted to obtain the minimum value,

$$\frac{dV_{CC}}{db} = 0 \tag{9.39}$$

From Equation (9.39), the minimum amplitude of the third harmonic is obtained. When the determined b is substituted into Equation (9.37), θ_m can be determined. The value of V_{CC} then can be obtained by substituting b and θ_m again into Equation (9.38). Finally, the normalized amplitudes of the fundamental and third harmonic by the value of V_{CC} can be obtained. This process can be extended to higher-order harmonics. In the case of the current waveform, using the similar calculation, the maximum efficiency waveform can be calculated from Equation (9.34) for up to the harmonic order of m–th. The results are summarized in Tables 9.4 and 9.5 where n represents the harmonic number of the voltage, while m represents the harmonic number of the current. The maximum values of voltage and current waveforms v_{max} and i_{max} in Tables 9.4 and 9.5 are normalized values using the values of the DC components, and defined as shown in Equations (9.40a) and (9.40b).

$$\delta_V = \frac{v_{max}}{V_{CC}} \tag{9.40a}$$

$$\delta_I = \frac{i_{max}}{I_{DC}} \tag{9.40b}$$

In addition, the values of V_1 and I_1 at the fundamental frequency are also normalized using the DC values as expressed in Equations (9.41a) and (9.41b).

$$\gamma_V = \frac{V_1}{V_{CC}} \tag{9.41a}$$

$$\gamma_I = \frac{I_1}{I_{DC}} \tag{9.41b}$$

Table 9.4 Maximum efficiency voltage waveform

n	$\delta_V = v_{max}/V_{CC}$	$\gamma_V = V_1/V_{CC}$	V_3/V_1	V_5/V_1
1	2	1	–	–
3	2	1.1547	0.1667	–
5	2	1.2071	0.2323	0.0607
∞	2	$4/\pi = 1.273$	$4/(3\pi) = 0.424$	$4/(5\pi) = 0.255$

Table 9.5 Maximum efficiency current waveform

m	$\delta_I = i_{max}/I_{DC}$	$\gamma_I = I_1/I_{DC}$	I_2/I_1	I_4/I_1
1	2	1	–	–
2	2.9142	1.4142	0.3540	–
4	3.000	1.5000	0.3890	0.0556
∞	$\pi = 3.142$	$\pi/2 = 1.571$	$2/3 = 0.667$	$2/15 = 0.133$

Figure 9.35 shows the voltage and current waveforms plotted using the results of Tables 9.4 and 9.5. In Figure 9.35, as the number of harmonics increases, the waveform is found to approach that of the ideal class-F power amplifier. In addition, slight ripples are observed to occur around the maximum and minimum values of the voltage in Figure 9.35. However, the ripples do not occur in the maximally flat waveforms.

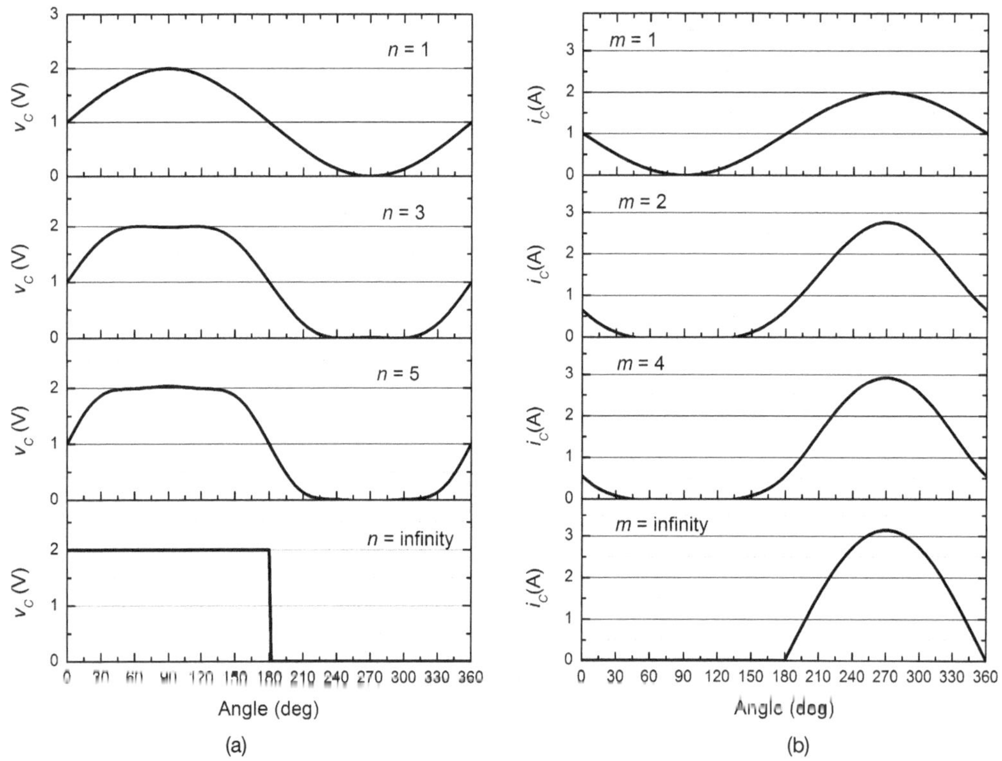

Figure 9.35 Maximum efficiency (a) voltage and (b) current waveforms of class-F power amplifier with the increase in the number of harmonics

Using the values obtained in Tables 9.4 and 9.5, the efficiency, optimum impedance, and output power capability can be calculated. First, the optimum impedance R_{opt} at the fundamental frequency is computed as shown in Equation (9.42).

$$R_{opt} = \frac{V_1}{I_1} \tag{9.42}$$

Then, the optimum impedance of the ideal class-F amplifier R_{opt} at the fundamental frequency can be expressed as Equation (9.43).

$$R_F = \frac{V_1}{I_1} = \frac{4}{\pi}\frac{V_{CC}}{\frac{\pi}{2}I_{DC}} = \frac{4}{\pi}\frac{V_{CC}}{\frac{\pi}{2}\times\frac{1}{\pi}I_{max}} = \frac{8}{\pi}\frac{V_{CC}}{I_{max}} \tag{9.43}$$

With the optimum impedance R_{opt}, the output power delivered to the load is shown in Equation (9.44).

$$P_o = \frac{1}{2}\frac{V_1^2}{R_{opt}} = \frac{1}{2}\frac{(\gamma_V V_{CC})^2}{R_{opt}} \tag{9.44}$$

In addition, since the DC power consumption is represented by Equation (9.45),

$$P_{dc} = V_{CC} I_{DC} = \frac{\gamma_V V_{CC}^2}{\gamma_I R_{opt}} \tag{9.45}$$

the drain efficiency is computed as Equation (9.46).

$$\eta_D = \frac{\gamma_V \gamma_I}{2} \tag{9.46}$$

This computation is summarized in Table 9.6.

Table 9.6 Efficiency according to the number of harmonics

Current Harmonics (m)	Voltage Harmonic (n)			
	1	**1,3**	**1,3,5**	**1,3,5,...,∞**
1	0.5	0.5774	0.6033	$2/\pi = 0.637$
1,2	0.7071	0.8165	0.8532	0.9003
1,2,4	0.7497	0.8656	0.9045	0.9545
1,2,4,...,∞	$\pi/4 = 0.785$	0.9069	0.9477	1

Finally, the power capability of the active device can be assessed using $v_{max}i_{max}$. The output power of the class-F amplifier given in Equation (9.44) cannot fully utilize $v_{max}i_{max}$, which leads to the definition of the output power capability M_p as

$$M_p = \frac{P_o}{v_{max}i_{max}} = \frac{\gamma_V \gamma_I}{\delta_V \delta_I} \tag{9.47}$$

The output power capability given by Equation (9.47) can generally be applied to various types of power amplifiers and can be used as the criteria for assessing the effectiveness of their operations in exploiting an active device's given output power capability. The M_p of the class-E amplifier is generally low compared to that of the class-F amplifier.

Example 9.8

For the amplifier shown in Figure 9E.15, the supply voltage is $V_{CC} = 10$ V and the maximum current is $I_{max} = 1$ A. Using Tables 9.4, 9.5, and 9.6, determine the optimum impedance R_{opt}, maximum output power, and efficiency.

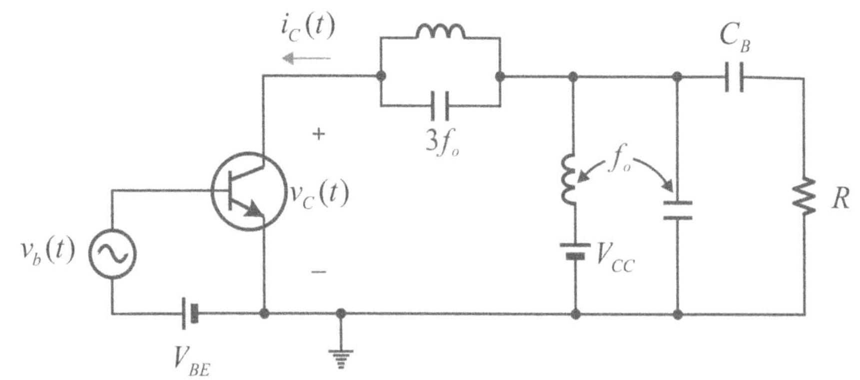

Figure 9E.15 A class-F power amplifier built from a class-B power amplifier through the third-order harmonic peaking circuit

Solution

From Figure 9E15, the impedance seen from the collector at f_o is found to be R, and at $3f_o$ it is found to be ∞, but at other harmonics the impedance is 0. Therefore, only the voltage harmonic 1, 3 can appear at the collector. In contrast, the fundamental and all even-harmonic currents can flow toward the load, which corresponds to the case of $m = 1, 2, 4, \ldots,$ and ∞. Thus, the maximum efficiency from Table 9.6 is about 90.7%. Also, in this case the ratio of the fundamental voltage to the DC voltage from Table 9.4 is

$$\frac{V_1}{V_{CC}} = 1.1547$$

Thus, $V_1 = 11.55$ V and, in the case of the current, since $\delta_I = \pi$, $I_{DC} = 1/\pi$ A. Therefore, the fundamental current is

$$I_1 = \gamma_I I_{DC} = \frac{1}{\pi} \times \frac{\pi}{2} = 0.5 \text{ A}$$

Thus,

$$R_{opt} = \frac{11.55}{0.5} = 23.1 \text{ ohm}$$

$$P_{max} = \frac{1}{2}\frac{V_1^2}{R_{opt}} = \frac{(11.55)^2}{2 \times 23.1} = 2.89 \text{ W}$$

In Example 9.7, the ideal class-F waveforms were explained by modeling the output of the active device as a switch. However, the switch model of the active device cannot explain the class-F waveforms of limited harmonics shown in Figure 9.35. The non-zero voltage appears at the time when the switch is on, and the non-zero current appears at the time when the switch is off. Thus, a reasonable model for the active device will be the transconductance-type current source model employed in the explanation of the class-B operation. The increased peak value of the sinusoidal-tip-shaped current will drive the active device output into saturation, and the clipped collector voltage appears as a result. Due to clipping, the class-F waveforms of limited harmonics can appear according to the class-F load circuit.

Example 9.9

For the power amplifier circuit in Example 9.8, set up the output circuit of the active device as a half-wave current source. For V_{CC} and I_{DC}, as in Example 9.8, plot the collector voltage and current waveforms. In addition, calculate the efficiency and compare it with the previously obtained results. Also, plot the load line and discuss the problems associated with assuming a half-wave current source.

Solution

Figure 9E.16 is a schematic for the class-F load circuit with the third harmonic peaking in Example 9.8. The n-Tone source **I_nTone** is the half-wave current source. Here, the Q of the parallel resonant circuit at the fundamental frequency is set to 20 and that of the third harmonic is set to 30. In addition, the load impedance is set to **Ropt** = 23.1 Ω. The diode in the circuit is inserted to prevent the collector voltage from becoming negative. In addition, the diode ideality factor is set to 0.1 so as to be as close to an ideal diode as possible.

The simulated waveforms are shown in Figure 9E.17. From that figure, the collector voltage waveform around the maximum and minimum values is flattened as expected, and the collector current can also be seen to maintain a half-wave current shape. In addition, the voltage waveform across the load is shown in the figure. Although the collector voltage and current show significant distortions, the load voltage is almost close to a sine wave.

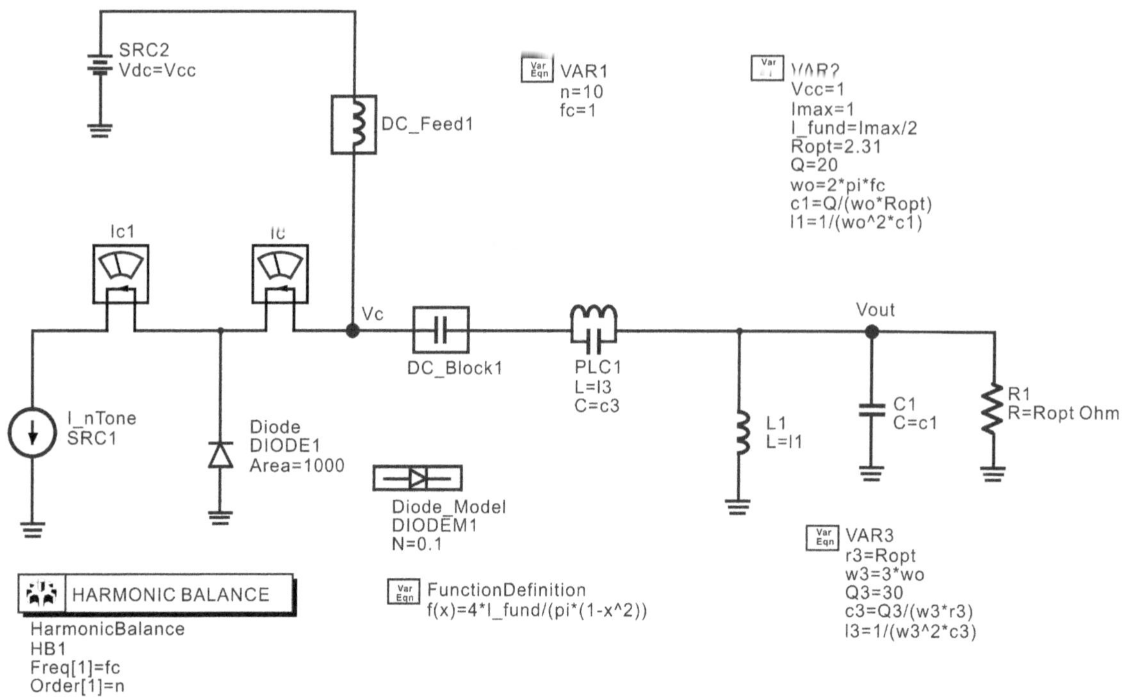

Figure 9E.16 Circuit for the simulation of the third harmonic peaking in the class-F power amplifier. **SRC1** is a half-wave current source. The diode will approximately model an ideal diode and it prevents the voltage of **SRC1** from being negative.

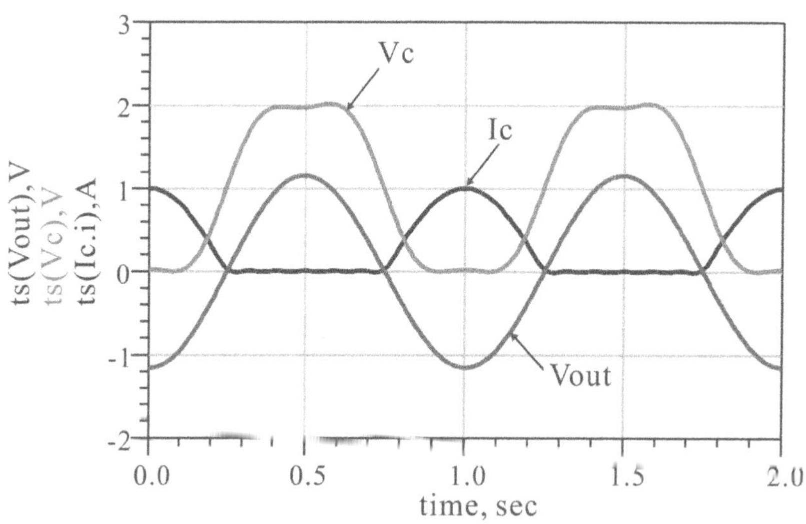

Figure 9E.17 Simulated collector voltage **Vc**, current **Ic**, and voltage across the load **Vout**

The equations shown in Measurement Expression 9E.2 are entered in the display window to calculate the collector efficiency and the delivered power to the load.

```
Eqn Vcc=10
Eqn Ropt=23.1
Eqn Idc=real(Ic.i[0])
Eqn eff=(mag(Vout[1]))**2/(2*Ropt)/(Vcc*Idc)*100
Eqn PL=0.5*mag(Vout[1])**2/Ropt
```

Measurement Expression 9E.2 Equations for calculating the drain efficiency and output power

The calculated collector efficiency is 90.7% and the delivered power is 2.89 W. These are the same results as those calculated in Example 9.8. Figure 9E.18 shows the plot of the load line.

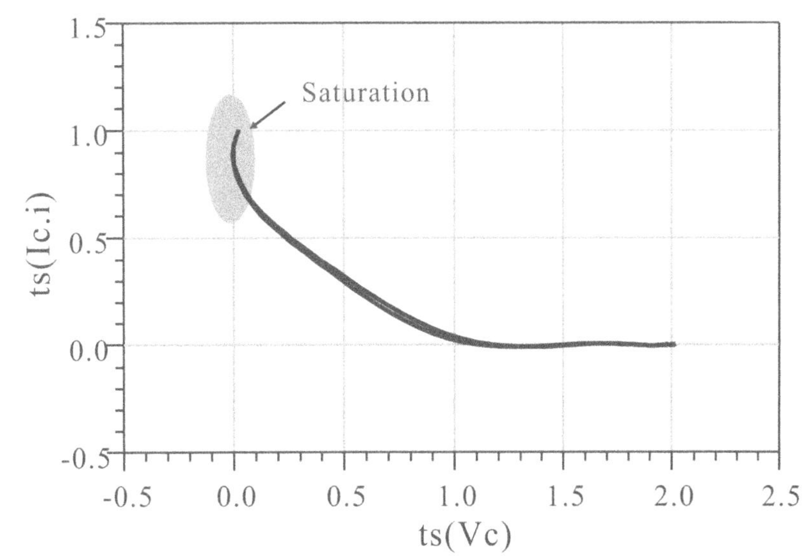

Figure 9E.18 Load line of the third harmonic peaking in a class-F power amplifier output circuit

The shaded area shown in Figure 9E.18 is where the collector voltage is close to 0. In this case, it is difficult, in reality, for the half-wave current source of Figure 9E.17 to maintain the half-wave shape. As the collector voltage is close to 0, the active device is in the saturation region, and the collector current cannot maintain its half-wave shape but instead drops to zero. Therefore, the calculated collector efficiency may not be achieved. This problem is not unique to this case, and it appears for most of the class-F waveforms shown in Figure 9.35. Therefore, in reality, implementing the class-F waveform with a finite number of harmonics may lead to problems.

9.5 DESIGN EXAMPLE

In this section, we demonstrate a design example for a compact 4-W class-F power amplifier module that operates in the WiMAX frequency band of 2.3–2.7 GHz.[7] The selected active device is a TGF2023-01 GaN HEMT chip device from TriQuint with a gate length and width of 0.25 μm and 1.25 mm, respectively. The shape of the active device is shown in Figure 9.36, where the source terminal of the GaN HEMT is connected to the ground at the bottom and the gate and drain terminals appear at the top. The drain supply voltage is selected as 28 V.

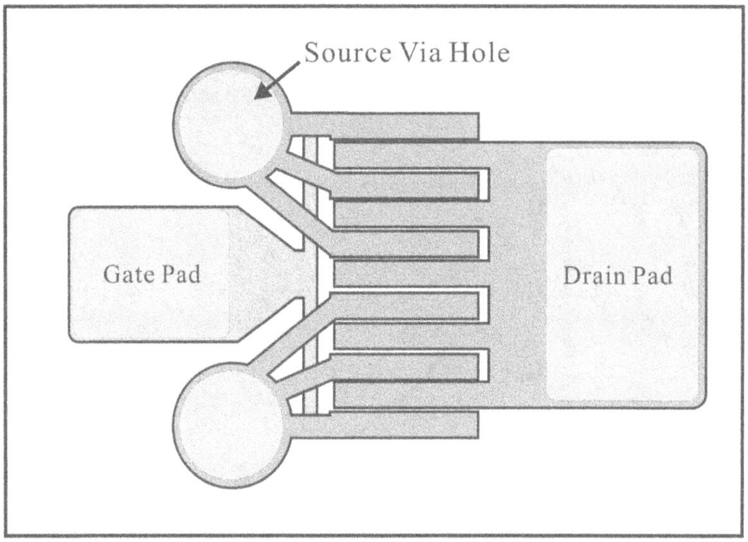

Figure 9.36 TriQuint's GaN HEMT TGF2023-01[8]

The characteristics of TGF2023-01 are shown in Table 9.7. Since the saturated output power is 38 dBm at a drain voltage of 28 V with a quiescent drain current of 125 mA, the device is considered to be adequate for the design goal of the 4 W above.

TriQuint provides the large-signal model of the TGF2023-01, which is shown in Figure 9.37. The S-parameter data components at the gate, source, and drain terminals in the large-signal equivalent circuit are those from the EM-simulated data of the source via, drain, and gate pads supplied by TriQuint. The large-signal equivalent circuit shown in Figure 9.37 is configured using the EEHEMT model whose required values are provided in the ADS discrete file format from TriQuint since the values to be entered are so many. The discrete file format consists of variable names and their corresponding values. The variable names of the discrete format file are meant to be the same as those in the EEHEMT model. The DAC (data access component) named **eehemt_dac** and shown in Figure 9.37 is

7. H-C Jeong, H-S Oh, and K-W Yeom, "A Miniaturized WiMAX Band 4-W Class-F GaN HEMT Power Amplifier Module," *IEEE Transactions on Microwave Theory and Techniques* 59, no. 12 (December 2012): 3184–3194.

8. TriQuint semiconductor, TGF2023-01 6 Watt Discrete Power GaN on SiC HEMT, November 2009; available at www.triquint.com/products/p/TGF2023-2-01.

Table 9.7 Performances of TGF2023-01

Item	TGF2023-01
Frequency range	DC–18 GHz
DC bias	Vd = 28–35 V, Vg = –3.6 V
PAE (Max)	55%
Technology	0.25 μm power GaN on SiC
Saturated power	38 dBm (6 W)
Power gain	10 dB @ 10 GHz
Quiescent DC current (Idq)	125 mA
DC current at saturated power (Idrive)	400 mA
Chip dimension	0.82 × 0.66 × 0.10

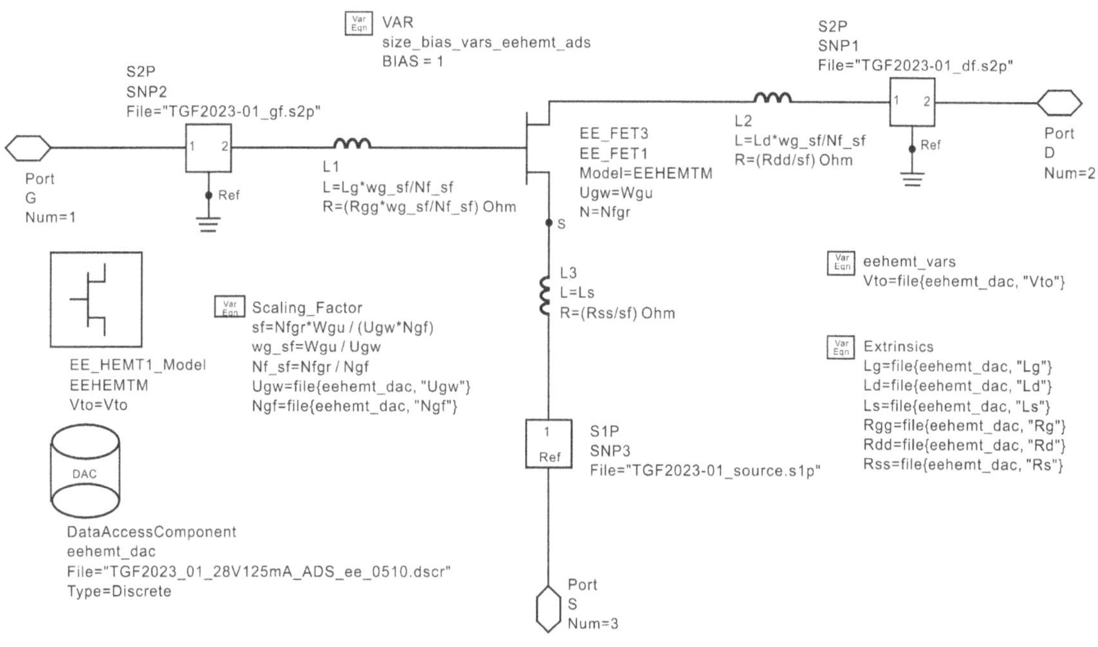

Figure 9.37 Large-signal **EEHEMT** model of TGF2023-01.[9] S-parameter data components at the gate, source, and drain are EM-simulated data for the gate, source, and drain pads. The EEHEMT model values in Table 9.8 are read into DAC **eehemt_dac** and those read values are again used to define the variables with the same names using the variable block named **eehemt_vars**. The resistances and inductances at the gate, drain, and source are separately scaled through the variable block named **Scaling_Factor**.

9. TriQuint Semiconductor, TriQuint EEHEMT Model Implemented in ADS and AWR for TQT 0.25 μm 3MI GaN on SiC Process 1.25 mm Discrete FET: 30 V @ 100 mA/mm @ 10 GHz, December 2009.

used to read the discrete file format. Then, the variable values of **eehemt_dac** are read into the variable block named **eehemt_vars** using the expression **Vto=file{eehemt_dac, "Vto"}**. Now the variable values in **eehemt_dac** are represented by the schematic window variables that have the same name as those in the discrete format file. Only one equation of **Vto** is shown in **eehemt_vars** and other variables are hidden to avoid congestion. Then, the variables defined by **eehemt_vars** are again used as the input for the EEHEMT model named **EEHEMTM**, in which all the variables are similarly hidden except one, **Vto**. The extrinsic inductors and resistors appearing at the gate, drain, and source are appropriately scaled using the number of fingers and gate width. The component **Scaling_Factor** specifies the scaling. Table 9.8 lists the values of the EEHEMT model parameters.

Table 9.8 Performances of TGF2023-01

Variable	Value	Variable	Value	Variable	Value	Variable	Value	Variable	Value
Vto	-4.623	Vtso	-10.0	Gammaac	0.002	C11th	1.497p	Idsoc	1.834
Gamma	1.03 m	Is	83.2 f	Vdeltac	0.0	Vinfl	-3.8938	Ugw	125
Vgo	-3.114	N	4.210	Gmmaxac	0.3077	Deltgs	0.6863	Ngf	10
Vdelt	0.0	Ris	0.9969	Kapaac	0.0248	Deltds	0.2636	Vco	-1.146
Vch	7.8966	Rid	1.6957	Peffac	23.7	Lambda	600u	Vba	0.1424
Gmmax	0.2688	Tau	4.049p	Vtsoac	-10.0	C12sat	58..9f	Vbc	0.2587
Vdso	28	Cdso	0.262p	Gdbm	25.0u	Cgdsat	73.6f	Mu	7.23u
Vsat	7.8770	Rdb	75.6M	Kdb	39.2m	Kbk	1.0m	Deltgm	0.2254
Kapa	0.0164	Cbs	35.7p	Vdsm	60.0	Vbr	125	Deltgmac	0.3961
Peff	337.13	Vtoac	-4.52	C11o	1.98p	Nbr	2	Alpha	0.201
Rd	0.711	Rs	0.097	Rg	0.432				
Ld	14.0p	Ls	4.56p	Lg	-9.38p				

9.5.1 Optimum Input and Output Impedances

Prior to the load-pull simulation, the gate voltage giving the quiescent drain current of 125 mA at the drain voltage of 28 V in Table 9.7 is determined from a DC simulation using the large signal equivalent circuit. The drain current versus gate voltage is plotted as shown in Figure 9.38. A DC drain current of approximately 125 mA is obtained at a gate voltage of $V_{GS} = -3.57$ V.

The gate voltage is thus determined as −3.57 V, and a drain voltage of 28 V is applied. A problem arises because the circuit is first configured to be stable before the load-pull simulation. The stability investigation requires S-parameter simulation. But before the S-parameter simulation, the S-parameters of an appropriate bonding ribbon for the connection of the active device's gate and drain should be predetermined because they appear as fixed components in assembly. After the 3D-simulation of the bonding ribbons with a 3-mil width using AnSoft HFSS, the bonding ribbon S-parameters are determined. They

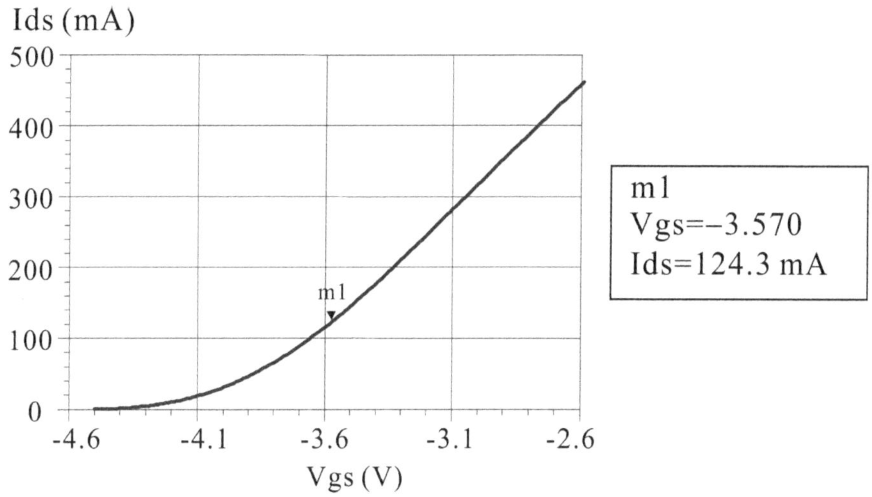

Figure 9.38 Simulated I_{DS}–V_{GS} characteristic at a drain voltage of 28 V. The quiescent current of 125 mA is achieved at **Vgs** = –3.57 V.

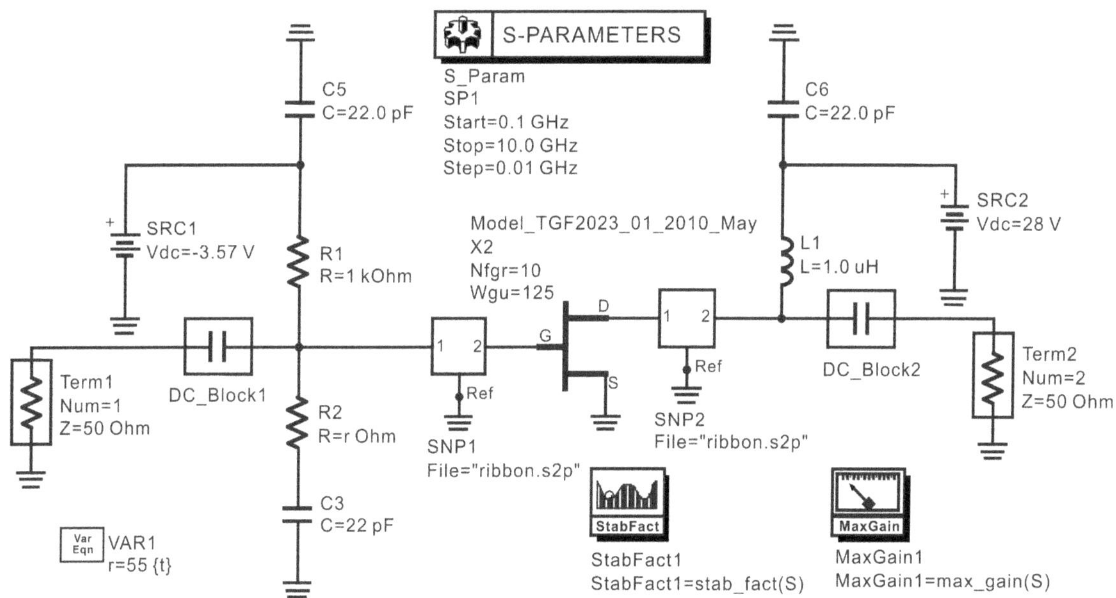

Figure 9.39 Schematic for a stability simulation. The two S-parameter components at the gate and drain represent those of bonding ribbons simulated using HFSS. The value **r** of resistor is tuned to stabilize the TGF2023.

appear as the S-parameter data components in Figure 9.39. The *RC* circuit connected in shunt to the gate is to ensure the stabilization of the active device. By adjusting the resistance value **r**, the stability factor can be adjusted to be greater than 1 in the full frequency band. For the value **r** = 55 Ω, the circuit is stabilized for a full frequency band of up to 10 GHz. Figure 9.40 shows the simulated stability and maximum gain. Since the stability

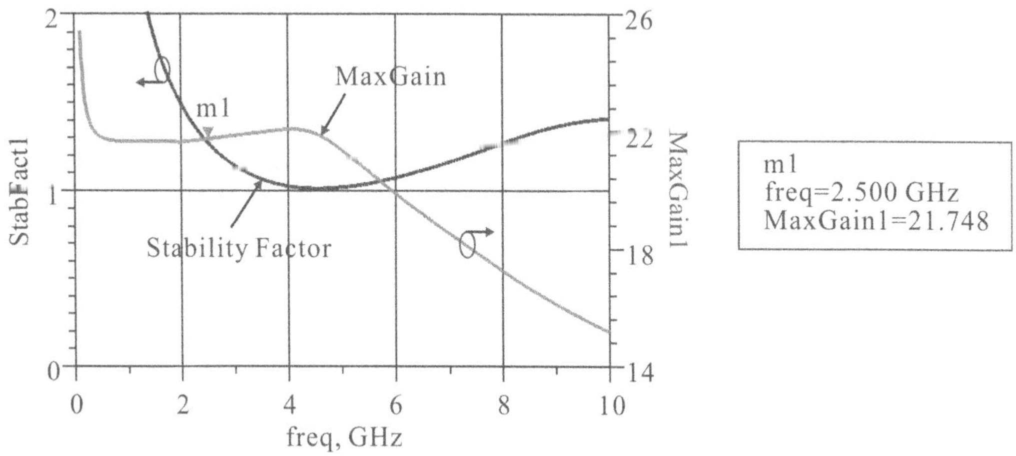

Figure 9.40 Stability simulation results at **r** = 55 Ω. The stability factor is above 1 up to 10 GHz and **MaxGain**1 at 2.5 GHz is about 21.7 dB.

factor is larger than 1 up to a frequency of 10 GHz, the device is found to be stable and the maximum gain at the center frequency of 2.5 GHz is approximately 21 dB.

After the stability circuit is thus configured, the load-pull simulation shown in Figure 9.41 is set up and the simulation is carried out to obtain the source and load impedances giving maximum output power. Here, the sweep center of the load reflection coefficient is set using the maximum output power impedance of the TGF2023-01 in the datasheet as **s11_center** = polar (0.32, 130°). The optimum source impedance is set to the conjugate of S_{11} of the small-signal S-parameters at 2.5 GHz in the datasheet. The optimum source impedance is implemented in the variable block named **Initial_S**11 in Figure 9.41. The circuit is thus set up and the initial load-pull simulation is carried out. Next, using the resulting initial load impedance values, the source-pull simulation is performed. This process is repeated three to four times to obtain the final results. The source impedance value that is eventually obtained is shown as **z1** in the variable block named **VAR**4 of Figure 9.41. It must be noted that the impedance of the load in **VAR**2 of Figure 9.41 is set to operate in class-F up to the third harmonic. The second harmonic impedance was set to 0 and that of the third harmonic to 1 kΩ, while the rest of the harmonic impedances were set to 50 Ω. On the other hand, all the harmonic impedances of the source were set to 50 Ω. The reason for setting the source impedance is that matching at the input is performed only at the center frequency. In addition, the number of harmonics in the HB simulator is set to 3.

The simulated load-pull contour plot is as shown in Figure 9.42. The maximum output power in that figure shows an output power of 37 dBm. The load reflection coefficient is 0.220∠55°, which can be seen to differ from the maximum output power reflection coefficient 0.320∠130° provided by TriQuint in the datasheet. The PAE can be seen to be approximately 60%. The difference in the maximum output power reflection coefficient is considered to be due to the addition of the stabilization circuit.

Figure 9.43 shows the source-pull simulation; see reference 1 at the end of this chapter for more information. This simulation is similar to the load-pull simulation. The load impedance value of 59.587 + *j*22.702 determined from the load-pull simulation is used as the load impedance, which is assigned up to the third harmonic for a class-F operation. The variable block named **Impedance Equations** is used to assign the sweep variable

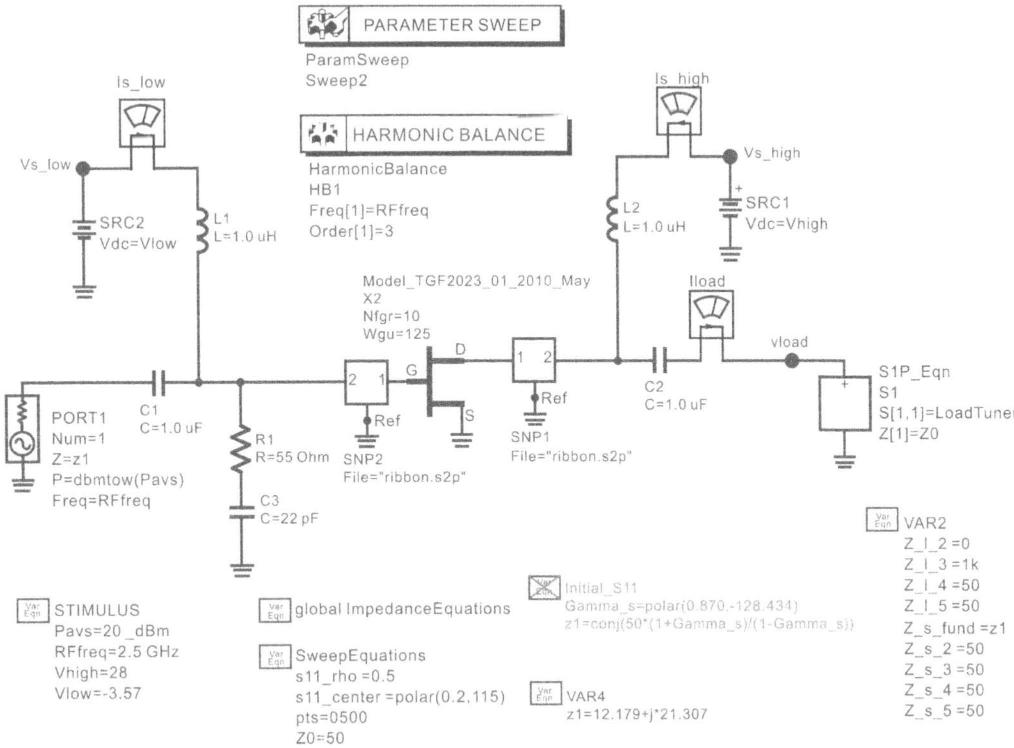

Figure 9.41 Load-pull simulation schematic. The available power **Pavs** is set to 20 dBm. The load circuit is a class-F load up to the third harmonics. The source impedance **z1** is initially set to the conjugate S_{11}. The source-pull and load-pull simulations are iterated to obtain the optimum load and source impedances.

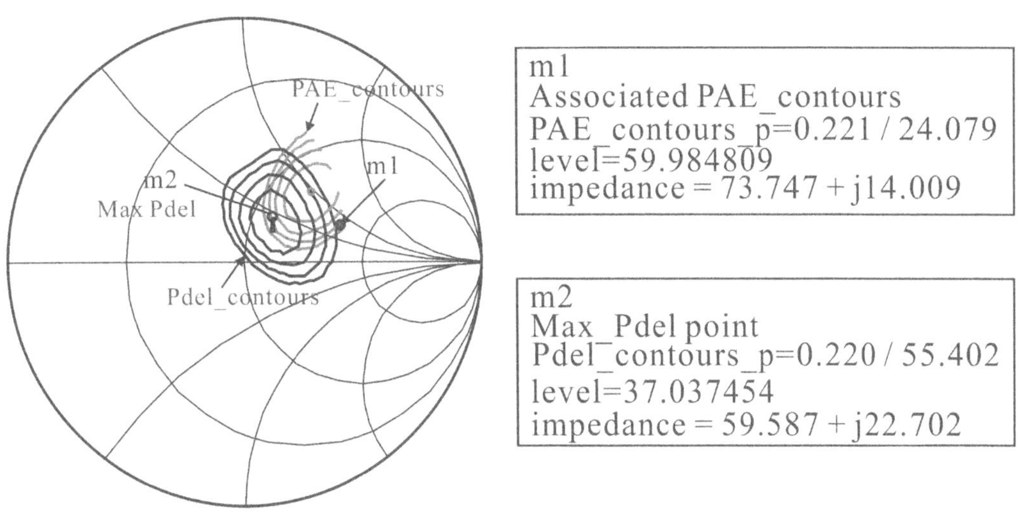

Figure 9.42 Simulated contour plots for the delivered power and PAE. The maximum output power is about 37 dBm at the load impedance of 59.587 + j22.702, which is the selected design point. The associated PAE is about 60%.

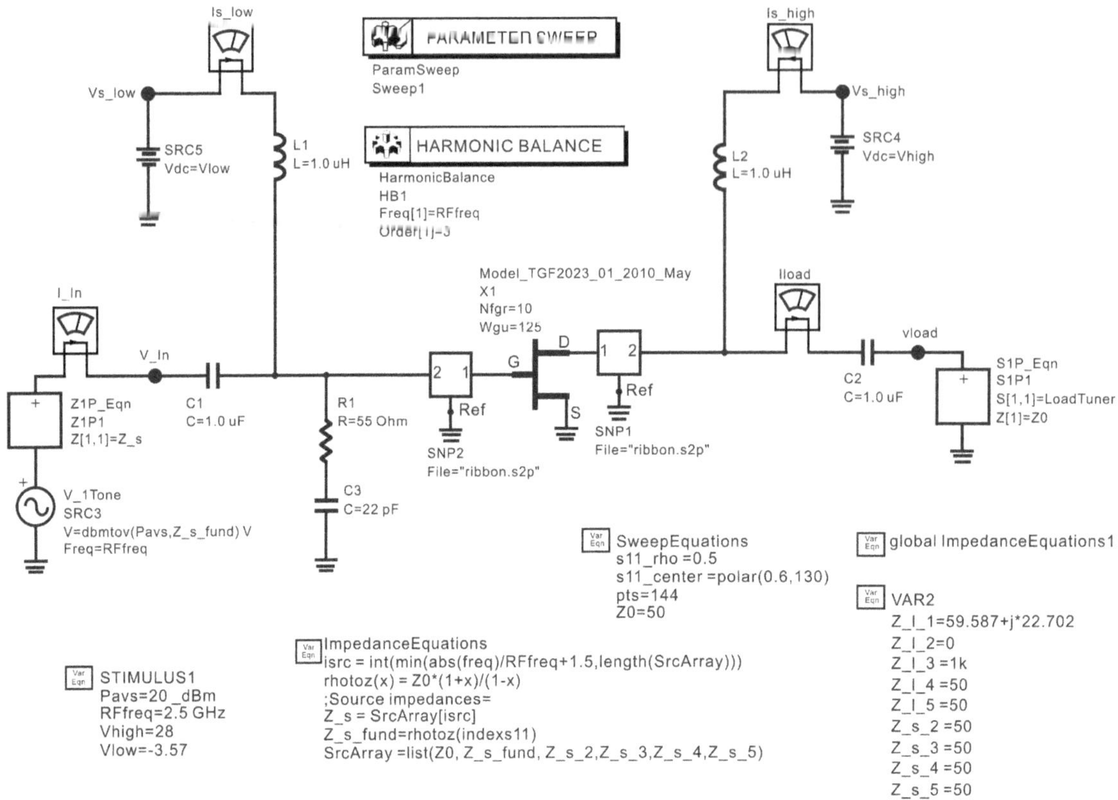

Figure 9.43 Source-pull simulation. The harmonic balance and parameter sweep are set to sweep for the source impedances, while the value of **LoadTuner** is fixed as $59.587 + j22.702$. The source impedance has 50 Ω except at the fundamental frequency. The sweep of the source impedance is the same as the load-pull simulation and is set in the variable block named **ImpedanceEquations**.

indexs11 to the source impedance. The sweep method of **indexs**11 is the same as the load-pull simulation, which uniformly samples the region of the circle. In addition, to supply the specified available power **Pavs**, the one-tone source **V_1Tone** is used. The value of the **V_1Tone** source is set using the dbmtov(·) function. Note that the impedance of the **V_1Tone** source is set using the impedance at the fundamental frequency **Z_s_fund**. The source-pull simulation results are as shown in Figure 9.44. Only the contours for the delivered power are plotted. The optimum source reflection coefficient providing the maximum delivered power can be seen to be $0.660\angle 132°$.

In summary, the simulated optimum source and load impedances are listed in Table 9.9. The output power is 37 dBm and the PAE is about 60% for the source and load impedances in Table 9.9.

Table 9.9 Source and load impedances determined from load-pull and source-pull simulations

Source impedance	$12.2 + j21.3\Omega$	$0.660\angle 132°$
Load impedance	$59.6 + j2.7\Omega$	$0.32\angle 130°$

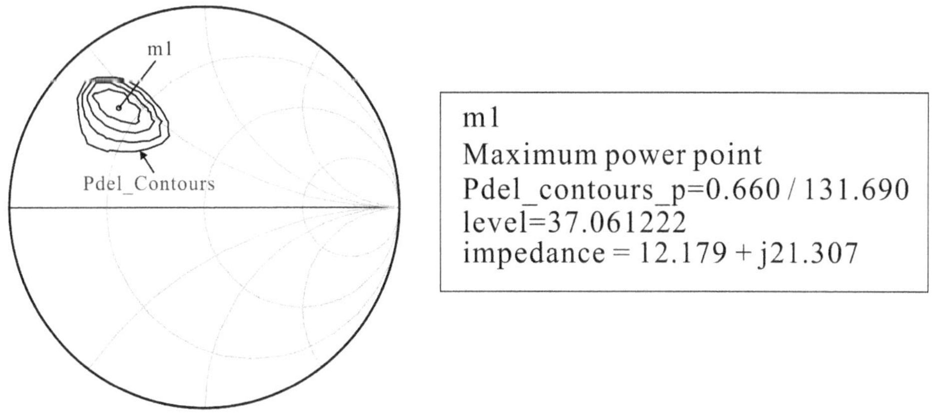

Figure 9.44 Source-pull simulation results. The maximum power is delivered at the source impedance of 12.179 + j21.307.

The simulation circuit shown in Figure 9.45 is set up to verify the obtained source and load impedances in Table 9.9. The source and load impedances at the fundamental frequency are declared using the variable block named **VAR**4 and are used to set the input and output impedance tuners. After the simulation, the equations in Measurement Expression 9.8 are entered in the display window to calculate the output power and the PAE. In addition, the equation for calculating the power gain is also entered.

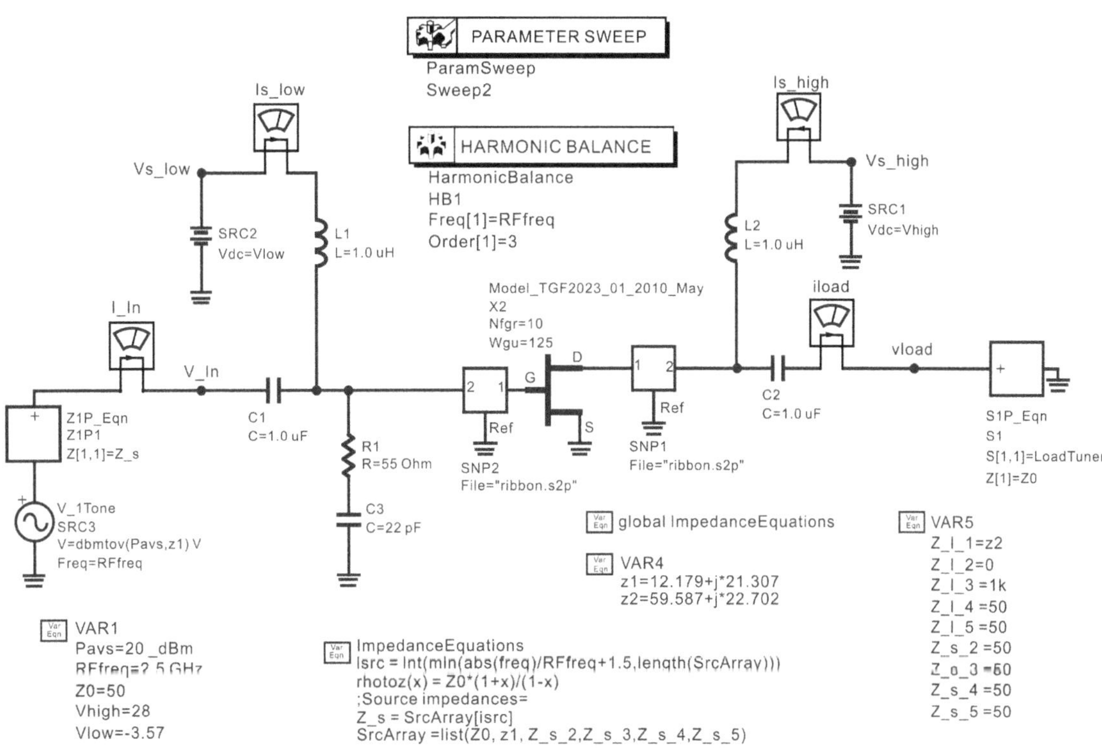

Figure 9.45 Simulation circuit for verifying the source and load impedances. These impedances obtained from the source and load pulls are in the variable block **VAR**4 and they are used to set the input and output impedance tuners.

```
Eqn Pdc=28*real(Id.i[0])
Eqn Pout_dbm=10*log10(0.5*real(vload[1]*conj(iload.i[1])))+30
Eqn Prf=10**(Pout_dbm/10-3)-10**(Pavs/10-3)
Eqn PAE=Prf/Pdc*100
Eqn gain=Pout_dbm-Pavs
```

Measurement Expression 9.8 Equation for calculating the output power, gain, and PAE

Figure 9.46 shows the calculated output power, gain, and PAE. As expected, the output power is 37 dBm at an input power of 20 dBm, and the PAE is about 62% at an output power of 37 dBm. In addition, the power gain can be seen to be approximately 20 dB.

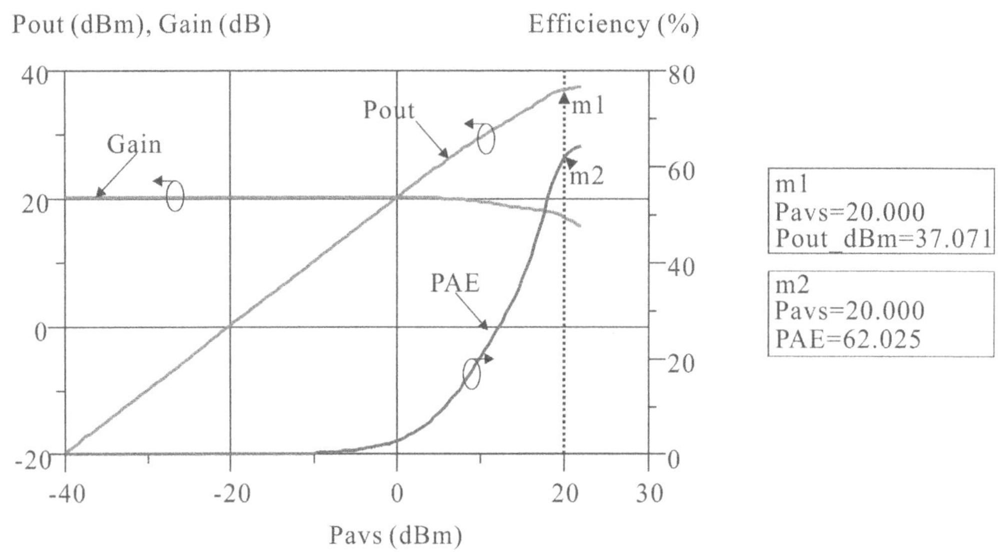

Figure 9.46 Gain, output power, and PAE for the selected impedance. The output power and PAE are about 37 dBm and 62%, respectively.

9.5.2 Input and Output Matching Circuits

A general class-F power amplifier requires an output matching circuit that yields the load impedance obtained from the load-pull simulation at the fundamental frequency and shows open for odd harmonics and short for even harmonics. However, in reality it is difficult to implement the circuit that satisfies the class-F conditions for an infinite number of harmonics. Thus, the circuit is implemented to satisfy class-F operation for a limited number of harmonics. Such circuits have been proposed by many researchers; see reference 3 at the end of this chapter for more information.

In this section, a simple implementation of the class-F output matching circuit for up to the third harmonic using lumped components will be discussed. Figure 9.47(a) shows the class-F output matching circuit that can yield the determined load impedance from the load-pull simulation. The capacitor C_p represents the drain capacitance, which is computed from the series-to-parallel conversion of the load impedance in Table 9.9. The value is fixed

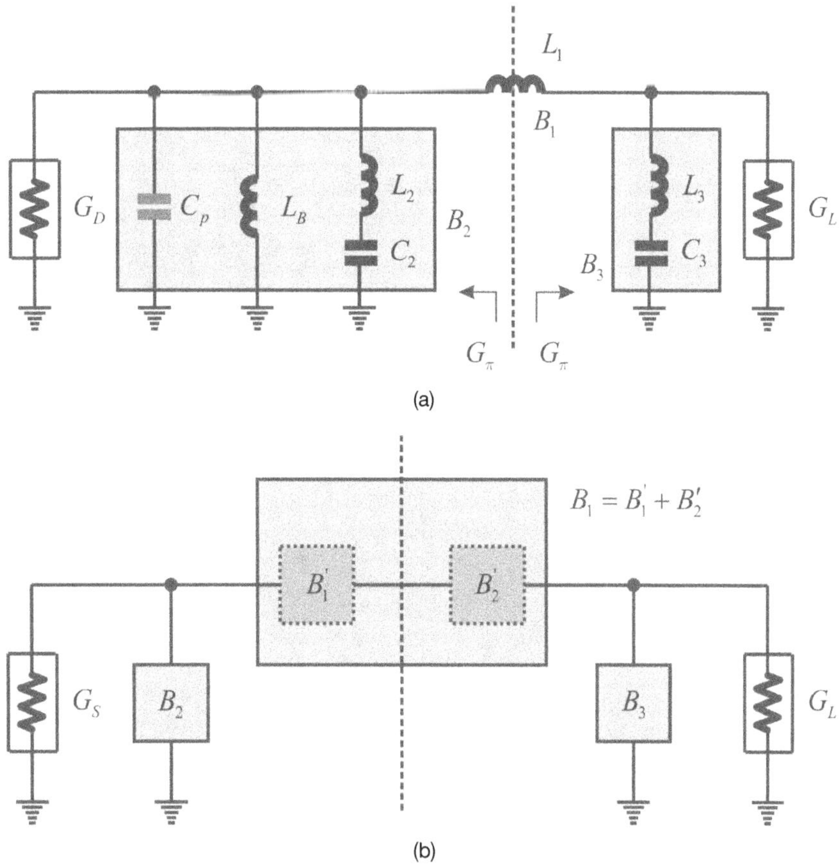

Figure 9.47 (a) Class-F output matching circuit and (b) equivalent circuit at the fundamental frequency. The matching circuit is the π-type matching circuit presented in Chapter 6.

as $C_p = 0.35$ pF. In addition, G_D and G_L represent the drain conductance and load conductance, respectively.

Basically, the output matching circuit is a kind of low-pass filter that provides a short for the second harmonic frequency and an open for the third harmonic frequency. Series-connected branches L_2–C_2 and L_3–C_3 are established to be series resonant and become short at the second and third harmonic frequencies, respectively. Therefore, the circuit in Figure 9.47(a) provides a short at the second harmonic due to L_2–C_2, independent of the rest of the load circuit. In order for the amplifier to operate in a class-F, the impedance seen from the drain should be open at the third harmonic. Since the impedance of L_3–C_3 is 0 at the third harmonic, the circuit seen from the drain with the branch L_3–C_3 shorted should be a parallel resonant circuit. Consequently, the impedance seen from the drain terminal can be made open at the third harmonic. Thus, the circuit can be set to operate in class-F up to the third harmonic.

In addition, the circuit should have the desired load impedance at the fundamental frequency. The circuit in Figure 9.47(a) becomes a π-type matching circuit at the fundamental frequency and the equivalent circuit of Figure 9.47(a) at the fundamental frequency is shown in Figure 9.47(b). Here, the drain-side admittance at the fundamental frequency is denoted as B_2, the 50-Ω load-side admittance is denoted as B_3, and the series admittance is denoted as B_1. Also, the ratios of the drain and load admittances to the π-matching admittance denoted as n_1 and n_2 are respectively defined as expressed in Equation (9.48).

$$n_1 = \frac{G_D}{G_\pi},\ n_2 = \frac{G_L}{G_\pi} \tag{9.48}$$

Note that $G_\pi > G_D, G_L$. Given G_π, the values of B_1, B_2, and B_3 can be calculated at the fundamental frequency ω_1 following the method explained in Chapter 6 that can be expressed with Equations (9.49a)–(9.49c).

$$B_1 = -\left(\frac{G_D}{n_1\sqrt{(1-n_1)}} \parallel \frac{G_L}{n_2\sqrt{(1-n_2)}}\right) \tag{9.49a}$$

$$B_2 = G_D\sqrt{\frac{1-n_1}{n_1}} \tag{9.49b}$$

$$B_3 = G_D\sqrt{\frac{1-n_2}{n_2}} \tag{9.49c}$$

The value of B_1 in Equation (9.49a) is selected to be negative because B_1 is the admittance of inductor L_1. From the calculated value of B_1, the value of L_1 can be computed as shown in Equation (9.50).

$$L_1 = -\frac{1}{\omega_1 B_1} \tag{9.50}$$

The admittance of the branch L_3–C_3 at the fundamental frequency is B_3, which is short at the third harmonic frequency, ω_3. Using the relationship $(\omega_3)^2 = (L_3C_3)^{-1}$,

$$B_3 = \omega_1 C_3 \frac{1}{\left(1-\frac{\omega_1^2}{\omega_3^2}\right)} = \frac{9}{8}\omega_1 C_3 \tag{9.51}$$

Thus, from B_3, the value of capacitor C_3 can be determined using Equation (9.51) and inductor L_3 can be subsequently determined using $(\omega_3)^2 = (L_3C_3)^{-1}$. In addition, the drain-side branch admittance at the fundamental frequency is B_2, and we obtain Equation (9.52).

$$B_2 = \omega_1 C_p - \frac{1}{\omega_1 L_B} + \frac{4}{3}\omega_1 C_2 \tag{9.52}$$

At the same time, the admittance seen from the drain at the third harmonic frequencies is 0 and this can be expressed as

$$\omega_3 C_p - \frac{1}{\omega_3 L_B} - \frac{1}{\omega_3 L_1} + \omega_3 C_2 \frac{1}{\left(1-\frac{\omega_3^2}{\omega_2^2}\right)} = 0 \tag{9.53}$$

As a result, Equations (9.52) and (9.53) become simultaneous equations for C_2 and L_B. Thus, by solving the simultaneous equations, C_2 and L_B can be determined as shown in Equations (9.54) and (9.55).

$$L_B = \frac{L_1}{-\frac{5}{32} - \frac{9}{32}\omega_3 L_1 B_2 + \frac{1}{4}\omega_3^2 L_1 C_p} \tag{9.54}$$

$$C_2 = \frac{3}{4}\left(\frac{B_2}{\omega_1} - C_p + \frac{1}{\omega_1^2 L_B}\right) \tag{9.55}$$

Thus, the values of all the elements in Figure 9.47(a) can be determined to satisfy the class-F operation up to the three harmonics.

The values of all the elements were calculated by setting $G_\pi = 2G_D$ and using Equations (9.50)–(9.55). The calculated results give a value of L_B too large to implement. Therefore, the value of C_p was increased by 0.5 pF and the values of all the elements were recalculated. Here, the increased value 0.5 pF can be implemented using a parasitic capacitor that appears together with L_B. The calculated results are shown in Table 9.10.

Table 9.10 Values of the output matching circuit elements

Output	L_1 (nH)	C_2 (pF)	L_2 (nH)	C_3 (pF)	L_3 (nH)	L_B (nH)
	3.654	0.7813	1.2968	0.7722	0.5832	4.203

In order to verify the results above, the simulation circuit shown in Figure 9.48 is set up and the simulation is carried out. The transmission characteristic of the designed class-F load circuit is shown in Figure 9.49(a). As expected, the transmission characteristic shows a match at the fundamental frequency with the transmission zeroes at the second and third harmonics. In addition, in order to plot the impedance seen from the drain, S_{11}, from the S-parameters renormalized by the reference, 50 Ω is plotted as shown in Figure 9.49(b). From that figure, the impedance provides an exact match at the fundamental frequency and provides a short and an open at the second and third harmonics, respectively.

The input matching circuit is designed to produce the source impedance value of $12.2 + j21.3\ \Omega$ in Table 9.9. The input matching circuit can be implemented using the L-type matching circuit described in Chapter 6. Figure 9.50 is a simulation schematic for computing the element values of the L-type matching circuit through the optimization. The L-type matching circuit is a low-pass filter type. After the optimization, $C_i = 2.2445$ pF and $L_i = 2.722$ nH are obtained.

The entire amplifier circuit is shown in Figure 9.51. Here, the input and output matching circuits are configured as subcircuits. It should be noted that since the active device includes the drain capacitor C_p in the output matching circuit of Figure 9.48, that drain capacitor must be deactivated when the output matching circuit is configured into the output matching subcircuit. The input matching subcircuit represents the L-type matching circuit in Figure 9.50. The gate DC voltage is supplied through a 1-kΩ resistor. The input and output DC block capacitors in Figure 9.51 are set to 22 pF.

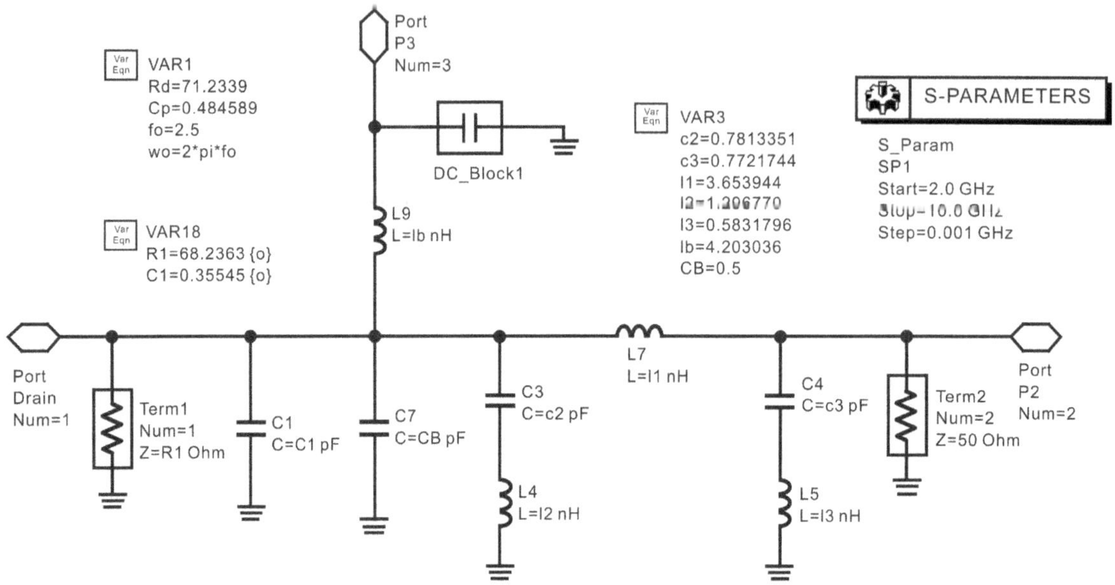

Figure 9.48 Output matching circuit simulation (C_B in the figure represents the parasitic capacitor due to the RFC inductor **L**9, which was set to 0.5 pF)

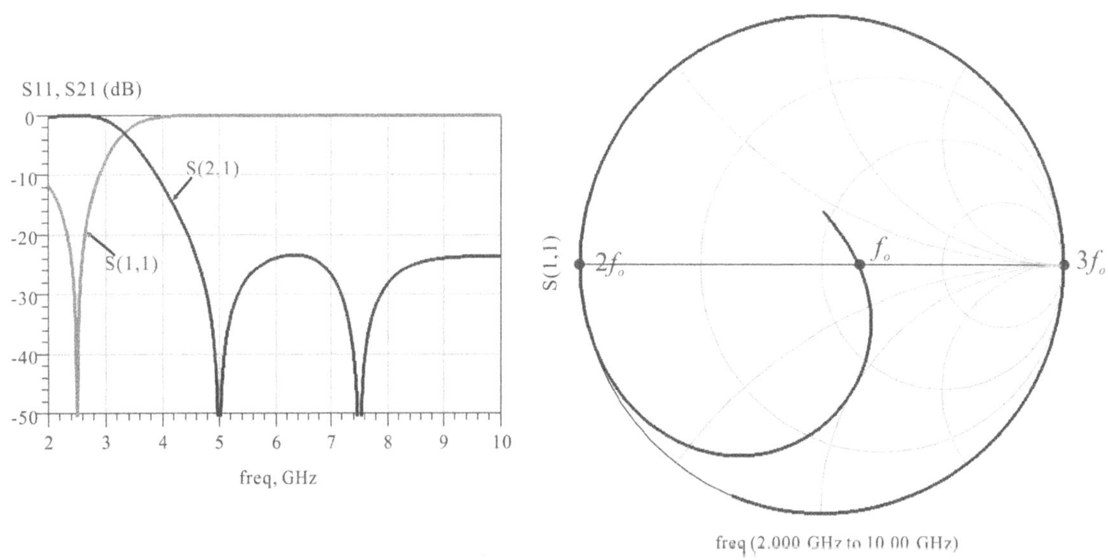

(a) (b)

Figure 9.49 Output matching circuit simulation results; (a) transmission and reflection characteristics and (b) impedance seen from the drain. The matching is achieved at the fundamental frequency and the transmission zeroes appear at the second and third harmonic frequencies.

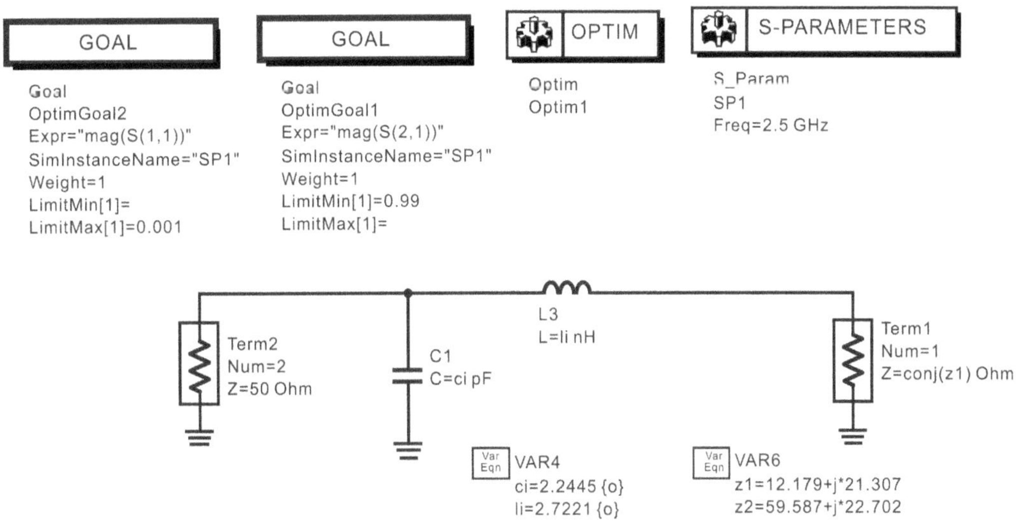

Figure 9.50 Input matching circuit simulation. The port 1 impedance is set to the conjugate of the desired input impedance and optimized for the maximum power transfer.

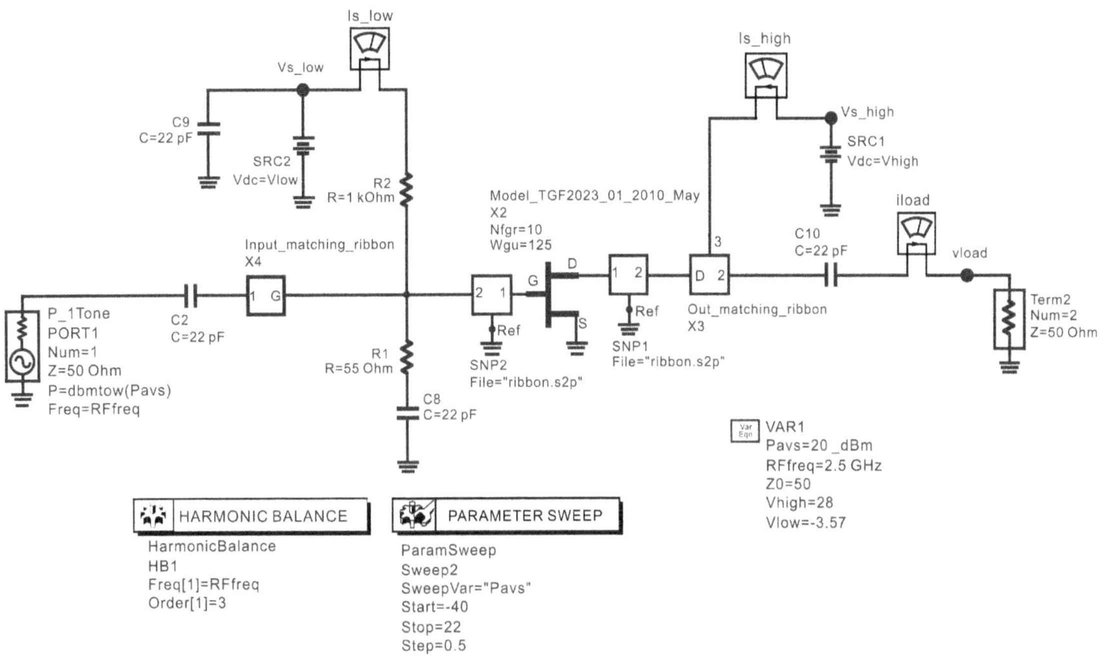

Figure 9.51 Simulation schematic for verifying the PA with the designed input and output matching circuits

The simulated output power and the PAE are shown in Figure 9.52. As in the previous simulation, the output power is 37 dBm, the PAE is about 60%, and the gain can be seen to be about 20 dB. Thus, the designed matching circuits can be seen to reproduce the characteristics obtained from the load-pull simulation results.

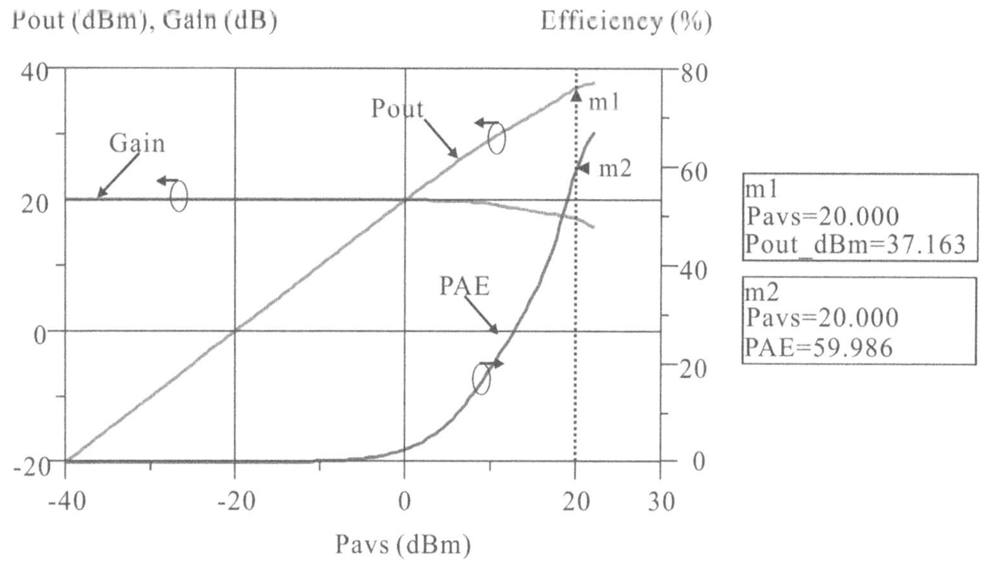

Figure 9.52 Simulated output power and PAE of the power amplifier with the lumped-element matching circuits. The small-signal gain is about 20 dB. The output power and PAE are about 37 dBm and 59%, respectively, and are exactly equal to the load-pull simulation results.

9.5.3 Design of Matching Circuits Using EM Simulation

In order to miniaturize the designed power amplifier's size, the inductors and capacitors in the matching circuits are implemented using spiral inductors and interdigital capacitors, respectively. The selected substrate here is a 5-mil-thick alumina with a dielectric constant of 9.9. The geometrical parameters of the spiral inductor include the number of turns, the diameter, the spacing, and the line width. In the case of the spiral inductor discussed in Chapter 5, for every 0.25 increase in the number of turns, the output terminal rotates with an angle of 90° to the input. Thus, the increase in the number of turns is set to 0.25. In addition, the line width and spacing of the matching circuit's spiral inductor are fixed at 30 mm and 10 mm, respectively. Thus, the spiral inductor value can be increased continuously by changing the diameter. In the case of the RFC inductor L_B for the DC power supply, line width and spacing are fixed as 60 μm and 30 μm, respectively, in consideration of the *fusing current*.

Figure 9.53 shows a simulation schematic for determining the initial dimension of the spiral inductor. Using the simulation, the value of inductance can be calculated by varying the inner diameter. The determined diameters are summarized in Table 9.11. In the case of inductors L_2 and L_3, their values were found to be too small and so were implemented using microstrip lines.

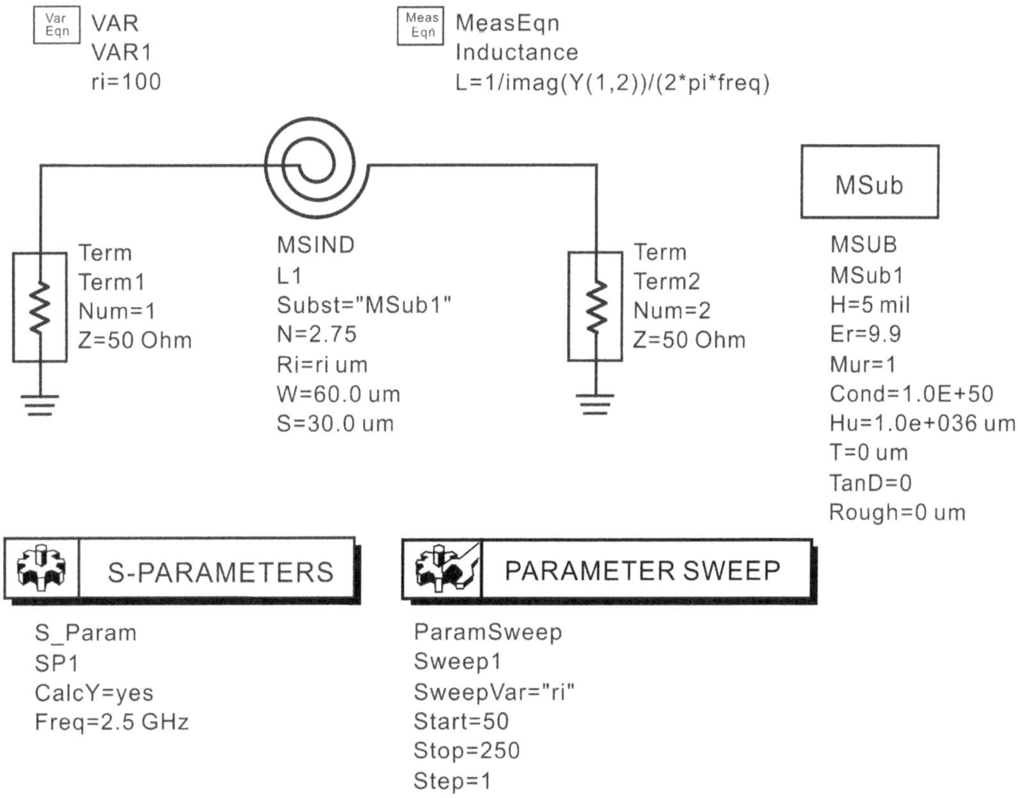

Figure 9.53 Simulation schematic for calculating the inductance of a spiral inductor. The number of turns and the width and spacing are fixed. To achieve the desired inductance, the internal diameter of the spiral inductor is adjusted.

Table 9.11 Inductance calculation results

Element	Design Value	Turns	Width	Spacing	Diameter	Initial Value
L_1	3.654 nH	2.5	30 mm	10 mm	187 mm	3.64 nH
L_2	1.2968 nH	—	—	—	—	—
L_3	0.5832 nH	—	—	—	—	—
L_B	4.203 nH	2.75	60 mm	30 mm	219 mm	4.19 nH
L_i	2.722 nH	2.5	30 mm	10 mm	134 mm	2.715 nH

The initial dimension of the interdigital capacitor can also be configured in a similar way. The line width, spacing, length, and number of fingers of the interdigital capacitor can be adjusted; among them, only the length is varied with other parameters fixed. The continuous increase of capacitance values can be obtained by changing the length. The line width, spacing, and number of fingers are fixed at 20 μm, 10 μm, and 10, respectively, while the length is adjusted. Figure 9.54 is the simulation schematic for calculating the capacitance.

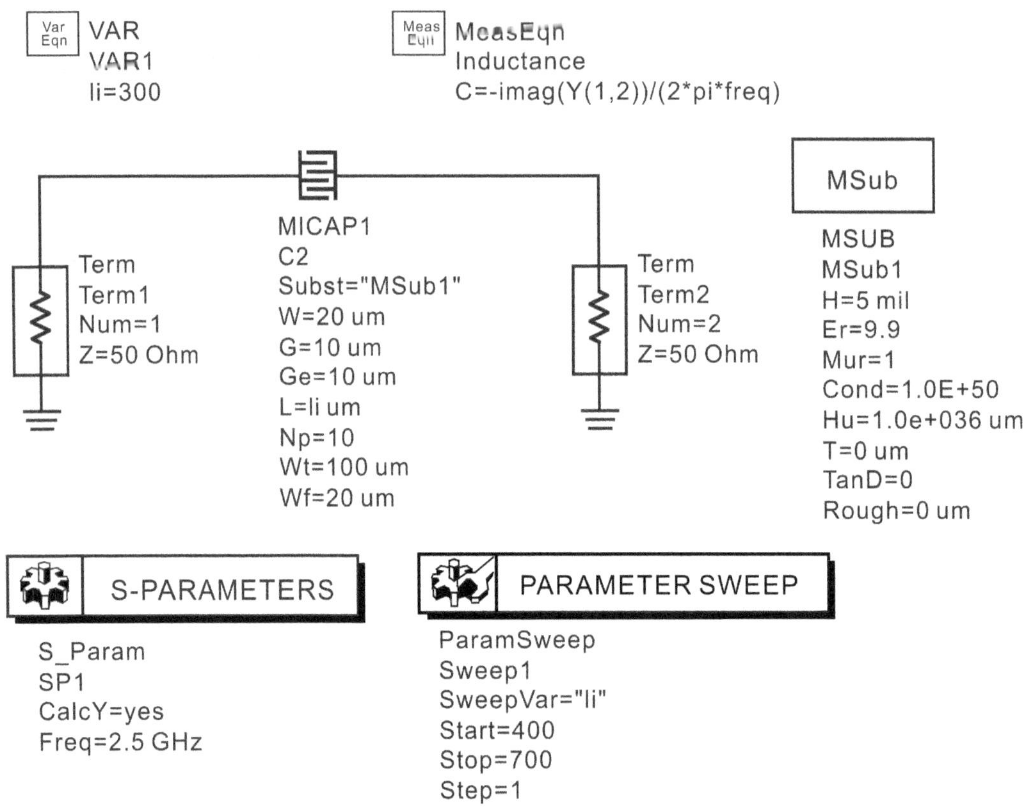

Figure 9.54 Simulation schematic for determining the initial dimensions of an interdigital capacitor. The width, spacing, and number of fingers are set to 20 μm, 10 μm, and 10, respectively. The desired capacitance is achieved by varying the length **li**.

The calculated dimensions are shown in Table 9.12. The value of C_i in the input matching circuit was found to be too large to be implemented as an interdigital capacitor, and so was implemented using a single layer capacitor (SLC). In addition, because the values of the DC block and bypass capacitors are also large, SLCs are used. To connect the SLC to the substrate, bonding wires are needed. The S-parameters of the bonding wire are required during simulation and have been precomputed using Ansoft's HFSS.

Table 9.12 Calculation results of interdigital capacitors

Element	Design Value	Fingers	Width	Spacing	Length	Initial Value
C_2	0.7813 pF	10	20 mm	10 mm	678 mm	0.780 pF
C_3	0.7722 pF	10	20 mm	10 mm	671 mm	0.771 pF
C_i	2.2445 PF	—	—	—	—	—

Using the initial dimensions of the inductors and capacitors, an initial layout of the power amplifier can be drawn and is shown in Figure 9.55. The patterns marked 1–1′, 2–2′, 3–3′, 4–4′ in the figure are inserted for on-wafer probing of the substrate. Using the on-wafer probing patterns, the substrate can be inspected using S-parameter measurements before the assembly of the power amplifier.

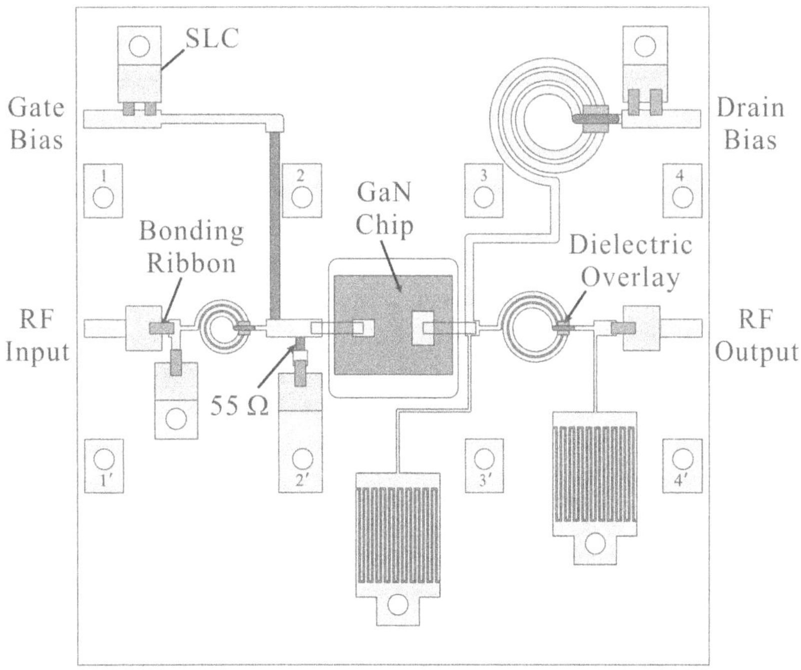

Figure 9.55 Power amplifier initial layout (4.4 × 4.4 μm²). The pads 1–1′, 2–2′, 3–3′, and 4–4′ are on wafer-probe pads for inspecting the matching circuits. The substrate is a 5-mil-thick alumina with a permittivity of 9.9. The dielectric overlay has a thickness of 4 μm. The sheet resistivity is 50 Ω/square. All bonding ribbons are 3 mil.

The initial power amplifier layout shown in Figure 9.55 should be tuned using EM simulation to obtain the desired input and output impedances. Figure 9.56 shows the schematic of the co-simulation with EM simulation for the output matching circuit. The substrate in the EM simulation, as previously described, is a 5-mil-thick alumina of dielectric constant 9.9 and the conductor thickness is set to 10 μm while the conductivity is set to 4.1×10^7 siemens/m. As mentioned earlier, the inductance of the bonding wire needed for connecting the DC block SLC was determined using Ansoft HFSS simulation, and the value is about 0.2 nH. The 0.2 nH inductors in Figure 9.56 represent the bonding-wire inductances. The effects of landing patterns for SLC capacitors should also be considered; in Figure 9.56, they are represented as data items that were obtained through a separate S-parameter simulation.

If co-simulation is not involved, the layout must be redrawn each time to reflect the changes in the layout dimensions, which is inconvenient. To work around the co-simulation, dimensions such as the inner diameter of the spiral inductors and length of the interdigital capacitors are first declared as variables in the layout. Then, those declared variables can

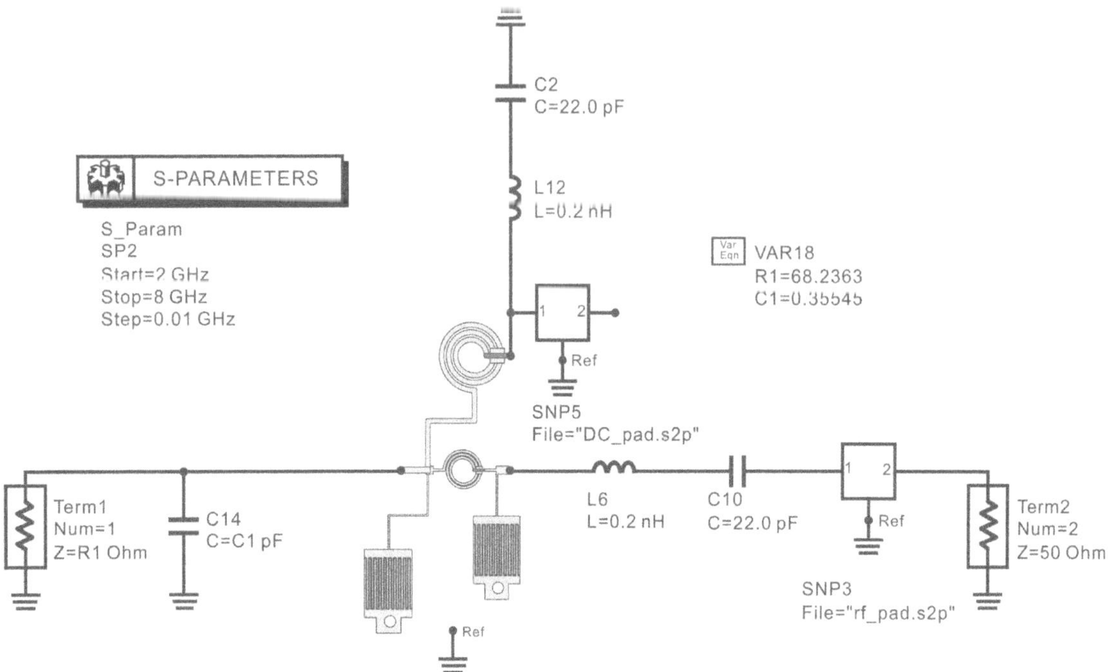

Figure 9.56 EM co-simulation of the output matching circuit. The 22-pF capacitors are SLC capacitors. The 0.2-nH inductors represent bonding-wire inductances. The **C**14 capacitor with a value of **C**1 pF is the drain capacitance. The layout component is the output matching circuit captured from the amplifier layout in Figure 9.55. To match, the inner diameters of the spiral inductors and the lengths of the interdigital capacitors are adjusted. The other dimensions are automatically adjusted according to the dimension changes of the spiral inductors and interdigital capacitors to preserve anchor points such as RF input and output, and drain bias points.

be used as variables in the circuit simulation window. The impedance can thus be obtained from the schematic co-simulation without the necessity of separate EM simulations for the layout with the given dimensions. Using the co-simulation, the layout tuning can be carried out repeatedly until the dimensions are close to the goal.

As mentioned previously, only the inner diameter of the spiral inductor and the length of the interdigital capacitors are selected as the adjustable dimensions. Certain dimensions should be fixed even when the selected dimensions are adjusted. For example, the distance between the drain and the 50-Ω load should be fixed and should not be changed by the adjustments of the inner diameter of the spiral inductor. Also, the location of the DC supplying pad in the substrate should be fixed irrespective of the adjustments. Thus, the length of the transmission line used for connection must also be specified as a variable to prevent changes to the previously fixed dimensions in the layout. In addition, the drain capacitor is increased by 0.5 pF in the output matching circuit. The increased capacitor is implemented by the parasitic elements arising from the spiral inductor for L_B and L_1, and from the interdigital capacitors C_2. Thus, a separate capacitor for the implementation of 0.5 pF is not used. The value of the parasitic capacitor is approximately 0.2–0.6 pF, which is somewhat dependent on the patterns of L_B, L_1, and C_2. The output matching circuit, including the parasitic capacitor, is thus optimized in the course of the co-simulation.

The desired matching can be achieved through optimization. However, because the optimization requires significant computation time, the matching is obtained by trial-and-error adjustment in the co-simulation. The transmission characteristics shown in Figure 9.49(a) help in the adjustment technique. The second harmonic-transmission zero depends on capacitor C_2 and the third harmonic-transmission zero depends on C_3. Thus, the desired transmission zeroes at the second and third harmonics can be achieved by adjusting C_2 and C_3. The matching at the fundamental frequency can be achieved by tuning inductor L_1. Thus, by adjusting capacitors C_2, C_3, and inductor L_1, the desired output matching circuit can be achieved.

Figure 9.57 shows the EM simulation results after completing the final adjustments for the output matching circuit. As the impedance is not the 50-Ω reference impedance, the S-parameters are first converted into the 50-Ω reference S-parameters and plotted in Figure 9.57(b). In that figure, the EM-simulated impedance is found to be well-matched to the desired impedance at the fundamental frequency. In addition, the short and open impedances at the second and third harmonics, respectively, were approximately achieved. However, significant losses appear at the second and third harmonics compared to the loss results in Figure 9.49(b). Despite these losses, looking at the transmission characteristics of Figure 9.57(a), there appear to be no problems eliminating the second and third harmonics. The insertion loss at the fundamental frequency can be seen to be about 0.6 dB.

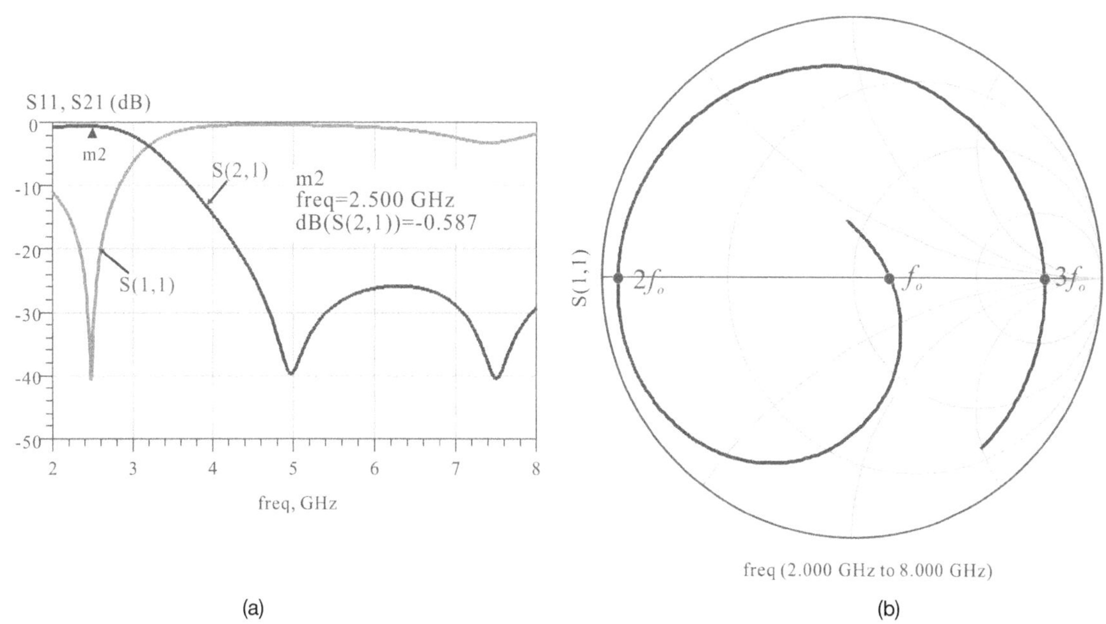

Figure 9.57 (a) Transmission characteristics of the drain to the load and (b) the EM-simulated impedance of the output matching circuit seen from the drain. The impedance seen from the drain shows a match to 68 Ω at the fundamental frequency and a short and open at the second and third harmonic frequencies, respectively. The insertion loss is about 0.6 dB at the fundamental frequency. Note that the impedances at the second and third harmonic frequencies are not exactly short or open due to the matching circuit loss.

The EM co-simulation schematic of the input matching circuit is shown in Figure 9.58. The input matching can be achieved by adjusting the value of the SLC C_i and the diameter of the spiral inductor L_i. Here, the inner diameter of the spiral inductor is related to the connection of the device and input port. Thus, the length of the connected transmission line is set to compensate for the change in the inner diameter of the spiral inductor so that an assembly problem does not arise in the connection. In addition, the bonding wire of the DC block SLC at the input that cannot be simultaneously simulated in ADS Momentum is pre-simulated using Ansoft's HFSS. This is represented by a 0.2-nH inductance, which is a fitted value that uses the HFSS simulation results. Similarly, the input-pad pattern is pre-simulated and represented by a data item for the co-simulation.

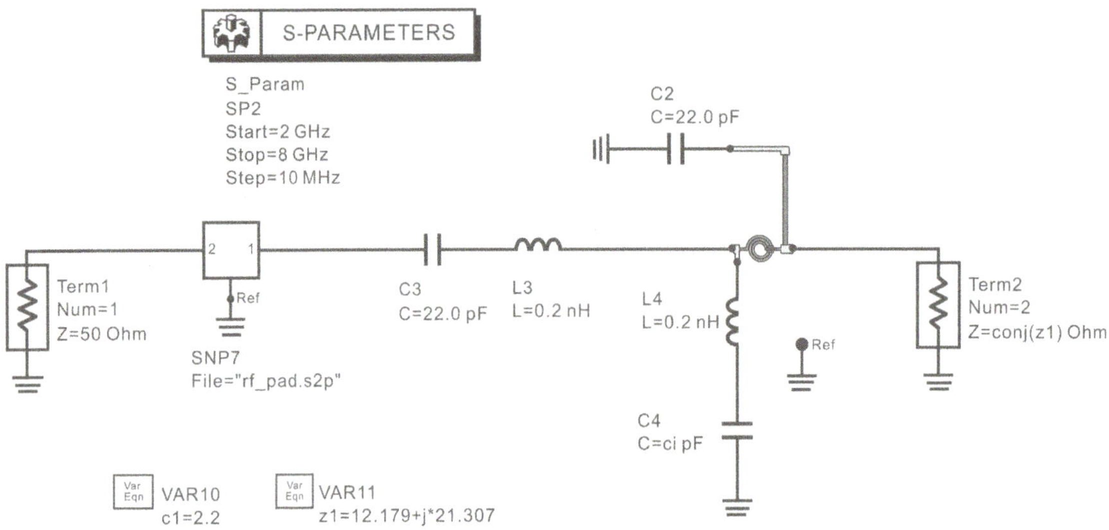

Figure 9.58 EM co-simulation schematic of the input matching circuit. **C2** and **C3** are SLC capacitors. The input matching is achieved by the spiral inductor and **C4** with the value of **ci**. The inner diameter of the spiral inductor and **ci** are adjusted for the input impedance match. The 0.2 nH inductors represent 3-mil bonding-wire inductances. The effect of the RF pad is simulated separately and represented by the data component **SNP7**.

It should be noted that because the value of the capacitor C_i is not continuous, it is adjusted using fixed discrete values available from the datasheet. The calculated value is 1.9 pF. The element values of the input matching circuit can be obtained using the properties of the L-type matching circuit, as explained in Chapter 6. The value of L_i is adjusted when the real part of the impedance seen from port 1 differs from 50 Ω. If the imaginary part is not matched, it is matched by adjusting the value of C_i.

The simulated impedance of the input matching circuit at port 1 is shown in Figure 9.59. In Figure 9.59(a), it can be seen that the input part is closely matched to 50 Ω at

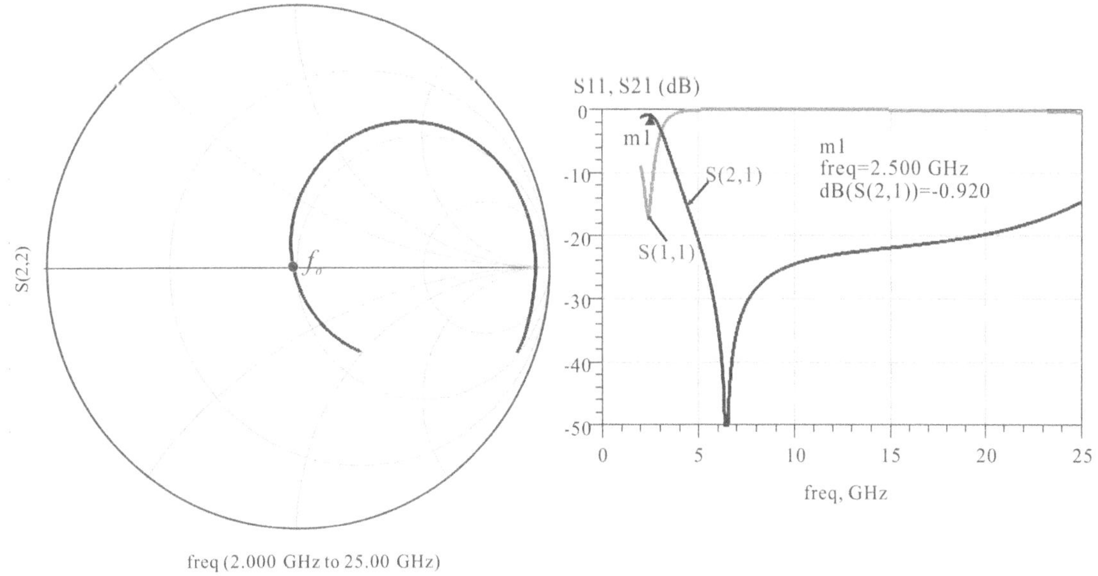

(a) (b)

Figure 9.59 Simulated (a) S_{11} and (b) transmission characteristics of the input matching circuit. The input matching circuit has a loss of about –0.920 dB and transmission zero is about 6.5 GHz.

2.5 GHz. However, looking at Figure 9.59(b), the matching circuit shows a loss of approximately 1 dB. This is considered to be caused by losses in the input matching circuit. It must be noted that although this loss is high, it does not reduce the output power. Therefore, in order to obtain the same output power, an input power greater by 1 dB must be applied to the input.

Finally, using the EM-simulated results thus obtained, the P_{out}–P_{in} characteristics of the power amplifier can be examined. The simulation schematic configured by replacing the EM simulation results by S-parameter data items is shown in Figure 9.60.

The output power, gain, and PAE are shown in Figure 9.61. Due to the losses at the input, the results at an input power of 21 dBm will correspond to the lossless circuit simulation results at an input power of 20 dBm. At 21-dBm input power, the output power is decreased by approximately 0.6 dB compared to the circuit simulation results due to the approximately 0.6-dB loss occurring in the output matching circuit. In addition, due to the same losses, the PAE is 53% and can be seen to have also decreased by about 7%.

The small-signal gain characteristics obtained from the EM simulation are shown in Figure 9.62. The gain in that figure is approximately 20 dB and the return losses for both input and output can be seen to be approximately 10 dB. The reason for the poor return losses is that the power amplifier is matched at large signals. This causes a mismatch at the small-signal level and results in poor return losses.

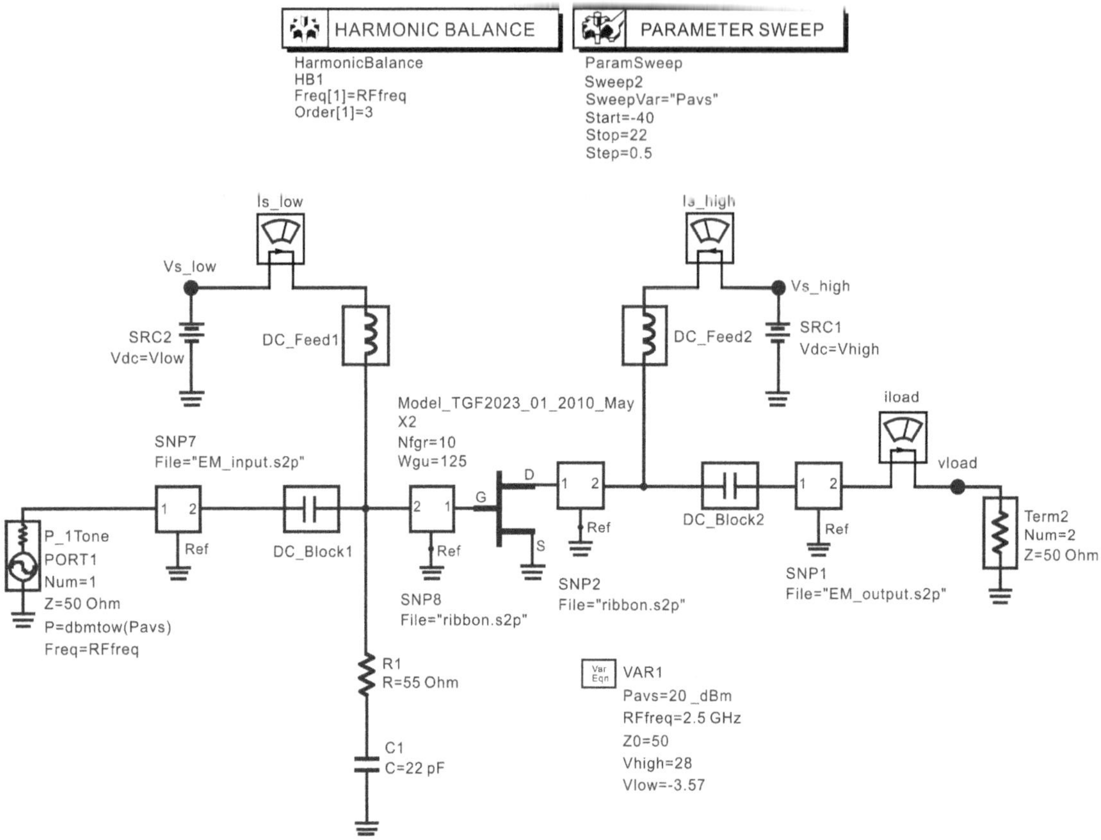

Figure 9.60 Simulation schematic for the verification of the power amplifier with the EM-simulated matching circuits. The data components **SNP7** and **SNP1** represent the EM-simulated S-parameters of the input and output matching circuits, respectively.

The transmission characteristics for the small signal are shown in Figure 9.63. As expected, the transmission zeroes can be seen to occur at the second and third harmonics. Another transmission zero exists between the two transmission zeroes as can be seen in Figure 9.63. This is the transmission zero caused by the input matching circuit, which was shown in Figure 9.59(a).

9.6 POWER AMPLIFIER LINEARITY

The previously discussed class-D, -E, and -F power amplifiers can significantly distort the input waveform, which can cause several problems in applications of power amplifiers to communication systems. The distortions in a power amplifier degrade the quality of communication, and cause problems for a receiver when demodulating a received signal. Therefore, when a power amplifier is used in a communication system, it is critical to determine whether or not the modulated input signal can be reliably demodulated beyond

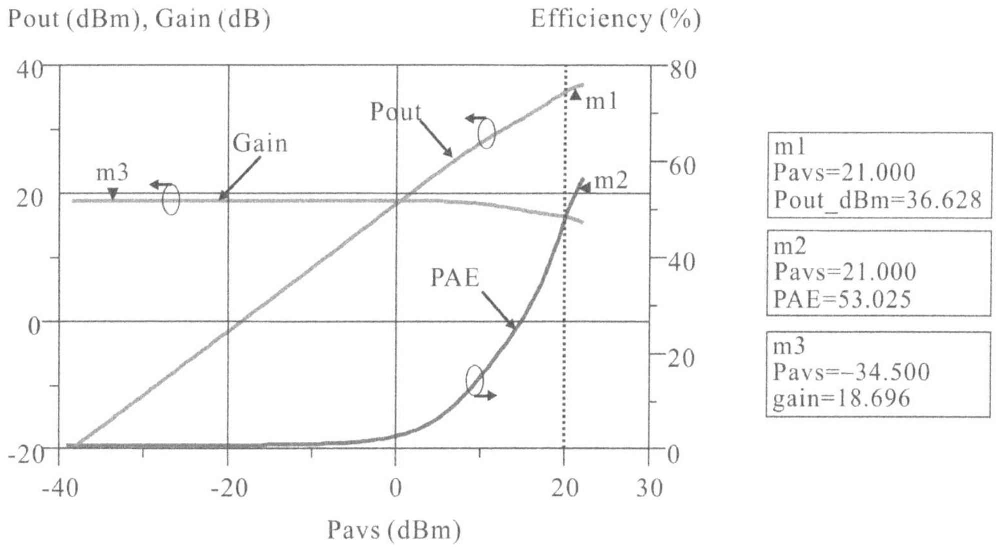

Figure 9.61 Output characteristics of the power amplifier with the EM-simulated matching circuits. The small-signal gain is reduced by about 1.5 dB compared with that of the circuit simulation due to the matching circuit loss. The input matching network has a loss of about 0.9 dB. Thus, the output power and the PAE are read at a 1-dB increased input power of 21 dBm. The output power and PAE are about 36.6 dBm and 53% at the input power of 21 dBm.

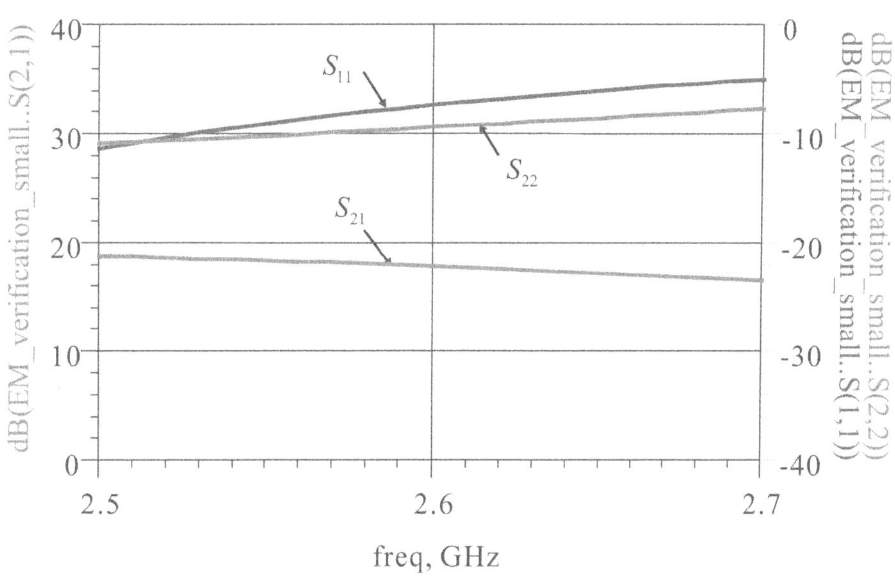

Figure 9.62 Small-signal in-band characteristics of the EM simulated power amplifier. Because the power amplifier is matched at large signals, the input and output return losses are poor.

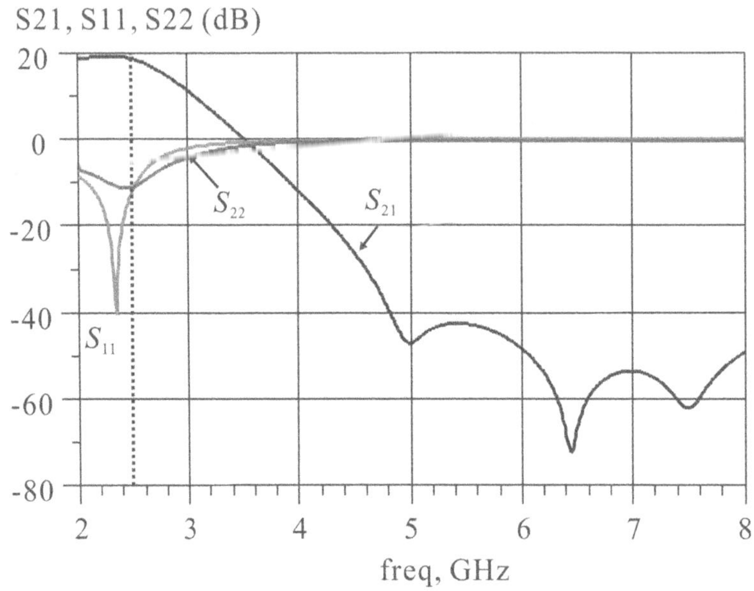

Figure 9.63 Transmission characteristic of the power amplifier with the EM-simulated matching circuits. The transmission zeroes appear at the second (about 5 GHz) and the third harmonic (7.5 GHz) due to the class-F design. The transmission zero between 6–7 GHz is due to the input matching circuit.

a certain limit. In the past, a two-tone test was the primary tool for assessing the distortion in a communication system that employed frequency division multiplexing (FDM/FDMA). In the two-tone test, a two-tone signal is applied to the input and, by measuring the amount of the third-order intermodulation distortion (IMD3), the applicability of the power amplifier can be determined. This way, the two-tone test was used as the main criterion for the applicability because the effect of intermodulation distortion on the other adjacent channels could be determined. However, with recent advances in digital communication systems, a power amplifier must transmit a digital modulation signal and the evaluation of the degree of distortion becomes a problem of significant concern.

Basically, the evaluation of a power amplifier for a digital modulation signal is, to some extent, related to the previously described two-tone test; however, the two-tone test results provide indirect evaluation for the power amplifier and a new, direct assessment method is required. This has led to the development of new test methods for assessing the degree of signal distortion in the digital modulated signal; they include the BER (bit error rate), the ACPR (adjacent channel power ratio), and the EVM (error vector magnitude) methods. These test methods are widely accepted as new standards. In this section, these methods of evaluating distortions in power amplifiers will be discussed together with simulation techniques in ADS.

9.6.1 Baseband Signal Modulation

An example of a digital bit stream is shown in Figure 9.64. The transmission rate of this signal is represented by the number of bits transmitted per unit time, and is defined as the *bit rate*. Its unit is bps (bit per second) and the bit rate of the signal in Figure 9.64 is expressed in Equation (9.56).

$$\text{Bit rate} = \frac{1}{T_b} \tag{9.56}$$

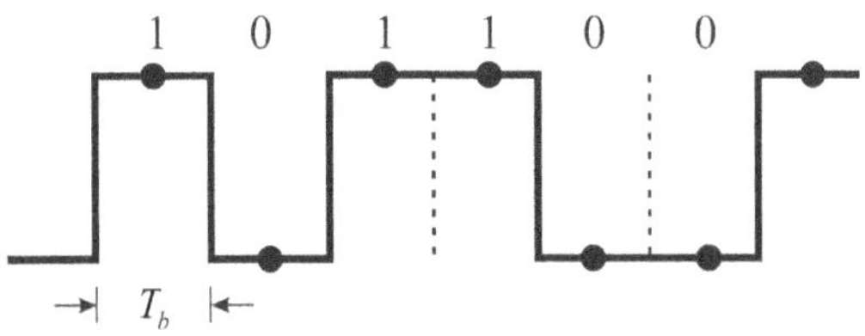

Figure 9.64 Digital signal waveform

However, when the digital signal in Figure 9.64 is transmitted directly, the bandwidth required for the transmission is wider than the bandwidth given by Equation (9.56), which is $1/T_b$ because the waveform in Figure 9.64 contains a lot of harmonics. To avoid this situation, two transmission techniques are used. One technique uses a digital filter that removes the harmonics; the other transmits a collection of multiple bits using IQ (in-phase and quadrature-phase) modulation. Denoting the modulation carrier frequency as ω_c, the IQ modulation uses two orthogonal carriers, $\cos(\omega_c t)$ and $\sin(\omega_c t)$ as shown in Equation (9.57).

$$x(t) = I(t)\cos\omega_c t + Q(t)\sin\omega_c t \tag{9.57}$$

When the odd bits shown in Figure 9.64 are assigned to $I(t)$ and the even bits are assigned to $Q(t)$, two bits can be transmitted simultaneously instead of only a single bit. Using the waveform in Figure 9.64 as an example, paired bits of (1, 0), (1, 1), (0, 0), ... are transmitted. A pair of bits is usually called a *symbol*. The number of transmitted symbols per unit time is defined as a *symbol rate*. In the example just explained above, two bits are transmitted as a symbol and the symbol rate can be seen to be one-half of the bit rate. Therefore, the transmission bandwidth can be reduced by half.

Extending this concept further, two odd bits can be assigned to $I(t)$ and two even bits can be assigned to $Q(t)$. This way, $I(t)$ and $Q(t)$ will each have four discrete levels. In the previous example, one bit is assigned to $I(t)$ and $Q(t)$, and the resulting $I(t)$ and $Q(t)$ will have two levels, 0 or 1. However, increasing the bit allocation to n, the levels of $I(t)$ and $Q(t)$ will increase to 2^n. The symbol rate then is determined as

$$\text{Symbol rate} = \frac{\text{Bit rate}}{2^n}$$

Thus, the transmission bandwidth required is reduced by 2^n. In the case where one bit is assigned to $I(t)$ and $Q(t)$, the modulation is called QPSK (quadrature phase shift keying) or 4 QAM (quadrature amplitude modulation). The modulation is called 16 QAM in the case of 2 bits, 64 QAM in the case of 3 bits, and 256 QAM in the case of 4 bits.

When an IQ-modulated signal is demodulated, $I(t)$ and $Q(t)$ appear at the output. The demodulated $I(t)$ and $Q(t)$ can be plotted in the I–Q plane. Figure 9.65 shows the I–Q plots for QPSK and 16 QAM. This is referred to as a constellation plot. By repeatedly plotting each demodulated signal at each symbol time, all possible symbols appear at the I–Q plane and the constellation plot will be similar to the plots shown in Figure 9.65. The

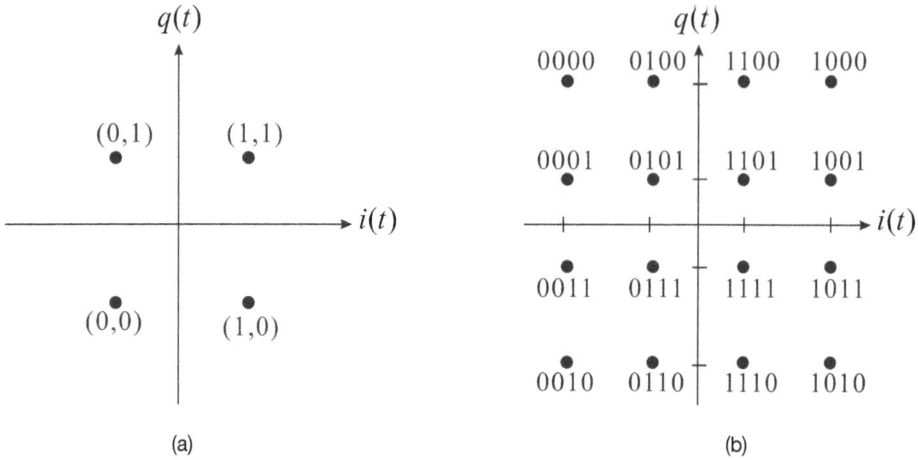

Figure 9.65 Constellation plot: (a) QPSK and (b) 16 QAM (gray coding)

constellation plot can be plotted using commercially available instruments. The lower the distortions in the channel of the communication system, the more likely it is that the constellation point or symbol point will appear at a single point, as shown in Figure 9.65. However, in a significantly distorted channel, jitters of the constellation points appear, and by evaluating the jitters, the degree of distortion in a power amplifier can be evaluated.

The standard way to evaluate the jitters of the constellation points is defined as EVM (error vector magnitude) measurement. Figure 9.66 illustrates the concept of EVM measurement.

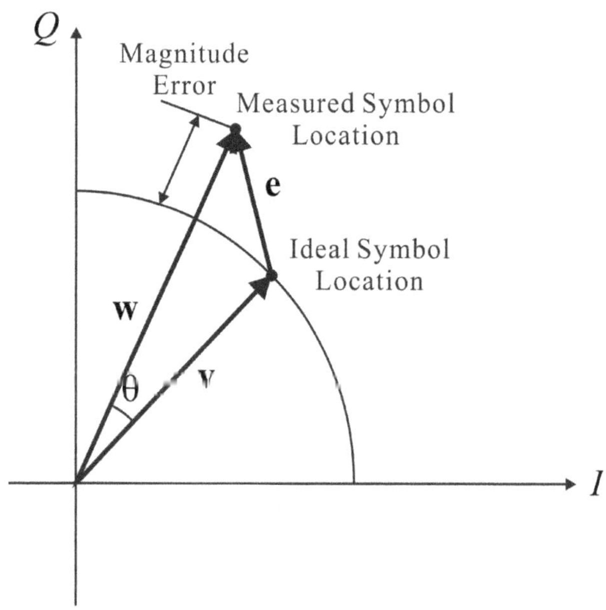

Figure 9.66 EVM (error vector magnitude)

The EVM can be measured for every symbol time, as shown in Figure 9.66, and calculating the RMS (root mean square) value enables the evaluation of the degree of distortion. Thus, EVM is defined as shown in Equation (9.58).

$$EVM = \frac{\sqrt{\frac{1}{N}\sum_{j=1}^{N}\left\{\left(I_j - I_{o,j}\right)^2 + \left(Q_j - Q_{o,j}\right)^2\right\}}}{\left|v_{\max}\right|} \tag{9.58}$$

Here, I_j and Q_j are the I and Q values measured at the j–th symbol time and $I_{o,j}$ and $Q_{o,j}$ are the ideal I and Q values. Also, $v_{\max}$ represents the value of the maximum vector magnitude of the ideal constellation point.

Figure 9.67 shows the measurement of the EVM. A spectrum analyzer with the EVM measurement utility is necessary for the EVM measurement. First, the reference constellation points are identified from the input reference signal, and then the EVM obtained with Equation (9.58) is measured by connecting the power amplifier's output to the spectrum analyzer.

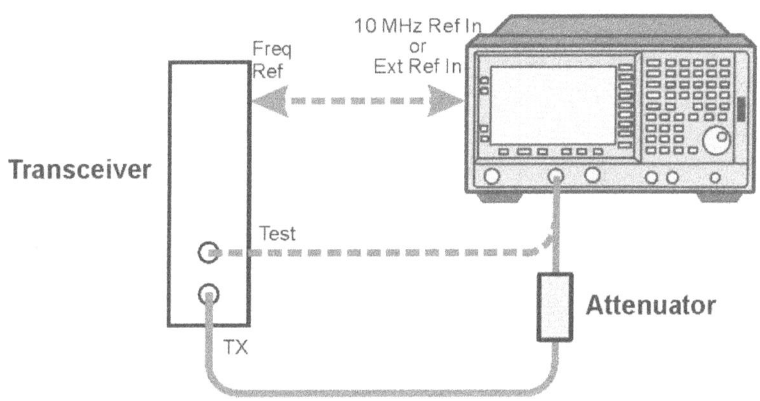

Figure 9.67 EVM measurement.[10] The test signal is measured first and becomes the reference of the constellation plot. Then, the power amplifier's signal is connected to the spectrum analyzer, which then evaluates the EVM with the built-in utility.

We have discussed that the required transmission bandwidth can be reduced using the IQ modulation. Another method to reduce the bandwidth is to use a low-pass filter, which converts the harmonic-rich bit waveform shown in Figure 9.64 into a smooth waveform. However, a simple analog lowpass filter generates ISI (intersymbol interference). When the bit waveform in Figure 9.64 is viewed as a superposition of independent square waveform bits, the analog low-pass filter's output waveform for a bit overlaps with that of another bit, which causes problems in the decision of bits. Usually, the impulse response

10. Agilent Technologies, Inc., Agilent Technologies ESA-E Series Spectrum Analyzers Modulation Analysis Measurement Personality, E4402-90071, 2002.

waveform of an analog low-pass filter disappears at $t = \infty$. Thus, the response of each bit overlaps, which results in an error in bit value at the sample time or the decision time, denoted as • in Figure 9.64.

Thus, a filter giving the impulse response

$$h(nT_b) = 0 \quad n \neq 0$$

is required. As an example, the following impulse-response waveform in Equation (9.59) gives a value of 0 at $t = nT_b$ and it does not affect other adjacent bits.

$$h(t) = \text{sinc}\left(\frac{t}{T_b}\right)\frac{\cos\left(\alpha\frac{\pi t}{T_b}\right)}{1-\left(2\alpha\frac{t}{T_b}\right)^2} \tag{9.59}$$

The spectrum of this impulse response $H(f)$ is given by Equation (9.60) and is called a *raised-cosine filter*. The α here is called the roll-off factor; its value is selected between $0 < \alpha < 1$, and 0.5 is usually used.

$$H(f) = \begin{cases} 1 & f < \dfrac{1-\alpha}{2T_b} \\ \dfrac{1}{2}\left\{1 + \cos\dfrac{\pi T_b}{\alpha}\left(f - \dfrac{1-\alpha}{2T_b}\right)\right\} & \dfrac{1-\alpha}{2T_b} < f < \dfrac{1+\alpha}{2T_b} \\ 0 & f > \dfrac{1+\alpha}{2T_b} \end{cases} \tag{9.60}$$

Figure 9.68 shows the plots of $h(t)$ and its frequency response $H(f)$ given by Equations (9.59) and (9.60). In that figure, when a digital-bit waveform is passed through a raised-cosine filter, the bandwidth of the transmitted signal is determined as approximately $1/T_b$.

9.6.2 Envelope Simulation

Most digital modulated signals in digital communication systems can be expressed as shown in Equation (9.61).

$$x(t) = \text{Re}\left(A(t)e^{j\omega_c t}\right) \tag{9.61}$$

The complex amplitude or phasor $A(t)$ is called an *envelope* and is assumed to be a slowly varying waveform compared to the carrier, while signal $x(t)$ can be viewed as a sine wave with the time-varying envelope as its amplitude. An example of the waveforms is shown in Figure 9.69.

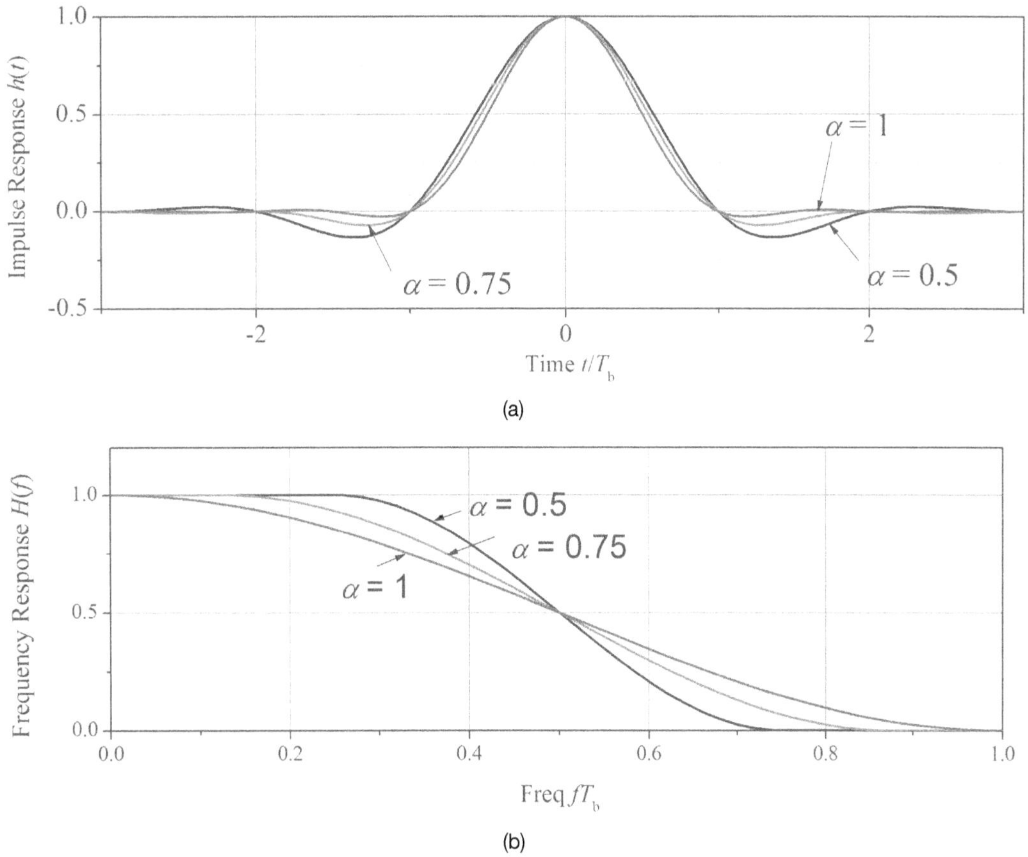

Figure 9.68 Raised-cosine digital filter: (a) impulse response, and (b) frequency response. Note that $h(t)$ is zero for $t = nT_b$. The bandwidth of the low-pass raised-cosine filter is about $1/(2T_b)$.

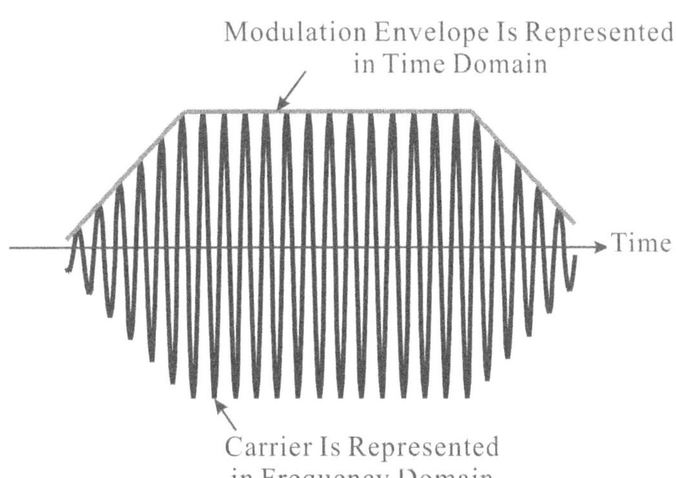

Figure 9.69 Example of a modulated signal in the time domain. The signal can be represented by the product of the time-varying envelope and carrier.

Figure 9.70 shows this signal behavior in the frequency domain. Here, the sine-wave envelope signal is chosen. The signal can be considered as a sine wave having amplitude A_1 at time t_1; the spectrum for the signal at time t_1 appears on the right side of Figure 9.70. Similarly, at time t_2, the signal can be considered as a sine wave having amplitude A_2. Thus, the modulated digital signal given by Equation (9.61) can be seen as a carrier whose amplitude is slowly varying with time in the frequency domain.

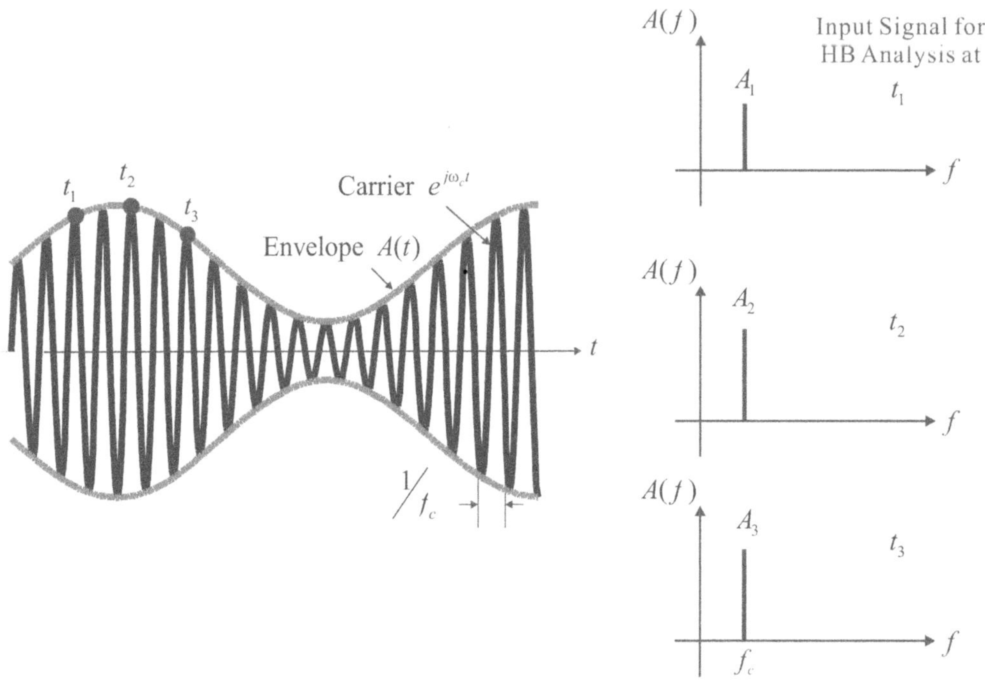

Figure 9.70 Spectrum of an amplitude-modulated waveform in the frequency domain. The carrier power in the frequency domain slowly varies with time.

Figure 9.71 shows the concept of the approximate circuit analysis for the envelope-modulated input signal. First, harmonic balance simulation is performed for the input sine wave with constant amplitude $V(t_1)$ at sample time t_1. The resulting output can be represented by harmonics and the amplitude of the fundamental frequency becomes $A(t_1)$, as shown in Figure 9.71. Here, $A(t_1)$ is called the envelope for the fundamental carrier and other envelopes for other harmonics can be similarly defined. The amplitude $A(t_1)$ is the approximate envelope of the time-varying output signal that corresponds to the time-varying input envelope signal. Through sequential harmonic balance simulations (also called envelope simulation), the envelope of each harmonic is obtained from the simulation results. Thus, the time-domain waveform of the envelope can be observed. Note that the envelope can be interpreted as a time-varying phasor. Furthermore, by expanding the envelope waveform in a Fourier series, the spectrum of each envelope can be seen in the frequency domain. Envelope simulation is used for analyses such as determining the ACPR of a power amplifier, the transient response of an oscillator, the tracking response of a PLL (phase locked loop), and so on. The details of the envelope simulations can be found in the ADS manual.

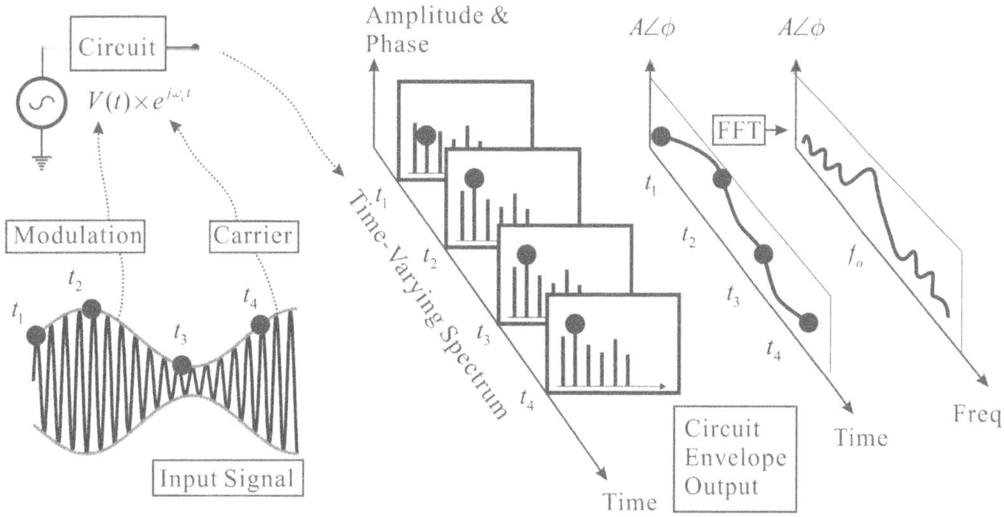

Figure 9.71 Envelope simulation concepts. The spectrum of the time-varying envelope can be obtained through FFT.

Example 9.10

Open the **QAM_16** in the ADS **examples/Tutorial/ModSources_prj** directory and plot the $I(t)$ and $Q(t)$ waveforms. Also, plot the waveform after it passes through the raised-cosine filter. In addition, plot the constellation as well as the spectrum of the modulated waveform. Finally, plot the probability distribution of the time-varying output power of the modulated signal.

Solution

Figure 9E.19 shows the opened view of the **QAM_16** file.

In practice, the baseband $I(t)$ and $Q(t)$ signals for a 16 QAM signal is obtained by serial-to-parallel data conversion followed by digital-to-analog conversion. For a serial digital-bit stream input, two serial bits are paired into a two-bit parallel data by a serial-to-parallel data converter. Through the serial-to-parallel data conversion, a pair of two-bit parallel data corresponding to the $I(t)$ and $Q(t)$ signals can be generated. Each of the two-bit parallel data are then converted to an analog signal through a digital-to-analog converter (DAC). However, baseband $I(t)$ and $Q(t)$ signals are directly generated by removing the serial-to-parallel converters and the DACs in **QAM_16** in Figure 9E.19. The values of MSB (most significant bit) and LSB (least significant bit) are specified as

$$V_{\text{MSB}} = \pm\frac{\mathbf{2V\_peak}}{3\sqrt{2}}, \quad V_{\text{LSB}} = \pm\frac{\mathbf{V\_peak}}{3\sqrt{2}}$$

The factor $3(2)^{1/2}$ is for normalization. When the maximum vector magnitude of the constellation is set to 1, the projection of the MSB onto the I-axis has a magnitude of $1/3(2)^{1/2}$. Therefore, **I1** in Figure 9E.19 represents the sum of the two signals that correspond to the MSB and LSB. The $Q(t)$ signal **Q1** is similarly constructed. As the MSB and LSB signals must have random distributions, a digital feedback method called **Tap and Seed** is used to generate a random distribution.

The reason for setting the variable **V_peak** is to set the average power of the IQ-modulated signal to 1. Since the modulated signal $x(t)$ from Equation (9.57) can be expressed as Equation (9.62),

$$x(t) = \sqrt{I^2(t) + Q^2(t)}\cos(\omega_c t + \phi) \tag{9.62}$$

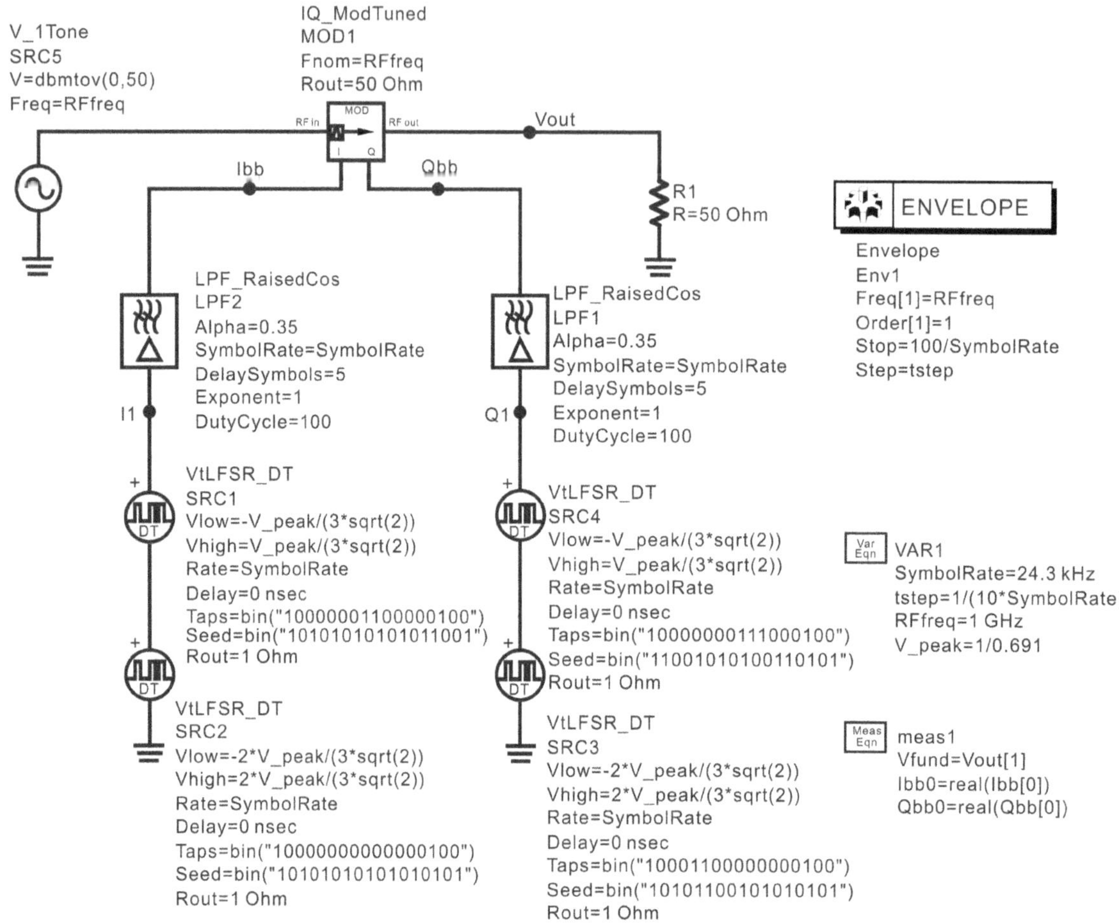

Figure 9E.19 A 16-QAM signal generation. The baseband sources **SRC**1, **SRC**2, **SRC**3, and **SRC**4 generate random bipolar digital signals. **SRC**1 and **SRC**4 are for LSB while **SRC**2 and **SRC**3 are for MSB. The two-sum digital signals in the I and Q channels are filtered by the low-pass raised-cosine filter with a bandwidth of the symbol rate. The IQ modulator **MOD**1 modulates the filtered digital signals using RF carrier **V_1Tone** of power 0 dBm. Actually, the IQ digital signals should be generated by serial-to-parallel data conversion and DAC; however, for simplicity in simulation, they are generated by the sum of the MSB and LSB digital signals. Although the power of the RF carrier is set to 0 dBm, the average modulated power is not 0 dBm. To make the power level 0 dBm, the digital signal amplitude **V_peak** is adjusted for 0 dBm average power.

the power of signal $x(t)$ is proportional to $(I(t) + Q(t))^{½}$. The power varies according to the random IQ modulation signals. The average power of an IQ-modulated signal must be set to 1. The value of **V_peak** is set to 1/0.691 for this purpose. The value of **V_peak** must be readjusted when a different random signal, say PRBS (pseudo-random binary sequence), is employed.

The generated baseband IQ signals pass through the raised-cosine digital filters and are converted into smooth waveforms in the time domain. Here, the exponent of the raised-cosine filter is 1 and the roll-off factor is set at 0.35, giving a 0.35 roll-off factor raised-cosine filter. Next, both smoothed $I(t)$ and $Q(t)$ signals are applied to the modulation inputs of an IQ modulator. The RF input signal is a sinusoidal voltage source. The source voltage corresponding to a power of 0 dBm under a 50-Ω load condition is set using the function **dbmtov**(·). The symbol rate in the variable block named **VAR**1 is set to 24.3 kHz, which is equal to the symbol rate of the raised-cosine filter.

The envelope simulation setting is similar to the transient simulation setting. The time step **tstep** is set to 10 times the symbol rate. Stop time is set to 100/**SymbolRate**, which corresponds to 100 symbols. The voltages defined as **Ibb** and **Qbb** are the filtered baseband signal, and a

Measurement Expression is used to view **Ibb** and **Qbb** at the output. In the case of HB simulation, since baseband signals correspond to DC signals, the outputs are obtained by setting **real**(**Ibb**[0]) and **real**(**Qbb**[0]). Figure 9E.20 shows the waveform before and after it passes through the raised-cosine filter. In Figure 9E.20, the waveform after passing through the filter is delayed by 5 symbols and it has become considerably distorted compared to the input waveform. However, the values at a sample time, which are marked as •, have not been affected.

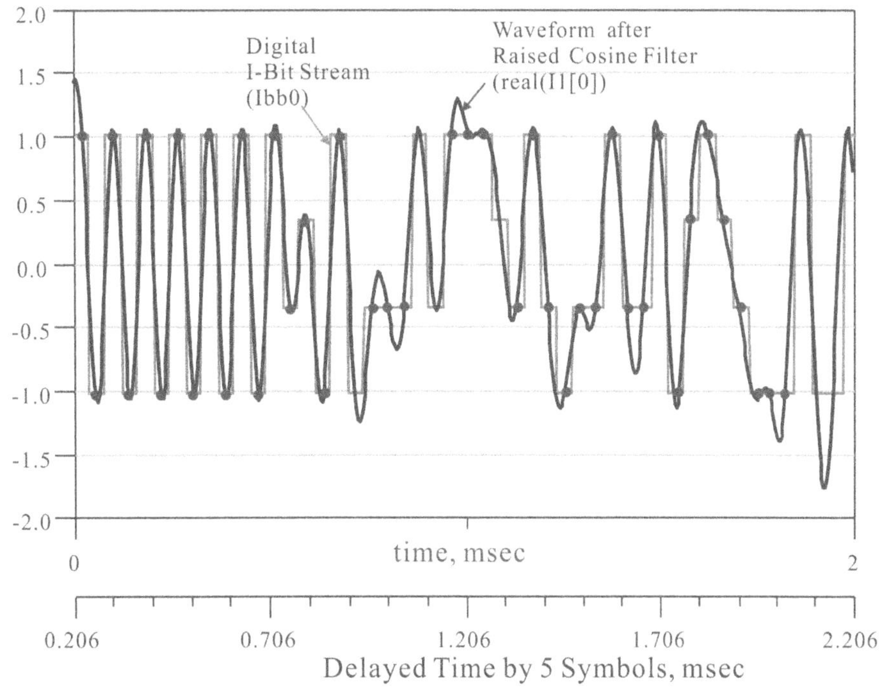

Figure 9E.20 Waveforms before and after passing through the raised-cosine filter (RCF). Although the RCF-filtered waveform is significantly distorted, it has exactly same value as the $I(t)$ at the sample time. The RCF-filtered waveform is drawn along the bottom time axis and is delayed by 5 symbols, 2.056 msec.

To draw the constellations, the equations shown in Measurement Expression 9E.3 are entered in the display window. The **Rotation** here is entered to rotate the constellation points and, as the value is currently set to 0, it does not affect the constellation plot.

```
Eqn time_pts=indep(Vfund)
Eqn tstep=time_pts[1]-time_pts[0]
Eqn Rotation=0
Eqn delay=0.5/Symbol_Rate[0]-1.5*tstep
Eqn Vrotated=Vfund*exp(j*Rotation)
Eqn Vreal=real(Vrotated)
Eqn Vimag=imag(Vrotated)
Eqn Traj=vs(Vimag,Vreal)
Eqn Const=constellation(Vreal, Vimag, Symbol_Rate[0], delay)
```

Measurement Expression 9E.3 Equations inserted to obtain the constellations

The variable **Vfund** in Figure 9E.19 is defined as the envelope of the fundamental carrier frequency. Thus, the real part of **Vfund** corresponds to the $I(t)$ signal while the imaginary part corresponds to the $Q(t)$ signal. The trajectory is drawn using the variable **Traj** shown in Figure 9E.21. The trajectory is a continuous locus of time between $I(t)$ and $Q(t)$, including I and Q values at sample times. To draw the constellation plot, the sample time must be determined, which will result in the constellation points showing minimized jitters. The sample rate is basically equal to the symbol rate but the starting sample time should be determined for minimum jitters in the constellation points. The delay variable in Measurement Expression 9E.3 is inserted in order to optimally set the sample time. For the selected **delay**, the jitters of the constellation points are minimized and will appear as shown in Figure 9E.21. For the QPSK, the function for determining the optimum sample time exists in ADS, but because the optimum sample time is usually not given, it is determined by trial and error. It can be seen in Figure 9E.21 that 16 constellation points appear at the normal positions.

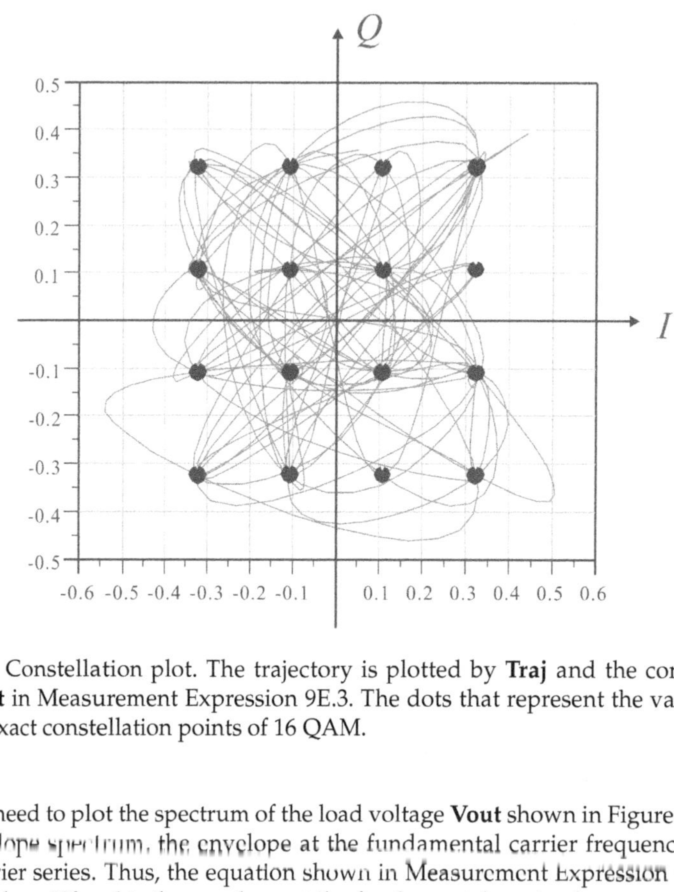

Figure 9E.21 Constellation plot. The trajectory is plotted by **Traj** and the constellation plot is drawn by **Const** in Measurement Expression 9E.3. The dots that represent the value at the sample time are at the exact constellation points of 16 QAM.

Next, we need to plot the spectrum of the load voltage **Vout** shown in Figure 9E.19. In order to obtain the envelope spectrum, the envelope at the fundamental carrier frequency should be converted in a Fourier series. Thus, the equation shown in Measurement Expression 9E.4 is entered in the display window. **Vfund** is the envelope at the fundamental carrier frequency and the function **fs**(·) converts **Vfund** into a Fourier series.

Eqn Spectrum=dBm(fs(Vfund,,,,,"Kaiser"))

Measurement Expression 9E.4 Equation for obtaining the spectrum of the output voltage

The spectrum is shown in Figure 9E.22. Since the symbol rate in Figure 9E.19 has been set to 24.3 kHz, the spectrum is almost 0 at frequencies beyond the symbol rate. Most of the spectral powers can be

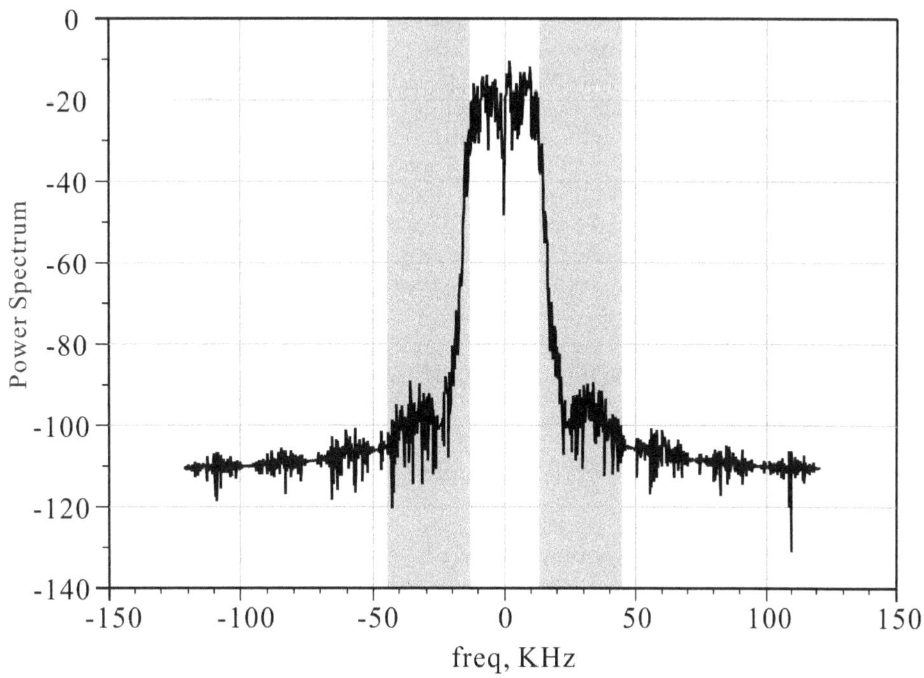

Figure 9E.22 Spectrum of the fundamental frequency envelope. Most of the spectral powers appear within the bandwidth determined by the symbol rate. The shaded frequency bands are the adjacent upper and lower channels. The leakage power to adjacent channels appears, although it is small, and it is specified using the ACPR metric, which calculates for the selected high- and low-side bandwidths.

observed to be gathered within ± 24.3 kHz. An ACPR calculation is often required for the envelope spectrum, for which the equations shown in Measurement Expression 9E.5 are entered in the display window.

Eqn mainlimits={-16.4 kHz, 16.4 kHz}

Eqn UpChlimits={mainlimits+30 kHz}

Eqn LoChlimits={mainlimits-30 kHz}

Eqn Main_Channel_Power = channel_power_vr(Vfund, 50, mainlimits, "Kaiser")

Eqn TransACPR=acpr_vr(V_fund, 50, mainlimits, Lochlimits, UpChlimits, "Kaiser"}

Measurement Expression 9E.5 Equations entered in the display window to calculate ACPR

The variable **mainlimits** in Measurement Expression 9E.5 represents the bandwidth. The variable **UpChlimits** represents the bandwidth moved up by 30 kHz, while the variable **LoChlimits** represents the bandwidth moved down by 30 kHz. The upper and lower channels are also shown in Figure 9E.22. The function that computes the power within the given bandwidth is **channel_power_vr**(·) and is based on the voltage and resistance. Thus, **Main_Channel_Power** represents the power within the **mainlimits** bandwidth.

The ACPR, the ratio of the power in the main bandwidth to the powers in the adjacent bandwidths, can be determined by the repeated use of the **channel_power_vr**(·)function described above. However, as there is a simple function **acpr_vr**(·) in ADS, the ACPR can be determined by using the **acpr_vr**(·) function shown in Measurement Expression 9E.5. The **Trans_ACPR** shown in that expression compares the power in the **mainlimits** to the powers in the **LoChlimits** and **UpChlimits**, and stores the results as **Trans_ACPR**(1) and **Trans_ACPR**(2), respectively, in dB. The values of **Trans_ACPR**(1) and **Trans_ACPR**(2) are −28.821 dB and −27.016 dB.

Finally, the power of the modulated signal varies with time and this leads to a probability distribution, which can be plotted using the **pdf**(·) function of ADS. Here, the output power and the probability distribution are set as shown in the following equations. The resulting probability distribution is shown in Figure 9E.23.

```
Eqn PdBm=10*log10(mag(Vfund)**2/(2*50))+30
Eqn y=pdf(PdBm)
```

Measurement Expression 9E.6 Equations for computation of the probability density function

In Figure 9E.23, the average output power is set to 0 dBm, which can be seen to show the highest probability.

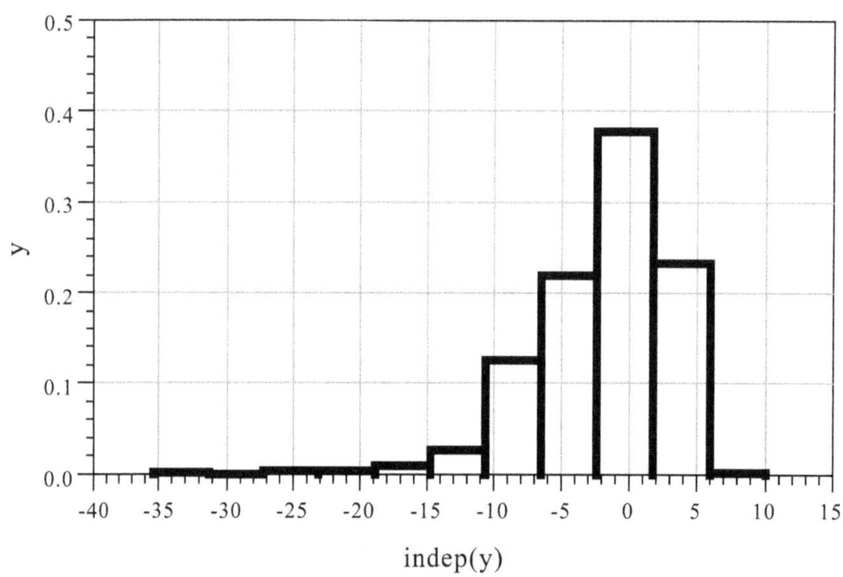

Figure 9E.23 Probability distribution of the modulated signal's power. Since the average power level is set to 0 dBm using the variable **V_peak**, the power 0 dBm shows the highest probability.

9.6.3 Two-Tone and ACPR Measurements

In the past, the two-tone test was the most commonly used method for evaluating the linearity of power amplifiers. In this method, two CW inputs with the specified frequency spacing are applied to the power amplifier and the output spectrum of the amplifier is then measured. Figure 9.72 shows the measurement method conceptually. When a two-tone signal with a specified frequency spacing is applied to a power amplifier, two

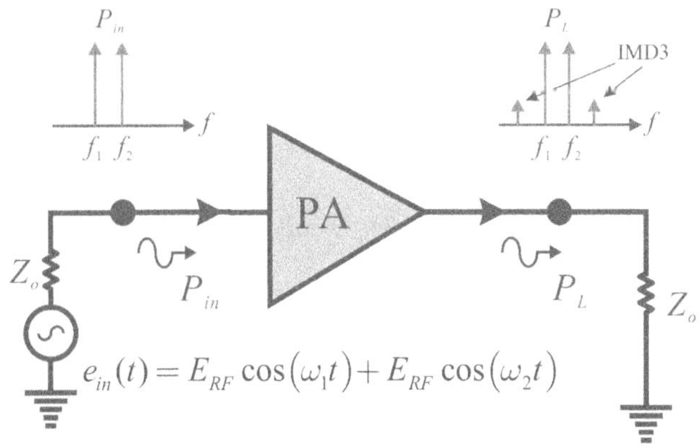

Figure 9.72 Two-tone measurement. Due to third-order nonlinearity, the IMD3 appears at the power amplifier output.

distorted signals with the same frequency spacing above and below the two input frequencies appear at the power amplifier's output. In a communication system using frequency division multiplexing, the frequency spacing corresponds to the channel spacing. The distorted signals appear at the adjacent channels as a result of the power amplifier distortion and they become interference signals to users communicating on the adjacent channels. The distorted signals generally show a third-order relationship to the input power. Therefore, a distorted signal usually increases in proportion to the cube of the input power and has a slope of 3 in a log-log plot. The amplified signal, on the other hand, increases in proportion to the input power. Due to the difference in slope, the extensions of these two lines intersect at a point that is referred to as a TOI (third-order intercept). As an alternative description, the ratio of the amplified signal to the distorted signal is used and is referred to as an IMD3 (third-order intermodulation distortion). The definitions of TOI and IMD3 are shown in Figure 9.73.

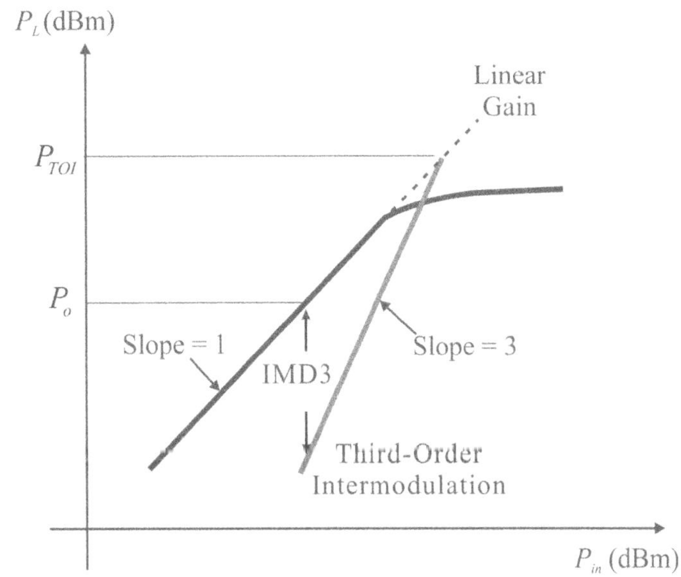

Figure 9.73 Definitions of IMD3 and TOI

From Figure 9.73, the TOI and IMD3 are not independent and measuring the IMD3 determines the TOI as expressed in Equation (9.63).

$$\text{TOI} = \frac{\text{IMD3}}{2} + P_o \tag{9.63}$$

Here, P_o as a tone output power represents the power at which the IMD3 is measured.

Example 9.11

For the power amplifier circuit designed in section 9.5, set the center frequency to 2.5 GHz, the input power per tone to 18 dBm, and the frequency spacing to 10 MHz. Calculate the IMD3 and the TOI. The reason for setting the power per tone to 18 dBm is to set the input power to 21 dBm because the sum of the two input powers of 18 dBm is 21 dBm. Note that the previously designed power amplifier has an output power of 36 dBm when the input power is 21 dBm. Plot the output tone power and the third-order distorted power for the input power change per tone from –10 dBm to 18 dBm.

Solution

After configuring the designed power amplifier circuit as a subcircuit, the ADS **P_nTone** signal source is used as the input source and the circuit is set up as shown in Figure 9E.24 to observe the two-tone response. In order to view the output tone power and third-order distorted power, after opening the Sweep tab of the Harmonic Balance simulation controller shown in Figure 9E.24, **pwr_in** is specified as the sweep parameter and adjusted from –10 dBm to 18 dBm in steps of 0.5 dB.

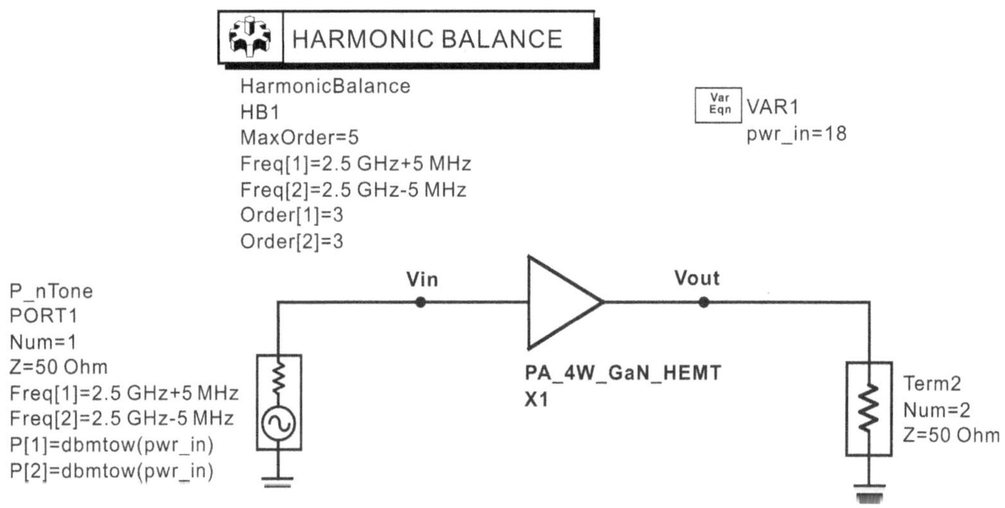

Figure 9E.24 Simulation schematic for the IMD3. Frequency spacing for two-tone simulation is set to 10 MHz. The tone power level is set to 18 dBm because the PA yields 36 dBm at a 21-dBm input power.

The calculated power spectrum at 18-dBm input power is shown in Figure 9E.25. **Vout**[56,::] corresponds to the output voltage for the input tone power of 18 dBm. The IMD3 in Figure 9E.25 can be seen to be approximately 16 dB. The combined output power of the two-tone signals is computed to be 35.343 (=32.343 + 3) dBm, which is about 1 dB less than the expected power. This is because the output power corresponding to the 1-dB difference is distributed into many spurious components.

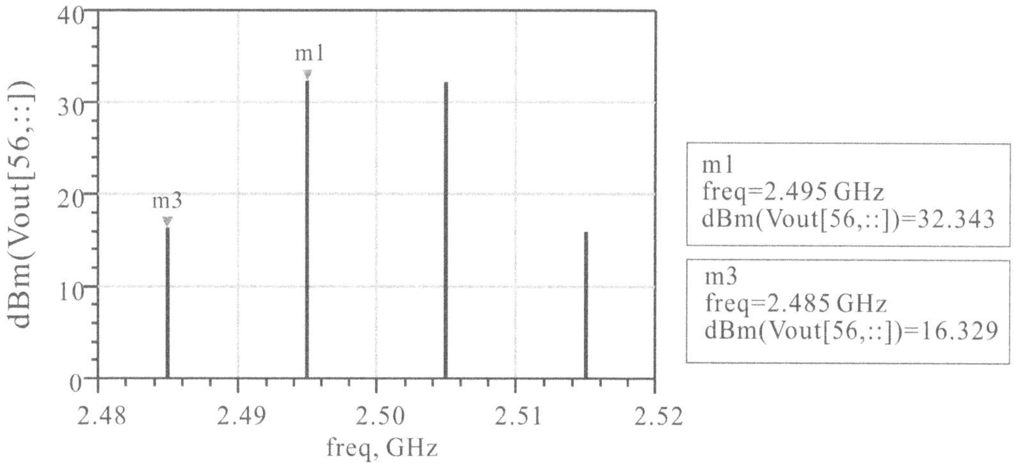

Figure 9E.25 Simulated output spectrum at input power 18 dBm. The 18-dBm input power that corresponds to the sweep index 56 IMD3 is about 16.329 dBm.

In addition, in order to calculate the TOI in accordance with Equation (9.63), the equation in Measurement Expression 9E.7 is entered in the display window, which gives a TOI = 40.35 dBm.

Eqn TOI=(3*m1-m3)/2

Measurement Expression 9E.7

The output tone power and the IMD3 power are shown in Figure 9E.26.

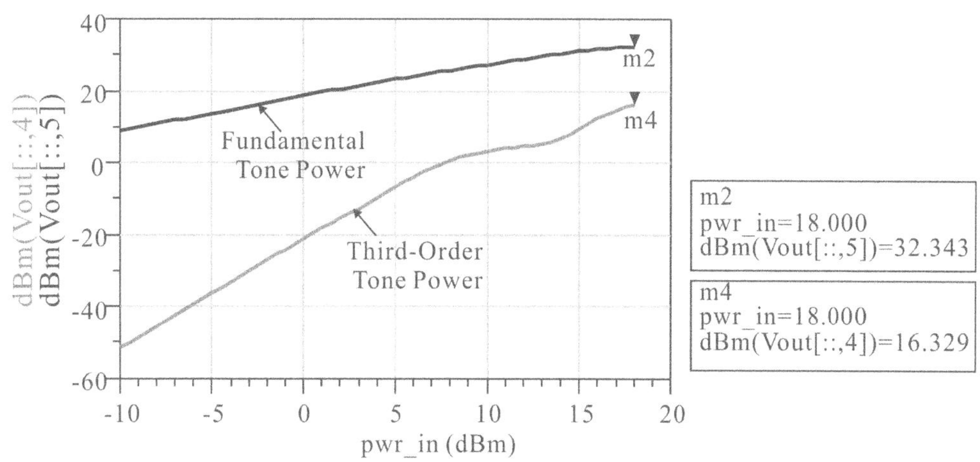

Figure 9E.26 Output tone power and third-order distorted power for the two-tone input power **pwr_in**

Here, **freq**[::, 4] corresponds to the frequency of the third-order distortion of 2.485 GHz and **freq**[::, 5] corresponds to the input frequency of 2.495 GHz. For an 18-dBm tone input power, the tone output power is 32.3 dBm and corresponds to the combined power of the two-tone output power, which is

35.3 dBm. As mentioned earlier, this is lower than the expected output power because a portion of the output power is distributed into many spurious components.

The previously explained two-tone method is the appropriate evaluation tool for the metric of power amplifier linearity used in communication systems employing frequency-division multiplexing. However, the two-tone method does not give direct criteria for power amplifier linearity in digital communication systems. The spectrum of digitally modulated signals is typically spread over a wide frequency range and the two-tone method does not provide a direct evaluation for the interference effects on other channels. As a result, in digital communication systems employing CDMA (code division multiple access), power amplifier linearity is evaluated by ACPR (adjacent channel leakage power ratio). ACPR can generally be defined as

$$\text{ACPR} = \frac{\text{Leakage power to adjacent channel}}{\text{Main channel power}}$$

However, although conceptually the same, the ACPR definition may depend on the type of communication system. In addition, various communication systems' standards are usually referred to in the evaluation of the ACPR of a power amplifier.

Example 9.12

The CDMA forward-link signal source in ADS is applied to the input of the power amplifier designed in section 9.5, as shown in Figure 9E.27. Set the input power level of the CDMA forward-link signal source to 12 dBm, which corresponds to a 9-dB back-off from the input power of 21 dBm that yields a 36-dBm output power. Determine the ACPR.

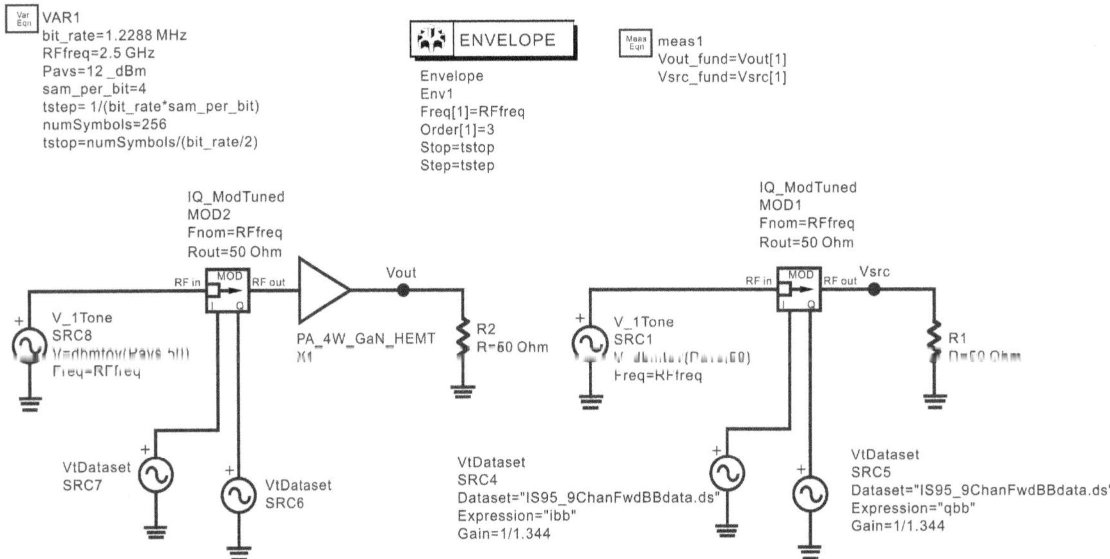

Figure 9E.27 Simulation schematic for calculating the ACPR. The right-hand side obtains the spectrum of the CDMA forward-link source.

Solution

Forward link refers to the link from the base station to the handset, while reverse link refers to the link from the handset to the base station. There is a difference between the two links because they use different modulation schemes. The left side of the schematic in Figure 9E.28 shows a CDMA forward-link signal source applied to the designed power amplifier circuit. The circuit at the right side of Figure 9E.27 is inserted for the analysis of the signal source.

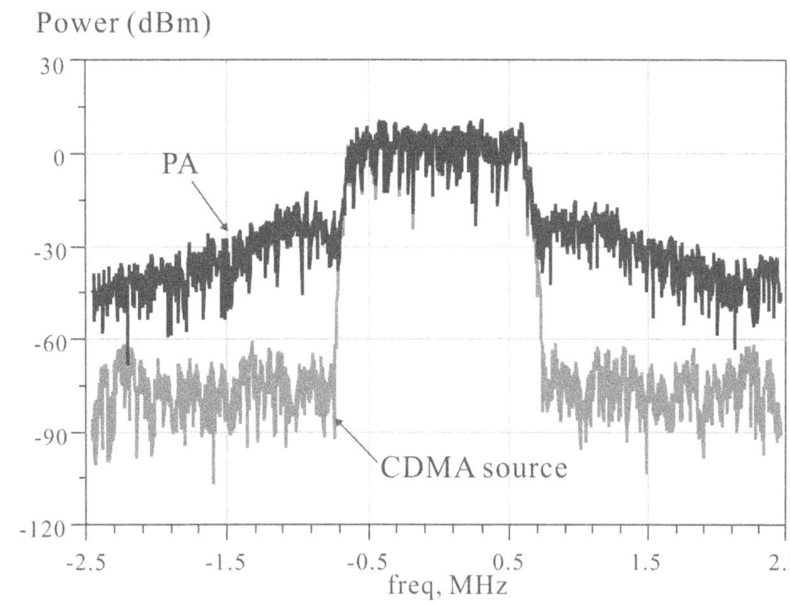

Figure 9E.28 Simulated output spectra. The upper trace is the PA output spectrum, while the lower trace shows the spectrum of the CDMA forward link source. The PA spectrum outside the bandwidth grows significantly.

In Korea, the center frequency of communication systems that employ a CDMA is 1.9 GHz but the center frequency of the designed amplifier is 2.5 GHz. Thus, the center frequency in this example is changed to 2.5 GHz. In addition, to improve the linearity in the designed power amplifier, the input power is set to a 12-dBm back-off of 9 dB from 21 dBm.

The signal shown in Figure 9E.27 represents the signal source modulated by a 1.288-MHz bit rate. The time step of the envelope simulation is defined to be one-quarter of the bit rate. This is the recommended parameter value in ADS. In addition, the number of symbols is set to 256 in order to be able to see all the variations of the signal source. Since the order is 3 in the Harmonic Balance controller, **Vout** in Figure 9E.27 shows three harmonics. To obtain the fundamental components of **Vout** and the signal source, **Vout_fund=Vout**[1], **Vsrc_fund=Vsrc**[1] are entered in the Measurement Expression shown in Figure 9E.27. **Vout_fund** and **Vsrc_fund** thus represent the time-varying envelopes of the fundamental carrier. The signal source and power amplifier output powers within the bandwidth can be calculated by summing the spectral powers within the bandwidth. The ADS function for this purpose is **channel_power_vr**(·). As explained earlier, this function calculates the power based on the voltage and resistance by summing the spectral powers within the bandwidth. The bandwidth **mainCh** in Measurement Expression 9E.8 is determined by the given **bit_rate**. The signal source and power amplifier output powers within the bandwidth can thus be calculated in dBm by entering the following equations in the display window:

Eqn mainCh={-(1.2288MHz/2),(1.2288MHz/2)}

Eqn PA_pwr=10*log10(channel_power_vr(Vout_fund,50, mainCh, "Kaiser")+30

Eqn Src_pwr=10*log10(channel_power_vr(Vsrc_fund,50, mainCh, "Kaiser")+30

Measurement Expression 9E.8

The calculated powers of the power amplifier output and signal source are 31.061 and 14.141 dBm, respectively. In the case of the signal source, the actual signal source power is approximately 2.141 dB larger than the input power set to 12 dBm. Therefore, the gain of the signal source must be reduced. Reducing the gain of the signal source by 1/1.72 results in a signal source power of 11.998 dBm, which is close to the desired input power level. Under this condition, the power amplifier shows an output power of 29.338 dBm. After making these settings and performing the simulation, the equations in Measurement Expression 9E.9 are entered in the display window to determine the signal source and the power amplifier's output spectrums.

Eqn Sp_src=dBm(fs(Vsrc_fund,,,,,"Kaiser"))

Eqn Sp_PA=dBm(fs(Vout _fund,,,,,"Kaiser"))

Measurement Expression 9E.9

Here, the function **fs**(·)converts a time-domain function into the frequency-domain spectrum using a Fourier series. Thus, these equations represent the spectra of **Vout_fund** and **Vsrc_fund**. Figure 9E.28 shows the spectra obtained from the simulation.

From Figure 9E.28, the adjacent channel leakage power of the signal source is found to be low, while that of the power amplifier output shows a significant increase. This degrades the ACPR performance. As a result, the following equations in Measurement Expression 9E.10 are entered in the display window to calculate the ACPR defined for CDMA communication systems:

Eqn UpCh={885 kHz, 915 kHz}

Eqn LoCh={-915 kHz, -885 kHz}

Eqn ACPR=acpr_vr(Vout_fund, 50, mainCh, LoCh, UpCh,"Kaiser")

Measurement Expression 9E.10

Function **acpr_vr**(·) determines the ratio of the powers within the bandwidth and in the upper and lower sidebands. The resulting ACPRs of the power amplifier are –45.96 dBc and –44.03 dBc for the upper and lower sidebands, respectively. It is worth noting that the bandwidths of the upper and lower sidebands are set narrower than the signal bandwidth. The ratio of the signal bandwidth to the bandwidths of the upper and lower sidebands is

$$\frac{1.2288\text{ MHz}}{(915-885)\text{kHz}} = 40.96$$

If the bandwidth ratio is normalized to 1, the ACPR will be degraded by 16.12 dB.

9.6.4 EVM Simulation

To some extent, the previously described ACPR can be used to assess whether a power amplifier is adequate for a digital communication system; however, the ACPR is still an indirect measure. The most direct method is to measure the BER (bit error rate) when the power amplifier is used in a digital communication system. However, the BER may be affected by the variety of components that constitute the digital communication system, making it difficult to estimate the degradation of the BER coming from the power amplifier alone. Another problem is that BER requires the analysis of more than 10^6 symbols to find the BER's degree of degradation. Thus, the computation time for BER simulation increases to an intolerable level. In order to simulate BER efficiently, a DSP (digital signal processing) simulator, which is not discussed in this book, is used together with the envelope simulation. Co-simulation with DSP is an efficient method for BER analysis. Refer to the ADS documentation in reference 6 at the end of this chapter for details. An alternative method is to perform the EVM analysis discussed in section 9.6.1. After the demodulation of the received signal is completed, plotting $I(t)$ and $Q(t)$ at sampling times gives the constellation plot, which jitters around the normal constellation points of $I(t)$ and $Q(t)$ due to distortions in the power amplifier. The power amplifier's suitability can be determined by looking at the degree of jitter in the constellation points. Of course, in this method as well, the contributions of other components in the communication system can appear, but only the power amplifier's contribution can be found. In the measurement given by Equation (9.58), the EVM shows the relative jitters of a constellation reference point; thus, the degree of jitter in the constellation points due to the power amplifier can be found using EVM and thereby the suitability of the power amplifier can be determined.

Example 9.13

Using the simulation results of the power amplifier circuit in the previous Example, obtain the EVM.

Solution

The same simulation schematic in Figure 9E.27 can be used. For the simulation results obtained from Example 9.12, the equations in Measurement Expression 9E.11 are entered in the display window to obtain the EVM.

```
Eqn time_pts=indep(Vsrc_fund)
Eqn tstep=time_pts[1]-time_pts[0]
Eqn delay = sample_delay_qpsk(Vout_fund, bit_rate[0], 0, tstep)
Eqn data = const_evm(Vsrc_fund, V_fund, bit_rate[0], delay, 0, 0, 0)
```

Measurement Expression 9E.11

The variable **time_pts** is defined using the independent variable of **Vsrc_fund**. Thus, **time_pts** becomes a time variable. Variable **tstep** is the time step in the envelope simulation. Entering the equations above can be avoided by specifying the output variable as **tstep** in the output tab of the envelope simulation controller. The optimum sample time may be between 0 and 1/bit_rate = 1/1.2288 MHz = 0.8138 μsec; however it is difficult to find the optimum sampling time manually, which minimizes the jitters of the constellation points. Thus, using the function **sample_delay_qpsk**(·), the optimum sampling time can be obtained. The optimum sampling time obtained is 0.2035 µsec. It is also noteworthy that the optimum sampling time depends on the modulation schemes. Since the CDMA forward-link modulation is a QPSK, the optimum sampling time can be calculated using the **sample_delay_qpsk**(·) function. The function **const_evm**(•) calculates the EVM by comparing the reference constellation points with the measured constellation points. The EVM results are stored in **data** in Measurement Expression 9E.11. The output **data**[0] and **data**[1] represent the constellation plot and trajectory of the reference signal at the given sample time, while **data**[2] and **data**[3] represent the constellation plot and trajectory of the signal to be compared. The constellation plots of the reference signal and those of the signal for comparison are shown in Figure 9E.29, respectively. Next, **data**[4] shows the error with respect to time, and **data**[6] shows the value of the EVM as a percentage. The resulting value of **data**[6] is 3.84%. Small jitters are found to occur at the constellation points compared with the reference QPSK constellation points.

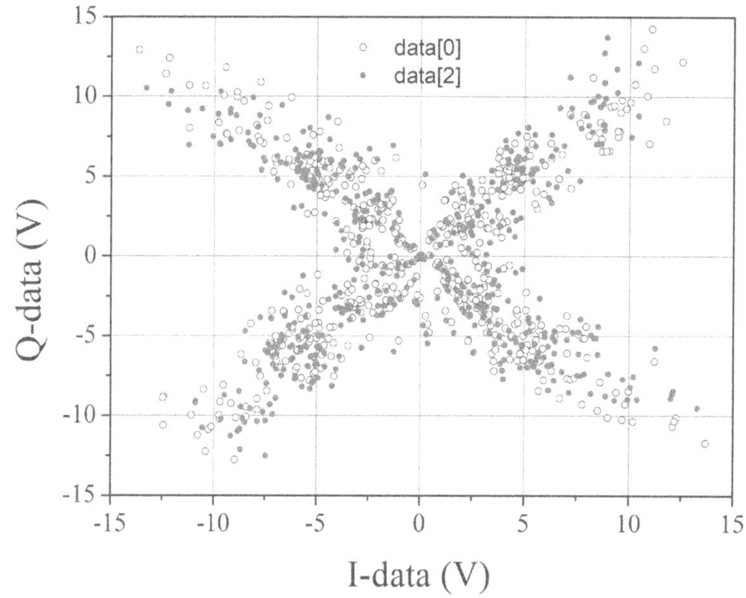

Figure 9E.29 Simulated reference (**data**[0]) and power amplifier output (**data**[2]) constellation plots. The function **const_evm**(·) gives the reference constellation plots as the output variable of index 0 and the constellation plots of comparison signal as the output variable of index 2.

The constellation points appearing at the origin of the IQ plane in Figure 9E.29 are due to QPSK modulation. In that modulation, transitions from (1,1) to (0,0) or from (0,1) to (1,0) cause the trajectory to pass through the origin in the IQ plane, which results in the constellation points near the origin in the plane. Such transitions make the value of the envelope 0 and cause a large amplitude change. Therefore, PAPR (peak-to-average power ratio) becomes larger, which causes problems in the efficient driving of the power amplifier. This can be solved using the **π/4-DQPSK** modulation scheme. Refer to references 2 and 5 at the end of this chapter for more details.

9.7 COMPOSITE POWER AMPLIFIERS

As mentioned previously, most power amplifiers distort the input signal to some degree. Thus, the linearity of a given power amplifier may be the key issue in determining its applicability in a communication system. Most power amplifiers have a satisfactory linearity at a low-output power level; however, significant distortion occurs as the output power approaches saturation. To avoid this problem in the past, it was common practice to use the power amplifier at an output power level backed off (by a certain amount) from the saturated power. Usually, a 3-dB back-off power level from the saturated power was widely used. However, with the use of digitally modulated signals today, a higher back-off output power level is required. The higher back-off output power level is very unsatisfactory, considering the cost-effectiveness of the active device. RF methods such as predistorter and feed-forward techniques are commonly used to improve the linearity. Another method is based on recent advances in digital signal processing techniques that can decompose the input signal to the power amplifier into envelope- and phase-modulated carriers. The power amplifier is then set to amplify only the phase-modulated carrier while the envelope controls the amplifier's power supply, thereby improving the amplifier's linearity and efficiency. This technique is called an EER (envelope elimination and restoration). This method will be summarized in this section.

In addition, the efficiency of the power amplifier generally improves as the output power level approaches saturation; however, communication systems employing digitally modulated signals are mostly operated at low power. In addition, the power amplifier is rarely operated at its maximum output power. In this case, as the power amplifier is used mainly at the output power level with low efficiency, the amplifier's efficiency significantly drops. Thus, a power amplifier with high efficiency at a wide input power range is required and one such amplifier is the Doherty. The operation of the Doherty amplifier will be briefly discussed in this section.

9.7.1 Predistorters

Figure 9.74 shows a power amplifier with a predistorter.

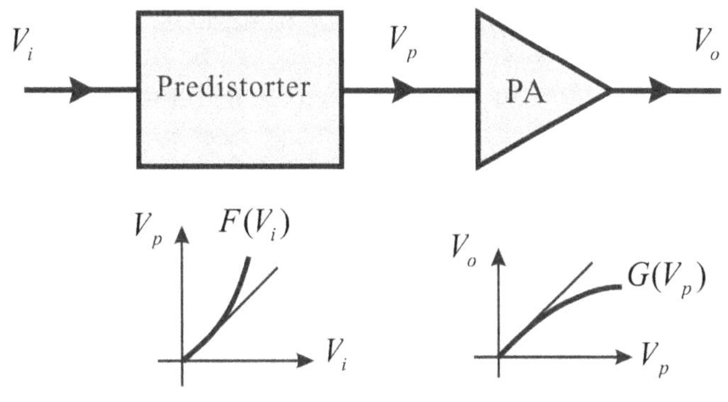

Figure 9.74 Concept of predistorter. The predistorter that compensates for the power amplifier's saturation helps to improve the amplifier's linearity

In Figure 9.74, V_i, V_p, and V_o represent the voltage amplitudes of the fundamental frequency. The typical V_o–V_p characteristic of a power amplifier is also shown in Figure 9.74. The output voltage departs from the straight line as the input voltage increases and enters saturation. If the output voltage V_p of the predistorter increases as V_i increases, as shown in Figure 9.74, the entire input-output characteristic of the system will show linearity. As a result, the linear characteristic can be improved. Figure 9.75 shows the method of obtaining the required input-output characteristic of the predistorter for a linear power amplifier. In Figure 9.75(a), the power amplifier has an output power $P_{L,1}$ for an input power $P_{in,1}$ (point A). In order to align this to the straight line, the output power $P_{L,2}$ corresponding to point B must appear at the output. The output power $P_{L,2}$ appears at the output for an input power $P_{in,2}$ instead of $P_{in,1}$, corresponding to point C. Therefore, for an input power of $P_{in,1}$, the predistorter should produce $P_{in,2}$. By repeating this process, the predistorter required for the linear power amplifier can be obtained and the linear output power range of the power amplifier can be increased up to the saturation point.

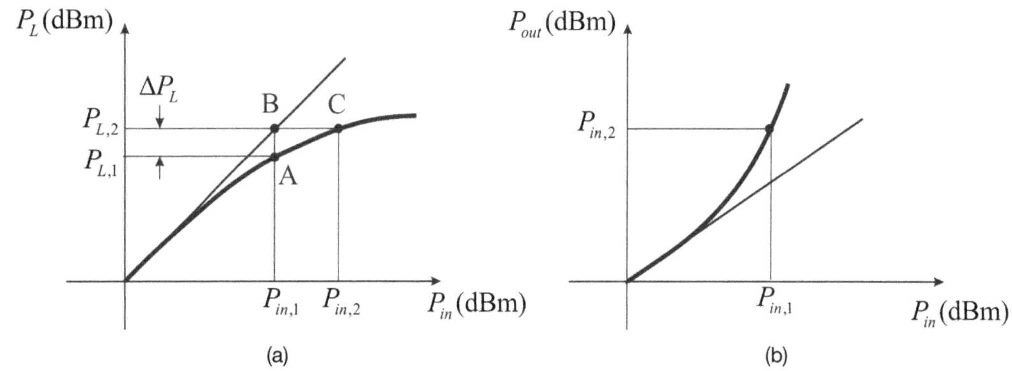

Figure 9.75 (a) Input and output characteristic of a power amplifier and (b) input and output characteristic of a predistorter. Using the P_L–P_{in} plot of the PA, the desired P_{out}–P_{in} characteristic of the predistorter can be computed.

In the previous explanation, three questions must be considered:

1. Is it possible to implement such a predistorter with the specified input-output characteristic?
2. The method in Figure 9.75 is possible only when the input-output characteristic of the power amplifier is assumed to be purely resistive. Can this method still be applied when the input output characteristic of the power amplifier is not purely resistive?
3. When configured in this way, is the linearity improvement truly achieved?

For the first question, various configurations for the predistorter are possible. For example, it can be implemented using a diode. Assuming the diode to be a purely nonlinear resistive device, it can be considered as an open circuit for a low-input power level, and as an approximate short circuit for a higher input power. Thus, using this property of the diode, the predistorter can be constructed using a well-known branch-line coupler, as shown in Figure 9.76.

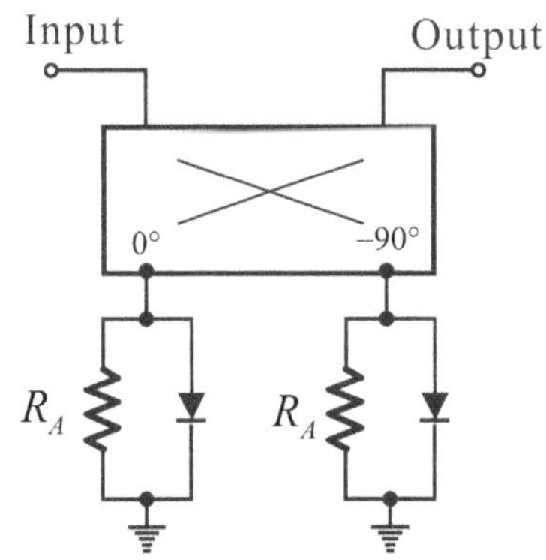

Figure 9.76 A predistorter using a branch-line coupler. For a small input power, the two diodes can be approximated as open and the small input power appears at the output due to mismatch by R_A. However, for a large input power, the two diodes approximately operate as short and the output power is almost equal to the input power due to reflection by the two diodes.

If R_A is close to the port impedance of the branch-line coupler, the predistorter provides a linear-attenuated output power due to a small mismatch by resistor R_A. Here, the diode is approximated as an open circuit because the RF input power level is low. Thus, the input-output characteristic is considered to be approximately linear. On the other hand, as the diode comes close to being a short circuit for a high RF input power, total reflection occurs from the diode and the sum of the two reflected powers appears at the output. This combined power will be close to the input power level. Thus, the attenuation is high when the input is low and the input-output characteristic is linear. On the other hand, when the input power level is high, the input-output characteristic shows a gain closer to 1. Thus, an appropriate input-output characteristic for the predistorter can be obtained. By adjusting the values of the devices in Figure 9.76, a predistorter having the required characteristic of Figure 9.75(b) can be designed. There are a variety of predistorter circuits and more details about them can be found in reference 2 at the end of this chapter.

For the second question, the distortion of power amplifiers requires further analysis. In the explanation above, by viewing the distortion of the power amplifier as purely resistive, the output voltage of the power amplifier v_L in terms of the input voltage can be written by Equation (9.64).

$$v_L = a_1 v_{in} + a_3 v_{in}^3 \tag{9.64}$$

When the input is a sinusoidal voltage given by $v_{in} = V_1 \cos\omega t$, by substituting this into Equation (9.64), the fundamental component of $V_L = \text{Fund}(v_L(t))$ can be rewritten as Equation (9.65).

$$\text{Fund}\left(v_L(t)\right) = \left(a_1 V_1 + \frac{3}{4} a_3 V_1^3\right)\cos\omega t \tag{9.65}$$

Here, depending on the sign of the coefficient a_3, the input-output characteristic shown in Figure 9.75(a) can be obtained. However, in general, the power amplifier does not take the form of Equation (9.64). It is commonly known that a time delay appears at the third-order term. Therefore, Equation (9.64) can be rewritten as Equation (9.66).

$$v_L = a_1 v_{in}(t) + a_3 v_{in}^3(t - \tau) \tag{9.66}$$

Thus, substituting the sinusoidal input $v_{in} = V_1 \cos\omega t$, and rearranging the fundamental component, it can be written as Equation (9.67).

$$\text{Fund}\left(v_L(t)\right) = a_1 V_1 \cos\omega t + \frac{3}{4} a_3 V_1^3 \cos(\omega t + \phi_3) \tag{9.67}$$

In this case, the previous resistive-predistorter design method may not work due to the differences in the phases of the fundamental and third-order distortion terms. However, the predistorter used in combination with a phase-shifter can improve the linearity of the power amplifier. This Solution is somewhat complicated, but is theoretically possible by expanding the input-output relation of the power amplifier in the form of Equation (9.66). Reference 2 at the end of this chapter can be consulted for more information.

The third question can be proved using the two-tone method. By substituting a two-tone signal into Equation (9.66) and expanding the equation, the third-order term can be removed and the linearity is clearly improved; as a result, the linearity for other modulation signals will also be improved. An ACPR improvement by 10–20 dB has reportedly been achieved, as shown in Figure 9.77. This method is simple but widely used for power amplifier linearization, and the literature currently describes several other methods.

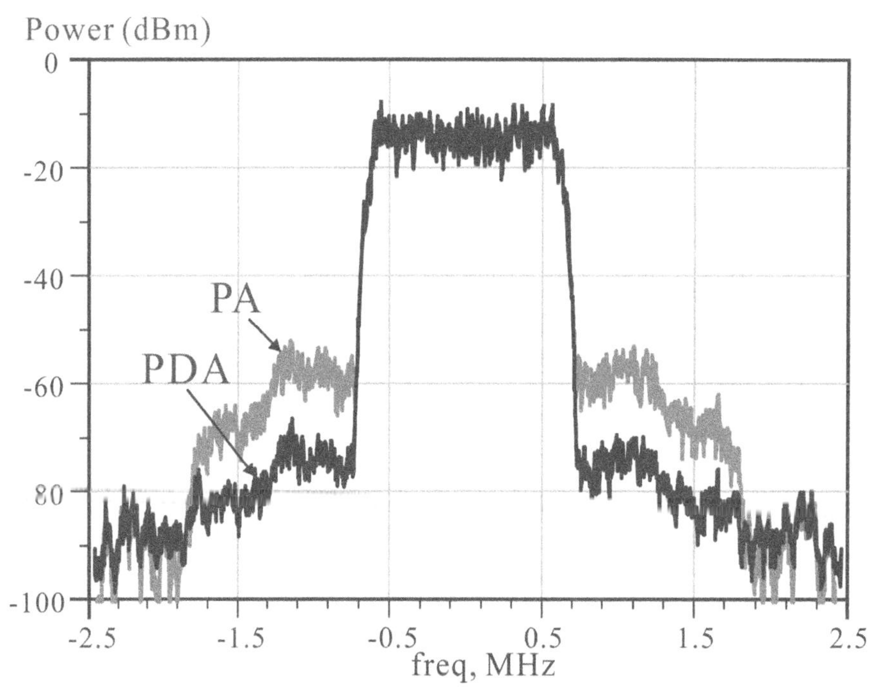

Figure 9.77 ACPR improvement of a power amplifier with a predistorter for digital modulation signal. The digital modulation signal is a IS95 forward-link signal. The spectra of the PDA (predistorted power amplifier) and power amplifier alone are compared. The adjacent power level is reduced by about 20 dB.

Example 9.14

Given that the input-output characteristic of a power amplifier is expressed as $v_L = a_1 v_i - a_3(v_i)^3$, where $a_1 = 10$ and $a_3 = 0.109a_1$, design an appropriate predistorter. In addition, compare the input-output characteristics of the power amplifiers with and without the designed predistorter. In addition, compare the output powers of the two power amplifiers for a two-tone input of a tone power $P_{in} = 10$ dB.

Solution

Suppose that the input-output characteristic of the predistorter is $v_p = v_i + b_3(v_i)^3$ and substituting this into v_i,

$$v_L = a_1 v_p - a_3 v_p^3 = a_1 v_i + b_3 a_1 v_i^3 - a_3\left(v_i + b_3 v_i^3\right)^3$$
$$= a_1 v_i + b_3 a_1 v_i^3 - a_3\left(v_i^3 + 3b_3 v_i^5 + 3b_3^2 v_i^7 + b_3^3 v_i^9\right)$$

The third-order term must be removed, which results in $b_3 = a_3/a_1$. This is configured through the VCVS (voltage-controlled voltage source) in ADS, as shown in Figure 9E.30. The simulated P_{out}–P_{in} characteristic is shown in Figure 9E.31. From Figure 9E.31, the power amplifier with the predistorter is found to show an improved 1-dB compression point compared with the original power amplifier.

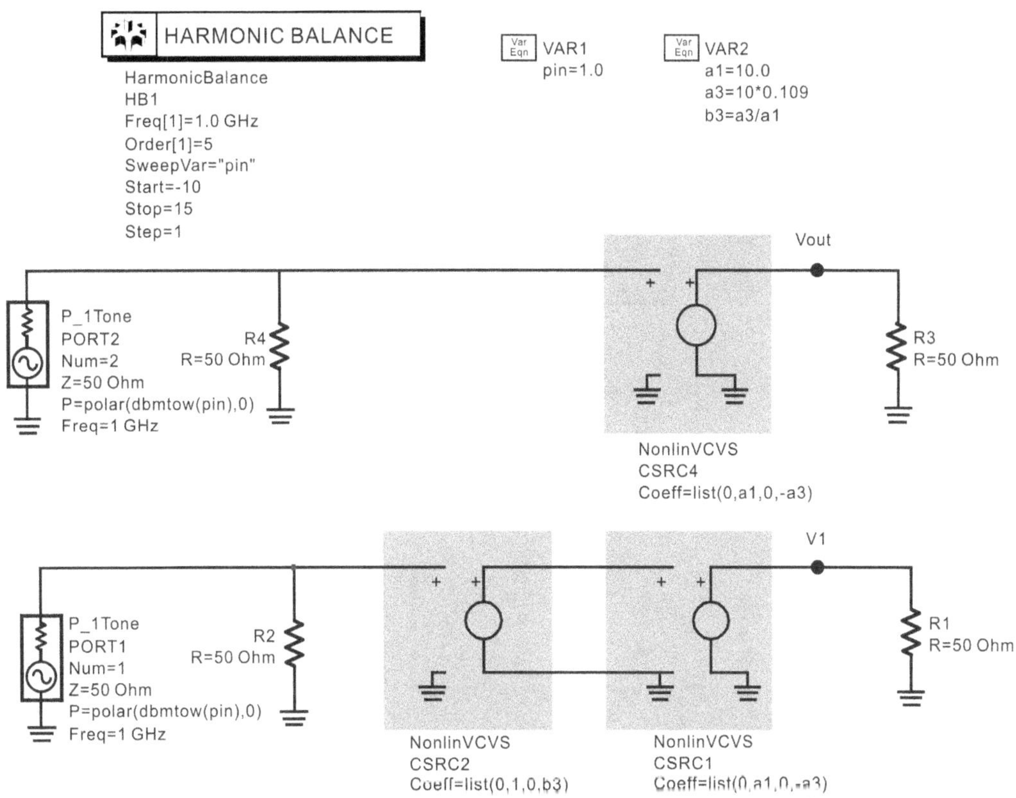

Figure 9E.30 Comparison of the input-output characteristics of the power amplifiers with and without predistorters. The top circuit models PA while the bottom circuit models PDA.

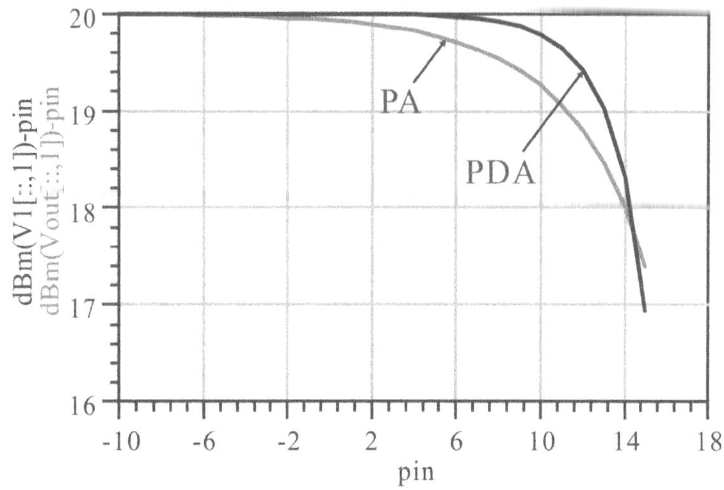

Figure 9E.31 Comparison of the simulated gain characteristics. From the plot, we can find the 1-dB compression point is increased in the case of the PDA.

Next, the simulation is carried out for the two-tone input signal with a center frequency of 2.5 GHz, a frequency spacing of 10 MHz, and a tone power of 10 dBm. The two simulated spectra for the power amplifier and the power amplifier with the predistorter are shown in Figure 9E.32. The IMD3 of the power amplifier with the predistorter in Figure 9E.32 is found to be improved by about 20 dB.

The linearity of the power amplifier is clearly improved in this example. However, it should be noted that the ACPR is somewhat poorer due to the contribution of the higher-order terms of the predistorter when the input power is not sufficiently backed off, especially in the case of digitally modulated signals such as CDMA.

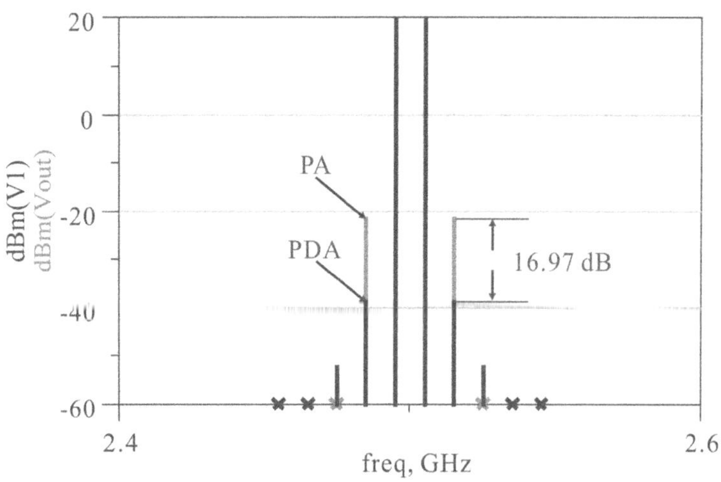

Figure 9E.32 Two-tone simulation results. The IMD3 of the PDA is reduced by about 20 dB.

9.7.2 Feedforward Power Amplifiers (FPA)

The basic concept of a feedforward power amplifier is to remove the distortion signal IMD3 that appears at the output of the power amplifier by adding an inverted distortion signal. When a two-tone signal is applied to the power amplifier's input, distorted signal components appear at the power amplifier's output due to the third-order distortion, as shown in Figure 9.78. By adding a signal whose magnitude is equal and the phase is 180° out of phase to the distortion signal, the distortion signal will disappear from the output and only the amplified components of the input signal will appear at the output.

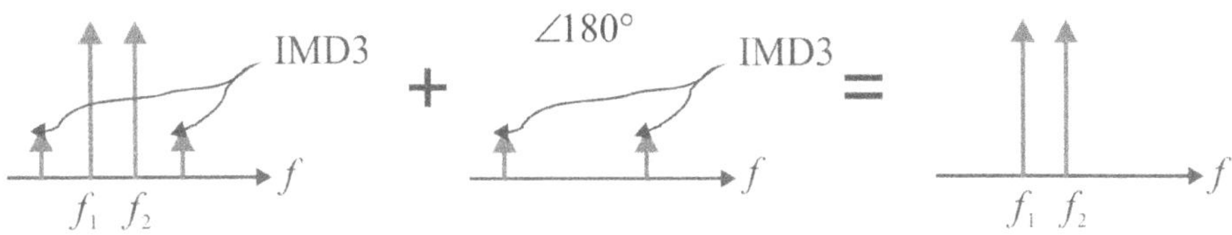

Figure 9.78 Concept of a feedforward power amplifier. The 180° out-of-phase IMD3 signal is generated and added to the PA output. Consequently, the IMD3 signal at the PA output disappears and linearity is improved.

The feedforward power amplifier is the implementation of this concept and is shown in Figure 9.79.

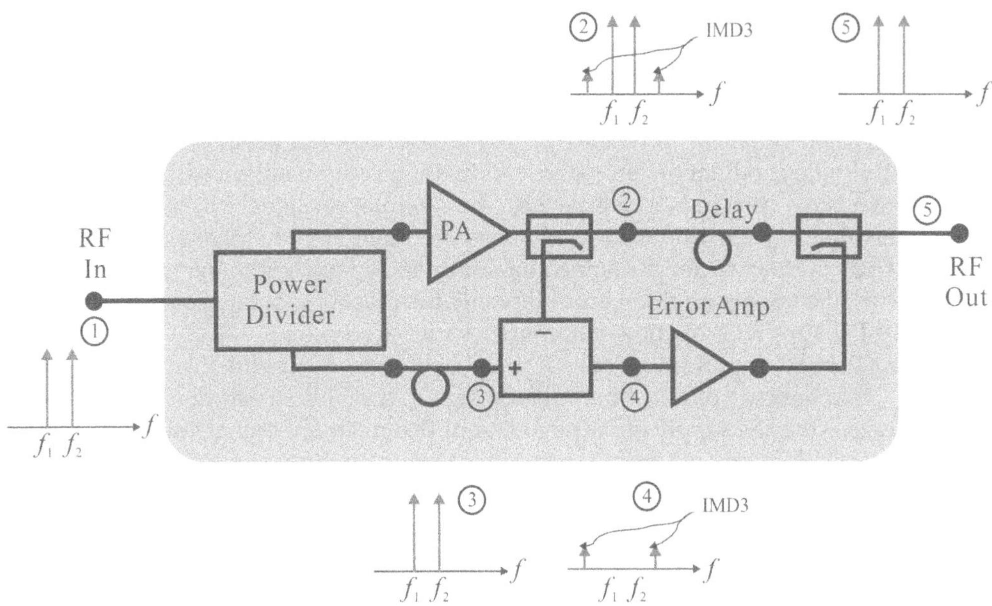

Figure 9.79 Feedforward power amplifier's structure and operation

In Figure 9.79, when a two-tone signal is applied to ①, the amplified two-tone signal with the IMD3 appears at the output of the power amplifier, ②. In contrast, a distortion-free two-tone signal with an appropriate delay appears at ③. Assuming the delay line offers a phase delay such that the two two-tone signals at ② and ③ have a phase difference of 180°, the resulting sum of the two signals appearing at ④ will be a pure IMD3 signal, as shown in the figure. Since the magnitude of the IMD3 signal is generally small, after being sufficiently amplified by the error amplifier, the two signals are combined at the power amplifier's output. Thus, an amplified two-tone signal alone appears at the output of the power amplifier at ⑤, as shown in the figure. In order to make the phase difference of 180° between the two IMD3 signals, a delay line is inserted at the output of the power amplifier.

Thus, the feedforward power amplifier can be experimentally designed using a two-tone input. There may be a disadvantage to this method in that the linearity performance is easily degraded by environmental changes because the adjusted condition to eliminate the distorted signal depends on the conditions of many components. In addition, since the condition is satisfied for selected two-tone input frequencies, the IMD3 elimination condition can also be broken when the frequency changes. Therefore, the bandwidth of the feedforward power amplifier can, in principle, only be a narrow one. In general, this type of amplifier is reported to show more improvement in linearity in a narrow bandwidth than does a predistorted power amplifier.

9.7.3 EER (Envelope Elimination and Restoration)

In 1950, the EER method was proposed by Kahn; in it, instead of directly amplifying the amplitude-modulated signals, the power amplifier amplifies a CW signal, and the power amplifier's DC supply varies according to the amplitude-modulation signal, thereby improving the amplifier's efficiency. Generally, the power amplifier shows better efficiency when a constant-amplitude square-wave input is applied. In addition, the higher the input power, the better the improvement in the efficiency. Therefore, by using EER, a dramatic improvement in the efficiency of the power amplifier can be achieved.

However, unlike conventional amplitude modulation, both the amplitude and phase of a carrier changes in modern digitally modulated signals. It is not easy to separate the amplitude-modulation and phase-modulation components. However, using digital signal processing, as shown in Figure 9.80, the modulation signal can be separated into the signals that require amplitude and phase modulations. The phase modulation is independently performed for the sinusoidal input and is applied to the power amplifier's input. Thus, the power amplifier amplifies only the phase-modulated signal. On the other hand, the DC supply of the power amplifier varies according to the envelope signal and, using this type of amplification, the efficiency of the power amplifier can be improved.

In the case of a fast modulation signal, the EER structure is not easy to implement because the DC supply of the power amplifier generally changes slowly. The power amplifier usually requires a large current, from a few amperes to several tens of amperes, and a DC supply voltage from as much as a few volts to typically several tens of kV. Therefore, varying this amount of power supply voltage according to the modulation signal is not an easy task. In addition, because the DC voltage for a power amplifier is generally supplied through a bypass capacitor, that capacitor significantly limits the high-frequency modulation signal.

9.7.4 Doherty Power Amplifier

Figure 9.81 shows the probability distribution of the output power of a power amplifier in a handset employing CDMA (code division multiple access). The power amplifier is driven

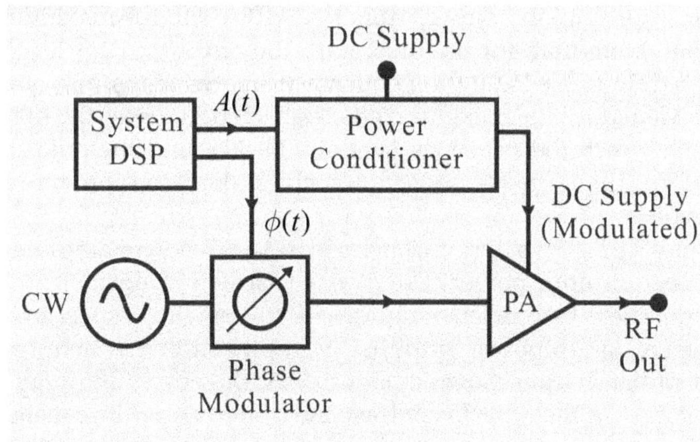

Figure 9.80 EER block diagram. The modulation signal is first decomposed into the amplitude and phase. The decomposed phase then modulates the CW signal and the phase-modulated signal is amplified by the PA. Thus, the PA operates in the high-efficiency region. The amplitude modulation is carried out using a DC power supply modulation.

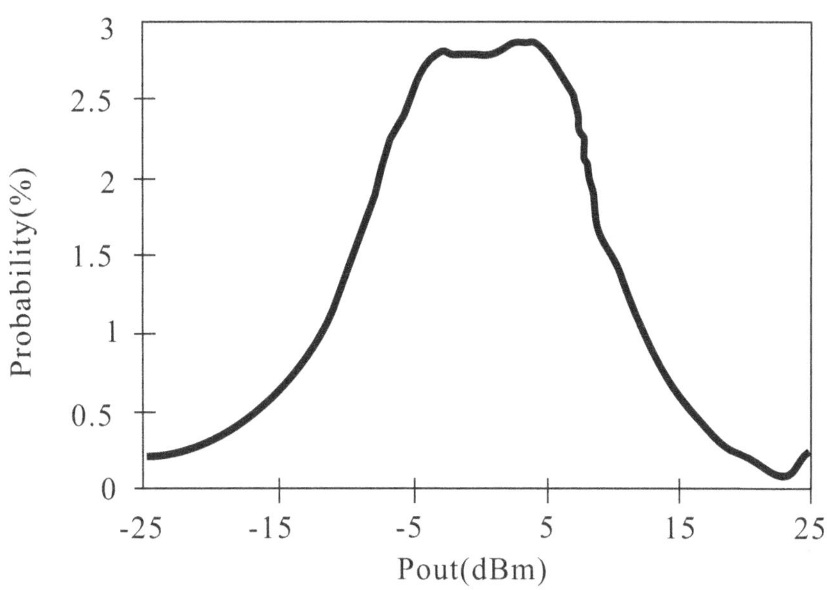

Figure 9.81 Probability distribution of CDMA (drawn after the reference).[11] The maximum power level of the PA is about 25 dBm but most of the PA's usage is at about a 0-dBm power level. As a result, the PA operates at low efficiency.

11. G. Hanington, P-F Chen, et al., "High-Efficiency Power Amplifier Using Dynamic Power-Supply Voltage for CDMA Applications," *IEEE Transactions on Microwave Theory and Techniques* 47, no. 8 (August 1999): 1471–1476.

for linear amplification from –25 dBm to 25 dBm, and the maximum output power is set to be 25 dBm. Note that, for the most part, the output power level in the power range of –5 dBm–5 dBm is used and the frequency using the maximum output power of 25 dBm is very low. Generally, the maximum efficiency is shown near the maximum output power of 25 dBm, at which the efficiency is about 30–50% and the efficiency rapidly falls with a decrease in the output power, which makes the power consumption increase considerably.

Figure 9.82 shows the efficiencies of a class-B power amplifier and Doherty amplifier with respect to input power. The x–axis is scaled in decibels. In the case of the class-B power amplifier, the efficiency rapidly increases with respect to the input power. Therefore, when a class-B power amplifier is employed in a system having the output-power probability distribution shown in Figure 9.81, the power amplifier will operate at low efficiency for most of the time. In contrast, the Doherty amplifier maintains efficiency at maximum output power within a significant input power range even when the input power is low. Thus, when the operation requires better efficiency for a wide range of input power, the Doherty amplifier appears to be the ideal choice. A multistage Doherty amplifier can be used to further improve the range of this efficiency. Refer to reference 2 at the end of this chapter for more information.

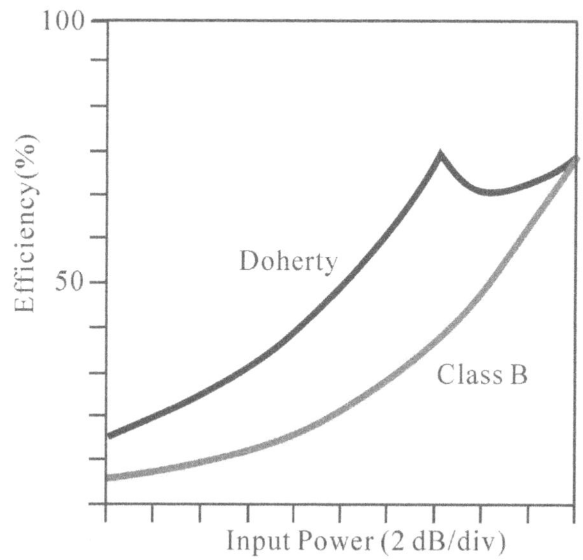

Figure 9.82 Comparison of the efficiencies for Doherty and class-B power amplifiers with respect to input power. The Doherty PA maintains high efficiency down to the 6-dB input power level.

Figure 9.83 shows the structure of the Doherty power amplifier. Two active devices are used in the power amplifier; the device used as "Main" is biased as class-B using the DC supply voltage V_{bm}; the device marked as "Peaking" is usually biased as class-C by adjusting the DC voltage V_{bp}. In addition, all the transmission lines used in the input and output are one-quarter-wavelength long. The two active devices used in the Doherty amplifiers may be different due to their costs or specifications; however, for this discussion it is assumed that both devices are the same.

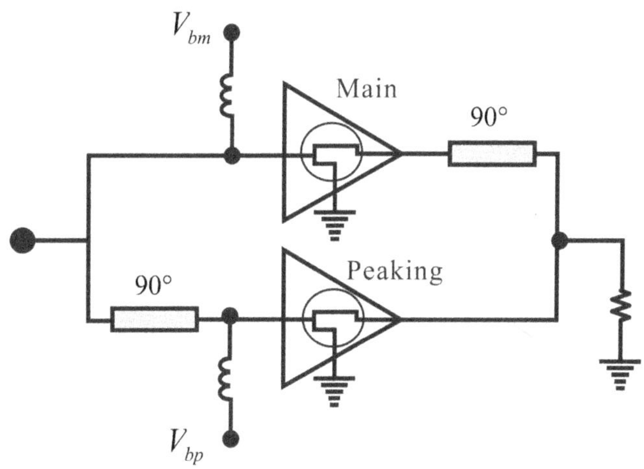

Figure 9.83 Doherty power amplifier. "Main" is a class-B power amplifier while "Peaking" is a class-C power amplifier.

Figure 9.84 shows the normalized drain currents versus input voltage for the devices "Main" and "Peaking," which operate in class-B and -C, respectively. The drain current waveform of the class-B power amplifier has a sinusoidal-tip shape. The drain current I_m in Figure 9.84 represents the fundamental component of the sinusoidal-tip-shaped waveform. A plot of I_p versus the input voltage can be similarly interpreted. The drain current I_m linearly increases from 0 with respect to the input voltage because the device "Main" is biased as class-B. However, in the case of I_p for the device "Peaking," the drain current I_p

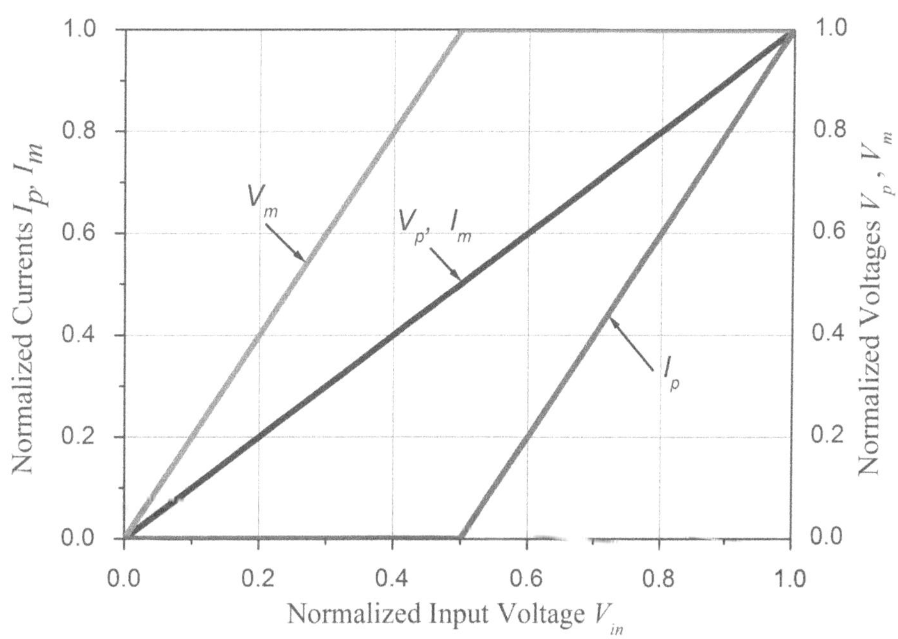

Figure 9.84 The "Main" and "Peaking" output voltages and currents versus the input voltage in the Doherty amplifier. This plot shows the typical characteristics of a 6-dB Doherty PA.

does not flow below a certain level of the input voltage because it is biased as class-C. The current I_p begins to increase with respect to the input voltage beyond the threshold voltage. In Figure 9.84, the "Peaking" device is so biased in class-C such that no output current flows below $V_{in} = 0.5$. In addition, denoting the maximum values of I_m and I_p as I_M and I_P, respectively, the normalized drain currents I_m and I_p by I_M and I_P are shown in Figure 9.84. It is also assumed that $I_M = I_P$. Thus, denoting the normalized input voltage V_{in} as $x(0 < x < 1)$, I_m and I_p can be expressed as shown in Equations (9.68) and (9.69).

$$I_m = x \tag{9.68}$$

$$I_p = \begin{cases} 0 & x < 0.5 \\ 2(x - 0.5) & x > 0.5 \end{cases} \tag{9.69}$$

In order to analyze the combined output of the "Main" and "Peaking" devices in Figure 9.84, the equivalent circuit of the Doherty amplifier's output from Figure 9.83 is drawn in Figure 9.85. The current sources I_m and $-jI_p$ in Figure 9.85 represent the currents arising from the drains of the "Main" and "Peaking" devices, respectively. The drain current of the "Peaking" device is phase-delayed by 90° due to the phase delay of the input voltage, as shown in Figure 9.85.

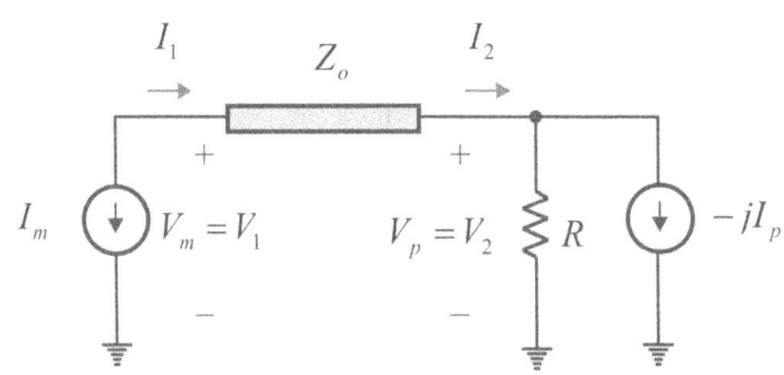

Figure 9.85 Model of the Doherty amplifier output. I_m represents the class-B half-wave current source of the "Main" device while I_p represents the class-C sinusoidal tip waveform of the "Peaking" device. The phase factor $-j$ appears as a result of the quarter-wavelength transmission line.

The transmission line is represented by ABCD parameters as expressed in Equation (9.70).

$$\begin{pmatrix} V_1 \\ I_1 \end{pmatrix} = \begin{pmatrix} 0 & jZ_o \\ jY_o & 0 \end{pmatrix} \begin{pmatrix} V_2 \\ I_2 \end{pmatrix} \tag{9.70}$$

From Equation (9.70), it can be seen that $V_p = -jZ_oI_m$. Thus, as shown in Figure 9.84, the drain voltage V_p of the "Peaking" device has the same dependence on the input voltage as I_m. The drain voltage V_p reaches the maximum value for $I_m = I_M$, and the maximum voltage of V_p is denoted as V_M. Then, the magnitude of V_p normalized by V_M can be plotted against the input voltage, which is the straight line of I_m shown in Figure 9.84. Using Equation (9.70) again, V_1 can be expressed as Equation (9.71).

$$V_1 = V_m = jZ_oI_2 = jZ_o\left(\frac{V_2}{R} - jI_p\right) = jZ_o\left(jZ_o\frac{I_m}{R} - jI_p\right) = -I_m\frac{Z_o^2}{R} + Z_oI_p \tag{9.71}$$

Because $I_p = 0$ for $x < 0.5$, there is no contribution of I_p up to $x = 0.5$ in Equation (9.71). Setting $(Z_o)^2/R$ to reach the maximum output voltage V_M at $x = 0.5$, the drain voltage V_m of the "Main" device will reach maximum voltage V_M at $x = 0.5$, as shown Figure 9.84. Since the "Peaking" device is in the off state, the efficiency of the "Main" device becomes the efficiency of the Doherty amplifier below $x < 0.5$. For $x > 0.5$, the "Peaking" device turns on and the drain current I_p flows. The voltage V_m changes due to the contribution of the current I_p and generally becomes a piecewise line. As the signs of the two terms in Equation (9.71) are different, they are found to cancel out each other. If the further increase by I_m is exactly cancelled by I_p for $x > 0.5$, the drain voltage V_m of the "Main" device maintains the maximum output voltage V_M at the normalized input voltage of 0.5. Thus, the plot of V_m appears, as shown in Figure 9.84. The cancellation of the two terms in Equation (9.71) is done to make the two terms have the same slopes for x. Since the slope of I_p is 2 for x, $(Z_o)^2/R$ should be equal to $2Z_o$ in order to cancel out each other. Thus, the relationship in Equation (9.72) is satisfied.

$$Z_o = 2R \tag{9.72}$$

Because $Z_o = 2R$ in Equation (9.72), the efficiency of the "Main" device when operating as a class-B amplifier decreases for $x > 0.5$; however, the efficiency of the "Peaking" device when operating as a class-C amplifier increases as x increases. As a result, the Doherty amplifier maintains an approximately constant efficiency in the range of $0.5 < x < 1$. The efficiency of the Doherty amplifier can be calculated from the Fourier series of class-B and class-C amplifiers explained in section 9.4.1. Since Figure 9.84 represents the amplitude of the fundamental component, the DC current component can be calculated using Equations (9.8) and (9.9a). The DC components of the "Main" and "Peaking" devices $I_{m,0}$ and $I_{p,0}$ are expressed in Equations (9.73) and (9.74).

$$I_{m,0} = \frac{2}{\pi} I_m \tag{9.73}$$

$$I_{p,0} = I_p \frac{\gamma_0(\theta)}{\gamma_1(\theta)} \tag{9.74}$$

Here, the conduction angle is $\theta = \cos^{-1}(0.5)$. Thus, setting the DC drain supply voltage to $V_{DD} = 1$, the normalized DC power consumption can be obtained with Equation (9.75).

$$P_{DC} = I_{p,0} + I_{m,0} \tag{9.75}$$

The RF output power is determined by computing the power supplied by the two current sources of Figure 9.85. The RF output power is given by

$$P_{RF} = \frac{1}{2}\left(I_p V_p + I_m V_m\right) = \frac{1}{2}\left(I_p V_{in} + V_m V_{in}\right) \tag{9.76}$$

From Equation (9.76), the efficiency is calculated as shown in Equation (9.77).

$$\eta(\%) = \frac{P_{RF}}{P_{DC}} \times 100 \tag{9.77}$$

The DC power consumption computed using the DC current calculated from Equations (9.73) and (9.74) and the RF power calculated from Equation (9.76) are shown in Table 9.13.

Table 9.13 Efficiency calculation of Doherty amplifier

V_p, I_m	V_p	I_p	$I_{m,0}$	$I_{p,0}$	P_{DC}	P_{RF}	η (%)
0	0	0	0.000	0.000	0.000	0.000	0
0.1	0.2	0	0.064	0.000	0.064	0.010	15.7
0.2	0.4	0	0.127	0.000	0.127	0.040	31.4
0.3	0.6	0	0.191	0.000	0.191	0.090	47.1
0.4	0.8	0	0.255	0.000	0.255	0.160	62.8
0.5	1	0	0.318	0.000	0.318	0.250	78.5
0.6	1	0.2	0.382	0.112	0.493	0.360	73.0
0.7	1	0.4	0.446	0.223	0.669	0.490	73.3
0.8	1	0.6	0.509	0.335	0.844	0.640	75.8
0.9	1	0.8	0.573	0.446	1.019	0.810	79.5
1	1	1	0.637	0.558	1.194	1.000	83.7

Figure 9.86 is a plot of the calculated efficiencies in Table 9.13. Considering that the Doherty amplifier is normally used in the high-efficiency range starting from the normalized input voltage $V_{in} = 0.5$, it means that the Doherty amplifier in Figure 9.86 provides high efficiency up to 6-dB back-off input power. The peak of the efficiency at $V_{in} = 0.5$ can be moved by adjusting the conduction angle; using this method, a Doherty power amplifier with various ranges of back-off input power can be achieved.

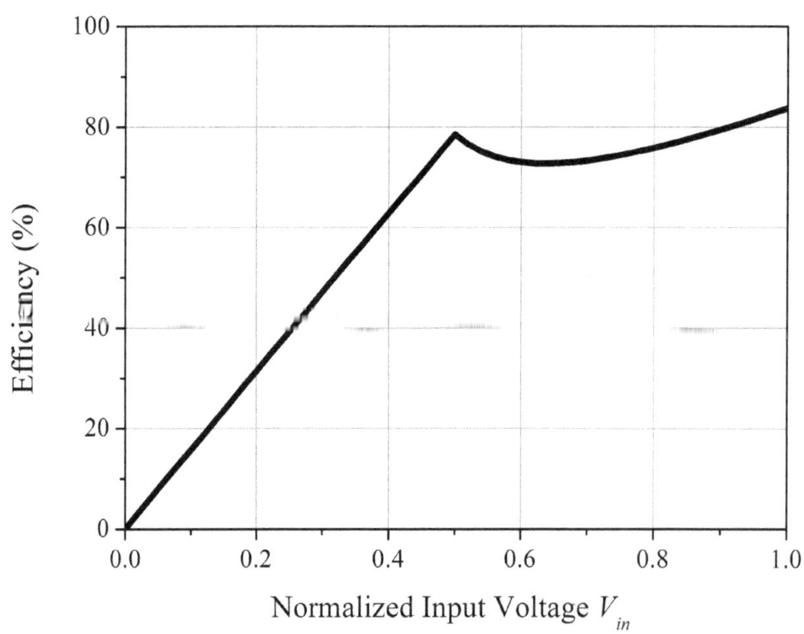

Figure 9.86 The calculated efficiency of the Doherty amplifier. High efficiency is maintained down to $V_{in} = 0.5$.

Example 9.15

Using the half-wave current source, calculate the efficiency and output power of the Doherty power amplifier in Figure 9.85 with ADS.

Solution

The equivalent circuit of the output in Figure 9.85 consists of two half-wave current sources, which can be implemented using Equations (9.8) and (9.9). Figure 9E.33 shows the "Main" and "Peaking" half-wave current sources implemented by using those equations. Since the output of the "Peaking" amplifier has a 90° phase delay at the fundamental frequency f_o, the phase delay is implemented using the **pd**(x) function that provides the phase delay of $n \times 90°$ at the n-th harmonic frequency. In addition, the two current sources are made to provide the same peak current values of 1 A and the conduction angle of the "Peaking" current source is set to $\theta = \cos^{-1}(0.5)$.

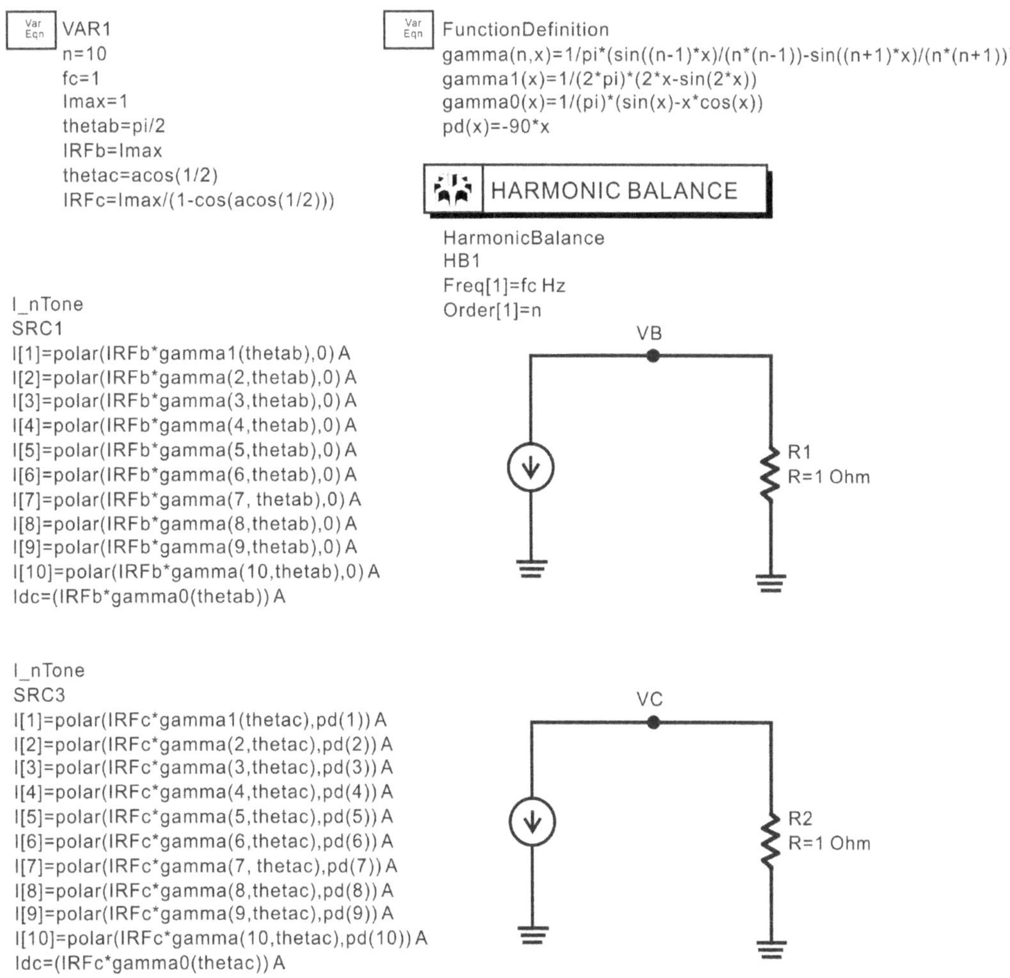

Figure 9E.33 Configuration of the output current sources for the Doherty amplifier

Figure 9E.34 shows the simulated waveforms of the current sources. The "Main" current source has the half-wave shape and its maximum value of 1 A is shown in Figure 9E.34. On the other hand, the "Peaking" current source has a phase delay of 90° compared with the "Main" current source, and its maximum value is also 1 A as expected. The conduction angle can be found from the width of the sinusoidal-tip-shaped waveform, and it is estimated to be about half of the "Main." Thus, the conduction angle can be found to be $\theta = \cos^{-1}(0.5)$.

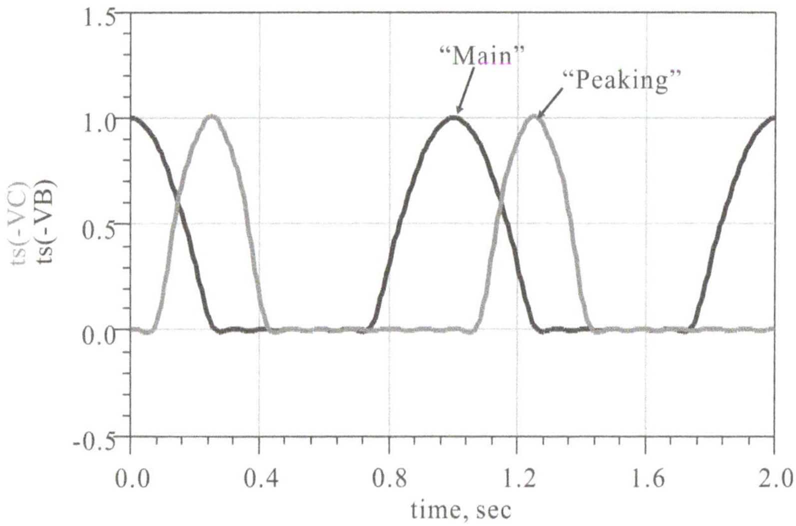

Figure 9E.34 Class-B and class-C simulated output current waveforms. The voltage **VC** is delayed by a phase of 90°.

Using the current sources thus obtained, the equivalent circuit of the Doherty amplifier's output can be configured as shown in Figure 9E.35.

First, in Figure 9E.35, **IRFb_max** is set to 2. **IRFb_max** corresponds to the maximum peak value of the "Main" current source that has the half-wave shape. The value of I_m in Figure 9.84 is the value of the fundamental amplitude when the half-wave-shape current is expanded into a Fourier series. The value of I_m is one-half of the half-wave peak value. Thus, since **IRFb_max** = 2, the maximum value of I_m is set to 1 A. Similarly, I_p in Figure 9.84 is the fundamental amplitude of the "Peaking" current source when the "Peaking" current is expanded into a Fourier series. The ratio of I_p to the peak value of the "Peaking" current source is $\gamma_1(\theta)$. To make the maximum value of I_p to 1 A, the peak value **IRFc_max** of the "Peaking" current source is set to **IRFc_max** = **1/gamma1(thetac)**.

The variable **x** represents V_{in} in Figure 9.84, which corresponds to the fundamental amplitude of the input voltage. The change of I_m for V_{in} is set to **IRFb_max*x**, by which the "Main" current varies in proportion to V_{in}. The "Peaking" current source can be also made to vary according to V_{in}. To implement I_p for V_{in} in Figure 9.84, I_p is set to **IRFc_max*(2*(x−0.5)*step(x−0.5))** because its slope is 2 and its non-zero value appears at the value of above **x** = 0.5.

L1 and **C1** in Figure 9E.35 constitute a parallel resonant circuit and its Q is set to 10. Using Equation (9.72), the transmission line impedance is set to **Z** = 2 * **Ropt**. The value of R_{opt} can be determined from the Doherty amplifier characteristics in Figure 9.84. When I_m is 0.5 A, the drain voltage of "Main", V_m, should be 1 V. Using $V_m = I_m(Z_o)^2/R_{opt}$ in Equation (9.71), $0.5 \times (Z_o)^2/R_{opt} = 4R_{opt} \times 0.5 = 1$. Thus, it can be found that **Ropt** = 0.5 Ω. After the settings, it must be verified whether or not the circuit of Figure 9E.36 has the Doherty amplifier characteristics shown in Figure 9.84. The values of I_m, I_p, V_m, and V_p are computed and plotted for V_{in} as shown in Figure 9E.36. The plot is found to show the Doherty amplifier characteristics in Figure 9.84.

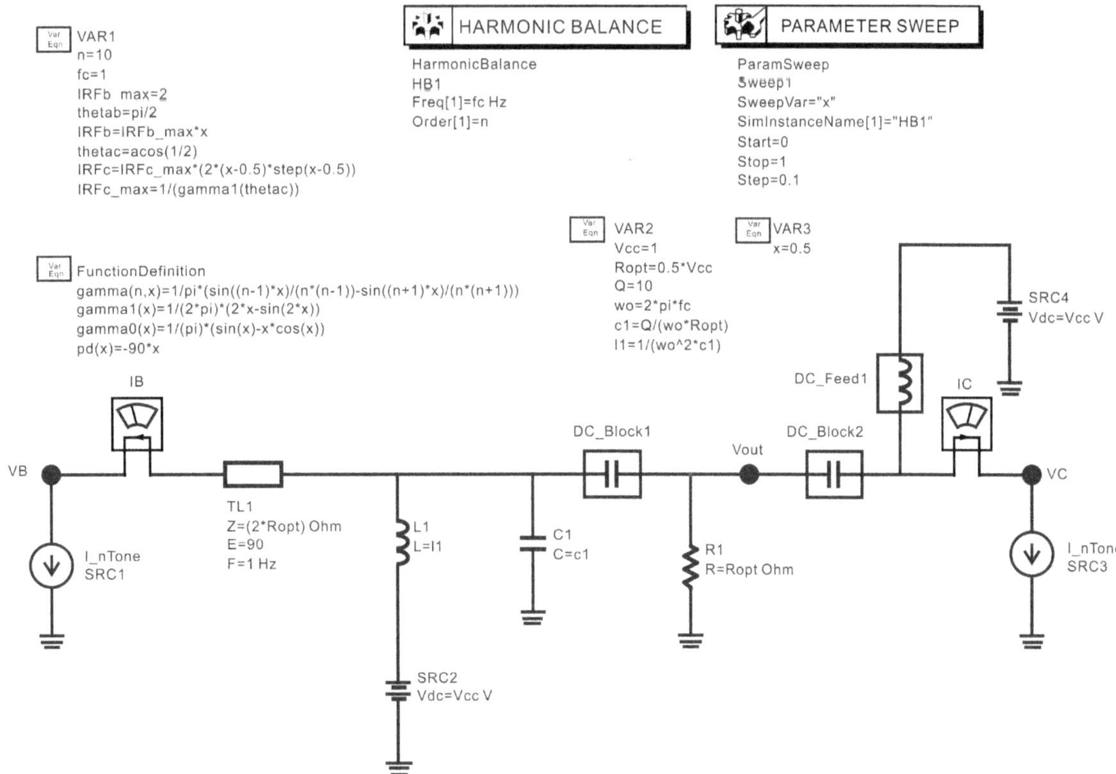

Figure 9E.35 Simulation schematic of the Doherty amplifier's output. **L**1 and **C**1 are parallel resonant at the fundamental frequency. **SRC**1 is the half-wave current source, while **SRC**3 is the current source of the sinusoidal-tip-shaped waveform.

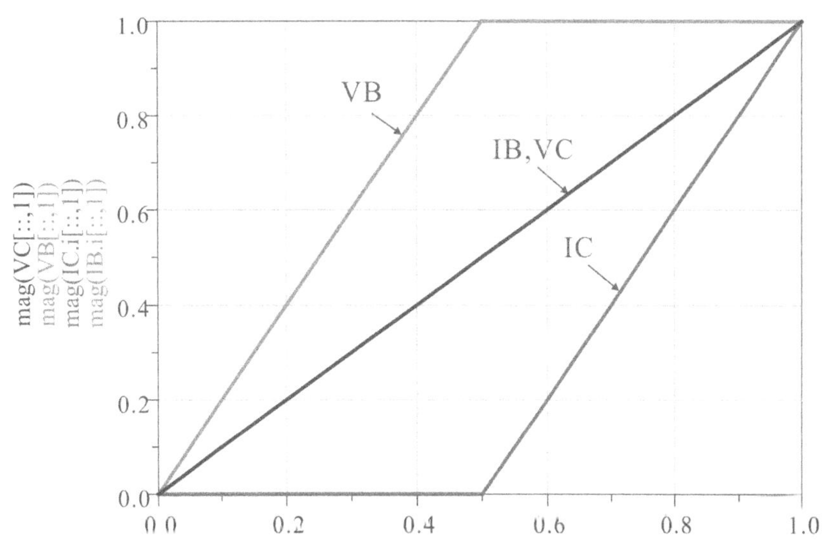

Figure 9E.36 Simulated output current and voltage characteristics of the Doherty amplifier. Here, B stands for class-B and C stands for class-C. **VC**[::,1], **VB**[::,1], **IC**.i[::,1], and **IB**.i[::,1] are the fundamental voltages and currents for the sweep variable **x**.

The following equations in Measurement Expression 9E.12 are entered in the display window to compute the efficiency, which is plotted as shown in Figure 9E.37. It can be seen that the efficiency in Figure 9E.37 is the same as that computed in Figure 9.86.

```
Eqn Pdc=mag(SRC2.i[::,0]+SRC4.i[::,0])
Eqn Pout=0.5*mag(Vout[::,1])**2/Ropt[0]
Eqn eff=Pout/Pdc*100
```

Measurement Expression 9E.12

In addition, in order to know the maximum current capability of the active devices, the output waveforms of the "Main" and "Peaking" current sources are plotted with V_{in} as a parameter. The variations of the currents are shown in Figure 9E.38. In the case of the "Main" amplifier in Figure 9E.38(a), a maximum current capability of 2 A is required, while in the case of the "Peaking" amplifier operating in class-C, a maximum current capability of 2.5 A is required. The waveform of the output voltage **Vout** with V_{in} as a parameter is shown in Figure 9E.38(b); the output voltage is close to a sinusoidal waveform due to the effect of the parallel resonant circuit, as expected.

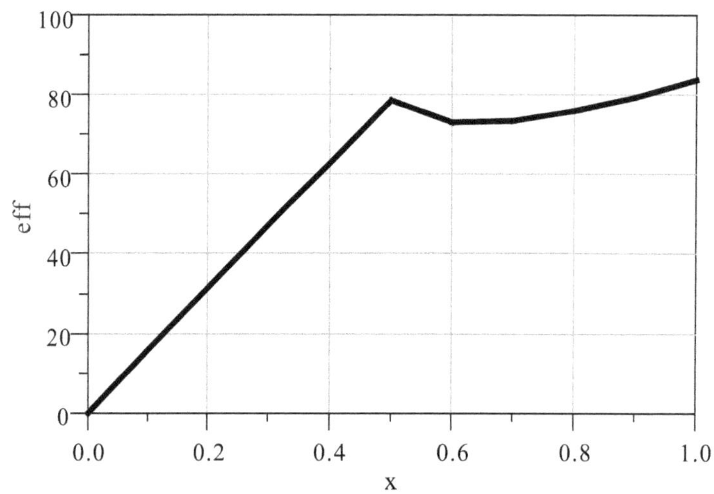

Figure 9E.37 Simulated efficiency for the sweep variable **x** of the Doherty amplifier. The simulated efficiency is found to be equal to that in Figure 9.86.

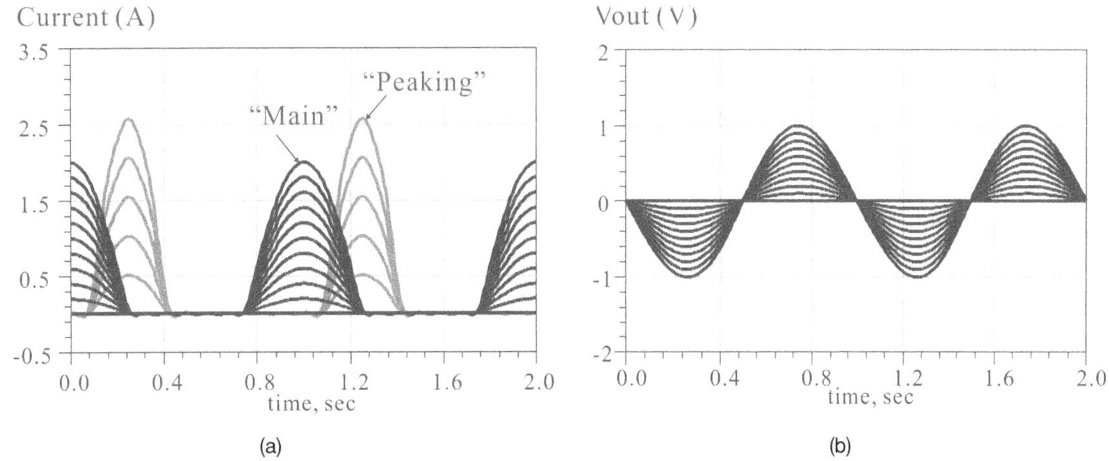

Figure 9E.38 (a) Class-B and class-C current waveforms and (b) output voltage **Vout** with respect to input

9.8 SUMMARY

- The power amplifier (PA) is a key component that determines the transmitting power level of a transmitter. The design of the PA is fundamentally different from that of an LNA. The load impedance is selected to meet the output power, efficiency, harmonics, and distortion.
- The active devices such as GaN HEMT and LDMOSFET that are widely used for PAs are introduced. GaN has a wide bandgap and good thermal conductivity, which leads to a high breakdown voltage and a high power dissipation capability. As a result, a GaN HEMT is an appropriate device for a high-power amplifier.
- The Si LDMOSFET is a device with a breakdown voltage that is improved through drain engineering. However, due to the low electron mobility of Si, its use is primarily limited to applications with frequencies below 4 GHz.
- The optimum load impedance of an active device is experimentally obtained from a load-pull measurement setup that employs impedance tuners. Alternatively, the optimum load impedance can be obtained through a load-pull simulation when the large-signal model of an active device is available. In the load-pull simulation, harmonic impedances should be taken into consideration.
- Power amplifiers can be classified into class-A, -B, -C, -D, -E, and -F. A class-A PA provides high linearity; however, its efficiency is limited to 50% or less. A class-B PA eliminates the stand-by power consumption of the class-A PA and its efficiency is below 78.5%. Moving the operating point further from the pinch-off results in the class-C PA provides a higher efficiency than does the class-B PA; however, from the output power point of view, the power capability of the active device is not fully exploited.

- Class-D and -E PAs operate with a transistor as a switch and their efficiency can reach 100%. However, they lose the amplitude information. A class-E PA employs the detuned resonator circuit. Using the class-E PA, the drain or collector voltage does not show an abrupt transition or possible device damage, and an efficiency drop can be avoided.
- A class-F PA uses multiple resonators in the load to flatten the drain voltage. The maximally flat and maximum efficiency waveforms for limited harmonics are derived. They can be used to estimate the efficiency and optimum resistance of a class-F PA.
- A design example of a class-F PA using the TriQuint GaN HEMT model is demonstrated. The design follows these steps:
  1. Stabilize the device
  2. Obtain the optimum source and load impedance extraction using load-pull simulation
  3. Implement the input matching circuit and a class-F load matching circuit using lumped-element matching circuits
  4. Replace the lumped-element matching circuits with the physical layout components
  5. Tune the implemented matching circuits through EM simulation
  6. Verify the EM-simulated matching circuits
- Power amplifier evaluation methods and specifications for linearity are presented. The two-tone, ACPR, and EVM methods are described.
- A power amplifier's linearity can be improved by building a composite PA. Predistorter, feedforward, and EER techniques are used to improve PA linearity.
- The high-efficiency range of a power amplifier can be extended by using a Doherty power amplifier. Using a simplified example, we demonstrate that the amplifier's high-efficiency range can be extended.
- The power amplifier is a key component of a communication system and it is constantly attracting the interest of many researchers. Because power amplifier design is a field in which new concepts are continuously emerging, the readers should keep abreast of recent research results.

REFERENCES

1. A. Howard, *Load Pull Simulation Using ADS*. Available at ADS directory: ADS_Home/examples/RF_Board/LodaPull_prj/LoadPullPres.pdf, 2002.
2. S. C. Cripps, *Advanced Techniques in RF Power Amplifier Design*, Boston: Artech House, Inc., 2002.
3. A. Grebennikov and N. O. Sokal, *Switchmode RF Power Amplifiers*, Amsterdam: Elsevier, 2007.
4. TriQuint semiconductor. *TriQuint EEHEMT model implemented in ADS and AWR for TQT 0.25 um 3MI GaN on SiC process 1.25 mm discrete FET*. April, 2009.
5. S.C. Cripps, *RF Power Amplifiers for Wireless Communications*, Boston: Artech House, Inc., 1999.
6. Agilent Technologies, *Advanced design system 2009U1*. Available at http://www.keysight.com/en/pd-1612226/ads-2009-update-1?cc=US&lc=eng, 2009.

PROBLEMS

9.1 Instead of the following variable declaration for the impedance tuner, use the **step(•)** function and implement the **LoadTuner**.

```
VAR global Impedance Equations
;Tuner reflection coefficient
LoadTuner=LoadArray[iload]
LoadArray=list(0,rho,fg(Z_l_2),fg(Z_l_3),fg(Z_l_4),fg(Z_l_5))
iload=int(min(abs(freq)/RFfreq+1.5,length(LoadArray)))
fg(x)=(x-Z0)/(x+Z0)
```

9.2 Prove that the optimum load resistance value of a class-C power amplifier is

$$R_C = \frac{V_{DD}\left(1-\cos\theta\right)}{I_{\max}\gamma_1} = R_A\frac{\left(1-\cos\theta\right)}{2\gamma_1}$$

Then, normalizing the output power by the maximum output power of a class-A amplifier, prove that the following ratio is obtained.

$$\frac{P}{P_A} = \frac{2\gamma_1}{1-\cos\theta} = \frac{2\theta-\sin 2\theta}{\pi\left(1-\cos\theta\right)}$$

9.3 **(ADS problem)** Using the output circuit of the class-B power amplifier shown in Example 9.5, and fixing the load resistance value as the resistance value that gives maximum output power, plot the output power versus the input power. In the plot, consider the input power as the output power of the fundamental frequency. Compare this result with the results of Figure 9.25 and discuss the reasons for the difference.

9.4 **(ADS problem)** Consider the BJT of the voltage-switching class-D amplifier shown in Figure 9P.1 as an ideal switch; using ADS, show that the output voltage V_L appears as a sinusoidal voltage. Set the Q of the series resonant circuit to be more than 10.

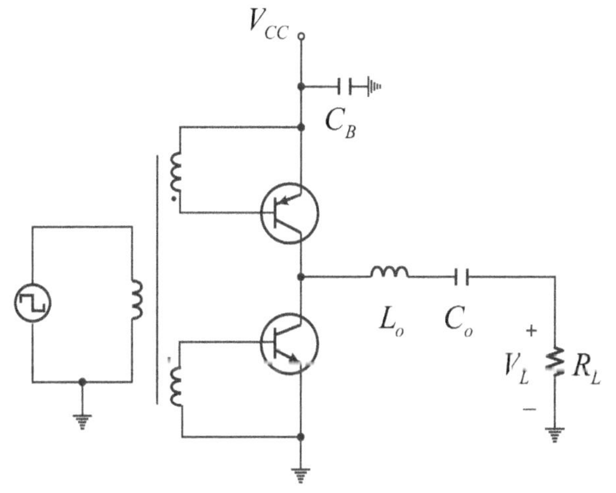

Figure 9P.1 Class-D push-pull amplifier for problem 9.4

9.5 (**ADS problem**) In the single-ended class-D amplifier circuit shown in Figure 9.26, simulate the circuit using ADS with the DC block capacitor replaced by a high Q-series resonant circuit. Explain why the result is different from that of problem 9.4.

9.6 Prove Equation (9.32).

9.7 The harmonic impedance Z_n of the class-E amplifier can be expressed by inspection as shown below. Explain why. Another way to express this is for Z_n to expand $v_n(t)$ and $i_n(t)$ into a Fourier series and Z_n can then be determined from the ratio of the Fourier series of $v_n(t)$ and $i_n(t)$ at the same frequency.

$$Z_1 = \left(R + j\omega_o L\right) \parallel \frac{1}{j\omega_o C}$$

$$Z_n = \frac{1}{jn\omega_o C} \quad n = 2,3,4,\cdots$$

9.8 Figure 9P.2 is a class-E amplifier. For operating in the optimum condition, the following relationship must be satisfied (see reference 3, above).

$$\omega L = 0.732R, \ \omega C = \frac{0.685}{R}, \quad R = 1.365\frac{V_{CC}^2}{P_{out}}$$

Then, setting $f_o = 1$, $P_{out} = 1$, and $V_{CC} = 1$, plot the waveforms of $v(t)$, $i(t)$, and $v_{out}(t)$.

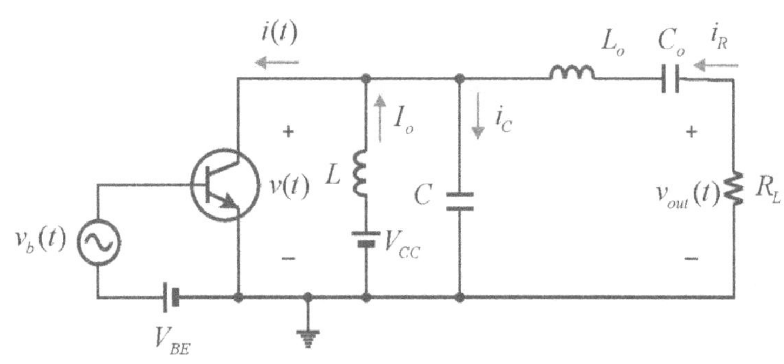

Figure 9P.2 Class-E power amplifier circuit

9.9 Prove that the value of b that satisfies Equation (9.39) is $b = 0.1667$.

9.10 Calculate the power capability, M_p, of the class-E amplifier in Figure 9.35. Compare this with the power capability of the ideal class-F amplifier.

9.11 For the following amplifier in Figure 9P.3, given that power supply voltage $V_{CC} = 10$ V, and the maximum current $I_{max} = 1$ A, determine the optimum impedance, R_{opt}, for maximum efficiency using Tables 9.4, 9.5, and 9.6.

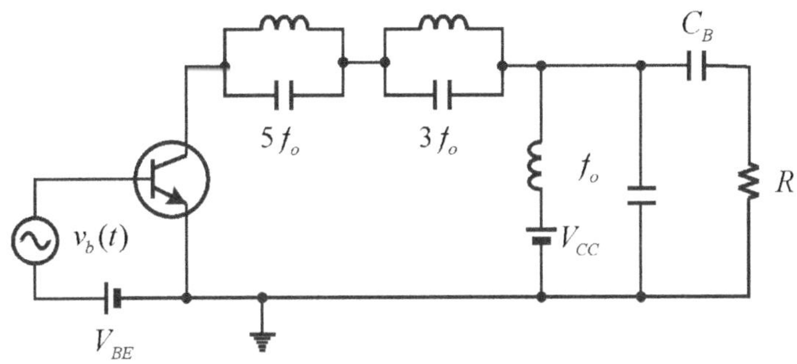

Figure 9P.3 Class-F power amplifier circuit

9.12 In the following circuit in Figure 9P.4, given that the L_o–C_o has a high-Q and resonates at the frequency ω_o and that it has the following settings for the output matching circuit,[12]

$$L_1 = \frac{1}{6\omega_o^2 C_{out}},\quad L_2 = \frac{5}{3}L_1,\quad C_2 = \frac{12}{5}C_{out}$$

Calculate the maximum efficiency.

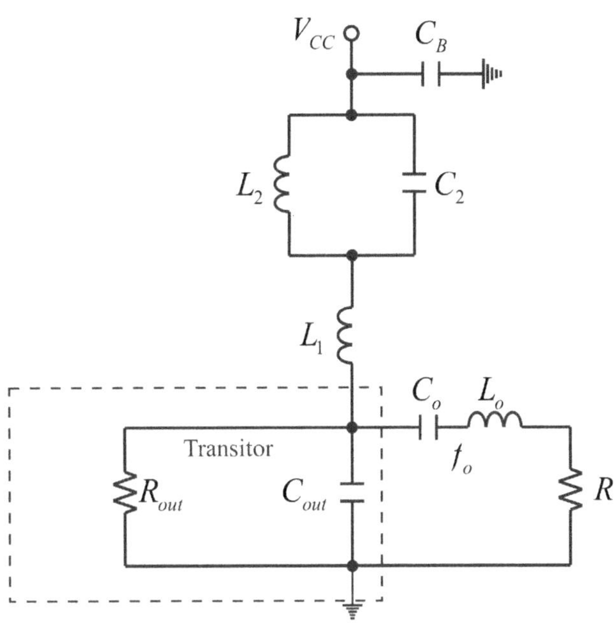

Figure 9P.4 Output equivalent circuit

12. A.V. Grebennikov, "Effective Circuit Design Techniques to Increase MOSFET Power Amplifier Efficiency," *Microwave Journal* 43, no. 7 (July 2000): 64–72.

9.13 **(ADS problem)** The circuit of Figure 9P.5 is an example of an ideal class-F amplifier. By considering the output of the active device as a switch, plot the waveforms of $v(t)$, $i(t)$, and $v_{out}(t)$. Now given that the $L_1 - C_1$ resonates at the fundamental frequency f_o, set a high value for Q and then simulate.

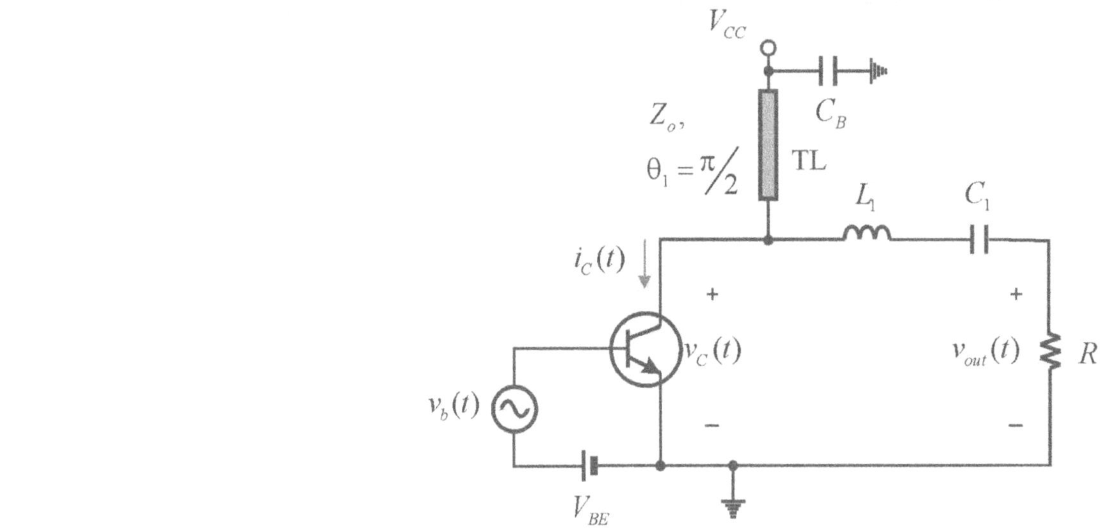

Figure 9P.5 Ideal class-F amplifier

9.14 Discuss the $\pi/4$-DQPSK modulation scheme, and explain why PAPR (Peak to Average Power Ratio) in this scheme can be improved.

9.15 A pre-distorter with the input-output characteristics given by $v_p = v_{in} + b_3(v_{in})^3$ is to be used to linearize the power amplifier represented by the input-output characteristics given by $v_L = a_1 v_p - a_3(v_p)^3$. In order to linearize the amplifier, prove that $b_3 = a_3/a_1$. With such a pre-distorter, the output voltage characteristic of the power amplifier is represented by the 5th, 7th, and 9th distorted terms; determine the coefficient of the 5th term.

9.16 For a power amplifier and a predistorter modeled by $v_L = a_1 v_p(t) - a_3(v_p(t-\tau))^3$, and $v_p = v_{in}(t) - b_3(v_{in}(t-\tau'))^3$ respectively, find a relation for a sinusoidal input to eliminate the third-order term at the amplifier output.

9.17 In the pre-distorter with $v_p = v_{in}(t) - b_3(v_{in}(t-\tau'))^3$, the third order term design requires the conditioning of amplitude and phase. When only the third-order term can be separately extracted, the design of the pre-distorter is made easy, which is called a "cuber" in reference 2. Propose the block diagram to extract the third-order term of the predistorter.

9.18 The most general approach for modeling the power amplifier will be the recently developed X-parameter method. Search X-parameters on the Web.

9.19 **(ADS problem)** In this chapter, the 6-dB back-off Doherty amplifier was explained, which maintains a high efficiency even at 6-dB back-off power. By a similar concept, a 10 dB back-off Doherty amplifier is also possible. Calculate the efficiency characteristics of this amplifier.

CHAPTER

10

Microwave Oscillators

CHAPTER OUTLINE

10.1 INTRODUCTION

An oscillator is a circuit that generates a high-frequency sinusoidal waveform by converting DC energy delivered from a DC supply. Figure 10.1 shows the formation of an oscillation waveform in the oscillator, which can be divided into two phases: transient and equilibrium states. The amplitude of the sinusoidal waveform with a specific frequency component grows exponentially in the transient state. Then, after passing through the transient state, the waveform reaches the equilibrium state where the sinusoidal waveform with constant amplitude appears.

Thus, an oscillator design must determine whether or not the sinusoidal waveform with the desired frequency grows exponentially for a given circuit, as shown in Figure 10.1; it must then calculate the amplitude and frequency of the oscillation waveform at the equilibrium. For a given circuit, the condition necessary for a sinusoidal waveform with a specific frequency that will grow exponentially is called an *oscillation start-up condition* or simply *oscillation condition*. Since the signal level in an oscillation start-up is low, small-signal analysis can be used to examine the oscillation condition that tells whether or not the sinusoidal waveform with a specific frequency can grow exponentially. In contrast,

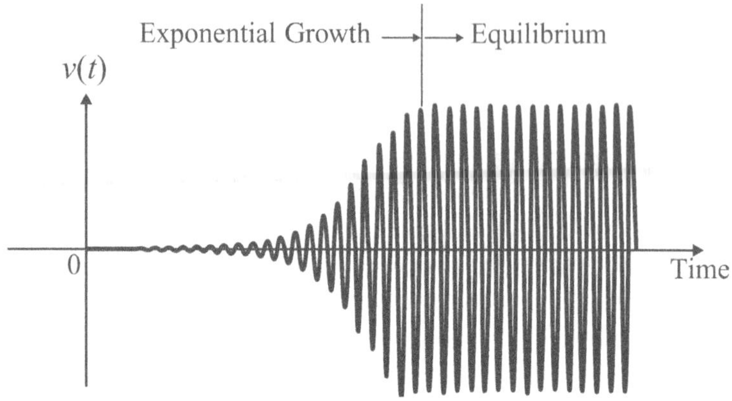

Figure 10.1 Formation of oscillation waveform

since the signal level is not low enough at the equilibrium, large-signal analysis must be carried out. Therefore, the oscillation condition at equilibrium or simply the *equilibrium condition* should be described using large-signal parameters. Active devices such as transistors are generally nonlinear, and it is not possible to directly apply a linear circuit analysis concept, such as impedance or reflection coefficient, as a way to describe the oscillation at equilibrium. However, the level of harmonics in an oscillator circuit is generally low and large-signal impedance or gain can be defined by extending the concept of small-signal impedance or gain to describe the oscillation condition at equilibrium. This is explained in Appendix D.

In this chapter, we will discuss not only the oscillation start-up but also the equilibrium conditions. The large-signal impedance and reflection coefficient will be used to describe the equilibrium oscillation condition. Next, we will learn about the transformation of oscillator circuits and their design. The oscillation waveform is not an ideal sinusoidal waveform and its amplitude and phase vary randomly with time. The measurement technique and modeling of the randomly varying amplitude and phase will also be discussed in this chapter.

10.2 OSCILLATION CONDITIONS

The small-signal oscillation (*start-up condition*) and large-signal equilibrium conditions of an oscillator circuit can be described in a number of ways. The reason for these various descriptions for the oscillation conditions is related to the ease of measuring the quantities that describe the oscillation conditions and it depends on the type of active devices that have evolved with advances in fabrication technologies. However, although the descriptions of the oscillation conditions may differ, they describe the same oscillation phenomena.

Earlier microwave oscillators were implemented using primarily one-port devices such as Gunn or IMPATT diodes. These oscillators can be decomposed into two parts, active device and load. The decomposed one-port network can easily be described using the impedance or reflection coefficient that can be measured directly. Thus, the one-port oscillation and equilibrium conditions that use the impedances or reflection coefficients are convenient for describing these one-port oscillators. As network analyzers are often used in microwave circuit measurements, the reflection coefficient is more direct and preferred for measurement than is the impedance. Thus, the one-port oscillation condition based on the reflection coefficient is a commonly used metric. The oscillation conditions in ADS are also based on the reflection coefficient. However, the impedance-based one-port oscillation conditions can still be applied to an oscillator circuit analysis after converting the reflection coefficient into the impedance.

Due to recent advances in microwave semiconductor technology, transistors such as the pHEMT or the HBT are primarily used in microwave applications instead of diodes such as the Gunn diode or the IMPATT diode. Thus, instead of the one-port oscillation condition, a technique based on an open-loop gain is more efficient. In other words, since the oscillator that uses transistors is, in general, implemented by a feedback network, the open-loop gain oscillation condition is obviously easy to apply. In this chapter, we will explain the open-loop gain conditions for the oscillator circuits that use transistors.

10.2.1 Oscillation Conditions Based on Impedance

10.2.1.1 Start-Up Conditions

In the oscillation condition analysis based on a one-port circuit, an oscillator can be viewed as a circuit composed of a one-port load and a one-port active part, as shown in Figure 10.2.

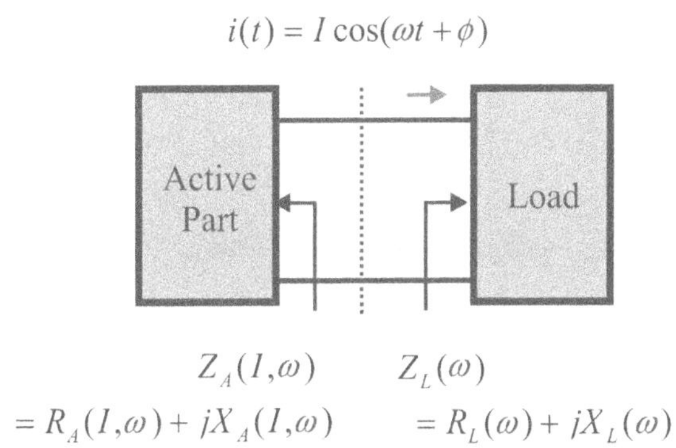

Figure 10.2 An oscillator circuit composed of a one-port load and the active part. The oscillator circuit is considered as the series connection of the one-port load and the active part.

In the oscillator circuit, the active part represents the one-port network that includes an active device, such as the Gunn diode or an IMPATT diode, and has a negative resistance. In general, the impedance of the active device depends on the amplitude of the RF current I, as shown in Figure 10.2. Thus, $Z_A(I,\omega)$ represents the large-signal impedance. The detailed explanation for obtaining the large-signal impedance, reflection coefficient, and gain using ADS or by measurement can be found in Appendix D. However, the load can be considered as a passive circuit that depends on frequency alone. The impedances of the active part and load can easily be measured using a network analyzer. Defining the impedance shown in Figure 10.2, the sum of the two impedances can be expressed as

$$\begin{aligned} R + jX &= Z_A(I,\omega) + Z_L(\omega) \\ &= R_A(I,\omega) + R_L(\omega) + j\left(X_A(I,\omega) + X_L(\omega)\right) \end{aligned} \tag{10.1}$$

Near the oscillation start-up point, the signal level is considered to be small, and since $I \cong 0$, Equation (10.1) can be rewritten as Equation 10.2.

$$R(0,\omega) + jX(0,\omega) = R_A(0,\omega) + R_L(\omega) + j\{X_A(0,\omega) + X_L(\omega)\} \tag{10.2}$$

Let the frequency at which the imaginary part $X(0, \omega) = 0$ be at ω_0. Also, when $R(0, \omega) < 0$ at ω_o, the current I grows exponentially with time, which can be seen to satisfy the oscillation start-up condition. That is, in order to start the oscillation, the following conditions in Equations (10.3a) and (10.3b) must be satisfied, and their frequency-dependent variations are shown in Figure 10.3.

$$R(0,\omega) < 0, \quad X(0,\omega) = 0 \tag{10.3a}$$

$$\left.\frac{\partial X(0,\omega)}{\partial \omega}\right|_{\omega=\omega_o} > 0 \tag{10.3b}$$

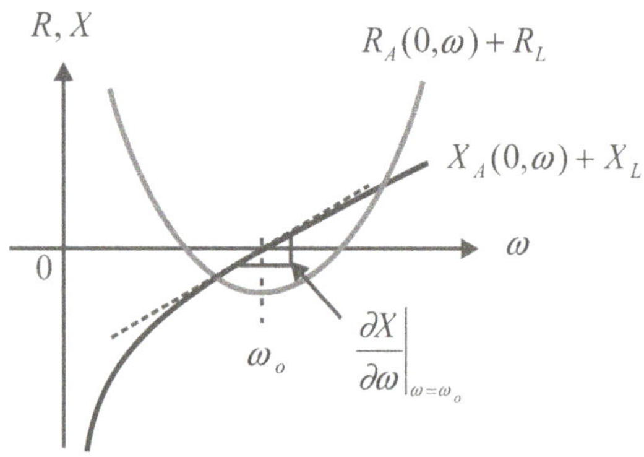

Figure 10.3 Series oscillation condition. The sum of the real part $R < 0$ and $X = 0$ at the oscillation frequency ω_0. In addition, the slope of X with respect to frequency should be positive.

It must be noted that Equation (10.3b) can be interpreted as a series resonant condition and oscillation does not occur when the condition in that equation is not satisfied. This is the description of the oscillation condition in terms of impedance. A similar description can be obtained in terms of admittance. The load and active part in Figure 10.2 can be considered as a series connection. However, the connection can also be considered as a parallel connection, as shown in Figure 10.4. When viewed as a parallel connection, the voltage is common to both of the one port circuits, whereas in the case of a series connection, the current is common.

The admittances of the active part and load are defined as follows:

$$Y_A(V,\omega) = G_A(V,\omega) + jB_A(V,\omega)$$
$$Y_L(\omega) = G_L(\omega) + jB_L(\omega)$$

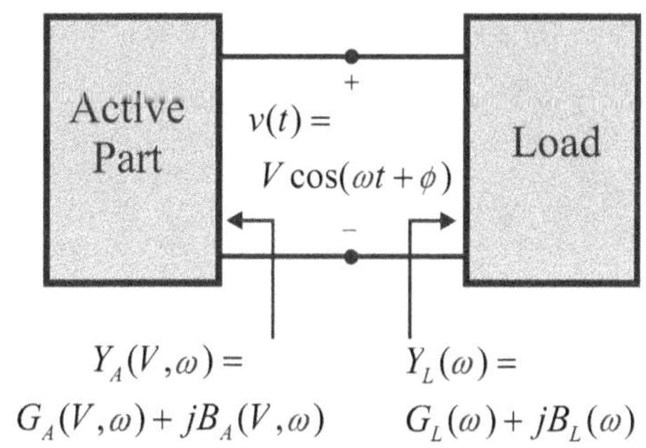

Figure 10.4 Oscillator circuit composed of a one-port load and the active part. The oscillator circuit is considered as the parallel connection of the one-port load and the active part.

Denoting the sum of the admittances as $Y = G + jB$, then Equations (10.4a) and (10.4b),

$$G(0,\omega) < 0, \quad B(0,\omega) = 0 \tag{10.4a}$$

$$\left.\frac{\partial B(0,\omega)}{\partial \omega}\right|_{\omega=\omega_o} > 0 \tag{10.4b}$$

express the similar start-up conditions that must be satisfied for the oscillation waveform to grow exponentially.

Note that the series oscillation condition in Equation (10.3) is the condition for the exponential growth of the RF current amplitude I, whereas the parallel oscillation condition in Equation (10.4) is the condition for the exponential growth of the RF voltage amplitude V when the active part and load are joined to form an oscillator. The series oscillation condition does not necessarily mean the exponential growth of V. Generally, the series oscillation condition described by the impedance does not satisfy the parallel oscillation condition described by admittance and vice versa. To investigate this, suppose that a fixed load (for example, $Z_o = 50\text{-}\Omega$ load) is connected to the series resonating active part, as shown in Figure 10.5(a). Suppose that the sum of the active part and load impedances satisfies the series oscillation condition near the frequency ω_o, $X(\omega_o) = 0$, $-r + Z_o < 0$, and the slope is positive.

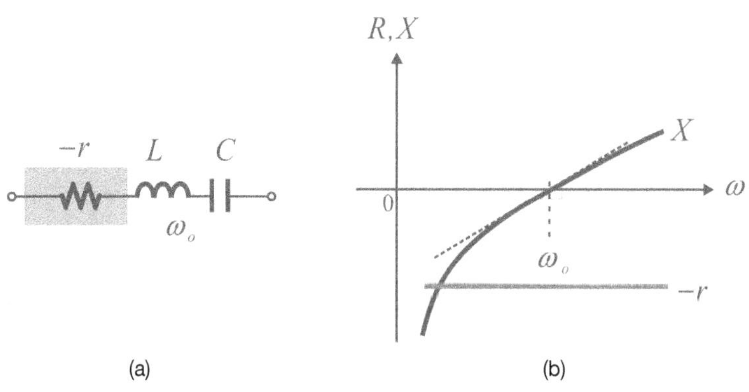

Figure 10.5 (a) Equivalent circuit of the active part and (b) the real and imaginary parts of the impedance for frequency

When the active part's impedance is converted into admittance, the admittance of the active part becomes

$$Y = \frac{1}{Z} = \frac{-r}{r^2 + \left(\omega L - \dfrac{1}{\omega C}\right)^2} + \frac{-j\left(\omega L - \dfrac{1}{\omega C}\right)}{r^2 + \left(\omega L - \dfrac{1}{\omega C}\right)^2}$$

The admittance Y can be drawn as shown in Figure 10.6.

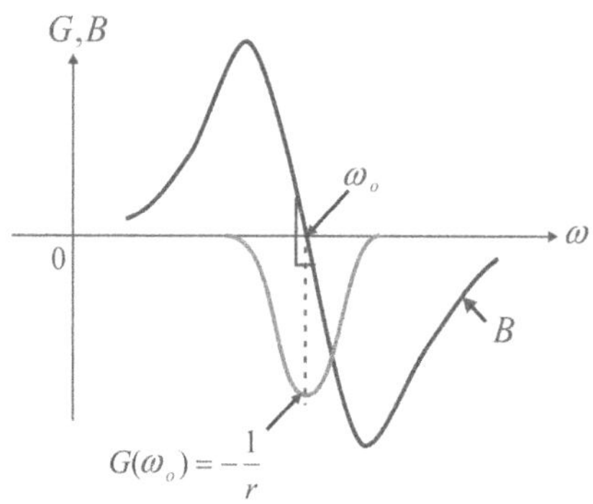

Figure 10.6 The admittance that corresponds to the impedance of Figure 10.5(b). Note that the slope of B with respect to frequency is negative at $B = 0$. Also, $|G| < Y_o$, and parallel oscillation conditions are not satisfied as a result.

The sum of the real part of Y at ω_o and the load conductance Y_o becomes

$$-\frac{1}{r} + Y_o = \frac{-Z_o + r}{rZ_o} > 0$$

That is, as the total conductance is positive and the result does not satisfy the parallel oscillation condition for which the RF voltage amplitude can grow exponentially. Thus, the series oscillation condition in Equation (10.3) does not provide the parallel oscillation condition given by Equation (10.4). Similarly, when the admittance that satisfies the parallel oscillation condition is converted into impedance, the converted impedance does not satisfy the series oscillation condition.

Note that the slope of B in Figure 10.6 can be seen to be negative. Thus, even when it shows positive conductance, if the slope of B at $B = 0$ is negative, it must be reinvestigated to determine whether or not the series oscillation condition is satisfied. Similarly, in terms of impedance, even though the total resistance R is positive, a point where X is 0 occurs, and if the slope of X is negative, it must be reinvestigated to determine whether or not the parallel oscillation condition is satisfied. Therefore, both oscillation conditions should be simultaneously investigated to check the oscillation condition. The reason for this is that the series oscillation condition describes the condition for the RF current amplitude I to grow exponentially, which is not the exponential growth condition of the RF voltage amplitude V.

Example 10.1

The active part of a series resonant oscillator connected to a load through a quarter-wavelength impedance inverter is shown in Figure 10E.1. To convert the load impedance of 50 Ω into 10 Ω, Z_o of the impedance inverter is selected as $Z_o = (10 \times 50)½ = 22.4$ Ω. Discuss the oscillation condition at the load plane and the oscillation condition at the device plane in Figure 10E.1.

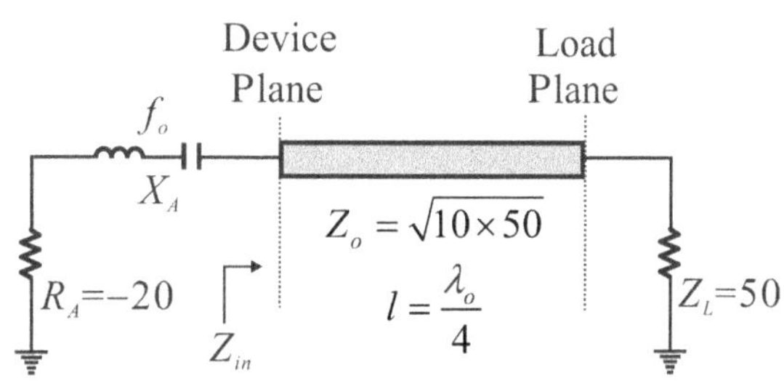

Figure 10E.1 Example of an oscillation circuit that uses a quarter-wavelength transmission line

Solution

The load impedance seen from the active part is 10 Ω. A series resonant circuit is formed at the device plane and a series oscillation then occurs. Thus, the RF current amplitude with the resonance frequency will grow. Conversely, when the active part is seen from the load, the resistance at the resonance becomes $Z_a = -(Z_o)^2/20 = -25$ Ω, which is smaller than the load resistance. However, the impedance of the active part at the load side is given by Equation (10E.1).

$$Z_{out} = \frac{Z_o^2}{R_A + jX_A} \tag{10E.1}$$

The equation above means that the series connection is transformed into a parallel connection through the impedance inverter and a parallel resonance appears at the load side. Thus, since the real part of the active part is smaller than the load, it can be found to satisfy the parallel oscillation condition. Therefore, at the load side, a growth condition of the RF voltage is satisfied. In conclusion, by changing the reference plane, the series oscillation condition can be changed into the parallel oscillation condition.

Example 10.2

The sum of the active part and load impedance $R(0,\omega) + jX(0,\omega)$ is computed in Figure 10E.2. Explain where oscillations can occur and also discuss if further investigation is required to check the possibility of oscillation.

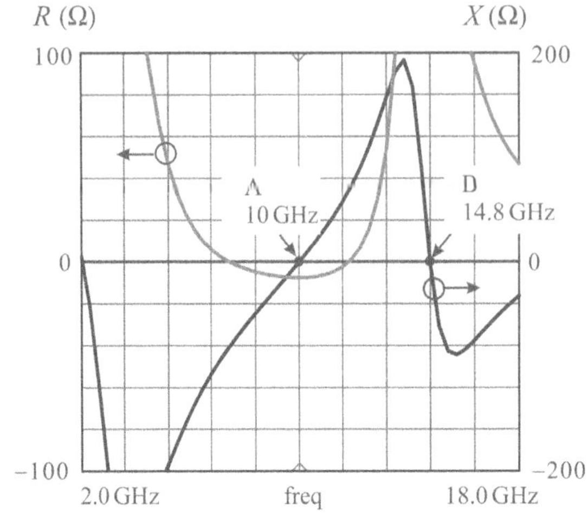

Figure 10E.2 Plot of the sum of the active part and load impedances. The series oscillation condition is satisfied at 10 GHz but the parallel oscillation condition may be satisfied at 14.8 GHz. The parallel oscillation condition should be checked again at 14.8 GHz.

Solution

From the figure, the possible point that satisfies the series oscillation condition described by the impedance is point A (10 GHz). However, oscillation is also possible at point B (14.8 GHz), which shows a parallel resonance. When the admittance satisfying the parallel oscillation condition is converted into an impedance, a positive resistance and negative slope appear at the parallel resonant frequency. Since the oscillation condition is investigated using the sum of the impedance alone, the possibility of satisfying the parallel oscillation condition at point B is not certain. Therefore, it is necessary to plot the sum of the admittance at 14.8 GHz (point B) to check the occurrence of a parallel oscillation.

10.2.1.2 Equilibrium Conditions

In Figure 10.2, when the equilibrium is reached, the amplitude of $i(t)$, I_o, becomes constant. Applying the KVL, we obtain Equation (10.5),

$$\left(Z_A(I_o,\omega)+Z_L(\omega)\right)I_o=\left\{R_A(I_o,\omega)+R_L(\omega)+j\left(X_A(I_o,\omega)+X_L(\omega)\right)\right\}I_o=0 \quad (10.5)$$

and since $I_o \neq 0$,

$$Z_A(I_o,\omega)+Z_L(\omega)=R_A(I_o,\omega)+R_L(\omega)+j\left(X_A(I_o,\omega)+X_L(\omega)\right)=0 \quad (10.6)$$

Rewriting Equation (10.6),

$$R(I_o,\omega)=R_A(I_o,\omega)+R_L(\omega)=0 \quad (10.7a)$$

$$X(I_o,\omega)=X_A(I_o,\omega)+X_L(\omega)=0 \quad (10.7b)$$

Thus, Equations (10.7a) and (10.7b) must be satisfied at equilibrium.

The ω_0 given by the oscillation start-up condition does not generally satisfy $X_A(I_o, \omega) + X_L(\omega) = 0$. Thus, the oscillation frequency at equilibrium can be slightly different from the frequency determined by the oscillation start-up condition. In general, for a high Q circuit, the oscillation frequency primarily appears near the frequency determined by the oscillation start-up condition. Assuming that $X_A(I, \omega_0)$ does not vary with I, the RF current amplitude I_o at equilibrium can be found by plotting the sum resistance R with respect to the RF current amplitude I. Figure 10.7 shows a plot of $R(I,\omega)$ for I. In Figure 10.7, the value of R at the small-signal $I \cong 0$ is found to be negative. The RF current amplitude will increase exponentially, thereby increasing along the x-axis. Conversely, when I is larger than I_o at the equilibrium point, since the value of R will be positive, the RF current amplitude decreases exponentially, thereby decreasing along the $-x$ direction. As a result, equilibrium is attained at the point where the value of R is 0, that is, the point where the RF current amplitude is I_o and thus a steady-state current amplitude appears.

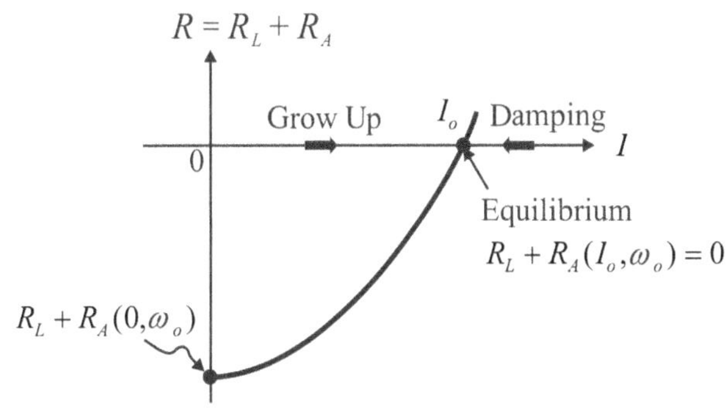

Figure 10.7 Plot of the series sum resistance with respect to the current amplitude I. For small I, I will grow exponentially because $R < 0$. In contrast, I will decrease because $R > 0$ for $I > I_0$. Eventually the equilibrium is formed at I_o.

In addition, the real and imaginary parts of the large-signal admittance at equilibrium condition, like the impedance equilibrium condition, are expressed in Equations (10.8a) and (10.8b), respectively.

$$G_A(V_o, \omega) + G_L(\omega) = 0 \tag{10.8a}$$

$$B_A(V_o, \omega) + B_L(\omega) = 0 \tag{10.8b}$$

Similarly, the oscillation frequency at the equilibrium given by Equation (10.8b) is slightly different from the oscillation frequency determined by the oscillation start-up condition. When $B_A(V, \omega_0)$ is assumed to be constant for the change of V, the equilibrium can be seen to be attained at an RF voltage amplitude V_o using reasoning similar to that for the impedance equilibrium condition.

10.2.1.3 Oscillation Start-Up and Equilibrium Condition Analysis Using ADS

Since it is necessary to calculate the sum of the series impedance in order to simulate a small-signal start-up oscillation condition using ADS, a port is inserted in series between the load and the active part, as shown in Figure 10.8(a). After the one-port S-parameter analysis, the impedance of $Z_A + Z_L$ at the small-signal start-up oscillation can be obtained. However, the admittance $Y_A + Y_L$ is obtained by inserting a port in parallel between the load and the active part, as shown in Figure 10.8(b), and then performing S-parameter analysis. Thus, to investigate the possibility of oscillation using the admittance or impedance obtained through S-parameter analysis, two S-parameter analyses should be simultaneously performed, as shown in Figure 10.8.

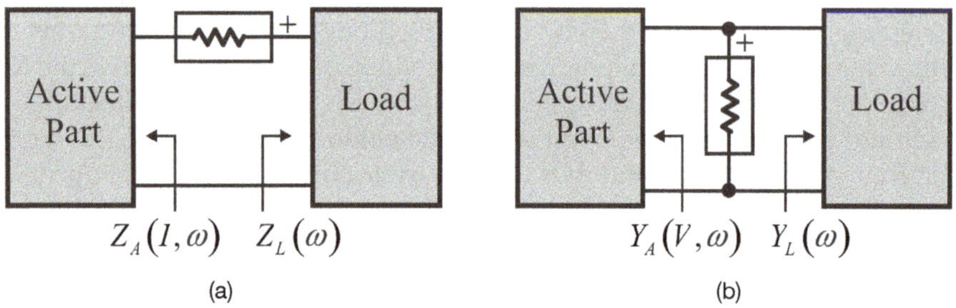

Figure 10.8 Simulation setup for checking the small-signal oscillation condition: (a) series and (b) parallel

Next, the large-signal equilibrium condition must be calculated in ADS to determine the exact oscillation frequency and amplitude. This is accomplished by connecting a large-signal port, as shown in Figure 10.9, and performing a harmonic balance simulation. For series oscillation, the circuit is connected, as shown in Figure 10.9(a), to determine the equilibrium oscillation point. For parallel oscillation, as shown in Figure 10.9(b), the circuit is connected to determine the equilibrium oscillation point.

From Figure 10.9(a), since the sum of the impedances at the series oscillation equilibrium point is 0, the voltage across the port becomes 0. In that case, the amplitude of the current flowing through the port will only be the RF current amplitude at equilibrium. In contrast, since the sum of the admittance becomes zero at the parallel-oscillation equilibrium point, the current flowing out from the port becomes 0 and thus the RF voltage amplitude across the port will be the amplitude of the voltage at equilibrium.

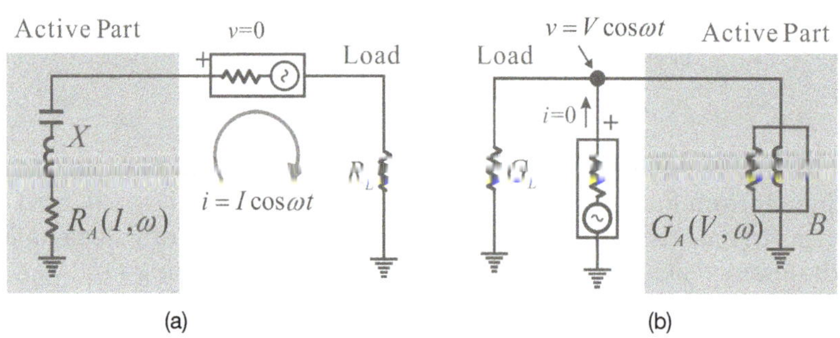

Figure 10.9 Simulation at large-signal equilibrium: (a) series and (b) parallel

Example 10.3

For the active device represented by the current–voltage relationship as

$$i = -v + \frac{1}{3}v^3$$

plot the conductance G_A of the active device; then, by simulation, plot the total conductance $G = G_A + G_L$ when $G_L = 0.2$. In addition, obtain the voltage when the total conductance is 0 and verify that the voltage is the oscillation output voltage at equilibrium.

Solution

When a sinusoidal voltage is applied to the active device, the current is

$$\begin{aligned} i(t) &= -V_p\cos\omega t + \frac{1}{3}\left(V_p\cos\omega t\right)^3 = -V_p\cos\omega t + \frac{1}{3}V_p^3\left(\frac{3}{4}\cos\omega t + \frac{1}{4}\cos 3\omega t\right) \\ &= -V_p\cos\omega t + \frac{1}{4}V_p^3\cos\omega t + \frac{1}{12}V_p^3\cos 3\omega t \end{aligned}$$

Thus, the large-signal conductance becomes

$$G_A\left(V_p\right) = \frac{-V_p + \frac{1}{4}V_p^3}{V_p} = -1 + \frac{1}{4}V_p^2$$

To confirm this by simulation, the given active device can be configured using an SDD (symbolically defined device), as shown in Figure 10E.3. In that figure, the parallel resonant circuit resonating at 1 GHz is connected in parallel with the SDD in order to set the oscillation frequency to 1 GHz. Since the parallel resonant circuit can cause problems at the DC operating point, a DC block is inserted in the parallel resonant circuit. After these settings and the simulation, the following equation in Measurement Expression 10E.1 is entered in the display window to plot the total conductance G, the result of which is shown in Figure 10E.4.

```
Eqn G=real(I_Probe.i[::,1])/Vac
```

Measurement Expression 10E.1 Conductance **G** calculation

In Figure 10E.4, **G** is the same as the theoretically calculated conductance, which is found to be a parabola. In addition, the value of **Vac** where **G** is 0 can be seen to be approximately 1.8 V. Thus at the oscillation equilibrium, the RF voltage amplitude of 1.8 V will appear. To confirm this, **OscPort**, which will be explained in the next section, can be used to verify the value of **Vac**. The oscillation output voltage can be obtained by setting up the schematic shown in Figure 10E.5 and simulating it. The simulated waveform is shown in Figure 10E.6. The waveform in that figure shows a significant amount of distortion due to harmonics, but the amplitude of the oscillation waveform can be seen to be approximately 1.8 V.

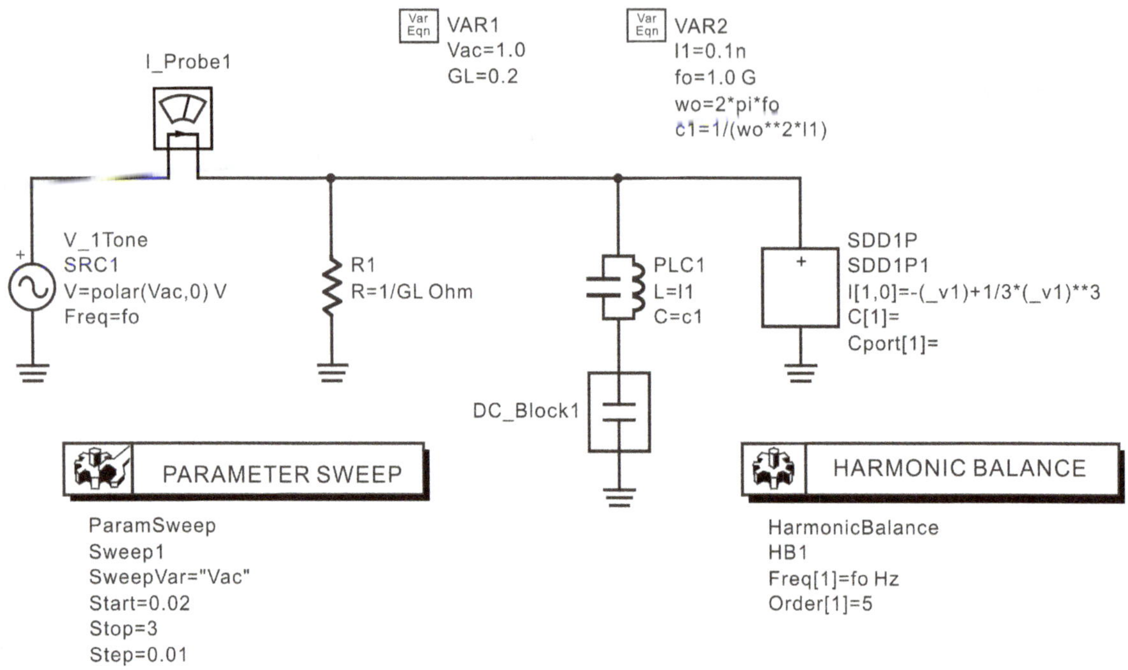

Figure 10E.3 Computation of the large-signal impedance for **Vac**. **DC_Block**1 is inserted so as not to disturb the DC bias.

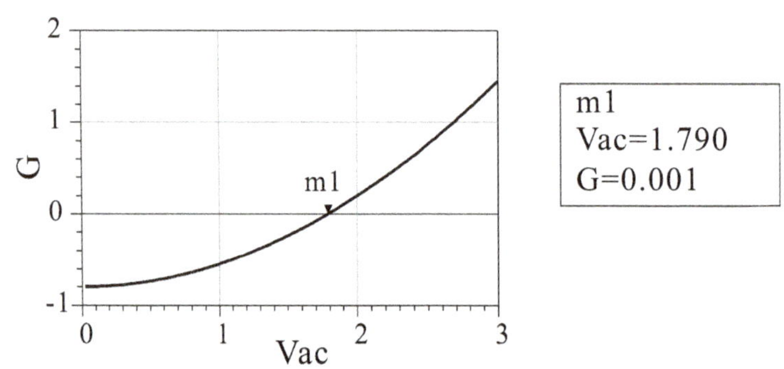

Figure 10E.4 Calculated total conductance

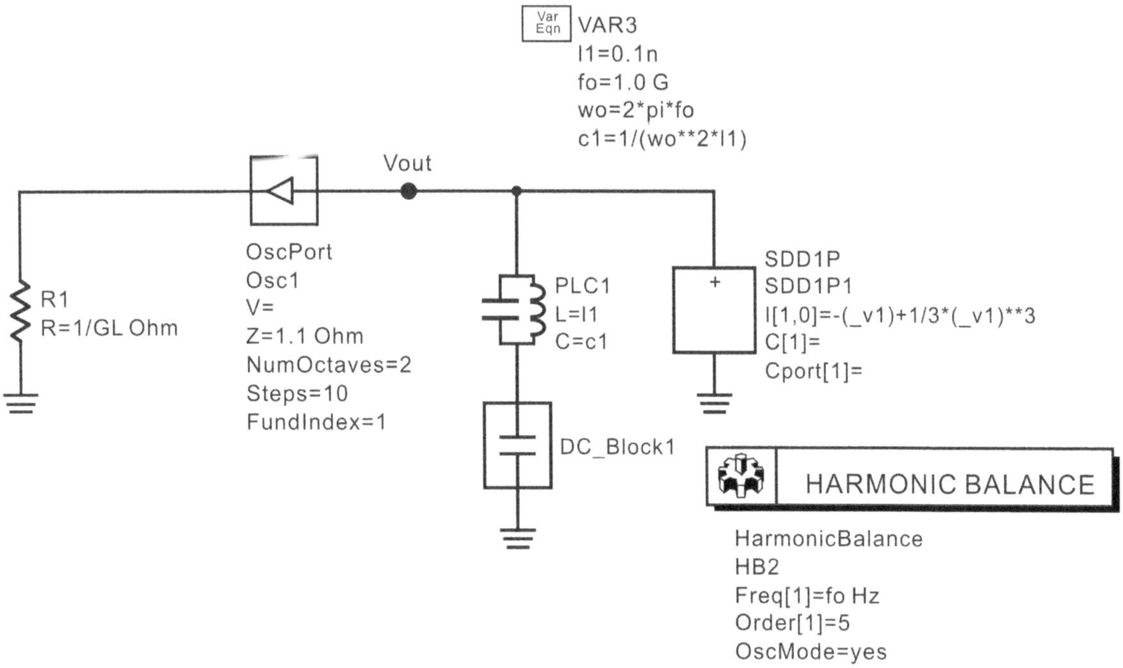

Figure 10E.5 Simulation schematic for the oscillation waveform. **OscPort** in the Harmonic Balance simulator is used to obtain the large-signal oscillation waveform of the oscillator circuit.

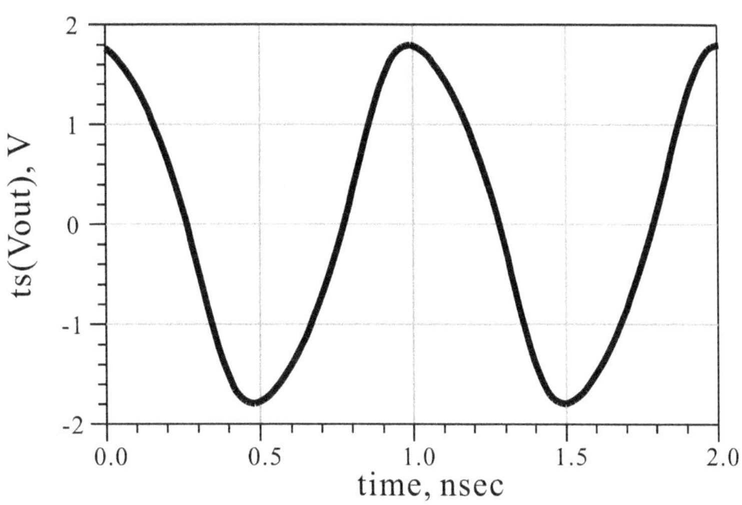

Figure 10E.6 Calculated oscillation waveform

10.2.2 Oscillation Conditions Based on the Reflection Coefficient

10.2.2.1 Start-Up and Equilibrium Conditions Based on the Reflection Coefficient

The oscillation condition based on the reflection coefficient is widely used because it is easier to measure the reflection coefficient at high frequencies compared to the measurement of impedance. The oscillation condition based on the reflection coefficient is similar to the oscillation condition based on impedance discussed earlier. In Figure 10.10, the reflection coefficients of the active part and load are defined for the same reference impedance Z_o, and their one-port reflection coefficients are Γ_A and Γ_L, respectively.

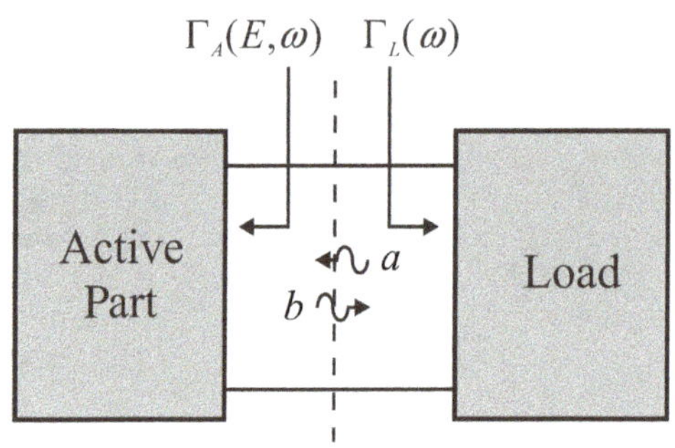

Figure 10.10 Oscillator circuit based on the reflection coefficient. Γ_A and Γ_L are the reflection coefficients defined with the same reference impedance. In the case of Γ_A, it is also a function of $|a| = E$.

When the incident voltage $a = E\cdot\cos\omega t$ corresponding to the available power of $P_A = ½E^2$ is applied from the load, the new reflected voltage from the load a' after a round-trip between the active part and the load is expressed in Equation (10.9).

$$a' = \Gamma_L(\omega)\Gamma_A(E,\omega)a \tag{10.9}$$

The reflected voltage a' becomes a new incident voltage that again makes a round-trip between the active part and the load. Thus, for exponential growth, the following start-up conditions must be satisfied, as expressed in Equations (10.10a) and (10.10b).

$$\left|\Gamma_L(\omega_o)\Gamma_A(0,\omega_o)\right| > 1 \tag{10.10a}$$

$$\angle\Gamma_L(\omega_o)\Gamma_A(0,\omega_o) = 0 \tag{10.10b}$$

Here, $\Gamma_A(0,\omega_o)$ represents the small-signal reflection coefficients of the active part. Equation (10.10a) is the condition that must be satisfied for the signal to grow through the repetition of round-trips, while Equation (10.10b) is the condition requiring that the phase remain unchanged when these round-trips repeat.

In addition, for the oscillation conditions given by Equations (10.10) to be stable, the slope of Equation (10.10b) for the frequency must be negative. This is expressed in Equation (10.11).

$$\frac{\partial\angle\Gamma_L(\omega_o)\Gamma_A(0,\omega_o)}{\partial\omega} < 0 \tag{10.11}$$

The example of the oscillation condition that satisfies Equations (10.10) and (10.11) is shown in Figure 10.11. The magnitude and phase of $\Gamma_A\Gamma_L$ that satisfy the oscillation condition are plotted in that figure, where it can be seen why the condition in Equation (10.11) is necessary. When the frequency is lower than the oscillation frequency, a positive phase occurs, and by repeated round-trips, the frequency increases as the phase continues to increase until it eventually approaches the oscillation frequency. In contrast, when the frequency is higher than the oscillation frequency, the phase becomes negative and, by repeated round-trips, the phase decreases continuously and eventually attains equilibrium at the frequency of oscillation. Thus, Equation (10.11) provides a stable oscillation.

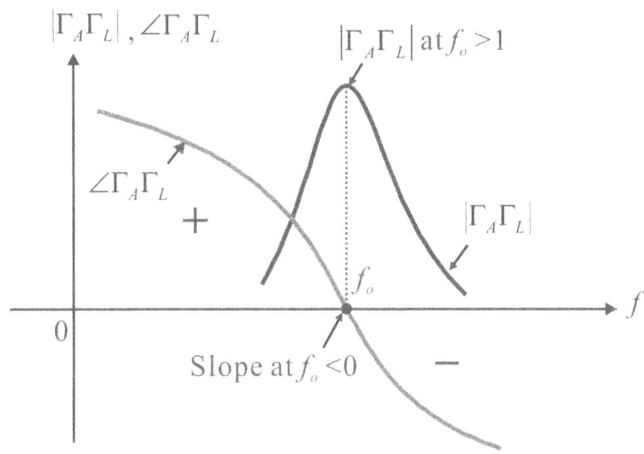

Figure 10.11 Plot of small-signal $\Gamma_L\Gamma_A$, which can be represented by $\Gamma_L\Gamma_A$ $(0,\omega)$. For oscillation at f_o, $|\Gamma_L\Gamma_A(0,\omega)| > 1$ and $\angle\Gamma_L\Gamma_A(0,\omega) = 0$ at f_o should be satisfied. Also, the slope of $\angle\Gamma_L\Gamma_A(0,\omega)$ with respect to frequency should be negative for stable oscillation.

It must be noted that the oscillation conditions given by Equation (10.10) obviously guarantee oscillation. However, oscillation can also occur under other conditions and thus the conditions given by Equation (10.10) are the conditions sufficient for oscillation. Furthermore, the conditions change when the reference impedance, used to measure the reflection coefficients of the active part and load, changes. This is summarized in Appendix F.

The plot of $\Gamma_L\Gamma_A(E,\omega_0)$ for E is shown in Figure 10.12. Similar to the impedance oscillation conditions described earlier, E will grow exponentially at the small-signal because $|\Gamma_L\Gamma_A(E,\omega_0)| > 1$ and increases along the x-axis, while E decreases when E becomes greater than E_0. Therefore, the equilibrium is achieved at E_0. That equilibrium is expressed in Equations (10.12a) and (10.12b).

$$|\Gamma_L(\omega_0)\Gamma_A(E_0,\omega_0)| = 1 \tag{10.12a}$$

$$\angle\Gamma_L(\omega_0)\Gamma_A(E_0,\omega_0) = 0 \tag{10.12b}$$

In other words, at the equilibrium point, the magnitude of the product of the reflection coefficients due to continuous round-trips is 1 and its phase is 0°.

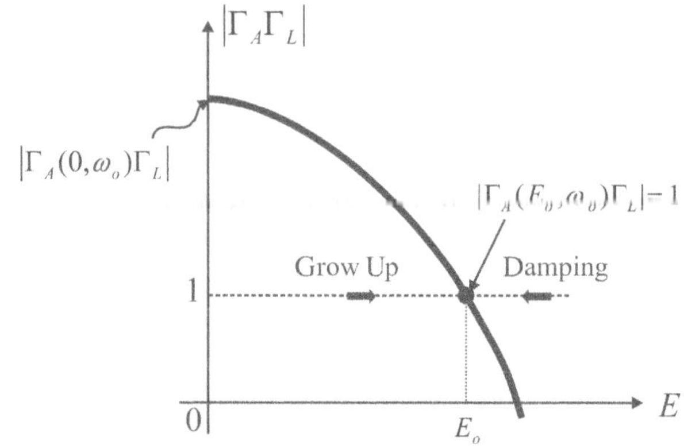

Figure 10.12 Plot of $\Gamma_A\Gamma_L$ with respect to the available power. For small E, E will grow through repeated round-trip because $|\Gamma_L\Gamma_A(0,\omega)| > 1$. In contrast, E will decrease because $|\Gamma_L\Gamma_A(0,\omega)| < 1$ for $E > E_o$. Eventually the equilibrium is formed at E_o.

10.2.2.2 Circuit Implementation

The previously described oscillation condition based on the reflection coefficient can be implemented using a circulator, as shown in Figure 10.13.

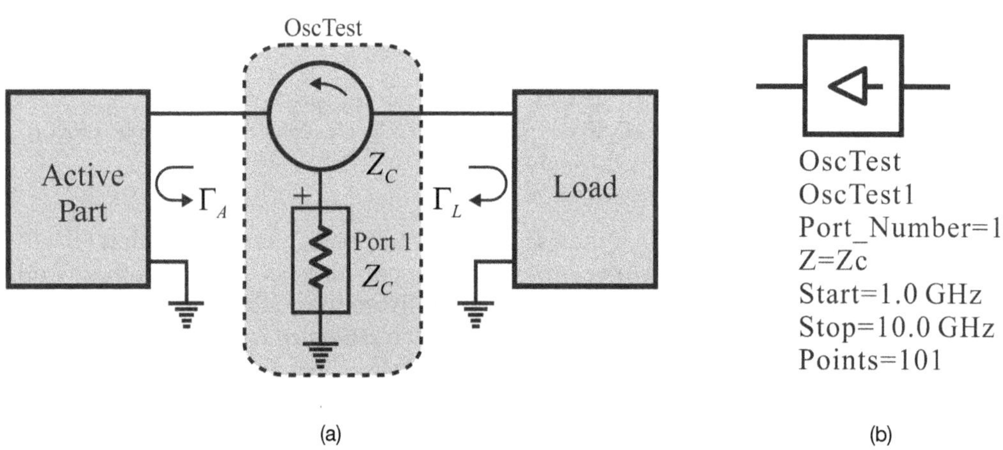

Figure 10.13 a) Measurement of $\Gamma_A\Gamma_L$. The reflection coefficient seen from the port becomes $\Gamma_A\Gamma_L$. (b) OscTest in ADS. Z represents the port impedance.

As shown in Figure 10.13(a), when the load and active part are connected to a broadband circulator with the reference impedance Z_C, the reflection coefficient S_{11} at the port can be expressed as $S_{11} = \Gamma_A\Gamma_L$. Note that the computed $\Gamma_A\Gamma_L$ depends on the reference impedance Z_C of the port. The different port impedance will result in a different value for $\Gamma_A\Gamma_L$. The oscillation condition based on the reflection coefficient is thus a function of the reference impedance Z_C. The circulator and port in the shaded box shown in Figure 10.13(a) are already implemented as **OscTest** in ADS, as shown in Figure 10.13(b). The variable **Z** of **OscTest** represents Z_C in Figure 10.13(a). The **OscTest** computes $\Gamma_A\Gamma_L$ for the frequency range that is specified as **Start** and **Stop**.

Example 10.4

This Example considers a small-signal series oscillation circuit shown in Figure 10E.7. For the reference impedance of 100 Ω, calculate $\Gamma_A\Gamma_L$ and verify that this is the same as computed using **OscTest** whose **Z** is set to 100 Ω.

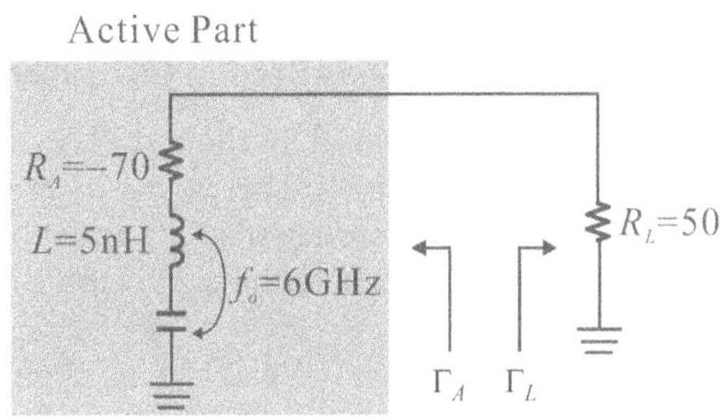

Figure 10E.7 Example of a small-signal series oscillation circuit

Solution

The magnitude and phase of S_{11} is computed using **OscTest** with **Z** = 100 Ω as shown in Figure 10E.8.

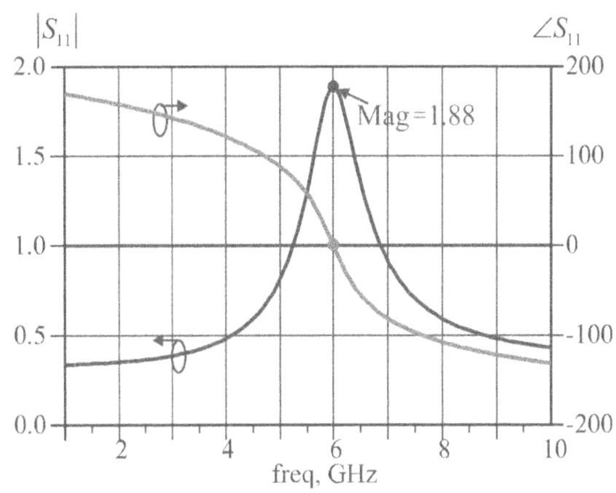

Figure 10E.8 S_{11} computed using OscTest with the reference impedance of 100 Ω

In that figure, $|S_{11}|$ at f_o = 6 GHz is greater than 1, and the phase is 0°. In addition, the phase slope with respect to frequency can be seen to be negative. Thus, the oscillation conditions based on the reflection coefficient are satisfied.

On the other hand, calculating $\Gamma_A\Gamma_L$ at the oscillation frequency of 6 GHz,

$$\Gamma_A\Gamma_L = \frac{-70-100}{100-70}\times\frac{50-100}{50+100} = 1.88$$

which is the same as the value of mag(S_{11}) shown in Figure 10E.8.

Notably, $\Gamma_A\Gamma_L = 0$ when Z_C is set at 50 Ω instead of 100 Ω, and the oscillation conditions in Equation (10.12) are not satisfied. However, this is the oscillation condition variation due to the change of Z_C. The oscillator circuit in this example clearly oscillates although $\Gamma_A\Gamma_L = 0$. Therefore, selecting the appropriate reference impedance makes it easy to check the oscillation condition. For more information, refer to Appendix F.

10.2.2.3 Equilibrium Based on the Reflection Coefficient

The oscillation start-up condition based on the reflection coefficient can be used to derive the large-signal equilibrium conditions. That is, irrespective of a parallel or series oscillation, the product of the reflection coefficients $\Gamma_A\Gamma_L$ must be 1 at equilibrium. Figure 10.14 shows the circuit measuring $\Gamma_A\Gamma_L$, where the port is replaced by the large-signal port.

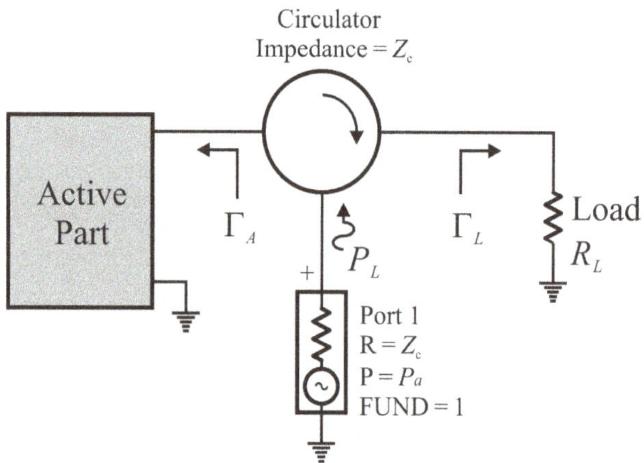

Figure 10.14 Computing method for a large-signal equilibrium in an oscillator. For a small signal, the oscillator circuit generates power and the port consumes the oscillation power; however, the port delivers the power to the oscillator circuit at a large-signal level. Thus, the equilibrium is formed when the dissipation power of the port is 0.

Denoting the available power from the port as P_a, the power delivered to the oscillator circuit from the port is $P_a(1 - |\Gamma_L\Gamma_A|^2)$. Thus, when P_a is small, the delivered power becomes negative because $|\Gamma_L\Gamma_A| > 1$. As a result, the port consumes the power rather than delivering it to the oscillator circuit. At equilibrium, the port is found to deliver no power to the oscillator circuit because $|\Gamma_L\Gamma_A| = 1$. Therefore, the equilibrium point can be found by determining when the delivered power from the port P_L becomes 0. Then, every voltage and current in the oscillator circuit at equilibrium can be obtained by calculating the currents and voltages of the oscillator circuit at the port power where delivered power from the port becomes 0.

Thus, to determine the large-signal equilibrium state using the reflection coefficient, the available power and frequency of the port is altered to yield $|\Gamma_A \Gamma_L| = 1$. Then, using the determined port power and the frequency at equilibrium, every current and voltage inside the oscillator circuit can be calculated; in turn, this calculation can be used to determine the waveforms of every node in the oscillator circuit at equilibrium. These operations can be performed automatically in ADS, and this is usually done using the **OscPort**. That is, by performing the simulation using the **OscPort** in conjunction with the **Harmonic Balance** simulator (i.e., a simulation in which the large-signal available power of the port is varied to determine the oscillation power and frequency at equilibrium), the oscillation at equilibrium can be found. The **OscPort** is shown in Figure 10.15(b). Figure 10.15(a) shows the concept of **OscPort** in ADS. The oscillation at equilibrium can be obtained by using the **HB simulator** together with the **OscPort** shown in Figure 10.15(b).

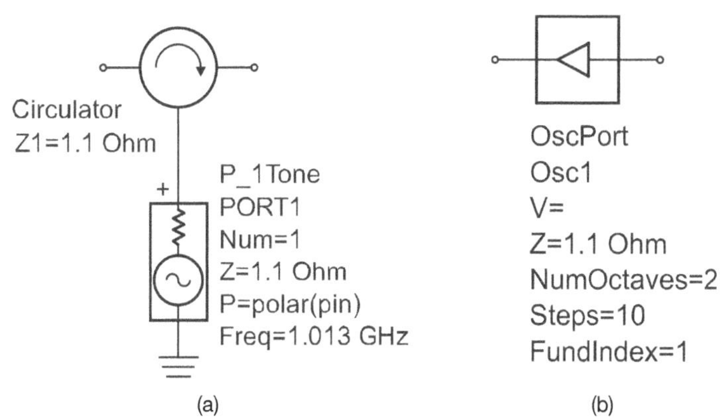

Figure 10.15 (a) Equivalent circuit of the **OscPort** and (b) **OscPort**

The **Z** of the **OscPort** in Figure 10.15(b) is the reference impedance of the circulator. **FundIndex** is the index of the estimated oscillation frequency in the **HB simulator**. It represents the simulation for the fundamental frequency, which in most cases is 1. Since the oscillation frequency at equilibrium differs from the estimated small-signal oscillation frequency, **NumOctaves** is specified to determine the frequency tuning range. When **NumOctaves** = 2, the oscillation frequency is sought for the frequency range that varies from 0.5 to 2 times the estimated frequency.

Example 10.5

The diode admittance is denoted as $Y_A = -G(A) + jB(A)$. Here, A represents the amplitude of the fundamental voltage. One diode has $G(A) = g(0) - k_1 A$ and the other has $G(A) = g(0) - k_2 A^2$, as shown in Figures 10E.9(a) and (b), respectively.

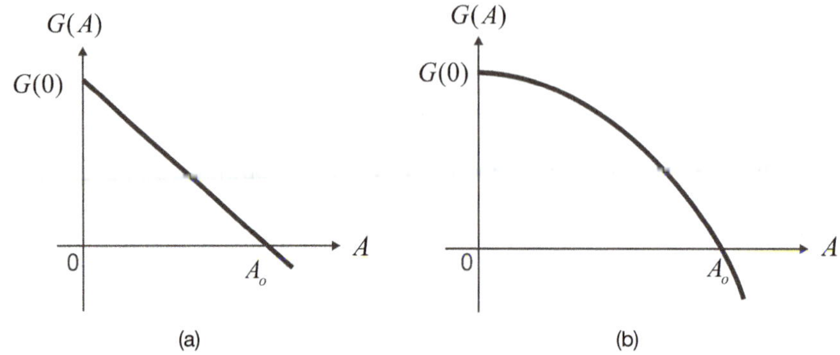

Figure 10E.9 $G(A)$ with respect to voltage amplitude A: (a) linear decrease and (b) quadratic decrease

The load values giving maximum oscillation output power are known to be $1/3 \cdot G(0)$ and $1/2 \cdot G(0)$, for $G(A)$ in Figures 10E.9(a) and (b), respectively. In Example 10.4, $-G(A) = 1 - 1/4 \cdot A^2$. Thus, $G(A)$ delivers the maximum oscillation output power to the load when the load conductance G_L is equal to 0.5. Confirm this through simulation and calculate the maximum oscillation output power.

Solution

As mentioned earlier, the maximum oscillation power occurs at $G_L = 0.5$ and since the amplitude at equilibrium must satisfy

$$-G(A) + G_L = -1 + \frac{1}{4}A^2 + \frac{1}{2} = 0$$

it can be seen that $A = (2)^{1/2}$. Thus, the maximum output power is

$$P_{L\max} = \frac{1}{2}G_L A^2 = \frac{1}{4} \times 2 = 0.5 \text{ (W)}$$

To confirm the oscillation output power for the change of G_L, the harmonic balance simulation is performed using **OscPort**, as shown in Figure 10E.10. After simulation, the equation shown in Measurement Expression 10E.2 is entered in the display window to calculate the output power P_L delivered to the load G_L.

```
Eqn PL=1/2*GL*mag(Vout[::,1])**2
```

Measurement Expression 10E.2 Equation for the delivered power to the load

The delivered power P_L is shown in Figure 10E.11. As expected, the maximum output power is 0.5 W at $G_L = 0.5$.

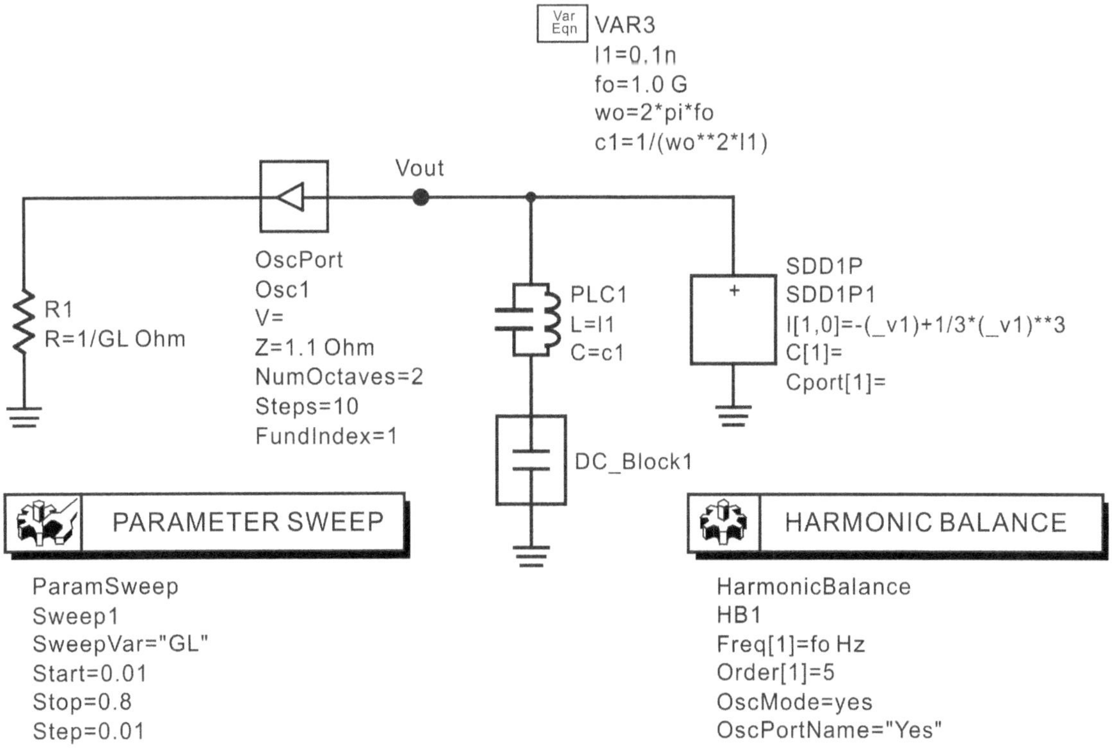

Figure 10E.10 Schematic for obtaining the oscillation output power for the variation of G_L

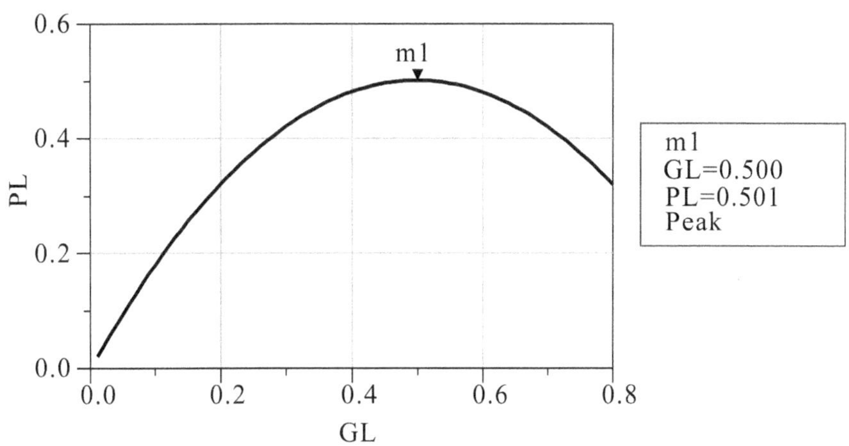

Figure 10E.11 Oscillation output power for the G_L change. As expected, the maximum oscillation power 0.5 W is achieved at $G_L = 0.5$.

10.2.3 Start-Up and Equilibrium Conditions Based on Open-Loop Gain

The one-port oscillation condition described earlier that uses the impedance or reflection coefficient is a direct method for analyzing oscillator circuits employing diodes, but not for analyzing oscillator circuits using transistors, which can be configured as oscillators using a feedback network. In the case of oscillators that use the feedback network, the description based on the one-port oscillation conditions generally makes it difficult to understand the role of the feedback network. Instead of the reflection coefficient or the impedance used in one-port oscillation conditions, the use of an *open-loop gain* makes it easy to understand oscillators that employ feedback. The open-loop gain also facilitates an understanding of the maximum output power condition and phase noise of oscillators using feedback. The open-loop gain is easily obtained when the reverse gain of a transistor is small, which enables the calculation of the open-loop gain in the direction of power delivery. However, at high frequency, the reverse gain is not small and it is not easy to calculate the open-loop gain due to the bidirectional properties of transistors. Recently published papers have revisited the calculation of the open-loop gain, taking the bidirectional properties into consideration.[1,2] Thus, the open-loop gain method will be widely used in the design of oscillators employing feedback.

10.2.3.1 Start-Up and Equilibrium Conditions Based on Open-Loop Gain

Figure 10.16 shows an oscillator configuration in which the output of the amplifier is fed back through a resonator. Here, the feedback loop is broken, which is done to obtain the open-loop gain.

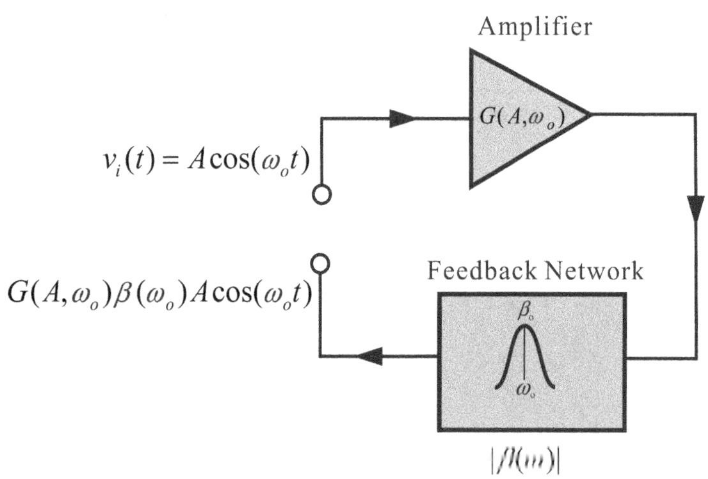

Figure 10.16 Oscillator configuration with feedback network. To measure the open-loop gain, the feedback loop is broken. The test signal is applied to one port where the loop is broken and the open-loop gain is obtained by measuring the output appearing at the other port.

1. M. Randall and T. Hock, "General Oscillator Characterization Using Linear Open-Loop S-Parameters," *IEEE Transactions on Microwave Theory and Techniques* 49, no. 6 (June 2001): 1094–110.

2. R. Rhea and B. Clausen, "Recent Trends in Oscillator Design," *Microwave Journal*, January 28, 2004.

In the figure, the gain $G(A,\omega)$ of the amplifier can be considered to vary with the input signal amplitude A, and the transfer function of the resonator can generally be expressed as shown in Equation (10.13),

$$\beta(\omega) = \frac{\beta_o}{1 + jQ_L\left(\frac{\omega}{\omega_o} - \frac{\omega_o}{\omega}\right)} \tag{10.13}$$

where β_o is the transmission coefficient at the resonant frequency ω_o, and Q_L represents the loaded Q of the resonator. The open-loop gain $L(A, \omega)$ is defined as Equation (10.14),

$$L(A,\omega) = G(A,\omega)\beta(\omega) \tag{10.14}$$

which is the output voltage per loop trip of the input signal. When a sinusoidal signal of $A{\cdot}\cos(\omega_o t)$ is applied to the input of Figure 10.16, a signal of $L(A,\omega_o){\cdot}A\cos(\omega_o t)$ appears at the loop output. It must be noted that the output signal has the same phase as the input signal and only the amplitude varies. The output signal $L(A,\omega_o){\cdot}A\cos(\omega_o t)$ is repeatedly applied to the input of Figure 10.16 when the loop is closed. If the loop gain $|L(A,\omega_o)| > 1$, the amplitude grows after every trip of the loop. Therefore, in order for the signal to grow as a small signal, $|L(A,\omega_o)| > 1$ at the frequency ω_o with the phase 0°. Then, oscillation can be formed. This is called the *Barkenhausen Criterion* and can be expressed as shown in Equations (10.15a)–(10.15c).

$$\angle L(0,\omega_o) = 0 \tag{10.15a}$$

$$|L(0,\omega_o)| > 1 \tag{10.15b}$$

$$\left.\frac{\partial \angle L(0,\omega_o)}{\partial \omega}\right|_{\omega=\omega_o} < 0 \tag{10.15c}$$

When the conditions above are satisfied, the amplitude grows by repeated feedback and oscillation can occur. The reason Equation (10.15c) must be satisfied is that when the frequency is lower than the resonant frequency, the phase of the open loop gain in Figure 10.17 becomes positive, and thus the phase grows positively for every trip of the loop. The continuous increase in the phase represents an increase in frequency, eventually approaching the resonant frequency. In contrast, when the frequency is higher than the resonant frequency, the phase of the open loop gain in Figure 10.17 becomes negative and decreases negatively for every trip of the loop, and the continuous decrease in phase represents a decrease in frequency, and thus approaches the resonant frequency. On the other hand, when the phase response of the open-loop gain is opposite of that given by Equation (10.15c), even when there is a small phase jitter, it will be away from the resonance frequency, and oscillation will not occur. Thus, Equation (10.15c) is an important criterion in determining the occurrence of oscillation.

The typical small-signal open-loop gain $L(0,\omega)$ is shown in Figure 10.17. The frequency response characteristics shown in Figure 10.17 satisfy the conditions of Equation (10.15). Because the gain $G(0,\omega)$ is almost constant, the shape of the open-loop gain with frequency generally resembles the frequency response of the resonator given by Equation (10.13). The maximum value of the open-loop gain occurs at the resonant frequency ω_o and decreases

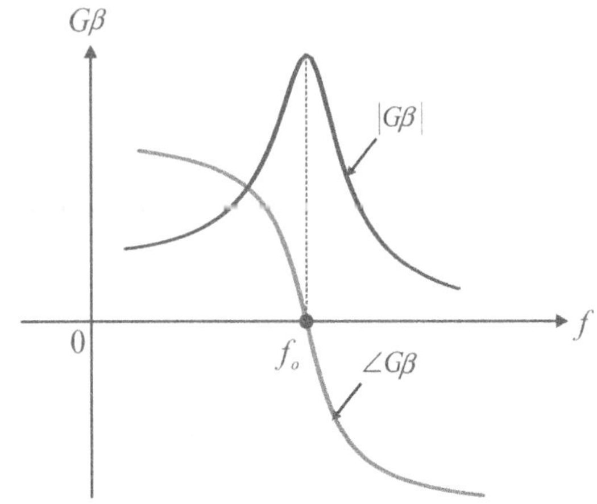

Figure 10.17 The frequency response example of the small-signal open-loop gain

below and above the resonant frequency, as shown in Figure 10.17. Furthermore, the phase is 0° at the resonant frequency; the phase approaches 90° below the resonant frequency, while it approaches –90° above the resonance frequency.

The change of the open-loop gain according to the amplitude is plotted in Figure 10.18. For small signals ($A \cong 0$), as the magnitude of the open-loop gain is greater than 1, the amplitude A increases exponentially. The open-loop gain is reduced due to the increase in A. In contrast, when the amplitude is greater than the equilibrium point, the amplitude decreases because the open-loop gain is less than 1. Eventually, the amplitude A reaches A_o at the point where the open loop gain is 1. Therefore, the oscillation frequency is determined by Equation (10.16)

$$\angle L(A_o, \omega_o) = 0 \tag{10.16}$$

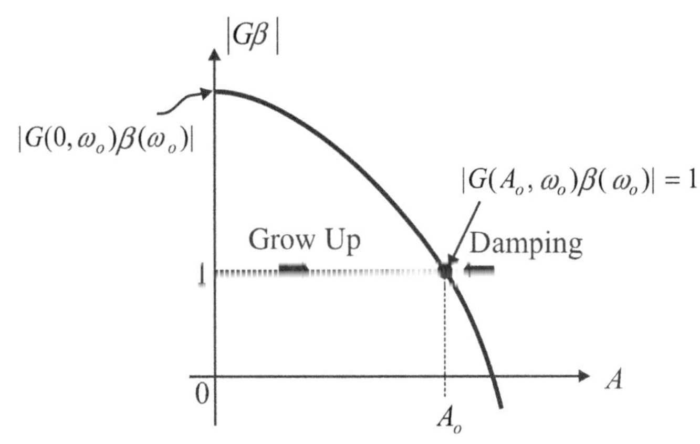

Figure 10.18 The open-loop gain for amplitude A. For small A, A will grow through repeated round-trips because $|G\beta| > 1$. In contrast, A will decrease because $|G\beta| < 1$ for $A > A_o$. Eventually, the equilibrium is formed at A_o.

and the oscillation amplitude A_o is determined by Equation (10.17).

$$\left|L(A_o, \omega_o)\right| = 1 \tag{10.17}$$

10.2.3.2 Open-Loop Gain

Compared to the one-port method, the open-loop gain method is obviously more intuitive and easier to apply when investigating the oscillation condition for an oscillator circuit using a feedback network. The oscillator circuit using feedback can be represented conceptually, as shown in Figure 10.19(a). In order to calculate the open-loop gain of the oscillator circuit, that circuit is cut to break the feedback, as shown in Figure 10.19(b).

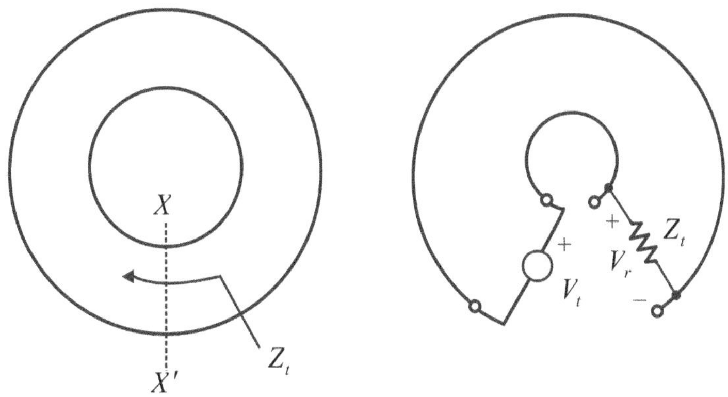

Figure 10.19 (a) Conceptual feedback network and (b) the open-loop network for calculating open-loop gain. The open-loop gain is equal to V_r/V_t.

The open-loop gain can be computed by applying a test source V_t and measuring the return voltage V_r at the load with the impedance Z_t, which is equal to that looking into the source side prior to breaking the loop. However, it is generally difficult to find the impedance Z_t at the cut plane. To this end, the feedback is considered as an infinite number of identical open-loop networks connected end to end. When the two-port S-parameters for the two-port open-loop network in Figure 10.19(b) are defined as S_{ij}, the open-loop gain L can be expressed as Equation (10.18).

$$L(\omega) = \frac{S_{21} - S_{12}}{1 - S_{11}S_{22} + S_{21}S_{12} - 2S_{12}} \tag{10.18}$$

For $L(\omega)$ to satisfy the oscillation conditions in Equation (10.15), the open-loop network can oscillate when it is closed. In that equation, S_{11}, S_{22}, and S_{12} are generally small, and $L(\omega)$ is simplified as

$$L(\omega) \cong S_{21} \tag{10.19}$$

From Equation (10.19), the oscillation condition is determined as

$$\mathrm{dB}(S_{21}) \geq 0 \tag{10.20a}$$

$$\mathrm{phase}(S_{21}) = 2n\pi \ (n=0,1,2,3\cdots) \tag{10.20b}$$

The feedback-type oscillator can easily be designed using Equations (10.20a) and (10.20b). The feedback-type oscillator is usually composed of cascaded two-port circuits. For example, the oscillator circuit in Figure 10.16 can be viewed as the cascaded connection of an amplifier and a resonator. Using the S-parameters of each block, the two-port S-parameters of the open-loop network can be obtained. Also, if each block is matched, the open-loop gain is simply the product of each block's S_{21}, and the oscillation frequency satisfying Equations (10.20) can be easily found. Once the open-loop gain satisfying Equation (10.20) at the oscillation frequency is obtained, the oscillator can be built up simply by closing the open loop. In addition, the large-signal equilibrium conditions can be obtained from the response of the open-loop gain for amplitude change, which can be computed using **OSCPort** in ADS.

Example 10.6

The circuit shown in Figure 10E.12 is a Colpitts oscillator. Calculate its oscillation frequency by the open-loop gain method using ADS.

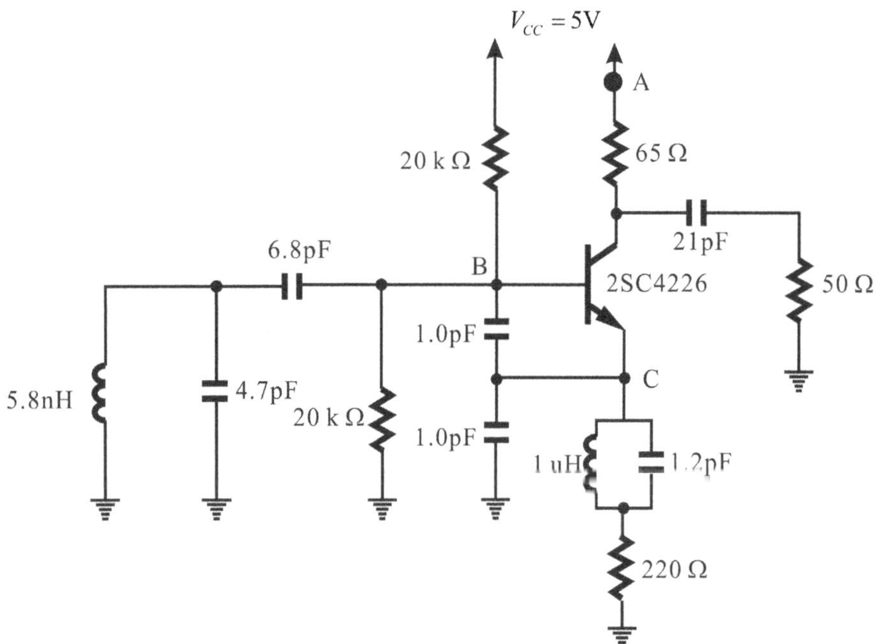

Figure 10E.12 Colpitts oscillator circuit

Solution

In the circuit shown in Figure 10E.12, the ground point is eliminated and a new ground point C is set at the transistor's emitter. With the changed ground point, the input of the open loop is defined by the base emitter (B–C plane) of the transistor and the open loop is formed by breaking the oscillator circuit at the B–C plane, as shown in Figure 10E.13. The collector-emitter voltage and base current are determined in advance through DC simulation for the circuit in Figure 10E.12. In order to maintain the DC operating point of the transistor, the determined collector-emitter voltage and base current are supplied by a new DC current source and a voltage source, as shown in Figure 10E.13. After setting up the circuit, the S-parameter simulation is carried out.

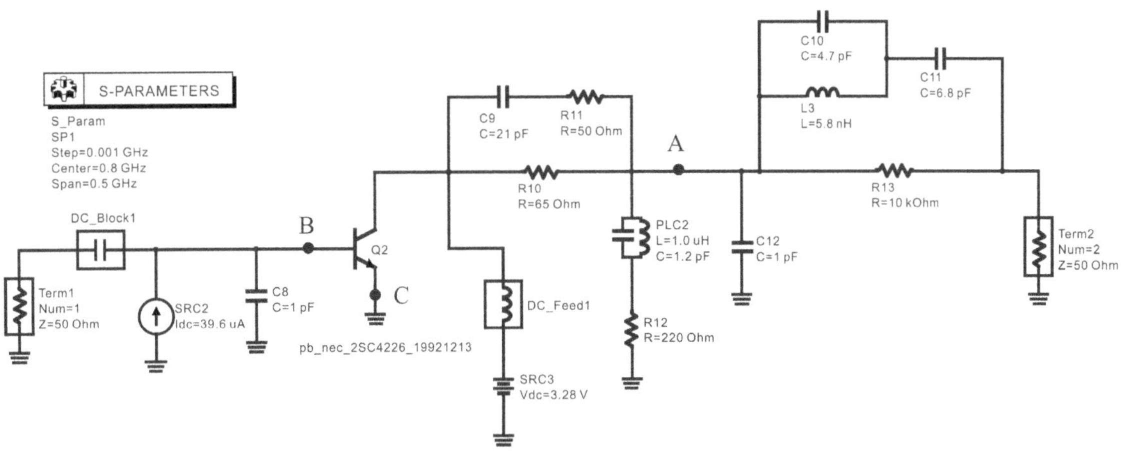

Figure 10E.13 Circuit for calculating the open-loop gain. First, DC analysis is carried out for the circuit shown here. Then, the computed base current and V_{CE} are applied to bias 2SC4226, as shown in this figure. Finally, the circuit, which is cut along points B and C, is redrawn and the new ground point is set to point C.

After simulation, the equation in Measurement Expression 10E.3 is entered in the display window to compute and plot the open-loop gain.

Eqn G=(S(2,1)-S(1,2))/(1-S(1,1)*S(2,2)+S(1,2)*S(2,1)-2*S(1,2))

Measurement Expression 10E.3 Open loop gain calculation using the simulated S-parameters

The magnitude and phase of the open-loop gain are shown in Figure 10E.14. The oscillation frequency is approximately 827 MHz. To compute the large-signal oscillation frequency, the previously described **OscPort** is inserted in the oscillator circuit in Figure 10E.12. The calculated oscillation frequency confirms the obtained 827 MHz shown in Figure 10E.14. Note that the ground point was moved to the transistor's emitter for calculating the open-loop gain, which made the calculation easy. This change to the ground point is frequently used in oscillator design and is called *virtual ground technique*.

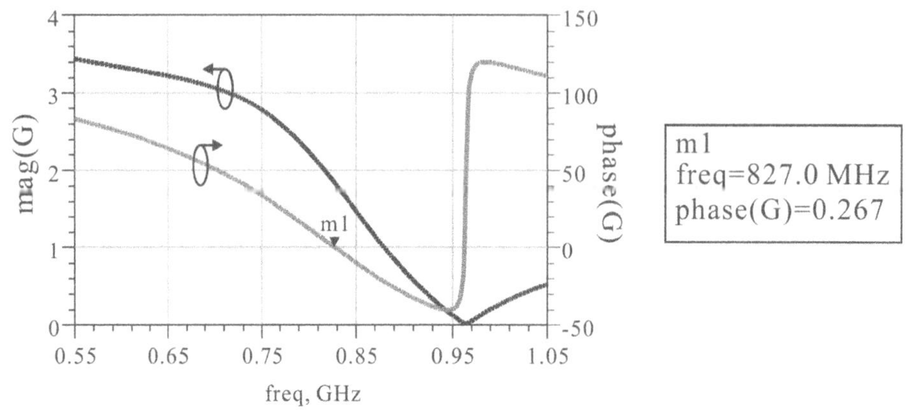

Figure 10E.14 Calculated open-loop gain. The phase of the open-loop gain crosses 0 at 827 MHz, at which the open-loop gain G > 1. Thus, the small-signal oscillation condition based on the open-loop gain is satisfied at 827 MHz.

10.3 PHASE NOISE

10.3.1 Spectrum of an Oscillation Waveform

The output waveform of an oscillator is not a pure sine wave and its amplitude and phase fluctuate with time. Thus, denoting the oscillation output power across a 1-Ω resistor as P, the waveform can be expressed in time domain as shown in Equation (10.21).

$$v(t) = \sqrt{2P} \cdot \left(1 + a(t)\right)\cos\left(\omega_o t + \phi(t)\right) \tag{10.21}$$

Here, we assumed the power P across the 1-Ω resistor but it does not lose generality. The $a(t)$ represents the fluctuation of the amplitude, and $\phi(t)$ represents the fluctuation of the phase in the time domain. The $a(t)$ is called the amplitude modulation (AM) noise while $\phi(t)$ is called the phase noise of the oscillator.

Two major issues are associated with understanding amplitude and phase noises. The first is the mathematical model or mechanism of the amplitude and phase noises and the second is how to measure these noises. We will first discuss the measurement method and then the mathematical model of the phase noise will be explained.

When the waveform represented by Equation (10.21) is observed on a spectrum analyzer, the spectrum is usually similar to that shown in Figure 10.20. The spectrum analyzer can be thought of simply as equipment showing the spectral power of an input signal that is the output power of a narrow-band filter. The center frequency of the narrow-band filter moves with time while maintaining the user-specified resolution bandwidth (RBW). Thus, the spectrum analyzer shows the power within the RBW on the axis of the frequency. In displaying the power with the RBW, the spectrum analyzer averages the measured power within the RBW in a given amount of time. The VBW (video bandwidth) is used as a measure of time averages. Usually, because VBW is expressed as a frequency, it is the reciprocal of the average time and so it is smaller than the RBW. The spectrum shown in Figure 10.20 is measured for a span of 1 MHz at the center frequency of 35.349 GHz. Also, RBW = 10 kHz and VBW = 3 kHz. Thus, the spectrum in Figure 10.20 represents average power in the 10-kHz bandwidth over 1/3 msec.

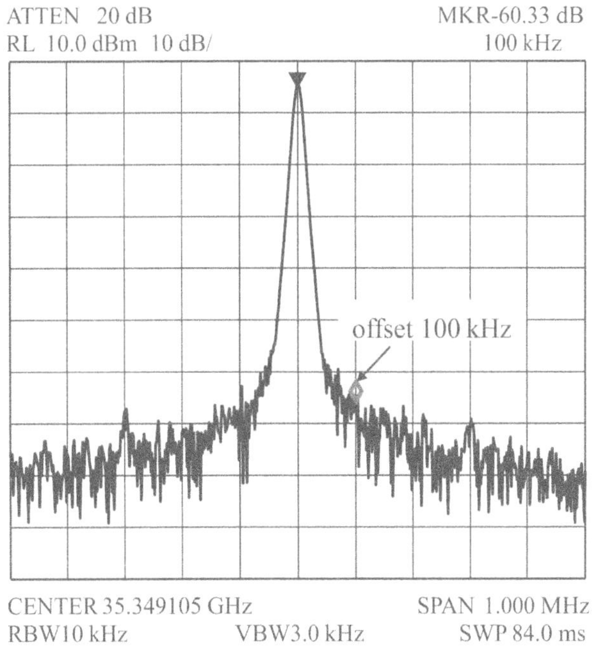

Figure 10.20 Example spectrum of an oscillator output waveform

Furthermore, when Equation (10.21) is expanded, it can be considered as the superposition of the sinusoidal component of frequency ω_0, whose power is P and a noise power. The sine wave power appears as the power of the center frequency component and the noise power has a distribution that is spread around the center frequency. Since the spectrum of the sinusoidal wave of frequency f_0 has the spectrum of $P\delta(f - f_0)$, a power P appears when f_0 is within the RBW, otherwise $P = 0$. In addition, the value of the sinusoidal power P does not change even when the RBW is changed; that is, whether the RBW is lowered or increased, the same power appears. However, it should be noted that noise density (noise power per bandwidth) is constant in the case of the noise. Thus, by lowering the RBW, the noise power measured within the RBW is lowered, whereas the noise power measured within the RBW is raised when the RBW increases.

Example 10.7

In Figure 10.20, the ratio of the center frequency power (or the carrier power) to the noise power at a frequency offset of 100 kHz from the center frequency is measured to be about –60.33 dBc. Calculate the carrier to noise power ratio measured at a 100-kHz offset when the RBW is changed to 1 Hz. Also calculate the power when the RBW is changed to 1 kHz.

Solution

As the power of the center frequency is the sine wave power, it does not change even if the RBW is changed. However, the power at the 100 kHz offset is a noise power and thus changes when the RBW is changed. Since the power is –60.33 dB when the RBW = 10 kHz, then –60.33 dB/10 kHz = –100.33 dB/Hz. In addition, when the RBW is changed to 1 kHz, the power at the marker can be measured by the same method to be –70.33 dB.

The spectrum in Figure 10.20 represents the contribution of a sine wave's output and noise. The noise power also comes from the combined effect of fluctuations in the amplitude and phase. However, for most oscillators, because the fluctuation effect coming from the amplitude is low compared to that coming from the phase, the spectrum noted above is generally known to occur due to phase fluctuation.

The following conceptual experiment can be thought of as the proof for the claim above of the phase noise dominance in the measured spectrum. That is, in order to eliminate the AM noise due to the amplitude fluctuation, the oscillator output is passed through a limiter and then the output is passed through a narrow bandwidth bandpass filter to remove harmonics. The resulting spectrum reflects only the phase fluctuation. However, in most cases, almost the same spectrum is obtained as a result of this experiment, which leads to the conclusion that the spectrum in Figure 10.20 is mostly due to the phase noise. In addition, because the AM noise can always be removed using the limiter and filter, the spectrum noted above is considered to represent the phase noise.

10.3.2 Relationship between Phase Noise Spectrum and Phase Jitter

To see the relationship between the spectrum shown in Figure 10.20 and the phase fluctuation, consider a part of the noise spectrum at a frequency offset of ω_m from the center frequency and the sine wave output shown in Figure 10.21. The waveforms in the time domain can be expressed as

$$v(t) = \sqrt{2P}\cos(\omega_o t) + \sqrt{2N}\cos((\omega_o + \omega_m)t + \varphi(t)) \tag{10.22}$$

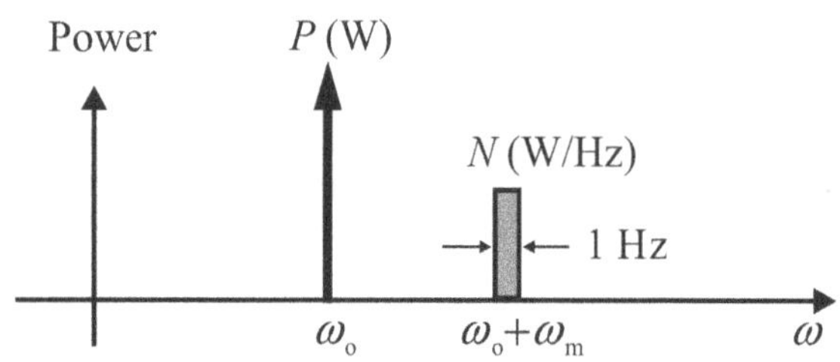

Figure 10.21 Carrier frequency component and noise in unit bandwidth

Expanding Equation (10.22) using the additive theorem of trigonometric functions and assuming that $P \gg N$, $v(t)$ can be written as

$$v(t) = \sqrt{2P}\cos\left(\omega_o t + \sqrt{\frac{N}{P}}\sin(\omega_m t + \varphi(t))\right) \tag{10.23}$$

Thus, the carrier is found to be phase-modulated by the noise signal and its power is almost equal to the carrier power. Since the maximum phase deviation is $(N/P)^{1/2}$, then the peak phase jitter becomes $(N/P)^{1/2}$. Alternatively, the phase jitter can be determined using a phasor diagram. The carrier becomes the phasor that rotates counter-clockwise and the noise phasor is placed at the end of the carrier phasor, which becomes a rotating phasor with angular velocity ω_m, as shown in Figure 10.22. Therefore, the maximum phase error for $P \gg N$ is obtained with Equation (10.24).

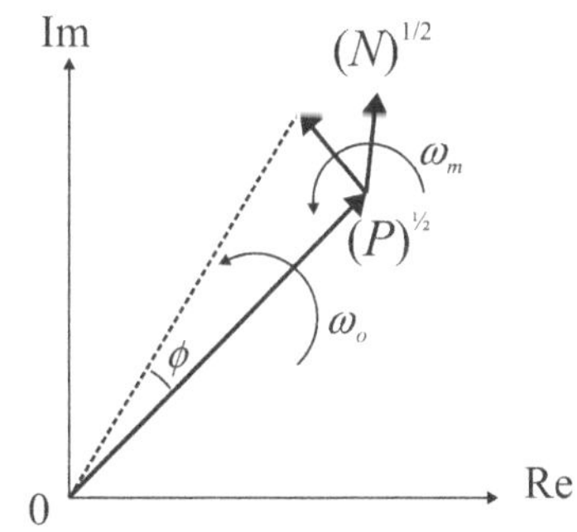

Figure 10.22 Phasor diagram of Equation (10.22). The carrier can be represented by the phasor that rotates counterclockwise with an angular velocity of ω_o, while the noise signal in Equation (10.22) can be represented by the phasor at the end of the carrier phasor rotating counterclockwise with an angular velocity of ω_m.

$$\overline{\phi^2}(t) = \frac{N}{P}\left[\text{rad}^2/\text{Hz}\right] \tag{10.24}$$

Thus, the phase fluctuation is a function of the offset frequency and the maximum phase jitter at the offset frequency of ω_m is expressed in Equation (10.25).

$$S_\phi(\omega_m) = \overline{\phi^2}(t) = \frac{N}{P} \tag{10.25}$$

Example 10.8

Calculate the peak phase jitter when the carrier to noise power at a 100-kHz offset is –100 dBc/Hz.

Solution

$$\frac{N}{P} = -100 \text{ dB/Hz}$$

Thus, the phase jitter = 10^{-5} rad.

10.3.3 Leeson's Phase Noise Model

The phase noise of an oscillator can be qualitatively explained using a simple oscillator model shown in Figure 10.23. In general, an oscillator can be represented as a circuit composed of an amplifier and a feedback network, as shown in Figure 10.23. Here, the frequency dependence of the amplifier gain is imposed on the feedback network and the amplifier is assumed to have a constant gain. In addition, the transfer characteristic of the feedback network can generally be expressed as shown in Equation (10.26),

$$S_{21} = \frac{\beta_o}{1 + jQ\left(\dfrac{\omega}{\omega_o} - \dfrac{\omega_o}{\omega}\right)} \cong \frac{\beta_o}{1 + j2Q\left(\dfrac{\omega_m}{\omega_o}\right)} \tag{10.26}$$

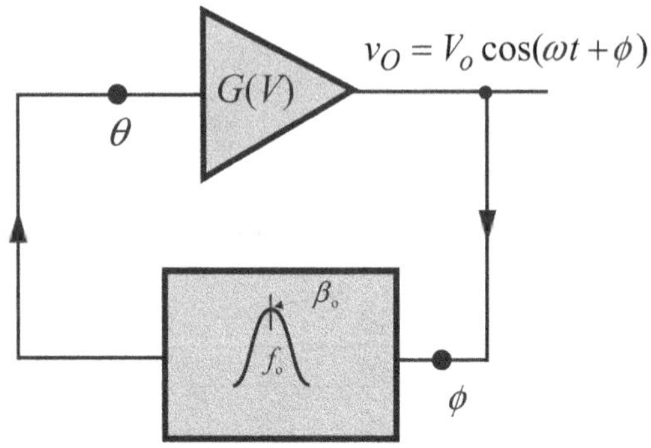

Figure 10.23 A simplified oscillator structure. The phase at the amplifier input is denoted by θ, while that of the amplifier output is denoted by ϕ.

where ω_o represents the oscillation frequency and $\omega_m = \omega - \omega_o$ represents the offset frequency. Generally, the magnitude of the frequency response in Equation (10.26) is approximately constant, while the phase is approximated as a straight line that decreases linearly within the 3-dB angular bandwidth of $BW = \omega_o/Q$.

In the oscillator structure shown in Figure 10.23, the equivalent noise source N can be placed at the amplifier input; its frequency characteristic is shown in Figure 10.24. The F in Figure 10.24 represents the noise factor, which will be added to the oscillator signal.

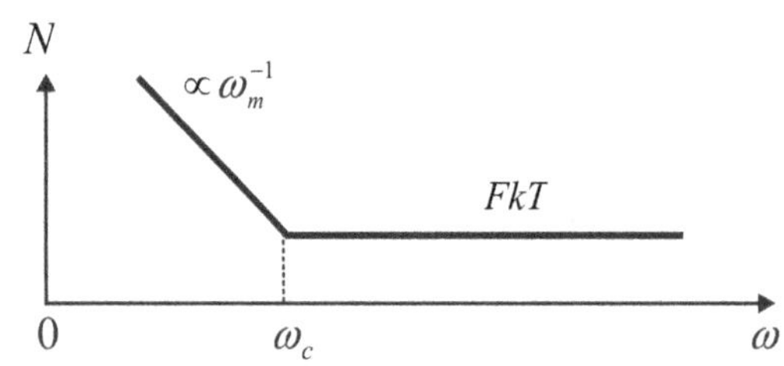

Figure 10.24 Noise at the amplifier input. The noise power at the amplifier input can be represented by FkT, which is explained in Chapter 4. At low frequency, the noise increases from FkT. The frequency $\omega_c = 2\pi f_c$ is called the corner frequency or flicker frequency.

Suppose that the oscillation signal is frequency-modulated by the noise signal shown in Figure 10.24 due to the nonlinearity of the amplifier. The frequency-modulated signal then appears at the output of the amplifier, which is the oscillator output signal. The peak frequency deviation of the frequency-modulated signal by the noise is denoted as $\Delta\omega$. Note that $\Delta\omega$ is proportional to N shown in Figure 10.24. Then, the frequency-modulated signal is applied to the input of the feedback network, and the output of feedback network appears again at the input of the amplifier. When the oscillator output, which is frequency-modulated by the noise, is applied to the input of the feedback network, the frequency-modulated signal is transformed into a phase-modulated signal by the feedback network.

The peak phase deviation $\Delta\theta$ of the phase-modulated signal is related to the peak frequency deviation as expressed in Equation (10.27).

$$\Delta\theta = -\frac{2Q}{\omega_o}\Delta\omega \tag{10.27}$$

Thus, the phase-modulated signal with the peak phase deviation given by Equation (10.27) will appear at the input of the amplifier. Note that both $\Delta\omega$ and $\Delta\theta$ are proportional to $(N)^{1/2}$, shown in Figure 10.24. Since the peak phase deviation $\Delta\theta$ is proportional to $(N)^{1/2}$, the single-sideband phase noise at the input of the amplifier can be plotted, as shown in Figure 10.25. Here, the carrier power at the amplifier input is denoted as P.

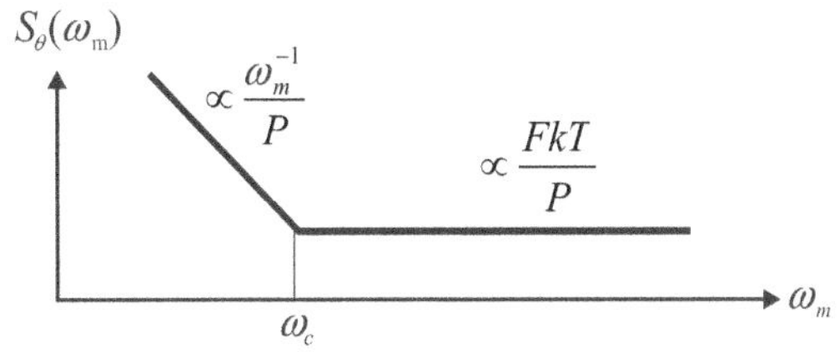

Figure 10.25 The phase noise at the amplifier input computed using the noise power shown in Figure 10.24

The phase noise of the amplifier input in Figure 10.25 can be written as Equation (10.28).

$$S_\theta(\omega_m) = \frac{FkT}{P}\left(1 + \frac{f_c}{f_m}\right) \tag{10.28}$$

In addition, at the oscillator output, the peak frequency deviation $\Delta\omega$ is related to the peak phase deviation $\Delta\phi$ by $\Delta\omega = \omega_m\Delta\phi$. Using Equation (10.27), the relationship between the phase noise appearing at the amplifier input and the phase noise appearing at the oscillator output is expressed in Equation (10.29).

$$S_\theta(\omega_m) = \left(\frac{2Q\omega_m}{\omega_o}\right)^2 S_\phi(\omega_m) \tag{10.29}$$

Also, outside the resonator bandwidth BW there is no such relationship, which is shown in Equation (10.30).

$$S_\theta(\omega_m) = S_\phi(\omega_m) \tag{10.30}$$

Therefore, combining Equations (10.29) and (10.30), the combined can be written as Equation (10.31).

$$S_\phi(\omega_m) = S_\theta(\omega_m)\left[1 + \left(\frac{\omega_o}{2Q\omega_m}\right)^2\right] \tag{10.31}$$

This is shown in Figure 10.26; that is, near the oscillation frequency, the phase noise decreases by ω^{-3} (30 dB/decade) and, after the 1/*f* noise disappears, the phase noise decreases by ω^{-2} (20 dB/decade). Then, outside the resonator's bandwidth, it is proportional to the noise figure and shows a constant phase noise. It should also be noted that the higher the *Q* of the feedback network, the lower the phase noise.

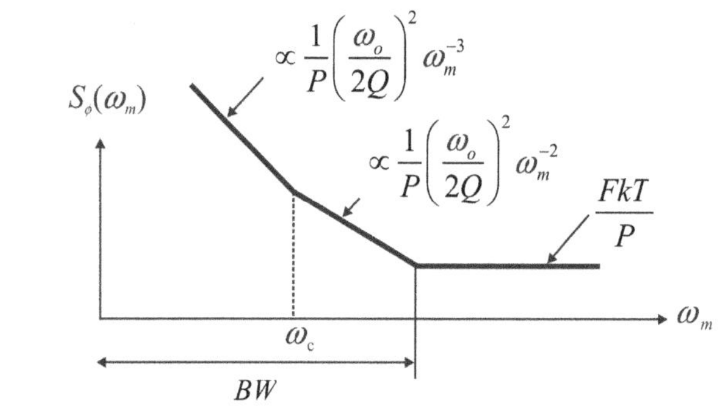

Figure 10.26 Phase noise of the oscillator. *BW* is the bandwidth of the feedback network. Outside the *BW*, the phase noise due to the white noise is given by FkT/P and it is increasing in proportion to ω_m^{-2} inside BW. The phase noise further increases due to the flicker noise in proportion to ω_m^{-3}.

In conclusion, the phase noise $S(f_m)$ based on Leeson's phase noise model can be expressed as

$$S(f_m) = -10\log\left\{\frac{1}{2}\frac{FkT}{P}\left(1+\frac{f_c}{f_m}\right)\left[1+\left(\frac{1}{f_m}\frac{f_o}{2Q}\right)^2\right]\right\} \tag{10.32}$$

The Leeson's phase noise model expressed in Equation (10.32) is an approximate description of phase noise. Note that the phase noise generation in an oscillator is basically a nonlinear phenomenon. However, the phase noise asymptotically approaches the Leeson's phase noise model for higher f_m. In addition, it should be noted that noise factor *F* of the amplifier in Equation (10.32) is seldom equal to the measured amplifier noise factor; for more information, see reference 3 at the end of this chapter. Noise factor *F* around the oscillation frequency is thought to be caused by the thermal noise and the DC bias-dependent shot noise. However, the DC bias-dependent shot noise generally becomes a function of the oscillating signal. As a result, it is modulated by the oscillating signal. Thus, it acts as a cyclostationary noise source in the oscillator, which makes *F* in Equation (10.32) differ from the measured amplifier noise figure. In addition, the flicker noise is also DC bias dependent and the flicker frequency f_c in Equation (10.32) may also differ from the measured flicker frequency due to flicker-noise conversion dynamics. Recently, significant published research has focused on cyclostationary noises. However, a design for a low-phase noise oscillator to meet a given phase noise specification is still only a theory despite recent research on oscillator phase noises and the emergence of modern CAD simulators. We will present the *experimental method to meet the design objective of the phase noise* in the design of a DRO (dielectric resonator oscillator) in section 10.7.

The Leeson model is based on deduction and should be proven experimentally.[3] The basic assumption is the frequency modulation by the noise at the amplifier input, namely the peak frequency deviation, is proportional to the noise frequency characteristic. This assumption has been experimentally verified by Pucel and Curtis, who measured the $1/f$ noise of the drain current of a GaAs FET under a given DC voltage. The fluctuation of the drain current ΔI_d^2 observed with a spectrum analyzer is shown in Figure 10.27. Using the drain current fluctuation, the peak frequency deviation Δf^2 of the oscillation output can be calculated. This peak frequency deviation is usually referred to as FM noise. In this case, the frequency dependence of the FM noise must be the same as that of the drain current noise since the FM noise is assumed to be proportional to the drain current noise. The measured and computed FM noises shown in Figure 10.27 are found to have the same frequency dependence as that of the drain current noise.

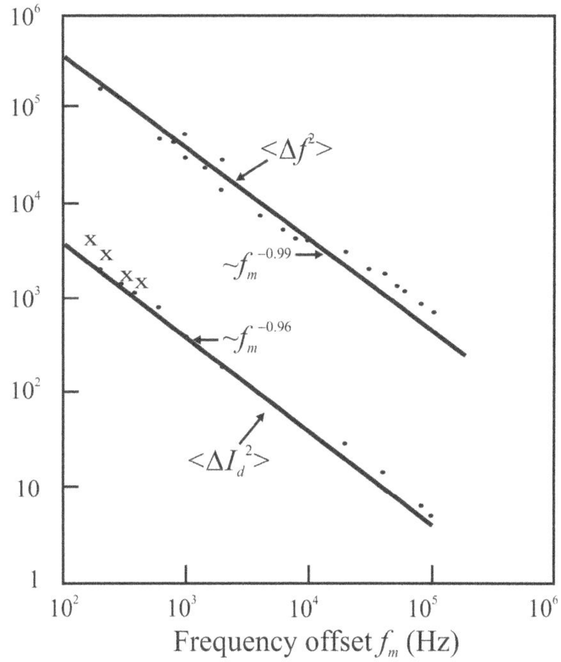

Figure 10.27 The Pucel experimental results of a GaAs FET 10-GHz oscillator phase noise.[4] Baseband noise ΔI_d^2 represents the measured drain current fluctuation in unit (nA^2/Hz), while FM noise is measured in the unit (Hz^2/Hz). Since the FM noise is directly proportional to the baseband noise as expressed in Equation (10.27), the FM noise should show the same dependence as in the baseband noise, and this clearly appears in the plot.

3. D. B. Leeson, "A Simple Model of Feedback Oscillator Noise Spectrum," *Proceedings of the IEEE* 54, no. 2 (February 1966): 329–330.

4. R. A. Pucel and J. Curtis, "Near-Carrier Noise in FET Oscillators," *IEEE MTT-S International Microwave Symposium Digest*, (May 31–June 3, 1983): 282–284.

10.3.4 Comparison of Oscillator Phase Noises

It is often necessary to compare the performance of oscillators in terms of phase noise even though the oscillators generally have different oscillation frequencies. To compare the phase noises of two oscillators with different oscillation frequencies, the frequency of one oscillator must first be made equal to that of the other using frequency division or multiplication. First, we will examine the changes in phase noise resulting from frequency multiplication or division.

Suppose that the time-domain waveform of an oscillator is given by Equation (10.33).

$$v(t) = \sqrt{2P}\left(1 + a(t)\right)\cos\left(\omega_o t + \phi(t)\right) \tag{10.33}$$

Then, after the frequency multiplication by n, the resulting output waveform can be expressed with Equation (10.34).

$$v(t) = \sqrt{2P}\left(1 + a(t)\right)\cos\left(n\omega_o t + n\phi(t)\right) \tag{10.34}$$

Thus, the phase noise of the multiplied waveform can be expressed as Equation (10.35),

$$S_\phi\left(\omega_m\right) = \overline{n^2\phi^2(t)} = n^2\overline{\phi^2}(t) \tag{10.35}$$

which represents a degradation of the phase noise by n^2. As an example, the frequency of a 10-MHz crystal oscillator is multiplied by $n = 1000$ to give the frequency of 10 GHz. Since $n = 1000$, the phase noise increases by 60 dB.

In this way, the phase noises of various oscillators can be compared, as shown in Figure 10.28, where a crystal oscillator, a DRO (dielectric resonator oscillator), and a

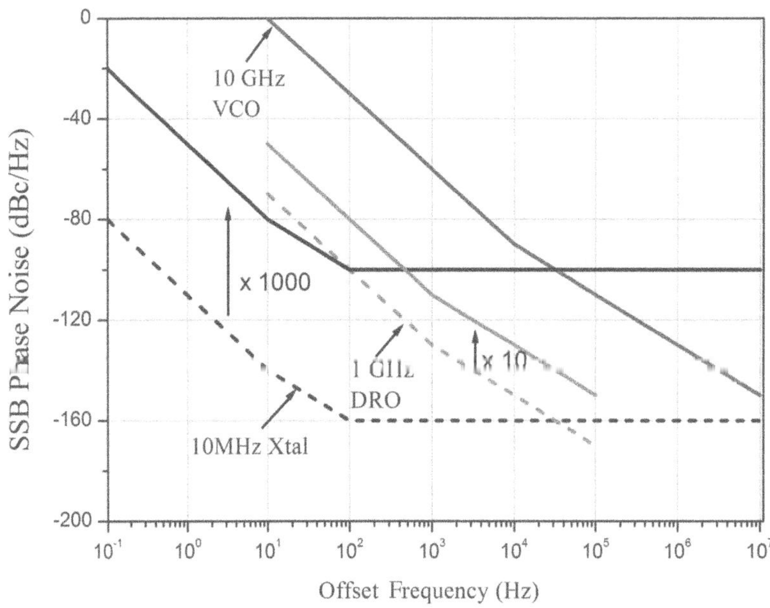

Figure 10.28 Comparison of the phase noises of various oscillators: a 10-MHz crystal oscillator, a 1-GHz DRO, and a 10-GHz VCO. All the frequencies are converted to the 10-GHz frequency.

microstrip VCO (voltage-controlled oscillator) are graphed. Their oscillation frequencies are 10 MHz for the crystal oscillator, 1 GHz for the DRO, and 10 GHz for the microstrip VCO. The frequencies of the oscillators are first set to 10 GHz. Thus, the frequencies of the crystal oscillator and the DRO are multiplied by factors $n = 1000$ and 10, respectively. After the frequency multiplication, the phase noises are compared, as shown in Figure 10.28. From this figure, although the noise floor of the crystal oscillator is higher (as a result of multiplying the frequency by a factor of 1000), the phase noise at low frequency can be found to be the lowest compared to the other oscillators. This is followed by the DRO, and then the microstrip VCO, which has the poorest phase noise.

Example 10.9

At a 100-kHz frequency offset, a VCO with a center frequency of 10 GHz has a phase noise of –100 dBc/Hz, while another VCO with a center frequency of 35 GHz has a phase noise of –96 dBc/Hz. Compare the phase noises of the two oscillators.

Solution

To set the frequency of the 10-GHz VCO to 35 GHz, the required frequency multiplication factor is

$$n = \frac{35}{10}$$

The phase noise of the 10-GHz VCO after its frequency multiplication by n equals

$$S_\phi = -100 + 10\log\left(\frac{35}{10}\right)^2 = -89 \text{ dBc/Hz}$$

Thus, the phase noise of the 10-GHz VCO is poorer than that of the 35-GHz VCO by 7 dB.

10.4 BASIC OSCILLATOR CIRCUITS

10.4.1 Basic Oscillator Circuits

The possibility of oscillation for a given oscillator circuit was investigated in section 10.2. Now we will present the design of oscillator circuits that oscillate at a specified frequency. The design of an oscillator circuit can be carried out using basic oscillator circuits.

First, after removing the DC bias circuits and all the elements that have no effects at the RF in a given oscillator circuit, most oscillator circuits can be categorized into two configurations: series feedback oscillators, as shown in Figure 10.29, or parallel feedback oscillators, as shown in Figure 10.30. In Figure 10.29, the series feedback is achieved by jy, which delivers the transistor DS output to the GS input. In contrast, the parallel feedback shown in Figure 10.30 is achieved by jy, which delivers the DS output to the GS input. Three types of series feedback oscillator configurations and three types of parallel feedback oscillator configurations are shown in Figures 10.29 and 10.30, respectively.

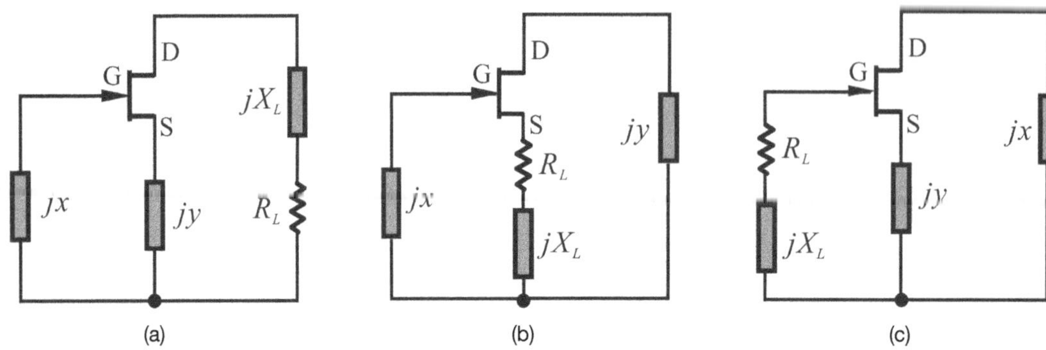

Figure 10.29 Three series feedback oscillators. (a) The load is connected to the drain, (b) to the source, and (c) to the gate terminals.

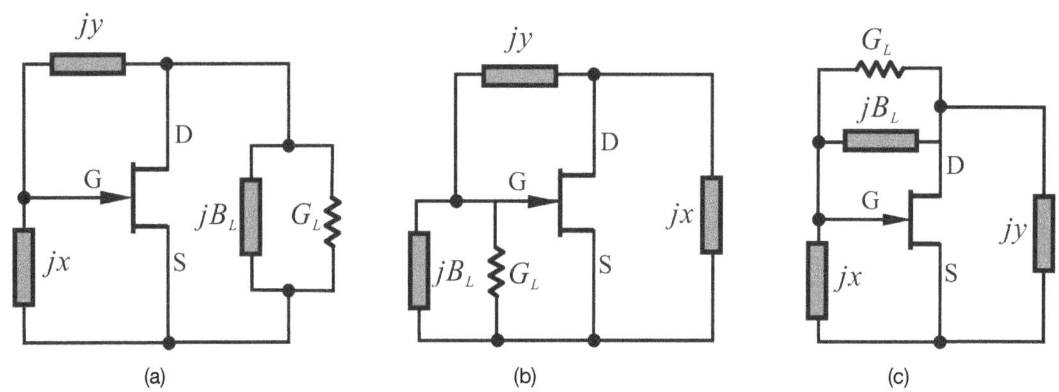

Figure 10.30 Three parallel feedback oscillators. (a) The load is connected to the drain, (b) to the gate terminals, and (c) to the feedback path.

The classifications for the three types of configurations in Figures 10.29 and 10.30 are based on where the loads are connected, whereas the feedback type can be found to be essentially the same for all three types. In addition, jx and jy in the figures represent the reactance of a capacitor or inductor in the series feedback configurations, while they represent the susceptance of a capacitor or inductor in the parallel feedback configurations.

The basic forms of the oscillator circuits in Figures 10.29 and 10.30 can be represented by an amplifier and a feedback network. The basic form of the series oscillator shown in Figure 10.29 can be converted into the T-type feedback network in Figure 10.31(a), and the basic form of the parallel feedback oscillator can be converted into the π-type feedback network in Figure 10.31(b).

The oscillation mechanism of the series or parallel configuration can be qualitatively understood by analyzing the feedback structure. First, for the series configuration of Figure 10.29(a), the reactance jx is replaced by capacitor C_2, the feedback reactance jy is replaced by inductor L, and the load is replaced by a capacitor C_1 and resistor r in series. In addition, since the input impedance of the transistor at high frequencies is generally low, by approximating it as short, the equivalent circuit of the transistor can be represented by a current-controlled voltage source, as shown in the shaded rectangle of Figure 10.32. The open-loop circuit can be obtained by cutting the FET input (the A–A′ reference plane in

Figure 10.31) and reconfiguring the circuit. As the input impedance of the transistor is approximated as short, a shorted load can be connected where the cut occurs. To obtain the open-loop gain, a unit current source is applied to the transistor input of the open-loop circuit, which is shown in Figure 10.32. The open-loop gain is then the $-I_r$ of Figure 10.32.

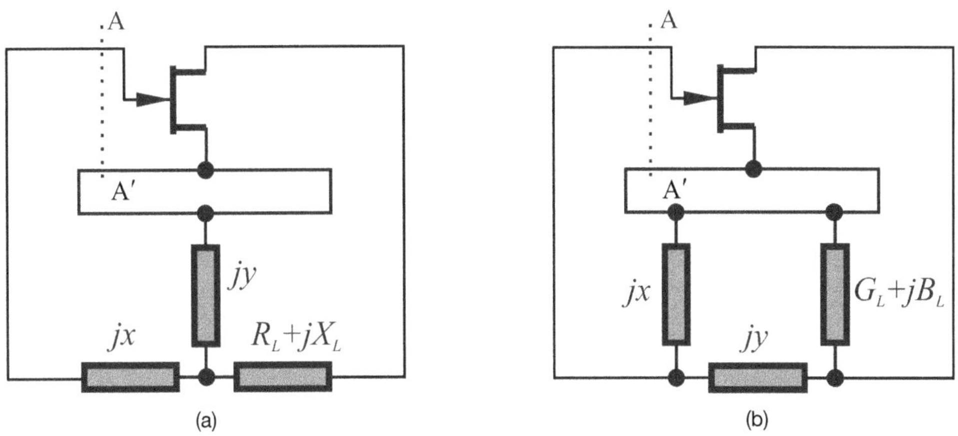

Figure 10.31 The conversion of the oscillator circuit into a feedback form. (a) T-type feedback network converted from the basic oscillator circuit in Figure 10.29(a) and (b) π-type feedback network converted from the parallel feedback oscillator in Figure 10.30(a)

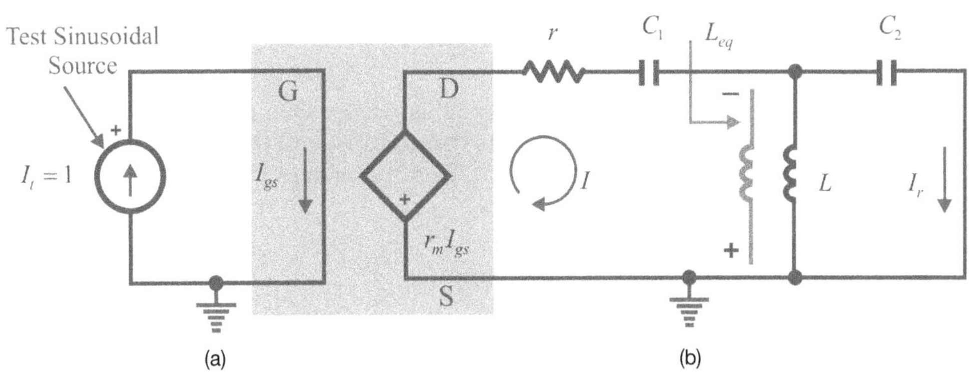

Figure 10.32 Open-loop circuit of the circuit in Figure 10.31(a). The gate-source plane is cut and the unit test current is applied.

Here, Equations (10.36a) and (10.36b) express the parallel impedance of L and C_2.

$$j\omega L_{eq} = \frac{j\omega L \dfrac{1}{j\omega C}}{j\omega L + \dfrac{1}{j\omega C}} = \frac{j\omega L}{1-\omega^2 LC_2} = \frac{j\omega L}{1-\dfrac{\omega^2}{\omega_r^2}} \tag{10.36a}$$

$$\omega_r^2 = \frac{1}{LC_2} \tag{10.36b}$$

Thus, at frequencies lower than ω_r, L and C_2 in parallel are equivalently treated as an inductor L_{eq}. In addition, since the current transfer function I_r/I becomes

$$k = \frac{I_r}{I} = \frac{j\omega L}{j\omega L + \dfrac{1}{j\omega C_2}} = \frac{1}{\left(1 - \dfrac{\omega_r^2}{\omega^2}\right)} \tag{10.37}$$

at frequencies lower than ω_r, the phase of k becomes 180° and a phase inversion occurs, as expressed in Equation (10.37). In contrast, at frequencies higher than ω_r, the phase becomes 0° and results in the disappearance of the phase inversion. It also is worth noting that k is a real number. In addition, the open-loop gain L is now expressed by Equations (10.38a) and (10.38b).

$$L = -k \cdot \frac{r_m}{r + \dfrac{1}{j\omega C_1} + j\omega L_{eq}} = -k \cdot \frac{r_m}{r + \dfrac{1}{j\omega C_1}\left(1 - \dfrac{\omega^2}{\omega_o^2}\right)} \tag{10.38a}$$

$$\omega_o^2 = \frac{1}{L\left(C_1 + C_2\right)} \tag{10.38b}$$

Since $\omega_o < \omega_r$, the open-loop gain is positive real at $\omega = \omega_o$ and therefore the phase of the open-loop gain is 0. The open-loop gain is

$$L\left(\omega_o\right) = \frac{C_2}{C_1}\frac{r_m}{r} \tag{10.39}$$

When the gain given by Equation (10.39) is greater than 1, oscillation can form. In the circuit, note that phase inversion occurs due to $L \,\|\, C_2$ and the oscillation frequency occurs at the resonant frequency of $L \,\|\, (C_1 + C_2)$. This is shown in Figure 10.33.

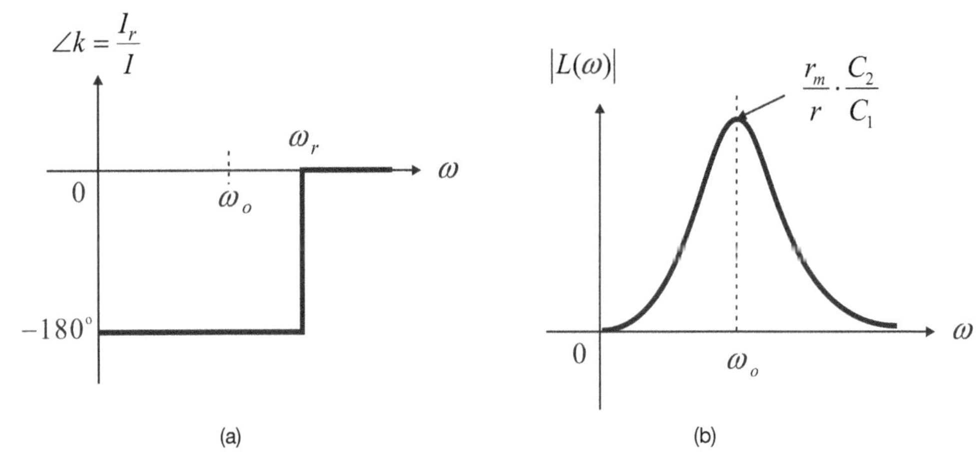

Figure 10.33 (a) Phase of k and (b) open-loop gain of the series feedback oscillator for the frequency

In the case of the parallel feedback oscillator in Figure 10.30(a), the oscillator circuit can be similarly implemented by replacing the reactance jx with a capacitor C_2, the feedback reactance jy is replaced with an inductor L, and the load is replaced with a capacitor C_1 and resistor R in parallel. Furthermore, the input impedance of the transistor at low frequencies is generally high; the equivalent circuit of the transistor can be approximately represented by a voltage-controlled current source, as shown in the shaded area of Figure 10.34. Similar to the previously discussed series-feedback-type oscillator, to obtain the open-loop gain, the circuit is cut at the transistor input and the open-loop circuit is drawn as shown in Figure 10.34, where a unit voltage source is applied to the input, and the open-loop gain is obtained by calculating the voltage V_r returning from the output to the input.

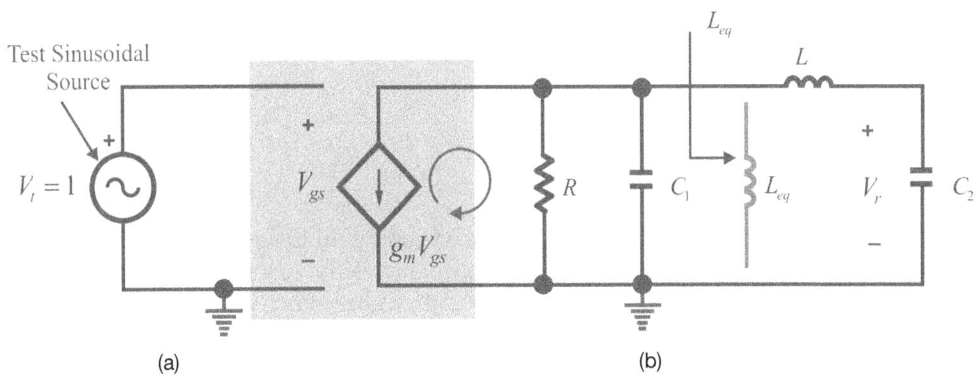

Figure 10.34 Circuit for calculating the open-loop gain of a parallel feedback oscillator circuit. The gate source in Figure 10.31(b) is cut and the test voltage is applied to calculate the open-loop gain.

From that figure, the voltage transfer function computed as $k = V_r/V_o$ is given by Equations (10.40a) and (10.40b).

$$k = \frac{\dfrac{1}{j\omega C_2}}{\dfrac{1}{j\omega C_2} + j\omega L} = \frac{1}{1 - \dfrac{\omega^2}{\omega_r^2}} \tag{10.40a}$$

$$\omega_r^2 = \frac{1}{LC_2} \tag{10.40b}$$

Thus, phase inversion appears for frequencies higher than ω_r and disappears for frequencies lower than ω_r. In addition, as in the case of series feedback, k is a real number. In addition, the open-loop gain is shown in Equations (10.41a) and (10.41b).

$$L = V_r = -g_m \frac{Z_t R}{R + Z_t} k \tag{10.41a}$$

$$Z_t = \frac{\dfrac{1}{j\omega C_1}\left(\dfrac{1}{j\omega C_2} + j\omega L\right)}{\dfrac{1}{j\omega C_1} + \dfrac{1}{j\omega C_2} + j\omega L} \tag{10.41b}$$

Thus, the open-loop gain is given by Equation (10.42).

$$L = V_r = -g_m \frac{1}{j\omega(C_1 + C_2)(1 - \omega^2 LC_1 \parallel C_2) + G(1 - \omega^2 LC_2)} \tag{10.42}$$

Since the imaginary part of the denominator of the expression above must be 0 for oscillation to occur, the oscillation frequency

$$\omega_o^2 = \frac{1}{L(C_1 \parallel C_2)} \tag{10.43}$$

It can also be seen that $\omega_o > \omega_r$ from Equations (10.40b) and (10.43). Substituting Equation (10.42) into the open-loop gain equation, the following condition must be satisfied for oscillation to occur:

$$L(\omega_o) = -g_m \frac{1}{1 - \frac{1}{LC_1 \parallel C_2} LC_2} = g_m R \frac{C_1}{C_2} > 1 \tag{10.44}$$

Thus, no oscillation occurs when the open-loop gain given by Equation (10.44) is less than 1. From that equation, the oscillation frequency is the resonant frequency of the overall *LC* resonant circuit seen from the output. Furthermore, since $\omega_o > \omega_r$, the value of *k* is found to be negative. That is, the oscillation frequency must always be higher than the frequency that causes the phase inversion. Because the amplifier is an inverting amplifier that has its own phase inversion of 180°, the *k* network should provide the phase inversion of 180° to restore the overall phase of the open-loop gain to 0°. This is shown in Figure 10.35.

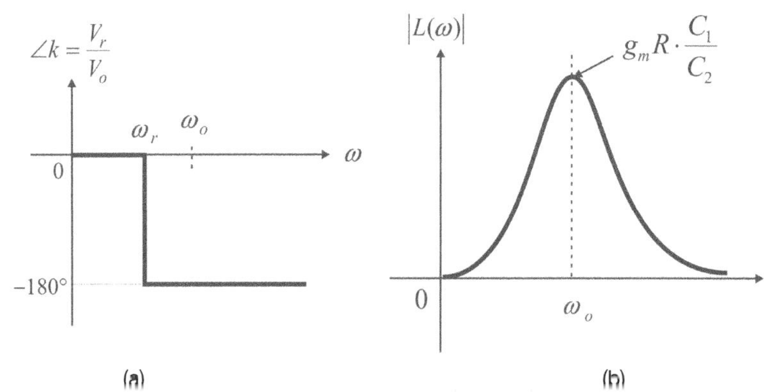

Figure 10.35 (a) Phase of *k* and (b) open-loop gain of the parallel feedback oscillator for the frequency

10.4.2 Conversion to Basic Forms

The actual oscillator circuit is realized by applying DC voltage to the basic-form oscillator circuit that represents the equivalent circuit at the RF frequency. The actual oscillator circuit sometimes looks slightly different from the basic forms. However, most oscillators when simplified can be converted to the previously mentioned basic forms of the oscillator. Thus, for a given oscillator circuit, the design tasks first require the conversion of the oscillator circuit to one of the basic forms and then DC voltage must be applied to the selected basic form of the oscillator circuit. In this section, we will examine these tasks through some examples.

Example 10.10

Figure 10E.15 represents a microstrip oscillator circuit. Simplify this circuit and convert it into the oscillator circuit's basic form.

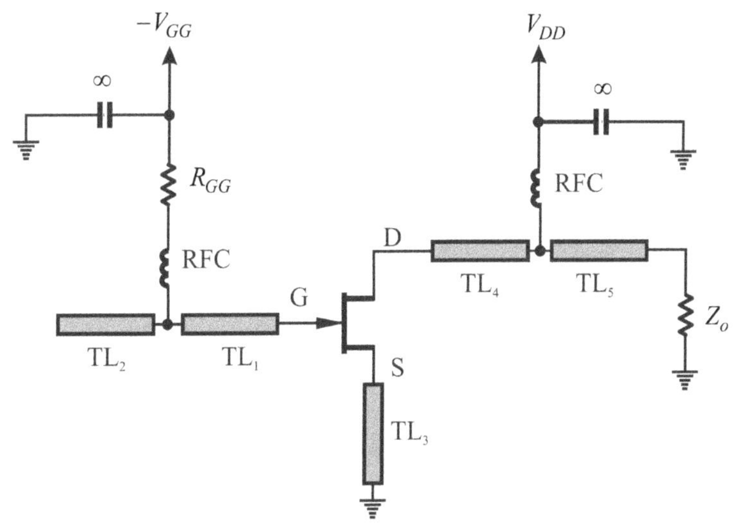

Figure 10E.15 Example of a microstrip oscillator circuit

Solution

Removing the DC bias circuit portion of Figure 10E.15 results in the bottom-right circuit in Figure 10E.16.

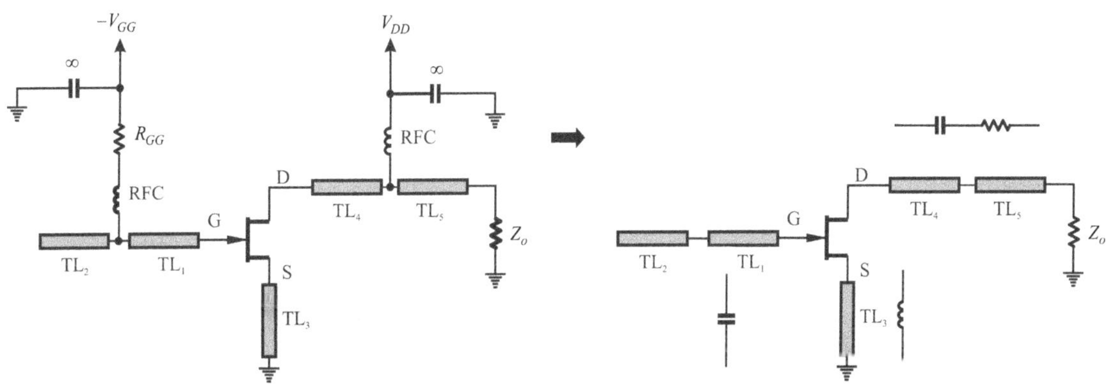

Figure 10E.16 Simplified oscillator circuit. RFCs are removed.

The circuit connected to the drain terminal can be seen as an *RC* series circuit at the oscillation frequency. Also, assuming the length of transmission line is short, the transmission line connected to the source can be seen as an inductor. In addition, the two transmission lines connected to the gate terminal can be considered as a capacitor and, by removing the ground, the microstrip oscillator circuit can be equivalently redrawn as the basic form of the series feedback oscillator in Figure 10.29(a).

Example 10.11

The circuit shown in Figure 10E.17 is a 200-MHz-band Colpitts oscillator.[5] The S-parameters of transistor NE85633 at Vce = 3.5 V and Ic = 10 mA are used. Show that the circuit oscillates at 200 MHz using the open-loop gain. Then, using ADS, convert the circuit into a basic parallel feedback oscillator and calculate the values of the resulting admittance jx, jy and the load admittance $G_L + jB_L$ at 200 MHz.

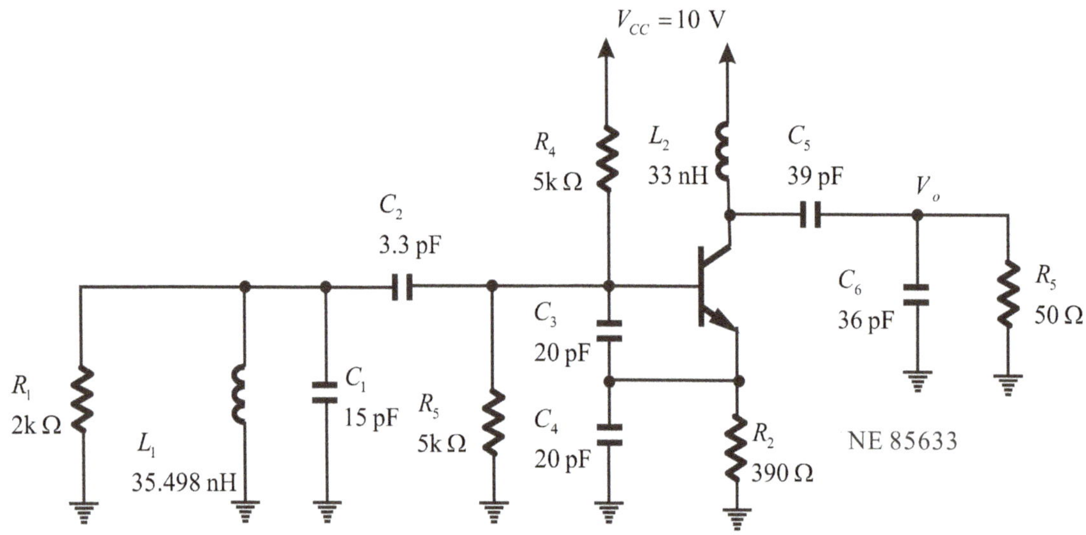

Figure 10E.17 A 200-MHz Colpitts oscillator circuit

Solution

After removing the ground point in Figure 10E.17 and moving the new ground to the transistor's emitter, the S-parameter of NE85633 is inserted and simulated in ADS, as shown in Figure 10E.18, to confirm the open-loop gain at 200 MHz. To compute the open-loop gain, the oscillator feedback is cut at the base-emitter plane. Two ports are connected where the feedback loop is cut, as shown in Figure 10E.18.

After the S-parameter simulation, the equation in Measurement Expression 10E.3 is similarly entered in the display window to compute the open-loop gain **G** using the simulated S-parameters. The simulated open-loop gain **G** is shown in Figure 10E.19. From the phase of **G**, the oscillation condition is found to be satisfied at a frequency of approximately 200.9 MHz.

5. M. Randall and T. Hock, "General Oscillator Characterization Using Linear Open-Loop S-Parameters," *IEEE Transactions on Microwave Theory and Techniques* 49, no. 6 (June 2001): 1094–1100.

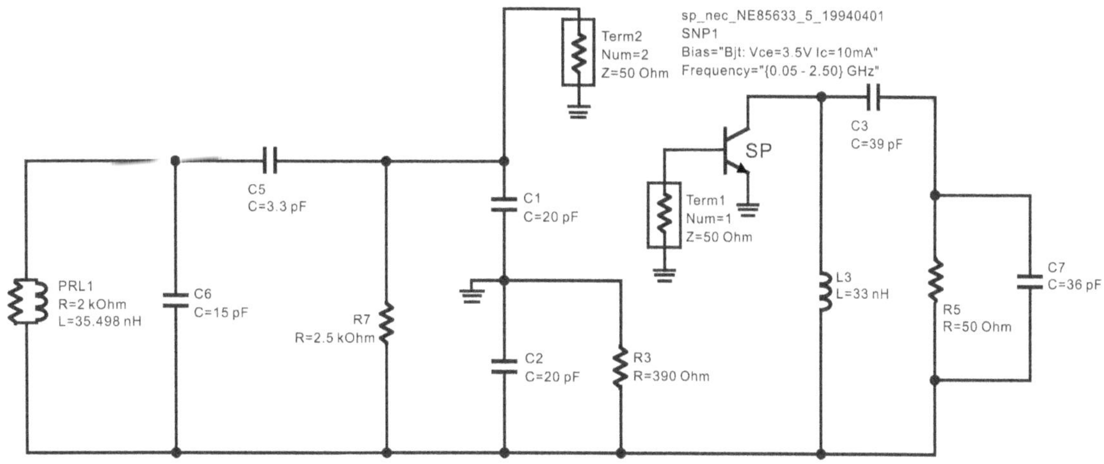

Figure 10E.18 Circuit for calculating the open-loop gain. The ground point is moved to the emitter and the oscillator is cut along the base emitter. After the breaking the loop, the oscillator circuit is redrawn. Note that the S-parameter data component for NE85633 is used.

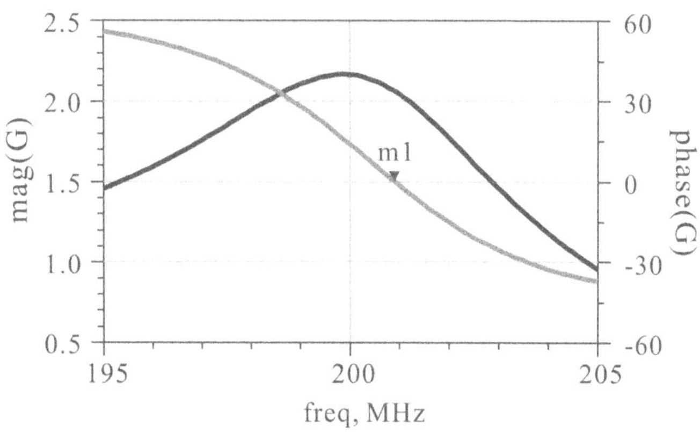

Figure 10E.19 Open-loop gain calculation result. The phase of the open-loop gain crosses 0 at about 200.9 MHz, at which $|G| > 1$ and the oscillation is possible.

The two-port parameter values of the oscillator circuit's feedback network can be obtained by removing the transistor, as shown in Figure 10E.20. The two-port Y-parameters of the circuit shown in that figure can now be obtained. These parameters can be represented by the π-type circuit shown in Figure 10.31(b). Since it is a passive network, $y_{11} + y_{12}$ corresponds to the admittance jx, and $y_{22} + y_{12}$ corresponds to $Y_L = G_L + jB_L$, while jy corresponds to $-y_{12}$ in the basic form of the parallel feedback oscillator. Therefore, the following equations shown in Measurement Expression 10E.4 are entered in the display window and the values of the admittances in Table 10E.1 are displayed in a list.

```
Eqn jx=Y(1,1)+Y(1,2)
Eqn jy=-Y(1,2)
Eqn YL=Y(2,2)+Y(1,2)
```

Measurement Expression 10E.4 Equations for the feedback parameters x, y, and Y_L

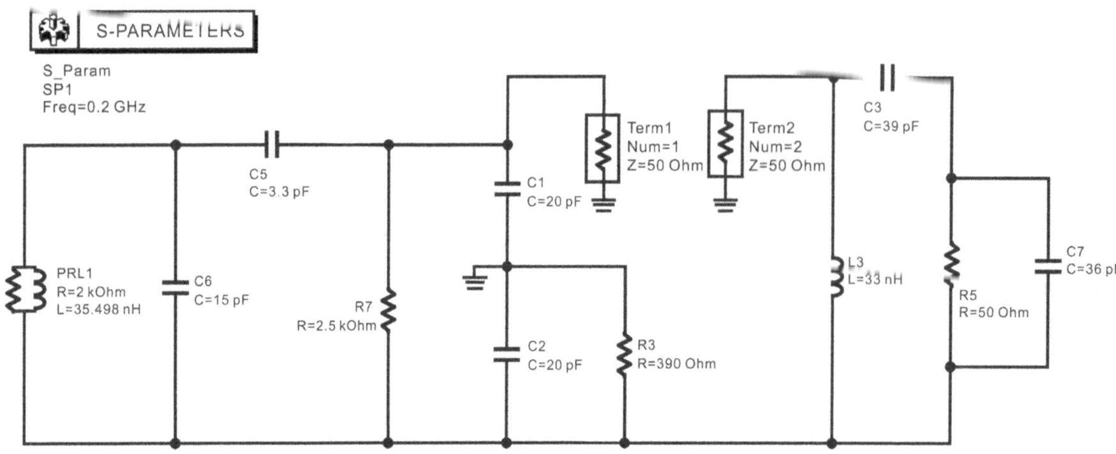

Figure 10E.20 Calculation of the two-port circuit parameters external to the oscillator. To obtain the Y-parameters of the feedback network, the BJT is removed and the Y-parameters are computed for the remaining network.

Table 10E.1 The values of *x*, *y*, and *jx*

jx	jy	Y_L
$1.8541 \times 10^{-2} + j2.9972 \times 10^{-2}$	$1.7236 \times 10^{-3} - j3.5450 \times 10^{-3}$	$2.4685 \times 10^{-3} + j4.3726 \times 10^{-3}$

In these equations, jy is an inductor and Y_L and jx are capacitors. However, note that jx and jy are not pure imaginary numbers because the collector terminal of the transistor in Figure 10E.18 is not connected to a ground but, instead, to a 33-nH inductor. Also note that the real part of jx is the largest and the load is connected to the base rather than the collector terminal.

Example 10.12

For the basic parallel-type oscillator shown in Figure 10E.21, put the ground point at the collector and then implement the oscillator circuit by adding a DC bias.

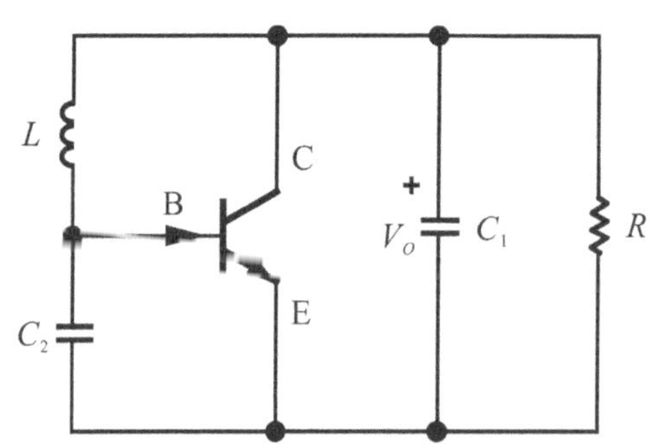

Figure 10E.21 Basic parallel feedback oscillator

Solution

The oscillator circuit with the collector as a ground is called a Colpitts oscillator and it can be implemented as shown in Figure 10E.22.

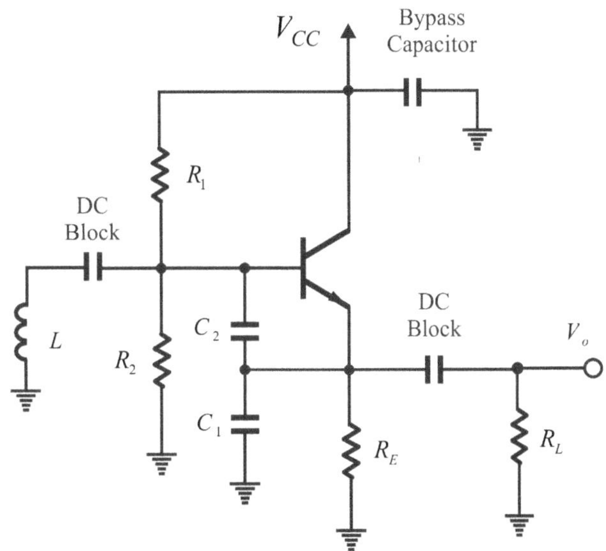

Figure 10E.22 An implemented Colpitts oscillator

The collector DC voltage is supplied through the bypass capacitor and the oscillation output is obtained from the emitter terminal. The DC voltage to the base is supplied using the bias resistors R_1 and R_2. Furthermore, a DC block capacitor is inserted between the inductor and base to prevent the base from being grounded. The emitter current of the BJT can then be set by the resistor R_E. In addition, the DC block capacitor is necessary to prevent the appearance of DC voltage at the output. The inserted resistors should provide higher impedances than those of the components around them at the oscillation frequency so as not to affect the RF signal at that oscillation frequency. The common collector implementation is easy and is thus widely used. The Colpitts oscillator circuits of Figures 10E.12 and 10E.17 are a type of common-collector oscillator circuit. In particular, in order to reduce the impact of the emitter bias resistor, the RFC may be used, as shown in Figure 10E.12. It should be noted that because the DC block capacitors and the RFCs inserted for the DC bias can be made to satisfy the oscillation conditions at other undesired frequencies, the appropriate values of the capacitors and the RFCs should be chosen so as not to satisfy the oscillation conditions at undesired frequencies. Instead of the common collector, the ground can be set as the emitter or the base. However, these kinds of configurations are not widely used due to the complexity of their implementations.

10.4.3 Design Method

Oscillator design from the impedance point of view is relatively simple; that is, the reference plane is set at the active part, which could be either Gunn or IMPATT diodes, and a series resonant load at the oscillation frequency is formed by adding a matching circuit to the 50-Ω load. The matching circuit must be designed such that the resistance looking into the load from the active part is smaller than the negative resistance of the active part. Alternatively, from the admittance point of view, a parallel resonant load is formed by adding a matching circuit to the 50-Ω load and the matching circuit must be designed such that the value of the parallel load is greater than the negative resistance of the active part.

The design concept from the impedance or admittance point of view can be similarly applied to the design of the series- or parallel-feedback-type oscillators. In the case of the series feedback type, the reference plane is set at the terminating reactance jx, as shown in Figure 10.36(a), and the active part is designed to satisfy the oscillation condition. For the purpose of simple design, the load $Z_L = R_L + jX_L$ is set as $X_L = 0$, $Z_L = Z_o$, and the series feedback reactance jy that gives the appropriate negative resistance value can be found by varying jy. Now, denoting the impedance looking into the active part from the reference plane as Z_{in}, the oscillation condition can be satisfied by setting the value of jx as $x = -\text{Im}(Z_{in})$. Next, adding the DC bias circuits for the transistor completes the oscillator design. This design method is simple; however, when the gain of the transistor is not high, the negative resistance is not induced at the reference plane, which causes problems in oscillator design. In that case, the design can be accomplished by trial-and-error adjustment of the load impedance value Z_L.

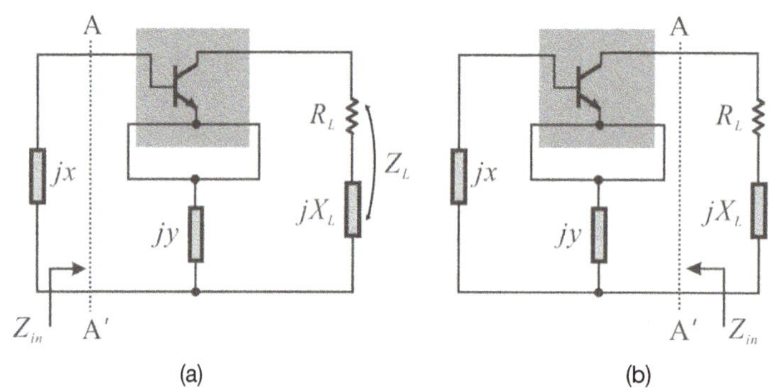

Figure 10.36 Design of a feedback-type oscillator: (a) jx reference plane and (b) load reference plane

Alternatively, the reference plane is set at the load, as shown in Figure 10.36(b).[6] In addition, the reactance pair jx and jy are set to make the real part of the impedance Z_{in} seen from the load Z_L negative. The selection of jx and jy is possible when the contour of the real part of Z_{in} is plotted in the (x, y) plane. The method of plotting these contours can be found in Appendix E and, by using this method, the (x, y) values giving negative resistance can be selected. Thus, for the selected (x, y) values, the impedance Z_{in} seen from the load Z_L can be calculated. The suitable load Z_L for this Z_{in} can then be synthesized using a matching network to satisfy the oscillation conditions. The real part of the load Z_L must be less than the negative resistance of Z_{in}, and $Z_L + Z_{in}$ must be also designed to be series resonant at the oscillation frequency. The basic form of the parallel oscillator can be designed using both the admittance condition and a method similar to the design of the series oscillator's basic form. This approach will be presented in the following section that deals with design examples. The mobile communication VCO will be designed following this approach.

6. M. Maeda, K. Kimura, and H. Kodera, "Design and Performance of X-Band Oscillators with GaAs Schottky-Gate Field-Effect Transistors," *IEEE Transactions on Microwave Theory and Techniques* 23, no. 8 (August, 1975): 661–667.

The oscillator design based on the previously explained one-port method can easily determine oscillation frequency, but the determination of the oscillation frequency alone is not enough in designing an oscillator with two-port devices such as a transistor. For example, it will be impossible to know whether the transistor in the oscillator is set to give maximum gain or is set to give maximum output power. The oscillator design based on the two-port method can provide an improved design even though it is more complex than the one-port method.[7]

First, consider the transistor shown in Figure 10.37 in an amplifier. For a given input power, the load impedance can be determined by the load-pull previously described in the power amplifier design in Chapter 9. The load impedance can be determined for maximum efficiency or for maximum output power. Then, the input voltage and current V_1 and I_1, and the output voltage and current V_2 and I_2 can also be determined. Using voltages and currents V_1, V_2 and I_1, I_2, the feedback network that yields V_1 and I_1 from the voltage and current V_2 and I_2 can be designed. The designed feedback network with the amplifier will form an oscillator that yields the designed frequency and output power of $P_{osc} = P_L - P_{in}$. Since a part of the output power P_L is fed back to supply input power P_{in} in the oscillator circuit, P_{osc} becomes $P_L - P_{in}$.

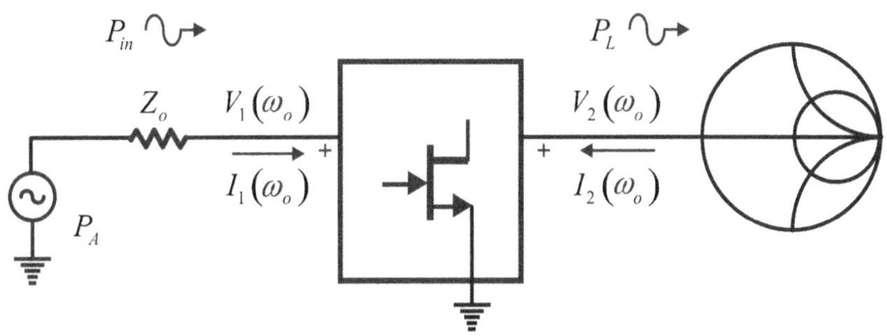

Figure 10.37 Determination of the input and output voltages and currents through the load-pull simulation

It is worth noting that the choice of the optimum source power level P_A to give the maximum oscillation power is unknown. This requires the following iteration: for the chosen P_A, the load impedance that maximizes P_L can be determined through the conventional load pull. For the determined load impedance, $P_L - P_{in}$ can be plotted for P_A. From the $P_L - P_{in}$ plot, the new optimum P_A can be obtained. For that new P_A, a new load impedance can be computed again through the load pull. Then, these steps are repeated to obtain the optimum load impedance that maximizes the oscillation output power.

Another problem can arise: In order to deliver the maximum input power to the input of the active device, the source and the input of the active device impedance must be conjugate matched. However, the large-signal input impedance of the active device is unknown until the load-pull simulation. Therefore, an initial value for the source impedance is determined using the small-signal S-parameters S_{11} and the load-pull simulation is then performed.

7. M. Q. Lee, S. J. Yi, S. Nam, Y. K. Kwon, and K. W. Yeom, "High-Efficiency Harmonic Loaded Oscillator with Low Bias Using a Nonlinear Design Approach," *IEEE Transactions on Microwave Theory and Techniques*, 47, no. 9 (September 1999): 1670–1679.

Using the determined voltages and currents V_1, V_2, I_1, and I_2, the feedback network should be designed to give the input and output voltages shown in Figure 10.38. With V_1, V_2, I_1, and I_2, the input power and oscillation output power are expressed in Equations (10.45a) and (10.45b).

$$P_{in} = \mathrm{Re}\left(V_1^* I_1\right) \tag{10.45a}$$

$$P_{osc} = P_L - P_{in} = -\mathrm{Re}\left(V_1^* I_1 + V_2^* I_2\right) \tag{10.45b}$$

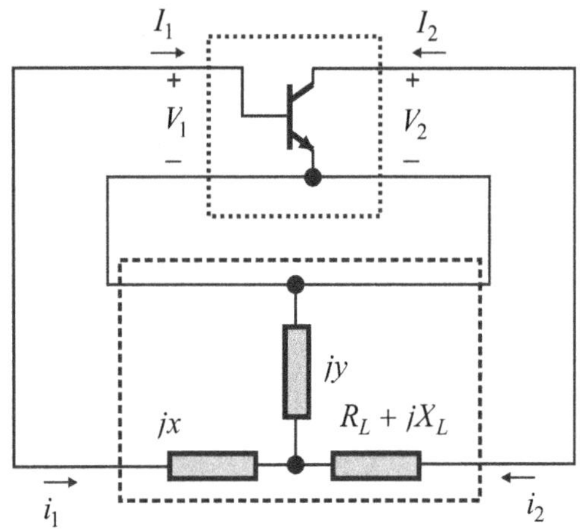

Figure 10.38 Definition of the voltages and currents in the oscillator

In addition, defining $i_1 = -I_1$ and $i_2 = -I_2$, as shown in the figure, and denoting the Z-parameters of the external feedback network as Z^e, the terminal voltages V_1, V_2 can be expressed as Equation (10.46).

$$\begin{pmatrix} V_1 \\ V_2 \end{pmatrix} = \begin{pmatrix} Z^e_{11} & Z^e_{12} \\ Z^e_{12} & Z^e_{22} \end{pmatrix} \begin{pmatrix} i_1 \\ i_2 \end{pmatrix} \tag{10.46}$$

Now, Z^e must be determined, because V_1, V_2, i_1, and i_2 have already been determined. The Z-parameters Z^e for the T-type circuit are shown in Equations (10.47a)–(10.47c).

$$Z^e_{12} = jy \tag{10.47a}$$

$$Z^e_{11} = jx + jy \tag{10.47b}$$

$$Z^e_{22} = Z_L + jy \tag{10.47c}$$

Thus, Z^e has four unknowns that take the real and imaginary parts of Z_L into consideration. Defining the new parameters expressed in Equations (10.48a)–(10.48d),

$$z_1 = -V_1/I_1 \tag{10.48a}$$

$$\beta_f = (1 + I_2/I_1) \tag{10.48b}$$

$$z_2 = -V_2/I_2 \tag{10.48c}$$

$$\beta_b = (1 + I_1/I_2) \tag{10.48d}$$

the values of x, y, and Z_L can be determined as shown in Equations (10.49a)–(10.49c).

$$Z_L = z_2 + j\beta_b \frac{\mathrm{Re}(z_1)}{\mathrm{Im}(\beta_f)} \tag{10.49a}$$

$$y = -\frac{\mathrm{Re}(z_1)}{\mathrm{Im}(\beta_f)} \tag{10.49b}$$

$$x = \mathrm{Im}(z_1) + \mathrm{Re}(\beta_f)\frac{\mathrm{Re}(z_1)}{\mathrm{Im}(\beta_f)} \tag{10.49c}$$

Similarly, in the case of a π-type feedback network, by defining the Y-parameters as Y^e and following a similar process, the following parameters can be defined for the π-type feedback network, as expressed in Equations (10.50a)–(10.50d).

$$y_1 = -I_1/V_1 \tag{10.50a}$$

$$\beta_f = (1 - V_2/V_1) \tag{10.50b}$$

$$y_2 = -I_2/V_2 \tag{10.50c}$$

$$\beta_b = (1 - V_1/V_2) \tag{10.50d}$$

From these equations, the values for the feedback network can be determined with Equations (10.51a)–(10.51c) as follows:

$$Y_L = y_2 + j\beta_b \frac{\mathrm{Re}(y_1)}{\mathrm{Im}(\beta_f)} \tag{10.51a}$$

$$y = -\frac{\mathrm{Re}(y_1)}{\mathrm{Im}(\beta_f)} \tag{10.51b}$$

$$x = \mathrm{Im}(y_1) + \mathrm{Re}(\beta_f)\frac{\mathrm{Re}(y_1)}{\mathrm{Im}(\beta_f)} \tag{10.51c}$$

Note that the solutions for x, y, and Z_L in Equation (10.49) make the open-loop gain equal to 1 at the large-signal condition. The gain at the maximum output power or large signal is generally lower than the small-signal gain. Thus, even when the open-loop gain is 1, the small-signal open-loop gain naturally becomes greater than 1 and does not cause major problems in oscillation start-up.

Example 10.13

FHX35LG is a pHEMT and its large-signal model is available in the ADS library. The drain current I_{DS} is about 4 mA at $V_{DS} = 2$ V, $V_{GS} = -0.5$ V. Design a self-bias circuit with a source resistor. Then, set up the load-pull simulation for the maximum oscillation power at the frequency of 2.5 GHz. Using the load-pull simulation results, determine the value of the series feedback network x, y, and Z_L using Equations (10.48) and (10.49). Verify the series feedback network by simulating the oscillator built with the designed series feedback network.

Solution

$V_{GS} = -0.5$ V is required for a drain current $I_{DS} = 4$ mA. From this, V_S should be 0.5 V when $V_G = 0$. The value of the source resistance is thus $R_s = 0.5/4 \times 1000 = 125\ \Omega$. This yields the source voltage of 0.5 V at a drain current of 4 mA. Since $V_{DS} = 2$ V, a drain supply voltage should be 2.5 V. The self-bias circuit is shown in Figure 10E.23.

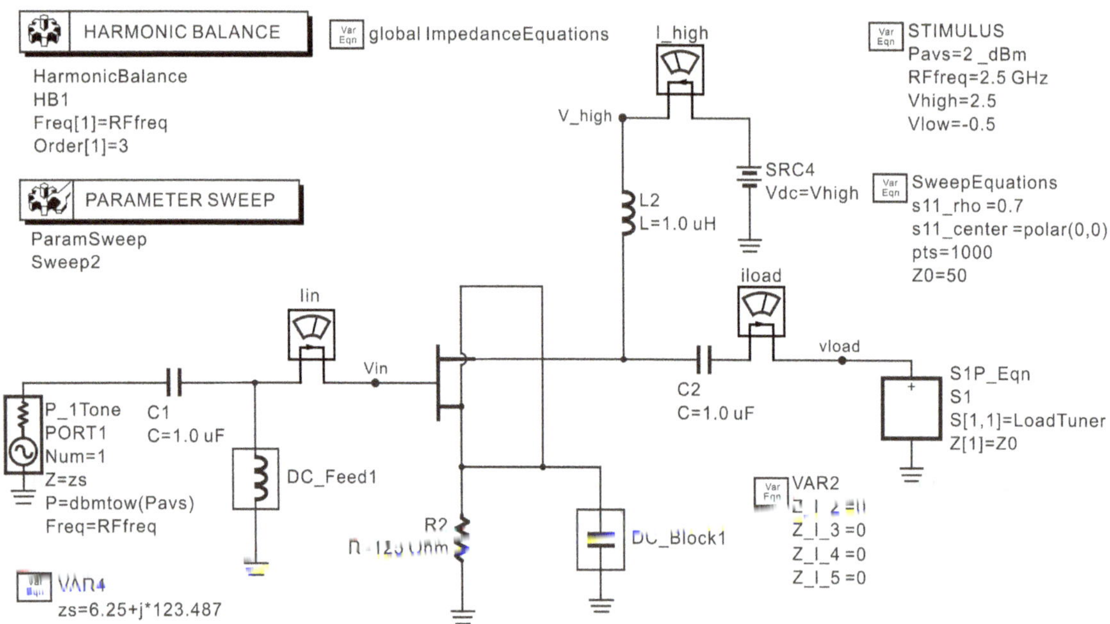

Figure 10E.23 Load-pull simulation schematic. The FET FHX35LG is self-biased to flow a drain current of 4 mA. The source impedance is determined to be the conjugate of S_{11}. The harmonic impedances of the load are unknown and they are all set to 0.

The load-pull simulation circuit can be set up by modifying the load-pull schematic in section 9.3.2 of Chapter 9. The source impedance is set to the conjugate of S_{11}. Its value is **zs** = 6.25 + *j*123.487 Ω. The harmonic impedances of the source should also be set to appropriate values. However, they are unknown and so all the harmonic impedance values are set to **zs**. The harmonic impedances of the load should also be appropriately set. All the harmonic impedances of the load from **Z_1_2** to **Z_1_5** are set to 0. Since the feedback network of the oscillator is a type of bandpass filter, the harmonic impedances can be approximated as short. Finally, the power level of the source must be set for the load-pull simulation. The source power level is initially set to 0 dBm. Figure 10E.24 shows the load-pull simulation results. From that figure, the maximum power level is about 7.95 dBm at the load impedance of 108.85 + *j*61.201 Ω.

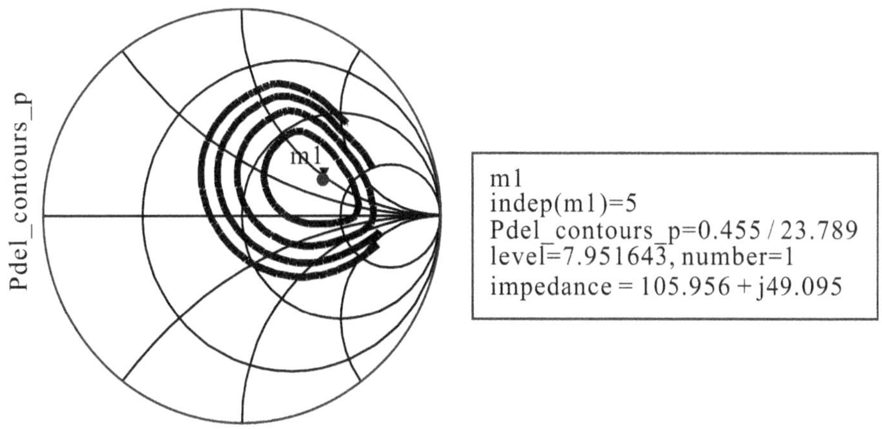

Figure 10E.24 Load-pull simulation results. At the load impedance of 105.956 + *j*49.095 Ω, the delivered power to the load is about 7.95 dBm.

Then, to plot $P_{osc} = P_L - P_{in}$, the simulation circuit shown in Figure 10E.25 is set up. To plot $P_{osc} = P_L - P_{in}$, the following equations in Measurement Expression 10E.5 are entered in the display window:

```
Eqn PL_dBm=10*log(0.5*real(conj(vload[1])*iload.i[1]))+30
Eqn Pin_Watt=0.5*real(conj(Vin[1])*Iin.i[1])
Eqn Posc_dBm=10* log(0.5*real(conj(vload[1])*iload.i[1])-Pin_Watt)+30
```

Measurement Expression 10E.5 Equations for $P_{osc} = P_L - P_{in}$

Using the plot of P_{osc}, the new source's power level can be found. The 0 dBm source power that was initially chosen is now found not to yield maximum oscillation power. After iteration, the source power level is determined to be 2 dBm. Generally, the load impedance changes according to the change in the source power level. However, the load impedance is almost the same as that at the source power level of 0 dBm. The load impedance at the source power level of 2 dBm is found to be 105.956 + *j*49.095 Ω, which is shown in Figure 10E.25. The plot of the delivered power and oscillation power is shown in Figure 10E.26. The maximum oscillation power appears at the source power of 2 dBm and is 7.314 dBm. The delivered power to the load for P_{osc} is 7.803 dBm.

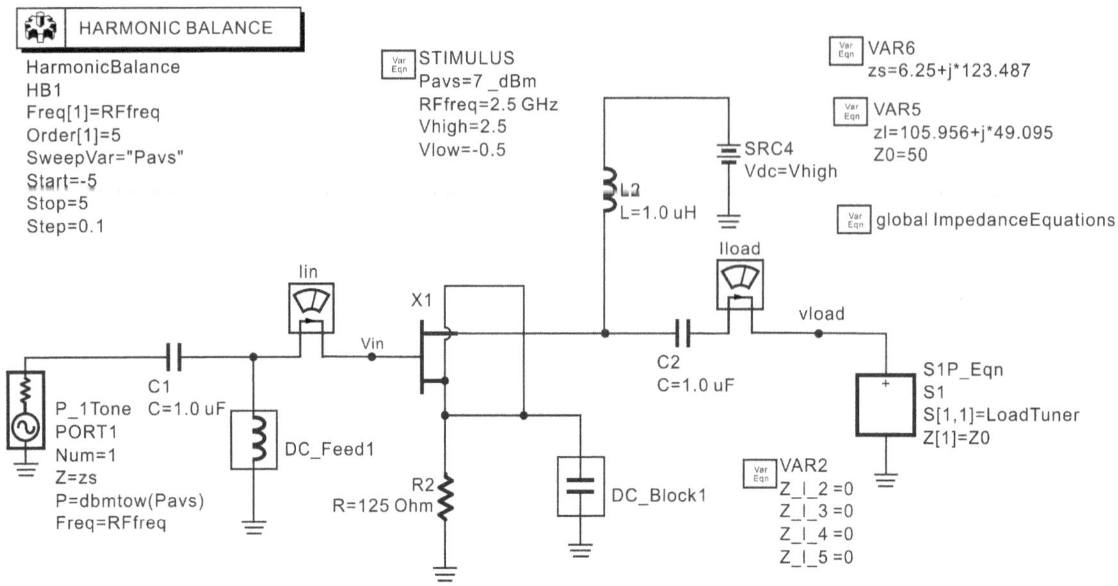

Figure 10E.25 Simulation schematic for determining the values of the oscillator's series feedback network. Setting the load impedance that is determined from the load-pull simulation, the source power is varied to find the optimum oscillation power. The input and output voltages and currents at the optimum power are also computed to determine the oscillator feedback network.

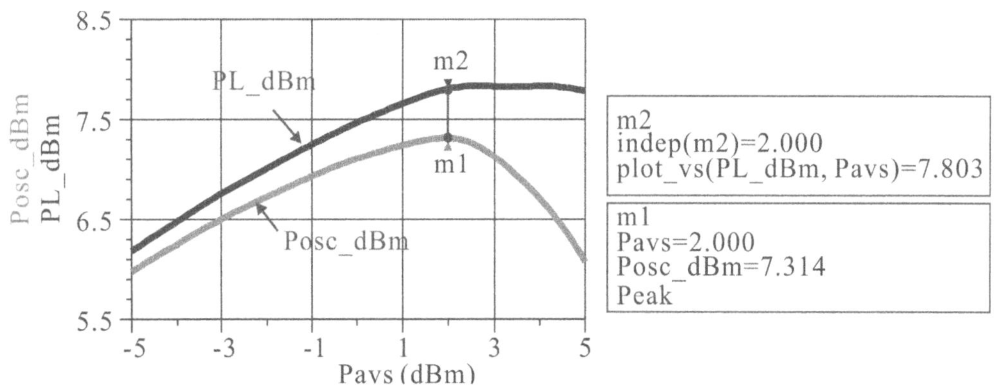

Figure 10E.26 Simulated P_L and P_{osc}. P_{osc} = 7.314 dBm. At this power level, the values of the oscillator feedback network are determined.

To determine the series feedback network element values x, y, and Z_L from the simulated output currents and voltages, the equations in Measurement Expressions 10E.6 and 10E.7 are entered in the display window. Measurement Expression 10E.6 is for the computation of the parameters z_1, z_2, β_f, and β_r in Equation (10.47). The index **n** in the first equation is the index of **Pavs**, which gives the maximum oscillation power. The outer index represents the harmonic number. Then, using the determined parameters z_1, z_2, β_f, and β_r from Measurement Expression 10E.6, the series feedback network element values of x, y, and Z_L in Equation (10.48) can be determined by Measurement Expression 10E.7.

```
Eqn n=find_index(HB.Pavs, indep(m1))
Eqn I2=-iload.i[n,1]
Eqn I1=iin.i[n,1]
Eqn V1=Vin[n,1]
Eqn V2=vload[n,1]
Eqn z1=-V1/I1
Eqn z2=-V2/I2
Eqn beta_f=(1+I2/I1)
Eqn beta_b=(1+I1/I2)
```

Measurement Expression 10E.6 Equations for β_f and β_b

```
Eqn ZL=z2+j*beta_b*real(z1)/imag(beta_f)
Eqn y=-real(z1)/imag(beta_f)
Eqn x=imag(z1)+real(beta_f)*real(z1)/imag(beta_f)
```

Measurement Expression 10E.7 Equations for the feedback parameters x, y, and Z_L

The computed results for x, y, and Z_L are 143.76 Ω, –11.594 Ω, and 94.662 + j49.194 Ω, respectively. Figure 10E.27 shows the simulation schematic to verify the computed x, y, and Z_L results.

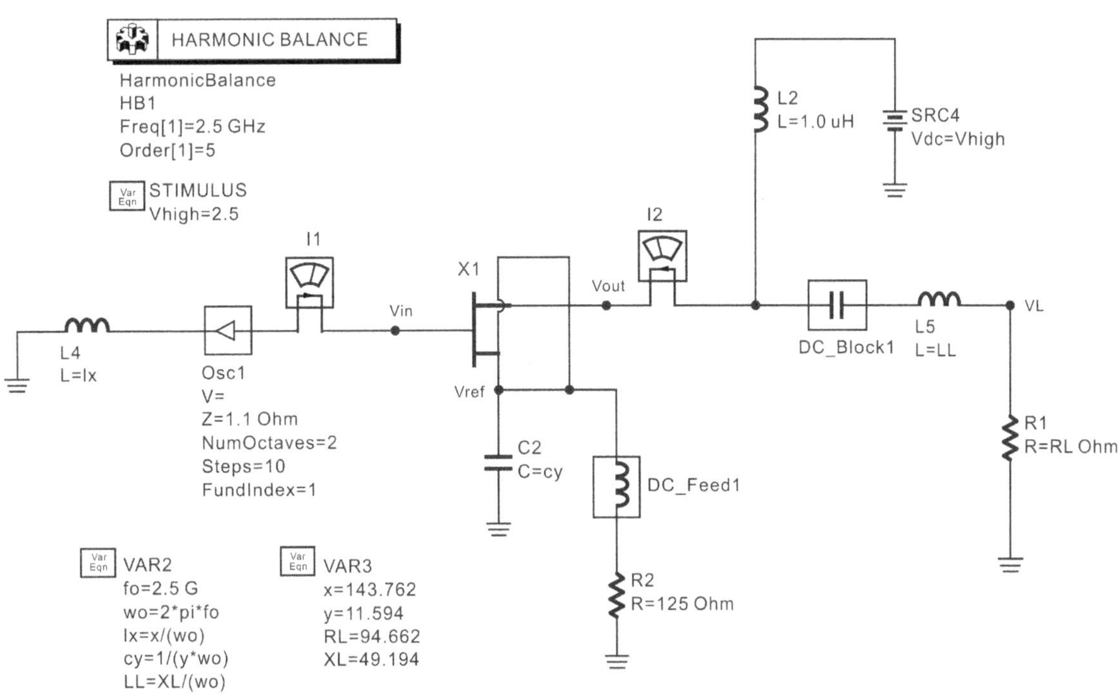

Figure 10E.27 Oscillator circuit for confirming the values of the series feedback network

The simulated waveform and spectrum are shown in Figures 10E.28(a) and 10E.28(b). Figure 10E.28(a) is the simulated time-domain waveform of **VL** and Figure 10E.28(b) is the spectrum of **VL**. The oscillation output power is about 6.3 dBm. The value is less than the 7.3 dBm in Figure 10E.26. In addition, the oscillation frequency is 2.483 GHz, although it is close to 2.5 GHz. The difference in the oscillation power and frequency is due to the difference in the load impedances. The load impedance seen from the drain can be computed using (**Vout** – **Vref**)/(–**I2**). The harmonic impedances up to the three harmonics are computed as 105.962 + j47.695 Ω, 239.971 + j391.488 Ω, and 107.703 – j783.136 Ω, which are different from 94.662 + j49.194 Ω, 0 Ω, and 0 Ω. The differences of the harmonic impedances also make the drain voltage V_2 in Figure 10.38 different. The ratio of V_2 in the two simulations shown in Figure 10E.25 and Figure 10E.27 is about 1 dB, which explains the power difference of 1 dB. However, it can be seen that the result is fairly close to the expected value. It is possible to tune the values of x and y in order to obtain the exact oscillation frequency, but this tuning is not performed in this example.

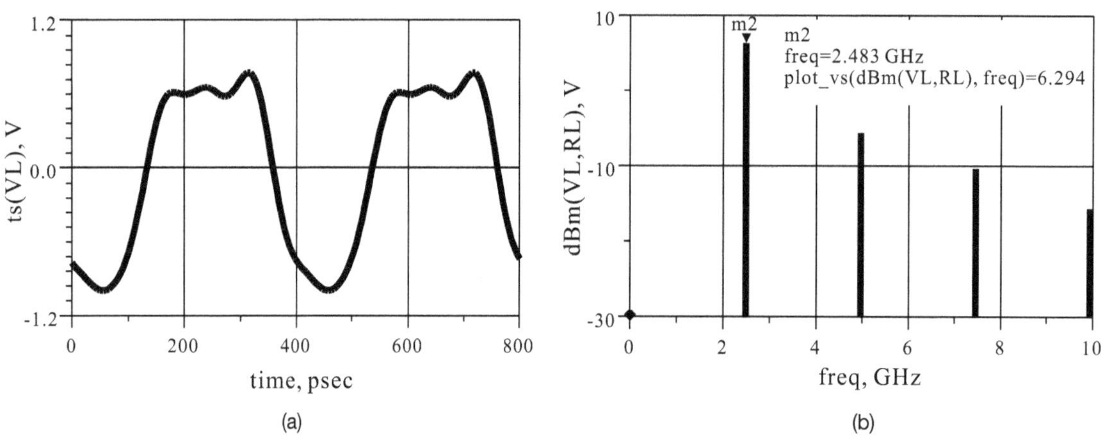

Figure 10E.28 (a) Oscillation waveform and (b) spectrum obtained through oscillator simulation. The oscillation power is smaller than the load-pull simulation given by 7.314 dBm. Note that the power is computed by **dBm** (**VL**, **RL**) because **RL** is not 50 Ω.

10.5 OSCILLATOR DESIGN EXAMPLES

10.5.1 VCO for Mobile Communications

Figure 10.39 shows the configuration of a voltage-controlled oscillator (VCO) used for mobile communications. The voltage-controlled oscillator has a size of 12 × 10 × 4 mm^3, and a volume of approximately 4.8 cc. The smaller-size VCOs are still fabricated using the same technology shown in Figure 10.39. As can be seen in that figure, the VCO is composed of a metallic cover and a multilayer printed circuit board that is used to mount chip components. The metallic cover provides both electromagnetic shielding and ground. Therefore, the total weight of the VCO is determined by the multilayer printed circuit board and the thin metal cover, which makes the VCO lightweight. In addition, the terminals for DC power supply, oscillator output, ground, and frequency tuning for the VCO are formed by cutting in half the center of the multilayer printed circuit board's through hole, which results in the half-plated cylindrical terminals. With the exception of the connection terminals, the bottom of the circuit board is coated with solder-resistant material. This provides electrical isolation and makes it possible for the PCB's lines to pass through the bottom of the VCO when the VCO is mounted on the PCB.

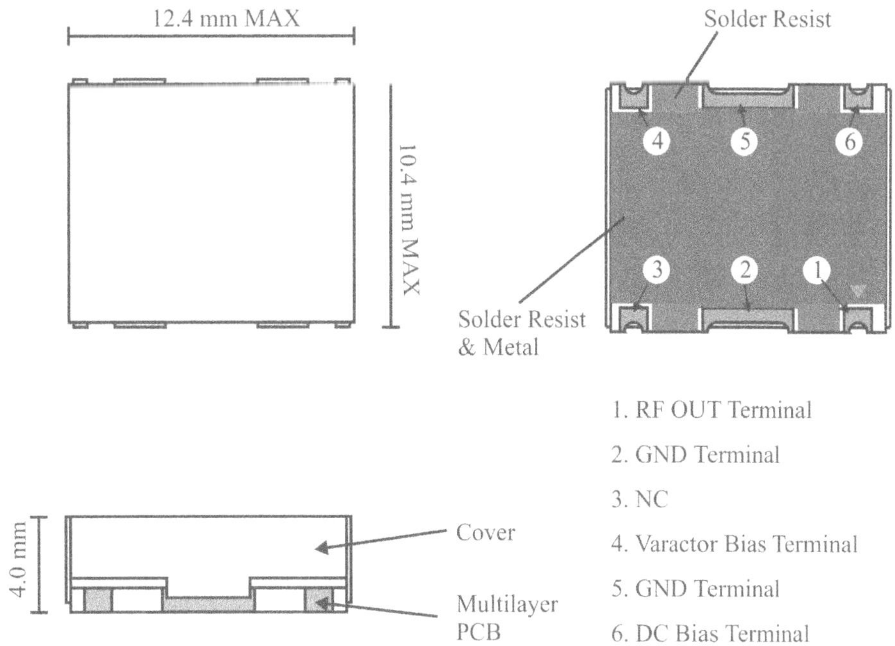

Figure 10.39 Configuration of a VCO for mobile communications

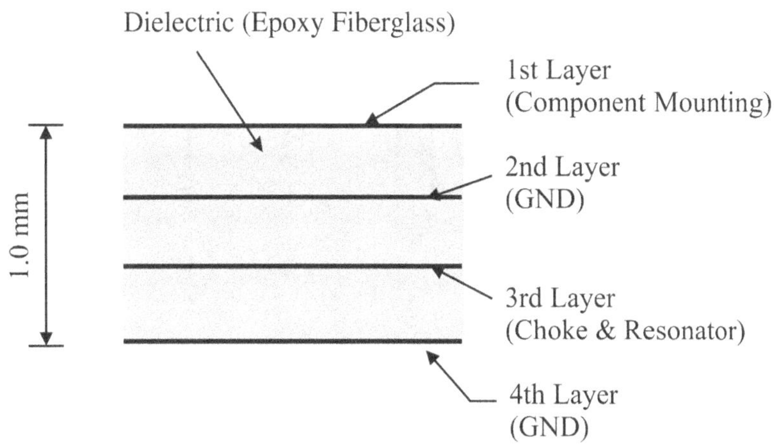

Figure 10.40 Configuration of a multilayer printed circuit board. Chip components are mounted on the first layer and the second layer acts as the ground for the first layer. Thus, the line on the first layer acts as a microstrip. The third layer is for the resonator and RFC, and the fourth layer is the ground. As a result, the lines on the third layer act as strip lines.

Figure 10.40 shows a cross-section of the multilayer PCB. The substrate is composed of three dielectric sheets of FR4 and the total thickness is 1.0 mm. Each dielectric has equal thickness. The first metal layer is for mounting components and the second metal layer is the ground for the first layer. The third metal layer is used for the RFC, strip-line resonator, and connection lines. The fourth metal layer is also the ground. Thus, the lines on the first layer are considered microstrip lines from the electromagnetic point of view, and the lines on the third layer become strip lines since they are surrounded by the ground planes. Note that except for the connecting holes, the lines on the first and third layers are electromagnetically isolated due to the second and fourth layers.

The VCO circuit used here is shown in Figure 10.41, where two transistors, Q_1 and Q_2, are connected in cascade. Transistors Q_1 and Q_2 share a common emitter current and the VCO consumes only about one-half of the DC current compared with other VCOs not using cascaded structures. However, a relatively low DC voltage is assigned between the collector emitter of transistors Q_1 and Q_2, and a low RF output power as well as significant distortions can occur, which is a disadvantage.

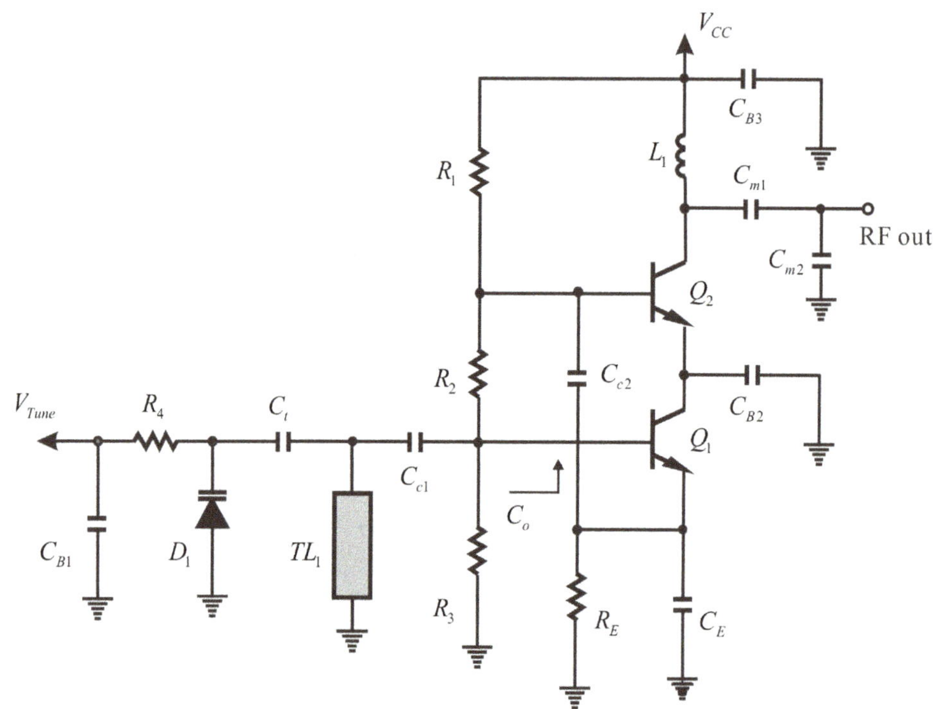

Figure 10.41 Schematic of a VCO (voltage-controlled oscillator)

In Figure 10.41, resistors R_1, R_2, and R_3 determine the DC base voltages of Q_1 and Q_2. Supply voltage is divided by resistors R_1, R_2, and R_3. The DC voltage across resistor R_3 is applied to the base of transistor Q_1. From this DC base voltage, the emitter current of transistor Q_1 can be controlled by varying the value of resistor R_E. In addition, the emitter current of Q_2 is equal to that of transistor Q_1. As the impedances of resistors R_1, R_2, and R_3 can be set higher than those of their surrounding components, their effects can be ignored at the oscillation frequency. Generally, in the frequency band of operation (800 MHz–2 GHz), the Q of a chip capacitor is generally higher than that of an inductor. Also, because of their small sizes, chip capacitors are primarily used to construct VCOs. An inductor is used for the RFC and a resonator is implemented using a shorted transmission line. The collector of transistor Q_1 is connected to the ground through a bypass capacitor C_{B2}, and the transistor Q_1 operates as a common collector. The capacitor C_E is a feedback capacitor that yields a negative resistance. When the value of the feedback capacitor C_E is small, only the resistor

R_E is left at the emitter of Q_1, which becomes a negative feedback circuit consisting of resistor R_E. As a result, the negative resistance disappears. In addition, when the value of C_E is too large, capacitor C_E operates as a short. Thus, transistor Q_1 operates as a common emitter and negative resistance is not induced. As a result, in order to induce negative resistance, the value of the feedback capacitor C_E must be appropriate at the oscillation frequency. To select an appropriate value for C_E, ignoring R_E and C_{c2}, the impedance Z_t looking into the base of the transistor Q_1 can be written as Equation (10.52).

$$Z_t = \frac{1}{j\omega\left(C_E \parallel C_{be1}\right)} - \frac{g_{m1}}{\omega^2 C_{be1} C_E} \tag{10.52}$$

Here, C_{be1} and g_{m1} represent the base-emitter capacitance and transconductance of Q_1. Oscillation is possible due to the induced negative resistance given in this equation.

The oscillation output appears across capacitor C_E, which is applied to the base of transistor Q_2 through capacitor C_{c2}. The transistor Q_2 operates as a common emitter amplifier due to the bypass capacitor C_{B2}. Thus, the oscillation output that appears across capacitor C_E is amplified by transistor Q_2, which is then delivered to the load. Chip inductor L_1 connected to the collector of the transistor Q_2 is an RFC and capacitors C_{m1}, C_{m2} are for matching, which aims the maximum power delivery to the load. Thus, C_E is set to generate a negative resistance at the operating frequency. Note that capacitor C_{c2}, connected to the input of transistor Q_2, appears in parallel to C_E and the contribution of C_{c2} should be considered in the determination of C_E. The lower the value of C_{c2}, the lower the power delivered to transistor Q_2. In contrast, the larger the value of C_{c2}, the larger the power delivered to Q_2; however, it causes distorted output to appear at the load.

The resonator in Figure 10.41 is composed of a shorted transmission line TL_1, a capacitor C_t, and a varactor diode. The resonant frequency of the resonator is tuned through a varactor diode, whose tuning range is limited by C_t in series. Thus, the oscillation frequency's tuning range can be adjusted by controlling capacitor C_t. The oscillation frequency tuning range is reduced compared with the direct oscillation frequency tuning using the varactor diode alone, but the Q of the resonator becomes higher due to C_t.

Capacitor C_{c1} is added in series to the base of the transistor Q_1, and the imaginary part of the impedance generated by the feedback capacitor C_E can be changed. This makes it easy to tune out the inductance of the resonator at the oscillation frequency. As a result, the resonator and active part can be in series resonance at the oscillation frequency.

In summary, the function of the capacitor C_t is related to the oscillation frequency's tuning range, the capacitors C_{c1} and C_E are associated with the oscillation formation, and C_{c2} is related to the coupling of the oscillation power and delivery to the amplifier. Capacitor C_{c1} and inductor TL_1 determine the oscillation frequency of the oscillator. In particular, by varying the value of inductor TL_1, it is possible, to some extent, to easily adjust the center of the frequency tuning range. Obviously, capacitor C_{c1} does not change the negative resistance as it is connected in series but, when the impedance of the active part is converted to the admittance through series-to-parallel conversion, C_{c1} gives a variation of negative conductance. Thus, by adjusting capacitor C_{c1}, the oscillation may disappear. Therefore, TL_1 is efficient for tuning the oscillation center frequency; however, it is difficult to tune once fabricated. Sometimes, by connecting a small tunable patch pattern in parallel with TL_1, an adjustment to the inductor value of TL_1 is possible and the center of the frequency tuning range can be tuned to some extent.

Most of the capacitor values explained above are implemented using chip capacitors that are mounted on the PCB during fabrication and it is easy to modify their values. Thus, the exact values of the length and width of the lines are not required as they would be in the design of a microstrip oscillator. Therefore, to design the VCO for mobile communications, the function of each component should be understood and the range of values that satisfy the given specifications should be found through design or simulation.

Table 10.1 provides brief specifications of the VCO for mobile communications. The transistor 2SC4226 from NEC is adopted.

Table 10.1 Design specification for voltage-controlled oscillator for mobile communications

Spec Item	Value
Frequency tuning range	1720–1780 MHz
DC supply voltage	3.3 V
DC current consumption	< 8 mA
RF output power	> −3 dBm
Active device	2SC4226

To design a VCO for mobile communications that satisfies the specifications above, the DC bias must first be set. The circuit of the active part is set up as shown in Figure 10.42 to determine the DC bias. All the capacitors used are implemented as chip capacitors and all the chip capacitors are replaced by a series *RLC*-equivalent circuit to include the parasitic components described in Chapter 2. In Figure 10.42, from a 3.3 V power supply, the base voltage of the oscillating transistor **Q2** is

$$V_B \cong \frac{R_3}{R_1 + R_2 + R_3} V_{CC} = 3.3 \frac{3.3}{2.7 + 2.7 + 3.3} = 1.25 \text{ V}$$

Thus, the emitter current is

$$I_E = \frac{V_B - 0.7}{R_E} = \frac{1.25 - 0.7}{50} = 11 \text{ mA}$$

The value of 11 mA is close to that of the DC current specification in Table 10.1. The exact value is determined through DC simulation. The calculated DC collector current is computed to be 6.45 mA at a supply voltage of 3.3 V. The resulting supply current is 6.83 mA, which satisfies the goal of 8 mA in Table 10.1.

Next, in order to operate the oscillating transistor (transistor **Q2**) in the common collector mode, the value of the DC block capacitor **SRLC3** is set to 100 pF. The value of the bypass capacitor **SRLC2** is set to 1000 pF. The value of the RF-output DC block capacitor **SRLC1** is set to 47 pF, and no output matching circuit is used. In addition, RFC **L1** is determined to be a 10-nH chip inductor (approximately 100 Ω at the oscillation frequency).

As mentioned earlier, the impedance $Z = R + jX$ seen from port 1 should have negative resistance, and it is determined by the feedback capacitor **Ce** and coupling capacitor **Cc2**. To determine their values, first **Cc2** is set to 0 and the impedance is calculated by varying **Ce**. The coupling capacitor **Cc1** is replaced by the DC block and Z is plotted with respect to **Ce**.

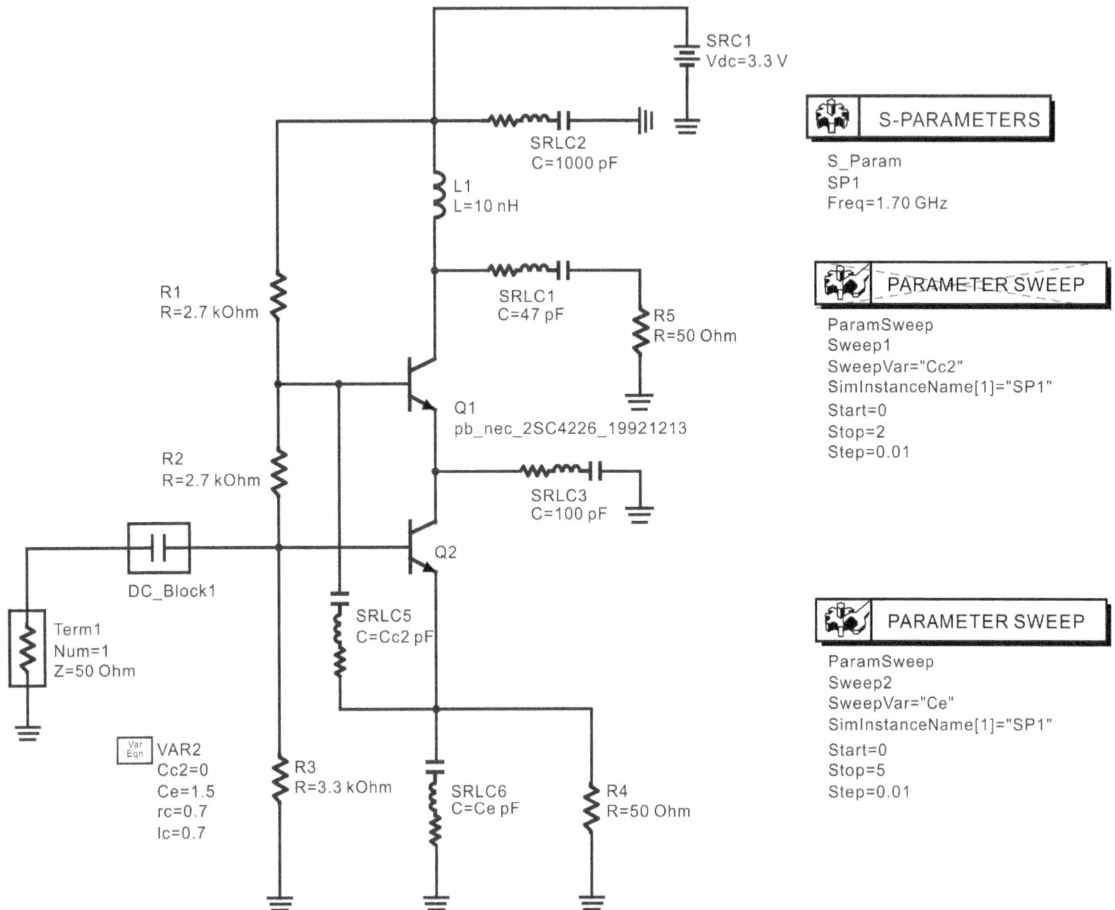

Figure 10.42 Simulation circuit for determining **Ce** and **Cc2**. First, the S-parameter simulation is performed at a fixed frequency of 1.7 GHz for the change of **Ce**, which is swept by **Sweep**2. The optimum value of **Ce** is determined from the **Ce** sweep. After the determination of the value of **Ce**, the value of **Cc2** is then swept by **Sweep**1.

The computed real and imaginary parts of Z are shown in Figure 10.43(a). The negative peak value of R occurs for **Ce** = 1.95 pF. To take the loading effect of **Cc2** into consideration, **Ce** is initially set to 1.5 pF smaller than 1.95 pF and **Cc2** is varied. As the value of **Cc2** increases, the magnitude of R is reduced, as shown in Figure 10.43(b). Thus, **Cc2** is set at 1.2 pF because the value of R begins to increase at the onset of 1.2 pF. The value of **Cc2** is the maximum value that preserves the value of R determined by **Ce**. The maximum value of **Cc2** is chosen to deliver the sufficient oscillation output power to the load. Note that the delivered power to the load is smaller for the smaller value of **Cc2**.

Next, a circuit must be set up to determine the oscillation frequency tuning range for the selected values of **Ce** and **Cc2**. The smv1235-079 varactor diode was used. The equivalent circuit of the smv1235-079 varactor diode is shown in Figure 10.44. Note that as the capacitance of the varactor diode at 0 V is large, the varactor diode will actually operate as a variable inductor. The equivalent circuit in Figure 10.44 is configured as a subcircuit. The oscillator circuit shown in Figure 10.45 is set up to determine the oscillation frequency tuning range.

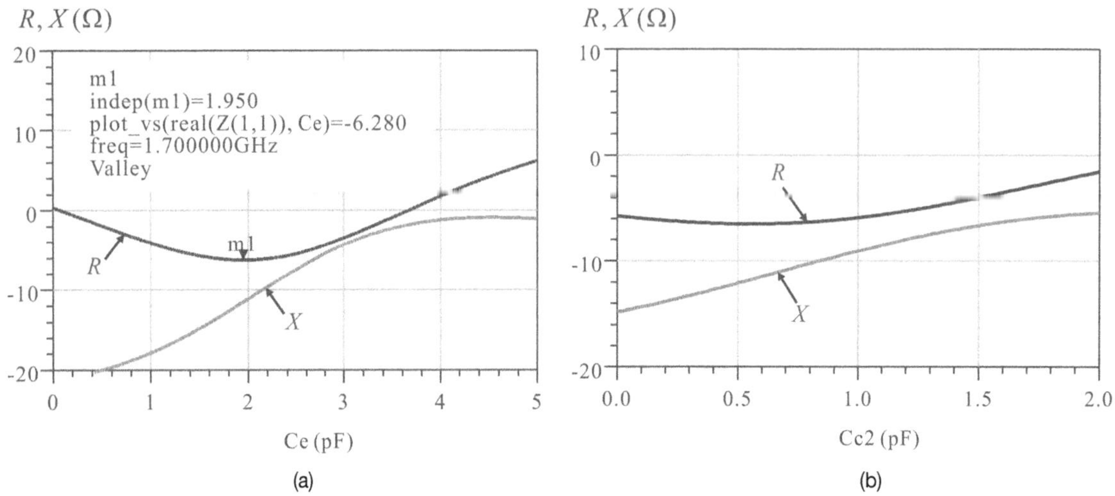

Figure 10.43 Simulated input impedance of the active part with C_e and C_{c2} values as parameters. (a) The input impedance for **Ce** sweep; the negative resistance is maximum at **Ce** = 1.95 pF. (b) The input impedance for **Cc2** sweep with fixed **Ce**; the value of **Ce** is determined to be 1.5 pF and **Cc2** is swept. The value of **Cc2** is determined to be 1.2 pF at the onset of the increasing resistance.

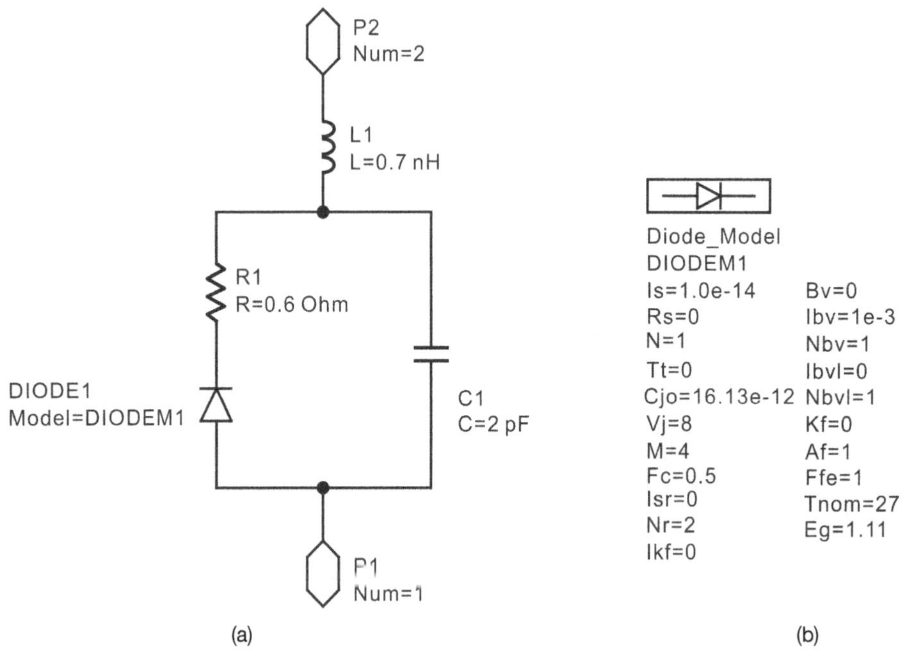

Figure 10.44 The equivalent circuit of the smv1235-079 varactor diode. Other diode parameters are not given and they are hidden.

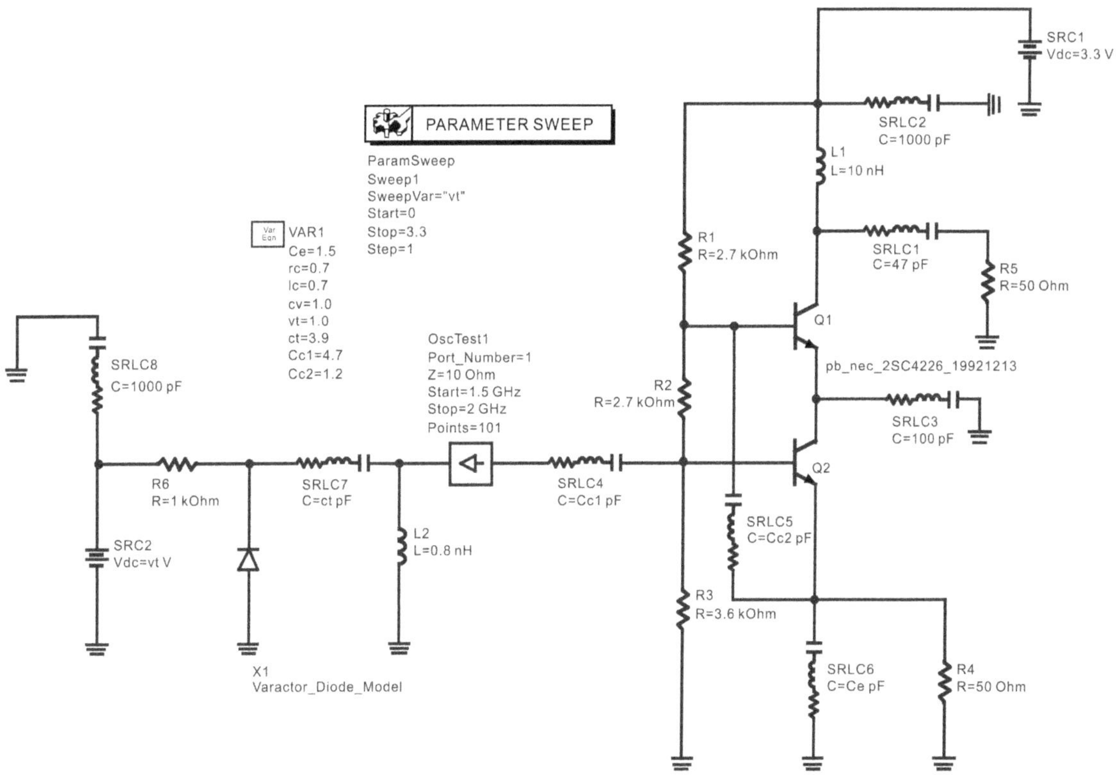

Figure 10.45 A VCO circuit for the simulation of the oscillation frequency's tuning range. **L2** is the resonator and is set 0.8 nH. The subcircuit **X1** is the varactor diode that is shown in Figure 10.44. The capacitor **Ct** is used to tune the oscillation frequency's tuning range.

As shown in Figure 10.45, the inductor in the resonator is set to 0.8 nH. The tuning voltage of the varactor diode is changed from 0 to 3.3 V, and the values of capacitors **Cc1** and **Ct** are then adjusted. The oscillation center frequency is moved using **Cc1** and the oscillation frequency tuning range is adjusted using **Ct**. Using the adjustment, the value of **Cc1** is set to 4.7 pF and **Ct** is set to 3.9 pF. Figure 10.46 shows the simulated oscillation conditions using **OscTest** as well as the magnitude and phase changes of S_{11} with respect to frequency. From S_{11}, the frequency tuning range is about 1.70–1.77 GHz, which is found to meet the specifications.

Using the small-signal simulation, oscillation is found to occur and it covers the specified oscillation frequency's tuning range. Next, the circuit shown in Figure 10.47 is built up and large-signal simulation is performed in order to determine the output power and frequency tuning range. The simulated oscillation frequency and power of the circuit are shown in Figure 10.48, where the output power range of –1.698 to –0.690 dBm is observed, which meets the desired specifications. The tuning voltage range of **Vt** that covers the desired oscillation frequency tuning range is approximately 0.6–2.8 V, as shown in Figure 10.48.

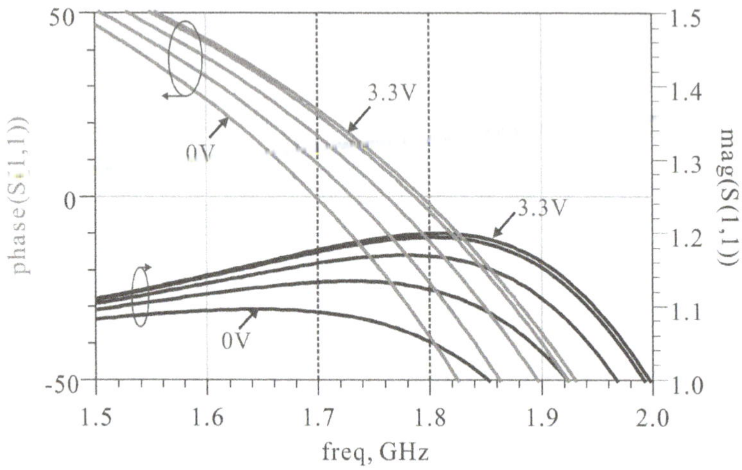

Figure 10.46 Frequency tuning range of the VCO circuit. The phase of S_{11} crosses 0 at a frequency range of 1.7–1.8 GHz and its $|S_{11}| > 1$, which provides the oscillation frequency tuning range of 1.7–1.8 GHz.

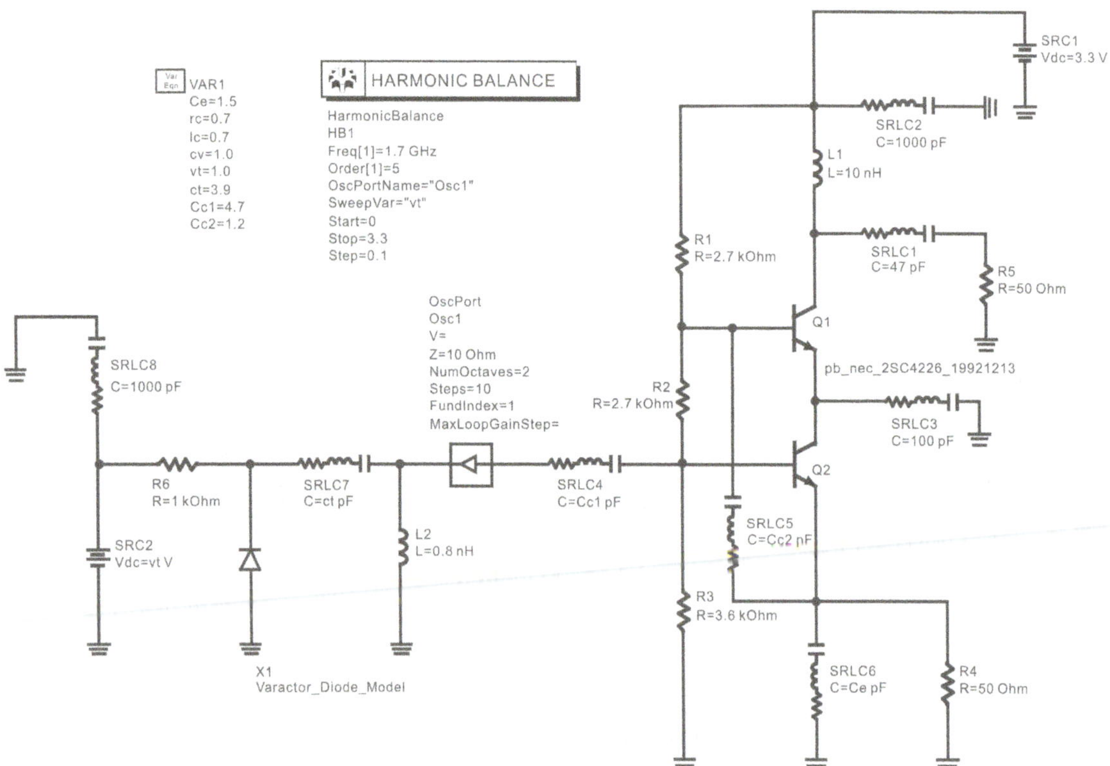

Figure 10.47 Circuit for a large-signal simulation. The **OscTest** in Figure 10.45 is replaced by **OscPort** for large-signal simulation.

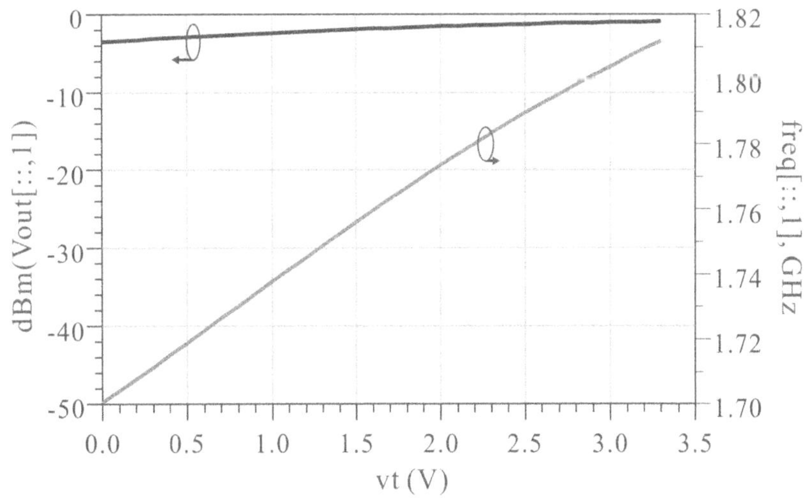

Figure 10.48 Oscillation frequency and output power variation with respect to tuning voltage **Vt**. The output power is about –3 dBm–0 dBm for the **Vt** change. The oscillation frequency changes from 1.70 to above 1.80 GHz for the **Vt** change.

In Figure 10.49(a), the time-domain waveforms are shown with the tuning voltage **Vt** as a parameter and the output waveform can be seen to be close to a sinusoidal waveform. The spectra of the oscillation waveforms are shown in Figure 10.49(b) and the second harmonic is found to be suppressed by approximately –20 dBc compared with the fundamental frequency.

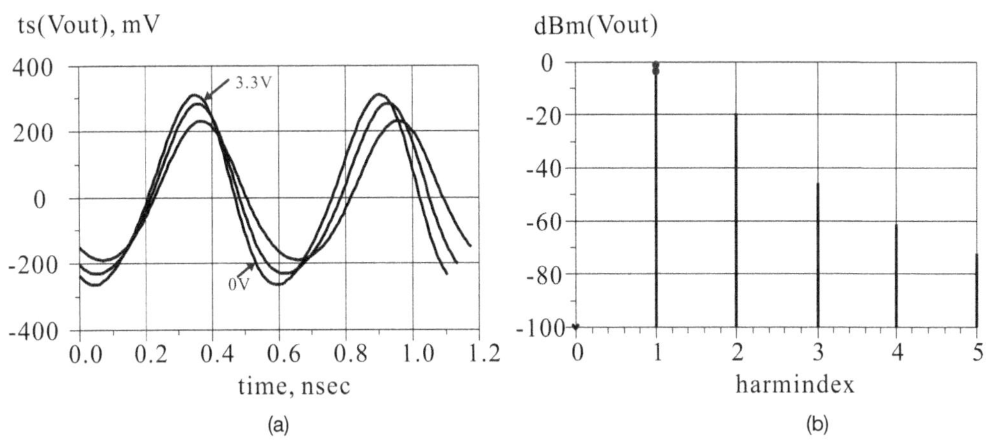

Figure 10.49 Output (a) waveforms and (b) spectra

10.5.2 Microstrip Oscillator

The technique used in the design of the VCO for mobile communications described earlier is useful when components can be easily connected and removed by soldering. Since soldering can readily adjust the values of the components, the exact values of the components are not required in the design stage. Throughout the design stage, the designer should identify the role of each component and confirm that the oscillator's fabrication will meet specifications. However, since the microstrip oscillator we will describe in this section is primarily composed of microstrip lines, it is not as easy to adjust those lines as it is in the mobile communications VCO. Thus, the exact dimensions of the microstrip lines are necessary in the design of a microstrip oscillator.

The microstrip oscillator can be designed using the impedance method in the design of the VCO for mobile communications described in the previous section. However, in this section we will demonstrate the design using the method from Example 10.13. The transistor employed for the design is a packaged pHEMT, FHX35LG (LG means low parasitic, hermetically sealed metal-ceramic package) shown in Figure 10.50. The large-signal model of the transistor is available in the ADS library. This section addresses the design of the oscillator using the large-signal model of the FHX35LG. Thus, the design can include both the oscillation frequency and the power. Typically, the large-signal model shows some errors, as discussed in Chapter 5, and these errors become larger as the frequency increases. Taking into consideration the error of the large-signal model, the oscillation frequency is set at 2.5 GHz, which is relatively low.

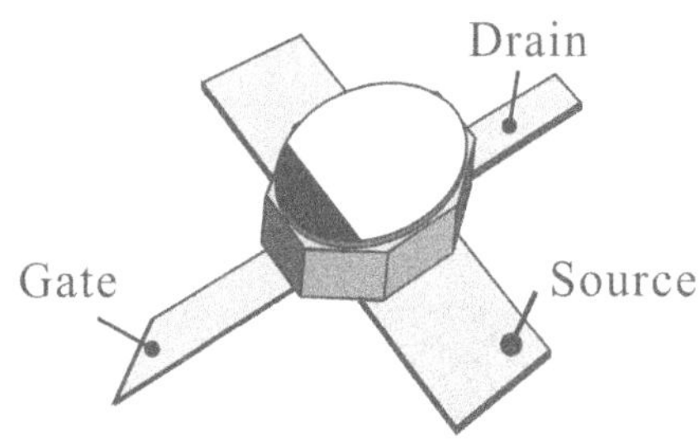

Figure 10.50 FHX35LG transistor package[8]

10.5.2.1 Implementation of Microstrip Oscillator

The oscillator circuit shown in Figure 10E.27 in Example 10.13 is implemented using lumped elements such as inductors and capacitors. Therefore, to complete the microstrip oscillator, the lumped elements of Figure 10E.27 must be replaced with microstrip circuits that give the same impedance values at the oscillation frequency.

Figure 10.51(a) shows a microstrip circuit for replacing the inductor connected to the gate. The selected substrate has a thickness of 20 mil, a dielectric constant of 2.5, a loss tangent 0.0019, and a conductor thickness of 17.5 μm; the diameter of the via is fixed at 0.5 mm. To replace the inductor with a microstrip yielding the same impedance, the length

8. Fujitsu, FHX35LG/LP Super Low Noise HEMT, July, 1999.

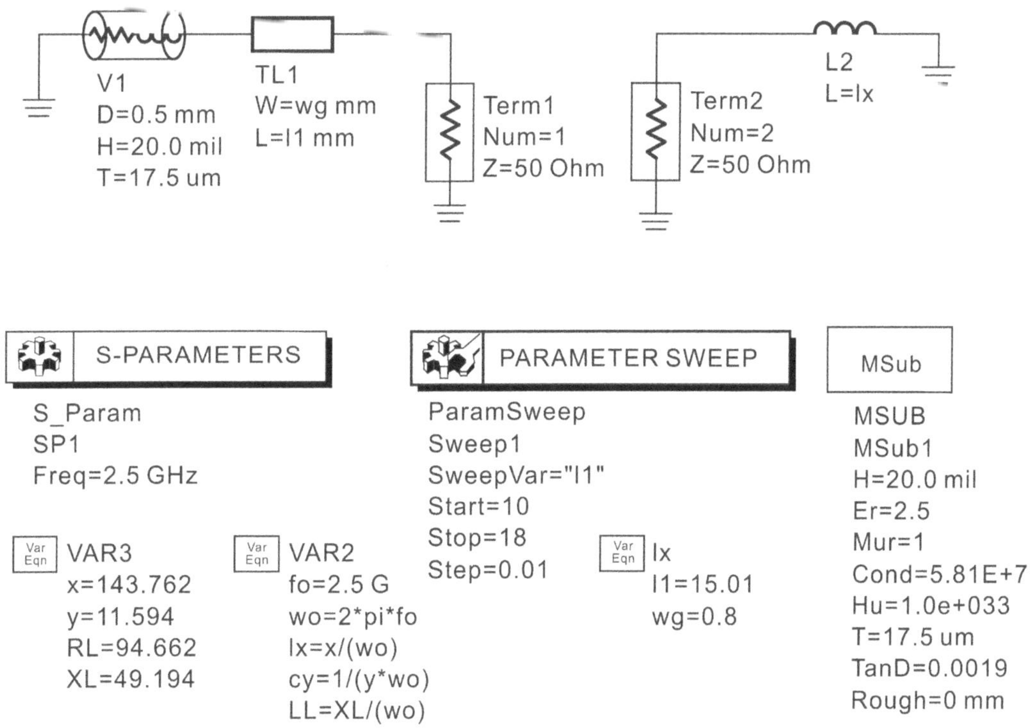

(a)

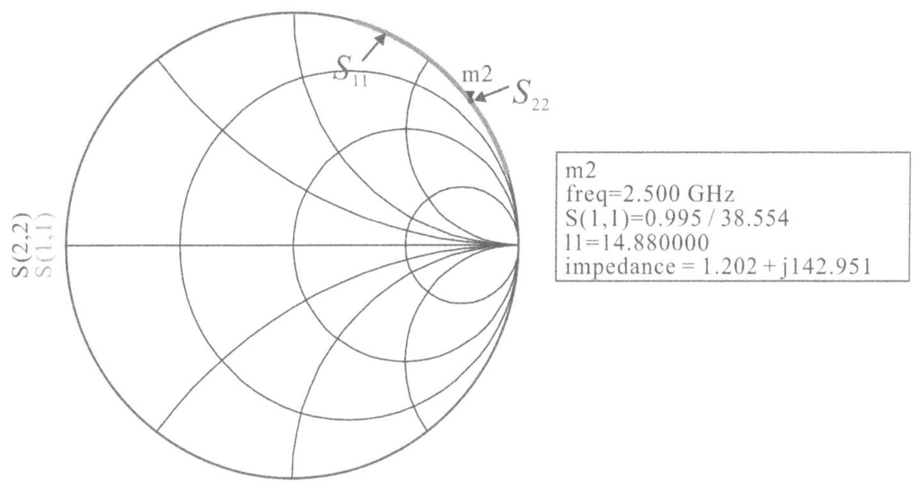

(b)

Figure 10.51 (a) The schematic for determining the microstrip length that gives the same impedance as the inductor and (b) simulation results. The length of about 14.88 mm gives the same impedance.

of **TL1**, l1 is varied to obtain the same inductance value as that of the inductor. Taking the gate lead width of the package in Figure 10.50 into consideration, the width of **TL1** is fixed at 0.8 mm. Figure 10.51(b) shows the result of the simulation. S_{11} represents the reflection coefficient with respect to the change of l1, and S_{22} is the reflection coefficient of the gate inductor **lx**. Thus, in Figure 10.51(b), the length l1 that gives the same value of reflection coefficient can be seen to be l1 = 14.88 mm.

Figure 10.52(a) is a microstrip circuit implementation of the capacitor **cy** that is connected to the source terminal. The selected FHX35LG device has two source terminals. Thus, the capacitor **cy** in Figure 10.52(a) is configured as the parallel combination of the two microstrip circuits shown in that figure. In addition, the microstrip width is set wider than the width of the source terminal whose value is set to 1.2 mm. A self-bias circuit must also be included, but only in one circuit of the parallel combination. The point where the self-bias is connected is set 3.0 mm away from the source terminal. Note that the impedance seen from port 1 is not affected by the self-bias circuit at the oscillation frequency of 2.5 GHz. The two open microstrip stubs have almost the same impedances at 2.5 GHz because the lengths of those stubs are set to have equal lengths, l2. The length l2 is adjusted to make the impedance of the parallel combination equal to the impedance of **cy**.

The RFC connected to the source terminal for the DC bias consists of a high-impedance microstrip with a length of about a quarter-wavelength and a radial stub. The width of the input terminal and angle of the radial stub are initially fixed at 0.8 mm and 70°, respectively. With the fixed width and angle, the length of the radial stub is set by separately simulating the radial stub's impedance to make the input impedance 0 at 2.5 GHz. The length is calculated to be 12.3 mm. The width of the high-impedance microstrip operating as the RFC is set to 0.2 mm, and the length is set such that its electrical length is 90° at 2.5 GHz. The value is the l1_90 in Figure 10.52(a), which when computed with **LineCalc** is determined to be approximately 22 mm. The RFC microstrip line is bent at 90° to make the DC biasing easy while maintaining the value of the length l1_90. The shape is shown in Figure 10.52(a). In addition, the resistor used in the DC bias is implemented with a chip resistor from the ADS library, and the 120-Ω value is selected because the determined value of 125 Ω is not available in the ADS library. The simulation is performed by varying the length l2 and its results are shown in Figure 10.52(b). From that figure, when l2 = 15.42 mm, the impedance of the parallel combination is the same as that of the capacitor **cy**.

In the case of the load circuit, the impedance seen from the drain terminal should be implemented to have the impedance $Z_L = 94.662 + j49.194\ \Omega$. This requires the synthesis of a matching network to transform the 50-Ω load into Z_L. In addition, the load circuit must be implemented to include the DC block and DC supply circuit shown in Figure 10.38. Conceptually, this circuit can be configured by inserting a shunt inductor, as shown in Figure 10.53(a), where the inductor can be formed by connecting a bypass capacitor at the end of the high-impedance transmission line as shown in that figure. Thus, the DC supply can be applied through the bypass capacitor and the design of a separate DC supply circuit is not required. The DC block capacitor is implemented with a 100-pF chip capacitor. As explained in Chapter 2, the chip capacitor is not a pure capacitor. However, in order to show the operation of the circuit in Figure 10.53(a), the impedance of the chip capacitor is approximated to 0. When the impedance of the 100-pF DC block chip capacitor is considered to be 0, the shunt inductor moves the 50-Ω load at the origin of point B, as shown in Figure 10.53(b). The impedance at position B is then moved into the point of Z_L through the clockwise rotation of the 50-Ω transmission line. The load circuit can then be implemented using the circuit shown in Figure 10.53(a).

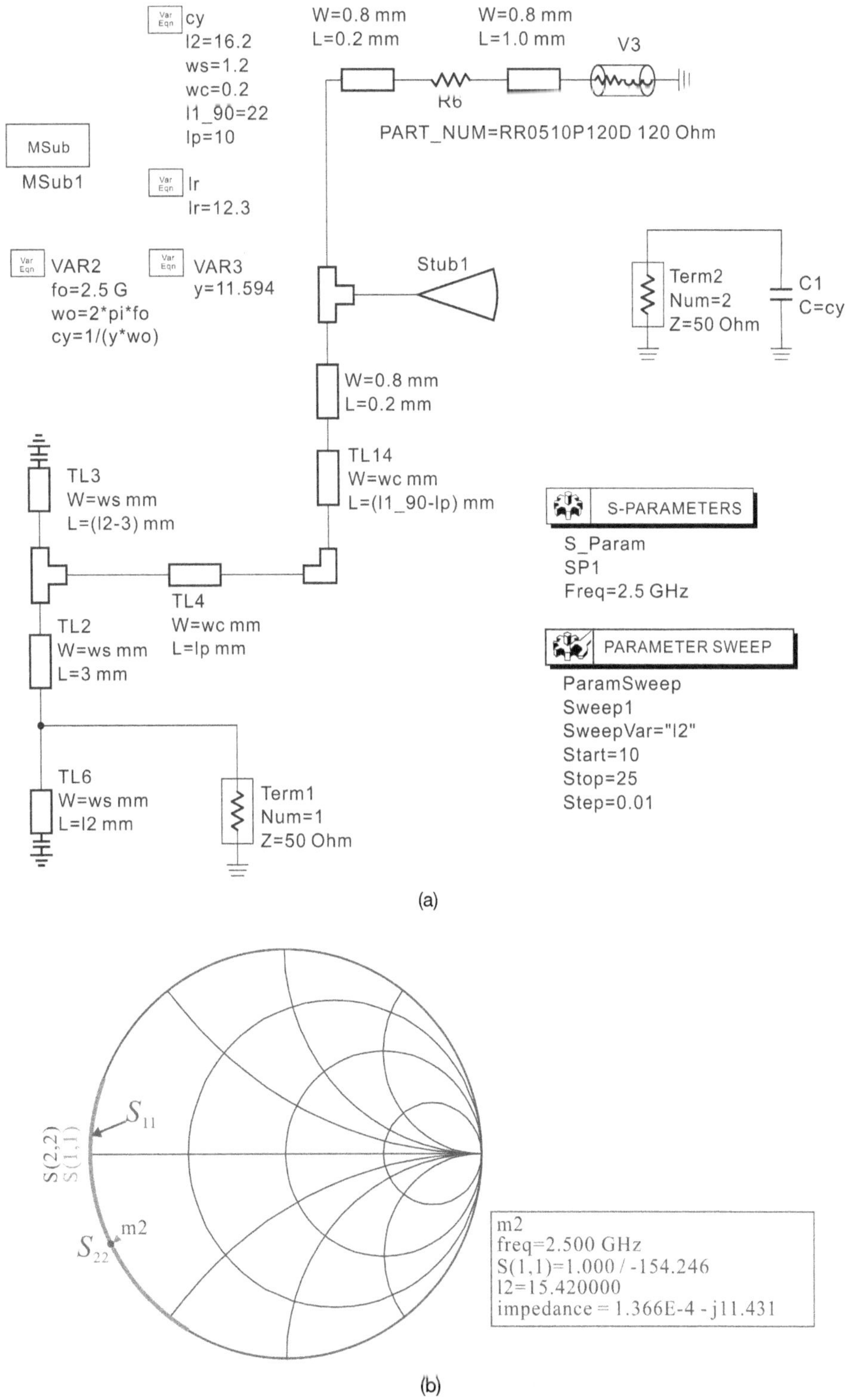

Figure 10.52 (a) Microstrip circuit implementation of the feedback capacitor and (b) simulation results. The length 15.42 mm gives the impedance of **cy**. In (a), **TL2**, **TL3**, and **TL6** are the two stubs to implement **cy**, while other elements are the RFC and bias circuit elements. The radial stub length **lr** = 12.3 mm is determined using the separate simulation.

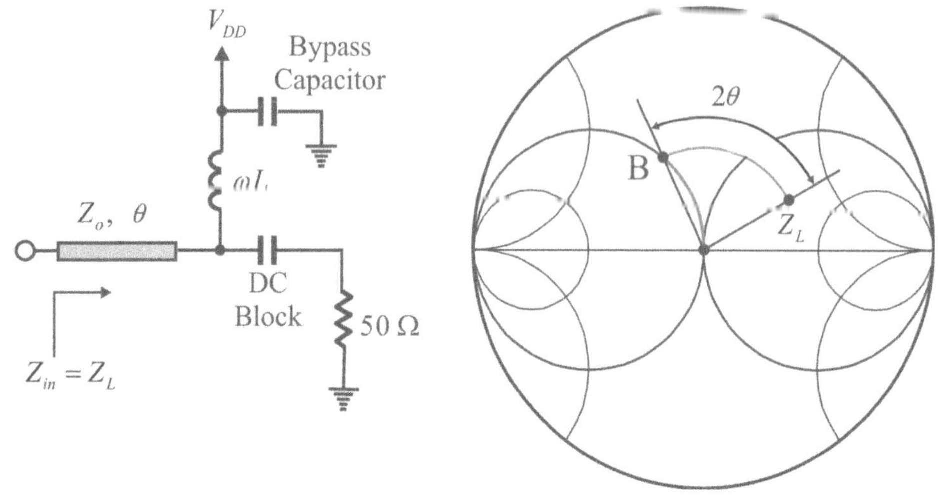

Figure 10.53 (a) Implementation of the load circuit and (b) matching in the Smith chart. Inductor L is used as a matching circuit element and it can also supply drain current at the same time. Z_L is the desired impedance and it can be obtained by moving 50 Ω to point B by ωL and rotating B to Z_L by the 50-Ω transmission line. In the matching, the DC block and bypass capacitors are assumed to be ideal.

Figure 10.54(a) shows the load circuit implemented with microstrip lines. The bypass capacitor for the DC power supply is implemented with a radial stub and a 100-pF chip capacitor. The radial stub has the dimensions used in the implementation of **cy**. In addition, a 100pF-chip capacitor is connected in parallel for the DC power supply. A microstrip with a width of 0.2 mm is used for the shunt inductor with a length of **ll3** that is varied to provide the appropriate inductor value. The microstrip line is bent to minimize possible unwanted coupling to the **cy** circuit. The width of the 50-Ω microstrip is approximately 1.4 mm and its length is set to **ll1** for optimization. A variable-length 50-Ω microstrip can also be inserted in front of the DC block chip capacitor. The length of this microstrip **ll2** is optimized together with **ll1** and **ll3**. Figure 10.54(b) shows the optimized impedance for the optimized dimensions shown in Figure 10.54(a). The impedance can be seen to be close to the desired design value of $Z_L = 94.662 + j49.194\ \Omega$.

A large-signal oscillator simulation is performed using the **lx** and **cy**, and the load circuit is implemented with the microstrip lines. The simulation shows an oscillation frequency of 2.439 GHz and an output power of 5.593 dBm. These values are slightly different from those obtained in the circuit simulation in Example 10.13. However, as mentioned previously, this can be due to the difference in harmonic impedances.

10.5.2.2 EM Simulation

The previously implemented microstrip oscillator circuit accurately predicts the behavior of the actual oscillator; however, more accurate values can be determined through EM simulation. Therefore, EM simulation can also further refine the values previously determined in circuit simulation. The procedure is the same as that described in the previous section. In the case of the gate inductor **lx**, the layout is generated from the previous

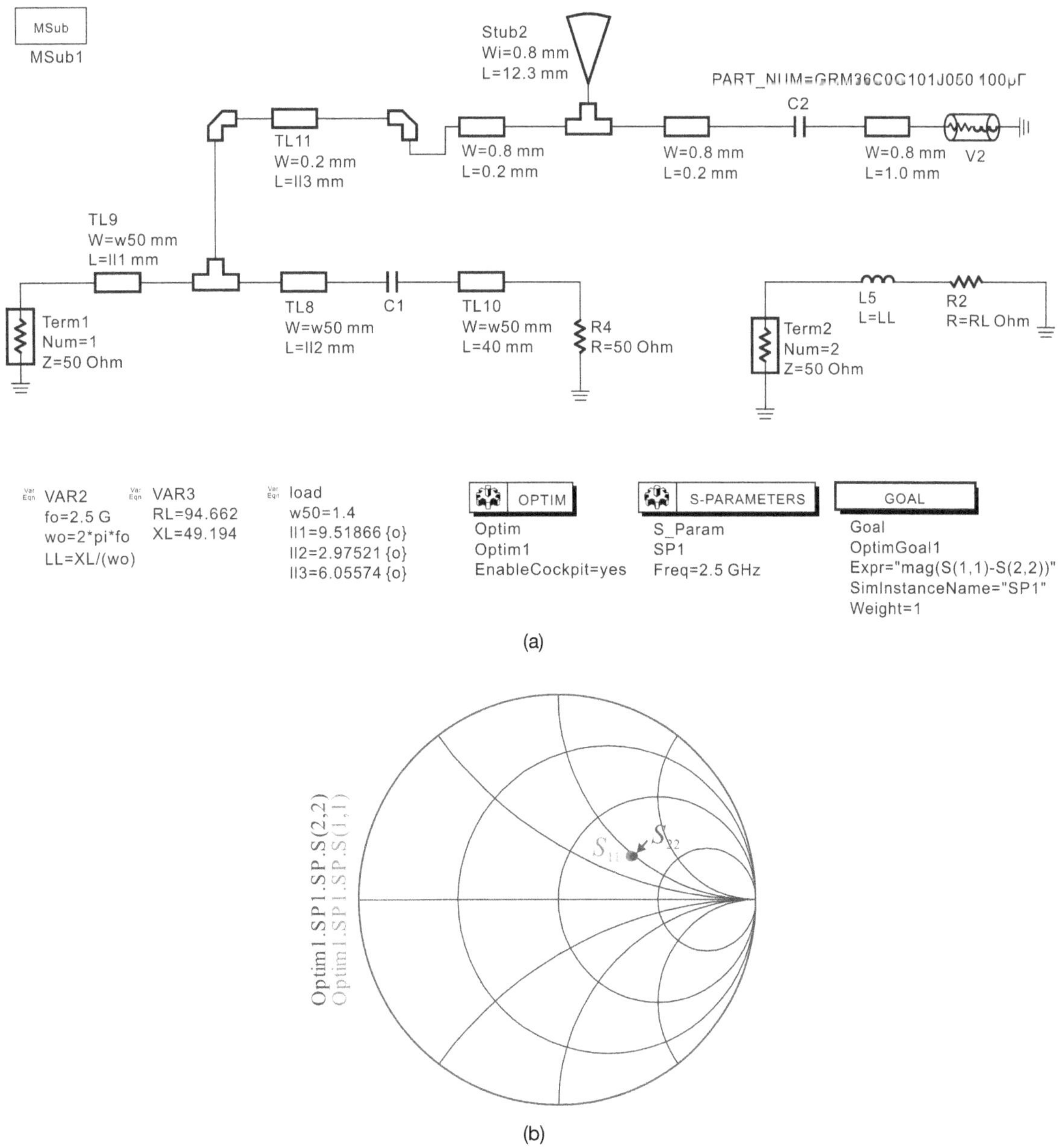

Figure 10.54 (a) Load configured with a microstrip circuit. The optimized values are in the variables of VAR named **load**. In (a), the line lengths **ll1**, **ll2**, and the stub length **ll3** are varied. The length **ll2** in the simulation is also set to vary taking into consideration the effect of the DC block capacitor **C1**. (b) The optimization results and S_{22} is the reference and S_{11} is optimized to match S_{22}.

microstrip circuit using the ADS autolayout utility and l1 in the generated layout is specified as a variable for a newly defined *layout component*. By entering the layout component in the schematic window, the simulation can be performed in the schematic window similar to that shown in Figure 10.51(a). The value of l1 changes slightly and is found to be l1 = 14.9 mm.

Internal ports in Momentum are used for the simulation of the **cy** circuit connected to the source terminal. These ports can cause changes in the ports' locations when the lengths of the transmission lines defined as variables change. There are several techniques for solving the port location changes. In this design, a fixed-length microstrip in a layer that has no function is used to connect the internal ports, thereby eliminating the changes in the locations of the ports due to the length changes. Also note that the accuracy of the ADS internal port is currently not well known. Using Sonnet for the EM simulation is recommended for greater accuracy. A length l2 = 15.5 mm is obtained through the EM simulation.

The lengths of the microstrip lines in the load circuit connected to the drain terminal are also determined in a similar way through optimization in the EM simulation. The computed values are obtained as ll1 = 9.0 mm, ll2 = 3.13 mm, and ll3 = 5.46 mm. The values obtained in the EM simulation can be seen to be significantly different from those obtained in the circuit simulation.

Figure 10.55 is the schematic of the oscillator whose dimensions are tuned through the separate EM simulations described above. The RFCs **DCFEED**1 and **DCFEED**2, and the DC block **DC_Block**1 in Figure 10.55 are used for computational efficiency. Without these RFCs or the DC block, the frequency of the EM simulation should be extended up to the DC, which significantly increases the computation time beyond what is acceptable and it also degrades the accuracy of the computation in the DC. However, it must be noted that these RFCs or the DC block inserted in the circuit have no effect on the computation at the DC. To this end, the frequency range of the layout components, such as **lx** and the load circuit, is set to 1–10 GHz. The simulation results are shown in Figure 10.56. Figure 10.56(a) shows the oscillation waveform obtained from the EM simulation, while the spectrum of the oscillation output is shown in Figure 10.56(b). The oscillation frequency and output power in Figure 10.56(b) are 2.461 GHz and 6.825 dBm, respectively. The values are close to those in the circuit design.

The oscillation frequency can be tuned to the design frequency of 2.5 GHz by varying **cy** or **lx**. When **lx** connected to the gate is varied, the oscillation frequency close to the desired oscillation frequency of 2.5 GHz can be obtained. Figure 10.57 shows the layout of the designed microstrip oscillator.

Using independent EM simulations, the layout of the microstrip oscillator shown here is determined by computing the variables of the layout components **lx**, **cy**, and the load circuit. This is done to simplify the calculation. Undesired coupling may exist between **lx**, **cy**, and the load circuit. The radial stubs of the load circuit and **cy** circuit in Figure 10.57 are quite close, which requires more accurate calculation. In addition, it can be seen that the bias circuit of the **cy** circuit and **lx** circuit are close and this can also cause unwanted coupling. To account for this coupling, the gate, drain, and source terminals of the FET must be set as internal ports, and the EM simulation of the external circuits **lx** and **cy**, and the load circuit together should be performed for those defined internal ports. However, since this simulation is not a new process, it can be done by following the previous EM simulation procedure.

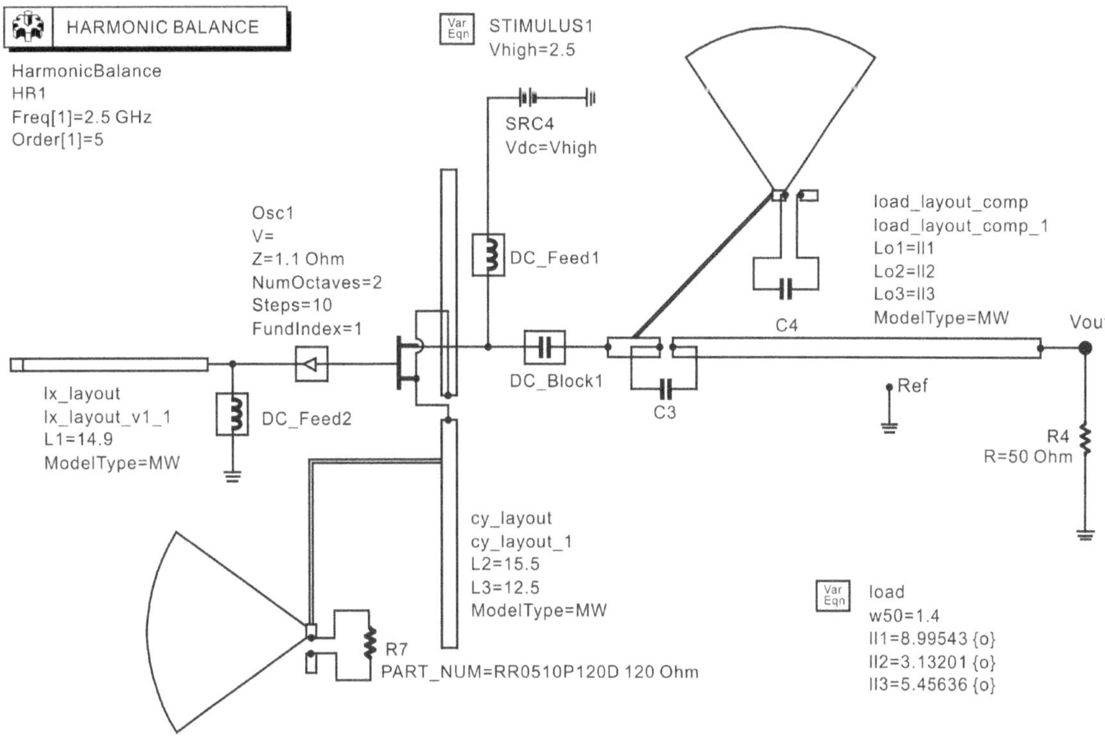

Figure 10.55 Large-signal simulation of a microstrip oscillator circuit using EM simulation. The variables in the layout components are separately determined through the EM simulation and their values are in the schematic. The circuit components such as **DC_Feed**s and **DC_block**s are inserted for fast simulation.

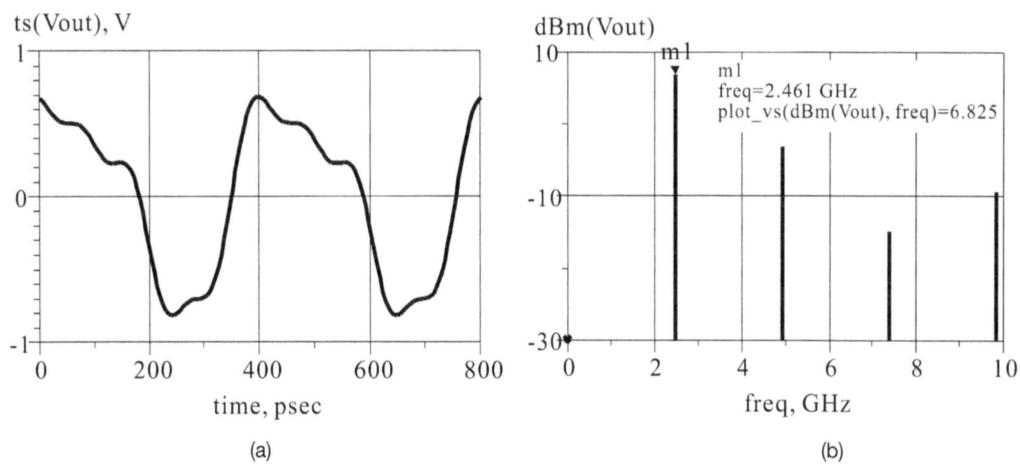

Figure 10.56 (a) Waveform and (b) spectrum from the EM simulation. Oscillation frequency and power are about 2.461 GHz and 6.825 dBm, respectively.

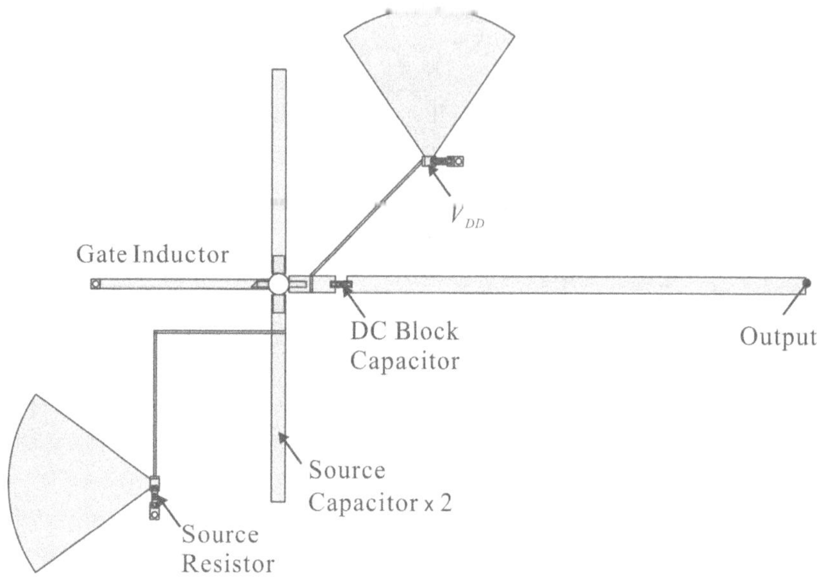

Figure 10.57 Microstrip oscillator circuit layout. V_{DD} represent DC bias supply point. In the layout the coupling between the lines are not considered.

10.6 DIELECTRIC RESONATORS

10.6.1 Operation of Dielectric Resonator (DR)

A dielectric resonator (DR) consists of a dielectric material with a cylindrical form, as shown in Figure 10.58, that has a typical relative permittivity of $\varepsilon_r = 30\text{–}100$ and acts as a resonator due to the high dielectric constant. The dielectric material used for a DR generally has a low loss and the dielectric resonator usually has a Q higher than 10^3. The size of the dielectric resonator generally expands or shrinks according to temperature change. The size of the dielectric resonator is directly related to its resonant frequency and the temperature change causes the resonant frequency change. The dielectric constant also changes with temperature, which contributes to the change in the resonant frequency as well. Defining the thermal expansion coefficient and temperature coefficient of the dielectric resonator's dielectric constant as α_L and τ_ε, respectively, the temperature drift in the resonant frequency, τ_f, is expressed in Equation (10.53).

Figure 10.58 Structure of a dielectric resonator

$$\tau_f \approx -\frac{1}{2}\tau_\varepsilon - \alpha_L \tag{10.53}$$

The temperature coefficient of the dielectric constant and the thermal expansion coefficient are dependent on the dielectric material used in the dielectric resonator, and by setting the temperature coefficient of the dielectric constant to compensate for the thermal expansion coefficient, the temperature drift of the dielectric resonator's resonant frequency can be minimized. As a result of recent studies of dielectric resonator materials, most commercially available dielectric resonators available today are designed to make the resonant frequency's temperature drift almost zero.

Similar to other resonators, several resonant modes are possible in a dielectric resonator and the fundamental mode with the lowest resonant frequency has the electric field distribution shown in Figure 10.59.

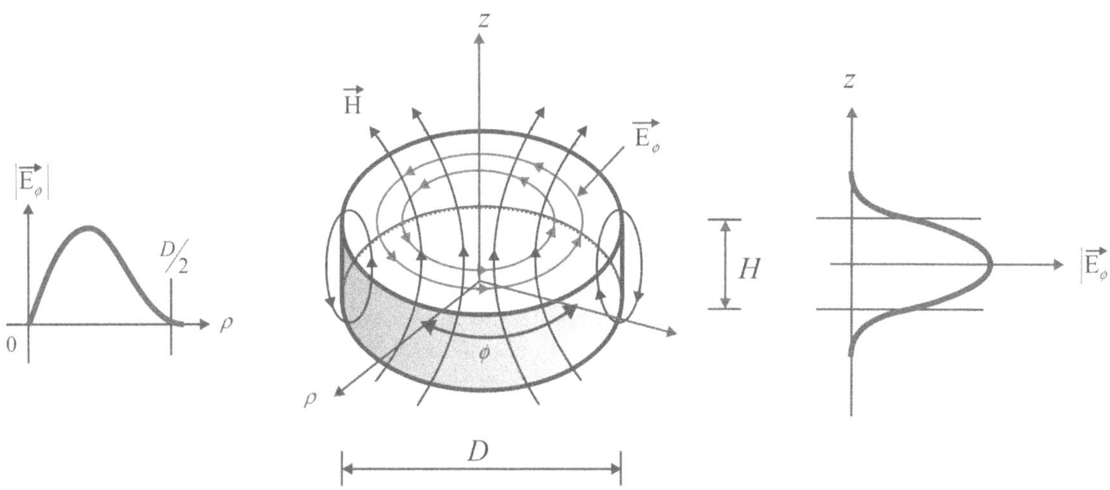

Figure 10.59 The electromagnetic field patterns in a dielectric resonator. The left plot shows the electric field intensity along the ρ-axis. The right plot shows the electric field intensity along the z-axis.

For the cylindrical $\rho\varphi z$ coordinate system shown in Figure 10.59, the circumferential electric field appears along the φ-direction and is denoted as E_φ. The distribution of E_φ for the z-axis shows a peak value at $z = 0$, and it rapidly decays, moving away from $z = 0$, as shown in the figure. The distribution of E_φ for ρ-axis is 0 at $\rho = 0$ and it shows a peak value inside the dielectric resonator. The field E_φ also rapidly decays outside of the dielectric resonator. Thus, when the z direction is taken as the propagation direction, the field is said to be in a transverse electric (TE) mode because the electric field exists only in the transverse plane of the propagation direction. The distribution of the electric field E_φ can be represented by a mode number that represents the standing waveform in each axis. The mode number in the ρ-axis is close to 1 and in the φ-axis it is 0 since there is no standing wave. The mode number in the z-axis is represented by δ since a complete standing wave is not established. Thus, the resonant mode in Figure 10.59 is called the $TE_{10\delta}$ mode.

On the other hand, the shape of the magnetic field primarily occurs as it passes through the dielectric resonator, as shown in Figure 10.59, which is generated by the time-varying electric field E_φ according to Maxwell's equations. Note that the electric field inside the dielectric resonator is small and can be approximated as 0 as $\varepsilon_r \to \infty$ according to

Maxwell's equations; otherwise, an infinite magnetic field occurs, which is impractical. Consider the case where the external magnetic field penetrates the dielectric resonator, as shown in Figure 10.59. The electric field E_φ is then generated according to Faraday's law of electromagnetism. The induced electric field E_φ again generates a magnetic field that occurs in a direction that cancels the incident magnetic field. Since the dielectric resonator is not a magnetic material, its relative permeability can be considered to be 1, and the finite electric field E_φ is induced. Then, the magnetic field generated by the finite induced electric field E_φ becomes ∞ due to $\varepsilon_r \rightarrow \infty$ as the generated magnetic field is proportional to the relative permittivity ε_r, which is unacceptable. As a result, the magnetic field inside the dielectric resonator is therefore close to 0. Otherwise, a small incident magnetic field can cause an infinite magnetic field in the opposite direction. This phenomenon of the electromagnetic field in the dielectric resonator is similar to that of an eddy current electromagnetically generated in a conductor. Therefore, the magnetic field with z-direction at the surface of the dielectric resonator can be approximated as zero.

The dielectric resonator with the properties previously described is usually coupled to the microstrip shown in Figure 10.60. The microstrip's magnetic field is incident to the dielectric resonator in the z-direction as shown in that figure and the magnetic field that penetrates the dielectric resonator becomes almost 0 at the resonant frequency of the dielectric resonator. In order to make the magnetic-field lines penetrate the dielectric resonator vertically, the resonator must be positioned higher than the substrate on which the microstrip is placed. To achieve this effectively, a separate spacer is inserted beneath the resonator, as shown in Figure 10.60, which makes the possible magnetic-field lines penetrate the resonator vertically. The magnetic field corresponds to the microstrip current and since the magnetic field at the resonant frequency of the dielectric resonator is 0, the microstrip current will be zero. As a result, the microstrip will be open circuited at the plane where the microstrip is coupled with the dielectric resonator.

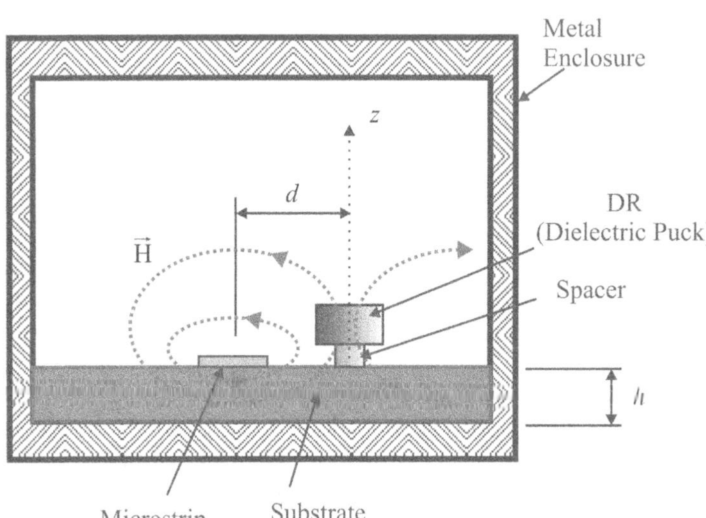

Figure 10.60 Coupling a dielectric resonator to a microstrip. The dielectric resonator is coupled by the magnetic field of the microstrip.

The circuit shown in Figure 10.61(a) represents the dielectric resonator coupled to a microstrip and its equivalent circuit can be represented as shown in Figure 10.61(b). When the distance d in Figure 10.61(a) is small, the coupling between the dielectric resonator and the microstrip becomes tight and the magnetic field of the microstrip is significantly affected by the resonator. On the other hand, when the distance is large, the coupling between the dielectric resonator and the microstrip becomes loose, and the effect of the resonator almost vanishes. Note that the length l in Figure 10.61(a) represents the plane where the microstrip is most tightly coupled to the dielectric resonator and the magnetic field is considered to be 0. Since the magnetic field is 0, the current is also 0 and the microstrip line is thus open circuited at the length l. The circuit in Figure 10.61(a) is represented by the equivalent circuit shown in Figure 10.61(b). Here, the transformer n represents the degree of coupling between the dielectric resonator and the microstrip, which is a function of d and the parallel resonant circuit L_r–C_r–R_r represents the dielectric resonator. The circuit in Figure 10.61(b) can be transformed into the circuit shown in Figure 10.61(c). The impedance seen from the transformer port connected to the microstrip line is a parallel resonant circuit and the circuit in Figure 10.61(c) can then be obtained. It should be noted that the resonance frequency of the circuit in Figure 10.61(c) is the same as that in Figure 10.61(b). However, the values of R, L, and C differ from those of R_r, L_r, and C_r. The value of resistor R reflects the degree of coupling between the dielectric resonator and the microstrip, which is a function of d.

Setting $l = 0$ in the circuit of Figure 10.61(c), the locus of S_{11} with frequency can be plotted as shown in Figure 10.62(a) with d as a parameter. The dielectric resonator has a

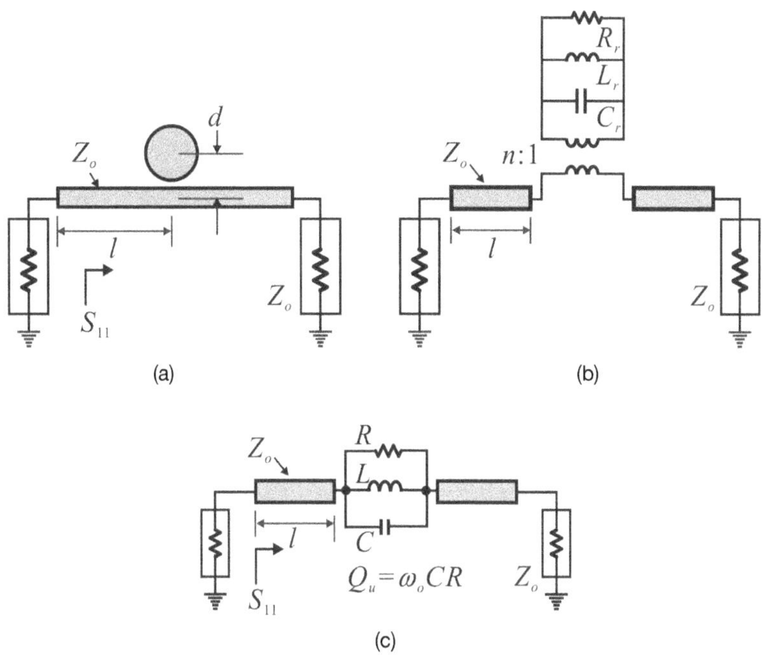

Figure 10.61 (a) Dielectric resonator circuit, (b) equivalent circuit, and (c) simplified equivalent circuit

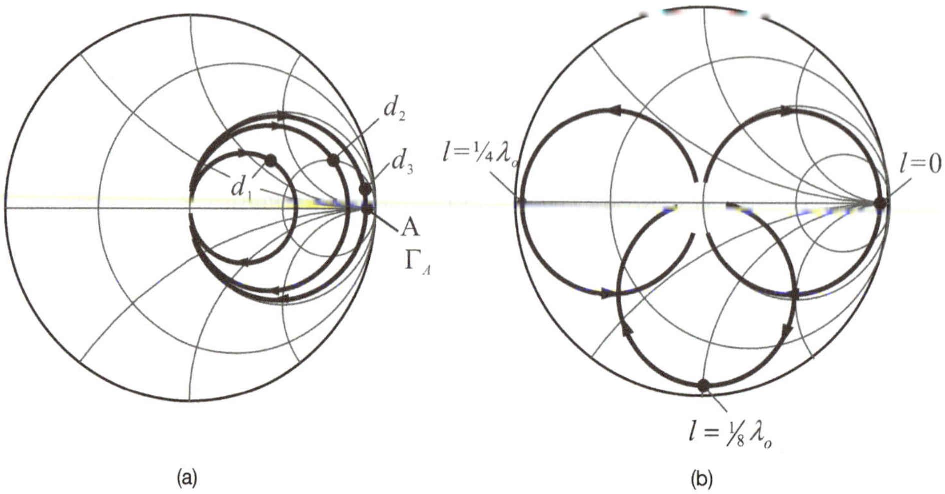

Figure 10.62 (a) Frequency response of S_{11} ($l = 0$) with respect to the change of d ($d_3 < d_2 < d_1$) and (b) the tightly coupled S_{11} (d_3) with respect to the change of l

negligible effect on the microstrip line for frequencies outside the resonant frequency; consequently, the impedance seen from the port is close to $Z_o = 50\ \Omega$ and appears at the origin of the Smith chart. However, the impedance seen from the input port at the resonant frequency becomes $Z_o + R$ and this is represented by point A on the Smith chart in Figure 10.62(a). In addition, this point depends on the coupling between the microstrip and the dielectric resonator. When the coupling is very tight, $R \to \infty$, and the impedance appears at a point ∞ of the Smith chart. Therefore, when the coupling is very tight, the impedance seen from the port at the resonant frequency is close to infinity.

For the tightly coupled dielectric resonator (d_3), Figure 10.62(b) shows the change of the S_{11} locus with the length l as a parameter. The locus rotates in a clockwise direction as l increases. Thus, denoting the electrical length at the resonant frequency as $\theta(l)$, the peak reflection coefficient Γ_{in} (l) of S_{11} in Figure 10.62(a) is obtained by rotating Γ_A by $2\theta(l)$, as expressed in Equation (10.54).

$$\Gamma_{in}(l) = \Gamma_A e^{-2j\theta} \tag{10.54}$$

Note that any reactance jX around the unit circle of the Smith chart can be implemented by varying the length l. In addition, the implemented jX that uses the dielectric resonator only appears near the resonant frequency and rapidly approaches Z_o as the frequency slightly away from the resonant frequency. The narrow-band implementation of jX is important in oscillator design. To emphasize its importance, the circuit is redrawn in Figure 10.63(a) where the dielectric resonator is placed at a distance θ away from the port and the S_{11} locus is shown in Figure 10.63(b). In the circuit configuration of Figure 10.63(a), the impedance seen from the port behaves as the open-circuited transmission line with an electrical length of θ near the resonant frequency, as shown in Figure 10.63(b), and the impedance becomes Z_0 at other frequencies. Thus, the oscillation condition can be satisfied only at the resonant frequency. At other frequencies, the oscillator circuit is generally stabilized because the impedance seen from the port is close to Z_0 and the oscillation condition is not satisfied except at the resonant frequency. As a result, the undesired oscillation condition at other frequencies can be easily avoided.

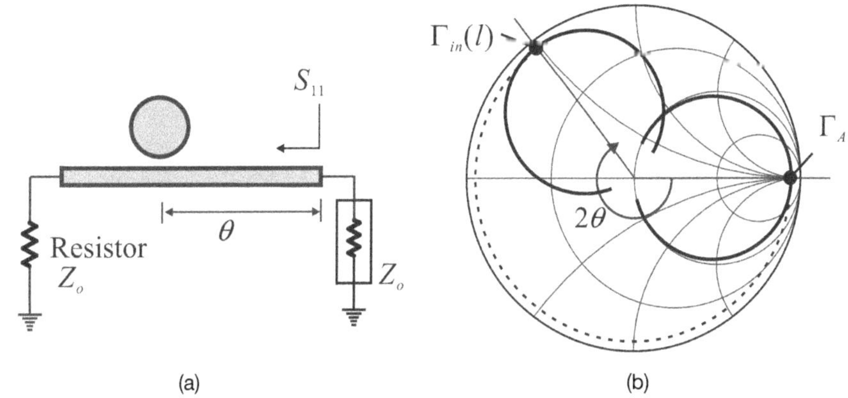

Figure 10.63 (a) A circuit of the DR coupled to microstrip and (b) its S_{11} locus apart from θ. By varying θ, all the reactance in the Smith chart can be implemented.

10.6.2 Extraction of the Equivalent Circuit of a DR Coupled to a Microstrip

In this section, we look at the method for extracting the values of the equivalent circuit for a dielectric resonator coupled to a microstrip. Denoting the impedance of the parallel resonant circuit in the circuit of Figure 10.61(c) as Z_p, S_{11} is expressed in Equation (10.55),

$$S_{11} = \frac{Z_p + Z_o - Z_o}{Z_p + Z_o + Z_o} e^{-2j\theta} = \frac{Z_p}{Z_p + 2Z_o} e^{-2j\theta} \tag{10.55}$$

where

$$Z_p = \frac{1}{\frac{1}{R} + j\omega C + \frac{1}{j\omega L}} \cong \frac{R}{1 + 2jQ_u \frac{\Delta\omega}{\omega_o}} \tag{10.56a}$$

$$f_o = \frac{1}{2\pi\sqrt{LC}} = \frac{\omega_o}{2\pi} \tag{10.56b}$$

$$Q_u = \omega_o CR \tag{10.56c}$$

In Equations (10.56a)–(10.56c), Q_u is the unloaded Q of the dielectric resonator. Substituting Z_p into Equation (10.55) and rewriting it, we obtain Equations (10.57a)–(10.57c).

$$S_{11} = \frac{\beta}{1 + \beta + 2jQ_u \frac{\Delta\omega}{\omega_o}} e^{-2j\theta} = \frac{\frac{\beta}{1+\beta}}{1 + 2jQ_L \frac{\Delta\omega}{\omega_o}} e^{-2j\theta} \tag{10.57a}$$

$$\beta = \frac{R}{2Z_o} \tag{10.57b}$$

$$Q_L = \frac{Q_u}{1 + \beta} \tag{10.57c}$$

Figure 10.64 shows the plot of $|S_{11}|$ with respect to frequency.

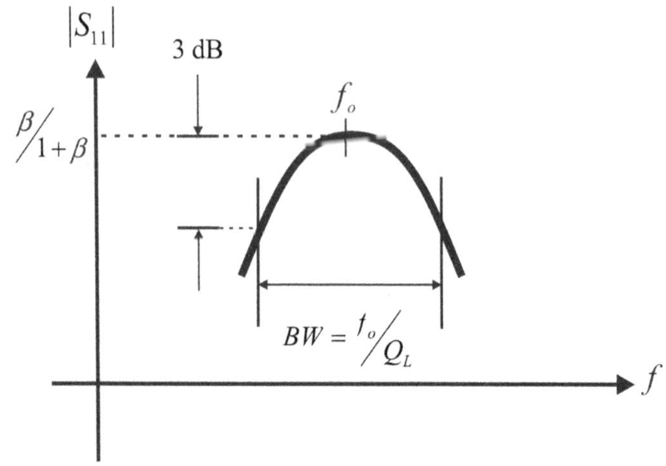

Figure 10.64 Sketch of the dielectric resonator's $|S_{11}|$ coupled to the microstrip for frequency

From Equation (10.57a), the peak value of $|S_{11}|$ occurs at the resonant frequency f_o. Since the peak value at the resonant frequency is $\beta/(1 + \beta)$, β can be obtained from the measured peak value. As a result, the value of R can be obtained using Equation (10.57b). Furthermore, from the measured 3-dB bandwidth in Figure 10.64, the loaded Q, Q_L, can be found as expressed in Equation (10.58).

$$Q_L = \frac{f_o}{BW} \tag{10.58}$$

Thus, Q_u, the unloaded Q, can be determined by substituting β and Q_L into Equation (10.57c). From the computed Q_u, the value of C can be determined using Equation (10.56c). Finally, the value of L can be obtained from the resonant frequency f_o and C using Equation (10.56b). Therefore, all the component values of the equivalent circuit can be determined using the frequency response of $|S_{11}|$.

Example 10.14

Figure 10E.29 shows the frequency response of $|S_{11}|$ for a dielectric resonator coupled to a microstrip. In that figure, $f_o = 10$ GHz, $|S_{11}|_{max} = -7.06$ dB, and $BW = 90$ MHz. Calculate the values of the microstrip circuit's equivalent circuit that is coupled to the dielectric resonator in Figure 10.61(c).

Solution

From $|S_{11}|_{max} = -7.06$ dB, β can be obtained as

$$\frac{\beta}{1+\beta} = |S_{11}|_{max} = 10^{-7.06/20} = 0.444$$

$$\beta = \frac{0.444}{1-0.444} = 0.799 = \frac{R}{2Z_o}$$

From β, the value of R can be obtained as

$$R = 80\Omega$$

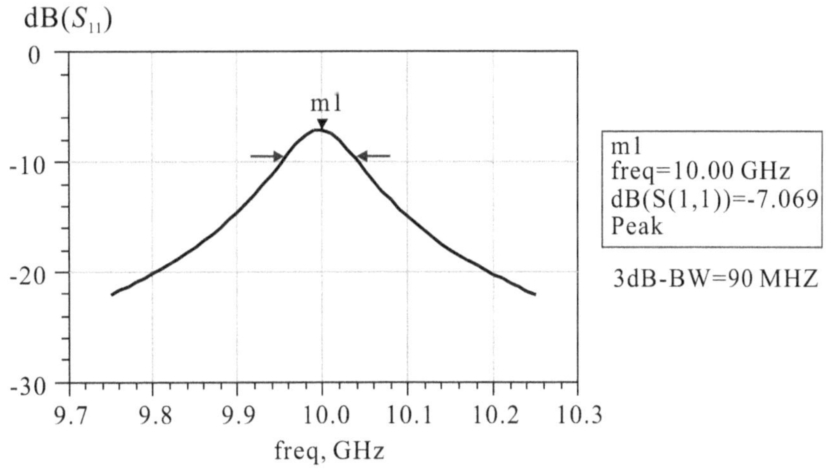

Figure 10E.29 Frequency response example of the dielectric resonator coupled to the microstrip circuit

From $BW = 90$ MHz and resonant frequency $f_o = 10$ GHz, Q_L and Q_u can be obtained as

$$Q_L = \frac{f_o}{BW} = 111 = \frac{Q_u}{1+\beta}$$

$$Q_u = 199.8 = \omega_o CR$$

Thus,

$$C = \frac{199.8}{2\pi f_o R} = 39.73 \text{ pF}$$

The value of L is also obtained from ω_0 as

$$L = \frac{1}{\omega_o^2 C} = \frac{1}{(2\pi f_o)^2 C} = 6.38 \text{ pH}$$

Therefore, the values of the components in the equivalent circuit that yield the given frequency response are as follows:

$$R = 80\Omega,\ C = 39.73 \text{ pF},\ L = 6.38 \text{ pH}$$

In Example 10.14, the equivalent circuit of the dielectric resonator coupled to microstrip is extracted using its S-parameters. The S-parameters can be obtained by measurement or from the 3D-simulation of the dielectric resonator coupled to a microstrip. The programs developed by vendors of dielectric resonators can also be used to compute the values of the equivalent circuit in Example 10.14.[9]

9. One of these programs can be found on the website for Trans-Tech, www.trans-techinc.com.

10.7 DIELECTRIC RESONATOR OSCILLATORS (DRO)

We saw in the preceding sections that a microstrip oscillator can be designed by replacing the capacitors and inductors of the basic series feedback oscillator by open- or short-circuited microstrip lines. There are several ways of designing a DRO (dielectric resonator oscillator) but the simplest method is to replace one of the capacitors and one of the inductors with a dielectric resonator coupled to a microstrip circuit as previously described. Another method is to use the dielectric resonator as a feedback resonator circuit. This section will discuss these two design methods.

In addition, we will present an oscillator design that meets the given phase noise design objective. The phase noise generation in an oscillator is basically a nonlinear phenomenon. The near carrier phase noise is primarily generated as a result of a nonlinear conversion of the low-frequency noise sources in an oscillator's components. This has been confirmed by significant recent research; see reference 3 at the end of this chapter for more information. However, the design of a low-phase noise oscillator to meet a given phase noise specification is still theoretical despite current research and the emergence of modern CAD simulators. In addition, the low-frequency noise models of active devices for phase noise simulation are not available for most CAD simulators and they should be prepared by designers. Rather than the design that uses phase noise simulation, we will present an experimental design method that meets the phase noise design's objective based on Leeson's formula. The measured phase noise of the prototype oscillator makes it possible to estimate the Q of the resonator which meets the phase noise design objective using Leeson's formula. In this section, we will show that the desired phase noise can be achieved through the reconfiguration of the resonator to yield the estimated Q.

10.7.1 DRO Design Based on Replacement

In Example 10.13, we showed how to compute the feedback circuit element values x, y, and Z_L. The circuit at which the dielectric resonator is coupled to a microstrip, shown in Figure 10.63, can be used to replace x or y in the feedback circuit elements, which enables us to build the DRO. This type of DRO is called a *reflection-type* DRO.

Example 10.15

Using the feedback circuit element values x, y, and Z_L obtained in Example 10.13, design a 2.5-GHz DRO. The dielectric resonator coupled to a microstrip circuit (DR circuit) has $Q_u = 3000$ and $Q_L = 40$.

Solution

In Example 10.13, the values of x and y can be implemented by an inductor and a capacitor, respectively. Note that the gate should be grounded at the DC. When x is implemented using the DR circuit, the gate terminal is connected to the ground through the 50-Ω resistor. Since the gate current is 0, the gate voltage is still 0 and the same self-bias circuit can be used without change. In contrast, when y is implemented by the DR circuit, a DC block capacitor is necessary so as not to disturb the self-bias circuit set by the source resistor R_S. In addition, since y should be implemented using two parallel connections, it is difficult to implement y using the DR circuit. Thus, the replacement of x by the DR circuit is the preferred method.

Figure 10E.30 is the simulation schematic for determining the length θ in Figure 10.63.

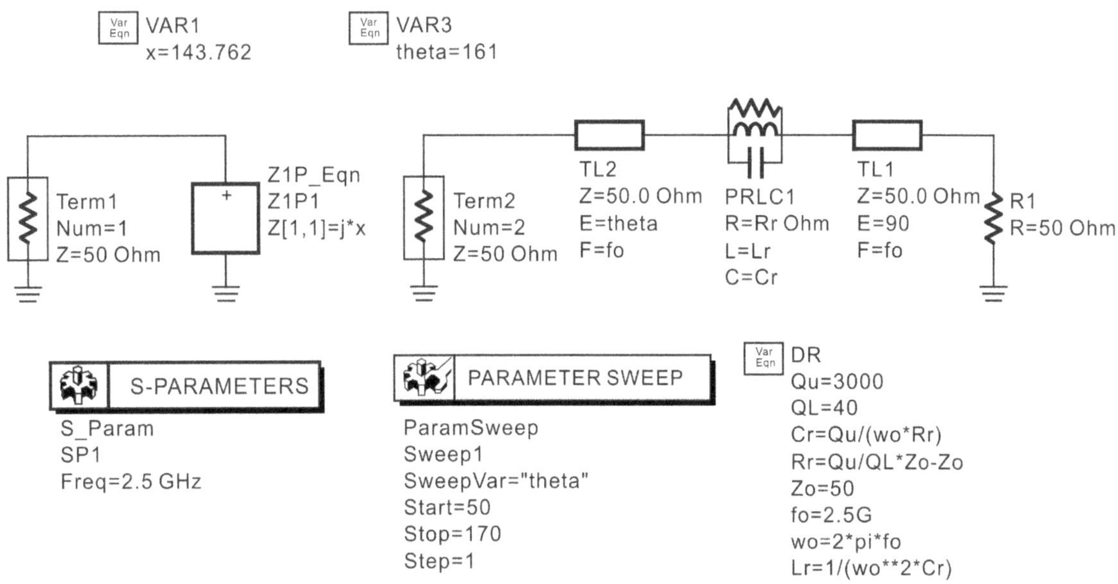

Figure 10E.30 Implementation of x using a dielectric resonator coupled to a microstrip circuit

First, using the given Q_u and Q_L, the values of R, L, and C can be computed using

$$R = \left(\frac{Q_u}{Q_L} - 1\right)Z_o, \quad C = \frac{Q_u}{\omega_o R}, \quad \text{and } L = \frac{1}{\omega_o^2 C}$$

These equations are included in a **VAR** named **DR**. The x is implemented using the equation component in ADS shown at the left of Figure 10E.30. The length θ can be found by sweeping **theta**, which gives the same value of x. Figure 10E.31 shows S_{11} and S_{22}. From that figure, the value of theta that gives the value of x is 161°. Now, replacing **lx** by the DR circuit in Figure 10E.30, the simulation can be performed. The simulated voltage waveform and spectral power are shown in Figures 10E.32(a)

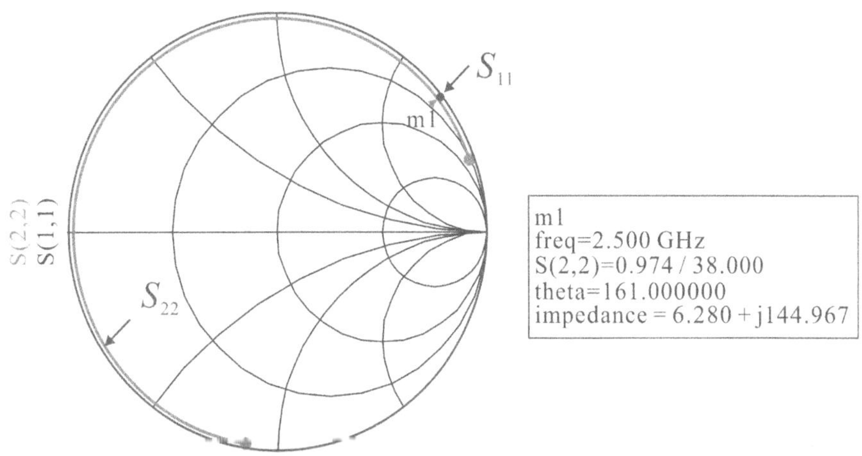

Figure 10E.31 Simulated S_{11} and S_{22}

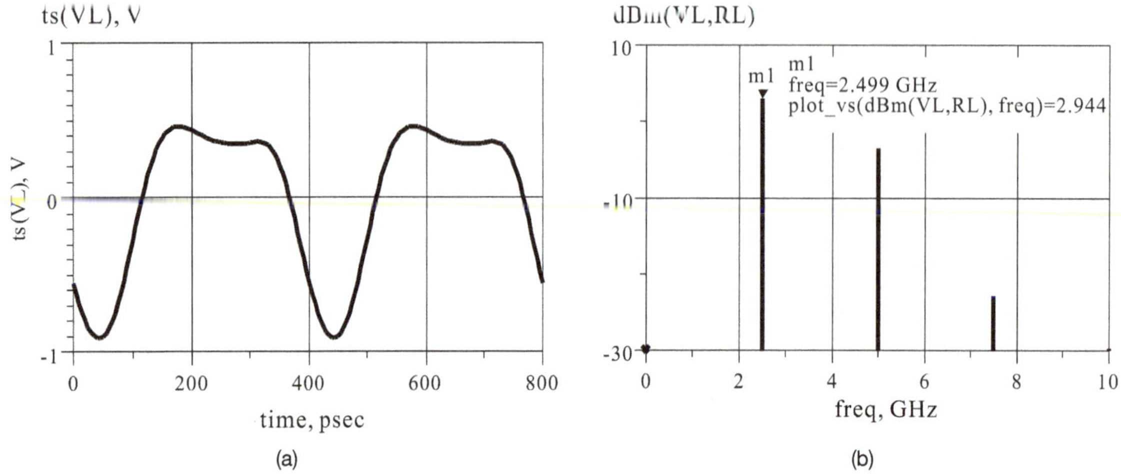

Figure 10E.32 (a) Simulated waveform and (b) spectral power of the DRO that is implemented by replacing x with the DR-coupled microstrip circuit

and (b), respectively. The oscillation frequency and power are 2.499 GHz and about 2.9 dBm, respectively. Although the oscillation close to 2.5 GHz is achieved, the oscillation power is significantly reduced from 6.3 dBm.

The reason for the reduced oscillation power is due to the loss of the DR circuit. This circuit cannot be considered as pure reactance but one that has loss. The loss can be found from the S-parameter shown in Figure 10E.31. The real part of the DR circuit is 6.28 Ω.

The disadvantage of this DRO design is that it may not take full advantage of the high Q of the dielectric resonator. Note that the loss of the dielectric resonator coupled to the microstrip circuit is small for a large value of β according to Equation (10.57a). This large value of β lowers the value of Q_L in Equation (10.57c). Consequently, the phase noise performance of the DRO that is related to Q_L is diminished. Thus, the design that takes advantage of the high Q of a dielectric resonator requires that the loss of the DR circuit be taken into account.

10.7.2 Dielectric Resonator Oscillator Design Using Feedback

10.7.2.1 Design Theory

An alternate way to design a DRO is to use the closed-loop configuration shown in Figure 10.65.[10] The dielectric resonator is used in the resonator shown in that figure. Here, the amplifier provides a sufficient loop gain, the oscillation frequency is adjusted using the phase shifter, and the output power is obtained by coupling a part of the oscillation power in the loop.

---

10. B. I. Son, H. C. Jeong, and K. W. Yeom, "Design of a Low Phase Noise Voltage Tuned DRO Based on Improved Dielectric Resonator Coupling Structure," *Asia-Pacific Microwave Conference Proceedings*, Kaohsiung, Taiwan (December 4–7, 2012): 1121–1123.

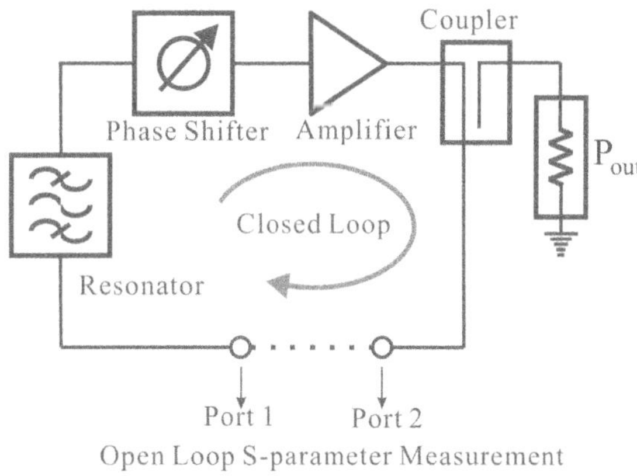

Figure 10.65 Closed-loop dielectric resonator oscillator circuit

The open-loop gain $L(\omega)$ given by Equation (10.18) can be obtained from the open-loop S-parameters of the cascade chains of the resonator, the phase shifter, and the amplifier. Usually, each component in Figure 10.65 is designed to have a small reflection in order to avoid a mismatch between the components. Thus, S_{11} and S_{22} of the open-loop S-parameters can be assumed to be small. In addition, the reverse gain S_{12} is small due to the amplifier's properties. Assuming S_{11}, S_{22}, and S_{12} are small, Equation (10.59) can be derived.

$$L(\omega) \cong S_{21} \tag{10.59}$$

From Equation (10.59), the oscillation condition can be rewritten as Equations (10.60a) and (10.60b).

$$\text{dB}(S_{21}) \geq 0 \tag{10.60a}$$

$$\text{phase}(S_{21}) = 0 \tag{10.60b}$$

To design an oscillator based on the open-loop gain method, the S-parameters for the open loop are measured. Then, when Equations (10.60a) and (10.60b) are satisfied at a desired oscillation frequency, the oscillator can simply be formed by closing the open loop.

From a design point of view, the phase shifter, amplifier, and resonator are independently designed in advance and the designed components are connected in cascade to form the oscillator. Denoting the S-parameters of the phase shifter, amplifier, and resonator as S_{φ}, S_A, and S_R, respectively, the open-loop gain S_{21} can be expressed as Equation (10.61).

$$S_{21} = S_{R,21} S_{\phi,21} S_{A,21} \tag{10.61}$$

Therefore, in order to satisfy the oscillation conditions given by Equation (10.60), $|S_{21}|$ has to be set greater than 1 and the phase of S_{21} should satisfy Equation (10.62).

$$\phi = \phi\big|_{resonator} + \phi\big|_{amplifier} + \phi\big|_{shifter} = 0 \tag{10.62}$$

Typically, the sum of the phases does not satisfy Equation (10.62). In this case, the oscillator can be designed by adding a 50-Ω transmission line with the appropriate electrical length to satisfy Equation (10.62). In addition, the group delay of the open-loop gain can be obtained by differentiating Equation (10.62) with respect to frequency, and the group delay of the open-loop gain becomes the sum of each component's group delays. However, as the group delay of the resonator is dominant, the group delay is determined by that of the resonator. Thus, the resonator must be designed to have the group delay that satisfies the design goal of the oscillator. This method is useful and efficient not only for theoretical design but also for the experimental design of an oscillator.

The oscillation frequency can be tuned using the phase shifter. The oscillation frequency occurs where the phase of S_{21} is 0 and the zero of the phase S_{21} changes by the phase shifter. Typically, the amplifier and phase shifter in Figure 10.65 have broadband characteristics and only the resonator has narrowband characteristics. Thus, the frequency response of the open-loop gain appears similar to that of the resonator, as shown in Figure 10.66. Given that a phase shifter has a maximum phase shift of $\pm\Delta\theta$, the phase shift of $\pm\Delta\theta$ is found to correspond to frequencies f_1 and f_2 from Figure 10.66. Thus, an electrical frequency tuning range of $\Delta f = f_2 - f_1$ is obtained by tuning the phase shifter. The slope of the phase of the open-loop gain at the center frequency f_o becomes the group delay and is defined by

$$t_g = -\left.\frac{\partial\phi}{\partial\omega}\right|_{\omega=\omega_o} \tag{10.63}$$

Using Equation (10.63), the electrical frequency tuning range Δf becomes Equation (10.64).

$$\Delta f \cong \frac{1}{180}\frac{\Delta\theta}{t_g}, \tag{10.64}$$

Here, $\Delta\theta$ is expressed in degrees. Thus, the higher the group delay, the narrower the electrical frequency tuning range becomes.

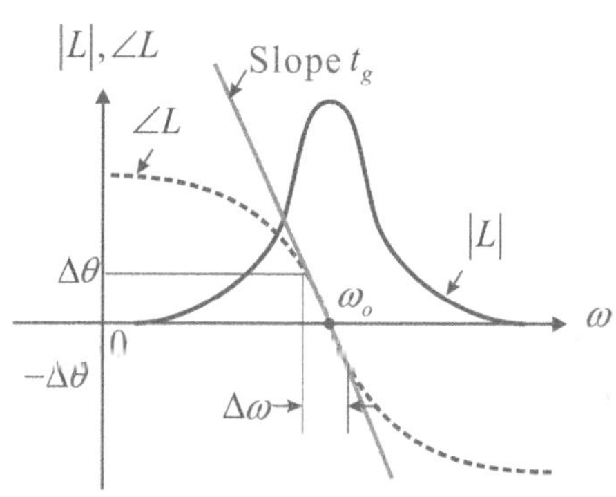

Figure 10.66 The frequency response of the open-loop gain. The dotted line represents the phase and solid line represents the magnitude of the open-loop gain. The slope of the phase at the center frequency is the group delay.

However, the Leeson's formula for the phase noise $S(f_m)$ is rewritten as Equation (10.65).

$$S(f_m) = -10\log\left\{\frac{1}{2}\frac{FkT}{P}\left(1+\frac{f_c}{f_m}\right)\left[1+\left(\frac{1}{f_m}\frac{f_o}{2Q}\right)^2\right]\right\} \tag{10.65}$$

Here, f_m and f_c represent the offset and flicker frequencies, and P and F represent the oscillation output power and noise figure of the amplifier, respectively. The Q in Equation (10.65) becomes the Q_L of the open loop gain, and Q_L is expressed in terms of the group delay as Equation (10.66).

$$Q_L = \pi f_o t_g \tag{10.66}$$

Therefore, the phase noise is improved as the group delay increases and this improvement is quite obvious from Equation (10.65).

Suppose that a phase noise of S_1 (dBc/Hz) is obtained when the oscillator shown in Figure 10.65 is formed with a group delay of t_{g1}. Assume that the group delay of the resonator is improved to a larger value t_{g2}, but other oscillator components remain the same. Then, the electrical frequency tuning range that results will be reduced by t_{g2}/t_{g1}, while the phase noise of the newly designed oscillator S_2 (dBc/Hz) will be improved, as expressed in Equation (10.67).

$$S_2 = S_1 - 20\log\frac{t_{g2}}{t_{g1}} \tag{10.67}$$

Thus, the phase noise can be systematically improved using the open-loop method. In conclusion, from Equations (10.64) and (10.65), the electrical frequency tuning range and the phase noise are found to have a trade-off relationship and the group delay of the open-loop gain is found to be a key parameter of that trade-off relationship.

In the derivation of Equation (10.67), we assumed that f_c, P, and F are invariant for the resonator change. However, the two values of F and f_c in Equation (10.65) for the resonator change may not be exactly equal. Although the identical insertion loss is assumed for the two resonators, the harmonics of the two DROs differ and this results in different noise conversions. However, assuming that the harmonic signals are small and ignoring the harmonic contributions in the noise conversion, the two values of F and f_c in Equation (10.65) can be approximated to be equal.

10.7.2.2 DRO Components

An amplifier layout for the HMC313 from Hittite is shown in Figure 10.67.[11] The schematic shown in Figure 10.67(a) is the recommended circuit in the datasheet and Figure 10.67(b) shows the layout. The points marked with • in Figure 10.67(b) represent the wafer-probe contact pads that enable two-port S-parameter measurement using on-wafer probing. All the chip DC block capacitors in Figure 10.67(b) have a value of 100 nF and the bypass capacitor C_{T2} is a 0.33-uF-chip tantal capacitor.[12] From the datasheet, the magnitude and phase of S_{21} are 17 dB and –75°, respectively, for a DC bias of 5 V and DC current of 47 mA.

11. Hittite, HMC313, GaAs InGaP HBT MMIC amplifier, available at www.hittite.com.

12. American Technical Ceramics, ATC 500S, ATC 545L capacitor, available at www.atceramics.com.

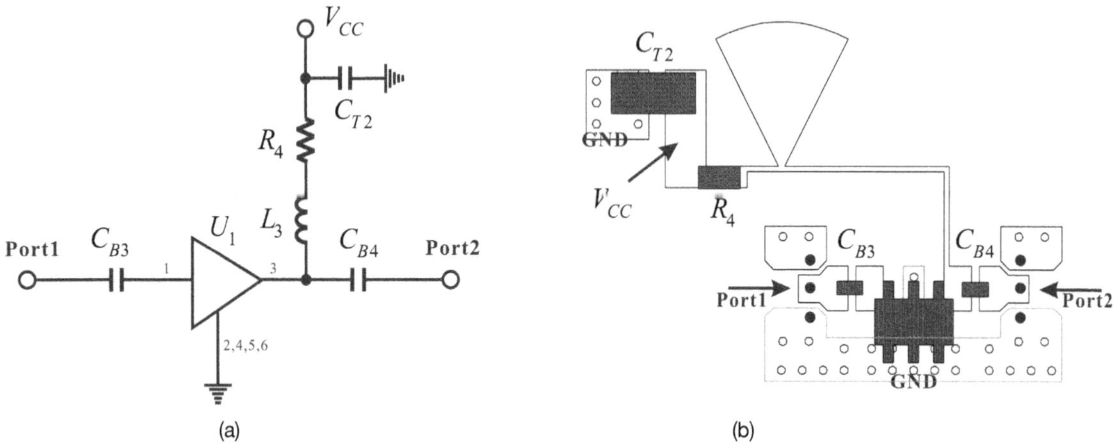

Figure 10.67 (a) Amplifier schematic where U_1 = HMC323, R_4 = 0, C_{T3} = 0.33 μF, L_3 is a one-quarter-wavelength choke, and C_{B3} and C_{B4} are 100-nF high-frequency DC block capacitors; (b) amplifier layout.

Figure 10.68 shows the measured results and datasheet values of S_{21}. The measured results can be seen to have a gain of 16.1 dB and a phase of approximately 52.5° within the frequency band of interest. The measured gain, S_{21}, is close to the value from the datasheet, but the phase is found to show a significant difference. The phase difference is due to the effects of the 50-Ω microstrip line lengths inserted for mounting the DC block capacitors and the HMC313. However, it can be seen that the gain of the amplifier is sufficiently high to compensate for the insertion losses of the resonator, phase shifter, and output coupler.

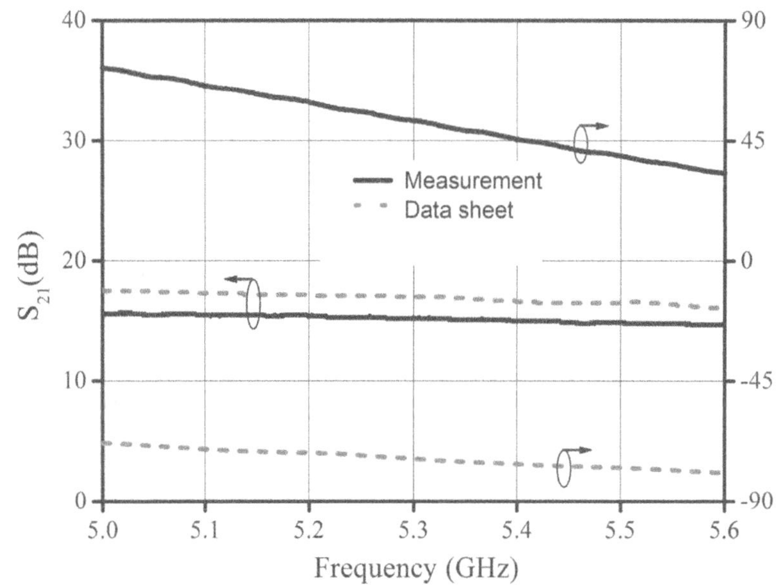

Figure 10.68 Comparison of the measured amplifier gain with that given in the datasheet for the HMC313. The dotted line is also from the datasheet. The two gains are close but a significant difference appears in the phase.

Figure 10.69(a) shows the phase shifter circuit.[13] In that figure, L_1 and L_2 are implemented using the identical varactor diodes, the SMV1245 from Skyworks, Inc.[14] This diode is series resonant due to packaging inductance and the series resonant frequency is approximately 1.5 GHz. Therefore, when used near the oscillation frequency of 5.3 GHz, the varactor diode acts as a variable inductor. The two varactor diodes are biased by resistors R_1, R_2, and R_3. The reason for using these resistors in the DC bias is that they can remove the parasitic resonance phenomenon that occurs when using an RF choke. Capacitors C_{B1} and C_{B2} are 100-nF high-frequency DC block capacitors. Thus, when the DC block and the DC bias resistor are removed, this circuit becomes a traditional all-pass filter and the capacitors C_1 and C_2 can be determined from Equations (10.68a) and (10.68b).

$$C_1 = \frac{1}{2\omega_o Z_o} \tag{10.68a}$$

$$C_2 = \frac{1}{\omega_o Z_o} \tag{10.68b}$$

Here, ω_o and Z_o represent the oscillation frequency and 50 Ω, respectively.

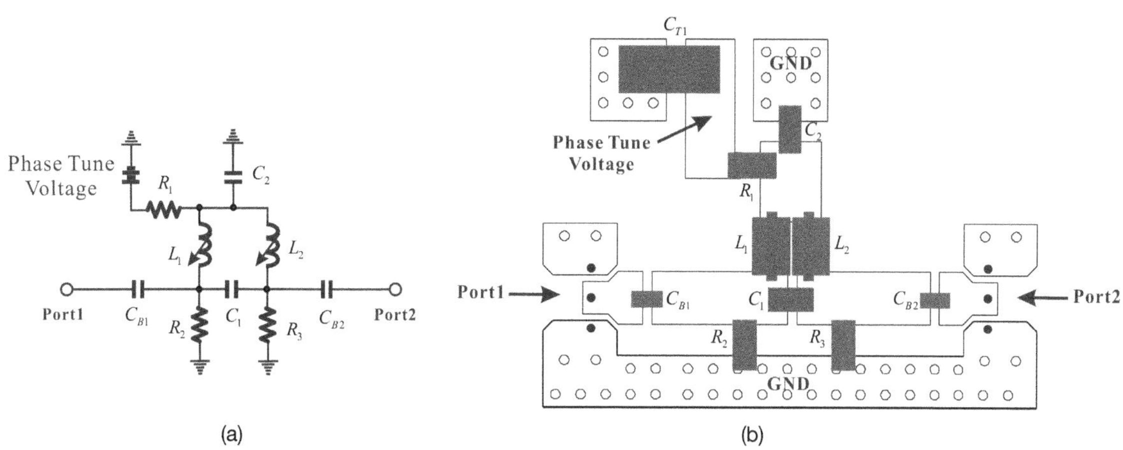

Figure 10.69 (a) Phase shifter circuit: L_1 and L_2 are varactor diodes, resistor $R_1 = R_2 = R_3 = 1$ kΩ, capacitor values are $C_1 = 0.3$ pF, $C_2 = 0.4$ pF, $C_{T1} = 0.33$ μF, and C_{B1} and C_{B2} are high-frequency 100-nF DC block capacitors; (b) phase shifter layout.

The computed values of C_1 and C_2 are approximately 0.3 pF and 1.2 pF, respectively, and based on these values, the capacitor values can be tuned in ADS simulation to achieve a phase shift of 70°. The value of C_2 is adjusted to 0.4 pF, while the value of C_1 is similar to the computed value 0.3 pF. Figure 10.69(b) shows the layout of the phase shifter circuit. Wafer-probe pads are inserted in the phase shifter layout to enable the independent S-parameter measurement of the phase shifter as in the case of the amplifier. The wafer-probe pads are shown as • in Figure 10.69(b). The phase shift is measured to be about 72°

13. L. Chen, R. Forse, A. H. Cardona, T. C. Watson, and R. York, "Compact Analog Phase Shifters Using Thin-Film (Ba,Sr) TiO3 Varactors," *IEEE MTT-S International Microwave Symposium*, Honolulu, HI, (June 3–8, 2007): 67–670.

14. Skyworks, SMV1245-011 varactor diode, available at: www.skyworksinc.com.

with respect to the varactor diode tuning voltage. The varactor diode tuning voltage varies between 0–10 V. The return loss varies according to the varactor diode DC tuning voltages. All the return losses are below 10 dB within the oscillation frequency band, which does not cause a serious mismatch in a cascade connection to other components.

10.7.2.3 Prototype Resonator

Figure 10.70 shows a resonator configured using a dielectric resonator, which is Murata's DRD107UC048.[15] At the frequency of 7 GHz, the resonator has parameters such as $\varepsilon_r = 37.7$ and $Q_u = 6800$. The resonant frequency of the dielectric resonator with a metal-box enclosure can be adjusted from 4.96 to 5.40 GHz. The substrate is RO4350, which has a permittivity of $\varepsilon_r = 3.66$ and a thickness $H = 30$ mil.[16] The spacer in Figure 10.70 is composed of a ceramic material and has a thickness of 1.016 mm and a relative permittivity of $\varepsilon_r = 4.5$. The resonance frequency can be adjusted within a frequency range of 100 MHz using the tuning screw that is fixed on the ceiling of the metal-box enclosure. The structure shown in Figure 10.70 is analyzed using Ansoft's HFSS to obtain the desired resonant frequency of 5.3 GHz by adjusting the tuning screw.[17] The detailed dimensions that yield the resonant frequency of 5.3 GHz are shown in the caption of Figure 10.70.

The designed resonator is fabricated and measured. The measured group delay can be derived using the slope of the phase against frequency and the value is measured to be 2.3 nsec as in the simulation. Using the group delay, Q_L can be computed from Equation (10.64) and the value is calculated to be 38.3. In addition, the electrical frequency tuning of about 84.5 MHz is estimated from Equation (10.62) when the phase shifter with a phase shift of $\pm 35°$ is employed in the oscillator.

10.7.2.4 Prototype DRO

Figure 10.71 shows the layout of the prototype oscillator that includes each of the separately designed and measured components. It can be seen from Figure 10.71 that the layout is designed to allow the independent S-parameter measurement of each block such as the resonator, amplifier, and phase shifter using a wafer probe. After measurement is carried out for each component, the components are connected in cascade by soldering 0-Ω resistors for R_5, R_7, R_8, and R_9 in Figure 10.71. Using this method, a closed-loop oscillator can be formed. The 50-Ω microstrip lines l_1 and l_2 in Figure 10.71 are deliberately inserted to make the phase of the open-loop gain to be an integer multiple of 360° at the oscillation frequency of 5.3 GHz. The oscillation output is obtained through capacitor C_3. After disconnecting one of the resistors R_5, R_7, R_8, or R_9 in Figure 10.71, the open-loop gain can be measured using wafer probes by placing those probes on the disconnected wafer-probe pads. In the open-loop gain measurement, a 50-Ω coaxial termination is connected to the oscillation output terminal.

Figure 10.72 shows the measured open-loop gain using an Agilent E8358A network analyzer. The phase tune voltage is set to 6 V, which corresponds to about the center value of the DC phase tune voltage range, which also corresponds to about the center frequency

15. Murata, DRD107UC048 dielectric resonator available at: www.murata.com.

16. Rogers Corp. RO4000R series high-frequency circuit materials available at: www.rogerscorp.com.

17. SpragueGoodman, GRRB70504SN10 dielectric resonator tuners available at: www.spraguegoodman.com.

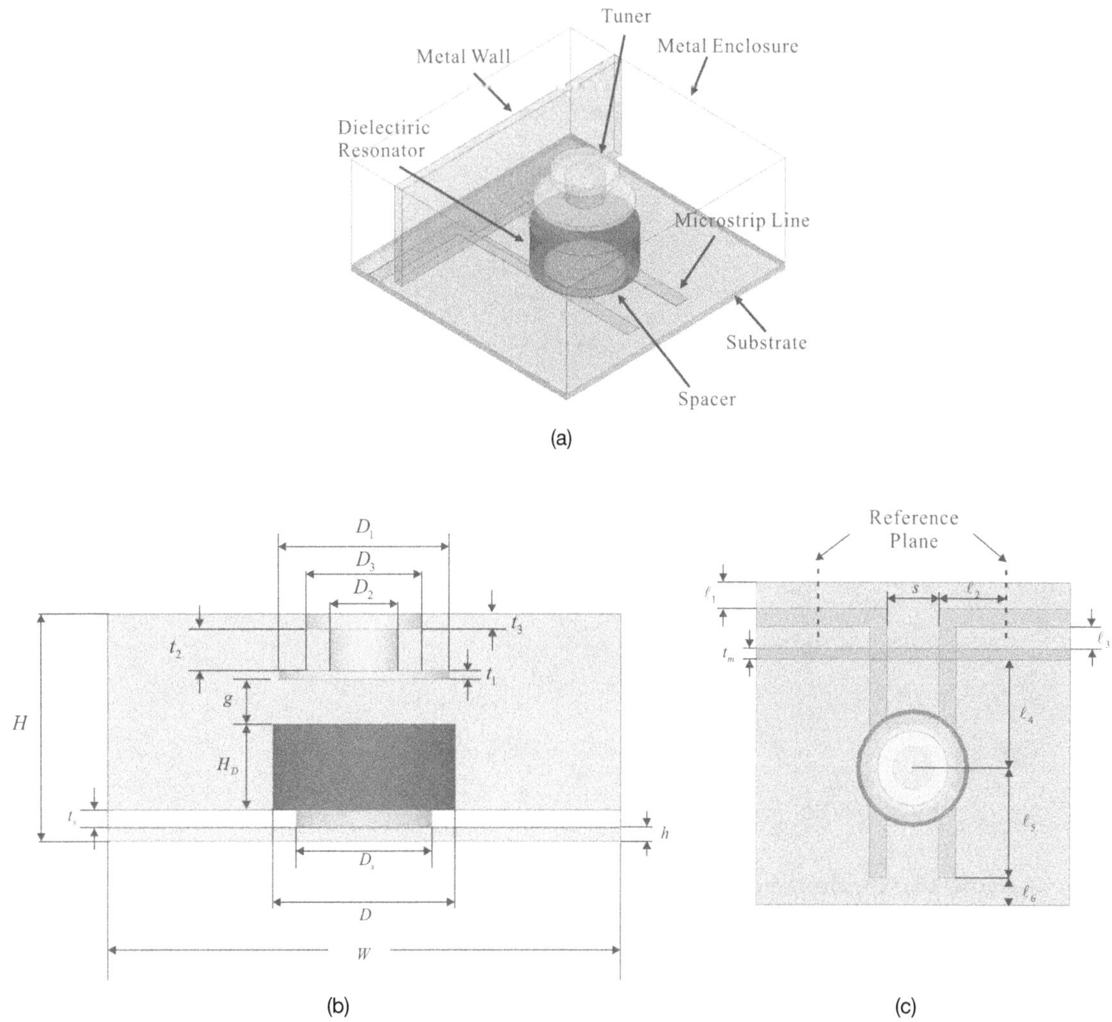

Figure 10.70 (a) Resonator structure; (b) front view dimensions: $W = 30$, $H = 12.762$, $h = 0.762$, $D_s = 8.001$, $t_s = 1.016$, $D = 10.75$, $H_D = 4.77$, $g = 2.514$, $t_1 = 0.5$, $t_2 = 2.35$, $t_3 = 0.85$, $D_1 = 10$, $D_2 = 4$, $D_3 = 6.8$; and (c) top view dimensions: $l_1 = 2.36$, $l_2 = 6.5$, $l_3 = 2.0$, $l_4 = 10.0$, $l_6 = 2.5$, $d = 5.6$, $w = 1.64$ (all units are in mm).

of the oscillator. The magnitude of the open-loop gain, S_{21}, is greater than 0 dB at the center frequency of 5.3 GHz, and the phase can be seen to be 0° at a frequency of 5.313 GHz. Therefore, the oscillation frequency can be expected to occur at 5.313 GHz, which is close to 5.3 GHz. The group delay can be obtained by calculating the slope at the frequency where the phase is 0°. Thus, the group delay is

$$t_g = \left.\frac{\partial \phi}{\partial \omega}\right|_{\omega=\omega_o} = -\frac{4.106-(-11.298)}{2\pi(5.31-5.318)\times 10^9} \times \frac{\pi}{180} \cong 2.5 \text{ nsec}$$

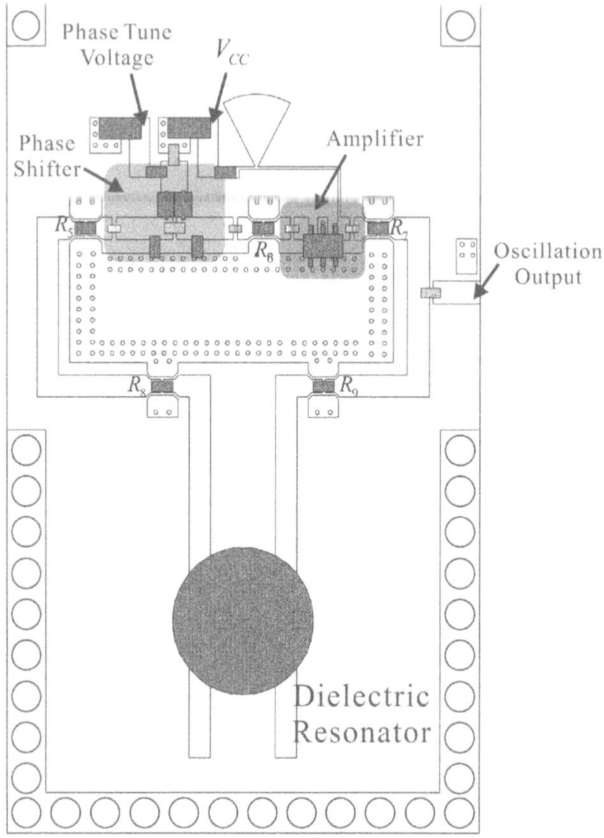

Figure 10.71 The prototype dielectric resonator oscillator layout (44×72 mm^2). Resistors R_5, R_6, R_7, R_8, and R_9 are 0-Ω resistors for connection.

The characteristics of the DRO can be measured with a spectrum analyzer; however, for more precise measurements, an Agilent E5052A signal source analyzer is used for the measurement. The electrical frequency tuning range is measured to be 99 MHz (5.264–5.363 GHz) at the phase tune voltage change of 0–10 V. The electrical frequency tuning range can be computed from the group delay, which is obtained from the open-loop gain measurement. From Equation (10.62), the electrical frequency tuning range is calculated to be

$$\Delta f = \frac{\theta}{\pi t_g} = \frac{36}{2.5n} \times \frac{1}{180} = 0.08 = 80\ \text{MHz}$$

Note that the electrical frequency tuning characteristic appears to be the same as the phase shifter's phase tune characteristic from Equation (10.62). Thus, the closer that phase tune characteristic is to a straight line, the more linear the electrical frequency tuning characteristic appears to be. Therefore, in order to have a linear frequency tuning characteristic, a phase shifter with a linear phase tune characteristic is required.

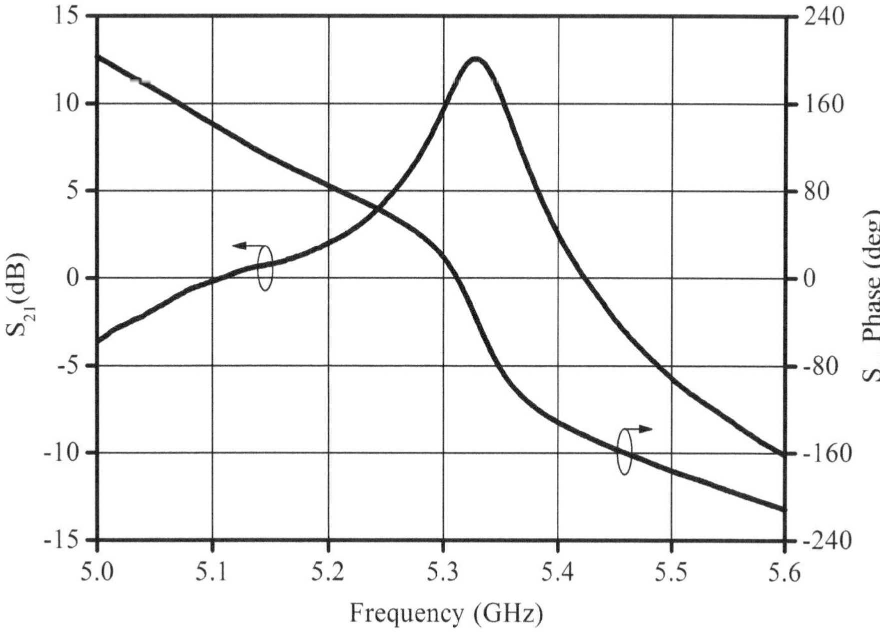

Figure 10.72 The measured open-loop gain (the DC phase tune voltage of the phase shifter is set to 6 V).

Figure 10.73 shows the phase noise characteristics measured with the E5052A signal analyzer from Agilent Technologies.[18] Figure 10.73(a) shows the phase noise characteristic at the center frequency of 5.3 GHz when the phase tune voltage is about 6 V. In that figure, the phase noise is about −111 dBc/Hz at the offset frequency of 100 kHz. Figure 10.73(b) shows the phase noise when the phase tune voltage is varied with the offset frequency fixed to 100 kHz. From the figure, even though the oscillation frequency is changed, the phase noise remains constant. Note that when the phase tune voltage is changed, the phase of the open-loop gain changes; however, the group delay does not change because it is chiefly determined by the resonator. As a result, regardless of the change in the phase tune voltage, a constant phase noise is displayed due to an approximately constant group delay. Usually, when the resonant frequency of the resonator is tuned using a frequency tuning device, such as a varactor diode, the Q of the resonator varies according to the resonant frequency change. Therefore, when the resonant frequency of the resonator is directly tuned using a frequency tuning device, the phase noise varies according to the variation in the oscillation frequency. However, when the oscillation frequency is tuned using a phase shifter, the Q of the resonator does not change and a constant phase noise is obtained regardless of the oscillation frequency tuning.

18. Agilent Technologies, 5989-6389EN, Agilent E5052B signal source analyzer, 10 MHz to 7, 26.5, or 110 GHz, 2007 available at: www.agilent.com.

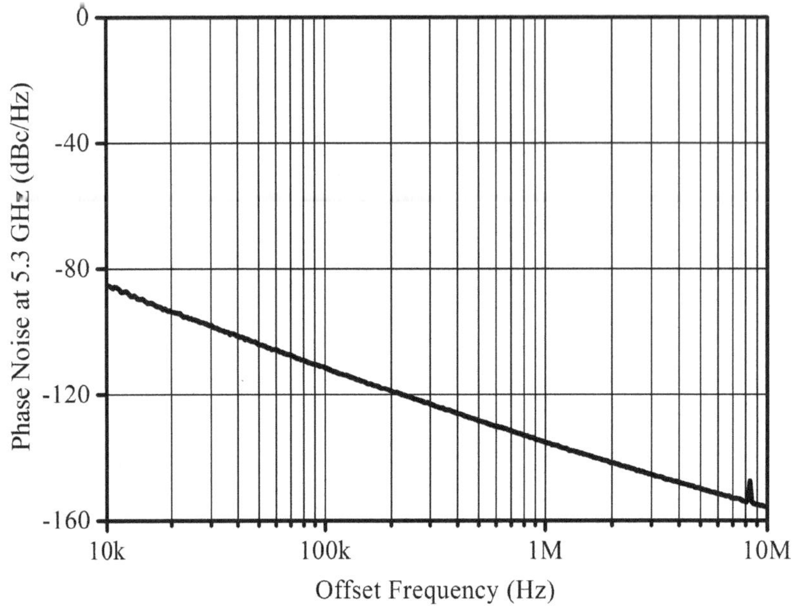

(a)

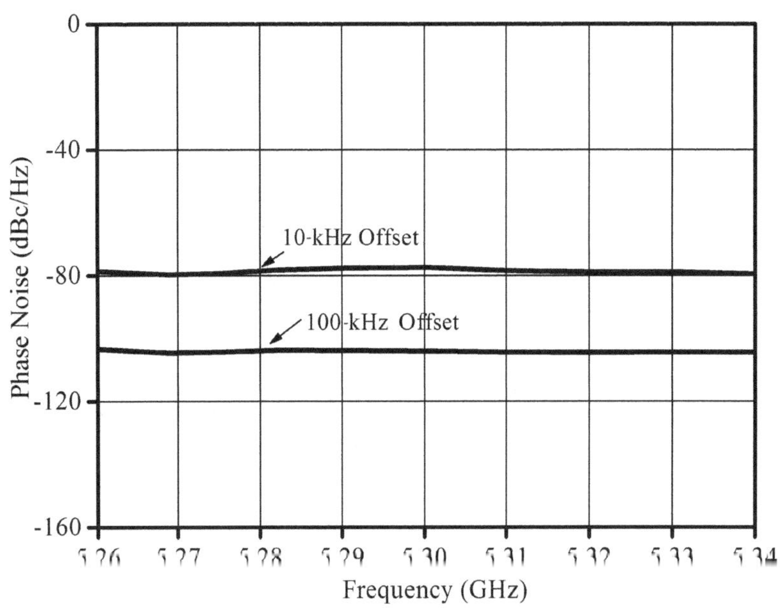

(b)

Figure 10.73 Frequency tuning characteristics of the fabricated DRO: (a) phase noise characteristics at an oscillation frequency of 5.3 GHz, and (b) phase noise at an offset frequency of 10 kHz and 100 kHz when the phase tune voltage is varied

10.7.2.5 Low Phase Noise DRO

There are often many applications that require further improvement in phase noise rather than in the wide electrical frequency tuning range. Suppose that the basic configuration of the oscillator is unchanged. Then, from Equations (10.63) and (10.64), it can be found that the increase in group delay leads to a narrower electrical frequency tuning range and improved phase noise. The improvement in phase noise due to an increased group delay can be estimated using Equation (10.67). Thus, when the design objective is set to a phase noise of −130 dBc/Hz at a 100-kHz offset frequency, the group delay of a resonator with more than 50 nsec (approximately 20 times improvement in the group delay of the previous DRO) is based on Equation (10.67).

Figure 10.74 shows a resonator configured with a dielectric resonator identical to that shown in Figure 10.70. The group delay of the resonator primarily depends on the distance (D) between the microstrip and the center of the resonator, and the length extension (L) of the microstrip from the center of the dielectric resonator, as shown in Figure 10.74. The width of the microstrip line can also be a variable but the width is fixed to that of the 50-Ω microstrip shown in Figure 10.70.

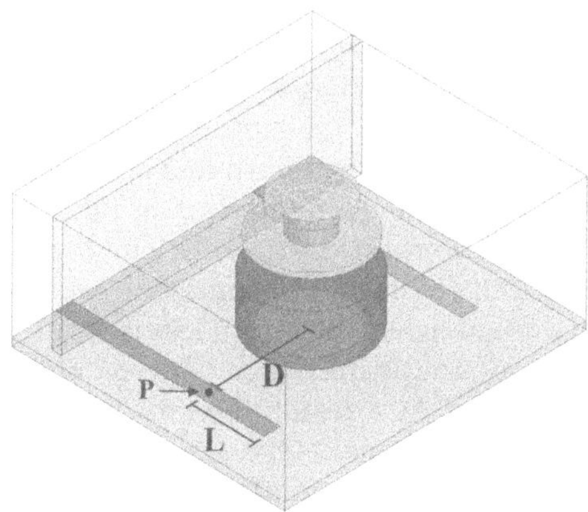

Figure 10.74 Definition of the parameters for the resonator that improves the group delay

Figure 10.75(a) shows the change of group delay with respect to D and L. The group delay of Figure 10.75(a) is related to the coupling between the dielectric resonator and the microstrip. Note that the coupling between them is magnetic. For a strong magnetic-field incident on the dielectric resonator from the microstrip, the coupling is high, while it is low for a weak magnetic-field incident. As D increases, the magnetic-field incident becomes weaker and the coupling becomes low. Note that the high coupling of the microstrip makes the loaded Q, Q_L decrease from the Q_u of the dielectric resonator. Thus, Q_L increases as the coupling becomes lower, thereby increasing group delay because $Q_L = \pi f_o t_g$. As a result, the group delay increases with the increase of D. The group delay variation with respect to L can be understood from the current standing wave pattern. Since the end of the microstrip line is open, the current is a maximum at point P when point P is placed away from the microstrip end by a quarter wavelength ($L = 7.4$ mm). Thus, the coupling by the microstrip's magnetic field is maximum and the group delay is minimum at $L = 7.4$ mm. The group delay increases when L increases or decreases from a quarter wavelength because the

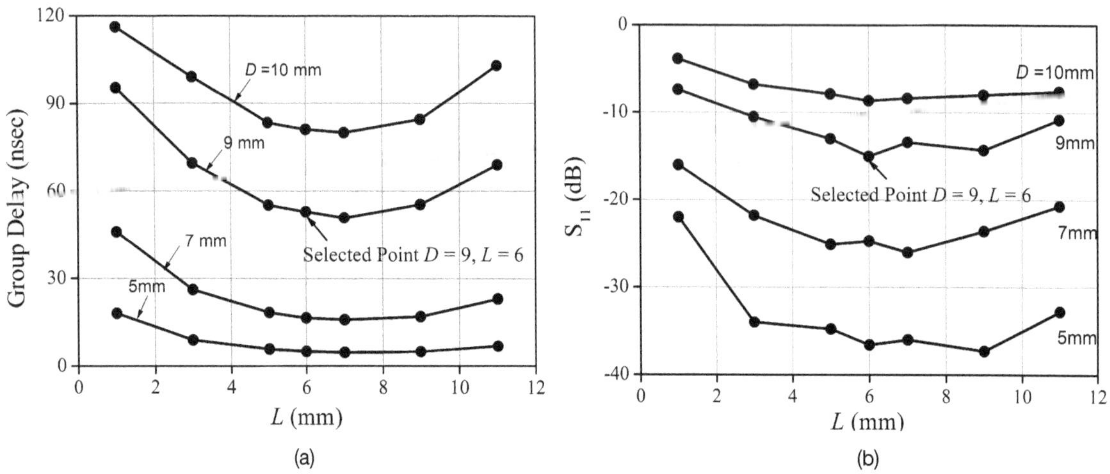

Figure 10.75 Simulation results of a resonator: (a) group delay and (b) return loss. The results are computed using AnSoft's HFSS.

current corresponding to the magnetic field at point P decreases when L is away from a quarter wavelength. Consequently, the group delay increases with the increase of D and is lowest near $L = 7$ mm, as shown in Figure 10.75(a).

In the DRO design, since the return loss and insertion loss depend on D and L, not only the group delay but also the return loss and the insertion loss should be considered simultaneously. A high insertion loss requires a high-gain amplifier. In that case, an amplifier with an impractically high gain is required to cause oscillation even though the group delay is high. In addition, when the return loss is small, the open-loop gain cannot simply be expressed as S_{21} due to the mismatch between other components such as the phase-shifter and the amplifier. As a result, the mismatch must be taken into account in the design process, which in turn makes the design difficult.

Figure 10.75(b) shows the variation of $|S_{11}|$ that is similar to the variation of the group delay in Figure 10.75(a). As D increases, a smaller power from the microstrip is delivered to the dielectric resonator. Thus, most of the power is not delivered to the other port but rather is reflected. This makes $|S_{11}|$ high as D increases. Since $|S_{11}|$ is high, the insertion loss increases. Therefore, the increased group delay necessarily accompanies both a larger insertion loss and smaller return loss. The distance $D = 9$ mm and the length $L = 6$ mm are chosen. The resulting group delay is 53 nsec, $\text{dB}(S_{11}) < -10$ dB, and $\text{dB}(S_{21}) = -2$ dB. Using Equation (10.65), the expected phase noise for the determined D and L is −132 dBc/Hz, and the electrical frequency tuning range is calculated to be 5 MHz. Compared to the previous DRO design, the phase noise at an offset frequency of 100 kHz is computed to be improved by approximately 26 dB and the frequency tuning range is reduced by about 1/20.

Figure 10.76 is a photograph of the DRO with the newly designed resonator. In this photograph, only the resonator has been changed and the remaining parts are basically the same as in the previous DRO circuit.

Figure 10.77 shows the measured results for the newly designed DRO using Agilent's E5052A signal analyzer. The measured electrical frequency tuning range is found to be approximately 5 MHz at a phase tune voltage of 0–10 V. An output power of approximately 4.5 dBm is measured at a phase tune voltage of 6 V. The phase noise against the

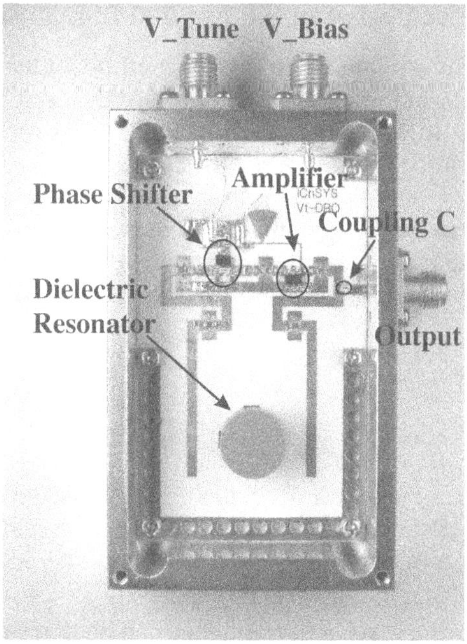

Figure 10.76 Fabricated DRO ($44 \times 72 \times 22$ mm^3) with the improved resonator

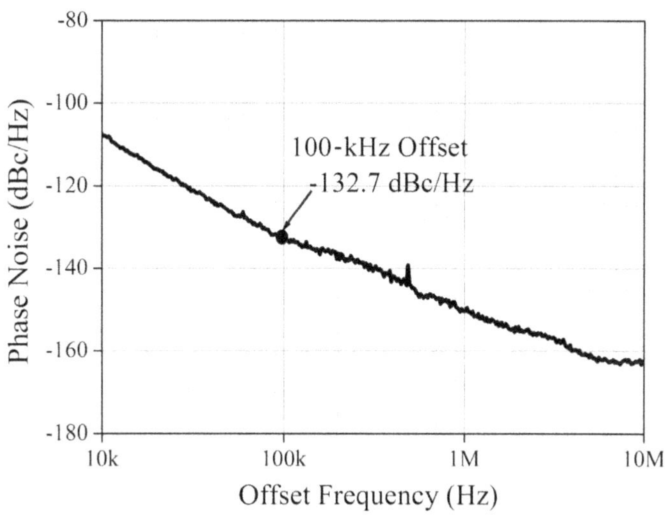

Figure 10.77 Measured phase noise of the fabricated low phase noise DRO employing the improved resonator at a phase tuning voltage of 6 V

frequency offset is shown in Figure 10.77 and the phase noise at a frequency offset of 100 kHz is measured to be –132.7 dBc/Hz. This result can be seen to be close to the phase noise predicted using the group delay. Similar to the prototype DRO, the phase noise with respect to oscillation frequency at a constant frequency offset is observed to be almost flat. This is the advantage of the loop-type DRO where the oscillation frequency is controlled using the phase shifter.

10.7.3 Comparison between the Two DRO Design Methods

We have so far investigated two design methods. One design is for a DRO that uses a dielectric resonator in the feedback network of the oscillator; the other design is the reflection-type method (discussed in section 10.7.1) that replaces the reactance x or y with a dielectric resonator coupled to a microstrip circuit. In terms of phase noise, one does not have an advantage over the other. From the design point of view, the phase noise of the reflection type may not be systematically improved and this type also does not provide a systematic way to tune an oscillation frequency. On the other hand, the feedback type is easy to design and its phase noise and electrical frequency tuning range can be systematically improved.

10.8 SUMMARY

- Whether or not a circuit oscillates can be checked using the small-signal oscillation conditions. Oscillation conditions can be described using impedances, reflection coefficients, and open-loop gain; however, they all give the same results for oscillation start-up.
- The oscillation power and frequency can be determined using the large-signal oscillation condition or the equilibrium conditions described using large-signal impedances, reflection coefficients, or open-loop gain.
- The amplitude and phase of an oscillation waveform fluctuates with time. The amplitude fluctuation is specified using AM noise and that of the phase is specified using phase noise.
- The phase noise spectrum can be determined using the ratio of noise to carrier power measured at a given frequency offset. The spectrum analyzer or other devices such as a signal analyzer or specially built equipment can be used to measure the phase noise.
- Phase noise is basically a nonlinear phenomenon; however, phase noise can be empirically described by Leeson's phase noise model. The near carrier phase noise is primarily dependent on the flicker noise of an active device in an oscillator.
- The simplified oscillator circuits at the RF are generally converted into two types of basic oscillator circuits: the series-feedback- and the parallel-feedback-type of oscillators. An oscillator circuit can be designed by modifying the basic oscillator circuits and adding DC bias circuits.
- An oscillator circuit yielding the desired oscillation frequency can be designed using the oscillation conditions. The design of an oscillator circuit for mobile communication is demonstrated as an example.
- The design of a microstrip oscillator circuit that provides optimum output power and the desired oscillation frequency is demonstrated. The microstrip oscillator that provides optimum power is designed using the synthesis of the feedback network derived from the load-pull simulation results.
- The operation of a dielectric resonator is introduced. In addition, the extraction of the equivalent circuit for the dielectric resonator coupled to a microstrip line is explained.

- The design of a DRO that meets the phase noise objective is demonstrated. First, the prototype DRO is designed and its phase noise performance is evaluated. Then, from the evaluated phase noise, the Q of the resonator in the DRO can be determined. Once the resonator is determined, the desired phase noise can be achieved.

REFERENCES

1. G. D.Vendelin, A. M. Pavio, and U. L. Rohde, *Microwave Circuit Design Using Linear and Nonlinear Techniques*, New York: John Wiley & Sons, Inc., 1990.
2. W. P. Robins, *Phase Noise in Signal Sources*, London: IEEE Press, 1984.
3. A. K. Poddar, U. L. Rhode, and A. M. Apte, "How Low Can They Go? Oscillator Phase Noise Model, Theoretical, Experimental Validation, and Phase Noise Measurements," *IEEE Microwave Magazine*, vol. 14, no. 6, pp. 50–72, September–October 2013.
4. D. Kajfez and P. Guillon, *Dielectric Resonators*, Dedham, MA: Artech House, Inc., 1986.
5. U. L. Rhode, A. K. Poddar, and G. Bock, *The Design of Modern Microwave Oscillators for Wireless Applications: Theory and Optimization*, New York: John Wiley & Sons, Inc., 2005.

PROBLEMS

10.1 Using the circuit shown in Figure 10P.1 and given that the oscillation equilibrium conditions are satisfied at the reference plane A–A′, prove that the oscillation equilibrium conditions are satisfied even when the reference plane is changed to B–B′. In other words, demonstrate that the oscillation equilibrium is established everywhere in the circuit regardless of the reference plane.

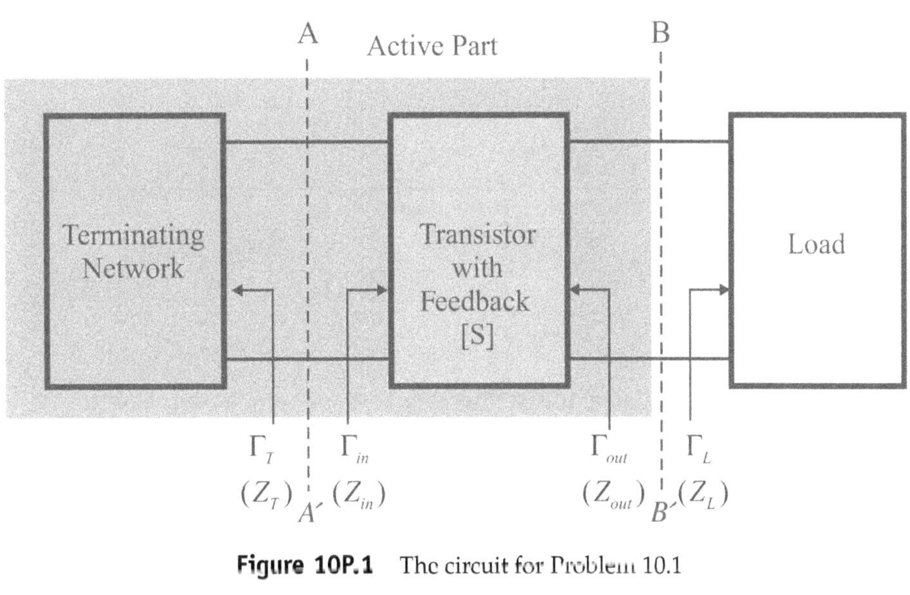

Figure 10P.1 The circuit for Problem 10.1

10.2 Using the circuit shown in Figure 10P.2, determine S_{11} in a polar format for the circuit at a frequency of 10 GHz.

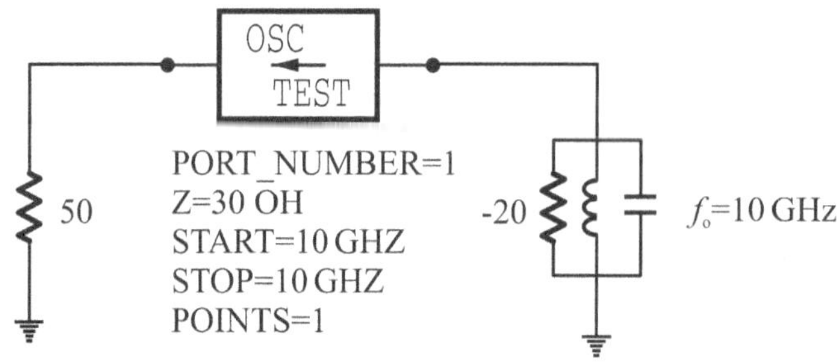

Figure 10P.2 The circuit for Problem 10.2

10.3 **(ADS problem)** Configure the series oscillating circuit of Example 10.4 in ADS and investigate the effect of the reference impedance on the oscillation condition by varying the reference impedance **Z** of **OscTest** for 10, 60, and 110 Ω.

10.4 From the measured spectrum shown in Figure 10P.3, in the following measurement determine the phase noise at a 100-kHz offset from the center frequency; also, calculate the maximum phase deviation φ from this offset frequency.

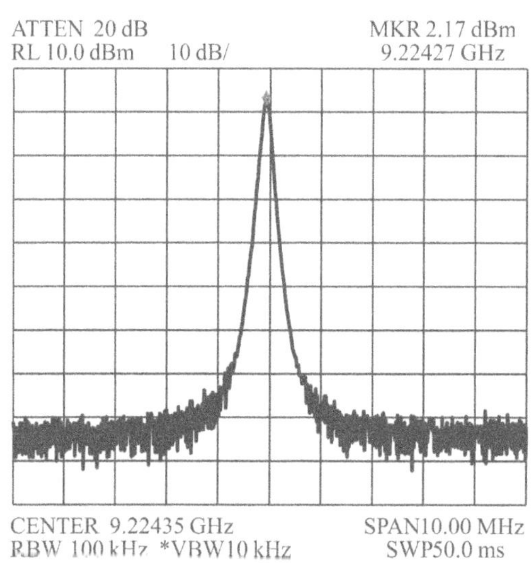

Figure 10P.3 The measured spectrum for Problem 10.4

10.5 Figure 10P.4 shows an oscillator circuit operating at 500 MHz–2 GHz. Simplify this circuit to the basic parallel feedback oscillator circuit and calculate the oscillation frequency and the values x, y, and Y_L of this parallel feedback oscillator circuit

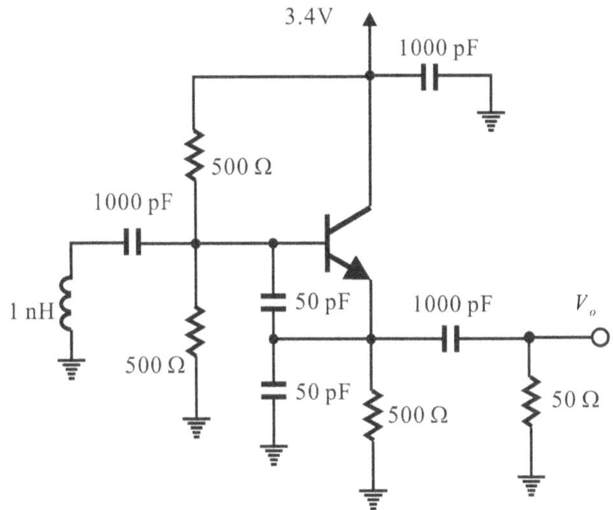

Figure 10P.4 The circuit for Problem 10.5

10.6 A dielectric resonator is used as the bandpass structure shown in Figure 10.76. When the measured Q_L for $|S_{21}|$ has the insertion loss of IL (dB) at the resonance frequency, prove that the unloaded Q, Q_u can be computed as

$$Q_u = \frac{Q_L}{1 - 10^{-\frac{IL}{20}}}$$

10.7 In the circuit shown in Figure 10P.5, assuming that the impedance of the active part changes linearly with the amplitude of RF current, perform the following tasks:

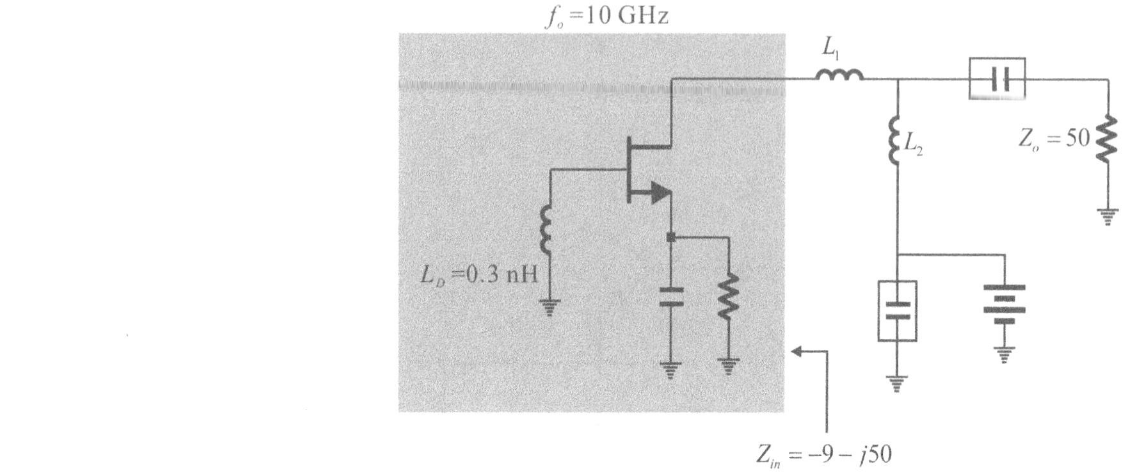

Figure 10P.5 The circuit for problem 10.7

(1) Set the load impedance value to give maximum output power and calculate the values of L_1 and L_2 to give this value.

(2) Given that the L_D of the active part in Figure 10P.5 is to be replaced using a dielectric resonator coupled to a microstrip, as shown in Figure 10P.6 below, determine the value of θ. Assume the DR is strongly coupled to the microstrip.

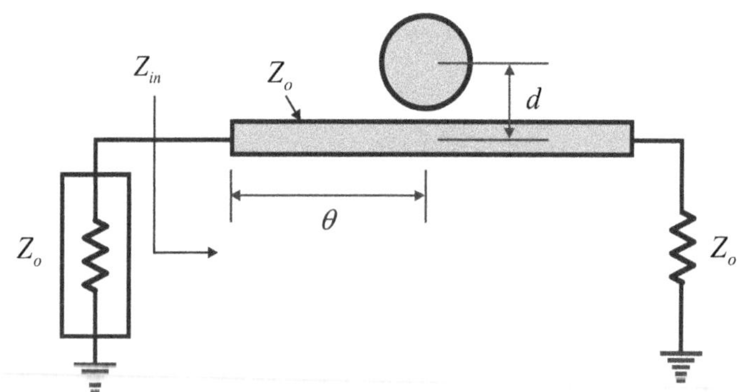

Figure 10P.6 Dielectric resonator coupled to a microstrip circuit

10.8 **(ADS problem)** Using the large-signal model of transistor NE42484, perform the load-pull simulation at 5 GHz at V_{DS} = 2 V and V_{GS} = –0.45 V, which yields about I_{DS} = 20 mA. Determine the feedback circuit that gives optimum output power. Then, verify the output power using ADS.

CHAPTER 11

Phase-Locked Loops

CHAPTER OUTLINE

11.1 INTRODUCTION

The direct application of most microwave oscillators, previously described in Chapter 10, to a communication system is seldom possible because their phase noises are generally not low enough. The phase noise is associated with frequency jitters in the time domain. As the phase noise of the microwave oscillator is typically high, the *frequency jitter* also increases significantly with time. The large-frequency jitter prevents the direct use of the microwave oscillator in a communication system. In addition, the microwave oscillator's phase or frequency in a communication system is often modulated and then that modulated signal is transmitted or received. Thus, the microwave oscillator's phase noise also limits the phase or frequency modulations necessary for communication.

A crystal oscillator is known to be the most stable oscillator to date and most communication systems require a phase noise as low as that provided by the crystal oscillator. However, it is not possible to fabricate crystal oscillators beyond a few hundred MHz. In contrast, microwave oscillators can be fabricated for high-frequency applications albeit with poor phase noise. The comparison of the phase noises of a 10-GHz VCO (voltage-controlled oscillator) and 10-MHz crystal oscillator (its frequency is multiplied by a factor of 1000 to obtain 10 GHz) is shown in Figure 11.1.

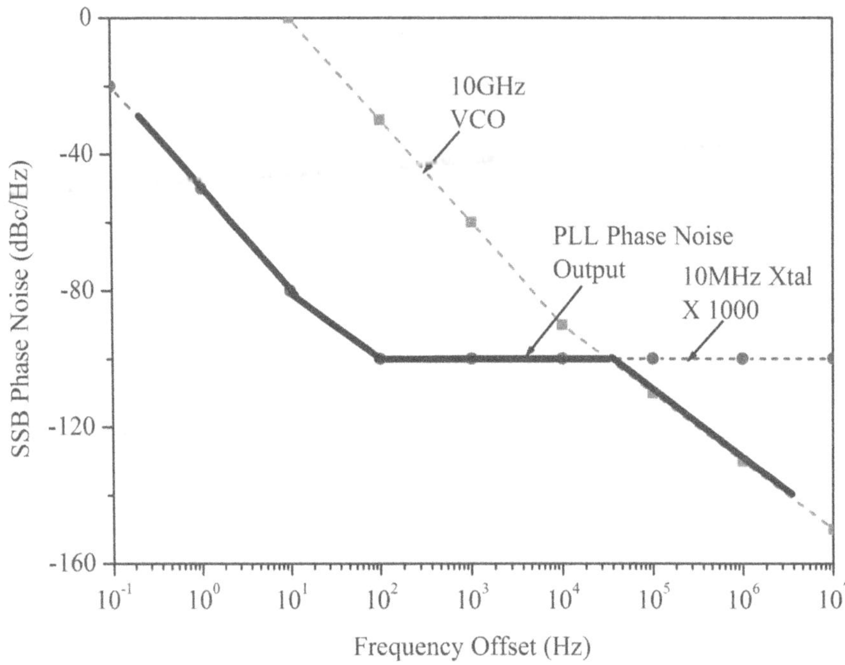

Figure 11.1 Phase noise of a PLL. The phase noise output for an optimally set loop bandwidth follows the minimum between the phase noise of a 10-GHz VCO and that of the crystal oscillator whose frequency is multiplied by a factor of 1000 to attain a frequency 10 GHz.

In that figure, the optimum frequency source may be one that has the phase noise of the crystal oscillator for a low offset frequency, while it follows the phase noise of the VCO for a high offset frequency. The optimum phase noise characteristic is shown by a solid line in the figure. The frequency source with the optimum phase noise characteristic in Figure 11.1 can be implemented using a PLL (phase-locked loop). In addition, a frequency source whose frequency increases or decreases with a specified frequency step (usually referred to as a channel) is usually required in communication systems. This type of frequency source is referred to a *frequency synthesizer* that can be configured using a PLL. In this chapter, the configuration and basic operation of a PLL will be discussed.

11.2 CONFIGURATION AND OPERATION OF A PLL

The configuration of a PLL is shown in Figure 11.2. The reference oscillator is usually a crystal oscillator that oscillates at a fixed frequency and provides the reference frequency. The phase of the reference oscillator is represented by θ_i, while that of the VCO is represented by θ_v. The phase noise of the VCO is represented by θ_n, which is shown as a separate noise source added to the phase of the VCO. Therefore, the output phase of the VCO is expressed in Equation (11.1).

$$\theta_o = \theta_v + \theta_n \tag{11.1}$$

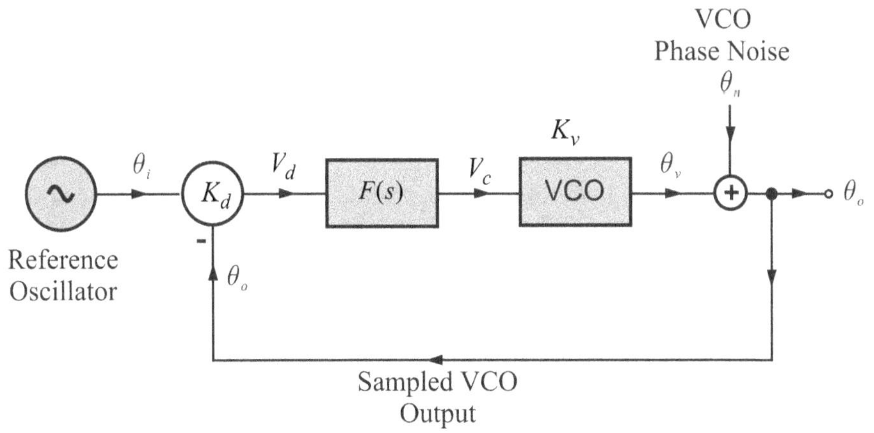

Figure 11.2 PLL configuration; the phase noise of the VCO is represented by an additive noise source θ_n. The PLL output is θ_o.

Since the oscillation frequency of the VCO, ω_v, varies according to the tuning voltage V_c, it can be expressed using θ_v with Equation (11.2).

$$\omega_v = \frac{d}{dt}\theta_v = K_v V_c \tag{11.2}$$

Expressing Equation (11.2) in a Laplace transform gives Equation (11.3).

$$\theta_v = K_v \frac{V_c}{s} \tag{11.3}$$

Here, K_v is termed the *tuning sensitivity* of the VCO, and its unit is [rad/sec·V]. The phase detector in Figure 11.2 is a device that converts the phase difference between the two input signals into a voltage and its proportionality constant is termed a *phase detector constant* K_d. Thus, denoting the output voltage of the phase detector as V_d, we obtain Equation (11.4).

$$V_d = K_d\left(\theta_i - \theta_o\right) \tag{11.4}$$

The dimension of K_d is thus [V/rad]. Note that the output voltage of the phase detector is a slow time-varying voltage close to the DC voltage. The low-pass filter connected to the phase detector is called the *loop filter* and its function is to remove the harmonics that appear in the phase detector. $F(s)$ represents the transfer function of the loop filter.

In order to examine the operation of the PLL in Figure 11.2, first consider the case of $F(s) = 1$, and $\theta_n = 0$, which results in $V_c = V_d$. Assuming that the steady state is reached, the tuning voltage V_c of the VCO is no longer considered to vary with time. If V_c varies with time, then the assumption of the steady state would be violated. Therefore, the output voltage of the phase detector V_d does not vary with time. Thus, the phase error θ_ε at the steady state is expressed in Equation (11.5).

$$\theta_\varepsilon = \theta_i - \theta_o = \frac{V_d}{K_d} \tag{11.5}$$

Under the steady state condition, since V_d is a DC voltage that no longer varies with time, then differentiating both sides of Equation (11.5) gives Equation (11.6).

$$\omega_i = \omega_o \tag{11.6}$$

Thus, when $F(s) = 1$, the frequency of the VCO is found to be equal to that of the reference oscillator; however, it should be noted that the reference oscillator and the VCO have a phase difference given by Equation (11.5). Initially, the frequencies of the VCO and reference oscillator are different prior to reaching the steady state and a tuning voltage proportional to the phase difference is then formed that drives the frequency of the VCO to a value equal to that of the reference oscillator. As a result, the tuning voltage of the VCO should be established to make the frequency of the VCO equal to the frequency of the reference oscillator and the phase error defined by Equation (11.5) inevitably occurs when $F(s) = 1$.

This phase error can be removed by implementing $F(s)$ with an integrator. However, even with the integrator, the frequency of the VCO should not be time-varying at the steady state and subsequently the tuning voltage, V_c, of the VCO does not become time-varying. Since the tuning voltage V_c in Figure 11.2 is the output of the integrator, that output is therefore constant and independent of time at the steady state. Thus, the input of the integrator must be zero. If that input is not zero, then by integrating the non-zero input with respect to time, the output will consequently vary with time.

When the input is positive, the output of the integrator increases with time while it decreases with time when the input is negative. As a result, the integrator output will vary with time. Since the integrator output is the tuning voltage V_c, that tuning voltage varies with time, which violates the assumption of a steady state. Therefore, by simply inserting an integrator into the PLL circuit, the phase error given by Equation (11.5) becomes 0, that is, $\theta_\varepsilon = 0$. As a result, when an integrator is used as a loop filter, not only the phases but also the frequencies of the two oscillators are made equal. The VCO is thus synchronized with the frequency and phase of the reference crystal oscillator.

Next, we examine the phase noise of the PLL's output (and also the VCO's output). Here, setting $K = K_d K_v$, the PLL output θ_o is computed as shown in Equation (11.7).

$$\begin{aligned} \theta_o &= \theta_i \frac{\dfrac{KF(s)}{s}}{1+\dfrac{KF(s)}{s}} + \theta_n \frac{1}{1+\dfrac{KF(s)}{s}} \\ &= H(s)\theta_i + \left(1 - H(s)\right)\theta_n \end{aligned} \tag{11.7}$$

Here, $H(s)$ represents the closed-loop gain of the PLL. Since the output phase noise consists of two independent noises θ_i and θ_n, the PLL output phase noise can be written as Equation (11.8).

$$\overline{\theta_o^2} = \left|H(s)\right|^2 \overline{\theta_i^2} + \left|1 - H(s)\right|^2 \overline{\theta_n^2} \tag{11.8}$$

In general, since the closed-loop transfer function $H(s) \to 1$ when $s \to 0$ and $H(s) \to 0$ when $s \to \infty$, $H(s)$ thus shows the characteristic of a low-pass filter. The phase noise of the PLL output from Equation (11.8) follows the phase noise of the reference oscillator within the bandwidth of $H(s)$, while it follows the phase noise of the VCO outside the bandwidth of $H(s)$. That is, the expression above can be approximately written as Equation (11.9).

$$\overline{\theta_o^2} = \begin{cases} \overline{\theta_i^2} & s \to 0 \\ \overline{\theta_n^2} & s \to \infty \end{cases} \tag{11.9}$$

Thus, the phase noise of the PLL's output can be determined using Equation (11.8). This is plotted as shown in Figure 11.3.

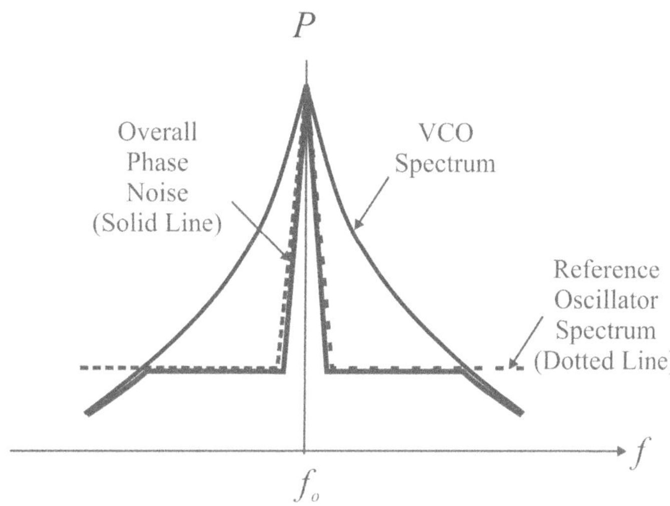

Figure 11.3 Phase noise of the PLL's output. Within the loop bandwidth, the phase noise of the PLL follows that of the reference oscillator, while it follows the VCO's phase noise outside the bandwidth.

Example 11.1

A PLL is configured using a 10-MHz crystal oscillator and a 10-GHz VCO, each having the phase noise shown in Table 11E.1.

The frequency of the 10-MHz crystal oscillator is multiplied by 1000 and it is used as the reference oscillator. The closed-loop transfer function is expressed as

$$H(s) = \frac{2\zeta\omega_n s + \omega_n^2}{s^2 + 2\zeta\omega_n s + \omega_n^2}$$

Plot the phase noise of the PLL output for $\zeta = 0.707$, $f_n = 1$ kHz, and 10 kHz ($\omega_n = 2\pi f_n$).

Table 11E.1 Phase noises of the oscillators

Frequency Offset	X-tal (10 MHz) Phase Noise (dBc/Hz)	VCO (10 GHz) Phase Noise (dBc/Hz)
0.1 Hz	80	—
1 Hz	–110	—
10 Hz	–140	0
100 Hz	–160	-30
1 kHz	–160	-60
10 kHz	–160	-90
100 kHz	–160	–110
1 MHz	–160	–130
10 MHz	–160	–150

Solution

The phase noise of the PLL output is computed using $\overline{\theta_o^2} = |H(s)|^2\,\overline{\theta_i^2} + |1 - H(s)|^2\,\overline{\theta_n^2}$ and is shown in Figure 11E.1.

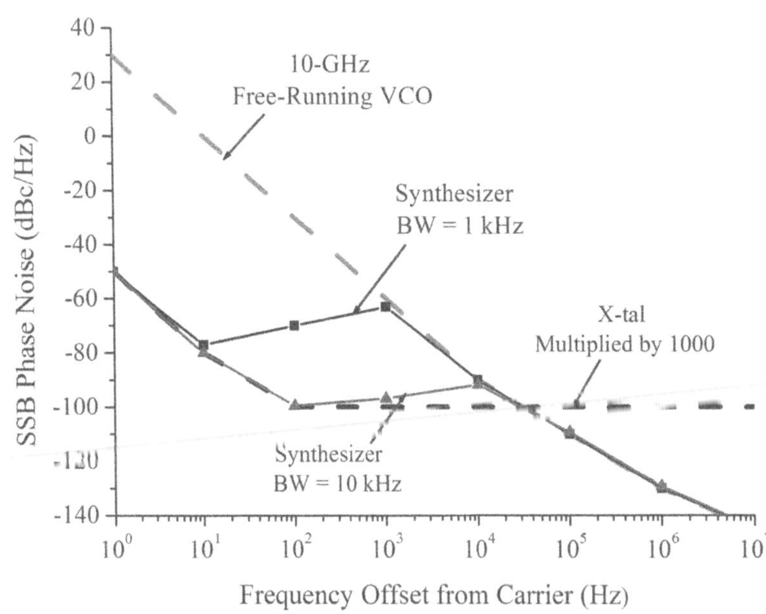

Figure 11E.1 PLL output phase noise. When the loop bandwidth is smaller than the intercept point of the two phase noises, then the PLL phase noise floor in the loop bandwidth is raised.

The bandwidth BW of $H(s)$ is defined by $(BW) = \omega_n/(2\pi)$. From Figure 11E.1, the phase noise of the PLL output follows the phase noise of the crystal oscillator below the bandwidth, as expected, and it follows the phase noise of the VCO above the bandwidth. Note that the phase noise floor rises for a small bandwidth such as 1 kHz. Therefore, a bandwidth of approximately 40 kHz, which is the intersection of the two phase noise curves, is appropriate for obtaining the optimum phase noise. As in the example, the PLL bandwidth (BW) can be selected for the optimum phase noise; however, the BW is sometimes determined by considering both spurious attenuation and *lock time*.

Example 11.2

The phase noise of a 10-MHz OCXO (oven-controlled crystal oscillator) and that of a 100-MHz VCXO (voltage-controlled crystal oscillator) are shown in Table 11E.2. It is necessary to phase-lock the VCXO to the OCXO (its frequency multiplied by 10) using a PLL. Determine the optimum bandwidth and the approximate phase noise characteristic of the phase-locked VCXO.

Table 11E.2 Phase noise of the oscillators

Frequency Offset	OCXO (10 MHz) Phase Noise (dBc/Hz)	VCXO (100 MHz) Phase Noise (dBc/Hz)
10 Hz	–115	–90
100 Hz	–135	–115
1 kHz	–145	–135
10 kHz	–150	–150
100 kHz	–150	–150
1 MHz	–150	–150

Solution

In order for the frequencies of the OCXO and VCXO to be equal, it is necessary to multiply the frequency of the OCXO by 10. Denoting the phase noise of the OCXO as $L_{OXCO}(f)$, then the phase noise multiplied by a factor of 10 is

$$L_{OCXO}(f) + 20\log 10 = L_{OCXO}(f) + 20$$

The phase noise of a 100-MHz OCXO is shown in the second column of Table 11E.3. In addition, for the purpose of comparison, the phase noise of the 100-MHz VCXO is shown in the third column. Then, using an ideal loop filter that yields $|H(s)| = 1$ inside the loop bandwidth while $|H(s)| = 0$ outside the loop bandwidth, the phase noise output of the phase-locked VCXO is found by selecting the minimum of the two phase noises in the second and third columns. A plot of the phase noises of the 100-MHz OCXO and of the VCXO in Table 11E.3 are shown in Figure 11E.2. Both the phase noises can be seen to be equal at an offset frequency of 100 Hz, which is thus the optimum loop bandwidth. In addition, the minimum of the two phase noise curves in the figure becomes the phase noise output of the phase-locked VCXO.

Table 11E.3 Phase noise worksheet

Frequency Offset	OCXO ×10 (dBc/Hz)	100-MHz VCXO (dBc/Hz)	PLL Output (dBc/Hz)
10 Hz	–95	–90	–95
100 Hz	–115	–115	–115
1 kHz	–125	–135	–135
10 kHz	–130	–150	–150
100 kHz	–130	–150	–150
1 MHz	–130	–150	–150

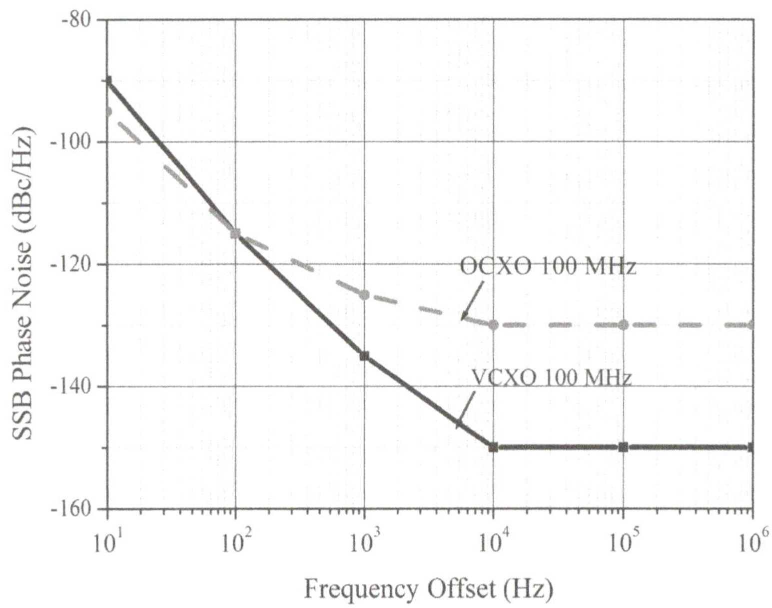

Figure 11E.2 Phase noises of a 100-MHz VCXO and a 100-MHz OCXO with respect to offset frequency. The optimal loop bandwidth for the PLL phase noise is about 100 Hz.

To configure the PLL in Figure 11.2, the frequencies of the VCO and of the reference oscillator must be equal. This is practically impossible to achieve as it is common for the reference oscillator's frequency to be low, while the VCO's is usually high. The problem can be solved by using a frequency multiplier or frequency divider in the PLL. The frequency divider is often used to configure the frequency synthesizer. Two configurations of the PLL using the frequency divider are shown in Figure 11.4.

The operations of the two PLLs can be explained using the steady-state concept. Assuming the loop filters in Figure 11.4(a) and 11.4(b) are implemented using a kind of integrator, the output voltage of the phase detector will be zero in the steady-state. Thus,

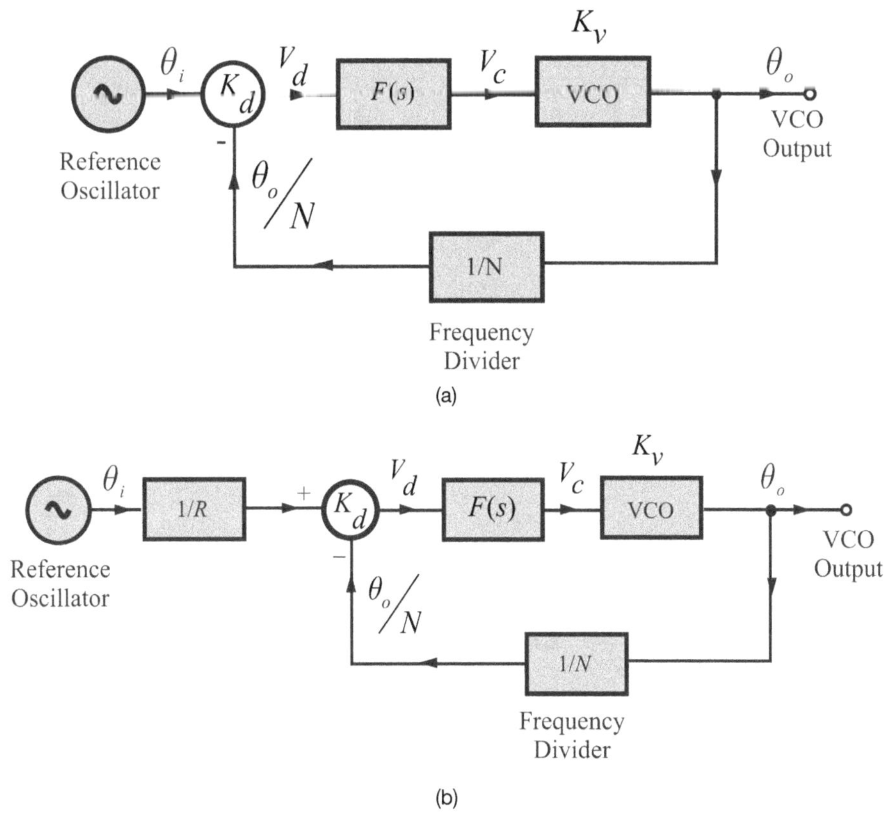

Figure 11.4 Frequency synthesizers; the frequency of the reference oscillator is (a) undivided and (b) divided by R

in the case of the frequency synthesizers shown in Figure 11.4(a), Equation (11.10) is satisfied.

$$\theta_i = \frac{\theta_o}{N} \tag{11.10}$$

The output frequency of the PLL can be obtained by differentiating Equation (11.10) with respect to time, which results in $\omega_o = N\omega_i$. Thus, the PLL output frequency can be obtained by multiplying the frequency of the reference oscillator by N. In addition, from Equation (11.9), the output phase noise is equal to the phase noise of the reference oscillator (multiplied by N) in the loop bandwidth, while it is equal to the phase noise of the VCO outside the loop bandwidth.

In the case of the reference oscillator shown in Figure 11.4(b), whose frequency is R-divided, Equation (11.11) is written by the same method as

$$\frac{\theta_i}{R} = \frac{\theta_o}{N} \rightarrow \theta_o = \theta_i \frac{N}{R} \tag{11.11}$$

Therefore, by varying the frequency division ratio, a frequency corresponding to integer multiples of the frequency step ω_i/R becomes the output frequency. At the same time, the phase noise of the reference oscillator multiplied by N/R is obtained inside the loop bandwidth, while outside the bandwidth, the phase noise of the VCO is obtained.

Example 11.3

Calculate the output frequency of the PLL with the configuration shown in Figure 11E.3.

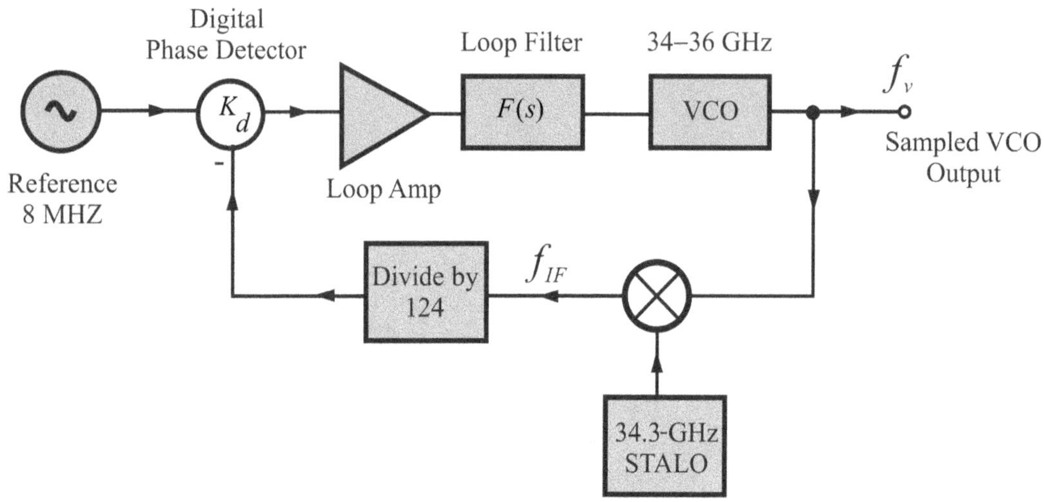

Figure 11E.3 Example of a PLL configuration

Solution

Assuming a steady state, the output of the phase detector is 0 and the frequencies of the two input signals of the phase detector are equal. Therefore, the output frequency of the mixer becomes

$$f_{IF} = 8 \times 124 \text{ MHz}$$

and the output frequency of the VCO is

$$f_v = f_{IF} + 34.3 = 35.292 \text{ GHz}$$

11.3 PLL COMPONENTS

The key components in a PLL include a VCO (voltage-controlled oscillator), a loop filter, a phase detector, and a frequency divider. Since we have already discussed the VCO in Chapter 10, in this section we will focus on the phase detector and frequency divider.

11.3.1 Phase Detector

In the past, due to the limitations in semiconductor manufacturing technology, it was not possible to fabricate high-speed logic circuits. DBMs (doubly balanced mixers) were widely used as high frequency phase detectors. However, thanks to recent technological breakthroughs, high-speed logic circuits can be fabricated for use as high-frequency phase detectors. The detailed configuration and characteristics of a DBM will be discussed in the next chapter. In this section, although the DBM is not widely used as a phase detector, we will briefly discuss the characteristics of a DBM phase detector and compare it with the high-speed phase detectors that use logic circuits. A DBM phase detector is shown in Figure 11.5.

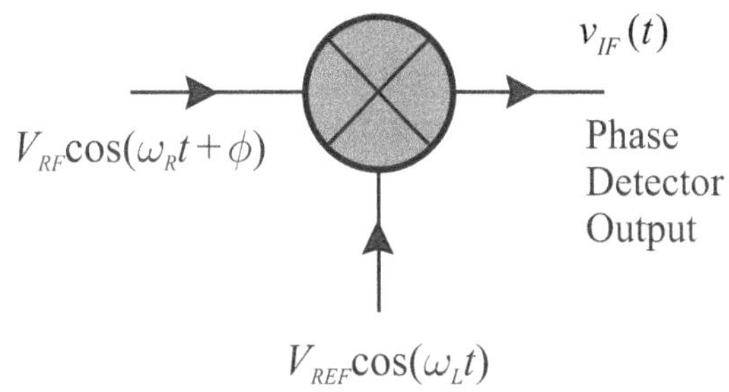

Figure 11.5 DBM phase detector

The output voltage $v_{IF}(t)$ of the DBM in Figure 11.5 can be written as Equation (11.12).

$$v_{IF}(t) = K_L V_{RF}\cos\big((\omega_R - \omega_L)t + \phi\big) + K_L V_{RF}\cos\big((\omega_R + \omega_L)t + \phi\big) \quad (11.12)$$

It should be noted here that $v_{IF}(t)$ in Equation (11.12) is independent of the LO signal amplitude, V_{REF}. This is possible when the amplitude V_{REF} of the LO signal in Figure 11.5 is sufficiently large. As a result, the output $v_{IF}(t)$ is proportional only to the RF signal amplitude, V_{RF}. Also note that $v_{IF}(t)$ is composed of two components: the frequency up-converted and down-converted components. However, only the down-converted components are used as the output of phase detector. Thus, when the frequencies of the LO and RF signals are equal, the output of the phase detector is expressed in Equation (11.13),

$$V_d = K_L V_{RF}\cos(\phi) = V_{bpeak}\cos(\phi) \quad (11.13)$$

which is proportional to the cosine of the phase difference between the two signals. Therefore, the output of the DBM phase detector acts as shown in Figure 11.6.

From Figure 11.6, the output of the phase detector has a period of 2π for the phase difference between the two signals, and it has a positive slope in the range of +180°–+360°. Therefore, phase-locking is possible for the phase difference in the range of +180°–+360°, but the linear range appears in a narrow range around 270°. The output voltage of the phase detector is 0 when the phase difference is 270°, and thus the reference signal and the VCO synchronize with the phase difference of 90° when a DBM phase detector is used in a

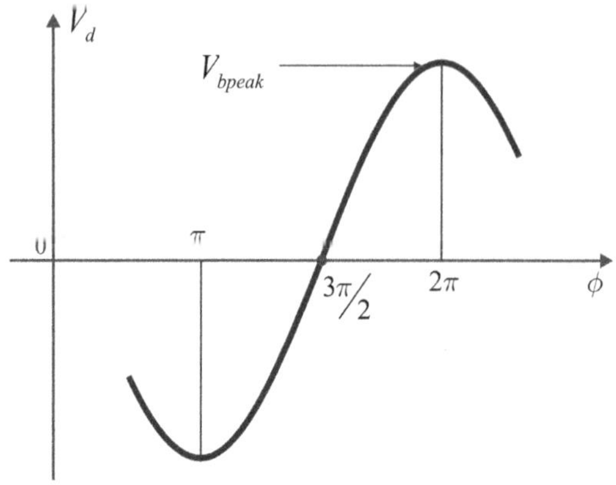

Figure 11.6 Output voltage of a DBM phase detector. The positive slope appears around $3\pi/2$ and the negative slope around $\pi/2$. Depending on the VCO's tuning characteristics, the slope of the phase detector can be chosen.

PLL. As is obvious from Equation (11.13), the disadvantage of the DBM phase detector is that the linear range occurs only in a limited range of the phase difference. Also, because the phase detector constant is proportional to the RF amplitude, the RF amplitude should be kept constant to preserve the phase detector constant given by the $K_v = K_L V_{RF}$ constant.

Example 11.4

A 500-MHz reference oscillator with a power of 10 dBm and a 500.010-MHz signal with a power of 0 dBm are applied to the LO and RF inputs of the DBM phase detector. At the phase detector's output, a peak-to-peak voltage of 100 mV and a 10-kHz sinusoidal waveform appear when measured with an oscilloscope. What is the value of the phase detector constant K_d?

Solution

A 10-kHz sinusoidal waveform and a 100-mV peak-to-peak voltage can be expressed as $V_p\cos(\omega t + \phi_1)$, where $V_p = 50$ mV. Thus, from Equation (11.13), $V_{bpeak} = V_p$, and $V_{bpeak} = 50$ mV. When the frequencies of the two signals are made equal, and the phase difference between the two signals is set to $90°+\phi$, the phase detector output becomes $V_d \cong V_{bpeak}\phi$. Thus, $K_d = V_{bpeak}$ and $K_d = 50$ mV/rad is obtained.

The DBM phase detector operates as a linear phase detector for a small phase difference. The linear range of the phase detector can be significantly increased by implementing the phase detector that uses logic circuits, and a typical example is the Exclusive OR (XOR) phase detector shown in Figure 11.7.

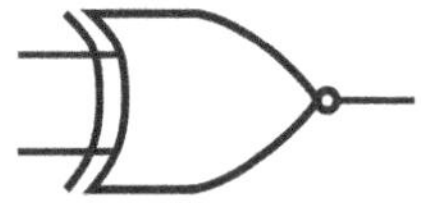

Figure 11.7 Exclusive OR phase detector

As shown in Table 11.1, the XOR gives the logical output for the two logical signal inputs. Consider the supply voltage V_{CC} and 0 (GND) as the logic values that correspond

Table 11.1 Exclusive OR truth table

SIG	REF	OUTPUT
L	L	L
H	L	H
L	H	H
H	H	L

to H (logic high) and L (logic low), respectively. Therefore, the XOR output yields H (V_{CC}) for the logic inputs pairs (H, L) or (L, H). When two waveforms with the same frequency are applied to the inputs of the XOR, the XOR output waveform appears, as shown in Figure 11.8. When the XOR is viewed as a phase detector, its output is the average DC voltage of the output waveform also shown in Figure 11.8.

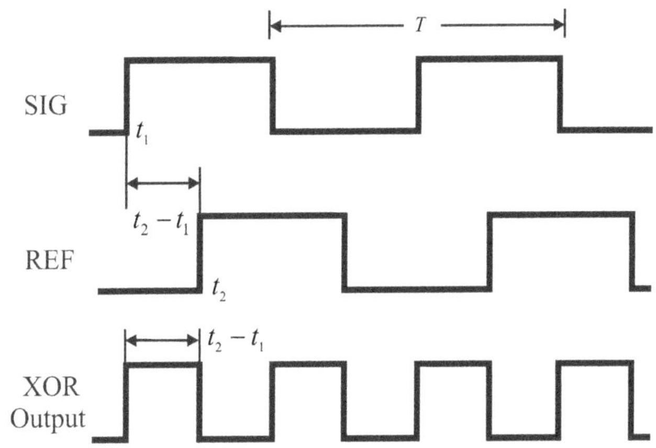

Figure 11.8 Output waveform of an XOR phase detector. The time average or DC component of the XOR output is proportional to the phase difference between SIG and REF.

From Figure 11.8, the output waveform shows two pulse tips per period T. The width and amplitude are $t_2 - t_1$ and V_{CC}, respectively. Thus, the average DC voltage can be expressed as Equation (11.14a).

$$V_d = 2V_{CC}\frac{t_2 - t_1}{T} = 2V_{CC}\frac{\omega(t_2 - t_1)}{\omega T} = V_{CC}\frac{\theta}{\pi} \tag{11.14a}$$

Here, θ corresponds to the phase difference between the two signals, and the phase detector constant is given by

$$K_d = \frac{V_{CC}}{\pi} \tag{11.14b}$$

Therefore, the XOR phase detector gives an output that is proportional to θ and has a phase detector constant given by Equation (11.14b).

When the two inputs applied to the XOR phase detector have a phase difference of less than 180°, the output voltage increases in proportion to the phase difference, as shown in Figure 11.9. Conversely, when the phase difference is greater than 180°, the phase detector output decreases as shown in Figure 11.9. From that figure, phase detector voltage $V_{d,1}$ arises for two phase difference values θ_1, and θ_2. An XOR phase detector does not have a one-to-one correspondence with the phase difference. The reason for this phenomenon is that the two pulse tips occur per period as a result of the XOR's operation. Therefore, the XOR phase detector can be used as a phase detector with a positive slope for a phase difference range of $0° < \theta < 180°$.

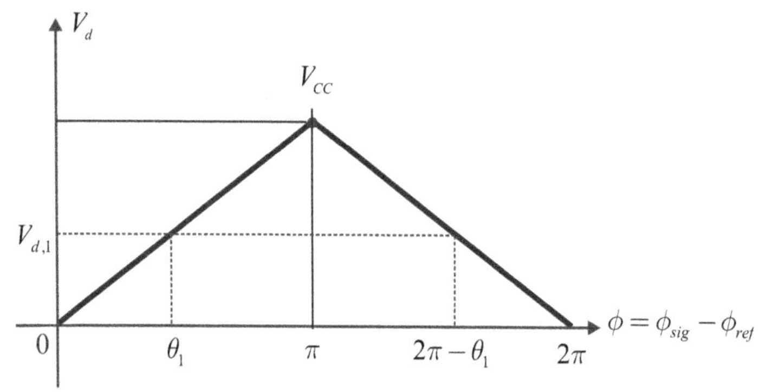

Figure 11.9 XOR phase detector voltage V_d with respect to phase difference

Another type of phase detector can be constructed that uses a sequential logic circuit, as shown in Figure 11.10. The phase detector in that figure is usually referred to as a PFD (phase frequency detector). The PFD can be configured with various sequential logic circuits.

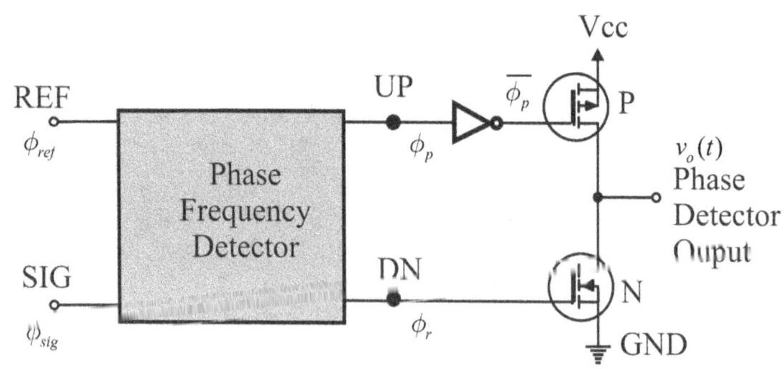

Figure 11.10 PFD (phase frequency detector) circuit. When the UP signal is high, the PMOS (*p*-type MOS) is on and the output voltage is high. When the DN is high, the NMOS (*n*-type MOS) is on and the output voltage is low. Otherwise, both MOSs are open and the output voltage is $V_{CC}/2$ (a high impedance state).

If the reference signal (REF) is leading when compared to the comparison signal (SIG), as shown in Figure 11.11(a), the UP output of the PFD in Figure 11.10 climbs to H at the rising edge of the REF signal and falls to L at the rising edge of the SIG. Conversely, if the comparison signal is leading when compared to the REF signal, as shown in Figure 11.11(b),

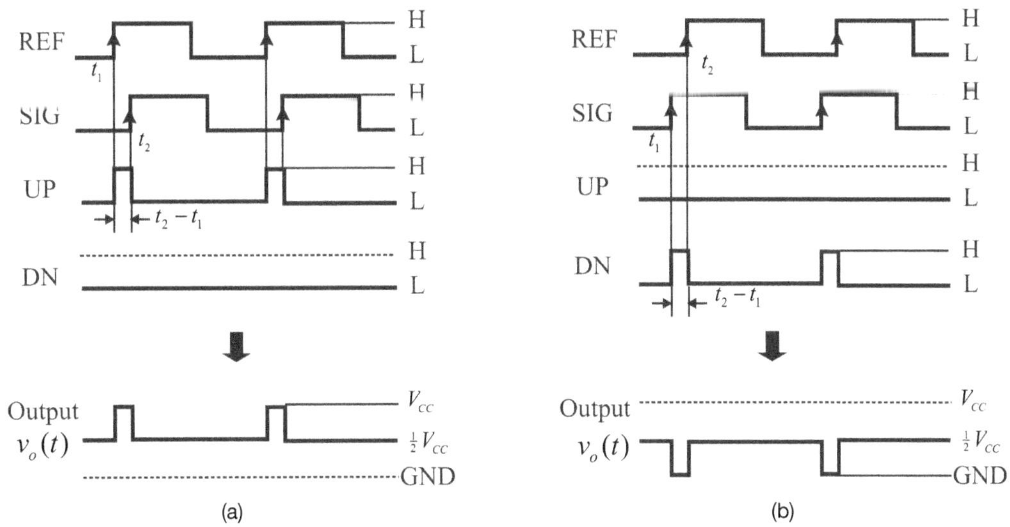

Figure 11.11 PFD waveforms: (a) UP and (b) DN

the DN output climbs to H at the rising edge of the SIG and falls to L at the rising edge of the REF signal. In addition, when the UP is in the H-state, the DN shows no change even at the rising edge of SIG. Similarly, when the DN is in the H-state, the UP shows no change even at the rising edge of REF.

Thus, for the UP and DN outputs, the phase detector output voltage $v_o(t)$ will appear as shown in Figs. 11.11(a) and (b), respectively. Note that in the absence of input, $v_o(t)$ stays at one-half of the supply voltage. In this state, the external circuit is also cut off from the PFD circuit and is said to be in a *high impedance state*.

Thus, for input waveforms of the same frequency, the PFD yields a pulse tip only once in a period, as shown in Figure 11.11. Since the phase detector output is the average DC voltage of $v_o(t)$, it can be expressed as Equation (11.15)

$$V_d = \frac{1}{2}V_{CC}\frac{t_2 - t_1}{T} = \frac{1}{2}V_{CC}\frac{\omega(t_2 - t_1)}{\omega T} = V_{CC}\frac{\theta}{4\pi} \tag{11.15}$$

and the output is proportional to θ. This is shown in Figure 11.12.

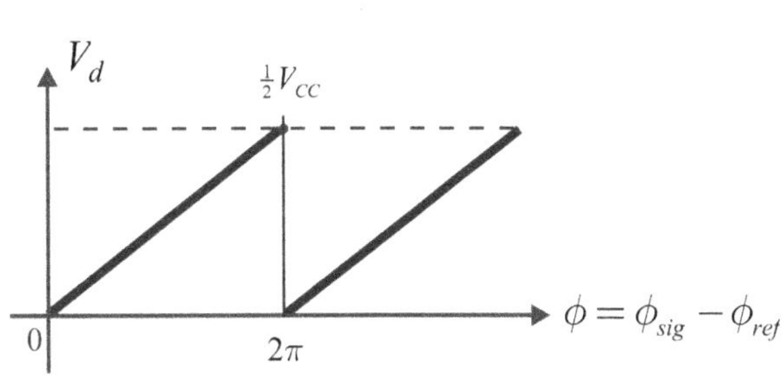

Figure 11.12 Phase detector voltage of PFD with respect to phase difference. The PFD has a one-to-one correspondence between the phase detector voltage output and the phase difference.

Note that the phase detector output is at its maximum when the phase difference between the two waveforms is 360°; also note that the phase detector voltage has a one-to-one relationship with the phase difference in the range of 0°–360°. In addition, compared with the other phase detectors previously discussed, the PFD is found to operate as a linear phase detector for a wide range of phase difference.

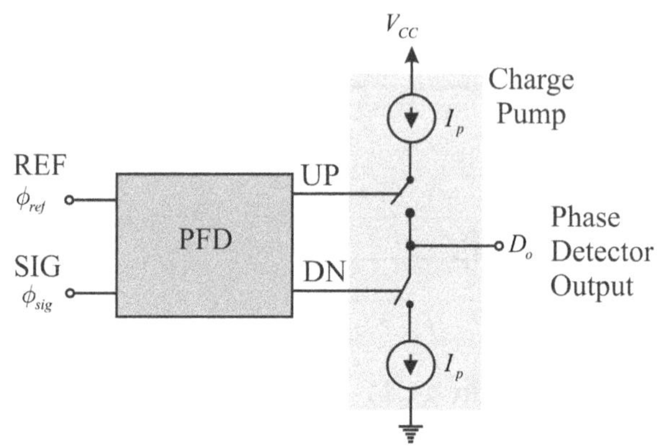

Figure 11.13 PFD circuit with a charge pump. The charge pump circuit makes the constant current flow out of the phase detector output when the UP is high. In contrast, the constant current flows into the phase detector output when the DN is high.

Note that the PFD output appears as a voltage. When the output of the phase detector is a voltage, an ideal integrator for the loop filter is difficult to implement with passive components. Thus, a current drive circuit is inserted at the end of the PFD shown in Figure 11.13. The circuit in that figure provides a constant current I_p flowing toward the load when the UP is H, whereas the constant current I_p flows from the load when the DN is H. With such a configuration, the phase detector voltage V_d of Equation (11.15) is converted to a current I_d and the relationship between the phase detector output current and the phase difference is written as Equation (11.16).

$$I_d = I_p \frac{\theta}{2\pi} \tag{11.16}$$

The current drive circuit is called a *charge pump*. Modern phase detectors are configured by connecting a charge pump to the PFD and the phase detector then provides a current proportional to the phase difference. With this type of configuration, the tuning voltage of the VCO becomes the phase detector current multiplied by the impedance of an external loop filter ($V_c = I_d Z(s)$), which drives the VCO. The integrator can thus be implemented with a simple capacitor. By easily connecting a capacitor to the phase detector in Figure 11.13, the tuning voltage becomes the integral of the phase detector current. However, in the case of the voltage-type phase detector, an integrator cannot be implemented with a simple capacitor.

Example 11.5

What is the phase detector constant when the output current I_p in the D_n output of Figure 11.13 is 5 mA?

Solution

$$K_d = \frac{I_p}{2\pi} = 5m/2\pi = 0.80 \times 10^{-3} \text{ [A/rad]}$$

As described earlier, the PFD provides a voltage output proportional to the phase difference and that voltage output is converted to a current output using a charge pump. However, the output of a typical phase detector is accompanied by a noise floor that can cause problems in the detection of the phase difference. Usually, the larger the frequency division ratio becomes, the smaller the phase noise of the frequency divided by the output of the VCO. The smaller divided phase noise is then applied to the input of a phase detector, which may be less than the noise floor of the phase detector. If the divided phase noise of the VCO is less than that noise floor, the output of the phase detector is simply its own background noise and the desired phase difference cannot be obtained.

The noise floor of a PFD phase detector is usually presented as a specification of the phase detector and an example of the PFD noise floor is shown in Figure 11.14. In that figure, when the comparison frequency increases, the noise floor of the phase detector also increases. The specification of the phase detector is usually defined as the phase detector noise at a comparison frequency of 1 Hz. Therefore, at a comparison frequency of f_r, the noise floor of the phase detector can be empirically expressed as

$$\begin{aligned}&\text{PD noise floor } (f_r) \text{ (dBc/Hz)}\\&= 1 \text{ Hz PD noise floor } + 10\log(\text{comparison frequency}) \text{ (dBc/Hz)}\end{aligned} \tag{11.17}$$

The empirical Equation (11.17) can be verified using Figure 11.14, in which the noise floor of the PFD at a comparison frequency of 10 kHz and 100 kHz is about –175 dBc/Hz and –165 dBc/Hz, respectively. Because of this, the PFD noise floor increases by 10 dB due to a decade change of the comparison frequency. Equation (11.17) clearly shows the computed decade change of the noise floor. Therefore, when the noise floor at an arbitrary comparison frequency is known, then the noise floor at all other comparison frequencies can be determined using Equation (11.17).

To examine the effect of the phase detector noise floor, consider the case in which the frequency of the VCO is divided by N in order to lower the VCO frequency to the comparison frequency. Denoting the phase noise of the VCO as $\phi_v(f)$, the divided phase noise of the VCO is $\phi_v(f)/N^2$. Also, denoting the noise floor of the phase detector as ϕ_{PD}, the phase detector provides an output proportional to the phase difference for the frequency-divided VCO input when the phase noise of the VCO is greater than ϕ_{PD}; this is given by

$$\frac{|\phi_v(f)|}{N^2} > |\phi_{PD}| \tag{11.18}$$

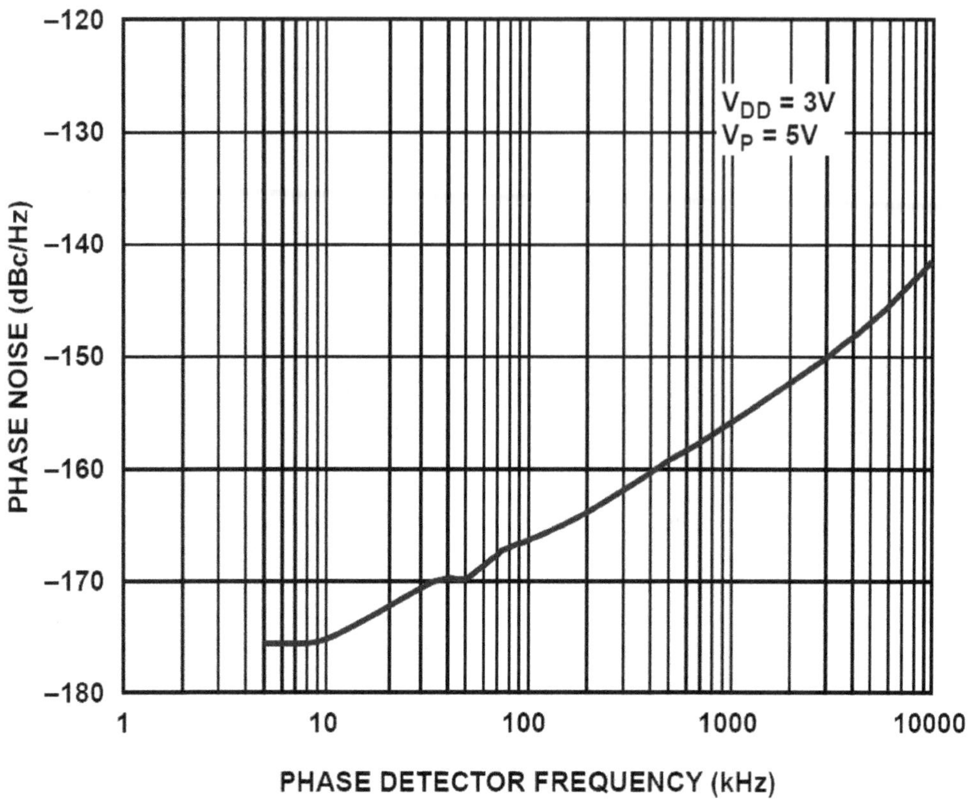

Figure 11.14 Noise floor of the phase detector. (*Source:* Analog Devices, RF PLL Frequency Synthesizers ADF4110/4111/4112/4113, August 2012.)

For noise levels lower than the value given by Equation (11.18), the phase detector comparison function is disabled. Equation (11.18) can then determine the offset frequency f_m of the VCO. Since the phase noise of the VCO increases as the frequency offset approaches 0 below f_m, the phase detector provides an output proportional to the VCO phase noise, while the phase detector provides an output corresponding to its noise floor above f_m. In addition, as the division ratio N in Equation (11.18) gets larger, the phase detector provides an output proportional to the VCO phase noise for only a narrow range of the offset frequency.

Example 11.6

A 900-MHz frequency synthesizer must be configured using a 900-MHz VCO and a 10-MHz crystal oscillator, both of which have the phase noise values shown in Table 11E.4. The normalized noise floor of the phase detector used in the PLL at 1 Hz is –210 dBc/Hz and the comparison frequency is 50 kHz. Calculate the bandwidth of a loop filter that is appropriate for the optimum phase noise output; also calculate the output phase noise of the 900-MHz frequency synthesizer. In addition, compare the output phase noise in this Example when the phase detector has no background noise.

Table 11E.4 Phase noise of the oscillators

Frequency Offset	10-MHz CXO (dBc/Hz)	900-MHz VCO (dBc/Hz)
10 Hz	−110	−20
100 Hz	−130	−55
1 kHz	−140	−85
10 kHz	−145	−110
100 kHz	−145	−130
1 MHz	−145	−150

Solution

Denoting the phase noise of the 10-MHz reference oscillator, which is a function of the offset frequency f, as $L(f)$, the phase noise at the comparison frequency of 50 kHz is $L(f) + 20\log(0.05/10)$. This is shown in the second column of Table 11E.5. Similarly, the phase noise of the VCO can be calculated by dividing its frequency by the comparison frequency of 50 kHz. The phase noise of a 50-kHz VCO is shown in the fifth column of Table 11E.5.

Table 11E.5 Phase noise worksheet

(1) Frequency Offset	(2) Reference Freq. 50 kHz	(3) PFD Noise Floor	(4) Max(2,3)	(5) VCO 50 kHz	(6) Synthesized Output 50 kHz	(7) 900-MHz Phase Noise
10 Hz	−156	−163	−156	−105	−156	−71
100 Hz	−176	−163	−163	−140	−163	−78
1 kHz	−186	−163	−163	−170	−170	−85
10 kHz	−191	−162	−162	−195	−195	−110
100 kHz	−191	−158	−158	−215	−215	−130
1 MHz	−191	−150	−150	−235	−235	−150

Also, since the noise floor of the given phase detector at 1 Hz is −210 dBc/Hz, the noise floor of the phase detector at the comparison frequency of 50 kHz can be calculated from Equation (11.17), from which a noise floor of −163 dBc/Hz is obtained. The noise floor value corresponds to the case in which the frequency offset is 0. The noise floor values are shown in the third column of Table 11E.5. The noise floor value appears to be constant up to the frequency offset of 1 kHz, which is not far from the comparison frequency, but the noise floor values significantly change above a frequency offset of 10 kHz.

The phase noise of the reference oscillator is independent of the phase detector noise floor. The maximum of the reference oscillator's phase noise and phase detector's noise floor becomes the new reference phase noise that is shown in the fourth column. Therefore, the phase noise synthesized by the PLL is obtained by comparing the newly defined reference phase noise and the VCO phase noise at 50 kHz. The minimum of the newly defined reference phase noise and the VCO phase noise becomes the phase noise of the frequency synthesizer. Comparing columns 4 and 5, the intersection is found to occur between 100 Hz–1 kHz. Thus, setting the loop bandwidth to approximately 100 Hz–1 kHz will give the optimum results. The synthesized phase noise at 900 MHz is thus calculated using the synthesized phase noise at 50 kHz by multiplying the latter by $n = 900/0.05 = 18{,}000$. This is shown in the last column of Table 11E.6.

In contrast, in the absence of the background noise in the phase detector, the values of the phase noise can be calculated in the same way as in Example 11.2. This is summarized in Table 11E.6. From that table, due to the effect of the phase detector noise floor on the frequency synthesizer, a difference of about 20 dB can be seen to occur between the frequency offset of 100 Hz–1 kHz.

Table 11E.6 Phase noise worksheet in the absence of the phase detector noise floor

(1) Frequency Offset	(2) Reference Frequency 50 kHz	(5) VCO 50 kHz	(6) Synthesized Output 50 kHz	(7) 900-MHz Phase Noise
10 Hz	−156	−105	−156	−71
100 Hz	−176	−140	−176	−91
1 kHz	−186	−170	−186	−101
10 kHz	−191	−195	−195	−110
100 kHz	−191	−215	−215	−130
1 MHz	−191	−235	−235	−150

11.3.2 Frequency Divider

Most frequency dividers are implemented using digital counters. In this section, we will explain a frequency divider that is based on the operation of digital counters; however, the discussion will not include details of the digital counter circuits. The operation of a frequency divider is shown in Figure 11.15. When the VCO input shown in that figure is applied to the frequency divider's input, the frequency divider shows changes only at the rising edges of the VCO waveform. When the rising edge is detected, *toggling* from a previous state to another state occurs. That is, the frequency divider changes to the "high" state if it was in the "low" state and vice versa. Through the toggling operation, the VCO frequency divided by 2 appears as an output, as in the second waveform shown in Figure 11.15. In addition, with continuous repetition of the process, waveforms whose frequencies are divided by 4, 8, 16, and so on, can be obtained. In this operation, the toggling is assumed to occur only at the rising edge of the VCO waveform. However, for some digital circuits the toggling can occur at the falling edges of the input and the frequency divider can be implemented with that falling-edge toggling.

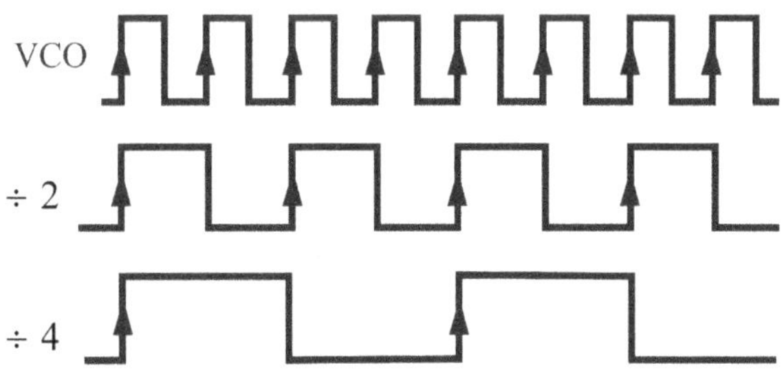

Figure 11.15 Operation of a divide-by-2 and divide-by-4

Generally, there are two types of frequency dividers: a fixed divider (with a fixed division ratio such as /2, /4, /8, and so on, as shown in Figure 11.15) known as a *prescaler* and a programmable divider whose division ratio can be determined by a program (i.e., a *programmable divider or counter*). The fixed divider generally operates up to high frequencies, while the programmable divider that is required for a PLL operates at low frequencies.

Thus, in order to divide the high-frequency output of the VCO, a high-speed fixed divider is first placed in the frequency divider's chains, as shown in Figure 11.16, and the programmable divider follows the fixed divider. However, as shown in that figure, due to the fixed divider, it is not possible to implement a frequency synthesizer that successively increases with a frequency step of the comparison frequency f_r. That is, when a frequency divider chain is built using the fixed P-divider and the programmable M-divider, the frequency that can be synthesized is $f_o = PMf_r$ ($M = 1,2, \ldots$). Thus, the frequency step of the frequency synthesizer becomes Pf_r.

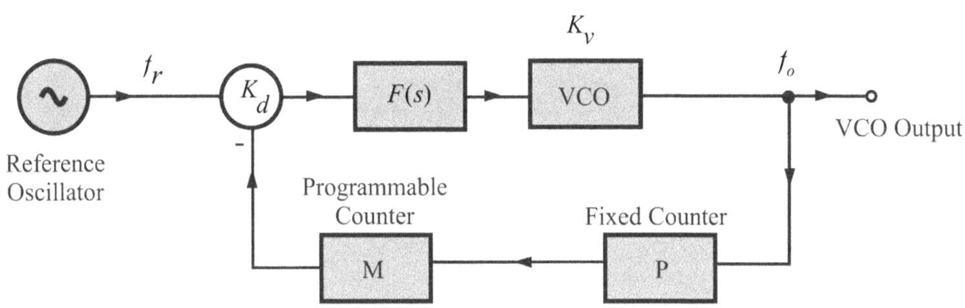

Figure 11.16 Configuration of a frequency synthesizer using a fixed divider

In contrast, for the high-speed fixed P-divider, it is possible to change the division ratio between P and $P + 1$ based on the control signal. This type of counter is called a *dual modulus counter*. Using a $P/P + 1$ dual modulus counter, the output frequency step of the frequency synthesizer can be lowered to f_r. Generally, the output frequency of the frequency synthesizer with a frequency step of f_r can be expressed as

$$f_o = Nf_r = (MP + A)f_r$$

where M and A are integers and represent the quotient and remainder, respectively, when N is divided by P. This can be rewritten as

$$N = MP + A = (M - A)P + A(P + 1) \tag{11.19}$$

Thus, based on Equation (11.19), the division ratio of N is accomplished by dividing the VCO signal A– times by $P + 1$, and then dividing $(M - A)$ times by P.

Example 11.7

For a 4/5 dual modulus counter, determine M and A for the consecutive numbers 960, 961, and 962. Also, determine the number that is divided by 4 and the number that is divided by 5.

Solution

$960 = 4 \times 240 + 0$, $M = 240$, $A = 0$; thus divided by 4, 240 times
$961 = 4 \times 240 + 1$, $M = 240$, $A = 1$; thus divided by 4, 239 times and by 5, 1 time
$962 = 4 \times 240 + 2$, $M = 240$, $A = 2$; thus divided by 4, 238 times and by 5, 2 times

---

To implement the right side of Equation (11.19), the divider chain is configured as shown in Figure 11.17. Then, one period of the square wave is generated after N cycles of the VCO input; that is, the frequency of the VCO can be divided by a general integer N, not by the integer multiples of P. The A-counter and M-counter of Figure 11.17 are programmable dividers, and the division ratios of M and A are initially determined by built-in registers. In addition, the division ratio of the dual modulus counter in Figure 11.17 is initially set to $P + 1$.

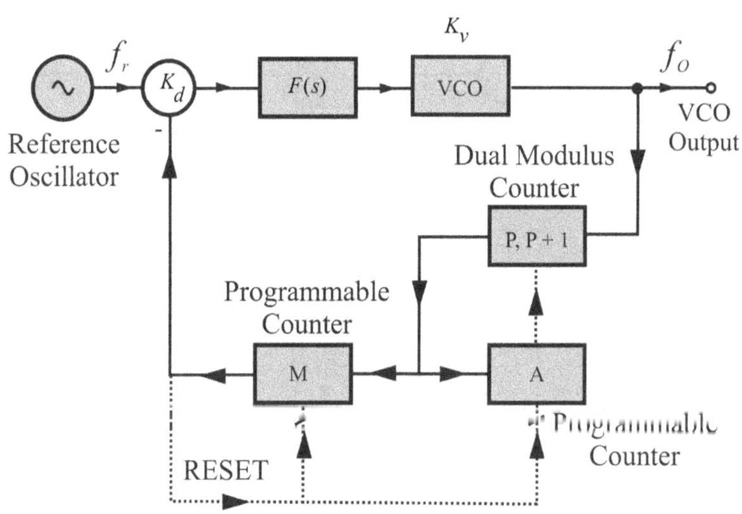

Figure 11.17 A PLL configuration using a dual modulus counter

From Figure 11.17, the VCO frequency divided by $P + 1$ is concurrently applied to the input of the A- and M-counters. When the A-counter completes A times of division, the A-counter generates the control signal (usually the rising edge occurring after counting is completed); the division ratio of the dual modulus counter is then set to P and no further division is performed in the A-counter. Thereafter, only the M-counter operates. Thus, for the rest of VCO cycle, the VCO frequency divided by P is the input to the M-counter, which then divides the input $(M - A)$ times. Similarly, the M-counter generates the "RESET" control signal when the division is completed and the M- and A-counters and the dual modulus counter return to their original states by the "RESET" control signal. After this, the operation explained above is repeated. As a result, the frequency of the VCO is divided by $N = (M - A)P + A(P + 1)$. That is, after N cycles of VCO output, one period of the square wave is generated. Therefore, the frequency divider then divides the VCO output frequency by an arbitrary integer N.

A fractional frequency divider can be configured by adding some structures to the architecture of the integer frequency divider shown in Figure 11.17. That is, when the division ratio is changed to N = 100, 100, 100, 101, it corresponds to an average division of 100.25. As a generalization, suppose that, among total F cycles, a signal frequency is divided by $N + 1$ for K cycles and then divided by N for the remaining $(F - K)$ cycles. The average division ratio M^* becomes,

$$M^* = \frac{K(N+1) + (F-K)N}{F} = N + \frac{K}{F}$$

which gives the same effect as dividing the signal frequency by a fraction. It should be noted that the integer K is less than F. The average frequency of the divided output obtained by this division is obviously equal to the VCO frequency divided by M^*, but the instantaneous frequency varies with time. The reason for implementing the fractional frequency divider in this way is that the frequency divider basically cannot perform fractional division and only integer division is possible. Thus, denoting the comparison frequency as f_r, the synthesized output frequency of the VCO f_o can be expressed in Equation (11.20).

$$f_o = \left(N + \frac{K}{F}\right) f_r \tag{11.20}$$

The implementation of Equation (11.12) depends on the periodic implementation of $N + 1$. Increasing the division ratio from N to $N + 1$ is done by increasing the division ratio of the A-counter by 1 from Equation (11.19). In addition, the periodic increment of the A-counter by 1 is accomplished by connecting the output of the M-counter (after the VCO input is divided by N or an $N + 1$ rising edge is generated) to the input of a separate counter (accumulator), which is made to generate K carries in F cycles. The carries then periodically increase the division ratio of the A-counter by 1. As described above, the implementation of 100,100,100,101 is done by implementing the accumulator as a 2-bit counter. The 2-bit counter generates a carry as

$$0, 1, 2, 3\uparrow 0, 1, 2, 3\uparrow \ldots \text{ (1 carry is generated every 4 cycles)}$$

When overflow occurs, the carry is generated. By passing it on to the A-counter, the A-counter is incremented by 1. Therefore, because the carry generation is accomplished with the accumulation of the rising edge outputs of the M-counter, the 2-bit counter is called the accumulator. The detailed architecture of the accumulator can be found in commercial application notes.[1]

1. Texas Instruments, Fractional/Integer-N PLL Basics, SWRA029, August 1999. Available online at www.ti.com/lit/an/swra029/swra029.pdf.

From the phase noise point of view, implementing the division ratio as a fraction has several advantages. First, it is not necessary to set the comparison frequency to be equal to the frequency step of the frequency synthesizer; the comparison frequency can be set to a higher frequency; for example, to make a frequency synthesizer of 890 MHz with a frequency step of 50 kHz using an integer divider, the comparison frequency must be f_r = 50 kHz. Thus, the division ratio becomes $N = 890/0.05 = 17{,}800$. In contrast, for $F = 16$, the comparison frequency $f_r = 50 \times 16 = 800$ kHz and the division ratio N_F becomes

$$N_F = 890 / 0.8 = \left(1112 + \frac{8}{16}\right)$$

In addition, note that the ratio between the integer and the fractional divider is $N/N_F = 16$. Due to the lowering of the division ratio by 20log(16) = 24 dB, the effect of the phase detector noise floor can be reduced by 24 dB according to Equation (11.18), assuming the noise floor of the phase detector is constant irrespective of the comparison frequency. Thus, the fractional frequency divider eliminates the need to divide the VCO frequency by a large number N and the *phase detector noise floor contribution is significantly reduced*. As a result, the phase detector noise floor may not be considered in the output phase noise of the frequency synthesizer. Also, because the comparison frequency is high, the loop filter bandwidth of the PLL can be made wider, giving the advantage of a faster lock time.

Example 11.8

A 900-MHz frequency synthesizer with a frequency step of 50 kHz must be configured using the 10-MHz reference oscillator given in Example 11.6. Compare the output phase noises of two frequency synthesizers: one uses the integer frequency divider and the other uses the fractional frequency divider with Fraction = 16.

Solution

The phase noises of the reference oscillator and VCO at the comparison frequency of 800 kHz (16 × 50 kHz) in the fractional frequency synthesizer can be computed in the same way as in Example 11.6. In addition, the noise floor of the phase detector at 800 kHz is similarly computed. This is shown in Table 11E.7, where the optimum loop bandwidth of the fractional frequency synthesizer appears to be between 1–10 kHz. The phase noise of the VCO at a frequency of 800 kHz is also calculated and is shown in the sixth column of Table 11E.7. By using the computed results, the phase noise of the fractional frequency synthesizer at 900 MHz is given in the last column.

N and K of the fractional frequency synthesizer can now be determined. Since

$$\frac{900\ \text{MHz}}{0.8\ \text{MHz}} = 1125 = \left(N + \frac{K}{F}\right)$$

then $N = 1125$ and $K = 0$. However, note that the frequency step becomes 0.8 MHz/16 = 0.05 MHz because $F = 16$. Thus, the frequency synthesizer can increase the frequency by a frequency step of 0.05 MHz (i.e., 50 kHz) when K is increased successively. The comparison of the phase noise of the fractional synthesizer with that of the integer frequency synthesizer in Example 11.6 is shown in Figure 11E.4. The phase noise of the fractional frequency synthesizer shows an improvement by approximately 10 dB in the offset frequency range of 10 Hz–1 kHz. The improvement is due to the reduced contribution of the phase detector noise floor compared with the integer frequency synthesizer.

Table 11E.7 Phase noise worksheet

(1) Frequency Offset	(2) Reference Frequency at 800 kHz	(3) PFD Noise Floor	(4) Max(2,3)	(5) VCO 800 kHz	(6) Synthesized Output at 800 kHz	(7) 900-MHz Phase Noise
10 Hz	−132	−151	−132	−81	−132	−71
100 Hz	−152	−151	−151	−116	−151	−90
1 kHz	−162	−151	−151	−146	−151	−90
10 kHz	−167	−151	−151	−171	−171	−110
100 kHz	−167	−150	−150	−191	−191	−130
1 MHz	−167	−147	−147	−211	−211	−150

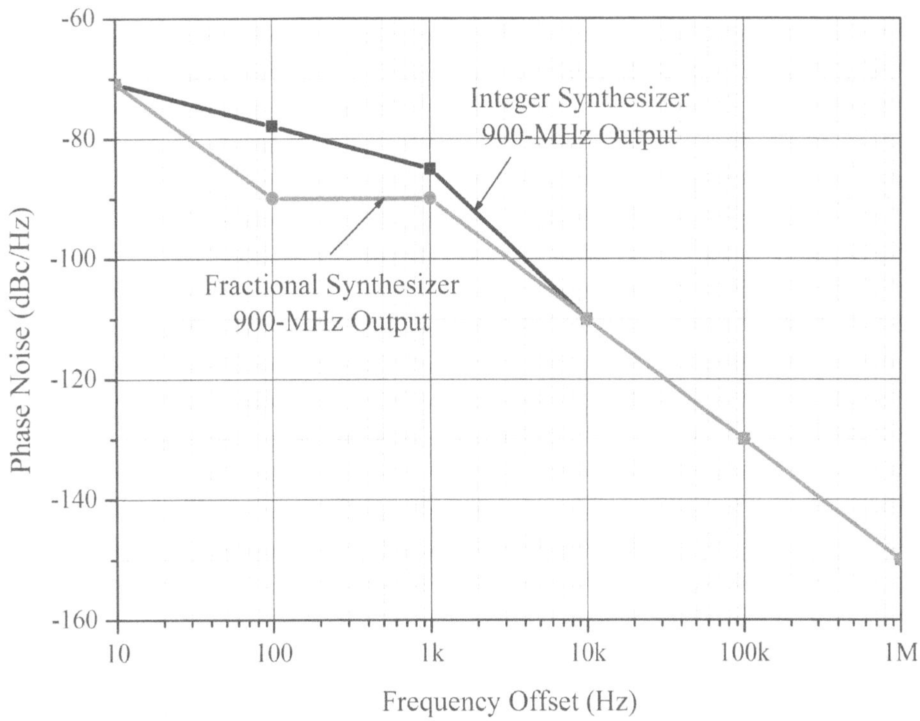

Figure 11E.4 Comparison of the phase noises of the fractional and integer frequency synthesizers. The fractional synthesizer provides a lower phase noise floor inside the loop bandwidth.

11.4 LOOP FILTERS

11.4.1 Loop Filter

A loop filter, unlike other components used in a PLL, determines the overall operation and characteristics of the PLL. Therefore, it is necessary to discuss the loop filter by considering the PLL's characteristics rather than by discussing the loop filter as a component. The loop filters are classified based on their component types and on the loop filters' transfer functions because those filters directly affect the transfer function of a PLL.

The architecture of a PLL is shown again in Figure 11.18. In that figure, because the PLL is structured as a feedback, it can be analyzed based on *open-loop gain* (simply referred to as *loop gain*). Denoting the transfer function of the loop filter by $F(s)$, the open-loop gain $L(s)$ is obtained as with Equation (11.21).

$$L(s) = \frac{KF(s)}{sN} = \frac{K(s^l + \alpha_1 s^{l-1} \cdots + \alpha_l)}{Ns^m(s^n + \beta_1 s^{n-1} + \beta_n)} \tag{11.21}$$

Note that $F(s)$ has the highest order of s^{m+n-1} in the denominator because the VCO contributes $1/s$ to the denominator.

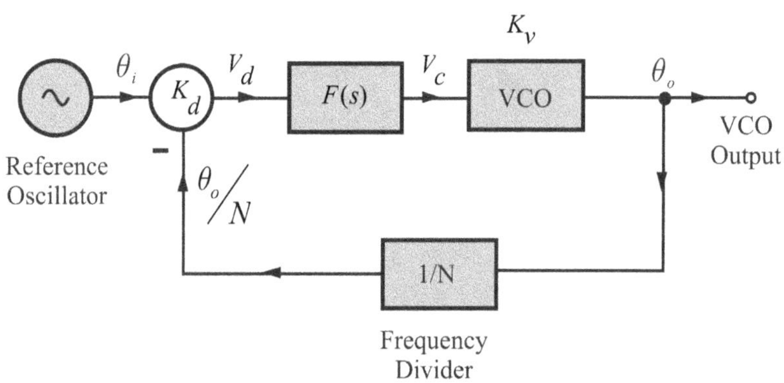

Figure 11.18 A PLL configuration

In addition, using the open-loop gain, the closed-loop gain can be expressed as Equation (11.22).

$$H(s) = N\frac{L(s)}{1 + L(s)} \tag{11.22}$$

The classification of the loop filter is based on the highest order of the denominator of $L(s)$, which is defined as the *order of the loop filter*. Therefore, for Equation (11.21), the order of the loop filter is $m + n$. In addition, even when the phase detector is directly connected to the VCO without the loop filter (in this case, $F(s) = 1$), the open-loop gain will have a term of s^{-1} due to the transfer function of the VCO. Thus, the order of the loop filter becomes a first order even when there is no loop filter in the PLL. Also, when a low-pass filter whose transfer function has a first-order denominator is included in the PLL, the order of the loop filter becomes a second order. The definition of the order may be different in other literature if the order is defined as the order of a loop filter alone.

In addition, the integration factor in Equation (11.21) is critical in determining the characteristics of the PLL, and its number is often used in the classification of loop filters. This is called the *type* of the loop filter. Equation (11.21) has the term s^m in the denominator and the loop filter is thus referred to as type m. That is, the type indicates the number of ideal integrators included in the transfer function. The closed-loop gain $H(s)$ of the PLL configuration in Figure 11.19 for a first-order loop filter is expressed as

$$H(s) = \frac{\theta_o}{\theta_i} = N\frac{L(s)}{1+L(s)} = \frac{N\frac{K}{sN}}{1+\frac{K}{sN}} = N\frac{\frac{K}{N}}{s+\frac{K}{N}} \tag{11.23}$$

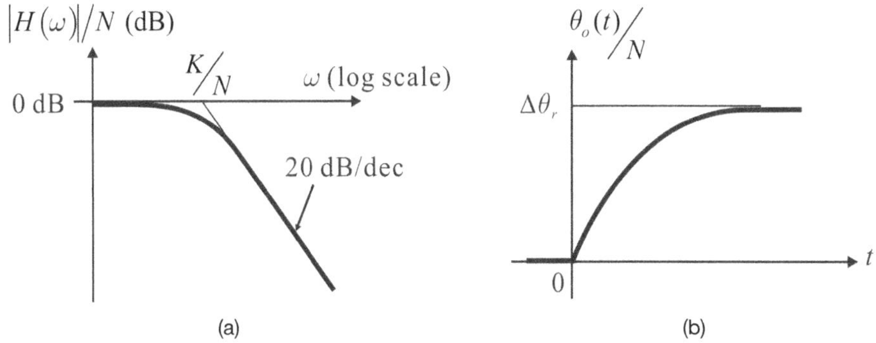

Figure 11.19 Responses of the first-order loop filter: (a) frequency response and (b) tracking for a step phase change

Therefore, from Equation (11.23), the 3-dB bandwidth (the frequency at which the magnitude of the open-loop gain is $1/2^{1/2}$) is K/N. As a result, the loop bandwidth is K/N.

When a step phase change ($\theta_i = \Delta\theta_r u(t)$) is applied to the input with the closed-loop transfer function in Equation (11.23), the output phase is

$$\theta_o = \theta_i H(s) = N\frac{\Delta\theta_r}{s}\frac{\frac{K}{N}}{s+\frac{K}{N}} \tag{11.24}$$

The inverse of the Laplace transform in Equation (11.24) can be written as

$$\theta_o = N\Delta\theta_r\left(1 - e^{-\frac{K}{N}t}\right) \tag{11.25}$$

From Equation (11.25), the output phase at the steady state can be seen to be $N\Delta\theta_r$ for a step phase change of $\Delta\theta_r$. Also note that the output frequency for a step frequency change follows the form of Equation (11.24) as expressed by Equation (11.26).

$$\omega_o = N\frac{\Delta\omega_r}{s}\frac{\frac{K}{N}}{s+\frac{K}{N}} \tag{11.26}$$

Thus, the output frequency tracks the step input frequency change as the output phase follows the step phase change. Figure 11.19 shows a first-order PLL frequency response and the output waveform for a step phase change.

Note that the phase detector output generates an error given by Equation (11.5) for tracking the step frequency change; however, the output frequency tracks the step frequency change without error. In the first-order PLL, the output is able to track the input frequency change that accompanies the phase error. In addition, as we saw in the previous explanation of phase detectors, not only the DC phase detector voltage, but also many of the comparison frequency's harmonic components are present in the actual phase detector output. Since the first-order loop filter applies the phase detector output to the VCO without filtering the harmonics, many frequency-modulated spurious components appear at the VCO's output due to those harmonics. Therefore, the first-order PLL is not commonly used due to its inability to remove the harmonics rather than because of the phase error problem.

11.4.2 Second-Order Loop Filters

The phase error and spurious components of the first-order PLL can, to some extent, be removed by configuring the PLL using a second-order loop filter. In order to remove the phase error, it is necessary to insert an integrator into the loop filter as previously explained. The second-order loop filter can be constructed by inserting an integrator in a first-order loop filter. The second-order loop filter transfer function can be expressed as

$$F(s) = \frac{1+s\tau}{s\tau_1} = \frac{1}{s\tau_1} + \frac{\tau_2}{\tau_1} \tag{11.27}$$

The first term in Equation (11.27) is the integration of the input and the second term is proportional to the input. Substituting Equation (11.27) into Equation (11.21), the open-loop gain can be obtained with Equation (11.28).

$$L(s) = \frac{KF(s)}{sN} = K\frac{1+s\tau_2}{s^2N\tau_1} \tag{11.28}$$

Thus, the closed-loop gain $H(s)$ is obtained as expressed in Equation (11.29).

$$H(s) = \frac{NL(s)}{1+L(s)} = \frac{K\frac{1+s\tau_2}{s^2N\tau_1}}{1+K\frac{1+s\tau_2}{s^2N\tau_1}} = \frac{N\left(2\zeta\omega_n s+\omega_n^2\right)}{s^2+2\zeta\omega_n s+\omega_n^2} \tag{11.29}$$

Here, ω_n and ζ are called the natural frequency and damping ratio, respectively, of the second-order transfer function, and ω_n and ζ are defined in Equations (11.30a) and (11.30b).

$$\omega_n = \left(\frac{K}{\tau_1 N}\right)^{1/2} \tag{11.30a}$$

$$\zeta = \left(\frac{K}{\tau_1 N}\right)^{1/2} \frac{\tau_2}{2} \tag{11.30b}$$

As $s \to 0$, $H(s)$ approaches N, and $H(s) \to 0$, for $s \to \infty$. Thus, the closed-loop gain $H(s)$ has the frequency response of a low-pass filter. Its frequency response is shown in Figure 11.20 with ζ as a parameter. The bandwidth is approximately ω_n, which varies slightly according to ζ. Calculating the 3-dB bandwidth results in Equation (11.31).

$$\omega_{3dB} = \omega_n \left[1 + 2\zeta^2 + \sqrt{\left(1 + 2\zeta^2\right)^2 + 1}\right]^{1/2} \tag{11.31}$$

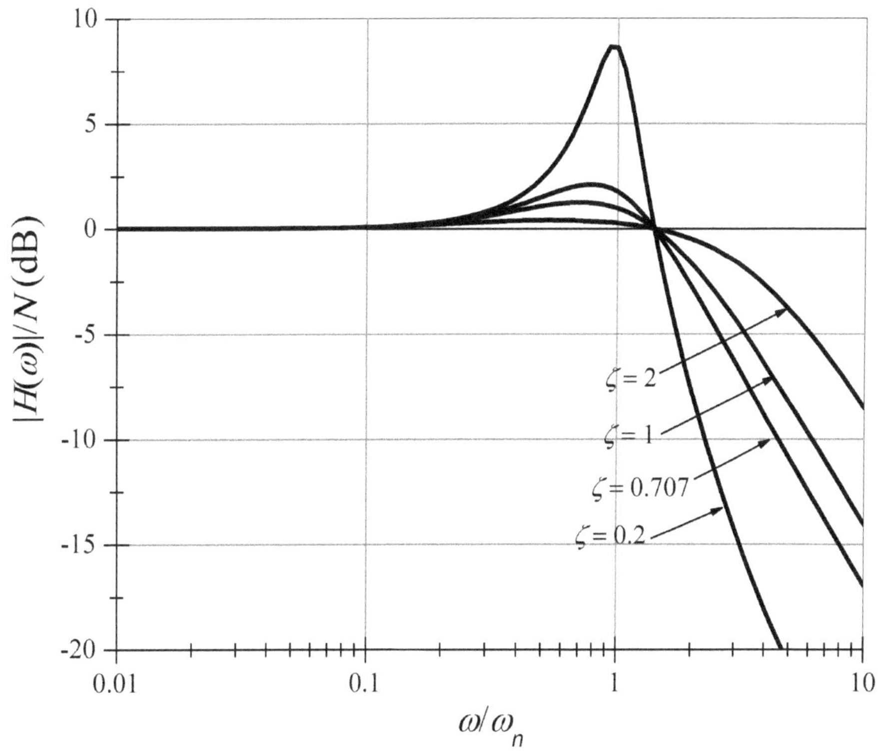

Figure 11.20 Frequency response of the closed-loop gain $H(s)$

From Equation (11.31), the loop bandwidth is close to the natural frequency and varies according to the damping ratio ζ. Therefore, the approximate loop bandwidth can be considered the natural frequency. From Equation (11.28), the magnitude of the open-loop gain $L(s)$ at the natural frequency ω_n can be approximated as 1 by ignoring the term $s\tau_2$, which is generally small at ω_n. Therefore, the natural frequency ω_n is *the frequency at which the magnitude of the open-loop gain becomes approximately* 1.

Using the transfer function of Equation (11.29), the output phase for a step phase change is found to be

$$\theta_o = \frac{\Delta\theta_r}{s} H(s) = N \frac{\Delta\theta_r}{s} \frac{\left(2\zeta\omega_n s + \omega_n^2\right)}{s^2 + 2\zeta\omega_n s + \omega_n^2}$$

and $\theta_o(t)$ can be obtained from an inverse Laplace transform expressed in Equation (11.32).

$$\frac{\theta_o(t)}{N} = 1 - e^{-\zeta\omega_n t}\left[\cos\left(\omega_n t\sqrt{1-\zeta^2}\right) - \frac{\zeta}{\sqrt{1-\zeta^2}}\sin\left(\omega_n t\sqrt{1-\zeta^2}\right)\right] \qquad (11.32)$$

The time response $\theta_o(t)$ is shown in Figure 11.21 with ζ as a parameter. Note that Equation (11.32) is the output phase change due to the step phase change, but as explained previously, the output phase change can be interpreted as the output frequency change for a step frequency change ($\omega_i = \Delta\omega_r u(t)$). The same response will be obtained for the output frequency $\omega_o(t)$ for a step frequency change $\omega_i = \Delta\omega_r u(t)$.

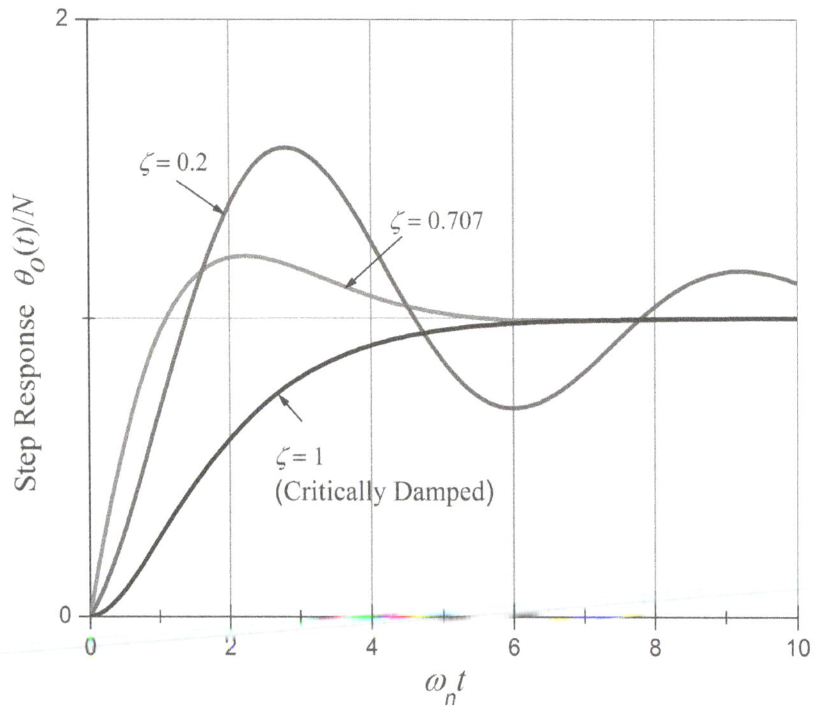

Figure 11.21 Step response of the second-order PLL

From Figure 11.21, at the steady state, the output phase tracks a step phase change that is irrespective of the damping ratio ζ. Thus, the output phase should track the input change as quickly as possible without overshooting it. From Figure 11.21, the optimal tracking can be observed for $\zeta = 0.707$. The time taken to reach the steady state is known as the lock time. The lock time is defined as the time typically taken to reach within 5% of the steady state's value. The lock time usually varies depending on ζ, but for $\zeta = 0.707$, Equation (11.33) gives

$$T_{lock} \sim \frac{1}{f_n} \tag{11.33}$$

Here $\omega_n = 2\pi f_n$. Note that in the case of the second-order loop filter, due to the role of the integrator, the phase and the frequency track the input without phase errors for a step frequency change. Therefore, beyond the second order, because the VCO tracks the phase and frequency of the reference oscillator without error, the VCO can be viewed as a phase lock state in the truest sense.

From the results of the second-order loop filter described above, the loop bandwidth is approximately the natural frequency ω_n and the optimal tracking for a step change can be achieved by selecting $\zeta = 0.707$. The lock time thus obtained is the same as that given by Equation (11.33). As a result, the narrower the loop bandwidth ω_n, the longer the lock time, whereas the wider the loop bandwidth ω_n, the faster the lock time. Therefore, once the loop bandwidth is determined in the second-order loop filter PLL, the optimal lock time can be automatically determined. However, in order to make the lock time faster, the loop bandwidth cannot be set to be infinitely large. The loop bandwidth should be set to be less than the comparison frequency. If the loop bandwidth is greater than the comparison frequency, the equations previously derived will not hold. In the derivation, the phase detector output was assumed to be DC. However, in such a case, the loop filter is unable to sufficiently filter out the harmonics of the phase detector and, just as in the first-order loop filter, the harmonics of the comparison frequencies are transferred directly to the VCO. The phase detector output waveform will then severely frequency-modulate a VCO frequency, which gives rise to many unwanted spurious characteristics.

11.4.3 Implementation of a Second-Order Loop Filter

The circuit implementations of the second-order loop filter described above are shown in Figure 11.22. The circuit in Figure 11.22(a) is a second-order loop filter for the phase detector with a charge pump and the circuits in Figures 11.22(b) and 11.22(c) are the active and passive loop filters for the voltage output phase detector.

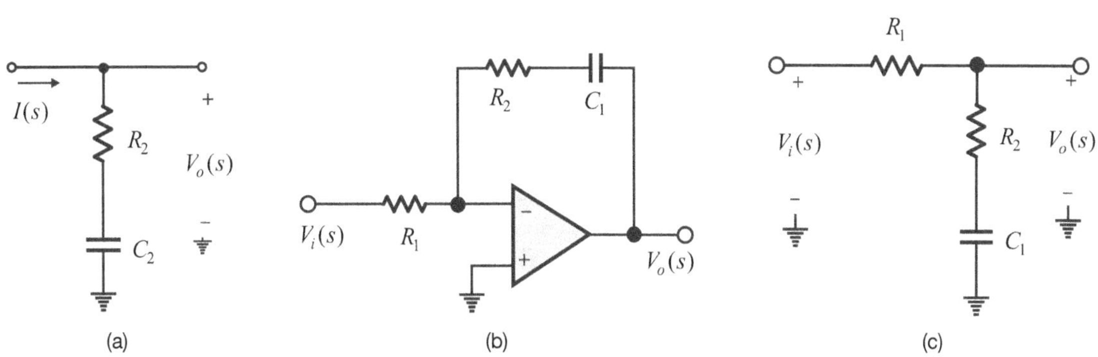

Figure 11.22 Second-order loop filters: (a) loop filter for the charge pump phase detector output and loop filters for the voltage phase detector output; (b) active and (c) passive loop filters

The loop gain of Figure 11.22(a) is expressed in Equation (11.34).

$$F(s) = Z(s) = \frac{V(s)}{I(s)} = R_2 + \frac{1}{sC_2} = \frac{1 + s\tau_2}{s\tau_1} \tag{11.34}$$

Thus, the ideal second-order transfer function can be implemented using only passive components. Also note that the transfer function of the loop filter $F(s)$ is the impedance $Z(s)$. In the case of Figure 11.22(b), assuming an ideal operational amplifier, Equation (11.35) gives us,

$$F(s) = \frac{V_2(s)}{V_1(s)} = \frac{R_2}{R_1} + \frac{1}{sR_1C_2} = \frac{1 + s\tau_2}{s\tau_1} \tag{11.35}$$

which also implements an ideal second-order transfer function. Since $F(s)$ has a voltage–voltage relationship, the loop filter in Figure 11.22(b) can be applied to the voltage–output phase detector without a charge pump. In addition, because the loop filter uses an active device, it is called an *active loop filter*. In the case of Figure 11.22(c), $F(s)$ is expressed in Equation (11.36)

$$F(s) = \frac{V_2(s)}{V_1(s)} = \frac{R_2 + \dfrac{1}{sC_2}}{R_1 + R_2 + \dfrac{1}{sC_2}} = \frac{1 + s\tau_n}{1 + s\tau_D} \cong \frac{1 + s\tau_n}{s\tau_D} \tag{11.36}$$

and can be used as an approximate second-order loop filter.

11.4.4 Measurement of a PLL

The characteristics of a PLL frequency synthesizer include *phase noise, spurious characteristics*, and *the transient response for a unit step frequency change*. The phase noise and spurious characteristics can be measured with a spectrum analyzer and the transient response for a unit step change can be measured with an oscilloscope or signal analyzer.

The spurious characteristics can be obtained by measuring the output spectrum of the VCO with the spectrum analyzer, as shown in Figure 11.23. To measure phase noise, a spectrum analyzer is used, as explained in Chapter 10. When the phase noise utility is installed in the spectrum analyzer, the phase noise can be conveniently measured without the computation of the N/C ratio as described in that chapter.

The unit step transient response can be obtained by measuring the tuning voltage V_c of the VCO with an oscilloscope, as shown in Figure 11.23. However, the tuning voltage is not a direct measurement of the VCO's frequency even though it is proportional to the VCO's frequency. The instrument that can directly measure the frequency change is the signal analyzer or modulation analyzer.[2] Alternatively, the frequency change with respect to time can be measured by setting the spectrum analyzer SPAN to 0 Hz, as shown in Figure 11.23, and applying the trigger pulse that is synchronized to the unit step frequency

2. Agilent Technologies, E5052A Signal Source Analyzer, Advanced Phase noise and Transient Measurement Techniques, Application Note 5989-1617EN, October 2004.

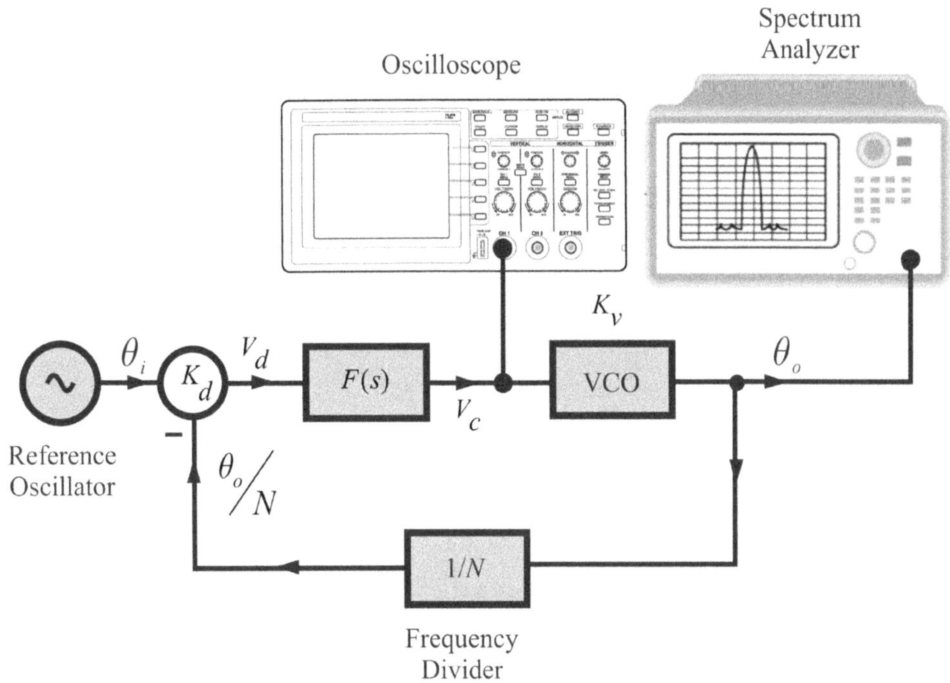

Figure 11.23 Measurement of a frequency synthesizer. The spectrum analyzer is used to measure the phase noise of a PLL and the oscilloscope is used to measure the tracking response.

change and to the external trigger input in the spectrum analyzer. With these settings, the spectrum analyzer shows the input power variation within the RBW (resolution bandwidth) of the specified center frequency versus time. Consequently, the measurement shows the frequency of the VCO with time and the lock time can be obtained by measuring the time taken to reach the steady-state frequency with an error of the RBW.

Some examples of measurement results for a frequency synthesizer are shown in Figure 11.24. The output phase noise of the frequency synthesizer and the transient frequency response for a unit step frequency change are shown in Figures 11.24(a) and 11.24(b), respectively. As expected, the output phase noise tracks the phase noise of the VCO outside the loop bandwidth, whereas it follows the phase noise of the reference oscillator within the loop bandwidth. In Figure 11.24(a), the phase noise approximated by two straight lines can be seen to intersect at about 15 kHz and the loop bandwidth can be estimated to be about 15 kHz. Figure 11.24(b) shows the time response to a unit step frequency change. This is measured using a modulation analyzer; the lock time in the figure can be seen to be about 500 μsec.

Next, Figure 11.25 shows the spectrum of the frequency synthesizer measured by expanding the SPAN of the spectrum analyzer. The SPAN is set wider than the comparison frequency. Spurious characteristics can be seen to occur in addition to the synthesized frequency components. As mentioned earlier, the phase detector includes many harmonic components of the comparison frequency. The loop filter removes these harmonics to some extent, but when the filtering is not sufficient, the remaining harmonics modulate the VCO and spurs occur.

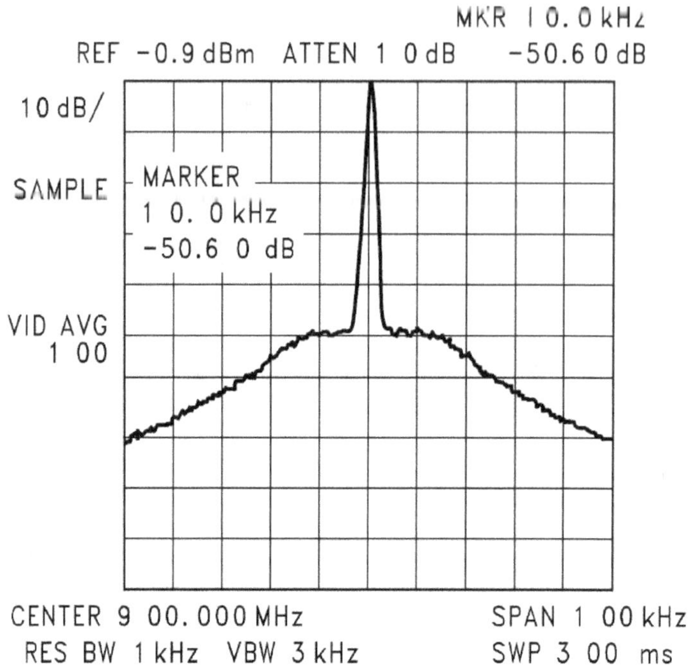

(a)

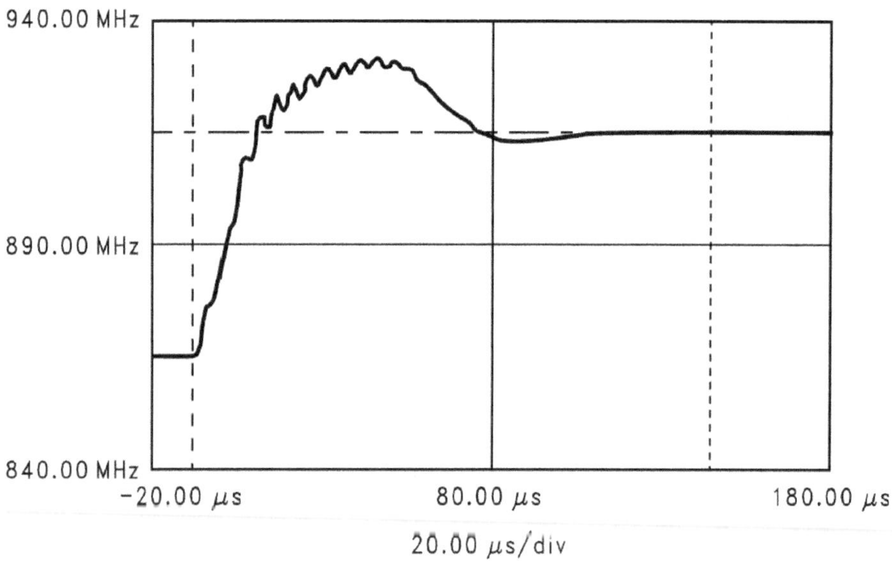

(b)

Figure 11.24 (a) Measured phase noise of the phase-locked VCO and (b) lock time for a frequency jump from 865 to 915 MHz. The PLL parameters are reference frequency $f_{ref} = 200$ kHz, $f_o = 900$ MHz, $N = 4500$, $K_v = 2\pi \times 20$ (Mrad/V), and $K_d = 5$ mA/(2π) (rad/V). The loop filter order is fourth and its loop bandwidth = 20 kHz, and phase margin is $\phi_p = 45°$.The lock time is about 153 μsec. Source: National Semiconductor Application Note 1001, *An Analysis and Performance Evaluation of a Passive Filter Design Technique for Charge Pump PLL's*, July 2001.

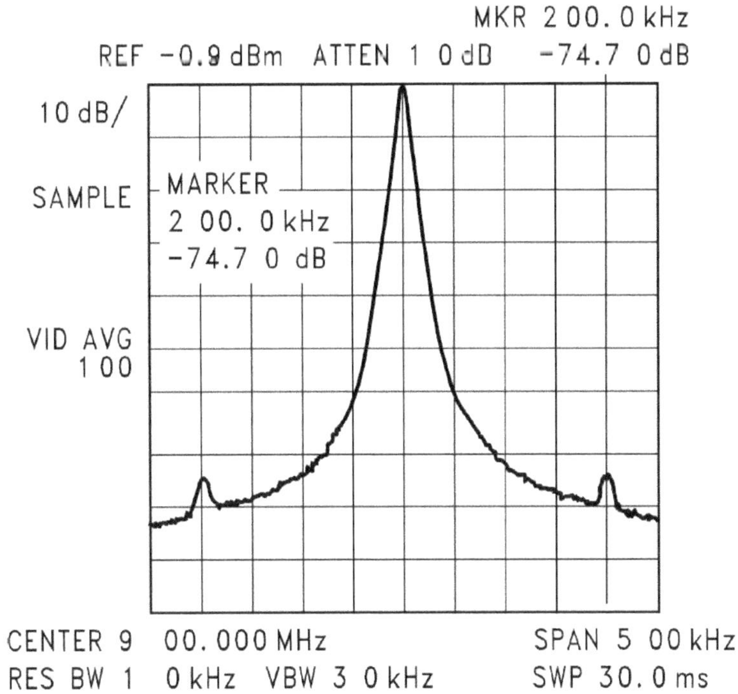

Figure 11.25 Spurs in the frequency synthesizer. The loop filter in Figure 11.24 is used. The spur level is measured to be about 74.7 dBc. Source: National Semiconductor Application Note 1001, *An Analysis and Performance Evaluation of a Passive Filter Design Technique for Charge Pump PLL's*, July 2001.

To see these effects, by denoting the leakage voltage of the comparison frequency harmonics at the output of the loop filter as $V_s \sin(\omega_m t)$, the VCO is FM modulated by this leakage voltage. This can be expressed with Equation (11.37).

$$v_o(t) = \cos\left(\omega_c t + K_v \frac{V_s}{\omega_m} \cos(\omega_m t)\right) \simeq \cos(\omega_c t) - \sin(\omega_c t) K_v \frac{V_s}{\omega_m} \cos(\omega_m t) \quad (11.37)$$

In addition to the synthesized frequency, the output, the two frequency components $\omega_c + \omega_m$ and $\omega_c - \omega_m$ from the right-hand term of Equation (11.37) appear. Thus, spurs occur as shown in Figure 11.25 and it is necessary to lower the spur level by narrowing the loop filter. In the case where a fast lock time is required, the bandwidth should be set wider and the loop filter is then unable to sufficiently attenuate the comparison frequency harmonics appearing at the phase detector's output. Consequently, this results in larger spurs at the VCO's output. However, when the bandwidth is overly reduced so as to suppress the spurs, the lock time becomes longer. Higher-order loop filters may be used in such a situation.

11.4.5 Higher-Order Loop Filters

The second-order loop filter can track the frequency and phase of the reference oscillator, and it can be used as a basic loop filter for PLL configuration. However, when spurs are above the specified level, the second-order loop filter does not provide an alternative way to suppress spurs other than to narrow the loop bandwidth at the expense of the lock time. To some extent, the problems that arise between the lock time and the spurs can be solved by using a higher-order loop filter.

Note that a higher-order loop filter yields the transfer function of an open-loop gain that is higher than a third order. Generally, the PLL can become unstable for the open-loop gain transfer function with the third order and above. One way of resolving the PLL instability is to provide a *phase margin* or *gain margin* in the open-loop gain. The phase margin method is commonly used in a PLL. The design of higher-order loop filters depends in part on the type of the loop filter. In this section, we will discuss a design that is limited to a charge pump PLL loop filter. The higher-order loop filters are shown in Figure 11.26 and they are formed by modifying the second-order loop filters to give more attenuation at the comparison frequency harmonics. Figure 11.26(a) shows a third-order loop filter and Figure 11.26(b) shows a fourth-order loop filter.

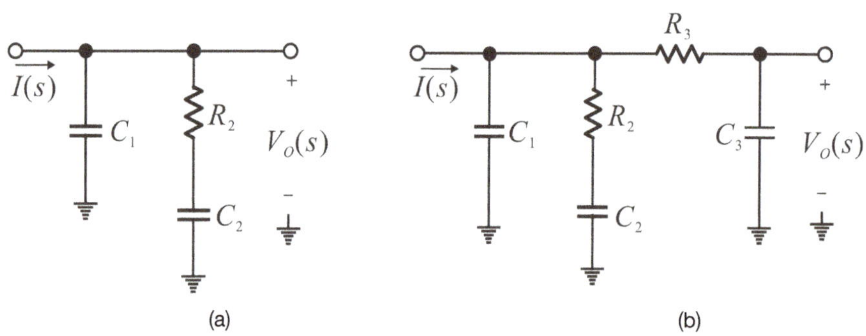

Figure 11.26 Circuit implementation of charge pump loop filters: (a) the third order and (b) the fourth order

In Figure 11.26(a), R_2 and C_2 comprise the second-order loop filter. The impedance $Z(s)$ of R_2–C_2 branch is expressed in Equation (11.38).

$$Z(s) = R_2 + \frac{1}{sC_2} \tag{11.38}$$

At lower frequencies, the impedance $Z(s)$ is approximated by $1/sC_2$, which decreases as the frequency increases. However, at higher frequencies, the impedance becomes $Z(s) \cong R_2$. Thus, there is no attenuation of the harmonic leakage for a frequency higher than $f_c = 1/2\pi R_2 C_2$. By adding C_1 in parallel to the R_2–C_2 branch, the harmonic leakage of higher frequency can be attenuated due to C_1. For the fourth-order loop filter in Figure 11.26(b), an R_3–C_3 stage is added to the third-order loop filter. The R_3–C_3 stage has no effect at low frequency. However, at high frequency the R_3–C_3 stage functions as a low-pass filter thereby severely attenuating the harmonic leakage of the phase detector output. Note that the values of C_1 and C_3 added in the higher-order loop filters are basically small and therefore have little or no effect on the basic form of the second-order loop filter within the loop bandwidth. However, they are inserted to give more harmonics attenuation outside the loop bandwidth. Thus, within the loop bandwidth, the added components in the third-order and fourth-order loop filters have minimum effect on the basic form of the second-order loop filter; this effect appears outside the loop bandwidth and attenuates the high-frequency harmonic leakage of the phase detector output. In reality, however, the added components somewhat distort the response of the second-order loop filter and the effects of the added components must be taken into account when determining their values.

When the third-order loop filter in Figure 11.26(a) is used in the PLL with the integer frequency divider shown in Figure 11.18, the loop gain $L(s)$ is expressed in Equations (11.39a)–(11.39c).

$$\text{Loop gain} = L(s) = \frac{KF(s)}{sN} = \frac{K(1+sT_2)}{s^2C_1N(1+sT_1)}\frac{T_1}{T_2} \tag{11.39a}$$

$$T_1 = R_2(C_1 \parallel C_2) \tag{11.39b}$$

$$T_2 = R_2C_2 \tag{11.39c}$$

Therefore, the frequency response characteristics can be computed using Equation (11.40).

$$L(s)\Big|_{s=j\omega} = \frac{-K(1+j\omega T_2)}{\omega^2C_1N(1+j\omega T_1)}\frac{T_1}{T_2} \tag{11.40}$$

Generally, the phase of $L(s)$ increases from –180° and then tends to decrease again as the frequency increases, as shown in Figure 11.27. Therefore, for stability, the loop gain must be 1 at the point where the phase is maximum, which then gives the maximum phase margin. The phase from Equation (11.40) is

$$\phi(\omega) = \tan^{-1}(\omega T_2) - \tan^{-1}(\omega T_1) + 180° \tag{11.41}$$

Differentiating Equation (11.41) to obtain the frequency ω_p at which $\phi(\omega)$ is maximum, we obtain Equation (11.42).

$$\frac{d\phi}{d\omega} = \frac{T_2}{1+\omega^2T_2^2} - \frac{T_1}{1+\omega^2T_1^2} = 0 \tag{11.42}$$

Then, the frequency for the maximum phase margin ω_p is given by

$$\omega_p = \frac{1}{\sqrt{T_2T_1}} \tag{11.43}$$

The frequency ω_p can be defined as the loop bandwidth. This definition is somewhat different from that of the loop bandwidth ω_{3dB} of the second-order loop filter, but it can be used to identify a newly defined loop bandwidth.

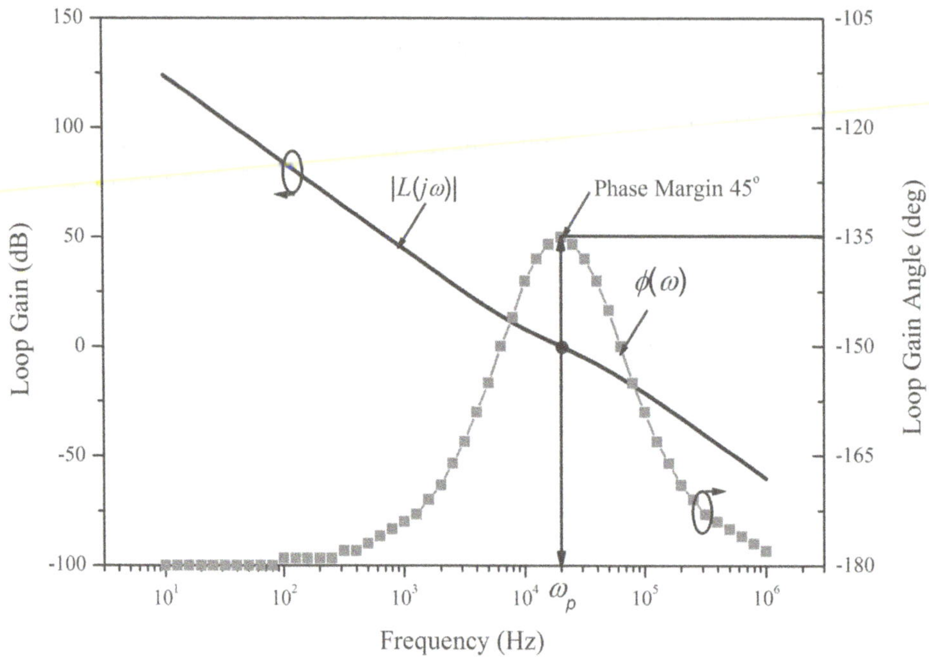

Figure 11.27 The open-loop gain of the third-order loop filter for frequency

From Equation (11.43), T_2 can be expressed in terms of ω_p and T_1 with Equation (11.44).

$$T_2 = \frac{1}{\omega_p^2 T_1} \tag{11.44}$$

Denoting the phase margin as ϕ_p, the phase margin from Equation (11.41) is $\phi_p = \phi(\omega_p) + 180°$. Thus, substituting the result of Equation (11.44) into Equation (11.41), T_1 can be obtained with Equation (11.45).

$$T_1 = \frac{\sec\phi_p - \tan\phi_p}{\omega_p} \tag{11.45}$$

Then, once the loop bandwidth ω_p and phase margin ϕ_p have been defined, the time constants T_1 and T_2 can be calculated from Equations (11.44) and (11.45).

In addition, in order to make ϕ_p of Equation (11.45) equal to the phase margin, for the determined T_1 and T_2, $|L(j\omega_p)| = 1$. Thus,

$$C_1 = \frac{KT_1}{\omega_p^2 N T_2}\left|\frac{(1 + j\omega_p T_2)}{(1 + j\omega_p T_1)}\right| = \frac{KT_1}{\omega_p^2 N T_2}\sqrt{\frac{1 + (j\omega_p T_2)^2}{1 + (j\omega_p T_1)^2}} \tag{11.46}$$

Since ω_p, T_1, and T_2 have already been determined, the value of C_1 can be calculated using Equation (11.46), from which R_2 and C_2 can be determined using Equations (11.39b) and (11.39c), and expressed in Equations (11.47) and (11.48).

$$C_2 = C_1\left(\frac{T_2}{T_1} - 1\right) \tag{11.47}$$

$$R_2 = \frac{T_2}{C_2} \tag{11.48}$$

In summary, for the given loop bandwidth ω_p and phase margin ϕ_p, the two time constants T_1 and T_2 can be calculated with Equations (11.44) and (11.45). As a result, all the element values of the third-order PLL loop filter can be determined. The lock time of the third-order loop filter is similar to that of the second-order loop filter, and $T_{lock} \sim 1/f_p$ ($\omega_p = 2\pi f_p$).

Example 11.9

The design of a frequency synthesizer of $f_o = 900$ MHz requires the use of a third-order loop filter. The third-order loop filter must have a loop bandwidth of $\omega_p = 2\pi \times 10$ kHz and a phase margin of $\phi_p = 45°$. The comparison (or reference) frequency is $f_{ref} = 200$ kHz. The tuning sensitivity of VCO is $K_v = (2\pi) \times 20$ Mrad/V. The phase detector constant is $K_p = 5/(2\pi)$ mA/rad. Compute the element values of the third-order loop filter.

Solution

From the given values,

$$N = \frac{900 \text{ MHz}}{200 \text{ kHz}} = 4500$$

By substituting the values of the loop bandwidth and phase margin into Equation (11.45),

$$T_1 = \frac{\sec\phi_p - \tan\phi_p}{\omega_p} = 6.592 \times 10^{-6} \text{ sec}$$

$$T_2 = \frac{1}{\omega_p^2 T_1} = 3.842 \times 10^{-5} \text{ sec}$$

Substituting the N and time constants T_1 and T_2 calculated above in Equation (11.46),

$$C_1 = \frac{KT_1}{\omega_p^2 N T_2}\sqrt{\frac{1+(\omega_p T_2)^2}{1+(\omega_p T_1)^2}} = 2.33 \text{ nF}$$

Using the value of C_1 and substituting C_1 into Equations (11.47) and (11.48), the third-order loop filter element values can be obtained as shown in Figure 11E.5.

$$C_2 = C_1\left(\frac{T_2}{T_1} - 1\right) = 11.26 \text{ nF}$$

$$R_2 = \frac{T_2}{C_2} = 3.413 \text{ k}\Omega$$

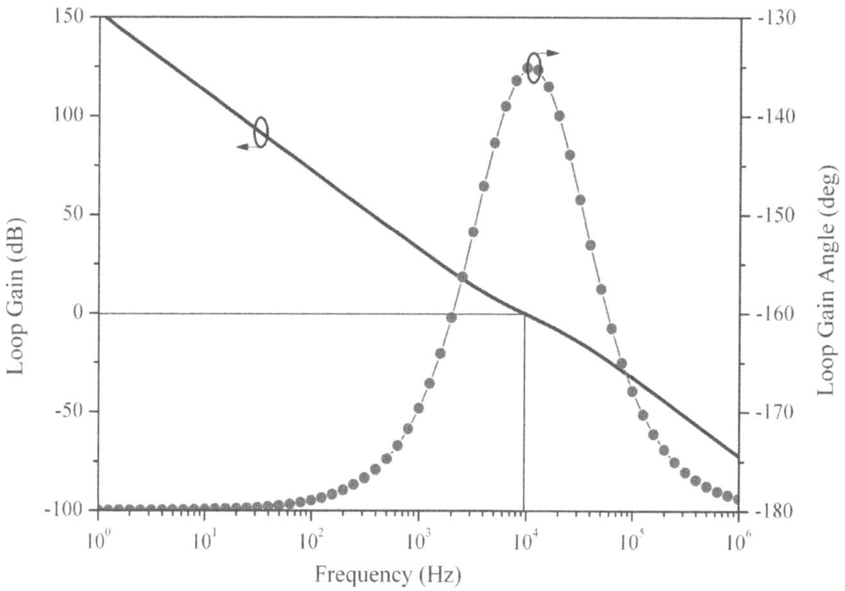

Figure 11E.5 Frequency response of the open-loop gain with the calculated element values

Substituting the calculated values into Equation (11.40), the frequency response of $L(s)$ can be computed and is shown in Figure 11E.5. In that figure, when the loop gain is 0 dB (i.e., the loop gain = 1), the phase is −135°, which is away from −180° by 45° and the phase margin can be seen to be 45°.

Example 11.10

For the parameters of the PLL given in Example 11.9, compute the third-order loop filter values using the equations in the ADS display window and graph its Bode plot for a frequency range of 10 Hz–10 MHz.

Solution

The PLL design in Example 11.9 occurs frequently in PLL design and requires only simple computations using a conventional calculator. This series of computations can be solved using the ADS display window without a calculator. Similar programs are available from the Web sites of PLL IC vendors.[3]

3. Refer to the Web site of Analog Devices at www.analog.com.

First, enter the equations that define the PLL parameters shown in Measurement Expression 11E.1 in the display window.

```
Eqn Kv=20 M  Eqn Kp=5m  Eqn fref=200k  Eqn fo=900 M  Eqn fp=10k
Eqn phi=45*pi/180
Eqn N=fo/fref  Eqn wp=2*pi*fp  Eqn K=Kv*Kp
```

Measurement Expression 11E.1 Definition of the PLL parameters in the display window

Note that expressions such as 20e6 and 5e-3 are not required and they can be stated using prefixes such as m, k, M, G, and so on. Next, determine the plot frequency range, which can be written as follows in Measurement Expression 11E.2:

```
Eqn fstart=10  Eqn fstop=10M  Eqn n=20  Eqn Dec=log(fstop/fstart)
Eqn x=generate(0, Dec,n*Dec)  Eqn f=fstart*10**(x)  Eqn w=2*pi*f  Eqn s=j*w
```

Measurement Expression 11E.2 Definition of the parameters for the Bode plot in the display window

Here, **n** sets the number of the computing points per decade of frequency change. The variable **Dec** is the decade number of the frequency range set by **fstart** and **fstop**, and thus **n*Dec** number of points are computed. The function **generate**() generates a sequence from 0 to **Dec** with points given by **n*Dec**.

The two sets of equations given above are commonly required in PLL loop filter design. Now, the third-order loop filter values can be computed using the equations derived from Equations (11.44) to (11.48) and can be written as follows in Measurement Expression 11E.3:

```
Eqn T1=(1/cos(phi)-tan(phi))/wp  Eqn T2=1/(wp**2*T1)
Eqn C1=K*T1/(wp**2*N*T2)*sqrt((1+(wp*T2)**2)/(1+(wp*T1)**2))
Eqn C2=C1*(T2/T1-1)  Eqn R2=T2/C2
Eqn L3=K*(1+s*T2)/(s**2*N*(C1+C2))*(1+s*T1))
```

Measurement Expression 11E.3 Loop filter element values and transfer function in the display window

The last equation is for the Bode plot. Using the listing column in ADS, the computed values can be found as

$$C_1 = 2.332\ \text{nF},\ C_2 = 11.26\ \text{nF, and}\ R_2 = 3.413\ \text{k}\Omega$$

These values are the same as those calculated in Example 11.9. In addition, the Bode plot of the open-loop gain **L3** with the calculated values can be obtained as shown in Figure 11E.6. From Figure 11E.6, the determined third-order loop filter yields the exact desired loop bandwidth (where the loop gain is 1) and phase margin of 10 kHz and 45°, respectively.

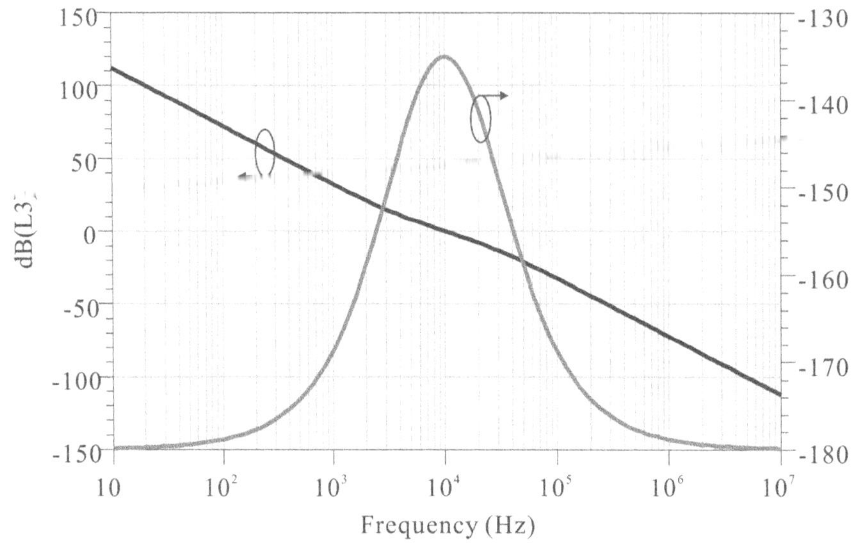

Figure 11E.6 Frequency response of the open-loop gain of the third-order loop filter with the calculated element values

With the loop filter thus configured, it is not possible to know absolutely the magnitudes of the spurs. However, they are more attenuated than in the case of the second-order loop filter and the relative degree of their attenuation can be determined by comparing the open-loop gain with that of the second-order loop filter. When the spurs of the configured third-order loop filter are measured and are still found to be greater than the specifications shown in Figure 11.25, the spurs' performance can be improved by employing a fourth-order loop filter, which is an extension of the third-order loop filter. In Figure 11.25, in order to make the spurs below the phase noise, an attenuation of approximately 10 dB is required. The 10-dB attenuation can be achieved by extending the third-order loop filter to the fourth-order loop filter in Figure 11.26(b). The transfer function of the additional R_3–C_3 stage is expressed in Equation (11.49).

$$A(s) = \frac{1}{1 + sC_3R_3} = \frac{1}{1 + sT_3} \tag{11.49}$$

Therefore, the required attenuation of the spurs at the comparison frequency f_{ref}, ATTN can be defined with Equation (11.50).

$$ATTN = 10\log\left[\left(2\pi f_{ref} T_3\right)^2 + 1\right] \tag{11.50}$$

Using Equation (11.50), the time constant T_3 can be calculated using Equation (11.51).

$$T_3 = \sqrt{\frac{10^{\frac{ATTN}{10}} - 1}{\left(2\pi f_{ref}\right)^2}} \tag{11.51}$$

Adding R_3–C_3 to the third-order loop filter causes a slight change in its element values. The loop gain of the fourth-order loop filter can be approximately written as Equation (11.52).

$$L(s) \simeq \frac{K(1+sT_2)}{s^2C_1N(1+sT_1)(1+sT_3)}\frac{T_1}{T_2} \simeq \frac{K(1+sT_2)}{s^2C_1N(1+sT_1+sT_3)}\frac{T_1}{T_2} \tag{11.52}$$

Here, the time constants T_1 and T_2 are the same as those in Equations (11.39b) and (11.39c). This is similar to the transfer function of the third-order loop filter given by Equation (11.39a). Therefore, Equation (11.44) can be rewritten as Equation (11.53)

$$T_2 = \frac{1}{\omega_p^2(T_1+T_3)} \tag{11.53}$$

and the relationship between the loop bandwidth and phase margin is obtained with Equation (11.54.)

$$T_1 + T_3 = \frac{\sec\phi_p - \tan\phi_p}{\omega_p} \tag{11.54}$$

Thus, T_3, T_1, and T_2 can be determined from Equations (11.51), (11.53), and (11.54). The calculated values T_1, T_2, and T_3 can be substituted into $L(s)$ of Equation (11.52). For the phase margin, $|L(s)|$ must be equal to 1 at ω_p. As in the third-order loop filter design, C_1 can be obtained with Equation (11.55).

$$C_1 = \frac{KT_1}{\omega_p^2NT_2}\sqrt{\frac{1+(\omega_pT_2)^2}{\left(1+(\omega_pT_1)^2\right)\left(1+(\omega_pT_3)^2\right)}} \tag{11.55}$$

In addition, C_2 and R_2 can be calculated from the time-constant relationships using Equations (11.56) and (11.57).

$$C_2 = C_1\left(\frac{T_2}{T_1} - 1\right) \tag{11.56}$$

$$R_2 = \frac{T_2}{C_2} \tag{11.57}$$

Using these formulas, all the fourth-order loop filter element values can be determined. However, as T_3 is expressed in terms of R_3C_3, there is a certain degree of freedom. The value of C_3 should be set 10 times smaller than C_1 because R_3C_3 must not affect C_1. Thus, all the element values are accurately calculated, and the reader may refer to reference 3 at the end of this chapter for the exact values.

Example 11.11

Using ADS, design the fourth-order loop filter for the PLL parameters given in Example 11.10, with an added parameter *ATTN* as

$$ATTN = 10 \text{ dB}$$

Also, compare the open-loop gain of the fourth-order loop filter with that of the third-order loop filter.

Solution

The first two sets of equations in the ADS display window are the same as those in the third-order loop filter design. The difference is in computations for the element values. Using Equations (11.50)–(11.57), the equations for the element values can be written as follows in Measurement Expression 11E.4:

```
Eqn ATTN=10   Eqn T3=sqrt((10**(ATTN/10)-1)/(2*pi*fref)**2)
Eqn T1=(1/cos(phi)-tan(phi))/wp-T3   Eqn T2=1/(wp**2*T1)
Eqn C1=K*T1/(wp**2*N*T2)*sqrt((1+(wp*T2)**2)/(1+(wp*T1)**2))
Eqn C2=C1*(T2/T1-1)   Eqn R2=T2/C2
Eqn C3=C1/10          Eqn R3=T3/C3
Eqn L4=K*(1+s*T2)/(s**2*N*(C1+C2)*(1+s*T1)*(1+s*T3))
```

Measurement Expression 11E.4 Loop filter element values and transfer function in the display window

Note that the open-loop gain **L4** obtained with Equation (11.52) is not an exact one. The open-loop gain of the fourth-order loop filter is somewhat more complex than that given by Equation (11.52). In the computation of **L4**, the element values of C_3 and R_3 are not required; only the value of **T3** is necessary. The value of C_3 is set to $C_1/10$ to make the effect of $R_3 - C_3$ branch small. The computed values of the elements are

$$C_1 = 1.539 \text{ nF}, C_2 = 12.52 \text{ nF}, R_2 = 3.067 \text{ kohm}, C_3 = 0.153 \text{ nF}, \text{and } R_3 = 15.5 \text{ kohm}$$

Figure 11E.7 shows the open-loop gain of the computed fourth-order loop filter. The open-loop gain magnitude is close to 0 dB and its phase margin is close to 45° at the loop bandwidth of 10 kHz. The Bode plot of the third-order loop filter open-loop gain is also included for comparison. The magnitude **L4** should be lower than that of **L3** by 10 dB at $f_{ref} = 200$ kHz. However, that difference is about 5 dB, which is smaller than 10 dB. The reason for the difference is because the loop filter calculation method relies on the previously explained approximation method. The method of finding exact values can be found in the end-of-chapter problems, or the programs that calculate the exact fourth-order loop filters are available on the Web sites of PLL IC vendors.

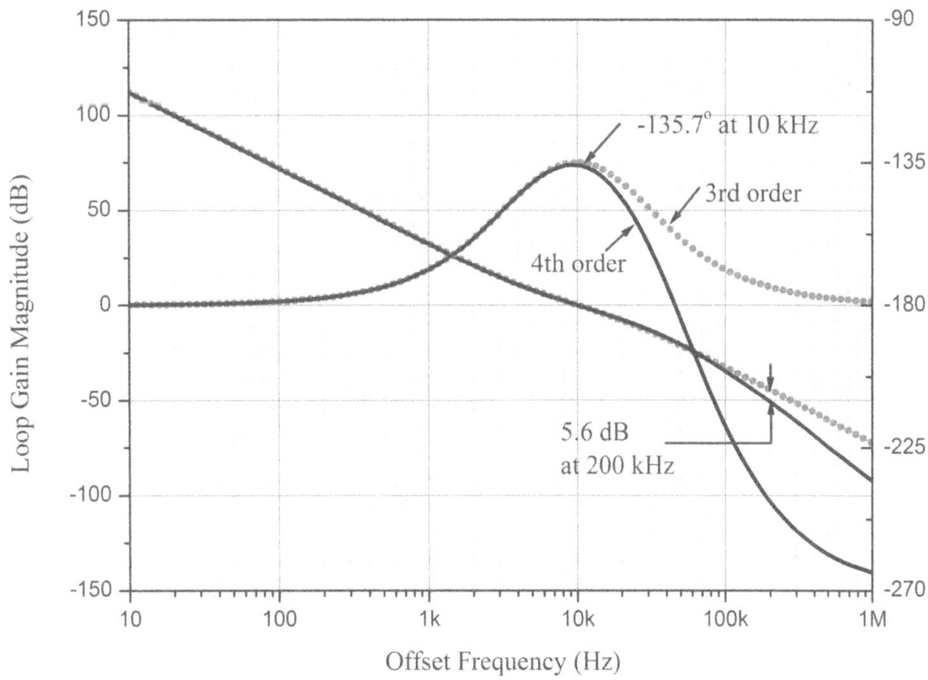

Figure 11E.7 Frequency response of the open-loop gain with the fourth-order loop filter

11.5 PLL SIMULATION IN ADS

A PLL can be simulated using ADS. The synthesis of a loop filter by optimization and the simulation of the phase noise and transient response for a unit step frequency change is also possible in ADS; however, ADS cannot simulate the spurious characteristics. For that simulation, a mathematical model of the phase detector's waveform at the steady state is required. The key factors of spur generation include the leakage current of the phase detector and the asymmetry of the charge pump current (see reference 3 at the end of this chapter). Currently, however, it is difficult to model these phenomena mathematically. As a result, accurately simulating the spur characteristics using ADS can be a problem. In this section, we will discuss loop filter synthesis and the simulation of phase noise and transient response in ADS.

As phase noise simulation is related to small-signal noise simulation, the phase noise simulation can be performed using an AC simulation, while the transient response requires an envelope simulation.

11.5.1 Loop Filter Synthesis

Examples of PLL simulation templates are in the **DesignGuide** of the ADS menu (DesignGuide/PLL). After the clicking the Menu command, the first pop-up window appears; then chose **Select PLL Configuration**. The second pop-up window for the PLL configuration will appear; in the **Type of Configuration** tab, choose **Frequency Synthesizer** and then select **Loop Frequency Response** in the **Simulation** tab. Next, choose **Charge Pump** in the **Phase Detector** tab and **Passive** 3 **Pole** in the **Loop Filter** tab. The design files named **SYN_CP_FQ_P3P**.dsn and **SYN_CP_FQ_P3P**.dds will be generated. Rename the generated design file as **SYN_third**. The design file contains three circuits: one for the closed-loop response, one for the open-loop gain, and one for the loop filter frequency response. Delete the two circuits for the closed-loop response and loop filter, and then modify the circuit as shown in Figure 11.28. There are two **MeasEqn**s that are initially hidden. These can be shown by setting **Parameter Visibility** to **Set All** in the **Component Options** of the pop-up window.

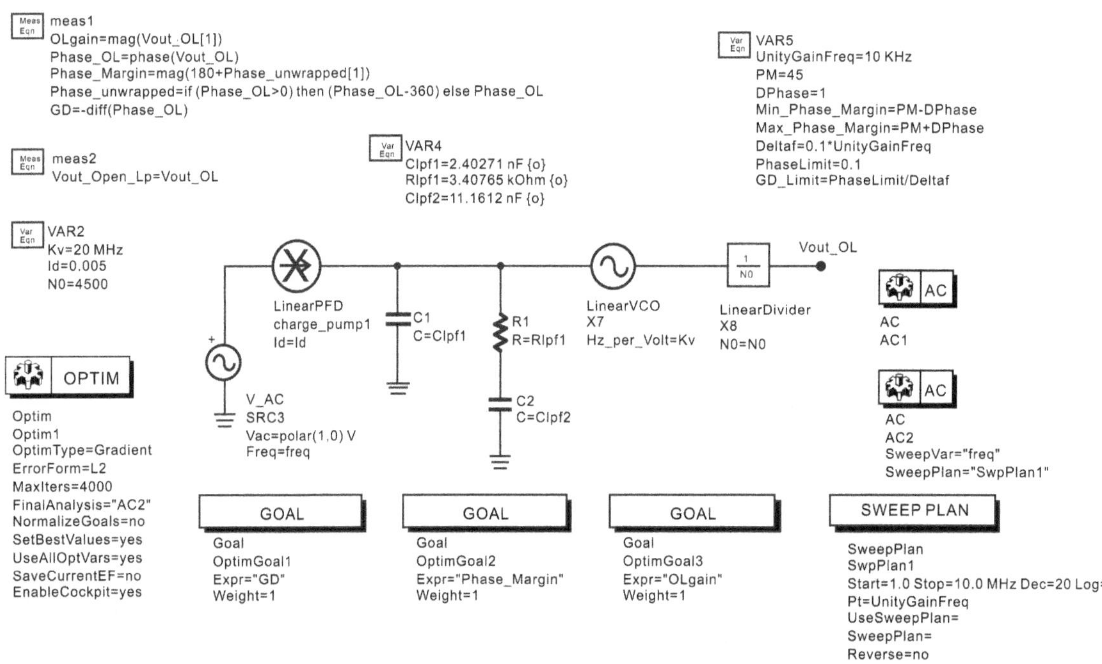

Figure 11.28 Loop filter design using optimization. The voltage **Vout_OL** corresponds to the open-loop gain of the PLL. **VAR2** sets the values of the PLL components and **VAR4** sets the values of the loop filter. VAR5 defines the parameters for the PLL design objectives. The measurement equation **meas**1 computes the phase and magnitude of **Vout_OL** and defines the phase margin. Then, **meas**2 defines the variable names for the display window. In **meas**2, the defined variables that correspond to the closed-loop filter and the loop filter output voltages are deleted. **Goal**s computes the values of the cost functions using the AC simulation result **AC**1. **AC**2 computes the frequency response of the open-loop gain for the frequencies given by **SwpPlan**1 when the optimization finishes.

In Figure 11.28, **LinearPFD** is a subcircuit built with a voltage-controlled current source (VCCS). The **LinearPFD** models the charge pump PFD explained in the previous section, which generates the current in proportion to the phase difference. Here, the phase difference is given by the voltage source. The **LinearVCO** is an integrator that is similarly modeled using two controlled sources. The **LinearDivider** also models the frequency divider using a VCVS (voltage-controlled voltage source). The previously mentioned component values of PLL are defined in **VAR2**. The element values of the loop filter are defined in **VAR4**.

In **VAR5**, the variable **UnityGainFreq** represents the loop bandwidth where the open-loop gain is 1. The variables **Min_Phase_Margin** and **Max_Phase_Margin** define the allowable values for the phase margin in optimization. The original schematic does not include the group delay. In that case, the phase of the open-loop gain may not be optimized at the phase peak frequency, which we do not want. Since the group delay is equal to 0 at the phase peak frequency, the allowable group delay limit is set using the variable **GD_Limit**, which is then used in **Goal1** to optimize the group delay to 0. The **meas1** defines the open-loop gain, phase margin, and group delay necessary for optimization. The **meas2** is a declaration of the output variable for the display window. The original schematic includes more variables but they are deleted for simplification.

The goals from **Goal1** to **Goal3** are connected to the AC simulation **AC1**. The **AC1** is modified to simulate three frequencies, from **UnityGainFreq-Deltaf** to **UnityGainFreq+Deltaf**. This is necessary to compute the group delay. The group delay **GD** defines the phase change with respect to frequency. Then, the loop filter can be designed using optimization to minimize the three goals. The optimization control **Optim1** optimizes the loop filter element values to fit the goals. The AC simulation **AC2** is not connected to the goals and it is used to compute the Bode plot of the open-loop gain after the optimization. The sweep frequency range of **AC2** is specified in **SwpPlan1**. After the optimization finished, the **Optim1** computes the Bode plot using **AC2**.

First, the loop filter element values are optimized for a 10-kHz loop bandwidth and a 45° phase margin using the PLL parameters given in Example 11.10. The phase margin is set to the range 45°– 46°. The **OLgain** is set to the range 0.99–1.01. The group delay **GD** is set to close to 0. The optimized element values are compared in Table 11.2.

Table 11.2 Comparison of the loop filter element values

	C_1 (nF)	C_2(nF)	R_2 (kΩ)
Calculation	2.332	11.26	3.413
Optimization	2.403	11.16	3.408

Figure 11.29 shows the magnitude and phase of the optimized open-loop gain. The phase margin is close to 45° and the open-loop gain is close to 1 at the frequency of the 10-kHz loop bandwidth. The higher-order loop filter, such as the fourth-order loop filter, can be obtained using a similar optimization procedure.

11.5.2 Phase Noise Simulation

The PLL phase noise simulation shown in Figure 11.30 can be set by selecting a PLL from the **DesignGuide**. Then, modify the loop filter as shown in Figure 11.30.

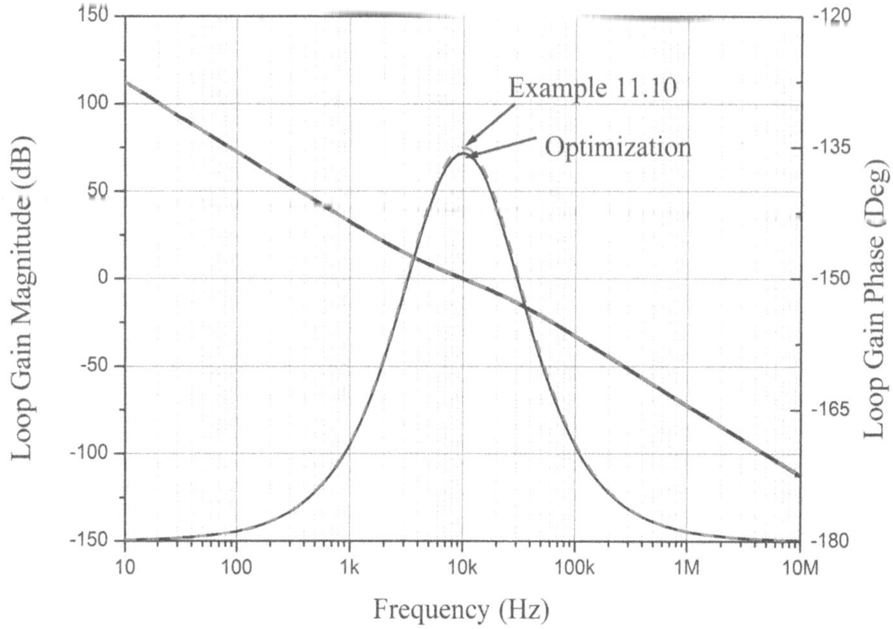

Figure 11.29 Comparison of the optimized open-loop gain with that computed in Example 11.10. The phase margin is close to 45° at 10 KHz and the difference is below 1°.

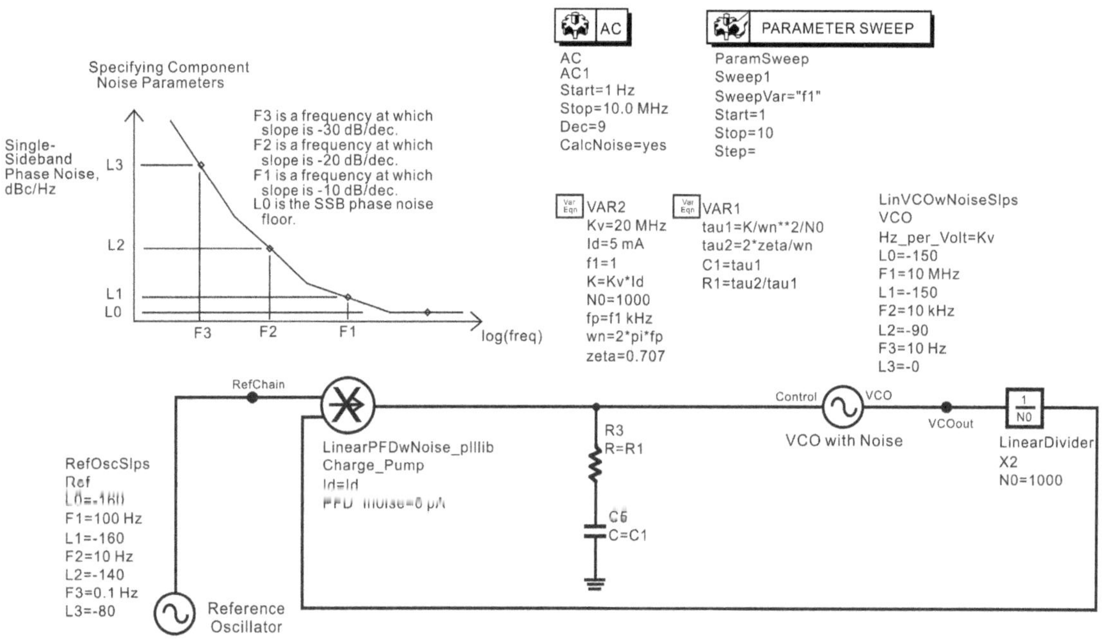

Figure 11.30 Phase noise simulation. In the simulation, voltages represent the phases of PLL components and the circuit is simulated using voltages. Reference oscillator is a noise voltage source whose spectrum is specified by the user as shown in the schematic. The component **Charge_Pump** generates the output current proportional to the difference of the two input voltages. The VCO produces the output voltage's integrating input current and then the spectral noise voltage is added. **LinearDivider** is modeled using a VCVS (voltage-controlled voltage source).

In order to compare the results of Example 11.1 with the ADS simulation, the loop filter is set to be a second-order loop filter, as shown in Figure 11.30. The values of C_1 and R_1 of the second-order loop filter can be determined using the given $f_n = 1$ kHz and $f_n = 10$ kHz, and the damping ratio $\zeta = 0.707$. From Equations (11.30a) and (11.30b), τ_1 and τ_2 can be determined as

$$\tau_1 = \frac{K_d K_v}{\omega_n^2 N}, \quad \tau_2 = \frac{2\varsigma}{\omega_n}$$

and from the values of τ_1 and τ_2, C_1 and R_1 can be computed as

$$C_1 = \tau_1, \quad R_1 = \frac{\tau_2}{\tau_1}$$

The values of C_1 and R_1 are computed using the equation in ADS.

Then, since the noise floor of the frequency divider and phase detector were not considered in Example 11.1, a frequency divider and a phase detector without the noise floor should be selected. However, the phase detector without the noise floor is not available in ADS and thus the noise current is set to 0 to eliminate the effect of the noise floor. The frequency divider without the noise floor is available in ADS and the frequency divider is replaced with the new frequency divider in the default PLL simulation schematic. Next, it is necessary to input the phase noises of the reference oscillator and VCO. Instead of a tabular form, the phase noise requires an input as shown in Figure 11.30 that is represented by the 30 dB, 20 dB, and 10 dB slopes, and the noise floor. The phase noise values in Table 11E.1 are thus entered as shown in Figure 11.30.

The simulated phase noise is shown in Figure 11.31. The phase noise simulation results with respect to the loop bandwidth change are the same as those in Example 11.1. The phase noises of the VCO and reference oscillator multiplied by 1000 are also shown in Figure 11.31. These correspond, respectively, to **PNTotal**, **PN_VCO_FreeRun**, and **PN_RefChain_FreeRun** of that figure. They are calculated in the display window using the following equations in Measurement Expression 11.1:

```
Eqn PN_VCO_FreeRun=10*log(0.5*VCO_FR.noise**2)
Eqn PN_RefChain_FreeRun=10*log(0.5*RefChain.noise**2)
Eqn PNTotal=10*log(0.5*VCOout.noise**2)
```

Measurement Expression 11.1 Definitions of phase noises in the display window

The calculated values in Figure 11.30 are the voltage values of the noise with respect to frequency. Note that because the simulation gives the noise simulation results, all the noise voltage variables in the previous equations have the **noise** extension. Of the computed noise voltages, **VCO_FR.noise** is not explicitly specified in the simulation schematic of Figure 11.30, but it is the value automatically calculated by the subcircuit VCO and stored in the dataset as **VCO_FR.noise**.

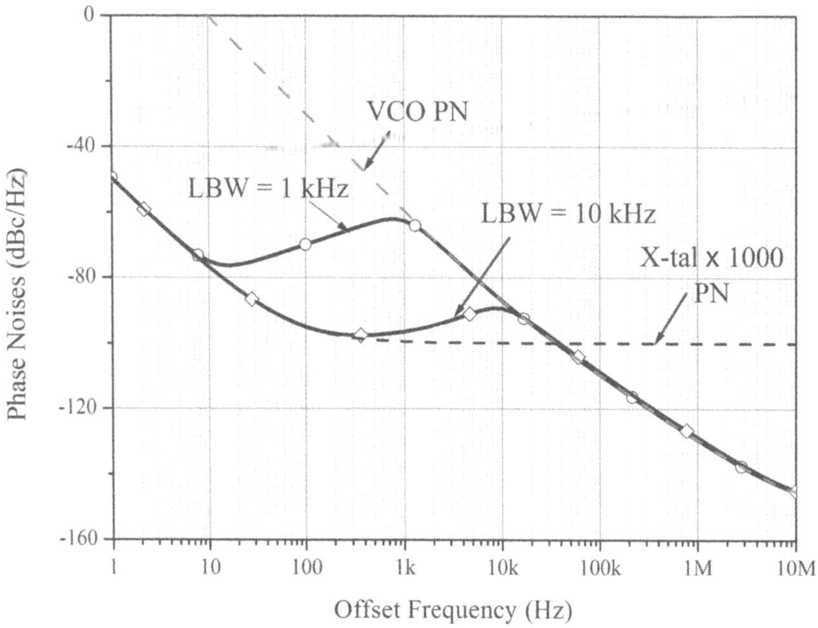

Figure 11.31 Phase noise simulation results

11.5.3 Transient Response Simulation

The simulation of a PLL's transient response can be performed by selecting the ADS **DesignGuide** menu and then **Transient Response Simulation** in the pop-up window that appears. Also, in the course of selecting the loop filter, **Passive** 3 **Pole** must also be selected for the transient simulation. This automatically generates the default schematic window for the transient response simulation shown in Figure 11.32. In the autogenerated circuit, the loop filter is configured as shown in Figure 11.32. The loop filter element values in the schematic window are replaced by those of the third-order loop filter of Example 11.9: K_v, K_d, N, and the comparison frequency f_{ref}, which correspond to **Kv**, **Id**, **N0**, and **Fref** in Figure 11.32.

In that figure, it is necessary to understand the two components, **VCO_DivideByN** and **PhaseFrequencyDetCP** in order to be able to simulate the PLL's transient response. **PhaseFrequencyDetCP** is the **PFD** with the charge pump explained earlier, and its input signal becomes the baseband signal, which is a time-varying waveform without a carrier. In Figure 11.32, the reference signal is set using a DC power supply and is made to vary with time using the following equation in Measurement Expression 11.2:

Eqn Vref=2*pi*Fref*rem(time+0.25/Fref, 1/Fref)-pi,

Measurement Expression 11.2 Reference signal definition in the schematic window

Here, the **rem**(x, y) function represents the remainder when x is divided by y. Denoting **1/Fref** = T, then $f(t)$ = **rem(time, 1/Fref)** is a sawtooth waveform with a period T. Thus, $0 \le f(t) < T$ and multiplying $f(t)$ by **2*pi*Fref** results in **Vref** having a value $0 \le V_{ref} < 2\pi$. As a result of subtracting the last π term, $f(t)$ becomes a sawtooth waveform having a value of

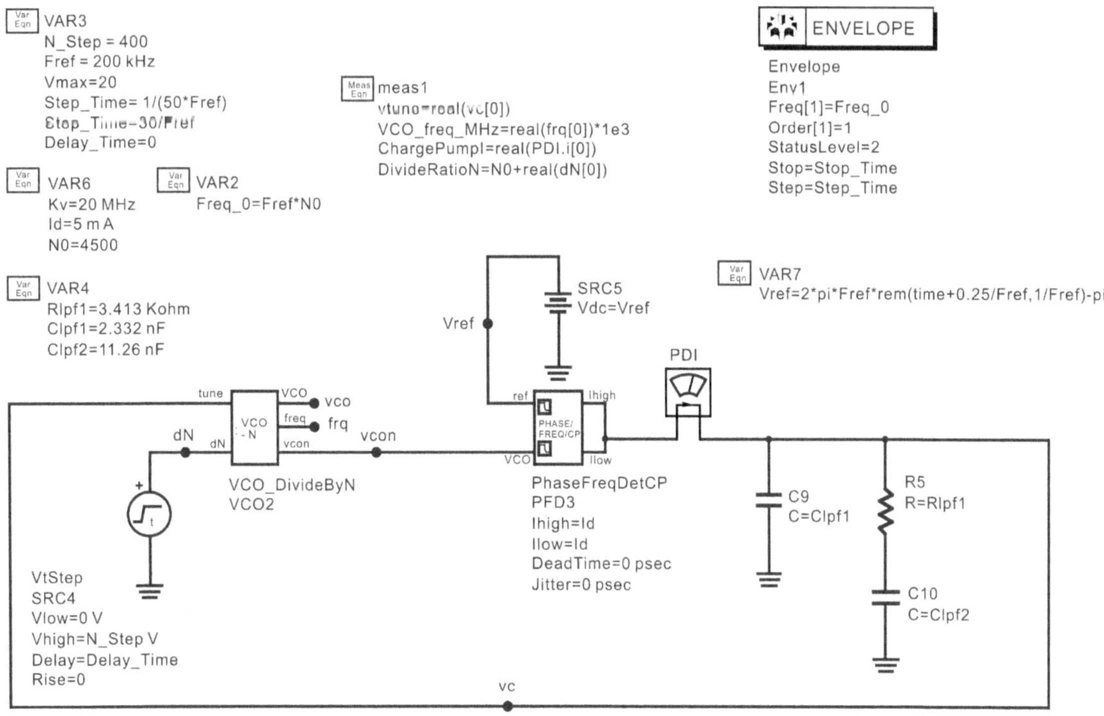

Figure 11.32 Transient response simulation. The terminal **VCO** is the undivided VCO output and the terminal **freq** outputs the undivided frequency values. The terminal **vcon** is the divided-by-(N + dN) output.

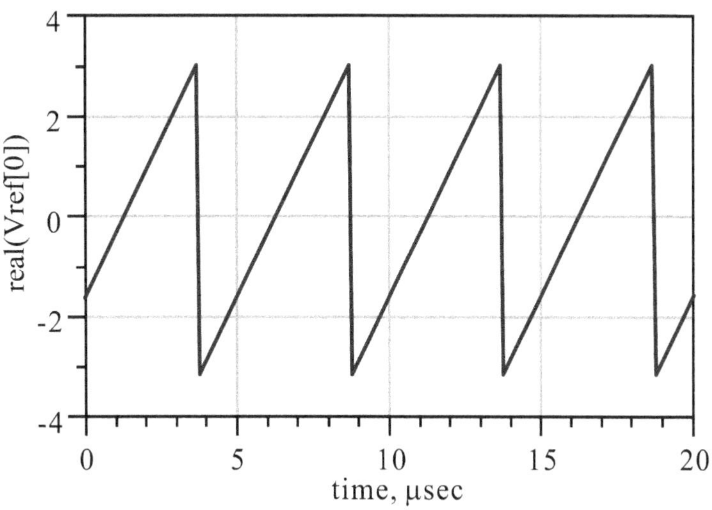

Figure 11.33 **Vref** waveform

$-\pi \le V_{ref} < \pi$. The function $f(t)$ also shifts to the left along the time axis by $0.25T$. The function $f(t)$ is shown in Figure 11.33. Therefore, despite being represented as a DC source, it is, in reality, a time-domain source.

The waveform in Figure 11.33 is fundamentally different from the **PFD** input waveform described earlier. However, because the **PFD** is a component that gives rise to a pulse-tip current proportional to the phase difference between the two input waveforms, the same result is obtained even though the input waveforms are different. In addition, the **PFD** changes at the rising edge of the input waveform. In the case of the sawtooth waveform shown in Figure 11.33, the rising edge corresponds to the time when the input sawtooth waveform crosses 0.5 V.

The output waveform of the VCO appears at the **VCO** terminal in the form of an envelope signal. The VCO frequency also appears at the **frq** terminal, which is the baseband signal. Note that the baseband signal's unit is GHz, while the following Measurement Expression 11.3 expresses it in MHz:

```
Eqn VCO_Freq_MHz=real(frq[0])*1e-3
```

Measurement Expression 11.3 VCO frequency definition in the MeasEqn

Therefore, **VCO_Freq_MHz** shows the change of the VCO's frequency (in MHz) with time. The output of the *N*–divided VCO in Figure 11.34 appears at the **vcon** terminal, which is also a baseband signal. The output waveform is also a sawtooth waveform that also represents a sawtooth waveform shifted to the left by 0.25*T* as in the case of the reference waveform. When the two inputs are applied to the input of the **PhaseFrequency DetCP**, the output current of the phase detector appears, as shown in Figure 11.34. In that figure, the output current begins to rise at the rising edge of the reference signal, **Vref** (the point where **Vref** crosses 0.5 V), and falls at the rising edge of the input VCO (the point where **vcon** crosses 0.5 V).

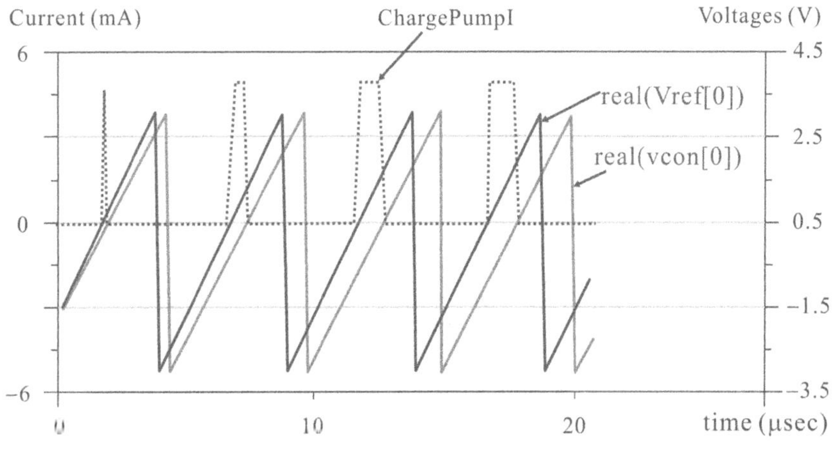

Figure 11.34 Input and output waveforms of the **PhaseFrequencyDetCP**

In addition, the VCO frequency division ratio can be made to change from **N** to **N + N_Step**, which is achieved by applying a step function with step height **N_step** to the **dN** terminal of the VCO. Here, **N_step** is set to **N_step** = 400, which allows for easy viewing of the transient waveform, but a similar waveform appears even when **N_step** = 1. The next issue is that the unit frequency step change represents the reference signal frequency step change in the previous explanation of section 11.4.2. However, here, the division ratio of the frequency divider is changed. This will give the same results as the unit step change of the reference signal; that is, when the division ratio of the frequency divider is changed from N to $N + 1$ at $t = 0$, the input signal to the phase detector becomes

$$\theta_i - \frac{\theta_o}{N+1} = \theta_i - \frac{N\theta_i}{N+1} = \frac{\theta_i}{N+1}$$

Thus, an increased step change of $\theta_i \times (N+1)^{-1}$ appears at the reference input of the phase detector. When the reference signal of the step change $\Delta\theta_i$ is applied, the phase detector experiences an increased step change of reference signal $\Delta\theta_i$. Thus, from a phase detector point of view, the step increase in the division ratio is the same as the reference signal step phase change.

The simulation result is shown in Figure 11.35. In that figure, because **N0** = 4500, the start frequency is **N0 × Fref** = 4500 × 200 kHz = 900 MHz. Also, since **N_Step** = 400, the frequency shift is **N_Step*Fref** = 400 × 200 kHz = 80 MHz. Figure 11.35 shows the VCO frequency settled at the steady state frequency of 980 MHz, which shows that the step change is tracked. In addition, because the bandwidth of the loop filter is set to 10 kHz, the lock time is

$$T_{lock} \sim \frac{1}{10} \times 1000\ \mu\text{sec} = 100\ \mu\text{sec}$$

From Figure 11.35, T_{lock} is found to closely predict the lock time of the PLL. Many ripples appear because the loop bandwidth is not sufficiently narrow compared with the comparison frequency.

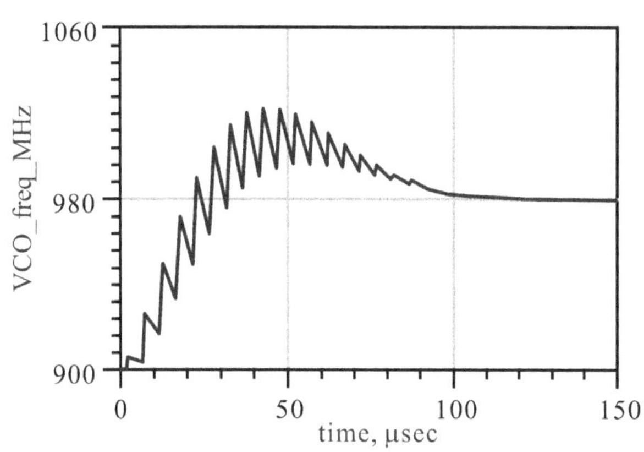

Figure 11.35 Time response of the VCO output frequency to the step frequency change

11.6 SUMMARY

- A PLL is composed of a phase detector, loop filter, VCO, and reference oscillator. The phase detector provides the output voltage proportional to the phase difference of the VCO and reference oscillator, which is characterized by the phase detector constant K_d (Volt/rad). The VCO in the PLL can be characterized by the tuning sensitivity K_v (rad/sec/V).
- At the PLL's steady state, the phase detector output voltage is constant and not time-varying. When the loop filter is above the second order, the phase detector output voltage is 0 at the steady state. This condition is used to find the VCO frequency at the steady state.
- Inside the loop bandwidth of the PLL, the phase noise of the VCO output follows that of the reference oscillator, while the phase noise of the VCO output as it is appears outside the loop bandwidth.
- Frequency dividers are used in a PLL to make the frequencies of the VCO and reference oscillator close; these frequencies are incident to the phase detector inputs. Frequency dividers are usually composed of a fixed divider and a programmable divider. Usually, the dual modulus counter is used as the fixed divider for consecutive dividing.
- Adding the accumulator to the divider composed of a dual modulus counter and a programmable divider, fractional dividing is possible and a fraction synthesizer can be constructed, which can reduce the in-band noise floor of the synthesizer.
- Phase detectors can be implemented using a DBM, Exclusive OR, and a phase-frequency detector (PFD). The phase detector characteristic of a PFD is linear and one-to-one for the phase difference 0–2π. When a charge pump circuit is added to the PFD, phase detector output becomes the current proportional to the phase difference, which makes for easy implementation of the loop filter that includes an ideal integrator.
- The loop filters can be designed using an open-loop gain. The second-order loop filter can be designed by setting ω_n to be equal to a loop bandwidth and damping ratio $\zeta = 0.707$. The open-loop gain of the third-order loop filter shows a phase peak. The third-order loop filter can be designed by setting the phase peak frequency to be equal to the loop bandwidth and setting the loop gain at this frequency to be equal to 1
- The fourth- and higher-order loop filters can provide spur attenuation. A higher-order loop filter can be designed by optimization.
- PLL simulation in ADS is possible and the process for simulating the phase noise and lock time of the PLL is presented.

REFERENCES

1. F. M. Gardner, *Phase-Locked Loop Techniques,* 2nd ed. New York: John Wiley & Sons, Inc., 1980.
2. U. L. Rohde, *Digital PLL Frequency Synthesizers: Theory and Design,* Englewood Cliffs, NJ: Prentice-Hall, Inc., 1983.
3. D. Barberjee, *PLL Performance, Simulation, and Design,* 4th ed. Indianapolis, IN: Dog Ear Publishing, 2006.

PROBLEMS

11.1 **(ADS Problem)** From Example 11.1, take f_n = 40 kHz and plot the output phase noise of the frequency synthesizer. Then, plot the phase noise using ADS.

11.2 Express the output phase noise, θ_o, of Figures 11.4(a) and 11.4(b) in terms of θ_i and θ_n.

11.3 Use Figure 11P.1 to solve the following problems:

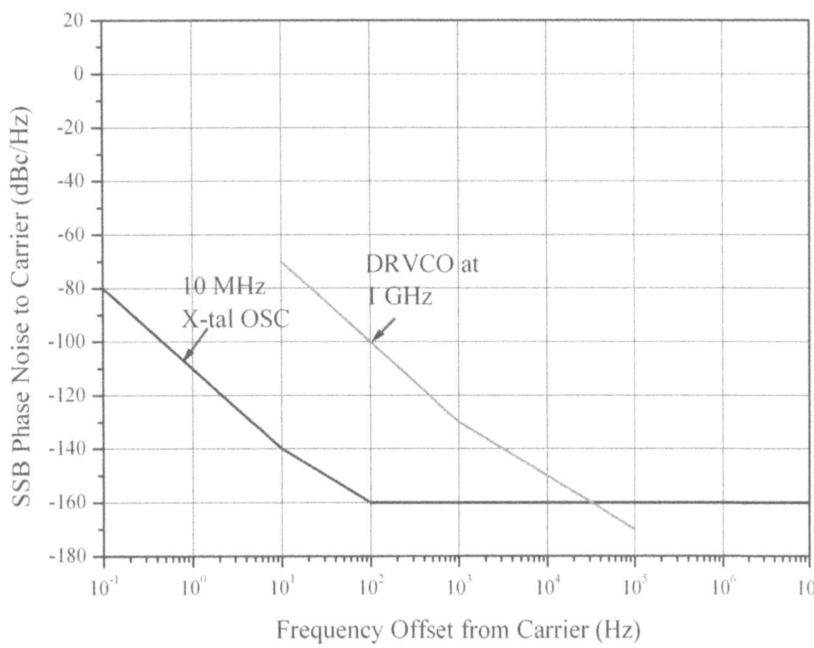

Figure 11P.1 Phase noises of crystal oscillator and DRVCO

(1) What is the phase noise at a 10-Hz offset frequency when a 10-MHz crystal oscillator is multiplied by 100? Compare this to the phase noise of 1-GHz DRO at a 10-Hz offset frequency.

(2) A 10-MHz crystal oscillator is used as the reference oscillator and it must phase lock at a 1-GHz DRVCO using a PLL. Obtain the optimum loop bandwidth BW. Then, plot the phase noise of the phase-locked DRVCO.

11.4 Use Figure 11P.2 to solve the following problems:

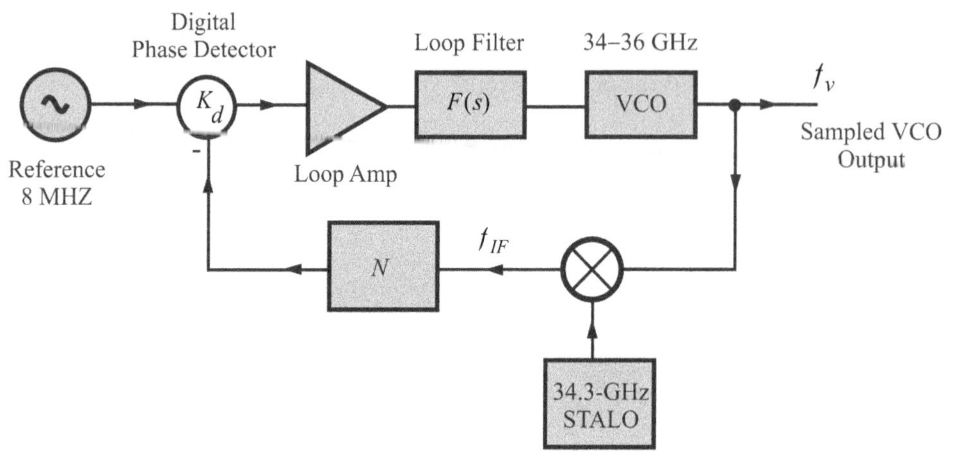

Figure 11P.2 PLL configuration

(1) From the figure, what is f_v when $N = 124$?
(2) Calculate the output frequency f_v when the frequency divider N is varied from 120–128.

11.5 Using Figure 11P.3, calculate the output frequency f_o.

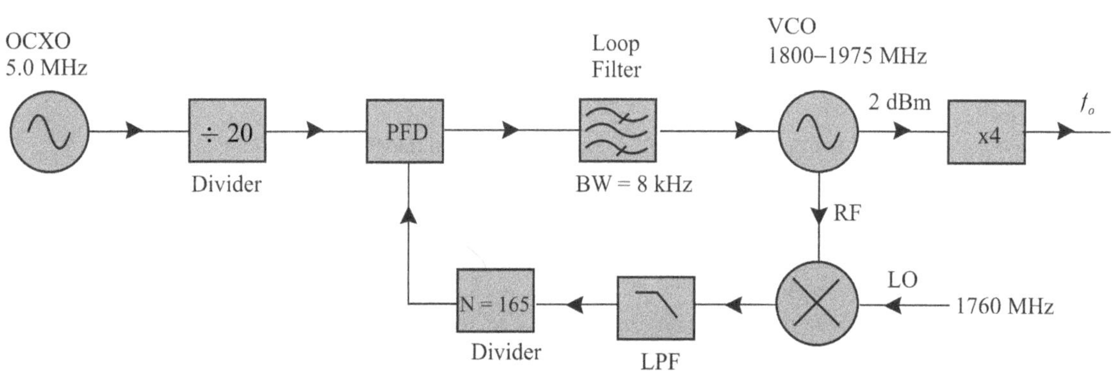

Figure 11P.3 PLL configuration

11.6 Using Figure 11P.4, solve the following problems:
(1) From the figure, what is f_o?
(2) Calculate the output frequency when the frequency divider n varies from 160–660.
(3) Given the phase noise characteristics of the OCXO and VCXO that appear in Table 11P.1, determine the optimum bandwidth of the F2 loop filter.

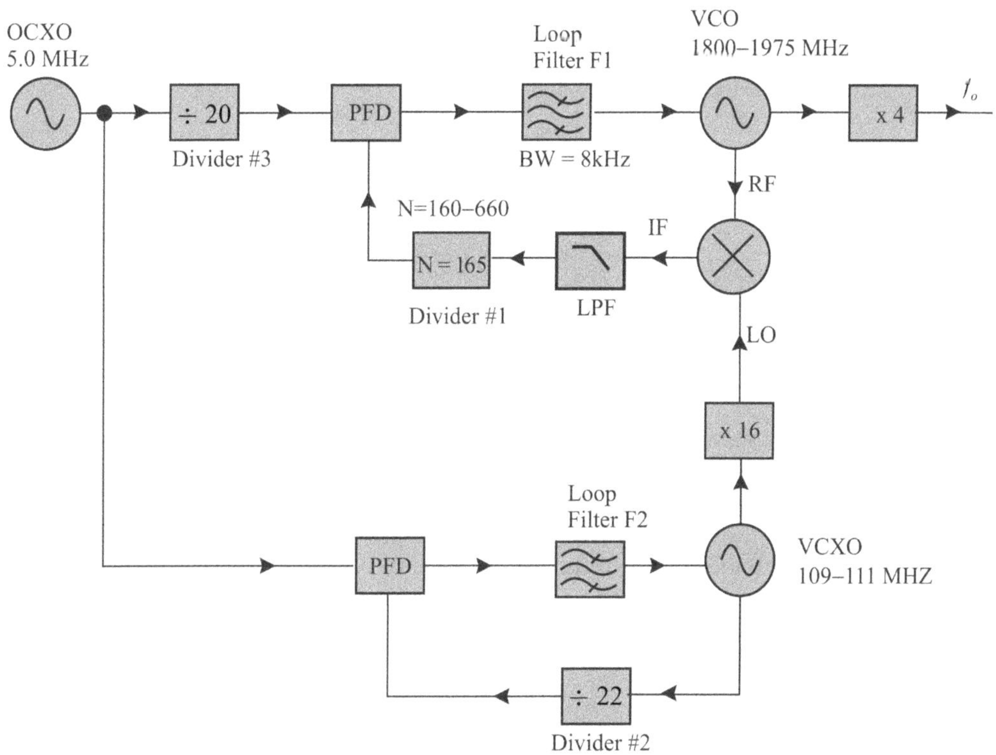

Figure 11P.4 PLL configuration

Table 11P.1 OCXO and VCXO phase noise

Frequency Offset	OCXO [dB/Hz]	VCXO [dB/Hz]
10 Hz	–115	–90
100 Hz	–135	–115
1 kHz	–145	–135
10 kHz	–150	–150
100 kHz	–150	–150
1 MHz	–150	–150

11.7 Explain why the characteristics of the XOR phase detector are as shown in Figure 11.9.

11.8 Solve the following problems using Figure 11P.5:

(1) Given that the frequency of the VCO varies from 1.5 to 1.6 GHz when the tuning voltage is varied from 0 to 3 V, calculate the tuning sensitivity K_v in [rad/sec/V].

(2) Given that the PFD flows a current of 1.52 mA when the phase signal leads that of the reference signal, and conversely, that the PFD flows a current of -1.52 mA when the phase signal lags, as shown in Figure 11P.5, what is the phase detector constant K_d [A/rad]?

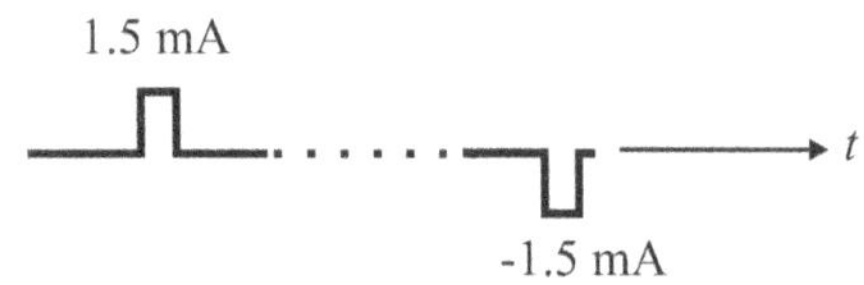

Figure 11P.5 The charge pump output waveform

(3) For a reference oscillator with a frequency of 250 kHz, a PLL is built using the PFD in problem 11.8.2, above, the VCO with a frequency of 1.5 GHz from problem 11.8.1, above, and a frequency divider with a ratio N ($N = 1500/0.25$). Determine the open-loop gain of this PLL. Determine C_2 and R_2, shown in Figure 11P.6, to yield the bandwidth for the PLL of 10 kHz.

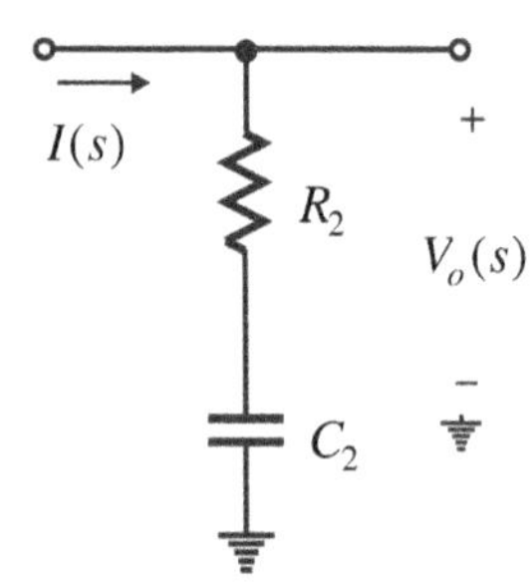

Figure 11P.6 The loop filter

11.9 By referring to the spectrum analyzer's manual and setting the SPAN to 0, explain the principles of measuring the time-varying power in an RBW.

11.10 For a given loop bandwidth ω_c and a phase margin ϕ, calculate the exact values of the fourth-order loop filter using the appropriate software. Confirm the results using a Bode plot.

11.11 **(ADS Problem)** Given that the loop bandwidth is 10 kHz, the phase margin is 45°, the comparison frequency is 200 kHz, and $f_o = 900$ MHz, after setting the phase detector constant for the tuning sensitivity of the VCO, as shown in Example 11.10, then calculate the element values of the second- and third-order loop filters and compare the loop gains.

11.12 **(ADS Problem)** In the simulation schematics shown in Figure 11.32, given that **N0** = 4500, **Fref** = 200 kHz, and **N_Step** = 10, simulate the transient response after configuring the third-order loop filter with a loop bandwidth of 1 kHz.

CHAPTER 12

Mixers

CHAPTER OUTLINE

12.1 INTRODUCTION

In communication systems, a mixer is frequently used as a building block for the frequency translation of an input signal. The mixer's basic operation can be best understood using a multiplier. Figure 12.1 shows a mixer based on the multiplier operation. The intermediate frequency (IF) output is obtained by multiplying two input signals, the RF and the local oscillator (LO), and it can be expanded as shown in Equation (12.1).

$$V_{IF} = KV_{RF}\cos\omega_{RF}t\cos\omega_{LO}t = KV_{RF}\left\{\cos\left(\omega_{RF} - \omega_{LO}\right)t + \cos\left(\omega_{RF} + \omega_{LO}\right)\right\} \quad (12.1)$$

The term $\cos(\omega_{RF} - \omega_{LO})t$ is the frequency down-converted term, while the term $\cos(\omega_{RF} + \omega_{LO})t$ is the frequency up-converted term. Thus, these two frequency components appear at the mixer output and the LO signal simply raises or lowers the RF input signal frequency. The down- or up-converted signal from the two frequency components can be selected using an appropriate filter.

In this chapter, we will examine the specifications of the mixer as well as their measurements, after which we will discuss the Schottky diode used in the mixer circuits. Transistors and diodes are the most widely used devices for mixers. Since transistors have been

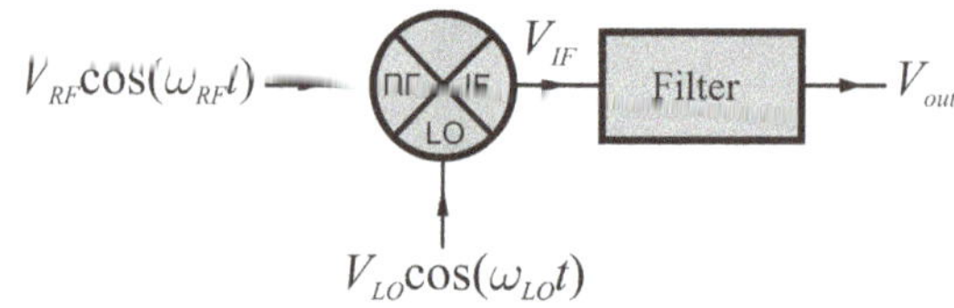

Figure 12.1 Mixer signals

covered extensively in Chapter 5, in this chapter, we will examine the physical properties of Schottky diodes that are mainly used in mixers.

In addition, there are various mixer configurations. These include the single-ended mixer (SEM), the single-balanced mixer (SBM), and the double-balanced mixer (DBM). We will examine the configurations of these mixers in this chapter and also explore their advantages and disadvantages.

12.2 SPECIFICATIONS

12.2.1 Conversion Loss and 1-dB Compression Point

As mentioned earlier, the mixer provides the frequency up- or down-converted signal to the IF output. Generally, the IF output power increases in proportion to the RF input power for a small RF input power. The ratio of the frequency up- or down-converted IF output power P_{IF} to the input RF power, P_{RF} is defined as the conversion loss, CL, which is expressed by Equation (12.2).

$$CL = \frac{\text{Available RF input power}}{\text{Delivered IF output power}} \tag{12.2}$$

The conversion loss can be measured using a spectrum analyzer, as shown in Figure 12.2. First, the available power of the signal generator with a single tone is measured by connecting the signal generator directly to the spectrum analyzer. Then, connecting the signal generator to the mixer input and connecting the spectrum analyzer to the mixer output, the mixer IF output signal appears at the spectrum analyzer. The conversion loss of the mixer can be obtained by measuring the mixer IF output power and calculating the ratio given by Equation (12.2). The procedure is shown at the top of Figure 12.2. Note that the mixer IF output power is generally proportional to the RF input power, as shown in Equation (12.1).

In addition, when the two-tone signals $f_{RF,1}$ and $f_{RF,2}$ (two sinusoidal inputs with different frequencies) are applied to the mixer input, as shown at the bottom of Figure 12.2, the mixer output presents two frequency-translated output signals that correspond to the down- (or up-) converted IF in Figure 12.2. Also, as a result of intermodulation, two spurious signals appear above and below the two frequency-translated IF signals. The frequency spacing between the spurious and the adjacent IF signal is the same as that of the two-tone signal. The spurious signals also generally increase with an increase in the RF input power. As the intermodulation usually occurs as a result of third-order nonlinearity in the mixer, the plot of the spurious signals for the RF input power generally shows a straight line whose slope is approximately 3 on a log-log plot, as shown in Figure 12.3. Since the slope is greater than that of the IF output, the two lines intersect, as shown in

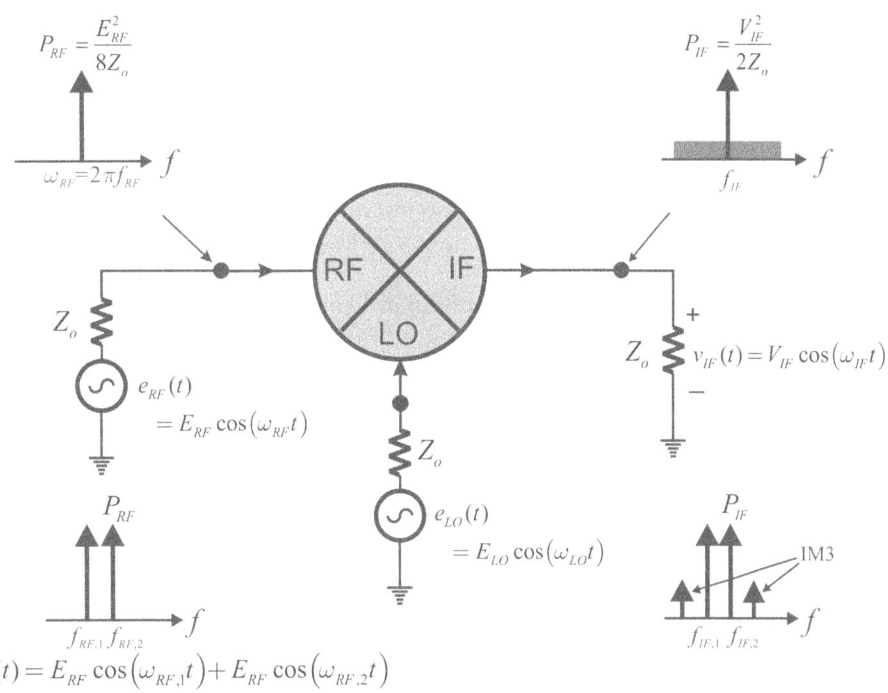

Figure 12.2 Measurement of mixer conversion loss; the upper portion of the figure illustrates the measurement of a mixer conversion loss and the lower portion shows the measurement of the mixer's IMD3.

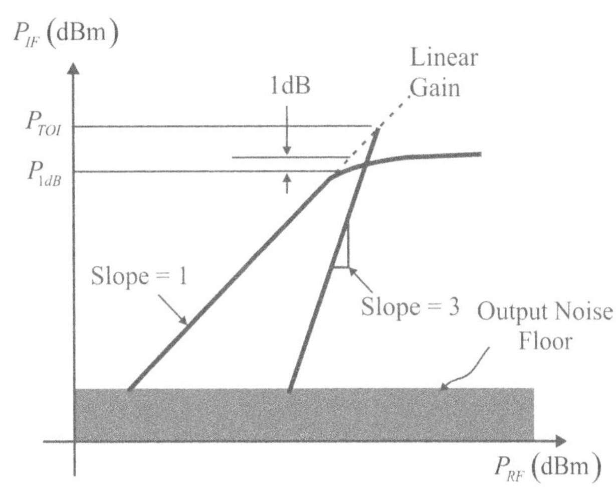

Figure 12.3 The IF output power and the third-order intermodulation frequency power of a mixer versus the RF input power. The intercept of the two extrapolated lines is defined as a TOI (third-order intercept).

Figure 12.3. The intercept point is known as a third order intercept (TOI). Rather than the TOI that is the power at the intersection point of the two linearly extrapolated lines, a third-order intermodulation distortion (IMD3) is often employed to specify the degree of the third-order distortion and is defined as the ratio of the intermodulation output power to the IF output power. IMD3 is thus expressed as Equation (12.3).

$$\text{IMD3} = \frac{\text{Spectral power of third-order tone}}{\text{IF output power}} \tag{12.3}$$

It is worth noting that the value of the IMD3 is dependent on the IF output power. Thus, it should be specified at what power level of the IF the IMD3 is measured.

Next, we consider the problem of LO power. The conversion loss in Equation (12.1) appears to be independent of the LO power. This is only partly correct. When the LO power is supplied above a certain level, an almost constant conversion loss is obtained irrespective of the LO power. However, the conversion loss varies significantly according to the LO power when the LO power is low. This is known as the LO characteristic of the conversion loss. Figure 12.4 shows an example of the LO characteristic; above an LO power of approximately 12 dBm, the conversion loss can be observed to be constant irrespective of the LO power.

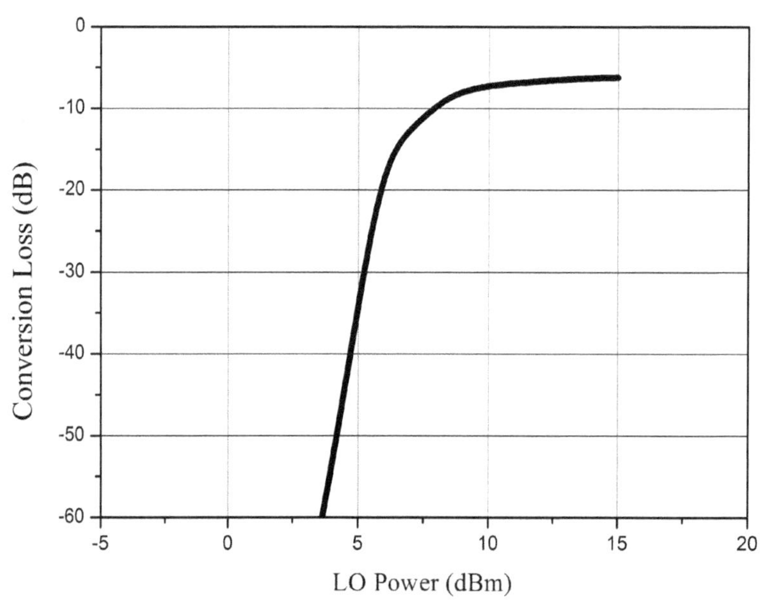

Figure 12.4 Example of the mixer's conversion loss with respect to the LO power

12.2.2 Mixer Isolation and VSWR

In an ideal mixer, no signal appears at the IF output except the frequency up- or down-converted components of the RF input. However, at the IF output of a real mixer, because the LO and RF signals leak, the RF and LO leakage signals also appear at the mixer's IF output. In addition, a leakage of the LO signal appears at the RF port. Mixers generally show different leakage powers, which also become one of the specifications for a mixer, usually referred to as the isolation between ports. The method for measuring the isolation is shown Figure 12.5. When a signal is applied to the LO port, it appears as a leakage at the RF and IF ports. The available power applied to the LO port is denoted as P_{LO} (the LO available power measurement method is the same as that of the RF available power measurement described in Figure 12.2), and then the leakage LO powers at the RF and IF ports are denoted as $P_{L\rightarrow R}$ and as $P_{L\rightarrow I}$, respectively. The isolation between the LO and RF ports $L\rightarrow R$ and the isolation between the LO and IF ports $L\rightarrow I$ are defined by Equations (12.4) and (12.5) as

$$L \rightarrow R = \frac{P_{L\rightarrow R}(\text{at RF port})}{P_{LO}} \tag{12.4}$$

$$L \rightarrow I = \frac{P_{L\rightarrow I}(\text{at IF port})}{P_{LO}} \tag{12.5}$$

Also, when an RF signal is applied, denoting the available power from the RF port P_{RF}, the isolation between RF and IF ports $R\rightarrow I$ is expressed by Equation (12.6).

$$R \rightarrow I = \frac{P_{R\rightarrow I}(\text{at IF port})}{P_{RF}} \tag{12.6}$$

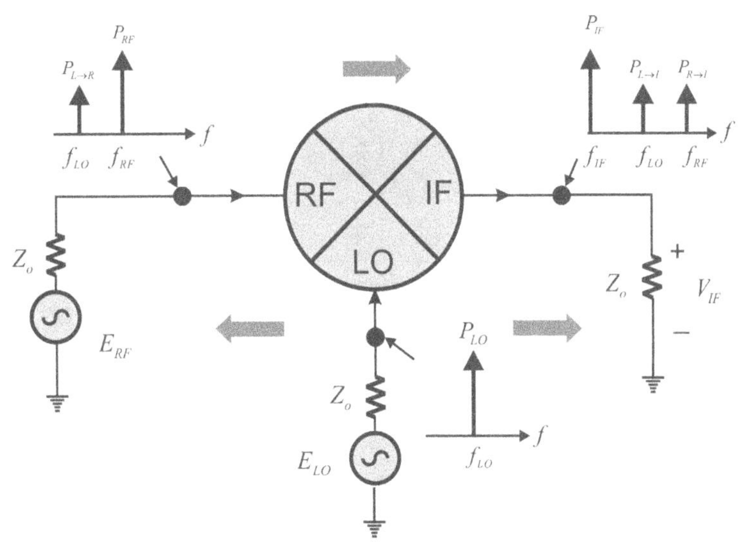

Figure 12.5 Measurement of mixer isolations

In addition, each port of the mixer is not perfectly matched, which often results in reflection problems when the mixer is applied in a system. A diagram for the respective definitions of these return losses is shown in Figure 12.6, and the return loss is usually defined for each frequency component and given by Equations (12.7a)–(12.7c).

$$\text{RF return loss} = \left.\frac{\text{reflected } P_{RF}}{\text{available } P_{RF}}\right|(\text{at RF port}) = \frac{P_{RF,r}}{P_{RF}} \tag{12.7a}$$

$$\text{IF return loss} = \left.\frac{\text{reflected } P_{IF}}{\text{available } P_{IF}}\right|(\text{at IF port}) = \frac{P_{IF,r}}{P_{IF}} \tag{12.7b}$$

$$\text{LO return loss} = \left.\frac{\text{reflected } P_{LO}}{\text{available } P_{LO}}\right|(\text{at LO port}) = \frac{P_{LO,r}}{P_{LO}} \tag{12.7c}$$

Note that the RF and IF return losses defined in the equations above should be measured after the LO signal of a specified power level is applied to mixer.

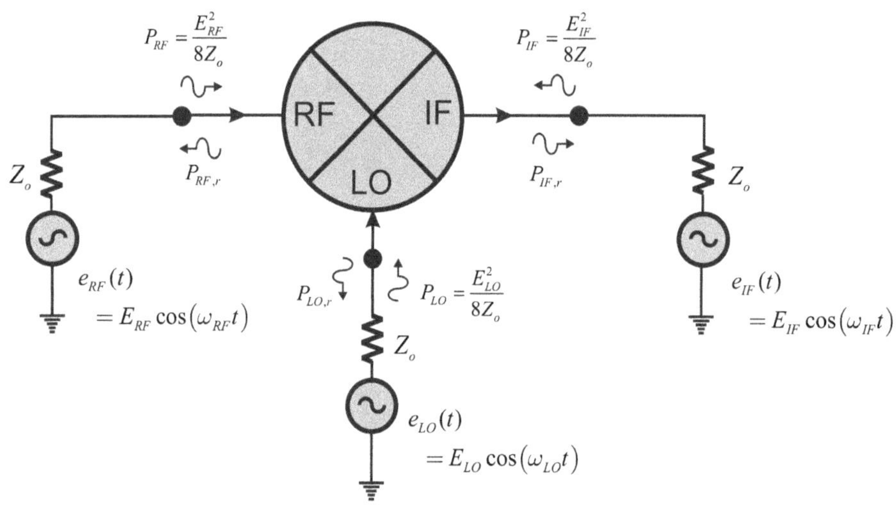

Figure 12.6 Measurement of a mixer's return losses. In the case of the return loss measurement for the LO port, both RF and IF powers should be set to 0. In contrast, the LO is supplied for the RF and IF ports' return loss measurements. In the case of the return loss measurement for the RF port, $E_{IF} = 0$, while that for the IF port return loss is $E_{RF} = 0$.

Example 12.1

In the ADS palette **System_Amp & Mixer**, search for a **Mixer** and set its parameters as shown in Figure 12E.1. Note that the mixer in ADS is modeled by polynomial expressions. The conversion loss is set to 6 dB. The TOI of the mixer is set to 12.5 dBm and the 1-dB compression point is set to 0 dBm. Note that the 1-dB compression and the TOI powers can be defined using the mixer's input or output powers. The variable **ReferToInput** is used to tell whether the input or output powers are used to specify the TOI and the 1-dB compression. In this example, the TOI and the 1-dB compression have been defined using the output power of the mixer as **ReferToInput** = OUTPUT. Also, **LO_Rej**1 represents the LO leakage into the RF port while **LO_Rej**2 represents the LO leakage into the IF port. The isolations of the LO signal to the IF and RF ports are set to –200 dB. **RF_Rej** represents the RF leakage into the IF port. The RF–IF isolation is also set to –200 dB.

In addition, the LO input power is set to 15 dBm and its frequency is set to 1.5 GHz; the RF is set to 1.15 GHz and its input power varies from –30 dBm to 20 dBm. The 1-dB compression point can be determined from the simulation. Using simulation, show that the conversion loss is 6 dB and the 1-dB compression point output power is 0 dBm.

Solution

As shown Figure 12E.1, the simulation is set up and run. The spectrum of the signal IF is shown in Figure 12E.2.

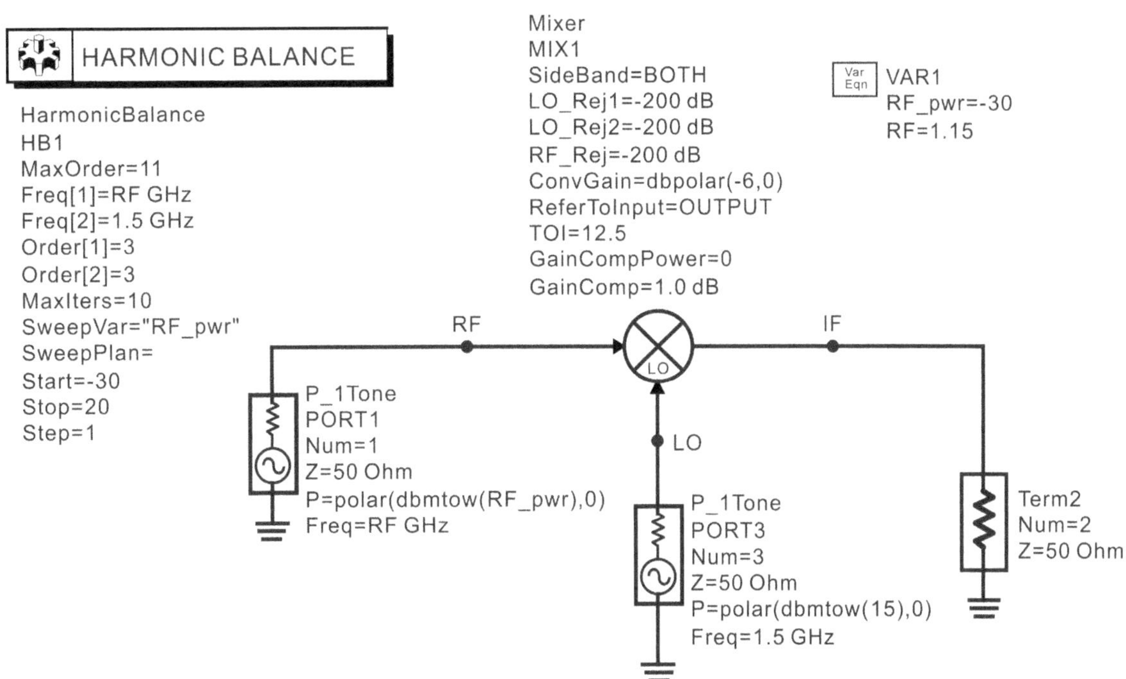

Figure 12E.1 Simulation of a mixer's conversion loss. **LO_Rej**1 and **LO_Rej**2 represent the LO-RF and LO-IF isolations, respectively. The variable **ReferToInput** asks whether the power levels of the TOI and the 1-dB compression points are measured at the mixer input or output. Here, they are measured at the output. The variable **GainComp** is used to specify the compression power levels; usually, the 1-dB compression is used but it can be 3 dB or other dB levels.

The IF frequency is (1.5 – 1.15) GHz = 350 MHz. The IF output power at 350 MHz can be read using a marker. Calculating the conversion loss at a small RF input power of –30 dBm gives –36 – (–30) = –6 dB, which yields the same results as the selected value of the conversion loss. Here, the inner array index 0 of the IF in Figure 12E.1 represents the sweep index of the RF input power, while the outer index corresponds to the harmonic index and **::** represents all indices for the swept variable. Thus, **IF**[0,::] represents all the harmonic spectra at the RF power level of -30 dBm.

Similarly, the array index [::,1] represents the IF frequency of 350 MHz for all the swept RF input power, **RF_pwr**. Thus, **IF**[::,1] is the set of output voltages with respect to the RF input power at a frequency of 350 MHz. To confirm this, the frequency values of the array index [::, 1] are listed using a listing box and the listings of the frequency and the mixer IF voltage are shown in Table 12E.1, where the IF voltage [::, 1] is found to correspond to 350 MHz.

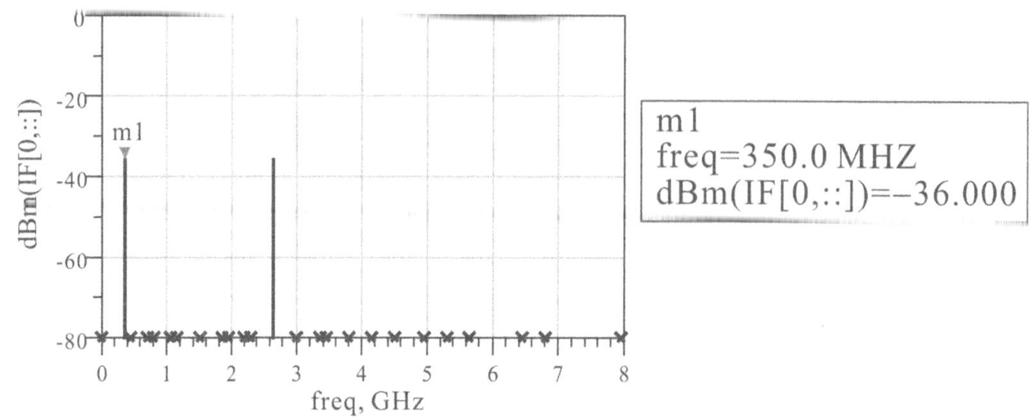

Figure 12E.2 Spectrum of IF output at -30-dBm RF input power. Since the RF input power sweeps from -30 to 20 dBm, IF[0,::] corresponds to all the harmonic spectra at an RF input power level of -30 dBm. At the RF power level of -30 dBm, the IF power level is -36 dBm and thus the conversion loss is 6 dB.

Table 12E.1 List of IF[::,1] with respect to RF_pwr

RF_pwr	IF[::,1]	freq[::,1]
-30.000	0.005 / -3.287...	350.0 MHz
-29.000	0.006 / -2.929...	350.0 MHz
-28.000	0.006 / 4.351...	350.0 MHz
-27.000	0.007 / -2.715...	350.0 MHz
-26.000	0.008 / -1.037...	350.0 MHz
-25.000	0.009 / -1.232...	350.0 MHz
-24.000	0.010 / -5.491...	350.0 MHz
-23.000	0.011 / -2.936...	350.0 MHz
-22.000	0.013 / -3.926...	350.0 MHz
-21.000	0.014 / -4.665...	350.0 MHz
-20.000	0.016 / -4.158...	350.0 MHz

Therefore, plotting the IF output power **IF**[::,1] versus **RF_Pwr** will find the 1-dB compression point. However, it is not easy to read the 1-dB compression point directly from the IF output power versus the RF input power plot.

The conversion loss can be found from the plot of dBm(**IF**[::,1])–**RF_pwr**, which also changes with respect to the RF input power. The conversion loss will have the value of –6 dB when the input RF power is low. Thus, the 1-dB compression point can be easily found by determining the –7dB point from the plot. A better way to find the 1-dB compression point is to plot dBm(**IF**[::, 1]) – **RF_pwr** + 6, which will give the curve decrease from 0 dB. Adding 6 dB to the value of the conversion loss will result in 0 dB at the low RF input power, which decreases as the RF input power increases, and passes through the point –1 dB. This is the 1-dB compression point. In Figure 12E.3, the 1-dB compression point occurs where the RF input power is 7 dBm. The corresponding 1-dB compression output power can be computed from the equation dBm(**IF**[::, 1]) – **RF_pwr** + 6 = –1. The output power dBm(**IF**[::, 1]) is computed to be 0 dBm. The value of 0 dBm is the 1-dB compression output power, which can be seen to be equal to the selected value.

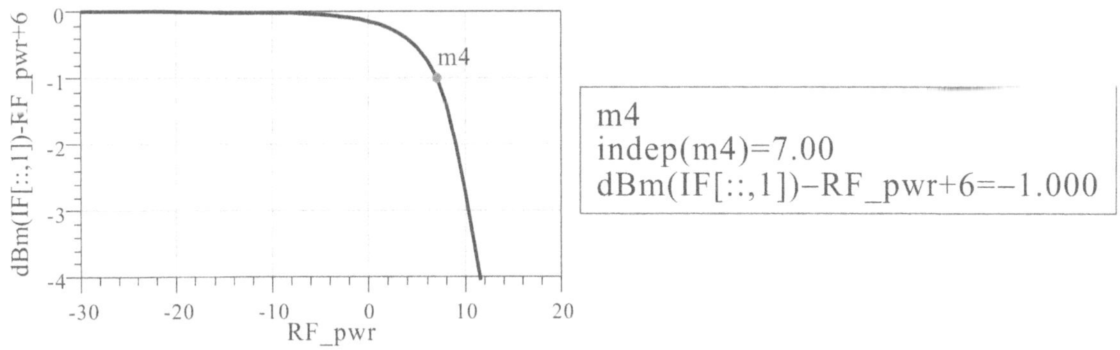

Figure 12E.3 The normalized conversion gain versus the RF input power. The 1-dB compression point can be easily read. Since the 1-dB compression occurs at the 7-dBm input power, the IF output power dBm(IF[::,1]) = 0 dBm at this input power. Thus, the 1-dB compression output power level is 0 dBm.

Example 12.2

Using the settings in Figure 12E.4, apply a two-tone input of 1.15 GHz and 1.25 GHz, and then verify the TOI.

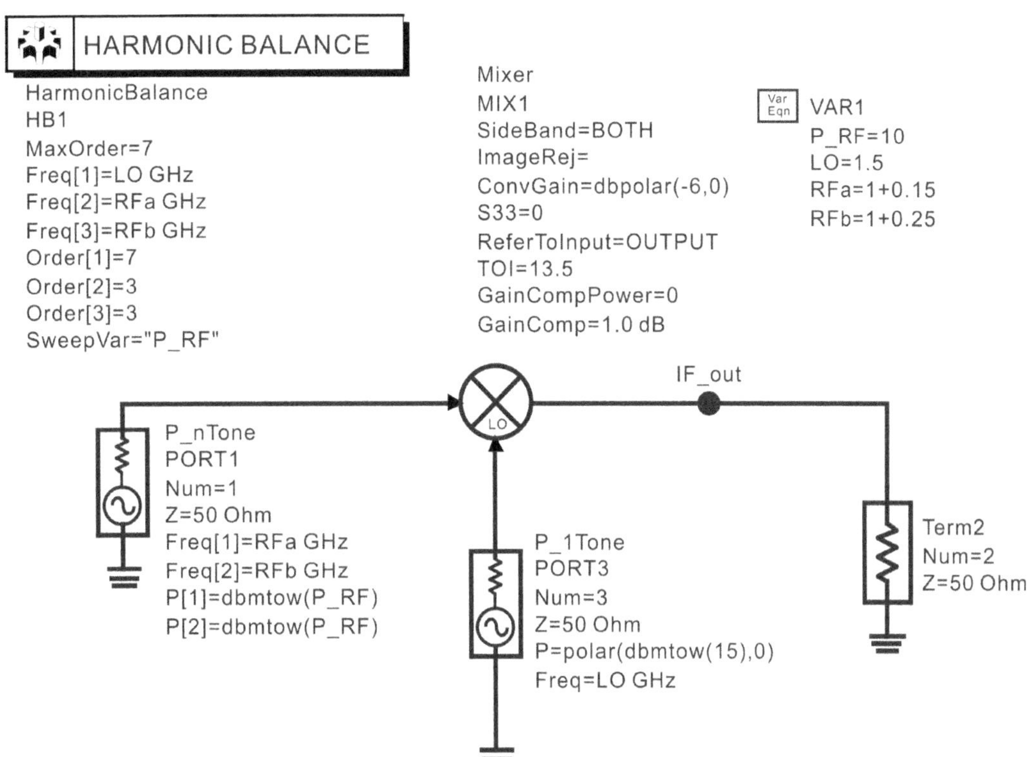

Figure 12E.4 Two-tone simulation schematic. The TOI can be measured using the two-tone input and the two-tone input signal is applied at the RF input.

Solution

The RF input power per tone, **P_RF**, varies from –30 to 20 dBm in a 1-dB step. The two-tone frequencies are set to 1.15 GHz and 1.25 GHz, respectively. Thus, the frequency spacing is 100 MHz. The plot of the mixer output at an input power of –10 dBm (corresponding to the **P_RF** sweep array index number =20) is shown in Figure 12E.5. The spurious outputs are also observed to appear with the two-tone output.

The third intermodulation frequency components appear at 150 MHz and 450 MHz in Figure 12E.5. The TOI can be determined using the extrapolated lines of the IF output and IMD3 powers. In order to determine the index of the intermodulation frequency components, **freq**[0,::] is listed using the listing box shown Table 12E.2. It can be seen from the listing box that the index of the lower intermodulation component for the 150-MHz frequency is 3. Similarly, the index of the 350-MHz IF frequency is 9.

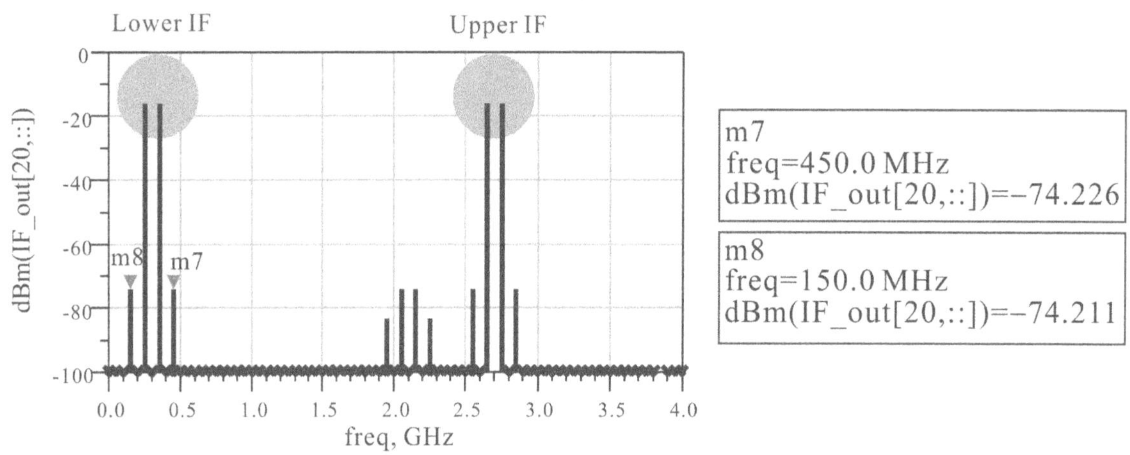

Figure 12E.5 The IF spectrum for the RF input of -10 dBm. Upper IF represents the up-converted signals and Lower IF represents the down-converted signals for the two-tone input.

Table 12E.2 List of variable frequencies

freq	freq[0,::]
0.0000 Hz	0.0000 MHz
50.00 MHz	50.00 MHz
100.0 MHz	100.0 MHz
150.0 MHz	150.0 MHz
200.0 MHz	200.0 MHz
200.0 MHz	200.0 MHz
250.0 MHz	250.0 MHz
300.0 MHz	300.0 MHz
300.0 MHz	300.0 MHz
350.0 MHz	350.0 MHz
400.0 MHz	400.0 MHz
400.0 MHz	400.0 MHz

Thus, when the **IF**[::, 3] and the **IF**[::,9] are plotted, they show the powers of the third-order intermodulation frequency and the IF frequency components of 150 MHz and 350 MHz with respect to the RF input power. In order to find the TOI, two extrapolated lines for the IF and the third-order intermodulation outputs for the small RF input power are necessary. In the case of the IF output power, the extrapolated line will be **P_RF** – 6. However, IMD3 power is –74.2 dBm at the RF power level –10 dBm from Figure 12E.5. Thus, this IMD3 power level of –74.2 dBm will grow by 30 dB at an RF power level of 0 dBm. As a result, the extrapolated line for the IMD3 power is 3(**P_RF**) – 44.2 = 3(**P_RF**—14.7). The two extrapolated lines are also plotted in Figure 12E.6. The intersection point of the extrapolated lines will yield the TOI. The two extrapolated lines are seen to intersect at the output power of 13 dBm. Even though this value of the TOI is slightly away from the selected value of the TOI, it is still relatively close to the selected TOI value.

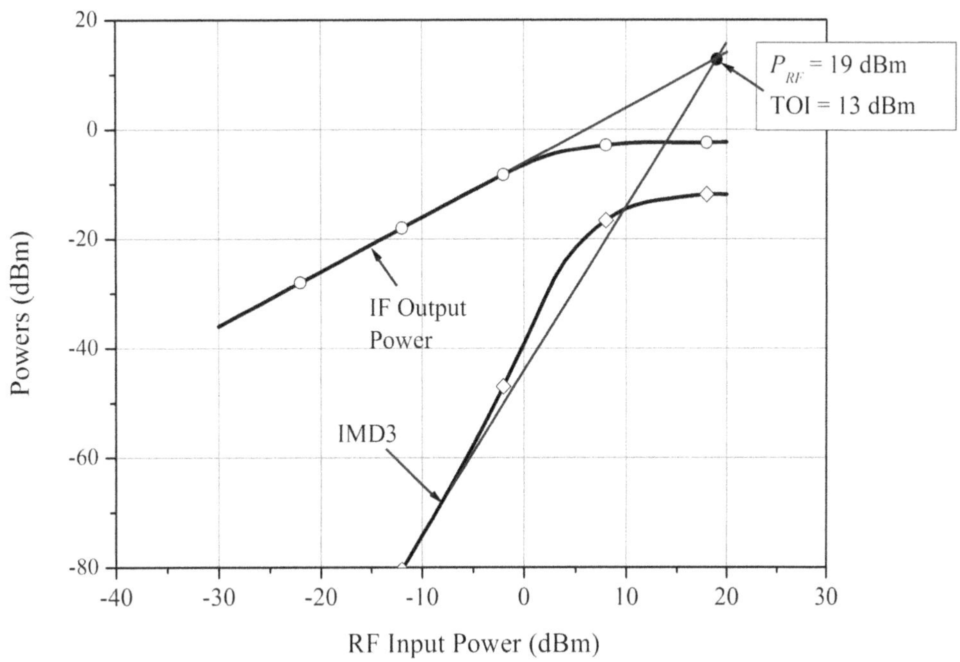

Figure 12E.6 Plot for the calculation of the TOI. The two extrapolated straight lines at the RF power level -10 dBm are also plotted to obtain the TOI. The computed value of the TOI is 13.0 dBm, which is smaller than the set TOI value of 13.5 dBm.

As an alternative to this cumbersome calculation, the TOI can be calculated by entering the following equation from Measurement Expression 12E.1 in the display window.

```
Eqn y=ip3_out(IF_out, {1,-1,0}, {1,-2,1},50)
```

Measurement Expression 12E.1 Calculation of the TOI

Since **freq**[1] = LO, **freq**[2] = **RFa**, and **freq**[3] = **RFb** in Figure 12E.4, the first index 1 in {1,–1,0} represents the LO harmonic index. Similarly, the second index –1 in {1,–1, 0} represents the **RFa** harmonic index. Thus, {1,–1, 0} represents the IF frequency of 350 MHz considering 1 × 1.5 + (–1) × 1.15 = 0.35 GHz, and {1,–2, 1} represents the frequency component of 1 × 1.5 + (–2) × 1.15 + 1 × 1.25 = 0.45 GHz.

Therefore, the TOI can be found by using the two frequency components. The listing box in Table 12E.3 shows the TOI value **y** for the RF input power. The value of the TOI varies according to the RF input power and is approximately 13.5 dBm at an input RF power of –30 dBm, which is the same as the selected TOI value. Thus, we can determine why the computed TOI value in Figure 12E.6 is smaller than the set TOI value of 13.5 dBm. Equation (9.63) in Chapter 9 is also used in **ip3_out** of Measurement Expression 12E.1 to compute the TOI.

Table 12E.3 List of the TOI values with respect to RF_pwr

P_RF	y
-30.000	13.496
-29.000	13.494
-28.000	13.493
-27.000	13.491
-26.000	13.489
-25.000	13.486
-24.000	13.482

Example 12.3

Set **LO_rej**1, **LO_rej**2, and **RF_rej** as shown in Figure 12E.7. The values of the TOI and the 1-dB compression point are the same as in Example 12.2. In addition, set the reflection coefficients of the RF and LO ports as shown in the figure. Note that because the LO port number is 3, **S**33 becomes the LO port's reflection coefficient. After the simulation, verify whether or not the selected isolations and reflection coefficients are obtained.

Solution

The IF output voltage's spectrum after the simulation is shown in Figure 12E.8.

From the reading of marker **m**1 in Figure 12E.8, the isolation between the LO and IF ports is computed to be –15 – 15 = –30 dB, and the isolation between the RF and IF ports is computed to be –70 – (–30) = –40 dB from the reading of marker **m**2. These can be seen to be the same as the selected values. From the reading of **m**3, the conversion gain is –36 – (–30) = –6 dB as was set in the simulation setup shown in Figure 12E.7. Figure 12E.9 shows the RF signal spectrum for confirming the L-R port isolation. The L-R isolation in Figure 12E.9 is –5 – 15 = –20 dB from the reading of the **m4** marker that was previously set.

Generally, the reflection coefficient can be found using a directional coupler. However, the reflection coefficient can also be determined from the simulated output voltage shown in Figure 12E.7. The reflection coefficient can be found using the following equation:

$$S_{11} = \frac{V_1 - \dfrac{E_{inc}}{2}}{\dfrac{E_{inc}}{2}}$$

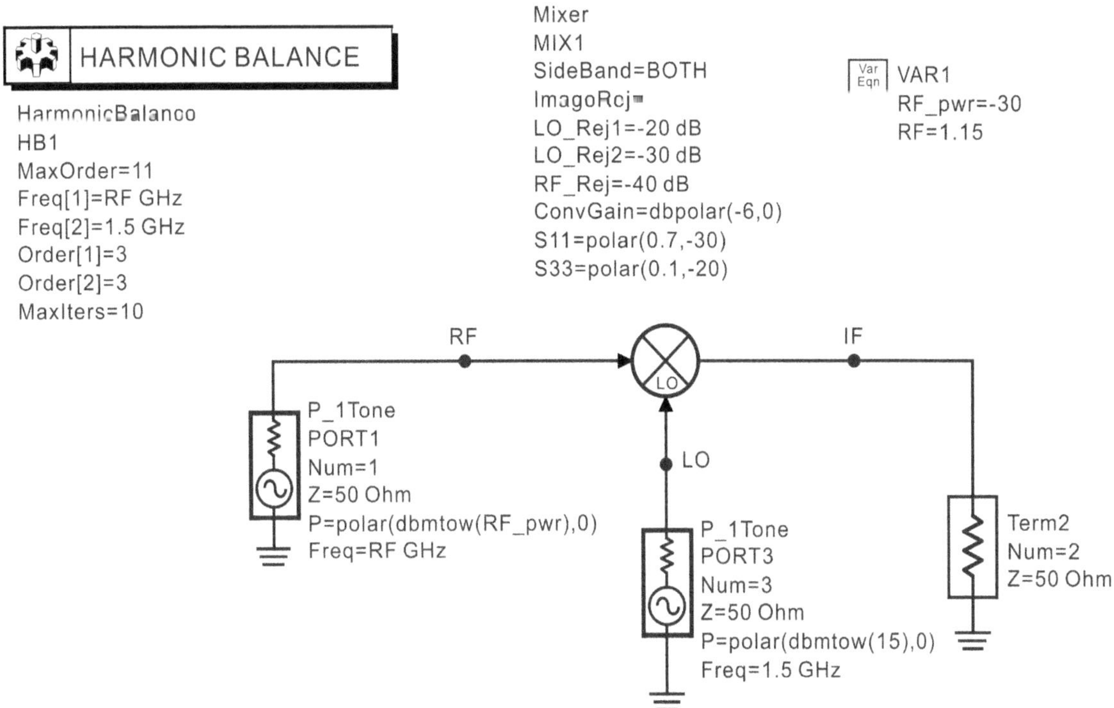

Figure 12E.7 Schematic for verifying port-to-port isolations and reflection coefficients. The LO-RF isolation is set to 20 dB and the LO–IF isolation is set to 30 dB. The variable **RF_Rej** represents the RF–IF isolation and is set to 40 dB. In addition, the reflection coefficients at the RF port and LO port are set to 0.7∠-30° and 0.1∠-20°, respectively.

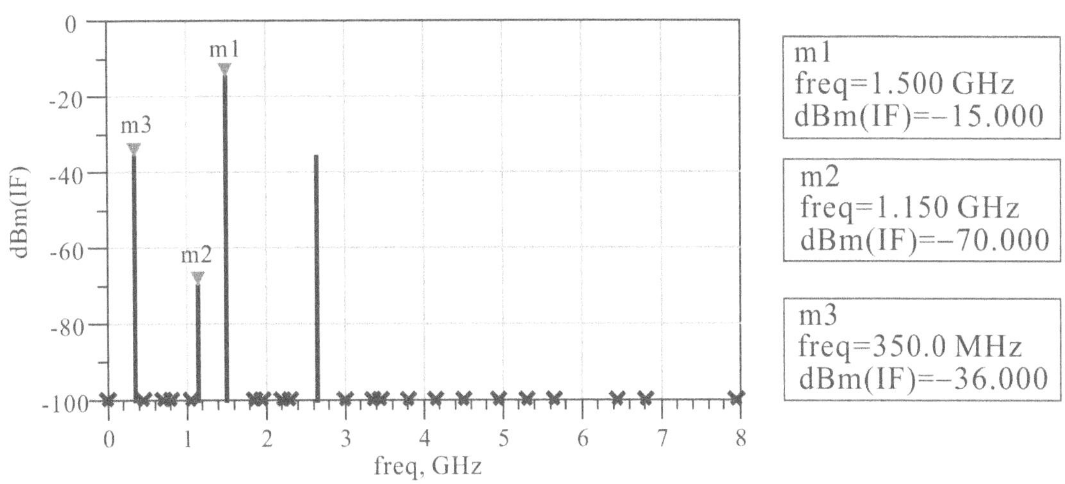

Figure 12E.8 Simulated IF output spectrum. Marker m3 represents the IF output power, marker m1 represents the LO leakage power, and marker m2 represents RF leakage power at the IF Port.

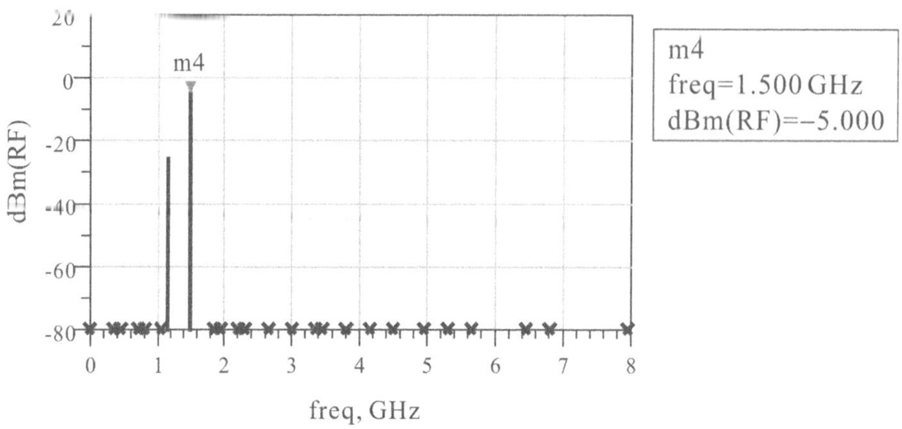

Figure 12E.9 Simulated power spectrum at the RF port. Marker m4 represents the LO leakage power at the RF port.

The voltage V_1 can be determined from the simulated RF voltage and E_{inc} can be determined from the RF input power. The equations shown in Measurement Expression 12E.2 are thus entered in the display window. The array index of the RF frequency is 6 and, as the array index of the LO frequency is 7, **RF**[6] and **LO**[7], respectively, are entered to obtain the RF and LO frequency components.

```
Eqn e_rf=sqrt(dbmtow(–30)*2*50)
Eqn e_lo=sqrt(dbmtow(15)*2*50)
Eqn ref1=(RF[6]-e_rf)/e_rf
Eqn ref2=(LO[7]-e_lo)/e_lo
```

Measurement Expression 12E.2 Equations for the LO and RF port reflections

The computed reflection coefficients **ref1** and **ref2** are shown in the listing box of Table 12E.4. The computed reflection coefficients are equal to those set in the simulation schematic.

Table 12E.4 List of the RF and LO reflection coefficients

ref2	ref1
0.100 / –20.000	0.700 / –30.000

12.3 SCHOTTKY DIODES

12.3.1 Structure of the Schottky Diode

Figure 12.7 is a top view of a typical Schottky diode. A cross-sectional view of the A–A′ plane in Figure 12.7 is shown in Figure 12.8. In order to form a Schottky diode, a thin n-type epitaxial layer is deposited on an electron-rich n^+ Si substrate, as shown in

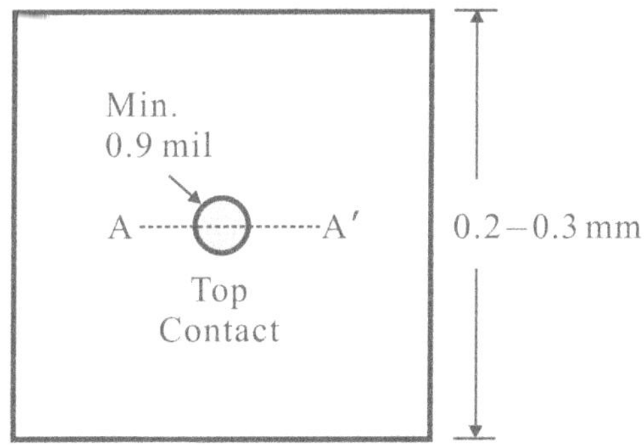

Figure 12.7 Top view of a Schottky diode

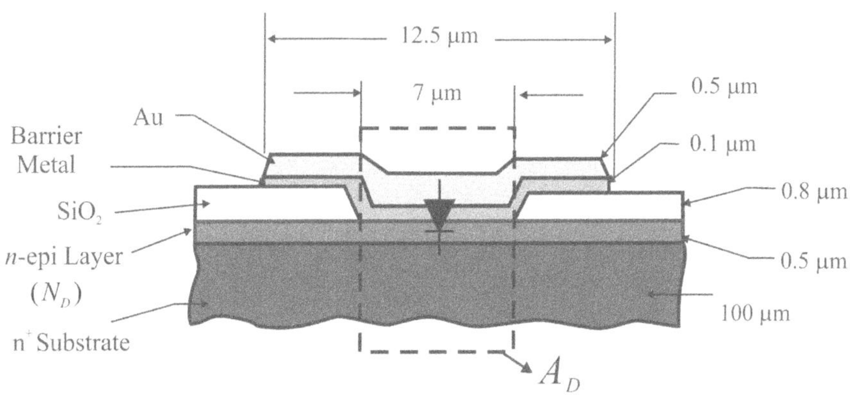

Figure 12.8 Cross-sectional view of a Schottky diode with its typical dimensions

Figure 12.8. The areas that do not form the diode are isolated using silicon oxide SiO_2. Conductive metal is then deposited on the open area to form the Schottky diode.

The equivalent circuit of the Schottky diode is shown in Figure 12.9. Resistor R_s represents the series resistance arising from the metal semiconductor junction. Capacitor C_j

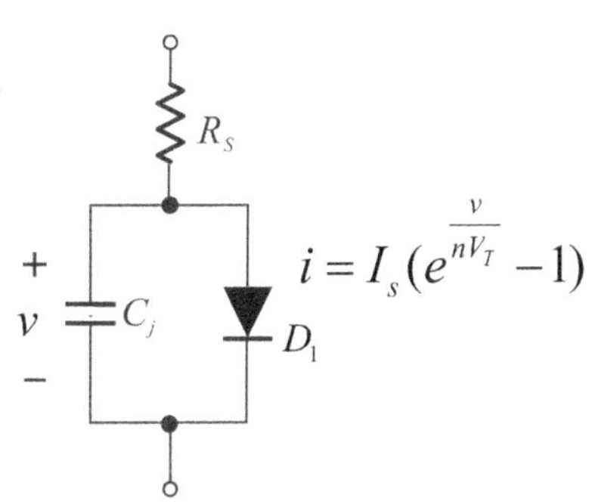

Figure 12.9 Equivalent circuit of a Schottky diode. C_j represents the depletion capacitance and R_s represents the series resistance.

represents the junction capacitance caused by the depletion region. The current–voltage relation arising from the Schottky junction is given by Equation (12.8).

$$i = I_s\left(e^{\frac{v}{nV_T}} - 1\right) \tag{12.8}$$

Here, I_s represents the saturation current, n is the ideality factor, and V_T is the thermal voltage constant.

The Schottky diode parameters given by Equation (12.8) can be found using DC measurements. When the diode is forward-biased, the term $e^{v/nVT}$ in the parentheses becomes dominant, and the current plotted against the voltage in the log-linear scale yields the plot shown in Figure 12.10; that is, for a forward-biased diode, the current in the log scale becomes a straight line as the voltage increases linearly. The slope of the line is related to the ideality factor n. Extrapolating the straight line to 0 V and reading the value of the current at 0 V gives the saturation current I_s. In addition, when the voltage is high, the diode current deviates from the straight line. Using that deviation, the value of R_s can be obtained. Since the deviation from the straight line for the appropriately selected current shown in Figure 12.10 is due to the voltage drop across R_s, the value of R_s can be determined by dividing the voltage difference between the two points by the selected current.

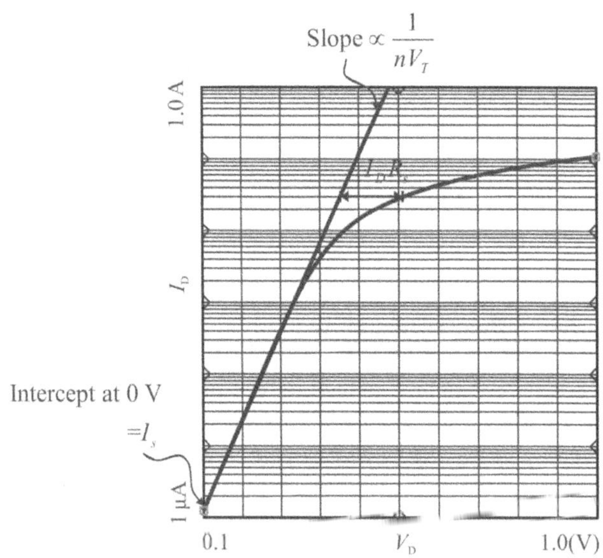

Figure 12.10 Schottky diode I–V characteristics. The saturation current can be obtained from the intercept at 0 V, the ideality factor from the slope, and R_s from the deviation from the straight line.

Example 12.4

Determine R_s, I_s, and n using the I–V characteristics of the Schottky diodes shown in Figure 12E.10.

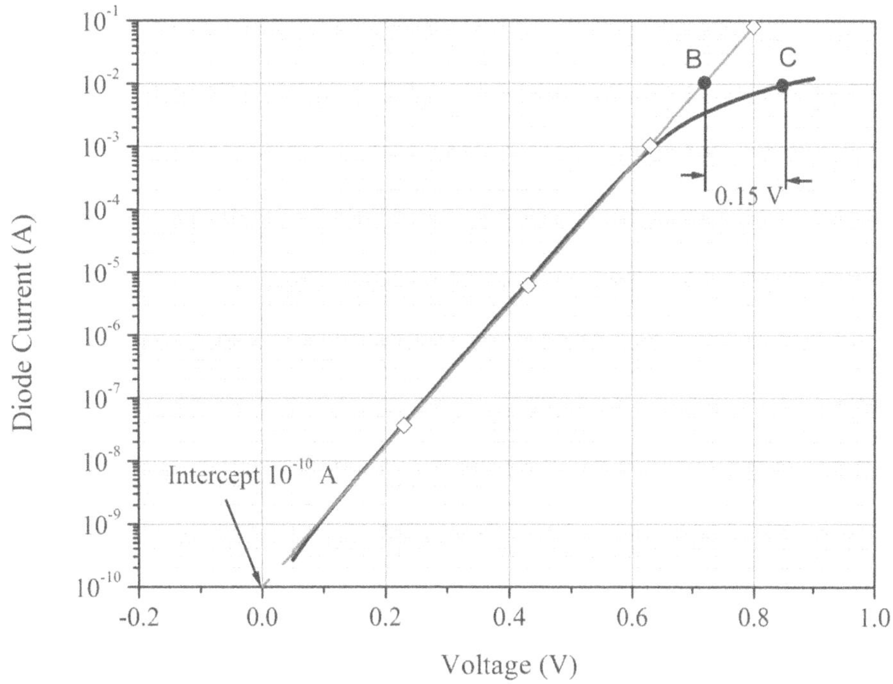

Figure 12E.10 Example of Schottky diode I–V characteristic

Solution

The current value at 0 V (point A) in Figure 12E.10 gives

$$I_s = 10^{-10}$$

Also, from the voltage difference at a current value of 0.01 mA (points B and C)

$$R_s = \frac{0.85 - 0.7}{0.01} = 15$$

The slope can be expressed as

$$\log_{10} i = \log_{10} I_s + \frac{v}{nV_T} \log_{10} e$$

Therefore, calculating the slope using points A and C gives

$$\frac{1}{nV_T} \log_{10} e = \frac{-2 + 10}{0.7} \rightarrow n \cong 1.52$$

As the area A_D of a typical Schottky junction (in Figure 12.8) becomes smaller, the capacitance C_j also becomes smaller and its application for a high frequency becomes possible. The area can be fabricated as small as is possible to allow for its application in a mixer circuit ranging up to hundreds of GHz; however, because of the very small area, it is not possible to assemble the device by wire bonding and the circuit is usually configured by a contacting wire called a *whisker*, as shown in Figure 12.11.

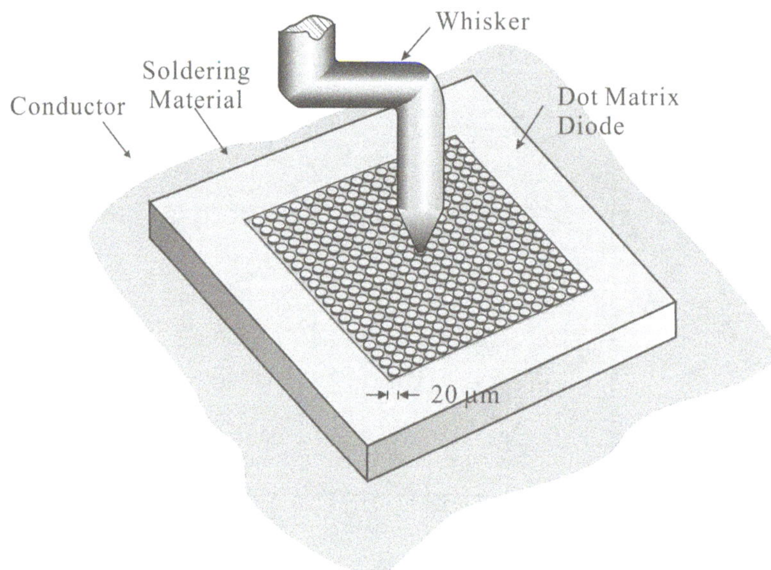

Figure 12.11 Multi-dot diode assembly. Since the dot size of the Schottky diode is too small to be wire-bonded, the whisker contact is used.

A dot in Figure 12.11 is one Schottky diode whose area is very small and each diode in the multidot diode is similarly fabricated as shown in Figure 12.8. One dot of the multidot diode is used in a mixer circuit after a whisker contact is formed. If the diode connected by the whisker contact is damaged, the circuit can be fixed by moving the whisker contact to another normal dot diode and reestablishing contact with the dot.

12.3.2 The Schottky Diode Package

The chip Schottky diodes described above can be used at higher frequencies because they have small parasitic circuit elements caused by packaging. However, their handling is difficult. As a compromise, a beam lead package, one of a variety of packages, has advantages for packaging. In contrast to other packages, since the beam lead package accompanies smaller parasitic circuit elements, it is frequently used for higher-frequency applications. The beam lead packaged diode is shown in Figure 12.12, where the package is made by forming the gold ribbons on the semiconductor substrate; the ribbons correspond to the lead terminals. Then, the back side of semiconductor substrate is etched to remove any unnecessary portions of the semiconductor substrate.

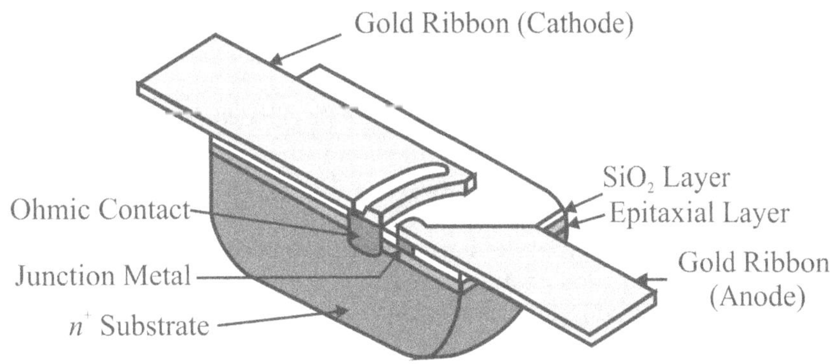

Figure 12.12 Beam-lead-packaged Schottky diode

Other types of packages are shown in Figure 12.13. A Schottky diode is assembled in a ceramic package for easy use in a waveguide or coaxial environment. The equivalent circuit of the packaged Schottky diode depends on the package. Generally, a bonding-wire inductance and a package capacitance are added to the equivalent circuit of the chip, as shown in Figure 12.13(b).

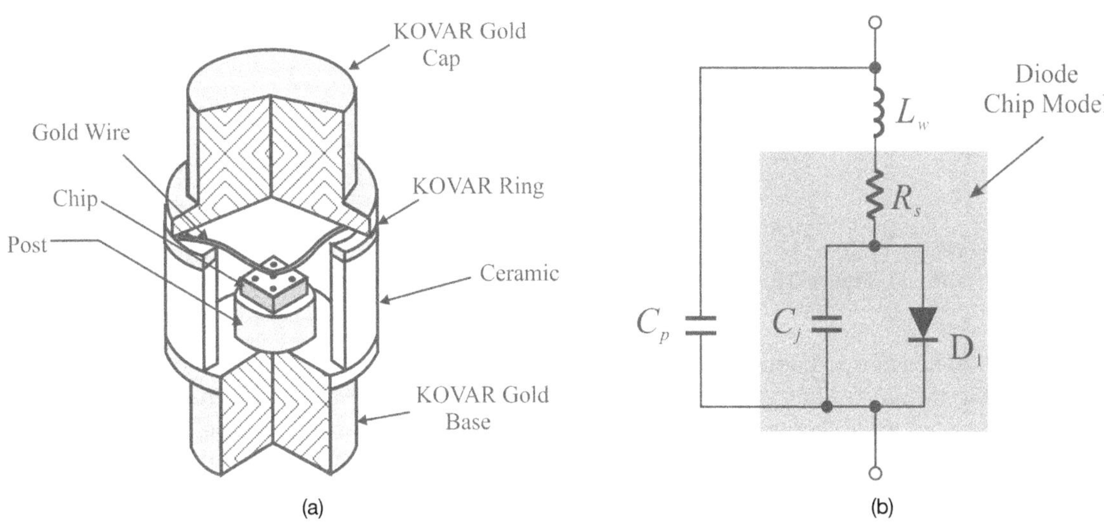

Figure 12.13 Ceramic-pill-packaged Schottky diode: (a) structure and (b) equivalent circuit. In (b), C_p and L_w represent the package capacitance and bonding-wire inductance, respectively. KOVAR is a nickel–cobalt ferrous alloy compositionally identical to Fernico 1, designed to be compatible with the thermal expansion characteristics of an alumina substrate.

12.3.3 Operating Principle of the Schottky Diode

12.3.3.1 Schottky Junction

The energy of electrons in a metal along the x-axis is shown on the left side of Figure 12.14. Most electrons have energies below the Fermi energy, which does not vary with respect to position in the metal along the x-axis. Here, E_O represents the energy of a free electron. The diagram is called an energy band diagram or simply an energy diagram. The energy required for an electron at the Fermi energy to become a free electron is defined as the work function $q\Phi_M$. Note that Φ_M represents the unit of voltage. Therefore, the work function can be expressed in Equation (12.9).

$$q\Phi_M = E_O - E_{FM} \tag{12.9}$$

In general, the value of the work function depends on the type of metal.

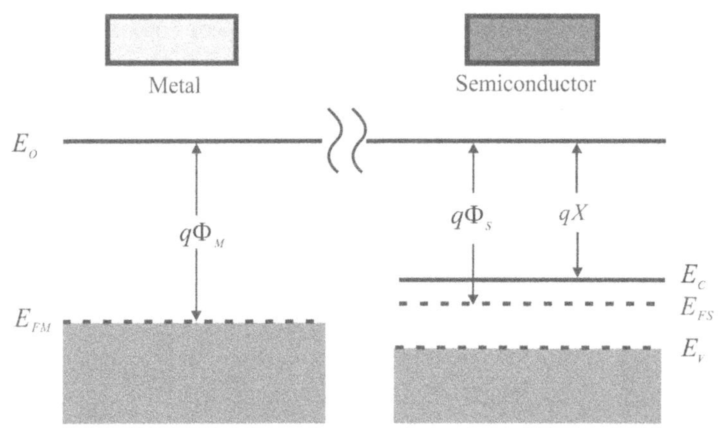

Figure 12.14 Energy band diagrams of a metal and a semiconductor

The energy diagram for electrons in a semiconductor is also drawn on the right side of Figure 12.14. At a sufficiently low temperature, most of the electrons have energy below the *valence band energy* E_V. However, as the temperature increases, some electrons achieve an energy value higher than the *conduction band energy* E_C and they populate in the conduction band. Therefore, the Fermi energy level is located between the energy values of the conduction and valence bands. Similar to conductors, the work function $q\Phi_S$ is the difference between the Fermi energy and the energy of a free electron. The difference between the conduction band energy and the energy of the free electron is termed the *electron affinity* qX and is defined as

$$qX = E_O - E_C \tag{12.10}$$

The value of the electron affinity defined in Equation (12.10) also depends on the type of semiconductor. We will discuss a case in which a junction is formed between a metal and a semiconductor having the energy distributions shown in Figure 12.14.

When a junction is formed between a metal and a semiconductor, the Fermi energies between the two materials become equal at a thermal equilibrium. In addition, there will be no difference in the energy gaps such as those between E_C and E_O. Also, the work function required to create the free electrons should show no difference because it only depends on the type of materials and does not change as a result of the junction.

Another constraint is that the energy of the free electrons is continuous. In the case of discontinuity at the junction, free electrons created on one side of the junction and the moving free electrons on the other side of the junction give rise to an energy difference. Therefore, applying all of the previous conditions, the energy diagram can be drawn as shown in Figure 12.15. The energy diagram of the semiconductor near the junction is bent but it has the same shape as that in the separated state shown in Figure 12.14 when it is sufficiently distant from the junction.

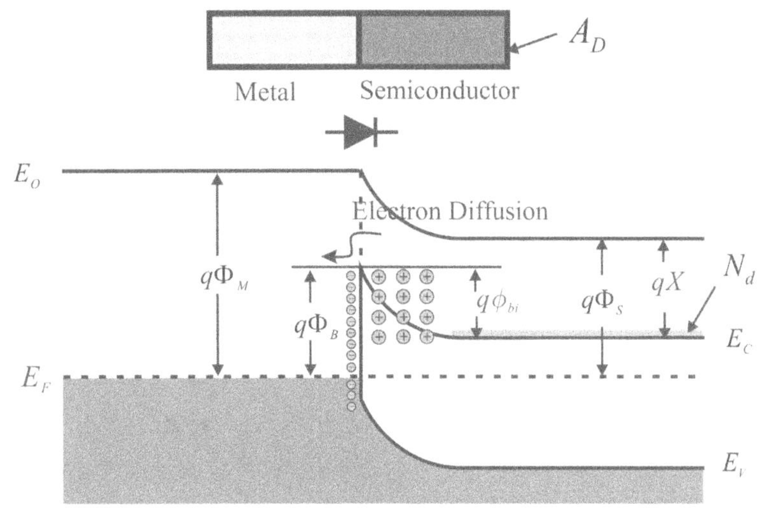

Figure 12.15 Energy diagram for the Schottky junction

Thus, a barrier arises between the metal and semiconductor junction given by Equation (12.11), which is usually called the Schottky barrier height.

$$q\Phi_B = q\Phi_M - qX \tag{12.11}$$

The difference between the energy of the electrons in the conduction band, which is sufficiently far away from the junction, and the energy of the electrons at the junction boundary is known as the built-in potential, and is given by Equation (12.12).

$$q\phi_{bi} = q\Phi_M - q\Phi_S \tag{12.12}$$

The occurrence of the Schottky barrier height and built-in potential can be explained in another way. When a junction is formed between a metal and a semiconductor, the electrons in the conduction band of the semiconductor diffuse across the junction into the metal. Since the electrons in the semiconductor's conduction band are formed due to a rise in temperature (in the case of a low temperature, most electrons exist in the valence band), the process is called thermionic emission. In contrast, most of the electrons in the metal, not having the appropriate energy in the semiconductor, are unable to diffuse into the semiconductor. As the electrons leave the semiconductor and move into the metal, positively charged bound atoms or uncovered atoms are left behind around the boundary. The uncovered atoms at the semiconductor junction again attract the diffused electrons into the metal. A layer of the uncovered atoms and electrons forms the junction and, due to the uncovered atoms and electrons, the electric field not only repels the electrons diffusing into the metal but also prevents the further diffusion of electrons from the semiconductor into the metal. In other words, equilibrium occurs between the electrons diffusing from the metal and the electrons diffusing from the semiconductor, with the result that the net movement of electrons is zero. The potential barrier formed between the electrons and the uncovered atoms in the semiconductor is called the built-in potential. This is the same as the previously mentioned induced built-in potential from the energy band diagram.

12.3.3.2 Depletion Capacitance

Suppose that the impurity atoms in the semiconductor are uniformly doped with a concentration N_d. The concentration of uncovered atoms per unit volume near the junction will be the same as the impurity concentration. Also, when the material is doped with N_d number of impurity atoms, the electrons N_d are mostly in the conduction band at room temperature. Since most electrons in the conduction band leave the junction when the junction is formed, the uncovered atoms of concentration N_d appear uniformly near the junction in the semiconductor (depletion approximation). In contrast, the distribution of electrons in the metal becomes a delta function distribution because the uncovered atoms in the semiconductor attract the diffused electrons near the metallic boundary. Due to neutrality, the charges will be distributed, as shown in Figure 12.16.

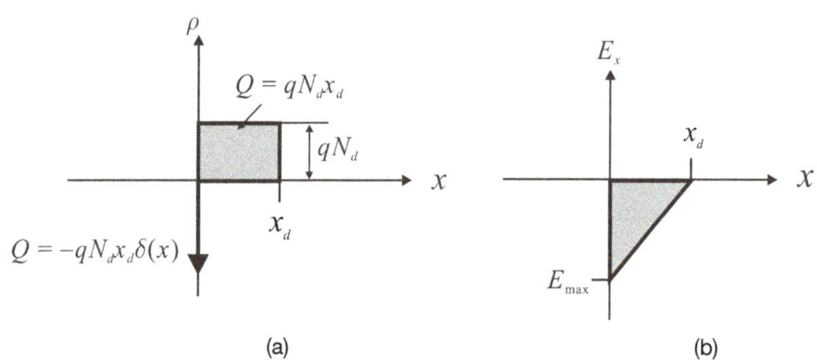

Figure 12.16 (a) Charge distribution and (b) electric field distribution near the junction

Applying the Gauss law, the electric field can be determined and will be spatially distributed as shown in Figure 12.16(b). The maximum value of the electric field is expressed as Equation (12.13),

$$E_{max} = -\frac{qN_d x_d}{\varepsilon_s} \tag{12.13}$$

where ε_s represents the permittivity of the semiconductor and the negative sign indicates the field is in the $-x$ direction. Therefore, the built-in potential ϕ_{bi} is obtained by integrating the spatially distributed electric field distribution, and can be expressed as

$$\phi_{bi} = \frac{1}{2}E_{max}x_d = \frac{1}{2}\frac{qN_d x_d^2}{\varepsilon_s} \tag{12.14}$$

Using Equation (12.14), the total charge Q_s in the semiconductor can be expressed as Equation (12.15).

$$Q_s = qN_d x_d = \sqrt{2q\varepsilon_s N_d \phi_{bi}} \tag{12.15}$$

When a positive external voltage V_a is applied to the metal side, the electric field direction of the external voltage will be opposite to that of the built-in potential. Thus, the ϕ_{bi} term in Equation (12.13) becomes $\phi_{bi} - V_a$ as there is no change in the impurity concentration N_d. Therefore, the total charge reflecting the external applied voltage is given by Equation (12.16).

$$Q_s = qN_d x_d = \sqrt{2qN_d\varepsilon_s(\phi_{bi} - V_a)} \tag{12.16}$$

The charge therefore varies according to the external applied voltage, which can be represented by a capacitance. The corresponding capacitance can be expressed with Equation (12.17).

$$C_j = A_D\left|\frac{\partial Q_s}{\partial V_a}\right| = A_D\sqrt{\frac{qN_d\varepsilon_s}{2(\phi_{bi} - V_a)}} = \frac{\varepsilon_s A_D}{x_d} \tag{12.17}$$

Thus, the width of the depletion layer x_d can be determined by measuring the depletion capacitance C_j and, as a result, the built-in potential can also be determined from Equation (12.14) when N_d is known.

12.3.3.3 Forward and Reverse Operation

When an external DC voltage is applied to a Schottky diode, as shown Figure 12.17, the applied voltage does not appear across the bulk region of the metal, which has a small resistance, or across the bulk region of the semiconductor, but appears primarily across the depletion region. As the direction of the voltage is opposite to the direction of the built-in potential, the barrier is lowered and the electrons in the semiconductor diffuse into the metal due to the lowered built-in potential. The Schottky diode thus begins conducting.

The concentration of electrons n_s in the semiconductor at the junction boundary without an applied voltage is given by Equation (12.18).

$$n_s = N_d \exp\left(-\frac{q\phi_{bi}}{kT}\right) \tag{12.18}$$

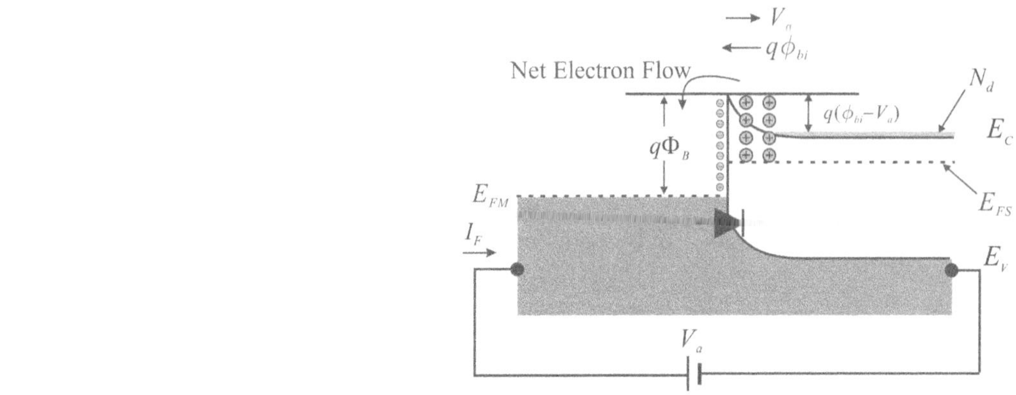

Figure 12.17 Electron flow in forward bias

Also, the current J_{SM} diffusing from the semiconductor into the metal is proportional to the electron density, which is expressed as Equation (12.19).

$$J_{SM} = KN_d \exp\left(-\frac{q\phi_{bi}}{kT}\right) \tag{12.19}$$

Since the net current is 0, J_{SM} will be the same current as the current J_{MS} diffusing from the metal into the semiconductor. Therefore, when the DC voltage is not applied, the currents diffusing from the metal into the semiconductor and vice versa are equal and given by Equation (12.20).

$$J_{MS} = J_{SM} = KN_d \exp\left(-\frac{q\phi_{bi}}{kT}\right) \tag{12.20}$$

When a DC voltage is applied, the electron density in the semiconductor is given by Equation (12.21).

$$n_s = N_d \exp\{-\frac{q(\phi_{bi} - V_a)}{kT}\} \tag{12.21}$$

The total current is expressed as $J = J_{MS} - J_{SM}$, and the total current J is given by Equation (12.22).

$$J = KN_d \exp\{-\frac{q(\phi_{bi} - V_a)}{kT}\} - KN_d \exp\left(-\frac{q\phi_{bi}}{kT}\right) = J_o\left[\exp(qV_a/kT) - 1\right] \tag{12.22}$$

In Equation (12.23),

$$J_o = KN_d \exp\left(-\frac{q\phi_{bi}}{kT}\right) \tag{12.23}$$

Note that the flow is due to the majority carriers, electrons. Unlike the *pn* junction diode, the flow is not due to minority carrier diffusion. Thus, the diffusion capacitance that appears as a result of the minority carrier storage is not present in the Schottky diodes. As a result, the Schottky diodes' switching speed is much faster than the switching speed in the *pn* junction diodes.

However, when a reverse-biased voltage is applied to a Schottky diode, as shown in Figure 12.18, the applied DC voltage and the built-in potential are in the same direction. The electrons in the semiconductor are attracted toward the terminals and more uncovered atoms appear near the junction. As a result, the junction widens and that leads to a higher built-in potential, which makes the diffusion of the electrons from the semiconductor into the metal almost zero. However, the same number of electrons in the metal when a voltage is not applied will diffuse toward the semiconductor. This current is J_o, as previously obtained in Equation (12.23), and it becomes the saturation current.

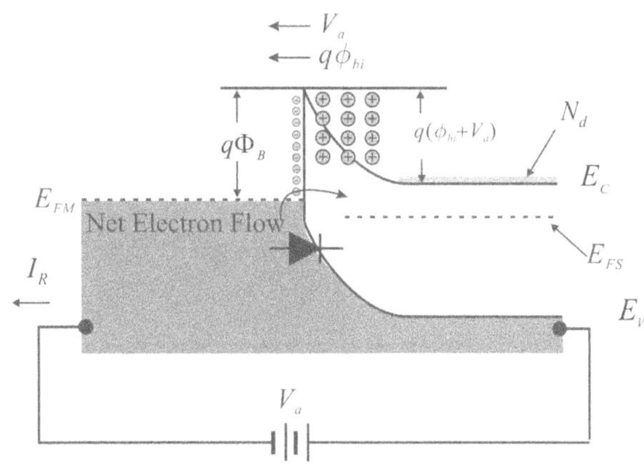

Figure 12.18 Flow of electrons in reverse bias

12.3.3.4 Schottky Barrier Height

The saturation current J_o of Equation (12.23) was previously derived using the uniform electron concentration at the boundary. However, assuming the electron concentration varies with respect to x, the current–voltage relationship can be recomputed using a distribution that closely approximates the actual distribution. As a result, the current–voltage relationship remains the same although the expression of J_o given by Equation (12.23) changes slightly, as expressed in Equation (12.24),

$$J_o = \frac{q^2 D_n N_c}{kT}\left[\frac{2q(\phi_{bi} - V_a)}{\varepsilon_s}\right]^{1/2} \exp\left(-\frac{q\Phi_B}{kT}\right) \tag{12.24}$$

where D_n represents the electron diffusion constant and N_c represents the electron density in the conduction band. From Equation (12.24), the value of the saturation current depends on the voltage applied to the diode. The square-root dependence is less significant than the dependence on the Schottky barrier height. The Schottky barrier height Φ_B depends on the work function of the metal and the electron affinity of the semiconductor as given by Equation (12.11). Note that the Schottky barrier height, unlike the built-in potential, does not depend on the number of impurity atoms in the semiconductor. The Schottky barrier height depends only on the work function and electron affinity, which in turn depend only on the material properties of the metal and semiconductor, and are therefore constant.

Once the metal and semiconductor properties are chosen, the Schottky barrier height can be considered to be constant.

The saturation current affects the turn-on voltage change of the diode. The turn-on voltage change then affects the LO power required for a diode mixer. The minimum point of the conversion loss of the mixer also depends on the LO power. This is shown in Figure 12.19, where the minimum conversion loss for the low barrier diode appears at approximately 0 dBm of the LO power, while the conversion loss of the high barrier diode can be seen to appear at about 3 dBm of the LO power.

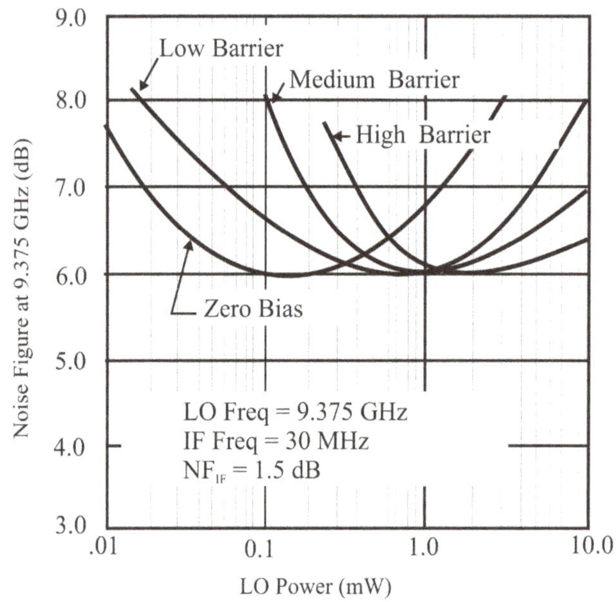

Figure 12.19 Mixer noise-figure dependence on the Schottky diode barrier height (here, the conversion loss and noise figure are considered to be almost equal).

12.4 QUALITATIVE ANALYSIS

In a mixer circuit using a diode, the diode can be approximated as an on-off switch, which operates according to the LO signal. This approximation holds since the LO signal in a mixer is generally a strong signal compared to the RF and IF signals. Therefore, when analyzing the mixer, as shown in Figure 12.20, the diode can be approximated as a switch by considering it to be a short circuit when the LO takes on a + polarity and an open-circuit when the LO takes on a – polarity. Thus, the diode can be viewed as a time-varying switch that operates according to the LO signal.

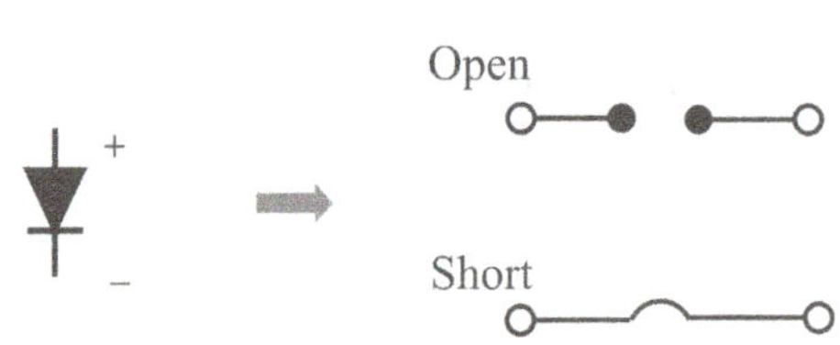

Figure 12.20 Diode approximation as a switch

A better method, since the diode equivalent circuit can be represented using R_s, C_j, and D as shown in Figure 12.21, would be to approximate the equivalent circuit by R_s when the diode is on, and by R_s in series with C_j when the diode is off.

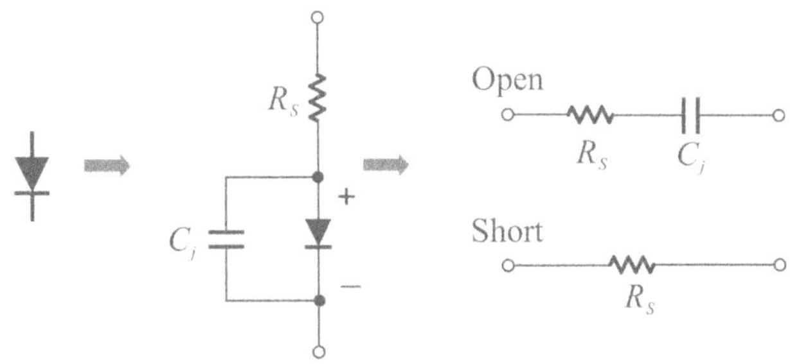

Figure 12.21 Impedance approximation based on a diode on/off switch

A mixer uses the impedance differences that result from the on-off switching of the diode. However, when the impedance difference from the on-off switching is not appreciable, the diode cannot be used as a mixer. In general, the frequency at which the impedance of R_s is equal to the impedance of C_j is known as the cut-off frequency f_c, beyond which the application of the diode to a mixer is considered unrealistic. Therefore, the cut-off frequency is expressed in Equation (12.25).

$$f_c = \frac{1}{2\pi R_s C_j} \tag{12.25}$$

The cut-off frequency f_c can be used as a figure of merit that indicates the high-frequency application limit of a diode to a mixer.

More operations are possible when a transistor is employed as a device for a mixer than would be available in the case of a diode. One of the operations is known as a transconductance operation and it utilizes a nonlinear relationship between the gate voltage and the drain (or collector) current, as shown in Figure 12.22. When the nonlinear relation is expanded in a polynomial, higher-order terms are generated. Denoting the second-order coefficient as a_2, the drain current due to the second-order term becomes $a_2(v_{GS})^2$. Thus, the intermodulation components of the LO and RF signals proportional to $2a_2\cos\omega_{RF}t\cos\omega_{LO}t$, can be obtained when the LO and RF signals are simultaneously applied to the gate (or base) of the transistor. This results in the up- or down-converted frequency component signals that can then be selected using an appropriate filter.

The operation of a transconductance mixer differs from that of an amplifier in the DC operating points. In the transconductance mixer operation, the gate DC bias voltage is usually set to the pinch-off voltage (V_p in Figure 12.22) or to the saturated region (I_{DSS} in Figure 12.22) in order to maximize the nonlinearity for an efficient frequency conversion. In contrast, in the amplifier, the gate DC bias voltage is selected to minimize the nonlinearity (generally in the middle of I_D–V_{GS} curve). In general, the transconductance mixer has a greater advantage in terms of gain over a diode mixer. However, the noise figure is not significantly improved and may even be inferior to a typical diode mixer.

Figure 12.22 Nonlinear I_D–V_{GS} characteristic of an FET

Another operation is referred to as a resistive operation, in which a transistor acts as an on-off switch, as in a diode mixer. Depending on the LO signal applied to the gate terminal, the drain source behaves as both a short and an open, as shown in Figure 12.23. Unlike the transconductance operation, the DC bias voltage is not applied to the drain terminal, $V_{DS} = 0$, and a weak RF signal is applied to the drain terminal. The drain source behaves like a variable resistor, varying according to the LO signal applied to the gate. The LO signal then switches the RF signal, which produces mixed-frequency components between the RF and LO signals. This kind of operation is close to an ideal multiplication between the LO and RF signals. As a result, spurious signals are generally small and the 1-dB compression point and TOI are higher than can be obtained with a diode mixer.

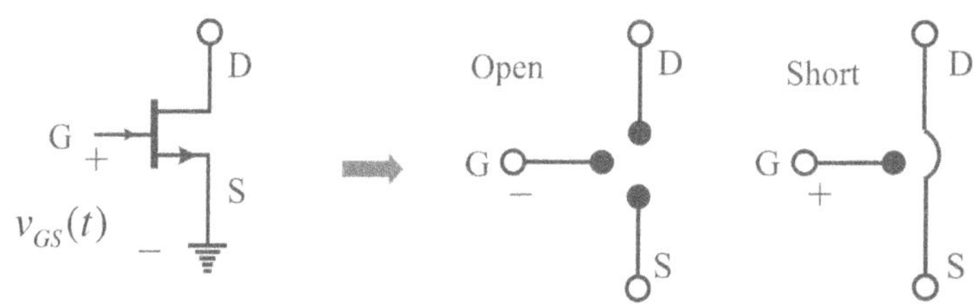

Figure 12.23 Switch approximation of an FET

Figure 12.24 represents a conceptual circuit to illustrate a mixer operation using an on-off switch that operates according to the applied LO signal. Figure 12.25 shows a time-varying function $S(t)$ that has the value 1 when the switch is closed and 0 when it is opened. Thus, the function $S(t)$ represents the on and off states of the switch.

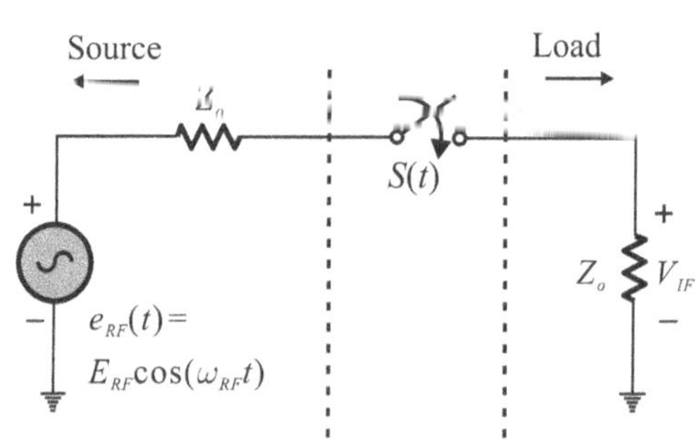

Figure 12.24 Mixer circuit implemented using a time-varying switch

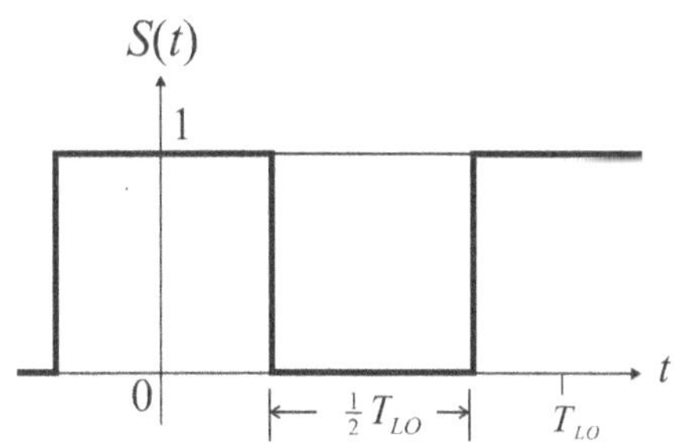

Figure 12.25 Diode switching function $S(t)$

The output signal $v_{IF}(t)$ can be written as Equation (12.26).

$$v_{IF}(t) = \frac{1}{2} E_{RF} \cos(\omega_{RF} t) \cdot S(t) \tag{12.26}$$

Expanding $S(t)$ in a Fourier series, it can be expressed as Equation (12.27),

$$S(t) = S_o + S_1 \cos(\omega_{LO} t) + S_2 \cos(2\omega_{LO} t) + S_3 \cos(3\omega_{LO} t) + \ldots \tag{12.27}$$

where S_n is given by Equation (12.28).

$$S_n = \frac{2}{T} \int_0^T S(t) \cos(\omega_{LO} t) dt = \text{sinc}\left(\frac{n}{2}\right) \tag{12.28}$$

The function sinc(x) is defined by sinc(x) = sin(πx)/(πx).

Therefore, the $n\omega_{LO} \pm \omega_{RF}$ frequency components are present at the output. The desired up- or down-converted frequency components are obtained by selecting the mixing term that corresponds to the $\cos(\omega_{RF} t)\cos(\omega_{LO} t)$ components in Equation (12.26). By selecting the down-converted frequency component, we obtain Equation (12.29).

$$v_{IF}(t)\Big|_{\omega = \omega_{LO} - \omega_{RF}} = \frac{1}{2\pi} E_{RF} \cos(\omega_{LO} - \omega_{RF}) t \tag{12.29}$$

Thus, the conversion loss defined by Equation (12.2) is given by Equation (12.30).

$$CL = \left(\frac{1}{2} \times \frac{\dfrac{E_{RF}^2}{4\pi^2 Z_o}}{\dfrac{E_{RF}^2}{8 Z_o}} \right)^{-1} = \pi^2 \approx 10 \text{ dB} \tag{12.30}$$

12.4.1 Single-Ended Mixer (SEM)

A single-ended mixer (SEM) is a mixer circuit that consists of one diode or one transistor and it can be constructed by replacing the switch with a diode or a transistor, as shown in Figure 12.24. A SEM circuit can be formed by adding an appropriate LO driving circuit for the diode or transistor. Also, in the case of a transistor, the use of a transconductance mixer is possible without relying on the switch-based circuit shown in Figure 12.24. In this section, we will discuss various types of SEM circuits.

12.4.1.1 Single-Ended Mixer (SEM) Using a Diode

Figure 12.26(a) shows an SEM circuit that uses a diode. In order to isolate the LO and RF circuits, a series resonant circuit is added to the LO drive circuits. The function of the series resonant circuit is to act as a short at the LO frequency and as an open at other frequencies. The resulting circuit for the RF signal is shown in Figure 12.26(b).

Another method for building an SEM is to input both the LO and RF signals simultaneously in a situation where RF and LO isolation is not required. The LO is used to drive the diode and the RF signal is either delivered to the output or not, depending on the LO signal. However, in this case, the resulting circuit for the RF signal is also the same as the circuit shown in Figure 12.26(b).

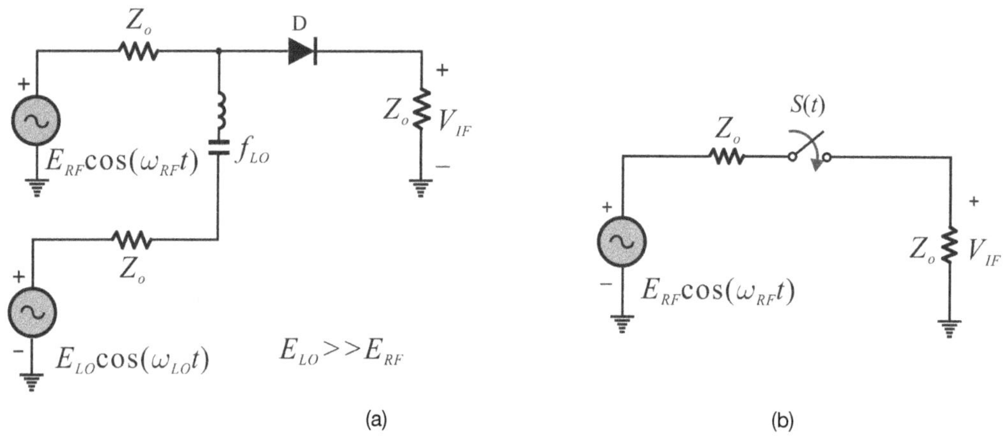

Figure 12.26 (a) Single-ended mixer circuit using a diode and (b) its equivalent circuit for an RF signal

Suppose that the diode is on during exactly half of the LO cycle, while it is off during the other half of the cycle. The RF signal will be delivered to the IF output, as shown in Figure 12.27. The conversion loss as found in Equation (12.30) is approximately 10 dB.

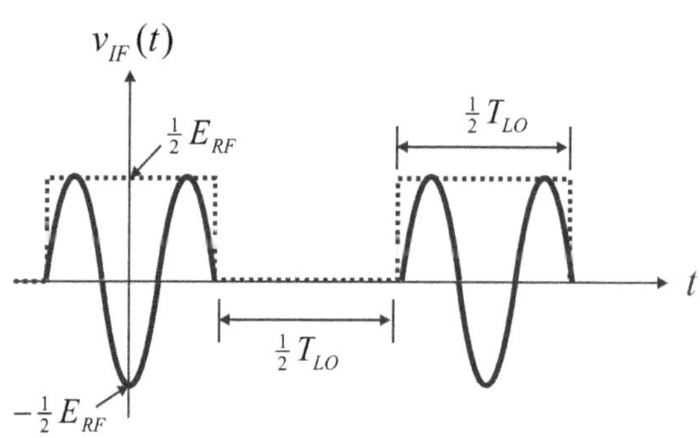

Figure 12.27 IF output waveform

Note that only the mixed frequency components such as $\omega_{LO} \pm \omega_{RF}$, $3\omega_{LO} \pm \omega_{RF}$, ... appear at the output because the even order of the Fourier series for the square wave in Figure 12.25 is 0. However, in the case of the real diode, the turn-on voltage is required and, consequently, the diode is turned on during a narrower interval, as shown in Figure 12.28. Expanding $S'(t)$ in the Fourier series, we obtain Equation (12.31).

$$S_n' = \frac{2}{T}\int_0^T S'(t)\cos(\omega_{LO}t)dt = \frac{2\tau}{T}\text{sinc}\left(\frac{n\tau}{T}\right) \tag{12.31}$$

In this case, all even-order frequency components are not zero and all frequency components of $n\omega_{LO} \pm \omega_{RF}$ will appear in the SEM output.

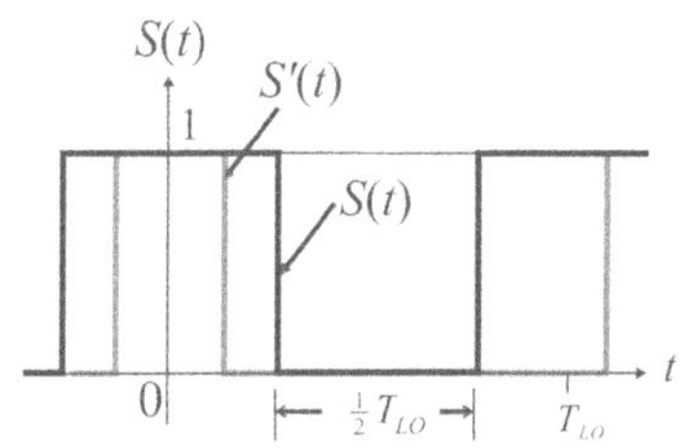

Figure 12.28 Comparison of the switching functions of a practical diode $S'(t)$ and an ideal diode $S(t)$

Another problem is that without filters, RF and LO signals appear directly as leakage signals in the IF output port of the SEM circuit. This results in poor isolation of the LO and RF signals from the IF port. In an application where the LO, RF, and IF frequencies are significantly apart, their isolations are primarily improved by including filters in the SEM circuit.

In addition, the circuit in Figure 12.26(a) does not provide efficient diode switching with respect to the LO due to the voltage drops across the RF and IF load resistors, which require higher LO power. To improve the LO driving, it is preferable that the IF port load resistance should be short at the LO signal frequency, while the RF port source resistance should be open. In this case, the LO signal can efficiently drive the diode to switch on and off.

Figure 12.29 shows the SEM circuit in which the filter circuits are inserted in order for the LO to drive the diode efficiently and improve the isolation. In Figure 12.29, the LO and RF signals are applied through the filter. The f_{RF} filter is short at the RF signal frequency, while it is open at other frequencies. Similarly, the f_{LO} filter is short only at the LO signal frequency, while it is open at other frequencies. Note that the RF and LO signals are shorted by inserting a bypass capacitor, as shown in Figure 12.29, for the efficient driving of the LO and for improved isolation. Thus, a weak LO signal can also easily drive the diode. The down converted IF voltage appears across the bypass capacitor. When the RF signal is applied and the diode turns on, the currents of all frequency components are generated by the diode, but only the IF current component is delivered to the IF port through the bypass capacitor. Thus, the impedance of the bypass capacitor at the IF frequency is sufficiently larger than the IF output resistance Z_o, which ensures that the IF signal can be transmitted without attenuation.

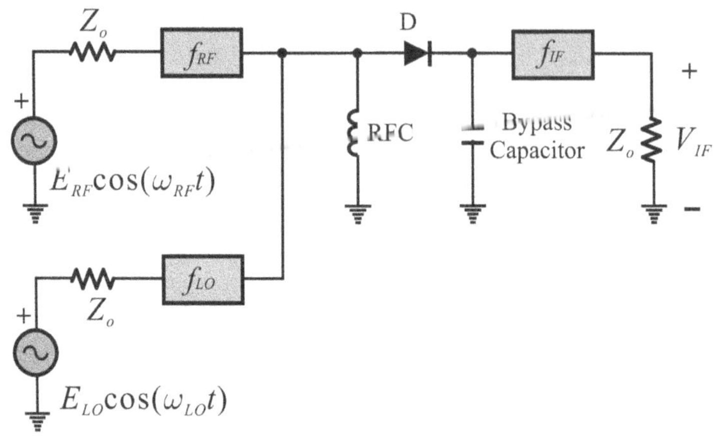

Figure 12.29 SEM mixer schematic with filters. The filters f_{RF}, f_{LO}, and f_{IF} are short for frequencies f_{RF}, f_{LO}, and f_{IF}, respectively, and open for other frequencies.

In addition, the diode produces the DC current when it is driven by the LO signal. The DC current is blocked by the f_{RF} and f_{LO} filters. Therefore, a DC current path for the diode is created by inserting an RFC inductor. Due to the RFC inductor, the DC voltage across the diode will be 0 V. The impedance value of the RFC must be set sufficiently large for it to operate as open at the RF and LO frequencies.

The addition of an RF matching circuit in the SEM circuit can improve the conversion loss and is usually placed after the f_{RF} filter. The maximum IF output is obtained when the RF signal is matched to the diode. The conversion loss of the SEM with the matching circuit can closely approach the theoretical conversion loss limit of 3 dB. The theoretical conversion loss of a mixer is limited to 3 dB. Generally, an ideal mixer converts an RF input signal into two frequency up- and down-converted components and one frequency-converted component is selected. Thus, the selected frequency-converted component has half of the RF input power and a minimum conversion loss of up to 3 dB can be achieved.

As an alternative mixer circuit, a mixer can be constructed using a shunt switch rather than the series switch shown in Figure 12.26. The circuit with the shunt switch is shown in Figure 12.30. In the shunt operation, the RF power is delivered to the IF when the switch is open, while the RF power is not delivered to the IF when the switch is shorted.

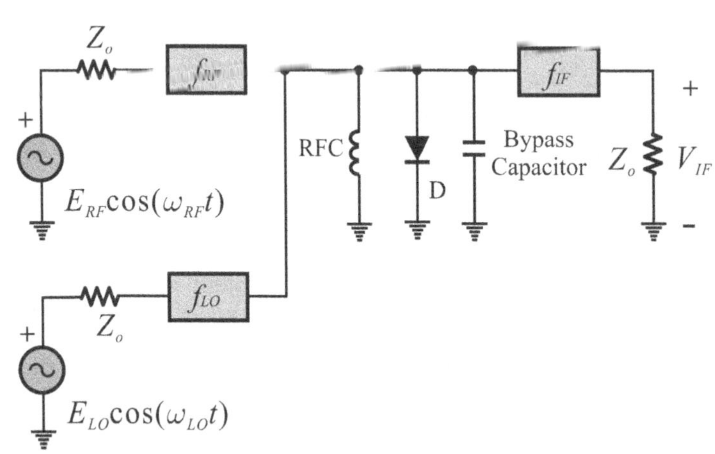

Figure 12.30 SEM schematic with the diode inserted in parallel

The general block diagram of an SEM circuit is shown in Figure 12.31. In this case, the coupler plays the function of mixing the RF and LO signals, and the matching circuit improves the conversion loss of the mixer. The RFC before the diode creates a path for the diode DC current generated by the LO drive. The capacitor behind the diode is a bypass capacitor and it prevents the leakage of the RF and LO signals into the IF output. The IMD3 performance is frequently used to compare mixers. The IMD3 of the SEM is poorer than that of any other type of mixer because the SEM does not provide any method for removing spurious signals generated by mixing the LO and RF signals.

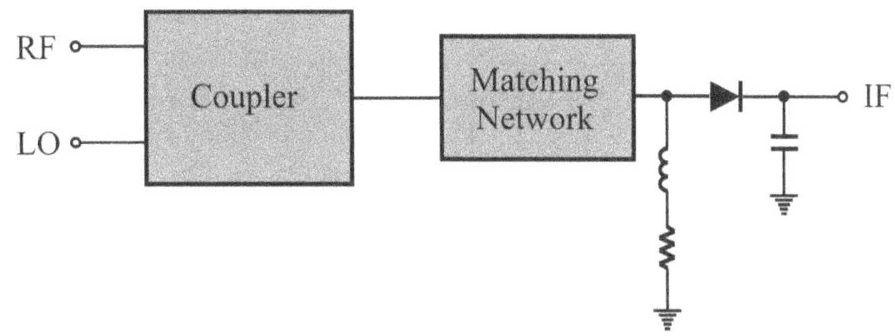

Figure 12.31 SEM mixer block diagram

Example 12.5

Using the default diode (a diode without series resistance and a junction capacitor) in ADS, verify the conversion loss of an SEM (single-ended mixer). Set up the SEM as shown in Figure 12E.11 and perform the simulation.

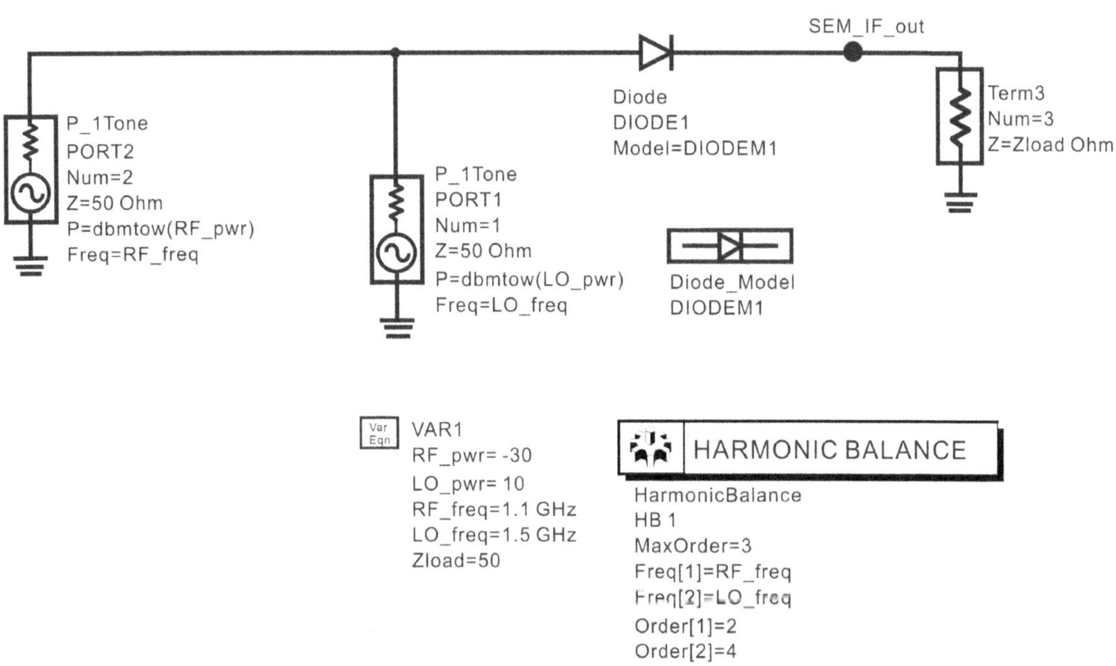

Figure 12E.11 An SEM circuit setup. The diode is the default diode that has the junction capacitance $C_j = 0$.

Solution

The simulation is performed and the spectrum output of the SEM is expressed in dBm.

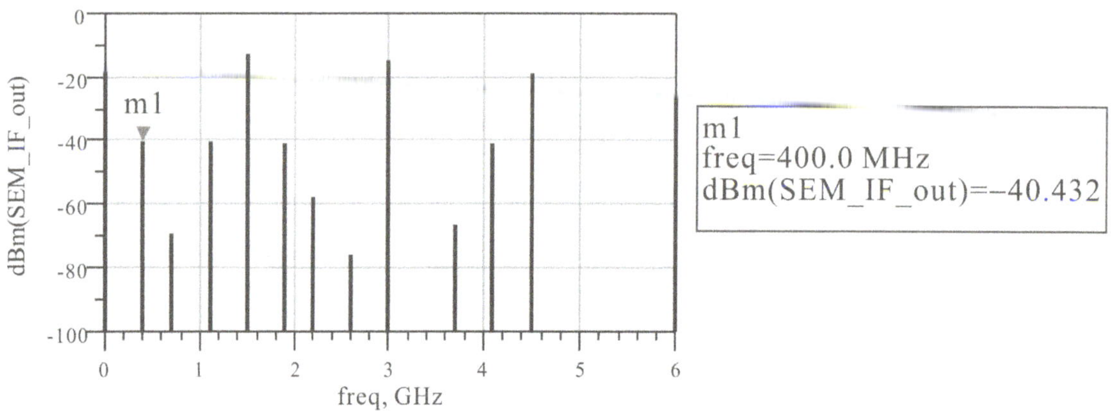

Figure 12E.12 The simulated IF output spectrum. The conversion loss is computed to be about 10.232 dB.

The simulated spectrum is shown in Figure 12E.12. Note that the RF and LO signals appear at the SEM output. The IF output power is –40.432 dBm. Since the RF input power is –30 dBm, the conversion loss is –30 – 40.432 = 10.232 dB. Therefore, the conversion loss of the SEM in Figure 12E.12 is 10.232 dB.

12.4.1.2 SEM Design Examples

An SEM constructed with a waveguide is shown in Figure 12.32. Here, both the RF and LO signals are applied through the common waveguide input. A waveguide step-impedance transformer is used to match the diode impedance at the RF frequency.

The chip diode is attached to the end plane of the coaxial low-pass filter (the IF Filter/Matching block in Figure 12.32) and its anode is connected to a whisker. The IF output is obtained through the output of the coaxial low-pass filter, which also provides a short to the RF and LO. The short appears just after the cathode of the diode. The waveguide back short of about a quarter wavelength provides an open circuit at the diode plane and thus avoids a possible waveguide loading effect. The LO signal must be sufficiently large to drive the diode. The goal of the design is to match the impedance of the diode block that includes the whisker at the RF frequency by using the step impedance transformer. The optimization of the step impedance transformer is necessary to obtain a minimum conversion loss. Although the RF and LO use the common step impedance transformer, that transformer is optimized for the RF matching. As a result, the LO signal may not be optimally matched.

As another design example, a planar mixer using microstrip lines is shown in Figure 12.33. In Figure 12.33, a beam lead package diode is used due to the planar-type mixer structure. The microstrip mixer structure basically resembles the waveguide mixer in Figure 12.32. A common input port is used for both the RF and LO signals, which are shorted just after the diode by the IF filter that acts as a coaxial low-pass filter in the waveguide mixer. The low-pass filter that starts with a bypass capacitor is used to prevent the RF and

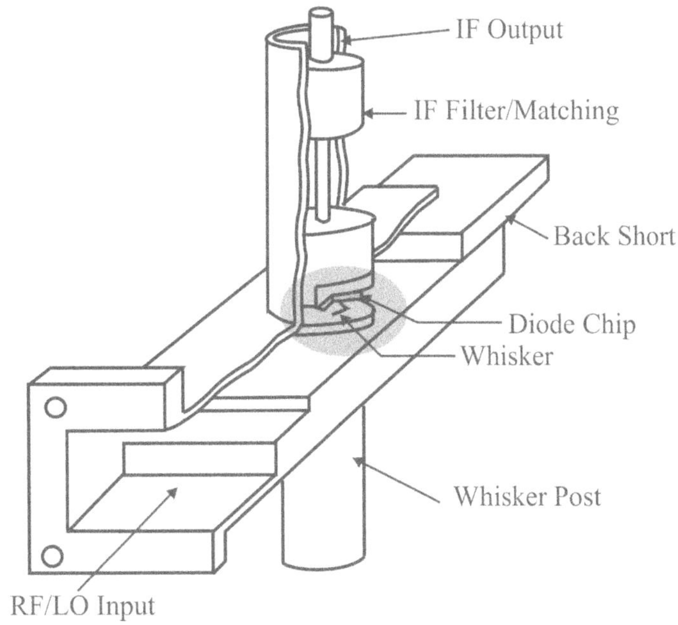

Figure 12.32 Waveguide SEM example

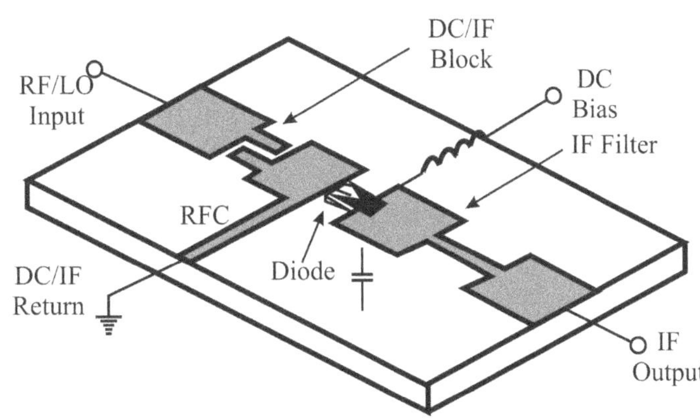

Figure 12.33 Microstrip SEM example

LO signals from leaking into the IF port. In addition, a microstrip RFC is used to create a path for the diode DC current generated by the LO driving. By making the RFC about a quarter wavelength at the RF and LO frequencies, it becomes open for both the RF and LO signals. The conversion loss of the mixer generally changes slightly when the DC voltage is adjusted; thus, a sufficiently large inductor is used to allow for the application of the DC voltage to adjust the conversion loss.

12.4.1.3 SEM Employing Transistors

A circuit for demonstrating the operation principle of an SEM that uses a transistor is shown in Figure 12.34. The gate voltage V_{GG} and drain voltage V_{DD} are applied for the DC

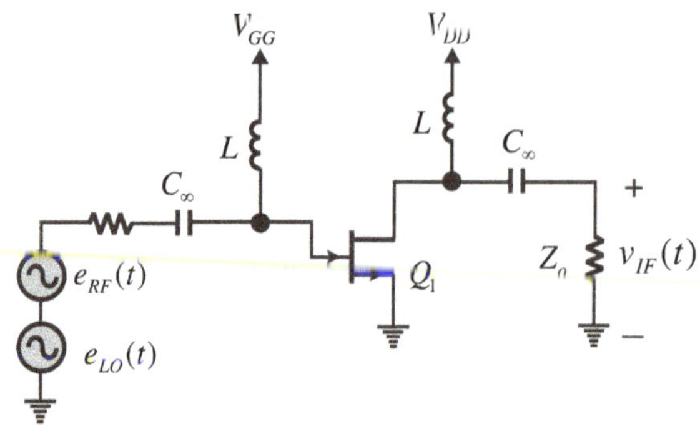

Figure 12.34 A transconductance SEM mixer. The DC bias of Q_1 is similar to that of the amplifier but V_{GG} is set near the pinch-off voltage.

biasing of the transistor. The gate voltage V_{GG} is usually selected near the pinch-off voltage of the transistor. The drain current i_D of the FET, which depends on the gate voltage v_{GS}, is expressed by Equation (12.32).

$$i_D = \begin{cases} I_{DSS}\left(1 - \dfrac{v_{GS}}{V_p}\right)^2 & v_{GS} > V_p \\ 0 & v_{GS} < V_p \end{cases} \tag{12.32}$$

Assuming that $V_{GG} = V_p$ and a small RF, the drain current gives rise to a squared sinusoidal-tip-shaped drain current similar to a half-wave rectifier. The drain current can be expressed in a Fourier series as Equations (12.33a)–(12.33c).

$$i_D(t) = I_{D0} + I_{D1}\cos\omega_{LO}t + I_{D2}\cos 2\omega_{LO}t + \ldots \tag{12.33a}$$

$$I_{Dn} = \frac{2}{T}\int_0^T i_D(t)\cos n\omega_{LO}t\,dt \tag{12.33b}$$

$$T = \frac{2\pi}{\omega_{LO}} \tag{12.33c}$$

The drain current in Equation (12.33) is disturbed by a small RF signal, $e_{RF}(t)$ and the resulting small-signal drain current $i(t)$ is given by Equation (12.34).

$$i(t) = \left.\frac{\partial i_D}{\partial v_{GS}}\right|_{V_{GG}+V_{LO}\cos\omega_{LO}t} e_{RF}(t) = g_m(t)e_{RF}(t) \tag{12.34}$$

The transconductance $g_m(t)$ that varies with respect to the LO signal can be similarly expressed in a Fourier series as Equation (12.35).

$$g_m(t) = g_{m0} + g_{m1}\cos\omega_{LO}t + g_{m2}\cos 2\omega_{LO}t + \ldots \tag{12.35}$$

Therefore, the frequency up- or down- converted components can be expressed with Equation (12.36).

$$v_{IF}(t) - g_{m1}Z_o \cos\omega_{LO}t \times E_{RF} \cos\omega_{RF}t \tag{12.36}$$

This operation is essentially based on the LO driving, which causes a time-varying drain current that produces the time-varying transconductance. As the small-signal RF is applied to the time-varying transconductance, the frequency-converted current flows through the load Z_o. As a result, the IF frequency component appears at the load. In contrast, in the case that $e_{LO}(t) = 0$, a constant drain current results, which in turn results in a constant transconductance and the amplified RF signal appears at $v_{IF}(t)$. Thus, the key difference between the amplifier and the transconductance mixer is in the transconductance. The amplifier has a constant transconductance, while the transconductance of the transconductance mixer is time-varying.

A mixer utilizing a transistor in a resistive operation is shown in Figure 12.35. Note that the DC bias voltage is not applied to the drain terminal and the impedance between the drain-source terminals becomes a variable resistor depending on the LO signal applied to the gate. The DC voltage V_{GG} to the gate is usually set to pinch off. Thus, when the sign of the LO voltage applied to the gate is negative, a high impedance appears between the drain source, and when the sign is positive, the drain source is approximately shorted. Therefore, the IF port is connected to the RF signal only in that time interval when the transistor is off. When it is on, the connection between the RF and IF ports is cut off and no signal appears at the IF port. The mixer can be simply illustrated as the circuit shown in Figure 12.35(b). The operation of the circuit in that figure is similar to the operation of the previous mixer circuit using a diode and the conversion loss is approximately 10 dB.

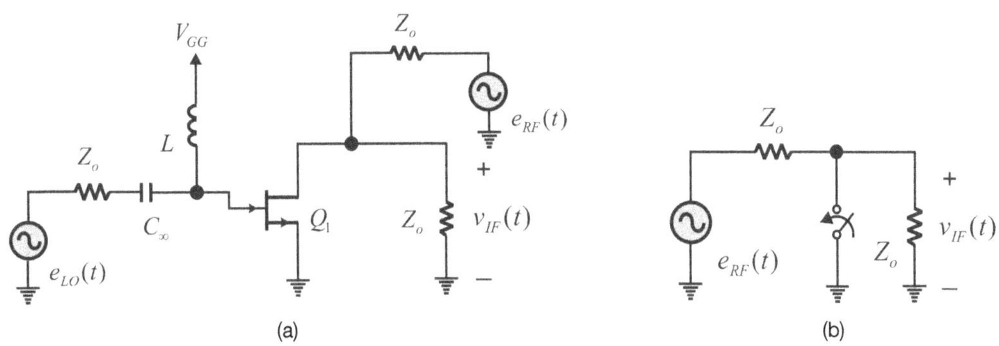

Figure 12.35 (a) An SEM circuit that uses the resistive operation of an FET and (b) its simplified equivalent circuit. V_{GG} is set to V_p.

The mixer in Figure 12.35 is called a resistive mixer and its operation is similar to that of the diode, but with an advantage over a diode mixer: First, the LO signal can be applied to a separate port, and consequently circuits for isolation, as in a diode mixer circuit (a series resonant circuit or filter), are not required. A second advantage of the resistive mixer is that the nonlinearity of the variable resistor of the drain source is lower compared to that in the diode, and the occurrence of spurious signals is also less than that in the diode. With fewer spurious signals, the 1-dB compression point is high and results in an improved TOI.

The dual-gate FET, which is an FET that has two gate terminals, is shown in Figure 12.36(a) and its equivalent circuit is shown in Figure 12.36(b). The fabrication of the dual-gate FET shown in Figure 12.36(a) is easy to process and it is not a new concept. A dual-gate MOSFET was developed in the 1960s with an existing silicon technology, and it is widely used in automatic gain control (AGC) amplifiers and mixers. In microwave applications, research on a mixer that utilizes a dual-gate GaAs MESFET was published in the 1980s. The advantage of the dual-gate FET is that because LO and RF signals can be applied to each of the gate terminals, isolation circuits for the LO and RF signals are not necessary. The LO–RF isolation between the ports is determined by the gate-drain capacitance of Q_1. The isolation is good when the gate-drain capacitance is small.

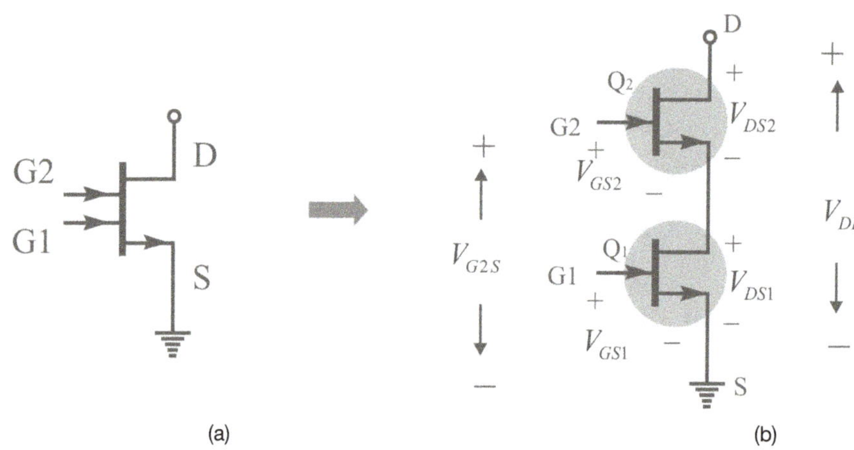

Figure 12.36 (a) A dual-gate FET and (b) its equivalent circuit[1]

The analysis of the dual-gate FET DC characteristics is basic to an understanding of the dual-gate FET mixer. Using the dual-gate FET DC characteristics, the appropriate operating DC bias point can be selected for a mixer application. Its DC characteristics can be derived from the characteristics of a single-gate FET. From Figure 12.36(b), the drain of Q_2, which is the only accessible drain terminal, is connected to V_{DD}. Thus, from Figure 12.36(b),

$$V_{DD} = V_{DS1} + V_{DS2} \tag{12.37}$$

Using Equation (12.37), the DC characteristics of the two FETs can be superimposed, as shown in Figure 12.37(a), where the DC characteristics of the two FETs are superimposed when $V_{DD} = 5$ V. From that figure, when $V_{GS1} = V_{GS2} = -1$ V, the drain-source voltages of the two FETs are found to be $V_{DS1} = V_{DS2} = 2.5$ V. The current flowing through the two FETs can be found to be approximately 30 mA. As another example, when $V_{GS1} = -1$ V and $V_{GS2} = -2$ V, the drain-source voltages of Q_1 and Q_2 are approximately $V_{DS1} = 0.2$ V and $V_{DS2} = 5$ V. Conversely, when $V_{GS1} = -2$ V and $V_{GS2} = -1$ V, the drain-source voltages of Q_1 and Q_2 are

1. C. Tsironis, R. Meierer, and R. Stahlmann, "Dual-Gate MESFET Mixers," *IEEE Transactions on Microwave Theory and Techniques* 32, no. 3 (March 1984): 248–255.

approximately $V_{DS1} = 5$ V and $V_{DS2} = 0.2$ V. Thus, from this observation, both FETs are in the saturation region (which corresponds to the active region in the BJT) when $V_{GS1} \cong V_{GS2}$. A slight difference between V_{GS1} and V_{GS2} drives one of the FETs and it is primarily in the current saturation region, while the other is in the triode region. The triode region is the region in which the current increases linearly with respect to V_{DS}.

Note that V_{GS2} is a floating voltage and it is easier to supply V_{G2S},

$$V_{G2S} = V_{GS2} + V_{DS1} \tag{12.38}$$

to the gate of Q_2. Plotting the DC characteristics defined by Equation (12.38) gives the plot shown in Figure 12.37(b).

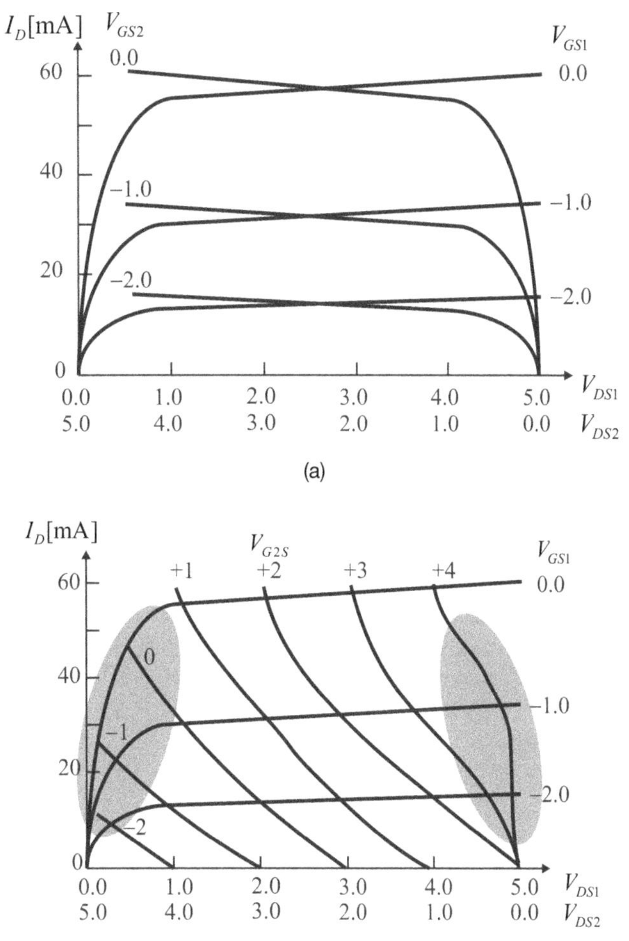

Figure 12.37 The characteristics of a dual-gate FET DC: (a) DC characteristics with respect to V_{GS1} and V_{GS2}; (b) DC characteristics with respect to V_{GS1} and V_{G2S}

From the discussion above about the DC characteristics of Figure 12.37(a), the region where both FETs are in the saturation region is considered to be inappropriate for a mixer application. In this case, when the LO signal drives one of the two FETs, a slight change in the LO signal makes the LO-driven FET alternate between the saturation region and the triode region. The other FET also alternates in the opposite region of the LO-driven FET. Thus, it is difficult to employ this region for a mixer. Therefore, the shaded regions of the FET in Figure 12.37(b) represent the appropriate operating regions. In this case, one of the FETs will be in the triode region, while the other will be in the saturation region.

Figure 12.38 shows the preferred method of applying signals to the dual-gate FET. The LO and RF signals are applied to the gates of Q_2 and Q_1, respectively, and the DC bias is necessary to drive Q_2 into the saturation region, while Q_1 is driven into the triode region; this is described in the left shaded region of Figure 12.37(b). Configured this way, the operation region of Q_2 does not change from the saturation region. In addition, because the RF signal is basically a small signal, the situation in which Q_1 moves out of the triode region does not occur.

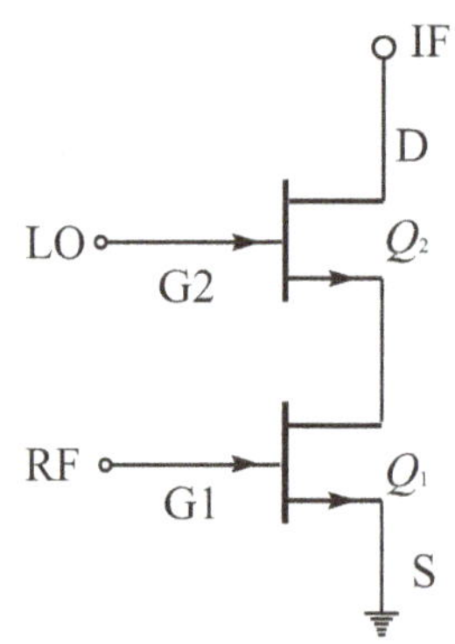

Figure 12.38 Method for applying LO and RF signals to a dual-gate FET

When the LO and RF signals are applied as shown in Figure 12.38, Q_1 becomes a variable resistor that varies according to the RF signal. The amplified LO signal that varies with respect to the variable resistor is delivered to the IF port. Therefore, denoting the transconductance of Q_2 as g_{m2} and the impedance of the IF port as Z_o, the voltage at the IF port, V_{IF} can be expressed as Equation (12.39).

$$v_{IF}(t) = v_{LO}(t)\frac{Z_o}{R_{ds}(t) + \dfrac{1}{g_{m2}}} \cong \frac{Z_o V_{LO}\cos\omega_{LO}t}{R_{ds}(t)} \tag{12.39}$$

Here, $g_{m2}R_{ds}(t) >> 1$ is assumed. Denoting $R_{ds}(t) = R_{ds0}(1 + kV_{RF}\cos\omega_{RF}t)$, since the RF signal is a small signal, Equation (12.39) can be rewritten as Equation (12.40),

$$v_{IF}(t) \cong \frac{Z_o V_{LO}\cos\omega_{LO}t}{R_{ds}(t)} \cong \frac{Z_o V_{LO}\cos\omega_{LO}t}{R_{ds0}}(1 - kV_{RF}\cos\omega_{RF}t) \tag{12.40}$$

and the desired IF signal is obtained by properly filtering the mixed signal given by Equation (12.40). Therefore, the IF output is obtained with the mixer circuit configuration using the dual-gate FET as shown in Figure 12.39.

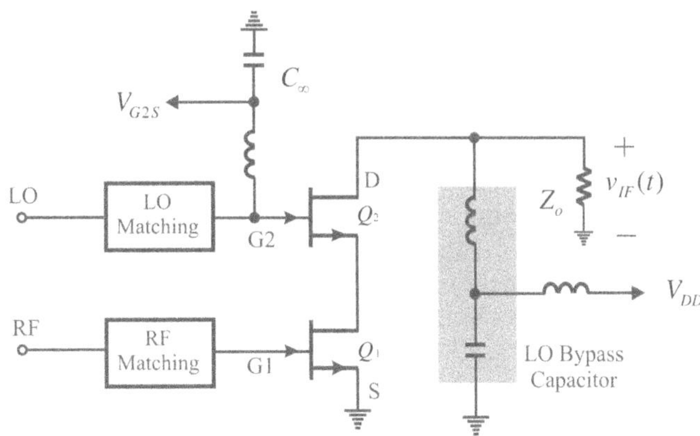

Figure 12.39 Mixer circuit using a dual-gate FET

12.4.1.4 SEM Design Examples Using a Transistor

An example of the transconductance SEM that uses a transistor is shown in Figure 12.40. The LO frequency is 7 GHz and the RF frequency is set to 7.9–8.4 GHz.

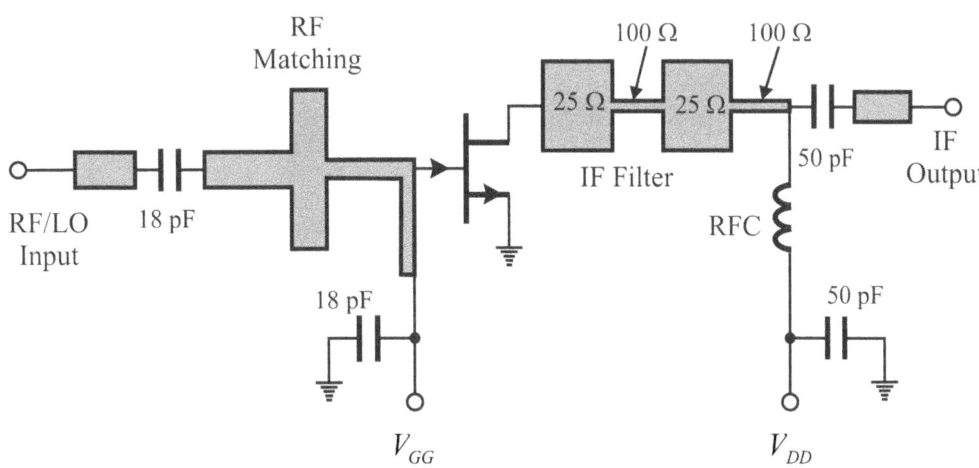

Figure 12.40 Transconductance SEM circuit example

As previously explained, the operation of the FET uses the nonlinear transconductance that depends on the gate voltage. The series 18-pF capacitor in the RF/LO signal input in Figure 12.40 is a DC block, and a quarter-wavelength high-impedance line connected to the gate operates as the RFC. The capacitor connected in shunt to the DC gate voltage V_{GG} is a bypass capacitor and the DC gate voltage must be biased near the pinch-off of the FET. The two-stage cascade connection of 25- and 100-Ω transmission lines connected to the drain is a low-pass filter and it passes only the IF frequency. Also note that the low-pass filter provides the short to the drain at the RF and LO frequencies. The DC drain voltage V_{DD} can be supplied to the drain through the inductor, which acts as the RFC at the IF frequency. The drain voltage V_{DD} should be selected in order for the FET to operate in the saturation region. The 50-pF capacitor connected in series to the IF port acts as a short at the IF frequency (the maximum IF frequency is 1.4 GHz) and serves as the DC blocking capacitor. Similarly, the shunt 50-pF capacitor at the V_{DD} supply acts as bypass capacitor.

A resistive SEM using an FET is shown in Figure 12.41. The DC gate voltage V_{GG} biases the FET near the pinch-off and is applied through the RFC that is made from two quarter-wavelength transmission lines as explained in Chapter 3. The series 51-pF chip capacitor is used as a DC block and also prevents the IF signal from leaking into the LO port. The LO frequency is 8.8 GHz and the LO filter matches the LO signal at the gate of the FET, while simultaneously isolating the RF signal.

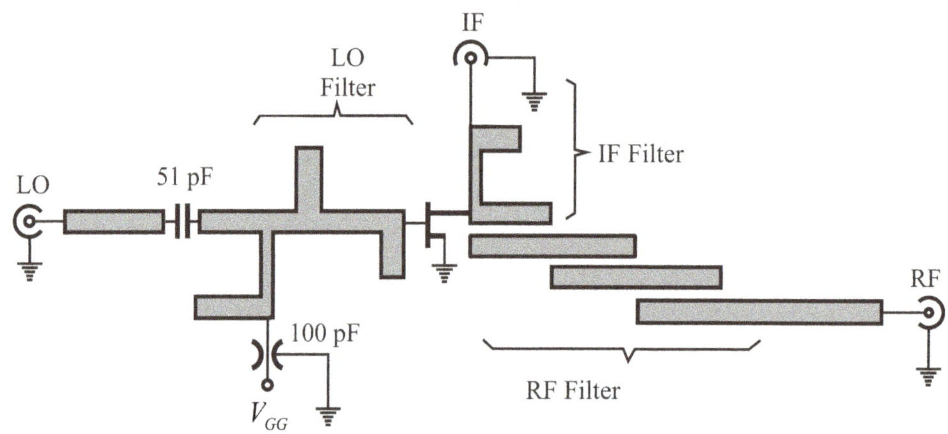

Figure 12.41 Example of a resistive mixer using an FET[2]

Note that as the DC drain voltage is not applied, the FET operates as a variable resistor. The outputs of the RF and IF ports are connected through the RF and IF filters. The frequency range of the RF is 10–10.5 GHz. The RF filter is open at other frequencies and prevents the RF signal from affecting the signals of those frequencies. The IF filter provides an open at the RF frequency, which minimizes the effect on the RF filter. The IF filter also prevents the leakage of the LO and RF signals to the IF port.

When the LO signal is applied to the gate, the resistance of the drain source is almost shorted for the positive half cycle of the LO signal. Consequently, the RF and IF ports will be disconnected. In contrast, for a negative half cycle, the drain source becomes open and so the RF voltage modulated by the LO is delivered to the IF port. The desired IF signal

2. S. A. Maas, "A GaAs MESFET Mixer with Very Low Intermodulation," *IEEE Transactions on Microwave Theory and Techniques* 35, no. 4 (April 1987): 425–429.

appears as the output through the IF filter. The resistive mixer has an improved 1-dB compression point and TOI. In this design, a 1-dB compression point of P_{1dB} 9.1 dBm and $IP_3 = 21.5$ dBm are reported for an LO power of 10 dBm.

12.4.2 Single-Balanced Mixer

The SEM previously described requires filters for the isolation of the RF and LO signals. The dynamic range of the SEM is narrow because it has no capability for suppressing spurious signals due to the mixing of the RF and LO harmonics, which is a disadvantage. The dynamic range can be improved by using hybrids in the mixer's circuit. In this section, we will discuss the hybrids and then examine the principles of operation for an SBM (single-balanced mixer). We will also consider some design examples for an SBM.

12.4.2.1 Hybrids

The most commonly used low-frequency hybrid is probably a triple-wound transformer that can be used up to a frequency of about several GHz. Its construction is shown in Figure 12.42(a). The triple-wound transformer is constructed by simultaneously winding three wires around a high-permeability ferrite core, as shown in Figure 12.42(a). The equivalent circuit of A–A′, B–B′, and C–C′ is shown in Figure 12.42(b) with B′ and C connected as a common terminal.

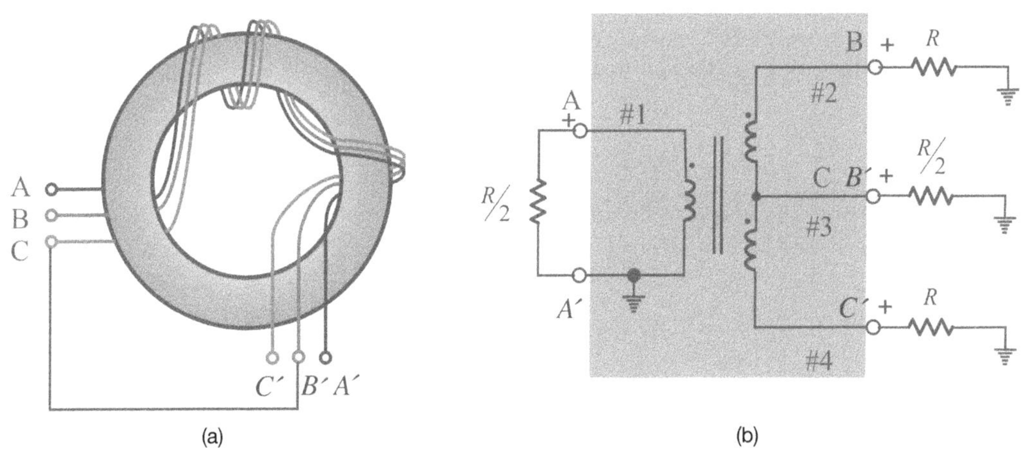

Figure 12.42 (a) Toroidal core transformer and (b) its equivalent circuit

Figure 12.43 shows the currents and voltages that result when a voltage source is applied to the input of port 1 (A–A′). Denoting the voltage across port 1 as V, the voltages equal to V appear in both the windings B–B′ and C–C′ due to transformer operation, as shown in Figure 12.43. Also, denoting the current flowing toward the dot at port 1 as I, currents equal to $\frac{1}{2}I$ flow out from the dot through both the B–B′ and C–C′ windings. Since the transformer is lossless, the power delivered to the input winding should be conserved and the two output windings should deliver power equal to that delivered to port 1. Thus, the current flowing from the two output windings is equal to $\frac{1}{2}I$. As a result, the current flowing through the resistor connected to port 3 becomes zero and the voltage at port 3 becomes zero. Therefore, the input power will be equally split between ports 2 and 4.

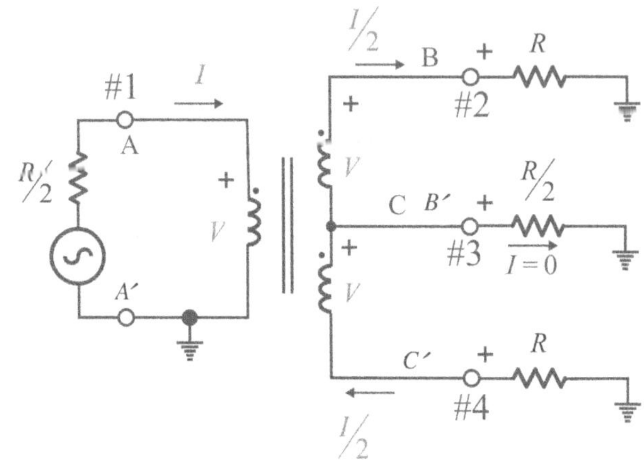

Figure 12.43 The voltages and currents that result when a voltage source is applied to port 1 of the transformer

However, the current direction through the resistors connected to ports 2 and 4 is opposite, and the port 2 and 4 voltages have a 180° phase difference. The operation is known as a Δ- or difference operation.

In addition, a total resistance of $2R$ is connected to the output windings of the transformer. Since the $2R$ resistance is connected through a 1:2 transformer, the impedance seen from the input can be expressed by Equation (12.41).

$$\frac{2R}{n^2} = \frac{R}{2} \tag{12.41}$$

Thus, port 1 is found to be matched. The S-parameters normalized by the port resistance are expressed by Equations (12.42a)–(12.42c).

$$S_{21} = \frac{1}{\sqrt{2}} \tag{12.42a}$$

$$S_{41} = -\frac{1}{\sqrt{2}} \tag{12.42b}$$

$$S_{31} = 0 \tag{12.42c}$$

In contrast, when the voltage source is applied to port 3, as shown in Figure 12.44, the input current through port 3 is divided equally into ports 2 and 4, and an in-phase voltage appear at ports 2 and 4. As the currents flowing from the two output windings of the transformer are in opposite directions, the current of the input winding becomes 0 and the voltage at port 1 therefore becomes 0. The operation is called an Σ- or sum operation.

Thus, when an input is applied to port 1, the voltages appearing at ports 2 and 4 have the same magnitudes with a phase difference of 180° between them; however, when an input is applied to port 3, in-phase voltages appear in ports 1 and 3. In addition, ports 1 and 3 become isolated.

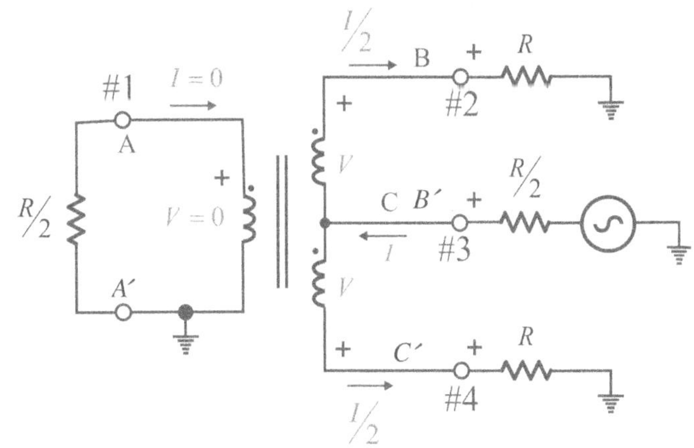

Figure 12.44 The voltages and currents produced when a voltage source is applied to port 3

The triple-wound transformer does not generally operate at microwave frequencies. Devices performing similar functions are constructed using waveguides or transmission lines. Figures 12.45(a) and (b) show a magic-T and a rat-race ring hybrid (or, for short, a ring hybrid) that perform the same function.

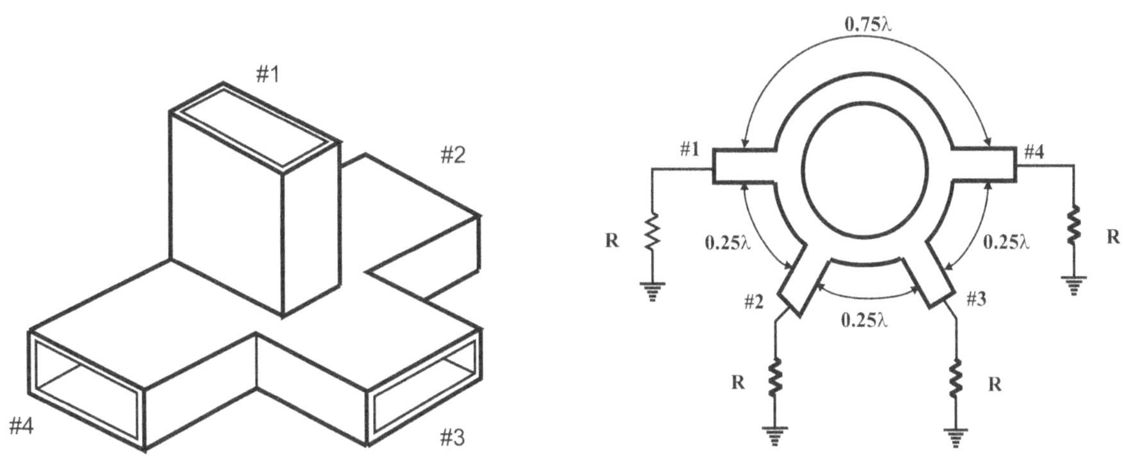

Figure 12.45 (a) Magic-T and (b) rat-race ring hybrid

In the magic-T shown in Figure 12.45(a), the input electric field into port 1 is divided between ports 2 and 4. Due to the shape of the electric field at the junction, the electric fields with a 180° phase difference appear at ports 2 and 4. In contrast, for the input applied to port 3, the in-phase electric field output appears at ports 2 and 4.

A device that performs the same function as the magic-T, not by using a junction but rather depending on the length of the transmission line, is called a rat-race ring hybrid, which is shown in Figure 12.45(b). When an input is applied to port 1, the output voltage at port 3 is zero. The applied input voltage at port 1 is divided into two voltage waves; one traverses the upper path of the ring and the other passes along the lower path of the ring. The two voltage waves arriving at port 3 have the 180° phase difference due to the difference of a half-wavelength phase delay and consequently the sum of the two voltage waves is 0 at port 3. Thus, ports 1 and 3 are isolated. Similarly, the voltage at port 2 has a phase delay of –90° with respect to the input at port 1. The voltage at port 2 is the sum of the two voltage waves with the one-quarter and five-quarter wavelength phase delays that are in phase. Similarly, at port 4, the output voltage with a phase delay of a three-quarter wavelength (–270°) appears. Note that the output voltages appearing at ports 2 and 4 have the same magnitudes but with a 180° phase difference. Using the same reasoning, when an input is applied to port 3, the output to port 1 will be 0, while the output voltages at ports 2 and 4 will have the same magnitudes and be in phase.

When port 3 in Figure 12.42 is shorted, a three-port circuit is formed, as shown in Figure 12.46. As mentioned earlier, when an input is applied to port 1, voltages equal in magnitude but opposite in phase will appear at ports 2 and 3. The signal at ports 2 and 3 is called a *balanced signal*, whereas the input signal at port 1 is called an *unbalanced signal* and it is defined with respect to the ground. Therefore, as the circuit converts an unbalanced signal into a balanced signal, it is called a *balun*.

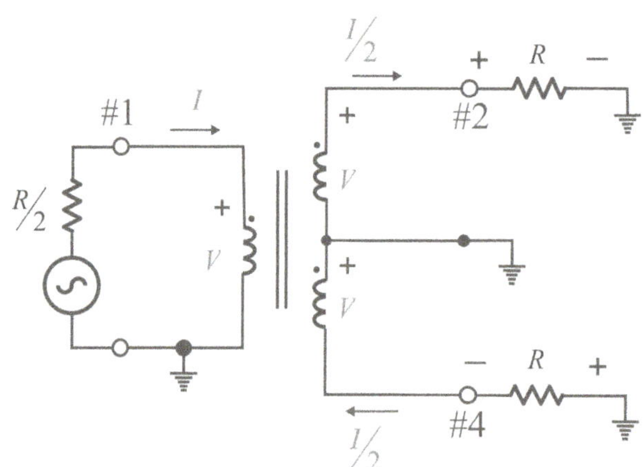

Figure 12.46 A balun

The balun can be naturally configured using the triple-wound transformer described previously. The triple-wound transformer balun is limited to a low-frequency application, and the balun for higher frequencies can be configured using a CPS (coplanar strip) or a coupled line, as shown in Figure 12.47. The even-mode impedances of the coupled lines appear between the housing and the strips, while the odd-mode impedances appear between the strips. The CPS transmission line in Figure 12.47(a) can be viewed as a coupled line whose even-mode impedance is ∞. When a quarter-wavelength CPS is used, as shown in Figure 12.47(a), a balanced signal can be obtained at the end of the CPS. Instead

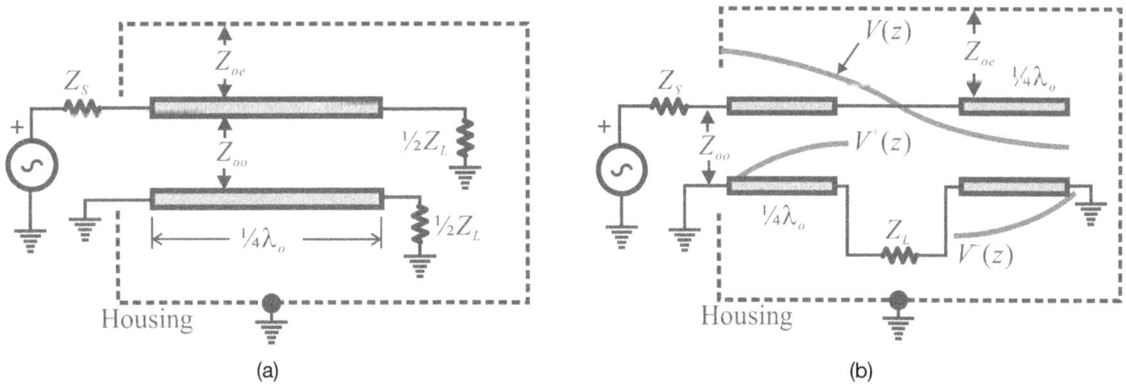

Figure 12.47 (a) A CPS balun and (b) a Marchand balun. Even-mode impedances appear between the housing and the strips, while odd-mode impedances appear between the strips. In (b), $V(z)$ is the standing wave voltage pattern of the upper strip, while $V^+(z)$ and $V^-(z)$ represent the voltages of the two bottom coupled lines.

of the CPS balun, removing the ground plane of the microstrip line and configuring it with the symmetrically balanced up and down microstrip lines achieves the same purpose; this is called a *parallel-plate balun* (refer to Figure 12.65).

Figure 12.47(b) is known as a Marchand balun; the exact description of its principles of operation is rather complicated, but a simple explanation can be given as follows: As this is an open-ended, half-wavelength line when viewed from the upper line of Figure 12.47(b), a voltage standing wave pattern $V(z)$ will occur, as shown in the figure. The standing wave voltage $V(z)$ is maximum at the end of the line and is 0 at the quarter wavelength from the end. Note that the polarity changes at the quarter wavelength and results in the standing wave as shown in the figure. The voltage at the bottom of the coupled transmission line is 0 at both ends and becomes larger as it progresses toward the center. Note that the bottom coupled-line voltages $V^+(z)$ and $V^-(z)$ have opposite polarity according to the polarity of the standing wave of the upper transmission line. Thus, a balanced signal appears across resistor Z_L. The disadvantage of the baluns in Figure 12.47(a) and (b) is that they do not provide a ground for the balanced signal as in the transformer balun. Therefore, they present a problem when an application utilizing the ground is required.

12.4.2.2 SBM Circuit Using Diodes

A single-balanced mixer (SBM) circuit is shown in Figure 12.48. In that figure, the RF and LO signals are applied to isolated ports 1 and 3 of the transformer in Figure 12.42, respectively. The LO and RF signals are thus found to be naturally isolated. Note that the LO is applied to the in-phase distribution port of the transformer (usually referred to as the Σ-port), whereas the RF is applied to the 180° out-of-phase distribution port (usually referred to as the Δ-port). Because the LO signal is distributed in phase, when one of the diodes is on, the other diode is automatically turned off.

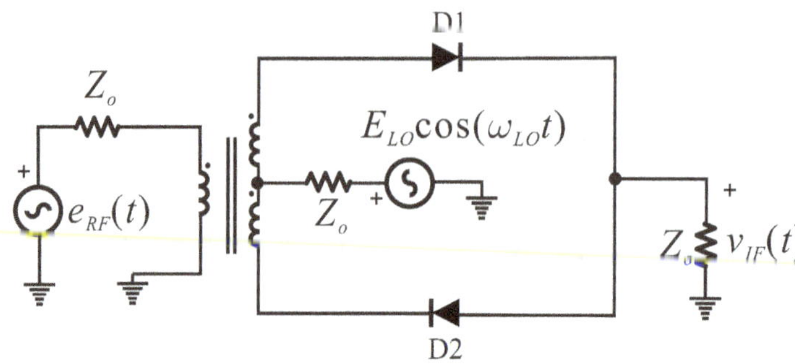

Figure 12.48 An SBM circuit

To understand the SBM circuit's operation, the equivalent circuit of the positive and negative half cycles of the LO are illustrated in Figures 12.49(a) and (b), respectively. In Figure 12.49(a), the voltage appearing at the IF port is expressed by Equation (12.43).

$$v_{IF} = \frac{1}{3}e_{RF}(t) \tag{12.43}$$

The negative half cycle is given by Equation (12.44).

$$v_{IF} = -\frac{1}{3}e_{RF}(t) \tag{12.44}$$

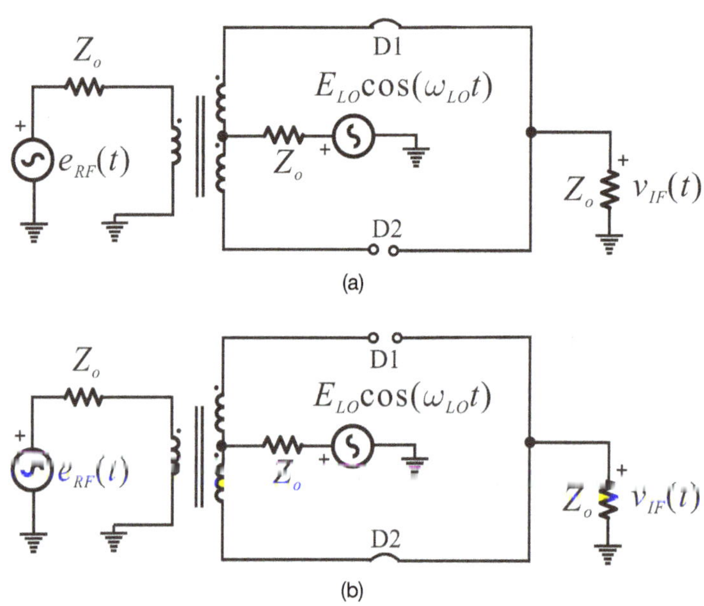

Figure 12.49 Equivalent circuit of SBM depending on LO signal: (a) the positive half cycle and (b) the negative half cycle of LO

The IF output waveform given by Equations (12.43) and (12.44) is illustrated in Figure 12.50.

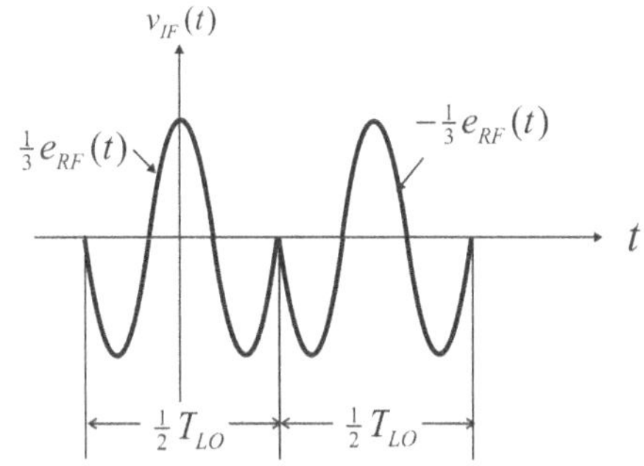

Figure 12.50 The IF output waveform of an SBM

The IF output waveform can be represented by the product of the switching function $S(t)$ and the RF signal, as shown in Figure 12.51, and can thus be expressed as Equation (12.45).

$$v_{IF} = \frac{1}{3}e_{RF}(t)S(t) \tag{12.45}$$

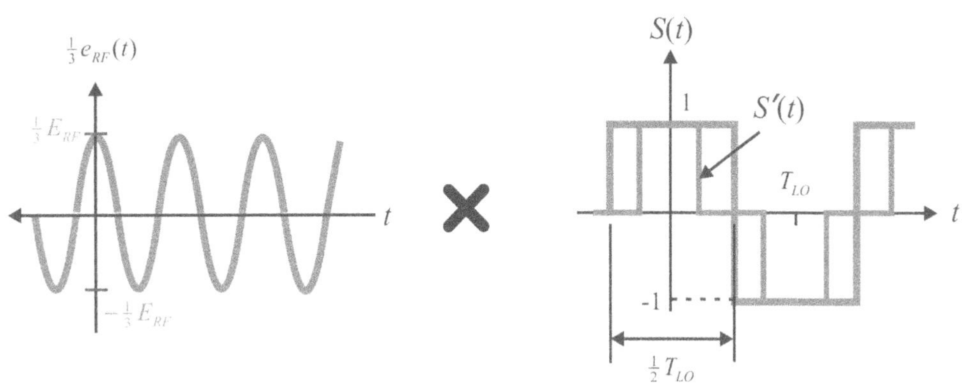

Figure 12.51 A representation of the IF output waveform

The $S(t)$ can be expanded in a Fourier series and it becomes Equation (12.46).

$$S(t) = \frac{4}{\pi}\left\{\cos\omega_{LO}t + \frac{1}{3}\cos 3\omega_{LO}t + \cdots\right\} \tag{12.46}$$

It must be noted that there are no even harmonics in the $S(t)$. This is true even when the diode does not turn on at exactly a half cycle of the LO. A real diode requires a small turn-on voltage during the positive half cycle of the LO, which narrows the duration of the turn-on time of the diode D_1. However, this is also true for diode D_2 during the negative half cycle of the LO. The turn-on time of diode D_2 also becomes narrow. Thus, for a low LO power, the turn-off time interval for both diodes D_1 and D_2 becomes larger. Consequently, a more realistic $S(t)$ for an SBM using real diodes may be $S'(t)$, where the durations of both the positive and negative turn-on times are reduced but $S'(t)$ is still found to have no even-order harmonics. Thus, most of the even-order harmonics of $S(t)$ vanish and, as a result, the mixed-frequency components due to the even-order LO harmonics disappear. The elimination of many of the spurious signals caused by the LO even-order harmonics in the SBM leads to a wider dynamic range than in the SEM. Thus, the SBM generally provides an important advantage over the SEM in mixer applications. The approximate conversion loss of an SBM can be computed as with Equation (12.47).

$$CL = \left(\frac{1}{2} \times \frac{\dfrac{4E_{RF}^2}{9\pi^2 Z_o}}{\dfrac{E_{RF}^2}{8Z_o}} \right)^{-1} = \frac{9\pi^2}{16} \approx 7.5 \text{ dB} \tag{12.47}$$

Figure 12.52 shows the typical structure of an SBM. The LO and RF signals are applied, respectively, to each of the two matched diodes through the 180° hybrid. The nature of the hybrid provides isolation between the RF and LO. In addition, due to the 180° hybrid, the equally split and in-phase LO signals appear at the two diodes, while the 180° out-of-phase RF signals also appear at the two diodes. The IF output signal is obtained through the bypass capacitor as in the SEM.

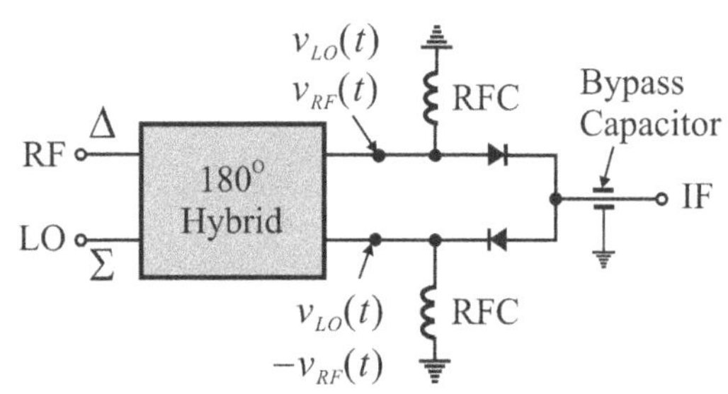

Figure 12.52 Block diagram of an SBM

In terms of the conversion loss, the SBM has a disadvantage compared to the SEM. Improving the conversion loss by matching is rather difficult in the case of the SBM and so matching is commonly not done while in the case of the SEM, the conversion loss can be improved through matching. However, the spurious signals due to the even-order LO harmonics are eliminated, which is the main characteristic of the SBM and this results in an improvement in the dynamic range.

Example 12.6

Use the MBD101 diode in ADS and calculate the conversion loss of the single-balanced mixer (SBM)

Solution

The SBM circuit is set up as shown in Figure 12E.13 and the simulation is performed.

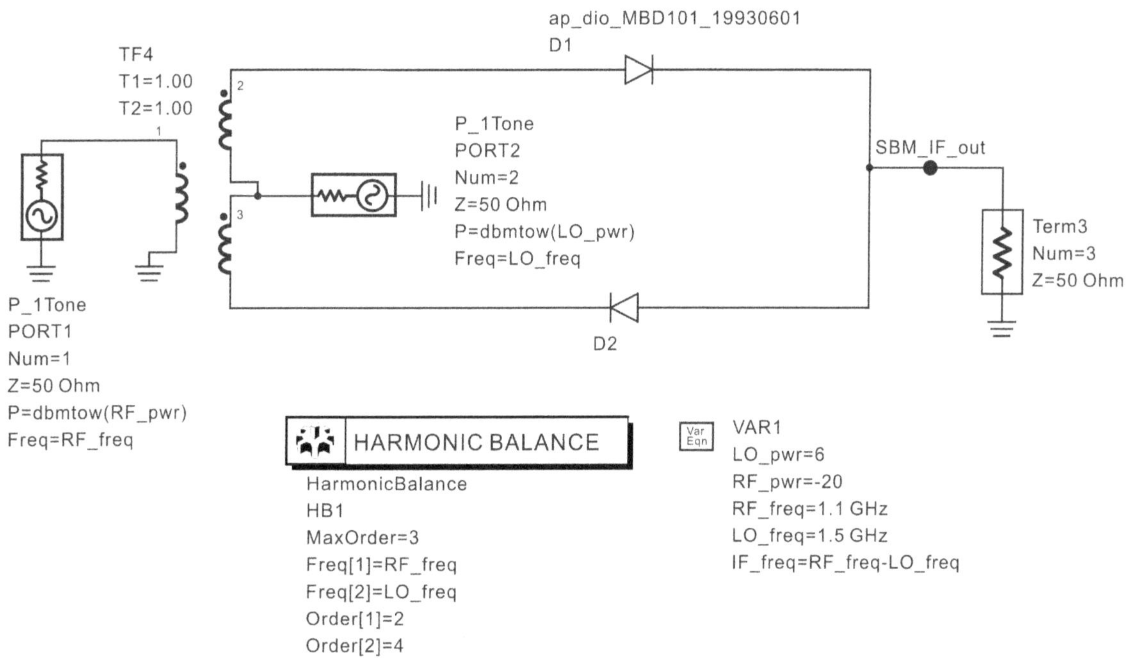

Figure 12E.13 SBM simulation circuit. The diode MBD101 is in the ADS library.

After the simulation, the mixer output voltage **SBM_IF_out** is expressed in dBm as shown in Figure 12E.14. In that figure, the IF output is –28.705 dBm. Since the value of the RF input power is –20 dBm, the conversion loss is –20 – (–28.705) = 8.705 dB. Thus, the SBM has a conversion loss of 8.705 dB.

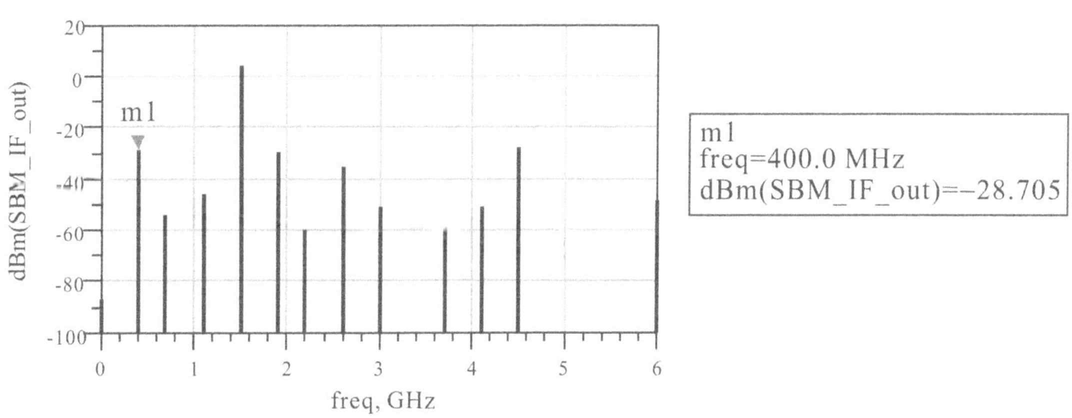

Figure 12E.14 The simulated IF output voltage spectrum

12.4.2.3 Design Examples of an SBM Using Diodes

Figure 12.53 is an SBM circuit consisting of a microstrip rat-race ring hybrid.

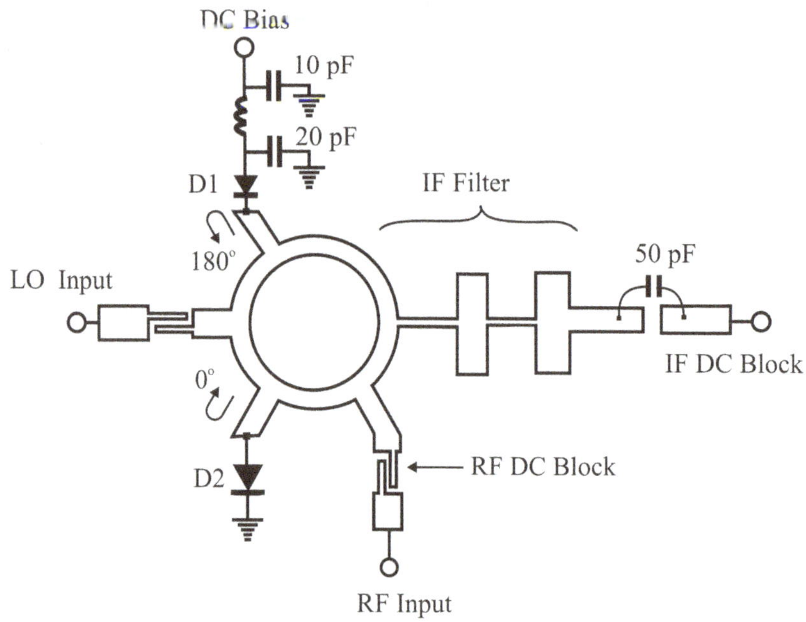

Figure 12.53 SBM circuit using rat-race ring hybrid

The LO signal is applied to the Σ-port of the rat-race ring hybrid and the LO signal is equally divided in phase into the two diodes. Thus, whenever one of the diodes is on, the other diode is turned off as the two diodes are connected to the microstrip in opposite directions. In contrast, the RF signal is applied to the Δ-port of the rat-race ring hybrid and the equally divided RF signals are delivered to the two diodes with a phase difference of 180°. Both of the RF signals incident on the two diodes with a phase difference of 180° are totally reflected and those reflected RF signals from the two diodes are delivered to the IF output port. As the IF output port is equidistant from the two diodes, the reflected signals will have the same phase delay. A microstrip low-pass filter is also connected to the IF port in order to block the leakage of the RF and LO frequency signals. In addition, the capacitor of the IF port is used as a DC block and two diode DC current paths are formed through the DC source and ground.

For the half cycle of the LO signal, diode D_2 is off and D_1 is on. The reflection at D_2 changes the phase of the incident RF signal by 0° because D_2 is open, as shown in Figure 12.53. However, the reflection at D_1 will change the phase of the incident RF signal by 180°, as shown in the figure. In addition, since the incident RF input signal to D_1 has a 180° phase and diode D_1 is shorted (reflection by –180°), the resulting reflected RF signal from D_1 has a 0° phase. Similarly, the incident RF signal to D_2 has a 0° phase and is reflected from diode D_2, which is open. Thus, the reflected RF signal from diode D_2 also has a 0° phase. Consequently, the two reflected signals from the two diodes will have the same phase of 0° and these signals are delivered to the IF output port. Conversely, when D_2 is on, D_1 is off by the LO signal and the two reflected signals with a 180° phase will be delivered to the IF output port. Thus, the SBM operation is completed and the frequency mixing by LO even harmonics will disappear at the IF output as previously explained.

Figure 12.54 is a single-balanced mixer that uses a waveguide.

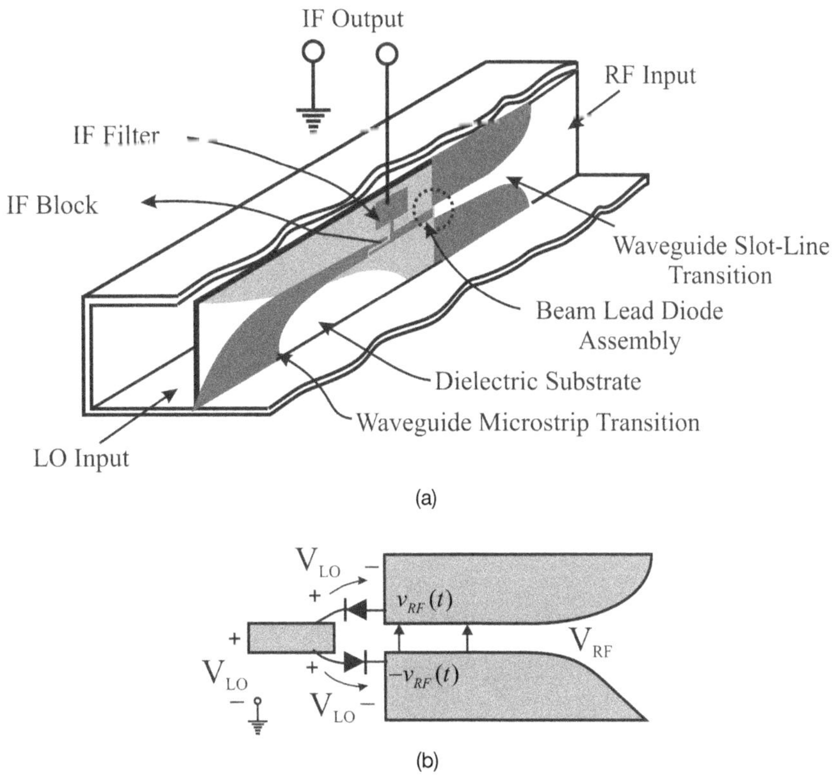

Figure 12.54 Fin-line SBM: (a) the structure and (b) electric fields near the diode

The LO signal that is incident from the waveguide is converted to the microstrip mode through the antipodal fin-line mode converter. In contrast, the RF signal that is incident from the other side of the waveguide input is converted to the slot-line mode using the slot-line mode converter. The circled area in Figure 12.54(a) is enlarged in Figure 12.54(b). Since the LO signal is converted to the microstrip mode, the LO electric field appears in phase at the two diodes, as shown in Figure 12.54(b). However, the RF signal appearing at the end of the slot line is a balanced signal and the RF voltages across the two diodes are 180° out of phase. Thus, the two diodes turn on and off alternately according to the LO signal polarity because the diodes are connected to the microstrip in opposite directions. The RF output voltage is alternately connected to the IF output through the diode that is in the on state. As a result, an opposite polarity RF voltage appears at the IF port according to the LO signal polarity. The circuit then becomes a single-balanced mixer. Here, the IF filter suppresses high-frequency signals such as LO and RF, and the coupled-line IF block acts as a DC block capacitor and prevents the IF output from being shorted.

Figure 12.55 is a single-balanced mixer using a waveguide magic-T. The LO signal is applied to the two diodes in-phase while the RF signal is applied to the two diodes 180° out-of-phase. The waveguide backshorts are used to provide an open circuit at the plane where the diode is connected. When one of the two diodes is turned off by the LO signal, the RF signal connected to the coaxial IF port through the diode opens. Thus, the RF signal that is 180° out of phase alternately comes out through the IF port constructed from a coaxial line. The operation also provides an SBM operation.

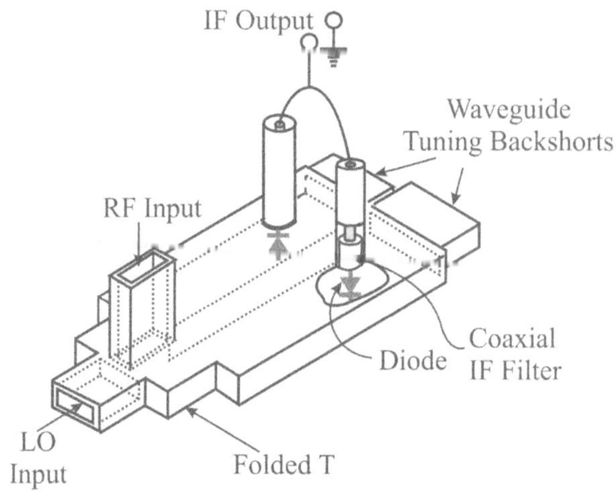

Figure 12.55 An SBM using a waveguide. RF and LO signals are applied to Δ and Σ ports of the magic-T for an SBM operation.

12.4.2.4 Design Examples of an SBM Using Transistors

In order to construct an SBM using transistors, it should be noted that *n*-type and *p*-type transistors are required. However, a matched *p*-type transistor is not commonly available at high frequencies. As a result, two baluns will be required. Figure 12.56 shows the structure of an SBM that uses resistively operated FETs. The transistors Q_1 and Q_2 are alternately turned on and off by the LO signal. Note that the RF signal is applied in phase to the two transistors that are alternately switched on and off by the LO. When Q_1 is on, $i_{RF}(t)$ flows into the drain of Q_1 and the current flowing into Q_2 is 0. Due to the IF balun operation, the voltage $-i_{RF}(t)Z_o$ appears across the IF output load Z_o. Similarly, when Q_2 is on, the IF output voltage of $i_{RF}(t)Z_o$ appears across the load. Thus, at the IF port, a balanced output appears. The configuration can also be applied to construct a transconductance SBM.

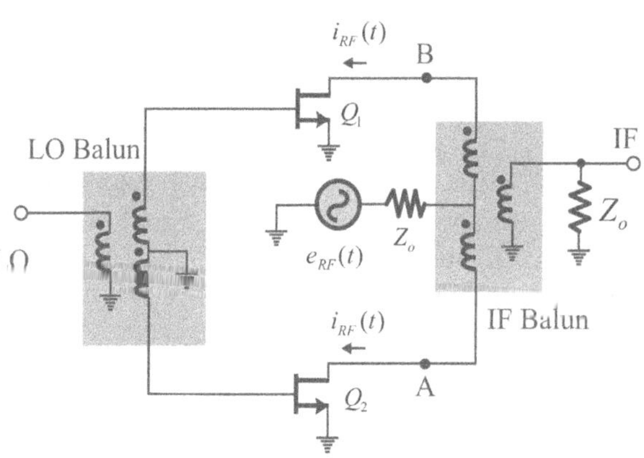

Figure 12.56 An SBM using resistively operated FETs

An SBM constructed using a differential amplifier is shown in Figure 12.57. In that figure, the LO signal is applied through the LO balun to the base terminals of transistors Q_1 and Q_2 with a phase difference of 180°. Thus, transistors Q_1 and Q_2 alternately turn on and off according to the LO signal. In contrast, as the RF signal is input to the base of the transistor Q_3, a current $i_C(t)$ that is proportional to the RF input voltage flows into the Q_3 collector. Thus, the in-phase RF current $i_C(t)$ flows through transistors Q_1 and Q_2 alternately, depending on the LO signal. When the transistor Q_1 turns on, a voltage of $-i_C(t)Z_o$ appears at the IF port. When the transistor Q_2 turns on, a voltage of $i_C(t)Z_o$ appears at the IF port. Thus, for each half cycle of the LO, an RF voltage with changed polarity appears at the IF port, which is an SBM operation.

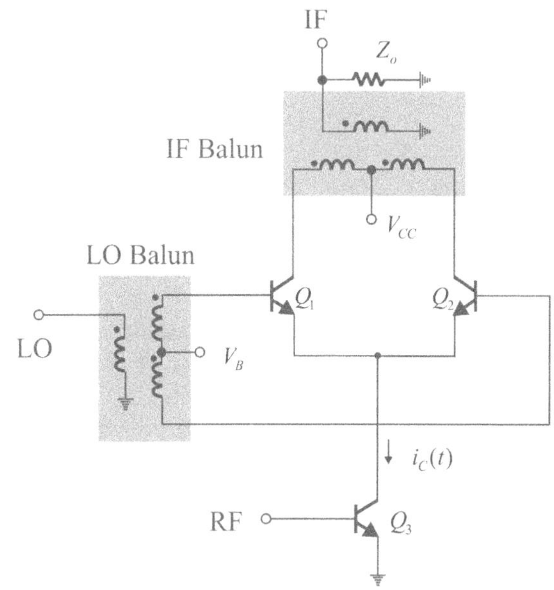

Figure 12.57 An SBM using a differential amplifier

When the frequency of the LO is low, the LO balun can be removed, as shown in Figure 12.58. At low frequencies, the collector impedance of transistor Q_3 is generally high and larger than the input impedances of the two transistors Q_1 and Q_2 seen from the LO terminal, and the applied LO voltage can be equally divided into the base terminals of transistors Q_1 and Q_2. As those divided voltages in the input of transistors Q_1 and Q_2 have a phase difference of 180°, the LO voltage can drive transistors Q_1 and Q_2 alternately without the LO balun.

In addition, a DC current flows through the IF balun but it can be avoided by using resistors. DC biasing can be accomplished by inserting resistors between the collectors and the DC supply, and the voltage between the collectors can be used as the input for the balun. Note that in a case where the output is taken from only one side of the resistor, the output will not be balanced and it becomes the same as the SEM output. As a result, the SBM's advantage cannot be realized.

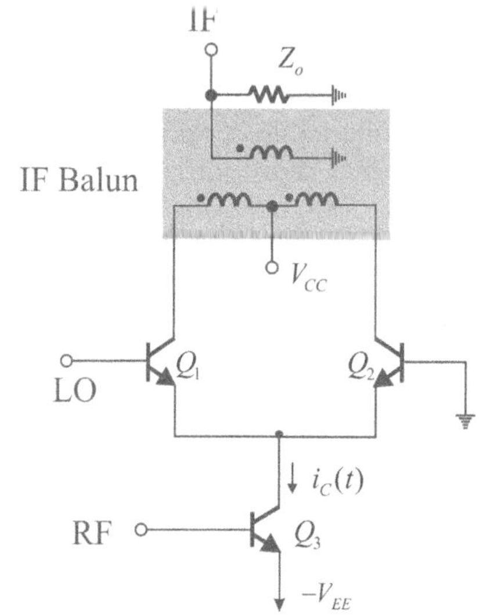

Figure 12.58 An SBM with the LO balun removed

12.4.3 Double-Balanced Mixer (DBM)

A double-balanced mixer (DBM) circuit that uses transformers is shown in Figure 12.59. The four diodes of Figure 12.59 are assumed to be completely matched. Since a balanced signal appears at the secondary winding of the LO balun and the diodes are identical, points A and A′ become the LO signal ground points. Thus, no LO voltage is developed across points A and A′, and they become the secondary winding of the RF balun. This means the LO signal is not delivered to the RF port. Similarly, the secondary winding of

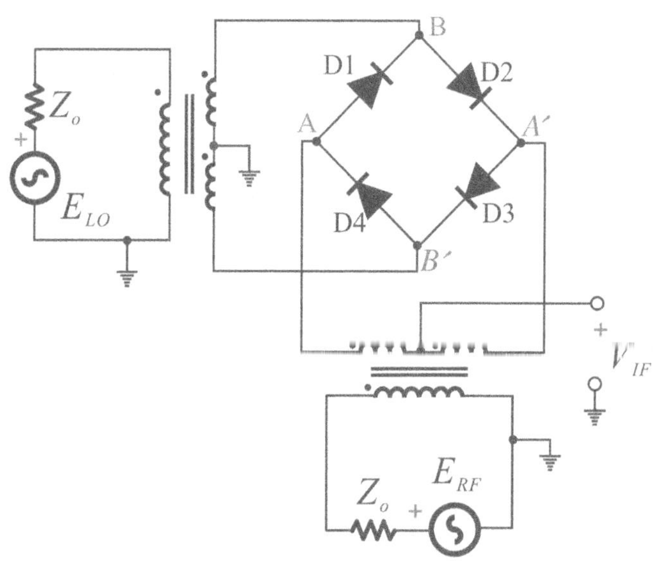

Figure 12.59 A double-balanced mixer circuit

the RF balun is connected to points B and B′ that then become the RF signal ground. Thus, the RF signal does not appear at the LO port. As a result, the RF and LO signals are naturally isolated by the DBM's circuit configuration.

In addition, because the RF and LO grounds are used as the IF port, that port is isolated from both the RF port and the LO port. Therefore, the RF and LO signal outputs do not appear at the IF port even when a low-pass filter is not used. As a result, the RF, LO, and IF ports are isolated from each other and, unlike the SEM, filters are not required to isolate each signal.

In order to analyze the DBM, the equivalent circuit seen from the LO secondary winding is shown in Figure 12.60. Since the ratio of the primary winding to the secondary winding is 1:2, the voltage across the secondary winding is $2E_{LO}$, and half of $2E_{LO}$ is assigned to each winding. In addition, the impedance seen from the secondary winding will be $4Z_o$ and the impedance $2Z_o$ can be similarly assigned to each winding. Therefore, the equivalent circuit becomes that shown in Figure 12.60. In that figure, during the positive half cycle of the LO, diodes D_2 and D_3 are conducting, while the diodes D_1 and D_4 will be turned on during the negative half cycle.

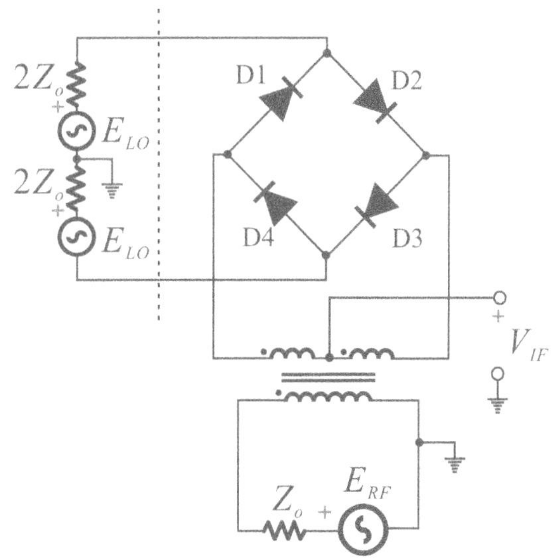

Figure 12.60 Equivalent circuit seen from the secondary winding of the LO

In Figure 12.61(a), the circuit at the RF frequency is shown during the positive half cycle of the LO. Since one side of the secondary winding of the RF transformer is open, that transformer will behave like a simple 1:1 transformer. Thus, the equivalent circuit seen from the secondary winding of the RF transformer will be further simplified, as shown in Figure 12.61(a). Therefore, the voltage at the IF port can be expressed as Equation (12.48).

$$v_{IF}(t) = \frac{Z_o}{Z_o + 2Z_o \| 2Z_o + Z_o} e_{RF} = \frac{1}{3} e_{RF}(t) \tag{12.48}$$

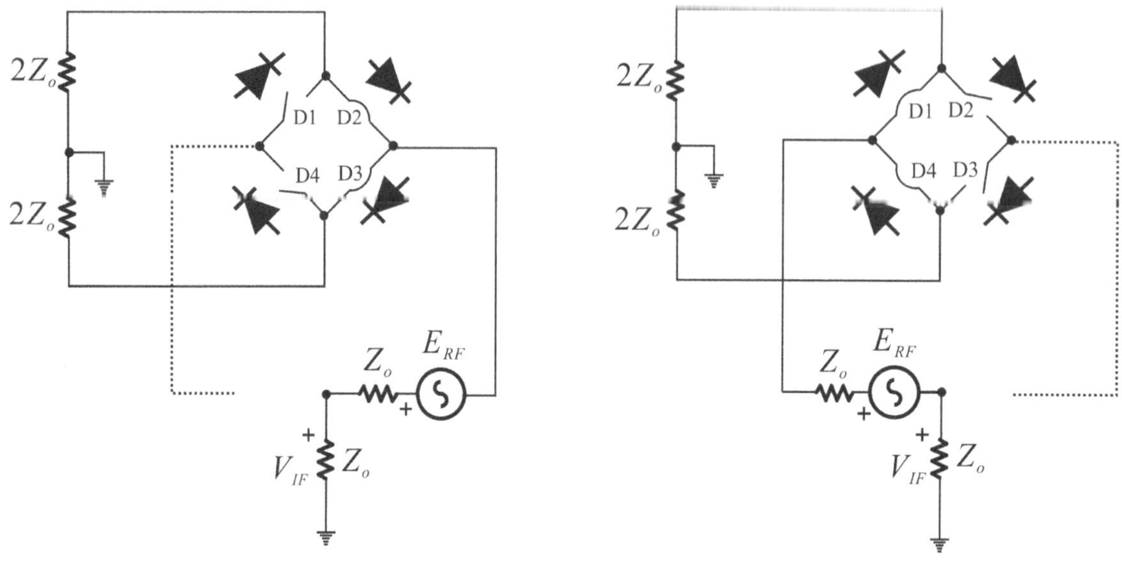

Figure 12.61 (a) The equivalent circuit of the positive half cycle of the LO signal; (b) the equivalent circuit of the negative half cycle of the LO signal

In contrast, during the negative half cycle of the LO, the equivalent circuit at the RF frequency can be simplified, as shown in Figure 12.61(b); therefore, the IF output is given by Equation (12.49).

$$v_{IF}(t) = -\frac{Z_o}{Z_o + 2Z_o \| 2Z_o + Z_o} e_{RF} = -\frac{1}{3} e_{RF}(t) \tag{12.49}$$

This is the same result that would be obtained for a single-balanced mixer and the mixing frequency components by the LO's even-order harmonics vanish. The conversion loss from Equation (12.47) is given by Equation (12.50).

$$CL \approx 7.5 \text{ dB} \tag{12.50}$$

The SBM differs from the DBM when the LO and RF signals are interchanged. The SBM becomes an SEM when the LO and RF signals are interchanged, while the DBM remains an SBM irrespective of the interchanged LO and RF signals. Figure 12.62 is the equivalent circuit at the RF frequency when the LO and RF signals are interchanged.

Analyzing the circuit in Figure 12.62, the circuit behavior is the same as that in Figure 12.61; a balanced RF signal appears at the IF output. Generally, when the RF signal is not low, the output frequency components of a mixer are given by $m\omega_{LO} \pm n\omega_{RF}$. In the case of an SBM, only the even harmonics of the LO signal are eliminated. However, analysis reveals that in the case of the DBM, all the RF and LO even harmonics are eliminated.

In order to show the difference between an SBM and a DBM, the output spectra of the SBM and the DBM for a two-tone input are simulated. Figure 12.63 shows the output spectra of the SBM and the DBM for a two-tone input. Figure 12.63(a) shows the output spectrum of the SBM while 12.63(b) shows that of the DBM. These figures show that more spurious frequency components are removed in the case of the DBM than in the SBM, which consequently leads to an improvement in the dynamic range.

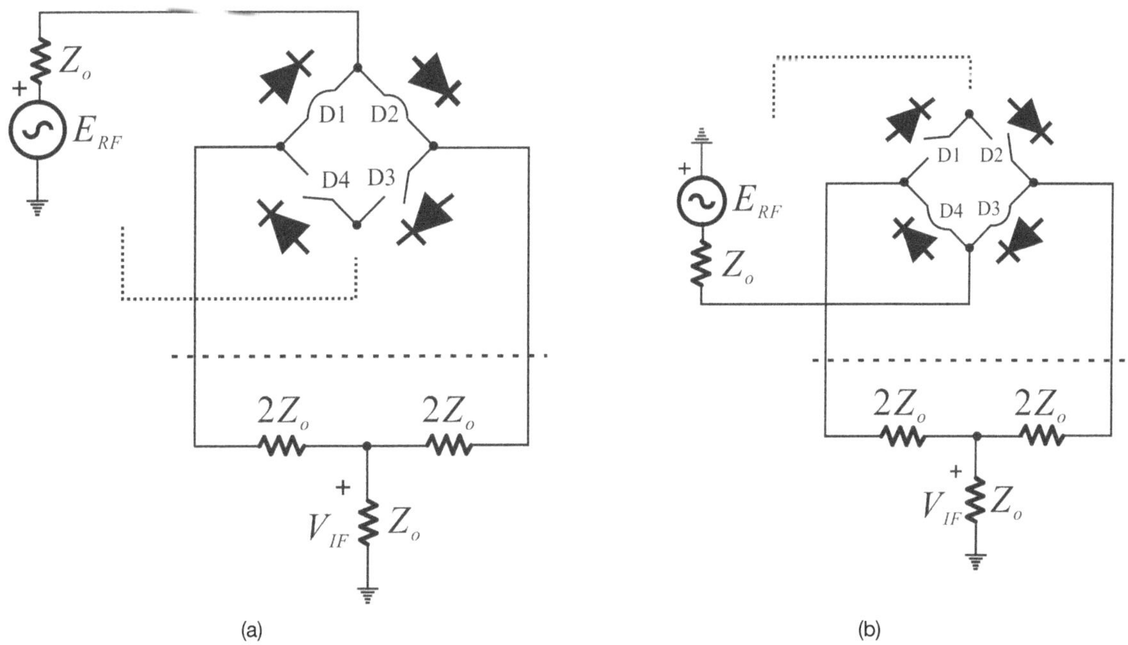

Figure 12.62 Interchanging the LO and RF signals: (a) the equivalent circuit of the positive half cycle of the LO signal and (b) the equivalent circuit of the negative half cycle of the LO signal

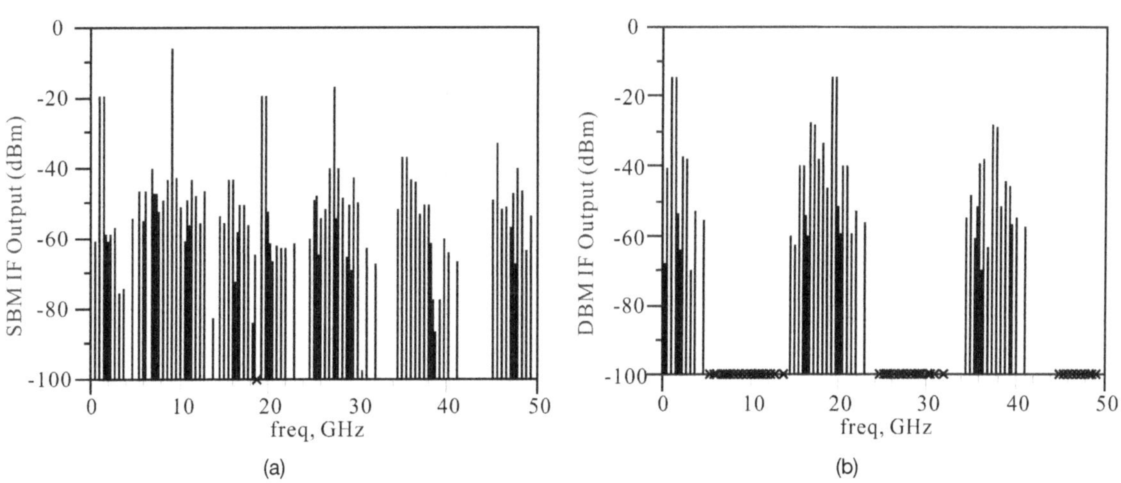

Figure 12.63 (a) Spurious characteristics of an SBM and (b) a DBM for a two-tone input. The SBM and DBM in Figures 12.48 and 12.59 are used in the simulation. The two-tone RF frequencies are 10.0 and 10.5 GHz, and their tone power is set to -10 dBm. The LO powers for the SBM and DBM are 6 and 10 dBm, and their frequency is set to 9.1 GHz. The diode model is the ADS default diode model.

Example 12.7

Using an MBD 101 diode, obtain the conversion loss of a DBM.

Solution

The DBM circuit is set up as shown in Figure 12E.15 and the simulation is performed.

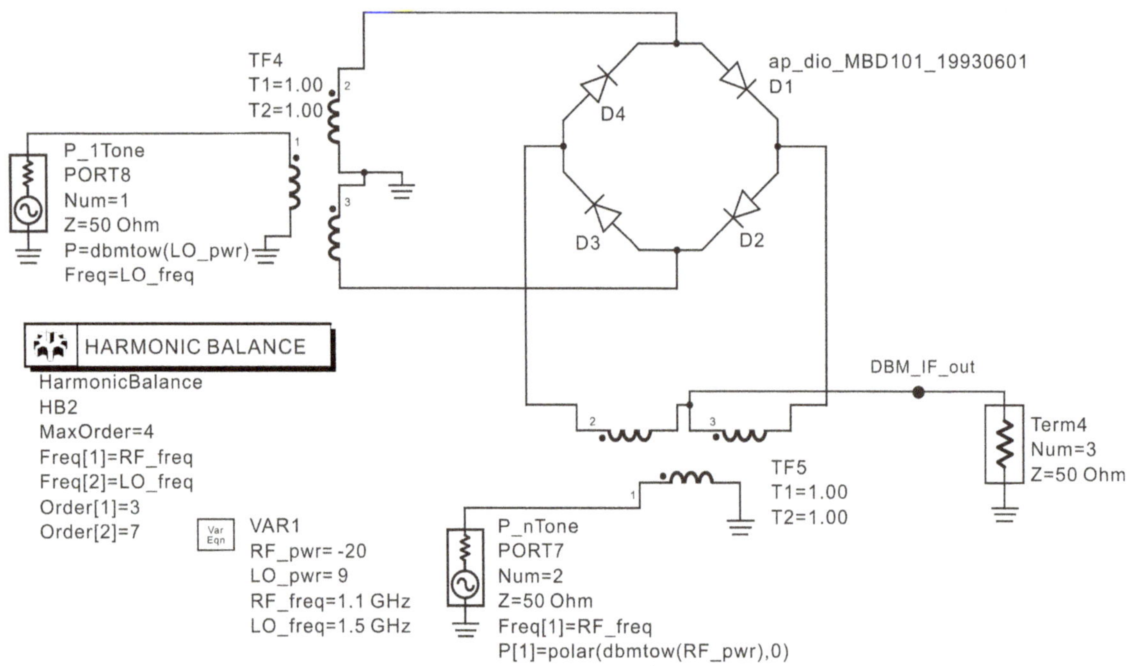

Figure 12E.15 DBM simulation circuit

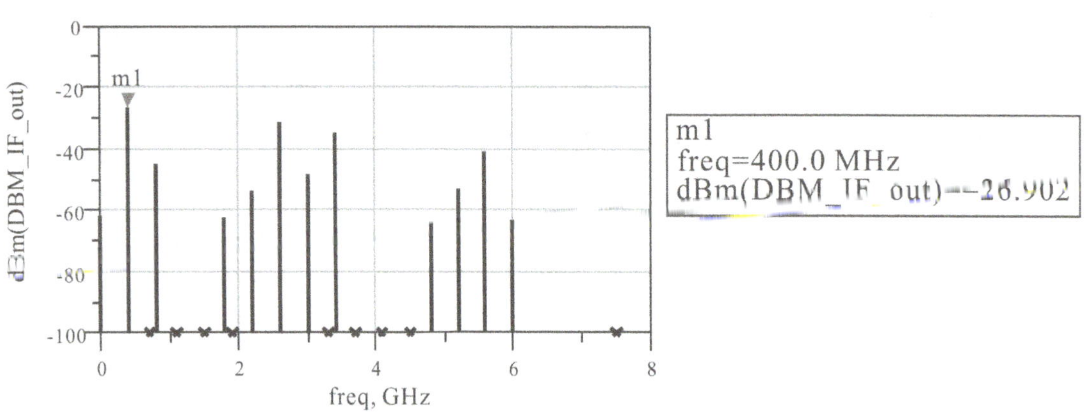

Figure 12E.16 Simulated IF output spectrum of the DBM

To calculate the conversion loss, the output voltage **DBM_IF_out** is plotted in dBm, as shown in Figure 12E.16. In the simulated spectrum, the value of the IF output power is – 26.902 dBm. Since the RF input power value is –20 dBm, calculating the conversion loss of the DBM gives –20 – (–26.902) = 6.902 dB. Thus, the conversion loss of the DBM is 6.902 dB.

12.4.3.1 Double-Balanced Mixer Design Examples

The configuration of a DBM using toroidal transformers is shown in Figure 12.64. The diode quad (four-diode pair) in the figure is fabricated on a single semiconductor substrate and is packaged. The balance characteristic of the DBM largely depends on the degree of the match between the toroidal transformers and the diode quad. This type of DBM is widely used and commercially popular at frequencies below 1 GHz.

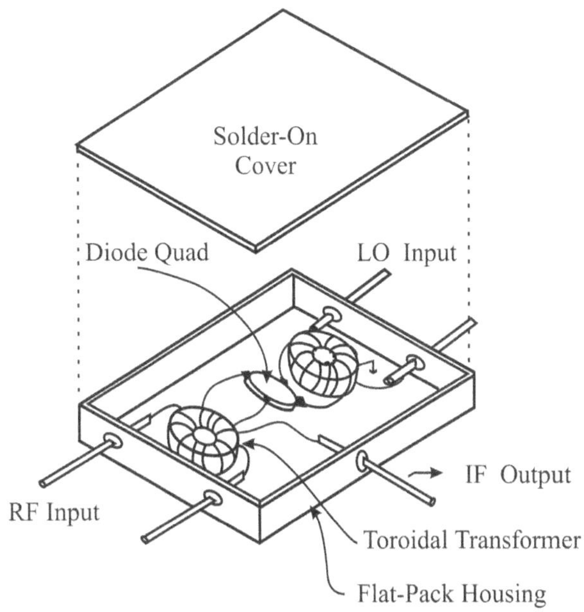

Figure 12.64 DBM using toroidal transformers

As the frequency increases, the toroidal transformer cannot be used to supply the balanced signal to the four diodes of the DBM. Thus, an alternative circuit is required at high frequencies, such as the parallel-line balun shown in Figure 12.65. The balanced microstrip line can be used to support the balanced signal. The balanced microstrip line mode is thus obtained by converting the microstrip mode using the transition shown in Figure 12.65. The electric-field shape of the microstrip is different from that of the balanced microstrip, and the microstrip mode is smoothly converted to the balanced microstrip mode using the tapered ground plane shown in Figure 12.65. The transition is called the parallel-line balun. Using this balun, the balanced signal can be supplied to the four ring-shaped diodes (diode quad).

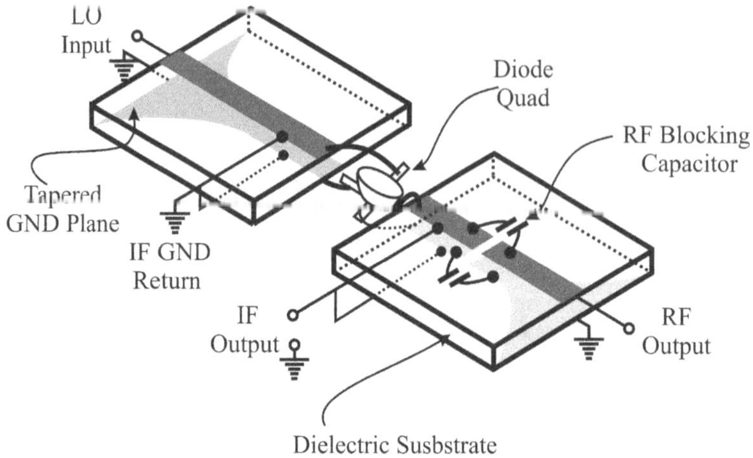

Figure 12.65 A DBM using parallel-line baluns

Although the balanced LO and RF signals can be supplied to the four diodes, the parallel-line balun provides no center tap as the toroidal transformer, which is used as the IF output terminal. Note that the center-tap grounds are also used as the diodes' DC current path. Thus, a diode DC current path must be created for the four diodes by inserting a pair of top and bottom RFCs in the LO and RF baluns, as shown in Figure 12.65. In addition, when the pair of top and bottom RFCs are connected, the center will be the ground for the LO and RF signals. Thus, the IF output can be formed as in the mixer using toroidal transformers. Note that the LO and RF signals are isolated from the IF output because the IF output and IF GND return are the ground points of the LO and RF signals. The potential IF current flow out toward the RF port is prevented by using DC block capacitors inserted in the RF balun, as shown in the figure. Note that these RFCs and DC block capacitors will obviously limit the bandwidth of the IF signal. In addition, there are many other types of DBMs, details of which can be found in reference 2 at the end of this chapter.

12.4.3.2 Double-Balanced Mixer Using Transistors

A DBM using transistors is shown in Figure 12.66. This is also known as the Gilbert cell and it is widely used in integrated circuits at low frequencies. The Gilbert cell's circuit acts as a DBM and is similar to a diode DBM.

Denoting the collector currents of transistors Q_5 and Q_6 as $i_{C,5}(t)$ and $i_{C,6}(t)$, for the applied RF signal, collector current $i_{C,5}(t)$ can be expressed as Equations (12.51a) and (12.51b).

$$i_{c,5}(t) = i_0(t) + i_1(t) + i_2(t) + i_3(t) + \dots \tag{12.51a}$$

$$i_n(t) = I_n \cos n\omega_{RF} t \tag{12.51b}$$

However, the RF signal to applied transistor Q_6 is 180° out of phase with transistor Q_5 and the collector current $i_{C,6}(t)$ can be expressed as Equation (12.52).

$$i_{c,6}(t) = i_0(t) - i_1(t) + i_2(t) - i_3(t) + \dots \tag{12.52}$$

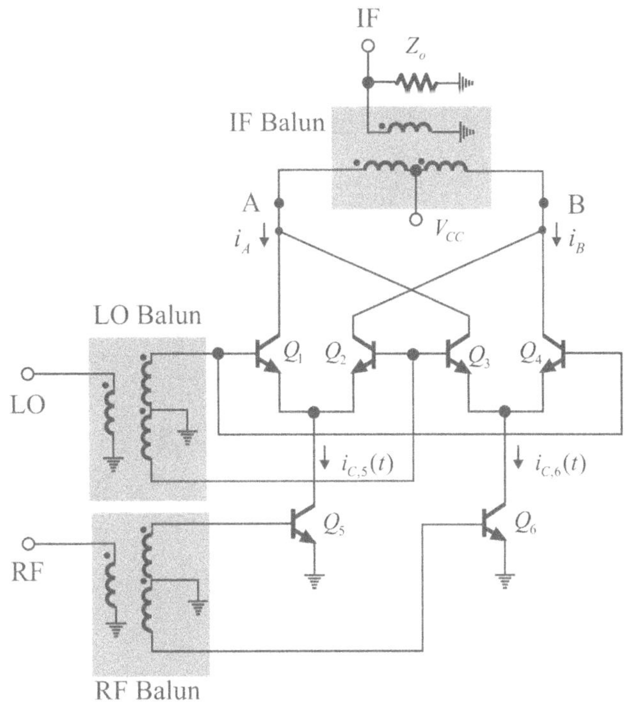

Figure 12.66 A Gilbert cell DBM

Note that in the case of even-order harmonics, the collector currents $i_{C,5}(t)$ and $i_{C,6}(t)$ are in phase, while for odd-order harmonics, they appear 180° out of phase. This is because the even-order harmonic currents are generated by the even-order nonlinearities, while the odd-order harmonic currents are generated by the odd-order nonlinearities.

Among the odd-order harmonic currents, we will first discuss the fundamental collector current due to the RF. Transistors Q_1 and Q_4 will be on during the positive half cycle of the LO signal, while transistors Q_2 and Q_3 will be in the off state and vice-versa. During the positive half cycle of the LO signal, $i_A = i_1(t)$ at point A, while $i_B = -i_1(t)$ at point B. As a result, during the positive half cycle of the LO signal, a voltage of $-i_1(t)Z_o$ appears at the IF port. However, during the negative half cycle of the LO signal, $i_A = -i_1(t)$ at point A, while $i_B = i_1(t)$ at point B. A voltage of $i_1(t)Z_o$ appears at the IF port. Thus, an alternating RF voltage appears at the IF port for every half cycle of the LO signal; that is, the LO output's even-order harmonics will not appear in the IF port.

In the case of the even-order RF harmonics, in-phase, even-order RF harmonic currents flow at points A and B irrespective of the polarity of the LO signal. Due to the operation of the IF balun, the IF output voltage due to these currents disappears. Therefore, in the case of the Gilbert cell, even-order LO and RF signals are not present at the IF output and the circuit acts as a double-balanced mixer.

However, the Gilbert cell mixer in Figure 12.66 uses three triple-wound transformers that are difficult to implement in an IC form. At low frequency, the triple-wound transformer can be removed using the property of differential amplifiers and this is shown in Figure 12.67. This figure is an example of the Gilbert cell DBM with the transformer baluns removed and many other forms are also possible.

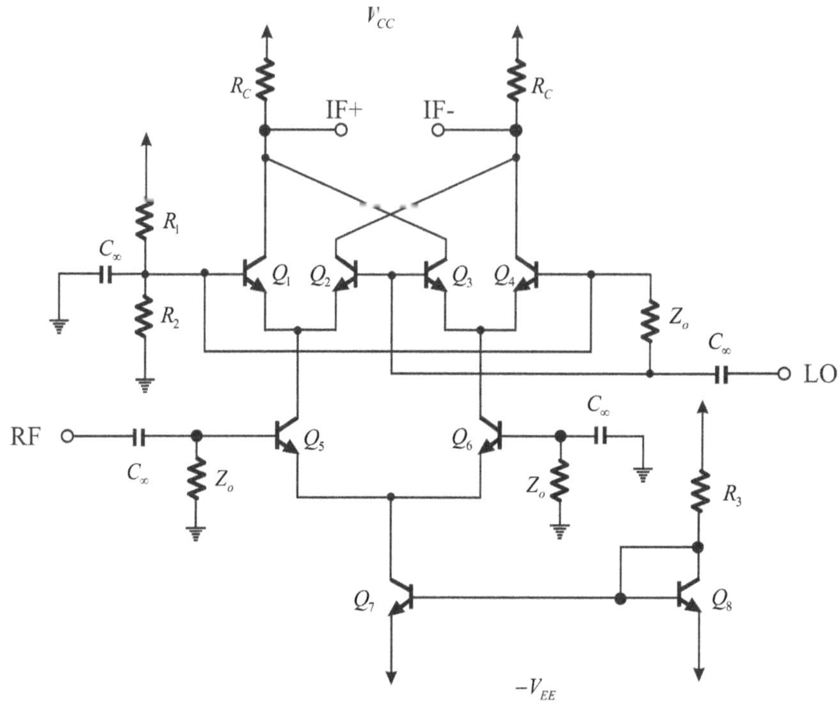

Figure 12.67 A Gilbert cell DBM with LO baluns removed

The DC bias is implemented using a constant current source composed of transistors Q_7 and Q_8. In addition, the LO and RF baluns are implemented by grounding one input end of the differential amplifier as in the differential amplifier SBM. This application of the RF and LO signals can approximately replace the triple-wound baluns because the collector impedance is generally higher than the base–emitter impedance at low frequency. For the triple-wound transformer IF-balun, a differential output IF^+ and IF^- is taken to replace the IF balun. The differential output can be connected to the back-stage differential amplifier. Therefore, integrated circuit implementation of the double-balanced mixer without three baluns is possible and the DBM in Figure 12.67 is widely used at low frequencies.

Another type of double-balanced mixer that uses a transistor is the ring-type resistive DBM shown in Figure 12.68. All the FETs in Figure 12.68 are assumed to be DC biased near the pinch-off, which is not shown in the figure. During the positive half cycle of the LO, the transistors Q_1 and Q_3 are turned on, while the transistors Q_2 and Q_4 are turned off. Therefore, the voltages $v_{RF}(t)$ and $-v_{RF}(t)$ will appear at points A and B, respectively. The resulting balun output voltage of $v_{RF}(t)$ will appear at the IF port. Conversely, during the negative half cycle of the LO, transistors Q_2 and Q_4 are turned on, while transistors Q_1 and Q_3 are turned off. Therefore, the voltages $-v_{RF}(t)$ and $v_{RF}(t)$ will appear at points A and B, respectively. Thus, a voltage of $-v_{RF}(t)$ will appear at the IF port and this represents the balanced output at the IF port. Also note that the RF and IF signals are isolated by the balun. Assuming all the transistors are identical, since points A and B are the RF signal ground, the RF signal does not leak into the IF port. In addition, due to the symmetry of the circuit, the LO signal also does not leak into the IF and RF ports.

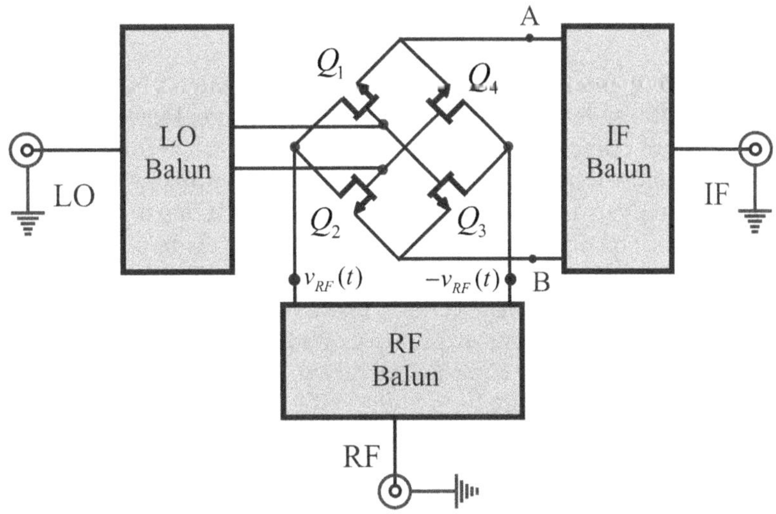

Figure 12.68 A resistive DBM structure; in the case where Q_1 and Q_3 are on, the voltages of points A and B are equal to v_{RF} and $-v_{RF}$, whereas the voltages are $-v_{RF}$ and v_{RF} when Q_2 and Q_4 are on. Thus, the balanced output appears at the IF output. Even-order harmonics disappear due to the operation of the IF balun as they would do in the Gilbert cell mixer. As a result, the mixer operates as a DBM.

12.4.4 Comparison of Mixers

The comparison of the mixers discussed so far is summarized in Table 12.1. The conversion loss for an SEM can be improved with a matching circuit, but its isolation and VSWR depends on the filter. In the case of an SBM, the conversion loss can also be improved with a matching circuit, but it is unrealistic to expect as much improvement as in an SEM. However, because the even-order harmonics of the LO can be eliminated, the SBM has improved spurious characteristics compared to the SEM. In the case of the DBM, the even-order harmonics of the RF and LO can be removed and, in terms of spurious response, the DBM has more advantages than any other mixer. However, more LO power is consumed in driving the DBM because it uses four diodes.

Table 12.1 Performance comparison among mixers

Type	Conversion Loss	Isolation	VSWR	Spurious
Single-ended	GOOD (Depends on matching)	Depends on filter	Depends on matching	BAD
Single-balanced (90°)	MEDIUM (Depends on matching)	POOR	GOOD	As SEM
Single-balanced (180°)	MEDIUM (Depends on matching)	GOOD	POOR	GOOD
Double-balanced mixer	MEDIUM	EXCELLENT	MEDIUM	EXCELLENT

12.5 QUANTITATIVE ANALYSIS OF THE SEM

The qualitative characteristics of a mixer were derived by approximating a diode as a switch, and the switching depended on the LO drive. However, because the diode in the mixer does not operate exactly as an on-and-off switch according to the LO signal, quantitative analysis is required in order to calculate the mixer's characteristics accurately. A numerical analysis method for a diode mixer includes a harmonic balance simulation method in which both the RF and LO signals are considered as large signals. However, in the small-signal harmonic balance simulation method, all signals except the LO signal are considered as small signals. Accurate results can be obtained from the harmonic balance method, but the calculations are complex and quite challenging when performing a theoretical analysis of a mixer circuit's operation. In this section, we will introduce the small-signal harmonic balance analysis of the SEM and we will discuss how to analyze a mixer circuit.

12.5.1 LO Analysis of a Mixer

In the case of an SEM circuit, the only nonlinear element is the diode and the external circuit connected to the diode is a linear circuit. Thus, the SEM can be considered as a four-port linear network with LO, RF, and IF ports connected to the diode, as shown in Figure 12.69. The S-parameters of the four-port linear network are defined in that figure.

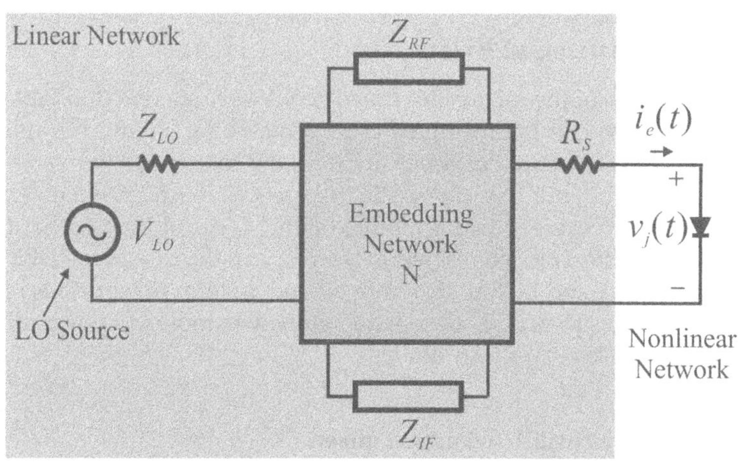

Figure 12.69 The circuit configuration of the SEM

The LO signal is treated as the only large signal while all other signals are treated as small signals. First, the voltage and current of the LO drive's diode is computed at the steady state with all small-signal sources turned off; this is the LO analysis, which is a kind of nonlinear circuit analysis and thus iteration is necessary. The resulting voltage and current of the diode will be periodically time-varying with the period of the LO signal. Next, using the LO analysis results, the time-varying small-signal equivalent circuit can be derived for small-signal RF excitation. Although the small-signal equivalent circuit for RF signal excitation is time-varying, the analysis is obviously a kind of linear circuit analysis. Thus, the RF signal analysis of the mixer circuit can be described as a small-signal, time-varying circuit analysis.

The small-signal mixer analysis resembles the small-signal amplifier analysis performed for electronic circuits. In the small-signal amplifier analysis, the DC operating point for an amplifier circuit is first obtained through nonlinear analysis with other AC signals turned off. Then, the small-signal equivalent circuit at the DC operating point is determined and the AC analysis for a small AC input signal is performed. From the small-signal circuit analysis, the performance characteristics of the amplifier, such as amplification, input, and output impedances, can be determined. The difference between the small-signal amplifier and the mixer is that the small-signal equivalent circuit of the mixer is a linear time-varying circuit, whereas that of the small-signal amplifier is a linear time-invariant circuit. Thus, in order to determine the time-varying operating point in the mixer, the LO analysis must first be performed, which is the nonlinear circuit analysis that determines the time-varying operating point of the nonlinear element with other small-signal sources turned off.

The diode in the circuit shown in Figure 12.69 can be represented by a Schottky diode and a depletion capacitance connected in parallel. Since series resistance R_s of the diode is linear, it is included in the four-port linear network. Thus, the circuit in Figure 12.69 can be represented as the LO signal shown in Figure 12.70. The DC voltage shown in that figure represents a possible DC source that is sometimes inserted for adjusting the conversion loss of the SEM. For this reason, it is also included in the LO analysis. The relationship of the diode and depletion capacitance is expressed in Equations (12.53) and (12.54),

$$i_d = I_o\left(\exp\left(\frac{v_j}{nV_T}\right) - 1\right) \tag{12.53}$$

$$C_j = C_{jo}\left(1 - \frac{v_j}{\phi}\right)^{-m} \tag{12.54}$$

where ϕ represents the built-in potential and v_j represents the voltage across the diode.

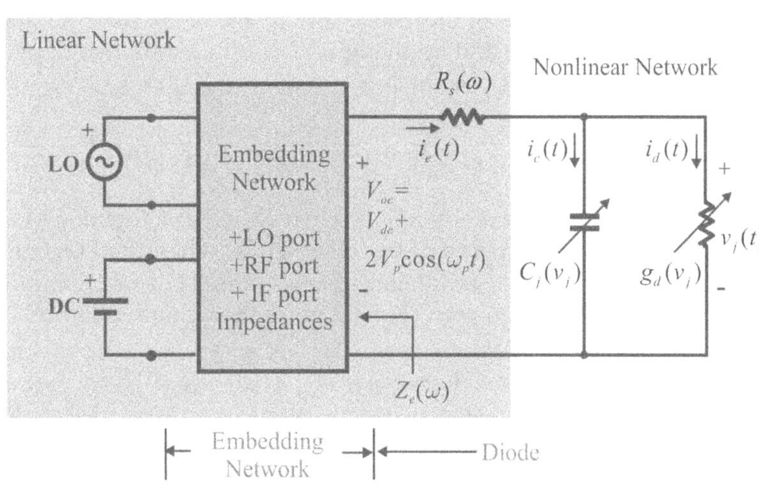

Figure 12.70 The SEM equivalent circuit with respect to the LO signal; here, $\omega_p = \omega_{LO}$.

Denoting the angular frequency of the LO signal as ω_p, the voltage across the diode can be expressed in a Fourier series as shown in Equation (12.55)

$$v_j(t) = \sum_{k=-\infty}^{\infty} V_k e^{jk\omega_p t} \tag{12.55}$$

where V_k is the Fourier series coefficient. Note that $V_k = (V_{-k})^*$ holds since $v_j(t)$ is a real waveform. Denoting the total current as $i_e(t)$ and the currents of the diode and depletion capacitance as $i_d(t)$ and $i_c(t)$, respectively, then $i_d(t)$ and $i_c(t)$ are given by Equations (12.56a) and (12.56b).

$$i_d = I_o\left(\exp\left(\frac{v_j(t)}{\eta V_T}\right) - 1\right) \tag{12.56a}$$

$$i_c(t) = \frac{\partial Q}{\partial v_j} = C_j \frac{dv_j(t)}{dt} \tag{12.56b}$$

The total current $i_e(t)$ can also be expressed in a Fourier series as Equation (12.57)

$$i_e(t) = i_c(t) + i_d(t) = \sum_{k=-\infty}^{\infty} I_{e,k} e^{jk\omega_p t} \tag{12.57}$$

where, similarly, $I_{e,k}$, is the Fourier series coefficient of $i_e(t)$, and $I_{e,k} = (I_{e,-k})^*$ again holds since $i_e(t)$ is a real waveform.

The relationship between the voltage V_k and the current $I_{e,k}$ given by Equations (12.55) and (12.57), respectively, can be obtained from the circuit in Figure 12.70. First, the LO embedding network seen from the diode is a one-port circuit and it can be represented by the Thevenin equivalent circuit. The Thevenin impedance $Z_e(\omega)$ is the impedance of the LO embedding network seen from the diode with the LO and DC sources killed. The Thevenin equivalent voltage V_{oc} is an open-circuit voltage and has the DC and LO frequency components of $\pm\omega_p$. Denoting the Thevenin voltage of DC and LO frequency components as V_{dc} and V_p, respectively, and applying KVL for the fundamental frequency, the relationship in Equation (12.58a) is obtained.

$$V_{\pm 1} = V_p - I_{e,\pm 1}\left[Z_e(\pm\omega_p) + R_s(\pm\omega_p)\right] \tag{12.58a}$$

The series resistance of the diode R_s is also expressed as a function of frequency by assuming that it varies with respect to frequency. Thus, when the LO signal is expressed as a sinusoidal waveform, the LO waveform has an amplitude of $2V_p$ but the analysis does not lose its generality. Also, Equation (12.58b) gives the value for the DC.

$$V_0 = V_{dc} - I_{e,0}\left[Z_e(0) + R_s(0)\right] \tag{12.58b}$$

Since no sources appear at the diode for the rest of the harmonics, the current and voltage relationship for those harmonics is given by Equation (12.58c).

$$V_k = -I_{e,k}\left[Z_e(k\omega_p) + R_s(k\omega_p)\right], \qquad k = \pm 2, \pm 3, \ldots, \pm\infty \tag{12.58c}$$

Thus, the voltage $v_j(t)$ can be computed as follows: When an initial voltage estimate is appropriately chosen as $v_j(t)$, current $i_e(t)$ can be directly calculated using Equation (12.56). Then, the transformation of $i_e(t)$ into the Fourier series $I_{e,0}$, $I_{e,\pm1}$, $I_{e,\pm2}$,... can be computed. Again, substituting $I_{e,k}$ into Equations (12.58a)–(12.58c) gives V_0, $V_{\pm1}$, $V_{\pm2}$,... and a new $v_j(t)$ can be obtained by transforming the Fourier series V_k into time-domain signal. In a case where the new voltage $v_j(t)$ is different from the initial estimate, then the newly obtained $v_j(t)$ is used to replace the old estimate. Through the repetition of the procedure above, $v_j(t)$ will converge to a certain value and, as a result, the $v_j(t)$ that satisfies the equations can be obtained. The computed waveforms for $i_e(t)$ and $v_j(t)$ are shown in Figure 12.71. The waveforms will vary slightly depending on the diode parameters, but they generally represent the voltage and current waveforms of the diode. When the diode is forward-biased, a constant voltage of approximately 0.7 V appears across the diode and the resulting current of the diode increases exponentially, whereas the current is almost 0 in the reverse bias.

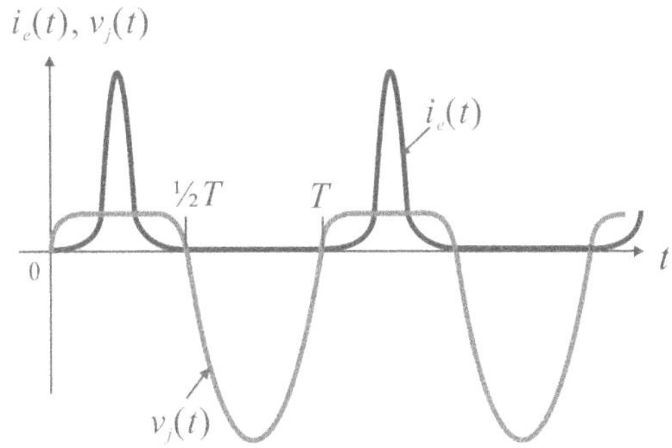

Figure 12.71 Typical $i_e(t)$ and $v_j(t)$ obtained through iteration; here, $T = (2\pi)/\omega_p$.

12.5.2 Small-Signal Analysis

When a time-varying small RF signal is applied to the mixer, a time-varying small-signal voltage $\delta v(t)$ is superimposed on the diode voltage $v_j(t)$ previously determined at a large-signal LO drive. Due to $\delta v(t)$, the diode current $i(t)$ determined at a large-signal LO drive will also change by $\delta i(t)$. Thus, the current $i(t) + \delta i(t)$ flows through the diode. The notation that adds a δ in front of the voltage or current indicates a small signal; that is, $\delta i(t)$ and $\delta v(t)$ represent the small-signal current and voltage, respectively, and also represent time-varying signals. We will consider a device with the general characteristic $i = f(v)$. For the device having this characteristic, the current $i(t) + \delta i(t)$ can be written as Equation (12.59).

$$i(t) + \delta i(t) = f\left(v_j(t) + \delta v(t)\right) = f\left(v_j(t)\right) + \left.\frac{\partial f}{\partial v}\right|_{v=v_j(t)} \delta v(t) \tag{12.59}$$

The first term on the right-hand side of the equation represents the time-varying current determined in the LO analysis, while the second term represents the small-signal current generated by $\delta v(t)$. The coefficient of $\delta v(t)$ corresponds to the slope of $f(v)$ whose value is time-varying according to $v_j(t)$ determined from the LO analysis. The slope can be interpreted as a small-signal conductance.

The frequency of the small-signal RF is denoted as $\omega_o + \omega_p$. Here, ω_o represents the offset frequency from the LO and it can be interpreted as the IF frequency. When the RF signal with $\omega_o + \omega_p$ is applied to the mixer, the frequency components in the mixer can generally be expressed as $n\omega_p \pm m(\omega_o + \omega_p)$. However, since the RF signal is a small signal, it will have the two values $m = 1$ or 0. In the case of $m = 0$, the frequency becomes $n\omega_p$ and these frequency components are harmonics of the LO signal. Since the LO harmonic signals are strong signals, after removing them (the voltages and currents for these frequency components are already obtained in the LO signal analysis), the small signals have the frequency components $m\omega_p \pm \omega_o$. However, the $-\omega_o$ term is not necessary and this will be explained later. Thus, the small-signal voltage can generally be expressed as Equation (12.60).

$$\delta v(t) = \sum_{m=-\infty}^{\infty} \delta V_m e^{j(m\omega_p + \omega_o)t} \tag{12.60}$$

The signal that has frequency components $m\omega_p + \omega_o$ is denoted as $\delta v_m(t)$ in Equation (12.61).

$$\delta v_m(t) = \delta V_m e^{j(m\omega_p + \omega_o)t} \tag{12.61}$$

Then, δV_m can be interpreted as the phasor of the signal with the frequency $m\omega_p + \omega_o$. In addition, $\delta \mathbf{v}$ is a column vector signal defined in Equation (12.62).

$$\delta \mathbf{v} =^t \left[\ldots \delta v_1(t), \delta v_0(t), \delta v_{-1}(t), \ldots \right] \tag{12.62}$$

Similarly, $\delta \mathbf{V}$ is also a column vector that consists of phasor δV_m, which is defined in Equation (12.63).

$$\delta \mathbf{V} = \left[\ldots, \delta V_1, \delta V_0, \delta V_{-1}, \ldots \right] \tag{12.63}$$

With these definitions, the small-signal RF voltage can be represented in Equation (12.64).

$$\delta v_1(t) = \delta V_1 e^{j(\omega_p + \omega_o)t} \tag{12.64}$$

Thus, $\delta v(t)$ can be expressed with Equation (12.65).

$$\delta v(t) = \sum_{m=-\infty}^{\infty} \delta v_m(t) = \sum_{m=-\infty}^{\infty} \delta V_m e^{j(m\omega_p + \omega_o)t} \tag{12.65}$$

Now, going back to Equation (12.59) and substituting the voltage $v_j(t)$ determined from the LO analysis into the small-signal conductance, the small-signal conductance waveform $g(t)$ can be obtained. The waveform $g(t)$ is also periodic with the LO period and can be expressed in a Fourier series as shown in Equation (12.66).

$$\left.\frac{\partial f}{\partial v}\right|_{v=v_j(t)} = g(t) = \sum_{k=-\infty}^{\infty} G_k e^{jk\omega_p t} \quad \left(G_k = G_{-k}^*\right) \tag{12.66}$$

The waveform $g(t)$ can be interpreted as the time-varying, small-signal conductance from the nonlinear resistor $f(v)$. Using $g(t)$, the time-varying small-signal current $\delta i(t)$ can be obtained as $g(t)\delta v(t)$, as shown in Figure 12.72.

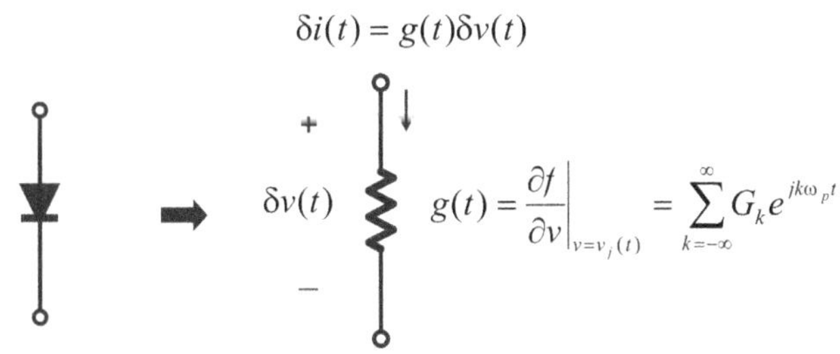

Figure 12.72 Small-signal current in a diode

Using Equations (12.65) and (12.66), the small-signal frequency components of $g(t)\delta v(t)$ are again found to be $m\omega_p + \omega_o$. Note that $\delta i(t) = g(t)\delta v(t)$ has the same frequency components as those of $\delta v(t)$. Thus, the time-varying resistor with the LO frequency does not generate new frequency components other than $m\omega_p + \omega_o$. The spectra of $\delta v(t)$ and $\delta i(t)$ are shown in Figure 12.73.

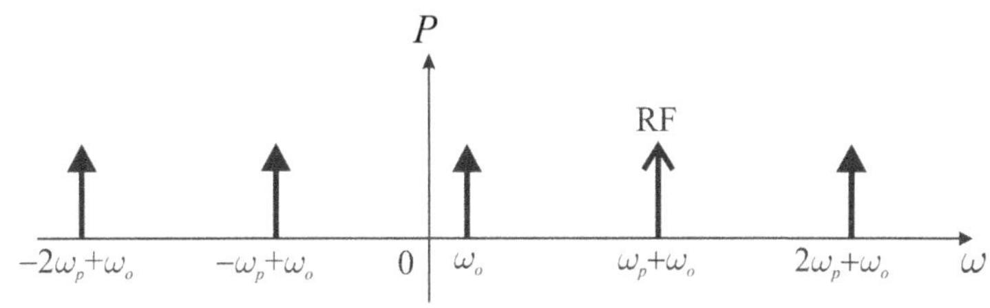

Figure 12.73 Small-signal spectra

However, the real spectrum for a small-signal excitation is shown in Figure 12.74. The spectra can be easily inferred. The harmonics of the LO will appear since the LO is a strong signal. In contrast, assuming the RF has an extremely small signal with frequency $\omega_o + \omega_p$, the first-order mixing terms will appear as shown in Figure 12.74.

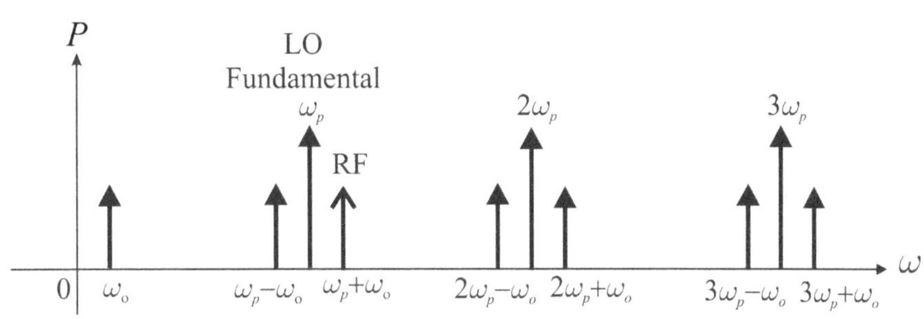

Figure 12.74 Spectra in a mixer

The spectra in Figure 12.73 do not show the LO harmonic components of $n\omega_p$ because these components are addressed in the first term of Equation (12.59) and they are not small signals. Excluding the LO signals that have the frequency component $n\omega_p$, there are still frequency components $n\omega_p - \omega_o$ that are unmatched to those in Figure 12.73. The frequency components $n\omega_p - \omega_o$ correspond to the negative frequency components $-n\omega_p + \omega_o$ in Figure 12.73. These negative frequency components are reflected back to the positive frequencies because no negative frequencies are permitted in real measurement. The concept of the negative frequency in Figure 12.73 is computationally easy; the spectra will be represented using the negative frequency. This is usually referred to as the *Saleh notation*.

By using the Saleh notation, the m-th harmonic current $\delta i_m(t)$ due to $\delta v(t)$ is expressed in Equation (12.67).

$$\delta i_m(t) = m\left(g(t)\delta v(t)\right) = m\left(\sum_{k=-\infty}^{\infty} G_k e^{jk\omega_p t} \sum_{m=-\infty}^{\infty} \delta V_m e^{j(\omega_o + m\omega_p)t}\right) \tag{12.67}$$

Here, $m(\bullet)$ represents the m-th harmonic component. Thus, we obtain Equation (12.68).

$$\delta i_m(t) = \left(\sum_{k=-\infty}^{\infty} G_{m-k}\delta V_k\right) e^{j(\omega_o + m\omega_p)t} \tag{12.68}$$

Therefore, δI_m can be expressed as Equation (12.69).

$$\delta I_m = \left(\sum_{k=-\infty}^{\infty} G_{m-k}\delta V_k\right) \tag{12.69}$$

The depletion capacitance $C_j(t)$ in Equation (12.56b) can be considered a time-varying capacitance that depends on the LO signal. Expressing the time-varying depletion capacitance as a Fourier series results in

$$C_j(t) = \sum_{k=-\infty}^{\infty} C_k e^{jk\omega_p t} \quad \left(C_k = C^*_{-k}\right)$$

The small-signal current of the time-varying depletion capacitance has the relationship shown in Figure 12.75.

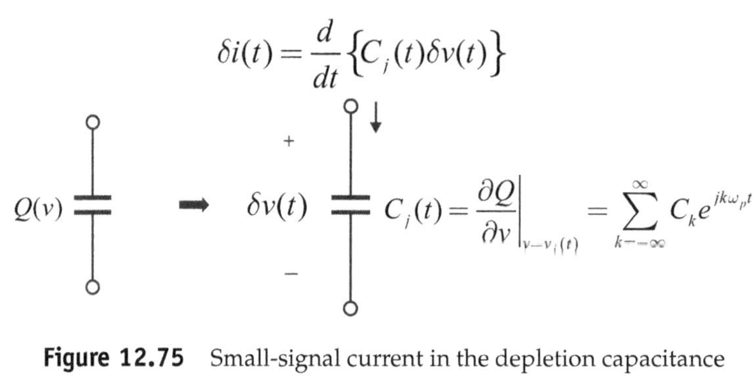

Figure 12.75 Small-signal current in the depletion capacitance

Thus, the m-th harmonic current $\delta i_m(t)$ is given by Equation (12.70).

$$\delta i_m(t) = m\left(\frac{d}{dt}\left[C(t)\delta v(t)\right]\right) = \left(\sum_{k=-\infty}^{\infty} j\left(\omega_o + m\omega_p\right)C_{m-n}\delta V_n\right)e^{j\left(\omega_o + m\omega_p\right)t} \tag{12.70}$$

The small-signal voltage applied to the diode can be represented by $\delta \mathbf{V} = {}^t(\ldots, \delta V_1, \delta V_0, \delta V_{-1}, \ldots)$ and gives rise to the small-signal current of all the frequency components $\delta \mathbf{I} = {}^t(\ldots, \delta I_1, \delta I_0, \delta I_{-1}, \ldots)$. Therefore, the diode can be represented by the following admittance matrix shown in Equation (12.71).

$$\delta \mathbf{I} = \mathbf{Y}\delta \mathbf{V} \tag{12.71}$$

The admittance matrix can be expressed with Equation (12.72),

$$Y = \begin{pmatrix} & \vdots & \vdots & \vdots & \\ \cdots & Y_{11} & Y_{10} & Y_{1-1} & \cdots \\ \cdots & Y_{01} & Y_{00} & Y_{0-1} & \cdots \\ \cdots & Y_{-11} & Y_{-10} & Y_{-1-1} & \cdots \\ & \vdots & \vdots & \vdots & \end{pmatrix} \tag{12.72}$$

where Y_{ij} is interpreted with Equation (12.73).

$$Y_{ij} = \frac{\text{Resulting current of frequency component } \omega_o + i\omega_p}{\text{Exciting voltage of frequency component } \omega_o + j\omega_p} \tag{12.73}$$

Using Equations (12.69) and (12.70), the value of Y_{mn} is given by Equation (12.74).

$$Y_{mn} = G_{m-n} + j\left(\omega_o + m\omega_p\right)C_{m-n} \tag{12.74}$$

Now, it is necessary to use an interpretation of Equation (12.71) as a circuit. The matrix $[Y_{ij}]$ represents the small-signal admittance matrix of the diode driven by the LO signal, which relates the small-signal current of the frequency component $\omega_o + m\omega_p$ to the small-signal voltage of the frequency component $\omega_o + n\omega_p$. Thus, the diode can then be represented as a multiport device, as shown in Figure 12.76.

Here, δV_m is an independent variable and δI_m is the dependent variable resulting from the port voltage excitation δV_m. Thus, δI_m is dependent on all the frequency components of the voltage δV_m. This multiport representation of the diode according to the small-signal frequencies is somewhat strange. However, considering the connection of the diode to the external linear circuit, the diode multiport representation is understandable. Note that the external circuit connected to the diode is considered to be linear and thus the voltage and current of the same small-signal frequency components are allowed. As a result, the external linear circuit is treated as multiple separate circuits when, depending on their small-signal frequencies, those circuits are separated. For example, a linear circuit with

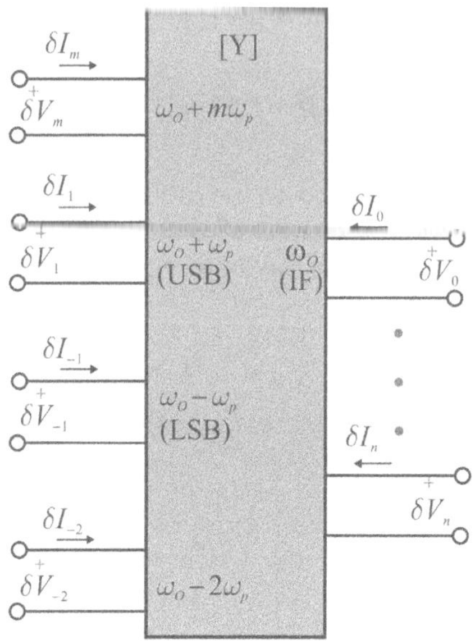

Figure 12.76 Diode multiport representation

multiple sinusoidal sources of different frequencies can be considered as separate linear circuits with different frequencies. Since the diode generates multiple small-signal frequencies, the external circuit is considered to be composed of separate circuits that depend on small-signal frequencies. Thus, the diode multiport representation provides an easy connection between the diode and the external linear circuit.

When the diode is viewed as a multiport device, the mixer circuit can be described using the diode multiport representation. Since the external circuit is a linear circuit, it can be represented using the Norton equivalent circuit at each $m\omega_p + \omega_o$ frequency. This is shown in Figure 12.77.

Note that there are no other sources apart from the RF signal with the frequency $\omega_p + \omega_o$. Thus, the only Norton current source appears at the RF frequency shown in Figure 12.77. Also, since the impedance of the external circuit including R_s seen from the diode is defined as $Z_e(\omega) + R_s(\omega)$, as shown in Figure 12.70, the impedance at $m\omega_p + \omega_o$ will be $Z_e(m\omega_p + \omega_o) + R_s(m\omega_p + \omega_o)$. The impedance $Z_e(m\omega_p + \omega_o)$ is denoted as $Z_{e,m}$. Similarly, $R_{s,m}$ represents $R_s(m\omega_p + \omega_o)$. Note that $\delta I'_m$ is the equivalent Norton current source of the external circuit that includes R_s. Thus, the port currents and voltages are denoted as $\delta I'_m$ and $\delta V'_m$. Then, the augmented admittance matrix that includes the external circuit impedance is obtained with Equation (12.75).

$$\mathbf{Y}' = \mathbf{Y} + \mathrm{diag}\left[\frac{1}{Z_{e,m} + R_{e,m}}\right] \tag{12.75}$$

Since $\delta\mathbf{V}' = \delta\mathbf{V}$, the resulting small-signal current and voltage relationship is given Equation (12.76).

$$\delta\mathbf{I}' = \mathbf{Y}'\delta\mathbf{V} \tag{12.76}$$

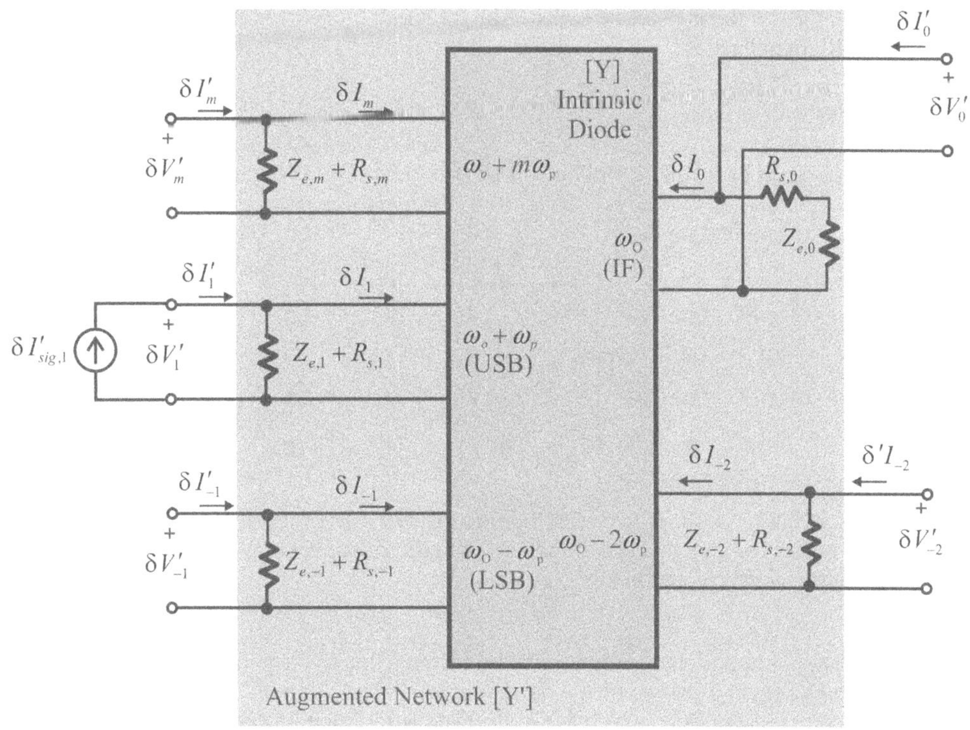

Figure 12.77 A representation of an SEM circuit that uses a diode multiport (an intrinsic diode is a diode without series resistance R_s).

There is no current source except the RF frequency component, and $\delta\mathbf{I}' = {}^t(\ldots, 0, \delta I'_1, 0, \ldots)$. Solving Equation (12.76) for the small-signal voltage, we obtain Equation (12.77),

$$\delta\mathbf{V}' = \mathbf{Z}'\delta\mathbf{I}' \qquad (12.77)$$

which is the small-signal voltage of all the frequency components that result from the small-signal current input of the RF component. Here, The $\mathbf{Z}'$ is termed as the *conversion impedance matrix*. The element Z'_{01} of the conversion matrix represents the IF output voltage due to the small-signal input current of the RF frequency; that is, it represents a conversion from the RF to the IF. Similarly, the element Z'_{21} can be interpreted as a conversion from the RF to the small-signal voltage at the frequency component of $2\omega_p + \omega_o$.

Example 12.8

Using the diode multiport shown in Figure 12.76, draw the circuit at the RF frequency for the SEM circuit given in Figure 12.26(a) assuming $R_s = 0$.

Solution

Since the LO filter is approximated as open for frequencies $\omega = \omega_p$, all the ports in the diode multiport see the same impedance as $2Z_o$. The resulting network is similar to the network shown in Figure 12.77

where all $Z_{e,m} + R_{s,m}$ is replaced by $2Z_o$. The source $\delta I'_{sig,1} = E_{RF}/(2Z_o)$. Note that the diode admittance at the RF frequency is

$$Y_{11} = G_0 + j\left(\omega_o + \omega_p\right)C_0$$

The circuit can be further simplified, as shown in Figure 12E.17.

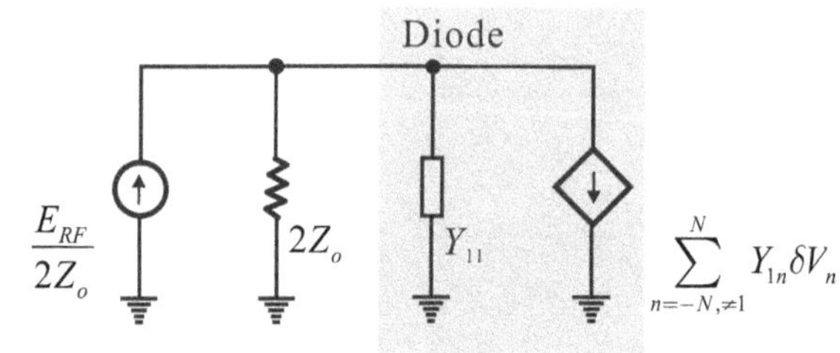

Figure 12E.17 A simplified SEM circuit at the RF frequency; the controlled source represents the RF frequency component current that is due to the voltage of the other frequency components developed across $m\omega_p \pm \omega_o$ and $m \neq 1$.

The controlled source represents the currents due to the voltage of the other frequency components $m\omega_p \pm \omega_o$, $m \neq 1$. The voltage developed across $2Z_o$ for $m\omega_p \pm \omega_o$, $m \neq 1$ makes the RF frequency component current flow, which can be found using Equation (12.71). Similarly, the circuit for each frequency component can be drawn. In conclusion, at the RF frequency, the diode can be thought of as the parallel connection of the admittance determined by the LO driving and the RF current source due to other mixing frequency components.

12.5.3 Calculation of Mixer Parameters

As mentioned above, when the RF input is applied, the voltage of all the frequency components that occur across the diode was determined by viewing the diode as a multiport. However, the voltages $\delta \mathbf{V}'$ are not the voltages at the real ports and the voltages appearing at the real ports must be obtained from the voltages $\delta \mathbf{V}'$. Now, we determine the conversion loss from the conversion impedance matrix. The conversion loss is defined as shown in Equation (12.78).

$$L = \frac{\text{Power available from source } Z_{e,1}}{\text{Power delivered to load } Z_{e,0}} \tag{12.78}$$

The element Z'_{01} of the conversion matrix represents the power ratio delivered to $Z_{e,0} + R_{s,0}$ but the real delivered power is that which is delivered to the load $Z_{e,0}$. The ratio of the power $P'_{e,0}$ delivered to $Z_{e,0} + R_{s,0}$ to the power $P_{e,0}$ delivered to the load $Z_{e,0}$ is given by Equation (12.79).

$$K_0 = \frac{P'_{e,0}}{P_{e,0}} = \frac{\text{Re}\left(Z_{e,0} + R_{s,0}\right)}{\text{Re}\left(Z_{e,0}\right)} \tag{12.79}$$

In addition, the real available power from the RF source is the power available from the source resistance $Z_{e,1}$, not from that of $Z_{e,1} + R_{s,1}$. The Norton current source $\delta I'_{sig,1}$ is converted from the Thevenin equivalent circuit in Figure 12.78.

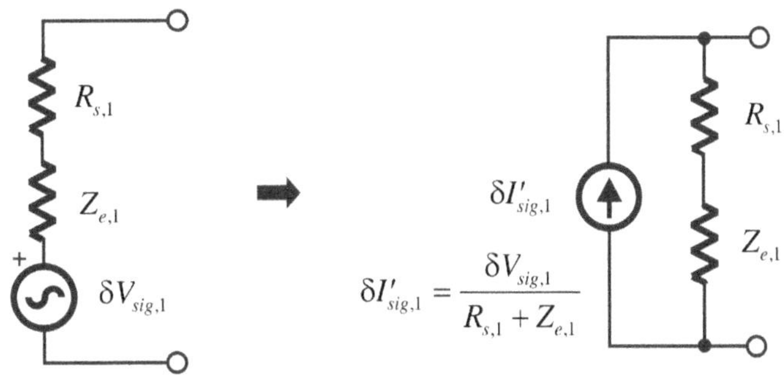

Figure 12.78 Thevenin to Norton conversion

Thus, the ratio of the available power $P_{e,1}$ from $Z_{e,1}$ to the available power from $Z_{e,1} + R_{s,1}$ is expressed in Equation (12.80).

$$K_1 = \frac{P'_{e,1}}{P_{e,1}} = \frac{\mathrm{Re}\left(Z_{e,1} + R_{s,1}\right)}{\mathrm{Re}\left(Z_{e,1}\right)} \tag{12.80}$$

Now, using $\delta V'_0 = Z'_{0,1}\delta I'_1$, and using Equations (12.79) and (12.80), the conversion loss L is computed with Equation (12.81).

$$L = \frac{1}{4\left|Z'_{0,1}\right|^2} \frac{\left|Z_{e,0} + R_{s,0}\right|^2}{\mathrm{Re}\left(Z_{e,0}\right)} \frac{\left|Z_{e,1} + R_{s,1}\right|^2}{\mathrm{Re}\left(Z_{e,1}\right)} \tag{12.81}$$

The other parameters, such as the impedances seeing into the RF and IF ports and the noise figure, can be obtained through the conversion impedance matrix. For further reading, see reference 4 at the end of this chapter.

Example 12.9

Set up the schematic as shown in Figure 12E.18 for a small-signal harmonic balance simulation. The mixer in Figure 12E.18 is the same as that in Example 12.1. Verify the small-signal frequency (**SM.freq**) and large-signal LO frequency (**HB.freq**) of the mixer and calculate the conversion loss for the RF input power swept from −30 dBm to 20 dBm.

Solution

SS_FREQ specifies the small-signal frequency in the Harmonic Balance simulator with the entry of the large-signal frequency **Freq**[1]. **SS_FREQ** corresponds to the difference between the RF and LO frequencies. Using the variable **IF** in **VAR1**, **SS_FREQ** is specified as 350 MHz. The power that corresponds to the small-signal frequency is specified as **P_LSB** or **P_USB**, which represents the

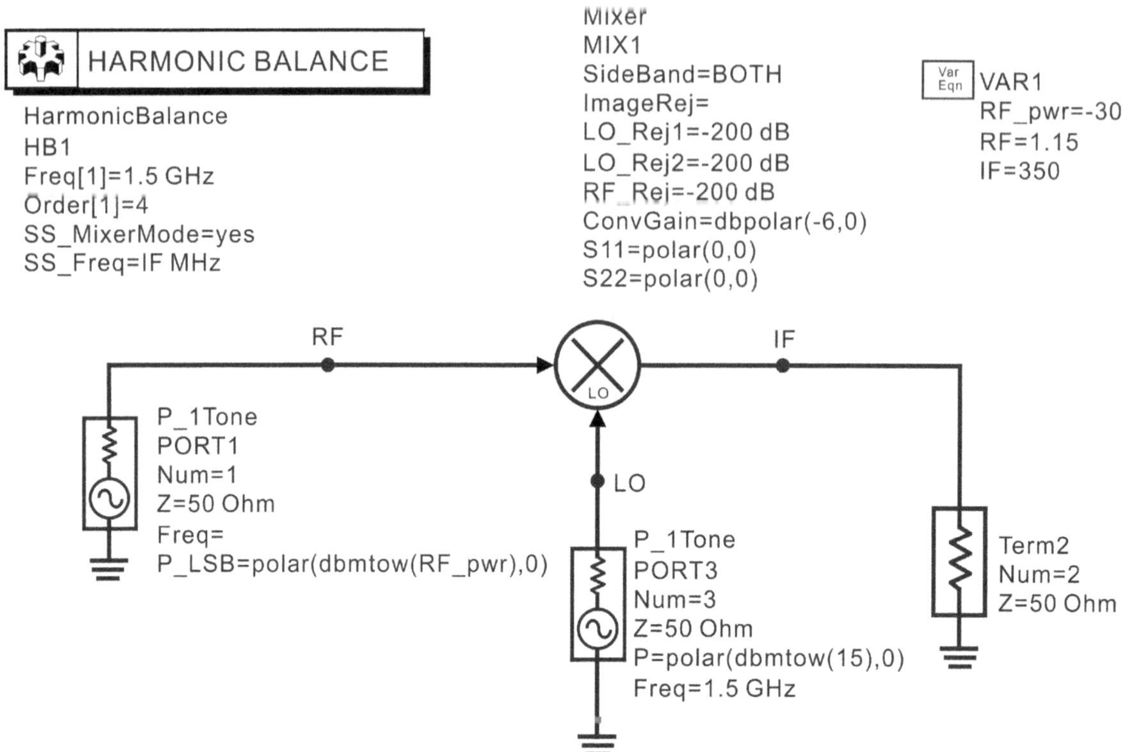

Figure 12E.18 Small-signal harmonic balance simulation. Note that the LO signal is the only large signal. **SS_FREQ** represents the IF frequency. Since the RF port is specified **P_LSB** (lower side band), the RF frequency is lower than the LO frequency by **SS_FREQ.**

small-signal power. Note that in the case of the small-signal frequency RF source, the **Freq** entry is specified as blank. Also note that the small-signal frequency is below the LO frequency because the RF power is specified using the **P_LSB** of the RF source. Thus, the RF frequency is 1.5 – 0.355 = 1.15 GHz. The simulation circuit is set up to reflect this and the simulation is performed. The frequency variables **HB.freq** and **SM.freq** appear in the dataset and are shown in Table 12E.5.

Table 12E.5 SM.freq and HB.freq

SM.freq	HB.freq
350.0 MHz	0.0000 MHz
1.150 GHz	1.500 GHz
1.850 GHz	3.000 GHz
3.350 GHz	4.500 GHz
4.150 GHz	6.000 GHz
4.850 GHz	
5.650 GHz	

Here, **HB.freq** represents the harmonics of the LO and the list of **SM.freq** shows the small-signal frequencies generated by the RF input.

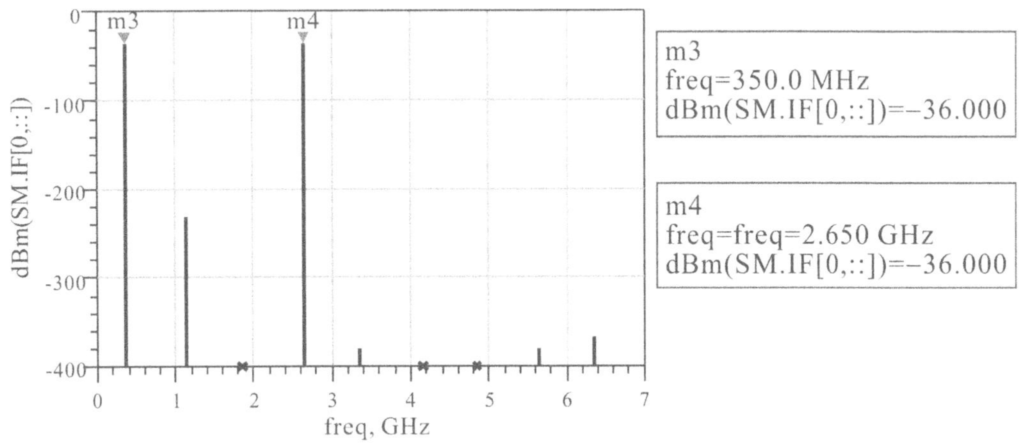

Figure 12E.19 Small-signal IF output spectrum

The small-signal output spectrum is shown in Figure 12E.19. To represent the small-signal output, **SM** is attached to the front of the variable name. Note that the spectrum of a variable with **SM** shows only the small-signal frequency components. Considering that the RF input power is set to −30 dBm, it can be seen that the conversion loss of the up- and down-converted IF output is −6 dB. Next, the RF input is swept from −30 dBm to 20 dBm and the down-converted IF output power is shown in Figure 12E.20. However, that figure does not show the 1-dB compression point at 0 dBm as in Example 12.1 because this was basically a small-signal simulation. Thus, the 1-dB compression point and the TOI cannot be found from the small-signal harmonic balance simulation.

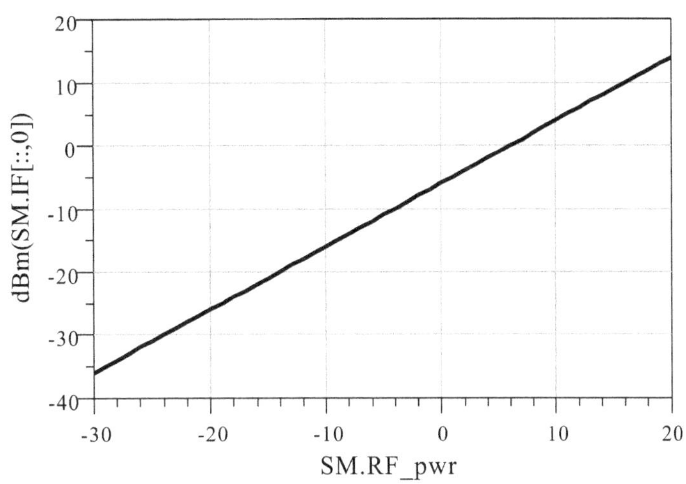

Figure 12E.20 The small-signal IF output power versus the RF input power

Example 12.10

For the SEM circuit in Example 12.5, determine the two-port S-parameters for the RF input (1.1-GHz frequency) and the IF output (400-MHz frequency) signals using the small-signal harmonic balance simulation.

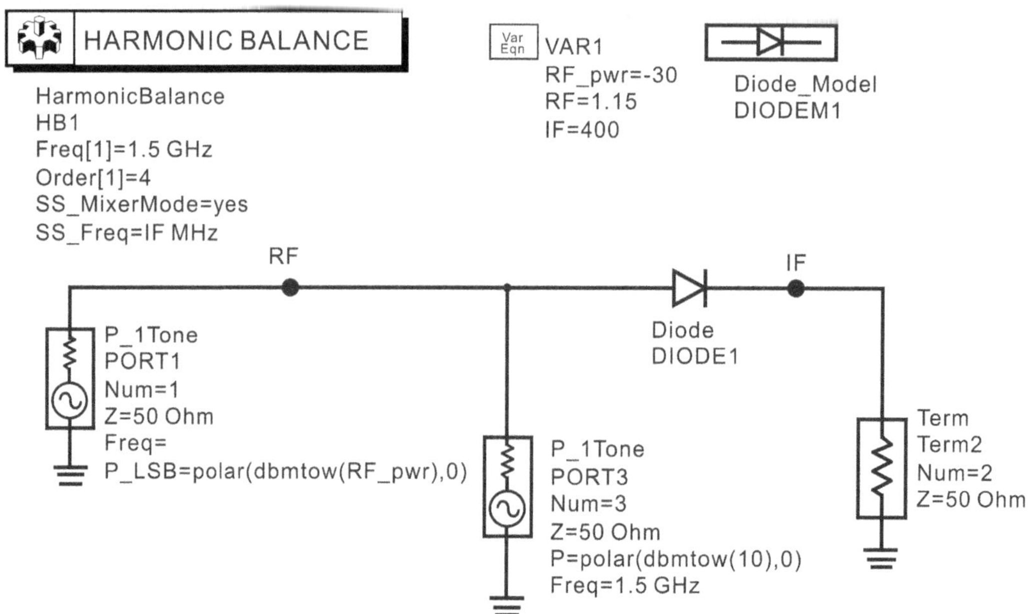

Figure 12E.21 Small-signal SEM simulation circuit

Solution

In Figure 12E.21, the frequency of the RF port is set to a 400-MHz frequency away from the LO frequency, and the RF input power is set to −30 dBm. Although the RF power is set to −30 dBm, the simulation results are independent of the RF input power. After the simulation, the small-signal IF output power is plotted as shown in Figure 12E.22. It can be seen that the conversion loss is approximately −15 dB.

In addition, the equations in Measurement Expression 12E.3 are inserted in order to calculate the S-parameters. Note that **SM.RF**[1] is the voltage corresponding to the RF frequency at the RF port, while **SM.IF**[0] represents the voltage of the IF frequency components at the IF port. Therefore, these equations will yield S_{11} and S_{21} of the two-port circuit composed of the RF and IF ports. The first equation computes the incident voltage from the RF port, while the second and third equations are for the computation of S_{11} and S_{21}.

```
Eqn e_rf=sqrt(dbmtow(-30)*2*50)
Eqn s_11=(SM.RF[1]-e_rf)/e_rf
Eqn s_21=(SM.IF[0])/e_rf
```

Measurement Expression 12E.3 Equations for calculating S-parameters

Entering these equations in the display window gives the results in Table 12E.6. Note that dB($|S_{21}|$) is equal to the conversion loss.

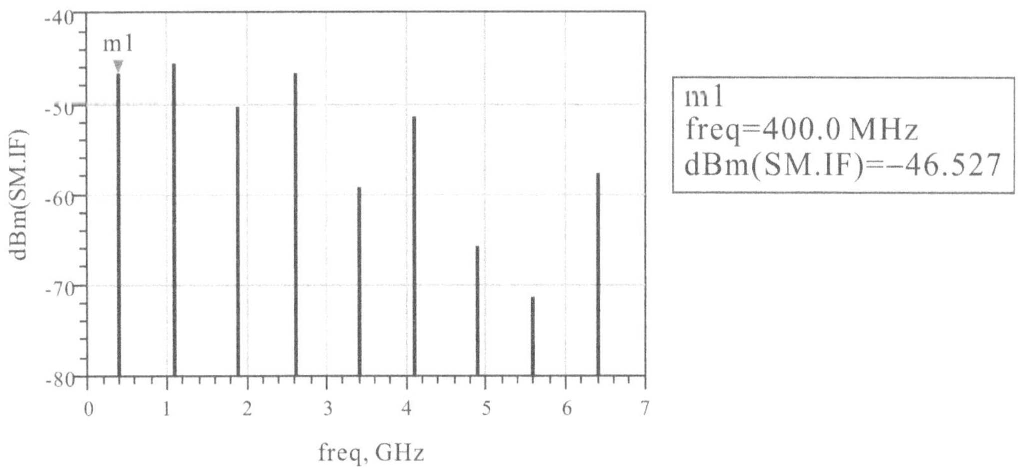

Figure 12E.22 Small-signal IF output spectrum

Table 12E.6 Computed results of S11 and S21

S_{11}	S_{21}
0.084 / 180.000	0.149 / 4.808E-15

Next, in order to determine S_{12} and S_{22}, a source is supplied to the IF port and the schematic is set up as shown in Figure 12E.23. Thus, S_{12} and S_{22} can be determined through the simulation. It must be noted that **SS_Freq** is set to 1100 MHz in order to set the IF frequency to 400 MHz.

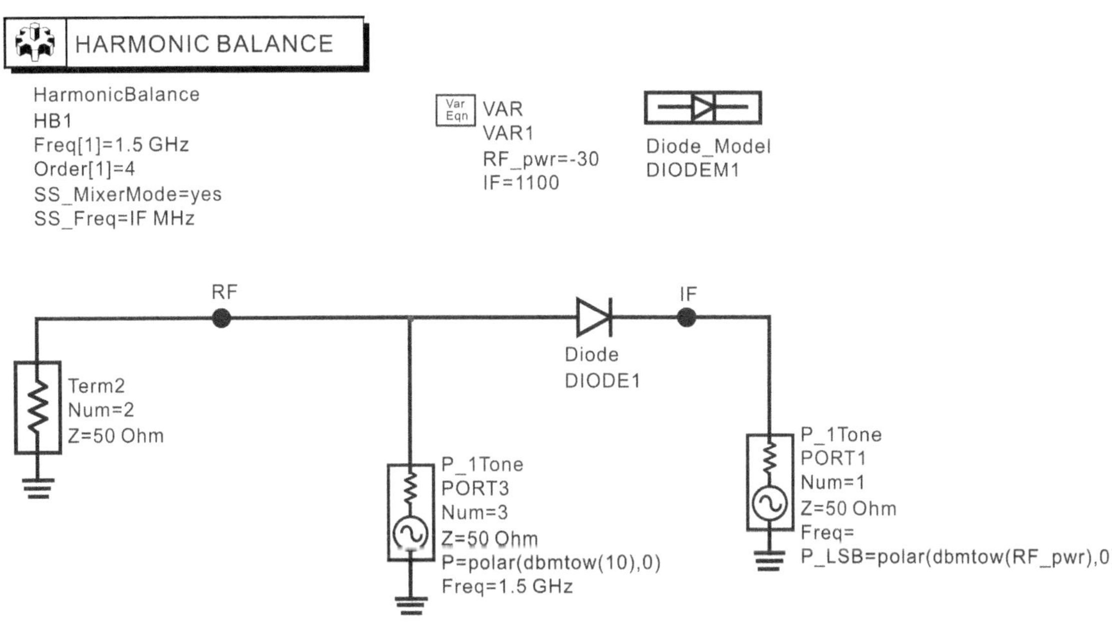

Figure 12E.23 Small-signal harmonic balance simulation; a source is applied to the IF port.

After the simulation, the equations shown in Measurement Expression 12E.4 are similarly entered in the display window to compute the S-parameters as in the computation of S_{11} and S_{21}.

```
Eqn e_if=sqrt(dbmtow(-30)*2*50)
Eqn s_22=(SM.RF[1]-e_rf)/e_if
Eqn s_12=(SM.IF[0])/e_if
```

Measurement Expression 12E.4 Equations for calculating S-parameters

The results are shown in Table 12E.7. Note that $dB(|S_{21}|) = dB(|S_{12}|)$, which corresponds to the conversion loss. In addition, since $S_{12} = S_{21}$, the mixer is found to be a passive network.

Table 12E.7 Computed results of S_{22} and S_{12}

S_{22}	S_{12}
0.662 / 4.791E-30	0.149 / 4.808E-15

As a result, the S-parameter of the mixer is

$$S = \begin{pmatrix} 0.084\angle 180^\circ & 0.149\angle 0^\circ \\ 0.149\angle 0^\circ & 0.662\angle 0^\circ \end{pmatrix}$$

Using the two-port S-parameters, complex matching may give the minimum insertion loss, but it must be noted that when the complex matching circuit is inserted, the given S-parameters change due to the loading effects of embedding impedances. Thus, it is not easy to improve the conversion loss using S-parameters.

12.6 SUMMARY

- An ideal mixer is a component that translates the RF input frequency up or down by the LO frequency. The mixer has three ports: RF, LO, and IF. The input is an RF port and the output is an IF port. The LO port is for the LO signal application.
- The mixer specifications are conversion loss, 1-dB compression, TOI, isolations, and port return losses. The conversion loss is defined as the ratio of the available power at the RF port to the delivered power of an up- or down-converted frequency component at the IF output port. Isolation is defined as the ratio of the available power at the RF and LO ports to the leakage power at the other ports. There are L–R, L–I, and R–I isolations. Since the mixer is a nonlinear component, port return losses are defined using the reflected powers of the same frequency components as the incident signals.
- A Schottky diode is composed of a metal semiconductor contact. The current characteristic of a Schottky diode is based on the majority electron diffusion in the semiconductor depending on the applied voltage, which is different from the *pn* junction

diode based on the minority carrier diffusion. As a result, the minority carrier diffusion capacitance that degrades high-frequency switching performance does not appear in a Schottky diode. As the area of a Schottky diode becomes smaller, the depletion capacitance becomes smaller and thus a Schottky diode with a small area can be applied to high-frequency mixer.

- Schottky diodes and transistors are widely used in mixers. A Schottky diode in a mixer operates as a switch that turns on or off according to the LO signal polarity. However, both operations are possible with a transistor, which can be used as a switch or a nonlinear device that mixes two applied signals.
- Three types of mixers, SEMs, SBMs, and DBMs, are introduced. The SEM employs one mixing device such as a diode or transistor. The conversion loss of the SEM can be improved by appropriate matching; however, it has no capabilities for the elimination of spurious signals. As a result, its dynamic range is narrower than other types of mixers. Isolation depends on using filters in SEMs.
- A triple-wound toroidal transformer is introduced. Its operation is similar to that of a magic-T or rat-race ring hybrid. The use of a balun is also explained.
- In an SBM, a balanced RF signal that is a result of the switching by the LO signal appears at the IF output. Thus, the mixing terms due to the even-order harmonics of the LO disappears at the IF output. Due to the disappearance of the mixing terms of the LO's even-order harmonics, the dynamic range is improved compared with an SEM. Due to the hybrids used in SBMs, good isolation between the LO and RF ports in the SBM is achieved.
- A DBM has good isolation between ports. The LO–RF, LO–IF, and RF–IF ports are naturally isolated without employing filters. In addition, the spurious signals occurring from the LO and RF even-order harmonic mixing terms do not appear at the IF output. As a result, a DBM has the widest dynamic range when compared with other types of mixers. However, due to its larger number of diodes, increased LO power is necessary for driving the diodes.
- Since the diode in a mixer does not exactly operate as a switch, the exact analysis of a mixer requires harmonic balance simulation. Since the RF signal power is small compared with the LO power, the small-signal harmonic balance is usually an efficient method for computing the mixer performances. LO analysis with other small-signal RF and IF sources set to zero gives a time-varying operating point for a diode. When a small-signal RF is applied to the LO-driven time-varying network, the frequency components generated in the mixer are $m\omega_p \pm \omega_o$. The LO frequencies ω_p and ω_o are the IF frequency.
- The diode can be seen as a diode multiport circuit with a frequency component $m\omega_p \pm \omega_o$. Using the diode multiport representation, the mixer circuit becomes a linear circuit and the conversion matrix can be formed. Using the conversion matrix, the performance of the mixer can be computed.
- The diode multiport analysis can be carried out using a small-signal harmonic balance simulation in ADS. The small-signal harmonic balance simulation gives a faster and more accurate solution for the mixer circuit than does a harmonic balance simulation.

REFERENCES

1. R. S. Muller and T. I. Kamins, *Device Electronics for Integrated Circuits*, 2nd ed. New York: John Wiley & Sons, Inc., 1986.
2. S. A. Maas, *Microwave Mixers*, Dedham, MA: Artech House, Inc., 1993.
3. H. C. Torrey and C. A. Whitmer, *Crystal Rectifiers*, New York: McGraw Hill, 1948.
4. D. N. Held and A. R. Kerr, "Conversion Loss and Noise of Microwave and Millimeter Wave Mixers: Part 1-Theory," *IEEE Transactions on Microwave Theory and Techniques*, vol. 26, no. 2, pp. 49–55, February 1978.
5. D. N. Held and A. R. Kerr, "Conversion Loss and Noise of Microwave and Millimeter Wave Mixers: Part 2-Experiment," *IEEE Transactions on Microwave Theory and Techniques*, vol. 26, no. 2, pp. 55–61, February 1978.

PROBLEMS

12.1 A two-tone signal with a –20-dBm power per tone, 200-kHz spacing, and a 2.4-GHz center frequency is applied to the RF port and, given that the observed IF output is as shown in Figure 12P.1, answer the following questions:

(1) What is the conversion loss?

(2) What is the third-order intercept (TOI) point?

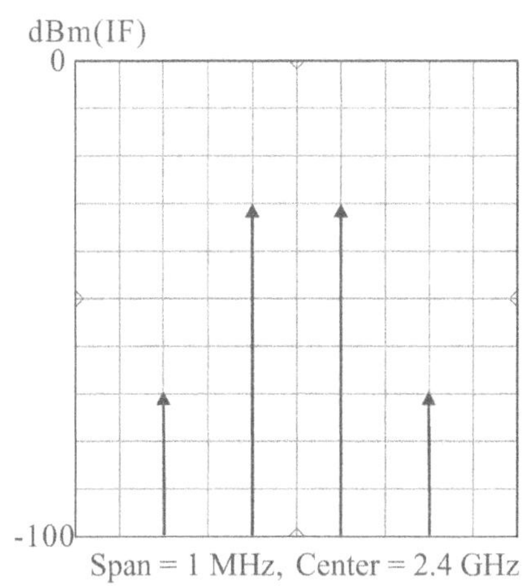

Figure 12P.1 IF output spectrum for problem 12.1

12.2 Figure 12P.2 shows the SEM where the two-diode polarity is changed. Assuming the diode turns on and off exactly for a half cycle of the LO, draw the two switching functions S_P and S_N. Also prove that

$$S_N(t) = S_p(t - \tfrac{1}{2}T)$$

Note that the Fourier series coefficients of $S_N(t)$ and $S_P(t)$, $S_{N,n}$ and $S_{P,n}$ have the following relationship:

$$S_{N,n} = \begin{cases} -S_{P,n} & n = \text{odd number} \\ S_{P,n} & n = \text{even number} \end{cases}$$

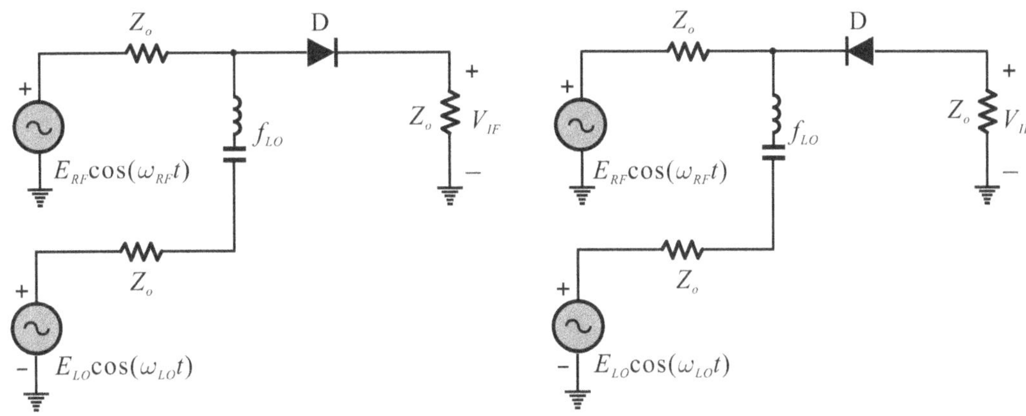

Figure 12P.2 Circuits for problem 12.2

12.3 Using the circuit shown in Figure 12P.3, compute S_{32}, S_{12}, and S_{42}.

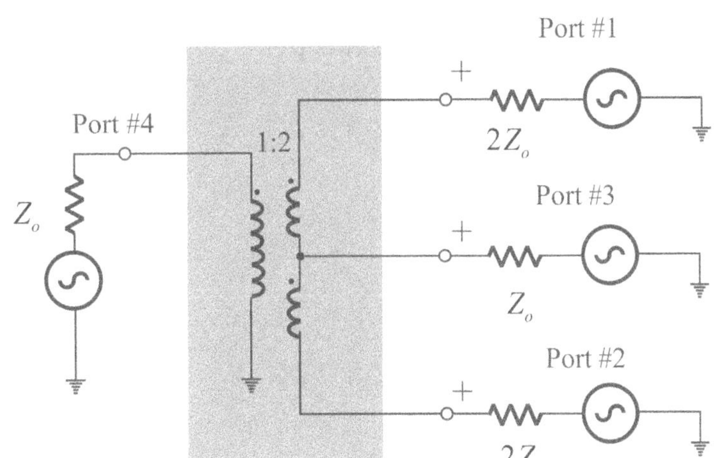

Figure 12P.3 Circuit for problem 12.3

12.4 The circuit shown in Figure 12P.4 must be operated as an SBM. Connect the RF and LO signals to the circuit and place two identical diodes, taking into consideration their directions. In addition, explain whether the circuit would be an SEM when the diodes' directions are specified differently.

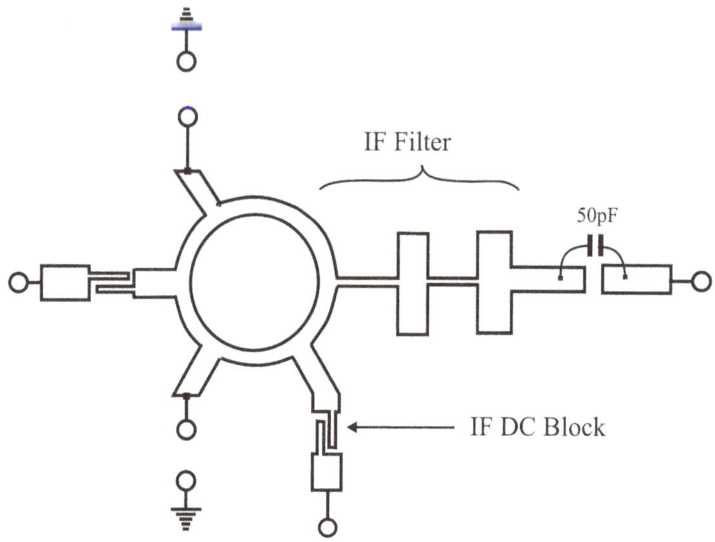

Figure 12P.4 Circuit for problem 12.4

12.5 Figure 12P.5 shows a waveguide single-balanced mixer. Explain its operation and also explain why it is a single-balanced mixer.

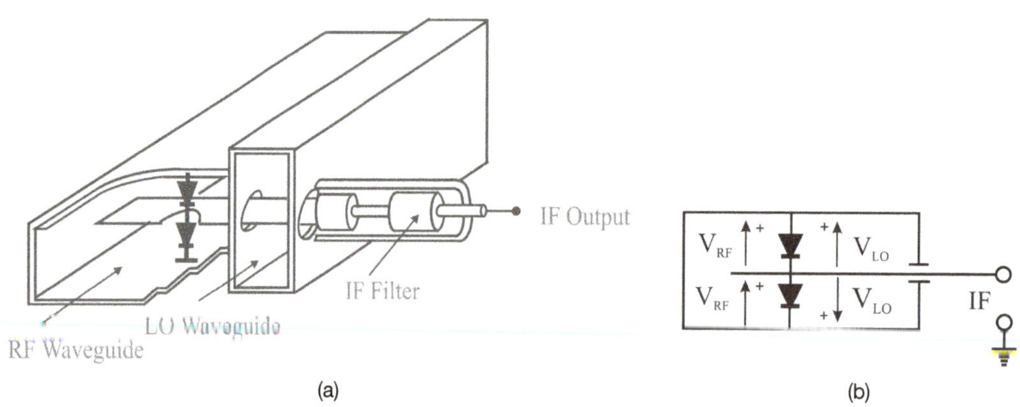

Figure 12P.5 A cross bar SBM: (a) structure, (b) electric field near the diode

12.6 For a low-frequency mixer, the differential amplifier mixer shown in Figure 12P.6 is frequently used and is much easier to implement in the form of integrated circuit. Explain how this differential amplifier becomes an SBM when the LO and RF signals are applied as shown below. In addition, when the LO and RF inputs are interchanged, is the mixer still an SBM?

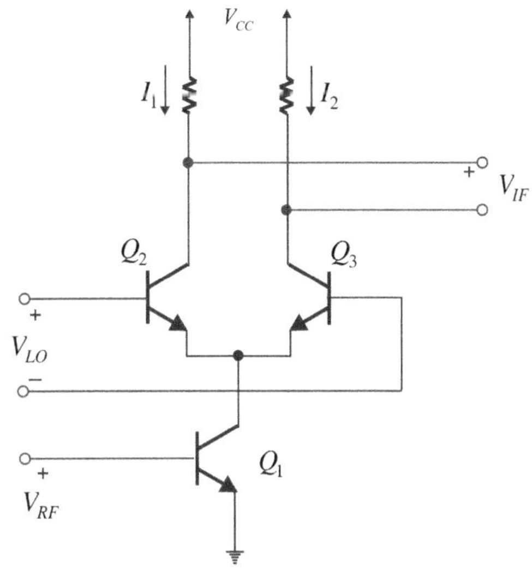

Figure 12P.6 Differential amplifier mixer for problem 12.6

12.7 Figure 12P.7 shows a double-balanced mixer that uses a differential amplifier called a Gilbert cell. Explain why this is a double-balanced mixer.

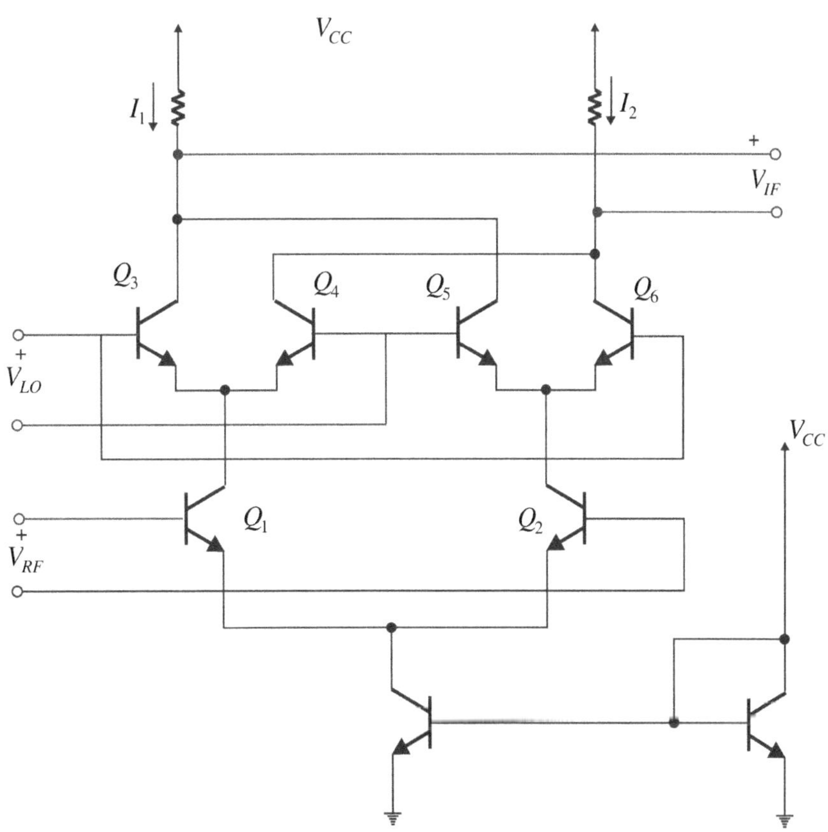

Figure 12P.7 Double-balanced mixer for problem 12.7

12.8 Figure 12P.8 shows a mixer that uses a dual-gate FET. Describe the operation of this mixer and explain why it is an SEM.

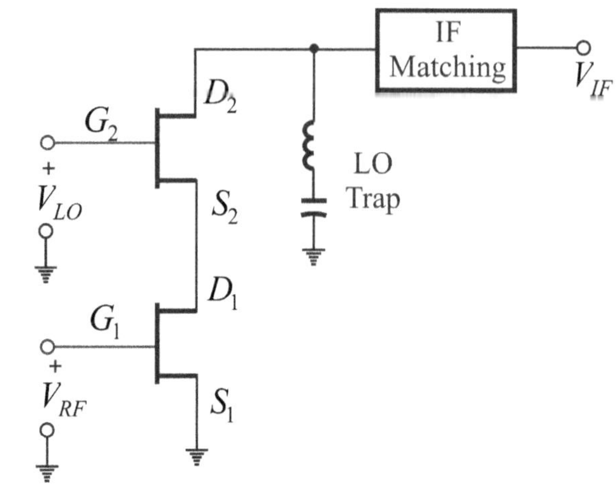

Figure 12P.8 Dual-gate FET for problem 12.8

12.9 Derive Equation (12.81).

A
Appendix

CHAPTER OUTLINE

A. Units
B. Cascaded Structure
C. Half-Wave Rectifier Analysis Using Mathcad
D. Large-Signal Impedance and Reflection Coefficient
E. Mathematical Analysis of Negative Resistance
F. Oscillation Conditions Based on Reflection Coefficients

A. UNITS

The most common measurement units in microwave engineering are decibels (dB) and decibel-related units. Most microwave components are commonly specified using powers, the level of which varies from 10^{-13}–10^{4} W. The change in the power level is found to be very large. The power change becomes arithmetic when expressed using decibels. In addition, a power level measured in decibels can be conveniently plotted on a graph. Also, by expressing power levels in decibels, the multiplication operation is converted into addition or subtraction, which is very convenient. We will explain the definitions of the decibel (dB) and its related units in this section. The decibel as a comparison of two power levels of the same frequency is defined as shown in Equation (A.1).

$$10\log\left(\frac{P_2}{P_1}\right)(\text{dB}) \tag{A.1}$$

Using this decibel definition, the insertion gain expressed in decibels is $20\log|S_{21}|$, which is expressed in Equation (A.2)

$$|S_{21}|^2 = \frac{P_L}{P_{av}} \tag{A.2}$$

where P_{av} and P_L are the available power from the source and the delivered power to the load, respectively. A decibel as a comparison of power levels has no dimension. However, a decibel can be used to express power; thus, dB-milliwatt (dBm) is defined in Equation (A.3).

$$\text{dBm} = 10\log\left(\frac{P}{1\text{mW}}\right) \quad \text{(A.3)}$$

Therefore, 0 dBm and 30 dBm represent 1 mW and 1 W, respectively.

Example A.1

Express the powers of –10 dBm and 10 dBm using power in watts.

Solution

Substituting the given values into Equation (A.3), these powers are 0.1 mW and 10 mW, respectively.

Similarly, voltage V in decibels can be defined as shown in Equation (A.4).

$$\text{dB}\mu\text{V} = 20\log\left(\frac{V}{1\mu\text{V}}\right) \quad \text{(A.4)}$$

As the quantity of the comparison is voltage, it is expressed as 20log(•). When the value of the resistance across which the voltage appears is given, the unit dBμV can be converted to dBm or dBW. In addition, it must be noted that V and 1 μV in Equation (A.4) represent root mean square (RMS) values.

Example A.2

Change –10 dBm and 10 dBm into dBμV quantities across a 50-Ω load.

Solution

–10 dBm, which is 0.1 mW, corresponds to 10^{-4} W. As a result, calculating the RMS voltage,

$$V = \sqrt{RP} = \sqrt{50 \times 10^{-4}} = 0.07$$

which, when expressed in dBμV, is 20log(0.07 × 10^6) = 97 dBμV.

In addition, power with different frequency components appears in a spectrum, the ratios of which are also expressed in decibels. However, because decibels are used to compare the power levels of the same frequency, this is expressed in Equation (A.5) using dBc.

$$10\log\left(\frac{P_2}{P_1}\right)\ (\text{dBc}) \quad \text{(A.5)}$$

It should be noted that here, P_1 and P_2 have different frequencies.

Example A.3

From the spectrum shown in Figure A.1, express in decibels the ratio of the center frequency power to the power of the adjacent sideband on the right.

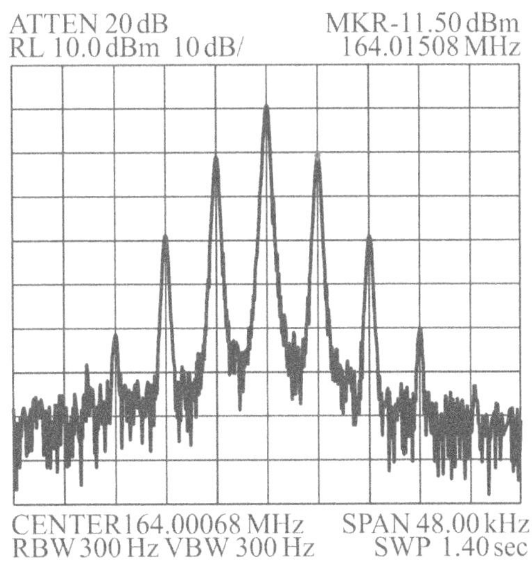

Figure A.1 Spectrum

Solution

Since one division is 10 dB, the ratio can be seen to be about 10 dBc.

B. CASCADED STRUCTURE

A cascaded structure is commonly used when constructing a microwave system. As the processed output of a front stage is carried over to the next stage as input, a cascaded structure allows separate designs for each component, which is easier than designing one big system. Commonly evaluated terms in a cascaded system include the gain, noise figure, 1-dB compression point, and third-order intercept (TOI). When individual blocks are connected in cascade to form a system, the system parameters previously mentioned are determined by those of the individual blocks. In this section, we will look at how the system parameters are determined by those of the individual blocks.

The total gain G_t for an amplifier composed of the n-stage amplifiers connected in cascade as shown in Figure B.1 becomes Equation (B.1).

$$G_t = G_1 G_2 \cdots G_n \tag{B.1}$$

Expressed in a decibel scale, the gain is given by Equation (B.2).

$$G_t(\text{dB}) = G_1(\text{dB}) + \cdots + G_n(\text{dB}) \tag{B.2}$$

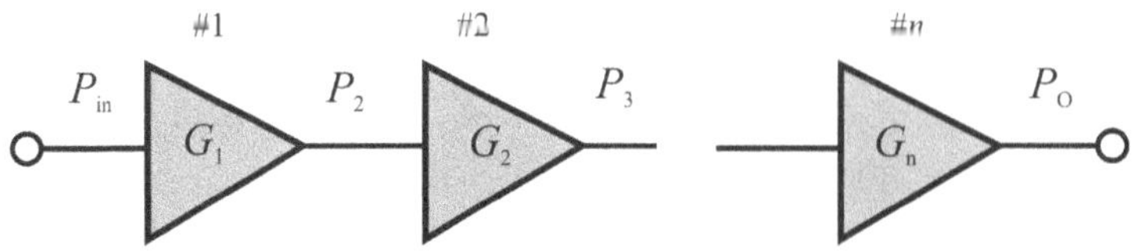

Figure B.1 An n-stage cascaded amplifier

The total gain is simply determined by addition where the output power P_o, when an input power P_{in} is applied to the input of the amplifier, is expressed in Equation (B.3),

$$P_o = G_t P_{in} = G_1 G_2 \cdots G_n P_{in} \tag{B.3}$$

which is expressed in a decibel scale and given by Equation (B.4).

$$\begin{aligned} P_o(\mathrm{dBm}) &= G_t(\mathrm{dB}) + P_{in}(\mathrm{dBm}) \\ &= G_1(\mathrm{dB}) + \cdots + G_n(\mathrm{dB}) + P_{in}(\mathrm{dBm}) \end{aligned} \tag{B.4}$$

Thus, the power levels and gains expressed in dBm at each stage can be determined from simple addition.

Example B.1

For an input power of –100 dBm, calculate the output power for the cascaded structure shown in Figure B.2.

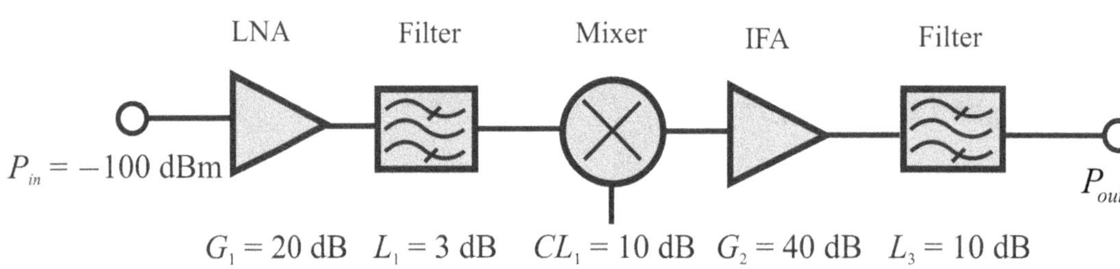

Figure B.2 Example of a cascaded structure

Solution

The total gain is $G_t = G_1 - L_1 - CL_1 + G_2 - L_3 = 20 - 3 - 10 + 40 - 10 = 37$ dB. Thus, $P_{out} = P_{in} + 37 -$ –63 dBm.

The total noise figure of the amplifier composed of the cascaded amplifiers shown in Figure B.3 has already been discussed in Chapter 4. The total noise figure F_t is given by Equation (B.5).

$$F_t = F_1 + \frac{F_2 - 1}{G_1} + \frac{F_3 - 1}{G_1 G_2} + \cdots \tag{B.5}$$

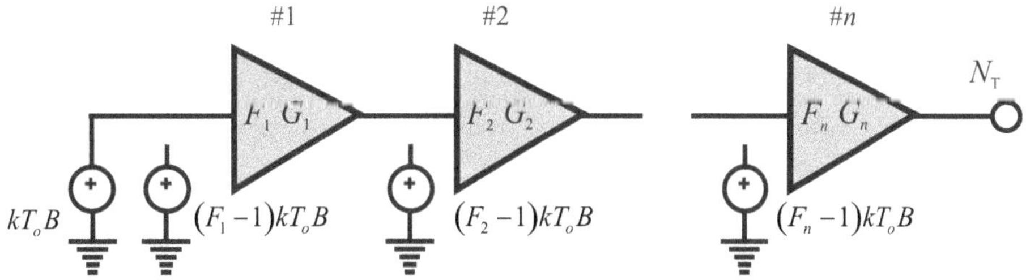

Figure B.3 An n-stage cascaded amplifier

It should be noted that because the gain and noise figure of Equation (B.5) are not expressed in the decibel scale, the gain and noise figure given in decibels must first be converted to a linear scale before using Equation (B.5) for the calculation.

A similar relationship can be derived for the 1-dB compression point. First, plotting a normalized output power with respect to the input power gives the plot shown in Figure B.4. That is, the output power decreases as the input power increases, which can be expressed as

$$\frac{P_o}{GP_i} = 1 - \delta P_i = 1 - \varepsilon P_o \tag{B.6}$$

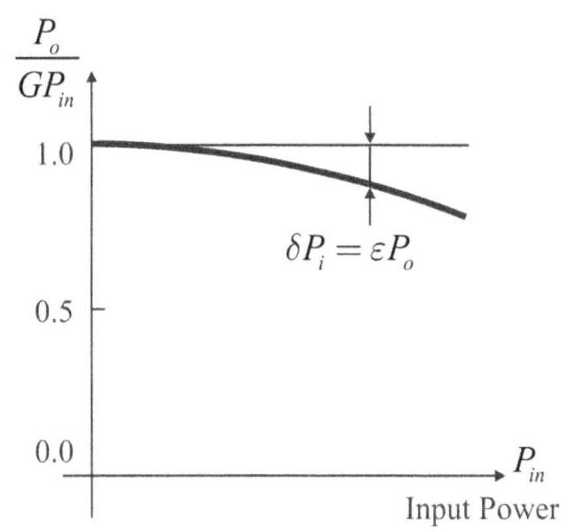

Figure B.4 The normalized output power for an input power

The left-hand term of Equation (B.6) represents the normalized output power. This is expressed in decibels by taking the logarithm of both sides of this equation, which gives the relationship shown in Equation (B.7).

$$10\log\frac{P_o}{GP_i} = 10\log(1 - \varepsilon P_o) \tag{B.7}$$

Using $\log(1 - x) = 1 - (\log_{10} e)x = 1 - kx$, the 1-dB compression point is given by Equation (B.8).

$$-1 = -k\varepsilon P_o = -k\varepsilon P_{1dB} \tag{B.8}$$

Thus, ε in equation (B.6) is shown in Equation (B.9).

$$\varepsilon = \frac{1}{kP_{1dB}} \tag{B.9}$$

Substituting Equation (B.9) into Equation (B.6), the change of the normalized output power with respect to the output power can be expressed in terms of P_{1dB} with Equation (B.10).

$$\frac{P_o}{GP_i} = 1 - \frac{P_o}{kP_{1dB}} \tag{B.10}$$

The overall 1-dB compression point in a cascaded structure can be determined using Equation (B.10). The gain and the 1-dB compression point defined for the n-cascaded amplifiers are shown in Figure B.5, where each n-th stage input power is denoted as P_n and thus the final-stage output power becomes P_{n+1}. Applying Equation (B.10) from stage 1 to the n-stage,

$$\begin{aligned} \frac{P_2}{G_1P_1} &= 1 - \frac{P_2}{kP_{1dB,1}} \\ \frac{P_3}{G_2P_2} &= 1 - \frac{P_3}{kP_{1dB,2}} \\ &\vdots \\ \frac{P_3}{G_nP_n} &= 1 - \frac{P_{n+1}}{kP_{1dB,n}} \end{aligned} \tag{B.11}$$

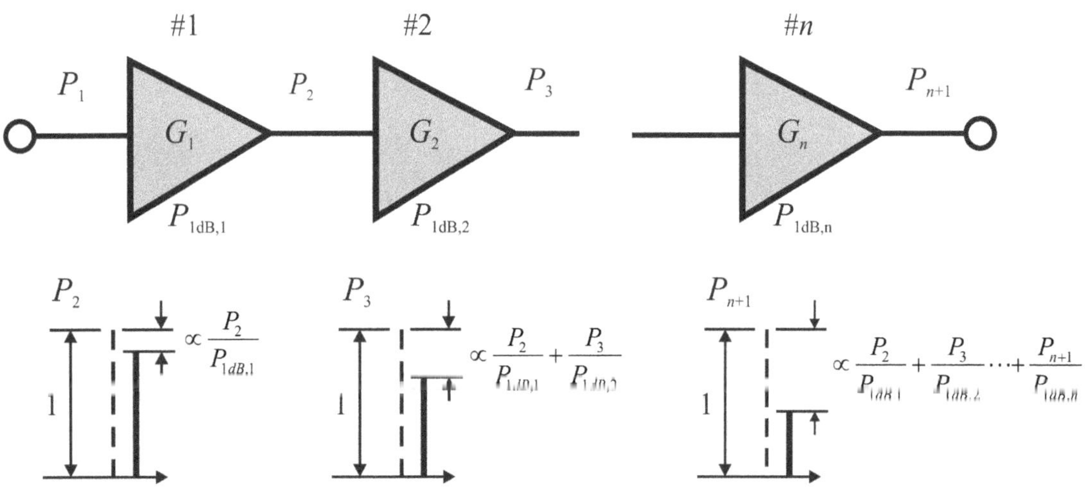

Figure B.5 Output power at each stage in an n-stage cascaded amplifier

Multiplying both sides of the Equation (B.11), we obtain Equation (B.12).

$$\frac{P_{n+1}}{G_1G_2\cdots G_nP_1} = \left(1-\frac{P_2}{kP_{1dB,1}}\right)\left(1-\frac{P_3}{kP_{1dB,2}}\right)\cdots\left(1-\frac{P_{n+1}}{kP_{1dB,n}}\right)$$
$$\cong 1-\frac{1}{k}\left\{\frac{P_2}{P_{1dB,1}}+\frac{P_3}{P_{1dB,2}}\cdots\frac{P_{n+1}}{P_{1dB,n}}\right\} \tag{B.12}$$

Replacing all P_2, P_3, ... on the right-hand side of the equation by P_{n+1}, we obtain Equation (B.13).

$$\frac{P_{n+1}}{G_1G_2\cdots G_nP_1} \cong 1-\frac{P_{n+1}}{k}\left\{\frac{1}{G_2\cdots G_nP_{1dB,1}}+\frac{1}{G_3\cdots G_nP_{1dB,2}}\cdots+\frac{1}{P_{1dB,n}}\right\} \tag{B.13}$$

Thus, comparing Equations (B.10) and (B.13), we obtain the result expressed in Equation (B.14).

$$\frac{1}{P_{1dB,t}} = \frac{1}{G_2\cdots G_nP_{1dB,1}}+\frac{P_3}{G_3\cdots G_nP_{1dB,2}}\cdots+\frac{1}{P_{1dB,n}} \tag{B.14}$$

From Equation (B.14), the total 1-dB compression point of the cascaded structure is dominantly determined by the final stage P_{1dB}. The 1-dB compression points of the front stages slightly degrade the total 1-dB compression point determined by the final-stage P_{1dB}. The total 1-dB compression point is most seriously degraded by the stage just before the final stage. Therefore, in order to preserve the final stage, P_{1dB} should satisfy the relation $G_nP_{1dB,(n-1)} >> P_{1dB,n}$.

Equation (B.14) can be interpreted differently using Figure B.5. If the first-stage amplifier is linear, P_2/G_1P_1 can be viewed as 1. Due to the 1-dB compression point, the ratio of P_2/G_1P_1 is reduced from 1 by a factor that is proportional to $P_2/P_{1dB,1}$. This is shown in the lower side of Figure B.5. After this reduced power passes through the second stage, the normalized output power is further reduced by $P_3/P_{1dB,2}$. Therefore, by repetition, the reduction of the normalized output power at the final stage is

$$\frac{P_2}{P_{1dB,1}}+\frac{P_3}{P_{1dB,2}}\cdots\frac{P_{n+1}}{P_{1dB,n}}$$

Considering the whole amplifier as a one-stage equivalent amplifier, this corresponds to the reduction of the normalized output power of the whole amplifier that is proportional to P_{n+1}/P_{1dB}. Equating P_{n+1}/P_{1dB} to the reduction factor above, Equation (B.14) can be obtained. In addition, it must also be noted that similar to the Frii's formula, all the quantities in Equation (B.14) are expressed in a linear scale rather than in a decibel scale.

Next, in Figure B.6 a cascaded structure for the calculation of the total TOI is shown. Similar to the calculation of P_{1dB}, applying input power to the first-stage amplifier with $P_{TOI,1}$ results in an IMD3 output occurring with adjacent sidebands, as shown in Figure B.6. When the output power P_2 is normalized to 1, as shown in the figure, the output power of the third-order intermodulation distortion (IMD3) is proportional to $P_2/P_{TOI,1}$. When the distorted output power is applied to the next stage, the IMD3 power proportional to $P_3/P_{TOI,2}$ is added by the second stage. Therefore, the total IMD3 at the second stage is proportional to

$$\frac{P_2}{P_{TOI,1}} + \frac{P_3}{P_{TOI,2}}$$

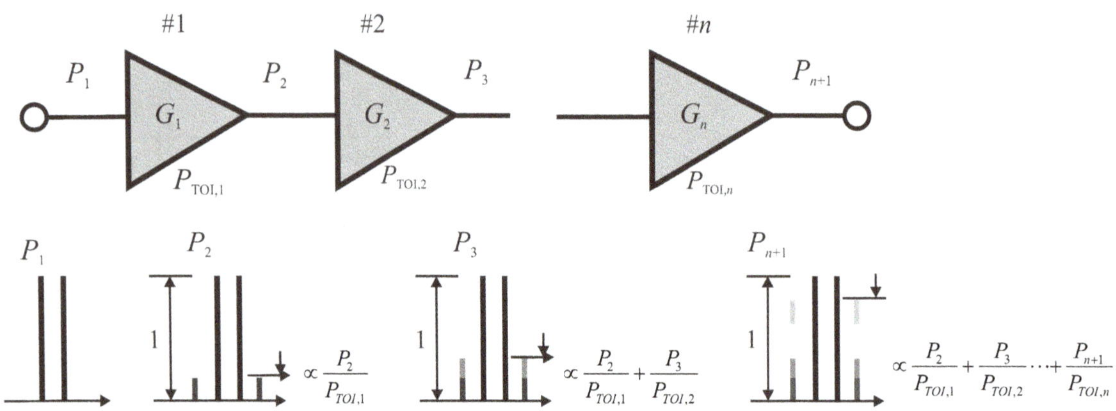

Figure B.6 A TOI in a cascaded structure

As a result, the total intermodulation output appearing after passing through the n-th stage is proportional to

$$\frac{P_2}{P_{TOI,1}} + \frac{P_3}{P_{TOI,2}} \cdots \frac{P_{n+1}}{P_{TOI,n}}$$

Converting this to an equivalent single-stage amplifier, the total TOI ($P_{TOI,t}$) obtained is

$$\frac{1}{P_{TOI,t}} = \frac{1}{G_2 \cdots G_n P_{TOI,1}} + \frac{1}{G_3 \cdots G_n P_{TOI,2}} \cdots + \frac{1}{P_{TOI,n}}$$

It can be seen that the obtained TOI has a similar property to 1-dB compression point.

Using the formulas presented above, the power, gain, noise figure, P_{1dB}, and TOI of each stage can be calculated using Microsoft Excel. However, there are numerous programs available online that can also perform these calculations.

Example B.2

Download Cascade101.xls from www.microwaves101.com and use it to calculate the total noise figure, gain, and P_{1dB} of the system shown in Figure B.7.

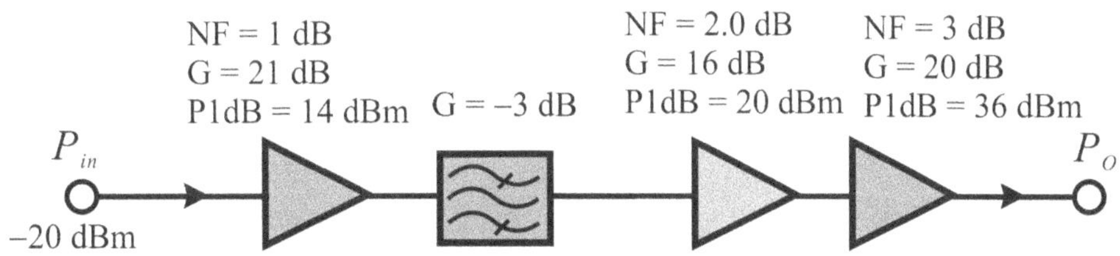

Figure B.7 Example of a cascaded structure

Solution

Enter the input power of –20 dBm in the first stage of the control sheet of the downloaded file; next, substitute the names of the components and their respective given values from the problem in the second sheet named **Component**, as shown in Table B.1.

Table B.1 Components specification

Stage	Stage Description	SSG (dB)	P1dBout (dBm)	NF(dB)
1	Amp	21.00	14.00	1.00
2	Filter	-3.00	100	3.00
3	Amp	16.00	20.00	2.00
4	Amp	20.00	36.00	3.00

Then, the results in the Cascade sheet are automatically calculated and shown in Table B.2. The cumulative gain is shown in the **SSG** column where total gain is seen to be 54 dB. The output power at each stage appears in the **Pout** column and the final output power level is seen to be 33.26 dBm. P_{1dB} at the input and output appears at the **P1dBin** and **P1dBout** columns, respectively, which represents the resulting cumulative P_{1dB} appearing at each stage. The total output P_{1dB} in this example is 34.30 dBm. Finally, the noise figure is the overall noise figure accumulating at each stage, the total of which is 1.06 dB.

Table B.2 Calculation results

Stage	Stage Description	SSG (dB)	Pout (dBm)	P1dBin (dBm)	P1dBout (dBm)	NF(dB)
1	Amp	21.00	0.96	-6.00	14.00	1.00
2	Filter	18.00	-2.04	-6.00	11.00	1.03
3	Amp	34.00	13.75	-13.79	19.21	1.06
4	Amp	54.00	33.26	-18.70	34.30	1.06

C. HALF WAVE RECTIFIER ANALYSIS USING MATHCAD

Harmonic Balance Calculation

Parameter Definition

$f := 1 \cdot 10^{3}$	Frequency
$Ip := 5a \cdot 10^{-3}$	Supply Current
$R := 1 \cdot 10^{3}$	Value of Parallel Resistor
$h := 1 \cdot 3$	Harmonic Number

Angular Frequency

$$\omega_O := 2 \cdot \pi \cdot f \quad \mathbf{T} := \frac{1}{f}$$

Function Definition

$N := 1$	Ideality factor
$Is := 10^{-14}$	Saturation current
$error := 0.01$	Tolerance
$M := 2^{5}$	Sample number
$V_T := 0.025$	Thermal voltage

$$\mathbf{f}(x) := \left| \begin{array}{l} a \leftarrow \text{Is} \cdot \left(e^{\frac{x}{N \cdot V_T}} - 1 \right) \\ a \leftarrow \text{Is} \cdot \left(e^{\frac{1}{N \cdot V_T}} - 1 \right) + \dfrac{\text{Is}}{N \cdot V_T} \cdot e^{\frac{1}{N \cdot V_T}} \cdot (x - 1) \text{ if } x \geq 1 \end{array} \right.$$

Diode current equation

$$df(x) := \left| \begin{array}{l} a \leftarrow \dfrac{\text{Is}}{N \cdot V_T} \cdot e^{\frac{x}{N \cdot V_T}} \\ a \leftarrow \dfrac{\text{Is}}{N \cdot V_T} \cdot e^{\frac{1}{N \cdot V_T}} \text{ if } x \geq 1 \end{array} \right.$$

Equation for derivative of diode current

$$g(x) := \left| \begin{array}{ll} 1 & \text{if } x = 0 \\ 2 & \text{otherwise} \end{array} \right. \qquad A1 := \left| \begin{array}{l} M \leftarrow 0 \\ \text{while } M \leq 100 \\ M \leftarrow M + 1 \\ \text{return } M \end{array} \right.$$

$V = $ $m \leftarrow 0$ — Initial definition

$a \leftarrow (m\ {-1}\ 5\ 0\ 0)$

while $m < 200$ — FFT function definition

$\quad k \leftarrow 0$

$\quad$ while $k < M - 1$

$$\qquad v_k \leftarrow \sum_{z=0}^{3} \mathrm{Re}\left(a_{m,z+1} \cdot e^{i \cdot z \cdot w_O \cdot \frac{T \cdot k}{M}} \right)$$

$\qquad V \leftarrow v_k$

$\qquad id_k \leftarrow f(V)$

$\qquad id\_diff_k \leftarrow df(V)$

$\qquad k \leftarrow k + 1$ — Definition of each harmonic current

$\quad Id \leftarrow CFFT(id)$

$\quad Id\_diff \leftarrow CFFT(id\_diff)$

$\quad n \leftarrow 0$

$\quad$ while $n \le h$

$\qquad i_n \leftarrow id_n \cdot \varepsilon(n)$

$\qquad n \leftarrow n + 1$

$\quad j \leftarrow 0$

$\quad k \leftarrow 0$

$\quad$ while $k \le h$ — Definition of the matrix F

$\qquad$ while $j \le h$

$$\qquad\quad F_{k,j} \leftarrow \left(\frac{1}{2} \cdot Id\_diff_{k+j} + \frac{1}{2} \cdot Id\_diff_{|k-j|} \right) \cdot \varepsilon(k)$$

$$\qquad\quad F_{k,j} \leftarrow \left(\frac{1}{2} \cdot Id\_diff_{k+j} + \frac{1}{2} \cdot \overline{Id\_diff_{|k-j|}} \right) \cdot \varepsilon(k) \text{ if } k - j < 0$$

$\qquad\quad j \leftarrow j + 1$ — Definition of the matrices I and G

$\qquad j \leftarrow 0$

$\qquad k \leftarrow k + 1$

$\quad n \leftarrow 0$

$\quad$ while $n \le h$

$$\qquad I_n \leftarrow -\left(i_n - F_{n,0} \cdot a_{m,1} - F_{n,1} \cdot a_{m,2} - F_{n,2} \cdot a_{m,3} - F_{n,3} \cdot a_{m,4} \right)$$

$$\qquad G_{n,n} \leftarrow \frac{1}{R}$$

$\qquad n \leftarrow n + 1$

$\quad I_1 \leftarrow I_1 + Ip$

$\quad n \leftarrow 0$

$\quad m \leftarrow m + 1$

$\quad a_{m,0} \leftarrow m$

$\quad$ while $n \le h$

$\qquad a_{m,n+1} \leftarrow 1solve(F + G, I)_n$

$\qquad n \leftarrow n + 1$

$$\quad \text{break if error} > \frac{\left(\left| Id_0 + \frac{a_{m,1}}{R} \right| \right)^2 + \left(\left| Id_1 + \frac{a_{m,2}}{R} \right| \right)^2 + \left(\left| Id_2 + \frac{a_{m,3}}{R} \right| \right)^2 + \left(\left| Id_3 + \frac{a_{m,4}}{R} \right| \right)^2}{Ip}$$

return a

Results of Calculation

$\text{Rows} := \max\left(V^{\langle 0 \rangle}\right)$

$\text{Rows} = 14$

$k := 0..2^5 \quad t_k := \frac{k}{M} \cdot T$

$$vD_K := \sum_{z=0}^{3} \text{Re}\left(V_{\text{Rows},z+1} \cdot e^{i \cdot z \cdot \omega_O \cdot \frac{T \cdot k}{M}}\right)$$

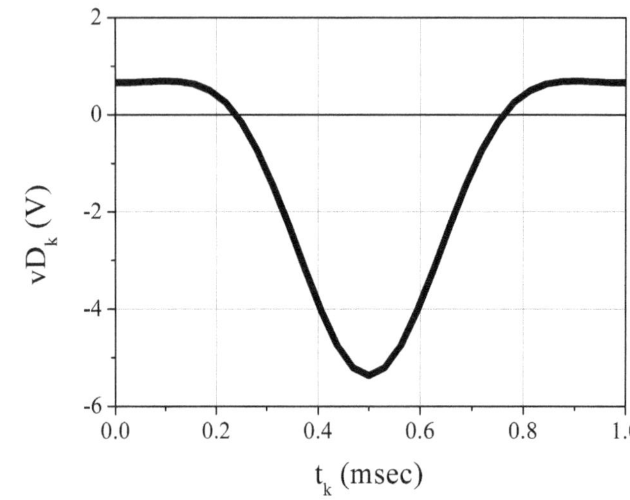

$$iD_k := Ip \cdot \cos\left(\omega_O \cdot \frac{T \cdot k}{M}\right) - \frac{vD_k}{R}$$

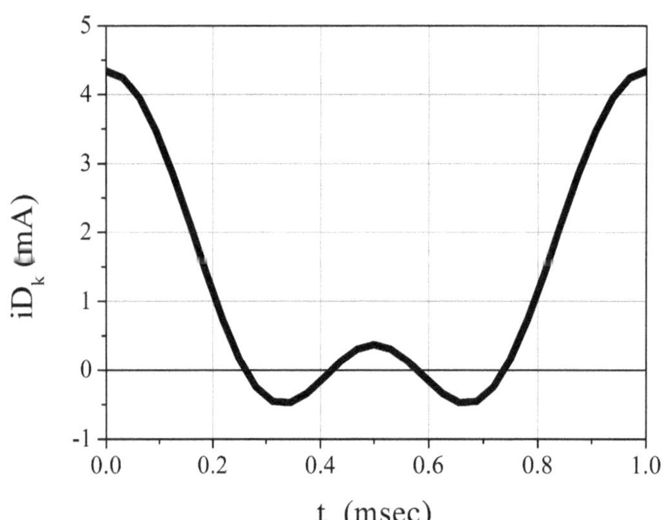

D. LARGE-SIGNAL IMPEDANCE AND REFLECTION COEFFICIENT

Large-signal impedance or admittance can be defined by extending the concept of a small-signal impedance and admittance, respectively. For the one-port nonlinear device shown in Figure D.1, a sinusoidal voltage source is applied to the input, as shown in the figure, to calculate or measure the large-signal admittance. When the input voltage expressed in Equation (D.1) is applied,

$$v(t) = V\cos(\omega t) \tag{D.1}$$

harmonic currents including DC flow through the nonlinear device as

$$i = I_{DC} + I_1\cos(\omega t + \phi_1) + I_2\cos(2\omega t + \phi_2)\cdots \tag{D.2}$$

The phase and amplitude of the fundamental current in Equation (D.2) vary as V increases. Representing the $V\cos(\omega t)$ and $I_1\cos(\omega t + \phi_1)$ by a voltage and current phasors, respectively, the large-signal admittance is defined with Equation (D.3).

$$Y_A(V) = \frac{I_1 e^{j\phi_1}}{V} \tag{D.3}$$

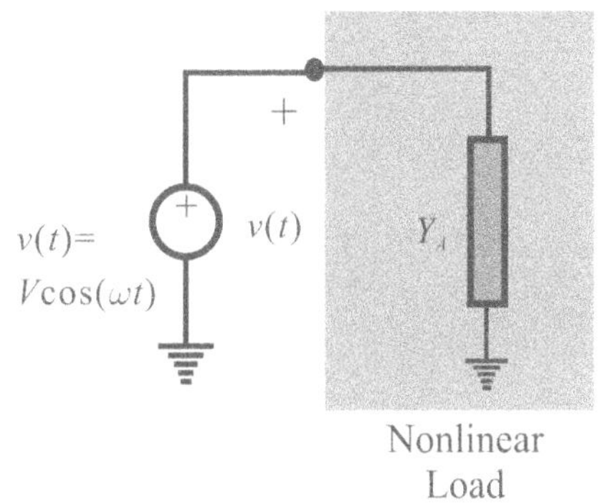

Figure D.1 Large-signal admittance measurement

It should be noted that when V is small, $Y_A(V)$ is the same as a typical small-signal admittance, which is a constant. However I_1 and ϕ_1 vary as V increases and the large-signal admittance $Y_A(V)$ depends on V, in contrast to a small-signal admittance.

Similarly, when a current source $i(t) = I\cdot\cos(\omega t)$ is applied to the input shown in Figure D.2, the harmonic voltages including the DC voltage appear across the nonlinear device as expressed in Equation (D.4).

$$v = V_{DC} + V_1\cos(\omega t + \phi_1) + V_2\cos(2\omega t + \phi_2) + \cdots \tag{D.4}$$

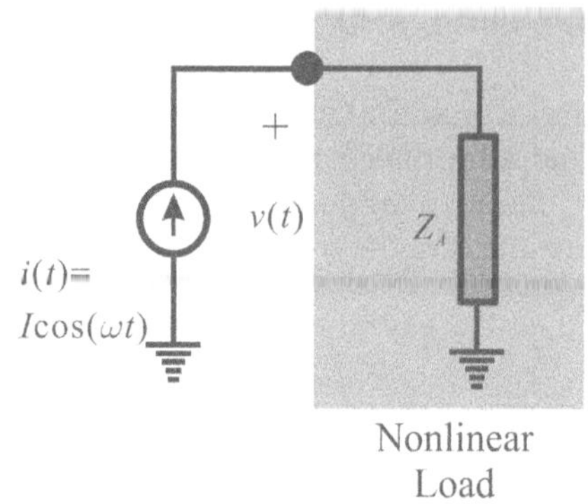

Figure D.2 Large-signal impedance measurement

The amplitude and phase of the fundamental voltage varies as I increases and the large-signal impedance is defined as the ratio of the fundamental voltage phasor to the input current phasor that is given by Equation (D.5).

$$Z(I) = \frac{V_1 e^{j\phi_1}}{I} \tag{D.5}$$

It is also worth noting that this impedance also depends on the amplitude of the current source.

Examples D.1

A diode has the following current–voltage relationship:

$$i = I_s(e^{\frac{v}{V_T}} - 1)$$

Here, V_T and I_s are constants. Calculate the large-signal admittance of this diode.

Solution

Substituting $v(t) = V{\cdot}\cos(\omega t)$ into the current–voltage relation,

$$i = I_s\left(e^{\frac{V\cos\omega t}{V_T}} - 1\right)$$

where

$$e^{\frac{V\cos\omega t}{V_T}} = I_o\left(\frac{V}{V_T}\right) + 2\sum_{n=1}^{\infty} I_n\left(\frac{V}{V_T}\right)\cos n\omega t$$

In this case, $I_n(\cdot)$represents the n-th order modified Bessel function. It can thus be found that the large-signal admittance is the conductance as

$$Y_A = G_A = \frac{2I_s}{V} I_1\left(\frac{V}{V_T}\right)$$

This can also be computed using ADS and the simulation circuit is set up as shown in Figure D.3. The amplitude of the sinusoidal source **Vac** is set to vary 0.01–0.8 V. The diode maximum current I_{max} is set to $I_{max} = 100$ A to preserve the diode exponential I–V characteristic at higher voltages. The result is shown in Figure D.4 and it can be seen that a difference arises between the theoretical and the calculated results at the lower voltages noted above. The reason is that the diode model in ADS is different from the simple exponential model expressed above.

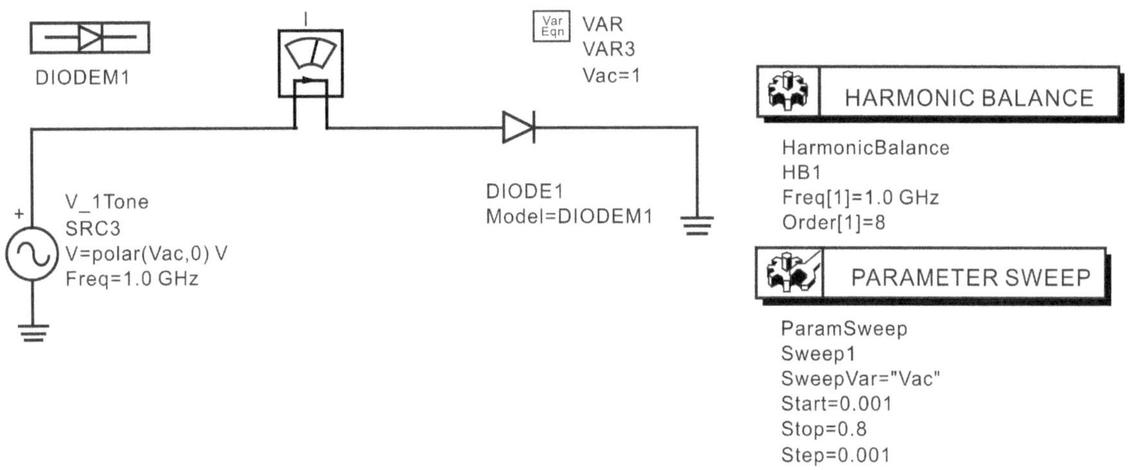

Figure D.3 Schematic for obtaining a large-signal admittance of the diode

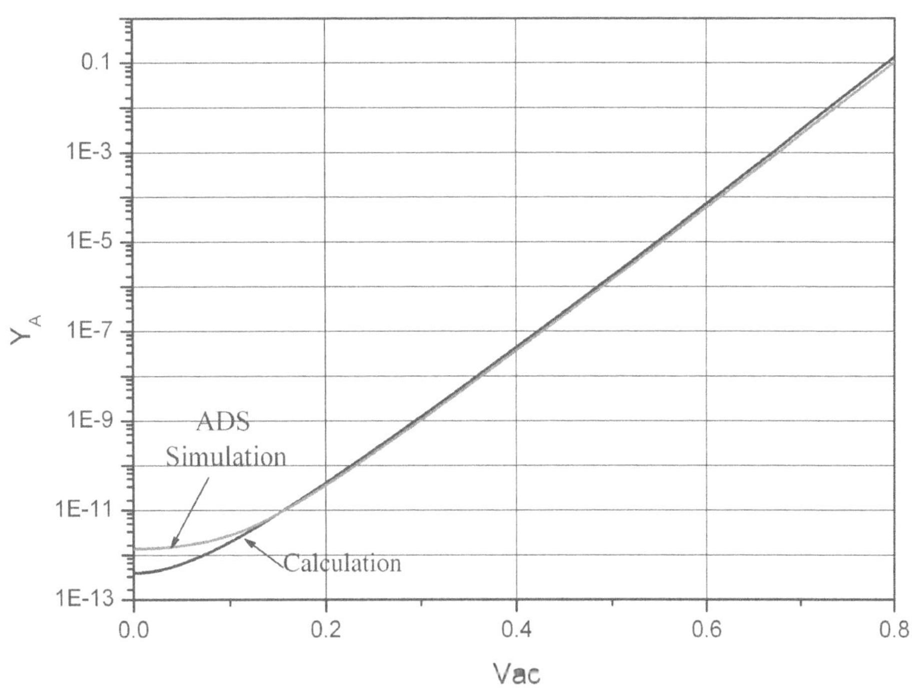

Figure D.4 Comparison of the calculated large-signal admittances of the diode

The large-signal reflection coefficient can be similarly defined with the small signal reflection coefficient. A port with internal impedance Z_o is connected to the nonlinear element in Figure D.5. By varying the magnitude of the incident power, various harmonic components of the reflected voltage, including DC, appear across the nonlinear device as expressed in Equation (D.6).

$$v = V_{DC} + V_1\cos(\omega t + \phi_1) + V_2\cos(2\omega t + \phi_2) + \cdots \tag{D.6}$$

The incident voltage is obtained with Equation (D.7).

$$v_{inc}(t) = \frac{e(t)}{2} \tag{D.7}$$

Since Equation (D.8) gives

$$v_{ref}(t) = v(t) - v_{inc}(t) = v(t) - \frac{e(t)}{2} \tag{D.8}$$

by taking the phasor of the reflected voltage, the reflection coefficient can be expressed by Equation (D.9).

$$\Gamma = \frac{V_1e^{j\phi_1} - \frac{E}{2}}{\frac{E}{2}} \tag{D.9}$$

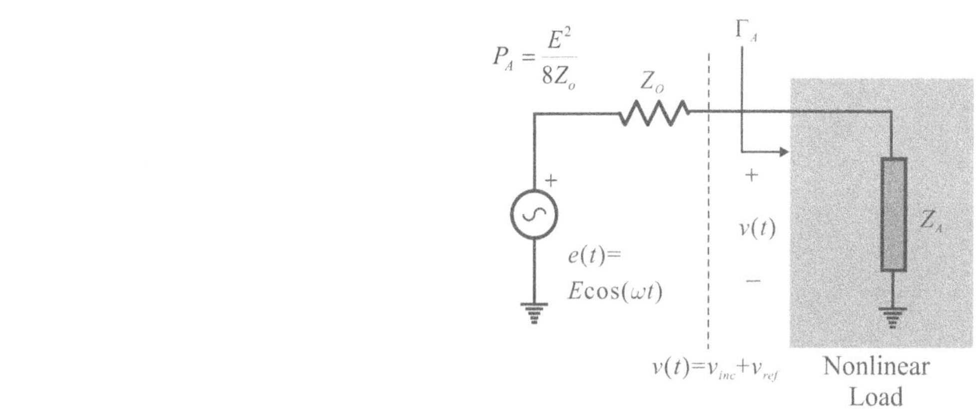

Figure D.5 Measurement of the large-signal reflection coefficient

It should be noted that the conversion of the large-signal impedance into an admittance or reflection coefficient is generally not possible. In the case of small signals, when one is known from measurement, the others can be found through conversion. However, this is only possible with small signals and not so in the case of large signals. The conversion can be examined by simply comparing the large-signal impedance and the admittance. In order to measure the admittance, a sinusoidal voltage with an amplitude V is applied to the input of the nonlinear device and the resulting fundamental current phasor is denoted as $I_1e^{j\varphi_1}$. It should be noted that when a sinusoidal current $I_1\cos(\omega t + \varphi_1)$ is applied to the input of the same device in order to determine the impedance, the voltage appearing across the nonlinear device is seen to be different from $V_1\cos(\omega t)$. In addition, the voltage includes many harmonics that do not exist in the admittance measurement. Thus, the conversion from one parameter to the other is valid only when harmonics do not occur.

Examples D.2

Set up the following simulation circuit and calculate the large-signal reflection coefficient for the input powers of −10–20 dBm using ADS.

Solution

The large-signal reflection coefficient can be calculated using an AC simulation and applying Equation (D.9) to the simulated results, but the reflection coefficient can also be calculated by performing a large-signal S-parameters simulation, as shown in Figure D.6. Since the diode becomes open-circuited at a lower input power, the impedance of the circuit in Figure D.6 is close to 50 Ω and the large-signal reflection coefficient is 0. However, when the input power is high, the diode becomes short-circuited and the large-signal reflection coefficient approaches 1. This is shown in Figure D.7.

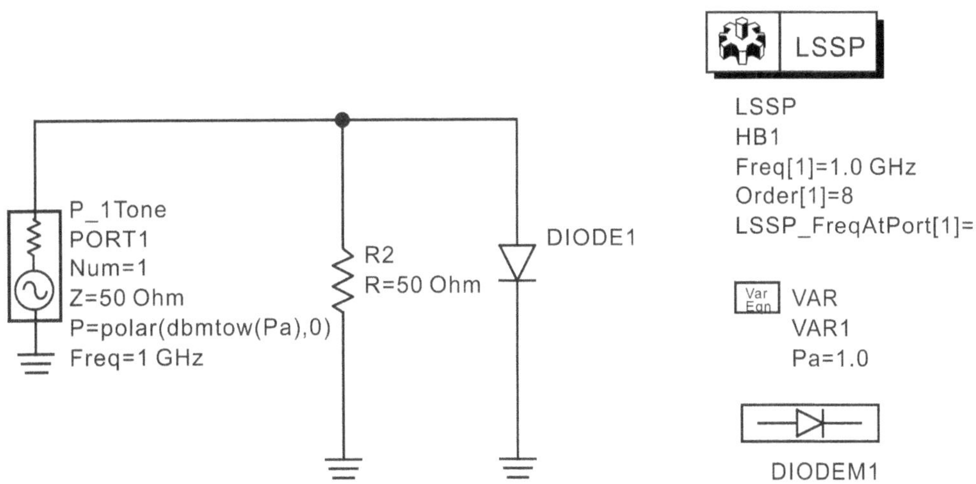

Figure D.6 Schematic for computing the large-signal reflection coefficient

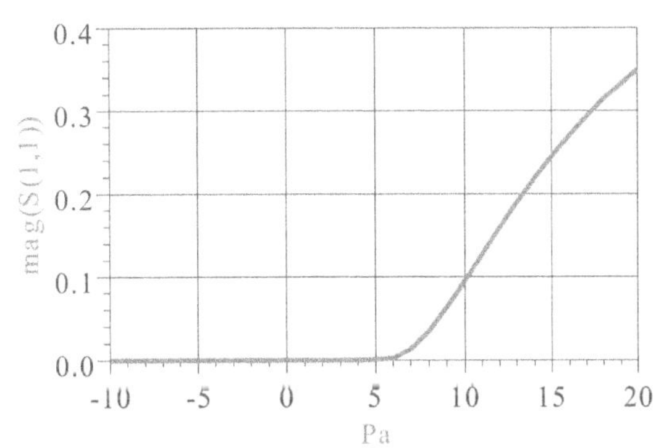

Figure D.7 Computed large-signal reflection coefficient

E. MATHEMATICAL ANALYSIS OF NEGATIVE RESISTANCE

The active device employed in an oscillator at high frequency is significantly more complex than the typical active device employed in the Colpitts oscillator described in Chapter 10. Thus, it is difficult to estimate the range of the element values in the feedback circuit that provides sufficient gain; that is, negative resistance. In addition, due to the significant differences between the simplified active device model and the measured results, oscillators at high frequencies are usually designed to use measured two-port parameters.

Figure E.1 shows a basic series feedback oscillator circuit. The first task in the design of the oscillator circuit in that figure is the selection of an appropriate (x, y) pair that yields Z_{out} with a negative resistance. In the case of a Colpitts oscillator with the simplified active device model described in Chapter 10, selecting this type of (x, y) pair is easy. However, the active device's equivalent circuit is not a simple one, as noted in the previous explanation of the Colpitts oscillator's operation. Since the input of the transistor cannot be approximated as an open or short, the (x, y) pair can be selected using a trial-and-error method. When the range of (x, y) pairs having negative resistance is known, the number of these trial-and-error selections can be minimized.

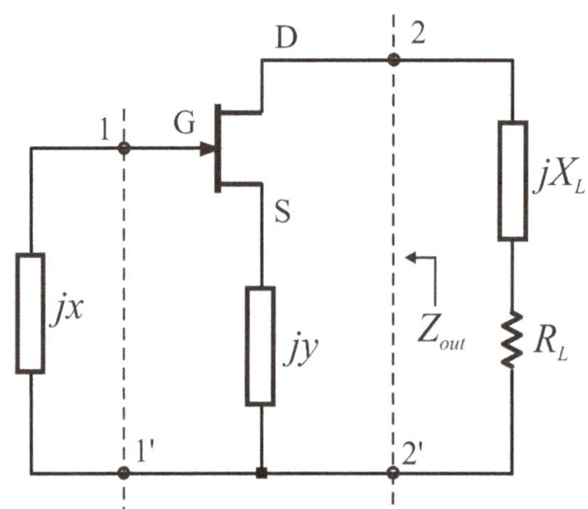

Figure E.1 Impedance seen from the load side

For the (x, y) pair thus selected, where the active part shows a negative resistance, the oscillator can be designed by synthesizing the load. The load should be synthesized so that R_L is set lower than the resulting negative resistance value of the active part and for X_L to cancel out the active part reactance. The series oscillation condition is then satisfied and the oscillation waveform can be formed.

The impedance seen from the load can be obtained in terms of the Z-parameters of the transistor, which are defined as

$$Z_t = \begin{pmatrix} z_i & z_r \\ z_f & z_o \end{pmatrix}$$

and the two-port parameters arising from jy are given by

$$Z_y = \begin{pmatrix} jy & jy \\ jy & jy \end{pmatrix}$$

In addition, as these two two-port parameters are connected in series, the Z-parameters for them are defined by dashed lines 1–1′ and 2–2′, as shown in Figure E.1, and they are expressed as

$$Z = Z_t + Z_y$$

Using those Z-parameters, since jx is a termination for the newly defined two-port parameters, the impedance seen from the load Z_{out} can be obtained as follows with Equation (E1).

$$Z_{out} = z_o + jy - \frac{(z_r + jy)(z_f + jy)}{z_i + jx + jy} \tag{E.1}$$

Then, the contour (x, y) that gives the same value of the real part $\mathrm{Re}(Z_{out}) = R$ satisfies the following quadratic equation of x, y with R as a parameter,

$$\alpha x^2 + 2\beta xy + \gamma y^2 + \delta x + \varepsilon y + K = 0 \tag{E.2}$$

where $\alpha, \ldots, K$ are functions of R. Equation (E.2) is generally a locus of conic sections. Translation of the (x, y) coordinate followed by a rotation results in a newly defined coordinate, (ξ, η). In the (ξ, η) coordinate, Equation (E.2) can be converted into a well-known form as expressed in Equation (E.3).

$$\frac{\xi^2}{a^2} + \frac{\eta^2}{b^2} = 1 \tag{E.3}$$

Thus, the locus of Equation (E.2) represents a translated, rotated hyperbola, ellipse, parabola, or a pair of straight lines in the (x, y) coordinate system. The discriminant D can be used to classify the conic sections given by Equation (E.2) and the discriminate is computed with Equations (E.4a)–(E.4c).

$$D = r_i(R - R_c) = \begin{cases} > 0 & \text{hyperbola} \\ = 0 & \text{parabola} \\ < 0 & \text{ellipse} \end{cases} \tag{E.4a}$$

$$R_c = R_\tau \left\{ 1 - \frac{(r_f - r_r)}{4(r_i r_o - r_f r_r)} \right\} \tag{E.4b}$$

$$R_\tau = \frac{r_i r_o - r_f r_r}{r_i} \tag{E.4c}$$

D can be seen to be a linear function of R. When r_i is positive, the line has a positive slope. Thus, since D varies from negative to positive as R increases, it can be seen that the contour changes from an ellipse to a hyperbola through a parabola. In addition, there is no locus of (x, y) when the value of R becomes less than a certain value R_m. Thus, as R increases, the locus changes from point → ellipse → parabola → hyperbola. In contrast, when r_i is negative, this trend is reversed and thus the locus exists for all negative values of R, (i.e., the negative resistance can become $-\infty$). As R increases, the change is from hyperbola → parabola → ellipse → point and thus shows a positive maximum-resistance value.

For convenience, the new parameters are defined with Equations (E.5a) and (E.5b).

$$R_q = r_o + \frac{r_i - (r_f + r_r)}{2} \tag{E.5a}$$

$$Q = -r_i^3 (R - R_\tau)(R - R_m) \tag{E.5b}$$

To plot the contour with a constant R in the (x, y) coordinate system, the values of the major axis a and the minor axis b, (x_c, y_c) correspond to the origin of the (ξ, η) coordinate system, and the rotation angle θ_r between the ξ- and x-axes is necessary. After extensive manipulation, the parameters for the conic sections can be obtained with Equations (E.6)–(E.10).

$$y_c = \frac{r_i}{4D}\left\{2(x_f + x_r)(r_o - R) - \frac{r_f + r_r}{r_i}(r_f x_r + r_r x_f)\right\} \tag{E.6}$$

$$x_c = \frac{2r_i - (r_f + r_r)}{(r_f + r_r)} y_c - x_i + \frac{r_i(x_f + x_r)}{r_r + r_f} \tag{E.7}$$

$$\theta_r = \frac{1}{2}\tan^{-1}\left(\frac{2r_o - 2R - r_f - r_r}{r_f + r_r - r_i}\right) + \frac{\pi}{2}\left\{H(Q) + H(R_q - r_o)\right\} \tag{E.8}$$

$$a^2 = \frac{Q}{D^2}\left\{R_q - R + \operatorname{sgn}(Q)\sqrt{(R_q - R)^2 + D}\right\} \tag{E.9}$$

$$b^2 = \frac{Q}{D^2}\left\{R_q - R - \operatorname{sgn}(Q)\sqrt{(R_q - R)^2 + D}\right\} \tag{E.10}$$

Using these equations, the conic sections can be plotted in the x–y plane based on the parameter method. That is, in the case of the ellipse, by setting

$$\xi = a\cos t$$

$$\eta = b\sin t$$

and then substituting t whose range is $0 \leqq t \leqq 2\pi$ and plotting, an ellipse with a major axis a and a minor axis b is obtained. For the hyperbola, by setting

$$\xi = a \cosh t$$

$$\eta = b \sinh t$$

since $\xi \geqq 1$. When both a and b are positive in the equations above, the locus becomes a hyperbola that exists in the positive direction. Thus, the remaining half is obtained by substituting $\xi = -a{\cdot}\cosh t$ instead of ξ. The conic section loci can be classified using the discriminant and (ξ, η) can be parameterized to include the ellipse and hyperbola using parameter t as

$$\xi = \frac{1}{2}\left(a \cosh t + a \cos t\right) + \frac{1}{2}\operatorname{sgn}(D)\left(a \cosh t - a \cos t\right)$$

$$\eta = \frac{1}{2}\left(b \sinh t + b \sin t\right) + \frac{1}{2}\operatorname{sgn}(D)\left(b \sinh t - b \sin t\right)$$

Using the parameterization above, a half hyperbola appears when D is positive and an ellipse appears when D is negative. The (x, y) trace on the x–y plane can be obtained by rotating the locus in (ξ, η) obtained above, which is followed by translation as in the following expression:

$$\begin{pmatrix} x \\ y \end{pmatrix} = \begin{pmatrix} x_c \\ y_c \end{pmatrix} + \begin{pmatrix} \cos\theta_r & -\sin\theta_r \\ \sin\theta_r & \sin\theta_r \end{pmatrix} \begin{pmatrix} \xi \\ \eta \end{pmatrix}$$

Denoting the range of the axis for the plot as α, the range of t becomes $t = 1/b{\cdot}\sinh^{-1}\alpha$ and the contour of $R = 0$ can be plotted on the x–y plane. Only one side of the hyperbola appears, in which case the entire hyperbolic pair can be drawn again for $-\xi$.

Then, by plotting the point in this region where the maximum occurs, it will be easy to determine the region of negative resistance. Mathematically, the maxima R_m that gives the maximum negative resistance is expressed below, where (x_m, y_m) represents the x, y coordinate that gives this value and is expressed by Equations (E.11)–(E.13).

$$R_m = R_\tau \left\{1 - \frac{\left|z_f - z_r\right|}{4\left(r_i r_o - r_f r_r\right)}\right\} \tag{E.11}$$

$$y_m = -\frac{\left|z_f\right|^2 - \left|z_r\right|^2}{2\left(x_f - x_r\right)} \tag{E.12}$$

$$x_m = -\frac{\left|z_{fg}\right|^2 - \left|z_{rg}\right|^2}{2\left(x_{fg} - x_{rg}\right)} \tag{E.13}$$

The subscript g indicates a common gate. Note that x_m and y_m vary according to frequency and the locus of (x_m, y_m) with frequency can be plotted using the equations given above.

Figure E.2 illustrates a typical case of the locus for the change of R in the x–y plane, and the maximum point that gives the maximum negative resistance is also plotted. As can be seen in that figure, at the value of R where the negative resistance is maximum, $-R_m$ is one point and when the value of R is increased in the positive direction, the locus varies from an ellipse ($R = -4/5R_m$) to a hyperbola ($R = -2/5R_m$).

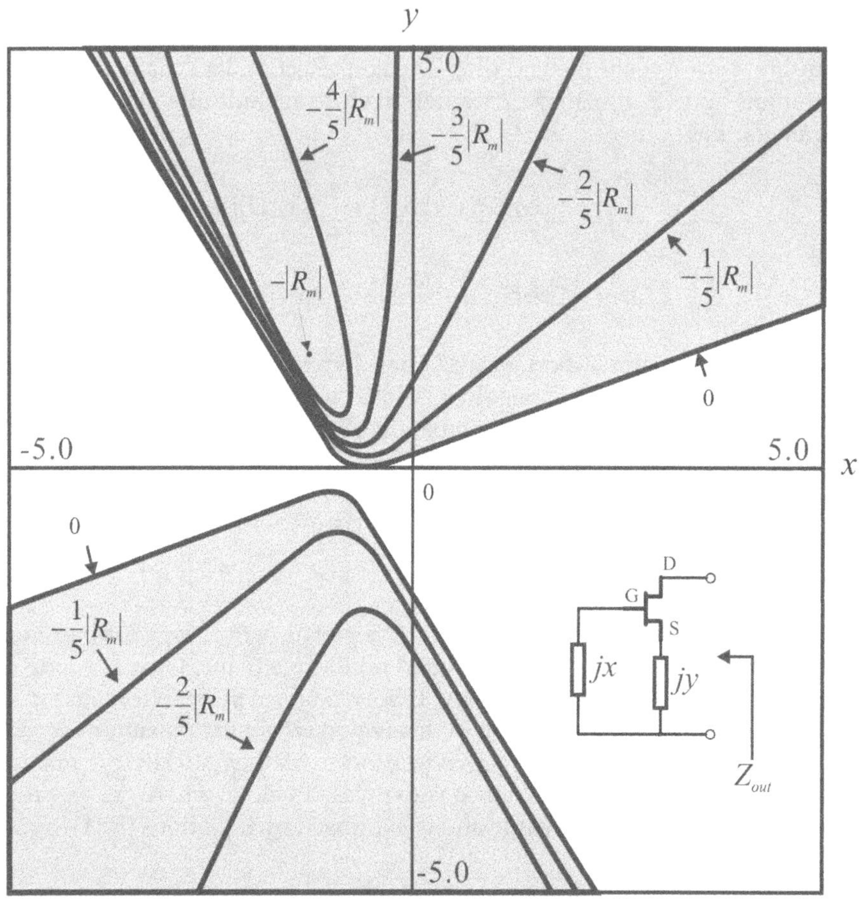

Figure E.2 Typical contour plot of negative resistance

In addition, for the remaining six cases of basic series and parallel feedback oscillators, the contours can be plotted in the same way. First, for the case where the gate and drain are changed, the port numbers are changed when obtaining the S-parameters. In addition, when the output is taken from the source terminal, the common gate S-parameters computed using ADS can be used for the plot. This calculation can be done analytically, but the current version of ADS makes it significantly easier. In the case of a parallel feedback, the same results can be obtained by using Y-parameters instead of Z-parameters. Therefore, when conductance is taken instead of resistance, the same results will be obtained in the plot.

In the case of series feedback oscillators, the change in the gate's output terminals from the drain leads to the same results shown in Figure E.2. However, when the output is taken from the source terminal, r_i becomes negative as does the slope of D and thus the magnitude of the negative resistance obtained becomes ∞. It should be noted that this case is often applied in YIG oscillators and in certain other cases. The contour plot of the negative resistance for this case is shown in Figure E.3, where R_m is positive and the contour does not exist above R_m. When R is varied from positive to negative, the locus varies from a point to an ellipse and then a hyperbola, and the feedback element value y exists for all negative values of R.

Therefore, in order to design an oscillator at a desired frequency, the oscillator circuit's form is first selected, which allows the locus for $R = 0$ and the region giving negative resistance to be determined using the S-parameters. The (x, y) point that gives the appropriate negative resistance is then selected, and the load circuit can be synthesized to form an oscillator. The real part of the load is set smaller than the obtained negative resistance (usually by one-half or one-third) and the imaginary part is set to cancel the imaginary part of the active part, which completes the RF design of the small-signal oscillator. Adding the appropriate DC supply circuit to this RF circuit completes the oscillator design with the desired oscillation frequency in the small-signal domain.

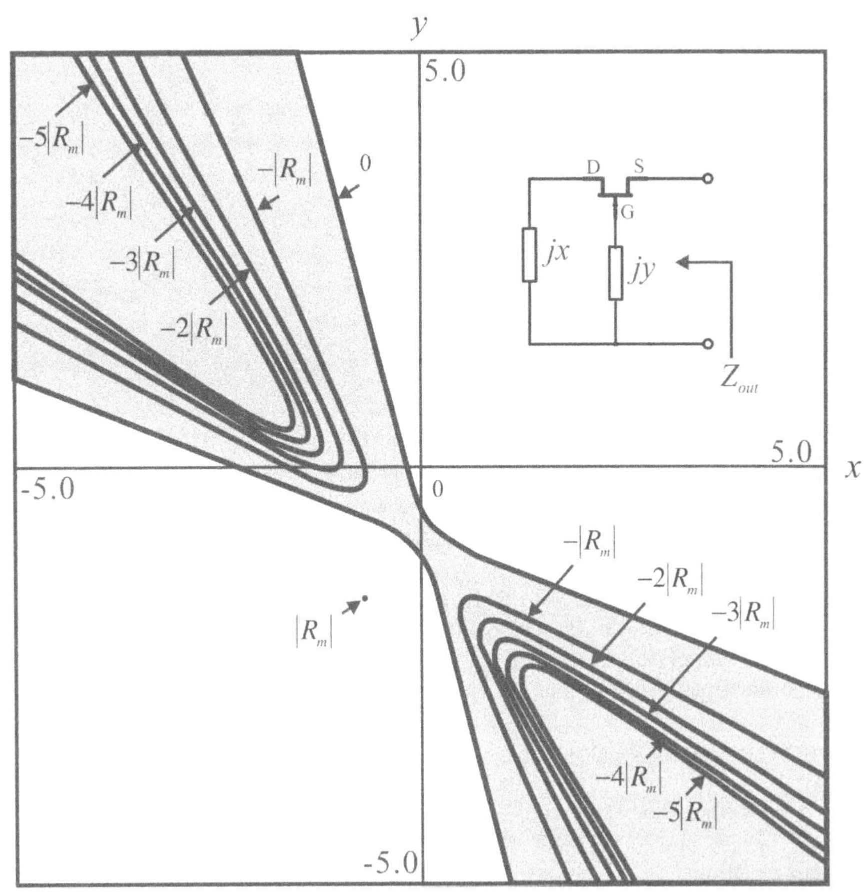

Figure E.3 Contour of the negative resistance seen from the source terminal

Example E.1

Plot the $R = 0$ contour and the negative resistance maximum point with respect to the frequency for NE32484 using MDS.

Solution

First, the Z-parameters of the NE32484 are obtained from the S-parameters, after which each parameter is calculated using the equations above and the contours can then be plotted using the parameter method.

```
n = 17; this is a number that represents frequency
R = 0; the resistance value for the contour to be plotted
z = Stoz (S); the S–Z parameter conversion
```

[Parameter definitions]

```
zi=z[1,1,n]
zf=z[2,1,n]
zr=z[1,2,n]
zo=z[2,2, n]
ri=real(zi),
xi=imag(zi)
rf=real(zf)
xf=imag(zf)
ro==real(zo)
xo=imag(zo)
```

[Parameter calculations]

```
Rt=(ri*ro-rf*rr)/ri
Rm=Rt* (1-mag(zf-zr)^2/(4*(ri*ro-rf*rr)))
Rc=Rt*(1-(rf-rr)^2/(4*(ri*ro-rf*rr)))
Rq=ro + (ri-(rf+rr))/2
D=-ri(R-Rc)
Q=-ri^3*(R-Rt)*(R-Rm)
xc=(2*ri-rf-rr)/(rf+rr)*yc-xi+ri*(xf+xr)/(rf+rr)
yc=ri/(4*D)(2*(xf+xr)*(ro-R)-(rf+rr)/ri*(rf*xr+rr*xf))
theta=1/2*atan((2*ro-2*R-rf-rr)/(rf+rr-ri))+pi/2*(step(Q)+step(Rq-ro))
asq=Q/D^2*(Rq-R+sign(Q)*sqrt((Rq-R)^2+D))
bsq=Q/D^2*(Rq-R-sign(Q)*sqrt((Rq-R)^2+D))
a=sqrt(mag(asq))
b=sqrt(mag(bsq))
```

[Running parameter definition]

```
t=generate(-5,5,51)
```

[Locus equations]

```
xi=1/2*(a*cosh(t)+a*cos(t))+1/2*sign(D)*(a*cosh(t)-a*cos(t))
eta=1/2*(b*sinh(t)+b*sin(t))+1/2*sign(D)*(b*cosh(t)-b*cos(t))
x1=xc+cos(theta)*xi-sin(theta)*eta
```

```
y1=xc+sin(theta)*xi+cos(theta)*eta
x2=xc-cos(theta)*xi-sin(theta)*eta
y2=xc-sin(theta)*xi+cos(theta)*eta
```

After this, **y1** is plotted against **x1** in the display window and, in case of a hyperbola, a plot of **y2** against **x2** should be added. In addition, to add the locus of the maximum point (x, y) when the value of R is the maxima R_m for the frequency, the following equations are established:

```
ym=-(mag(zf)^2-mag(zr)^2)/2/(xf-xr)
xm=(2*ri-rf-rr)/(rf+rr)*ym-xi+ri*(xf+xr)/(rf+rr)
```

Similarly, **ym** is plotted against **xm**. The curves plotted as described above are shown in Figure E.4.

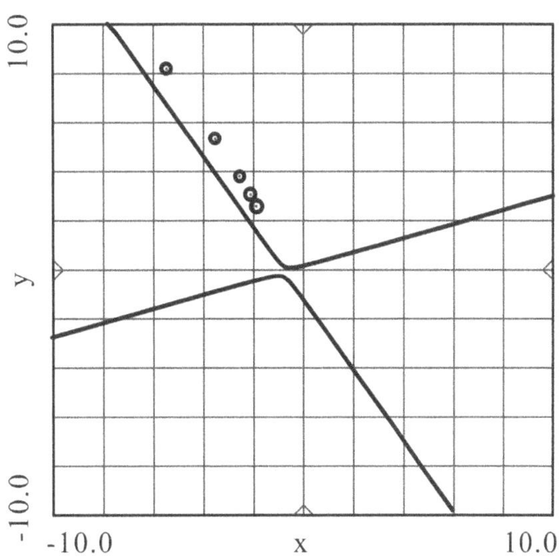

Figure E.4 Contour of R = 0 for NE32484 plotted on MDS

F. OSCILLATION CONDITIONS BASED ON REFLECTION COEFFICIENTS

Unlike the impedance, the reflection coefficient changes when its reference impedance changes. Thus, the oscillation conditions based on the reflection coefficients shown below in Equations (F.1a)–(F.1c) also vary with the change of the reference impedance.

$$\left|\Gamma_L(\omega_o)\Gamma_A(0,\omega_o)\right| > 1 \tag{F.1a}$$

$$\angle\Gamma_L(\omega_o)\Gamma_A(0,\omega_o) = 0 \tag{F.1b}$$

$$\frac{\partial\angle\Gamma_L(\omega_o)\Gamma_A(0,\omega_o)}{\partial\omega} < 0 \tag{F.1c}$$

In other words, it is often possible for oscillation to occur even when Equation (F.1) is not satisfied. As a simple example, consider the oscillator circuit that consists of a 50-Ω load and a series resonant active part. The real part of the active part is negative and its magnitude is greater than the 50 Ω. The oscillation condition of the oscillator circuit based on the reflection coefficient can be obtained using the circuit shown in Figure F.1. By selecting $Z_c = 50\ \Omega$, $\Gamma_A\Gamma_L = 0$ irrespective of frequency. This corresponds to setting the reference impedance of **OscTest** in ADS to 50 Ω. Thus, the oscillation conditions given by Equation (F.1) seem not to be satisfied. However, when the oscillation condition is determined from the previously discussed impedance-based method, oscillation is possible although $\Gamma_A\Gamma_L = 0$. Obviously, the example circuit oscillates at the series resonant frequency.

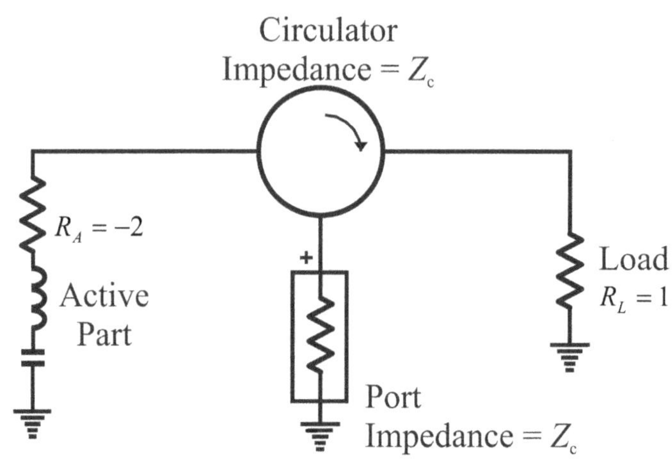

Figure F.1 Simulation schematic for the variation of $\Gamma_A\Gamma_L$ according to the circulator reference impedance

Since it is generally difficult to show the variation of oscillation conditions with respect to the change of reference impedance, we will consider the oscillator circuit in which a simple resistive load is connected to a series resonant active part, as shown in Figure F.1.

The load is normalized by its resistance of 50 Ω and set to 1. Let the real part of the active part impedance normalized by 50 Ω be –2 and the active part be series resonant at the oscillation frequency. Thus, the series oscillation condition based on the impedance criteria is satisfied. The oscillation conditions based on the reflection coefficient for a selected reference impedance Z_c at the series resonant frequency are expressed as

$$\Gamma_L\Gamma_A = \left(\frac{1 - Z_c}{1 + Z_c}\right)\left(\frac{-2 - Z_c}{-2 + Z_c}\right) \tag{F.2}$$

$\Gamma_A\Gamma_L$ given by Equation (F.2) is plotted against the reference impedance Z_c and is shown in Figure F.2.

In Figure F.2, $\Gamma_A\Gamma_L$ approaches $\pm\infty$ at $Z_c = -R_A = 2$ at point B because the reflection coefficient for the active part, Γ_A is computed to be ∞. At point A ($Z_c = 1$), the load and reference impedances are equal, which results in $\Gamma_L = 0$ and leads to $\Gamma_A\Gamma_L = 0$. Note that when $Z_c > -R_A = 2$ in Figure F.2, $\Gamma_A\Gamma_L$ becomes greater than 1. In addition, since the sign is positive, the phase of $\Gamma_A\Gamma_L$ becomes 0°. Therefore, in the case of $Z_c > R_L$ and $Z_c > -R_A$, the proper oscillation conditions in Equation (F.1) are satisfied. This case is shown in Figure F.3(a).

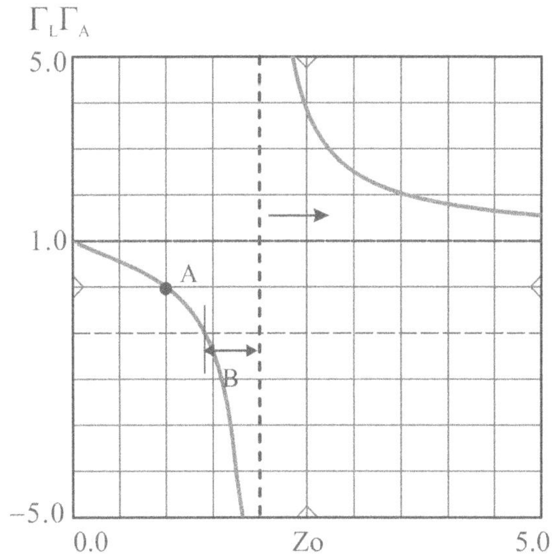

Figure F.2 $\Gamma_A\Gamma_L$ with respect to reference impedance

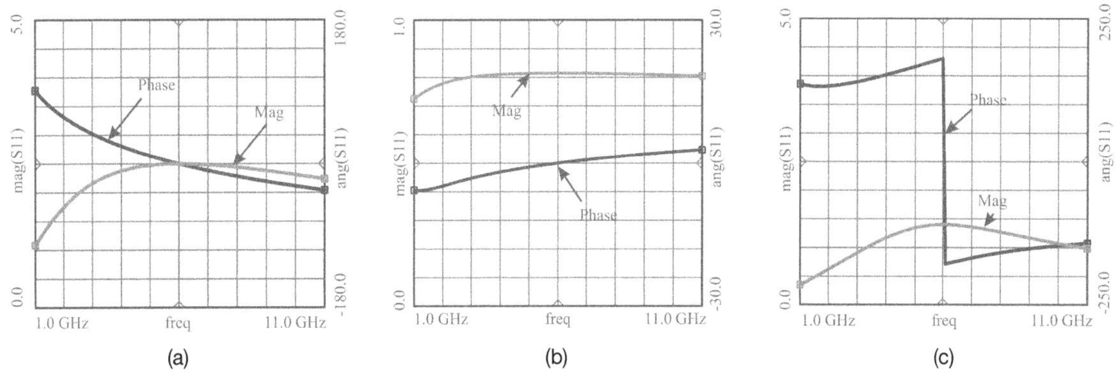

Figure F.3 (a) Proper oscillation condition; (b) $|R_A|$, $R_L > Z_c$; (c) $R_L < Z_c < |R_A|$

However, when $Z_c < R_L$ (the reference impedance is lower than point A), the phase of $\Gamma_A\Gamma_L$ becomes 0° and the magnitude of $\Gamma_A\Gamma_L$ can be seen to be less than 1. In this case, the phase slope of $\Gamma_A\Gamma_L$ with respect to frequency is positive and appears to be different from that of the proper oscillation condition. This case is shown in Figure F.3(b).

In addition, when $R_L < Z_c < -R_A$ (in the range between points A and B), the phase of $\Gamma_A\Gamma_L$ becomes 180° because the sign is negative. Note that $|\Gamma_A\Gamma_L|$ may be less than 1 or greater than 1. This case of the oscillation condition is shown in Figure F.3(c). However, all the oscillation conditions shown in Figure F.3 provide the oscillation start up. The conditions show the changes according to the changes of the reference impedance. Thus, the proper oscillation condition can be demonstrated by changing the reference impedance.

Example F.1

In the circuit shown in Figure F.4(a), Z_c is set to 100 Ω in order to investigate the possibility of series oscillation at a frequency of 10 GHz. The result $S_{11} = \Gamma_A \Gamma_L$ is shown in Figure F.4(b). Discuss the possibilities of oscillation for this circuit at 10 GHz.

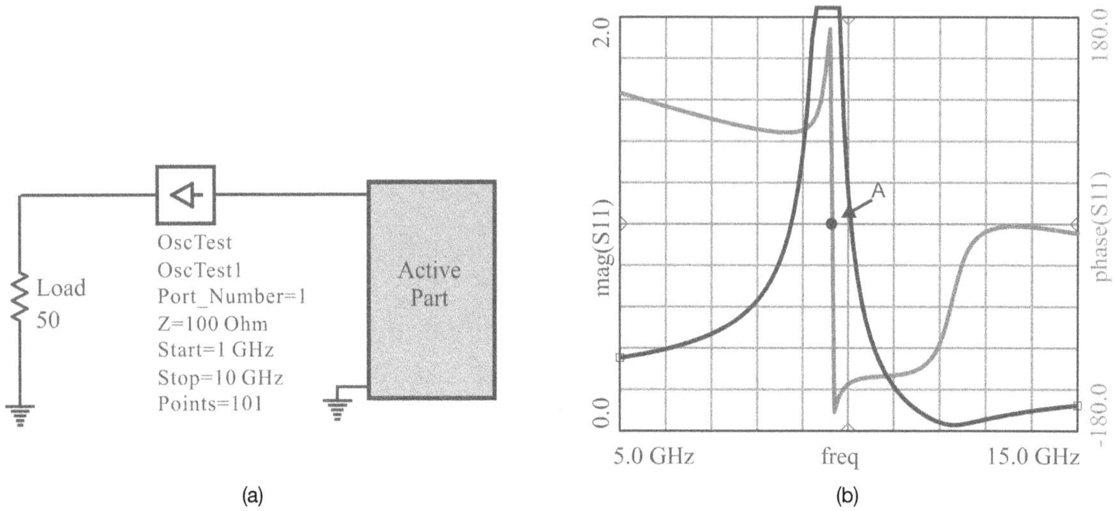

Figure F.4 Oscillation condition for $Z_c = 100$ ohm

Solution

In Figure F.4(b), the phase of $\Gamma_A\Gamma_L$ is 180°at the frequency indicated by point A. Also, the magnitude of $\Gamma_A\Gamma_L$ is greater than 1. It can be concluded that the value $Z_c = 100$ Ω of **OscTest** is smaller than $-R_A$ even though greater than R_L. The proper oscillation condition can be obtained by making $Z_c > 100$ Ω. Thus, Z_c was changed to $Z_c = 1000$ Ω, and the recalculation of S11 gave the results shown in Figure F.5. It can be found that at about 10 GHz, $Z_c = 1000$ Ω yields mag($\Gamma_A\Gamma_L$) greater than 1 and the phase of $\Gamma_A\Gamma_L$ is 0°. In addition, from Figure F.5, since a negative phase slope of $\Gamma_A\Gamma_L$ is obtained, the proper oscillation condition is also obtained.

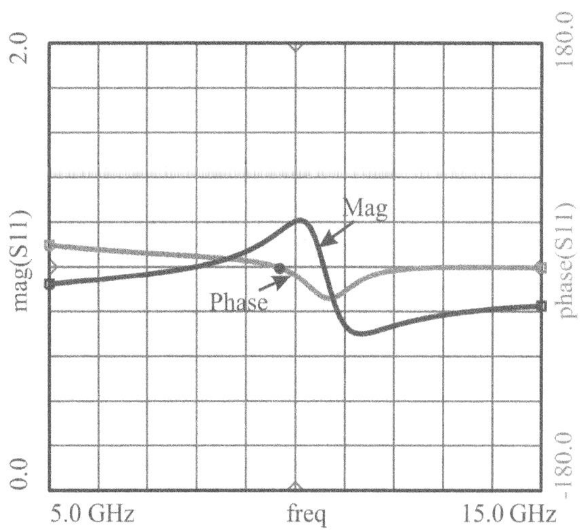

Figure F.5 Oscillation condition for $Z_c = 1$ kΩ

The example above shows the variation of the series oscillation condition with respect to the change of the reference impedance. Similarly, in the case of parallel oscillation, we look at the variation of the oscillation condition with respect to the change of the circulator's reference admittance. First, suppose that the normalized real part of the parallel resonant active part by the reference admittance is –2 and the normalized load admittance is 1. The product of the reflection coefficients is thus

$$\Gamma_L\Gamma_A = \frac{1-Y_c}{1+Y_c}\,\frac{-2-Y_c}{-2+Y_c} \tag{F.3}$$

The result given by Equation (F.3) is the same as that in Equation (F.2). The plot of $\Gamma_A\Gamma_L$ with respect to the reference admittance results in the same plot as in Figure F.2. Only the x-axis changes with respect to the reference admittance. Thus, when viewed as impedance, the proper parallel oscillation condition is obtained when the reference impedance is lower than both the load impedance and the real part of the active part, but when the reference impedance is higher than these two impedances, an inverse oscillation condition is obtained. These are summarized in Table F.1. Thus, in the parallel oscillation condition, the reference impedance must be lower than the load impedance and the real part of the active part, whereas for the series case, the reference impedance must be set to have a value greater than the two impedances.

Table F.1 Variation of the oscillation condition according to the settings of *OSCTEST* reference impedance

Condition	$Z_c < -R_A$ **and** R_L	$-R_A < Z_c < R_L$ $-R_A > Z_c > R_L$	$Z_c >- R_A$ **and** R_L
Series oscillation	Inverse	180° phase change	Proper
Parallel oscillation	Proper	180° phase change	Inverse

In conclusion, to view the proper oscillation condition, in series oscillation the reference impedance must be set higher than both side impedances, but in parallel oscillation it must be set lower than both side impedances. It is easier to determine the oscillation frequency based on the phase because oscillation is possible where the phase is 0° and 180°. The proper oscillation condition can be obtained at the frequency where the phase is 0° and 180° by sweeping the reference impedance. Using the obtained proper oscillation condition, we can clearly determine the oscillation start-up. For sweeping convenience, inserting **OscTest** on the load side provides an easier way to perform the oscillation start-up check because the value of the load impedance is known. Then, only unknown impedance is the active part's impedance.

Example F.2

Figure F.6 shows a Colpitts oscillator circuit, whose output is obtained from the emitter terminals of the transistor. To investigate the oscillation condition of this circuit, the **OscTest** impedance is set at 1.1 Ω to be lower than the load impedance. Look at this result and verify through simulation the possibility of oscillation at other frequencies.

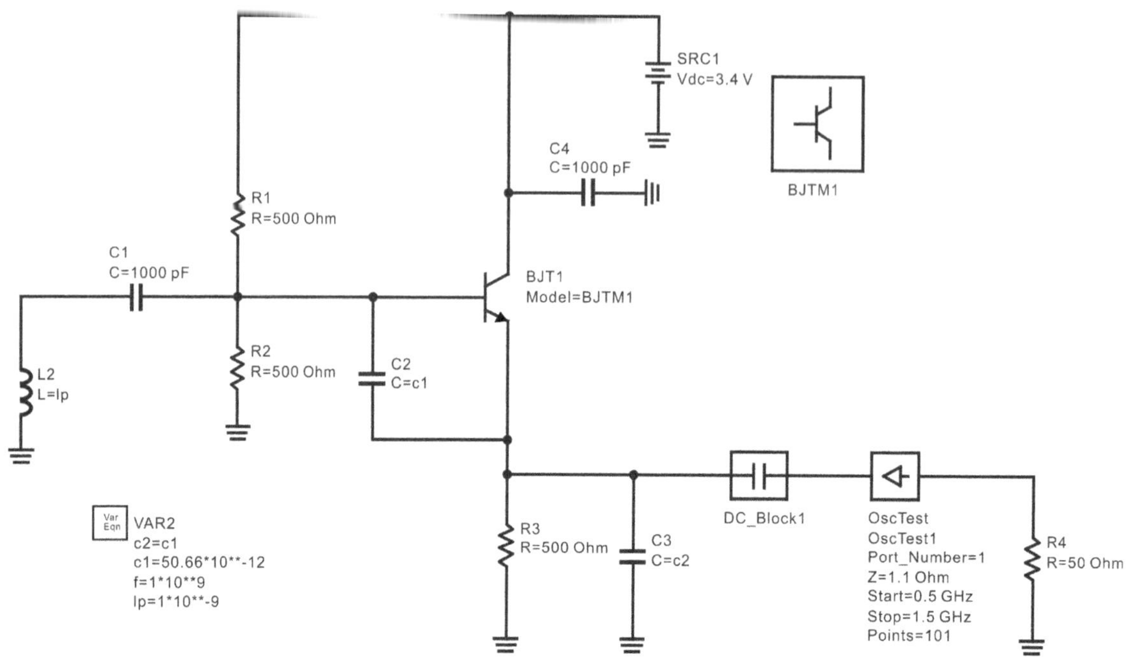

Figure F.6 Colpitts oscillator (output extracted from emitter terminals)

Solution

The oscillation condition for the reference impedance of 1.1 Ω gives the result shown in Figure F.7. The magnitude of $\Gamma_A\Gamma_L$ at about 1 GHz (point A) is greater than 1 and the phase is 0°, and since the phase slope is negative, it can be seen that parallel oscillation is possible. It can also be concluded that the active part impedance is higher than the reference impedance of 1.1 Ω.

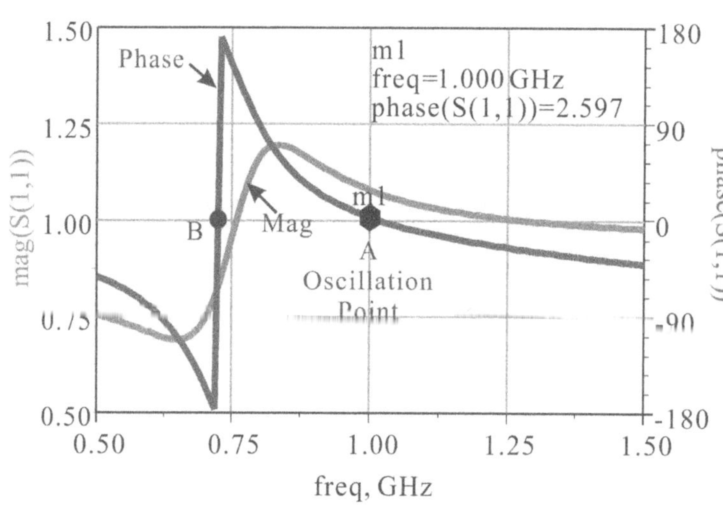

Figure F.7 Oscillation investigation of Colpitts oscillator, $Z_c = 1.1\ \Omega$

In the case of point B, since the phase is 180°, the oscillation start-up may be possible at this point. From Table F.1, it can be concluded that $-R_A < Z_c < R_L$ for the parallel oscillation condition start-up. However, taking the selected reference impedance Z_c =1.1 into consideration, the case $-R_A < Z_c < R_L$ is hardly expected. Therefore, point B is caused by a series resonant active part. To see a proper oscillation condition at a series resonance, the reference impedance is set to 1 kΩ and the simulation is performed. The results are shown in Figure F.8.

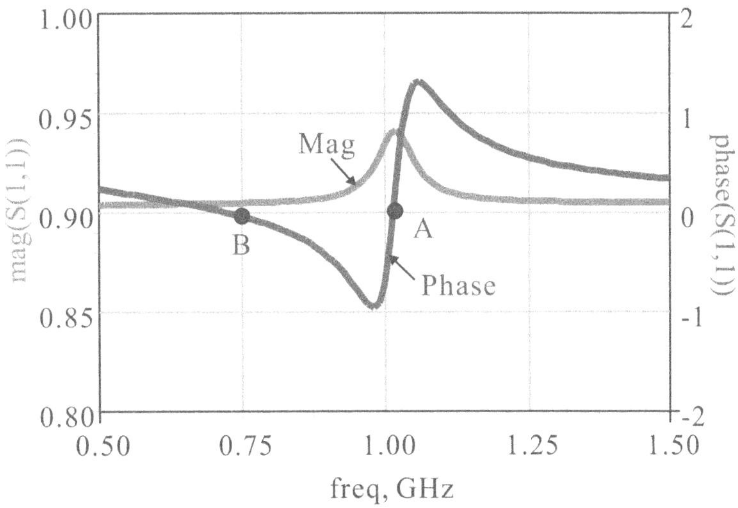

Figure F.8 Investigation of oscillation possibility at point B

For Z_c =1 k Ω, since point A is a point satisfying the parallel oscillation start-up, an inverse oscillation should occur. Looking at point A, since the phase slope is positive and the magnitude is less than 1, point A definitely shows an inverse oscillation condition. On the other hand, in the case of point B, it shows a proper oscillation condition because the phase slope is negative. However, because the magnitude is less than 1, the oscillation condition is not satisfied. Thus, the oscillation is not formed at point B.

Index

D

www.ingramcontent.com/pod-product-compliance
Lightning Source LLC
LaVergne TN
LVHW081959200926
848481LV00010B/2771